AF520224

GEOGRAPHY OF RAJASTHAN

Dr. T. S. Chouhan
Former Professor
Department of Geography
University of Rajasthan
Jaipur (India)

Published by

SCIENTIFIC PUBLISHERS (INDIA)
5 A, New Pali Road
P.O. Box 91
Jodhpur - 342 001 INDIA
E-mail: info@scientificpub.com
Website: http://www.scientificpubonline.com

ISBN: 978-93-89184-28-0
eISBN: 978-93-89184-30-3

Printed in India

Dedicated to my Life Partner

Smt Vidya Chouhan

for

Constant Inspiration and Support

JAI NARAIN VYAS UNIVERSITY
JODHPUR, RAJASTHAN (INDIA)

DEPARTMENT OF GEOGRAPHY

Dr. J. K. Jain

(Former) Senior Faculty Member and Head

US Fulbright Post-Doctroral Fellowship Awardee

FOREWORD

Rajasthan is endowed with diversified natural resources with a unique combination of Thar Desert, Aravali Hill, Hadoti Plateau and Eastern Alluvial Plain. To meet the increasing demands of burgeoning human and livestock population, improved technology has to be employed in order to obtain optimum yields from our natural resources. However, exploitation of nature cannot be reckless because the agro ecosystem is fragile and prone to degradation. Land resources assessment and management is thus absolutely essential in order to conserve our natural resources and sustain their productivity.

The book is an endeavour to develop resource development plan for sustainable development through the use of latest technology of Remote Sensing and GIS. Resources and infrastructure scenario has also changed. The book is therefore a revised and updated version using latest inputs. I hope the book will be useful for planners at government level as also researchers and teachers in the Universities, NGOs and all those interested in developmental activities.

All issues of concern have effectively been addressed and synthesized well in this book. It will serve as an useful document. In fact, the author of this book has long experience of working in regional geography. He has done a commendable job which I am sure will be useful to researchers, extension workers, students and policy planners alike.

(J. K. Jain)

Jodhpur

JAI NARAIN VYAS UNIVERSITY
JODHPUR, RAJASTHAN (INDIA)

DEPARTMENT OF GEOGRAPHY

Dr. J. K. Jain

(Former) Senior Faculty Member and Head

US Fulbright Post-Doctoral Fellowship Awardee

FOREWORD

Rajasthan is endowed with diversified natural resources with a unique combination of Thar Desert, Aravali Hill, Hadoti Plateau and Eastern Alluvial Plain. To meet the increasing demands of burgeoning human and livestock population, improved technology has to be employed in order to obtain optimum yields from our natural resources. However, exploitation of nature cannot be reckless because the agro-ecosystem is fragile and prone to degradation. Land resources assessment and management is thus absolutely essential in order to conserve our natural resources and sustain their productivity.

The book is an endeavour to develop resource development plan for sustainable development through the use of latest technology of Remote Sensing and GIS. Resources and infrastructure scenario has also changed. The book is therefore a revised and updated version using latest inputs. I hope the book will be useful for planners at government level as also researchers and teachers in the Universities, NGOs and all those interested in developmental activities.

All issues of concern have effectively been addressed and synthesized well in this book. It will serve as an useful document. In fact, the author of this book has long experience of working in regional geography. He has done a commendable job which I am sure will be useful to researchers, extension workers, students and policy planners alike.

(J. K. Jain)

Jodhpur

The Author

Dr. T.S. Chouhan is former Professor in the Department of Geography, University of Rajasthan, Jaipur (India). He has to his credit over 35 years of experience in teaching and research in interdisciplinary subject.

Dr. Chouhan has participated in number of International Conferences and Presented Research Paper and Chaired Technical Sessions. He has visited Brazil, Morocco, Saudi Arabia, Tunisia, Argentina, Germany and France.

Dr. Chouhan has worked intensively on the following projects:

(i) Ecology of Tropical Deserts in Special Reference to Arid Land Physiology (UNESCO Project Encyclopaedia of Life Support System).

(ii) Desert Eco System in India (UNESCO Project Encyclopaedia of Life Support System).

(iii) Ecology of Tropical Deserts in Special Reference to Biogeography and Evolution of Desert Animals (UNESCO Project Encyclopaedia of Life Support System).

He has also completed one UGC major Research Project on "Geo-informatics for Combating Land Degradation and Desertification in Fragile Aravalli Foothills Eco System of Rajasthan".

He has to his credit more than 70 research papers published in reputed journals in India and abroad and more than two dozen well acclaimed books. He has been guiding research work for the last more than two decades and supervised more than fifty doctoral theses.

Dr. Chouhan is recognized consultant/expert to the Government of Rajasthan, Planning Department (Gr. IV), Jaipur in various fields like Wasteland Development, Forestry, Land Management, Water Conservation, Remote Sensing and Cartography.

Dr. Chouhan holds active membership of various International, National and Regional Academic Societies.

Preface

Rajasthan state has been a wide vistas, diversified natural and cultural spectrum and changing pattern of progress and development in various fields like agriculture, forestry, mining, industries, economic and social way of life since historical past. The development of the state is considered in harmony with nature with certain exceptions like environmental hazards and resource exploitation.

Rajasthan is endowed with diversified natural resources. The survival of mankind has always been supported by nature and its resources. For centuries man has been exploiting these resources for his survival and development. In the rapid march of progress, the balance that needs to be maintained between exploitation and regeneration was however largely ignored. With the increase in population there has thus been great pressure on these resources leading to their scarcity and degradation. Proper Planning for management of growth is possible only when relevant information of resource, population, availability of social and industrial infrastructure and the trends in growth are known.

Rajasthan is a land of extremes and the present book indicates the great variety in relief, climate, soil, vegetation, agricultural activity and mineral deposits.

The aim of the book is comprehension and understanding through functional correlation. Detailed information is furnished for the purpose of illustrating a point and general functional relationship.

The subject has been treated in its regional aspects and an attempt has been made to preserve a balance between the physical, human and economic aspects. This book provides updated information regarding land, people, available resources and infrastructural facilities mainly through maps, graphs, charts and tables. I hope the book will be useful for planners, researchers, teachers, students, NGO and all those interested in developmental activities in Rajasthan.

Jaipur **T.S. Chouhan**

Acknowledgement

The author is grateful to the officials of the various department of the Government of India as well as State Government of Rajasthan especially to Dr. J.K. Jain, former Head, Department of Geography, Jai Narayan Vyas University, Jodhpur, the Director, Department of Economics and Statistics, Jaipur; the Director, Board of Revenue (Land Records), Ajmer; Director, Department of Mines and Geology, Udaipur; Director, Indian Meteorological Department, Poona; Director, Census Operations Rajasthan, Jaipur; Director, Agricultural Census Commissioner, Jaipur; Chairman, Indian Space Research Organisation, Bangalore; Director, National Remote Sensing Centre, Hyderabad, Department of Land Resources Ministry of Rural Development, Govt. of India, New Delhi; Land Use and Cover Monitoring Division (LRUMG), Remote Sensing & GIS Applications Area, National Remote Sensing Centre, Indian Space Research Organisation, Department of Space, Government of India, Hyderabad; Ministry of Environment, Forest and Climate Change, Government of India; Director, Space Applications Centre, Indian Space Research Organisation, Department of Space, Government of India, Ahmedabad; Chief Engineer, Department of Irrigation, Jaipur; Director, Central Arid Zone Research Institute, Jodhpur; Chief Engineer, Ground Water Department, Jodhpur; Director Central Ground Water Board Western Region Jaipur; Director Regional Remote Sensing Service Centre, Jodhpur; Director State Remote Sensing Application Centre, Jodhpur; Forest Resource Survey of India, Dehra Dun, Ministry of Environmet, Forest and Climate Change, New Delhi; Director, Geological Survey of India, Calcutta; Director General, Survey of India, Dehra Dun; Director, National Atlas Organisation, Calcutta; National Dairy Development Board, Anand for their cooperation in providing me the relevant literature and data required in the writing of this book.

I shall be failing in my duty if I do not express my sincere thanks to my wife Vidya who strived hard during the preparation of this book.

T.S. Chouhan

Contents

Foreword v
Preface ix
Acknowledgements x

1. INTRODUCTION 1–9
2. GEOLOGY 10–32
3. PHYSIOGRAPHY AND GEOMORPHOLOGY 33–46
4. DRAINAGE SYSTEM AND LAKES 47–105
5. CLIMATE 106–153
6. SOIL 154–170
7. ENVIRONMENTAL POLLUTION 171–193
8. NATURAL VEGITATION AND FORESTS 194–219
9. DISASTERS MANAGEMENT 220–274
10. SURFACE WATER RESOURCES 275–289
11. GROUND WATER RESOURCES 290–336
12. LAND USE AND AGRICULTURE DEVELOPMENT 337–409
13. WATERSHED DEVELOPMENT 410–418
14. IRRIGATION 419–434
15. LIVESTOCK AND DAIRY DEVELOPMENT 435–487
16. MINERAL RESOURCES 488–513
17. WASTELANDS 514–530
18. ENERGY DEVELOPMENT 531–549
19. TRANSPORTATION AND COMMUNICATION 550–566
20. TOURISM AND CULTURAL HERITAGE 567–604
21. INDURISTAL DEVELOPMENT 605–634
22. POPULATION 635–662
23. GEOGRAPHICAL REGION 663–671

REFERENCE 672–675

Chapter 1

INTRODUCTION

Rajasthan, the largest state of India (3,42,239 sq. km.) situated in the north-western part of the Indian union (23° 3' and 30° 11' North latitude and 69° 29' and 78° 17' East longitude) is largely an arid state for most of its part. The Tropic of Cancer passes through south of Banswara town. Presenting an irregular rhomboid shape, the state has a maximum length of 869 km. from west to east and 826 km. from north to south. The western boundary of the state is part of the Indo-Pak international boundary, running to an extent of 1,070 km. It touches four main districts of the region, namely, Barmer, Jaisalmer, Bikaner and Ganganagar. The state is girdled by Punjab and Haryana states in the north, Uttar Pradesh in the east, Madhya Pradesh in south east and Gujarat in the south west (Figure 1.1).

Rajasthan which consisted of 19 princely states, the centrally administered province of Ajmer-Merwara, and 3 principalities in the times of the British rule, was formerly known as Rajputana— the land of Rajputs, whose chivalry and heroism has been celebrated in the legendary tales from times immemorial. The. formation of Rajasthan state in its present form started in 1948 when the states Reorganization Commission reconstituted the various provinces.

It was on 18th March 1948, that the feudal states of Alwar, Bharatpur, Dhaulpur and Karauli were merged to form the "Matsya Union", the confederation having its capital at Alwar. Only about a week later, on 25th March, 1948, other ten states viz. Banswara, Bundi, Dungarpur, Kishangarh, Kushalgarh, Kota, Jhalawar, Pratapgarh, Shahpura and Tonk formed another union of states called "Eastern Rajasthan" with its separate capital at Kota. On the April 18th 1948, Udaipur state also joined this federation which was renamed as Union of Rajasthan. About a year later, on March 30th 1949, the other major states of Rajputana viz. Bikaner, Jaipur, Jodhpur and Jaisalmer also joined the federation. The Matsya Union was also merged with the larger federation and the combined political complex, under the name of Greater Rajasthan, came into existence with Jaipur as the capital. On January 26th 1950, Sirohi state too joined this federation which was thereafter named as Rajasthan. The centrally administered area of Ajmer Merwara was merged with Rajasthan on November 1, 1956, when the recommendations of the State Reorgani-zation Commission were accepted, and the new state of India came into existence (Figure 1.2).

The area of Rajasthan is nearly equivalent to some of the urbanized countries of the western world like Norway (3,24,200 sq. km.), Poland (3,12,600 sq. km.) and Italy (3,01,200 sq. km.), The desert area of Rajasthan are among the few tropical deserts of the world which have the highest population density. Dating back to the old archeological history of the area, the artefacts and stone age tools, discovered at a few sites of the Luni Basin, the Buddha Pushkar, Ahar valley, Kalibanga, Pilibanga and Sanchore reveal that the area was inhabited by early man, 6 to 10 thousand years ago, leading to the surmise that the region had a climate favourable for human habitation in those times. It was only in the later ages, more especially in the last thousand years the climatic conditions gradually became adverse and the degradation of natural ecosystem was accentuated by the indiscriminate economic activities of man.

Source: Survey of India; GOI.

Figure 1.1 Location of Rajasthan

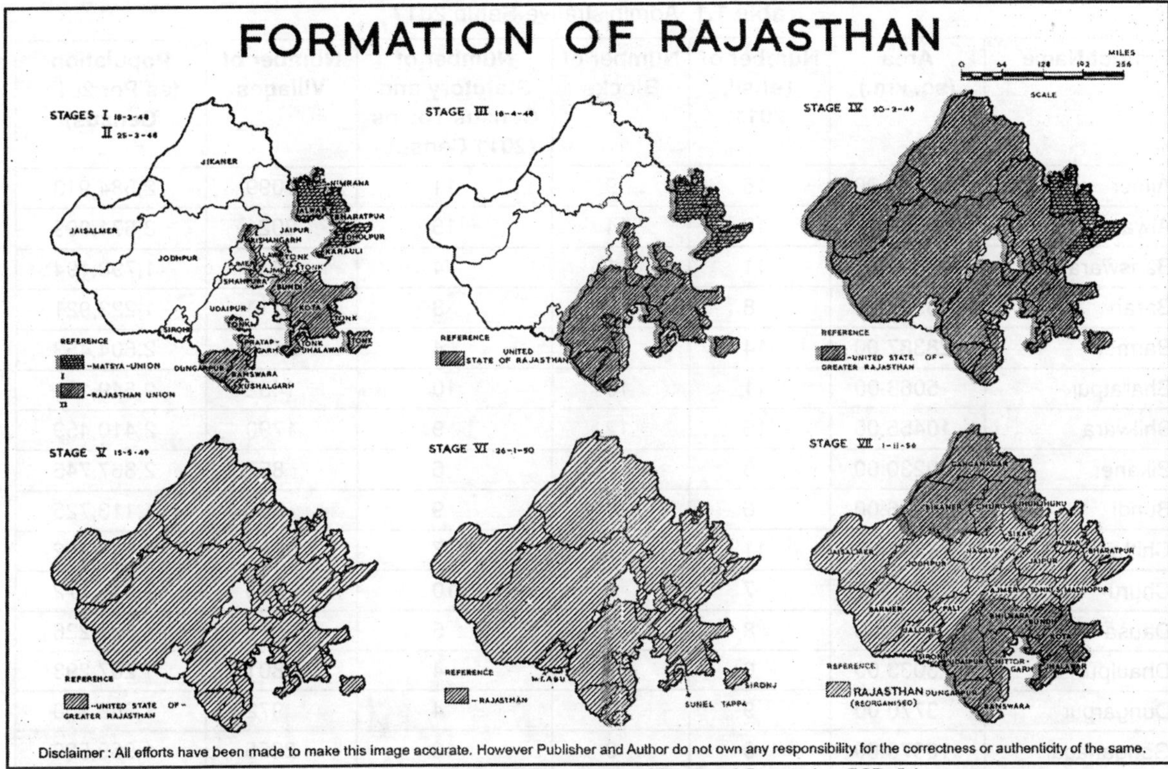

Source: Revenue Board, Government of Rajasthan, Ajmer and Department of Personnel Secretariat, GOR, Jaipur

Figure 1.2 Formation of Rajasthan

The exploitation of land resources at an accelerated rate is a recent phenomenon promoted by the -diffusion of technology to meet human needs in recent decades, and though limited to certain favourable areas, it protends an un-equitable distribution and consumption pattern of resources in the state. Exploitation of non renewable resources at a faster rate, merely to meet immediate necessities and to alleviate man-made maladies, reveals a myopic view. A closer examination needs be made of the state's resources with due consideration of their rate of exploitation.

Rajasthan has exhibited spectacular progress in several areas like agricultural production, harnessing of mineral resources, development of means of transport and communication, and the production of energy resources but the rate of progress and plans of economic development have been slowed to a large extent by a parallel growth of human population and livestock. Human resources are, therefore, to be geared into the channel of economic progress while taking care to arrest their future growth-rate.

The rich wealth of non-renewable resources is yet to be explored and exploited. Their judicious exploitation can make the state economically self-sufficient. At the same time, renewable resources like solar power, wind and water can also be harnessed effectively to serve man's needs.

ADMINISTRATIVE SETUP

For administrative purpose the state has been divided in 33 districts, 314 tehsils and 297 Panchayat Samiti (Development blocks). District wise list of Panchayat Samiti is presented in Table 1.1 and Figure 1.3.

Table 1.1 Administrative Setup 2017

District Name	Area (sq. km.)	Number of Tehsil, 2011	Number of Blocks	Number of Statutory and Census Towns (2011 Census)	Number of Villages	Population (as Per 2011 Census)
Ajmer	8481.00	16	9	11	1099	2,584,913
Alwar	8380.00	16	14	16	2021	3,671,999
Banswara	4522.00	11	11	4	1485	1,798,194
Baran	6922.00	8	7	8	1114	1,223,921
Barmer	28387.00	14	17	4	2457	2,604,453
Bharatpur	5066.00	11	10	10	1432	2,549,121
Bhilwara	10455.00	16	12	9	1790	2,410,459
Bikaner	30230.00	8	7	6	857	2,367,745
Bundi	5776.00	6	5	9	867	1,113,725
Chittaurgarh	7822.00	11	11	7	1581	1,544,392
Churu	13835.00	7	7	10	873	2,041,172
Dausa	3432.00	8	6	5	1079	1,637,226
Dhaulpur	3033.00	6	5	4	801	1,207,293
Dungarpur	3770.00	9	10	4	972	1,388,906
Ganganagar	10978.00	9	9	16	2855	1,969,520
Hanumangarh	9656.00	7	7	6	1831	1,779,650
Jaipur	11143.00	16	15	19	2126	6,663,971
Jaisalmer	38401.00	4	3	2	756	672,008
Jalor	10640.00	9	8	4	791	1,830,151
Jhalawar	6219.00	8	8	9	1488	1,411,327
Jhunjhunun	5928.00	8	8	18	923	2,139,658
Jodhpur	22850.00	13	16	7	1836	3,685,681
Karauli	5524.00	7	6	4	851	1,458,459
Kota	5217.00	6	5	11	805	1,950,491
Nagaur	17718.00	13	14	18	1575	3,309,234
Pali	12387.00	10	10	11	1017	2,038,533
Pratapgarh	4449.00	5	5	3	970	868,231
Rajsamand	4655.00	9	7	9	1037	1,158,283
Sawai Madhopur	4498.00	8	6	4	758	1,338,114
Sikar	7732.00	9	9	15	1162	2,677,737
Sirohi	5136.00	5	5	9	470	1,037,185
Tonk	7194.00	8	6	8	1116	1,421,711
Udaipur	11724.00	13	17	17	471	3,067,549
RAJASTHAN	**342239.00**	**314**	**297**	**297**	**43264**	**68,621,012**

Source: Census of India, 2011 and Statistical Year Book of Rajasthan, 2017

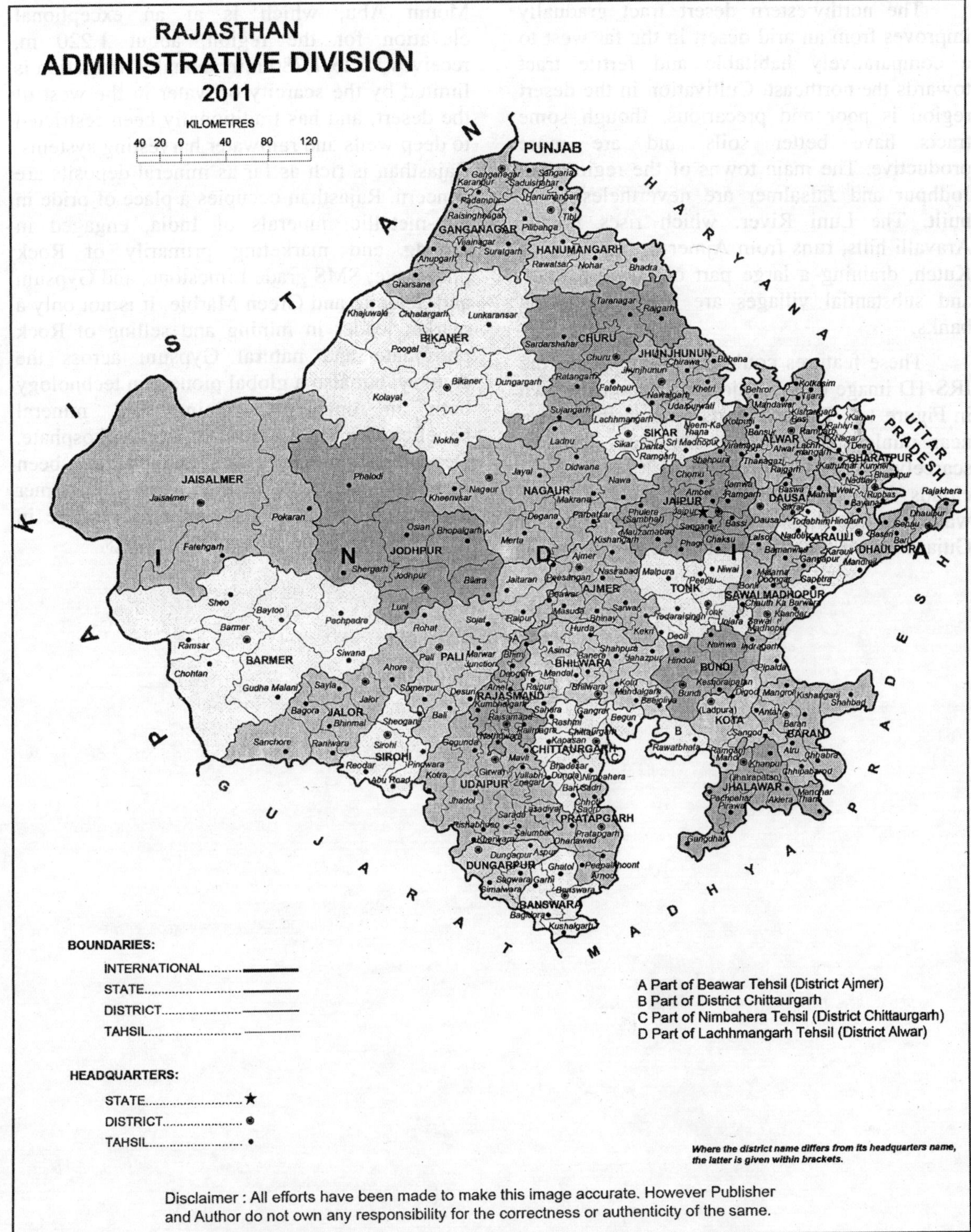

Source: Census of India, 2011 and Statistical Year Book of Rajasthan, 2017

Figure 1.3 Rajasthan : Administrartive Division, 2011

The northwestern desert tract gradually improves from an arid desert in the far west to a comparatively habitable and fertile tract towards the northeast. Cultivation in the desert region is poor and precarious, though some tracts have better soils and are more productive. The main towns of the region like Jodhpur and Jaisalmer are nevertheless well built. The Luni River, which rises in the Aravalli hills, runs from Ajmer to the Rann of Kutch, draining a large part of the arid tract, and substantial villages are found along its banks.

These features could be visualized in the IRS-1D image of the Rajasthan State as shown in Figure 1.4. The arid parts of Rajasthan are a near rainless desert. In some areas, rainfall scarcely averages more than 120 mm. Rain clouds from the oceans empty most of their water on the high ranges in Kathiawar in Gujarat or on the slopes of the Aravali range. Mount Abu, which is at an exceptional elevation for the region, about 1,220 m, receives 1,525-1,780 mm annually. Irrigation is limited by the scarcity of water in the west of the desert, and has traditionally been restricted to deep wells and rainwater harvesting systems. Rajasthan is rich as far as mineral deposits are concern. Rajasthan occupies a place of pride in non-metallic minerals of India, engaged in mining and marketing primarily of Rock phosphate; SMS grade Limestone, and Gypsum and Selenite and Green Marble. It is not only a market leader in mining and selling of Rock phosphate and natural Gypsum across the country, but also a global pioneer in technology both in open-cast mining and mineral beneficiation of Carbonate Rock phosphate. Depositions of natural gas and oil have been identified recently in the areas of Barmer district making Rajasthan as a focus state in terms of mines and geological settings.

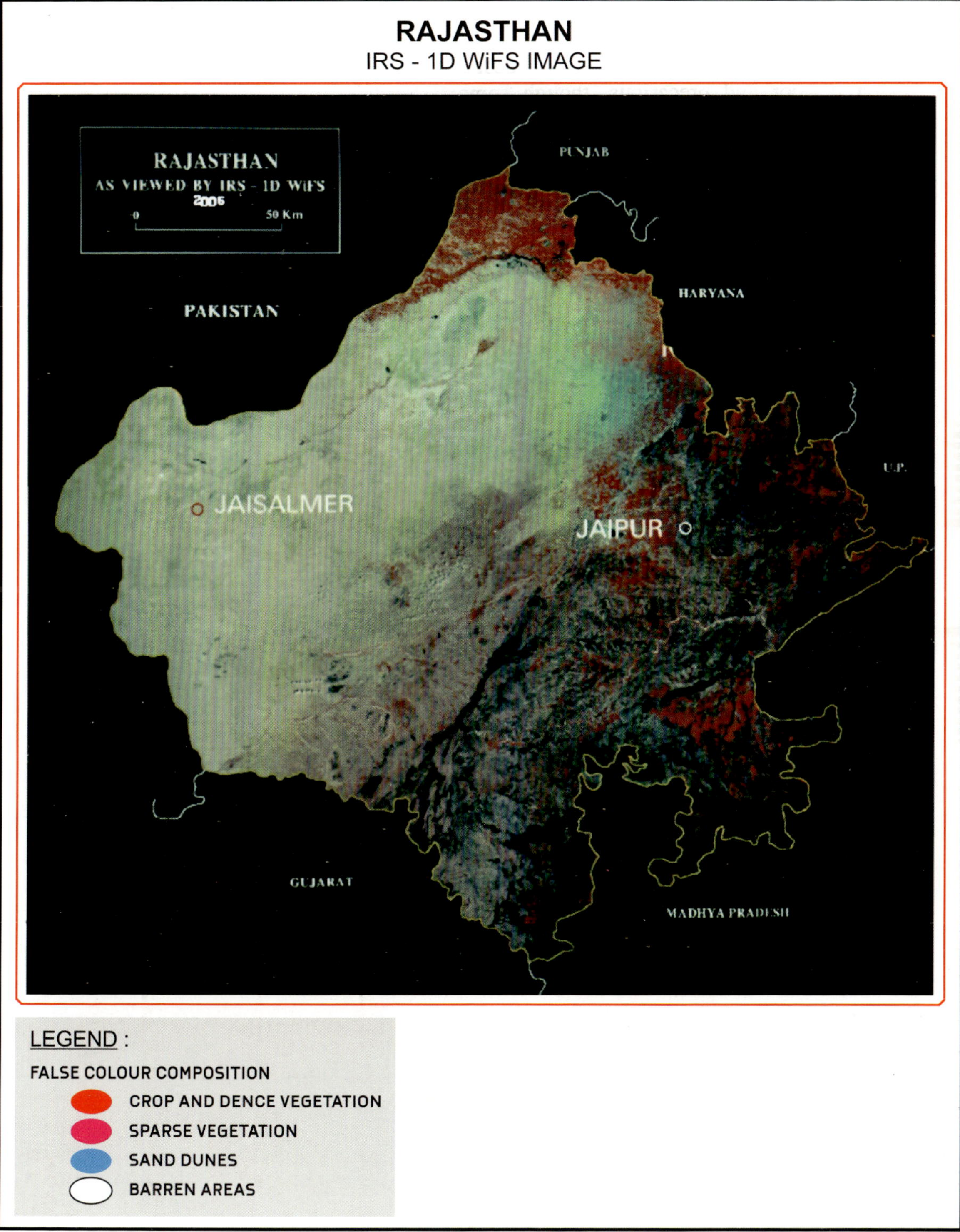

Source: NRSA, Hyderabad

Figure 1.4 Rajasthan Image Map

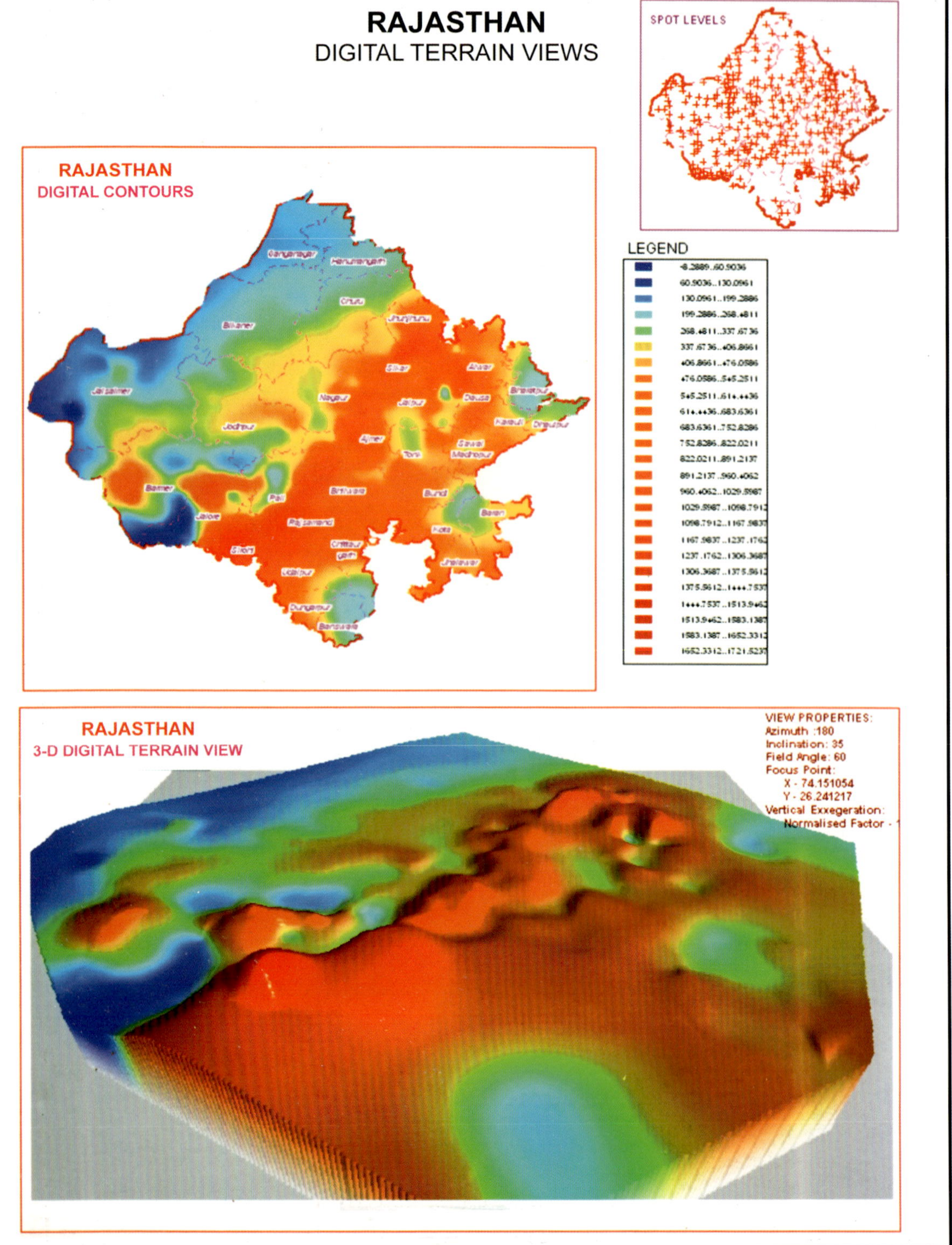

Source: Survey of India (SOI).

Figure 1.5 Digital Terrain Model of Rajasthan

A pictorial 3D view of generated from Digital Terrain Model of Rajasthan from different angles is shown in Figure 1.6. The cross-sectional profiles generated from the DTM along major directions are shown in Figure 1.6.

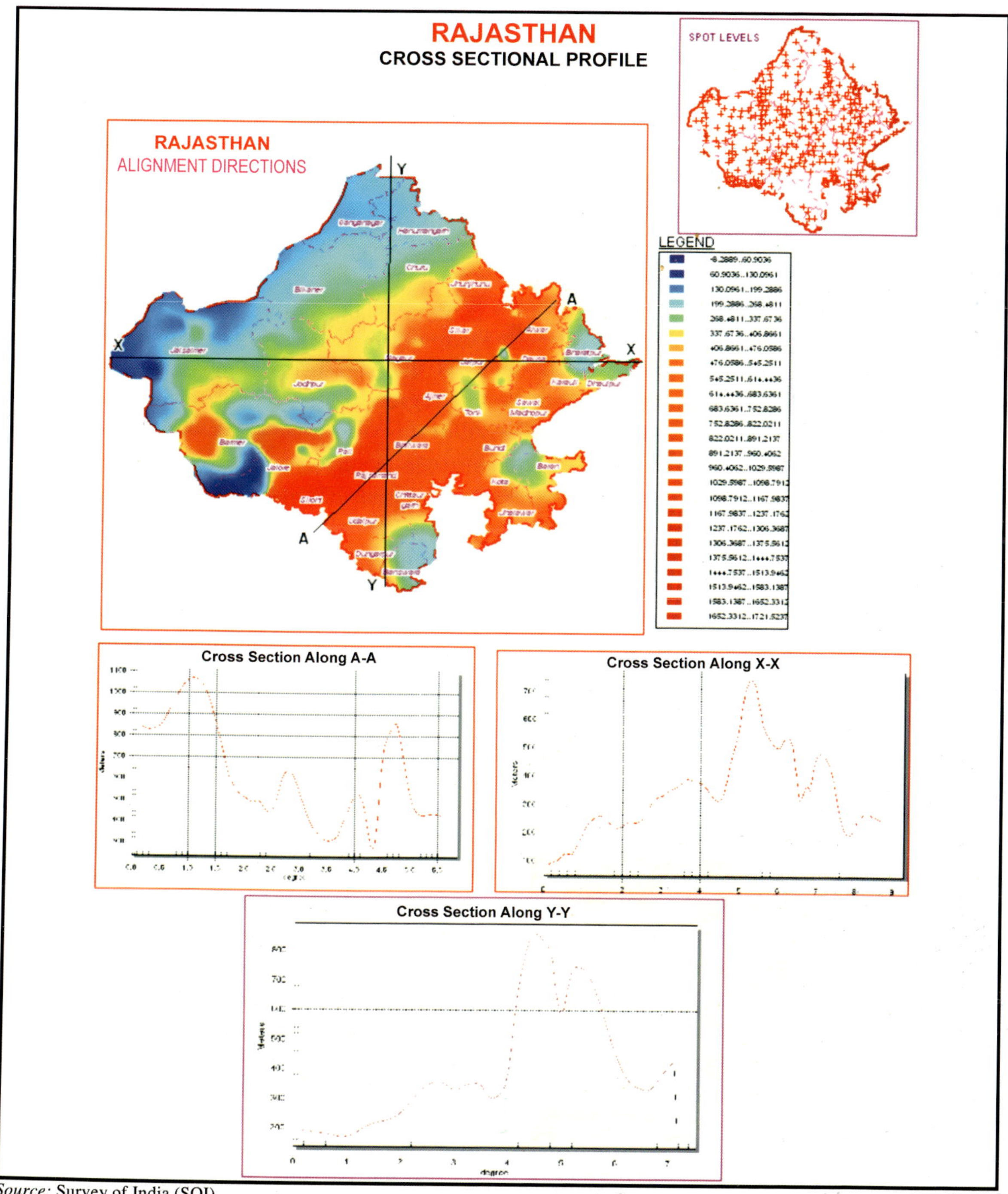

Source: Survey of India (SOI).

Figure 1.6 X-Sectional Profile of Rajasthan

Chapter 2

GEOLOGY

Rajasthan is endowed with a continuous geological sequence of rocks from the oldest Archaean Metamorphites, represented by Bhilwara Super Group (more than 2,500 million years old) to sub-recent alluvium and wind blown sand. The western and north-western parts of the state are covered by vast blanket of young unconsolidated deposits ineluding the blown sand of the Thar Desert (Marusthal) of western Rajasthan. The remaining area exposes wide variety of hard rocks which include various types of metamorphic schists, quartzites, marbles and gneisses of Pre-Cambrain age with associated acid and basic intrusive rocks. The sedimentaries include the rocks of Aravalli Super group, Delhi Super group, upper Precambrian Vindhyan Super group and of Cambrian to Jurassic, Cretaceous and Tertiary ages. south-eastern extremity of the state is occupied by a pile of basaltic flows of Deccan Traps of Cretaceous age. Several mineral deposits of economic importance occur in association with the above rock units.

The geological sequence of the state is highly varied and complex, revealing the co-existence of the most ancient rocks of Pre-Cambrian age and the most recent alluvium as well as wind-blown sand. The Aravallis, one of the most ancient mountains in the world, have the oldest granitic and gneissic rocks at their base, overlain by the rocks of the Aravalli Super group, Delhi Super Group, the Vindhyan Super group and younger rocks. These rocks are highly metamorphosed at certain places and show rich occurrences of minerals of great commercial importance. The characteristic feature of the geology of The characteristic feature of the geology of Rajasthan is the presence of several groups of rocks belonging to Archaean and Pre-Cambrian ages. They form the Aravalli mountain system which runs across the state from the north of Delhi in the north-east to the Gulf of Cambay in the south-west. The central part of the Aravalli ranges is occupied by a great synlinorium composed of Aravalli and Delhi rocks. Because of the thin deposits of sand in this region, the rock exposures are good but in the alluvium and desert sands.

The Archaeans consist of the Bhilwara : Super Group (Bundelkhand Gneiss and the Banded Gneissic complex). The Aravallis, an enormously thick series ofargillaceous rocks, came into existence at the close of the Archean era when the sediments which were deposited in the seas of that age, underwent an upheaval by orogenic activities. These vast mountains were peneplaned in pre-Cretaceous time but were again rejuvenated in later ages. The Aravalli Super group is a vast formation composed of basal quartzites, shales, conglomerates, composite gneisses and slates.

The Delhi Super group overlies the Aravallis. Delhi Super group is divided into lower Railo group, middle Alwar group and upper Ajabgarh group. Railo group is rich in crystalline limestones, grits, schistose There was an encroachment of an arm of the sea from the south-Westernly direction into western Rajasthan during the Jurassic period. Jurassic formations are distinctly noticeable in a vast area around Jaisalmer and some of the fossils, of this age are found in these rocks. There Was an encroachment of an arm of the sea from the south-was ternly direction in to Western Rajasthan during the jurassic period. Jurassic formation are distinctly noticeable in a vast area around Jaisalmer and some of the fossils, of this age are some of the found in these rocks.

The outcrops of these rocks are, partly, covered by wind-blown desert sands. Of special interest are the Bap (Jodhpur district) and Pokran (Jaisalmer district) beds of upper Carboniferous age which have now been exploited for ground water. They are composed of boulders of Malani rhyolites showing effects of glaciation. Violent volcanic activity in the form of fissure eruptions marked the close of

Mesozoic era in the lower Cretaceous age. The main characteristic of this was a stupendous outburst of covered volcanic energy, resulting into the eruption of thick streams of lava over vast areas. These rock known as Deccan traps, are found in southern and south-eastern Rajasthan. The Deccan trap, extending over a vast area in southern Jhalawar and in the eastern parts of Chittaurgarh and Banswara districts, are notable formations of Upper Cretaceous to Lower Eocene age when a large area of peninsular India was also covered with fissure eruptions of During Eocene times, marine transgression seems to have inundated a large part of western Rajasthan with the deposition of thick beds of fossilliferous limestone. To the north of Jaisalmer, the Jurassics are overlapped by nummulitic limestone. Pleistocene sandy alluvium, blown sand, kankar (calcium nodules), carbonate bed and evaporite deposits ofirecent and sub-recent age are found over a large area of western and eastern Rajasthan. The Great Boundary Fault through which the River Chambal has carved its course, passes through south-eastern parts of the state. This fault is visible in Begun (Chittaurgarh district) and northern parts of Kota. It reappears again in Sawai Madhopur, Karaulli and Dhaulpur districts. Besides this, several mega lineaments also traverse in the state (Figure 2.1 and Table 2.1).

Table 2.1 Generalized geologic succession (after Heron, 1953) and lithology of Rajasthan

Recent and Sub recent		Alluvium and blown sand
Vindhyan System		Sandstone, limestone & boulders
	Malani Rhyolites	Granite, Ultrabasic rock, Erinpura granite, pergmatite, aplite Epidiorite and hornblende schist
Delhi System	Ajabgarh Series	Upper phyllite, Limestone, Biotitic limestone and calc-schist, Phyllites, biotite schist and composite gneiss
	Alwar Series	Quartzites, Arkose, grit and conglomerates
	Raialo Series	Garnetiferous biotite schist, Limestone (marble), Local basal grit
Aravali System		Impure limestone, quartzites, phyllites, biotite-schist, composite gneiss, Quartzites, grits and local soda-syenites, conglomerate, Local amygdaloids and tuffs
	Banded Gneissic Complex (BGC)	Schists, gneisses and composite gneiss, Quartzites

Source: Geological Survey of India and Ground Water Department, GOR, Jaipur

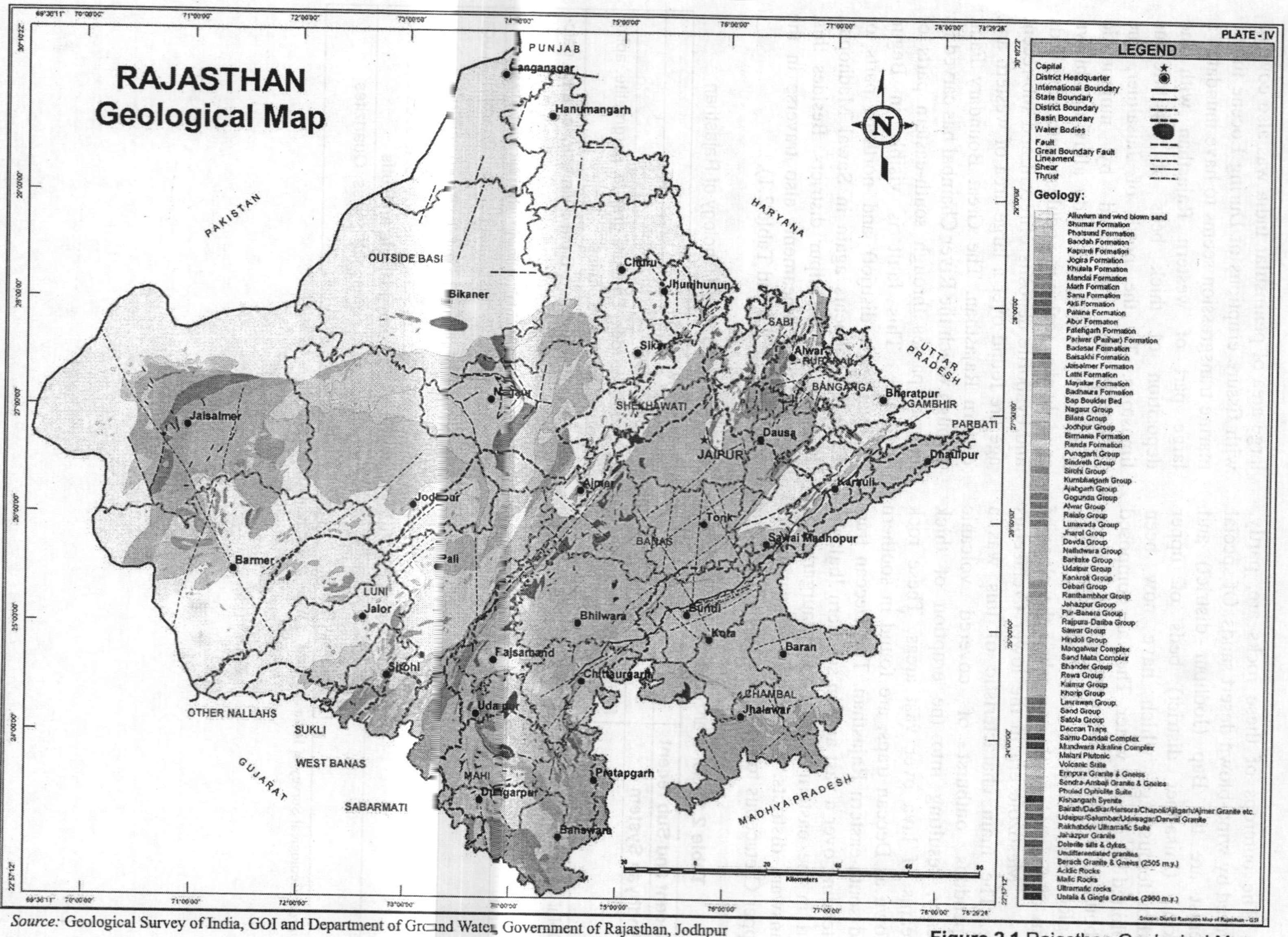

Source: Geological Survey of India, GOI and Department of Ground Water, Government of Rajasthan, Jodhpur

Figure 2.1 Rajasthan Geological Map

ARCHAEAN

The Precambrian rocks of Rajasthan were studied in the initial stages by Hacket (1881), Heron (1917, 1936, 1953), Coulson (1927, 1928) and Gupta (1934). Heron (1936) and Gupta (*et. al.*) classified the Precambrian sedimentary sequences of Rajasthan in the order of increasing antiquity as the Vindhyan System, Delhi System, Aravalli System and Banded Gneissic Complex (pre-Aravallis).

Heron (1953) considered the Banded Gneissic Complex (BGC) as representing the remnants of primordial crust on which the Aravalli sediments, equivalent to the Lower Dharwars of South India, were deposited. However, these pre-Aravalli rocks were not differentiated by the earlier workers. Kurien (1950) advocated that the older schists were granitised to form the gneisses and migmatites (pre-Aravalli) within which occur the quartz-sillimanite schists and calc-granulites as unassimilated lenses of older schists.

Raja Rao (1967) and Raja Rao et al. (1971) attempted for the first time to classify the BGC and designated these rocks as the Bhilwara Supergroup. Gupta *et al.* (1980) synthesised the voluminous ground data generated over decades of pioneering work in Rajasthan and parts of Gujarat and designated pre-Aravalli rocks as the Bhilwara Supergroup.

BHILWARA SUPER GROUP

The pre-Aravallis in Rajasthan, hitherto known as the Banded Gneissic Complex, have been regrouped into the Bhilwara Supergroup which form the oldest stratigraphical horizon around/ upon which the entire geological succession of Rajasthan has developed. On the basis of tectonoen-vironmental setting, magmatic events, lithological homogeneity and order of superposition, the rocks of Bhilwara Supergroup have been classified into the Sand Mata Complex, Mangalwar Complex and Hindoli Group (Archaean); Rajpura-Dariba, Pur Banera, Jahajpur, Sawar and Ranthambhor Groups (Lower Proterozoic). Rocky regions of central Mewar, to the east and southeast, comprise essentially igneous rocks of different types and periods of intrusion which have been welded into a streaky and banded composite gneiss, having wide variations of texture and composition. Banded biotite gneiss and foliated gneissic granites are the most common members of the complex which have been intruded by basic dykes and acid sills of widely variable composition and texture. The basement granite gneisses are intruded by many younger granites, basic and ultrabasic rocks, and support many sedimentary cover sequences. Their geology and age relations are complex and the entire assemblage has been further differentiated into the Sand Mata Complex (granulite facies) and the Mangalwar Complex (amphibolite facies) on metamorphic consideration.

Sand Mata Complex

The rocks of the Sand Mata Complex occur in a belt (200 km long and 50 km wide) from Lambori and Amet in the south to Kishangarh and Patan in the north. The western boundary is marked by the Delhi Supergroup of rocks which overlie the Sand Mata Complex with an overlap. The major part of the contact is, however, faulted. The eastern boundary with the Mangalwar Complex lying to the east of Parbati and passing through Agucha and Jatan, described as a faulted contact, has been rendered obscure by high-grade metamorphic effects and migmatisation. The metamorphic contact between the Sand Mata Complex and the Mangalwar Complex practically coincides with the Delwara Lineament. The present spatial disposition of the rocks of Sand Mata Complex may be due to vertical tectonics. On the southeast, the rocks of the Sand Mata Complex are bordered by the rocks of the Aravalli Supergroup. The metasedimentary assemblage of the Sand Mata Complex is heterogeneous in composition, comprising migmatite, composite gneiss, calc-gneiss, biotite schist, garnet-sillimanite schist, sillimanite-staurolitegarnetiferous biotite schist, hornblende schist, biotitechlorite schist, dolomitic marble, conglomerate, quartzite, para-amphibolite, epidiorite and pyroxene granulite and has been regionally metamorphosed to high-pressure amphibolite to granulite facies, and migmatised to different degrees. The group consists of varied igneous suites like ultramafics, mafics and felsics.

Mangalwar Complex

The heterogeneous assemblage containing pelitic rocks deposited during the orogenic phase of the Bhilwara geological cycle, regionally metamorphosed to amphibolite facies and syntectonically migmatised have been included in the Mangalwar Complex. The complex occurs to the east of Delwara Lineament in three main stretches, the first one extending as a 12 to 48 km wide belt from Pipalkhunt to Mangalwar for over 350 km; the second one occurs from west of Amrita to Dadala over 50 km (Sarara-ki-Pal inlier), and the third one extends from Kunda to Pratappur for over 30 km (Mando-ki-Pal inlier) having a NE-SW trend. The eastern boundary of the complex is marked by the Hindoli Group of rocks with an uncertain stratigraphic relationship. The Mangalwar rocks are unconformably overlain by the Aravalli Supergroup rocks to the west and by the Deccan Traps in the south. The Mangalwar Complex is considered representing an Archaean primary granite-greenstone belt comprising of migmatite, gneiss, feldspathic mica schist, garnetiferous mica schist, sillimanite-mica schist, hornblende schist, quartz-feldspar gneiss, granite gneiss, calc-silicate rock, dolomite, quartzite, impure marble and para-amphibolite. These rocks have been intruded by the Untala-Gingla Granites, Giyangarh-Asind acidic rocks, RaipurJalayan mafic rocks and Ultramafic rocks. The complex has been subdivided into the Lasaria, Kekri, Sarara, Mando-ki-Pal, Suwana, Potla and Rajmahal Formations.

Hindoli Group

The Hindoli Group named after Hindoli village, located 20 km northwest of Bundi, comprises a lowgrade (greenschist), predominantly argillaceous flyschlike sequence deposited during geosynclinal phase of the Bhilwara geological cycle. The rocks of the group occur as a 10 to 50 km wide arcuate belt extending from Gyaspur in the south to Deyi in the north and further north-eastward for 400 km distance, trending NNW-SSE in the southern part, becoming N-S near Chittaurgarh, and further northwards veers to ENEWSW in Hindoli-Deyi area. The group represents a secondary granite-greenstone belt and is bounded by the Vindhyans in the east and by the Deccan Traps in the south. In the west its contact with the Mangalwar Complex is not clear. The group comprises shale, slate, phyllite, mica schist with interstratified sequence of metagreywackes, metalvolca-nics, quartzite, dolomite and limestone. The grade of metamorphism increases gradually towards west, and greenschist fades rocks grade into intermediate pressure amphibolite fades at the contact with the Mangalwar Complex. The rocks are migmatised in Mangalwar-Deoli area.

PROTEROZOIC

The Proterozoic rocks of Rajasthan are included in the Rajpura-Dariba, Pur-Banera, Jahazpur and Sawar Groups (Bhilwara Supergroup); the Aravalli, Delhi, Vindhyan and Marwar Supergroups.

Overlying the Hindoli Group of rocks and the Mangalwar Complex with -an unconformity occur the next younger groups of rocks classified as the RajpuraDariba, Pur-Banera, Jahazpur and Sawar Groups which are exposed in a series of isolated linear belts. All these synformal metasedimentary basins/ structures of Lower Proterozoic age occur in disjointed belts as outliers. They are mostly composed of dolomite, marble, calc-gneiss, calc-schist, and calcbiotite schist; graphite-kyanite- staurolite schist, garnetiferous mica schist; banded chert, banded ferruginous chert and quartzite. These groups of rocks form important repositories for deposits of copper, zinc, lead, silver, iron ore, marble, kyanite, steatite, quartz, corundum and garnet.

Rajpura-Dariba Group

The sulphide-bearing predominantly chemogenic rocks, admixed with minor elastics, lying unconformably over the rocks of the Mangalwar Complex and occurring from Bhinder in the south to northwest of Dhor, over a strike length of 100 km as a 2.5 km wide belt have been assigned to the RajpuraDariba Group. The chief lithounits of the group are sulphide-bearing graphitic mica schist, dolomitic marble, calc-biotite schist, actinolite schist, garnetiferous mica schist and quartzites. The metasediments exhibit polyphase deformation and regional metamorphism corresponding to almandineamphibolite fades. The sulphides of zinc, lead, copper and iron occur as thin bands and layers exhibiting characteristic syngenetic bedded geometry and show relict sedimentary structures. The lithounits have been classified into the Bhinder, Malikhera, Dariba, Sindesar and Satdudhia Formations.

Pur-Banera Group

The dominantly chemogenic sequence with bands of elastics occurs from south of Banera to Samodi in the west, for over 80 km in a 3 km to 12 km wide belt. These rocks overlie unconformably the rocks of the Potla Formation of the Mangalwar Complex and have been included in the Pur-Banera Group. The sequence comprises conglomerate, garnetiferous mica schist, calc-schist, amphibolite schist, calc-gneiss, dolomitic marble, calc-silicate marble, magnetite quartzite (a persistant marker horizon), banded magnetite chert and carbonate rock with associated sulphide mineralization. These rocks exhibiting superposed deformation and regional metamorphism corresponding to amphibolite fades, have been grouped into the Pur, Pansal, Rewara, Tiranga and Samodi Formations.

Jahazpur Group

Jahazpur Group comprises volcanic, elastic and chemogenic package occurring 2 km southeast of Nandrai in Bhilwara district to Naenwa in Tonk district running over 110 km as a 0.5 km to 8 km wide linear belt. The different lithounits occur in three subparallel stretches: (1) extending from Nandrai to Jhikri, (2) from Katralia to Jahazpur and (3) from Utrana to Naenwa, and have been grouped under the Chuleswarji, Jhikri and Umar coeval Formations. The rock types of the group comprise conglomerate, quartzite, gritty quartzite, mica schist, ferruginous chert and breccia, dolomitic marble, carbonaceous phyllite and metavolcanics. The rocks show low-grade metamorphism (greenschist fades), and unconformably overlie the Sujanpura Formation of the Hindoli Group.

Sawar Group

The chemogenic and elastic rocks exposed as isolated outcrops over 13 km long and 3 to 5 km wide area between Bajata in the north and Sawar in the south, have been designated as the Sawar Group. It comprises quartzite, sulphide-bearing dolomitic marble, calc-silicate marble, mica schist, biotite schist, phyllite and carbonaceous phyllite. These rocks unconformably overlie the Kekri Formation of the Mangalwar Complex. The group has been divided into the Morhi and the Ghatiali Formations.

Ranthambor Group

It comprises quartzite and shale with intercalations of slate and phyllite and is considered to be the youngest member of the Bhilwara Supergroup. The molasse-like sediments exhibiting low-grade metamorphism occurring as isolated outcrops over a distance of 200 km from southeast of Bari Sadri to north of Hora and Mandalgarh have been included in the Ranthambhor Group. It comprises quartzite, slate and phyllite intruded by concordant dolerite sills. The rocks occur as synformal hills and overlie the Bhadesar Formation of the Hindoli Group with structural discordance. Three coeval formations viz., The Bari Sadri, Hora and Mandalgarh have been recognised on the basis of their spatial disposition in isolated basins of deposition.

ARAVALLI SUPER GROUP

A thick pile, chiefly comprising metamorphosed and complexly folded clastic sediments with minor chemogenic and organogenic assemblages and interlayered basic volcanics, overlying the Mangalwar Complex and the Sand Mata Complex of the Bhilwara Supergroup with an erosional unconformity, has been assigned to the Aravalli Supergroup. This assemblage of stratified metasediments and interlayered extrusives, together with synorogenic and late- to postorogenic acidic, basic and ultrabasic rocks, covers a time span from 2,500 Ma to 2,000 Ma.

The rocks of the Aravalli Supergroup are wellexposed over a stretch of 350 km. The width of the belt in the north is about 40 km gradually fanning out to 150 km in the south. The eastern margin, marking the first order erosional unconformity with the underlying Bhilwara Supergroup, represents one of the most ancient strand line. The present disposition of the strand line is arcuate in shape, with a NW-SE trend in the south, gradually veering to NE-SW in the north. On the west, the Aravalli Supergroup is overlain by the rocks of the Delhi Supergroup and towards south, its continuity is lost below the cover of Deccan Traps and Tertiary sediments. The rocks have undergone polyphase deformation and attained progressively higher grade of regional metamorphism reaching up to amphibolite facies. The rocks also show migmatisation in the vicinity of the synorogenic plutonic activity.

The boundary between the Aravalli and the preAravalli rocks is marked by the development of a thin quartz-arenite bed, overlain by a suite of synsedimentational basic metavolcanics followed up in the sequence by a coarse pebble polymict metaconglomerate, coarse meta-arkose, feldspathic quartzite, quartzite, dolomite, calcarenite and carbonaceous phyllite. This basal sequence, indicating a near-shore shelf sedimentation along the cratonic border, has been designated to as the *Debari Group.* The basal sequence is followed upward by a flysch (proximal turbidite) sequence designated as the *Udaipur Group,* and this is overlain by a distal shaly flysch-like sequence comprising chloritic phyllite and quartzite, representing the depositional episode in the Aravalli geosynclinal trough, named as the *Jharol Group.* Between the turbidite sequence and shaly-flysch again synsedimentational basic volcanics associated with epiclast are seen along a N-S trending lineament and are designated as the *Bari Lake Group.* A sequence of phyllite, mica schist, metasubgreywacke, quartzchlorite schist, metasemipelite minor dolomitic lenses, manganiferous phyllite, exposed in southern Rajasthan, has been included in the *Lunavada Group.* This assemblage represents a molasse facies developed after the main paroxysm of the Aravalli orogeny when the deeper trough was transformed to shallow-basins.

The dominantly calcareous rocks to the north of the Banas Lineament, representing a shallow-marine environment (epicontinental realm) of deposition and migmatites are included into the *Kankroli Group* and the *Dovda Group.* These two groups, considered to be coeval with the Udaipur and Bari Lake Groups, and the Jharol Group respectively, are exposed in the area south of Banas Lineament. Dolomitic marble and other calcareous units along with conglomerate, quartzite and carbonaceous phulite of ghasar-rama-gurali-nathwara area, representing shallow-water facies, have been desigenated to as the nathwara group. The rocks of the aravalli Supergroups are intruded by the synorogenic (aravalli orogeny) granites in Udipur, Udaidagar, Darwal and Jaisamand areas, in Jaisamand and Banswara areas the rocks of the Udaipur Group are migmatised. Similarly, the rocks of the Lunavada Group also show migmatisation.

Debari Group

Clastic, chemogenic and biogenic coastal and shelfsediments deposited marginal to Mangalwar, Sararaki-Pal and Mando-ki-Pal cratonic masses with first order erosional unconformity and associated synsedimentational shore-line basic volcanics corresponding to an early depositional and volcanic episodes of the Aravalli geological cycle, have been included in the Debari Group. It comprises a sequence of conglomerate, meta-arkose, quartzite, phyllite, mica schist, basic metavolcanics with associated pyroclastics, calcareous quartzite, dolomitic limestone, dolomite, calcitic marble, ferruginous chert, algal phosphatic dolomite and chert, and carbonaceous and manganiferous phyllite. The entire sequence has undergone polyphase deformation and regional metamorphism to greenschist facies.

The rocks of the Debari Group forming a NW-SE trending chain of high hills in sharp contrast to the peneplained tract of Mewar to the east, can be traced between Pakhand to the north and Waori Khera to the south over a distance of 160 km. Beyond Waori Khera the continuity of these rocks is lost below the Deccan Traps. The Debari Group is further exposed in the Sarara-ki-Pal and Mando-ki-Pal areas. Lithostratigraphy of the Debari Group has been worked out separately in Debari, Jaisamand, Ghatol and Sarada sectors.

Udaipur Group

A thick sequence of flysch-like coarse- to finegrained rocks associated with minor bands of chemogenic/biogenic rocks representing the main depositional episode during the geosynclinal phase of the Aravalli geological cycle, has been included in the Udaipur Group.

Lithologically, the Udaipur Group comprises a thick pile of phyllite, metagreywacke, feldspathic mica schist and migmatite with intercalatory bands of greywacke, conglomerate, quartzite, dolomite, gritty dolomite, dolomitic marble, oligomictic conglomerate, phosphatic and stromatolitic dolomite, carbonaceous, manganiferous and ferruginous phyllite, chert and amphibole schists. The rocks show frequent interfingering and gradational relationships of different litho-units. The argillaceous metasediments constitute the bulk of the group. The rocks have undergone superimposed deformation and show a grade of metamorphism ranging from greenschist fades to amphibolite fades associated with syntectonic migmatisation. The rocks of the Udaipur Group can be traced from northwest of Eklingjee in the north to Santrampur in south, over a distance of 180 km in a NNW-SSE trending broad belt. The outcrop of the group is narrow between Eklingjee and Sapetia from where it widens to 15 km in the Udaipur valley and extends southwards up to Sarara-ki-Pal Inlier where the outcrop bifurcates. The eastern one extends southeast-wards through Salumber, Ganora and Talwara to Banswara where its continuity is lost below the Deccan Traps and the Lunavada Group. The western outcrop follows a N-S trend and extends southwards through Paduan, Rakhabdev, Dungarpur, Peit and Vereth to Santrampur where it is overlapped by the rocks of the Lunavada Group.

The rocks of the Udaipur Group have been divided into the Sabina Formation, Eklinggarh Formation, Balicha Formation, Neemach Mata Formation and Banswara Formation in the Debari sector while Mandli, Baroi Magra and Zawar Formations are recognised in the Sarada sector.

Bari Lake Group

An assemblage of coarse- to fine-grained elastic sediments, synsedimen-tational basic volcanics, pyroclastic and volcaniclastic rocks overlying the Udaipur Group has been designated as the Bari Lake Group. This volcano-sedimentary association marks the local upwarping and shallowing of the trough and contemporaneous activation of a N-S trending deepseated fracture (Rakhabdev Lineament) resulting in the outpouring of basic lavas.

The Bari Lake Group is principally composed of basic metavolcancis viz., actinolite-chlorite schist and chlorite schist, metamorphosed pyroclastics, quartzpebbleconglomerate, meta-arkose, quartzite, chloritephyllite and metasiltstone with minor bands of dolomite, chert and carbonaceous phyllite. The rocks have undergone polyphase deformation and lowgrade regional metamorphism of greenschist facies.

These rocks occur over a distance of 30 km between Kharpena in the south to Madar in the north through Bari Lake as a 3-5 km wide belt. Further eastwards and northeastwards the group gradually widens to 15 km and can be traced up to Khamnor where it is terminated against the E-W trending Banas Lineament.

The coarser clastic rocks have been included in the Sajjangarh Formation while the overlying volcanosedimentary rocks are included in the Varla Formation. A thick sequence of phyllite and mica schist, around Khamnor, Batia, Kadmas and north of Iswal occupying area in continuity of the Bari Lake metavolcanics, and representing the metamorphosed products of fine grained pyroclastic and volcanic rocks, has been designated as the Khamnor Formation and is included in the Bari Lake Group. The Formation is overlain by a persistent quartzite horizon considered to be the base of the overlying Jharol Group.

Kankroli Group

A complexly folded and metamorphosed assemblages of rocks of argillaceous and calcareous facies developed under shallow-marine conditions around Kankroli has been included in the Kankroli Group. The rocks appear to have deposited on a topographic high developed towards the northern extremity of the Udaipur-Zawar trench due to the reactivation of the Banas Lineament.

The bulk of the Kankroli Group consists of garnetiferous muscovite-biotite schist, hornblende schist, hornblende-biotite schist, hornblende gneiss and dolomitic marble with minor intercalations of quartzite. These rocks have undergone polyphase deformation and regional metamorphism to amphibolite facies conditions.

The rocks of the Kankroli Group are exposed as a hammer-head shaped arcuate outcrop between Sangat on the west and Pipli on the east. The Kankroli sequence is surrounded on all sides by the younger rocks of the Dovda Group except along its southern margin where it is in contact with the older members of the Debari Group. Upper units of the Kankroli Group show interfingering relationships northeast of Nathdwara with the rocks of the Bari Lake Group.

The rocks of the group are divided into the Madra, Morchana, Rajnagar, Puthol and Sangat Formations. The Madra, Morchana and Rajnagar Formations have been considered coeval to the Udaipur Group while the Puthol and Sangat Formations are equivalent to the Bari Lake Group.

Jharol Group

A thick sequence of almost carbonate-free, shaly, flysch-like elastic sequence representing the orogenic phase of the Aravalli geological cycle has been assigned to the Jharol Group. This predominantly argillaceous suite with minor arenites, deposited in the distal axial portion of the Aravalli geosyncline, is designated as the Jharol Group and considered to have developed, perhaps, consequent to the reactivation of the Rakhabdev Lineament. The bulk of the group is made up of phyllite of different types, chlorite schist, garnetiferous mica schist, intercalatory quartzite and minor lenses and bands of calc-schist and marble. Intercalations of quartzite are frequent and thicker in the upper portion of the group. The rocks have undergone polyphase deformation and show a progressive increase in the grade of metamorphism from lower greenschist fades in the east to lower amphibolite fades in the west. Towards the east, the Jharol Group overlies volcanosedimentary association of the Bari Lake Group and is overlain by the rocks of the Delhi Supergroup in the west.

The rocks of the Jharol Group are exposed in a N-S trending linear belt, extending from Modasa in the south to Paraoli in the north, over a distance of 190 km. The belt is about 40 km wide in the south gradually decreasing to 10 km near Butala from where it runs as a 2-5 km wide strip up to Paraoli.

The Goran Formation, predominantly argillaceous in character, and the overlying Samlaji Formation which is also argillaceous but with frequent thicker intercalatory sandy layers and minor calcareous layers, have been identified in the Jharol Group.

Dovda Group

An association of chemogenic and elastic rocks showing attributes of shelf-type sedimentation in a shallow-epicontinental basin has been included in the Dovda Group. This group is chiefly composed of an interbanded assemblage of hornblende schist, hornblende-biotite schist, biotite schist, garnetiferous biotite schist, amphibolite gneiss, quartz-biotite gneiss and migmatite with numerous intercalatory bands of calc-silicate rock, dolomitic marble and quartzite. The assemblage as a whole has undergone polyphase deformation producing complex fold patterns and regional metamorphism under amphibolite facies conditions accompanying syntectonic migmatisation.

The rocks of the Dovda Group are exposed north of the Banas Lineament surrounding the rocks of the Kankroli Group. The outcrop extends from Masind in the southwest to Khandel in the northeast. The Dovda Group overlies the Kankroli Group in the central part, the Bari Lake Group in the southern part and shows interfingering and gradation relationship with the Jharol Group (at Paroti). It is overlain by the rocks of the Delhi Supergroup in Richer and Kuanthal areas. Depti and Devthari Formations have been recognised in the Dovda Group.

Nathdwara Group

Predominantly chemogenic rocks associated with minor clastic sediments deposited in shallow-inland basins have been assigned to the Nathdwara Group. The group is temporally correlated with the Jharol and Dovda Groups and rests on the Debari, Udaipur and Bari Lake Groups.

The bulk of Nathdwara Group in the eastern tract comprises phyllite and calcareous phyllite with numerous intercalatory bands of dolomite and calcitic marble. Thick bands of quartzite occur towards the base. The Nathdwara sequence has undergone polyphase deformation and regional metamorphism to greenschist facies.

The rocks of the Nathdwara Group occur in two separate areas. The eastern tract lies between Rama in the south and Berach in the north while the western tract extends from Madar in the south to Kotharia in the northwest through Iswal and Nathdwara.

The Kadmal and the Rama Formations have been identified in the eastern tract. The most abundant rock type in the western tract is dolomite/ dolomitic marble with thin intercalatory bands of ferruginous chert, quartzite and mica schist, referred to as the Haldighati Formation.

Lunavada Group

A characteristic assemblage comprising thick accumulation of clastic metasediments with minor bands of chemogenic and biogenic rocks developed in southern Rajasthan during late to post-orogenic phase of the Aravalli geological cycle has been referred to as the Lunavada Group. With excellent preservation of sedimentary structures the rocks represent a shallow-marine environment of deposition perhaps in the intertidal, beach and shelf zones with a regression of coastline from NE to SW. Structural discordance and typical coarser lithology separate this mollassetype assemblage from the flysch-like and volcanosedimentary asemblages o f the Udaipur, Jharol and Bari Lake Groups. The bulk of the Lunavada Group is composed of a sequence of phyllite, mica schist, quartz-chlorite schist, metasubgreywacke and quartzite with minor bands and lenses of dolomitic limestone, conglomerate, manganiferous phyllite and phosphatic algal dolomite. At least two phases of deformation have affected the Lunavada rocks resulting in the complex folding of the quartzite bands and formation of dome and basin structures. The rocks generally show greenschist facies of metamorphism. The continuity of the group is either terminated by the intrusive Godhra Granite or lost below the Deccan Traps. The rocks of the Lunavada Group have been divided into the Kalinjara, Wagidora, Bhawanpura, Chandanwara, Bhukia and Kadana Formations.

The notable igneous events of this period are the intrusion of the Udaipur, Salumber, Udaisagar and Darwal granites. The geosynclinal phase of basic and ultrabasic intrusion include metabasalts and ultrabasics. The Rakhabdev ultrabasic suite comprising serpentinite, talc-carbonate rocks and the talc-chlorite schist represents the geosynclinal phase.

DELHI SUPER GROUP

The metamorphosed terrigenous and marine clastics with subordinate chemogenic sediments and temporally related concordant and discordant intrusives and their extrusive phases unconformably resting over the rocks of the Bhilwara Supergroup in northeast Rajasthan have been included into the Delhi Supergroup.

The rocks of the Delhi Supergroup are traceable as a NE-SW trending rectilinear belt from Delhi in the northeast to Himmatnagar in the southwest over a strike distance of about 850 km. These form the Aravalli orographic axis separating the plains of Marwar from the hilly tract of Mewar in southwest Rajasthan. The continuity of the rocks in the northeast is punctuated by the sandy tract of Rajasthan desert where to the north of Sambhar lake these continue as ridges into the main expanse of Delhis in northeast Rajasthan. In the southwest, beyond Himmatnagar, the solid geology is lost under the cover of alluvial blanket.

The Delhi Supergroup comprises dominantly quartzite, biotite schist, calc-schist, calc-gneiss, and marble with subordinate conglomerate and chert. Metavolcanics, granite, ultramafics, basic flows, epidiorite, hornblende schist, amphibolite and gabbro occur as associated intrusives and extrusives. These rocks exhibit polyphase deformation and regional metamorphism varying from greenschist facies to amphibolite facies.

In Rajasthan and in contiguous parts of North Gujarat the Delhi sedimentation took place in two depositional regimes. The deposits in the north of the Rajgarh-Pisangan line (Ajmer and northeast Rajasthan) represent beach and shelf-type sedimentation which has been classified into the Raialo, Alwar and Ajabgarh Groups. South of Rajgarh-Pisangan line the deposition took place on shelf-margin to shelf-interior, main geosynclinal trough and in the post-orogenic successor basins. The dominantly arenaceous and argillaceous facies deposited on shelf-margin has been referred to as the Gogunda Group, the mixed calcareous and argillaceous facies of the shelf-interior have been included in the Kumbhalgarh Group. The clastic sediments of trough-fill have been included in the Sirohi Group and the molasse sediments of the successor basins have been accommodated in the Punagarh and Sindreth Groups. Besides the sediments, the magmatic phases of Phulad Ophiolite Suite, synsedimentational flows and tuffs and synorogenic granite, have also been included in the stratigraphy.

Raialo Group

The name 'Raialo Series' was introduced by Heron (1917) to designate a predominantly calcareous sequence with subordinate quartzite occurring between the Aravalli and the Delhi Supergroups. Heron (1953, p.22) correlated these calcareous formations occurring in widely separated outcrops in parts of Ajmer, Alwar, Udaipur and Bhilwara districts. He assumed that the Raialo rocks, which had no equivalent anywhere else in India, were deposited during the 'Eparchaean Interval'. The correlation of geographically isolated outcrops of the so called 'Raialos' of Heron (*op. cit.*) is hardly tenable and the type Raialo Group having a restricted geographic spread is developed only in Jaipur, Dausa and Alwar districts. The entire sequence of the Raialos, mainly developed around Baldeogarh and Tehla, gradually thins out and disappears in strike extension. This suggests its sedimentation in a localised basin during the 'Eparchaean Interval'. The Raialo Group, comprising a lower calcareous and an upper arenaceous sequence (viz., the Dogeta and the Tehla Formations respectively) with marble, phyllite and schist is best exposed in the northeast sector around Dogeta, Gola-ka-Bas, Kharar, Jhiri, Tilwarhi, Todi and Kho-Dariba.

Alwar Group

The metamorphosed dominantly arenaceous clastics with subordinate argillaceous and calcareous rocks and interlayered basic volcanics resting unconformably over the Raialo Group have been included in the Alwar Group. Heron (1917) described these rocks as the Alwar Series. The type sections can be examined near Jaipur and at many places in Alwar and Ajmer districts. The generalised sequence of the Alwar Group is a basal conglomerate, arkose and quartzite followed by a sequence of argillaceous and impure calcareous rocks represented by pelitic schists containing different mineral assemblages, impure marble and para-amphibolite which in turn is followed by a sequence of mainly arenaceous rocks with some amount of pelitic sediments. The rocks of the group have been classified into the Rajgarh, Kankwarhi, Pratapgarh, Nithar, Badalgarh and Bayana Formations in northeastern Rajasthan and into the Srinagar and the Naulakha Formations in the Ajmer sector respectively.

Ajabgarh Group

The metamorphosed argillites with intercalated arenites and subordinate carbonates have been included in the Ajabgarh Group. The 'Kushalgarh Limestone' and 'Hornstone Breccia' horizons of Heron (op. cit.) have been grouped under the Ajabgarh Group. These assume the status of formation and have been named as the Kushalgarh and the Sariska Formations respectively. Other rock units of this group include carbonaceous phyllite, interlayered sequence of quartzite, phyllite and schists containing chlorites, garnet, staurolite and andalusite. These are included under the Thanagazi, Bhakrol and Arauli Formations. The rocks of the Ajabgarh Group in the Ajmer area have been given independent formational status and described as the Ajmer Formation.

Gogunda Group

The metamorphosed terrigenous clastics comprising quartzite and interbanded schist with subordinate impure calcareous metasediments deposited mainly in the shelf zones of the Delhi trough have been included in the Gogunda Group. The quartzite and intimately associated schists of the Gogunda area and its strike extension in Mewar were included in the Alwar Series of the Delhi System by Heron (1953). Even though the continuity of these quartzites is punctuated in the northeast near Badnor, an emperical correlation was made by Heron (op. cit.) between the Delhi rocks of southwest Rajasthan (Mewar) with those of Jaipur and Alwar in northeast Rajasthan.

The rocks which were earlier mapped as Lower Delhi Group by Arora and Ramchandran (1971) together with the outliers of calc-schist of Ajabgarh (cf. Heron) have been grouped together under the Gogunda Group. The rocks of the Gogunda Group are traceable from Kishangarh to Richer. These rocks take a sinistral shift near Antalia and then continue southwards (up to Himmatnagar) in Gujarat. The unit attains its maximum width around Gogunda. The rocks of the Gogunda Group have been separated into three depositionally related litho- associations which have been designated as the Richer, Antalia and Kelwara Formations in descending order of antiquity.

Kumbhalgarh Group

The metamorphosed, marine, calcareous sediments with subordinate argillaceous and arenaceous sediments deposited in the shelf interior of the main rift-controlled Delhi geosyncline have been included in this group.

The rocks of this group are traceable as a broad, linear belt occupying the western part of almost the entire length of the main Aravalli range. It runs NESW to NNE-SSW from south of Rajgarh-Pisangan area in the north up to Khed Brahma in the south, over a distance of about 290 km.

Lithologically, this group is represented by a sequence of calc-gneiss and calc-schist, pure and impure calcitic marble, muscovite-biotite schist, calcbiotite schist, migmatite and gneiss, quartzite, conglomerate and amphibolite. The sequence has a general sharp contact with the rocks of the underlying Gogunda Group and at places with the overlying Sirohi Group. The rocks of the Kumbhalgarh Group have been regionally metamorphosed to amphibolite facies and the effects of contact metamorphism are well-marked towards the western side where they are intruded by acidic and basic rocks. The rocks of the Kumbhalgarh Group have been divided into the Todgarh, Sendra, Basantgarh, Beawar, Kotra, Barr, Ras and Kalakot Formations which do not always indicate the order of superposition. The rocks show a general younging towards west, and the Barr and the Ras Formations may represent the younger part of the group.

Sirohi Group

The flysch-like thick deposits of fine argillaceous sediments viz., phyllite, mica schist and biotite schist along with intercalations of arenaceous and calcareous sediments (quartzite and marble) deposited during or terminal orogenic phase of the Delhi geological cycle, have been designated as the Sirohi Group. The rocks occur in the form of a broad belt occupying the plains of Marwar lying to the west of Aravalli range. They have been traced as outcrops from Reodar Chandrawati area in southwest to Gandoch and Kot in the northeast for a strike length of about 160 km over a width of about 30 km. The continuity of the belt is either lost under a thick soil/sand cover in the area or interrupted by the intrusive Erinpura Granite. The rocks have been deformed and have been regionally metamorphosed to greenschist facies. Lithologically, the group comprises a thick pile of metasediments consisting predominantly of phyllite, mica schist and biotite schist with intercalated bands of calcitic marble and quartzite. The metasediments of the Sirohi Group have been subdivided, in order of antiquity, into the Jiyapura, Reodar, Ambeshwar and Khiwandi Formations.

Punagarh Group

A thick clastic sequence representing a mollasse like association accompanied with basaltic volcanics deposited in a continental basin and developed during the late to post-orogenic phase of the Delhi geological cycle in the Pali district of Rajasthan has been assigned to the Punagarh Group. The rocks of the Punagarh Group occur in roughly oval-shaped outcrop from Sheopura in the northeast to Bhumadra in the southwest and from Jadan in the east to Nimal in the west. Lithologically, the group consists of a sequence of shale, slate, phyllite, mica schist and quartzite, with synsedimentational basic volcanics containing pillow lava, metatuff, bedded chert. The entire sequence is folded to form doubly plunging syncline. The Sojat, Bombolai, Khambal and Sowania Formations have been recognised in the Punagarh Group.

Sindreth Group

Clastic sequence representing molasse -like sediments accompanied with basaltic volcanics, deposited in an isolated successor basin, developed during the late to post-orogenic phase of the Delhi geological cycle in Sirohi district of Rajasthan, has been assigned into the Sindreth Group. The rocks of the Sindreth Group are traceable on a roughly linear outcrop from northeast of Padia in the northeast to north of Mirpur in the south and Gotali in east to Angor in the west over a distance of 20 km. The entire sequence of metasediments is enveloped by intrusive granite and migmatite. Lithologi-cally, the group consists of a sequence of conglomerate, gritty quartzite, micaceous quartzite, shale, phyllite and synsedimen-tational basic flows. The entire sequence appears to have been folded into a synform. The Angor and the Goyali Formations have been recognised within the group.

VINDHYAN SUPERGROUP

The western segment of the Vindhyan basin occupies an area of about 26,000 km' in the south-eastern Rajasthan. The Vindhyan sediments overlie the metamorphites of the Bhilwara Supergroup and the Berach Granite with a sharp, angular unconformity (Great Eparchaean Unconformity) defined by synsedimentary volcanics (The Khairmalia Andesite) and discontinuous bodies of conglomerate of the basal Khardeola Sandstone and Bhagwanpura Limestone Formations.

The Vindhyan Supergroup comprises an alternating sequence of sandstone, shale and limestone having a total thickness of about 3,200 m. They are divisible into two broad subdivisions, namely the Lower and the Upper Vindhyans. The Lower Vindhyans are predominantly calc-argillaceous while the Upper Vindhyans are mainly arenaceous.

The Vindhyan sedimentation in Rajasthan took place in two different sub-basins, the ChittaurgarhJhalawar sub-basin in the southwest having a full development of the Lower Vindhyan stratigraphyfurther classified, in an ascending order, into the Satola, Sand, Lasrawan and Khorip Groups and the Sapotra-Karauli sub-basin in the northeast, with only a partial representation of the lower horizons of the Lower Vindhyan sequence equivalent to the Satola and the Sand Groups of Chittaurgarh area. The two sub-basins which later came under one cover of the sediments were separated by a median upland near Bundi where the Lower Vindhyans are absent.

Lower Vindhyan

Chittaurgarh-Jhalawar sub-basin : The initial phase of the Satola Group sedimentation started with volcanic activity of andesitic affinity (Khairmalia andesite, pyroclastic and tuffs) followed by conglomerate, cross-bedded sandstone and shale of the basal Khardeola Sandstone overlying the metasediments of the Bhilwara Supergroup and the Berach Granite. This is followed by the tidal-flat deposits of the Bhagwanpura Limestone with thin, impersistant conglomerate beds at the base signifying deposition in the coastal regime. The prolific growth of stromatolites belonging to the genera *Collenia, Cryptozoon, Oncolites, Conoplayton* and *Indophyton* in the Bhawanpura Limestone indicate their deposition under stable platform condition. The succeeding Sand Group comprising the Sawa Sandstone and the Palri Shale with procellanite, was deposited in quiet water when there was an incursion of the sea casusing overlaps. The overlying Lasrawan Group of sediments comprising the Kalmia Sandstone and the Binota Shale with glauconitic siltstone interbeds were deposited in deeper water under neritic conditions. The calcargillaceous sediments of the overlying Khorip Group comprising Khori Malan Conglomerate (at the base) grading into the Jiran Sandstone followed by the Bari Shale, the Nimbahera Limestone and the Suket Shale with limestone were deposited in an unstable shelf.

Sapotra-Karauli sub-basin : Sedimentation, contemporary to the Khardeola and Bhagwanpura Formations, also commenced in this sub-basin with the basal conglomerate, glauconitic sandstone and shale followed by the Tirohan Limestone with prolific algal growth of genera *Collenia, Conophyton and Cryptozoon.* The Tirohan Breccia with chert and procellanite, deposited conformably over the Tirohan Limestone, is equated with the Palri Shale of the Sand Group.

The bouldery nature of the Basal Conglomerate suggests proximity of the source area and the rate of sedimentation exceeding subsidence of the basin. The succeeding

Tirohan Limestone containing algal structures is indicative of its deposition under littoral marine environment. The overlying Tirohan Breccia and porcellanite is autoclastic in origin. The presence of Tirohan Breccia in the vicinity of the GBF zone has been attributed to synsedimentational tectonism. Yonger units of the Lower Vindhyans (the Lasrawan and the Khorip Groups) are missing in this sub-basin. Sedimentological studies of the Lower Vindhyan sediments indicate beach to near-shore conditions. Sub-rounded to poorly-sorted arkosic sediments of the Khardeola Sandstone, Bhagwanpura Limestone and Sawa Sandstones indicate little transportation compared to rounded and well-sorted sediments of the Jiran Sandstone. The palaeocurrent direction during the Lower Vindhyan time indicates an easterly transport of sediments in the Chattaurgarh sub-basin, and NE to ENE in the Sapotra-Karauli sub-basin. The Bhilwara, Aravalli and Delhi Supergroups acted as provenance.

Upper Vindhyan

A well-marked change in the sedimentary environment from predominatly calc-argillaceous (littoral shallow-marine) to clastic (continental fluviatile) took place during the Upper Vindhyan time. During the Kaimur, Rewa and major part of the Bhander time, the two sub-basins in Rajasthan remained unconnected. In the Sapotra- Karauli subbasin the Tirohan Breccia is unconformably overlain by the Badanpur (Kaimur) Conglomerate and the Akoda Mahadev Sandstone suggesting a sharp erosional hiatus between the Lower and the Upper Vindhyans. On the contrary, deposition continued uninterruptedly in the Chittaurgarh- Jhalawar subbasin where the Chittaurgarh Fort Sandstone (Lower Kaimur Sandstone) conformably overlies the Suket Shale (Khorip Group). Some more part of the basement came under sedimentation and the strand line shifted further to the north and the Akoda Mahadev Sandstone (Upper Kaimur Sandstone) was directly deposited over the older rocks northwest of the GBF, with a distinct basal conglomerate between Akoda and Indargarh. The absence of the Lower Vindhyans in Bundi-Indargarh area is considered to be due to the basement-high known as the Bundi Upland.

The Kaimur basin was not so deep initially and it gradually subsided with sedimentation. The orthoquartzitic nature, textural maturity, good sorting and prolific development of crosss-beds and ripple marks at all levels in the Kaimur Sandstone are indicative of its deposition under shallow- marine shelf regime. The presence of glauconite in the upper levels (around Rawatbhata) suggests its disposition under littoral marine condition. Restricted ferruginisation, presence of sun cracks, rain prints and intraformational conglomerate suggest development of local sub-aerial conditions towards the close of the Kaimur sedimentation.

The succeeding Rewa Group of sediments representing alternating argillaceous and arenaceous facies suggest periodic instability of the basin floor. The extremely fine-grained Panna Shale with fine parallel-laminations was locally deposited in a shorelagoon environment. The presence of glauconitic siltstone beds at the base in Rawatbhata-Javda area indicates local reducing condition. The overlying Indargarh Sandstone (Lower Rewa Sandstone) having a restricted development is orthoquartzitic, texturally matured, profusely cross-bedded and rippled, and was deposited in littoral to neritic zones under a shallowmarine environment. The Jhiri Shale, covering the older formations, was deposited when there was transgression of the sea. The dominant reddishpurplish colour glauconitic siltstone intercalations and proliferation of stromatolitic colonies in the limestone member of the Jhiri Shale indicate that the depositional environments were dominated by warm-water conditions. The overlying orthoquartzitic, texturallymatured Taragarh Fort Sandstone (Upper Rewa Sandstone) showing prolific development of ripples and cross-beds, occurring in a blanket-type disposition was deposited under shallow-marine tidal-flat conditions. The presence of glauconite in sandstone and its ferruginous nature locally, suggests deposition under reducing environment. The presence of thin, discontinuous bodies of conglomerate in the upper parts indicates local high-energy conditions.

The orthoquartzite-carbonate-shale association of the Bhander Group of sediments is broadly attributed to stable shelf regime. The reddish-purplish ferruginous Ganurgarh Shale with frequent glauconitic siltstone beds showing ripples, cross-laminations, sun cracks and intraformational conglomerate at the base suggest a shore-lagoon tidal-flat deposition where as the argillaceous limestone and intraformational conglomerate and breccia at the top indicate supratidal conditions. The intraformational conglomerate and breccia, sun cracks and ripples in the lower beds of the Lakheri Limestone suggest their deposition under subaerial conditions. The rest of the formation was deposited under more or less chemogenic environment as suggested by the chert and anhydrite intercalations and almost general absence of detritus. The corresponding limestone unit is missing in the Sapotra-Karauli sub-basin. This condition continued during the Samria Shale. The prolific growth of stromatolites in the limestone interbeds of the Samria Shale favours their deposition in an open, intertidal mudflats, mainly in the protected lagoon. The orthoquartzitic nature, high maturity and good sorting together with profuse crossbeds, ripple marks and current lineations suggest deposition of the Bundi Hill Sandstone under shallow, stable-shelf conditions.

During the deposition of the Sirbu Shale both the sub-basins got united into a single repository wherein the Sirbu Shale and other younger formations viz., the Maihar Sandstone, Balwan Limestone and Dholpura Shale in Bundi-Indargarh area and Upper Bhander Sandstone in Sapotra-Karauli area were deposited in a shallow-marine conditions.

The principal mode of palaeocurrent direction in the Maihar Sandstone in Sapotra-Karauli sub-basin is towards north. Uniformity and consistency of the direction of sediment transport signifies stable palaeoslope and general tectonic stability of the basin, especially during the Upper Bhander time. The Balwan Limestone which is stromatolitic limestone, indicates shallow-marine condition of deposition. The youngest Dholpura Shale indicates restricted condition. Thus, it appears that throughout the Vindhyan sedimentation the conditions of deposition were fluctuating due to subsidence and elevation of the basin and consequent transgressions and regressions of the sea.

Several evidences of organic life in the form of stromatolite, trace, macro and micro fossils have been reported from the Vindhyans. The stromatolites preserved in the Vindhyans (similar to the Ripheans of the former USSR) indicate an age ranging from over 1,300 Ma to less than 600 Ma. Radiometric dating of basal Khairmalia Andesite (1,250 Ma) and the Kaimur glauconite (940-910 Ma) is in fair agreement with the results arrived at from the stromatolites.

MALANI IGNEOUS SUITE

The Malani Igneous suite encompasses Late Proterozoic plutonic and volcanic rocks. The suite covers an area of about 51,000 km² in parts of Jaisalmer, Jodhpur, Barmer, Churu, Pali, Sirohi and Jalor districts of western Rajasthan and extends further in Sind province of Pakistan.

Malani magmatism, the largest anorogenic acid volcanism in India showing little metamorphic effect, is associated with post-Erinpura granitic activity and pre-Marwar sedimentation and covers a span of 850550 Ma. Three phases of igneous activity have been identified viz., eruption of basic flows, followed by voluminous acid rhyolite flows culminating with ashflow deposits (first phase); intrusion of subjacent discordant granites as plutons, ring dykes, bosses and plugs (second phase) and the indiscriminate intrusion of mafic and felsic, coarse- to fine-grained dykeswarms (third phase).

Malani volcanics have erupted in three stages. The volcanism commenced with eruption of basalt with occasional andesite flows (first stage) followed and covered by the voluminous outpouring of peralkaline and peraluminous rhyolite, rhyodacite and trachyte flows (second stage) and ceased with the outburst of ash-flow deposit (third stage).

The dominant felsic volcanics are rhyolites and rhyodacites having a spread of about 31,000 kin z. The other rock types associated with rhyolite are trachytes, dacites, pitchstone, welded-tuff, vitric, lithic and crystal-ash, ignimbrite, obsidian, pyroclastic slates, agglomerate volcanic breccia and volcanic conglomerates. The cumulative thickness of felsic volcanics exceeds 3-5 km in 95 cumulative flows Bhushan (2001). The final eruptive stage witnessed ash-flow deposits near Pokaran. These ash beds, 50 m in thickness, are composed essentially of light yellow and light brown lithic-, crystal- and vitric-ash having microscopic xenoliths of basalt and rhyolite embedded in it. The mafic-felsic volcanic suite (first phase) is intruded by post-tectonic, anorogenic Siwana and Jalor Granites during the second phase. The Siwana Granite occurs as ring dykes while Jalor Granite attains batholithic dimension near Jalor and also occurs as small bosses near Salawas. The hornblende granite (Malani Granite) occurs as porphyry dykes in Barmer area. The third phase of igneous cycle experienced intrusion of felsic and mafic dykes trending either NNW-SSE or NE-SW direction and the dolerite forming the predominant dyke phase. Five subphases in a dyke-swarm near Sankra have been recorded commencing with felsic intrusions and culminating with the basic intrusions.

The Jalor Granite plots in the granodiorite field. The main difference between the Siwana and the Jalor Granites is the presence of normative acmite and absence of normative anorthite in the former. The Jalor Granite is metaluminous. It has higher SiO_2 and Al_2O_3. The Siwana Granite which is peralkaline has very high TiO_2 alkalies, REE and iron due to presence of alkali pyroxene and alkali amphiboles. At Siwana, the alkali granites as well as the associated rhyolites (comendites and pantellerites) and trachytes are characterised by the presence of riebeckite, aegirine and arfvedsonite indicating a co-magmatic relationship.

Malani volcanics occur towards the western side of tightly compressed Delhi- Aravalli fold belts. A possible sudden release of pressure, causing rifting may have triggered the eruption. NW-SE trending fissures and several calderas and cones were the possible mode of eruptions.

MARWAR SUPER GROUP

The trans-Aravalli sediments, equivalent of Vindhyans deposited on the NW flank of the Aravalli hill range, occurring as numerous scattered flat topped hills and covered under the sands of the Thar desert towards north, have been grouped under the Marwar Supergroup. The general resemblance to the Vindhyan sediments notwithstanding the Marwar beds having deposited in a separate basin, are considered correlatable to the Upper Bhander Group. Laid down in continental seas, upon the Malani volcanics, at a time just before the life forms appeared in abundance, the Marwars are of great interest. Apart from *chordophyceous* tracks and fusiform bodies, bearing some relation of the *Cruziana furcifera,* no determinable organic remains have been found in the Marwar beds.

The rocks of the Marwar Supergroup, over 1,000 m thick, are divided into three groups, namely the Jodhpur, Bilara and Nagaur Groups.

The Pokaran Boulder Bed, Sonia and Girbhakar Formations constitute the *Jodlipur Group.* The Pokaran Boulder Bed, about 3 m thick, extending southward from Pokaran in the north to Hariyoli *Nadi* in the south, over a distance of 42 km consists of pebbles and cobbles of granite, syenite, basic rock and quartzite in a red-brown sandy matrix (near Karewa). Cobbles of granite set in a red arenaceous matrix (near Baratka-Gaon). The overlying Sonia Formation, horizontally disposed with rolling dips, commences with a maroon shale with mega ripples which alternates with creamish-buff to reddish, fine- to medium-grained sandstones interbedded with chert, cherty dolomite and jasper. The contact with the overlying Girbhakar Formation is marked by a short break in sedimentation indicated by the presence of pebbly horizon containing pebbles of the Sonia sandstone and chert. The Girbhakar sandstone is reddish and consists of coarse to gritty micaceous sandstone alternating with red and light green shale with pebbly and gritty layers. The Jodhpur Group of rocks shows NW to W palaeocurrent direction. The well-sorted, mature sandstones were deposited in near-shore environment. Heavy mineral analysis suggests that the materials were received from acid plutonic volcanics and low-grade contact metamorphic rocks. The beds are, for most of the part, unfossiliferous.

The calcareous *Bilara Group* which is 300 m thick comprises Dhanapa Formation-chert at the base, followed by well-bedded to laminated cherty dolomite, siliceous dolomitic limestone and stromatolitic limestone; Gotan Formation-dark grey laminated limestone, containing bands of clay; and the Pondlo Formation comprising dolomite, cherty dolomite, claystone, siltstone and reddish sandstone. It is stromatolitic towards top near Shiv Sar and contains siliceous oolites grading into pisolites near Lordiyan. *Weedia, Collenia, Cryptozoon* and algal nodules, illpreserved lamellibranchs and brachiopod shells are reported from the Bilaras.

The *Nagaur Group* unconformably overlying the Pondlo dolomite (the Bilara Group) commences north of Khichan with a 5 m thick conglomerate bed, containing cobbles, pebbles fragments and pieces of dolomite, chert, cherty dolomite, feldspar, quartz and sandstone cemented in a pink-brown calcareous sandy matrix. These are overlain by red, mottled, claystone, siltstone and fine-grained friable sandstone. The Nagaur sandstones have been divided into two formationsthe older unit, designated as the Nagaur Formation, comprises a sequence of brick-red to red claystone, siltstone and sandstone with bands of clay and blotches of green clay and the younger unit, designated as the Tunklian Formation, comprises brick-red claystone, siltstone, calcareous clay and gritty to pebbly sandstone containing pebbles of Malani rhyolite-granite, chert, dolomite, quartz and quartzite near Gudarli.

PALAEOZOIC

After the deposition of the Nagaur Group (Proterozoic) there was a break in sedimentation as a result of which sediments of most of Cambrian, Ordovician, Silurian, Devonian and Lower Carboniferous periods are not represented in Rajasthan.

Upper Carboniferous glaciation is represented in the Nagaur basin by the NE-SW trending remnants of the Bap Boulder Bed-an assortment of pebbles and cobbles with occasional boulders and a few erratics exhibiting well-marked glacial striations, spread over 68 km z from Khirwa to Nokhra. The rock types include gravel/boulder, 5 cm to 25 cm across in size, of rhyolite, tuff, granite, basalt and dolerite of the Malani Igneous Suite ; granite, syenite, amphibolite and basic rocks, quartzite, phyllite and slate of pre-Delhi and Delhi age and limestone, dolomitic limestone and sandstone of the Marwar Supergroup.

This is overlain by a very small remnant of the marine Badhaura sandstone of 30 m computed thickness. Extending from northwest of Badhaura up to Harbans, these marine sediments, designated as the Badhaura Formation, comprise medium- to coarsegrained brownish-yellowish and ferruginous sandstone, clay, shale and siltstone containing Permian fauna and flora. The subsurface continuation of the formation at Pugal between depths 500 and 590 m, and at Shumarwali *Talai* has been established by drilling. The Permian sediments thus have thickened dipwards and represent the post-glacial sediments. The Badhaura sandstone which can be correlated with the sequence of the Indus Basin, implies that the Indus Basin (Salt Range) was connected with the north and northwest Rajasthan.

MESOZOIC AND CENOZOIC

Deccan Traps

The Deccan Traps (Cretaceous to Eocene) covering an area of about 10,435 km² Z in southeast Rajasthan have developed in two separate geographic sectors, namely, the Kota-Jhalawar sector in the southeast and the Chittaurgarh-Banswara sector in the southsoutheast.

Kota-Jhalawar sector : In this sector the Deccan Traps overlying the Vindhyan sediments have attained a cumulative thickness of over 180 m. Twelve flows, varying in thickness from 7 to 56 m are present between 280 and 481 m RL. Out of the 12 flows, five are Pahoehoe-type and the rest are Aa-type. The individual flows are separated either by impersistent red bole bed or by Inter-trappean Beds comprising conglomerate, clay and chert. Both fossiliferous and unfossiliferous clay/chert beds are present among the older flows. *Physa prinsepii* and ostracodes have been recorded near Parbati (24°23' N : 76°31' E), Sohankheri (24°30' : 76°05') and Gadiyo (24°31' : 76°03'). Around Dag (23°56' : 75°50') and Chaumahala (23°57' : 76°36') the flows have undergone widespread laterisation. A gritty, sandy and cherty unfossiliferous friable Infra-trappean clay bed containing angular pieces of sandstone occurs widely in the area around Chippabarod (24°33' : 76° 43'), Chabra (24°40' : 76°51') and Chandloi (24°30' : 76°10').

Chittaurgarh-Banswara sector : The area is characterised by the N-S trending Deccan Traps plateau with prominent scarp on the west. The Deccan Traps overlie the Lower Vindhyan rocks in the north and northeast, but towards the west and south they rest over the rocks of the Mangalwar Complex and the Aravalli Supergroup. In all, sixteen flows varying in thickness from 11 to 55 m have been identified between 226 and 606 m RL. Out of these, four flows have Pahoehoe characters and the rest Aa-type characters.

Fairly widespread unfossiliferous clay bed, as Infratrappean, is present around Kushalgarh (23°12' : 74°27') and Barodia whereas fossiliferous Physa *(Bullinus) prinsepii* Infra-trappean limestone bed is reported from Ram-ka-Munna in Banswara district. Unfossiliferous thin Inter-trappean clay beds are noticed around Arnaud (23°53' : 74°49') Dhamotar (24°06' : 74°45'), and Kulmipura (24°07' : 74°45') whereas unfossiliferous chert bed, about 1 m in thickness, is noticed near Acheri (24°26' : 74°50') and Semara (24°26' : 74°42').

Tertiary Alkaline Complex

The Tertiary magmatic event of the western Rajasthan is characterised by the mildly alkaline and hyperalkaline plutonic volcanic suites of Sarnu Dandali, Karara-Tavidar and Mundwara areas. The hyperalkaline suite of rocks clearly traverses the Malani rhyolites and Cretaceous clastics and, therefore, represents a major Tertiary alkaline magmatic event. The Sarnu-Dandali suite is dominantly nephelinitic in composition, various members being mafic alkaline rocks, ijolites, foidal syenites, phonolites and carbonatites belonging to a carbonatite-nephelinite association. The Mundwara suite is mostly basanitic comprising picrites, foidal gabbros and foidal syenite with carbonatites which intrude the Erinpura Granite.

Sedimentaries

Triassic rocks do not outcrop anywhere in west and northwest Rajasthan but its subsurface presence, identified on pollen-spore assemblage, at Shumarwali *Talai* borehole, has been established. Land conditions existed during post-Permian to pre-Jurassic period and extensive erosional agencies reigned supreme leaving only traces of these Permian deposits. The Jurassic period commenced with a fluviatile environment which gave rise to the Lathi sandstone.

The area north, west and south of Lathi was filled up by the sediments of the Lathi Formation, the southern continuity being checked by the Jodhpur-Pokaran-Chotan-

Malani ridge. The outcrops of the Lathi sandstone are seen intermittently, separated by the desert sands. The Lathi sandstone, 360 m thick, comprises a sequence of coarse-grained sandstone, gritty sandstone, pebbly sandstone, gritty and pebbly sandstone, conglomerate, arkose, lithic-arenite, siltstone, shale, ferruginous shale, ferruginous sandstone and clay.

The Lathis overlie the Malani Igneous Suite of rocks, the Pondlo dolomite, the Nagaur sandstone, the Bap Boulder Bed and the Badhaura sandstone. The contact with the Malanis is marked by a conglomerate bed, grading into pebbly sandstone and grit. A friable, weathered, dirty-white to yellowish sandstone comes in contact with the other older underlying rocks.

The Lathi sandstone has been subdivided into lower and upper members. The lower member comprises a sequence of conglomerate, arkose, lithicarenites and coarse-grained sandstone, indicative of terrestrial deposition under stable delta in fresh water conditions. The upper member comprises bands of sandstone and siltstone, grading into limestone, indicating shallow-marine shelf-sedimentation.

The abundance of fossil wood logs is so conspicuous that an area near Akal has been converted into a Fossil Wood *Park*. The fossil wood is of gymnospermous origin and belongs to dicotyledonous stem. The Lathi sandstone has leaf impressions of *Petrophyllum* sp. *Ptillophyllitm acutifolium, Equisctites* sp. and preserves silicified dicotyledonous wood and silicified gastropod shells. The Lathi sandstone contains thin bands of lignite as noted in dug wells near Devikot, Nar Singh-ki-Dhani, Jhinjhinyali and south of Unror. The Mayakor Formation, which overlies the Badhaura Formation with an unconformity has been equated with the Lower Lathis. The deposition of Lathi/ Mayakar Formation culminated with the initiation of the marine transgressions in different basins. These include the Jaisalmer basin in the west, the Barmer basin in the south-central and the Palana-Ganganagar Shelf in the north and northwest of Rajasthan.

Jaisalmer Basin

The Jaisalmer basin deepened towards northwest extending into Pakistan. The Jaisalmer-Mari arch, the raised platform, exposes the entire sequence. The Mesozoics are preserved in NE-SW trending strip with their northwesterly boundary being located near Niba, Sam, Kanoi, Mahmud-ki-Dhani, Jaidesar, Sanu and Lila; the Lower Tertiaries are exposed towards northwest of these localities. The Mesozoics are represented by the Lathi sandstone, Jaisalmer limestone, Baisakhi shale, Bedesar sandstone, Parihar (Pariwar) sandstone and Abur limestone, while the Lower Tertiary includes Sanu sandstone, Khuiala limestone and Bandah limestone. Pleistocene Shumar grits overlie the Lower Tertiary sequence as capping of ferruginous grit and laterites. All these units have been assigned formational status.

The Jaisalmer limestone has intertonguing contact with the underlying Lathis and represents the first marine transgression of the Mesozoic in Rajasthan.

Subsurface thickness of this formation is ca 600 m in the Jaisalmer basin. The Baisakhi shale and the Bedesar sandstone occur as mappable rock forma lions of 165 m and 65 m thickness respectively, but in subsurface they occur as intergrading sequences. Each of these formations has been subdivided into units A, B and C. The overlying Pariwar sandstone which represents continental deposition contains angiospermic plant remains and is 300 m thick. The overlying Abur limestone marks the end of the Mesozoic Era in the Jaisalmer basin. The Abur Formation is 66 m thick and in subsurface it is divisible into the Goru Formation and the Parh Formation of younger ages showing continuity of the marine environments in the deeper parts of the Jaisalmer basin.

The Palaeogene succession is exposed in detached outcrops near Mahmud-ki-Dhani, Sanu, Shera and Lila. It commences with the Sanu sandstone which is 75 m thick and consists of fresh-water arenaceous and glauconitic sandstone, clay, shale and ferruginous shale showing current bedding. The

Sanu sandstone appears to grade in subsurface into non-marine and marine bands showing fish teeth. The Khuiala limestone of a total thickness of 100 m overlies the Sanu sandstone and is overlain by the Bandha limestone of 75 m computed thickness marking the close of the sedimentary cycle in the Jaisalmer basin. The Tertiaries occur oblique to the Mesozoic strike and find representation in the Barmer basin and in the Palana-Ganganagar shelf area.

Barmer Basin

Southeast of Jaisalmer a sequence of the Mesozoic and Lower Tertiary rock formations occupies a N-S trending linear tract for about 100 km from Mandai and Fatehgarh in the north to Barmer in the south. This synclinal exposure having a maximum width of 50 km at Akli is known as the Barmer basin. The configuration of the basin is related to Meso-Cenozoic tectonic framework of the Cambay Graben in Gujarat. The rock formation of the Barmer basin include the Fatehgarh sandstone, Akli bentonite, Mandai sandstone and Kapurdi fuller's earth-each unit enjoying a formational status.

The Fatehgarh sandstone overlies the Lathi sandstone in the Barmer basin and represents the Cretaceous marine trans-gression. It occurs along the northern half of the synclinal depression of the Barmer basin and comprises conglomerate, phosphatic sandstone, highly fossiliferous sandstone, ferruginous sandstone and clay in a 50 m thick sequene. These beds are overlain by the Akli bentonite which contains 265 m thick sequence of the bentonite, bentonitic clays, siliceous earth containing phosphatic beds. The continental Mandai sandstone (100 m) overlies the Aklis and is overlain by the Kapurdi fuller's earth (30 m) which marks the closure of Tertiary cycle in the Barmer basin.

Quaternary

A large track of western Rajasthan constitute the Thar desert. The NE-SW trending Aravalli hill range define its eastern boundary while the flood plains of the Ghaggar-Sutlaj rivers represent its northern periphery. In the southwest the Great Rann of Kachchh is its limit. Average elevation of this undulating desertic terrain varies from about 200 m to 300 m MSL with a few isolated hilly tracts rising up to 500 m above sea level.

The Quaternary continental sediments in the Thar desert comprise a multiple episodic succession of fluvial, fluvio-lacustrine and aeolian deposits. A complete succession (reaching up to 300 m) is not developed in any one section. Although the lithofacies show variations in their intensity of development in different sectors, their geological evolution seems to be conformable. The neotectonic movements, forming a series of NE-SW trending horst and graben structures, are considered to have controlled the origin, configuration and developement of depositional basins and their gradual thickening westward. Change in climate from semi-arid to arid resulted in widespread formation of calcrete.

Fluvial and aeolian agencies are the major presentday processes which also dominated throughout the Quaternary Era. The fluvial activity prevalent in the eastern margin while the aeolian processes in the west - the relative spatial dominance being a function of the distribution of rainfall. Numerous stable aeolian bedforms inherited from an earlier dry climatic phase have been vigourously acted upon by fluvial processes creating a fine network of rills and gullies along the eastern margins of the desert.

The most common landform sequence produced by the fluvial processes within the desert are the desert hill slope, rocky/ gravelly pediment, flat buried pediment, older and younger alluvial plains and palaeo-channel signatures. In the Jaisalmer area where near-horizontal, high-level, structural plains (hamada, bounded by steep escarpment) are exposed, the sequence consist of serir, gravel-leg, sandy loam, calcretised conglomerates etc.

A number of saline depressions, ranns or playas, occur within the desert. They are considered to have formed as remnants of a receding sea; as aeolian segmentation of former streams during a drier climate; due to deflation in the windshadow zone of a hill and stream-trapping by lee-longitudinal dunes etc.

The spatial extent of the fluvial landforms has reduced over time due to their progressive burial by the aeolian sands. New units like sand dunes, interdune plains and sandy undulating plains (buried pediment and alluvial plains) have been recognised.

Sand dunes are classified on the basis of their age of formation and stability, and the types recognised are the clustered parabolic, transverse, reticulate, longitudinal and barchans etc. Interdune plains, in between the dune ridges, occur as narrow to wide sandy plain which are flat, sandy arid undulatinghosting gypsite/calcrete beds at places. The calcretes show a large range in the degree of development and morphological appearance depending upon the age of the landform. Soft nodular calcretes have formed within aeolian sandy plains in the 250-400 mm rainfall zone. Nodules, 1-4 cm across, constitute 3-15% of the bulk and occur randomly distributed amidst mildly calcareous sandy mass. Hard nodular calcretes occur extensively in the aggraded older alluvial (Pleistocene) plains or on more ancient surfaces. With depth, the nodules get larger and irregular in shape and are cemented together. These deposits locally called *murnd* find extensive use in construction of roads in sandy dry land terrain.

STRUCTURE

The Precambrian rocks of Rajasthan exhibit widely varying structural patterns resulting from repeated periods of deformation and accompanying emplacement of igneous rocks. The regional strike of foliation in the Bhilwara and the Delhi Supergroups of rocks is dominantly NNE-SSW to NE-SW, while it is N-S and NW-SE in the Aravallis.

The foliation dips dominantly towards the west at a steep to moderate angle. Local variations in dips are seen towards the east in the rocks of the Hindoli Group. The regional structure is defined by the presence of a series of isoclinal overturned folds and by normal upright folds as in the Delhis of Alwar region. The fold axes in the Mangalwar and the Hindoli Groups of eastern Rajasthan trend dominantly NE-SW swerving to N-S in Rajpura-Dariba and Pur Banera-Bhinder belts of southern Rajasthan. The fold pattern in these rocks indicates en echelon shifts of the fold axes presumably indicating later cross-folding. On the other hand, the fold axes in the Aravallis of Udaipur-Banswara region in southern Rajasthan trend NW-SE, swerving gradually to N-S and NE-SW trends near Udaipur and Nathdwara. There are a series of culminations and depressions, anticlinal culminations in Sarada dome and the Hammer-head syncline of Kankroli as well as other structural complications along the Aravalli belt in comparison with the divergent structural trends in the Bhilwara and the Aravalli Supergroups, the fold axes in the Delhi Supergroup along the main Aravalli mountain range and in Alwar-Jaipur-Sikar region trends consistently in NNE-SSW and NE-SW directions. In the southern part of the Aravalli range, south of Beawar there is a main synclinorium flanked by a series of anticlines and synclines, all plunging towards southwest. Towards the north of Beawar, however, there is a plunge reversal beyond which a series of doubly plunging linear folds (anticlines and synclines) extend towards the northeast as far as the Khetri Copper Belt. The anticlinal culminations within these fold systems are often marked by Alwar quartzites, flanked all around by Ajabgarh schists, phyllites, calcgneiss, calc-schists etc. In additiion to this, the linear anticlinal axes in the Alwar quartzites, in Khetri region of Khetri Copper Belt, also reveal a series of north to south plunging en echelon dextral- and sinistral-type of later sheer folding. After a tight anticlinal arcuate swerve, in the Neem-ka-Thana-Babai region, the rocks display a major synclinal structure in the Sikar-JaipurAlwar region. This is marked by a series of sub-parallel upright synclines and anticlines plunging dominantly towards northeast and marked by alternating cluminations and depressions of the axes as one proceeds northward.

Precambrian rocks in Rajasthan have been subjected to many periods of superposed deformation resulting in varying structural trends. The earliest deformation has given rise to tight, isoclinals, reclined folds plunging at about 20°-40° towards WNW. The second deformation which is more intense and prominent has given rise to open to tight upright folds with their plunge varying from 10° to 50° towards SW or SSW in the southern part and NE or NNE in the northern part. The third deformation exhibits NE or SW plunging open, upright folds with sub-vertical axial planes striking WNW-ESE.

The first two generation of folds are noticed in the Bhilwara and the Aravalli Supergroups. The superposition of these two generations of fold has led to different structural trends along NW-SE, E-W and NE-SW axes. The rocks of the Delhi Supergroup, on the other hand, contain the second, third and fourth generation of folds resulting in the formation of tight longitudinal and sometimes isoclinals overturned folds having a dominantly NE-SW trend. The absence of the first generation of folds in the Delhi Supergroup marks a major structural break between the Delhis and the pre-Aravallis/Aravallis of Rajasthan. The other major tectonic element in these rocks is a series of sub-parallel NE-SW trending faults and presumably thrusts in the Delhi and other groups of rock, specially those within the Delhis along the Aravalli range and in Alwar region.

The rocks of the Vindhyan Supergroup are very little disturbed and have gentle dips towards south and south-southeast. The basement rocks beyond Chittaurgarh in the west, Bundi-Indargarh in the northwest, and Shivpuri in the east acted as rigid block. The Ramgarh Dome acted as a center from which the stresses were centrifugally radiated resulting in the formation of Bundi-Indargarh and Phalodi anticlines in the northwest; causing a series of anticlines and synclines in Sawai Madhopur, Sapotra and Karauli areas in the north-northwest; generating open rolls in Shabad area in the east; forming Jhalawar anticline in the southeast and developing a series of anticlines and synclines in Chittaurgarh in the west. The rejuvenation of adjoining Aravalli-Delhi mobile fold-belt in the west resulted in the Great Boundary Fault (GBF) which trends roughly NE-SW for a distance of 400 km between Chittaurgarh and Agra. It started as a gravity-fault along which gradual sinking of the basin floor took place. The present reverse geometry with low-angle thrusting has brought the older basement rocks in juxtaposition with the Vindhyan sequence. The GBF is not a single plane of disruption but it evolved in stages and represented a 10-20 km wide zone (Balmiki Prasad, 1984) comprising a number of parallel and oblique faults showing horst and graben features. Recent studies along the corridor of GBF carried out in Sawai Madhopur and Chittaurgarh sectors suggest that it is not a marginal basin fault, but formed during syn- to post-deformation F1 folds. The Marwar Supergroup of western Rajasthan are practically undisturbed except for the presence of a north plunging main basinal synclinal axes trending N-S to NNW-SSE from Sojat to Bilara and Nagaur.

The Malanis were emplaced dominantly as fissure eruptions and accompanying sheet-like intrusions along NW-SE and N-S faults and fissures of post-Delhi age.

The rocks of the Mesozoic and Tertiary Era in the western Rajasthan are under formed and constitute very low-dipping strata marked, at places, by gravity faults. The area being mostly covered by desert sand, the subsurface anomalies/ disconti-nuities, revealed by geophysical and geological exploration by GSI, ONGC and other State agencies, indicate presence of a number of faults which have shaped the present disposition of the various litho units in the area.

Chapter 3

PHYSIOGRAPHY AND GEOMORPHOLOGY

The physiography of Rajasthan is the product of long years of erosional and depositional processes. The present landforms and drainage systems have been greatly influenced and determined by the geological formation and structures.

The land forms of Rajasthan are the outcome of sub-aerial processes both by fluvial and aeolian aqueous. The long years of denudation and accumulation of blown sand have given its present physical form. The most prominent-physiographic feature of Rajasthan is the layout of hill ranges that extend south-west to north-east diagonally. This system of hills, called as the Aravallis, form the watershed boundary that divides the State virtually into two broad physiclimatic zones, i.e. arid and semi-arid. The arid zone that lies to the west of the Aravalli axis is covered by vast stretch of sand, and at places interrupted by numerous hillocks of low elevations. The Great Indian Desert or the 'Thar Desert' lies further west of this axis. The north-eastern part of the range shows discontinuity and is full of wind gaps. It is through these wind-gaps that the western sand-dunes penetrate into the east.

The area to the south-east of the Aravalis is a plateau which is dissected by numerous south-easterly flowing streams, most of them draining into the Chambal. Except for the south-southeast scraps (Vindhyan Scarplands) running parallel to the Chambal at its left bank, the rest of the area, lying east of the Aravallis, is a vast covered by a thick mantle of alluvium. (Figure 3.1).

There are thus variations in the physical relief of Rajasthan. It ranges from 30 metres in the vicinity of Rann of Kutch to 500 metres and even over 750 above of the mean sea level at a few places. The major part, of the state, however, lies in the relief range of 100 to 350 metres above mean sea level as well as their precentage coverage in the total area of the state are given Table 3.1.

Table 3.1 Area under various ranges of Relief.

Relief Range (m)	Coverage	
	Area (in'000 sq.kms.)	Percent
Under 50	1.0	03
50 to 100	1.5	0.5
100 to 150	19.0	5.6
150 to 200	38.0	11.2
200 to 250	45.0	13.1
250 to 350	86.0	25.2
350 to 400	18.5	5.3
400 to 450	32.0	9.4
450 to 500	40.0	11.6
500 to 600	27.5	8.0
600 to 750	24.5	7.1
750 & over	9.0	2.7

Source: Topography Sheets, Survey of India.

The map alongside shows the relative relief position in the state. The relief values viz., 150,300,600 and 900 metres above the sea level by various relative relief lines and their respective values have been shown at the end of each such line. The intervening spaces between these lines showing relative relief are shaded with high to low intensity as indicated in the legend of the map.

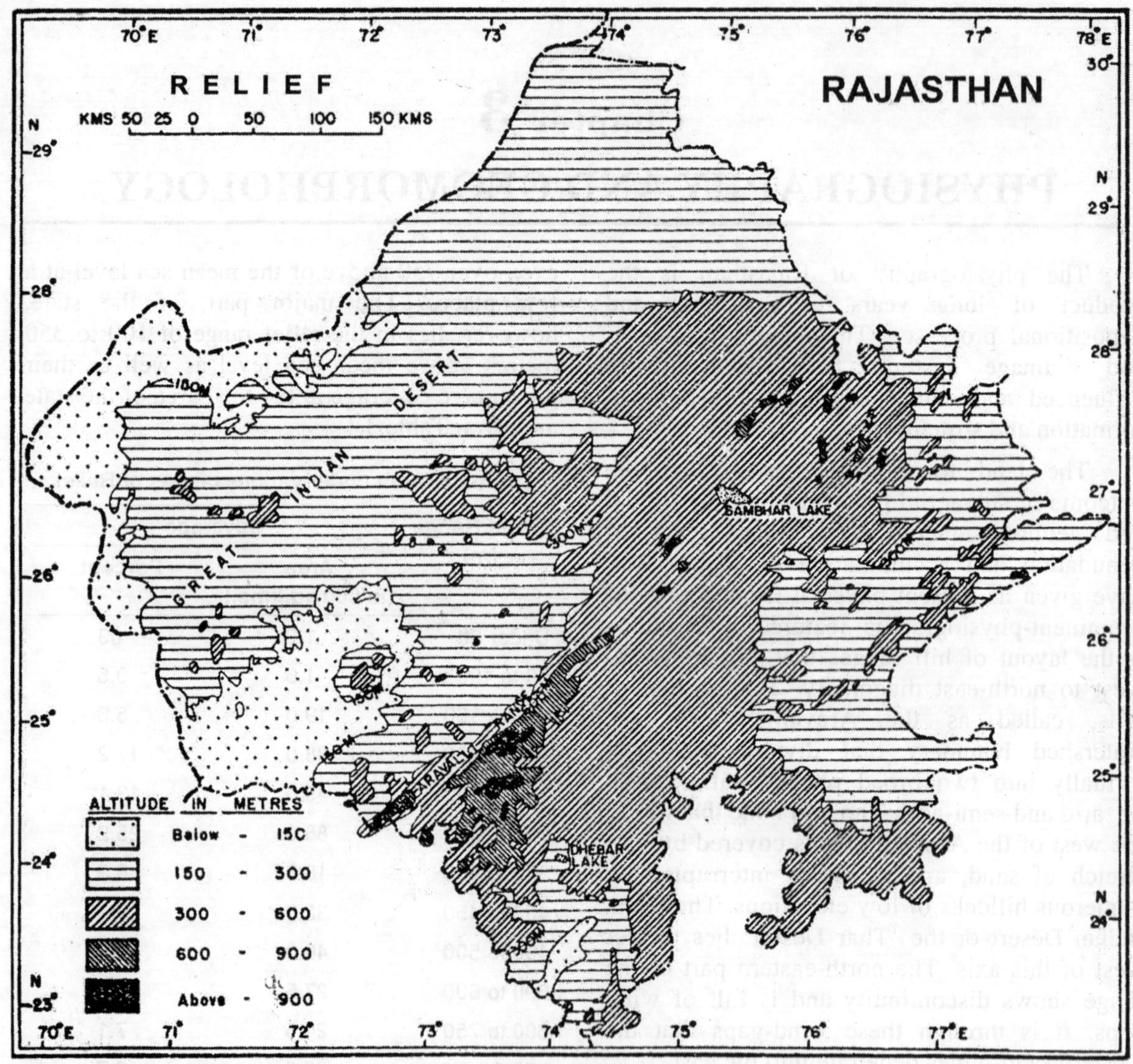

Source: Million Map of Rajasthan and Administrative Map, published by Census Commission of India.

Note : The external boundary and the coastlines of India appearing on the map are neither verified nor authenticated by the Survey of India.

Figure 3.1 Relief

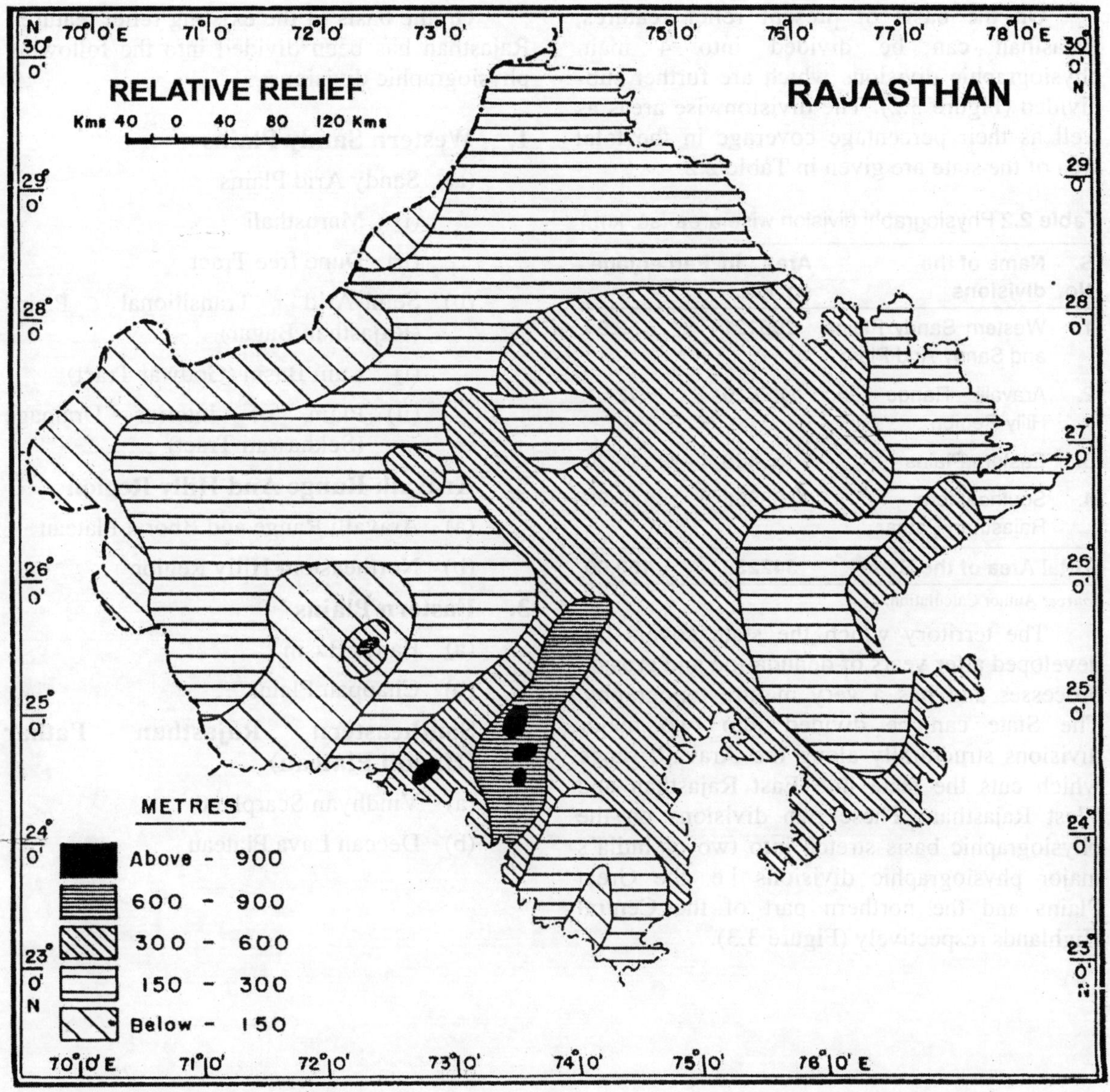

Source: Administrative Map, published by Census Commission of India.

Note: The external boundary and the coastlines of India appearing on the map are neither verified nor authenticated by the Survey of India.

Figure 3.2 Relative Relief

On the basis of present relief features, Rajasthan can be divided into 4 main physiographic divisions which are further sub-divided (Figure 3.2). The divisionwise areas as well as their percentage coverage in the total area of the state are given in Table 2.2.

Table 2.2 Physiographi division wise area (sq. km.)

S. No.	Name of the divisions	Area in sq. km.	Percentage
1.	Western Sandy Plains and Sandy Arid Plain	205121	59.94
2.	Aravalli Range and Hilly Region	42826	12.51
3.	Eastern Plains	55426	16.20
4.	Southeastern Rajasthan Pathar	38866	11.35
Total Area of the State		**342239**	**100.00**

Source: Author Calculation.

The territory which the state covers has developed after years of denudation and erosion processes and has a very mature topography. The State can be divided into two major divisions structurally along the Aravalli range which cuts the state into East Rajasthan and West Rajasthan. These two divisions on the physiographic basis stretch into two of India's major physiographic divisions i.e. the Great Plains and the northern part of the Central Highlands respectively (Figure 3.3).

On the basis of the existing relief features, Rajasthan has been divided into the following physiographic divisions:

1. Western Sandy Plains

(a) Sandy Arid Plains

(i) Marusthali

(ii) Dune free Tract

(b) Semi-Arid Transitional Plains (Rajasthan Bagar)

(i) Luni Basin (Godwar Tract)

(ii) Plain of Interior Drainage (Sekhawati Tract)

2. Aravalli Range And Hilly Region

(a) Aravalli Range and Bhorat Plateau

(b) Northeastern Hilly Region

3. Eastern Plains

(a) Banas Basin

(b) Chappan Plains

4. Southeastern Rajasthan Pathar (Hadoti Plateau)

(a) Vindhyan Scarpland

(b) Deccan Lava Plateau

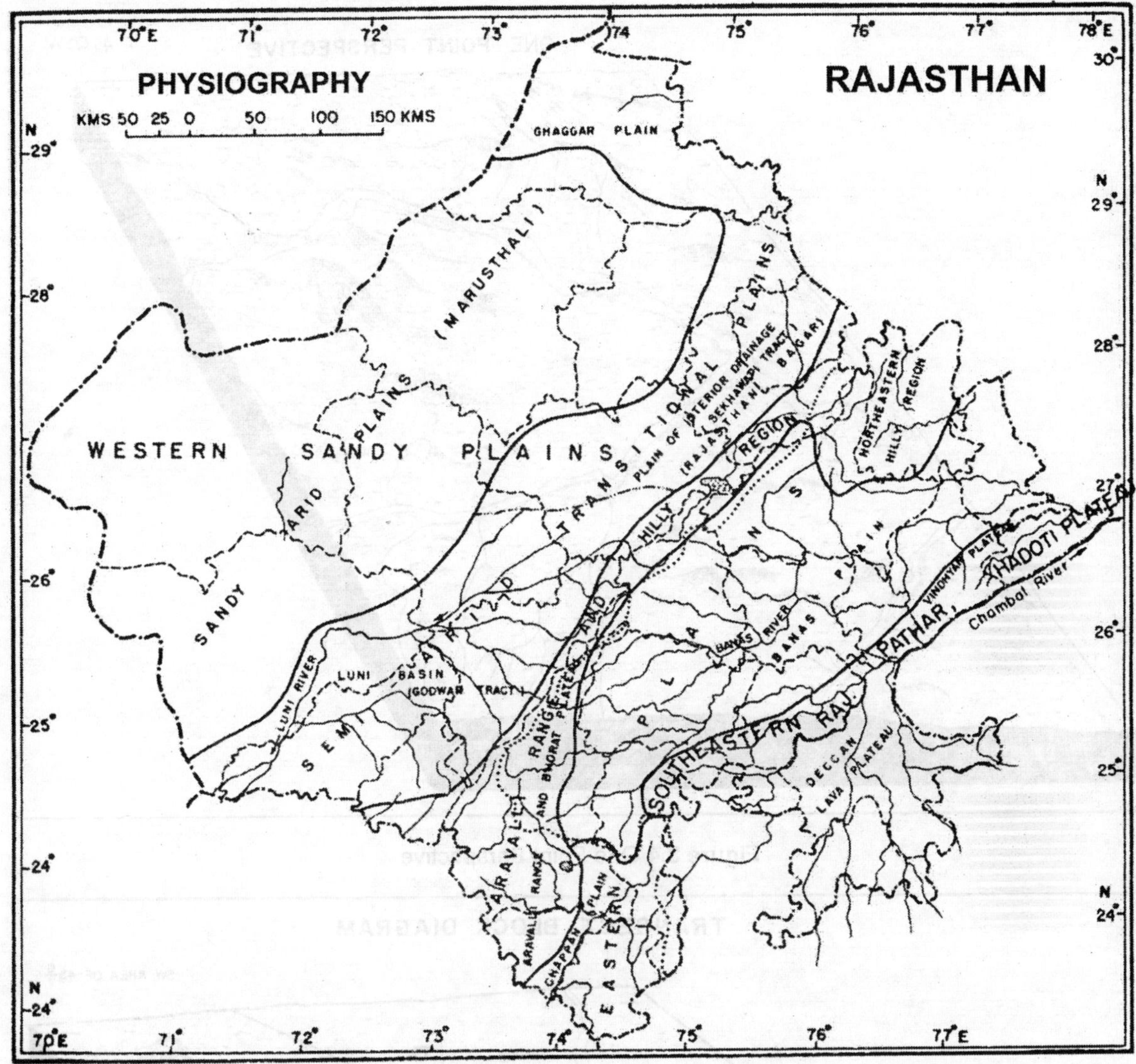

Source: Million Map of Rajasthan. Administrative Map, published by Census Commission of India.

Note: The external boundary and the coastlines of India appearing on the map are neither verified nor authenticated by the Survey of India,

Figure 3.3 Rajasthan Physiography Map

1. WESTERN SANDY PLAINS

The western Plain covers a large part of Rajasthan in the west and the northwest of the Aravalli axis. Its northern boundary is marked by Punjab and the southwest by Gujarat and the western boundary is the international boundary between India and Pakistan. The eastern boundary of the western sandy plain is marked by the Aravalli range's western sub-montane zone up to the northern point of Udaipur district. Beyond which the boundary is marked by the great Indian watershed and 50 cm isohyet (Figure 3.4 and 3.5).

The surface of western Rajasthan is covered with vast stretches of sand occasionally interrupted by rocky protruding right up to the Pakistan border. The Northwestern division is sandy, poorly watered and sterile. The Rajasthan desert in the west has been subdivided into several natural regions because of the Aravalli hills and the rocky terrain of Barmer-Jaisalmer-Bikaner tract.

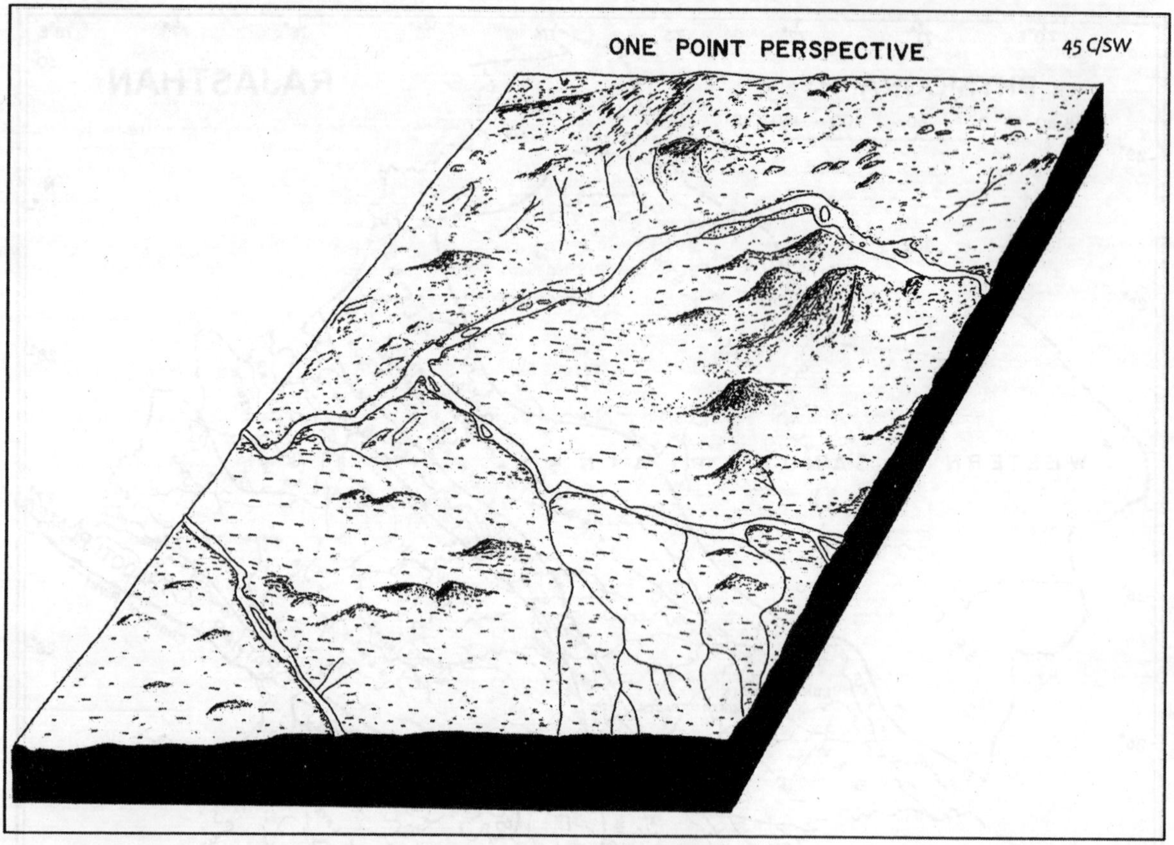

ONE POINT PERSPECTIVE

45 C/SW

Figure 3.4 One Point Perspective

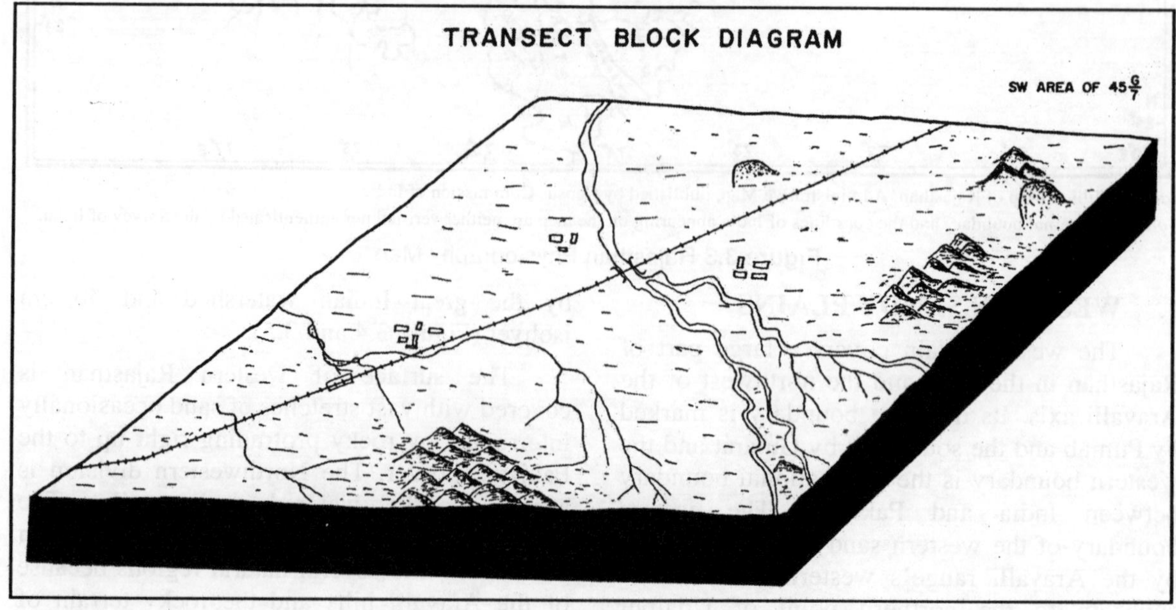

Figure 3.5 Transect Block Diagram

The Western Sandy Plains are subdivided into two units:

(a) Sandy Arid Plains (Marushthali), and

(b) Semi-Arid Transitional Plain (Rajajsthan Bagar) divided into

(i) Luni Basin (Godwar Tract)

(ii) Plain of Interior Drainage (Shekhawati Tract) and

(iii) Ghaggar Plain

The line dividing the Sandy Arid Plains and the Semi-Arid Transitional Plain is climatic, i.e. 25 cm isohyte. The western-most belt which is 'The Great Desert' is covered by sand dunes, stretches from the Great Rann along the Pakistan border to Punjab. The degree and extent of sand dunes greatly influence the economic activity in this area. 63% and dune area of Rajasthan is concentrated in the desert districts of Barmer, Jaisalmer and Bikaner.

2. ARAVALLI RANGE AND HILLY REGION

The principal and most dominant mountain range of Rajasthan is the Aravalli range. The geological history of the Aravalli range can be traced back probably close of the dharwar times. This range runs diagonally across the State from northeast near Delhi and extends to southwest up to the Plains of Gujarat, for a distance of about 692 kilometers., the range runs from Khetri in the northeast to Khed Brahma in the southwest Within Rajasthan for a length of about 550 kilometers. It is composed of rocks belonging originally to the Delhi system, folded in a synclinorium occupying the site of the geosynclines which have been deeply eroded.

The whole of this Aravalli range and Hilly Region has been further sub-divided into two smaller physiographic units:

(a) Aravalli Range and Bhorat Plateau

This division constitutes east Sirohi, nearly the whole of Udaipur except a narrow belt in the east and whole of Durgarpur district.

The highest section of the Aravalli range known as 'Bhorat' Plateau lies northwest of Udaipur between Kumbhalgarh-Gogunda and the average elevation of this plateau is 1,225 m.

(b) North-Eastern Hilly Tract

Northern edge of Bhorat Plateau marks the boundary of north eastern hilly tract in the southwest. North of Kumbhalgarh, especially at the Alwar hills the elevation varies between 550 m to 670 m. Further north and northeast, the hills are broken and their elevation decreases in the final extension of hills south of Delhi to about 306 m above sea level. This level is about 60 m to 90 m above the surrounding plain.

3. EASTERN PLAINS

The area on the northeast, east and southeast of the Aravalli range is known as the Eastern Plain. The Vindhyan Plateau marks the southeastern limit of the Plain. The western boundary is demarcated by the eastern edge of the Aravalli up to north of Udaipur. This Plain is further subdivided into two physiographic units-

(i) The Banas Basin

(ii) The Chappan Plain

(i) Banas Basin

The great watershed of India runs in an easterly direction starting From Udai Sagar, east of Udaipur. The watershed acts as the southern boundary of the Mewar Plains and south of this watershed is the Chappan Plain. The region is marked by various types of erosional features, produced in the granite and gneiss rocks mark the topography of the area. The soil is stony and the annual rainfall is about 73 cm. From the foot of the Aravalli range the plain slopes gradually towards northeast. River Banas and its tributaries drain the Mewar plain.

(ii) The Chappan Plains

This plain drained by the tributaries of the Mahi River lies south of the great Indian watershed in southeastern Udaipur, Banswara and the southern part of Chittorgarh district.

The Chappan area is deeply and intricately eroded resulting in the formation of separate hillocks, which is not identical to the Mewar plain in the north. This deeply dissected area is locally known as 'Begaar' and includes the hilly tracts of Banswara and Dungarpur.

4. SOUTHEASTERN RAJASTHAN PATHAR (HADOTI PLATEAU)

The eastern part along the Chambal River is covered by the Hadoti Plateau. The Great Boundary Fault of the Aravallis forms its northwest boundary which extends eastward across the Rajasthan border. River Chambal drains the large part of this area. The Plateau has been divided into two smaller units:

(a) Vindhyan Scarp Land

The scarp lands formed by massive sandstones separated by shale have an average elevation between 350m to 580m. The scarps are facing towards the south-southeast between the Banas and the Chambal and extend towards the east over Bundelkhand. A scarp block occupies the areas of Dholpur and Karauli.

(b) Deccan Lava Plateau

The Deccan Lava Plateau is also known as 'pathar' (stony) or 'uparmal' plateau. It is a wide stony upland, including Kota-Bundi plateau section. The Chambal along with its tributaries Kali Sindh and Parbati form a triangular alluvial basin of 210m-275m at Kota.

GEOMORPHOLOGY

The geomorphology of the state is quite varied. Broadly three types of units occupy major part of the state, viz, Aeolian, Fluvial and to some extent denudational. Aravallis form central NE-SW trending axis occurring as structural ridges and linear ridges in the state and the eastern/southeastern part represented by Vindhyans. (Figure 3.6, Table 3.3 and Figure 3.7)

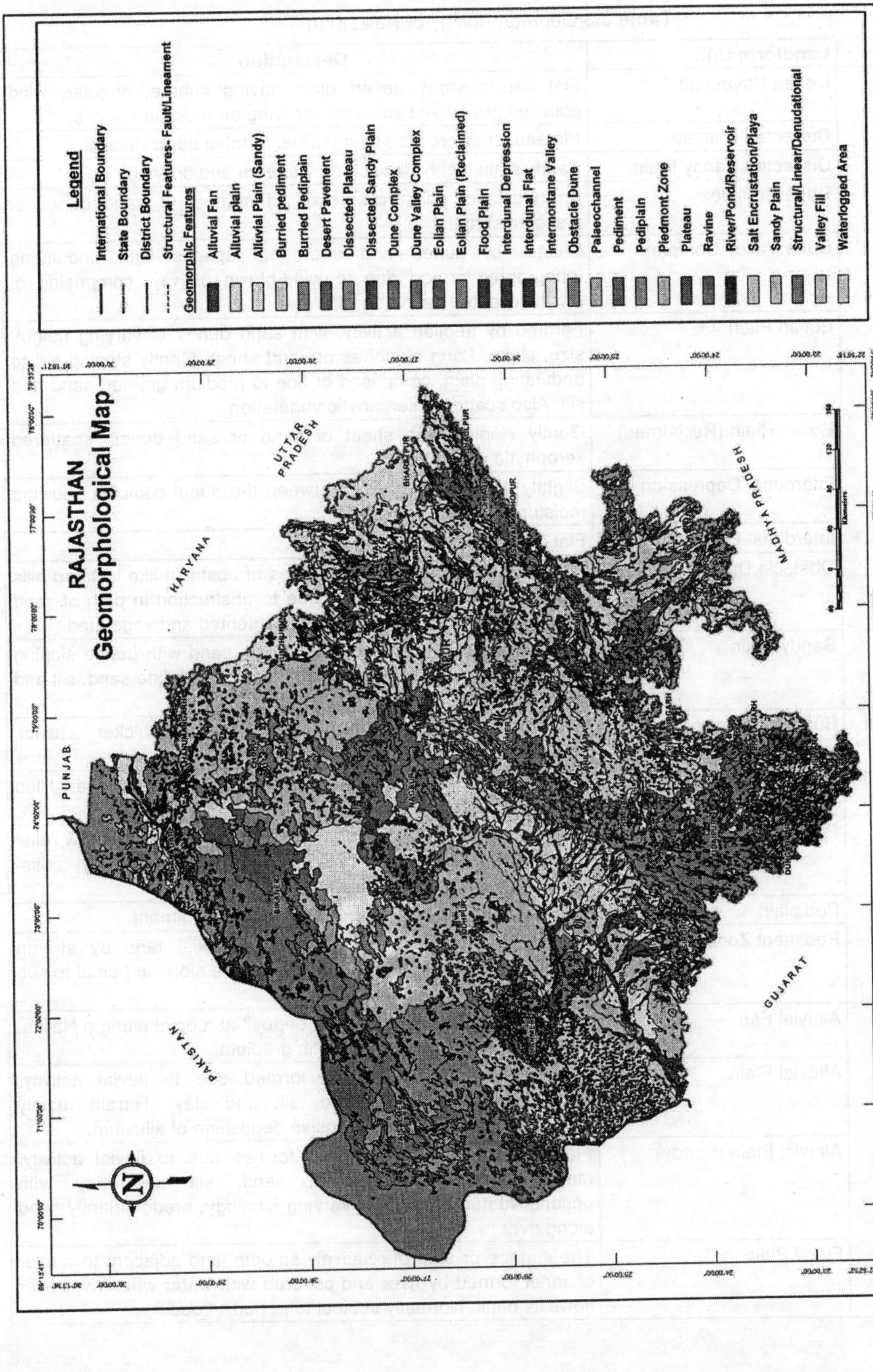

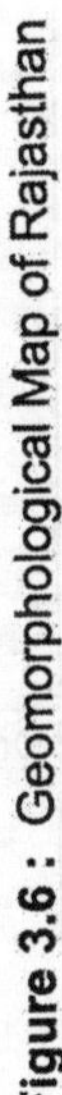

Source: Department of Ground Water, Government of Rajasthan, Jodhpur

Figure 3.6 : Geomorphological Map of Rajasthan

Table 3.3 Geomorphology of Rajasthan

Origin	Landform Unit	Description
Aeolian	Desert Pavement	Flat barren stony desert plain having coarse, angular, wind polished gravel and small stones lying on rock cut surface.
	Dissected Plateau	Plateau, criss-crossed by fractures forming deep valleys.
	Dissected Sandy Plain	Sandy plain highly dissected by stream and drainage
	Dune Complex	An undulating plain composed of number of sand dunes of cressent shape.
	Dune Valley Complex	Cluster of dunes and interdunal spaces with undulating topography formed due to wind-blown activity, comprising of unconsolidated sand and silt.
	Eolian Plain	Formed by aeolian activity, with sand dunes of varying height, size, slope. Long stretches of sand sheet. Gently sloping flat to undulating plain, comprised of fine to medium grained sand and silt. Also scattered xerophytic vegetation.
	Eolian Plain (Reclaimed)	Gently sloping with sheet of sand or sand dunes, scattered xerophytic vegetation.
	Interdunal Depression	Slightly depressed area in between the dunal complex showing moisture and fine sediments.
	Interdunal Flat	Flat, narrow land between dunes.
	Obstacle Dune	Formed on windward/leeward sides of obstacle like isolated hills or continuous chain of hills, dune to obstruction in path of sand laden winds. Badly dissected well cemented and vegetated.
	Sandy Plain	Formed of aeolian activity, wind-blown sand with gentle sloping to undulating plain, comprising of coarse sand, fine sand, silt and clay.
Denudational	Burried Pediment	Pediment covers essentially with relatively thicker alluvial, colluvial or weathered materials.
	Intermontane Valley	Depression between mountains, generally broad & linear, filled with colluvial deposits.
	Pediment	Broad gently sloping rock flooring, erosional surface of low relief between hill and plain, comprised of varied lithology, criss-crossed by fractures and faults.
	Pediplain	Coalescence and extensive occurrence of pediment.
	Pediment Zone	Formed by coalescence of several alluvial fans by stream covering large area at foot hills, with gentle slope in humid to sub humid region.
Fluvial	Alluvial Fan	A fan shaped mass of sediment deposit at a point along a Nallah, river where there is a decrease in gradient.
	Alluvial Plain	Mainly undulating landscape formed due to fluvial activity, comprising of gravels, sand, silt and clay. Terrain mainly undulating, produced by extensive deposition of alluvium.
	Alluvial Plain (Sandy)	Flat to gentle undulating plain formed due to fluvial activity, mainly consists of gravels, sand, silt and clay with unconsolidated material of varying lithology, predominantly sand along river.
	Flood Plain	The surface or strip of relatively smooth land adjacent to a river channel formed by river and covered with water when river over flows its bank. Normally subject to periodic flooding.

Origin	Landform Unit	Description
	Paleochannel	Mainly buried on abandoned stream/river courses, comprising of coarse textured material of variable sizes.
	Valley Fill	Formed by fluvial activity, usually at lower topographic locations, comprising of boulders, cobbles, pebbles, gravels, sand, silt and clay. The unit has consolidated sediment deposits.
	Ravine	Small, narrow, deep, depression, smaller than gorges, larger than gulley, usually carved by running water.
	Salt Encrustation/Playa	Topographical depression comprising of clay, silt, sand and soluble salts, usually undrained and devoid of vegetation.
	Water logged/ Wetland	Area submerged in water or area having very shallow water table. So that it submerges in water during rainy season.
Hills	Denudational Hill	Steep sided, relict hills undergone denudation, comprising of varying lithology with joints, fractures and lineaments.
	Linear Ridge	Long narrow low-lying ridge usually barren, having high run off may form over varying lithology with controlled strike
	Plateau	Formed over varying lithology with extensive, flat, landscapes, bordered by escarpment on all sides. Essentially formed horizontally layered rocky marked by extensive flat top and steep slopes. It may be criss crossed by lineament.
	Structural Hill	Linear to arcuate hills showing definite trend-lines with varying lithology associated with folding, faulting etc.

Source: Department of Ground Water, Government of Rajasthan, Jodhpur

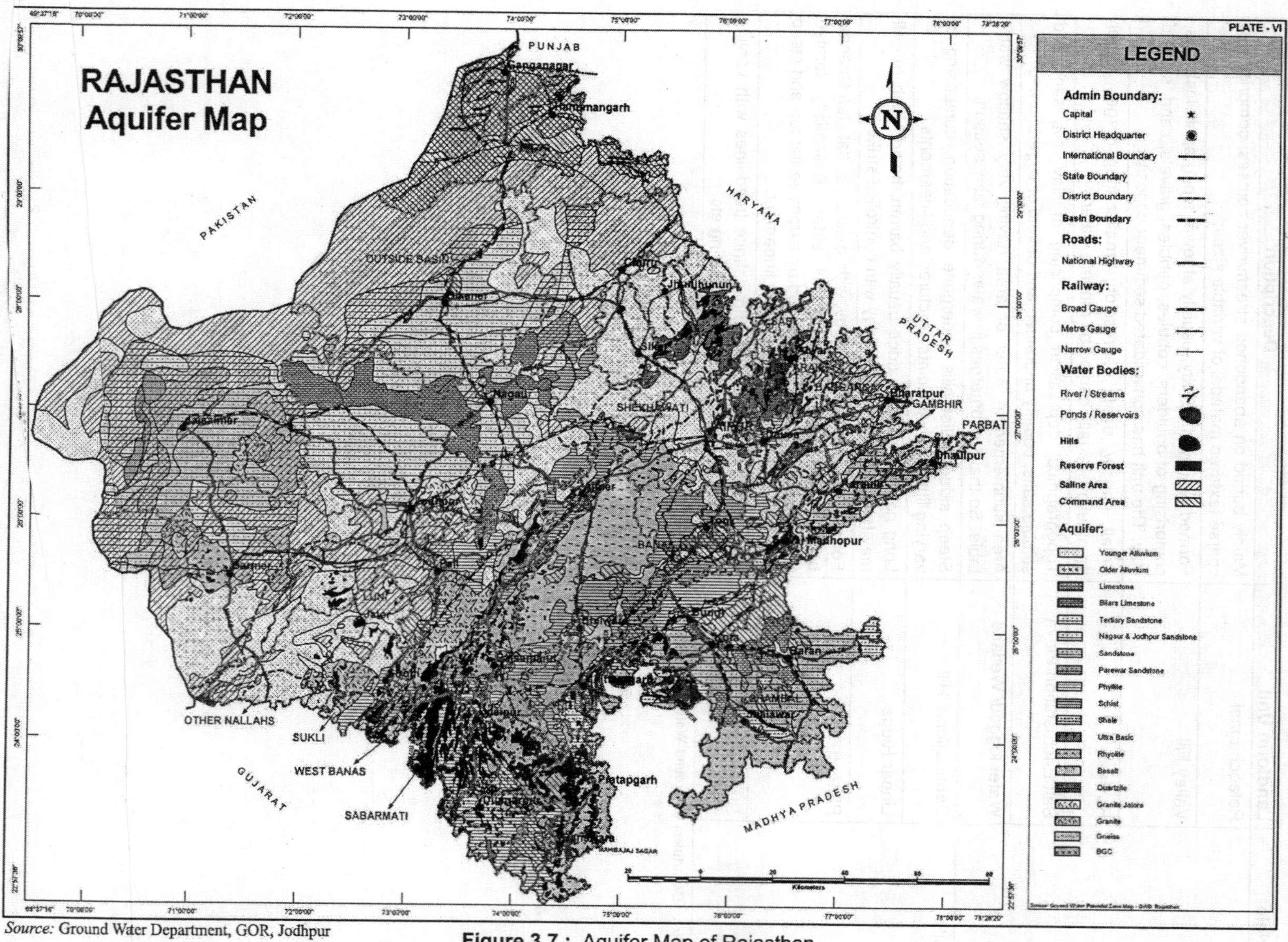

Source: Ground Water Department, GOR, Jodhpur

Figure 3.7 : Aquifer Map of Rajasthan

AQUIFERS

Standard Aquifer Type	Area (sq km)	% of total	Occurrence in District	Description of aquifer
Younger Alluvium	47,662.80	13.92	Ajmer, Baran, Barmer, Bundi, Churu, Dausa, Ganganagar, Hanumangarh, Jaipur, Jaisalmer, Jalor, Jhunjhunun, Jodhpur, Kota, Pali, Sikar, Sirohi.	It is largely constituted of Aeolian and Fluvial sand, silt, clay, gravel and pebbles in varying proportions.
Older Alluvium	95,134.95	27.78	Alwar, Barmer, Bharatpur, Bikaner, Churu, Dausa, Dhaulpur, Ganganagar, Hanumangarh, Jaipur, Jalor, Jhunjhunun, Jodhpur, Karauli, Nagaur, Pali, Sawai Madhopur, Sikar, Tonk.	This litho unit comprises of mixture of heterogeneous fine to medium grained sand, silt and kankar.
Limestone	4,136.09	1.21	Banswara, Baran, Bhilwara, Bundi, Chittorgarh, Kota, Pali, Sawai Madhopur.	In general, it is fine to medium grained, grey, red yellowish, pink or buff in colour.
Bilara Limestone	9,102.33	2.66	Bikaner, Churu, Jodhpur, Nagaur.	It is grey to buff coloured hard and compact.
Tertiary Sandstone	24,272.32	7.09	Barmer, Bikaner, Jaisalmer, Nagaur.	Medium to coarse grained, consolidated to semi consolidated sandstone.
Nagaur & Jodhpur Sandstone	23,920.14	6.99	Bikaner, Churu, Jaisalmer, Jodhpur, Nagaur.	Buff to reddish brown in colour, fine to medium grained hard and compact sandstone.
Vindhyan Sandstone	26,930.64	7.87	Baran, Bhilwara, Bundi, Chittorgarh, Dhaulpur, Jhalawar, Karauli, Kota.	Fine to medium grained, red colour and compact and at places.
Parewar Sandstone	1,285.44	0.38	Jaisalmer.	It is feldspathic ferruginous sandstone.
Gneiss	13,776.11	4.02	Ajmer, Bhilwara, Dausa, Jaipur, Pali, Tonk.	Comprises of porphyritic and non porphyritic gneissic complex.
Phyllite	14,456.52	4.22	Banswara, Bundi, Dausa, Dungarpur, Pali, Rajsamand, Sawai Madhopur, Sirohi, Udaipur.	These include meta sediments and represented by carbonaceous phyllite.
Schist	21,438.73	6.26	Ajmer, Bhilwara, Chittorgarh, Churu, Dungarpur, Jaipur, Jodhpur, Nagaur, Pratapgarh, Rajsamand, Sirohi, Tonk, Udaipur.	Medium to fine grained compact rock. The ltho units are soft, friable and have closely spaced cleavage.
Shale	4,574.00	1.34	Baran, Bundi, Chittorgarh, Jhalawar, Kota, Pratapgarh, Sawai Madhopur.	Grey, light green and purple in colour and mostly splintery in nature.

Standard Aquifer Type	Area (sq km)	% of total	Occurrence in District	Description of aquifer
Ultra Basic	103.28	0.03	Dungarpur.	This comprises serpentinite, hyperstinite and amphibolite.
Rhyolite	6,588.26	1.92	Barmer, Jaisalmer, Jalor, Jodhpur.	Rhyolite is porphyritic and has phenocryst of quartz and feldspar.
Basalt	8,969.64	2.62	Banswara, Baran, Jhalawar, Pratapgarh.	Dark grey, olive green and green colour, compact, vesicular, amygdaloidal and weathered.
Quartzite	4,998.73	1.46	Alwar, Bharatpur, Dausa, Jaipur, Jhunjhunun, Karauli, Sawai Madhopur, Sikar, Udaipur.	Medium to coarse grained and varies from feldspathic grit to sericitic quartzite.
Granite Jalore	1,852.64	0.54	Jodhpur.	Grey to pink in colour, medium to coarse grained, and non porphyritic.
Granite	9,431.08	2.75	Barmer, Jaipur, Jaisalmer, Jalor, Pali, Sirohi, Udaipur.	Light gey to pink colour, medium to coarse grained, and characteristically have porphyritic texture.
BGC	8,324.08	2.43	Banswara, Chittorgarh, Dungarpur, Pratapgarh, Rajsamand, Udaipur.	Grey to dark coloured, medium to coarse grained rocks.
Hills	14,058.86	4.11	Ajmer, Alwar, Banswara, Baran, Barmer, Bharatpur, Bhilwara, Bundi, Chittorgarh, Dausa, Dungarpur, Hanumangarh, Jaipur, Jalor, Jhalawar, Jhunjhunun, Jodhpur, Karauli, Nagaur, Pali, Pratapgarh, Rajsamand, Sawai Madhopur, Sikar, Sirohi, Tonk, Udaipur.	
Reserve Forest	1,387.87	0.41	Chittorgarh, Jalor.	–
TOTAL	**342,404.50**	**100.00**		

Source: Department of Ground Water, Government of Rajasthan, Jodhpur

Chapter 4

DRAINAGE SYSTEM AND LAKES

The rivers of Rajasthan, except for the Chambal, are ephemeral and flow only during the rainy season. The river catchment areas, which are largely influenced by local topography and amount of precipitation, vary largely showing an indelible impact upon the drainage system. In the absence of any formidable mountain barrier in the state and the low quantum of annual rainfall, the river catchment areas are mostly small. The flat, rugged and undulating rolling lands of the low hills and plateaus, which are the origins of the rivers and streams, are highly dissected and devoid of deep channels to add stream density and so the catchments are of low stream density.

The Aravalli ranges which run across the state from the south-west to the north-east direction largely influence the drainage system of the state. While a major part of western Rajasthan has an inland drainage system, the southern, the south-eastern and the eastern parts have a well developed drainage system. The river system of the state can be conveniently put into two major groups (i) draining its run off ultimately to the Bay of Bengal, and (ii) flowing to reach the Arabian Sea. While all the rivers flowing on the eastern side of the Aravalli ranges fall in the first group, rivers flowing in the southern and the western part of the state belong to the second group.

The river Chambal, the only perennial river of the state, which forms the boundary with Madhya pradesh, has its origin south of Mhow, near Manpur form the northern parts of Vindhyans. It enters Rajasthan near Chaurasigarh Fort flows through Kota district. Later, it flows to form the district boundary between Kota and Bundi and also between Kota and Sawai Madhopur. This river covers a distance of about 135 kms in Rajasthan out of its total length of about 965 kms. and has therefore, a smaller share of drainage system in the state. It has a catchment of 75,680 sq. km. (including Banas river) in Rajasthan. The river collects the major part of drainage discharge from Jaipur, Dausa, Sawai Madhopur, Karaulli Ajmer, Tonk, Bhilwara, Bundi, Kota, Baran, Jhalawar, Chittaurgarh, Udaipur and Rajsamand districts. Its main tributaries are Banas, Kali Sindh, Parvati etc. The river Chambal joins the river Jamuna near Etawah in U.P. (Figure 4.1).

The river Banas originating from the Khamnor hills of the Aravalli ranges (about 5 km from Kumbhalgarh) is one of the major rivers of the state which, in its entire course, flows through Rajasthan. It flows from Kumbhalgarh towards the south upto the Gogunda plateau and after cutting the Aravalli ranges at right angles, it flows through Nathdwara, Rajsamand and Railmagra. The river collects the major run-off of Udaipur, Rajsamand, Chittaurgarh and Bhilwara districts by the tributary, Berach. The Banas meets the river Chambal near Khandar. This total length of this river is about 480 kms and its main tributaries are Berach. Kothari, Khari, Daj, Mashi, Dundh and Morel.

The river Banganga which originates from the low hills of Bairath (Jaipur district), flows towards the east, entering Sawai Madhopur district and finally in Bharatpur district where its water spreads over a large area. The total length of this river is about 380 kms. The river joins the river Jamuna near Fatehabad of Agra district.

The Gambhiri which originates form Karaulli hills in Karaulli and Sawai Madhopur districts, flows through these ephemeral and flow only during the rainy season. The river catchment areas, which are largely influenced by local topography and amount of precipitation, vary largely showing an indelible impact upon the drainage system. In the absence of any formidable mountain barrier in the state and the low quantum of annual rainfall, the river catchment areas are mostly small. The flat, rugged and undulating rolling lands of the low hills and plateaus, which are the origins of the rivers and streams, are highly dissected and devoid of deep channels to add stream density and so the catchments are of low stream density.

Luni is the only major stream on the west of the Aravalli ranges which meets the sea in the Rann of Kutch. It originates in Nag Pahar in Ajmer and flows south-west through Jodhpur, Banner and Jalor over a distance of about 320 kms, covering a total catchment area of 34,250 sq. km. Its catchment area has spread over some parts of Nagaur district also. It flows only during the rainy season and even at this time, the river is choked with advancing sands at many places. Even in the monsoon months when the water flow rate is high and at its peak, the flowing water is not able to cut these sand deposits on the river bed, as a result of which, the run off spreads to the adjoining fields on both the banks of the river braiding it. This results in sand deposition on arable and non-arable lands, thus, adversely affecting their productivity. A unique characteristic of this river is that its water is good upto Balotra, but it becomes brackish thereafter till it drains in the Rann of Kutch. A number streams and tributaries like Bandi, Sukri, Mithri, Jawai, Sagi, Jojri, Lilri, Guhiya etc., flows on the western side of the Aravalli range and drain the run off of some parts of Pali, Jalor, Sirohi and Banner district areas (Table 4.1).

The river Mahi, which originates form the Mhow ranges of Madhya Pradesh, and after flowing for some distance towards noth, enters Rajasthan state near Khandu village of Banswara district. Its catchment area in Rajasthan largely spreads over the districts of Udaipur, Banswara, Chittaurgarh, Pratapgarh and Dungarpur, covering a total geographical area of about 16,030 sq. kms. Important tributaries of this river are Som, Jakham, Anas, Chap and Moren.

The west Banas originates from the western slope of the Aravalli ranges and drains some parts of Sirohi district, finally draining its ran off in the Rann of Kutch.

The Sabarmati, originating form the low hills of Gogunda (Udaipur district) and called Wakal in its initial course, covers a catchment are of 4,300 sq. km. In the state and drains part of its water from Udaipur and Sirhoi district before finally traversing across Gujarat to meet the Gulf of Cambay.

The major part of western Rajasthan has sandy terrain with relatively low rainfall. There are a number of small rivers/nallahs and seasonal streams which flow in this area for some distance, and then disappear in the vast expanse of sandy terrain.

The river Kantli originates from the Khandela hills of Sikar district. It flows during the rainy season only. After flowing over a distance of 100 kms. towards the north through Sikar and Jhunjhunun districts, it enters Churu district where its water is lost in the sand terrain. The river may flow with gushing waters during heavy rainfall, causing lateral cutting of the banks and a braided channel in some parts during its course.

Sahibi river has its origin from the Sewar hills (Jaipur district). It flows through Bansur, Bahrod, Mandawar and Tijara tehsils to enter Gurgaon district of Haryana state where, after flowing for some distance, its water is either lost in the adjacent areas or spread over vast stretches of land to cause flood havoc (Figure 4.2).

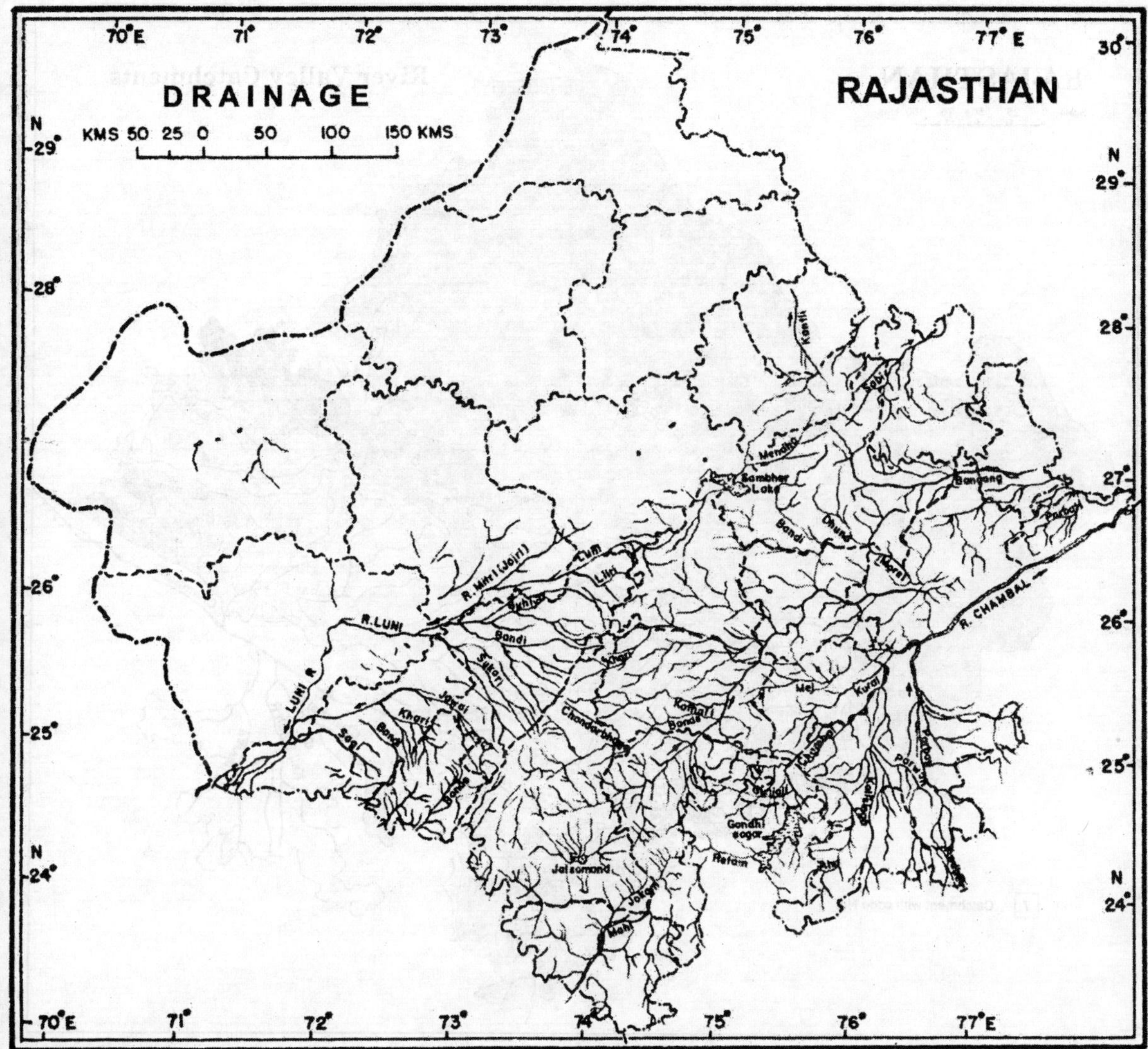

Source: Million Map of Rajasthan, Administrative Map, published by Census Commission ot India and Irrigation, Department of Rajasthan, Jaipur.

Note : The external boundary and the coastlines of India appearing on the map are neither verified nor authenticated by the Survey of India.

Figure 4.1 Drainage

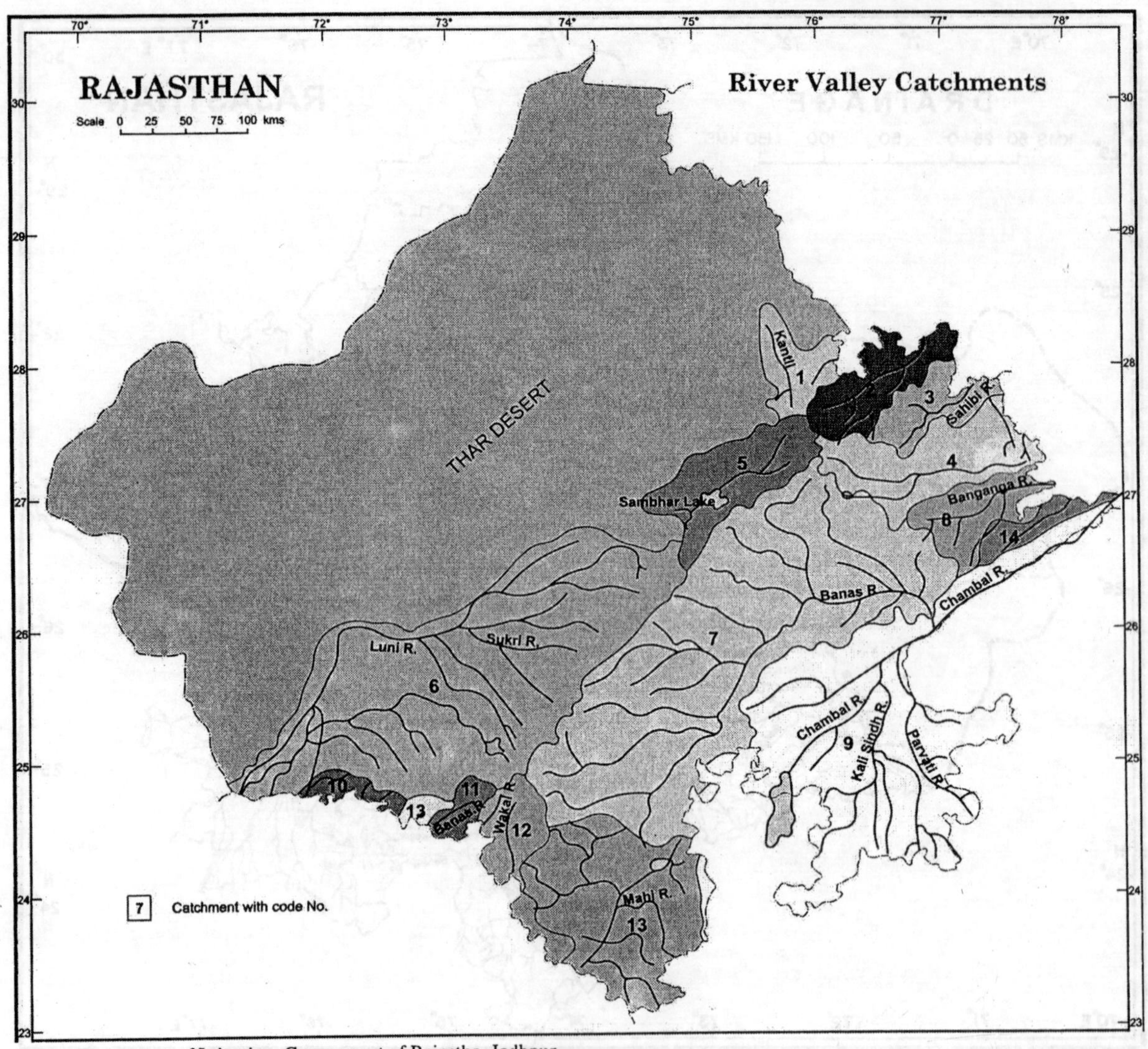

Source: Department of Irrigation, Government of Rajastha, Jodhpur

Note : The external boundary and the coastlines of India appearing on the map are neither verified nor authenticated by the Survey of India.

Figure 4.2 River Valley Catchments of Rajasthan

Table 4.1 : Catchment Area of Major River System of Rajasthan

S. No.	River System	Catchment Code No.	Catchment Area (sq.km.)	Percentage of Total Catchment Area
1	Kantli	1	2570	1.52
2	Dohan	2	1400	0.83
3	Sahibi and Sota	3	4480	2.64
4	Bar	4	4520	2.67
5	Mendha and Rupangarh (Sambhar Lake)	5	6940	4.10
6	Luni	6	34250	20.21

S. No.	River System	Catchment Code No.	Catchment Area (sq.km.)	Percentage of Total Catchment Area
7	Banas	7	46570	27.48
8	Banganga and Gambhiri	8	14360	8.47
9	Chambal	9	29110	17.18
10	Sukal	10	1940	1.14
11	Banas (West)	11	3000	1.77
12	Sabarmati (Wakal)	12	4300	2.53
13	Mahi	13	16030	9.46
	TOTAL		169470	100.00

Source: Department of Irrigation, Government of Rajastha, Jodhpur

Location of the great Indian watershed and the existence of the Aravalli axis greatly influence the drainage system of Rajasthan. The drainage to the Bay of Bengal and the Arabian Sea gets divide due to the Great Indian watershed which runs along the Aravalli axis from the Sambhar Lake southward to Ajmer. From here before proceeding towards the southwest of Udaipur city the line runs to southwest, a few kilometers east of Beawar and to Deogarh and Kumbhalgarh further extending to in the west, past Udaisagar and runs to southeast to Bari Sadri, from Choti Sadri to Pratapgarh. Table 4.2 shows the river basin area (Figure 4.3).

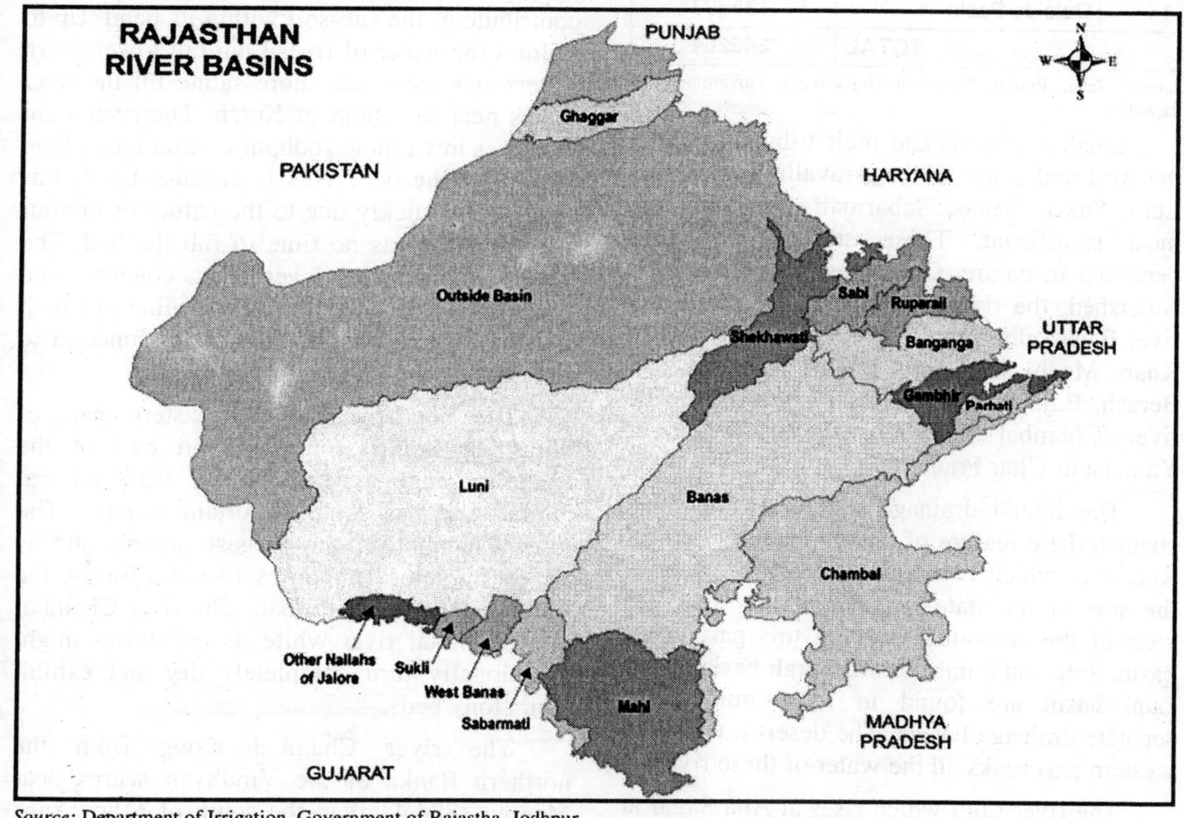

Source: Department of Irrigation, Government of Rajastha, Jodhpur.

Figure 4.3 River Basins Area in Rajasthan

Table 4.2 River Basin Areas in Rajasthan State (2012)

S.No.	Name	Basin Area (Sq.Km.)
1	Shekhawati	9691
2	Ruparail	2550
3	Banganga	9949
4	Gambhir	4934
5	Parvati	1891
6	Sabi	4615
7	Banas	46902
8	Chambal	31229
9	Mahi	16598
10	Sabarmati	4118
11	Luni	69580
12	West Banas	1835
13	Sukli	994
14	Other Nallah of Jalore	1775
15	Outside Basin	135603
	TOTAL	**342264**

Source: Tahal Report, Irrigation Department, Government of Rajasthan

Smaller streams and their tributaries drain the west and south of the Aravalli axis. Rivers Luni, Sukri, Banas, Sabarmati and Mahi are most significant. These streams are non-perennial in nature. On the eastern side of the watershed, the river Chambal is joined by the river Banas along with its main tributaries like Khari, Moshi and Morel on the left bank and Berach, Bajasen and Golwa on the right. The river Chambal ultimately joins the river Yamuna in Uttar Pradesh.

The inland drainage system is the most characteristic feature of the drainage system of Rajasthan which is that nearly 60.2 per cent of the area of the state. Nearly all this area lies west of the Aravalli range. In this part Kanti basin, Sota and Sahibi basin, Barah basin of the Luni basin are found in large number of separate drainage basins. The desert tract in the western part soaks all the water of these rivers.

The river Luni which rises at Ana Sagar at Ajmer is the only significant water course in this area is and flows towards the southwest for a distance of about 320 km through the districts of Jodhpur, Barmer and Jalor in the semi-arid tract west of the Aravalli range., the river has a small catchment area of about 32 sq km at the source at Talod Road. A small tributary joins from the Pushkar valley and the basin of the river widens. Near Ajmer, the river flows down the Aravalli slope and after 10 km flows towards the southwest. This river drains the total catchment area of about 34,866.40 sq kilometers.

As it is a rain-fed stream the river is choked with advancing sands at many places during the dry season. When the river is carrying maximum water during the monsoon season it is not able to cut the Aeolian deposits. It recieves many smaller hill torrents from the western slope of the Aravalli range, like Lalri, Ghuhia, Bandi, Sukri, Jawai, Jojri and Sagai, all joining on the left bank. All these streams contribute to the sub-soil within its bend. Up to Balotra the water of river Luni but lower down it becomes more and more saline till the river drains near the Rann of Kutch. The river Luni increases in width at Jodhpur district rather than deepening the bed. This is because the floods develop so quickly due to the nature of rainfall that the river has no time to rub the bed. The River Luni spills over the country and occasionally damages the railway line to which it actually runs parallel from Luni Junction to Gole during the rainy season.

The southern and the eastern part of Rajasthan, south, southeast and east of the Aravalli range receives more than 80 cm rainfall and has some important streams. The river Chambal is the largest stream and is joined by some tributaries like the Banas, the Kali Sindh, and the Parbati. The river Chambal is a perennial river while its tributaries might occasionally turn completely dry and exhibit their stony beds.

The river Chambal rising from the northern flanks of the Vindhyan scarps near Manpur (884.4 m) in the south of Mhow runs for about 325 km through a long narrow and steep gorge which overhangs the valley on both the sides rising about 60 m to 90 m above the valley floor. The river falls at 505 m near Chaurasigarh to Kota. The river Chambal is joined by its first major tributary - river Kali Sindh near Monera village. Another tributary Parbati joins about 48 km downstream. Taking a straight course for about 212 km, it bends southeast at Pinahat and flows to join the river Yamuna near Muradganj, after a total run of about 965 kilometres. For a length of about 153 km the river flows entirely in Rajasthan. the river forms the boundary between Rajasthan and Madhya Pradesh from Palia to Pinahat for about 241 km. it forms the boundary between Madhya Pradesh and Uttar Pradesh Before joining the river Yamuna in Uttar Pradesh. In Rajasthan two other tributaries Kurai and Banas join the Chambal river on the left bank (Table 4.3).

The Banas river flows through the heart of the Mewar Plains. Its chief tributaries are Berach, Kothari, Khari, Dhoond and Morel. The river Banas rises from the catchment area lying between Kankroli and Nathdwara and flows towards the east as far as Mandalgarh and further it flows towards the northeast up to Tonk where it again turns towards the east and finally this river turns at right angles and flows south to join the river Chambal. The upper reaches of this stream are hilly and have good rainfall.

Table 4.3 Districtwise Ravinous Area (in sq. km.)

S.No.	District	Total Geographical Area	Ravinous Area	%
1	Alwar	8380	206 .07	2.46
2	Baran	6992	225 .50	3.23
3	Bharatpur	5066	276.42	5.46
4	Bundi	5776	785.45	13.60
5	Dausa	3432	758.53	20.10
6	Dhaulpur	3033	726.40	23.95
7	Jaipur	11143	455 .22	4.08
8	Jhalawar	6219	708.34	11.39
9	Jhunjhunun	5928	299 .11	5.04
10	Karauli	5524	704.32	12.75
11	Kota	5217	573.44	10.99
12	Sawai Madhopur	4498	950.27	21. 13
13	Tonk	7194	375.00	5.21

Source: Ground Water Department, Government of Rajasthan, Jodhpur.

Ravine is an advanced stage of soil erosion by water. Prolonged inundation and impact of Seven eastern districts, viz. Baran, Kota, Bundi, Sawai Madhopur, Katauli, Dhaulpur and back water during high floods result in destruction Bharatpur have the bulk of the ravine areas. Other of protective vegetation and are the main reasons districts affected are Jaipur, Dausa, Jhalawar, for ravine formation. Depending upon the type of Alwar, Tonk, Bhilwara, and to some extent Sirohi, soil, geology, degree of vegetation cover, Ajmer, Jhunjhunun Dungarpur and Banswara. topography, rainfall pattern and the run-off from One or both banks of the river Chambal and its the catchment area, ravines of different sizes, main tributaries, namely, Kali Sindh, Parvati, shapes and depth are formed. Total area Parwan, Mej, Banas, Morel, Ghambhiri, gullied under and ravine land in Rajasthan, as estimated Banganga, etc. are affected with this problem of by digital analysis of satellite data on 1: 50,000 land degradation. scale, is about 7044.07 sq. km. About 49 per cent of the ravme area 1s Ravines have a very adverse effect on the owned by the Government, whereas private agro-economical activities of the area. Conversion ownership is about 40 percent. The remaining 11 of fertile cultivated land into unproductive ravine percent is the property of the Gram Panchayats. land is also mainly associated with many social problems. These are known to provide safe hiding places for dacoits. Due to lack of effective measures of management and reclaimation of these areas, their spread had been encroaching adjoining good arable lands and thus, the area under the ravines has increased considerably over the years. It has been estimated that the rate of encroachment of arable lands by ravines ranges from 0.5 to 1.07 percent per year in different situations.

Vindhyans. It enters Rajasthan near Chaurasigarh Fort flows through Kota district. Later, it flows to form the district boundary between Kota and Bundi and also between Kota and Sawai Madhopur. This river covers a distance of about 135 kms in Rajasthan out of its total length of about 965 kms. and has therefore, a smaller share of drainage s'ystem in the state. It has a catchment of 75,680 sq. km. (including Banas river) in Rajasthan. The river collects the major part of drainage discharge from Jaipur, Dausa, Sawai Madhopur, Karaulli Ajmer, Tonk, Bhilwara, Bundi, Kota, Baran, Jhalawar, Chittaurgarh, Udaipur and Rajsamand districts. Its main tributaries are Banas, Kali Sindh, Parvati etc. The river Chambal towards noth, enters Rajasthan state near Khandu village of Banswara district. Its catchment area in Rajasthan largely spreads over the districts of Udaipur, Banswara, Chittaurgarh, Pratapgarh and Dungarpur, covering a total geographical area of about 16,030 sq. kms. Important tributaries of this river are Som, Jakham, Anas, Chap and Moren.

The west Banas originates from the western slope of the Aravalli ranges and drains some parts of Sirohi district, finally draining its run off in the Rann of Kutch.

CHAMBAL RIVER BASIN

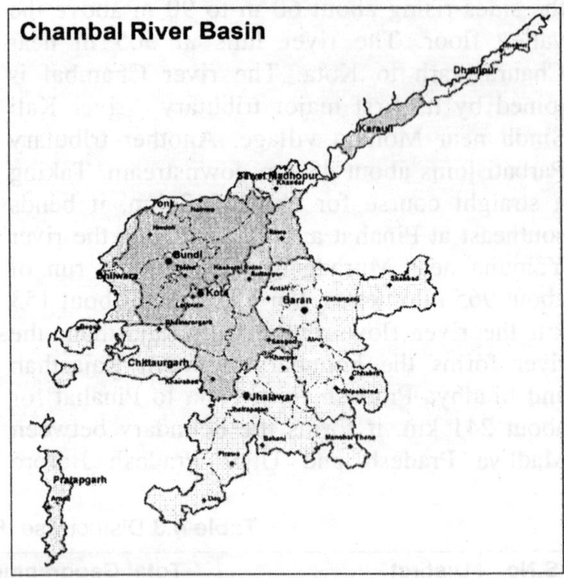

Source: Ground Water Department, GoR, Jodhpur

Figure 4.4 Chambal River Basin

Location

Chambal River Basin is located in the south-eastern part of Rajasthan. It stretches between 23° 31' : 03.91' to 26° 56' 03.72' North latitudes and between 74° 45' 15.00' to 78° 13' 33.86' East longitudes. It is bounded primarily by the Banas river basin in the west, partly by Mahi, Gambhir and Parbati river Basins. The eastern boundary is defined by Madhya Pradesh State where the rest of the catchment area of Chambal river lies (Figure 4.4). The Chambal basin extends over parts of Baran, Bhilwara, Bundi, Chittaurgarh, Dhaulpur, Jhalawar, Karauli, Kota, Pratapgarh, Sawai Madhopur and Tonk Districts. The total catchment area of the basin is approximately 31,239 km2 within the state of Rajasthan (excluding Banas river basin which is tributary to Chambal River. River Chambal is a principal tributary of river Yamuna and originates in the Vindhyan ranges near Mhow in Indore District of Madhya Pradesh, entering through a deep gorge in Rajasthan at Chourasigarh, about 96 km upstream of Kota. The deep gorge extends up to Kota and the river then flows for about 226 km within Rajasthan in a north-easterly direction, and then forms the boundary between MP and Rajasthan for about 252 km.

Administrative Set-up

Administratively, Chambal river basin extends over parts of Baran, Bhilwara, Bundi, Chittaurgarh, Dhaulpur, Jhalawar, Karauli, Kota, Pratapgarh, Sawai Madhopur and Tonk districts encompassing 41 Blocks and 5,786 towns and villages (Figure 4.5).

Table 4.3 Administrative Set-up of Chambal River Basin

S. No.	District Name	Area (sq km)	% of Basin Area	Total Number of Blocks	Total Number of Towns and Villages
1	Baran	6,993.9	22.4	7	1,213
2	Bhilwara	1,045.3	3.4	2	219
3	Bundi	5,610.2	18.0	5	827
4	Chittaurgarh	2,197.9	7.0	3	359
5	Dhaulpur	844.9	2.6	4	151
6	Jhalawar	6,315.3	20.2	6	1,608
7	Karauli	816.0	2.6	2	82
8	Kota	5,122.3	16.4	5	903
9	Pratapgarh	895.2	2.9	3	218
10	Sawai Madhopur	952.2	3.1	2	121
11	Tonk	446.0	1.4	2	85
	Total	**31,239.2**	**100.0**	**41**	**5,786**

Source: Ground Water Department, Government of Rajasthan, Jodhpur.

Climate

Chambal river basin falls within Semi-arid to Sub-humid climatic regions. It is very cold from November to February while turning hot from March to June when the maximum temperature rises up to 48°C. The mean annual rainfall over basin is approximately 697 mm (year 2010). Most of the rainfall is received during the Monsoon months between June/July to September. Months of September and October are very pleasant before onset of winters.

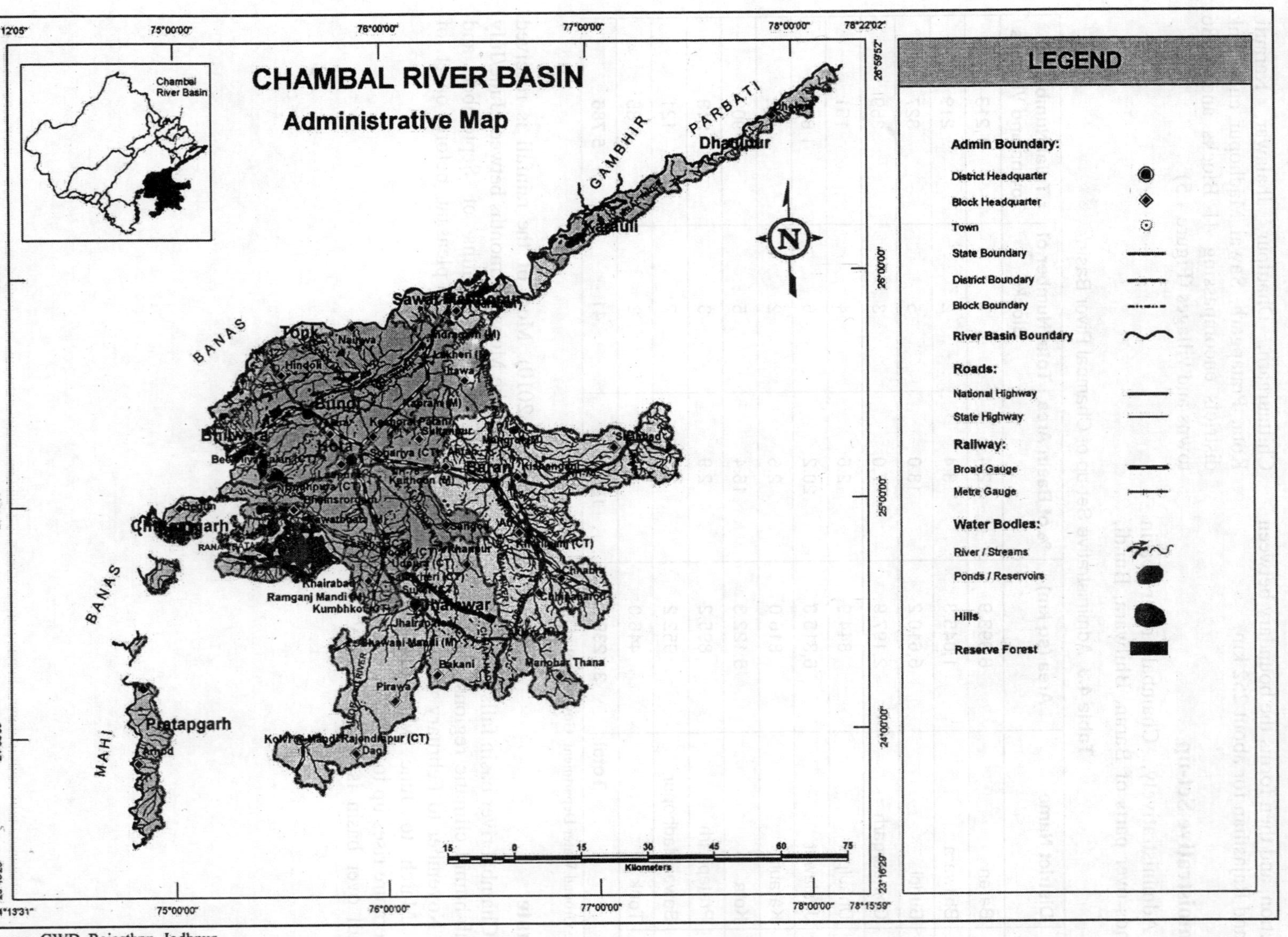

Source: GWD, Rajasthan, Jodhpur

Figure 4.5 : Administrative Map of Chambal River Basin

Topography

High elevation areas surround the central low lying parts of basin from east, south and western sides. The river flows in the north eastern direction and a narrow part extends through Karauli and Dhaulpur districts. The minimum elevation in the basin is 111 m amsl in Dhaulpur district where the river drains out of the basin and the maximum elevation is approximately 605 m amsl in Bhilwara district. Nearly half of the basin is hilly while rest is undulating plains (Figure 4.6 and Table 4.4).

Table 4.4 District wise minimum and maximum elevation

District Name	Min. Elevation (m amsl)	Max. Elevation (m amsl)
Baran	196.0	547.3
Bhilwara	304.3	605.3
Bundi	186.6	547.0
Chittaurgarh	293.6	604.1
Dhaulpur	111.3	341.2
Jhalawar	245.9	523.9
Karauli	131.0	469.5
Kota	177.0	517.0
Pratapgarh	437.0	575.0
Sawai Madhopur	164.7	502.7
Tonk	217.1	473.4

Source: Ground Water Department, GoR, Rajashan

Rainfall

The general distribution of rainfall across the Chambal river basin can be visualized from isohyets presented in the Plate III where most of the hilly areas seem to receive higher rainfalls (total annual rainfall in the range of 700-900mm) whereas, the central part received lower rainfall (between 400 to 700mm). The average annual rainfall for the year 2010, computed based on available station data is about 697 mm and Begun block had received highest rainfall measuring about 1,790 mm followed by Bijolia, 880 mm.

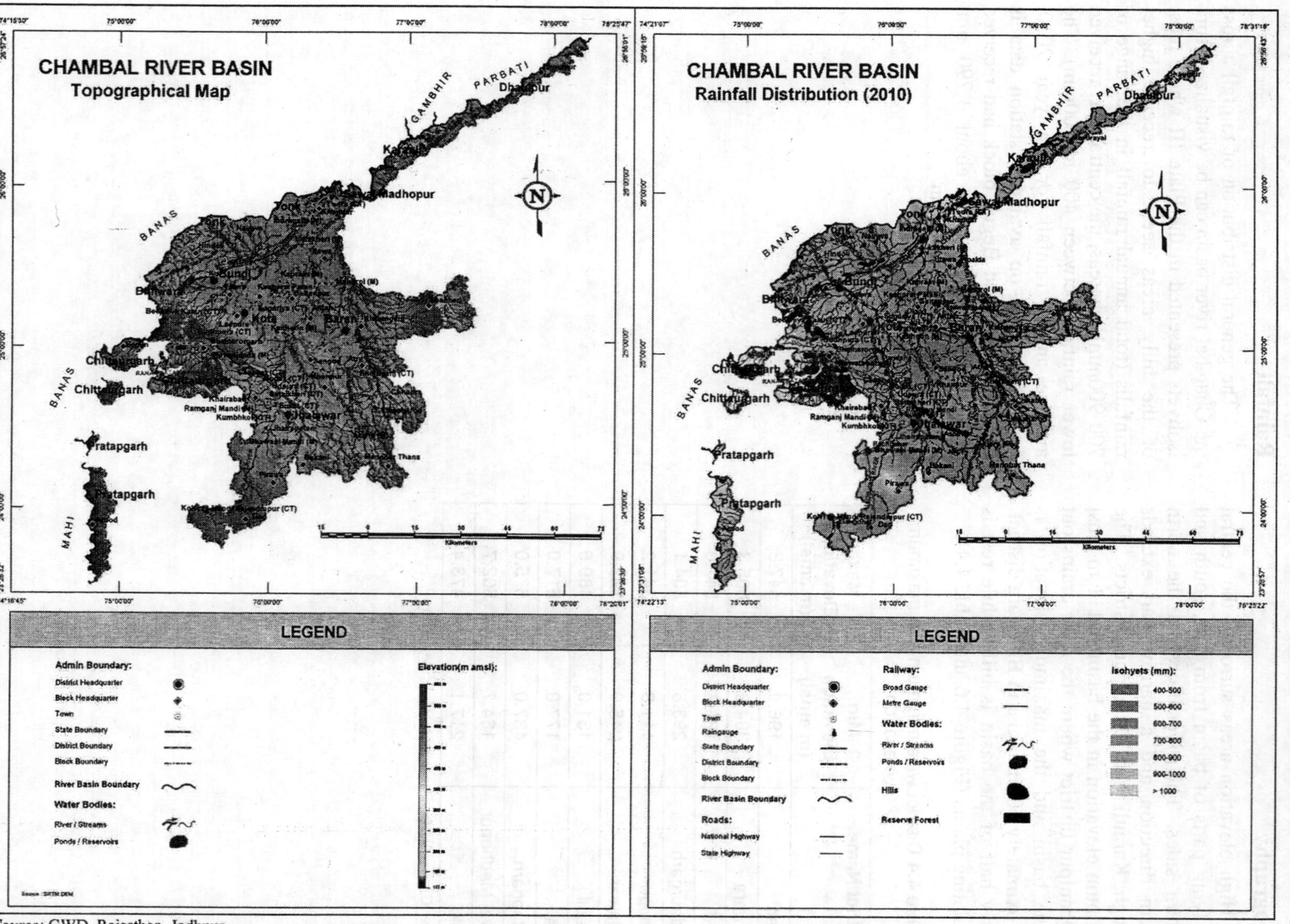

Source: GWD, Rajasthan, Jodhpur

Figure 4.6 : Chambal River Basin (Topographical and Rainfall Distribution)

BANAS RIVER BASIN

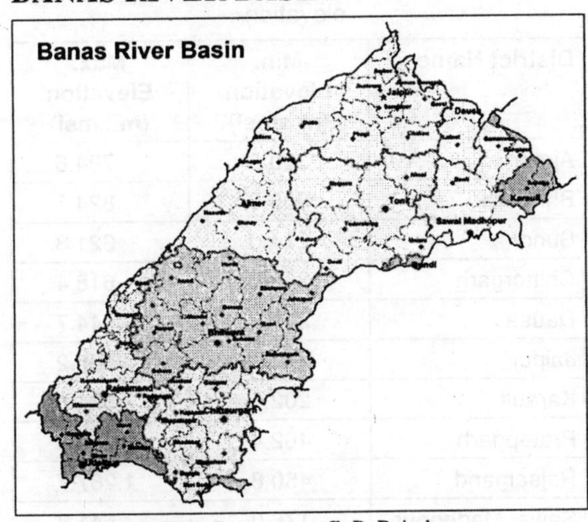

Source: Ground Water Department, GoR, Rajashan

Figure 4.7 Banas River Basin

Location

Banas River Basin is located in eastern part of Rajasthan and occupies significant area in the east of Aravali mountain range. It stretches between 24° 17' 14.22" to 27° 18' 15.24" North latitude and 73° 20' 54.84'' to 77° 00' 36.49''East longitudes (Figure 4.7). It is bounded in the east by Chambal river basin, in the north by Gambhir and Banganga river basins, in the west by Shekhawati and Luni river basins and in the south by Sabarmati and Mahi river basins. It is a tributary to Chambal River, which in turn flows into Yamuna River. Originating in the Khamnor Hills of the Aravali Range, about 5 km from Kumbhalgarh in Rajsamand District the Banas River flows for its entire length through Rajasthan only. It flows northeastwards through Mewar region of Rajasthan, meets the Chambal river near the village of Rameshwar in Khandar Block of Sawai Madhopur District. Major right bank tributaries include Berach and Menali whereas the left bank tributaries are Kothari, Khari, Dai, Dheel, Sohadra, Morel and Kalisel. Total length of the river is approximately 512 kilometers.

Administrative Set-up

Administratively, the basin extends over parts of Ajmer, Bhilwara, Bundi, Chittorgarh, Dausa, Jaipur, Karauli, Pratapgarh, Rajsamand, Sawai Madhopur, Tonk and Udaipur districts encompassing 78 Blocks and 8,511 towns and villages (Figure 4.8 and Table 4.6).

Table 4.6 Administrative Set-up of Banas River Basin

S. No.	District Name	Area (sq. km)	% of Basin Area	Total Number of Blocks	Total Number of Towns and Villages
1	Ajmer	5,547.4	11.8	9	625
2	Bhilwara	9,397.2	20.0	12	1,534
3	Bundi	166.3	0.4	1	29
4	Chittorgarh	5,450.8	11.5	13	1,269
5	Dausa	1,185.7	2.5	3	431
6	Jaipur	6,504.8	13.8	11	1,314
7	Karauli	1,113.2	2.4	3	161
8	Pratapgarh	177.3	0.4	1	29
9	Rajsamand	4,208.7	9.0	8	950
10	Sawai Madhopur	3,834.6	8.2	5	642
11	Tonk	6,744.4	14.3	6	1,015
12	Udaipur	2,669.0	5.7	6	512
	Total	**46,999.4**	**100.0**	**78**	**8,511**

Source: Ground Water Department, GoR, Jodhpur

Climate

Climate of Banas river basin is by and large, sub-humid and had received about 803 mm of average total annual rainfall in the year 2010. Winter sets in in the month of October and lasts till February while warmer period turning hot extends from March to July. The basin receives good rainfall during the four Monsoon months (July-September). Large tracts of the basin are suitable for agriculture practices.

Topography

The western part of the Basin is marked by hilly terrain belonging to the Aravali chain. East of the hills lies an alluvial plain with a gentle eastward slope. Ground elevations in the western hilly part range approximately from about 850 m above mean sea level (m amsl) to about 1,291 m amsl, while the alluvial plain elevations range approximately from 450 m amsl to 176 m amsl where river meets the river Chambal (Figure 4.9 and Table 4.7).

Table 4.7 Districtwise minimum and maximum elevation

District Name	Min. Elevation (m amsl)	Max. Elevation (m amsl)
Ajmer	300.6	794.6
Bhilwara	306.6	824.5
Bundi	274.0	321.8
Chittorgarh	352.0	616.4
Dausa	238.0	544.7
Jaipur	271.1	786.2
Karauli	202.1	521.8
Pratapgarh	462.4	566.8
Rajsamand	450.8	1,287.7
Sawai Madhopur	176.0	541.8
Tonk	238.8	608.3
Udaipur	450.5	1,291.1
Ajmer	300.6	794.6
Bhilwara	306.6	824.5

Source: Ground Water Department, GoR, Jodhpur

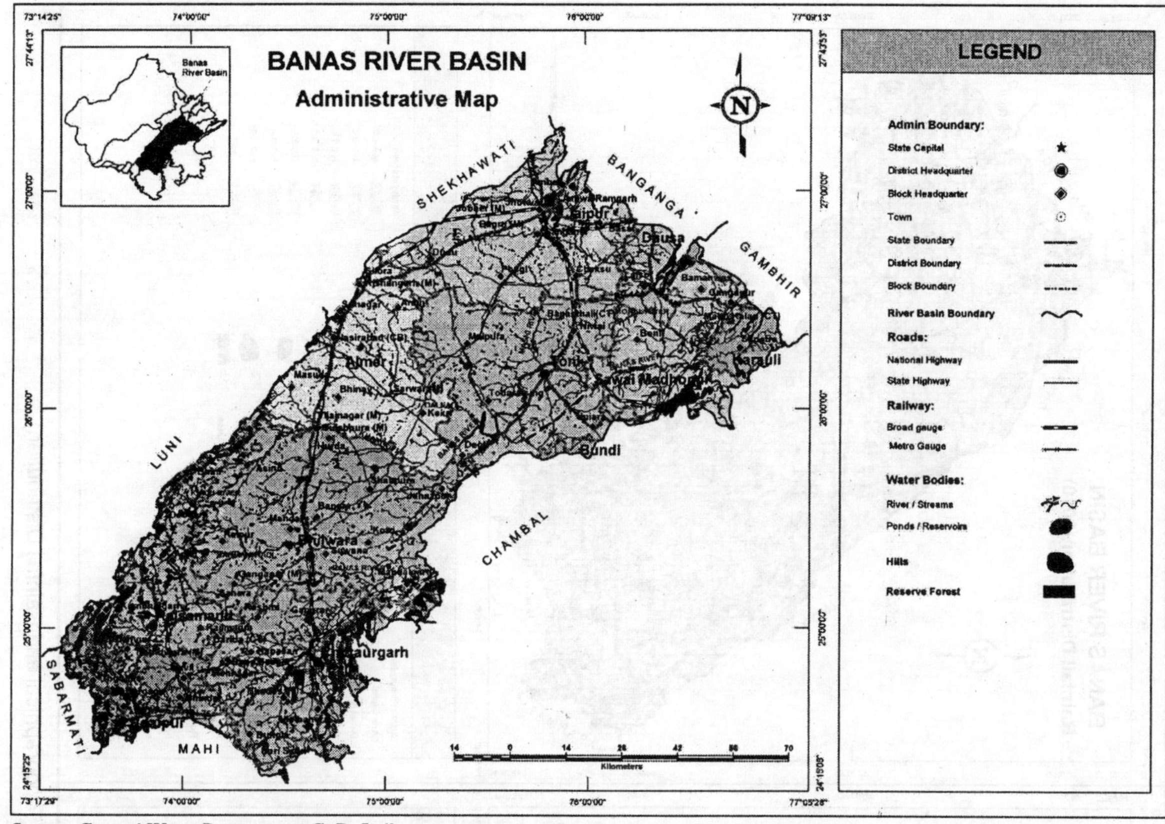

Source: Ground Water Department, GoR, Jodhpur

Figure 4.8 Administrative Map of Banas River Basin

Rainfall

The general distribution of rainfall across the Banas River Basin can be visualized from isohyets presented in the Figure 4.9, where most of the basin appears to receive good rainfall that is more than 1000 mm of total annual rainfall in northeastern and southwestern parts. Other areas to the west of Bhilwara and east of Tonk receive less rainfall.

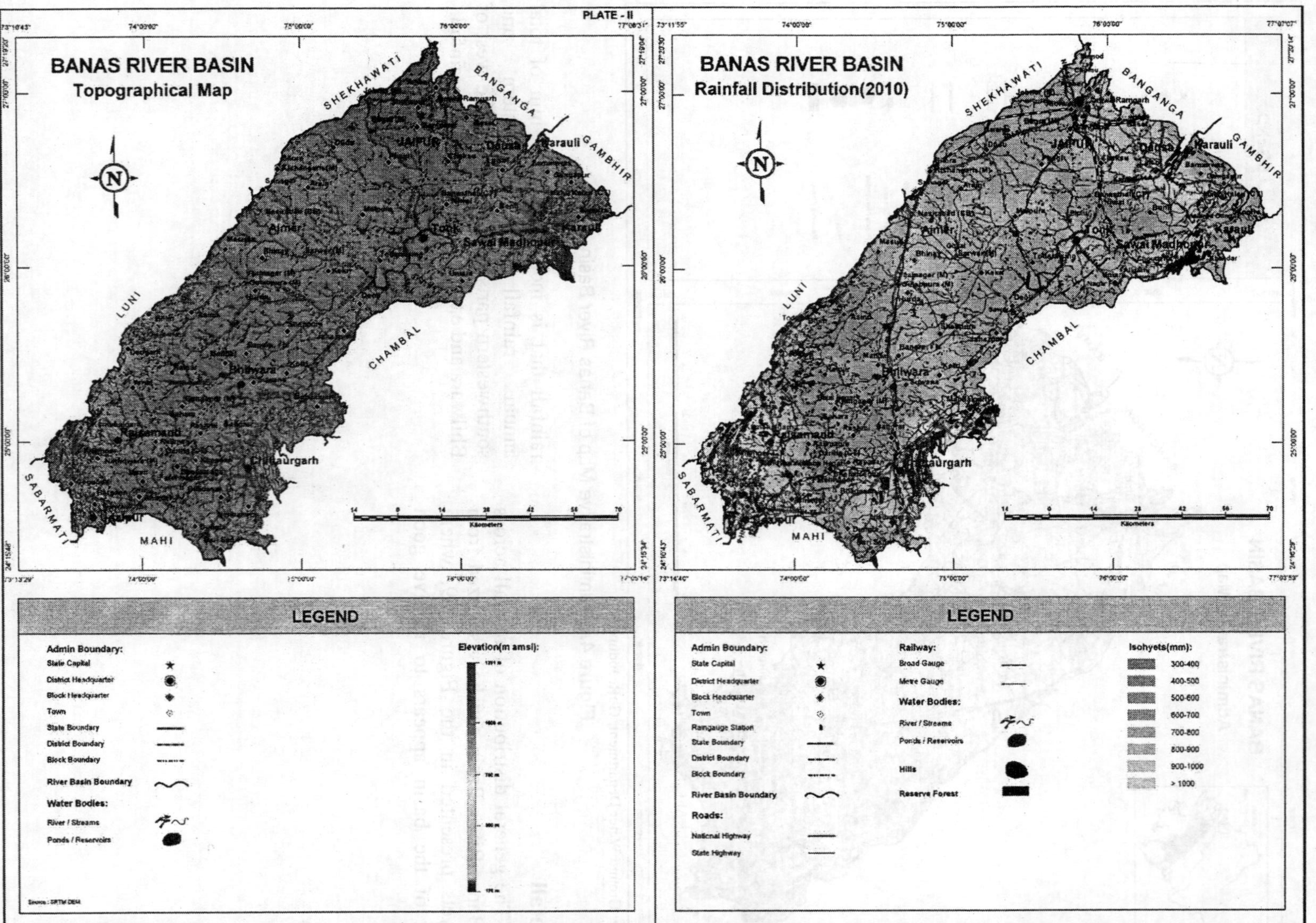

Source: GWD, GoR, Jodhpur

Figure 4.9 : Banas River Basin (Topographical and Rainfall Distribution)

MAHI RIVER BASIN

Source: Ground Water Department, GoR, Jodhpur

Figure 4.10 Mahi River Basin

Location

Mahi River Basin is located in south-eastern part of Rajasthan extending between 23° 03' 03.45" and 24° 35' 46.46" North latitudes and longitudes 73° 18' 29.57'' to 74° 51' 11.73'' East. It lies south of Banas Basin, its eastern edge borders Chambal Basin, and its western edge borders Sabarmati Basin whereas its south edge creates the state border with Gujarat. Mahi River Basin extends over parts of Banswara, Chittorgarh, Dungarpur, Pratapgarh and Udaipur Districts. The total catchment area of the Basin is 16,630 sq km approximately (Figure 4.10).

Mahi River originates in the Mahi Kanta hills in the Vindhyachal range, in the western part of Madhya Pradesh, and enters Rajasthan in Banswara District, near Chandangarh. It leaves the State at Salakari village. On an average the river is about 100 - 130 m wide and it flows mostly through rocky terrain. Its banks may be steep, though not very high. The main tributaries of the Mahi River are the Anas, Hiran, Eru and Chap Rivers, in Banswara District. Of these, only the Anas River is perennial. The Jakam and Gomti Rivers are the next most important downstream tributaries of the Mahi River, originating from Chittorgarh and Udaipur Districts, respectively. In Dungarpur District, the last lap of the Mahi River in Rajasthan, the main tributary is the Som River. Another tributary, the Moran, a seasonal river, also flows through this District.

Administrative Set-up

Administrative, Mahi River Basin extends over parts of Banswara. Pratapgarh, Dungarpur and Chittaurgarh and Udaipur Districts and encompasses 3,979 towns and villages (Figure 4.11 and Table 4.9).

Table 4.9 Administrative Set-up of Mahi River Basin

S. No.	District Name	Area (sq. km)	% of Basin Area	Total Number of Blocks	Total Number of Towns and Villages
1	Banswara	4,493.1	27.0	8	1,413
2	Chittaurgarh	188.4	1.1	2	85
3	Dungarpur	3,164.8	19.0	5	724
4	Pratapgarh	3,328.2	20.0	5	699
5	Udaipur	5,455.1	32.9	8	1,058
	Total	**16,629.6**	**100.0**	**28**	**3,979**

Source: Ground Water Department, GoR, Jodhpur

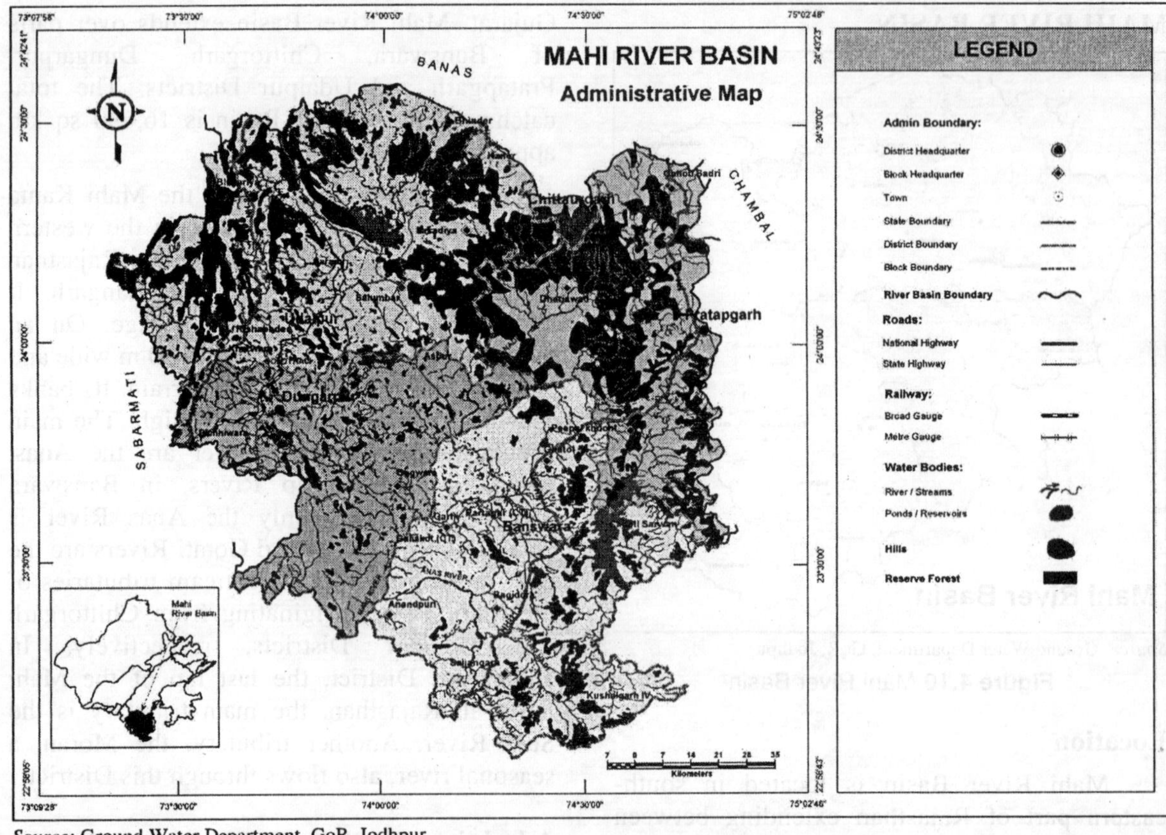

Source: Ground Water Department, GoR, Jodhpur

Figure 4.11 Administrative Map of Mahi River Basin

Climate

Mahi River basin largely falls within humid and sub-humid to even semi-arid. It is extremely cold from October to February while turning hot from March to September. Eastern part in nature and receives good rainfall. The mean annual rainfall over Mahi River Basin was 690 mm, of which about 94% is received during the four Monsoon months (June-September) (Figure 4.12).

Topography

Geographically, the Mahi basin is marked by hilly terrain belonging to the Aravali chain in the north western part of the basin. Southern hilly part of the basin less elevated as compare to north western part of the basin. The alluvial plain elevations above mean sea level (amsl) range from 112 meter to 262 meter approximately. The upstream part of the Basin lies outside Rajasthan, extending over 8,864 km², in Madhya Pradesh and Gujarat States. The topography levels in the basin are ranging from 112 meter to 1009 meter, highest being in the northwestern part and lowest in the southern region (Figure 4.12 and Table 4.10).

Table 4.10 : Districtwise minimum and maximum elevation

District Name	Min. Elevation (m amsl)	Max. Elevation (m amsl)
Banswara	112.6	582.3
Chittaurgarh	310.7	587.1
Dungarpur	112.0	540.8
Pratapgarh	162.7	576.8
Udaipur	159.3	1,009.0

Rainfall

The general distribution of rainfall across can be visualized from isohyets presented in the Figure 4.12 where most of the central part of the basin received total annual rainfall in the range of 500-700 mm in year 2010, and the hilly areas received highest rainfall reaching a maximum of 1,000 mm.

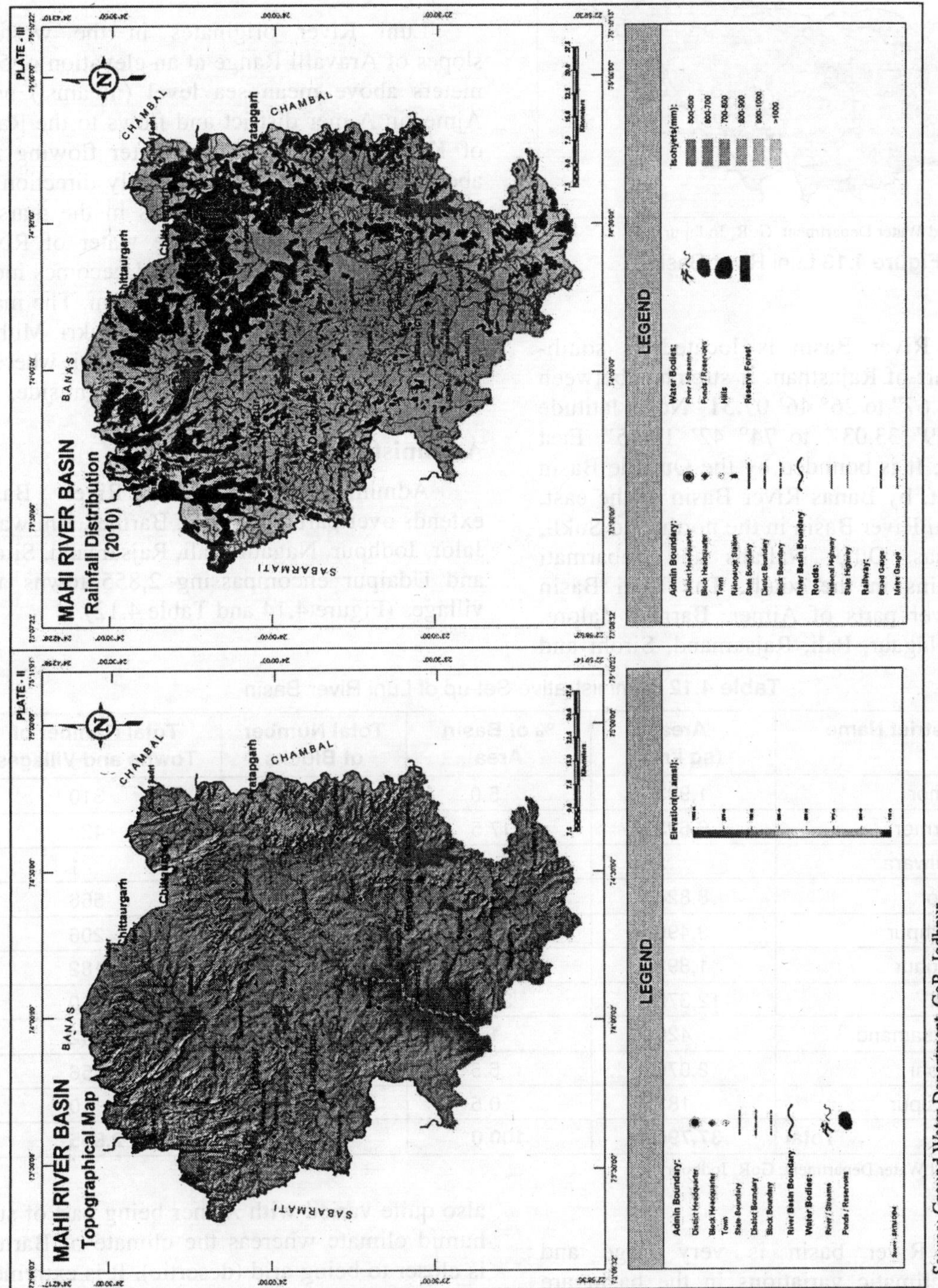

Source: Ground Water Department, GoR, Jodhpur

Figure 4.12 : Mahi River Basin (Topographical and Rainfall Distribution 2010)

LUNI RIVER BASIN

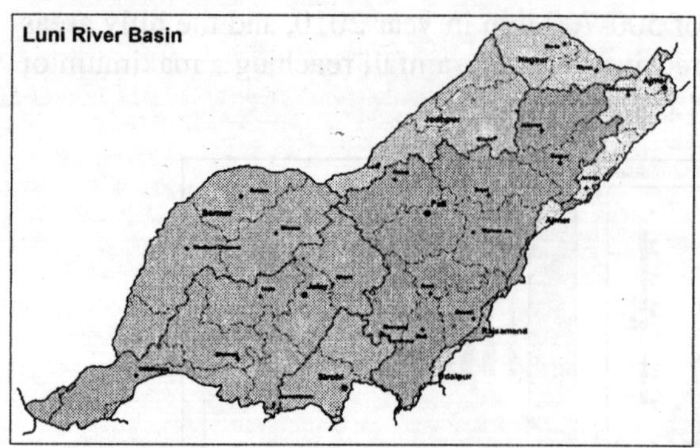

Source: Ground Water Department, GoR, Jodhpur

Figure 4.13 Luni River Basin

Location

Luni River Basin is located in south-western part of Rajasthan. It stretches between 24° 36' 35.67" to 26° 46' 07.31" North latitude and 70° 59' 33.03'' to 74° 42' 18.45'' East longitudes. It is bounded by the Outside Basin in the west, by Banas River Basin in the east, Shekhawati River Basin in the north, and Sukli, West Banas, Other Nallahs and Sabarmati River Basins in the south. The Luni Basin extends over parts of Ajmer, Barmer, Jalore, Jodhpur, Nagaur, Pali, Rajsamand, Sirohi and Udaipur Districts. The total catchment area of the Basin is 37,796.37 km². Luni River Basin lies to the west of the Aravali hills and forms part of the mid-west alluvial plain (Figure 4.13).

Luni River originates in the western slopes of Aravalli Range at an elevation of 550 meters above mean sea level (m amsl) near Ajmer in Ajmer district and flows to the Rann of Kutch in Gujarat State. After flowing for about 495 km in a south-westerly direction in Rajasthan, the river disappears in the marshy land of Rann of Kutch. The water of River Luni is sweet up to Balotra and becomes more and more saline further downstream. The main tributaries of Luni on the left are Sukri, Mithri, Bandi, Khari, Jawai, Guhiya and Sagi, whereas only Jojari River joins it from the right side.

Administrative Set-up

Administratively, Luni River Basin extends over parts of Ajmer, Barmer, Bhilwara, Jalor, Jodhpur, Nagaur, Pali, Rajsamand, Sirohi and Udaipur encompassing 2,855 towns and villages (Figure 4.14 and Table 4.12).

Table 4.12 Administrative Set-up of Luni River Basin

S.No.	District Name	Area (sq km)	% of Basin Area	Total Number of Blocks	Total Number of Towns and Villages
1	Ajmer	1,903.5	5.0	4	310
2	Barmer	6,623.3	17.5	5	422
3	Bhilwara	2.9	-	1	1
4	Jalor	8,824.2	23.3	8	566
5	Jodhpur	3,496.6	9.3	4	206
6	Nagaur	1,892.5	5.0	4	182
7	Pali	12,375.0	32.8	10	960
8	Rajsamand	420.7	1.1	3	42
9	Sirohi	2,075.1	5.5	5	156
10	Udaipur	182.7	0.5	2	10
	Total	**37,796.4**	**100.0**	**46**	**2,855**

Source: Ground Water Department, GoR, Jodhpur

Climate

Luni River basin is very large and therefore climatic variations in the basin are also quite varied with Ajmer being part of sub-humid climate whereas the climate of Barmer is closer to being arid (desertic). It is extremely cold from October to February while turning hot from March to September. Areas adjoining Aravalli range receive good rainfall and area east of Barmer receives very less to negligible rainfall. The mean annual rainfall over Luni River Basin was 643 mm, of which most of the rainfall is received during the four Monsoon months (June-September).

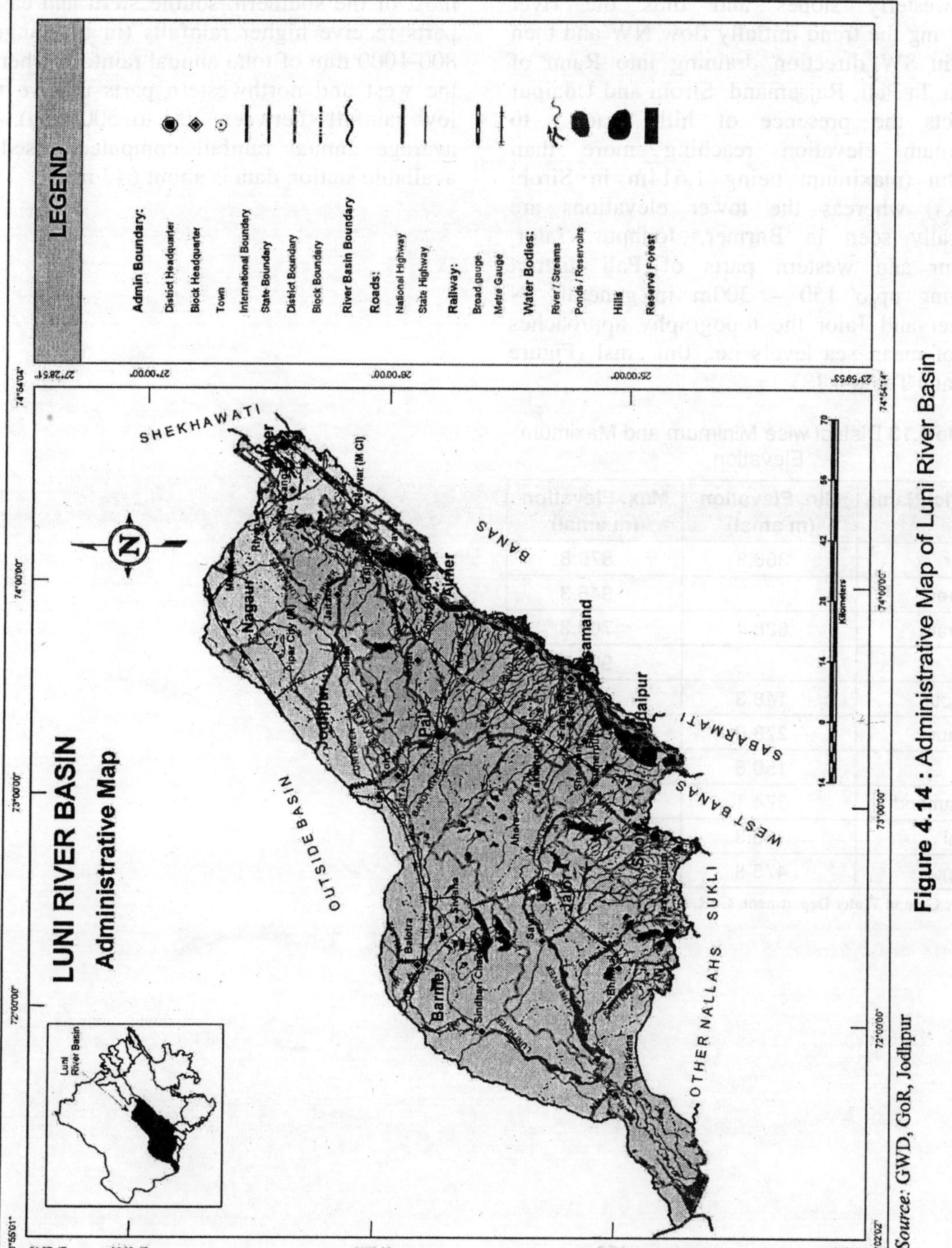

Figure 4.14 : Administrative Map of Luni River Basin

Source: GWD, GoR, Jodhpur

Topography

The Luni river flows from northeast to southwest following the general slope of the terrain within the basin,. Aravali ridges have a northwesterly slopes and thus the river following the trend initially flow NW and then turn in SW direction draining into Rann of Kutch. In Pali, Rajsamand, Sirohi and Udaipur districts the presence of hills leads to maximum elevation reaching more than 1,000m (maximum being 1,614m in Sirohi district) whereas the lower elevations are generally seen in Barmer, Jodhpur, Jalor, Nagaur and western parts of Pali district reaching upto 150 – 300m in general. IN Barmer and Jalor the topography approaches that of mean sea levels i.e., 0m amsl (Figure 4.15 and Table 4.13).

Table 4.13 District wise Minimum and Maximum Elevation

District Name	Min. Elevation (m amsl)	Max. Elevation (m amsl)
Ajmer	366.3	870.8
Barmer	-	946.3
Bhilwara	628.4	700.3
Jalor	-	969.0
Jodhpur	148.3	450.4
Nagaur	278.6	791.9
Pali	150.6	1,066.5
Rajsamand	374.1	1,173.4
Sirohi	183.3	1,614.8
Udaipur	475.8	1,126.0

Source: Ground Water Department, GoR, Jodhpur

Rainfall

The general distribution of rainfall across the Luni River Basin can be visualized from isohyets presented in the Figure 4.15 where most of the southern, southeastern and eastern parts receive higher rainfalls (in the range of 800-1000 mm of total annual rainfall) whereas; the west and northwestern parts receive very low rainfall (between 300 to 500 mm). The average annual rainfall computed based on available station data is about 643 mm.

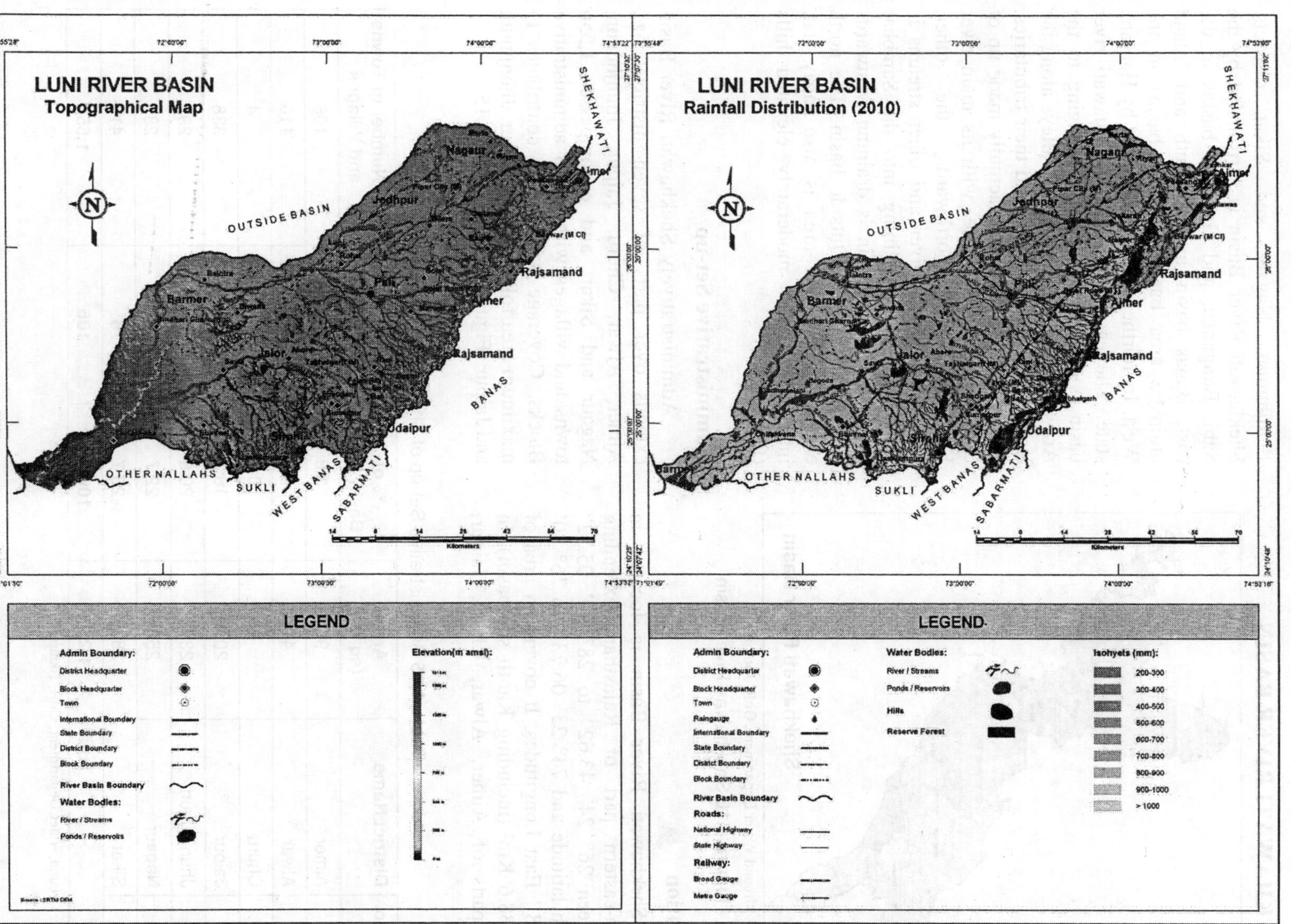

Source: GWD, GoR, Jodhpur

Figure 4.15 : Luni River Basin (Topographical and Rainfall Distribution)

SHEKHAWATI RIVER BASIN

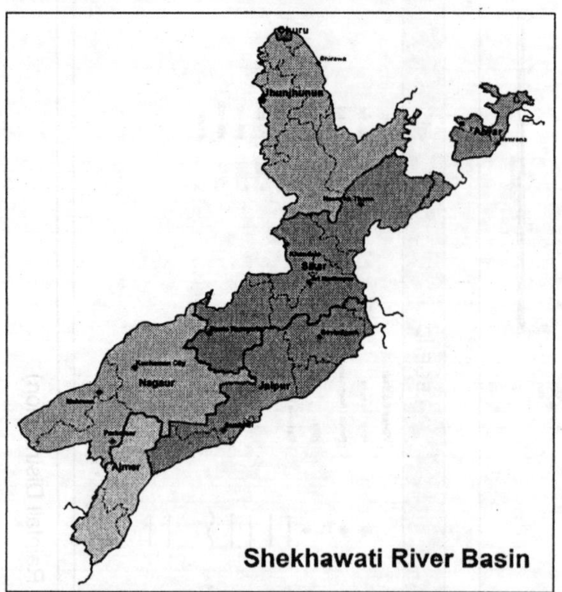

Source: Ground Water Department, GoR, Jodhpur

Figure 4.16 Shekhawati River Basin

Location

Shekhawati River Basin is located in north-eastern part of Rajasthan, extending between 26° 24' 43.62" to 28° 23' 05.52" North latitude and 74° 23' 06.73'' to 76° 30' 26.18'' East Longitudes. It covers an area of 11,238.6 Km² (including Kantli sub-basin), in the parts of Ajmer, Alwar, Churu, Jaipur, Jhunjhunun, Nagaur and Sikar districts. Shekhawati River Basin is surrounded by the Sabi, Banganga, and Gambhir basins in the east, Banas basin in the South and Inland drainage area basin (Outside basin) in the West. Its northern edge is bounded by Haryana state. The eastern parts of Shekhawati river basin is marked by terrain belonging to the Aravali chain, with fairly flat valleys along the Kantli and Mendha rivers and their tributaries. Shekhawati river basin is actually made up of two distinct units – one, with its main river Kantli draining northwards, the other, including Mendha river and other streams is landlocked basin draining into the Sambhar lake. Shekhawati basin is draining westwards and belongs to the Outside basin. The north eastern part of the area is also fairly flat, interspersed with moderately elevated hills (Figure 4.16).

Administrative Set-up

Administratively Shekhawati River Basin extends over parts of seven districts viz., Ajmer, Alwar, Churu, Jaipur, Jhunjhunun, Nagaur and Sikar; and encompasses 1,556 towns and villages within 106 administrative Blocks. Coverage wise, Sikar contributes to maximum area followed by Nagaur, Jhunjunun and Jaipur (Figure 4.17 and Table 4.15).

Table 4.15 Administrative Set-up of Shekhawati River Basin

S. No.	District Name	Area (sq km)	% of Basin Area	Total Number of Blocks	Total Number of Towns and Villages
1	Ajmer	984.4	8.8	4	106
2	Alwar	418.0	3.7	2	110
3	Churu	26.8	0.2	1	4
4	Jaipur	2056.7	18.3	7	338
5	Jhunjhunun	2313.2	20.6	8	299
6	Nagaur	2555.6	22.7	6	281
7	Sikar	2883.9	25.7	6	418
	Total	**11,238.6**	**100.0**	**106**	**1,556**

Source: Ground Water Department, GoR, Jodhpur

Climate

Shekhawati River basin largely falls within semi-arid region and has an internal drainage. Rains occur in the monsoon months between June-July and September (total annual rainfall ranging between 500-1000 mm in different Blocks). Temperatures remain relatively high throughout the year and especially higher in the summer months of April to July, whereas the average daily temperatures dip down significantly in winters. While the maximum temperatures during the months of May and June can reach close to 50°C, the winter months of November to February are pleasant, with average temperatures ranging from 15–18°C and temperatures in the night falling to 2-3°C.

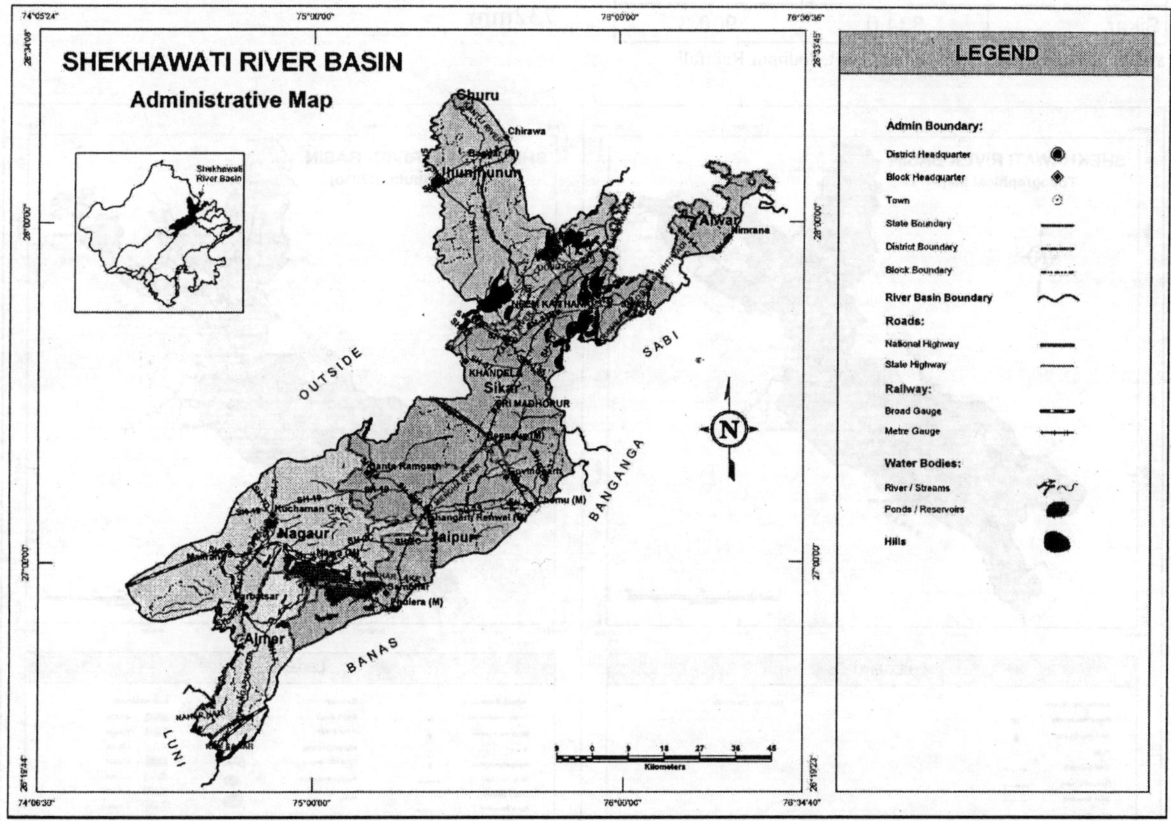

Source: Ground Water Department, GoR, Jodhpur

Figure 4.17 Administrative Map of Shekhawati River Basin

Topography

Isolated hills (part of Aravalli range) are seen around Neem ka Thana in the northern part of the basin and also around Makrana and Danta Ramgarh. Otherwise, in general, the topography of the basin is flat to undulating. The higher elevations (about 976m) have been observed in the northeastern portion and lower elevations (around 265m) are observed the southwestern and northern parts indicating general sloping towards north and into Sambhar Lake. Sambhar lake is the result of inland drainage and marks topographic low and accumulates drainage from areas all around it (Figure 4.18).

Table 4.16 District wise minimum and maximum elevation

District Name	Min. Elevation (m amsl)	Max. Elevation (m amsl)
Ajmer	360.6	802.7
Alwar	265.5	575.1
Churu	282.1	301.1
Jaipur	327.4	701.7
Jhunjhunun	283.6	976.5
Nagaur	344.5	793.2
Sikar	334.0	900.3

Source: Ground Water Department, GoR, Jodhpur Rainfall

The general distribution of rainfall across the basin can be understood from isohyets presented in the Figure 4.18 where the area around Kuchaman city (Nagaur) and that around Nimrana (Alwar) seem to receive the minimum rainfall of the basin that is between 400-600mm of Total annual rainfall. The hilly areas around Neem ka thana receives maximum rainfall in the basin reaching more than 1200mm in the year 2010. The annual average rainfall in the basin thus works out to 732mm.

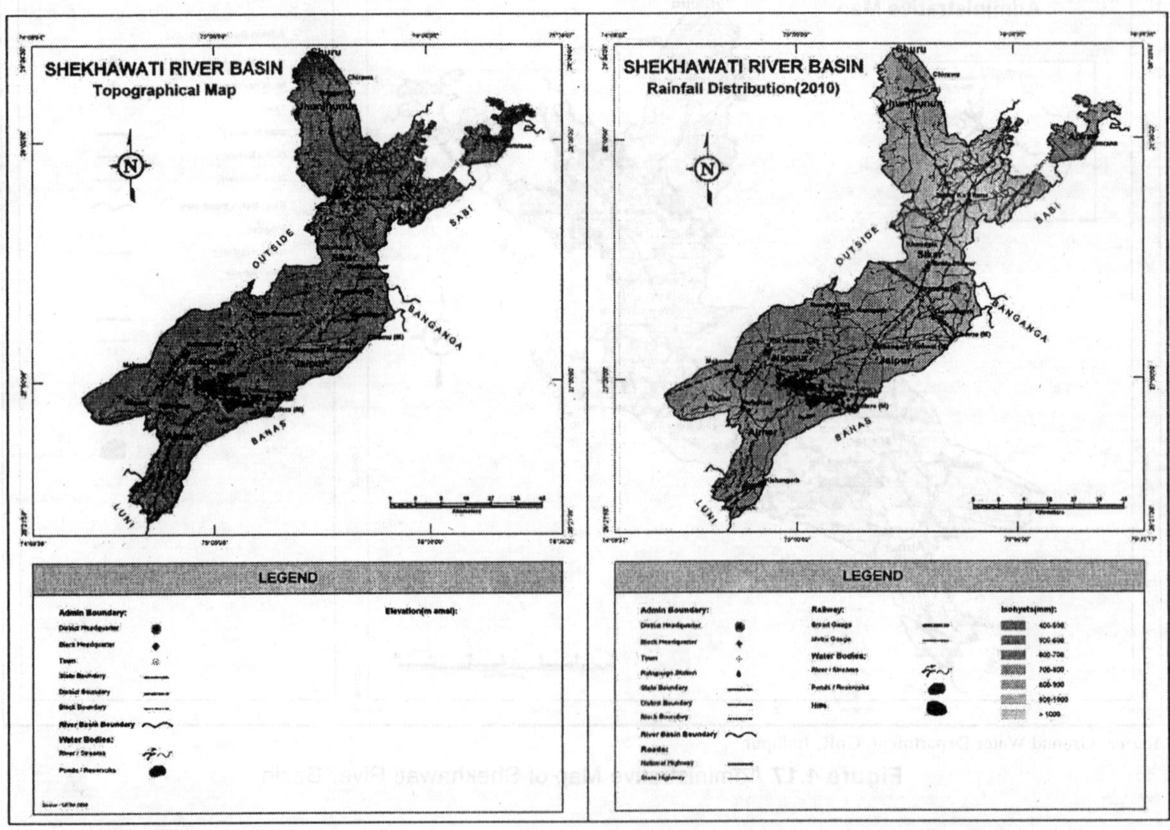

Source: GWD, GoR, Jodhpur

Figure 4.18 Shekhawati River Basin (Topographical and Rainfall Distribution)

SABI RIVER BASIN

Location

Sabi (also called Sahibi) River Basin is located in the mid north eastern part of Rajasthan. It stretches between 27° 18' 39.13" to 28° 13' 55.10" North latitude and 76° 58' 21.09" to 75° 45' 35.05" East longitude. It is bounded in the northwest by Shekhawati River Basin and Ruparail and Banganga River Basins in the southeast. The northern boundary is shared administratively with Haryana State.

The Basin extends over parts of Alwar, Jaipur and Sikar districts. The total catchment area of the Basin is 4607.9 km² (Figure 4.19 and Table 4.18).

The Sabi River originates in the eastern slopes of the Saiwar Protected Forest (PF) hills in Sikar District, enters Jaipur district near the foot of these hills, and after initially flowing southeast and east turns northeastwards near Shahpura and continues further till it exits the State to enter Haryana. Important left bank tributary is Sota River and there are numerous right bank tributaries that originate in the hills of Alwar district. It flows in Rajasthan for a distance of about 157 km before entering Haryana State.

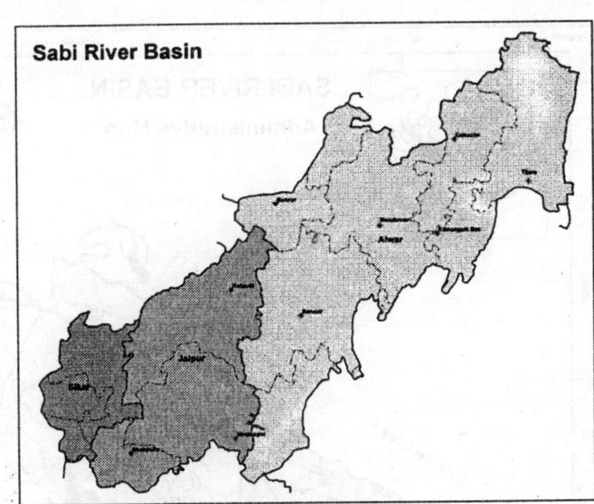

Source: GWD, GoR, Jodhpur

Figure 4.19 Sabi River Basin

Table 4.18 Administrative Set-up of Sabi River Basin

S. No.	District Name	Area (sq km)	% of Basin Area	Total Number of Blocks	Total Number of Towns and Villages
1	Alwar	2,937.8	63.8	10	745
2	Jaipur	1,249.2	27.1	3	225
3	Sikar	420.9	9.1	3	43
	Total	**4,607.9**	**100.0**	**16**	**1,013**

Source: GWD, GoR, Jodhpur

Administrative Set-up

Administratively, Sabi River Basin extends over parts of Alwar, Jaipur and Sikar districts through 16 blocks and encompassing 1,013 towns and villages. Major part of the basin area lies in Alwar district which accounts for about 64% of its catchment area and Sikar the least with just about 9% of the basin area (Figure 4.20).

Climate

The climate of the Sabi River Basin is semi-arid with very hot summers and extremely cold winters. The monsoon season is of very short duration. The cold season starts by the middle of November and continues up to the beginning of March. The summer season follows thereafter and extends up to the end of the June. The south-west monsoon continues from July to mid-September. The mean annual rainfall over Sabi basin is 616mm, of which about 91% falls during the four monsoon months. The period from mid-September to mid-November forms the post-monsoon season.

Topography

The general topographic map presented in Plate – II reveals that the southern, south eastern and eastern fringes are hilly whereas the general slope of the terrain is towards north to northwest. Aravali ridges constitute the hills and give rise to ridge and valley topography in hilly areas. The maximum elevation (791m amsl) is found in Sikar district where the Sabi River originates.

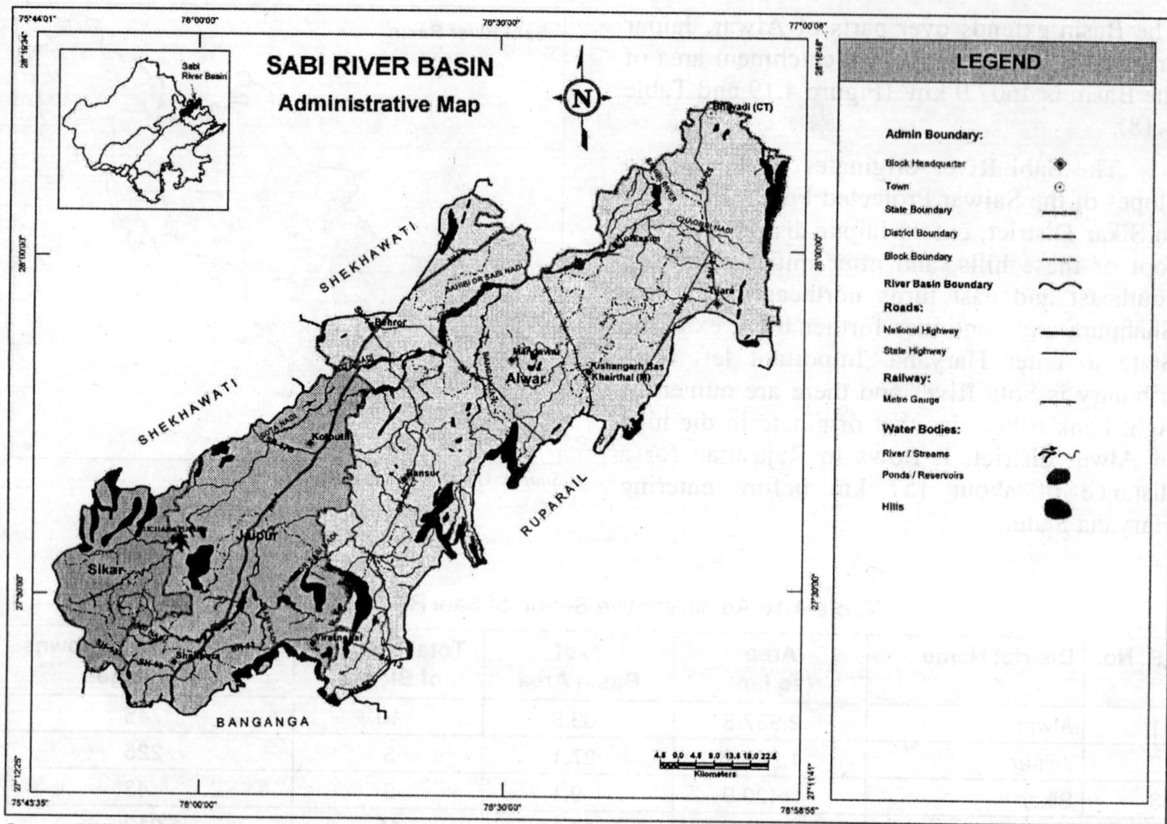

Source: GWD, GoR, Jodhpur

Figure 4.20 Administrative Map of Sabi River Basin

The maximum elevation found in other districts is also at comparable levels (760m in Jaipur district and 770m in Alwar district). The lowest elevation in the basin is 191m in Alwar district is at the point where the river Sabi exits Rajasthan and enters Haryana (Figure 4.21 and Table 4.19).

Table 4.19 : District wise minimum and maximum elevation

District Name	Min. Elevation (m amsl)	Max. Elevation (m amsl)
Alwar	190.8	769.7
Jaipur	317.6	760.4
Sikar	406.3	791.5

Source: GWD, GoR, Jodhpur

Rainfall

The general distribution of rainfall across the Luni River Basin can be visualized from isohyets presented in the Figure 4.21 where except for the area around Behror and in the Bhiwandi – Tapukara – Tijara region (total annual rainfall received is less than 600mm), most of the basin receives good rainfall (total annual rainfall >800mm) especially hills and in their vicinity. Highest rainfall was recorded by the Kotkasim rain gauge station which measured total annual rainfall of 1,013mm in the year 2010.

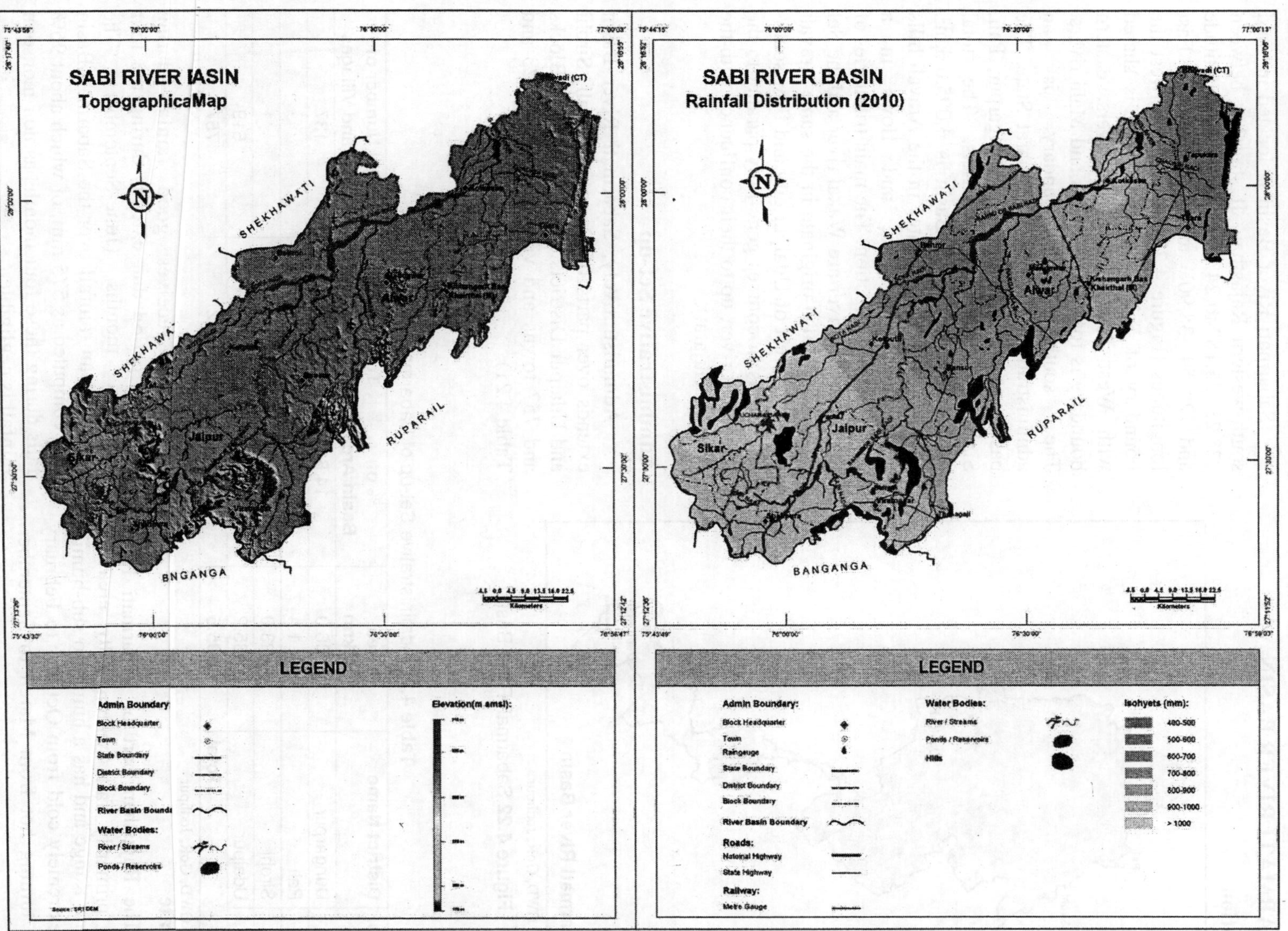

Source: IWD, GoR, Jaipur

Figure 4.21 : Sabi River Basin (Topographical and Rainfall Distribution)

SABARMATI RIVER BASIN

Location

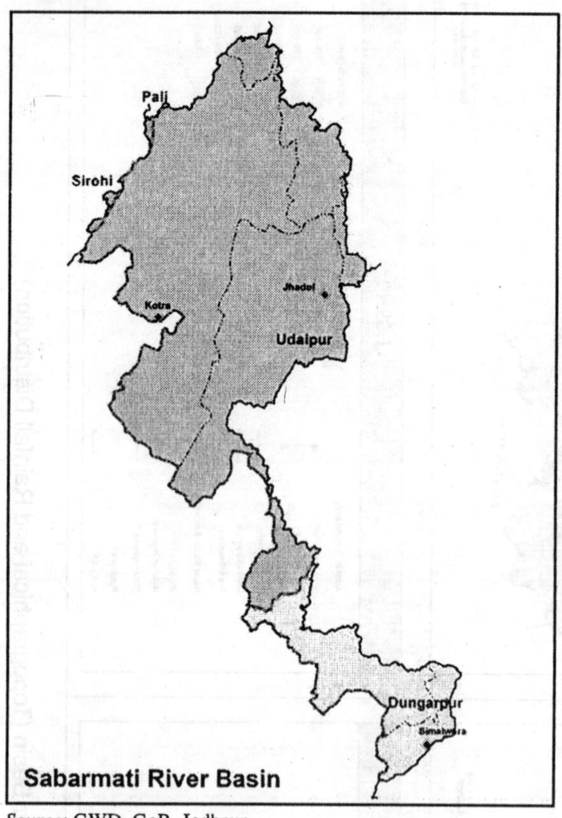

Source: GWD, GoR, Jodhpur

Figure 4.22 Sabarmati River Basin

Sabarmati River Basin is located in mid south-western Rajasthan. It stretches between 23° 28' 25.18" to 24° 55' 14.72" North latitude and 72° 59' 38.90'' to 73° 48' 29.62'' East longitudes (Figure 4.22). In the northwest the boundary of Sabarmati river basin is shared with West Banas and Luni rivers, eastern boundary is formed by Banas and Mahi rivers. The southwestern boundary is an administrative border with Gujarat State. The basin extends over parts of Dungarpur, Pali, Sirohi and Udaipur Districts. The total catchment area of the Basin is 4,095.6 km². Sabarmati River originates in the Aravali hills in Mount Abu area, and flows in a southwesterly direction. The main tributaries of the Sabarmati river are Wakal river and the Sel Nadi, which also originate in the same Aravali hill range west of Udaipur city and flow south-westwards in courses gene-rally parallel to the Sabarmati river, up to their confluence with the river (in Gujarat).

Administrative Set-up

Administratively, Sabarmati River Basin extends over parts of Dungarpur, Pali, Sirohi and Udaipur Districts encompassing 12 Blocks and 757 towns and villages (Figure 4.23 and Table 4.21).

Table 4.21 Administrative Set-up of Sabarmati River Basin

S. No.	District Name	Area (sq km)	% of Basin Area	Total Number of Blocks	Total Number of Towns and Villages
1	Dungarpur	606.0	14.8	3	137
2	Pali	1.4	-	1	-
3	Sirohi	23.0	0.6	2	1
4	Udaipur	3,465.2	84.6	6	619
	Total	**4,095.6**	**100.0**	**12**	**757**

Source: GWD, GoR, Jodhpur

Climate

The Rajasthan part of Sabarmati River basin primarily falls within hilly areas of Aravalli Range and has a humid to sub-humid. It is extremely cold from October to February while turning hot from March to September. Overall, it receives good rainfall during monsoon period that extends during the four Monsoon months (June-September). The annual mean rainfall over the Sabarmati Basin was computed as 575 mm, of which about 96% falls during these four months as non-monsoon rainfall is negligible.

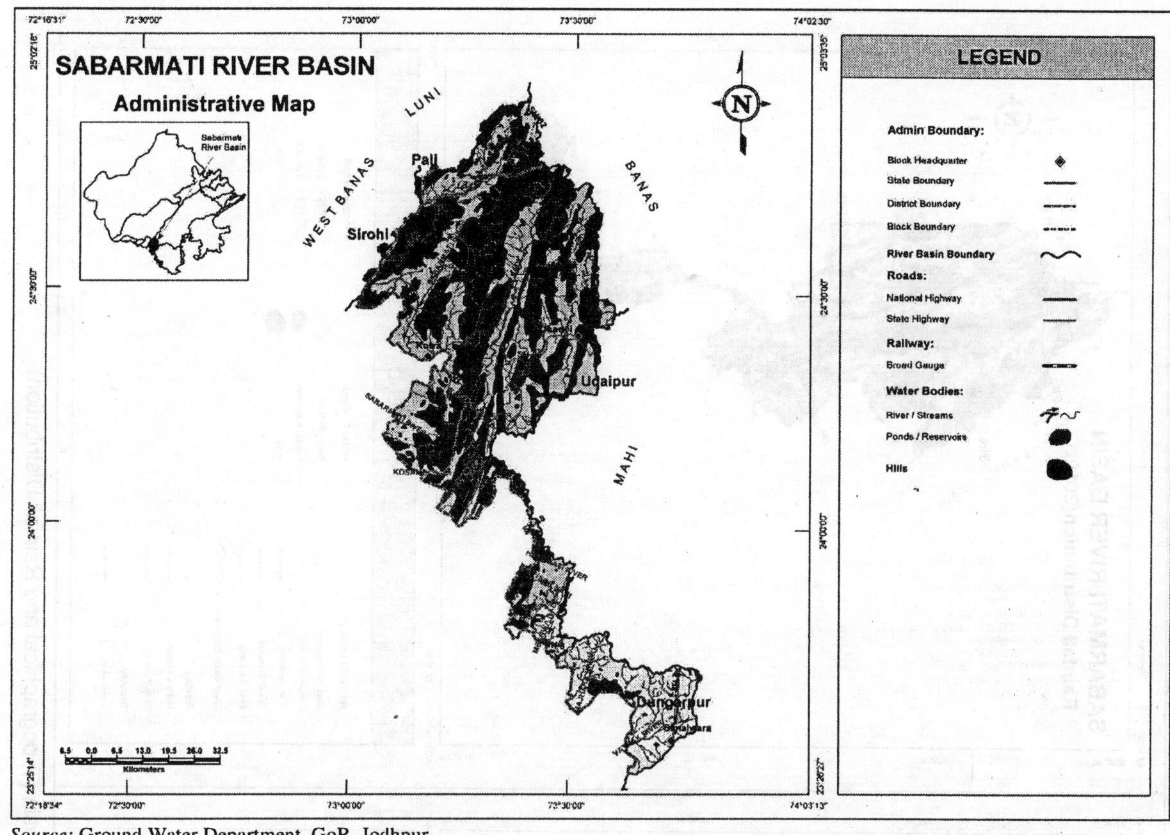

Source: Ground Water Department, GoR, Jodhpur

Figure 4.23 Administrative Map of Sabarmati River Basin

Topography

The major part of the Basin is marked by hilly terrain belonging to the Aravali chain marked by steep slopes and narrow valleys. The hills trend in north-east-southwest direction. The elevation variation within the part of basin is very high (to the extent of about 1000m). The higher elevations (about 1168m) have been observed in the northeastern portion and lower elevations (about 175m) are observed in the southern part of the basin indicating general sloping downwards towards south to southwestern direction. The river and its tributaries either flow parallel to the ridges in NW-SE direction or cut across them in a SE-NW or E-W directions (Figure 4.24 and Table 4.22).

Table 4.22 Districtwise minimum and maximum elevation

S. No.	District Name	Min Elevation (m amsl)	Max. Elevation (m amsl)
1	Dungarpur	174.6	496.9
2	Pali	683.8	952.2
3	Sirohi	434.3	901.7
4	Udaipur	242.1	1,167.0

Rainfall

The area of the basin being small, hill and less inhabited, has very few rain gauge stations. The rainfall distribution (Figure 4.24)) reveals high rainfall areas being located in the hilly areas in Udaipur district and reduces further southeastwards reaching minimum in Dungarpur area. The total average rainfall in the year 2010 had been around 1000mm.

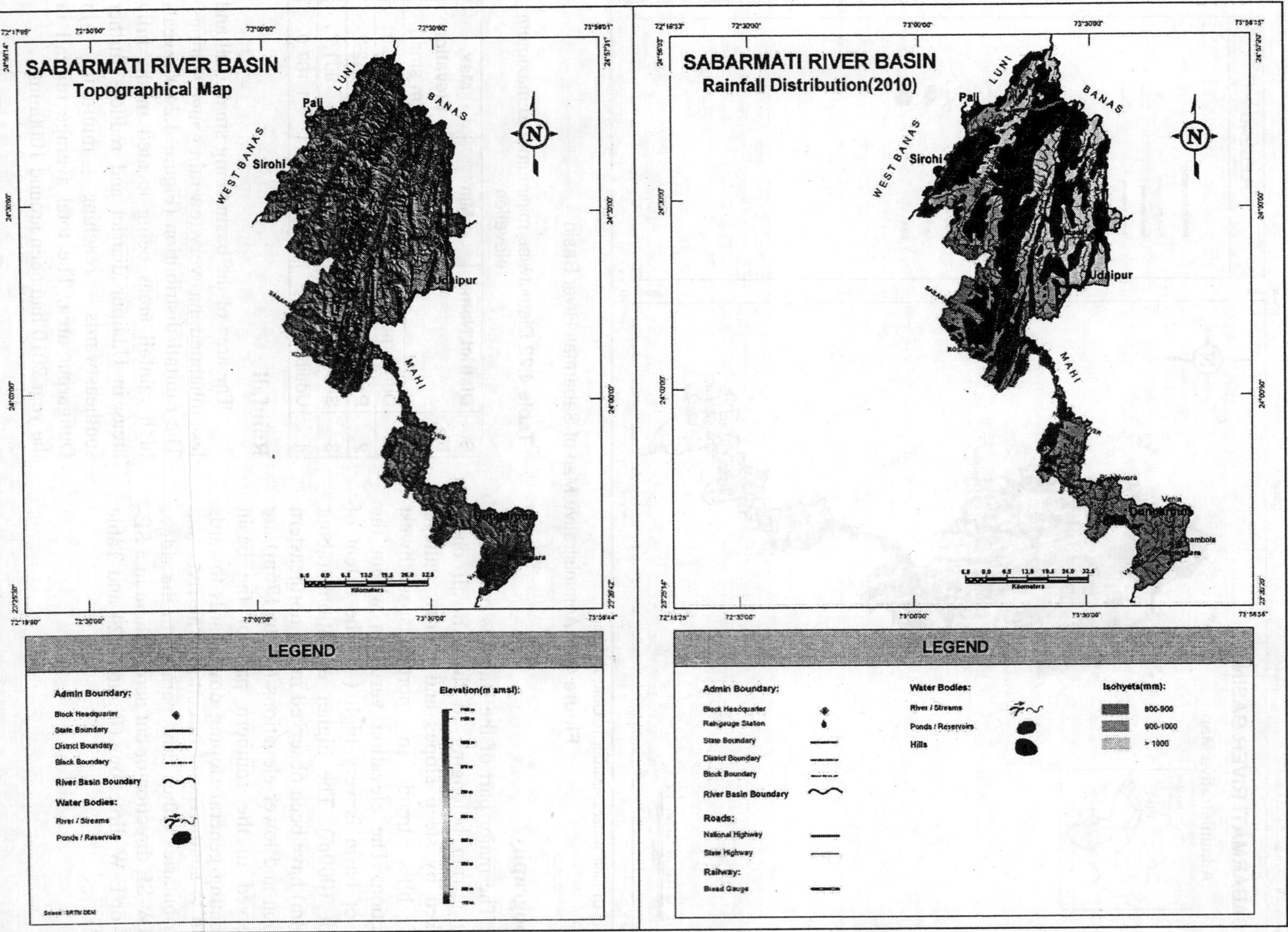

Source: GWD, GoR, Jodhpur

Figure 4.24 : Sabarmati River Basin (Topographical and Rainfall Distribution)

RUPARAIL RIVER BASIN

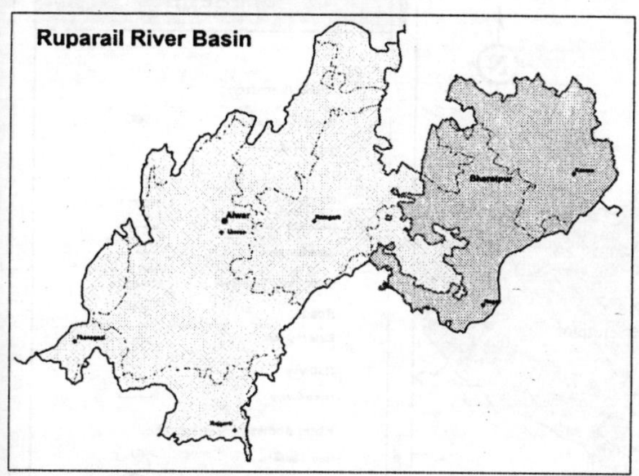

Source: GWD, GoR, Jodhpur

Figure 4.25 Ruparail River Basin

Location

The Ruparail River Basin is located in the northeastern part of Rajasthan. It stretches between 27° 09' 22.24" to 27° 55' 10.35" North latitude and 76° 14' 42.36'' to 77° 21' 34.47'' East longitudes. The basin extends in a broadly W-E direction and is bounded by the Sabi River Basins in the northwest side and the Banganga River Basin in the southeast. The northern border is shared with Haryana State. The basin extends over parts of Bharatpur and Alwar Districts. The total catchment area of the Basin is very small and extends over an approximate area of 3,809.6 km² (Figure 4.25).

River Ruparail, also known as Barah, rises in the Udainath hills of Thanagazi Reserved Forest (RF), Alwar District. It traverses these hills northwards, turning towards the east and northeast before disappearing in Bharatpur District. It flows first through hills and subsequently through plains nearly up to Kusalpur in Bharatpur district. It has a total length of about 104 km. A number of smaller streams rise from the various sub-ranges of the above-mentioned RF hills, e.g. the Narainpur, Golari, Sukri, Shanganga and Nalakroti Rivers and contribute to Ruparail flow.

Administrative Set-up

Administratively, Ruparail River Basin extends over parts of Bharatpur and Alwar districts encompassing 15 Blocks divided into 984 towns and villages. Major part of the river flows through Alwar district where its catchment is about 71% of basin area whereas the rest falls within the Bharatpur district (Figure 4.26 and Table 4.24).

Table 4.24 Administrative Set-up of Ruparial River Basin

S. No.	District Name	Area (sq km)	% of Basin Area	Total Number of Blocks	Total Number of Towns and Villages
1	Alwar	2,694.5	70.7	12	582
2	Bharatpur	1,115.1	29.3	3	402
	Total	**3,809.6**	**100.0**	**15**	**984**

Source: GWD, GoR, Jodhpur

Climate

The basin is located on the northeastern part of Rajasthan sharing its boundary with Haryana. It falls within the semi-arid to sub-humid regions but witnesses extreme temperatures during the hottest months which are May and June. Coldest months are December and January where temperatures sometimes reach near-zero levels. Incidentally, India's highest ever recorded temperature of 50.6°C was measured at Alwar on 10 May 1956. The rain are mostly received during monsoon season i.e., from July through September, and mean annual rainfall over the basin was computed as 576mm.

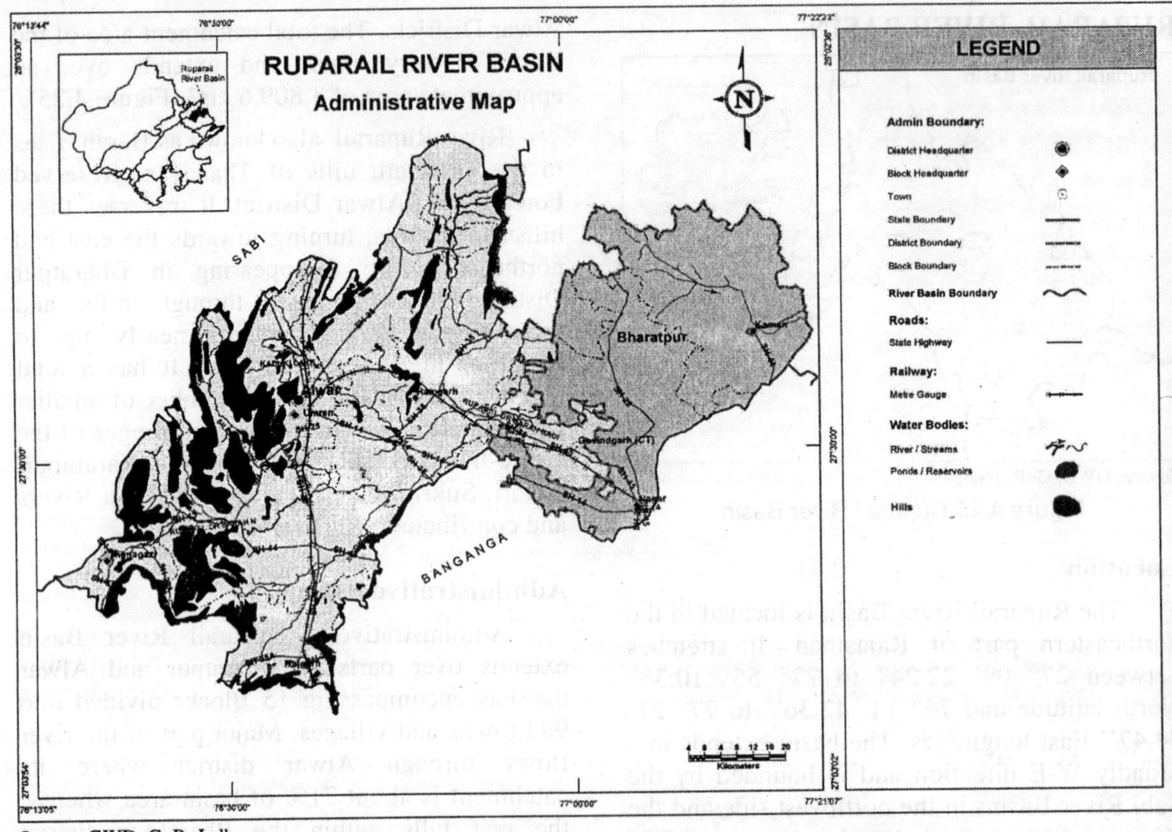

Source: GWD, GoR, Jodhpur

Figure 4.26 Administrative Map of Ruparail River Basin

Topography

Table 4.25 Districtwise minimum and maximum elevation

S. No.	District Name	Min Elevation (m amsl)	Max. Elevation (m amsl)
1	Alwar	204.0	716.4
2	Bharatpur	179.4	366.2

Source: Ground Water Department, GoR, Jodhpur

The western part is hilly where the river originates, and then the terrain slopes towards east and then to northeast. The hills mostly belong to the Aravali ridges are locally curvilinear representing dome and basin structures but follow a broad southwest – northeast trend. The river has an inland drainage system and ends up in the depression around Bharatpur city. The elevation ranges from a minimum of 179m in Bharatpur district to a maximum of about 716m in Alwar district. Ghana bird sanctuary area is actually a regional depression with no exit that leads to retention of surface water flows (Figure 4.27 and Table 4.25).

Rainfall

The general distribution of rainfall across the Ruparail River Basin can be visualized from isohyets presented in the (Figure 4.27) where the basin seems to have received good rainfall in the year 2010. The total annual rainfall was in excess of 1200mm in basin as recorded by the Govindgarh met station followed by Kaman. The lowest total annual rainfall in the basin was recorded at the Pahari station which accounted for 669mm. The average annual rainfall computed based on available station data is about 871 mm.

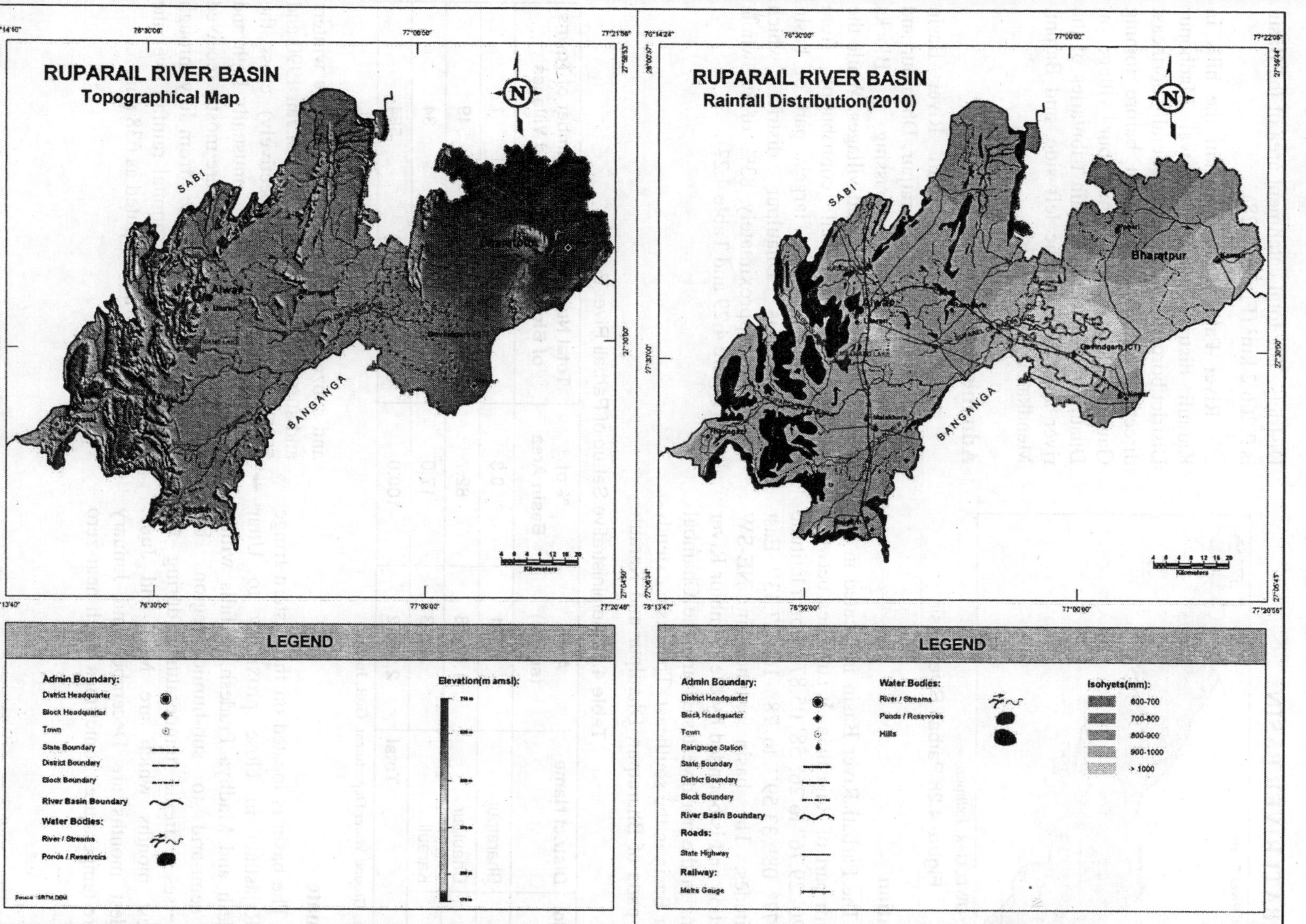

Source: GWD, GoR, Jodhpur

Figure 4.27 : Ruparail River Basin (Topographical and Rainfall Distribution)

PARBATI RIVER BASIN

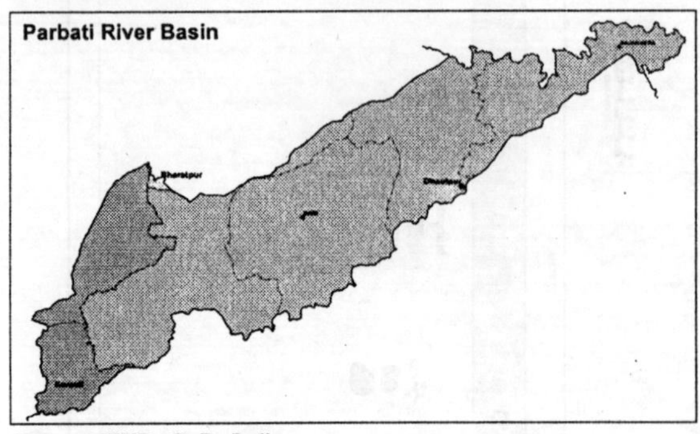

Source: GWD, GoR, Jodhpur

Figure 4.28 Parbati River Basin

Location

The Parbati River Basin is located in the eastern part of Rajasthan. It stretches between 26° 20' 29.76" to 26° 58' 05.87" North latitude and 77° 08' 33.59'' to 78° 16' 27.33'' East longitudes. The basin extends in NE-SW direction and is bounded by the Gambhir River Basins in the northwest side and the Chambal River Basin in the southeast. The basin extends over parts of Bharatpur, Dhaulpur and Karauli Districts. The total catchment area of the Basin is 2,266.2 km² (Figure 4.28).

River Parbati originates in the hills in Karauli district close to Sawai Madhopur District border. It flows in a generally northeast direction for about 123 km before joining Gambhir River near Kharagpur village in Dhaulpur District. The main tributaries of the river are Sairni on the left side, and Bamni, Mendka on the right side.

Administrative Set-up

Administratively, Parbati River Basin extends over parts of Bharatpur, Dhaulpur and Karauli districts encompassing 7 Blocks divided into 584 towns and villages. While the basin has very small contribution from Bharatpur district, the largest part of basin contributor is Dhaulpur district which contributes approximately 83% of its whole area (Figure 4.29 and Table 4.27).

Table 4.27 Administrative Set-up of Parbati River Basin

S. No.	District Name	Area (sq km)	% of Basin Area	Total Number of Blocks	Total Number of Towns and Villages
1	Bharatpur	7.1	0.3	1	1
2	Dhaulpur	1,873.8	82.7	4	549
3	Karauli	385.3	17.0	2	34
	Total	**2,266.2**	**100.0**	**7**	**584**

Source: Ground Water Department, GoR, Jodhpur

Climate

The basin is located on the eastern fringe of Rajasthan in close proximity to Uttar Pradesh and Madhya Pradesh. It falls within the semi-arid to sub-humid regions but witnesses extreme temperatures during the hottest months which are May and June. Coldest months are December and January where temperatures sometimes reach near-zero and subzero levels. In some years, the winters and summers (recorded -4.3°C in Jan 1990 and 50°C in June 1995 respectively) cross the extremes and record such unusually high and low temperatures. The rain are mostly received during monsoon season i.e., from July through September, and mean annual rainfall over the Parbati Basin was computed as 638 mm.

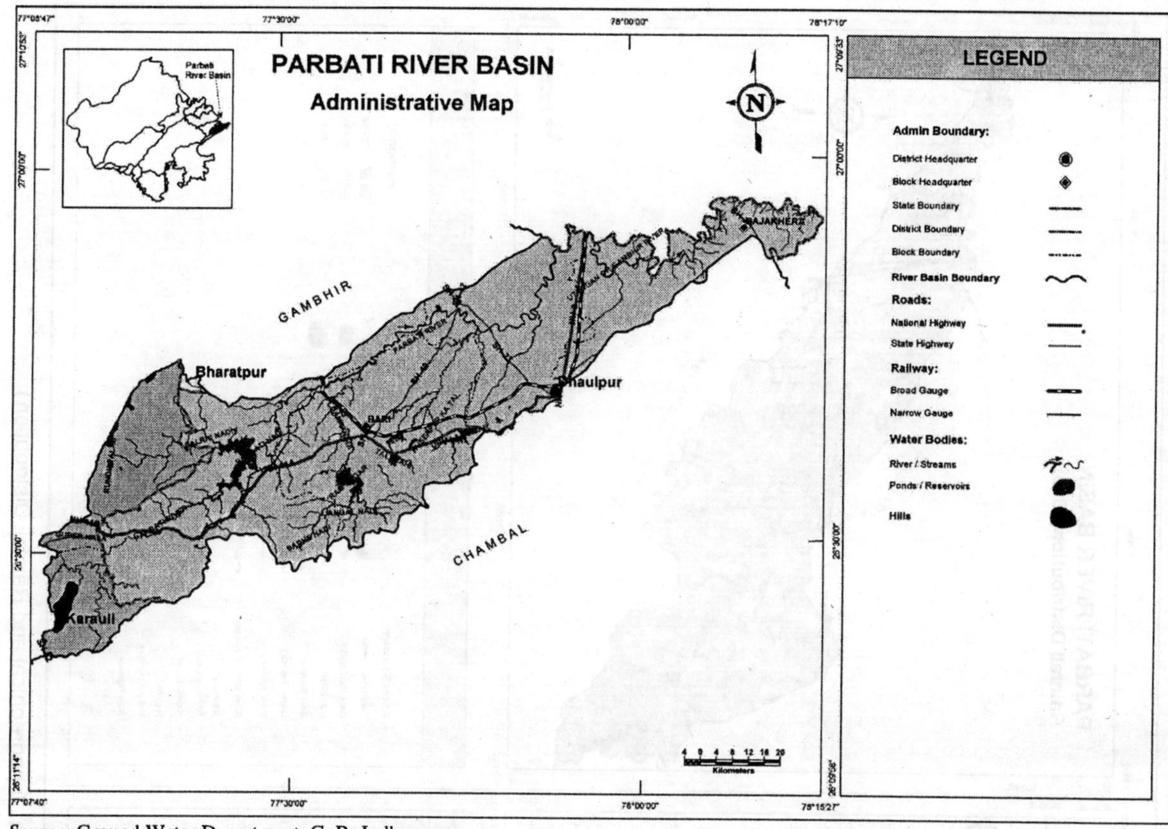

Source: Ground Water Department, GoR, Jodhpur

Figure 4.29 Administrative Map of Parbati River Basin

Topography

Table 4.28 Districtwise minimum and maximum elevation

S. No.	District Name	Min Elevation (m amsl)	Max. Elevation (m amsl)
1	Bharatpur	277.0	325.4
2	Dhaulpur	135.4	340.5
	Karauli	221.1	378.7

Source: Ground Water Department, GoR, Jodhpur

The western part is hilly which is the area where the Parbati river originates, and then the topography slopes towards northeast. The hills mostly belong to the Aravali ridges and have a southwest – northeast trend and the river flowing in the intervening valleys also follows the same trend till it drains out of the basin and enters Uttar Pradesh eventually to join the Yamuna river. The elevation ranges from a minimum of 135m in Dhaulpur district to a maximum of about 379m in Karauli district. The undulating plains between Bari and Rajkhera often show the development of ravines especially when closer to Uttar Pradesh border (Figure 4.30 and Table 4.28).

Rainfall

The general distribution of rainfall across the Parbati River Basin can be visualized from isohyets presented in the Figure 4.30 where the western part receives lesser rainfall (minimum total annual rainfall of 712mm) which seem to increase gradually towards east where it reaches a high of upto 975mm. The average annual rainfall computed based on available station data is about 829 mm.

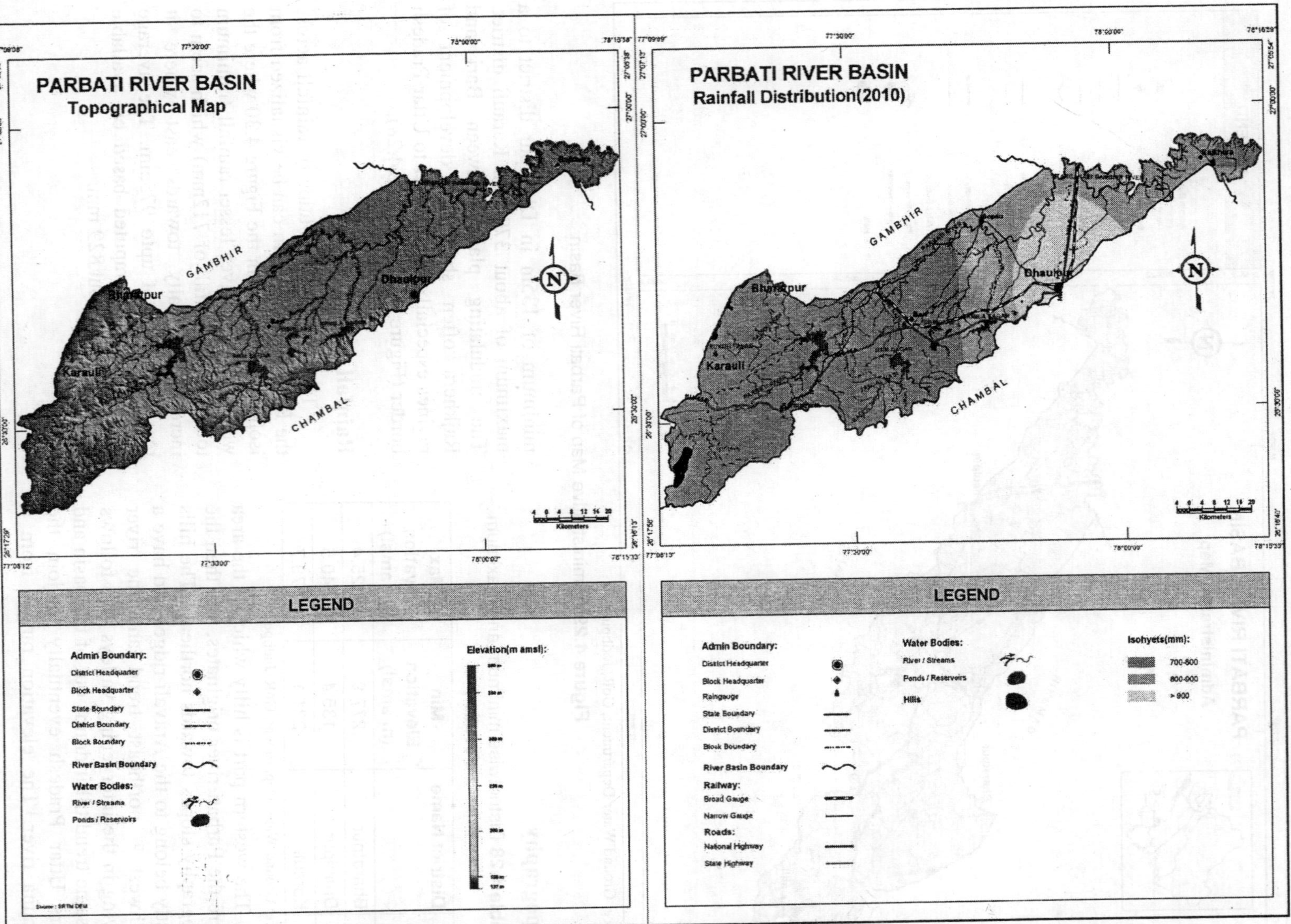

Source: GWD, GoR, Jodhpur

Figure 4.30 : Ruparail River Basin Topographical and Rainfall Distribution)

WEST BANAS RIVER BASIN

Location

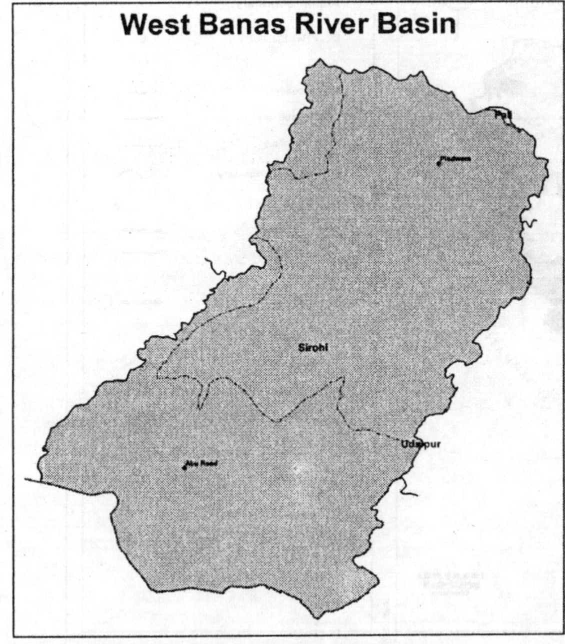

Source: GWD, GoR, Jodhpur

Figure 4.31 West Banas River Basin

The West Banas River Basin is located in the southwestern part of Rajasthan. It stretches between 24° 19' 55.98" to 24° 54' 51.02" North latitude and 72° 35' 22.07" to 73° 10' 03.50" East longitudes. The basin extends in a broadly N-S direction and is bounded by the Sabarmati River Basins in the eastern side, the Sukli River Basin in the west, Luni River Basin in the north and northwest; and the southern and southeastern part of the border is shared administratively with Gujarat State. The basin extends over parts of Sirohi, Pali and Udaipur Districts. The total catchment area of the basin is very small and extends over an approximate area of 1,876 sq km (Figure 4.31).

The West Banas River originates in the hills south of Sirohi and Pindwara towns. It initially flows through a valley near Sirohi and Pindwara in the Mount Abu range of Aravalli hills. It flows for a distance of 50 kms in Rajasthan state before entering Gujarat. The basin has a well-developed drainage system and the main tributaries of West Banas River are Kukli Nadi, Danot Nadi, Sangbaria Nala, Gangari Nala, Gabir Nala, Batriya Nadi, Kaleri Nadi, Gomati Nadi etc.

Administrative Set-up

Administratively, West Banas River Basin extends over parts of Pali, Sirohi and Udaipur districts encompassing 6 Blocks divided into 182 towns and villages. Major part of the river flows through Sirohi district where its catchment is about 99.8% implying that very minimal parts fall within the other two districts accounting for a total of just 0.2% of its basin area (Figure 4.32 and Table 4.30).

Table 4.30 Administrative Set-up of West Banas River Basin

S. No.	District Name	Area (sq km)	% of Basin Area	Total Number of Blocks	Total Number of Towns and Villages
1	Pali	2.5	0.2	1	1
2	Sirohi	1,873.0	99.8	4	180
3	Udaipur	0.8	-	1	1
	Total	**1,876.3**	**100.0**	**6**	**182**

Source: Groudn Water Department, GoR, Jodhpur

Climate

The West Banas River basin is very small and falls at the fringe of Arid climatic region. Bye and large it is a dry area however surrounded by hills. It is extremely cold from October to February while turning hot from March to September. Summer records very high temperatures in the range of 47 °C – 48 °C but winters are chilly when temperature falls to about 2 °C. The basin receives fairly low rainfall because the mean annual rainfall over West Banas River Basin was 767.7 mm, of which most of the rainfall is received during the four Monsoon months (June-September).

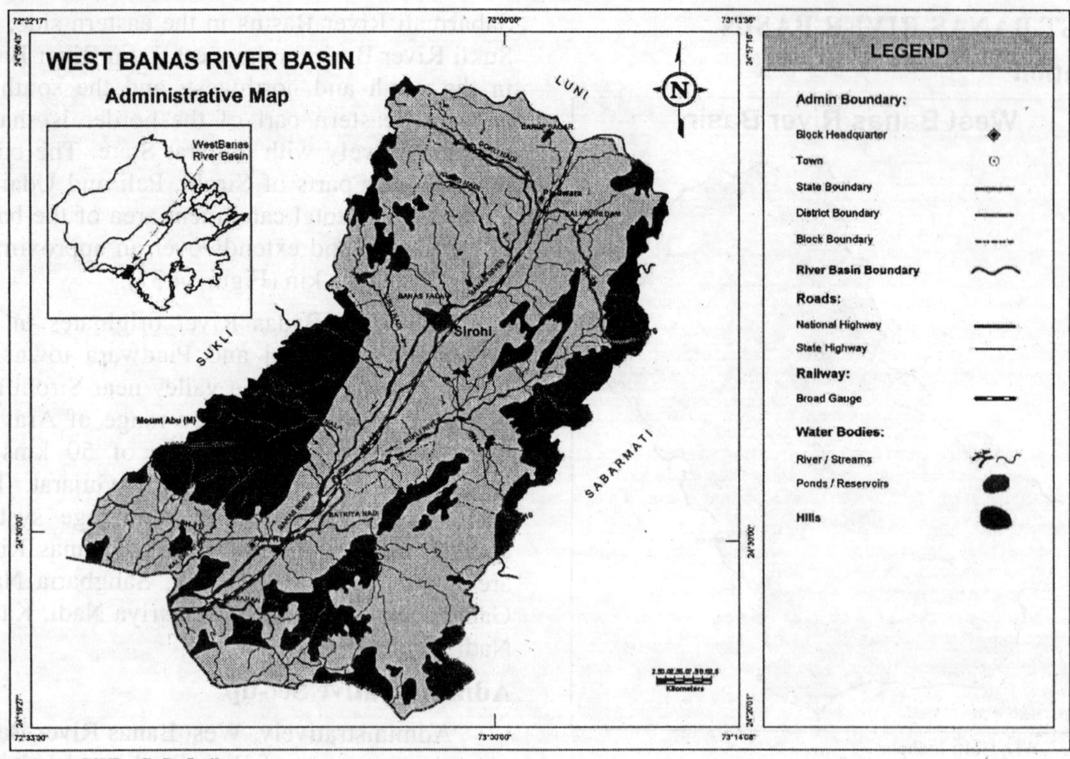

Source: GWD, GoR, Jodhpur

Figure 4.32 Administrative Map of West Banas River Basin

Topography

The basin is bounded both in the east and west by prominent hills of Aravali range. The hills on the western side are higher and are part of Mt. Abu range. The streams originating from hills flow in E-W directions as well as in south-southwestern to contribute to the drainage to West Banas River's main course flowing towards southwest. Being part of the Mt. Abu range, the maximum elevation seen within the basin is about 1695m amsl. While the general elevation within hilly part is in the 250-750m range, the same in the non-hilly part is generally under 500m amsl. The hills in the eastern part re not as high as those in the west. The overall slope of the terrain is from a broad north to south – southwest. The lowest point is also in Sirohi district itself and is located near the exit point southwest of Abu road stands at just less than 220m amsl which leaves a relief difference presence in the valley to 217.5m amsl (Figure 4.33 and Table 4.31).

Table 4.31 Districtwise minimum and maximum elevation

S. No.	District Name	Min Elevation (m amsl)	Max. Elevation (m amsl)
1	Pali	392.5	499.8
2	Sirohi	217.4	1,695.0
3	Udaipur	454.6	892.3

Source: Ground Water Department, GoR, Jodhpur

Rainfall

The general distribution of rainfall across the West Banas River Basin can be visualized from isohyets presented in the Figure 4.33 where, it is seen that the central part of the basin receives highest rainfall and also in in the western and southwestern part. There are two operational rain gauge stations operating within the basin which reveal a total average annual rainfall of 852 mm in the year 2010. Rainfall reduces as one moves towards northwest and southwest. The general rainfall range in the basin had been in the range of 800 to 1000mm.

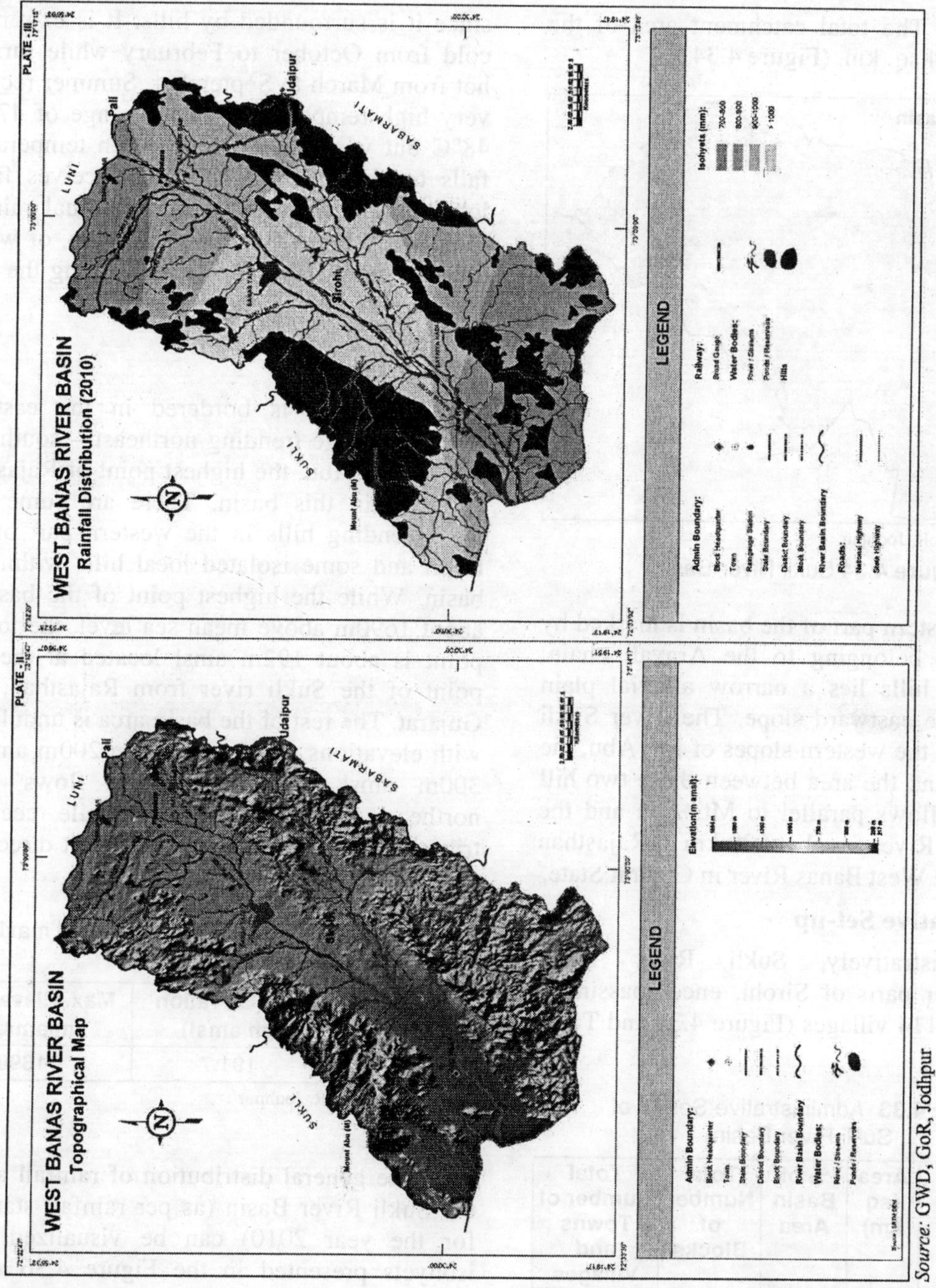

Figure 4.33 : West Banas River Basin (Topographical and Rainfall Distribution)

Source: GWD, GoR, Jodhpur

SUKLI RIVER BASIN

Location

River Sukli (also called Sipu River) flows in the south-western part of Rajasthan, and its basin is very small in geographic extend measuring less than 1,000 sq kms. It stretches between 24° 24' 42.01" to 24° 47' 22.30" North latitudes and 72° 19' 43.99'' to 72° 48' 32.40'' East longitudes. It is bounded on the north by the Luni River Basin, on the east by the West Banas River Basin and on the west by the area drained by Other Nallahs of Jalore. The southern border is shared with Gujarat State. The basin extends over parts of Sirohi district only. The total catchment area of the Basin is 999.4 sq. km. (Figure 4.34).

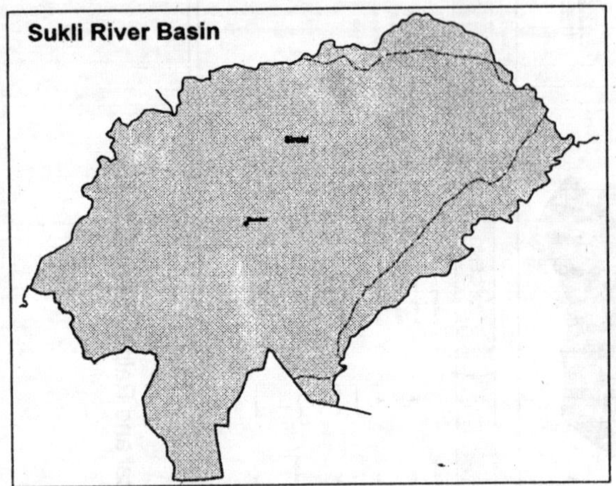

Source: GWD, GoR, Jodhpur

Figure 4.34 Sukli River Basin

The western part of the basin is marked by hilly terrain belonging to the Aravali chain. East of the hills lies a narrow alluvial plain with a gentle eastward slope. The River Sukli originates in the western slopes of Mt. Abu, the Silari hills and the area between these two hill features. It flows parallel to Mt. Abu and the West Banas River for about 29 km in Rajasthan and joins the West Banas River in Gujarat State.

Administrative Set-up

Administratively, Sukli River Basin extends over parts of Sirohi, encompassing 3 Blocks and 114 villages (Figure 4.35 and Table 4.33).

Table 4.33 Administrative Set-up of Sukli River Basin

S. No.	District Name	Area (sq km)	% of Basin Area	Total Number of Blocks	Total Number of Towns and Villages
1	Sirohi	999.4	100.00	3	114

Source: GWD, GoR, Jodhpur

Climate

The Sukli River basin is very small and falls at the edge of arid climatic region border. By and large it is a dry area but not much arid since it is surrounded by hills. It is extremely cold from October to February while turning hot from March to September. Summer records very high temperatures in the range of 47°C–48°C but winters are chilly when temperature falls to about 2°C. The basin receives fairly low rainfall because the mean annual rainfall over Sukly River Basin was 350mm, of which most of the rainfall is received during the four Monsoon months (June-September).

Topography

The basin is bordered in the east by Aravalli Range trending northeast – southwest and Mount Abu, the highest point of Rajasthan falls within this basin. There are some east west trending hills in the western part of the basin and some isolated local hills within the basin. While the highest point of the basin is about 1690m above mean sea level, the lowest point is about 192m amsl located at the exit point of the Sukli river from Rajasthan, into Gujarat. The rest of the basin area is undulating with elevations ranging between 200m amsl to 300m amsl. The Sukli river flows in a northeast–southwest direction, while receiving tributaries from east, north and west directions (Figure 4.36 and Table 4.34).

Table 4.34 : District wise minimum and maximum elevation

S. No.	District Name	Min Elevation (m amsl)	Max. Elevation (m amsl)
1	Sirohi	191.7	1690.2

Source: GWD, GoR, Jodhpur

Rainfall

The general distribution of rainfall across the Sukli River Basin (as per rainfall statistics for the year 2010) can be visualized from isohyets presented in the Figure 4.36 where most of the northern, eastern and western parts receive higher rainfalls (in the range of >1000mm of total annual rainfall) whereas; the southern parts received relatively low rainfall (between 800 to 1000mm). The average annual rainfall for the basin computed based on available station data is about 1281 mm.

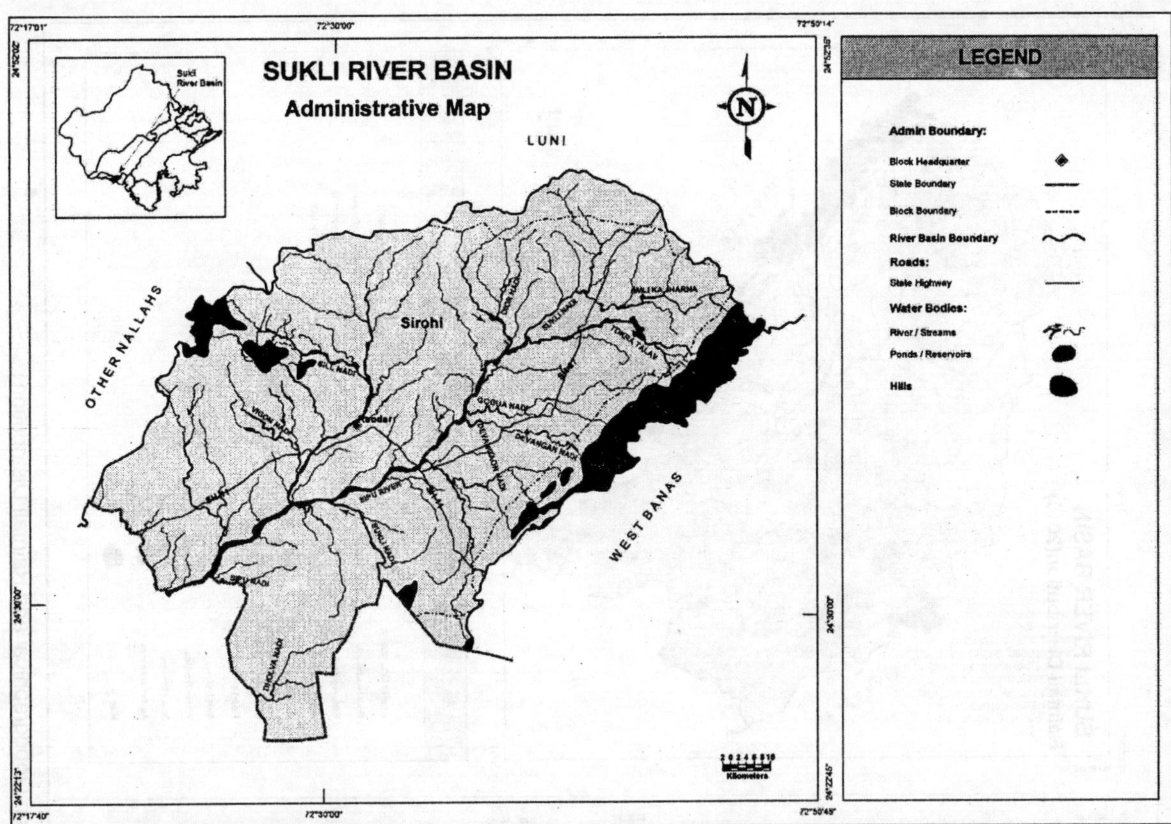

Source: Groudn Water Department, GoR, Jodhpur

Figure 4.35 Administrative Map of Sukli River Basin

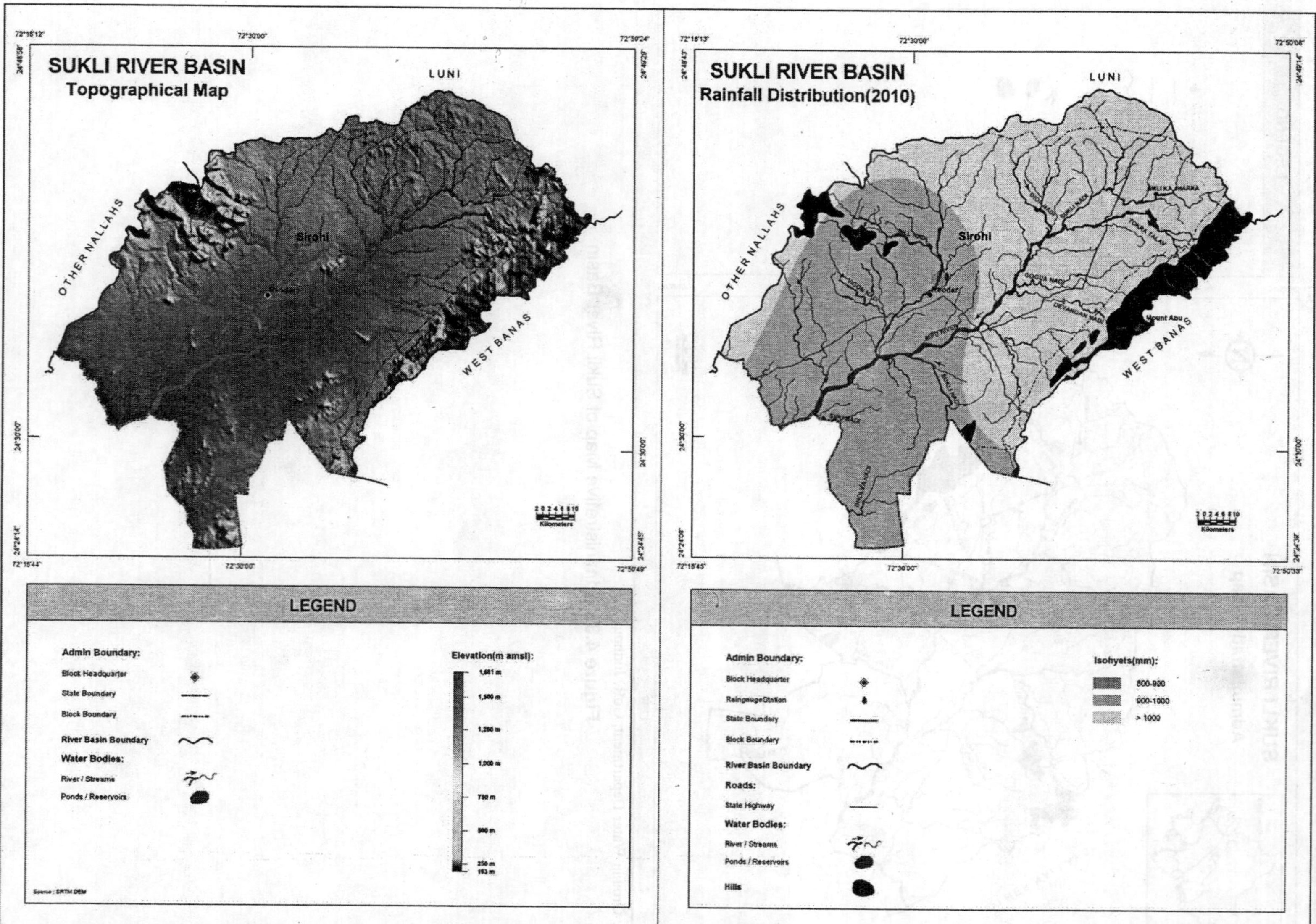

Source: GWD, Rajasthan, Jodhpur

Figure 4.36 : West Banas River Basin (Topographical and Rainfall Distribution)

GAMBHIR RIVER BASIN

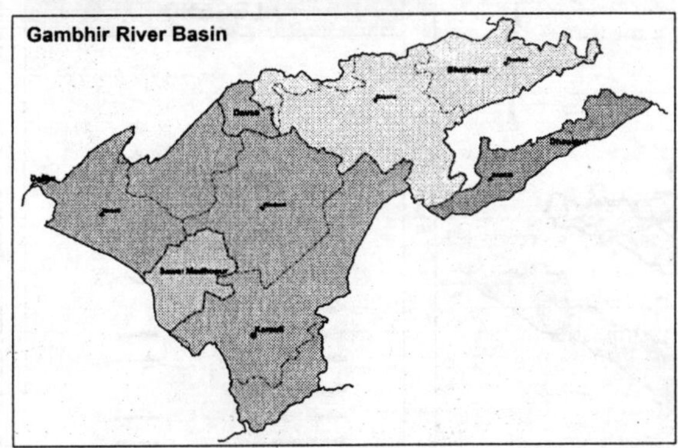

Source: GWD, GoR, Jodhpur

Figure 4.37 Gambhir River Basin

Location

Gambhir River Basin is located in northeastern part of Rajasthan. It stretches between 26° 16' 07.47" to 27° 05' 23.88" North latitude and 76° 31' 56.35'' to 77° 52' 25.31'' East longitudes. It is bounded by the Banganga river basin in the north, Banas river basin in the south west, Chambal and Parbati in the southeast; Uttar Pradesh state constitutes part of the boundary in the northeast. The Basin extends over parts of Bharatpur, Dausa, Dhaulpur, Karauli and Sawai Madhopur Districts. Gambhir is a small river basin and its approximate total catchment area is 4,316 km² (Figure 4.37).

River Gambhir originates in the hills near Karauli village in Sawai Madhopur District. It flows from south to north up to Kanjoli village (Toda Bhim), then towards northeast up to village Mertha of Roopbas Block, before entering Uttar Pradesh. The river again enters Rajasthan near Catchapaura village in Dhoulpur District and forms the boundary between UP and Rajasthan. It then enters Mainpuri District in UP to finally joins River Yamuna. The river is ephemeral, but becomes perennial after its confluence with the Parbati, outside Dhoulpur District. The total length of the river in Rajasthan is 288 km. important tributaries of the Gambhir are Sesa, and Kher.

Administrative Set-up

Gambhir river basin extends over parts of Bharatpur, Dausa, Dhaulpur, Karauli and Sawai Madhopur Districts and divided into 13 Blocks and encompassing 896 towns and villages (Figure 4.38 and Table 4.36).

Climate

Gambhir basin is small and falls within Sub-Humid climatic region receiving moderate to good rainfall resembling more to Uttar Pradesh than to semi-arid Rajasthan. It is very cold from November to February while warming up between March-April and the months of May-June are very hot when temperatures reaches high of 49-50°C. The mean annual rainfall over Gambhir river basin was 616 mm, of which around 95% is received during the Monsoon months of June-September.

Table 4.36 Administrative Set-up of Gambhir River Basin

S. No.	District Name	Area (sq km)	% of Basin Area	Total Number of Blocks	Total Number of Towns and Villages
1	Bharatpur	1,210.5	28.0	3	279
2	Dausa	100.1	2.3	2	27
3	Dhaulpur	321.6	7.5	2	105
4	Karauli	2,426.5	56.2	5	450
5	Sawai Madhopur	257.6	6.0	1	35
	Total	**4,316.3**	**100.0**	**13**	**896**

Source: Groudn Water Department, GoR, Jodhpur

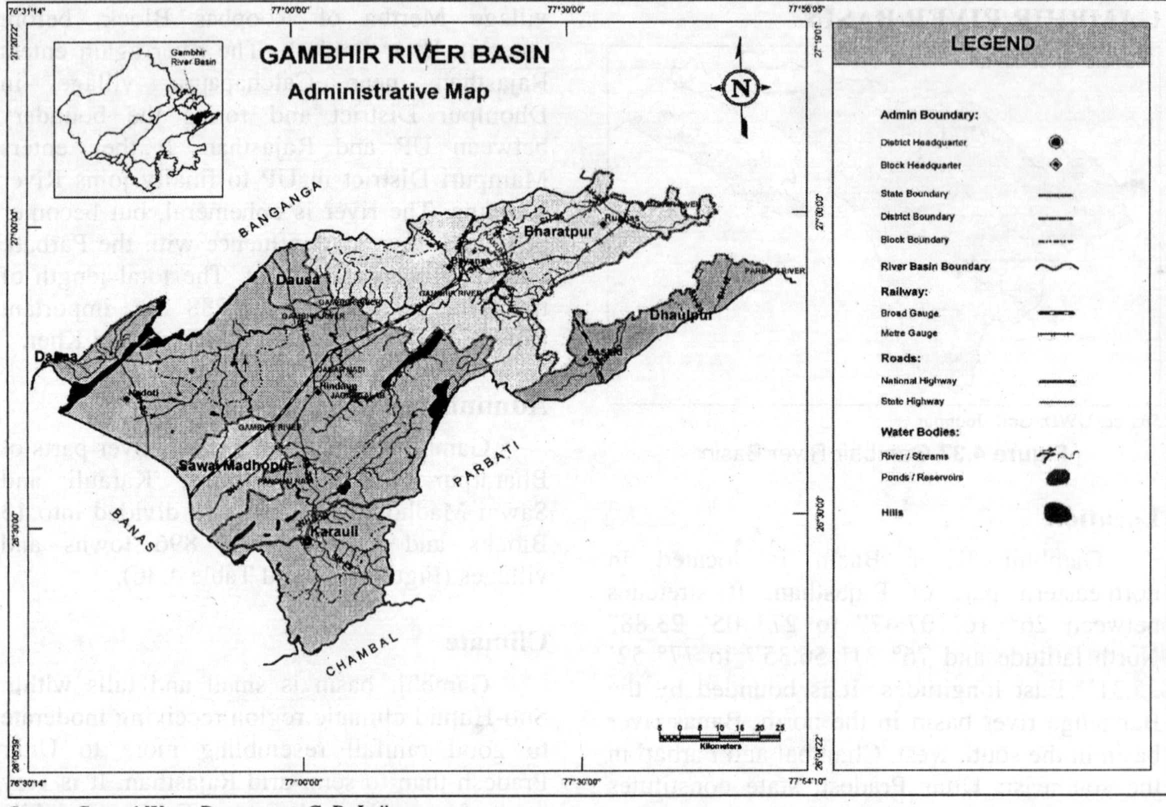

Source: Ground Water Department, GoR, Jodhpur

Figure 4.38 Administrative Map of Gambhir River Basin

Topography

The central and northeastern parts of the Basin falling within Sawai Madhopur, western parts of Karauli district and in Bharatpur and Dhaulpur districts is marked by undulating plains. Areas around Bayana, southwestern part of Dausa (within basin) and Karauli are hilly which belong to the Aravali range. Figure 4.39 depicts the topography of the basin (Table 4.38).

Table 4.38 District wise mini. and max. elevation

District Name	Min Elevation (m amsl)	Max. Elevation (m amsl)
Bharatpur	157.0	390.0
Dausa	203.5	491.4
Dhaulpur	156.5	295.3
Karauli	200.5	492.6
Sawai Madhopur	224.7	362.0

Source: Ground Water Department, GoR, Jodhpur

Rainfall

Perusal of Figure 4.39 reveals the distribution of total annual rainfall in the basin which seem to have ranged from 600-800mm in the year 2010. The northeastern parts of the basin adjoining Uttar Pradesh had received highest rainfall of the basin i.e., in the range of 900-100mm whereas western part around Nadbai in Karauli district had received least in the basin which ranged between 500-600mm of total annual rainfall.

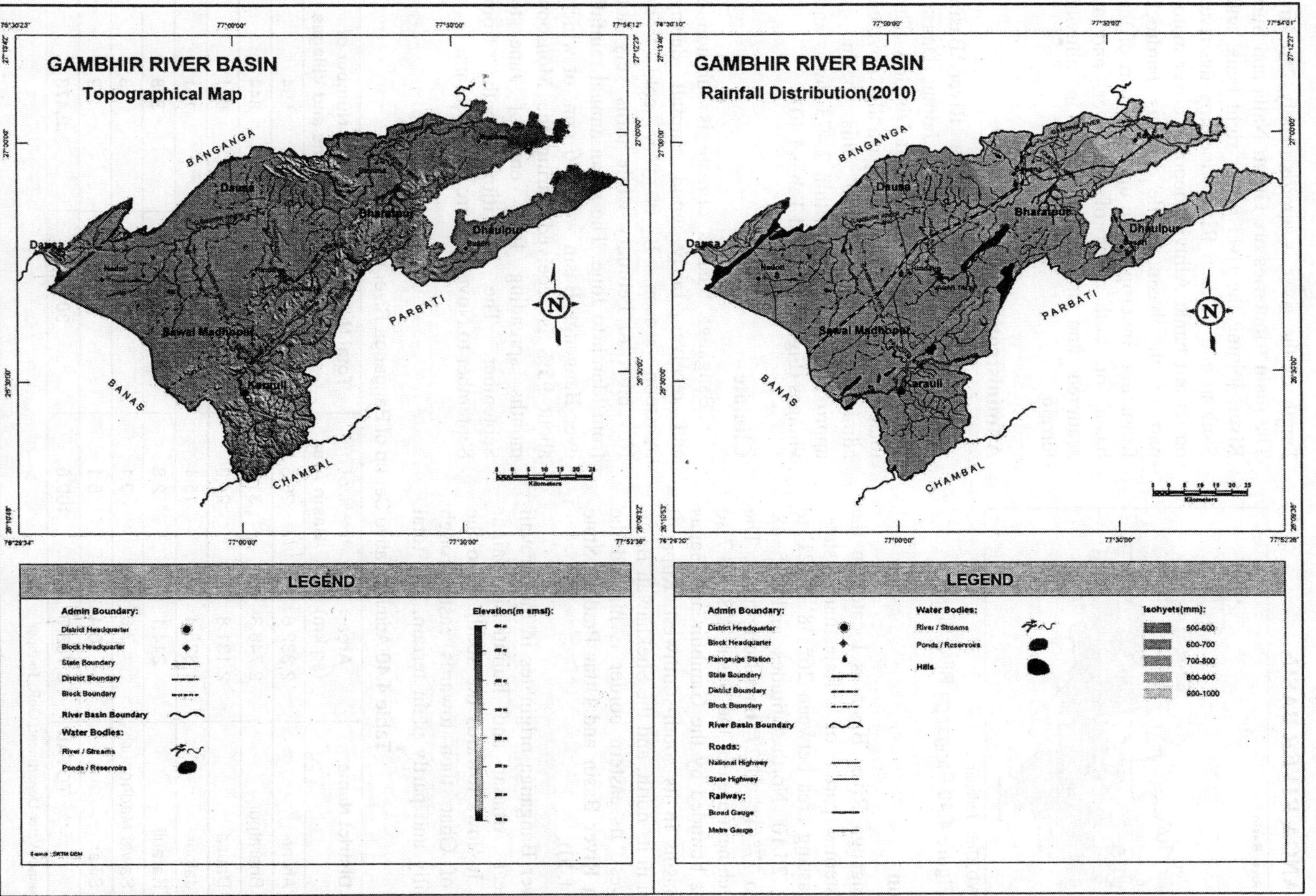

Source: Ground Water Department, GoR, Jodhpur

Figure 4.39 : Gambhir River Basin (Topographical and Rainfall Distribution)

BANGANGA RIVER BASIN

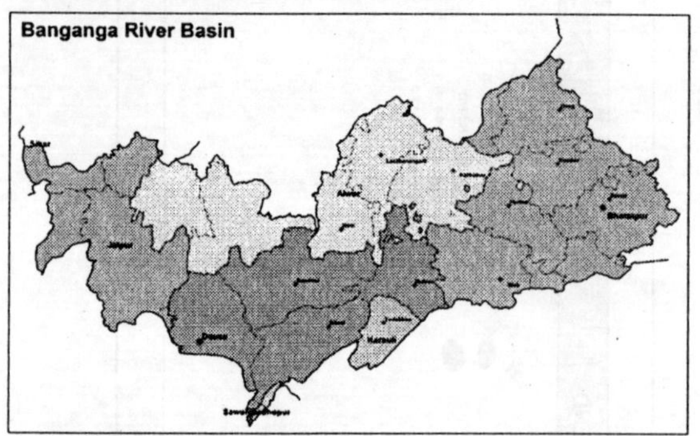

Source: GWD, GoR, Jodhpur

Figure 4.40 Banganga River Basin

Location

Banganga River Basin is located in the northeastern part of Rajasthan state, encompassing area between 26° 38' 57.21' to 27° 37' 25.10' North latitudes and 75° 48' 23.37' to 77° 41' 31.27' East Longitudes. The total catchment area of the Basin is 8,878.7 sq km. It is bounded by the Gambhir and Banas River Basins in its south-southwest; Ruparail and Sabi in its north; and the Shekhawati Basin in its west. Its eastern border is marked the Yamuna River Basin and Uttar Pradesh State (Figure 4.40).

River Banganga originates in the Aravali hills, near Arnasar and Bairath in Jaipur District. It flows towards the south up to the village of Ghat, then towards east through partly hilly and partly plain terrain. The total length of the river is approximately 240 km. The main tributaries are Gumti Nalla and Suri River, joining the river on its right bank, and Sanwan and Palasan Rivers meeting the river on its left bank. Alluvial deposits cover major part of the basin, while various hardrock formations outcrop in the upstream part of the basin, on both sides of the river, and as scattered and isolated hillocks in alluvial terrain.

Administrative Set-up

Administrative, Banganga River Basin extends over parts of Alwar, Jaipur, Dausa, Sawai Madhopur and Bharatpur Districts. The total number of fully or partially covered administrative blocks in this basin is approximately 30 and within 2,473 towns and villages (Figure 4.41 and Table 4.40).

Climate

Banganga basin's climate is sub-humid and receives fairly good rainfall during monsoon season. It is very cold from November to February while turns very hot from March to June. The mean annual rainfall over Banganga Basin was 596 mm, of which about 95% is received during the Monsoon months spreading from end of June to September. The transition months from September to November are very pleasant.

Table 4.40 Administrative Set-up of Banganga River Basin

S. No.	District Name	Area (sq km)	% of Basin Area	Total Number of Blocks	Total Number of Towns and Villages
1	Alwar	2,332.6	26.5	7	594
2	Bharatpur	2,746.8	31.3	9	844
3	Dausa	2,131.8	24.3	5	621
4	Jaipur	1,325.4	15.1	5	101
5	Karauli	244.1	2.8	2	308
6	Sawai Madhopur	7.6	0.1	1	3
7	Sikar	0.4	0.1	1	2
	TOTAL	**8,788.7**	**100.0**	**30**	**2,473**

Source: Ground Water Department, GoR, Jodhpur

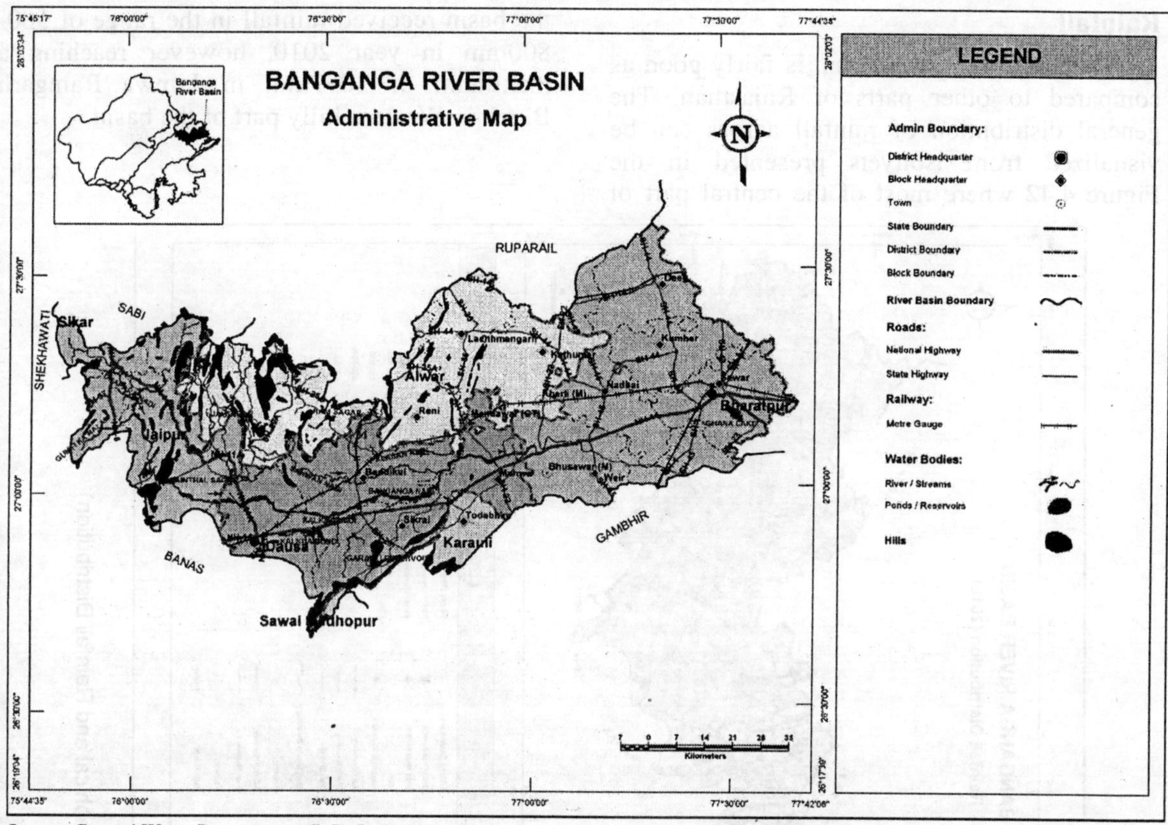

Source: Ground Water Department, GoR, Jodhpur

Figure 4.41 Administrative Map of Banganga River Basin

Topography

The western part of the Basin is marked by hilly terrain belonging to the Aravali range, with fairly flat valleys along the Banganga River and its tributaries. East of the Todabhim - Mandawar chain of hills, lies an extensive alluvial plain with gentle easterly slopes towards the Yamuna River in Uttar Pradesh. The northeastern part of the area is also rather flat, interspersed with moderately elevated hills (Figure 4.42).

Physiographically, the district can be divided into two distinct units, viz. Alluvial plain and Hilly areas. The Western part of the basin where the Banganga originates is largely hilly and further westwards and most of the eastern part is alluvial plain with gently undulating topography. The hills attain a maximum elevation of 727m in Jaipur district and topographically the minimum elevation of 160m is seen in Bharatpur district. The Banganga River is an east flowing river and after flowing through the basin it drains into Ghana Lake near Bharatpur city and thus forming an inland drainage system (Table 4.41).

Table 4.41 Districtwise minimum and maximum elevation

District Name	Min Elevation (m amsl)	Max. Elevation (m amsl)
Alwar	200	689
Bharatpur	160	416
Dausa	222	594
Karauli	233	438
Jaipur	322	727
Sawai Madhopur	430	526
Sikar	575	698

Source: Ground Water Department, GoR, Jodhpur

Rainfall

The rainfall in this basin is fairly good as compared to other parts of Rajasthan. The general distribution of rainfall across can be visualized from isohyets presented in the Figure 4.42 where most of the central part of the basin received rainfall in the range of 500-800mm in year 2010, however reaching a maximum of 1039mm in Jamwa Ramgarh Block in Western hilly part of the basin.

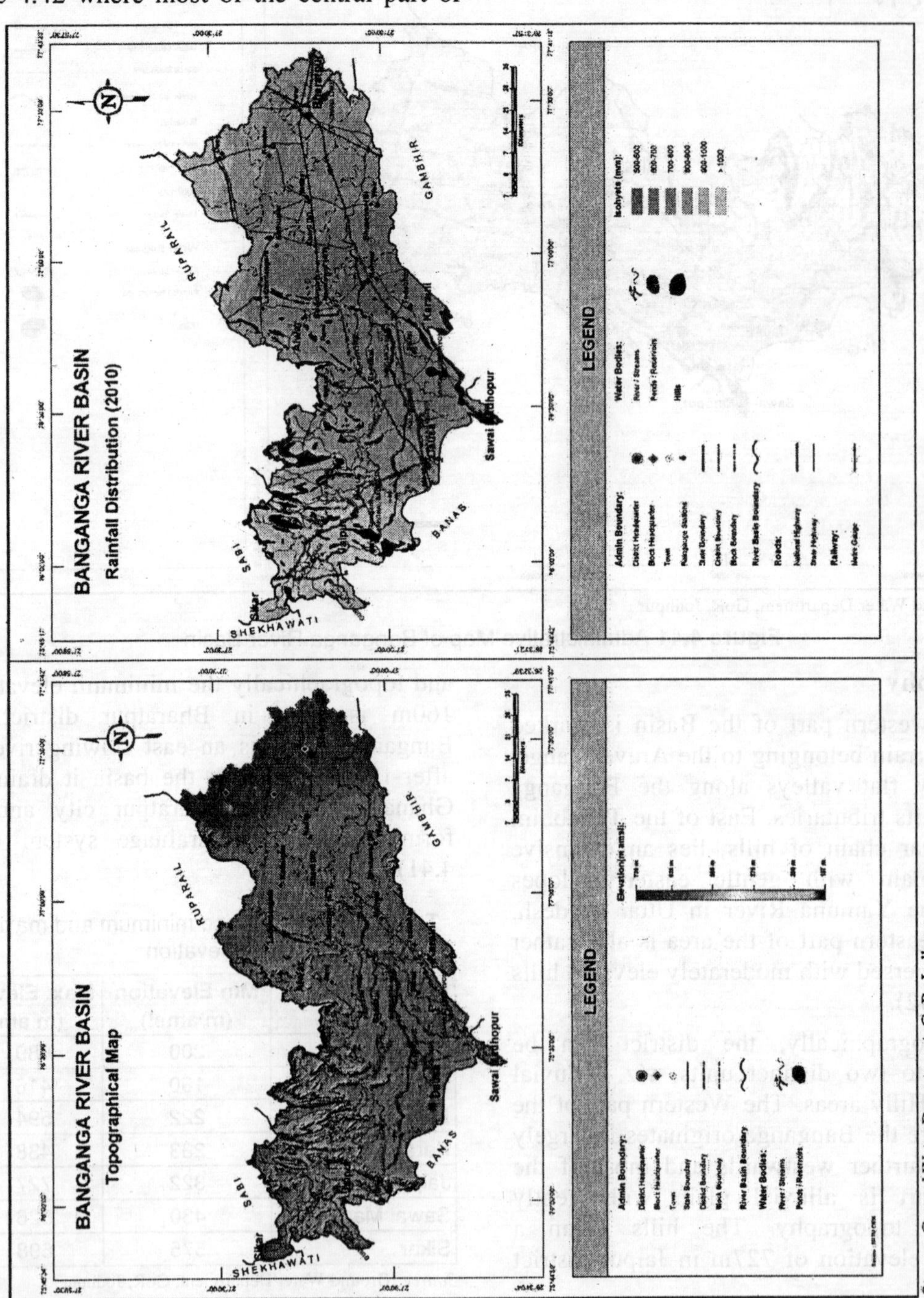

Source: Ground Water Department, GoR, Jodhpur

Figure 4.42 : Banganga River Basin (Topographical and Rainfall Distribution)

OTHER NALLAHS BASIN

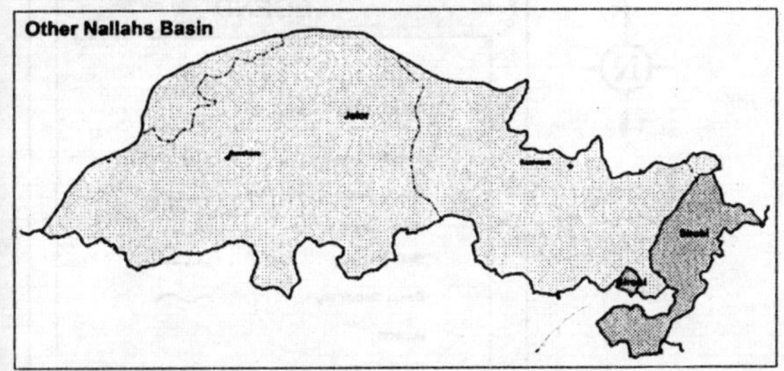

Source: GWD, GoR, Jodhpur

Figure 4.43 Other Nallahs Basin

Location

This is one of the very small basins in Rajasthan that measure less than 2,000 km2 in area (approximately 1900 km²). It is located in southwestern part of the state, bounded in the north and northwest by Luni river basin and in the east by Sukli river basin. The southern border is shared with administrative boundary of Gujarat state. It stretches between 24° 31' 50.08" to 24° 56" 27.96' North latitude and 71° 31' 26.59" to 72° 27' 32.25" East longitudes. Major part of the basin (>90%) falls within Jalor district whereas the remaining area falls in Sirohi district. The basin is a collection of small streams that drain the area but exit the border of Rajasthan State before contributing to any major river. Bandi river, Bargaon, Rip, and Sukal are some of the streams flowing in the area (Figure 4.43).

The general slope of the terrain is from east to south or west with hills appearing in the east which slowly roll into plains towards west in Sanchor.

Administrative Set-up

Administratively, Other Nallahs basin extends over parts of Jalor and Sirohi districts which are further divided into 5 blocks, encompassing 154 towns and villages (Figure 4.44 and Table 4.43).

Table 4.43 Administrative Set-up of Other Nallahs Basin

S. No.	District Name	Area (sq km)	% of Basin Area	Total Number of Blocks	Total Number of Towns and Villages
1	Jalor	1731.2	91.1	4	138
2	Sirohi	168.6	8.9	1	16
	TOTAL	**1899.8**	**100.0**	**5**	**154**

Source: Ground Water Department, GoR, Jodhpur

Climate

The basin falls in Semi-arid to Arid region and experiences extreme temperatures in winter (down to 4 °C) and summer (reaches high of 50 °C). Rainfall is scanty, with mean annual rainfall of about 353 mm only. As a result the rainy days are very few spread over the months of July to September and for most part of the year it remains dry. Winter ensues rainy season and till February the low temperatures can be felt. Summer sets in from April and May – June months are really hot.

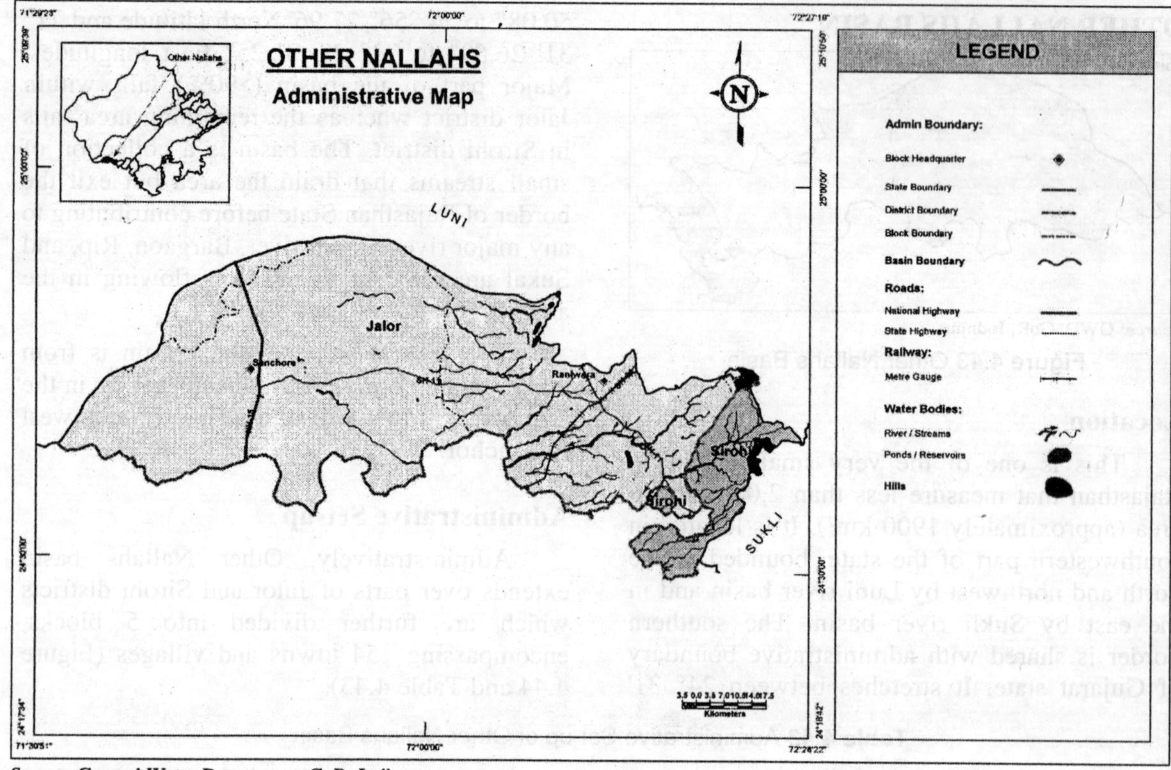

Source: Ground Water Department, GoR, Jodhpur

Figure 4.44 Administrative Map of Other Nallahs

Topography

Though the area that forms part of the basin is not large but still the elevation variation is significant, ranging from minimum of 21 m above man sea level in Jalor district (in the westernmost part) to a maximum of 977m amsl in the eastern hilly parts falling in Sirohi district. The eastern part is hilly and the general slope of the topography is westwards. The hills in Sirohi district and partly in Jalor district are major sources of drainage in the basin which flow westwards (Figure 4.45 and Table 4.44).

Table 4.44 : District wise minimum and maximum elevation

District Name	Min Elevation (m amsl)	Max. Elevation (m amsl)
Jalor	21	850.9
Sirohi	203.7	976.8

Source: Ground Water Department, GoR, Jodhpur

Rainfall

The general distribution of rainfall across the Other Nallah Basin can be visualized from isohyets presented in the Figure 4.45. Total Annual rainfall for the rain gauge stations has been interpolated to understand the distribution across the basin. The year seems to have received a higher rainfall than the other years since it varies from a high of about 1225mm to a low of 938mm indicated by the coloured regions. Higher rainfall is received in the hilly region of eastern part of the basin and reducing gradually towards west.

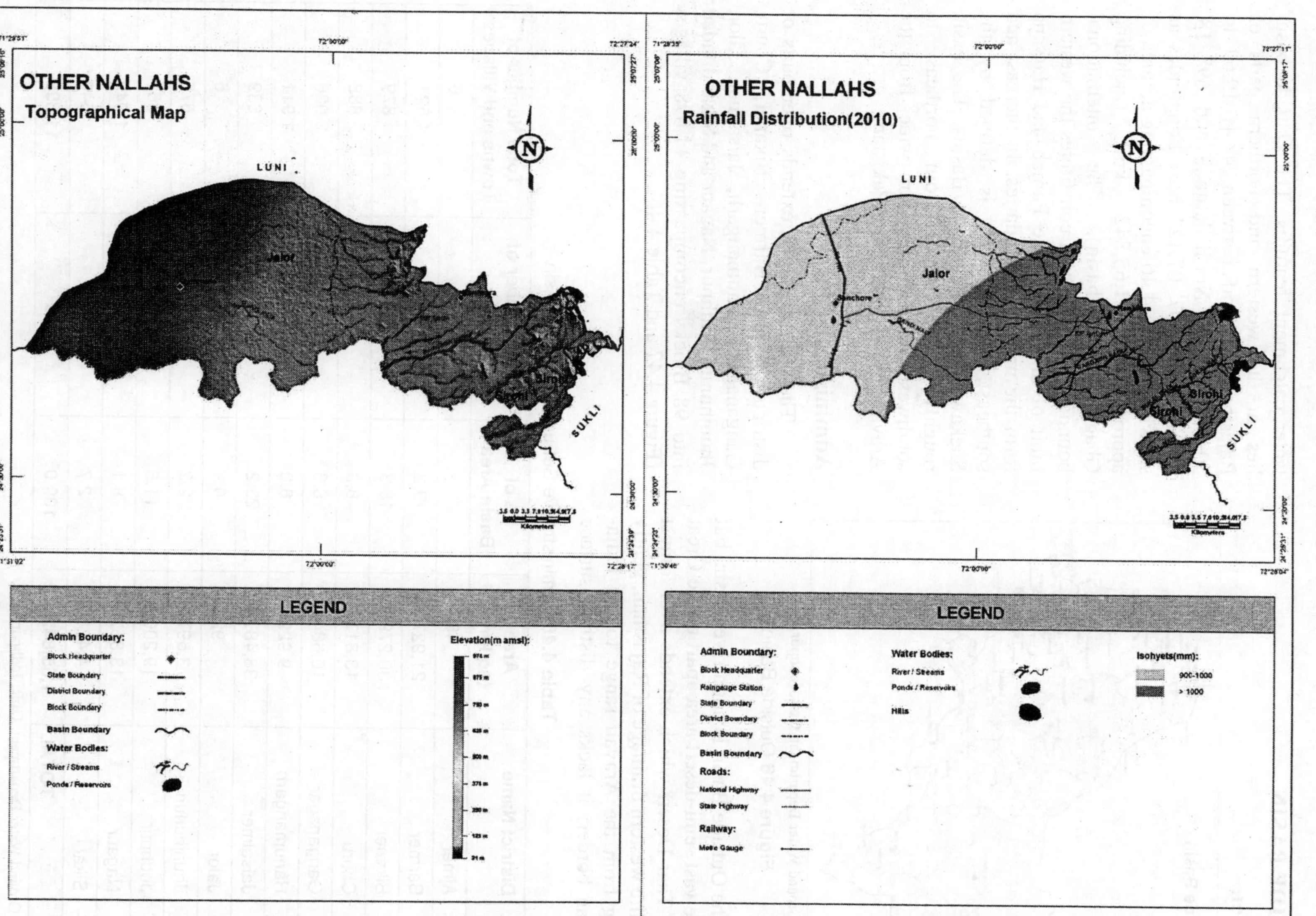

Source: GWD, GoR, Jodhpur

Figure 4.45 : Other Nallahs (Topographical and Rainfall Distribution)

OUTSIDE BASIN

Location

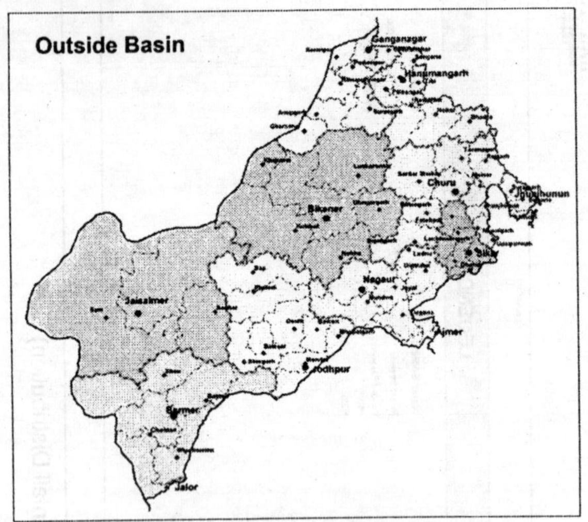

Source: Ground Water Department, GoR, Jodhpur

Figure 4.46 Outside Basin

The Outside Basin is not a river basin, but it is the vast semi-desert area apart of the Great Indian Thar Desert which extends over almost the entire western drainage of Rajasthan, gently sloping from the Aravali range to the India-Pakistan border; it lacks any distinct surface water catchment features. The Outside basin lies in the western and northern parts of Rajasthan, It extends between 24° 41' 19.69" to 30° 12' 08.60" North latitude and 69° 18' 41.91" to 76° 07' 07.12" East Longitudes as shown in Figure 4.46 encompassing an area of approximately 1,65,842 km² (including Ghaggar sub basin). The international boundary with Pakistan constitutes the western limit of the basin while Punjab and Haryana form the northern boundaries. In the east and southeast, the boundary is defined by the Shekawati and Luni River Basins. The basin roughly 600km long, from northeast to southwest and about 280 km wide, from the Aravalis to the border with Pakistan.

Administrative Set-up

The Outside Basin extends over parts of districts Ajmer, Barmer, Bikaner, Churu, Ganganagar, Hanumangarh, Jaisalmer, Jalor, Jhunjhunun, Jodhpur, Nagaur and Sikar divided into 98 Blocks encompassing 11,898 villages (Figure 4.47 and Table 4.46).

Table 4.46 Administrative Set-up of Outside Basin

S. No.	District Name	Area (sq km)	% of Basin Area	Total Number of Blocks	Total Number of Towns and Villages
1	Ajmer	48.7	-	3	6
2	Barmer	21,927.7	13.2	8	1,521
3	Bikaner	30,279.3	18.3	11	879
4	Churu	13,817.3	8.3	6	888
5	Ganganagar	10,684.0	6.3	19	3,026
6	Hanumangarh	9,929.2	6.0	13	1,911
7	Jaisalmer	38,487.2	23.2	3	639
8	Jalor	92.0	0.1	1	5
9	Jhunjhunun	3,597.8	2.2	8	573
10	Jodhpur	19,200.3	11.6	9	861
11	Nagaur	13,357.0	8.1	11	1,049
12	Sikar	4,421.6	2.7	6	540
	TOTAL	**1,65,842.1**	**100.0**	**98**	**11,898**

Source: Ground Water Department, GoR, Jodhpur

Climate

The basin is very large, covering almost half of Rajasthan in its western part but it falls into Arid climatic zone which is typically exemplified by dry conditions throughout the year and temperature extremes of very high (reaching upto 50 °C) in mid-summer to very low (down to 1 to 2 °C) in winter nights. Even the diurnal temperature variations are very high with hot days and cold nights for the most part of the year. Rainfall is scanty dipping to the lowest mean annual rainfall of about 100mm in the desert areas to high of upto 400mm in eastern parts adjoining Aravali hills. With almost negligible monsoon season, the two pronounced seasons are summer and winter.

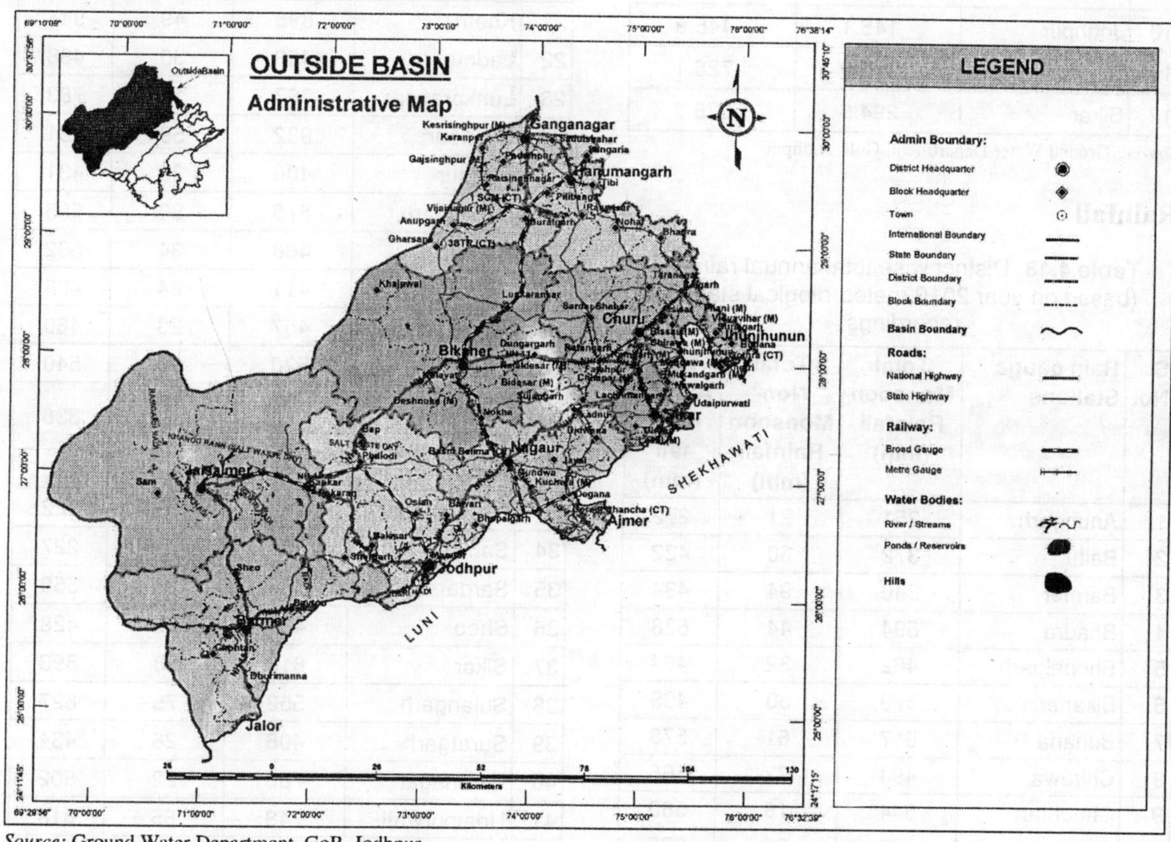

Source: Ground Water Department, GoR, Jodhpur

Figure 4.47 Administrative Map of Outside Basin

Topography

Most of the basin area is represented by low undulating stabilized and active dunal plains but bordering areas in the southernmost part of Barmer district has minimum elevation of 14m amsl and the hills in Jhunjhunun district have maximum elevation of about 1037 m amsl (Table 4.47).

Table 4.47 Districtwise minimum and maximum elevation

S. No.	District Name	Min Elevation (m amsl)	Max. Elevation (m amsl)
1	Ajmer	436.2	567.2
2	Barmer	14.1	622.6
3	Bikaner	105.6	348.2
4	Churu	192.6	478.0

S. No.	District Name	Min Elevation (m amsl)	Max. Elevation (m amsl)
5	Ganganagar	104.4	230.0
6	Hanumangarh	164.2	240.7
7	Jaisalmer	31.5	347.8
8	Jalor	16.2	90.4
9	Jhunjhunun	259.6	1,036.8
10	Jodhpur	145.1	445.3
11	Nagaur	273.9	726.1
12	Sikar	294.5	1,026.7

Source: Ground Water Department, GoR, Jodhpur

Rainfall

Table 4.48 District wise total annual rainfall (based on year 2010 meteorological station recordings

S. No.	Rain gauge Stations	Total Monsoon Rainfall (mm)	Total Non-Monsoon Rainfall (mm)	Total Annual Rain-fall (mm)
1	Anupgarh	251	21	272
2	Baitu	372	50	422
3	Barmer	340	94	434
4	Bhadra	594	44	638
5	Bhopalgarh	462	32	494
6	Bikaner	379	30	409
7	Buhana	517	61	578
8	Chirawa	494	72	566
9	Chouhtan	544	16	560
10	Churu	570	65	635
11	Degana	541	64	605
12	Didwana	592	10	602
13	Fatehgarh	362	34	396
14	Fatehpur	629	44	673
15	Ganganagar	351	15	366
16	Hanumangarh	519	28	547
17	Jaisalmer	285	24	309
18	Jayal	441	28	469
19	Jhunjhunu	775	53	828
20	Karanpur	282	31	313
21	Khetri	895	49	944
22	Ladnu	458	30	488
23	Lunkaransar	767	16	783
24	Malsisar	932	59	991
25	Nagaur	406	25	431
26	Nawalgarh	575	80	655
27	Nohar	468	34	502
28	Osian	411	24	435
29	Padampur	457	23	480
30	Pokaran	520	20	540
31	Raisingh Nagar	328	8	336
32	Ratangarh	666	47	713
33	Sadulshahar	286	36	322
34	Sam	225	2	227
35	Sardarshahar	536	20	556
36	Sheo	401	27	428
37	Sikar	818	78	896
38	Sujangarh	552	75	627
39	Suratgarh	406	28	434
40	Taranagar	780	22	802
41	Udaipurwati	748	65	813

Source: http://waterresources.rajasthan.gov.in

LAKES OF RAJASTHAN

There are two types of lakes in Rajasthan:

1. Saline Water Lakes

2. Sweet Water Lakes

Rajasthan is a state of 'Varieties of Lakes'. Some lakes are natural while others are artificial. Some of them are saline while others are sweet ones. The fresh water lakes have been created by damming the rivers and storing water for drinking and irrigation purposes. Some of these lakes are playing an important role in the economy of Rajasthan where as others adding to its beauty and natural resources (Table 4.49).

A brief description of lakes is given below:

1. Saline water lakes are found in the western part of Rajasthan.

Saline Water Lakes

(i) Sambhar Lake: It is situated at a distance of 65 km from Jaipur on Phulera rail-route. This is the largest lake of saline water in India. It has been famous for producing salt in India.

This is the largest lake of Rajasthan. Situated between 27°-29°N latitude and 74°-75°E longitude, this lake is on the border of Jaipur & Nagaur district. From north to south, the lake extends. to a length of 32 km while from west to east it extends to a breadth of 3 to 12 km. The maximum lake area during the monsoon season is 234 sq. km. whereas during summer, the area shrinks to 130 sq. km. only and water remains in small pools in different parts of the lake. It has an estimated quantity of 35 crore tons of common salt. The Hindustan Salt Ltd. is the main manufacturer of sodium chloride (common salt) and sodium sulphate which are produced here. The salt is manufactured by solar evaporation method from March to July.

(ii) Deedwana Lake: This salt-water lake is near Deedwana town (Nagaur district). It is about 4 km. long and 2 km. broad. The depression is surrounded by sand hills on all sides except in the west where an isolated spur of the Aravalli forms its western boundary. The bed of the depression shows the presence of black clay and is similar in appearance to that of the Sambhar lake.

(iii) Pachpadra Lake: This lake is about 25 sq. km in area and situated near Pachpadra (Barmer district) The salt of Pachpadra corresponds more to the sea salt. The salt pits in this lake are scattered over an area of about 5000 sq. metres.

It is situated near Balotra in Barmer district. It produces salt of high quality. It contains sodium chloride up to a level of 98%.

(iv) Lunkaransar Lake : It is about 6 sq. km in area and located near Lunkaransar town (Bikaner district) Mainly sodium chloride is produced.

Sweet Water Lakes

2. Sweet water lakes are as follows:

(i) Jai Samand: Jai Samand lake near Udaipur. It is the largest freshwater lake of Rajasthan. This lake is said to have been at one time the biggest artificial fresh water lake in the world. The lake was built by Maharana Jai Singh in the year 1691. it was created in 1685-91 by building a dam across the river Gomti near Salumbar (Udaipur district) The dam is 35 m. high, 375 m. long and 5 cm broad. This lake is situated at 73°55' to 74°43' east longitudes and 24° 12' to 24°18' north latitude. Its breadth varies from 1 km. to 8 km. with a circumference of about 50 km. It contains seven islands inhabited by Bhils who use rough boats, called Bhel for transport. After 1950, two canals viz. Shyampura and Bhat canals were built for irrigation purposes.

(ii) Raj Samand: Raj Samand lake near Kankroli in Udaipur district. This lake was built in 1662 A.D. by Maharaja Raj Singh of Udaipur. The lake is in Raj Samand district. It is 6.5 km. long and 3 km. broad. Today, the lake is being used for irrigation. It is fed by Banas feeder canal.

Raj Samand is famous for one of the longest literary works ever inscribed on stone. In nine marble pavilions from which the embarkment gets its name of Navchowki, are 25 marble slabs containing 1017 verses written in Sanskrit describing the early history of Mewar.

(iii) Pichhola: Pichhola and Fateh Sagar lakes in Udaipur. This lake took its name from the small Village of Pichhola nearly and was originally created by a Banjara in the 15th century during the reign of Maharana Lakha. The lake Pichhola is near the Udaipur city. It was said to be again renovated in the early 20th century by Maharaja Udai Singh. Apart from the two large island palaces-the present Lake Palace hotel and Jagmandir, there are several small islets. The lake is 7 km long and 2 km broad. Fateh Sagar and Swaroop Sagar are parts of the same lake.

(iv) Fateh Sagar: To the north of Pichhola lake and connected to it by a canal, is another artificial lake known as Fateh Sagar. The lake covers an area of about 3 sq. km. when full and has one small island which has been made a public park with a fountain.

(v) Udai Sagar: This picturesque lake 4 km. long and 3 km. wide situated 15 km. east of Udaipur was built by Maharana Udai Singh between 1559 and 1565. It is said to be the site of the famous meeting between Maharana Pratap and Kunwar Man Singh.

(vi) Ana Sagar: it was built by Ana ji, the Lord of Ajmer in 1137 A.D. It is a small lake in Ajmer city getting water from the nearby hills. Emperor Jahangir built Daulat Bagh (now called Subhash Garden) near this lake and Emperor Shahjahan built 378 metre long 'Baradari' on the lake to have scenic view.

(vii) Nakki Lake: It is a small natural lake near Mt. Abu which is surrounded by hills. The lake is about 35 m. deep and has an area of 9 sq. km.

(viii) Pushkar: This is a small natural lake with an area of 3 sq. km. It is located 11 km. west of Ajmer city. It is of an irregular elliptical shape, surrounded by hills except in the east. Around its bank are big buildings, temples, cenotaphs and bathing ghats. Hindu submerges the ashes of their departed relatives in the lake, as this lake is considered to be very holy.

(ix) Siliserh: This small natural lake, which is surrounded on all sides by low hills, is situated 12 km west of Alwar city. It has an area of about 10 km. King Jai Singh of Alwar built a system of canal irrigation from this lake in early 20th century.

(xi) Kolayat Lake: This lake is near Kolayat town about 48 km. south-west of Bikaner city. Here is the famous Kapil Muni Ashram. Every year on Kartik Pumima, there is a fair attended by Hindu worshippers.

(xii) Bal Samand: It is a large artificial lake situated about 6 km. from Jodhpur city. It has got a small palace and small pavilians living on its banks. It is just near the famous Mandore gardens.

Table 4.49 Districtwise Lakes, Ponds and Dams in Rajasthan

District	Lakes, Ponds and Dams
Ajmer	Anasagar, Foysagar, Pushkar, Narain Sagar Dam
Alwar	Mangalsar, Silised, Jaisagar, Devti, Vijaysagar
Bharatpur	Avar Dam, Band Baretha, Ajaan, Lalpur, Moti Lake
Bundi	Navalakhan lake, Navalsagar, Jaitsagar, Bhinrtal, Jawahar Sagar, Bardha Dam
Banswara	Bajaj Sagar Dam, Boravangadi,BKadana Dams near Banswara.
Bhilwara	Meja Dam, Sareri Dam, Khari Dam, Jaitpur Dam, Nahar Dam, Arwar, Parvatisagar
Bikaner	Loonkaransar, Anoop Sagar, Sursagar, Kolayat, Gajnei
Baran	Umed Sagar, Bilas, Kalisot, Banthali, Ratai
Barrner	Pachpadra
Chittorgarh	Bhupal Sagar, Gambhiri, Orai, Bassi, Vagan, Kundiya, Bankiya
Churu	Chhaparlai
Dholpur	Talabshahi, Ramsagar, Urmilasagar, Parbati
Dungarpur	Gabsagar, Lodisagar, Gajpur
Dausa	Madho Sagar Dam, Kalakh Sagar, Jhilmili, Devanchaii
Ganganagar	Taiwara Lake
Jhalawar	Bhimsagar Dam, HarishchandraSagar Dam, Chanwli Dam
Jhunjhunu	Pannalal shall Talab, Fateh Sagar Talab,Ajit Sagar Dam

District	Lakes, Ponds and Dams
Jalore	Balsamand, Bankali
Jodhpur	Kailana Dam or lake near Jodhpur, Bal Samand, Pratap Sagar, Ummed Sagar, Takhat Sagar, Piciiiyak Dam, Jaswantsagar, Kailiba
Jaipur	Ramgarh Dam of Jaipur (now lying dry), In the budget for 2013-14, it has been proposed that water would be brought from Meja river in Ramgarh Dam. Meja river originates from Bhilwara district and joins Chambal near Laklieri in Bundi.Gaita, Mansagar, Kanota Bandh, Deoyani, Ramgarh Dam, Chhaparvvara
Jaisalmer	Dharsi Sagar, Godisar, Amar Sagar, Bujh Lake
Kota	Jawahar Sagar Dam, Kota Dam, Kishorsagar
Karauli	Panchana, Kalisil, Khirkhiri, Bishan Samand
Nagaur	Sambhar, Kutchaman, Deedwana Lakes
Pratapgarh	Jakham Dam, Jai Sagar, Bhanwar Semla
Pali	Flemavas Dam, Jawai Dam, Bankali, Sardar Samand, Saei, Kharda, Raipur, Luni, Meethri
Rajsamand	Rajsamand, Nandsamand, Agria, Kundeli
Sirohi	Nakki Lake (Abu Mount), Nevara, Angor, W.Banas
Sawai Madhopur	Morel, Pancholas, Galai Sagar, Binori, Sagar Mansarovar
Tonk	Beesalpur Dam, Tordisagar, Galwa, Motisagar
Udaipur	Jai Samdnd, Raj Samand, Udai Sagar, Fateh Sagar, Swaroop Sagar, Pichola, Som Kagdar, Rohini

Source: District Gazetteer, GoR, Jaipur

Some lakes are natural and some are artificial or man-made. Sambhar lake is natural and Pushkar lake is also natural. Nakki lake of Mount Abu is quite beautiful and it is an attraction for the tourists.

Chapter 5

CLIMATE

The influence of climate is most conspicuous in traditional peasant agriculture of under-developed regions. It puts limit to cultivation and also manifest the type of soil. The characteristics of the soil are the product of present and past climate along with the vegetation that flourish in them.

The climate of any region can be considered as one of the most important geographical factors which influence the distribution of agricultural activities and type of crops. The study of the climate of a region is vast subject but, within the limited space and the scope of this book only major facts of the climate of Rajasthan have been considered. Climate is generally defined as the average state of the weather. The elements to be considered are the same while studying the climate or the weather conditions of a place, the difference lies in the length of the time considered. Weather refers mainly to short periods like a day, week, a month or longer, while climate is concerned with average conditions determined by observations made over extended period.

The study of the climate includes a number of elements or components like temperature and humidity of the air, the rainfall, the wind velocity, the duration of sunshine and many other elements of lesser significance. In their turn, these elements are products of several determining causes or factors. Among the important factors governing the climate of Rajasthan, as for any other region, are latitude, position relative to water, mountain barrier, altitude, prevailing winds, prevalence of cyclonic storms and contineotaliiy. Among various factors primarily, latitude is such, which greatly controls and determines the length of the day during the year. Since it is a mathematically controlled factor, it is considered as a reliable element. Latitude also influences the duration of the sunshine and the temperature variations.

The Rajasthan State stretches between 23°3'N to 30°12'N latitudes and 69'30'E to 78°17'E longitudes. Some other countries located between these latitudes are north Arabia, part of Egypt and northern Sahara and part of Mexico.

Although located in the same latitudes, the climatic conditions here are not so severe as occurring in some of these countries. The southern part of Rajasthan is about 225 km from the Gulf of Kutch and about 400 km from the Arabian Sea. The northern part extends up to Delhi. The elevation of land surface ranges between 214 m to 1375 m at few places in the Aravallis, while the larger part of the State is less than 370 m in altitude. The latitudinal location determines to a great extent the amount of insolation and the direction of the prevailing winds in the Tropics. The Aravallis stretching from north-east to south-west are the most conspicuous physiographic feature in the State. The eastern section of the Aravallis is the meeting place of Supan's broadly classified Indus and Indo-Australian Monsoon climatic provinces. The Indus Province is remarkable for great dryness and heat while the other in the east is characterised by strong periodical rains, temperature being fairly uniform. In general, the climate of Rajasthan west of the Aravallis, as of other desert and semi- desert regions, is characterised by great extremes of temperatures and long periods of severe drought accompanied by high wind velocity and low relative humidity to semi-arid conditions. The winter is quite cold and at many places the temperature sometimes falls below freezing point and frost occurs. On the other hand, the heat during the summers is intense and scorching. The arid region in western Rajasthan is the hottest region in India. East and south of the Aravallis, there is considerable variation in the amount of rainfall and temperature distribution.

India has a great variety of climatic variations but the influence of monsoonal changes in this sub-continent has resulted in some form of unity in the climatic conditions. The study of the climate of Rajasthan is based on the seasonal variations and the main features of the climatic changes have been followed for each month

Climatically, the year in Rajasthan has been divided into three major conventional seasons as follows:

1. The hot-weather season (March to mid-June)
2. The season of general rains (mid-June to September)
3. The cold weather season (October to February).

The Meteorological Department of India has again sub-divided the cold season into two divisions i.e. (a) the season of retreating monsoon (October to December) and (b) the cold season (January and February). These seasonal variations have been broadly based on the temperature and rainfall conditions in different months. Temperature and rainfall are the most important elements, which have been frequently observed in the study of the climate of any region. Other factors like the amount of sunshine, cloudiness and humidity have also their influence on the climate of this region.

Hot Weather Season

In Rajasthan, the hot weather season is initiated in the month of March and it prevails from April to June. The north-ward migration of the sun results in a rise of temperature, first on the Deccan plateau and then gradually in the north, north-west and western parts of India. The temperature rises during this month is somewhat uniform in the whole of Rajasthan. Due to increase in temperature, the atmospheric pressure falls subsequently over the heated land. In April, the winds are from west to east. As they come after crossing the heated Thar desert, they are characteristically dry and warm. In the months of April and May, the sun is almost overhead, the diurnal range of temperature increases more and days become hotter. In the western part of Rajasthan mainly at Bikaner, Phalodi, Jaisalmer and Barmer, the maximum daily temperature goes up to 40°C to 45°C during these months.

The Thar Desert in the west is one of the hottest regions in India and because the diurnal range is large, the minimum daily temperature drops down considerably. There is intense heat and glare during the day. The humidity drops down to about 1 per cent. The surface is covered with loose sand, occasionally level and mostly in the form of dunes. There is no live or green vegetation except cactii or only those plants capable of growth in the typical desert conditions. The annual range of temperature is between 14°C to 17°C but the diurnal range of temperature is much and in fact the difference between day and night temperatures at different stations in Rajasthan is so much that the mean values convey a very inadequate picture of the true climatic conditions. The rays of the sun reach the sandy and rocky surface without any obstacle since the air is dry and the skies are cloudless. The heat during the day increases rapidly due to the rising of the sun above the horizon. The increase is so rapid that by noon the temperature goes to about 38°C and in the afternoon it rises to about 49°C. On account of the dryness of the atmosphere, the clear skies and the sandy nature of the soil, there is rapid radiation of heat from the earth soon after sunset and the heat of the day rapidly dies down. The result is that when the maximum temperature during the day is very high, the minimum temperature at night is considerably lower. The day temperature increases considerably in association with the heat waves which generally develop during this period over north India. Ganganagar has thus recorded a maximum temperature of 50°C, Jodhpur, Bikaner and Barmer 49°C, Jaipur and Kota 47.7°C and Jhalawar 46.6°C. The high temperature during the day makes the weather very trying. Occasionally, dust-storms bring about a sudden fall in temperature and sometimes the storms are followed by rain which results in slightly lowering the temperature. The nights even in summers are cool and in winter sometimes frost occurs. The diurnal range of temperature is about 14°C to 17°C in the west of the Aravallis and in other parts of the State also, the heat of the sun during the day makes the sand and exposed rock surfaces so heated that it becomes well nigh impossible to walk bare footed. The local heating results in strong convection currents which in turn result in the formation of local swirls (dust devils). Sometimes they become associated with cyclonic storms which are more fearsome being full of sand. The visibility due to sand becomes so poor that sometimes it is difficult to see more than few meters ahead.

The mean isotherms of 32°C, 34°C and 36°C for the months of May and June are shown in Figure 5.1. There is constant increase in temperature in north and west of the Aravallis. Kota and Jhalawar also show an average temperature of about 36°C. The 36°C isotherm runs west of Bikaner and north of Jaisalmer district. In May, the maximum average monthly temperature of 36°C occurs at Kota. The trend of isothermal lines in this State is no indication of the amount and intensity of aridity.

Season of General Rains

The temperature, pressure and wind conditions in June indicate that the existing conditions have become more intensified. The northern, north-western and western India continue to get heated till the onset of the monsoon by the middle of June. The monsoon current coming across the Indian Ocean is divided in two branches namely Bay of Bengal Branch and the Arabian Sea Branch. Since their goal is towards the low pressure of Sind, Rajasthan, due to its location, experiences the indraft of both monsoon currents. The ascensional movement, however, does not lead to heavy rainfall. In spite of the monsoon currents towards the western part of Rajasthan, there is considerable decrease in rainfall This is mainly due to a variety of factors. The air currents from the south-east already give up their moisture in their passage up the Ganga plains. The currents coming from the southeast, after crossing the warm sea come over an intensely heated land; thus the relative humidity decreases from 90 per cent to 50 per cent. Ordinarily, even this percentage of relative humidity would cause rainfall if the air current rises to about 920 m but due to the existence of an upper current of hot dry air from further west, most of the moisture is absorbed by this wind and the skies, again become cloudless. Ultimately, the clear, cloudless air and the scorching rays of the vertical sun of the sandy desert conspire to maintain that dryness of the air which makes rainfall almost an impossibility (Miller, 1961). The Aravalli range lying in the north-east to south-west direction shows a marked influence on the rainfall during this period. The average elevation of these hills in the south-west is about 920 m which declines gradually till in the north-east, the hills peter out into scattered rocks around Delhi region. The result is that the onset of the monsoon showers brings relief after the prolonged heat of the summer season only to the south-west. But in the north-western part of the desert area, temperature, as high as 46°C has been recorded.

Cold Weather Season

The cold weather season has been further divided into two sub-divisions i.e. (a) the season of the retreating monsoon (October to December) and (b) the cold season (January and February).

(a) The season of the Retreating Monsoon (October to December): After the retreat of the

inter tropical front and southwest monsoon from the Indian climatic scene, there is slight increase in temperature for sometime and then there is gradual decrease to the winter minimum. In those areas where there has been rain, the high temperature combined with greater persisting humidity results in sultry weather. The temperatures in October in the whole of Rajasthan are fairly uniform, the maximum being between 33.3°C to 36.1 °C and minimum 17.7°C to 21. PC at various stations. November is slightly cooler, as maximum and minimum ranges of temperature at various stations indicate. Mt. Abu is considered as a separate case. The temperatures at Mt. Abu are much lower than in the surrounding areas since it is located at a much higher altitude. During these months due to the withdrawal of the monsoon, the winds die down considerably and are very light and highly variable.

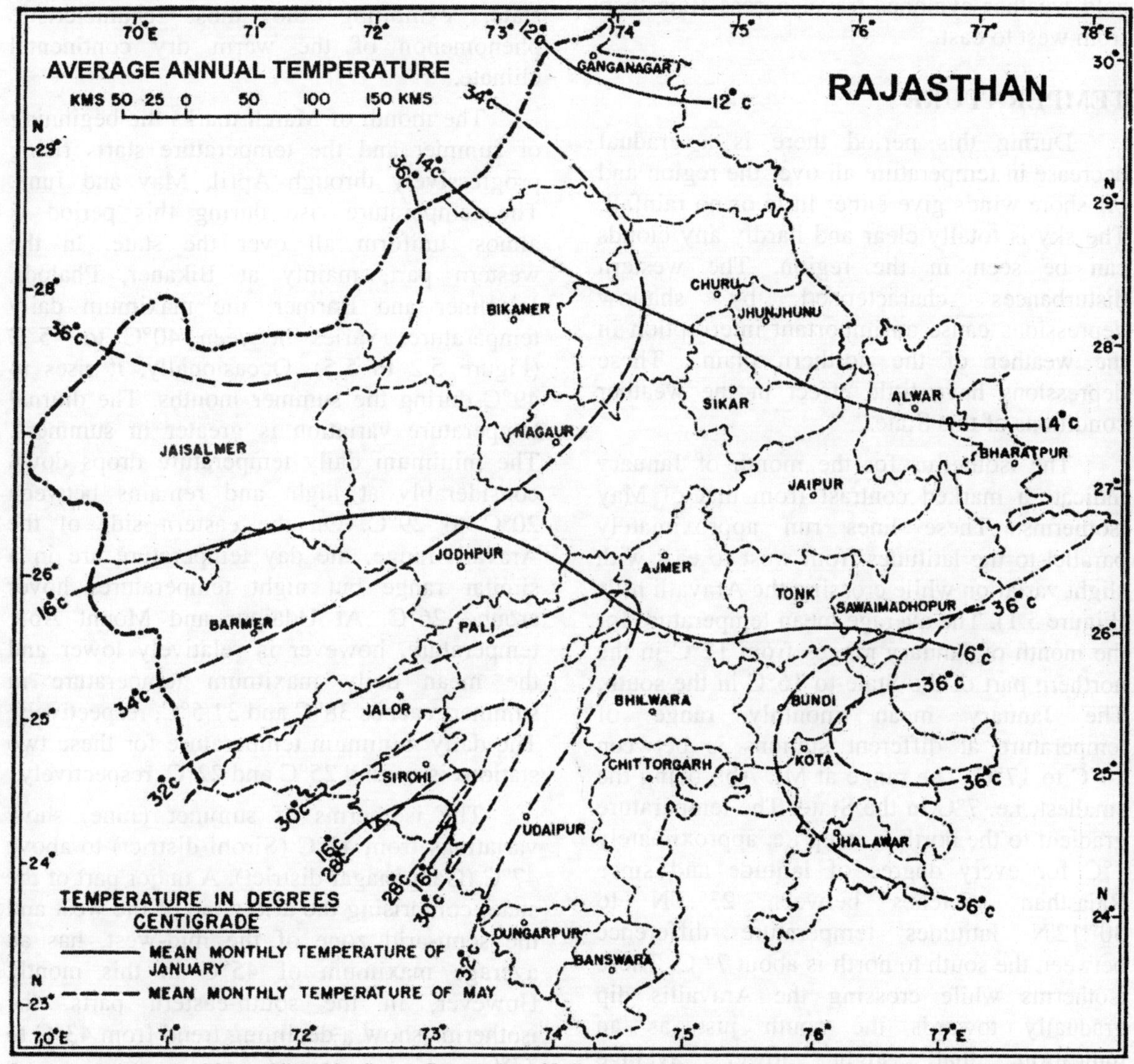

Source : Administrative Map, published by Census Commission of India and data obtained from India Meteorological Department, Poona.

Note : The external boundary and the coastlines of India appearing on the map are neither verified nor authenticated by the Survey of India.

Figure 5.1 Average Annual Temperature

(b) The Cold Weather Season (January and February): During these months, the sun migrates to the southern hemisphere and the sub-tropical anti- cyclonic cell extends from the west with its axis near the base of the Peninsula. This controls the pressure gradient, temperature and wind conditions over the Indian region. In the northern part of the State, light northwesterlies and in the south and south-east, northeasterlies are caused due to the anti-cyclonic cell of India. The result is that cold weather cyclones travel across Rajasthan from west to east.

TEMPERATURE

During this period there is a gradual decrease in temperature all over the region and off shore winds give either little or no rainfall. The sky is totally clear and hardly any clouds can be seen in the region. The western disturbances characterised by shallow depressions cause an important interruption in the weather of the northern plains. These depressions have little effect in the weather conditions of this State.

The isotherms for the month of January indicate a marked contrast from that of May isotherms. These lines run approximately parallel to the latitudes from west to east with slight variation while crossing the Aravalli hills (Figure 5.1). The average mean temperature for the month of January ranges from 12°C in the northern part of the State to 16°C in the south. The January mean monthly range of temperature at different stations is between 14°C to 17°C. The range at Mt. Abu being the smallest, i.e. 7°C in the State. The temperature gradient to the north is steep i.e. approximately 1°C for every degree of latitude and since Rajasthan stretches between 23°.3'N to 30°.12'N latitudes temperature difference between the south to north is about 7° C. These isotherms while crossing the Aravallis dip gradually towards the south just as an indication that slight lower average temperature is found over the Aravallis at a southern latitude. In the southern part of Rajasthan mainly in the districts of Kota, Jhalawar, Jodhpur, Barmer and part of Udaipur, the mean monthly January temperature is about 16°C. Mt. Abu, located on a hill, has the lowest average i.e. 14.6'C in January. In this cold weather month, winds are generally light and variable but northwesterly and northerly winds are more frequent than those from other directions.

Marked variations in diurnal and seasonal range of temperatures occur at all places in the state, exhibiting the most characteristic phenomenon of the warm dry continental climate.

The month of March marks the beginning of summer and the temperature starts rising progressively through April, May and June. The temperature rise during this period is almost uniform all over the state. In the western part, mainly at Bikaner, Phalodi, Jaisalmer and Barmer, the maximum daily temperatures varies between 40°C to 45°C (Figure 5.2 to 5.5). Occasionally, it rises to 49°C during the summer months. The diurnal temperature variation is greater in summers. The minimum daily temperature drops down considerably at night and remains between 20°C to 29°C. On the eastern side of the Aravalli range, the day temperature are in a similar range but night temperatures hover around 26°C. At Udaipur and Mount Abu, temperature, however is relatively lower and the mean daily maximum temperature in summer reaches 38°C and 31.5°C, respectively. The daily minimum temperature for these two stations is around 25°C and 22°C, respectively.

The isotherms of summer (June) show variations from 42°C (Sirohi district) to above 47°C (Ganganagar district). A major part of the state, comprising the arid zone of the west and the semi-arid zone of the mid-west, has an average maximum of 45°C in this month. However, in the south-eastern parts, the isotherms show a declining trend from 43°C to 37°C.

January is the coldest month of the year. The winter season remains from December through February with marked decline in minimum temperature in December and January. The minimum temperature may fall to minus 2°C in the night at some places like Sikar, Churu, Pilini and Bikaner. A sharp decrease in night temperatures is experien-ced throughout the arid and semi-arid zone of western Rajasthan on account of quick release of thermal radiations from the sandy soil soon after dusk. Occasional secondary Western disturbances which mostly cross months, causing light rainfall, may also lower the night temperatures and increase wind speed creating wind chill effect.

The isotherms of winter (January), ranging from less than 60°C in the north to above 10°C in the south-east, show few variations in temperatures. The greater part of Rajasthan, except the south-eastern part of Kota, Bundi and Baran and the western district of Barmer exhibits the mean temperatures of more than 10°C. The secondary depressions (western distur-bances) accompanied by the ice-cold winds, emanating from the higher latitudes, may bring entire Rajasthan under the spell of the cold wave for the periods of 2 to 5 days during the months.

A major feature of distribution in space and time of normal to maximum, minimum and mean temperatures is the small extent of variations over large areas in the monsoon season. One of the peculiar features of Rajasthan is the great extremes of temperature. Variation of as much as 22°C is noted between the mean minimum temperatures due to the sandy nature of the soil. During winter the north-western Rajasthan is colder than the south-eastern region, and at many places the temperature even falls below freezing point and frost occurs. During summer, the westerly or southwesterly winds raise up the temperature making the western region intensely hot and scorching. Due to the dryness of the atmosphere, the nature of the soil, and want of vegetation, the change in the temperature from day to night is sudden, very large and trying. However, in the southern and eastern parts and on the hill tops of Mt.Abu and Kumbhalgarh, the speed of the winds is very slow, temperature does not go very high and the nights are cool and pleasant.

The cool season lasts from December to February. January is the coldest month of the year. The Normals of daily maximum, minimum and mean temperatures for a dozen observation stations in the State. Monthly normals of daily minimum temperature vary from 5 to 7.5°C in January to 25 to 30°C in May and June and 25 to 27.5°C in July and August. The annual minimum temperatures vary between 17 to 20'C. Monthly normals of daily maximum, temperature are 20°C to 25°C in January to 40-42.5°C in May and June. The mean annual maximum temperature is 32 to 34°C. Mean daily temperature is 15±2°C in January and 33±2°C in May and June. The highest temperature recorded on any day during the past sixty years or more has exceeded 47.5°C over most of the State and reached as high as 50°C in some areas. The scorching heat and hot 100 winds make the weather during day very trying; while owing to dry atmosphere, clear sky and sandy soil, rapid radiation of heat lowers down the temperature considerably by might.

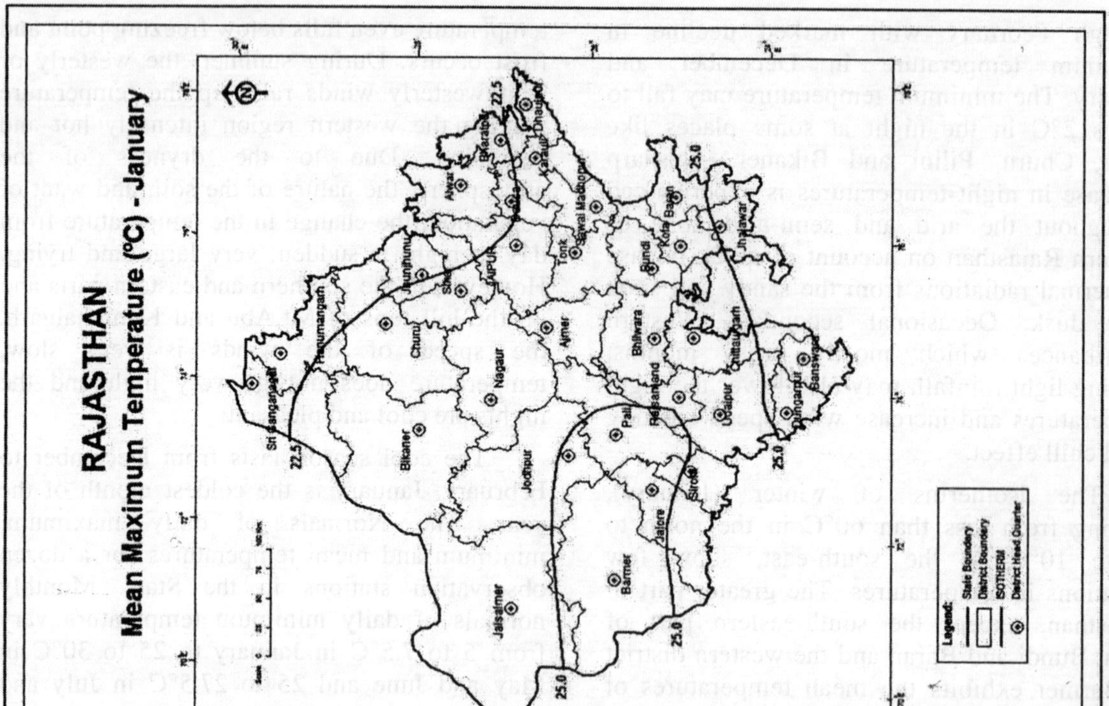

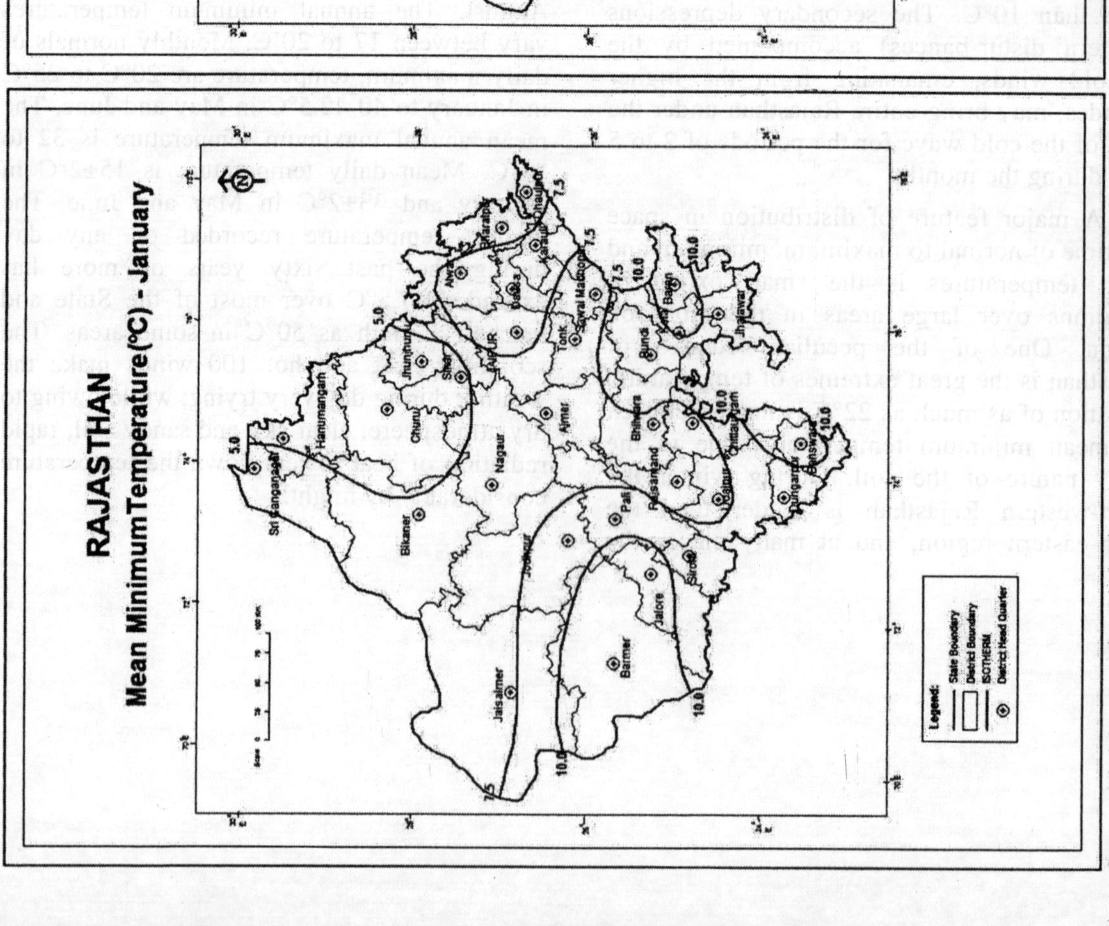

RAJASTHAN

Mean Minimum Temperature (°C) - January

Legend:
- State Boundary
- District Boundary
- ISOTHERM
- District Head Quarter

RAJASTHAN

Mean Maximum Temperature (°C) - January

Legend:
- State Boundary
- District Boundary
- ISOTHERM
- District Head Quarter

Source : Administrative Map, Published by Census Commission of India and data obtained from India Meteorological Department, Poona.

Figure 5.2

Figure 5.3

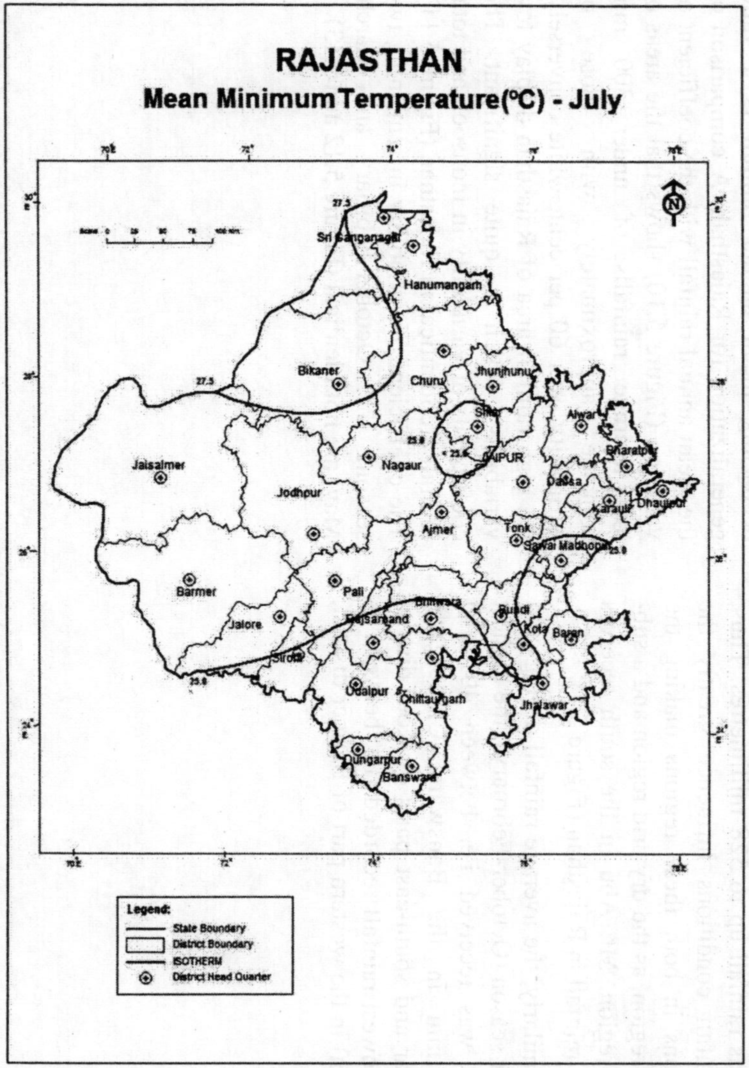

RAJASTHAN

Mean Minimum Temperature(°C) - July

Figure 5.4

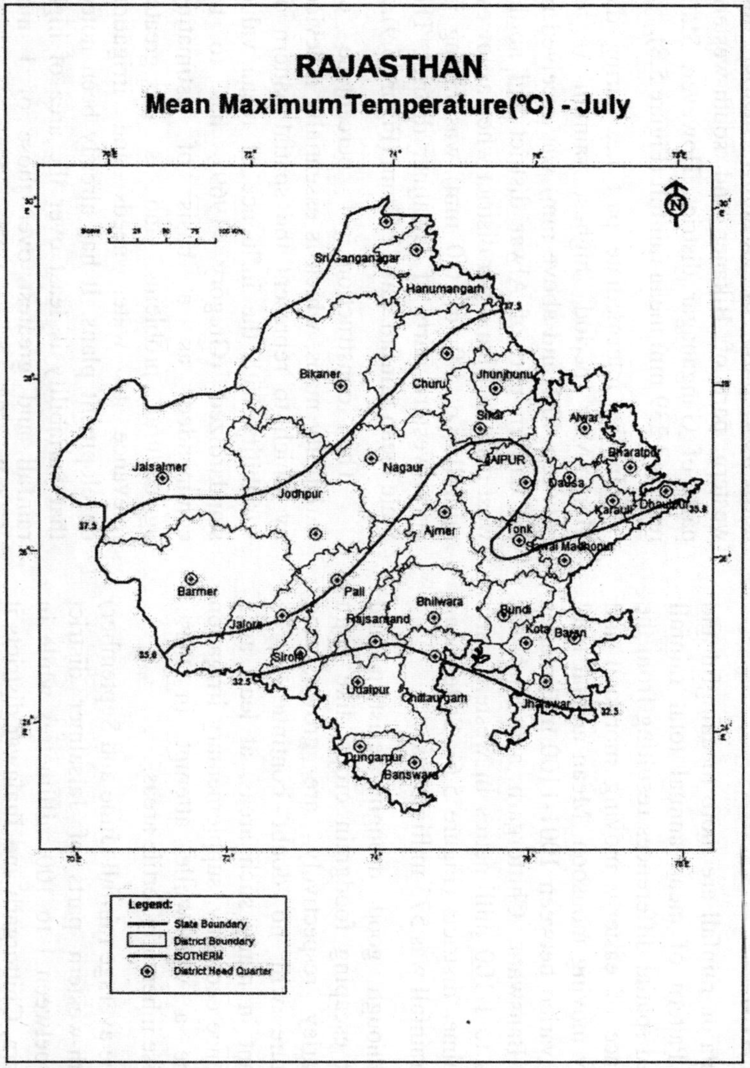

RAJASTHAN

Mean Maximum Temperature(°C) - July

Figure 5.5

Source : Administrative Map, Published by Census Commission of India and data obtained from India Meteorological Department, Poona.

RAINFALL

Data on rainfall are more meaningful and sound. Pattern of mean annual total rainfall indicated spatial differences resulting from the prevalence of easterly moving monsoon and westerly moving monsoon. Mean annual total rainfall varied between 1001-1100 millimetres in the Banswara, Chittorgarh and Jhalawar districts to 1-100 millimetres in western parts of Jaisalmer districts (Figure 5.6). State mean annual rainfall was 573 millimetres.

Although good drought resistant and drought escaping foodgrain crops, like bajra and barley respectively, are grown, and agriculture will no doubt continue to be important in many such areas, at least as a subsistence economy, supplementary irrigation may be a worthwhile attempt in these, otherwise inherently fertile areas.

The average rainfall (June and September) in north-western parts of Jaisalmer district ranges between 1 to 100 millimetres while in Banswara, Chittorgarh and Jhalawar districts it varies between 901 to 1000 millimetres. The State has rainfall up to 528 millimetres. Thus, the rainfall conditions influence greatly the conditions in both these regions making the former region, as the dry arid region and a sub-humid region. Mt. Abu in the south receives highest rainfall in Rajasthan (Figure 5.7).

Similarly, the average rainfall during cold weather season (October-February), the highest rainfall was received i.e. between 401 and above mm in the Banswara, Chittorgarh, Jhalawar and south-east parts of Kota district while lowest rainfall recorded (i.e. between 1-100 mm) in the western part of Barmer district, north-west parts of Jaisalmer district and western part of Bikaner and south-western parts of Ganganagar district. However, State received 239 mm mean rainfall (Figure 5.8).

In the intermediate part i.e. during the March-May period, highest rainfall (i. e. between 40,1 and above mm) was received in the eastern part of Alwar district and north-west parts of Bharat pur district whereas lowest rainfall (i.e. 1.0 to 10.0 mm) was falling in north-western parts of Jaisalmer district. The State mean rainfall was 22.0 mm (Figure 5.9).

The construction of percentage of variability maps, which is essentially a device by which to represent the spatial pattern of variability with the influence of mean value standardized (Gregory, 1969), has to be commended as a basis of estimating agricultural problems and is of greater relevance to water needs and irrigation development plans. It has already been noted that variability is least over the area of high rainfall and greatest over those of 1 mm rainfall. The variability increase with decreasing mean annual rainfall is a valid generalization for Rajasthah. A comparison of the mean annual rainfall with the coefficient of variation (Figure 5.10) shows that the areas of low average rainfalls of under 300 mm coincide approximately with those of variability over 60 per cent, while conversely, the high rainfall area of Rajasthan display low variability, which is quite significant. The coefficients of variability in monsoon and total rainfall are identical in Rajasthan (Figure 5.11). The coefficient of variability in remaining two seasons (i.e. October-February and March-May) are not identical (Figure 5.12 and 5.13).

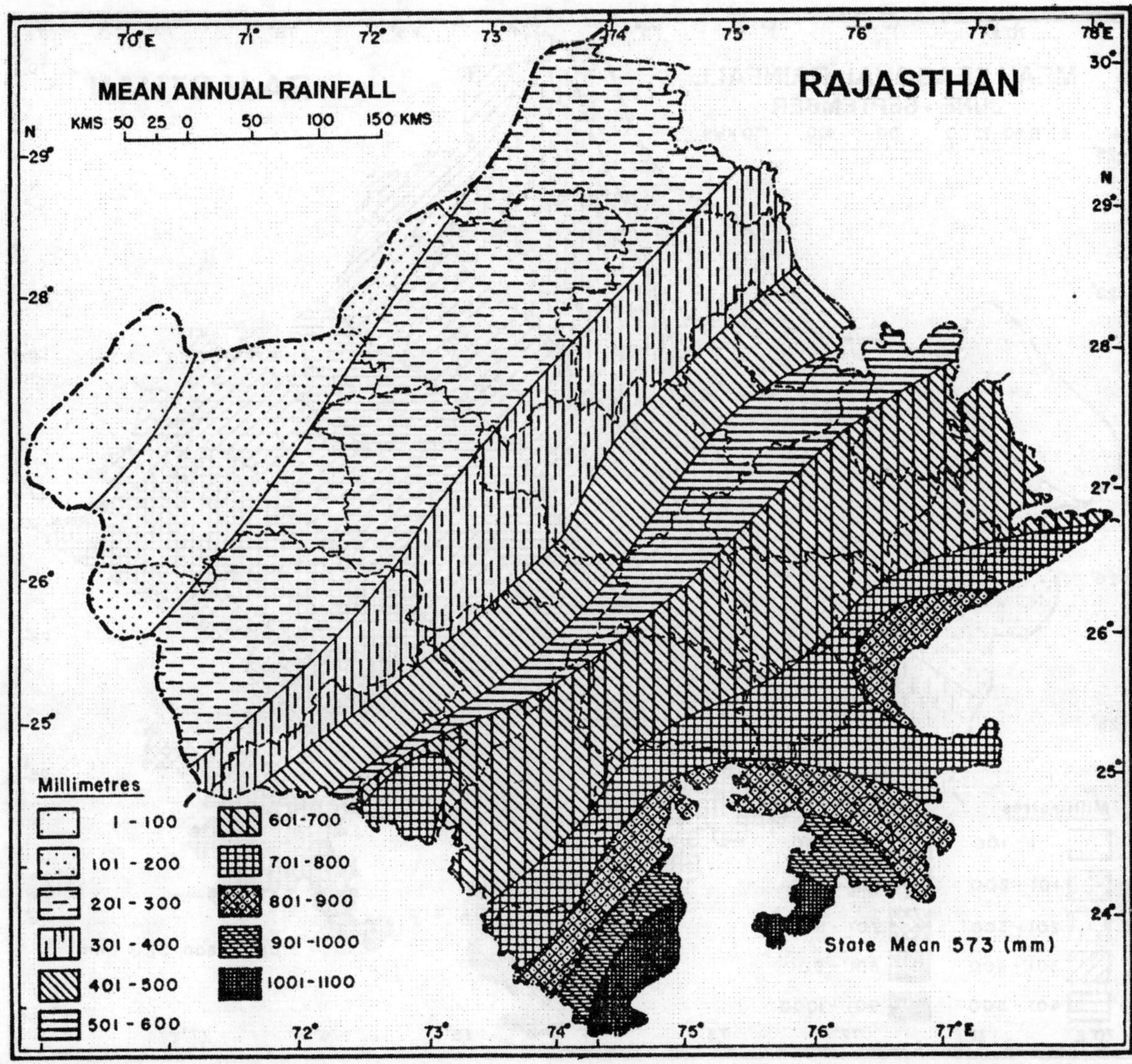

Source : Administrative Map, published by Census Commission of India and data obtained from India Meteorological Department, Poona.

Note : The external boundary and the coastlines of India appearing on the map are neither verified nor authenticated by the Survey of India.

Figure 5.6 Mean Annual Rainfall

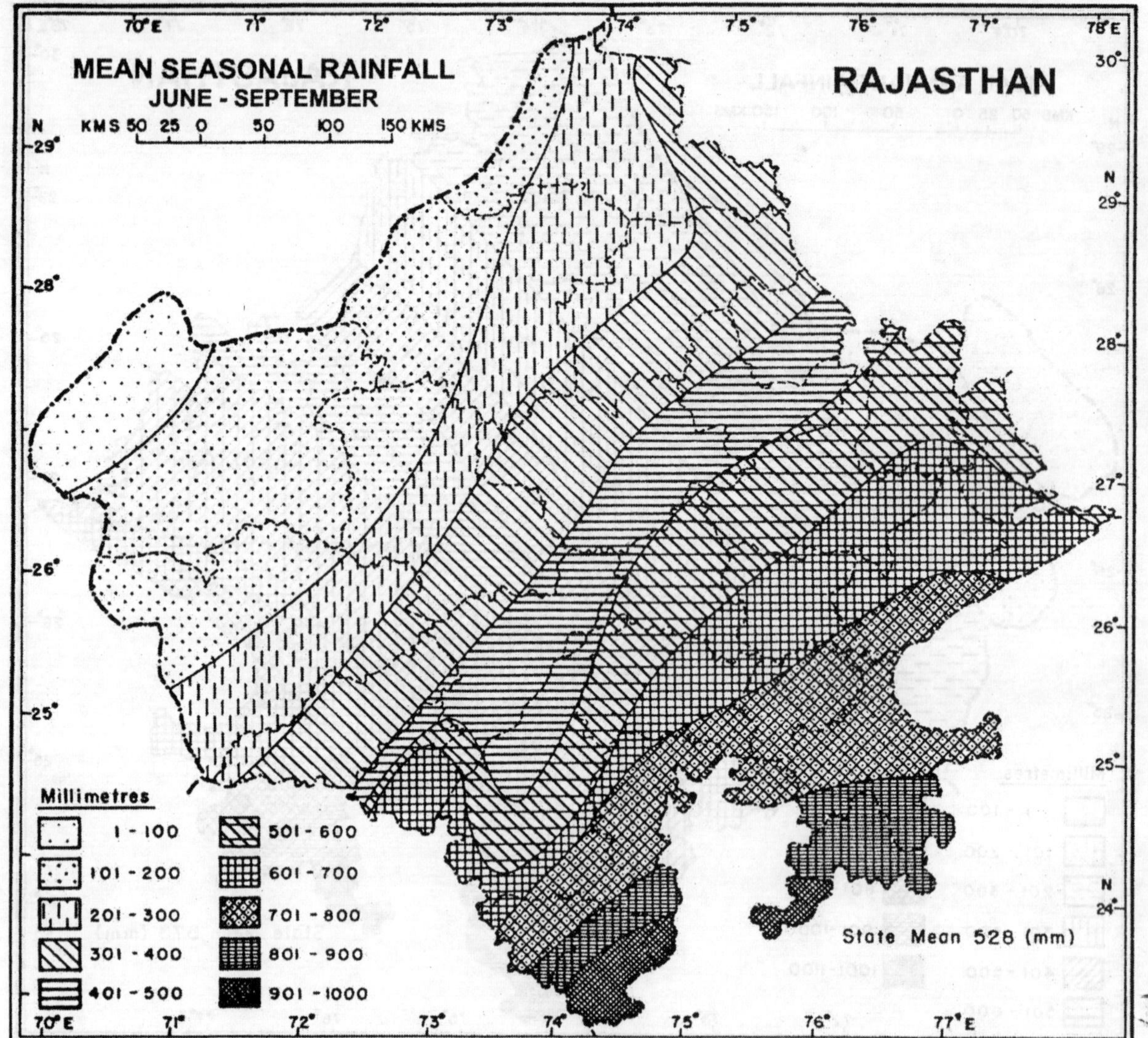

Source : Administrative Map, published by Census Commission of India and data obtained from India Meteorological Department, Poona.
Note : The external boundary and the coastlines of India appearing on the map are neither verified nor authenticated by the Survey of India.

Figure 5.7 Mean Seasonal Rainfall, June September

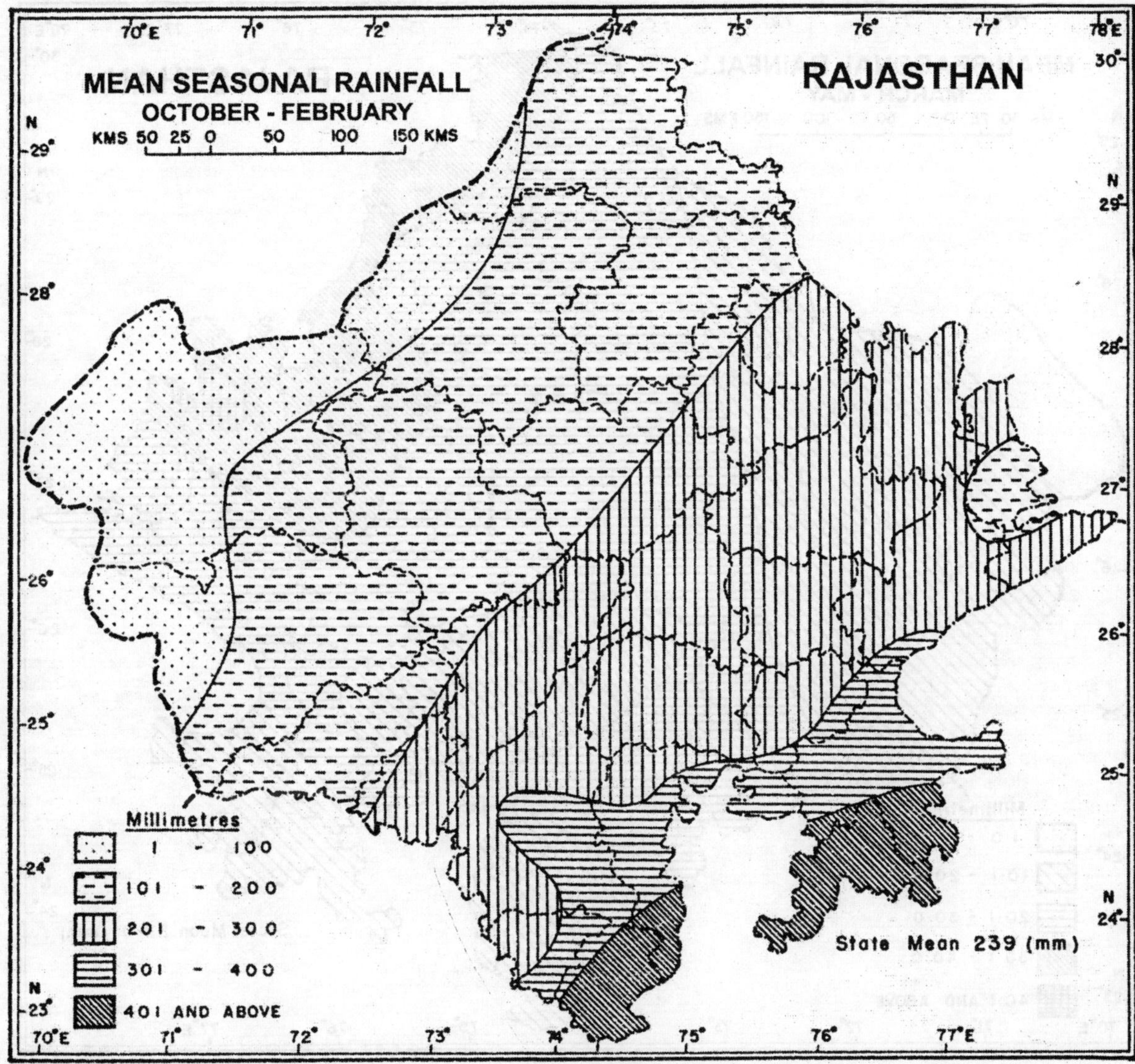

Source : Administrative Map, published by Census Commission of India and data obtained from India Meteorological Department, Poona.
Note : The external boundary and the coastlines of India appearing on the map are neither verified nor authenticated by the Survey of India.

Figure 5.8 Mean Seasonal Rianfall, October-February

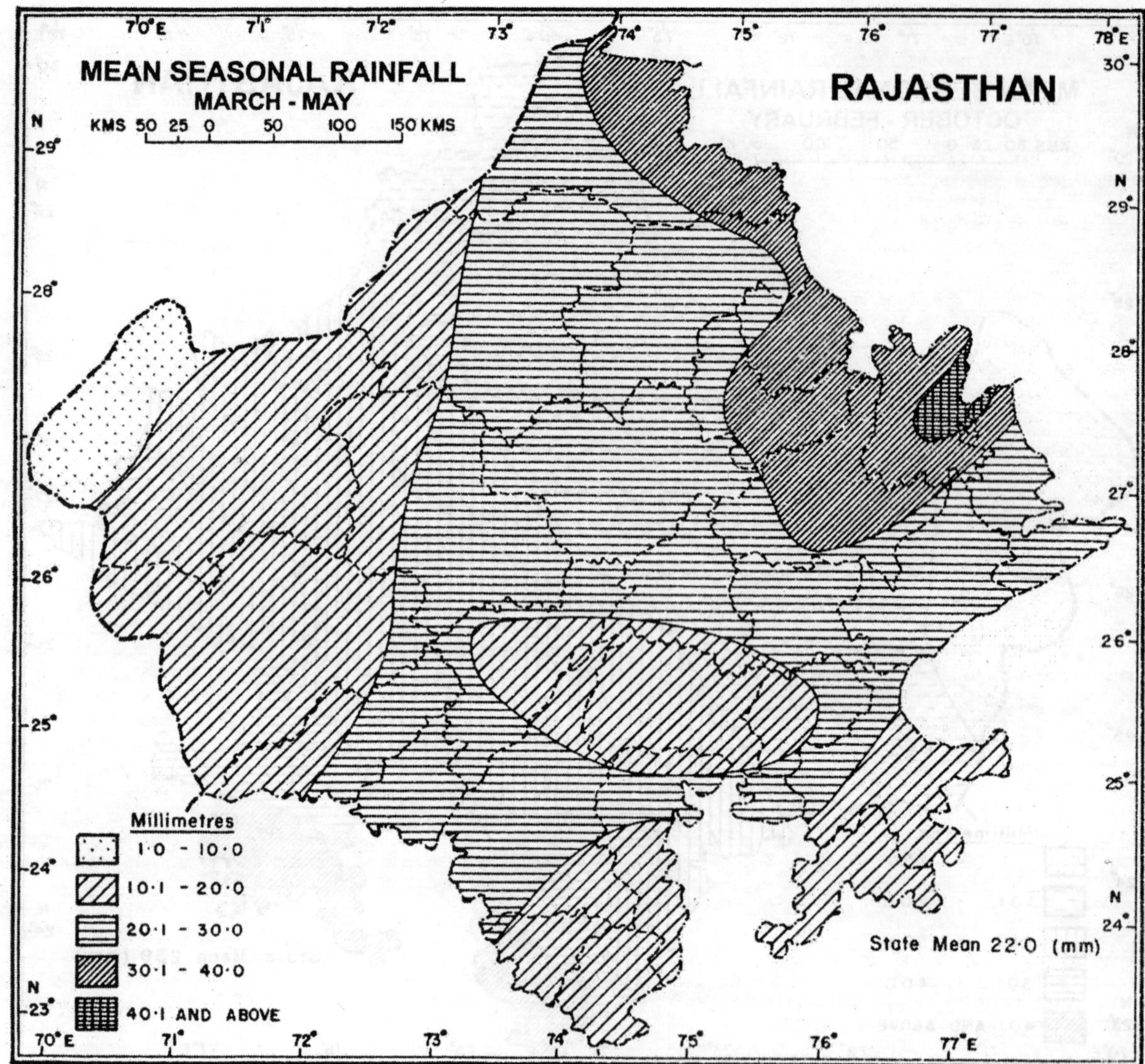

Source : Administrative Map, published by Census Commission of India and data obtained from India Meteorological Department, Poona.
Note : The external boundary and the coastlines of India appearing on the map are neither verified nor authenticated by the Survey of India.

Figure 5.9 Mean Seasonal Rainfall, March-May

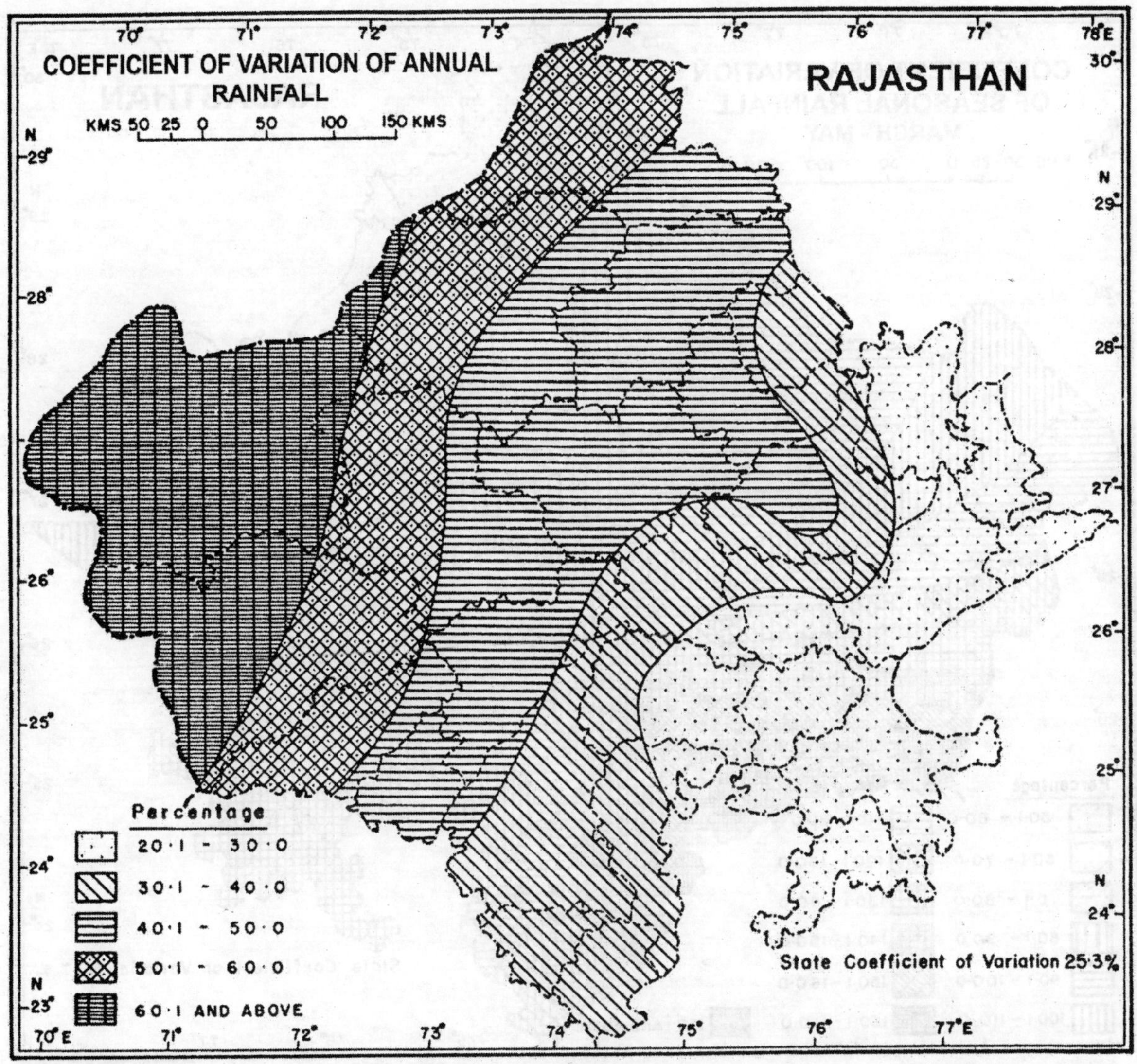

Source : Administrative Map, published by Census Commission of India and data obtained from India Meteorological Department, Poona.

Note : The external boundary and the coastlines of India appearing on the map are neither verified nor authenticated by the Survey of India.

Figure 5.10 Coefficient of Variation of Annual Rainfall

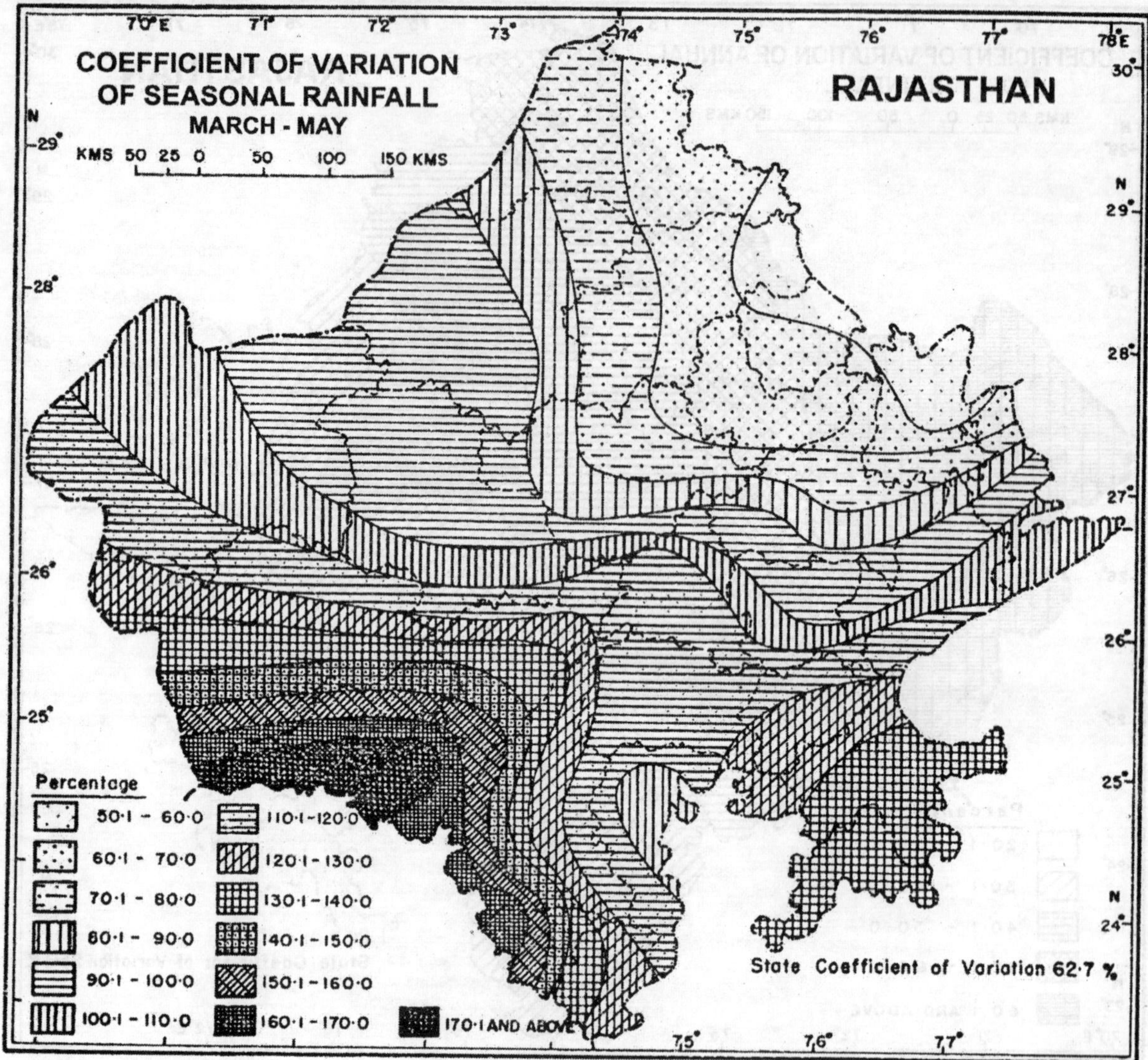

Source : Administrative Map, published by Census Commission of India and data obtained from India Meteorological Department, Poona.
Note : The external boundary and the coastlines of India appearing on the map are neither verified nor authenticated by the Survey of India.

Figure 5.11 Coefficient of Variation of Seasonal Rainfall; March-May

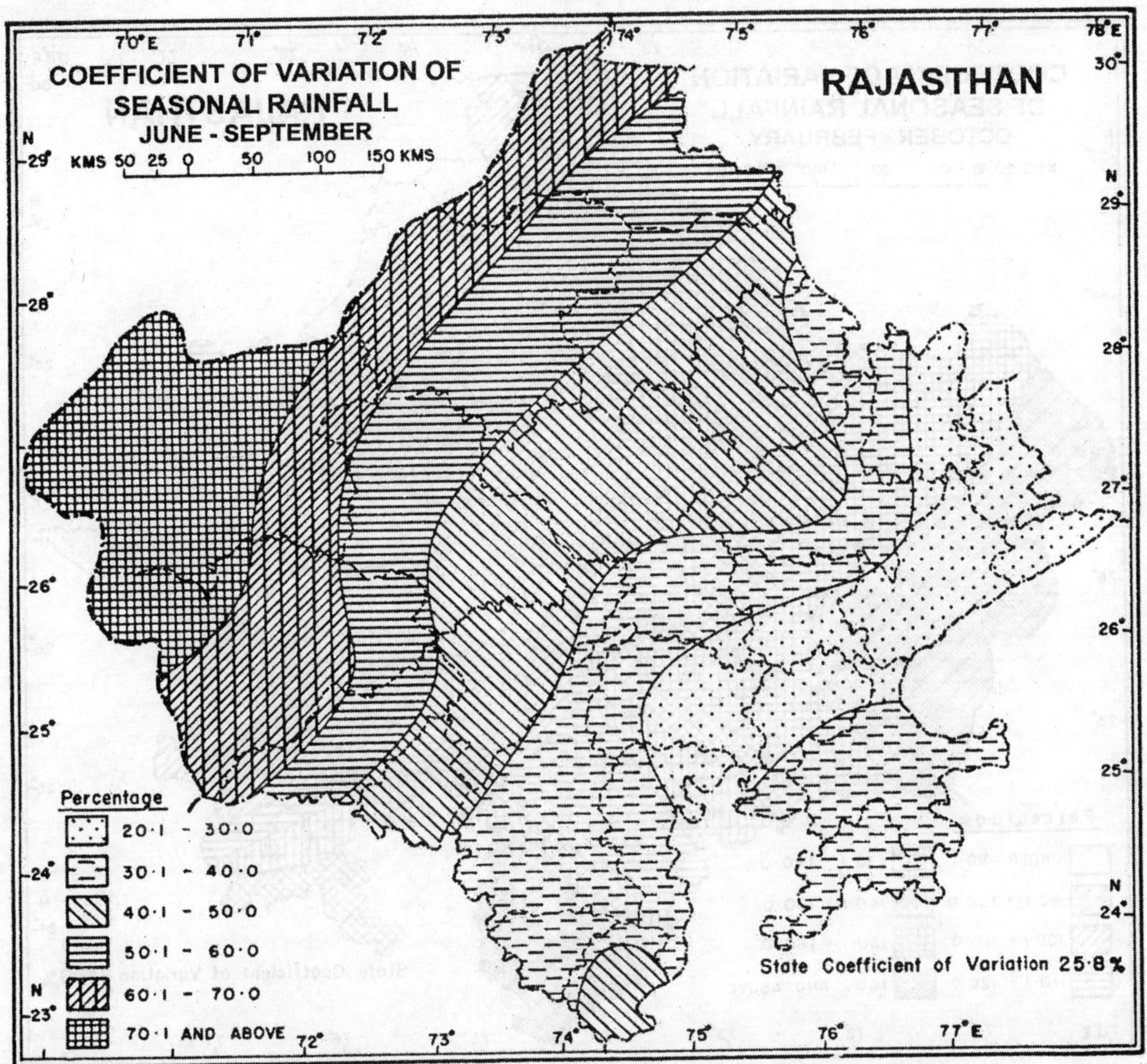

Source : Administrative Map, published by Census Commission of India and data obtained from India Meteorological Department, Poona.

Note : The external boundary and the coastlines of India appearing on the map are neither verified nor authenticated by the Survey of India.

Figure 5.12 Coefficient of Variation of Seasonal Rainfall; June-September

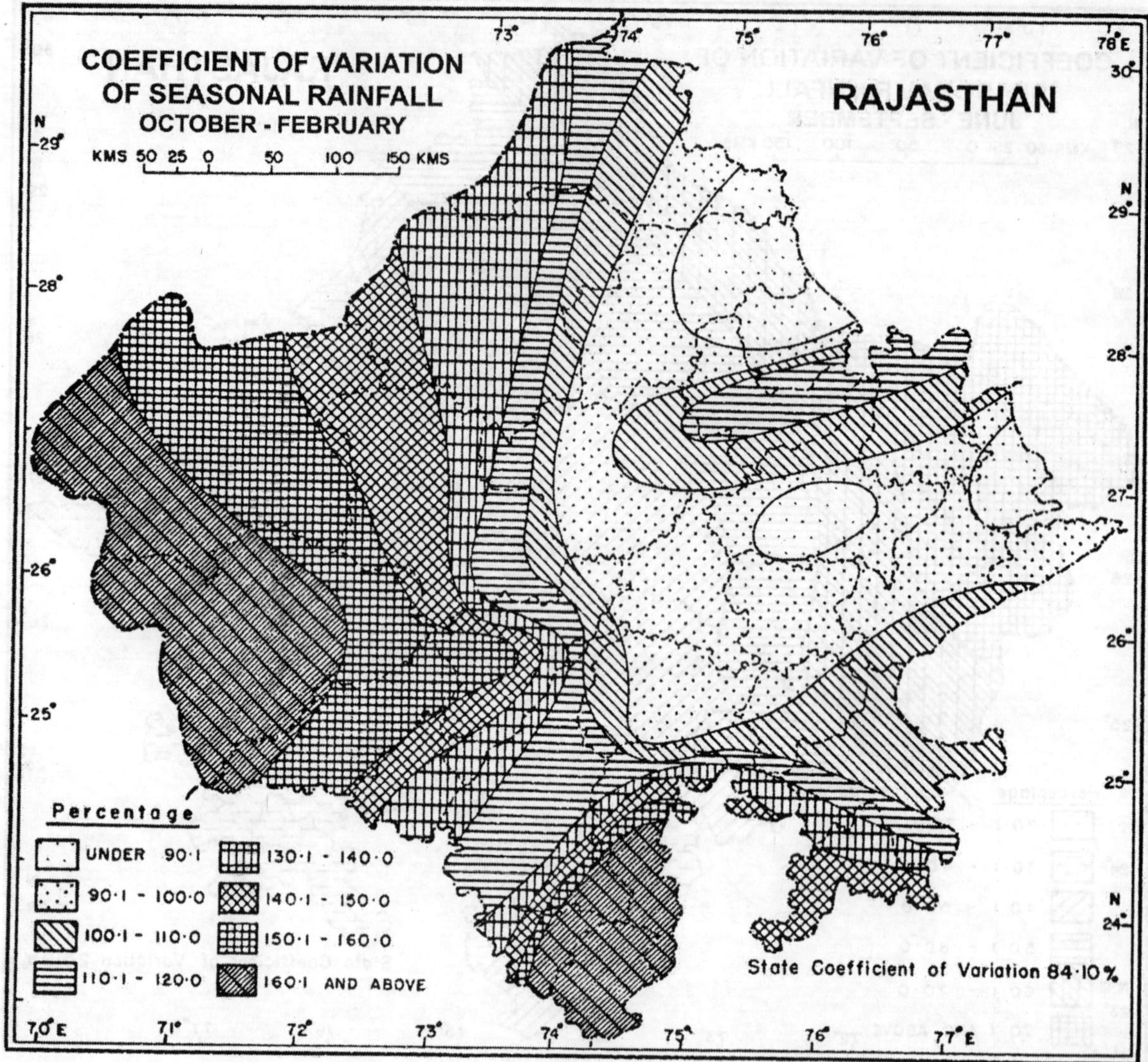

Source : Administrative Map, published by Census Commission of India and data obtained from India Meteorological Department, Poona.

Note : The external boundary and the coastlines of India appearing on the map are neither verified nor authenticated by the Survey of India.

Figure 5.13 Coefficient of Variation of Seasonal Rainfall; October-February

The rainfall is very scanty and erratic. The individual rain gauge station data of year 2010 shows very wide variation from less than 100mm to more than 1000mm of total annual rainfall in the state, whereas district level averages reveal that the minimum to be about 360mm in Ganganagar to a maximum of 948mm in Sirohi district. To the east of Aravalli Range, the annual total rainfall is, in general, higher and averages above 600mm whereas the districts in the west of Aravalli receive very less rainfall, i.e., in the range of less than 300mm to 400 mm annually minimum being in Ganganagar and Jaisalmer districts (Table 5.1).

Table 5.1 Monsoon, Non-Monsoon and Annual Rainfall in Rajasthan

District	Total Monsoon Rainfall (mm)	Total Non-Monsoon Rainfall (mm)	Total Annual Rainfall (mm)*
Ajmer	532	124	656
Alwar	732	51	783
Banswara	651	35	687
Baran	562	51	613
Barmer	475	48	524
Bharatpur	729	70	799
Bhilwara	610	156	767
Bikaner	573	23	596
Bundi	583	105	689
Chittaurgarh	729	99	828
Churu	621	46	667
Dausa	639	101	739
Dholpur	670	94	763
Dungarpur	605	61	665
Ganganagar	337	23	360
Hanumangarh	527	35	562
Jaipur	735	95	830
Jaisalmer	357	16	373
Jalore	658	125	783
Jhalawar	587	54	641
Jhunjhunu	705	63	768
Jodhpur	395	38	434
Karauli	628	114	742
Kota	563	55	618
Nagaur	531	40	571
Pali	552	123	675
Pratapgarh	707	40	747
Rajsamand	653	187	840
S. Madhopur	668	142	810
Sikar	763	74	836
Sirohi	784	164	948
Tonk	630	130	760
Udaipur	780	102	882
Statewide Average	**614**	**816**	**695**

* District level total rainfall is computed from 2010-2011 Monsoon season data of rainguage stations averaged at district level. (Source of data: http://waterresources.rajasthan.gov.in)

Rajasthan receives the rain from both the Arabian Sea and the Bay of Bengal system of the South-West monsoon and about 80% to 90% of the rain in the State occurs from mid-June to September. The 500 mm isohytel line which runs along the western edge of the Aravalli range divides the State into two regions. The rainfall gradually increases towards East and South of this line, and rapidly decreases towards the West. On an average, the State receives 536 mm of rainfall, but there are wide fluctuations in its distribution. Jaisalmer and Bikaner in the west receive hardly 170mm and 390mm of rain respectively where as Banswara and Jhalawar in the South-East receive about 930 mm of rain. On account of temporal and spatial variations in the occurrence of rainfall, a flood-drought syndrome afflicts the State (Figure 5.14).

The annual rainfall in the state varies significantly. The general trend of Isohyets is from north-west to south-east. There is a very rapid and marked decrease in rainfall west of the Aravalli range, making western Rajasthan the most arid part. The average annual rainfall in this part ranges from less than 10 cm in north-west part of Jaisalmer (lowest in the state), to 20 to 30 cm in Ganganagar, Bikaner and Barmer regions, 30 to 40 cm in Nagaur, Jodhpur, Churu and Jalore regions and more than 40 cm in Sikar, Jhunjhunun and Pali regions and along the western fringes of the Aravalli range. On the eastern side of the Aravalli range, the rainfall ranges from 55 cm in Ajmer to 102 cm in Jhalawar as shown in Table 5.2 and Figure 5.15. In plains, Banswara (92.0 cm) and Jhalawar (102 cm) districts receive the maximum annual rain. Mount Abu (Sirohi district) in the south west, however receives the highest rainfall in the state (163.8 cm). The yearly total rainfall is highly variable at different places all over the state and it is most erratic in the western half with frequent spells of drought, punctuated (Figure 5.15) occasionally by heavy downpour in some years, associated with the passing low pressure systems over the region.

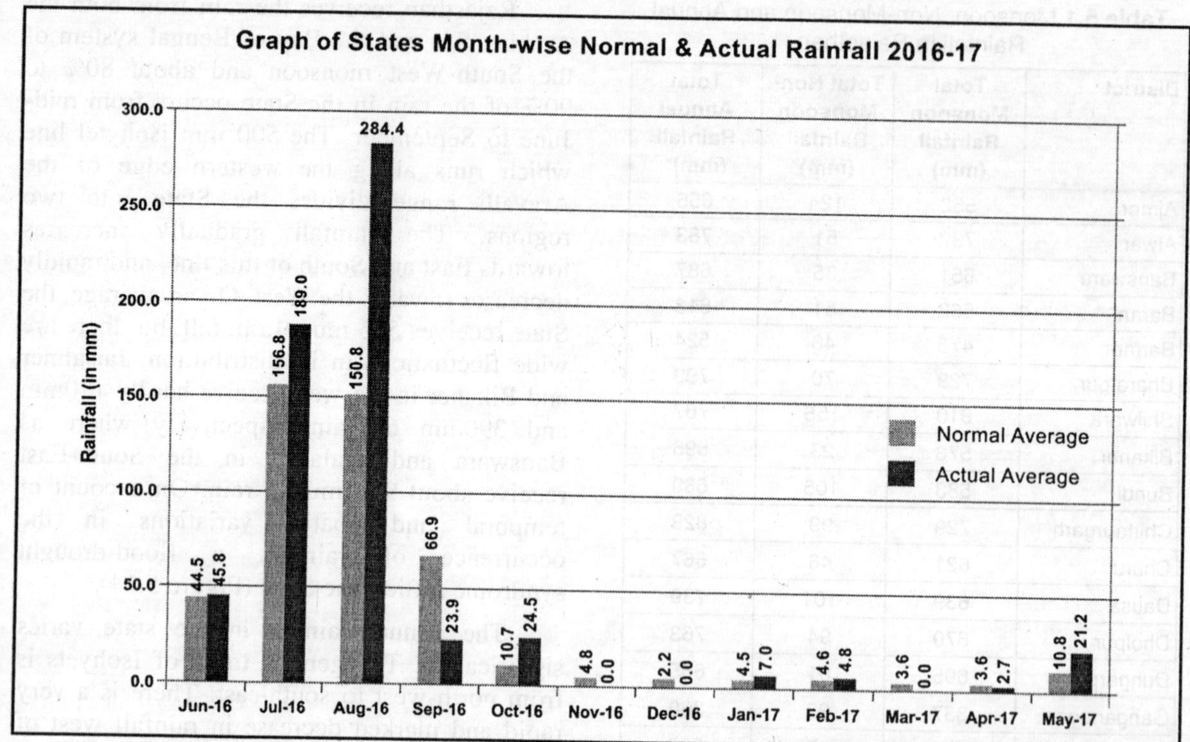

Source : Administrative Map, published by Census Commission of India and data obtained from India Meteorological Department, Poona.

Figure 5.14 Graph of States Month-wise Normal and Actual Rainfall 2016-17

The south-west monsoon which has its beginning in the last week of June in the eastern parts, may last till mid-September, Pre-monsoon showers begin towards the middle of June and post monsoon rains occasionally occur in October. In the winter season also, there is sometimes, a little rainfall associated with the passing western distribution over the region. At most places, the highest normal monthly rainfall is during July and August. The number of rainy days during this period varies widely in different places, ranging from 10 in Jaisalmer to 40 in Jhalawar and 48 in Mount Abu. Rainfall during the rest of the period ranges from 2.1 cm at Jaisalmer to 7.2 cm at Jaipur, distributed over 2.5 to 6 rainy days.

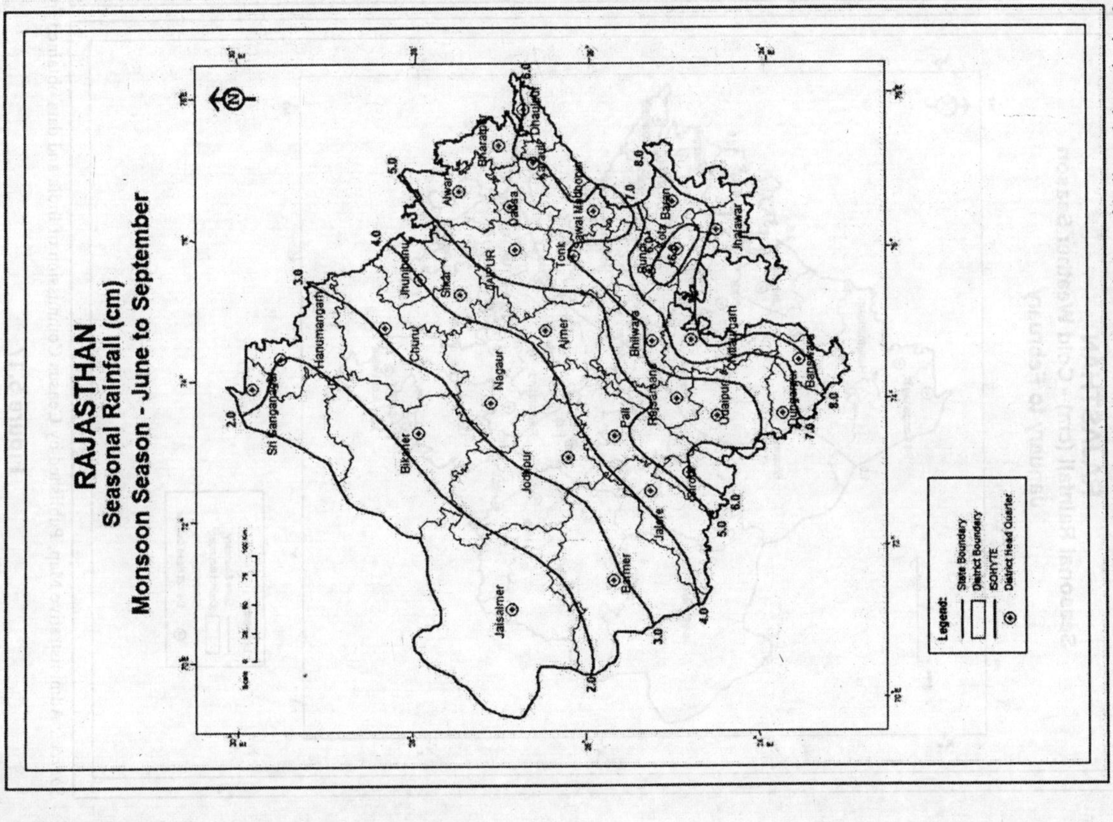

RAJASTHAN
Seasonal Rainfall (cm)
Monsoon Season - June to September

Legend:
State Boundary
District Boundary
ISOHYTE
District Head Quarter

Figure 5.15

RAJASTHAN
Annual Normal Rainfall (cm)

Legend:
State Boundary
District Boundary
ISOTHERM
District Head Quarter

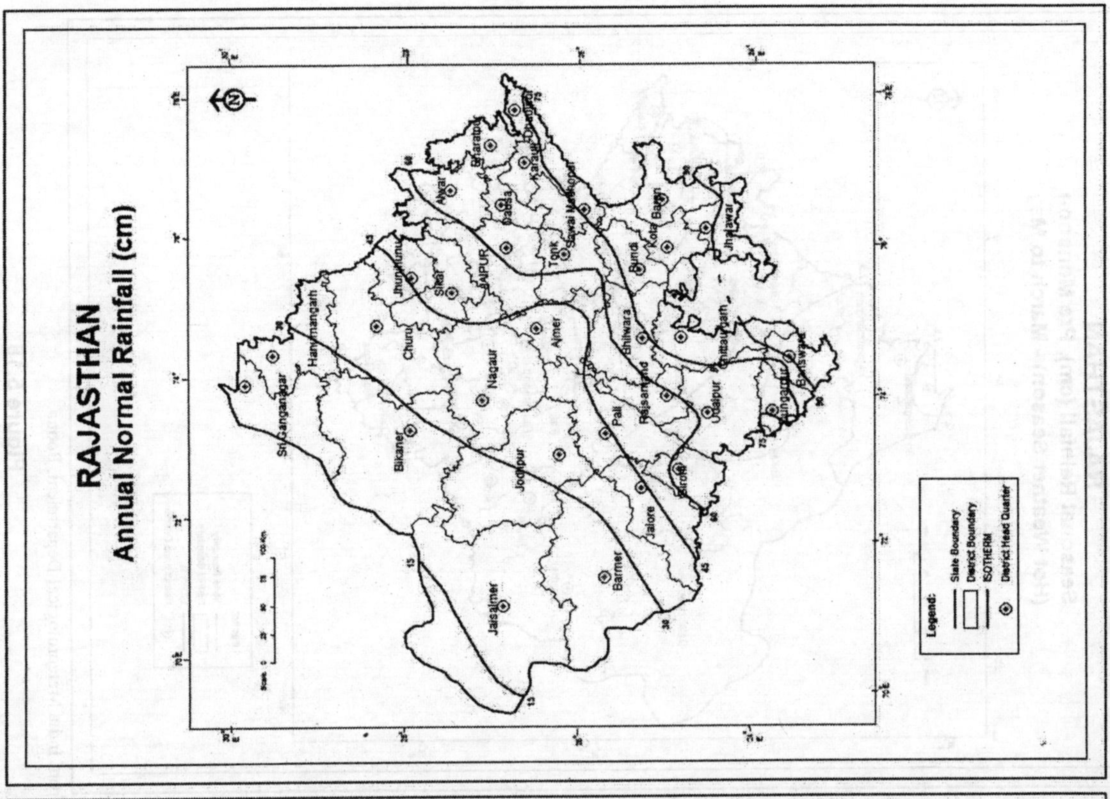

Figure 5.16

Source : Administrative Map, Published by Census Commission of India and data obtained from India Meteorological Department, Poona.

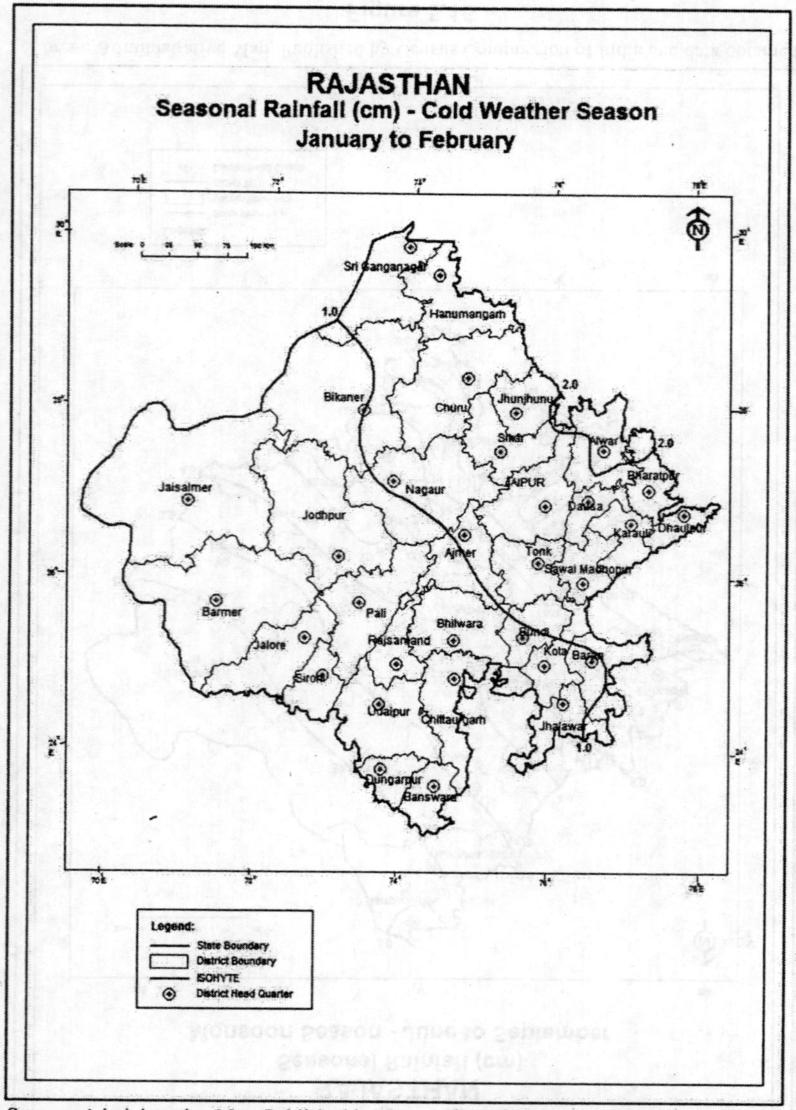

Figure 5.17

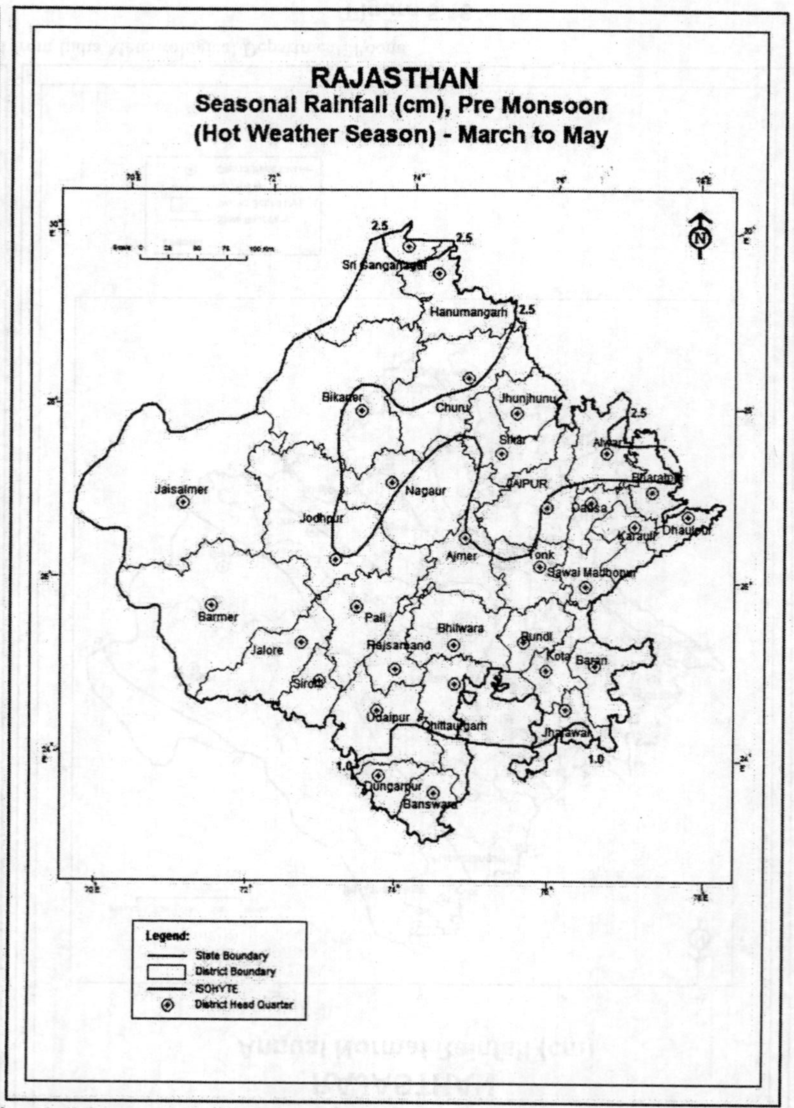

Figure 5.18

Source : Administrative Map, Published by Census Commission of India and data obtained from India Meteorological Department, Poona.

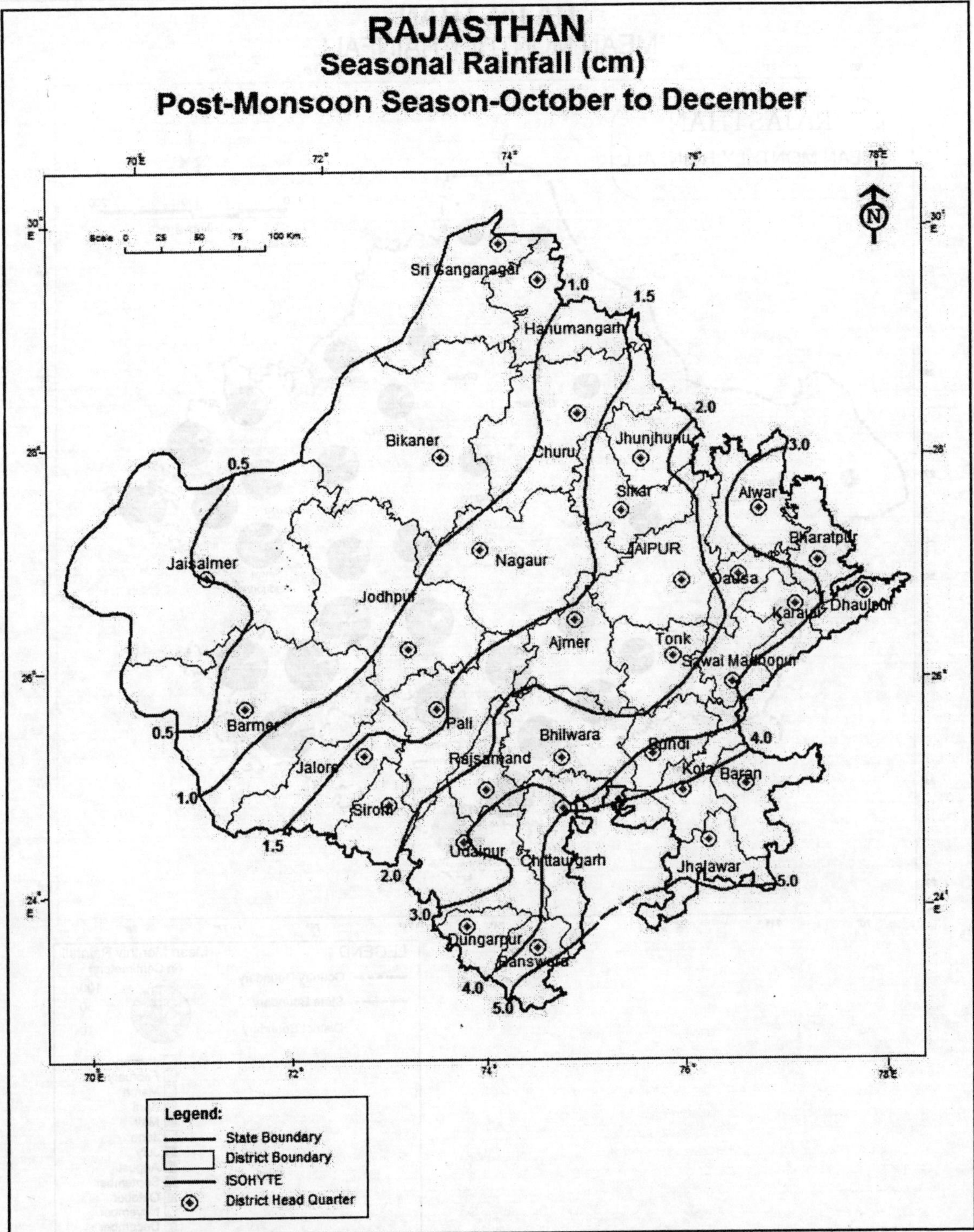

Source: Administrative Map, Published by Census Commission of India and data obtained from India Meteorological Department, Poona.

Figure 5.19 Seasonal Rainfall (cm) Post-Monsoon Season; October-December

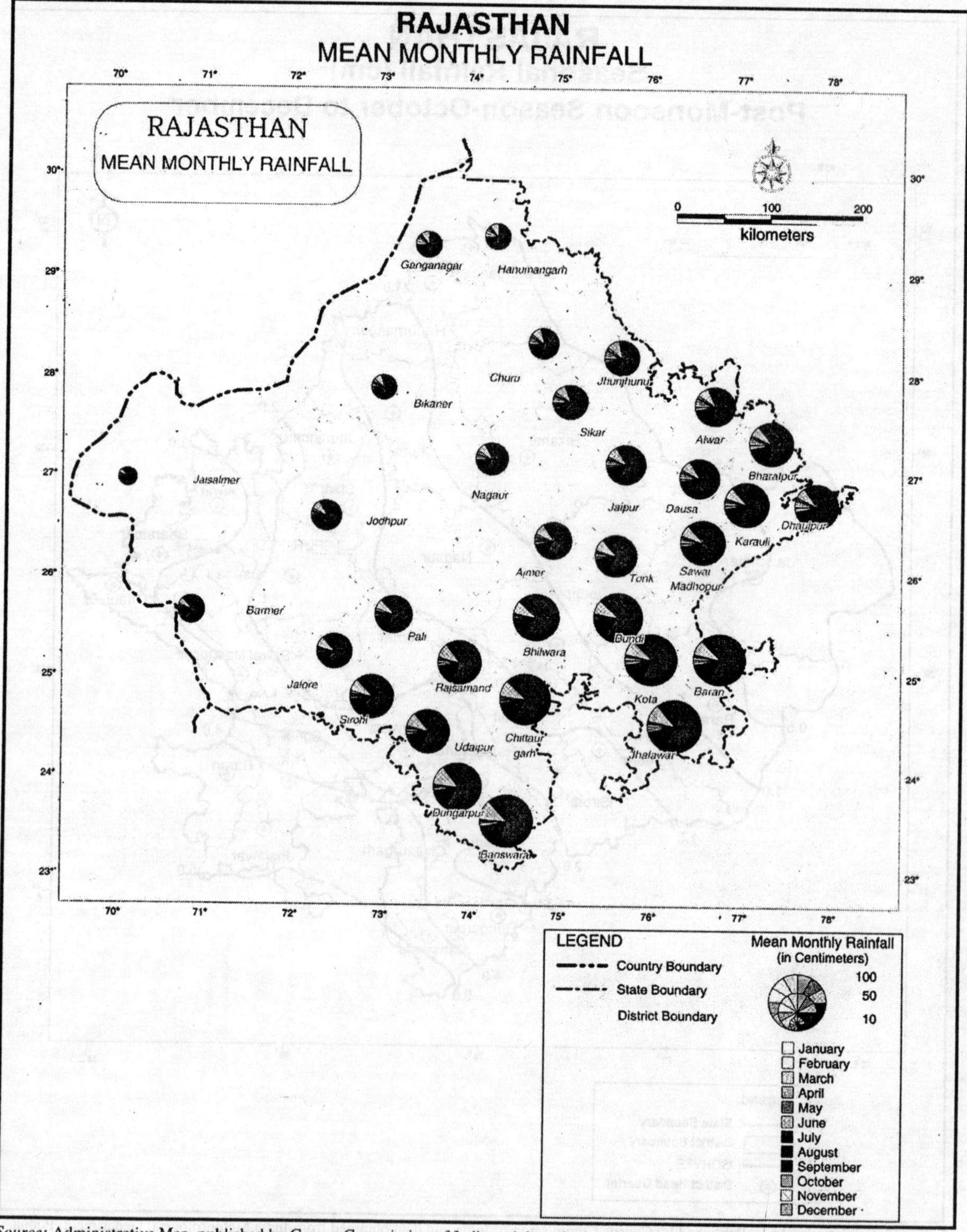

Source: Administrative Map, published by Census Commission of India and data obtained from India Meteorological Department, Poona.

Figure 5.20 Mean Monthly Rainfall in Rajasthan

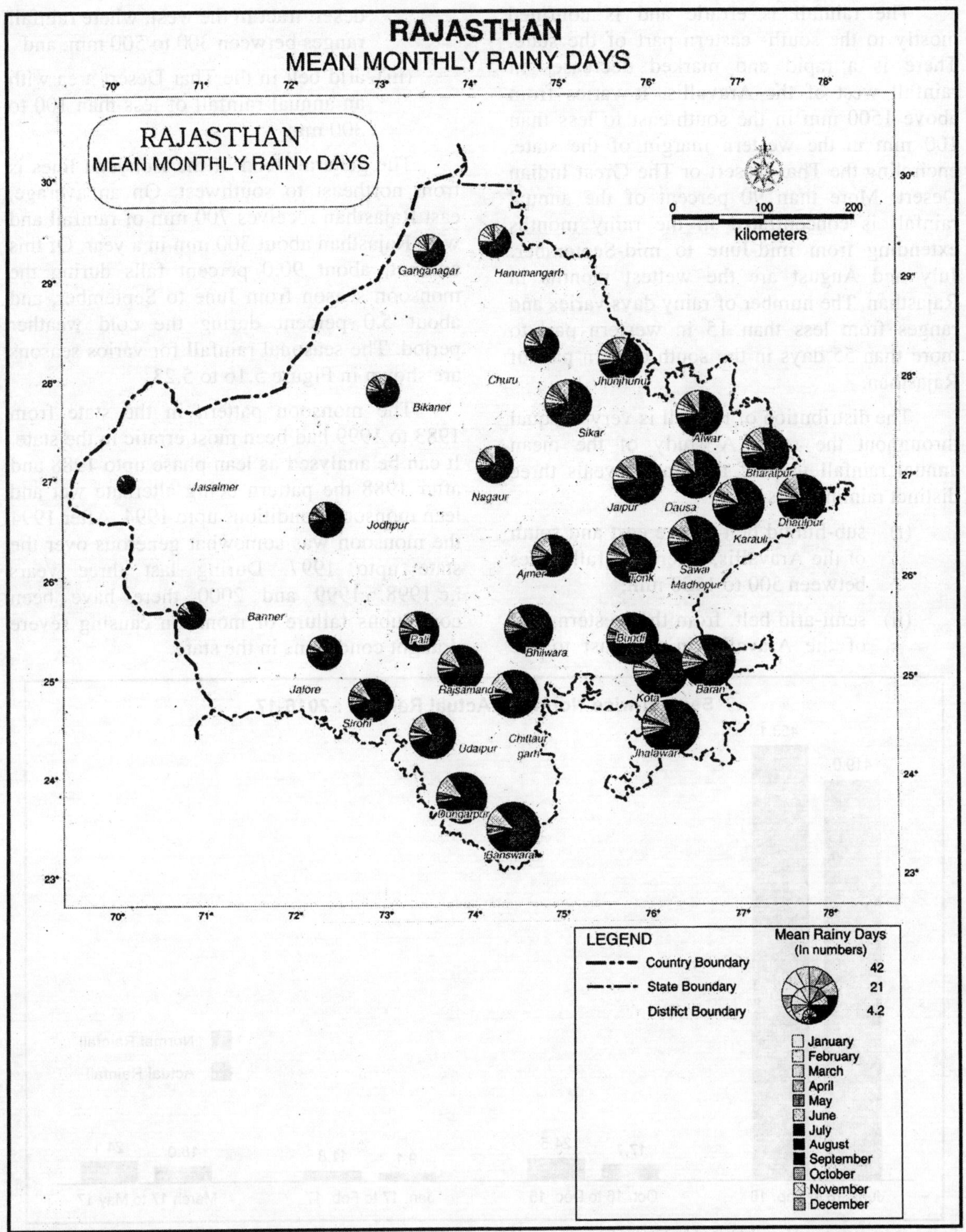

Source: Administrative Map, published by Census Commission of India and data obtained from India Meteorological Department, Poona.

Figure 5.21 Mean Monthly Rainy Days in Rajasthan

The rainfall is erratic and is confined mostly to the south- eastern part of the state. There is a rapid and marked decreases in rainfall west of the Aravallis. It varies from above 1500 mm in the south-east to less than 100 mm in the western margin of the state, encircling the Thar Desert or The Great Indian Desert. More than 90 percent of the annual rainfall is concentrated in the rainy months extending from mid-June to mid-September. July and August are the wettest months in Rajasthan. The number of rainy days varies and ranges from less than 15 in western part to more than 55 days in the south-eastern part of Rajasthan.

The distribution of rainfall is very unequal throughout the state. A study of the mean annual rainfall map of the state reveals three distinct rainfall belts :

(i) sub-humid belt on the east and south of the Aravallis, where rainfall varies between 500 to 1500 mm;

(ii) semi-arid belt, from the western hills of the Aravallis in the east to the desert tract in the west, where rainfall ranges between 300 to 500 mm; and

(iii) arid belt in the Thar Desert area with an annual rainfall of less than 100 to 300 mm.

The general trend of the isohyetal lines is from northeast to southwest. On an average, east Rajasthan receives 700 mm of rainfall and west Rajasthan about 300 mm in a year. Of this amount, about 90.0 percent falls during the monsoon season from June to September, and about 5.0 percent during the cold weather period. The seasonal rainfall for varios seasons are shown in Figure 5.16 to 5.23.

The monsoon pattern in the state from 1983 to 1999 had been most erratic in the state. It can be analysed as lean phase upto 1988 and after 1988 the pattern being alternate wet and lean monsoon conditions upto 1994. After 1994 the monsoon was somewhat generous over the state upto 1997. During last three years i.e.1998, 1999 and 2000 there have been continuous failure of monsoon causing severe draught conditions in the state.

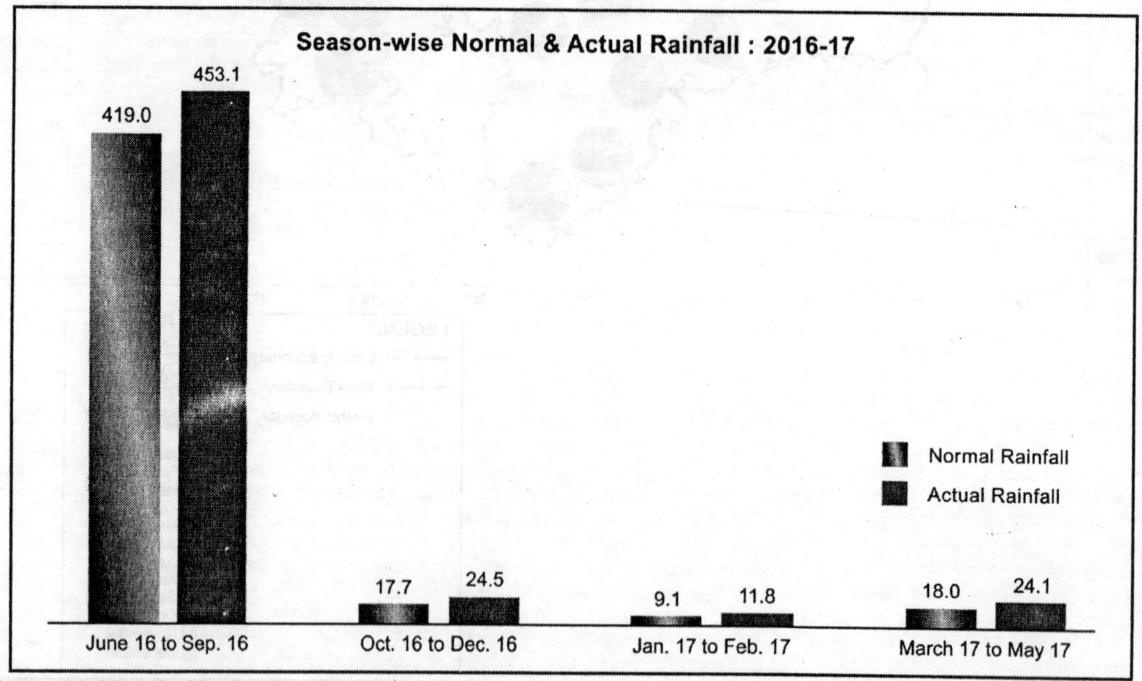

Source: Administrative Map, Published by Census Commission of India and data obtained from India Meteorological Department, Poona.

Figure 5.22 Seasonwise Norma and Actual Rainfall : 2016-17

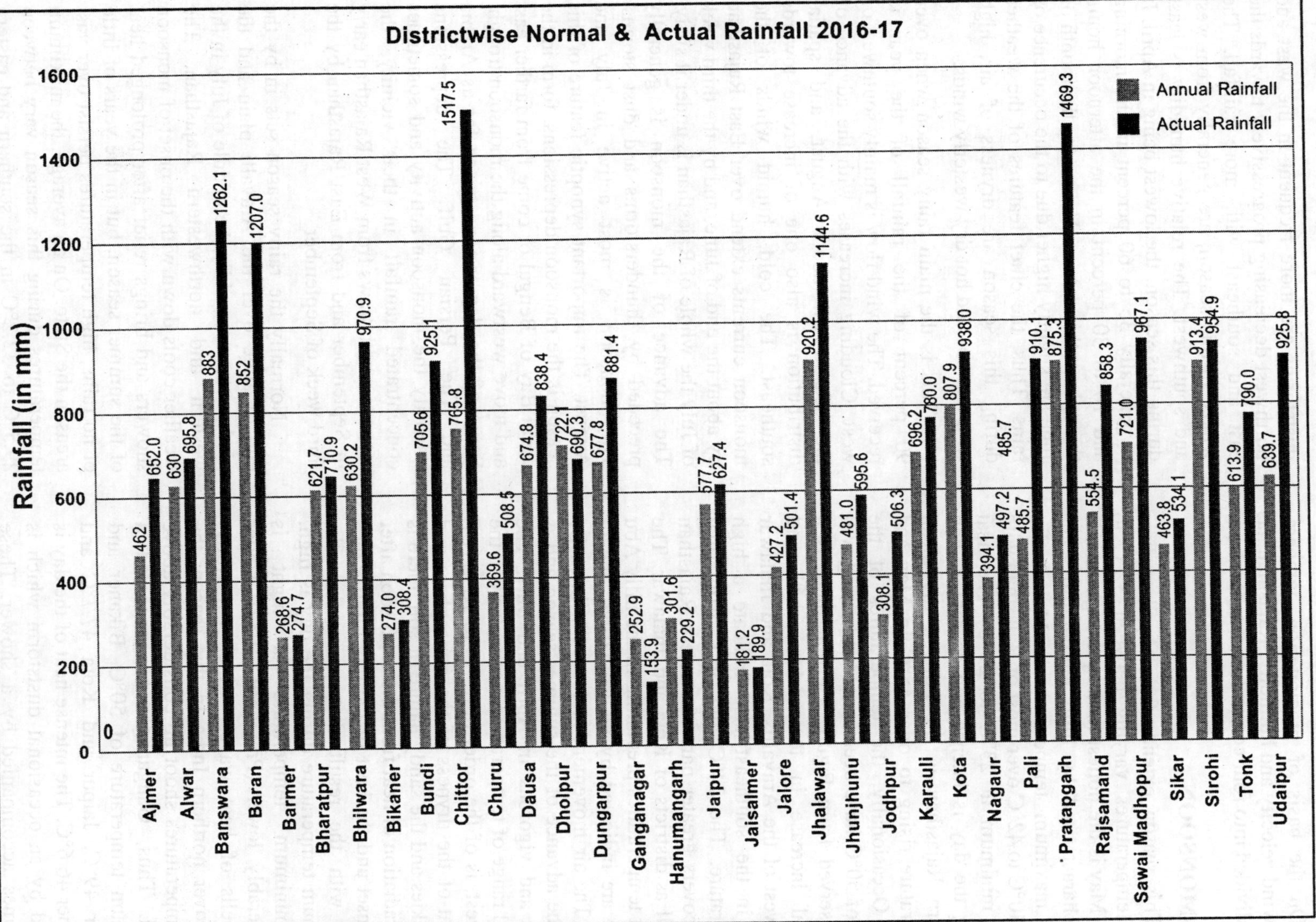

Source: Administrative Map, Published by Census Commission of India and data obtained from India Meteorological Department, Poona.

Figure 5.23 Districtwise Normal and Actual Rainfall 2016-17

On the basis of precipitation, temperature, wind velocity and humidity, Rajasthan has been divided into four conventional seasons:

PRE-MONSOON

This season extends from April to June with temperatures varying between 32°C to 45°C. May is the hottest month, though June is also characterised by high temperature. The mean maximum temperature in May varies from 40°C to 42°C over most of Rajasthan with mean minimum at 23°C to 27°C. The heat during the day is intense and scorching, and Western Rajasthan is the hottest, the mean temperature rising to over 45°C in May and June. Occasionally, the temperatures of the order of 50°C have been recorded here. It will be observed from Figure 5.1 that there is a gradual increase in temperature north and northwest of the Aravallis. Kota and Jhalawar areas in the southeast also indicate a high temperature. The 34°C and 36°C isothermal lines covers greater part of western Rajasthan as well as districts of Kota and Jhalawar. The lowest temperatures prevailing in the hilly Abu region are indicated by an isothermal line of 28°C. The air movement increases in intensity with the advance of the summer season and is strong and vigorous in April and May. The diurnal range of temperature, particularly in the arid west, is of the order of 15°C to 18°C. On account of the dryness of the atmosphere and clear skies and the sandy nature of soil, there is rapid radiation of heat from the earth soon after the sunset and the heat of the day rapidly dies down, with the result that even when the maximum temperature during the day is high, the minimum temperature at night is considerably lower (Pramanik, 1952). Under the spells of heat waves, which sometimes occur over northern India in this season, the day temperatures shoot up considerably above normal. Thus, Ganganagar has recorded a maximum temperature of 50°C, Bikaner and Barmer 49°C, Jaipur and Kota 47.7°C and Jaisalmer 46.6°C. The intense heat of the day is reduced by an occasional duststorm which is sometimes accompanied by a shower. These duststorms are more frequent in the west and northwest decreasing progressively towards the east and southeast with more rainfall. The winds, in this season, are generally from west and southwest. The relative humidity is least during this season, the lowest being in April. It is generally 35 to 60 percent in the morning and 10 to 30 percent in the afternoon hours throughout this season. Values for the month of June are slightly higher due to the occurence of rains. Thus, the chief features of the weather during this season are dryness of air, high temperatures and hot dry westerly winds.

This is the main rainy season when over 90 percent of the rainfall of the year is received. The wind flow is mainly southwest or west. Clouding increases with the advance of the season till about August, and spatial distribution is also one of increase towards southeast. The cold, humid winds of the monsoon currents extend over East Rajasthan by about the end of June, and by the third week of July the whole of Rajasthan is under it sway. The advance of the monsoon is generally preceded by thunderstorms and dust storms. The monsoon is most active in July and August. The important synoptic feature of this season is the monsoon depressions, form in the north Bay of Bengal or come from farther east and move westward along the monsoon trough. Some of the depressions go as far as Arabia and The Persian Gulf. The depressions intensity the monsoon activity and sometimes concentrated rainfall in their vicinity, the monsoon withdraws from West Rajasthan early in September and from East Rajasthan by the third week of September.

Normally the rainy season sets in by the end of June over the Aravalli plain and the piedmont zone and by the middle of July in the western and northwestern Rajasthan. The weather cools down with the onset of monsoon showers and brings relief after prolonged heat of the summer season, but in the years of little or no rains, high temperatures persist over vast areas in the State. On an average, the maximum temperatures during this season vary between 29.5°C to 32.2°C in the southern and eastern parts of the state to over 37.7°C in the northern and western parts.

The relative humidity is maximum during this season, being highest in August. But it is less marked in the areas to the west and northwest than in the eastern Rajasthan. Winds are generally from west and south west but whenever a depression or low approaches o enters the state; the winds change their direction so as get into the circulation of the disturbance.

MONSOON

The development of monsoon is directly related with the atmospheric activity governed by external forces. These forces include solar and terrestrial radiations, mountains and influence of ocean etc. The above factors however help in creation of monsoon. Its further migration from one place to another depends upon differential heating of region, which creates low pressure zone and invites air loaded with high proportion of humidity from area of high pressure. The South West Monsoon system developed in the Indian Ocean and touched Kerala Coast on date of 5th June. Since the onset took place over Kerala on 5th June, it rapidly covered the south peninsula and northeast India by 9th June and central, eastern parts by 12th June. It advanced further over some parts of SE Rajasthan on 14th June i.e. one day earlier than the normal date of 15 th June. Presence of low pressure area over east Rajasthan and neighborhood superposed with a trough in the mid and upper tropospheric westerlies provided conditions conducive for the large scale convection and wide spread Monsoon rains over northwest India during next 3 days. This helped Monsoon to advance further over the entire state by 26th June, about a month earlier than its normal date of 15th July. On the same day Monsoon covered the whole country also.

POST-MONSOON

With the withdrawal of the monsoon, clear autumn weather sets in over the area. The skies are generally clear, winds weak and variable over west Rajasthan and north over east Rajasthan. The monsoon is withdrawn from the area by the mid-September after which a second hot season follows which ultimately cools down and merges with the cold season by December. October and November are months of clear and light clouded skies. The temperatures during October, in the whole of Rajasthan, are fairly uniform, the maximum temperature ranging between 33°C to 38°C and minimum between 18°C to 20°C at various stations. The diurnal range of temperature is high; varying from 14°C to 16°C. The relative humidity is quite low, even tower than in the cold season at a number of stations. The winds die down considerably during this season as the anticyclonic conditions succeed to cyclonic, and are generally light and variable.

WINTER

This is a season of cool weather, clear skies, low humidity and low temperature with large diurnal variation. The wind flow is generally weak northerly. The fine weather is occasionally broken by brief spells of cloudy weather, caused by western disturbances which are attended by clouding and rise of temperature in the forward portion, while in its rear western and northwestern winds bring a spell of cold weather.

The winter season sets in by the month of December and is marked by a gradual decrease in temperature all over the region. January is the cold set month of the year. The mean maximum temperature during this month ranges from 20°C at Ganganagar to 22°C at Bikaner, 22.8°C at Jaipur and Ajmer 24.4°C at Barmer and Jodhpur to 25°C at Kota. The mean minimum temperature varies from 3.3°C at Ganganagar to 7.7°C at Ajmer, 8.3°C to 10°C at Jaipur, Bikaner, Jodhpur and Jhalawar to 10.5°C at Kota. The isotherms for the north of January show a marked contrast from those of the May isotherms. These isothermal lines approximately run from east to west, with slight variations while crossing the Aravalli hills where these lines dip gradually towards the south, an indication that slight lower average temperature is found over the Aravallis at southern latitude.

The incidence of western depressions is a special feature of the weather of the north Indian plains during this season but only a few of them affect and give rain Rajasthan. Only northern and northeastern portions of the state often come under the sway of these disturbances. They bring with them a little rainfall together with significant changes in temperature. The amount of winter rainfall is an insignificant portion of the annual total, hardly 5 to 6 percent. The relative humidity, in this season, is generally between 50 to 60 percent in the morning hours and 25 to 35 percent in the afternoon hours. The winds are generally light and variable, but northeasterly and northerly winds are more frequent than those from the other directions.

PROBABILITY DISTRIBUTION OF RAINFALL

In the present analysis daily rainfall data of 26 district headquarter stations of 75 years have been used. The statistical parameters e.g. mean and range in the rainfall on monthly and annual rainfall are worked out for all the stations. The contribution of monthly rainfall to annual rainfall were also worked out. The cumulative probability of occurrence of varying amount of rainfall during southwest monsoon period (June, July, August and September and probability distribution of annual rainfall were computed and presented. The different statistical parameters described above are discussed below :-

Mean-The simple arithmetic mean is defined as

$$X = \frac{\sum X}{N}$$

Where x is the variate and N is the number of observation. In meteorology if N is equal to or more than 30, the mean is known as "normal".

Range - The difference between the extreme maximum value and extreme minimum value of observation under analysis is known as range.

PROBABILITY DISTRIBUTION OF ANNUAL RAINFALL

The number of occurrence of a variate in a descrete number of variables is called frequency. The frequency divided by total number of occurrences is called probability (P). The total probability for all variates should be equal to unity, or EQUATION Distribution of probabilities of all variates is called probability distribution. The probability distribution analysis of the state are presented in map. (Figure 5.24). The selection of class interval is purely arbitrary.

PROBABILITY ANALYSIS OF SOUTHWEST MONSOON RAINFALL (during June, July, August and September)

The probability of occurrence of equal to or more than certain specified amount of rainfall during the months of southwest monsoon period (June, July, August and September) were computed and presented in map (Fig. 5.25). The specified amount of rainfall were chosen at an interval of 50 mm. The lowest and highest limit were selected depending upon the climate condition of the area. For arid region it is 50 to 500 or 550 mm and semi-arid and sub- humid condition from 100 to 600 mm. In humid climatic near Mount Abu region, the lowest and highest value selected were 100 to 1000 mm at the interval of 100 mm.

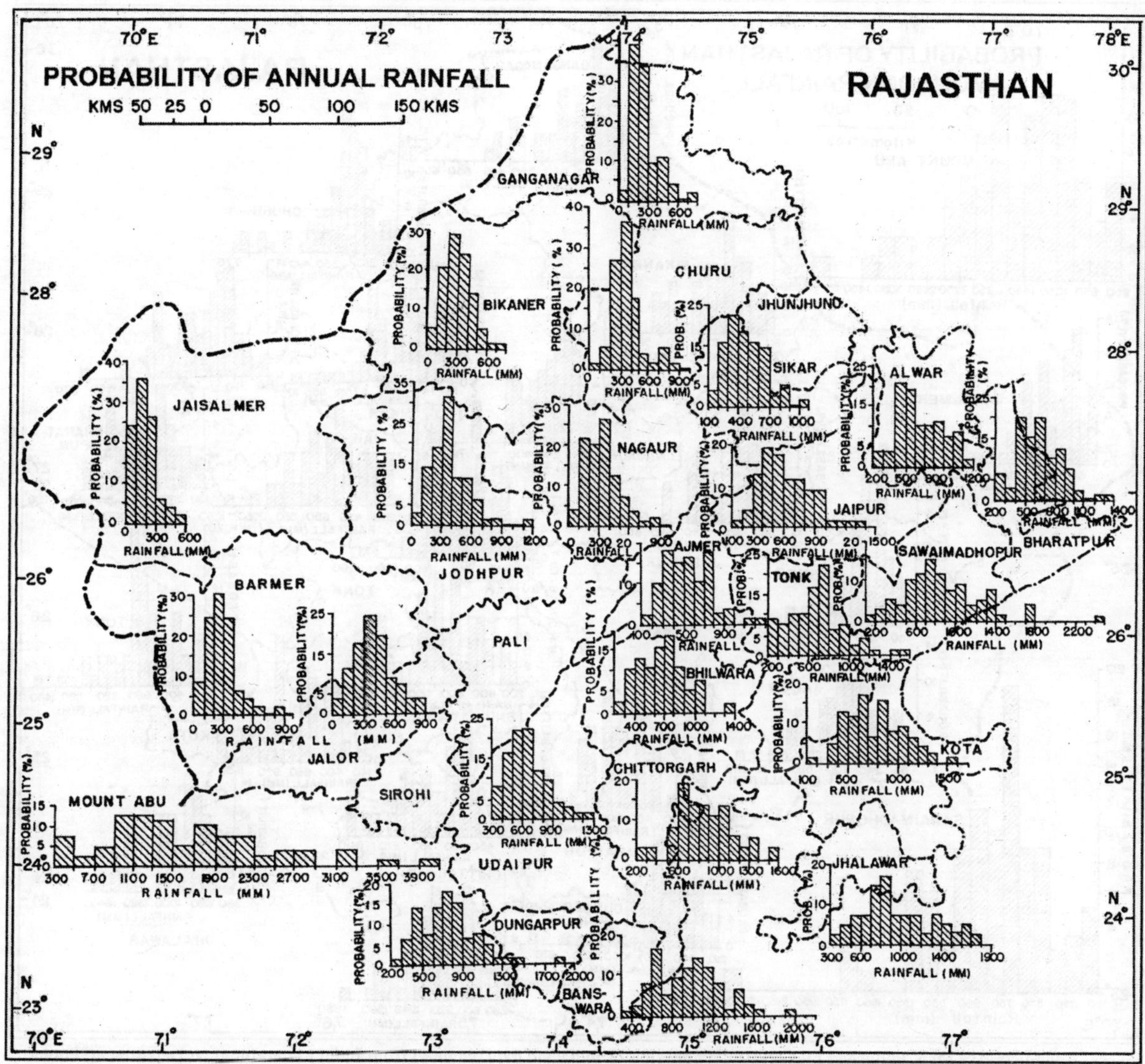

Source: Administrative Map, Published by Census Commission of India and data obtained from India Meteorological Department, Poona.

Note : The external boundary of India appearing on the map are neither verified nor authenticated by the survey of India. This is for academic purpose only.

Figure 5.24 Probability of Annual Rainfall

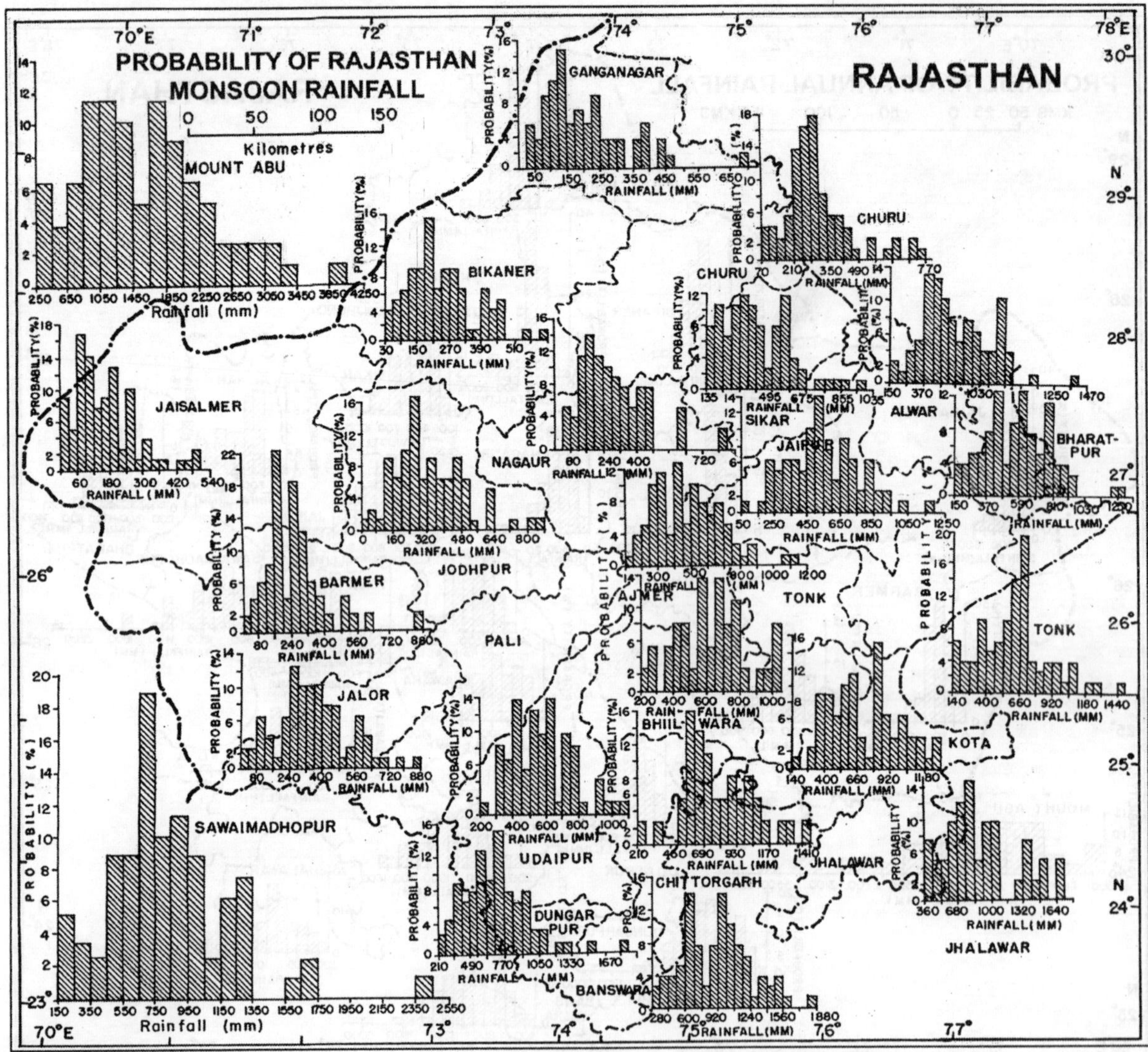

Source: Administrative Map, Published by Census Commission of India and data obtained from India Meteorological Department, Poona.
Note : The external boundary of India appearing on the map are neither verified nor authenticated by the survey of India.

Figure 5.25 Probability of Southwest Monsoon Rainfall

EVAPORATION

Evaporation closely follows the march of seasons. When the winds are slight and temperatures are low, evaporation is low. As temperature rises and wind velocity increases, evaporation also increases. Evaporation rates reach their peak in the summer months of April and May generally and display the higher evaporation rates during this period. With the advent of the moisture laden monsoon winds and precipitation, there is a marked fall in evaporation rates.

EVAPOTRANSPIRATION

Evapotranspiration is the combined loss of water vapour to atmosphere in the form of evaporation from soil, water surfaces and transpiration from the plants. This loss is controlled by two factors, namely the availability of moisture at the earth's surface and the ability of the atmosphere to supply energy to vaporise water and transport the vapour. If supply of water is unlimited, loss of water in the form of evaporation and transpiration will depend on the atmospheric conditions and this loss is termed as potential evapotranspiration (PE).

POTENTIAL EVAPOTRANSPIRATION

Potential evapotranspiration is the quantum of water vapour likely to be evaporated from the soil surface and transpired by the vegetation of a certain area under free water availability (moisture) conditions. Increase in evapotranspiration in the summer season is due to higher temperatures, low relative humidity and stronger wind regime. The higher the evapotranspiration, the lower is the spread of vegetal cover on the ground.

MEAN ANNUAL POTENTIAL EVAPOTRANS-PIRATION

Annual potential evapotranspiration is highest (206.32 cm) at Jaisalmer and the lowest (166.22cm) at Ganganagar in western part of Rajasthan. In eastern Rajasthan, it varies from 138.12 cm in Udaipur to 174.52 cm in Jaipur. Monthly variations in potential evapotranspiration are marked, making the mean annual values to a limited use as shown in Figure 5.26. Potential evapo-transpiration is lowest in the month of December when temperatures are low, but shows a gradual rise from January. From March onwards, this increase remarkably and reaches its highest degree in the month of May and thereafter with the onset of the monsoon, it starts decreasing till the month of August. Then again, there is a slight increase during September and October, owing to rise in day temperature, followed by a gradual decrease from November onwards as winter set in. In Jaisalmer and Bikaner, however, June records the highest rate of potential evapotranspiration.

Potential evapotranspiration is higher than precipitation even during rainy months in the western part of Rajasthan, making these regions arid to extremely arid in climate with perpetual water deficit in all the months. On the eastern side, however, the rainfall in much higher during the rainy season than the potential evapotranspiration demands during that time, thus providing water surplus during the rainy season.

The potential evapotranspiration isolines indicate that they are lowest in southern Rajasthan in Dungarpur and Banswara districts (130 cm). The values rise sharply in the western areas, more than 200 cm in Jaisalmer district. In the eastern areas, the values remain around 170 cm in Bharatpur, Dhaulpur, Jaipur and Tonk districts.

The root system of the plants utilises water. The water available in the root zone of the plants is all that is absorbed by the roots and subsequently transpirated through the leaf system. The amount of water that percolates below the root zone is the excess water over and above that utilised by the plant. The loss of rain water is mainly due to evapotranspiration besides that resulting from surface runoff, percolation and evaporation zone of the soil. The evapotranspiration depends on solar radiation, air temperature, wind velocity and humidity. Based on these factors, a formula to determine potential evapotranspiration is given by Penman. The mean annual potential evapotranspiration varies from a maximum of over 20001 mm and above in northwestern parts of Jaisalmer district to a minimum under 1301 mm in Banswara and Dungarpur districts. This formula was further modified to suit Indian conditions, and on the basis of modified formula, the monthly and annual values of evapotranspiration are worked out.

Potential Evapotranspiration (PE) is governed by precipitation, temperature and relative humidity, which in turn determine the water balance of a region. Owing to scanty rainfall, most of the districts in western Rajasthan have very high annual PE rate of above 1600 mm. Highest annual PE is observed in Jaisalmer (2063.2 mm), followed by Barmer (1857.7 mm), Jodhpur (1843.0 mm), Bikaner (1772.4 mm) and Ganganagar (1662.2 mm). In the eastern region, Jaipur district has the record of 1745.2 mm, followed by Ajmer (1566.3 mm), Jhalawar (555.5 mm), Kota (524.7 mm) and Sikar (1503.2 mm). The areas with comparatively higher rainfall such as the south and south-eastern part of the State have annual PE rate below 1400 mm. The annual PE in Udaipur (1381.2 mm) district is the lowest. Near a dam site, in most cases the water remains during the whole year, while in the canal conveyance area and in the command area, the water is made available only during the growing season for the crop. Water is lost by evaporation and percolation in the reservoir and also in are as of the conveyance canal.

The water lost through evaporation may have some effect, on micro climate, but the water that is lost through seepage from the reservoir, canal system and irrigation adds to the water table beneath the ground. The extent of seepage depends on the type of soil. In case of black clay soils, there may be very little seepage; but in case of sandy soils, seepage may account for a substantial part of water loss. Similarly, the lower the extent of cover of soil with vegetation, the lower the moisture content in the soil; and evaporation from such exposed surfaces takes place at a faster rate than from surfaces covered with vegetation. During periods of continuous rainfall, there is substantial loss of water through run off and percolation, while in case of intermittent rainfall there is very little loss due to run off. In case of deep sandy soils, the run off may be low even if the rain at high intensity. In case of clay-soils, where infiltration is poor, the losses due to run off are very high. The construction of proper bunds, may temporarily help storing water, thus reducing losses effected by run off.

The annual PE and its monthly variation for selected stations along with the monthly rainfall values are given in Figure 5.26. It is evident that PE is lowest in December, after which it gradually starts increasing. From March onwards, PE increases remarkably and reaches its highest in the month of May. Thereafter the PE decline during rainy season but once again resumes an upward trend as soon as the monsoon ends. However, in western part of the State the PE rate tends to be high in rainy months, owning to scanty precipitation. In the eastern part, the rainfall exceeds PE during rainy season.

Whenever the moisture available in the soil in any day is less than the PE value, the actual evapotranspiration will decrease depending upon the deficit, and will even cease completely if the moisture level in the top 20 cms is below wilting point. Thus during the years when the rainfall is above normal, there will be a further recharge of water table and increased runoff losses. During the years when the rainfall is below normal and distribution of precipitation is less than evapotranspiration, does the crops may suffer, thus reducing the productivity; but there will be not effect on ground water table level which will decrease only if water is pumped out and used for irrigation.

From a comparison of the monthly rainfall with potential evapotranspiration for the various stations in Rajasthan, it is possible to obtain a clear picture of the period of water surplus and deficiency in relation to monthly rainfall and to bring into perspective the nature of water problems and requirements in the area for successful crop husbandry. The mean annual potential evapotranspiration (Figure 5.26) varies from a maximum of over 2001 mm and above in north-western parts of Jaisalmer district to a minimum under 1301 mm in Banswara and Dungarpur districts.

In the present analysis of evapotranspiration and moisture index, 26 stations were selected which fall in different climatic zones of the State. Daily rainfall data from 1901 or from the year of inception of the observatory to 1979 have been converted to monthly rainfall and used. Since there is not significant variation in potential evapotranspiration from year to year in any meteorological periodicities, normal value of potential evapotranspiration based on Penman's formula was used.

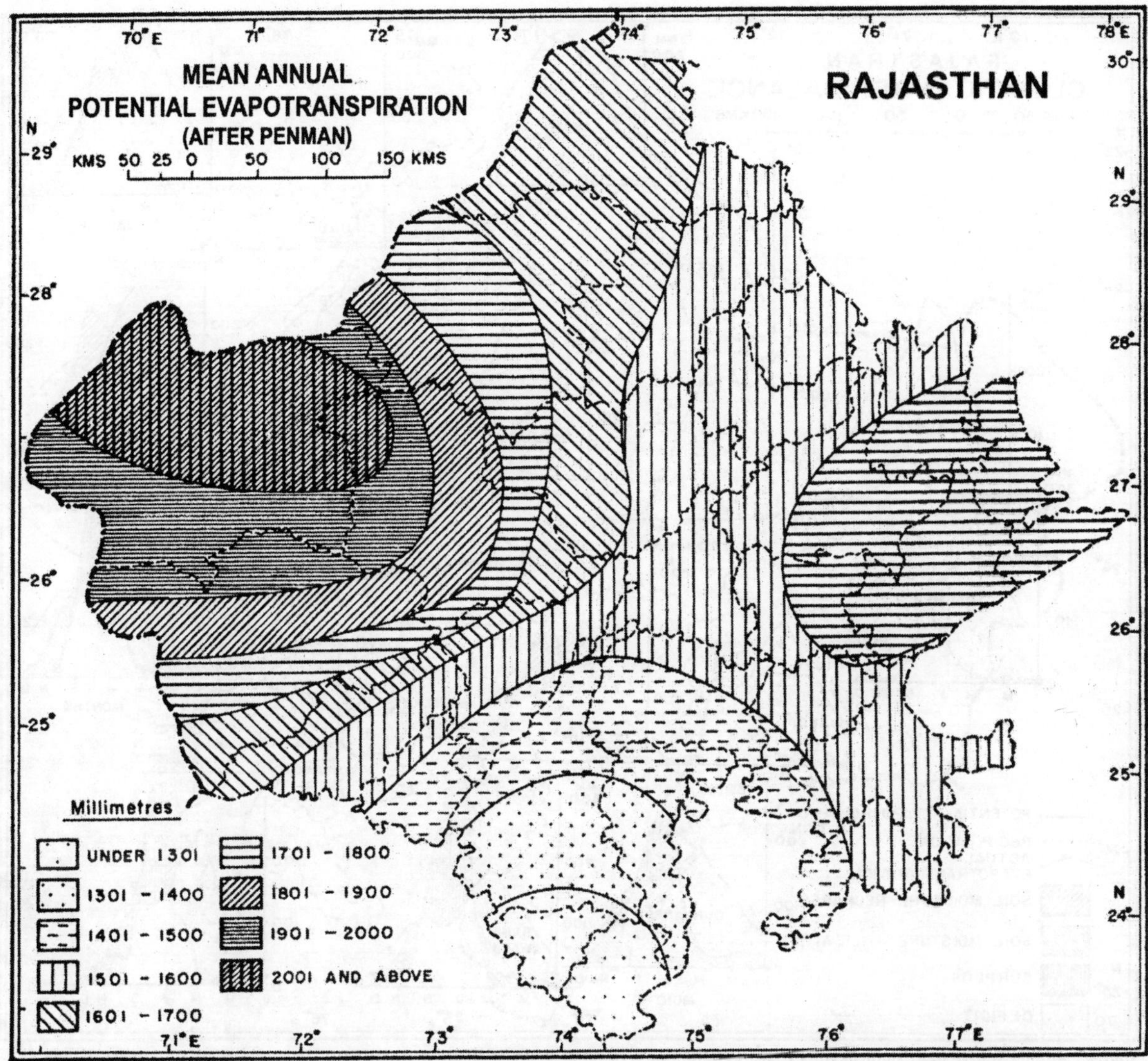

Source : Administrative Map, published by Census Commission of India and data obtained from India Meteorological Department, Poona.

Note : The external boundary and the coastlines of India appearing on the map are neither verified nor authenticated by the Survey of India.

Figure 5.26 Mean Annual Potential (After Penman)

During the monsoon season when the water table is close to the surface, water accumulates in the low lying areas, and where swamps and marshes exist, the computed PE tends to be lower than that would be expected of a moisture land surface under similar temperature conditions. The march of potential evapotranspiration generally follows a near symmetric bell-shaped curve. On an average, a maximum PE of 225 mm to 300 mm is obtained for April and May and a minimum of 50 mm to 80 mm for the month of December. As monthly PE and the mean monthly temperature are empirically related, the computed PE compares favourably with march of average monthly temperatures.

A comparison of seasonal course of rainfall with the course of monthly PE for a station provides much information on many aspects of water relations, such as actual

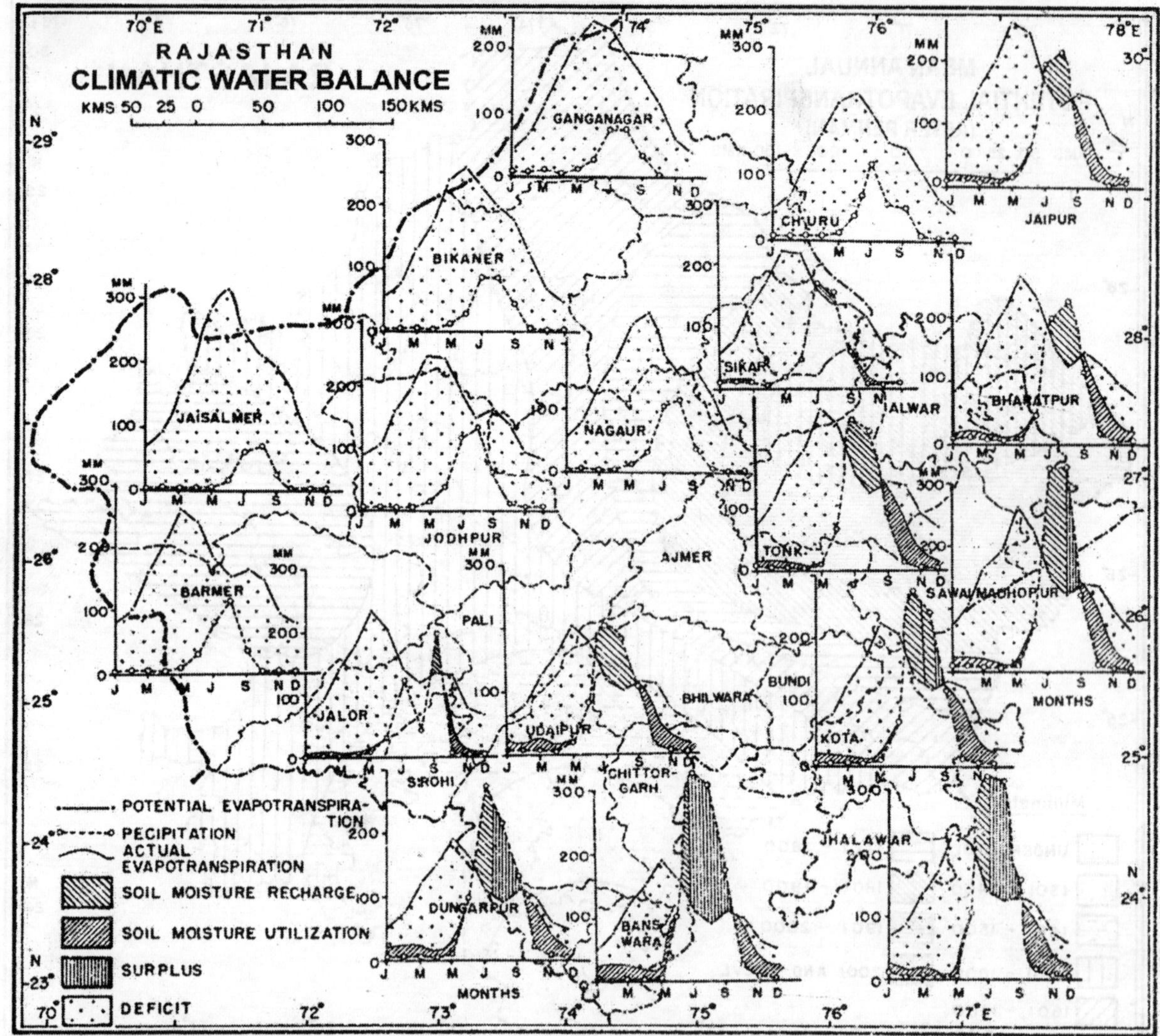

Source : Administrative Map, published by Census Commission of India and data obtained from India Meteorological Department, Poona.

Note : The external boundary and the coastlines of India appearing on the map are neither verified nor authenticated by the Survey of India.

Figure 5.27 Clmate Water Balance

evapotranspiration, periods of soil moisture recharge and soil moisture utilization, water deficit and water surplus. Of these, only the aspects of water deficit and water surplus have been discussed in detail. It was assumed that under the existing vegetation conditions, the soils of Bharatpur can on an average hold only 150 mm of moisture and the soils of Banswara at field capacity, can retain only 90 mm of moisture (Figure 5.27).

Generally, the water deficit occurs in two distinct phases, i.e. the pre-monsoon (March to June). Throughout Rajasthan, the imbalance between PE and rainfall is such that on an average more water is needed by plants than is actually obtained from rainfall. The average annual water deficit decreases from a little more than 2000 mm in the north-west to a little less than 400 mm in the south (Figure 5.28).

The gradient of change in water deficit aspect is relatively steep, which probably reflects the rainfall characteristics

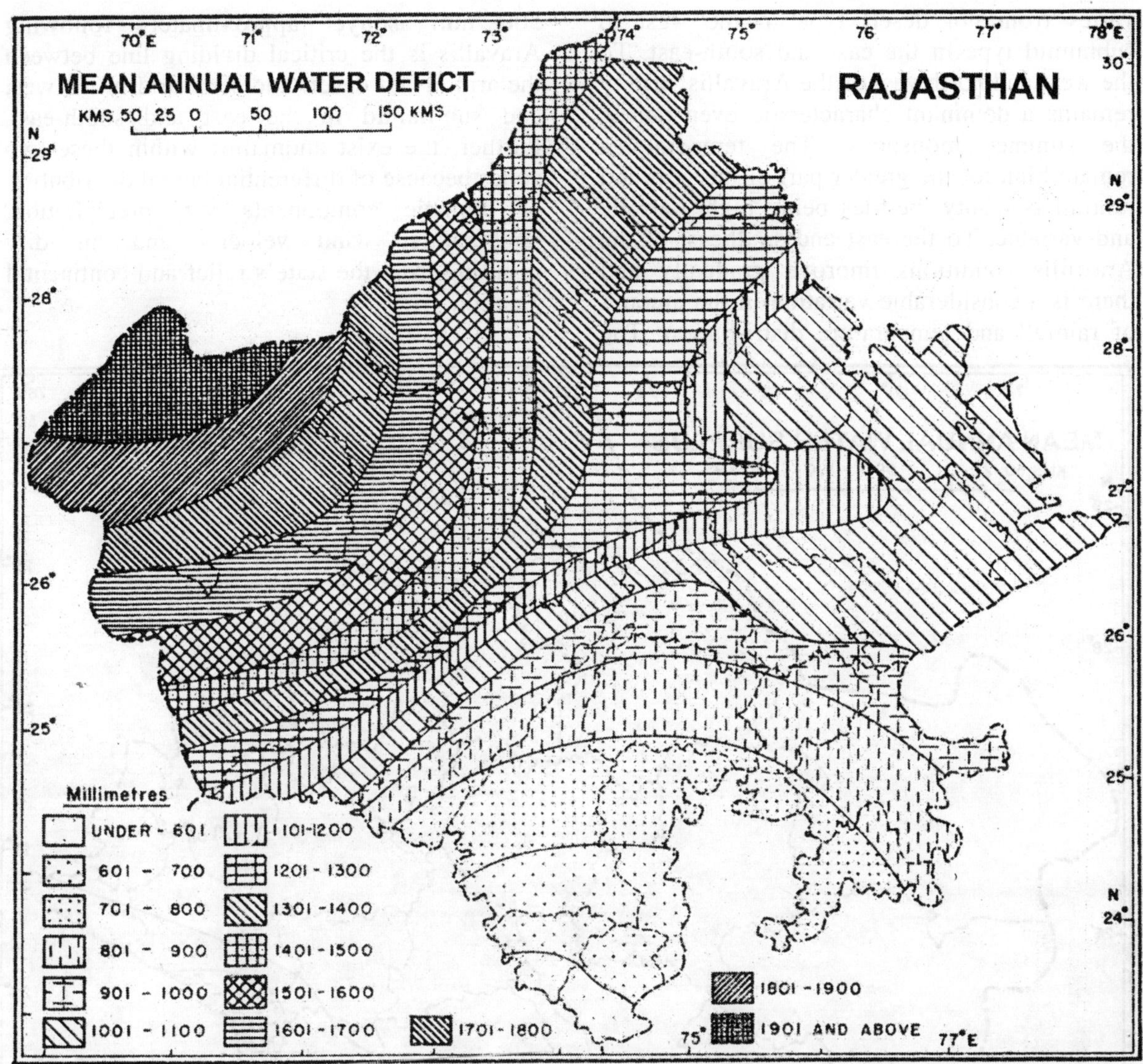

Source : Administrative Map, published by Census Commission of India and data obtained from India Meteorological Department, Poona.

Note : The external boundary and the coastlines of India appearing on the map are neither verified nor authenticated by the Survey of India.

Figure 5.28 Mean Annual Water Deficit

The distribution and pattern of average annual water deficit relate inversely to rainfall. It is also known that under normal rainfall conditions, repeated periods of soil moisture utilization and soil moisture recharge tend to minimize the total drought conditions so well characteristic of the dry years. Under such rainfall conditions, the onset of drought follows no particular rule.

As shown in Figure 5.29 most of the districts of Rajasthan suffer from deficiency in annual water surplus. The average annual water surplus varies from zero to a maximum of 450 mm in Mt. Abu area.

Rajasthan occupies a special position in the climatic history of the Indian sub-continent. The climate of Rajasthan plays a dominant role in structuring its physical as well as biotic environment. The climate is in general varies from hot desert type in the west to subhumid type in the east and south-east. To the west and northwest of the Aravallis, aridity remains a dominant characterstic even during the summer monsoons. The temperatures remain high for the greater part of the year and rainfall is scanty, besides being highly erratic and variable. To the east and southeast of the Aravallis conditions improve gradually and there is a considerable variation in the amount of rainfall and temperature distri-bution. The 500 mm isohyet approximately following Aravallis is the critical dividing line between the arid or hot desert and semi-arid in the west and sub-humid in the east and south-east. Further; the exist anomalies within these two zones because of differential lateral distribution of climatic components viz. precipitation, temperature, wind velocity and humidity resulting from the state's relief and continental location.

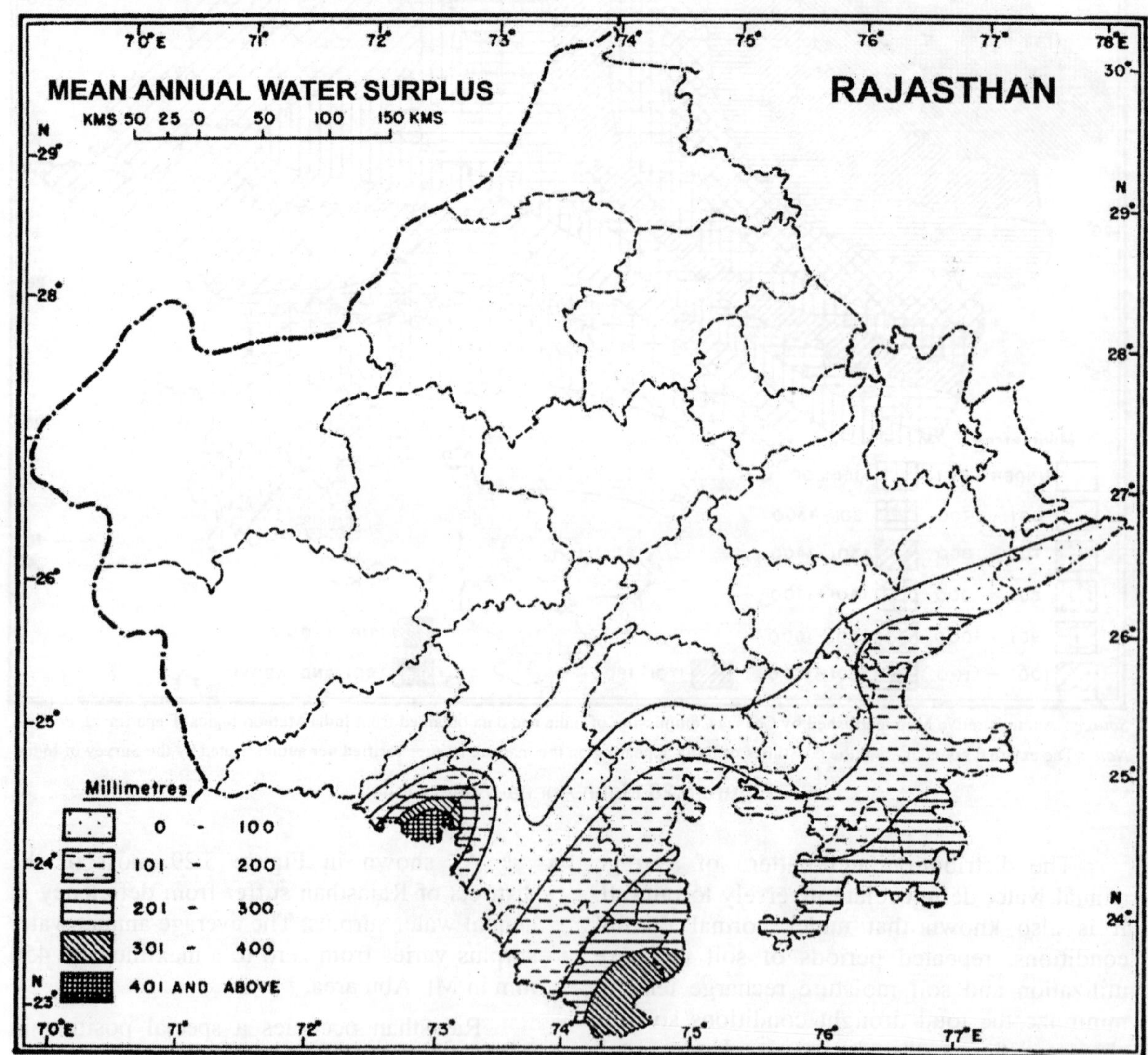

Source : Administrative Map, published by Census Commission of India and data obtained from India Meteorological Department, Poona.

Note : The external boundary and the coastlines of India appearing on the map are neither verified nor authenticated by the Survey of India.

Figure 5.29 Mean Annual Water Surplus

The climate of Rajasthan, West of Aravalli, is characterized by great extremes of temperatures, long periods of severe drought, accompanied by high winds and low to semi-arid conditions of humidity. The winter is severe and at places the temperature sometimes falls below freezing point. The heat during summer is intense touching 49°C in some parts of the State. In the east and south of the Aravalli there is a considerable variation in the amount of rainfall and range of temperature. Due to the sandy nature of soil, which gets rapidly heated during the day and cools down quickly after dark, difference of as much as 22°C have been recorded betweendaily maximum and minimum temperature. Based on climatic conditions, Rajasthan is the driest State of the country. Rajasthan is climatically devided in 4 zones based on Moisture Index (After Thornthwaite) as shown in Figure 5.30.

The climate of Rajasthan state has varied contrasts and the presence of Aravallis is the greatest influencing factor. The state can broadly be divided into Arid, Semi-Arid and Sub-Humid Regions, on the basis of rainfall intensities. The Western Rajasthan i.e. in the arid region consist of the districts of Hanumangarh, Jaisalmer, Barmer, Ganganagar, Churu, Jhunjhunu, Sikar, Nagaur, Jodhpur, Pali and Jalore covering an area of nearly 1,43,842 sq. km. The region is characterized by low and highly variable rainfall years creating inhospitable living condition to both human and livestock population. An area of 9,290 sq. km. in extreme western parts of the state has true desert conditions. With an improvement in rainfall pattern from the west towards the east Rajasthan semi-arid conditions are created in an area of about 66,830 sq. km. in the districts of Alwar, Jaipur, Bharatpur, Ajmer, Tonk, Sawai Madhopur, Bhilwara, Bundi, Kota, Chittorgarh, Udaipur, Sirohi, Dungarpur and parts of Jhalawar and Banswara.

ARIDIDY AND CLIMATIC ZONES

Aridity in the State is the root cause of a unique ground water and drought scenario, which calls for especial attention and should be studied and dealt with reoriented and renewed scientific approach. It is the aridity, which has resulted to water deficiency, deteriorated surface and ground water quality, origin of Thar Desert and incident of frequent drought spells.

Aridity signifies deficiency of moisture to support vegetation, water deficiency exceeds water demand i.e. evaporation exceed precipitation.

Several methods have been suggested for classification of climates and climatic zones.

KOPPEN'S METHOD (1923)

Sets out the limit between desert, steppe and dry and tree climate.

a) Boundary between dry and tree climate r = 0.44t-3

b) Boundary between steppe and desert

$$r = \frac{0.44t-3}{2}$$

Where t and r are mean annual temperature (°F) and mean annual precipitation (inches). Depending on values of r Western Rajasthan is in desert while eastern Rajasthan has been classified as steppe with the exception of Jhalawar, Dungarpur and Mount Abu regions.

DEMARTONNE'S METHOD (1928)

$$I = \frac{\text{Annual precipitation (mm)} + 10}{\text{Mean Annual temperature (°C)}}$$

Indices below 5 characterises true desert from botanical and hydrographical of view, above 10 corresponds to dry steppe, 20-30 to prairies and above 30 as forest vegetation. True desert is confined only to the westernmost parts of Jaisalmer and Bikaner districts. Rest of the areas of western Rajasthan is classified as semi-desert. Southern Rajasthan is classified as prairies, the remaining eastern Rajasthan as dry steppe and Mount Abu as forest vegetation.

GAUSSEN'S METHOD

The method takes in account factors of temperature, precipitation, number of rainy days, atmosphere humidity, mist and dew.

The climates are classified, as warm, warm temperate and temperate if annual mean temperature is positive. Among these regions, if xerothermic index exceeds 300, the climate is desert and 200-300 it is hot sub-desert, which can be subdivided as accentuated and attenuated. True desert is confined to small portions of Jaisalmer and Bikaner. Remaining State falls under sub-desert climate being accentuated in west Rajasthan and attenuated in east Rajasthan.

EMBERGER'S METHOD (1955)

Suggested the delimitation of bio-climates by means of the pluviothermic quotient (Q)

$$Q = \frac{100\,P}{M^2 - m^2}$$

Where, P= annual rainfall, M= average maximum temperature in the hottest month and m= average minimum temperature in the coldest month.

Values below 40 in west Rajasthan (arid), 60 to 100 from the north to south portions respectively (Semi-arid) of east Rajasthan and above 300 in Mount Abu.

THORNTHWAITE METHOD (1948)

Thornthwaite introduced the concept of potential evapotranspiration, which is amount of water that will lost from the surface completely covered with vegetation if there is sufficient water at all times for the use of vegetation. A moisture index has also been defined, which is amount of precipitation necessary to meet the demands of the evapotranspiration (water need) to the amount of water available stored as soil moisture.

$$\text{Moisture index } Im = \frac{100\,s - 60\,d}{n}$$

Where,

s = surplus moisture in the soil during humid season

d = lack of moisture during dry season

n = necessary quantity of moisture for the plant and is taken to be equal to P.E.

MOISTURE INDEX	CLIMATE TYPE
< - 40	Arid
-20 to -40	semi-arid
+20 to -20	Sub humid
> 20	Humid

According to this classification, entire Rajasthan falls under arid and semiarid category except Jhalawar and Mount Abu (moisture index +84), which are under sub humid and humid climates respectively. The line of separation passes through rann of kutch – Pali and Jhunjhunu along Aravalli hill range.

DRYNESS INDEX (DLJ)

Dryness index (Id) was developed by the Defence Laboratory, Jodhpur and mathematically expressed as:

$$Id = 56 \log_{10} \frac{120D}{R}$$

Where,

D = Annual average of diurnal range of temperature (°C)

R = Annual average precipitation (mm)

	CLIMATE TYPE	RANGE OF Id
Arid	Extreme	+ 71 to + 90
	Moderate	+51 to +70
	Mild	+ 31 to + 50
Semi-arid		+11 to +30
	Dry	0 to 10
Sub humid	Moist	- 1 to –10
Humid		-11 to –30
Per-humid	Moderate	-31 to – 50
	Extreme	- 51 to –70

Koppen's Classification

Koppen's classification for the World Climatic regions is totally based on the vegetation, as the effects of temperature and rainfall are directly evident and visible it. Here, the three categories are associated with Tropical/Mega-thermal climates, Dry (arid and semiarid) climates and Mild Temperate/ Meso-thermal climates respectively. The classification of Rajasthan according to Koppen is as follows (Figure 5.30 and 5.32):

(1) Aw or Tropical Humid Region: The southern parts of Dungarpur district and Banswara come under the region. Here, arid tropical grasslands and Savanna like region are found along with deciduous trees of Monsoon type. Winter season is arid and cool whereas summers experience scorching heat. Rainfall also mainly occurs in summer season. The temperature is more than 18°C in the coldest month records.

(2) Bshw Climatic Region: This climatic region is semi-arid, where winters are dry and even in summers there is no sufficient amount of rainfall. Vegetation is of steppe type, characterized with thorny bushes and grasses. This region comprises the districts of Barmer, Jalore, Jodhpur, Nagaur, Churu, Sikar, Jhunjhunu and Hanumangarh.

(3) Bwhw Climatic Region: The region has arid-hot desert climate with very scanty rainfall. On the contrary the process of evaporation is very active. Thus, these areas are known as desert region, limited to western parts of Thar Desert. North-western Jodhpur, Jaisalmer, western Bikaner and western parts of Ganganagar district are included in this category.

(4) CWG Climatic Region: The south-eastern areas of Aravalli are the part of the region. Rains are limited to few monsoon months only. Seasonal winds do not bring rains to this region during winters (Fig. 5.30).

Thorntwaite's Classification

The base of this classification Thornthwaite took into account the amount of rainfall and evaporation along with seasonal and monthly distribution of temperature and rainfall which made it popular and widely accepted. Rajasthan may broadly be divided into the following four regions (Figure 5.30 and 5.33):

(1) CA'w Climatic Region: this region is characterized by wet summers and dry winters. Savanna and Monsoon type of vegetation are found here. This region includes south-eastern parts of Udaipur and districts of Banswara, Dungarpur, Kota, and Jhalawar.

(2) DA'w Climatic Region: The region comprises south and eastern areas of Rajasthan, covering districts of Sirohi, east Jalor, Pali, Ajmer, Chittorgarh, Bundi, Sawai Madhopur, Tonk, Bhilwara, Bharatpur, Jaipur, Dausa, Alwar, Sikar and Jhunjhunu. Rainfall is meager and Semi-arid vegetation is found. Summers are marked by High temperatures.

(3) DB'w Climatic Region: It includes the northern areas of Rajasthan i.e. Ganganagar, Hanumangarh, Churu and Bikaner districts. The region has short and dry winters whereas summers are long and associated with good rains. Thorny bushes and semi-arid vegetation are found here.

(4) EA'd Tropical Desert Climatic Region: This region of Marusthali includes Barmer, Jaisalmer, western Jodhpur, south-western Bikaner district. The climate is very hot and arid. Rains are deficient in every season. Only Xerophytic vegetation grown only (Figure 5.30 and 5.33).

Based on moisture index (Im), climatic classification has been done for Rajasthan State Thornthwaite classification, 1955 (Figure 5.31). The values of Im may vary from—100% to more than + 100%. The different climatic zones for different Im ranges are as follows:

Moisture index	*Type of climate*
+ 100%	Per humid
+ 100% – +20%	Humid
+ 20% – 0%	Moist sub-humid
0 – -33.3%	Dry sub-humid
-33.3% – 66.6%	Semi-arid
-66.6% – 79 9%	Arid
-80.0% – -100%	Extremely arid

Moisture Index (Im) for all the stations has been computed and isoline have been drawn to know the climatic zones. It has been observed that four types of climatic zones exist in Rajasthan. The southeastern parts of the State, semi-arid type of climate, having a small area of Mt. Abu with humid climate. The extreme northwestern parts of the State falls under extremely aria type of climate while small area of central part has arid climate, ate (Figure 5.31).

These climatic types and their areal extent are of great significance in terms of agricultural operations as will be shown in the sections on variable cropping patterns and regional imbalances in levels of production. Further, the preceding climatic regionalization may be of immense value in future agricultural planning and irrigation developmental programmes. Additionally, it may help the agronomists in devising the dry farming programmes.

To sum up: While rainfall remains the chief factor limiting agricultural and livestock potentials, the picture of rainfall conditions, as presented, is only part of the story-Potentialities for irrigation to open up dry areas or to explore marginal lands or to supplement rainfall during dry weeks or months have been exploited only on a limited scale in the Bagar or Sandy Undulating Plain and Hills. The wide variations in rainfall amount, negligible favourable probability and acute soil moisture deficit make all aspects of water conservation and judicious water-resource planning of vital importance in the agricultural economy of Rajasthan.

Appreciation of the annual, seasonal, monthly, weekly and daily distribution of rainfall and its variability in amount and distribution is important not only for overall assessment of agricultural potentials, but also for improvements in farming practices as much in areas with good rainfall as in areas with marginal falls. Moreover, the estimates of rainfall probability by considering the consumptive use of primary and major commercial crops, average moisture balance patterns and soil moisture surplus or deficit would in fact provide an estimate of long-term crop risk and future irrigation planning. Till recently, mistakes have been made in pioneer developments through inadequate understanding of the climatic conditions.

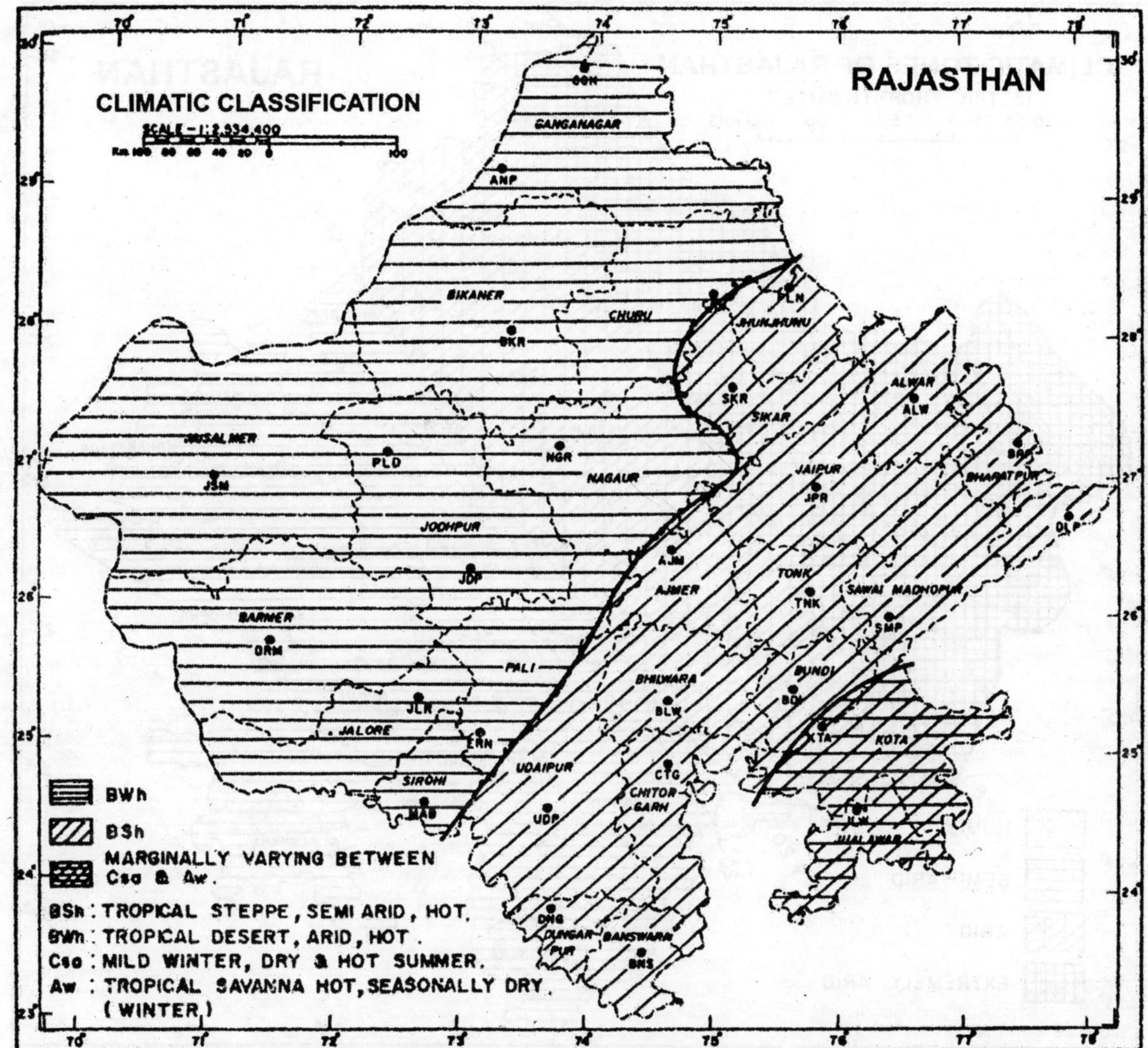

Source : Administrative Map, published by Census Commission of India and data obtained from India Meteorological Department, Poona.

Note : The external boundary and the coastlines of India appearing on the map are neither verified nor authenticated by the Survey of India.

Figure 5.30 Climatic Classification

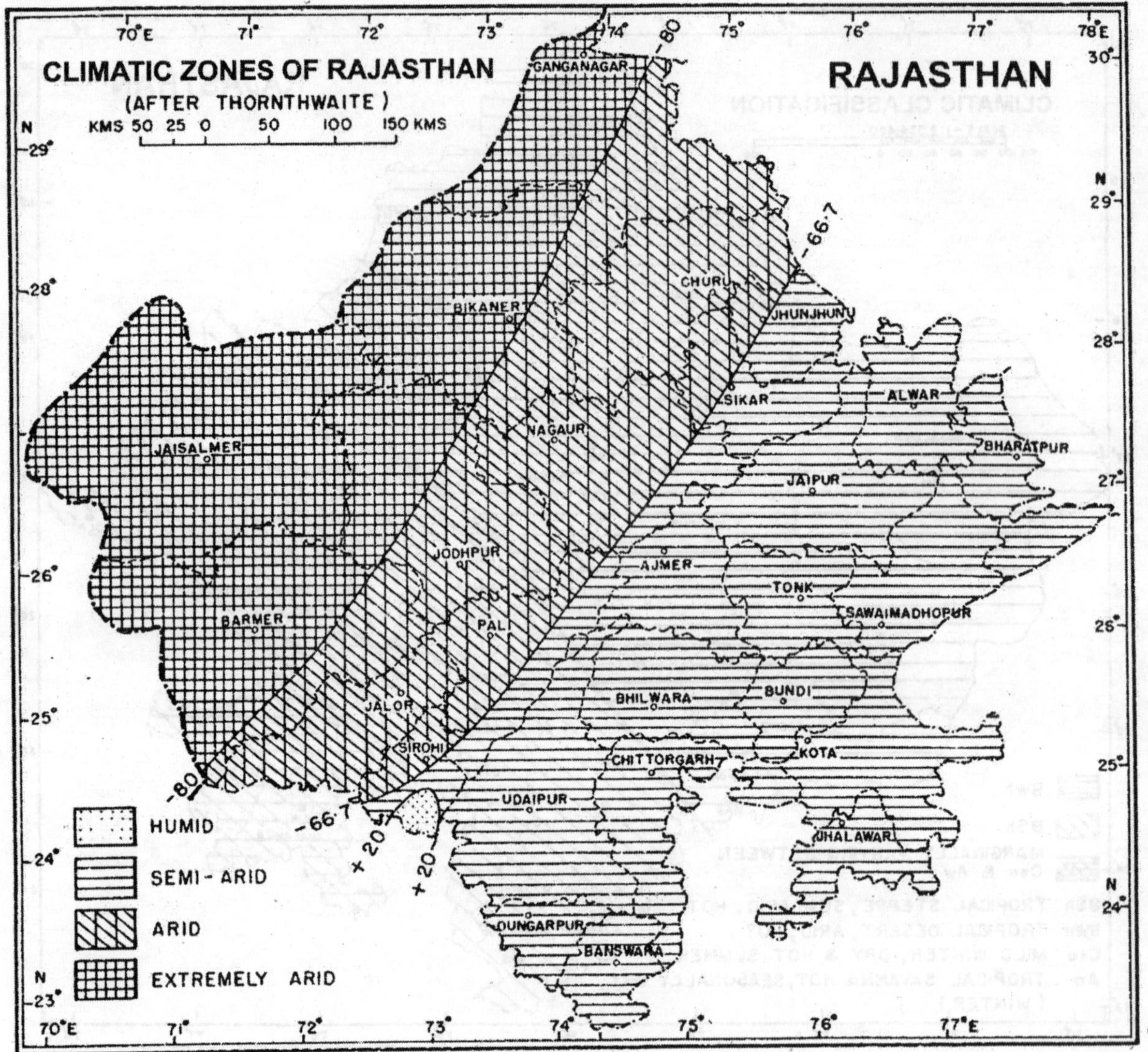

Source : Administrative Map, published by Census Commission of India and data obtained from India Meteorological Department, Poona.

Note : The external boundary and the coastlines of India appearing on the map are neither verified nor authenticated by the Survey of India.

Figure 5.31 Climatic Zones of Rajasthan

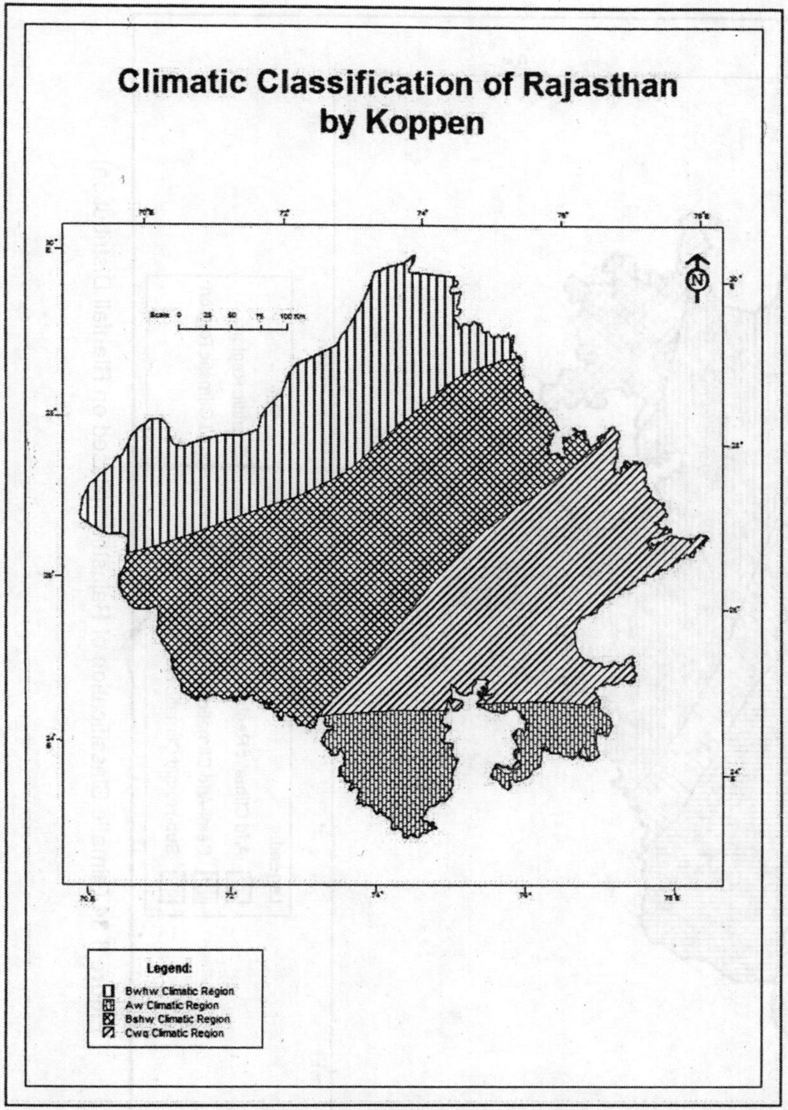

Climatic Classification of Rajasthan by Koppen

Scale 0 25 50 75 100 Km.

Legend:
- Bwhw Climatic Region
- Aw Climatic Region
- Bshw Climatic Region
- Cwg Climatic Region

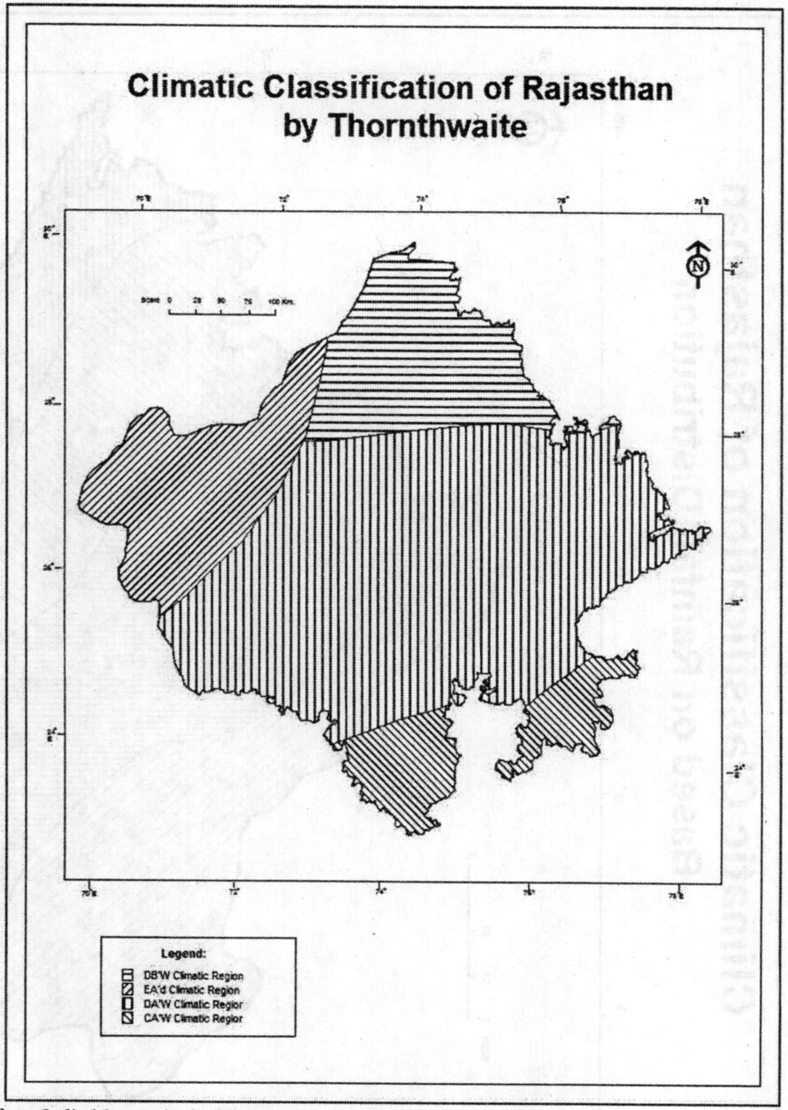

Climatic Classification of Rajasthan by Thornthwaite

Scale 0 25 50 75 100 Km.

Legend:
- DB'W Climatic Region
- EA'd Climatic Region
- DA'W Climatic Region
- CA'W Climatic Region

Source : Administrative Map, Published by Census Commission of India and data obtained from India Meteorological Department, Poona.

Figure 5.32

Figure 5.33

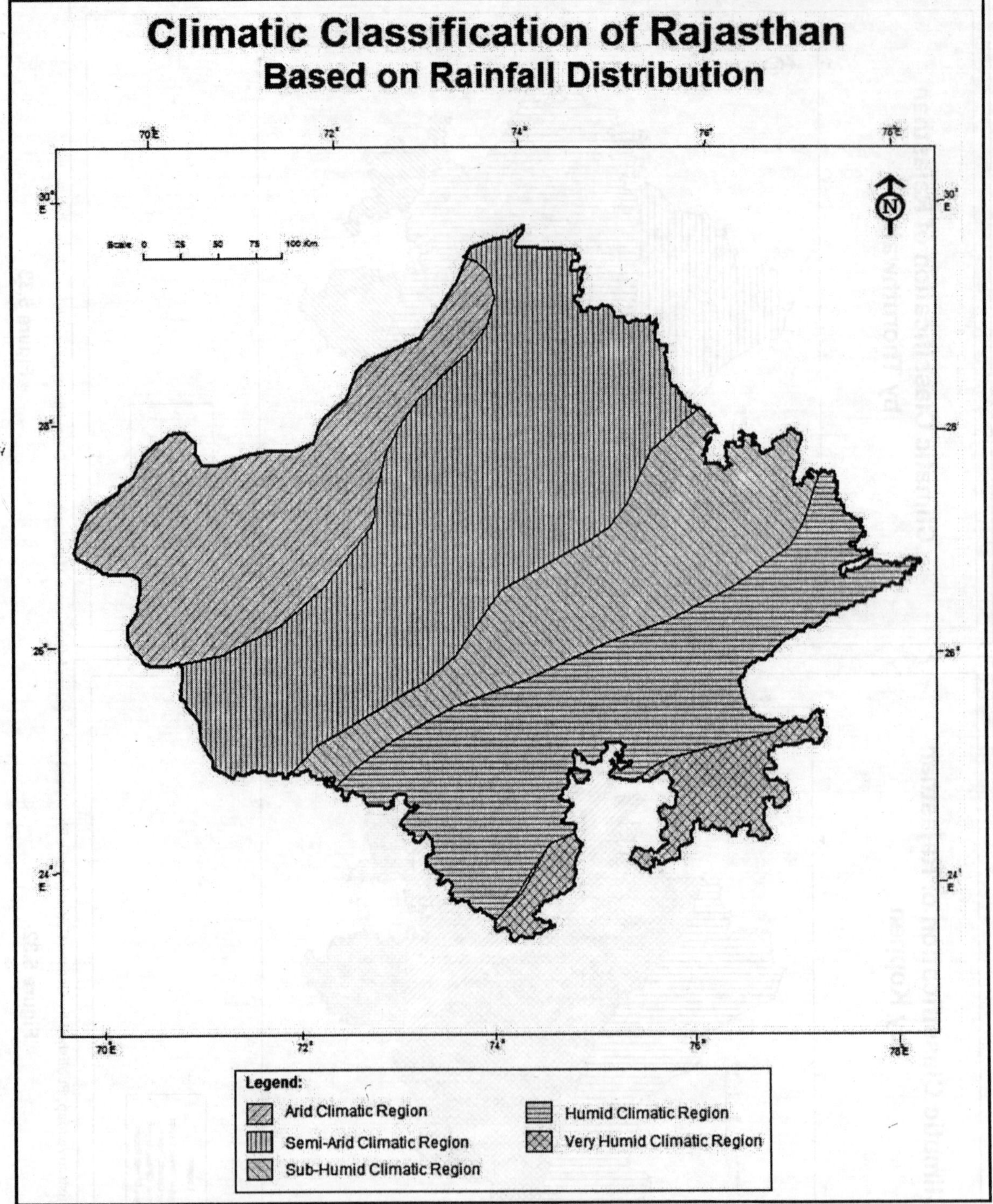

Figure 5.34 Climatic Classification of Rajasthan (Based on Rianfall Distribution)

The distribution of climatic regions of Rajasthan on the basis of rainfall and temperature variations is divided into the following categories (Figure 5.34):

(1). Arid Region: The Arid region includes Jaisalmer district, northern parts of Barmer, western of the Phalodi Tehsil of Jodhpur, western parts of Bikaner and southern parts of Ganganagar district. Climate of the region is very severe and arid. Rainfall less than 10 cm in extreme west parts of regions and rest areas record less than 20 cm rainfall. The average temperature during summer is recorded more than 34°C and during winters it ranges in between 12°C to 16°C.

(2) Semi-arid Region: The average temperature during winter season ranges between 10°C and 17°C and the summer season temperature range 32°C to 36°C. As the region has erratic as well as torrential rainfall it brings floods too each time. Rainfall ranges 20 to 40 cm. This region comprises the western parts of Ganganagar, Hanumangarh, Jodhpur and Barmer districts. The winter season is very short and arid in the northern parts of this region.

(3) Sub-humid Region: In the semi arid humid region, rainfall is meager and the amount of rainfall is limited to a few monsoon months only. The rainfall is between 40 to 60 cm and the average temperature during summer season ranges from 28°C to 34°C whereas it is recorded 12°C in northern parts and 18°C in the southern parts. Alwar, Jaipur, Dausa and Ajmer, eastern parts of Jhunjhunu, Sikar, Pali and Jalore districts, north-western parts of Tonk, Bhilwara and Sirohi districts are included in this category. This region has steppe type of vegetation.

(4) Humid Region: this region receives winter rainfall associated with cyclones along with monsoon season rainfall which varies from 60 to 80 cm. Deciduous trees dominate the region. Humid region is found at the districts of Bharatpur, Dholpur, Sawai Madhopur, Bundi, Kota, Barmer and Rajsamand and the north-eastern parts of Udaipur.

(5) Very Humid Region: Very Humid Region includes south-east Kota, Baran, Jhalawar, Banswara, south-west Udaipur and adjacent areas of Mt. Abu. Here, the summers are very hot and winters are cold and dry. Rainfall received is between 80 cm to 150 cm, which is mostly during the rainy season. Monsoon savanna type of vegetation is present in the region.

CLIMATIC PARAMETERS

Atmospheric Sea Level Pressure and Winds:

There is a very systematic manner in which the seasonal variations of atmospheric pressure take place over the state, with a maximum in the winter (January) and a minimum in the monsoon season (July). Except during the late summer and monsoon season the pressure gradient generally remains weak over the state. During winters the higher pressure is to the north and during summers the pressure decreases from south to north in Rajasthan.

The light winds are mainly from northwest-north-northeast over the state in January which gradually turn anti-clockwise replaced by light north-westerly to westerly or south-westerly winds in April.In July with the advance of the summers, the pressure gradient increases and correspondingly the winds from northwest to southwest also strengthen reaching their maximum strength. The pressure decreases from west-southwest to east-northeast over the state, in July. Having the weakest pressure gradient the month of October is the transition period. The changeover of the pressure and wind pattern commences and north-northeasterly winds appear October onwards.

Cloudiness

It is cloudless or slightly clouded during October to May but the part of the period from January to March is more clouded, especially during afternoons. The sky remains cloudless or lightly clouded over the state during April and May. During July and August, the monsoon skies are heavily clouded. The skies remains overcast for 5 days and clear on 7 days per month in West Rajasthan and for more than 8-9 days and clear on 4-5 days per month in East Rajasthan on an average during July and August. The cloud cover decreases to a great extent over the entire state during October.

Thunderstorms and Dust storms

Dust storms are more frequent over the western than the eastern parts. Their frequency is maximum in the month of June and it is 3 to 4 over the western parts and 1 to 2 over the eastern parts.

For the occurrence of thunderstorms and dust storms convective activity is essential. Thunder activity becomes pronounced due to ground heating with the advance of the summer and when the moisture is insufficient in the atmosphere, dry thunderstorms or dust storms occur. Maximum numbers of thunderstorms occur, with the approach of the monsoon current, while dust storms are mainly confined to the summer months of March-June. Hail is sometimes accompanied with Pre-monsoon and monsoon thunderstorms. The maximum occurrences of thunderstorms are during July in the state and minimal in the months from November to January. The average number of days of during the monsoon season is about 11 in West Rajasthan and 10 in East Rajasthan. The average annual number of thunderstorms in the West and East Rajasthan are 17.0 and 15.0.

Fog

Fog occurs occasionally due to lack of sufficient moisture in the region. The maximum frequency of occurrence of fog is during December and January in both Western and Eastern Rajasthan. The favorable Conditions for occurrence of radiation fog like light to calm winds, clear skies etc. exist after the withdrawal of the monsoon till the month of February. During the monsoon season hill fog occurs at few places, when air is almost saturated and is easily cooled below the dew point while rising over high elevations.

Air Temperature

The pleasant winters in the state witness mean night temperatures above 4.7°C to 10.6°C. During summers the western Rajasthan becomes the hottest place in the country with mean maximum air temperatures varying from 38.6°C to 42.6°C in May. The recorded extreme temperatures in the state were -4.4°C to 50.0°C in the western Rajasthan and -2.8°C to 47.8°C in the eastern Rajasthan (Indian Meteorological Department, 2010).

RELATIVE HUMIDITY

The relative humidity in the arid region of Western Rajasthan is also quite high because of the unfavorable circulations of the atmosphere and low precipitation that occurs in the arid region as compared to other semi-arid and sub-humid regions. Wind regime and associated phenomena.

The relative humidity is minimum in the summer months of March, April and May and maximum during the monsoon months of July, August and September, being lowest in April and highest in August. The moisture content of the air is lowest during the cold season. During the summer season, relative humidity ranges between 35 to 60 percent in the morning hours and 10 to 30 percent in the afternoon hours. In the winter seasons, mainly December to February, relative ranges between 50 to 60 percent in the morning and 25 to 35 percent in the afternoon hours. Ganganagar (68.5 percent) records a slightly higher relative humidity percentage, mainly due to irrigation and appreciably lower dry bulb temperature prevailing during these months of the year. The relative humidity is lowest at Jaisalmer (56.6 percent) and compared with the values of vapour pressure, it will be seen that Jaisalmer is drier than other stations. During the monsoon is less marked increase in relative humidity. This increase is less marked in the west and the northwest of the Aravallis as compared to the areas lying to the east and southeast where rainfall is much higher. After the withdrawal and sometimes the percentage is much lower than that of the winter season.

WIND VELOCITY, SPEED AND DIRECTION

The wind directions are mostly south-westerlies or west-south-westerlies during large part of the year whereas during winter they are northeast to north. During the winter season the Winds are light and variable but in summers especially around May to July strong winds prevail in the most parts of Rajasthan. Summers witness severe dust storm period too when the wind speed rises up to 60 kilometer per hour causing erosion from the dry soils.

The mean monthly and annually wind speed for the year as a whole, winds from south-west or west are most frequent in Rajasthan. Wind velocity is greatest in summer months and various from 10 to 20 km/hour. Wind speed starts increasing from April and reaches its highest in June, after which it gradually declines and comes to its lowest in November. The highest wind velocity is recorded in the month of June at Jaisalmer. (27.2). the minimum wind speed is recorded in the month of November at Udaipur (1.5 km/hour). The mean wind speed is highest at Phalodi station (15.3 km/hour) and minimum at Udaipur station (3.9 km/hour). In the month of April, the wind direction is from west to east. As they come after crossing the heated Thar Desert, they are characteristically dry and warm. During summer, hot and dust raising wind is a common phenomenon. Violent dust storms are most common in the western arid region. Numbers of such dust storms are very few in the eastern part of the state.

Chapter 6

SOIL

Despite the recent great strides in technology and mechanization, agriculture has remained the world's most important primary industry, in which the soil plays pivotal role. About 66 per cent of the global population, comprising of farmers derives its living directly from the soil. There are no natural resources more important than soil resource.

Whatever its production capacities, whether high in some places of irrigated pockets or low in major areas of dryland and desert, due to inherent limitations, the soil resource of Rajasthan, as a medium of growing crops, has furnished directly or indirectly, a significant share in the income of the state. Rajasthan, being geographically the largest state in India, has proportionately a greater soil resource. Therefore, the soil resource in the sate needs to be used extensively and efficiently so that the state finds an appropriate place in the national food, fibre and fodder production and the state economy is sustained without any depletion through erosion, degradation or overuse. The information and knowledge of soils of the state which could be gained through the study of their physical and chemical properties and their geographical distribution pattern, is an essential prerequi-site for their proper utilization, management and conservation. It also helps in proper selection of crops and better land use.

When seen in detail at the village level, the soils of Rajasthan are complex, and highly variable, reflecting a variety of differing parent materials, physiographic land features, range of distribution of rainfall and its effects, etc. However, broadly, the soils can be put in five major groups, based on the basic fabric of soils i.e. soil texture which governs its many other properties. They are, (1) sandy soils or light soils, (2) sandy loam or light medium soils, (3) loam or medium soils, (4) clay loam to clay or heavy soils and (5) skeletal soils or shallow rocky and hilly soils.

The soils are generally evaluated for their production capacity through study of their ability to supply plant growth requirements in terms of water, nutrient and rooting media. The capacity to retain, as also to conduct soil moisture in profile, depends upon the soil texture, soil porosity and soil structure. For example, heavy soils due to their fine texture and porosity, provide effective capillary tubes for moisture movement. In loose sandy soils, due to big particles and pores, with fewer fine capillary pores, there is lesser soil moisture movement. Further, heavy soils are more difficult to work with implements than in light soils. During dry hot periods, however, heavy soils that get compacted, resist wind erosion, but under heavy rainfall, due to their low infiltration rates, they generate high surface run off and thus cause extensive soil erosion through water. On the other hand, the sandy or light soils that are loose and single grained are more prone to wind erosion and are often subject to sand drift, but due to their coarse and open texture, coarse pores and resultant high infiltration capacity, they are not susceptible to water erosion, even during heavy cloud bursts. Medium textured soils have moderate condition between the above extremes. As such, these different soils create different types of habitats for plant growth, and therefore, the crop choice and cropping patterns on such kind of soils greatly vary.

Soils are thus, variable in their soil-water-plant relationship, conservation needs and production potentials. To assess the potential of agricultural development and to plan proper conservative land use, soil survey is the basic requisite. The knowledge of soils gathered through such surveys is not only useful in finding out agriculturally potential areas but also those areas which have remained unnoticed under one or the other soil degradation process for a long period and are now considered as problematic areas. Further, the agronomical and soil moisture conservation technologies, developed for crop production, are often more suitable, particularly for one or the other kinds of soils. Therefore, the distribution of such differing soils depicted through soil survey maps, is useful for easy and proper transfer of such technologies to suitable areas.

Soil resource is also a malleable environment moulded by the agriculturist in many ways e.g. its production capacity can be improved through improvements of its fertility by use of organic manures and fertilizers. Its sustainability can also be maintained by its conservation against damages through erosion, preventing salinization, alkalinization, compaction, depletion in nutrient level, etc. These all are known soil degradational processes which lessen the current or potential capability of soils to produce crops or biomass. Many technologies are available to check this and conserve soils for sustained productivity.

MAJOR SOIL GROUPS

The physiography-soil association relationship has been presented in Table 6.1 and Figure 6.1 (Joshi, D.C., 1993). The broad fertility characteristics are presented in Table 6.2.

Table 6.1 Physiography and Soil Association

I.	Sandy arid plain	(a) torripsamments/ calciorthids/cambort hids/ palcorthids
II.	Alluvial plain of mid west	(a) Camborithids/ calciorthids (b) Salorthids
III.	Plain of interior drainage	(a) Torripsamments (b) Salorthids
IV.	Gagger plain	(a) Torrifluvents/ torripsamments (b) Camborthids (c) Torripsamments
V.	Aravalli plain	(a) Haplustalfs (b) Torripsamments
VI	Eastern plain	(a) Haplustalfs (b) Ustifluvents
VII.	Chappen plain	(a) Ustochrepts (b) Chromuster
VIII.	North eastern hilly region	(a) Haplustalfs
IX.	South-eastern south- eastern hadoti plateau	(a) Haplustalfs (b) Chromustert (c) Ustifluvent

Salient characteristics of each of the major soil groups have been reported below.

Table 6.2 Soil Fertility Status

No.	Districts	Dominant soils	Fertility group		
			N	P	K
1	Ajmer, Jalore and Nagaur	Brown soil, gray brown loam and sandy plain soil	L	M	M
2	Banswara, Chittorgarh, Udaipur, Dungarpur	Medium black, red loam brown soil	M	M	M
3	Bharatpur, Alwar	Soils of younger alluvial plain	L	M	M
4	Barmer, Jaisalmer, Bikaner, Churu	Sandy plain and dune soils	VL	M	M

No.	Districts	Dominant soils	Fertility group		
			N	P	K
5	Bundi, Kota, Tonk, Bhilwara	Medium black, brown soil	M	M	M
6	Shri Ganganagar	Indus alluvial soil	VL	M	H
7	Jaipur, Sawai Madhopur	Brown soil	VL	H	M
8	Jhunjhunu, Sikar	Soils of sandy plain	VL	M	M
9	Jodhpur, Pali	Sandy plain, brown light loam, and gray brown loam	L	M	M

Dune and Inter-Dune Soils

Dunes are the spectacular features which can be seen scattered throughout arid region. These occur in continuous chain in two zones. One is in the extreme west in the Barmen Jaisalmer arid Bikaner districts. The another chain of sand dunes runs throughout the eastern part of Bikaner and Churu districts. A discontinuous zone of sand dunes runs from Sanchor, Saila, Phinch, Didwana and Lachmangarh. These three comprises of six types of sand dunes viz. obstacle, parabolic, coalesced parabolic, longitudinal transverse, barchans and shrub coppice (Ghose *et al*.,1977).

The soils associated with the dunes are very deep, line sandy and light yellowish brown to pale brown. The fine sand fraction constitute 88-92%, clay 2.0-5.0% and silt 0.4 to 1.37%. These are generally non calcareous but in Jaisalmer and Barmer districts dunes having 1-2% $CaCO_3$ in powder form or as root cast are also observed. In the associated inter-dunes the soils are similar to the dune except slightly higher amount of silt (2-3%), Clay (5-7%) and $CaCO_3$ (2-5%). The surface of inter-dunes is covered with loose sand and sand hummocks.

The dune soil profile does not show pedogenic manifestation. In the profile of inter-dune soil, accumulation of lime or gypsum concretions can be observed. The surface soils are loose and single grained but in the subsoil very weak aggregation is observed. These are coarse loamy, mixed, hyperthermic Typic Tarripsamments.

The organic matter content of these soils is very low (0.05- 0.10%). The available phosphorus is medium to low. The available potassium is medium to high (Dutta and Joshi, 1989). The available zinc and iron are adequate but the available manganese and copper are below the critical limits (Joshi and Dhir, 1983). The nutrient status is sufficient to meet the requirement of natural vegetation permitted by the low rainfall but upon availability of irrigation water from Indira Gandhi Canal these soils will need besides the macronutrients, also the micro nutrient application for raising the crops.

These soils have low water retention capacity and high water percolation rate, within one meter soil depth. 50-80mm of water is retained out of which 80% is available to the plants. Under extreme dry condition also these soils retain water in the subsoil which can meet the requirement of natural vegetation.

These soils are prone to wind erosion. Use of heavy machinery for soil operations and uncontrolled grazing result in the loosening of surface soils which lead to large scale sand movement. The dunes should be covered with the perennial grasses/shrubs and should not be cultivated. In the inter-dunes, pearl millet, moong and moth should be cultivated with strips of grasses.

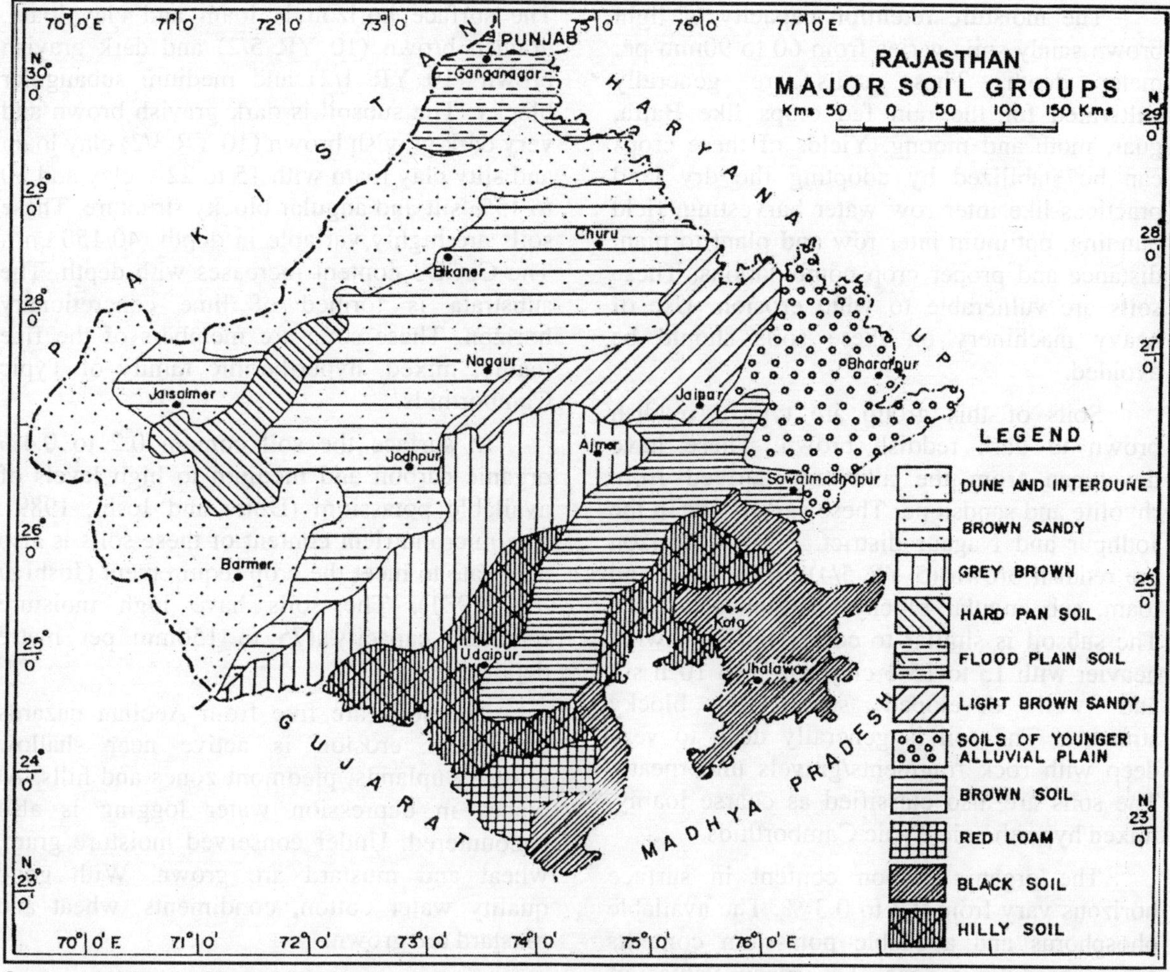

Source : Administrative Map, published by Census Commission of India and Joshi, D.C. (1993)

Note : The external boundary and the coastlines of India appearing on the map are neither verified nor authenticated by the Survey of India.

Figure 6.1 Major Soil Groups in Rajasthan

Brown Sandy Soils

These soils widely occur on sandy plains in Jodhpur, Barmen Nagaur, Bikaner, and Jaisalmer districts. The texture of the soil is fine sand to loamy sand. The surface of these soils is covered with loose sand sheeting with occasional hummocks. The surface soil is brown, yellowish brown and pale brown, single grain to granular and noncalcareous. The subsoil is loamy sand having 4 to 6% clay and 3 to 5% silt and weak subangular blocky structure. There is slight increase in $CaCO_3$ content which is present in powdery form. Below 60 to 90cm lime concretionary horizon is commonly encountered. These soils have been classified as coarse loamy, mixed, hyperthermic Typic Calciorthids. Similar soils occurring in Nagaur, Churu, Jhunjhunu and Sikar districts are non calcareous and have been classified as coarse loamy, mixed, hyperthermic Typic Catnborthids.

The organic matter content of these soils is low (0.1 to 0.2). The available potassium content is medium to high (Dutta and Joshi, 1988). These soils are fairly provided with the available iron, manganese, zinc and copper (Joshi *et al.*, 1982).

The moisture retention capacity of light brown sandy soils varies from 60 to 90mm per meter depth. These soils are generally cultivated for the rain fed crops like Bajra, guar, moth and moong. Yields of these crops can be stabilized by adopting the dry land practices like inter row water harvesting, field bunding, optimum inter row and plant to plant distance and proper crop combinations. These soils are vulnerable to wind erosion. Use of heavy machinery on these soils should be avoided.

Soils of this group are typical reddish brown to dark reddish brown. These have developed from the alluvium derived from rhyolite and sandstone. These sails occur in the Jodhpur and Nagaur district. Surface horizon are reddish brown (5 YR 5/4) sandy loam and loam, sub angular blocky and noncalcareous. The subsoil is similar to colour but somewhat heavier with 13 to 15% clay and 8 to 10°h silt and well developed subangular blocky structure. The soil is generally deep to very deep with rock fragments/gravels underneath. The soils are also classified as coarse loamy, mixed hyperthemic Typic Camborthids.

The organic carbon content in surface horizons vary from 0.2 to 0.3 %. The available phosphorus and available potassium contents arc in medium ranges. The mean values of available iron, manganese, copper and zinc are also fairly within the adequacy limits. The moisture retention capacity varies between 80-120 mm per meter depth. These soils, wherever provided with good quality irrigation water are very productive. High yields of wheat, mustered, cumin and vegetables are obtained on these soils. Under conserved moisture sorghum, pearl millet and pulses are grown in Kharif.

Gray Brown Soils

These grayish brown soils are loam, clay loam and silt clay loam in texture. The soils have developed from the fine textured alluvium and occur in south eastern part of Jodhpur and Nagaur and in Pali, Sirohi and Jalore districts. The surface horizon is loam and clay loam, grayish brown (10 YR 5/2) and dark grayish brown (10 YR 1/2) and medium subangular blocky. The subsoil is dark grayish brown and very dark grayish brown (10 YR 3/2) clay loam and silty clay loam with 15 to 22% clay and 10 to 15% silt and angular blocky structure. These soils are highly variable in depth (40/150 cm). The CaCO3 content increases with depth. The substrata is formed of lime concretionary horizon. These soils are members of the fine loamy, mixed, hyperthermic family of Typic Camborthids.

In surface the soil contain 0.2 to 0.4% organic carbon and medium to high levels of available potassium (Dutta and Joshi, 1989). The micronutrient content of these soils is also adequate to meet the crop requirement (Joshi *et al.*, 1981). The soils have high moisture retention capacity (75 to 150mm per metre depth).

The soils are free from Aeolian hazards but water erosion is active near shallow gravelly uplands, piedmont zones and hills. At places in depression water logging is also encountered. Under conserved moisture gram, wheat and mustard are grown. With good quality water cotton, condiments, wheat and mustard are grown.

Hard Pan Soils

These soils are 30 to 40 cm deep underlain by a strongly lime indurated horizon. Hard pan soils occur in Phalodi-Dhawa area of Jodhpur, Jayal-Katoti of Barmer and in scattered areas in Jaisalmer districts. This horizon is formed of lime concretions and gravels/rock fragments cemented together to form an indurated layer which is slight pervious to water but plant roots are unable to penetrate. Surface soils are brown (10 YR 5/3), light yellowish brown (10 YR 6/2) and yellowish brown (10 YR 5/4), fine sand to loamy sand and noncalcareous. The clay and silt content varies from 6 to 10 and 4 to 8%. These soils are the members of coarse loamy, mixed, hyperthermic family of Typic Paleorthids.

The content of organic carbon (0.1 to 0.2%), available phosphorus and potassium contents arc quit low.

Because of coarse texture and shallow depth the moisture retention capacity is only 40/70mm. These soils are generally unsuitable for rain fed cropping. During normal rainfall year moth and moong can be raise but these soils are better suited for grasses and shallow rooted shrubs. Wind erosion is very active and at places the calcareous substrata is exposed.

Flood Plain Soils

The landscape is the northern part of the State, covering Ganganagar and part of Bikaner districts is formed of flood sediments deposited by the river Ghaggar. The surface soil is brown to gray brown, very fine sand to loamy sand and sandy loam with granular and subangular blocky structure. The content of silt and clay range from 12 to 22 and 11 to 18 per cent respectively. There is wide variation in the morphological characteristics of the soils both within the profile and also between different profiles. These variations are attributed to the varied type of sediments deposited by the river.

The sand particles of the Ghaggar flood plain soils are much finer compared to the coarse textured soils in rest of the arid region. The moisture retention capacity of these soils is very high ranging from 180 to 231 mm per meter. The content of organic carbon is medium (0.21 to 0.45%). Available phosphorus, potassium and micro nutrient contents is better than other arid zone soils. These soils are fertile and because of availability of good quality water produce bumper crop yields. Sand hummocks and low dunes locally called 'Tibbis' are scattered on the surface of these soils. The 'tibbis' are 5 to 10 metre high and 2 to 3 meter long. These are members of loamy, mixed, hyperthermic, Typic uslifluvents.

The Ghaggar flood plain soils are often infested with high concentration of salt. The soil profile has fine textured horizon at some depth due to which drainage is impeded. These soils can be reclaimed with the good quality water available from the Indira Gandhi Canal. Because of good fertility and availability of canal water ibis region is changing in the green belt. High yields of cotton, sugarcane, paddy, wheat, oil seeds and pulses are being realised. With the optimisation of irrigation water and fertilizer application this region has good potentiality.

Light Brown Sandy Soils

These soils are typical light yellowish brown, fine sand, non calcareous and very deep. These extensively occur in Nagaur, Sikar, Churu, and Jhunjhunu districts. The surface is covered with sand hummocks, loose sand and low dunes. The soil profile is sandy, yellowish brown (10 YR S/4) and dark brown (7.5 YR 4/4) and many meters deep. The soils are free of $CaCO_3$. These soils are very pervious to water. Due to high susceptibility to wind erosion thick sand sheeting and hummocks at fence line and in field proper are spectacular and fields appear in bowl shape. The water retention capacity ranges between 60-70 mm per meter depth. These soils have been classified as coarse, loamy, mixed, hyperthermic Typic Torripsamments.

The organic matter content is medium (0.2-0.4%) and soils are also medium with respect to major and micro nutrient convent, but deficiency of Zn and Cu has been reported in some areas. These soils are responsive to fertilizer application and irrigation.

Soils of Younger Alluvial Plain

These are the soils of vast alluvial plain in the districts of Bharatpur, Alwar and parts of Jaipur district. Soils of these regions are very deep, pale brown and sandy loam to loam. In the subsoil the content of clay and silt ranges respectively from 8 to 15 and 6 to 10 per cent. The $CaCO_3$ content varies from 1 to 2.5% in different horizons which is present in powdery form. In lower horizons soft nodules of $CaCO_3$ may also occur. These soils are fairly resistant to wind erosion because of compact surface.

These are the members of loamy, mixed, hyperthermic family of uslifluvents/ napluslafs.

The organic carbon content of the soils varies from 0.3 to 0.4%. Other major and micro nutrients are fairly well provided. The moisture retention capacity ranges from 200 to 215 mm per meter. Sand hummocks and low dunes may be seen scattered on the fence line. With irrigation available from canal and dug well, winter crops like what, mustard, sorghum, pulses and groundnut are grown. The soils are also responsive to fertilizer application. The problem of salinity/ sodicity is wide spread due to water logging in low lying areas.

Brown Soils

These brown, medium to fine textured alluvial soils occur in Bhilwara, Tonk and Ajmer districts and also scattered in the Udaipur, Chittorgarh and Jaipur districts. These are strong brown (7.5 YR S/8), brown (10 YR S/3) and yellowish brown (10 YR S/4), sandy loam to clay loam at the surface and non-calcareous. The subsoil has clay loam and loam texture and sub angular and angular blocky structure. These soils are slightly calcareous and substrata consist of loosely packed weathered rock fragments. These soils have argillic horizon and exhibit moderate profile development. These are the members of loamy, mixed, Typic Hapluslaf.

The organic carbon content varies from 0.5 to 0.8%. The moisture retention capacity varies from 110 to 250 mm per meter. At places these soils are associated with rocky undulating uplands. In such situations water erosion is very active. Natural salinity in soil is the major problem. In canal command areas and saline water irrigation secondary salinisation dominates. Otherwise these soils are very productive both in the irrigated and rain fed areas.

Red Loam Soils

These soils generally occur in narrow valley fills or in the flat intermountain areas. These have developed from the sediments deposited from the adjoining hills. Such soils are widespread in southern part of Rajasthan covering Dungarpur, Udaipur and Chittorgarh districts. These soils are string brown (7.S YR S/6), yellowish brown (10 YR S/6) and pale brown (10 YR 6/3), sandy loam and loam devoid of $CaCO_3$. Such soil is slightly darker with somewhat more of clay and silt content. These are also member of the loamy mixed, family of Typic Ustalft (Joshi and Seth, 1972).

The organic carbon content in the surface soil is medium (0.4 to 0.6%). Deficiency particularly of Zn has been reported. The moisture retention capacity varies from 75 to 80 mm per meter. During rains water is pounded in low-lying areas where paddy is cultivated. on upper slopes soils are shallow and suffer from moderate to severe water erosion. Contour bunding, field bunding and terracing is required to reduce soil loss due to erosion. Perennial shrubs and trees also help in reducing soil losses. In the depression areas, the soils are very productive and cash crops like cotton sugarcane, wheat and grain are raised with very low irrigation requirement.

Black Soils

These dark grayish brown and very dark grayish brown, very deep soils occur in south and south eastern part of Rajasthan State. In the Kota, Bundi and Sawai Madhopur districts the black soils have developed from the Vindhyan rock and Deccan trap sediments deposited by the river Chambal and its tributaries, the Kali Sindh and the Parvati. In Banswara, Jhalawar and Pratapgarh - Dhariyavad area of Udaipur-Chittorgarh districts the black soils have developed from the Deccan trap. In this region the black and red soils occur at juxtaposition.

The black soils are clay loam, silty clay loam having angular blocky structure. The different horizons of the profile do not show variation in the silt and clay content 20-25% and 35-40% respectively. The substrata have lime concretionary. Sharma and Kant (1976) have reported slight translocation of finer particles and their subsequent concentration in alluvial horizons.

The soils have characteristic property of swelling when wetting and shrinking in dry condition. These soils develop cracks of 50-100 cm depth after drying. These soils have been classified as Typic Chromusterts and Typic Pellusterts.

The organic carbon content of the soil ranges between 0.4 to 0.6%. Deficiency of copper is common in these soils. In the black soils of Jhalawar deficiency of Zn has been reported. These soils have very high moisture retention capacity (145-200 mm/meter). Because of very low water percolation in these soils the rainwater remains ponded and water logging is common. In absence of good drainage system these soils develop problem of salinity. There are two big dams for irrigation on the Chambal and the Mahi rivers which supply water for irrigation. Higher yields of crops like cotton, sugarcane, sorghum, wheat, mustard, pulses and oil seeds can be obtained by adopting efficient water management systems.

Along the banks of the Chambal and its tributaries, very deep ravines have been formed. These are further extending towards the cultivated fields. Proper treatment in the catchment area, by adopting suitable mechanical means, soil conservation measures and growing perennials shrubs and grasses, is required.

Soils of Highly Region

The southern part of the State in Udaipur, Dungarpur, Banswara and Sirohi districts is occupied by the Aravalli hill range, similarly the south eastern part of Rajasthan covering Sawai Madhopur and Bundi districts is also hilly. The landscape of this region is very uneven, sloppy, covered with stones and rocks. The soils, which occur in pockets, are very shallow gravelly, coarse sandy, reddish brown to grayish brown and non calcareous. These suffer due to severe water erosion. Cultivation sloppy and undulating lands are not possible and good grasslands can be developed. In the depression valley areas where sediments are deposited soils are deep to very deep and fine textured. These are highly variable in texture (sandy loam to silty clay loam) and colour (reddish brown to dark grayish brown). Mehta *et al.* (1962) have observed that these soils have moderate illuviation and have monogenic character of the profile. These are very productive and with well irrigation good yields of paddy, cotton, sugarcane, wheat, pulses and oil seeds are harvested. Hilly soils are the member of Typic ustochreps.

The organic carbon content of soils in depressions varies from 0.8 to 1.5%. These soils also contain good amount of available phosphorus, potassium and micronutrients (Joshi *et al.*, 1973; Lodha and Baser, 1971). The shallow soils have very low moisture retention capacity but in deep soils 150-200 mm water can be retained per meter depth.

Salt Affected Soils

Salt affected soils are of common occurrence throughout the State. In Rajasthan 1 million hectare area is affected with salinity. Causes of high salinity in soil may be as follows:

Natural Salt Affected Soils

Because of arid climate in western Rajasthan, the salinity in sail and groundwater is quite high. Saline water lakes occur near Sambhar, Pachpadra and Didwana. In Pali, Jalore and south eastern part of Jodhpur and Nagaur districts vast stretches of natural salt affected soils occur. Due to weathering of rocks salts are released the catchments of ephemeral streams which is carried through water and salinity developed along the banks and in depressions where these streams are lost. Thus large area turns saline.

The profile of the salt affected soils is charged with very high concentration of salts. The pH and Ec values (1:2) respectively ranges from 8.2 to 8.7 and 20 to 82 dsm-1. The salinity decrease with depth. These soils are sandy loam to loam and clay loam, dark grayish brown (10 YR 4/2) to very dark grayish brown (10 YR 3/2) with angular blocky structure. The solum is very hard and compact and do not allow rainwater to percolate. These soils are devoid of vegetation except the salt tolerant bushes like Prosopis juliflora. The soils are moderately deep to deep (30-50 cm) underlain by a compact lime concretionary horizon. Inspite of high salinity these soils are well provided with the available iron, manganese, zinc and copper (Joshi, 1988a, 88b). In the flood plain of Ghaggar in Bikaner and Ganganagar district large natural saline areas occur. These are highly saline throughout the soil profile. Wherever, the canal water is available, these areas are being reclaimed and brought under cultivation. The problem of natural salinity is also common in parts of Bhilwara, Ajmer, Chittorgarh and Bharatpur districts. In low-lying areas salinity is commonly encountered. In the Shahpura and Banera tehsils of Bhilwara districts the problem of slick spots is common.

Anthropogenic Salinity

Irrigation with saline/high residual sodium carbonate water, construction of dams and in the command area of canal is resulting in the salinization of soils. The problem of brackish ground water is common in western part of the State but it is also encountered in Bhilwara, Ajmer, Bharatpur, Alwar, Jaipur and Chittorgarh districts. Due to increasing build up of salinity in soil the crop yields decrease. Farmers are left with the choice of salt tolerant crops and their varieties. The use of water containing high residual sodium carbonate has resulted in severe degradation of soil productivity (Figure 6.1). Due to development of high sodicity (Table 6.1) the soils become very hard and compact and nutrient availability has also been reduced (Joshi, 1988, 90). In such areas gypsum application is essential to rehabilitate the land productivity (Joshi and Dhir, 1988, 1992). The construction of irrigation dams results in the increased salinsation in command area of the Chambal and the Indira Gandhi Nahar Projects.. Similar problem has also developed with other irrigation dams in the semi arid region in Ajmer, Bharatpur, Alwar, Bhilwara and Pali districts. These salt affected soils can be managed by growing salt tolerant shrubs and grasses.

Soil is the natural body made up of minerals and organic constituents, differentiated in horizons of variable depth which differ from the material below in morphology, physical make up, chemical properties, composition and biological characteristics. Soil forms thin layer on the earth's surface and spatially differs in characteristics and properties. Virtually all human food is produces either directly from crops grown on soils or from animals which graze upon plants growing on the soils. Soils support plants, supply moisture and nutrients. Soils are not only important food producers but also integral part of the eco-system. As a constituent of environment, soils contribute in hydrological cycle; perform functions, like filtration, adsorption and storage of nutrients and moisture and buffering of certain chemical reaction, as well as cycling of carbon, nitrogen and sulphur.

Levelled land with deep, fertile and freely draining soil, is a highly valuable natural resource which is in great demand for production of crops. But with the increasing human and animal population and urbanization such good soils are going out of cultivation. Simultaneously, cultivation on marginal lands viz. upland with shallow soils, land with steep slope, sand dune flanks and soils with hard pan and over grazing on common land, have resulted in accelerating wind/water erosion, salinisation, destruction of soil structure and compaction. The manner in which the soils are used or misused is the subject of human decision. Many decisions about land use are taken without consideration of soil characteristics. This result in deterioration of soil productivity and land degradation. In Rajasthan alone, land have degraded due to wind erosion in 98.11 lakh ha, water erosion in 59.03 lakh ha, salinisation in 11.83 lakh ha, waterlogging in 1.46 lakh ha and ravines in 2.90 lakh ha (MOA, 1994). Therefore, soil resource should be used according to its capability. The constraints and potentials associated with soils should be ascertained through assessment of soil characteristics and mapping their geographical extent to achieve sustainable management.

The state is fortunate to have large variety of soils ranging from vast area of sandy soils covered with sand dunes, to medium and fine textured soils of alluvial plain. These have distinguishing characteristics important for biological productivity. Inspite of large variations, soil remain an over looked and under valued natural resource in the State planning and development programms. Therefore, efforts have been made to present the soil information in simpler form to highlight potentials and constraints of this resource for sustainable management and developmental activities. The State level scenario on physiography soil series relationship, Soil Taxonomy, land capability and, land irrigability have been discussed. Districtwise detailed soil series, land capability and land irrigability information along with thematic maps of soil series and land capability have been presented.

Soils are the products of natural surface processes. Five factors of soil formation climate, vegetation, relief, parent material and time are well known. When we consider factors affecting distribution of soils at local level, role of relief/ physiography dominates. Physiography is the relative configuration of landscape. The close relationship between soil and physiography/land form has been widely recognized. Factors responsible for evolution of landforms are causative factor of soil formation. The governing principle has been that the soils are the products of the same natural processes and conditions that sculpture the landscape they exist in.

The soils of the Rajasthan State have large variability in their characteristics including soil texture (fine sand to clayey), colour (Pale brown to very dark, grayish brown), calcareousness, salinity, depth, diagnostic horizons and substrata. These soils distinguish themselves from rest of the zonal soils in sandy texture, high salinity, segregation of lime and gypsum. The factors which have significantly contributed to soil variability are nature of parent material, history of landscape evolution, climate and physiography. Arid climate is responsible for weak chemical weathering and pedogenic processes. Once the landscape had stabilized the physiography has played significant role in development of soil characteristics.

Rajasthan stretches into two of India's major physiographic divisions viz. the great plain and the central highlands. The area lying west of Aravalli is the western sandy plain while the area east of Aravalli falls in the central high land. Both of these major divisions are marked by a variety of physiography and relief features and accordingly have been sub classified in nine physiographic zones viz.: 1. Arid north western sandy plain, 2. Alluvial plain of Luni basin, 3. Transitional plain of inland drainage, 4. Irrigated north western plain,
5. Semi arid eastern plain, 6. Flood prone eastern plain, 7. Sub humid southern plain and Aravalli hills, 8. Humid southern plain, 9. Humid south eastern plain.

Soil Irrigability Classification

Irrigation is applied to sustain crops and obtain higher yields. Water is a scarce commodity therefore specific management practices should be adopted for its efficient utilization. Information are needed for areas suitable for irrigation, crops that can be grown, expected yields, land development needs, areas with drainage problems and special reclamation measures for optimum utilization and also for commissioning irrigation projects. These information can be obtained through land irrigability classification of the command area.

Irrigation water is retained in soil pores and subsequently utilized by plants. Porosity of soil depends on soil texture and structure. Soil texture refers to the relative proportion of sand, silt and clay particles. The relative arrangement of soil particles/ aggregates is called structure. The structural pattern is controlled by clay and organic matter content. Porosity of sandy soils usually ranges between 35 to 50 per cent and that of clay soils between 40 to 60 per cent. Two classes of porosity are recognized viz. (i) capillary, and (ii) non-capillary. The capillary pores contain water which remains after free drainage. Non-capillary pores do not hold water after free drainge but these are filled with air. Large non-capillary porosity of sandy soil result in better drainage and aeration but in lower water holding capacity. An ideal soil has its pore spaces equally divided between capillary and non- capillary porosity. Therefore, while judging the soil for irrigation purpose, apart from other characteristic, soil texture and depth are given consideration.

Soil irrigability classes are groupmgs of soils according to their suitability for irrigation . The classes are defined in terms of degrees of soil limitations. Following five soil irrigability classes have been recognized for sustained use under irrigation.

Class A : None to slight soil limitations.

Class B : Moderate soil limitations.

Class C : Severe soil limitations

Class D : Very severe soil limitations.

Class E : Not suitable for irrigation.

Soil irrigability classes are established without regards to availability of irrigation water, water quality, land preparation cost and availability of drainage outlet. The most limiting property determines its classification.

Land irrigability

Land irrigability is the combination of soil irrigaility class with associated land features like topography, slope, surface restrictions for irrigation, suface drainage, depth to water table and climate. If any one of the parameter is limiting the lower irrigability class is assigned.

Following land irrigability classes are recognized.

Class 1 : Lands that have few limitations for sustained use under irrigation.

Class 2 : Lands that have moderate limitations for sustained use under irrigation like very gentle slope, un favorable topography, less than ideal depth, texture and permeability.

Class 3 : Lands that have severe limitations for sustained use under irrigation like gentle slope, unfavourable topography and/ or drainage, unfavourab le soil depth, texture, permeability, moderately severe salinity-alkalinity.

Class 4 : Lands that are marginal for sustained use under irrigation, have very severe limitations like very unfavourable topography or drainage conditions, moderately steep slopes, very unfavourable soil depth, texture, permeability, severe salinity- alkalinity,.

Class 5 : Lands that are temporarily classed as not suitable for sustained use under irrigation pending investigations .

Class 6 : Lands not suitable for sustained use under irrigation.

The soil series according to their suitability for irrigation have been grouped into six classes of progressively increasing difficulty to bring under irrigation. Land irrigability Class 1 can be relatively easily and inexpensively developed with good returns. Class 2 and 3 lands need increasing development and production cost while Class 4 land are marginal with high cost of development. Class 5 land are presently non-irrigable but might be proven otherwise by special studies. Class 6 land are non irrigable.

For classifying various soil series according to their suitability for irrigation they have been classified in six classes of progressively increasing difficulty to bring under irrigation. There are two steps (1) classification for soil irrigability, and (2) Classification for land irrigability. Class 1 land has no restrictions, but other classes (2-6) are classified in sub classes based on restrictions. Four sub classes are recognized indicating dominant limiting factors viz. topography (t), drainage (d), soil problem (s) and climate (c). The land irriga,bility classification given at the time of survey may change with any change in the irrigability parmeters made later on.

Class 1. Very good irrigable land

Soil series classified in land irrigability class 1 are presented below. These lands are highly suitable for irrigation farming, being capable of producing sustained and relatively high yields of wide range of climatically adaptd crops. The lands are level with gentle slope, soils are medium to fine texture, deep and moderately permeable, free from accumulation of soluble salt/ sodium, with good moisture holding capacity and well drained. Very good cultivable land include six soil series covering 4001.85 sq km (1.17 %) area.

Soil series : Anupgarh, Bhusawar, Bijaipur, Ladpura, Sarkholi, Sriganganagar

Class 2. Good irrigable land

Soil series classified in land irrigability class 2 are presented below. Land in the class 2 are moderately suitable for irrigation farming, require some expenses to bring under irrigation. The restrictions may be due to si;irface unevenness, slight salinity, soil texture and surface drainage. The restrictions may be one or more than one. Under good irrigable land class there are 118 soil series covering 73879 sq. km(21.58%) area of the State.

Soil series: Akhiyagarh, Aluda, Anadra, Anta, Arnaud, Bagoliya, Balapur, Baloda, Bamor, Banswara, Baran, Barwas, Bassi, Betia, Bhadrajun, Bhainsara, Bhansrorgarh, Bhatwara, Bhiladi, Bhinmal, Biraunda, Borawas, Borunda, Borwara, Chambal, Chandawal, Chaumahala, Chechat, Chhapiya, Dantrai, Datwas, Deodaspura, Deogarh, Devikot, Devli, Gajsinghpura, Giriyawas, Gopalpura, Gundawada, Hantra, Hindon, Hinglot, Islampurghata, Jaitaran, Jawaja, Jayal, Jogiwara, Jugaridurg, Kaman, Kapasan, Karanpur, Khajewala, Kharkhawa, Kharnnor, Koela, Kojra, Kota, Kotputali, Kukra, Kumela, Kundal, Kushalgarh, Lawan, Lipoli, Mahwa, Maloni, Malpura, Mandal, Mandar, Mandha, Mangrol, Marwa, Masalpur, Masitawali, Mau, Mewara, Mittarpura, Mokalsar, Nadol, Nagdi, Narwar, Nayagaon, Ninore, Niwari, Pal, Palri, Palsana, Parbatsar, Pipar, Rajakheda, Ramdeora, Rameara, Ranipura, Rawatsar, Relmagra, Roopsi, Saledipura, Sangod, Sangwas, Sankra, Saradhana, Sarol, Satu, Sayra, Sikrai, Silgaon, Simalwara, Sobhasar, Sojat, Soni, Surwal, Talwara, Taswaria, Tilwara, Ulela, Umar Baori, Virampura, Wasundra.

Class 3. Moderately good irrigable land

Soil series classified in land irrigability class 3 are listed below. Land is moderately suitable for irrigation because of restrictions due to topography, drainage, salinity, soil depth, and texture. Most of these restrictions are manageable but require more cost for development of irrigation than class 2 land. There may be one or more restrictions. Moderately good cultivable land include 120 soil series covering 103717 sq. km (30.30%) area.

Soil series: Adsar, Ajwa, Arnbapani, Arnet, Antralia, Asop, Asotra, Asotra sodic, Awan, Balia, Balotan, Bamanwas, Bant, Bap, Barwala, Bhaleri, Bhamatsar, Bhanpura, Bhanwarpura, Bhavi, Bhimpura, Bhimra, Bishanpura, Bochifala, Borwat, Chhinch, Chipparwada, Chirai, Chitalwana, Chohtan, Chomu, Danda, Dangabas, Danpur, Dhaber, Damari, Dhanoti, Dhingasari, Didwari, Dokua, Gamari Deval, Gapapura, Ghatmika, Gomati, Gundoj, Harjela, Harmara, Hultana, Jadan, Jairampura, Jasoti, Jethana, Jhalra, Jhankol, Jhawar, Jharwasa, Julmi, Kalyanpur, Kalyanpur sodic, Kamalia khedi, Kasti, Kesarpura, Khajwan, Khamanpura, Khiran, Kirwa, Kotri, Kuchaur, Kushalpura, Lalsot,

Lanela, Leta, Lunkaransar, Mahlan, Malkosani, Manda, Mangaliyawas, Manpur, Manpura, Mathania, Mehrana, Modasar, Moharnrnadpura, Mundiyar, Nadbai, Naldhara, Nandta, Nangal, Nimoda, Pali, Panchroli, Panditwali, Parbatsar sodic, Poharka, Pushkar, Raiki, Ramgarh, Rangmahal, Ranwa, Ratwai, Riched, Ruppu"ra, Rupra, Sahwa, Sakhi, Sam, Samota, Sanchor, Sareri, Sarkandi khurd, Sarwan, Seruna, Siyawa, Soyla, Sudasari, Sultanpur, Suratgarh, Taranagar, Thar, Vijaypura.

Class 4. Marginal irrigable lands

Lands with very severe limitations in soils, topography and/or drainage that may further reqmre special management / higher development for irrigation. Soil series classified in land irrigability class 4 are given below. There are 41 soil series covenng 55694 sq. km (16.27%) area which are marginally suitable for irrigation.

Soil series: Akodiya, Alniyawas, Anandgarh, Banedia, Bankoda, Baytu, Bhukia, Bhuttewala, Chandali, Chhau, Dadiya, Dhamand, Faloj, Fatehpur, Gajner, Ghatol, Jagakheri, Jajiwal, Jhamar, Jharoli, Kamli, Kanwat, Kesarbagh, Kolu, Kothri, Malkisar, Mokalsar sodic, Molasar, Nachna, Naglamai, Navalgarh, Palanheda, Panchdariya, Pichiyak, Raja ka karera, Rajela, Rozana, Runeja, Samdari, Sanchli, Siswali.

Class 6. Economically non-irrigable lands

Soil series under land irrigability Class 6 are economically not suitable for irrigation. These have topographic constraints like sand dunes, hills; shallow depth constraints like very shallow, rough and broken -lands, ravines, mining and playa. These series may have more than one constraints. There are 59 soil series in this categories covering 104946 sq. Jan (30.65%) area.

Land Inrrigability classification based on soil, land and climatic parameters, suggests guide lines for adopting practices and frequency of irrigation, and quantity of water to be applied for raising the crops. It also suggested precautionary measures to provide drainage etc. so that the irrigation can be sustained without causmg deterioration in soil quality.

SOIL FERTILITY

A perusal of the Table 6.3 indicating that 33 district in Rajasthan, the soil in 25 districts are deficent in nitrogen while phosphorus is low only in two distric namely Bundi and Jaisalmer. The potash ranges from medium to high. Need is thus apparent for further intensifying the use of nitrogenous fertilizers which is also exident from a large number of fertilizer experiments in the State. Applica-tion of phosphoric and potassic fertilizer is also required in are where P and K status is low or medium (Figure 6.2, 6.3 and 6.4).

Table 6.3 Distirctwise Soil Characteristics

Sr. No.	Name of the district	Fertility Status							Area of Problematic Soil [Hector]		Annual Normal Rainfall in mm
		Major Nutrients (LMH)			Micro Nutrient (deficient in per cent)						
		OC	P	K	Zn	Fe	Cu	Mn	Saline	Sodic / Alkali	
1	Ajmer	L	M	M	36.54	26.14	0.15	1.11	16712	19830	462.2
2	Alwar	L	M	M	54.67	64.08	1.19	2.97	15976	97625	630.9
3	Banswara	M	M	M	53.52	31.08	19.17	10.51	2131	2130	883
4	Baran	M	M	M	27.23	27.11	3.76	9.88	1008	1584	852.7
5	Barmer	L	M	H	55.58	55.97	4.69	4.03	-	1989	268.6
6	Bhartpur	L	M	M	57.49	38.92	4.58	8.16	32613	45217	621.1
7	Bhilwara	M	M	H	28.16	46.37	2.47	6.17	27950	13470	630.2
8	Bikaner	L	M	M	58.47	56.99	21.4	28.9	14134	14033	274
9	Bundi	M	M	H	2.89	19.39	1.02	7.36	6009	9229	705.6
10	Chittorgarh	M	M	M	25.45	53.15	2.42	14.65	17720	11397	765.3
11	Churu	L	L	M	48.64	63.47	14.06	25.01	-	250	369.6
12	Dausa	L	M	M	7.2	45.93	11.9	9.78	4056	38437	674.8
13	Dholpur	L	M	M	47.84	38.33	2.71	4.15	5373	20121	722.1
14	Dungarpur	M	M	M	31.21	38.89	2.59	17.98	2819	3928	677.8
15	Ganganagar	L	M	H	53.28	52.44	6.13	10.34	14000	5100	252.9
16	Hanumangarh	VL	M	H	49.07	40.09	3.63	10.34	14214	6517	301.6
17	Jaipur	L	M	M	20.51	35.21	6.96	8.18	74224	117474	577.2
18	Jaiselmer	L	M	M	9.04	98.6	0.84	3.53	-	460	181.2
19	Jalore	L	M	M	42.36	39.99	11.78	3.63	12500	23491	427.1
20	Jhalawar	L	M	H	35.12	30.84	0.01	0.16	7409	540	920.2
21	Jhunjhunu	L	M	M	61.89	59.38	5.55	13.52	1596	27612	481
22	Jodhpur	L	M	M	62.65	65.88	2.57	2.82	2902	9527	308.1
23	Karauli	L	M	M	46.52	40.03	13.66	2.97	7002	7200	696.2
24	Kota	M	M	H	27.36	29.38	0.24	0.46	2947	6223	807.9
25	Nagaur	L	M	M	52.14	64.51	1.36	4.84	3315	10658	394.1
26	Pali	L	M	M	58.87	41.36	1.54	4.86	26374	57962	485.7
27	Pratapgarh	L	M	M	17.48	45.44	0.22	2.26	-	-	915.3
28	Rajsamand	L	M	H	36.94	51.18	25.42	27.92	9922	6251	554.4
29	S. Madhopur	L	M	H	11.44	40.94	5.25	7.65	12530	20027	721
30	Sikar	L	M	M	43.83	50.39	1.19	3.9	59936	30036	463.8
31	Sirohi	L	M	M	54.44	45.44	9.05	13.15	5310	9035	913.4
32	Tonk	L	M	M	40.65	48.3	1.9	0.76	18014	16593	613.9
33	Udaipur	L	M	M	16.42	23.53	4.17	5.32	5636	9493	639.7

Source: Rajasthan Agricultural Statistics at a Glance Year 2016-17, Commissionerate of Agriculture, Rajasthan, Jaipur

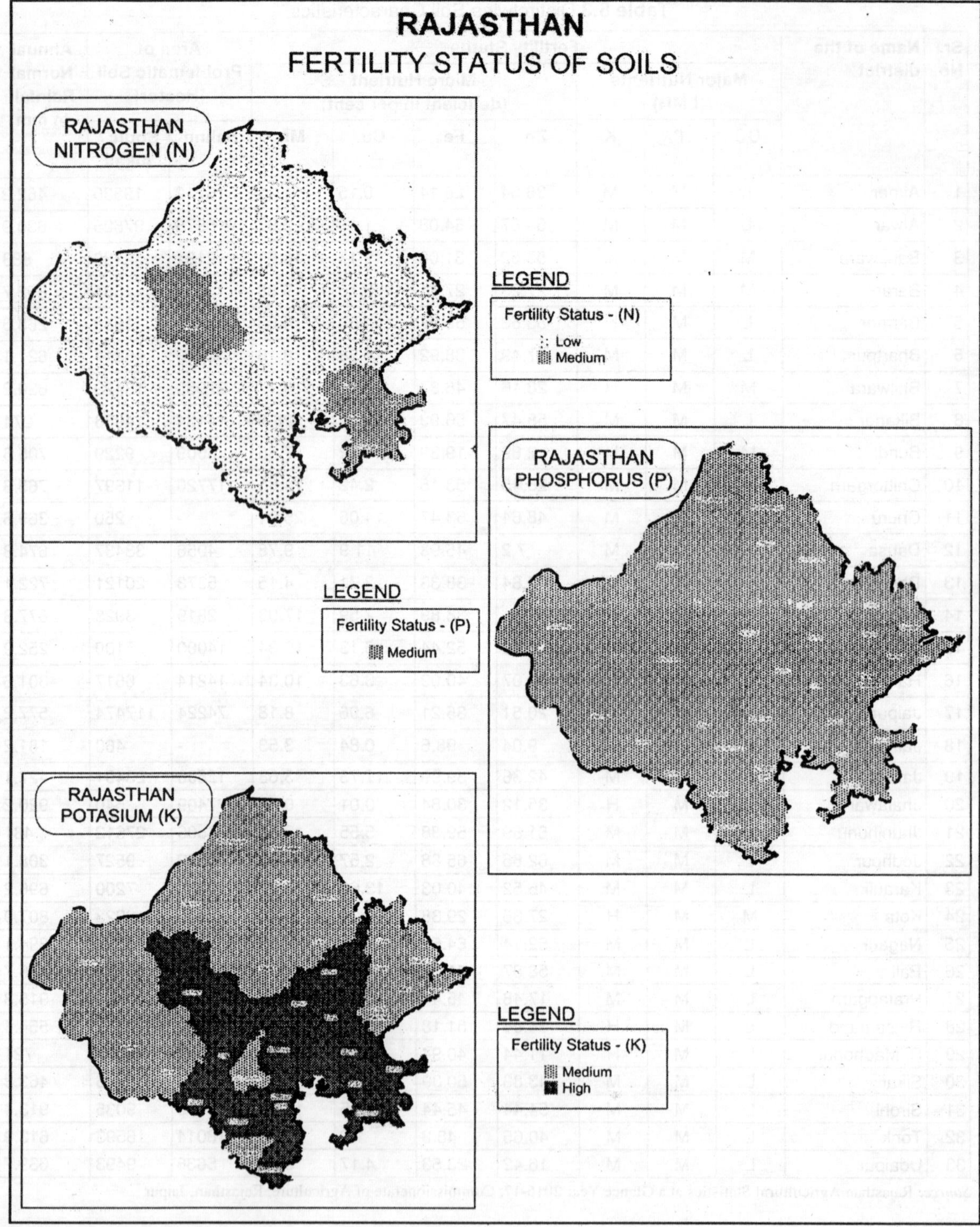

Source : Vital Agriculture Statistics

Figure 6.2 Fertility Status of Soils

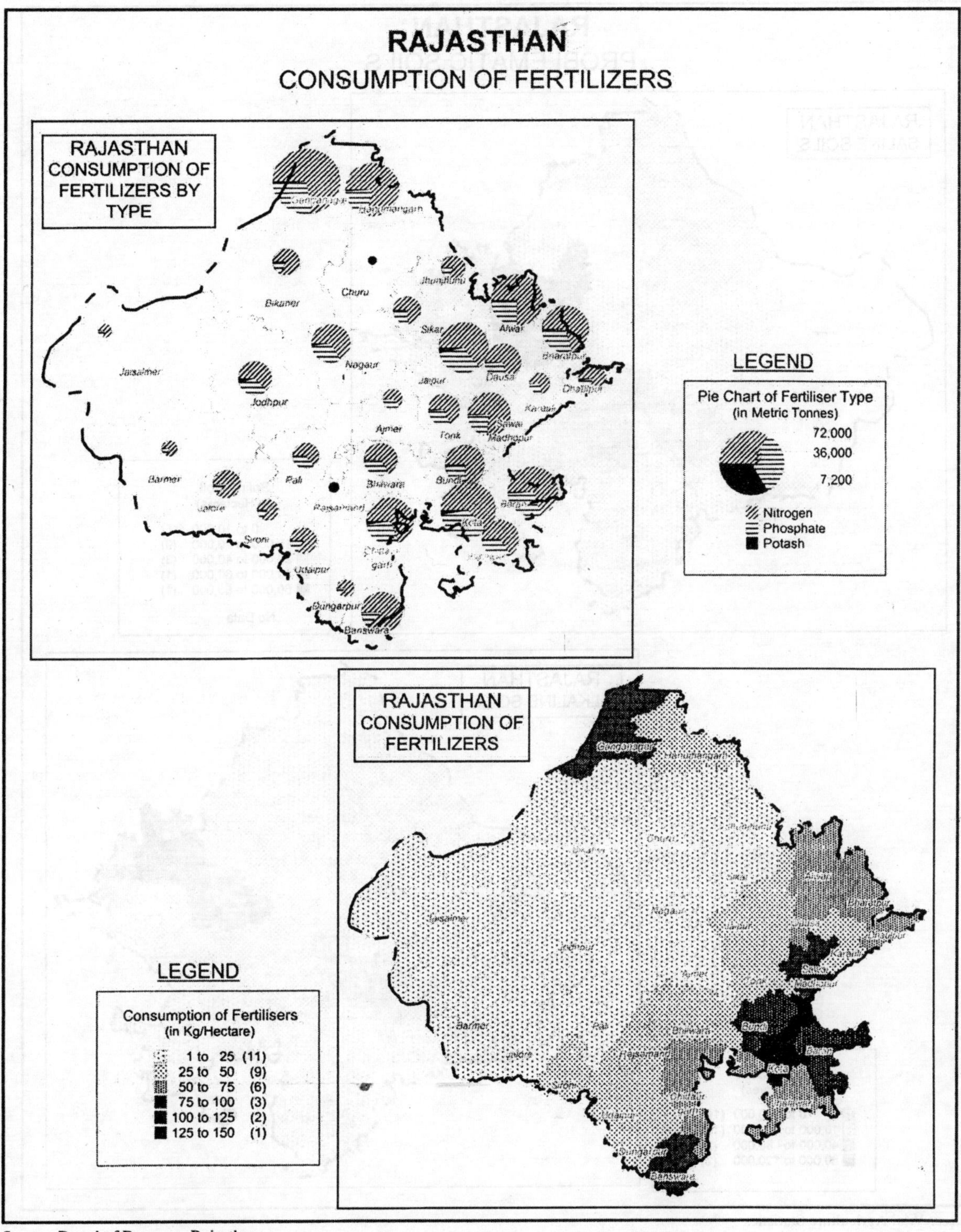

Source: Board of Revenue, Rajasthan

Figure 6.3 Consumption of Fertilizers

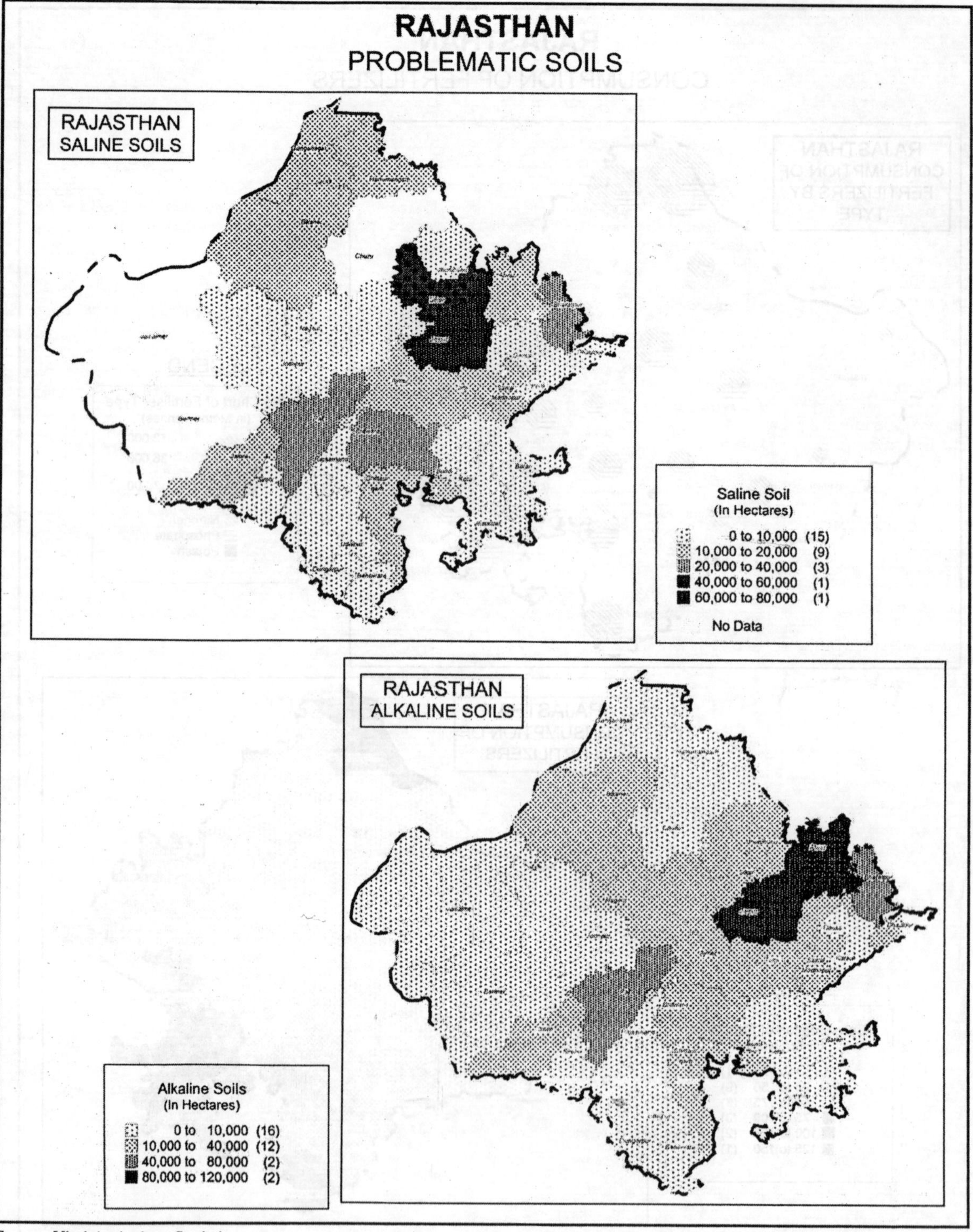

Source: Vital Agriculture Statistics

Figure 6.4 Problematic Soils

Chapter 7

ENVIRONMENTAL POLLUTION

With 10.41 per cent land mass of the country, Rajasthan has 5.67 per cent (2011) of India's population. It's large and mass has given it a variegated landscape ranging from sandy areas of Thar desert to green areas of eastern and south-eastern Rajasthan. The average rainfall varies from 130 mm in Jaisalmer to over 900 mm in Jhalawar and Banswara. Agricutlure is mostly rain-fed with canal irrigation system well developed in Chambal and Indira Gandhi Canal and Mahi Bajaj Sagar dam. Else-where irrigation is done largely from wells and tanks. Less than one third of the Sate's cultivated areas is under irrigation. Due to scanty nature of rainfall, the State is often visited by droughts. Of late, frequency of droughts has increased appreciably.

Rajasthan's economy is predominantly agrarian with animal husbandry as an important supplementary occupation. In desert areas animal husbandry provides main sustenance to the population. This area has the best breed of sheep and cattle. Despite its agrarian economy, agro-based industries have not well developed in the State. Notable exceptions being oil extraction plants in eastern Rajasthan. The pace of industrialization has been slow in Rajasthan despite its rich entrepreneurial heritage. Major impediments in rapid industrialisation of the State relate to inadequate communication net work, low literacy, endemic shortages of water and power, lack of infrastructure and prevalence of feudal culture. Rajasthan accounts for about 2 per cent of the total central investment in the public sector in the country. There has been no new central industrial undertaking in the Sate for the last 2 decades. As a result of efforts made by the state, Kota, Alwar, Udaipur, Bhilwara, Jodhpur, Bhiwadi and Abu Road have emerged as industrial centres.

Rajasthan has no coal or iron ore deposits of significance except lignitic deposits at Palana and Kapurdi. Mineral based industries like cement factories, Hindustan Zinc Limited, Hindustan Copper Limited, etc. dominated the industrial land scape. Rajasthan has emerged as a major producer of synthetic yarn in the country. Printing of cloth is done at places like Pali, Balotra and Sanganer. Chemical industries are located largely in Kota, Alwar and Udaipur districts.

Natural Geo-Environment Divisions

Here, in Rajasthan till recent years, the problems regarding environment were more concerned with Natural Environment than due to im-balances in environment created by human activities. (Now problems regarding im-bal- ances in environment due to human activities have not remained less important on any cunt than the problems due to Natural Environment gifted to Rajasthan). The vast stretches of desert, the ranges of hills, poor fertility of soil, soil erosion and dust storms due to strong winds, poor water availability, lack of potable water, scenty rainfall, etc, have been the negative factors of Rajasthan's Environment which the people of the State have faced for the uncounted years and are destined to face in future also. It can be devided to Geo-environmental Divisions.

Thar Desert Environment

To the west and north-west of the Aravalli Ranges lies the semi-desert and desert lands of the Thar Desert, occupying large areas of the Bikaner and Jodhpur Divisions. The area is dotted with small hillocks and isolated patches of Shrubs and stunted vegetation. But generally the entire area is avast sandy expanse with its ever-shifting sand dunes.

The Western Desert Plain is a sandy desert waste of ill-watered unpreductive area. The Luni and Sukri rivers form the main drainage of the area and flow7 in a south-westerly direction. A large part of Western Rajasthan is characterised by its inland drainage and carries some of the largest salt-lakes, viz., Sambhar, Didwana, Kuchaman and Talchapar.

Aravalli Hills Environment

The most conspicuous physiographic feature of the State is the Arvalli Range. It runs is the south-west-north-east direction from Palanpur in Gujrat to Delhi. It has a steep, discontinuous slope towards the north- west side facing the Thar desert and a relatively gentle slpe towards the south-east side facing the Alluvial Plains.

The Aravalli Mountain Range, composed principally of rocks of the Delhi Supergroup, comprises alternating hill ranges and vallyes extending from near Delhi on the north-east to the plains of North Gujarat on the southwest,-a distance of about 650 km. The highest peak, called Guru Sikhar, 1,700 m. above MSL is situated on the Mount Abu. The Aravalli range, partly forms one of the main watersheds of India, which divides drainage to the Bay of Bengal from the Arabian Sea. Ajmer city, lying almost at the centre of the range, is in fact on a tri-junction of the watershed. The watershed passes along the centre of the range from Ajmer to the south-west upto Kumbalgarh, where it swings to the south-east for some distance to pass ultimately through Udaipur. City. To the south of Udaipur city; the drainage of the range is to the south into the Gulf of Cambay through the Samaarmati river. The streams emanating from the north and norht-east of Udaipur city flew into the Banas through its tributaries. The Sabi river drains the northern parts and flew ultimately into the Jamuna at Delhi. The Banganga drains the areas lying in Alwar and Jaipur districts, flows through a large part of Bharatpur district, and reaches the Jamuna by indefinite channels.

Eastern Plains and Plateau Environment

The Easter Plains to the east of the Aravalli range constitute a vast undulating plain country occupied by the pre-Aravalli gneisses and schists as well as the phyllites and rocks of Aravalli Supergroup. The terrain of the pre-Aravalli Banded Gheissic Complex is covered by a thick mantle of alluvium and windblown sand.

The Vindhyan Plateau in south-eastern Rajasthan is an elevated hilly and plateau country' constituted by sandstones, shales and limestones forming the western extremity of the main Vindhyan Basin. The Vindhyans in this area comprise three plateaux, the Bhander plateau forming the uppermost group to the north-east, follow ed by the Rewa plateau surrounding the former to the south and west, and the Kaimur plateau encircling the Rewa and Bhander plateaux.

The Deccan trap plateau further is continuous with the basalt plateau of Madhya Pradesh. The junction between the Vindhyan plateau and the Eastern Plains of Rajasthan is marked by the Great Boundary' Fault. The Vindhyans here are continuous to the north-east upto Dholpur marked by a number of narrow; desconnected hill ranges, separated from the main Vindhyans in Gwalior by deeply dissected valleys of the Chambal river. The drainage in this part of Rajasthan is generally to the north and north-east. The Barach river which drains the south-western parts of the Vindhyan plateau runs northwards, turns to the north-east and flows parallel to and on the northern side of the Great Boundary Fault. Elsewhere in the Vindhyan plateau of south-east Rajasthan, the rivers and streams flow in different directions and join the Chambal river w hich flows in a north-easterly direction parallel to the same Fault through Kota city . The berach river flow's into the Banas which joins the Chambal river in Dholpur area. The Chambal drains into the Jamuna in the adjacent parts of Uttar Pradesh and is the only perennial river in the whole of Rajasthan.

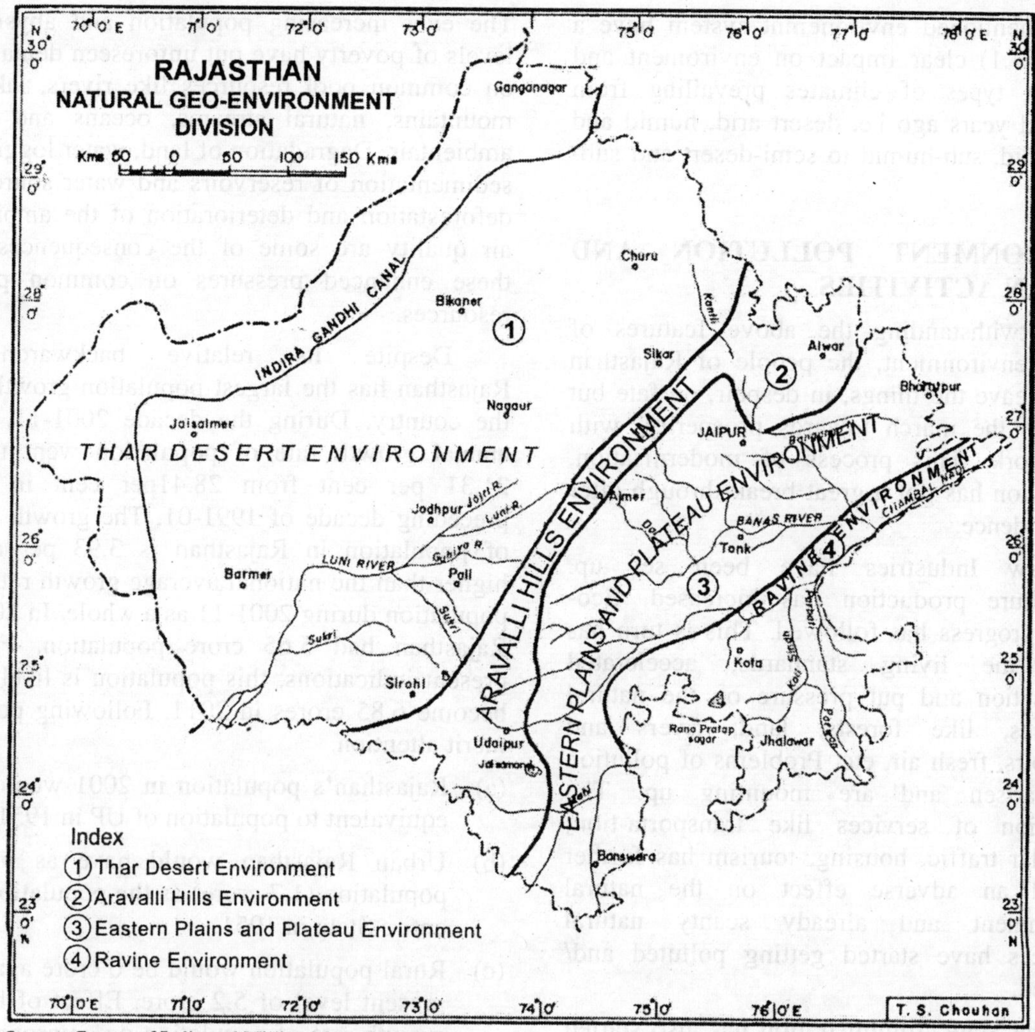

Source: Census of India and Million Map of Rajasthan.

Figure 7.1 Rajasthan: Natural Geo-Environment Division

The rainfall in the Aravalli hill range and further to the east is moderate, as much as 1500 mm per year. But to the west of the Aravalli range the rainfall is scanty, as much as 100 mm per year or even less, and most of the Western Rajasthan is often affected by drought.

From the environmental comparison point of view with other states of the country, the State of Rajasthan is on different footing because of its spatial, topographical and eco-climatic conditions. The land is full of contrasts with its desert dunes, saline lakes, jungles, valleys and ravines.

Rajasthan is a well known land of varied fragile environment. It covers following major environment (1) Thar Desert Environment by including active process of desertification mainly on the marginal areas. Indira Gandhi Canal commands area also lies in its between belt, (2) Aravalli Envrironment the area is stretching in the direction of NE to SW in Rajasthan State, (3) Eastern Plain and Plateu Environment- it lies in the eastern portion of Rajasthan State, including wetland and (4) Ravine Envrironment- it is deeply eroded extensive ravine areas in the basin of river Chambal having a national impact. These above mentioned environemnt system have a (Figure 7.1) clear impact on enviroment and different types of climates prevailing from thousand years ago i.e. desert arid, humid and sub-humid, sub-humid to semi-desert and sub-humid.

ENVIRONMENT POLLUTION AND HUMAN ACTIVITIES

Notwithstanding the above features of natural environment, the people of Rajasthan do not leave the things, in despair, on fate but continue the march towards prosperity' with hard work. This process of modernisation, civilisation has got a great break-through after Independence.

New Industries have been set up. Agriculture production has increased. Economic progress has followed. This is turn has raised the living standards, accelerated urbanisation and put pressure on the natural resources, like forests, land, rivers and reservoirs, fresh air, etc. Problems of pollution have arisen and are mounting up. The expansion of services like transporta-tion, vehicular traffic, housing, tourism has further exerted an adverse effect on the natural environment and already scanty natural resources have started getting polluted and/ depleted.

The State Administration has also started sensing the gravity of the problem of environment degradation and joined hands with the Centre in making necessary enactments for environment conservation and pollution abatement.

FACTORS AFFECTING ENVIRONMENT

(1) Population

Burgeoning population endemic poverty and uncontrolled industrialisation have been the major contributors in the degradation of environment in India. The population of India in 2011 was 121.09 crore, which is likely to touch the figure of 130 crore by the year 2021. The ever increasing population and abysmal levels of poverty have put unforeseen demands on common pool resources like rivers, lakes, mountains, natural streams, oceans and the ambientair. Degradation of land, water logging, sedimentation of reservoirs and water sources, deforestation and deterioration of the ambient air quality are some of the consequences of these enhanced pressures on common pool resources.

Despite its relative backwardness, Rajasthan has the largest population growth in the country. During the decade 2001-11, the annual growth rate of population went upto 21.31 per cent from 28.41per cent in the preceding decade of 1991-01. The growth rate of population in Rajasthan is 3.93 per cent higher than the national average growth rate of population during 2001-11 as a whole. In 2011, Rajasthan had 5.65 crore population. From present indications, this population is likely to become 6.85 crores in 2011. Following points merit attention:

(a) Rajasthan's population in 2001 would be equivalent to population of UP in 1951.

(b) Urban Rajasthan would have as much population (1.7 crore) as the population of entire State in 1951.

(c) Rural population would be 6 crore against present level of 5.2 crore. Effect of huge growth rate of population on environment would well be imagined particularly because high growth rate areas are those which have low' biotic reserves with low water endowment.

(2) Water Scarcity

Rajasthan has less than 1 per cent water of the entire country. With the rapid increase of population and increasing trend of urbanisation, frequent droughts and exploitation of water in desert areas for agriculture, water crises have become endemic. Deteriorating environment is telling upon the water availability in the state in a significant way. Emerging danger signals in this regard can be inferred from he following:

(a) Continuous lowering of water table throughout the State excepting in canal areas.

(b) Water earlier meant for irrigation is being utilised for consumption in urban areas. Even this is not proving adequate. Witness, the difficult situation of water availability in cities like Ajmer, Beawar and Jodhpur.

(c) Due to low rainfall in arid and semi-arid areas, re-charge is low. Due to large scale of drilling of tube wells, water which accumulated in 100 years in this areas is being exploited very rapidly for agriculture. It is a moot point whether over exploitation of under water which accumulated in the ground over centuries would not pose danger to existence of humna population in this area in few decades from now.

Water is one of the basic needs of the human beings. Only 0.3% of the total Water on earth is fit for human consumption and this small quantity' is also not distributed evenly. Rajasthan shares only small fraction of it, that is very' much less than the average. The rain fall, on an average, is only 150 mm and very irregular with no rians for years in certain areas. This results in severe condition of draught and consequently problems of water availability. The ground water is either too deep (even more than 200 m) or is foil of excessive salts, fluorides particularly in the western Rajasthan.

In Rajasthan, 42,952 villages out of the total 43,282 villages are problem villages with regard to drinking water availability and in these villages drinking water is provided for through deep wells or from regional water supply schemes.

The Government is taking necessary actions through its various agencies to control the adverse effects of negative aspects of the natural environment and to mitigate the sufferings of people of the State befalling on them time to time.

Soil Erosion and Desertification

Soil is structural component of human environment. Soil erosion is environmental destruction. The soil turns out to be a pollutant when not in its original home. Soil erosion is the cause of crises in physical, biotic and socioeconomic environments. After soil erosion due to water, it is only soil erision due to wind which very serious in the country, particulary in Rajasthan, on the whole affecting about 50 million hectares of land.

The process of desertification has assumed an alarming proportion. According to the First Five Year Plan Report, the Rajasthan Desert has been advancing at the rate of half a mile per year for the last 50 years. The climate has turned hostile and land resources have depleted below the bare minimum line. In desert areas, agriculture has become a gamble. Thus, the socio-economic environment has registered fantastic degradation resulting into nomadic way of life and migratory grazing in the vast affected areas.

(3) Deteriorating Quality of Soils

Due to massive felling of trees, high run-off and low investment on soil conservation the soils in Rajasthan have been deteriorating over time. Increase in bovine population particularly of sheep and goats and consequent heavy grazing is contributing to the depletion in the soil quality in the Sate. There has been high silting of dams and tanks. Massive expansion of cultivated areas even by bringing marginal lands under the plough is another contributory factor. The soil is the most precious resource of the mankind and, therefore, its depletion is going to effect Rajasthan's environment in a major way in coming years.

(4) Air Pollution

Fortunately Rajasthan's air is relatively free from pollution. One could see azure blue sky at any time of the day. Even in the night, one could see clear sky illuminated by stars except near locations of cement plants or near cities having chemical industries main sources of .Air pollution in Rajasthan.

(5) Water Pollution

Water pollution is contributed largely by industrial effluents and city waste disposal. At places like Pali, Balotra and Udaipur this has already reached alarming proportion. Attention is therefore now getting focused on chemical treatment of effluents before their release in the natural drainage system. In the absence of proper sewerage system, lakes which earlier stored water for the city are now becoming dumping ground for the sullage and solid wastes. Sursagar (Bikaner) and Pichhola lake (Udaipur) are ecellent examples of this category.

(6) Depleting Forest Wealth

Despite its large land mass and low density of population, Rajasthan does not have large well stocked forest areas. There has been massvie felling of trees caused by greed of men, demand of rapidly increasing population and acceptance of urban patterns of living requiring high consumption of timber. Aravali has been largely denuded of its forest cover. The denudation of Aravali which is the sky line of norht India is the single most contributory factor for deterioration of environment in the State. There has been growing imbalances between exploitation of forest wealth and restocking of forests. At many places people are now finding it difficult to procure fuel wood for burning the dead body at a very short notice. The growing imbalacne in supply and demand of fuel wood has already led to a situation where price of charcoal is approaching the price of food grains. In another 15 year's time charcoal would be costlier than food grain due to longer gestation period required for production of fuel wood.

Pollution Scenario : An Overview

The geographical area of Rajasthan constitutes 10.41% of the country and is inhabited by 5.67% of the country's population, but it has been endowed w ith only 1.15% of the total water resources of the nation, The scarcity of water in the State is matched by the abundance of its mineral resources, bulk of which are non-ferrous and non metallic.

Industrial Growth

Mining is prevalent in almost all parts of the State. Lime stone mining is carried out in Udaipur, Chittor, Kota, Bundi, Sawaimadhopur belts and also at Kotputli, Sikar, Alwar, Jodhpur, Nagaur (Gotan), Sirohi and Beawar. These mining operations are open cast in nature. There also exists great potentiality of lime stone exploration at Jaisalmer due to availability of steel grade limestone. Khetri in Jhunjhunu District and Dariba in Alwar are areas where copper mining takes place, which is open cast as well as underground in nature. Mining of Zinc and Rock phosphate is prevalent in Udaipur and mining of lead and zinc in Bhilwara.

Mining of building stone is predominant in Jodhpur and Jaipur regions whereas Bhilw ara is known for soap stone activities.

Majority of the industries in the State are mineral based, of which cement and cement products are the predominant ones. There are 14 major and around 80 small cement units in the State producing around 7-8 million tones of cement per annum and the production capacity7 is expected to be doubled in the near future. Similarly, Zinc Smelter at Debari (Udaipur), Copper Smelter at Khetri and Super lead zinc Smelter at Chittor are three major Smelter plants in the State. Rajasthan is generating around 1400 M. W. Power from Kota Thermal Power Plant, nuclear Power Plant, Rawatbhata, Gas based power plant at Anta and a few other captive power plants. Rock phosphate is another important mineral which makes use of Sulphuric acid to produce phosphatic fertilisers. There are 6 sulphuric acid plants and seven Single super phosphate plants in the State majority of them are at Udaipur. Besides, there is one major Nitrogeneous fertilizer plant at Kota, 2 Chlor-alkali plants, one at kota and another al Alwar, major tyre plant at Kankroli, man made Fibre plants at Kota and Jhalawar, major PVC plant at Kota and Freon gas manufacturing plant at Alwar along with major textile units at Bhilwara.

In addition to the above, there are clusters of SSI units in the State. Textile units are predominant at Pali, Jodhpur, Balotra, Sanganer and Bagru, whereas CPW units are confined to MIA, Alwar. Mineral grinding units, bricks/lime kilns, stone crushers etc. are spread all over the State.

Industrial water pollution is mainly confined to Jodhpur, Pali, Balotra, Bagru and Sanganer due to the existence of clusters of textile units; in Kota, it is on account of Thermal Power, Fertilizers and chemicals. Rayon units etc.; In Udaipur, due to Zinc Smelter, SSP, Sulphuric acid, chemical units, etc; and in Alwar on account of Chlor-Alkali, Sulphuric acid, SSP, Freon gas manufacturing, and chemical units. The predominant water pollutants are BOD, COD, Oil and Grease, mercury, TSS, TDS, PH, Fluorides and in specific cases, trace metals.

Water pollution is primarily affecting surface and ground water Udaipur, Kota, Alwar and Ajmer are prone to surface water pollution whereas Jodhpur, Pali, Balotra and Jaipur are the areas where possibility of ground water pollution is quite significant.

Industrial air pollution is prevalent in Kota due to coal/ gas based Thermal power plants, Calcium carbide, Nitrogenous fertilizer, steam generating boilers. Cement plants etc; in Udaipur on account of Zinc Smelter, SSP H_2SO_4 and Cement plants; in Chittor due to Cement, Captive power and Super Lead-Zinc Smelter plants: in Alwar on account of Chemicals, SSP. H_2SO_4, Mini-Steel plants etc. Besides air pollution is also predominant in areas like Gotan, Behror and others where clusters of mini-cement, brick/lime kilns, mineral grinding units, stone crushing etc. have developed. The predominant air pollutants that are released into the atmosphere are basically particulate matter, Sulphur-di-oxide and oxides of nitrogen all in the form of steady state conditions.

The status of sewerage facilities in the state is rather poor. Under the IDA assisted Water Supply and Sewerage Project, Jaipur, Jodhpur, Bikaner and Kota were partially sewered and an extended aeration treatment plant designed to cater approximately 4.5 lakh population of the walled city of Jaipur was built. In addition, Udaipur has also been partially sewered out of State pain funds.

None of the other cities has been benefitted with sewage treatment facility so far mainly because of resource constraint. During the Vth plan period a provision of Rs. 866.00 lacs has been kept for sewerage and sewerage treatment against which an However, the State has wide range of mineral deposits bulk of which are non-ferrous and non-metallic, next only to Bihar. A determined effort has been made by the State Government to provide the requisite infrastructure facilities for industrial development. This has led to a steady development of industries in general and small scale industries in particular. For instance, Rajasthan had 138 large/ medium water polluting industries as listed in 2010 and by 2018 this number has risen to about 302, corresponding to more than 100% increment. Presently there are 390 air polluting units identified in the State. The number of large/ medium water/ air polluting industries, according to 2018 statistics are given Table 7.1:

Table 7.1 Number of large/ medium water/ air polluting industries

S.No.	Nature of unit	Water	Air
1.	General Engineering	85	28
2.	Textile	67	55
3.	Man made fibre	5	2
4.	Cyrogenics	3	
5.	Chemical	15	35
6.	Ore Processing/ mining	9	10
7.	Dairy	11	9
8.	Edible oil	16	16
9.	Ceramic and refractory	8	23
10.	Paper and Pulp	13	8
11.	Pharmaceuticals	3	7
12.	Distillaries	5	5
13.	Sugar	3	4
14.	Fertiliser	4	11
15.	Asbestos fibre	-	3

S.No.	Nature of unit	Water	Air
16.	Tanneries	2	1
17.	Pesticide	2	1
18.	Rubber and allied products	4	10
19.	Iron and steel	-	23
20.	Chlor-alkali	1	2
21.	Hotels	2	2
22.	Thermal Power Plants	2	1 1
23.	Plastic	3	
24.	Cement	3	62
25.	Fllod (Breweries)	4	5
26.	Metallurgical	-	5
27.	Miscellaneous	10	52
	Total	280	390

(Rest 302-280=22 water polluting units are still to be classified)

Source: Pollution Control Board of Rajasthan, Jaipur

Industrial sectors such as textiles, woolen, cement, chemical and mineral grinding are some of the important industries of the State. Rajasthan's smelters produce the entire output of zinc and substanital copper production in the country. Besides, there are units for manufacture of Fertilisers, Sulphuric Acid, Synthetic fibre, insulated bricks, machine tools, instruments, electric meters, T. V. Sets, etc. As a result of efforts made by the State, Kota, Alwar, Bhiwadi, Udaipur, Bhilwara, Jodhpur, Pali, Balotra, Abu Road, Behror and Chittor have emerged as industrial centres.

Pollution Prone Areas

The problem of industrial water pollution in the State is mainly confined to Kota, Alwar, Udaipur, Jodhpur, Pali, Balotra, Sanganer, Bagru etc. and that of domestic waste water to Kota, Udaipur, Jaipur and other major towns of the state. It would be worthwhile to mention that Udaipur, Kota, Jaipur, Ajmer etc. are the areas where possibility of surface water pollution is always there whereas in Jodhpur, Pali, Balotra, Sanganer. Bagru, Khetri, Debari (HZL) etc. the possibility of ground water pollution is more significant.

The problem of industrial air pollution in the State is mainly confined to Kota, Alwar, Udaipur, Chittor, Gotan (Nagaur) and pockets where major cement units or cluster of tiny cement units, mineral grinding units, stone crushers, brick kilns, lime kilns etc. are in existance. Besides, Khetri (HCL) and Debari (HZL) are the significant pockets where problem of air pollution due to rel ease of sulphur-di-oxide is quite significant and its potentiality is further enhanced due to topographical conditions.

Hazardous chemicals and waste problems are mainly at Kota, Alwar and Udaipur due to the existance of chlor-alkali, fertiliser, chemical sulphiric acid, pesticides, chlorinated paraffin wax, nuclear power plants etc.

Compliance with Water and Air Emissions

As per the 2016-17 statistics, in Rajasthan, out of 301 Water Polluting major/ medium industries, 112 industries have installed complete treatment facilities and 35 industries have installed partial treatment facilities. Treatment plants in 14 industries are under construction and in 53 industries, the same are proposed. 42 industries have attained zero discharge status. In 3 industries treatment plants are not required and remaining industries are being persuaded to install treatment plants.

Out of 335 air polluting major/medium industries, 114 industries have installed air pollution control equipments; 99 industries do not require control equipments and 122 industries are being persuaded to install air pollution control equipments. Under Air (Prevention and Control of Pollution) Act, consent has been granted or extended to 140 industrial plants and 26 consents have been refused. In year 2016-17, four industries have installed treatment plants and 7 industries have attained zero discharge position.

A detailed breakup of number and status of ETP arrangement for large/ medium water polluting industries under each Regional Office (as on March 2018) is shown below. Activities on the inventorisation and control of hazardous waste have not yet begun (Table 7.2).

Table 7.2 Large/ Medium Water Polluting Industries

S. No.	District	Total number of industries	Pollution control measures taken			
			Adequate treatmtnet arrangement	Treatment arrange-ments are partial not required	Pollution Controal equipments	ETP Yet to be installed
1	2	3	4A	4B	4D	5
1.	R.O. Alwar					
	Alwar	112	28	8	-	15
	Dholpur	2	1	-	-	-
	Bharatpur	4	3	-	-	-
2.	R.O. Jaipur					
	Jaipur	47	12	5	-	9
	Ajmer	15	6	2	-	4
	Sikar	1	1	-	-	-
	Tonk	4	2	1	-	1
	Jhunjhunu	3	3	-	-	-
3.	R.O. Jodhpur					
	Jodhpur	10	2	3	-	3
	Bikaner	2	1	-	1	-
	Nagaur	-	-	-	-	-
	Sriganganagar	5	1	4	-	-
	Barmer	-	-	-	-	-
4.	R.O. Kota					
	Kota	20	13	-	1	1
	Bhilwara	21	15	1	1	-
	Sawai Madhopur	-	-	-	-	-
	Jhalawar	2	1	-	-	-
	Bundi	2	1	-	-	-
	Chittorgarh	6	4	2	-	-
5.	R.O. Pali					
	Pali	2	1	1	-	-
	Sirohi	7	3	2	-	-
	Jalore	1	-	-	-	-
6.	R.O. Udaipur					
	Udaipur	31	16	3	-	3
	Banswara	4	1	1	-	1
	Dungarpur	1	1	-	-	-
	Total	**302**	**116**	**33**	**3**	**37**

Source: Pollution Control Board of Rajasthan, Jaipur

A detailed breakup of number and status of pollution control (PC) arrangements for air pollution (AP) in large/medium air polluting industries under each regional office (as on March 2018) in given Table 7.3.

Table 7.3 Air Pollution in Large/ Medium Air Polluting Industries

S. No.	District	Industries which have installed camplete/ partial P. C. Plant.	Industries which do not require PC plant and/or have controlled AP by properly designed chimney	Industries which are yet to take steps for P.C.	Total
1	2	3	4	5	6
1.	Ajmer	1	9		11
2.	Alwar	25	54	58	13.7
3.	Banswara	2	2	1	5
4.	Bharatpur	2	2	1	5
5.	Bhilwara	4	27	3	34
6.	Bikaner	-	1	1	2
7.	Bundi	1	.	2	3
8.	Chittorgarh	8	-	2	10
9.	Churu	1	-	2	3
10.	Dholpur	1	1	1	3
11.	Jaipur	7	17	10	34
12.	Jodhpur	9	2	4	15
13.	Kota	16	10	2	28
14.	Nagaur	6	1	1	8
15.	Pali	1	1	1	3
16.	Sikar	6	1	1	8
17.	Sirohi	12	2	3	17
18.	Sri Ganganagar	-	3	2	5
19.	Udaipur	21	17	6	44
20.	Dungarpur	-	1	-	1
21.	Jalore	-	1	-	1
22.	Jhalawar	1	1	-	2
23.	Sawai Madhopur	2	1	-	3
24.	Tonk	-	2	-	2
25.	Jhunjhunu	3	2	-	5
	Total	**129**	**158**	**103**	**390**

Source: Pollution Control Board of Rajasthan, Jaipur

Automobile air pollution is predominant in major urban areas and is estimated to count for over 50% of the total air pollutional load. There were 11.4 lakh vehicles in the year 2017-18 in the State of Rajasthan which is an increase of 26.84% and 11.72% over the no. ofvehicles in 89-90 and 2016-17 respectively. Jaipur alone accounts for around 25% of the total vehicles of the State. As per rapid assessment made on theoretical assumption, the total pollutional load in the State on account of vehicular emissions may be around 1145 Tons Per Day (TPD) of which, Jaipur alone accounts for 280 TPD. Rajasthan State Pollution Control Board has been organising automobile emission camps from time to time in Jaipur and other towns and monitoring the vehicular emissions to generate the background data for evolving appropriate strategy for the control of automobile air pollution.

Pollution created by noise is prevalent mainly in urban areas, near the airports, industrial estates, construction sites, busy traffic junctures etc. Noise in urban areas is predominantly from automobiles, loud-speakers, construction activities, aircraft and railway operations etc. The Executive magistrates are authorised to pass regulatory orders under the Rajasthan Noise Control Act, 1963. The Rajasthan Pollution Control Board is empowered to regulate noise levels in industrial areas under the Air (Prevention and Control of Pollution) Act, 1981 through consent of administration. The Rajasthan Pollution Control Board also keeps on monitoring decible levels of noise in different zones periodically. Standards of ambient noise in sensitive, residential, commercial and industrial areas have been prescribed by the Central Government.

Mineral Resources Availability and potentiality of Industrialisation

The State of Rajasthan is rightly recognized as the mineral store house of India. Though it contributes on the small amount of mine production due to absence of Coal, Iron ore which together account for over 74% of the mining activity in India, while many important mineral deposits like Lead, Zinc, Copper etc. are available in the State and are being effectively mined. There are other minerals like Rock Phosphate, Fluorospar, Lime Stone, Gypsum, Bentonite, Quartz, Talk, Mica etc. which are also found in good quantities in different parts of the State.

It would be worthwhile to classify the different type of minerals under the following categories-

(i) Metallic minerals

(ii) Chemical Minerals

(iii) Salt and Salt Based chemicals

(iv) Fertilizer Minerals

(v) Insulants, Refractories and Ceramics

(vi) Miscellaneous Minerals.

The type of the minerals under the categories stated above have been dealt with in the following table where mineral reserves have been mentioned alongwith their occurence in the State of Rajasthan (Table 7.4).

Table 7.4 Mineral Reserves in Rajasthan

	Mineral	Occurence	Million Tonnes
	1	2	3
	Miscellaneous Minerals		
i)	Lignite	Bikaner	23.0
ii)	Garnet	Sawai Madhopur, Jaipur, Sikar, Tonk and Udaipur	Quantities not known
iii)	Emerald	Ajmer, Udaipur	not estimated
iv)	Bentonite	Barmer, Bikaner	20.0
v)	Calcite	Bundi, Sikar, Sirohi, Pali	not estimated
vi)	Slate	Udaipur, Alwar	- do -
vii)	Fuller's Earth	Bikaner, Jaisalmer	0.5
viii)	Tal	Bhilwara, Udaipur, Jaipur Dungarpur	not estimated
ix)	Ochre	Chittor, Bikaner, Dungarpur Bikaner, Jaisalmer, Nagaur	not estimated
x)	Silaceous Earth	Barmer	2.3
xi)	Building and Ornamental stones	Jaipur, Sirohi, Nagaur, Udaipur, Alwar, Kota	not estimated
xii)	Semi Precious and Precious stones	Jhalawar, Sawai Madhopur, Jodhpur, Bikaner, Chittorgarh Jaisalmer, Jalore	not estimated
I	**Metallic Minerals**		
i)	Iron	Udaipur, Jaipur and Bhilwara	10
ii)	Manganese	Banswara	0.34
iii)	Copper	Khetri, Alwar belt	70.5
iv)	Lead and Zinc	Zawar Belt	149
v)	Tungsten	Nagour	6.2

	Mineral	Occurence	Million Tonnes
II			
i)	Barytes	Alwar, Udaipur	124 (thousand tonnes)
ii)	Fluorospar	Dungarpur, Sikar	1.2
iii)	Limestone	Sirohi, Pali, Chittor Kota, Jaipur, Sikar, Nagour, Jodhpur, Banswara	2000
III	**Salt and Salt based chemicals**		
i)	Salt	Nagour, Jodhpur Jaisalmer, Barmer	50.8
IV	**Fertilizer Minerals**		
i)	Gypsum	Nagour, Bikaner, Jaisalmer Pali, Ganganagar, Churu	1000
ii)	Rock Phosphate	Udaipur, Jaisalmer	55
iii)	Pyrites	Sikar	1116
V	**Insulants, Refractories and Ceramics**		
i)	Asbestos	Ajmer, Bhilwara Jodhpur, Udaipur	27.6 (thousand tonnes)
ii)	Clay	Barmer, Bikaner, Jodhpur, Ajmer, Jalore, Sawai-madhopur, Sikar	Quantities not known
iii)	China Clay	Udaipur	10-12
iv)	Kyonite	Banswara, Ajmer, Bhilwara, Udaipur, Dungarpur	Quantities no known
v)	Falspar	Ajmer, Alwar, Pali, Sirohi Dungarpur, Bhilwara	- do -
vi)	Fire Clay	Bikaner, Jaisalmer	18.5
vii)	Graphite	Banswara	1.0
viii)	Magnesite	Ajmer	1.5
ix)	Mica	Bhilwara	Quantities not known
x)	Quartz Silicz and Glass Sand	Sawai Madhopur, Ajmer, Pali, Sirohi, Bundi	1.18
xi)	Verzmiculite	Ajmer	A few thousand tonnes

Source: Geological Survey of India, Jaipur and Department of Mines and Geology, GoR, Udaipur

Noise Pollution

It is a well recognized fact that the pollution created by noise is confined to Urban areas, near Airports, Industrial Estate, Construction activities etc. It has also been felt that noise pollution has assumed significant dimensions and consequently the noise has been included under section 2(a) of the Amended Air (Prevention and Control of Pollution) Act, 1987.

The State Pollution Control Board in collaboration with Central Pollution Control Board measure noise levels under a survey during 13th Feb. to 16 Feb. 90 in different parts of Jaipur city. The main purpose of this study was to assess the quality of ambient air noise lavels in different parts of Jaipur city for working out future strategies for the prevention and control of noise pollution. The observations recorded have been listed in Table 7.5.

Table 7.5 Prevention and Control of Noise Pollution

		8-11 AM	2-5 PM	11-2 AM
A.	**Residential Area**			
	a) Bani Park	45+	34+	
	b) Bapu Nagar	-	56+	36+

		8-11 AM	2-5 PM	11-2 AM
B.	**Sensitive Area**			
	a) Ram Niwas Park	65+	61 +	42+
	b) SMS Hospital	66+	68+	45+
	c) Durlabhji Hospital	-	59+	40+
	d) University of Raj.	-	55 +	43 +
	e) Birla Temple Circle	69+	72+	44+
C.	**Commercial Area**			
	a) Ajmeri Gate	68+	70+	76+
	b) Raja Park	67+	75 +	50+
	c) Rambagh	76+	74+	62+
D.	**Industrial Area**			
	a) Sudarshan Pura	55 +	59+	48+
	b) Vishwakarma	55+	66+	62+
	c) Bais Godawn	65+	66+	48+
E.	**Air Port Area**			
	a) Residential	66+(90+)*	62+(87)*	62+(78+)*
	b) On the Road	69+(98+)*	79 (104+)*	71 +

*db (A) during taking off

Source: Pollution Control Board of Rajasthan, Jaipur

The observation recorded above clearly reveals that the residential and industrial areas were found within the permissbile limits, whereas in commercial and sensitive areas, the observ ed values were found to exceed the permissbile levels during day and night hours. The higher values in these areas appear to have been mainly on account of vehicular traffic for which a well defined, traffic planning (As a short term measure) and urban planning (As a long term measure) shall have to be worked out in an integrated manner.

Automobile Air Pollution

There are more than 300 million passenger cars, trucks and buses the world over and their numbers are rapidly increasing, keeping pace with the industrial and social progress. The growing number of these vehicles is posing a serious threat to the urban air environment. An average car during a days driving discharges 0.50 to 1.5 kg. of pollutants depending on the condition of the car.

Vehicles in the major metropolitan cities of India are estimated to count for about 70% of all carbon monoxide, 5% of all hydro carbon, 30 to 40% of all Oxides of Nitrogen, 30% of suspended particulate matter and 10% of all sulphurdioxide pollution in these cities.

Rajasthan State accounts for around 7.68 lacs vehicles out of which two wheeleres are of the order of 5.47 lacs, three wheelers 0.95 lacs and four wheelers 1.26 lacs. Vehicles in Jaipur alone would be of the order of 25% of the total vehicle of the state. This would reveal that Jaipur would a cretical metropolitan city in times to come with respect to automobile air pollution. It is theoritically estimated that the state of Rajasthan contributes around 768 tonne of pollutents evety day into the atmosphere and Jaipur alone accounts for around 122 tonnes per day of pollutants due to automobile exhaust emissions.

Rajasthan Pollution Control Board has visualised the problem of automobile emissions at the initial stages and undertook extensive campaign for automobile exhaust emissions with the state department of environment during 22nd August to 5th Sept. 1988, During this compaign different categories of petrol and diesel driven vehicles were monitored with the main objective to evaluate and asses broadly the status of different category of vehicles with respect to air pollutants emitted out from these vehicles as a consequence of conbustion of fuel.

During the period of 9 days of automobile emission monitoring 1480 vehicles have been monitored out of which 1149 were petrol driven and 331 diesel driven vehicles. Compilation of observations have been done separately for petrol and diesel vehicles. It revealed that out of total 1149 petrol vehicles monitored 746 vehicles have been found within the prescribed limits which represents (64.92%). Out of 604 scooters, 140 mopeds, 170 motor cycles monitored; 66.22%, 58.57% and 78.23% respectively, of the vehicles have been found withing the permissible limits. This represents that out of 914 two-wheelers monitored 67.28% of the vehicles have been found within the permissible limits. As regards three wheelers in the form of Autorikshaw's are concerned 130 such vehicles have been monitored and 61.76% of which found within limits.

Four wheelers in the form of Ambassador, fiat, Maruti, Suzuki, Jeep etc. were monitored, out of which 55.74% have been found within the permissible limits.

The observations of diesel driven vehicles revealed that out of 331 vehicles monitored only 10.27% of the vehicles have been found within permissible limits of exhaust emissions. It was observed that 125 three wheelers have been monitored out of which only 1.58% of the vehicles have been found within limits. Out of 205 four wheelers monotored, 15.6% of the vehicles have been found within the prescribed limits.

It would thus be concluded that Diesel Driven Vehicles particularly three wheelers are the worst with respect to meeting of exhaust emission standards.

The State Pollution Control Board has again undertook the automobile emission monitoring campaign during 9-24 March, 1990 with State Transport Department herein 1151 vehicles were monitored. Out of which739 were petrol driven and 412 were diesel driven. Out of 739 petrol driven vehicles, 665 (90%) vehicles were found within the limits whereas in respect of 412 diesel driven vehicles, 115 (25.5%)vehicles were found within the limits it may be monitored here that the limits of pollutants from different categories of vehicles.

Automobile Emission Control Measures

The reduction of carbon monoxide, hydrocarbon and NOx emissions is generally achieved by air-fuel mixture gas control combustion control, exhaust gas treatment and fuel modification.

Regarding gasoline automobiles, the air fuel ratio will directly influence the emission components, fuel consumption and driving performance, so improvements are made in the air and fuel supply sy stems to achieve an optimum air fuel mixture. Meanwhile, cobustion control is being achieved by changing engine specification, optimization (delay) of ignition timing and exhaust gas recirculation. As for (exhaust treatment methods, thermal reactor has been developed first. Nowaday s, three-way catalyst system is being mounted for removing carbon monoxide, hydrocarbons and NO_x simultaneously.

Regarding diesel engines, the air is sucked into the cylinders and the fuel is injected in an automized state into the high-temperature, high pressure air that has been compressed adiabatically, to burn the fuel by auto-ignition. As a result a large quantity of oxygen remains in the exhaust gas, preventing the application of the three-way catalyst technique that is being applied to gasoline engines.

Regarding NO_x counter measure, the common methods which are being adopted include the delaying of fuel injection timing, decreasing of the swirl ratio and lowering of the compression ratio, but since these methods invariably deteriorate fuel economy and driving performance, technologies for the improvement of the engine performance are being introduced gradually; including turbo-charger introduction, combustion chamber shape modification and the adoption of a high fuel injection rate. Meanwhile, measure are attempted for switching over to alternative law pollution vehicles (electric automobiles, alcohol, fuel automobiles, hydrogen fuel automobiles etc.)

Regarding the use of alternative fuels, various ful sources have been attempted to use as automobile fuel after the oil crises. Nowadays, the use of clean energy such as natural gas, alcohol, hydrogen are expected to prevent urban air pollution. Further more, the use of fuel with high hydrogen to carbon ratio is effective for coping w ith the climate warming. Various countries are faced with the need to select optimum fuels as a part of their general energy policies.

Reducton of exhaust gaseous pollutants can be achieved by uxing an "after Treatment" device of some short, possible method include (a) Exhaust Manifold Reactor (b) Direct Flame After burner and (c) Catalytic Convertor.

The air pollution in kota has been attributed on account of the release of different types of air pollutants from Kota Thermal power Plant. (2×110 MW and 2×210 MW) cement, calcium carbide, chlor-alkali, amonical fertilizer and captive power plants (35 MW and 10 MW) of M/s. Shriram chemicals and fertilisers (SFS), bolier house of M/s. J.K. Synthetics, sulfuric acid, yarn dipping unit, boiler house, carbon tetra chloride and carbon-di-sulphide plants of M/s. Shriram Rayons; fumeces of M/s. Multimetals etc. These plants emit suspended particulate matters, sulfur-di-oxide, exides of nitrogen, ammonia, carbon-di-sulfide, hydrogen-sulfide, chlorine, mercury etc. into the atmosphere, or working environment.

All these units have installed air pollution control equipments and most of them are achieving the prescribed standards of the State Pollution Control Board. The Kota thermal power plant has installed ESP's of around 99.7% for which the unit is to increase the fields of ESP thereby providing more collecting area and increasing the efficiency of ESPs. The thermal power plants of M/s. SFC are equipped with ESPs of adequate capacity whereas Calcium Carbide plant is provided with fibre glass bag house, prilling tower with acid mist eleminator, fertilizer plant with ammonia desorption system, cement plant with ESPs, bag filters and cyclones of adequate capacities, similarly, M/s. Shriram Rayons have provided caustic scrubbing at carbon-tetra-chioride and carbon- di-sulphide, multiclones at boiler houses, which is to be replaced bu fibre glass bag filters or by providing other suitable control systems to achieve the permissible limits whereas M/s. J.K.Synthetics have provided fibre glass bag house at their boiler house and achieving the prescribed norms. It should be pertaining to note that around Rs. 17 crores have been spent on air pollution control equipments at Kota.

Ambient Air Quality Status at Kota

The State Polution Control board has installed 5 ambient air quality stations at following locations in Kota and they are in operation since February 1985.

(i) Vetenary Hospital

(ii) Borkhera

(iii) Akhelgarh (Now Regional Office of the Board)

(iv) Rajpura

(v) Anantpura.

These monitoring stations ate equipped with High volume sampler (HVS) and regular monitoring of suspended particulate matter (SPM), Sulfur dioxide (SO_2) and oxides of nitrogen (NO_2) is being done. The monitoring results reveal that there has been a marked improvement in the quality of ambient air with respect to SPM after the introduction of effective control measures on particulate matter emissions. However, the sulfurdioxide and oxides of Nitrogen in the ambient air have been found incresing partly due to the fact that no pollution control equipments have been provided for controlling the emissions and partly because of their cumulative effect.

The annual average concentration of SPM in the ambient air in the year 1985 has been observed as 313.47 mg/m^3 whereas in the year 1989 it has come down to the level of 223.12 mg/m^3 thereby reducing the concentration of SPM by 28.8% on the other hand; the SO_2 and Nox concentrations in the ambient air in the year 1985 have been observed as 17.78 mg/m^3 and 18.77 mg/m^3 whereas in the year 1989 it is observed as 20.90 and 81.90 mg/m^3 respectively which shows an increase of 17.5 and 336.3 percent respectively.

Air Polluting Industries

The State has significant amount of mineral resources and most of the air polluting industries are mineral based. There has been a great potentiality of setting up mineral based industries in the State and air pollution problem due to such industries would be max. around mineral rich areas. The State Board has till now identified 335 major/ medium air polluting industries. Out of which about 120 are located at Alwar and rest are scattered at Kota. Udaipur, Jaipur, Jodhpur, Sirohi, Chittorgarh etc.

Most of the Industries identified are Cement and cement product industries in nature. Such industries in the state are around 80 in number. Besides there are 2 chloralkali plants, one major thermal power plant at Kota, 6 Captive Thermal Power plants, at Chittorgarh and Kota. Seven Single super phospate plants, 6 sulphuric acid plants, one nitrogenous fertilizer plant, one gas based power plant, heavy water plant and nuclear power plant, 2 major smelter plants at Khetri and Udaipur and 6 mini steel plants.

The air pollution problem in the State can be characterised in the form of particulate matter, Sulphur dioxide and oxides of Nitrogen. There are few pockets where the possibilities of chlorine (Cl_2), Ammonia, fluorine, Hydrogen sulphide, Carbon-di-sulphide etc. exist in the ambient air in a localisd form. In addition there are no. of pockets, where brick, kiln lime kilns, stone crushers and different grinding units are in existence and causing air pollution in the nearby areas of Beawar, Ajmer, Sirohi, Sawai-Madhopur, Lakheri and Gotan in Nagaur.

The cluster of tiny cement plants in confined to Behror in Alwar, Abu Road in Sirohi, Jodhpur, Sikar etc. The Status of air pollution and its control for major and tiny cement plants is dealt separately.

Similarly a cluster of Chlorinated Paraffin Wax (CPW) is confined mostly to the industrial area of Alwar. This is so because chlorine is one the basic raw materials for CPW production, and it is easily available at Alwar from M/s. Modi Alkalis and chemicals Ltd., the largest manufacturer of chlorine in the state. Their air pollution control status is enumerated separately.

Major Cement Plants

There are 14 major cement industries in the State of Rajasthan out of which 10 industries have provided adequate pollution control measures in the form of Electro static precipitator, bag houses, multi clones and cyclones whereas 2 industries namely M/s. Jaipur Udyog. Ltd. at Sawai Madhopur and M/s. Swadeshi Cement Ltd. at Kotputli are being closed for the last couple of years. Out of the remamign 2 industries, M/s. Udaipur Cement Works, Udaipur is in the process of installing new ESP at one kiln and modifying/fortifying the existing ESP on kiln No. 2 so as to achieve the prescribed norms of the Pollution Control Board. The other units M/s. ACC Lakheri has installed one new ESP on one of the kilns and are planning to provide ESPs on remaining two kilns. They have submitted a new action plan for installing the ESPs by the end of 2019. Major cement units in the state have provided ESP, bag house, multi clones and cyclones. Inspite of the fact that these units have installed the pollution control equipments at various points but due to variation in coal quality, voltage fluctuation, power breakdown and variations in raw' materials received from mines, the units have not been able to stabilise the kiln conditions thereby emitting higher proportions of pollutants into the atmosphere. Besides, number of times the ESP is tripped as the Carbon monoxide levels exceed the permissible level for ESP operations. Under these circumstances the air pollutants are emitted out in the atmosphere without any control device. State Pollution Control Board had visualised these problems and discussed the issues with the representatives of various cement units.

In order to overcome above problems following suggestions have been considered and feasibility of technical, administrative and economic aspects is to be worked out. The variation in the quality of raw mix can be overcome by way of providing stacker-reclaimer so as to make the different grades of raw materials homogenised. The provision of on line X-ray analyser is also considered necessary' to regulate uniform compostion of raw mix to be fed into the kiln.

The variation in coal quality can be minimised or reduced either by way of processing the coal in coal washeries or blending the coal in blending silo to make it homogenised.

The voltage fluctuations, frequent power break down and erratic supply of kiln electricity' can be minimised in case of the operation of kiln is kept under captive power circuit of the unit. This will help in mantaining regular stablisation of the kiln thereby maintaining the emission levels in accordance with equipment performance.

In case the above points are implemented, it is quit possible that the run factor of electrostatic precipitator *v/s* operation of kiln would increase to a considerable extent.

Mini Cement Plants

There are around 36 tiny cement plants mainly of 20 TPD capacity, based on vertical shaft kiln (VSK) technology. These units are required to instal cyclones or wet scrubbers etc. on the stack attached with the kiln. About 35% industries have already installed cyclones or collection chamber on above source. Besides, all the tiny cement plants have hanging type bag filters on transfer points.

Chlorinated Paraffin Wax (CPW) Units

There are around 35 industries producing chlorinated paraffin wax using chlorine and normal paraffin wax as raw materials, The State Pollution Control Board has identified the problem of episodal air pollution alongwith fugitive type of chlorine and hydrogen chloride gas emissions in the early stages of the establishment of these units. The State Pollution Control Board in consultation with experts of Indian Institute of Technology (IIT) New Delhi has recommended the following measure to be adopted by the industrial units to minimise the problem of occurrence of above type of emissions.

- Replacement of existing HDPE exhanger by graphite heat exchanger system.
- Provision of emergency scrubbing system to take care of episodal pollution in an eventuality of any accidental situation along with exhaust arrangements to scrub chlorine.
- Provision of D.G. set and overhead water tank to take care of power breakdown situation. Besides, provision of dyke wall system for the storage of chlorine tonners with sloping floor toward the centre having a catch pit to reduce the evaporative effect due to spillage of chlorine.
- Provision of emergency alarming systems to take care abnormal conditions and initiating timely action.

As all these units are located near Ms/ Modi Alkali and Chemicals at Alwar, there is a strong need to prepare a disaster management plant of all the units combined together and working out the safe optimum number of units that can be allowed in that area.

Single Super Phosphate units

There are seven single super phosphate units in the state, which had provided two stage scrubbing systems. After the constant persuation of the State Pollution Control Board, tw o units have installed three stage scrubbing system whereas one unit is under the pipeline stage of installing the same. The remaining units have to instal three stage system for achieving the norms prescribed by the State Pollution Control Board. Such type of units emit fluorine based gases and particulate matter into the atmosphere.

Sulphuric Acid Plants

There are six sulphuric acid plants in the state, most of them have provided Double Conversion Double Absorption system (DCDA). The predominent air pollutant is in the form of sulphur dioxide and acid mist, which are released in the atmosphere.

Mini Steel Plants

There are six mini steel plants in the state, where the source of significant emissions is the electric arc furnaces. These furnaces emit substantial quantity of particulate matter which are quite reactive in nature. The pollution control in these electric are furnaces is quite complex partly due to the fact that the temperature of the gases are quite high (around 700 to 1000°C) and partly due to the fact that the emissions are of fugitive nature (not being channelised) and there is a big constraint to collect these gases and treat them effectively. In this context, the State Pollution Control Board has timely initiated the action and three units are in the stage of development of pollution control measures and for remaining units necessary action is being initiated.

Thermal Power Plants

There is a maj or thermal power plant at Kota having existing capacities of 2×110 MW and 2×210 MW (640 MW). The power plant has installed ESPs for arresting the particulate matter emissions. The efficiencies of the ESPs are not in accordance with the norms prescribed by the Pollution Control Board for which the unit is proposing to increase the fields of their ESPs thereby increasing the collecting area and consequently reducing the particulate matter emissions. Besides, there are two captive power plants of capacities 35MW and 10 MW at Shriram Chemicals and Fertilizers, Kota where the unit has provided ESPs of adequate capacities to achieve the norms. M/s. Birla Cement Works, Chittorgarh has three captive power plants of capacity 4,6 and 4.5 MW each, where the unit has provided adequate ESPs on 4.0 and 6.0 MW power plants and achieving the norms prescribed by the Pollution Control Board where as for 4.5 MW power plant the unit has provided multiclones, which is not enough to keep the emissions within the prescribed norms. However, they have now given an action plan for providing suitable pollution control system in order to acieve the norms prescribed by the Board.

These thermal pwer plants, whether major or small, contribute particulate matter and significant amount of SO_2, and NO_2 into the atmosphere. As there is no techno-economical solution for the control of SO_2 and NO_2, these emissions are of great concern as they are having commulative effect. The only regulating part is to have a better dispersion of the pollutants in the atmosphere thereby reducing the ground level concentrations.

Brick Kilns

There are hundreds of brick kilns in different parts of the state, which emit particulate matter into the atmosphere having localised effect in the nearby areas. No pollution control system has been adopted by these brick kilns. The main constraint appears to be that the location of the stack is at the kiln top where it may not be feasible to instal any control equipment due to weight factor. There is a possibility of providing down draft system and provide the stack at some suitable distance from the kiln to enable the control system to be installed in between. This may even reduce the fuel consumption and can be an economical solution of the problem. The feasibility of this sytems should be tried and if found successful, it should be implemented effectively.

Stone Crushers

There are hundreds of stone crushers in different parts of the state, where no control system has been provided as a result of which, significant amount of particulate matters are released into the atmosphere thereby causing localised problem of air pollution. The Central Pollution Control Board has studied this problem and prepared the document wherein measures have been suggested to minimised the dust levels. The measures to be provided are in the form of water spraying and providing containment system at all the transfer points.

Environmental Pollution and its Control in mining activities

Having significant mineral wealth in the State, the mining activities are consequently extensive and varied in nature. Almost all the part of the state is subjected to mining activities. Limestone mining is carried out in Udaipur, Chittor, Kota, Bundi, Sawai Madhopurbelt and at Kotputli, Sikar, Alwar, Jodhpur, Nagaur (Gotan), Sirohi etc. These mining operations are open cast in nature. Khetri in Jhunjhunu district and Dariba in Alwar district are the areas where copper mining takes place which is open cast as well as underground in nature, mining of zinc and rock-phosphate are prevalent in Udaipur whereas mining of building stone is predominant in Jodhpur region Bhilwara and is known for soap stone mining activities.

The mining operation in the State can be classfied in the following manner:

(a) Organised open cast mining without benefication like that of lime stone mining.

(b) Organised underground mining without beneficiation like that of tungstone, soap stone.

(c) Organised underground mining with beneficiation like that of copper mining etc.

(d) Non-organised open-cast mining like that of building stone, soap stone, mica, marble, felspar etc.

Water pollution

Mining operation in following ways contribute towards water pollution.

(a) Mine water pumped out during drainage operation.

(b) Spent water from handling plants, duest extraction and dust suppression systems.

(c) Effluents from preparation and beneficiation plants.

(d) Leachates/wash offs from waste/tailing dumps.

Acid mine drainage is serious problem which occurs wherever a mine of any type in permeable formations interacts with the water tank, aquifer, porched water body or where surface water finds its way into a mine in terrains where sulphides (particularly pyrites) are present in the ore or country rock. The acid mine drainage contributes to low pH and high level of sulphates, iron and total dissolved solids. These deplete oxygen level in water, increase toxicity by rendering heavy metals soluble and create corrosion problems.

Water Pollution Control Strategy

Some of the important control strategies have been listed hare under :

(i) Efforts should be made to preclude or reduce the dscharge of toxic and objectionable effluents into surface water bodies, ground water aquifers or usable lands to a minimum. Whereaever possible, reuse of waste water should be practised.

(ii) Wherever practicable, the mine water should be utilised for irrigation of the plantation raised to stabilize the mine waste dumps.

(iii) The control of acid mine drainage depends upon the prevalent situations.

The available techniques in this regard are as under

(i) To prevent the formation of acid mine drainage at source.

(ii) To control the acid mine drainage, the techniques include deep well injection, subsurface dams and grout curtains.

(iii) To comply standards, waste water treatment methods for enutralisation and removal of dissolved solids, the techniques include lime treatment accompanied by aeration or oxidation process to convert ferrous iron to ferric iron, neutralisation with soda ash, caustic soda, and anhydrous ammonia, reverse osmosis, ion exchange, desulfating, sulfide iron removal, microbiological control and permanent iron removal.

(iv) Waste with abnormally high concentrations of iron sulphides of other undesirable reactive elements should be disposed-of in sanitary landfills (solid waste).

(v) The tailing area decant ponds must be adequately deep to provide a quiscent settling zone for the particles. Wind and wave action would prevent settling of the pond which is too shallow1.

Air Pollution

The air pollution caused by mining and associated activities can be classified into :-

(a) Gaseous pollutants like sulfur-di-oxide, nitrogen-oxides, carbon mono-oxide and hydrocarbons.

(b) Suspended particulate matter like silica, fluorides, asbestos and metallic mineral dust.

Air pollutants from mining come from two types of sources, that is point and non-point sources. Point sources of air pollution are from drilling, crushing and screening, machinery, convey points etc. The non-point sources of air pollutants are from blasting (dust and combustion gases). The gaseous pollutants are released at the time of blasting and its effect is localised.

Air Pollution Control Strategy

The measure that can be used to control air pollution are:

(a) Dust suppression and extraction system should be provided with blowers, cyclones, multilcones and wet scrubbers depending upon the type of problem and prevailing situation of the ore handling plants i.e. crushing and grinding units.

(b) In case of open cast-mines, the haul roads are to be sprayed with water to suppress dust.

(c) Transportation of minerals from mines to plant is the main source of dust. Suitable chemicals or water maybe sprayed on the top of transport equipments to prevent dust effect. In the recent times wholly sealed containers have been used for transportation of metallic concentrates.

(d) The placement of hoods at transfer points to reduce the dust should be adopted.

(e) Dense belt of trees should be planted on road sides. These trees would serve as dust arresters. It has been reported that one hectare of trees can arrest 30-50 tonnes of dust. Dust generated from tailing dumps can be controlled by doing surface stabilisation by growing grass or herbs.

(f) The working environment level of pollutants should be monitored regularly or even continuously and in case the level exceeds the prescribed limits, the balsting or other related operation should be stopped till the pollutants are dispersed and the level of pollutants in the ambient air is brought down with in the acceptable limits.

Water Pollution its Control

The problem of industrial water pollution in the State is mainly confined to Kota. Alwar, Udaipur, Jodhpur, Pali, Balotra, Sanganer, Bagru etc. and that of domestic waste water to Kota, Udaipur, Jaipur and other major towns of the state. It would be worthwhile to mention that Udaipur, Kota, Jaipur, Ajmer etc. are the areas where possibility of surface water pollution is always there whereas in Jodhpur, Pali, Balotra, Sanganer, Bagru, Khetri, Deari (HZL) etc., the possibility of ground water pollution is quite significant.

There are around 300 major and medium scale water polluting units. Out of which 103 have installed the effluent treatment plants. In 20 such industries the effluent treatment plants are under construction and in 62 units the ETP is proposed as per approved feasibility report. 42 industires have made the arrangements for zero discharge by adopting cycling process or by changing their production process. 37 units have provided partial arrangements for water pollution control. The Board has been persuading 36 units to provide adeaute arrangement for the control of water pollution.

So far 228 cases against 157 water polluting industries have been filed under Water (Prevention and Contorl of Pollution) Act 1974; out of these 61 cases have been decided out. Of these cases, 27 cases have gone in favour of Board and 34 cases against the Board. Remaining cases against 167 industries are pending in different courts.

With the constant efforts and persuation of the Board the small scale textile industries have installed caustic recovery' paint, continuos mercerising plant and steam agers. In Pali 90 industries have adopted continuous mercerisation process instead of conventional double padding. For caustic recovery, 2 textile industries have installed caustic recovery plant. The main object of the arragements is to reduce the quantity' of caustic in the effluent.

There are 3 sugar industries in the state, one each at Chittorgarh, Sriganganagar, Bundi. Action under warer Act has been taken against 1 industry and 2 have been asked to submit action plant. There are 4 distilleries in the state namely at Udaipur, Sriganganagar, Kota and Chittorgarh. Only 1 unit has ETP and rest 3 are still to take effective measures. Besides, there are 4 tanneries namely one at Tonk, two at Bhiwadi and one at Sikar. Only one industry' has installed ETP closure order have been passed against one industry' and action is being taken against the others.

Rajasthan has 11 industries of paper and pulp of capacity, 3 TPD and above. All these units are lying closed at present. However, one unit has provided partial treatment plant and one unit is disposing the effluent in the factory premises itself.

There are 4 major fertilizer plants, 4 medium and small phosphatic fertilizer plants in the state. All the major plant have installed ETPs but they are not running satisfactorily at one place, Out 4 medium small scale plant, 2 are not discharging any waste water and selling the waste water to other units one is lying closed and one has installed ETP.

In view of above, it is evident that concerted efforts are required to contain water pollution in the state. Industries being quite small, financial constraint is being invariably felt in installing individual treatment plant. Therefore, State Board has perceived the problem of water pollution in the right and apt perspective and consequently it is proposed to instal common treatment facilities/ plants at Pali, Balotra, Jodhpur and Sanganer.

There are about 750 textile processing units in the SSI Sector in Pali, 159 in Jodhpur and 250 in Balotra. Processing of cotton fabric includes desizing, mercerising, kiering, bleaching, dyeing and printing. The combined effluent of these processes is alkaline with pH (9.12) and having hight TDS alongwith sizeable concentration of TSS, BOD and COD. Waste can be classfied as not easily bio-degradable because of its complexity and is potentially damaging the surface and ground water as well as the land and vegetation.

It is estimated that the combined industrial effluents, from tehse textile processing units at Pali, Jodhpur and Balotra are approximately 18,000 KLD, 15,000 KLD and 2500 KLD respectively. Since Pali is an unsewered city, entire domestic as well as industrial effluents of the city is discharged in Bandi river.

As regards Jodhpur and Balotra, the waste water is joining river Jojari and Luni respectively, which happen to be seasonal rivers. Seasonal nature of these rivers allows the waste water to flow without any dilution thereby polluting the subsoil water. With rainfall, the waste residues/ effluents flush out and pollute the down stream wells and the agricultural fields.

Since the units under reference are in SSI sector and the effluent is not easily biodegradable, it is difficult for each industrial unit to treat its effluent so as to attain the prescribed standards. This situtation has necessitated the concept of common effluent treatment Plant (CEPT), at the terminal point.

One such CETP is already in existence in the Mandia Road Industrial Area (MRIA) Pali to treat only 1.0 million gallons (MGPD) per day out of 1.6 MGPD generation is from MRIA. Looking to the quantity of waste water, two more CETPs have been proposed by NEERI. one of 6 MGPD capacity at the Juncture of Gandhi Nagar Nala and Bandi river and another of 3 MGPD at MRIA.

Physical work of one more CEPT at Pali and one CETP at Jodhpur is likely to be taken up early during the year 1993 -94. In case of Jodhpur, a feasibility report prepared by Small Industries Development Bank of India (SIDBI) has already been cleared by the Board, while NEERI has carried out detailed scientific investigation for additional CETP's at Pali. In both the cases, the sources of funds for the CETP have been identified, and land for construction is being set apart. NEERI has also been entmsted with the task of preparing a feasibility report for the textile waste at Balotra.

Environmental Management Policies

India has plethora of legislations which are directly or indirectly meant for the protection of environment. People at large asalso their representativ es believe almost at the level of conviction that introduction of a piece of legislation alone is capable to set the maladies in order, but inadequate infrastructure alongwith absence of integrated planning and mangement aspects lead to proliferation of such legislation.

The Pollution Control Boards in India have already com; leted their first decade of existence and perhaps it w ould not be out of place to suggest an objective evaluation of their performance to be able to answer the question: what effect have our management policies and their implementation had on our environment, our resources and our develop-ment so far. Evaluation could include some relevant parameters to see what effect we have had so far on the following:

1. The qualtiy staus of our water bodies and ambient air
2. Land use planning and planning for protection of water resources, fisheries and special eco-systems.
3. Mobilization of financial resources and use of finanacial incentives judiciously.
4. Choice of manufacturing technology, fuel cost effective waste treatment processes, optimisation in pollution control techniques, etc.
5. Improvements in operation, main-tenance and reliability of performance, etc. Handling and use of toxic chemicals, heavy metals, pesticides, etc.
7. Man-power development.
8. Functions enumerated u/s 17 of the water and air acts and other duties enumerated under different acts.
9. Public awareness system, procedure and approach.
10. Consulting practice and development of environmental research oriented programmes.
11. Environmental training and education programmes.

The pollution control authorities and the Central/State Department of Environment would have to critically evaluate and judge for themselves what effects their policies are having on each of the above mentioned aspects. It is presumed that earlier activities were by their veiy newness, adhoc in nature, but gradual experience has now led to review of initial programmes and new staffing should have brought in fresh blood with new ideas and better training in systems approach for effective planning and management.lt is the time right now where we should be able to evaluate and assess our earlier policies and programmes and thereafter formulate process of integrated planning and management and strategies for meeting the challenges likely to be confronted, over the coming years. However, following priority areas may be considered for effective integrated planning, management and implementation within a time bound frame work.

1. Strenthening of monitoring net-work to be able to establish periodically the quality status of our water bodies and ambient air alongwith regulating the compliance of emission norms from the polluting source.
2. River basin studies to be undertaken comprehensively with a view to have an integrated planning and management of various environ-mental domains.
3. Generation of base line data alongwith compilation and simulation of available data. This can provide useful imput to the process of planning.
4. Environmental information system be established whereall type of literature, data information, research work,etc. should be made available.
5. Mass awareness and campaign education system should be established with specific programmes coupled with evaluation system to assess the output.
6. Environmental research and develop-ment activities based on field applications should be encouraged at all levels.
7. Site selection of industries, industrial estates, mining areas, power projects, dams etc. should be made after undertaking comprehensive EIA study.
8. Man-power development and achieve-ment of professional competence, to takle environmental problems alongwith development of adequate infrastrusture should be undertaken.

Heritage for Environmental Protection

The roots of environmental conscious-ness are burried deep in the soil of Rajasthan. Maharaja Abhai Singh of Marwar (1724-49 A.D.) wanted to build a palace, for which timber was required. Village Khejarli located 25 kms. from Jodhpur and inhabited by the Bishnoi community, had Khejri trees in abundance. Maharaja's men went to cut the trees which was resisted by the Bishnois. Amrita Devi embraced a Khejri tree and challenged the soldiers that they could cut it only over her dead body. Amrita Devi, three of her daughters and his husband were killed by the solodiers. The other Bishnoi villagers followed suit, and met the same fate. In all 363 persons sacrificed their lives to protect the Khejri trees. Ultimately, the Maharaja himself ordered the stoppage of the carnage-both of men and the Khejri trees.

Chapter 8

NATURAL VEGITATION AND FORESTS

The state of Rajasthan is endowed with a wealth of a wide range of vegetaion. They can be discussed in two broad categories i.e. into two distinct groups–one comprising the arid vegetaion, falling into western part of the state while the other belonging to semi-arid to sub humid of eastern and southern Rajasthan.

A. VEGETATION OF WESTERN ZONE

Being largely influenced by edaphic and biotic factors, the western Rajasthan is characterized by sparse vegetation. Vegetation of this arid tract has well adapted to existing arid conditions. The potential vegetation of this region has been modified due to intense biotic factors. The vegetation of western Rajasthan has been grouped under the following categories based on two prominent indices i.e. density and frequency. These have been used to identify the dominant and co-dominant species of vegetation.

1. Calligonum-Haloxylon-Leptadenia Type

It is represented in the area of large stabilized longitudinal, parabolic and transverse dunes having the lowest rainfall zone (10 to 15cm). Only shrubs an under-shrubs which are well adapted to prevailing conditions are predominant in this tract. This region is represented by psammophytic Scrub Desert Vegetation. Phog is the major dominant shrub on the dunes in Barmer region while it is associated with Haloxylon in Jaisalmer tract.

2. Salvadoran Oleoides-Euphorbia Caducifolia Type

A distinct habitat of good vegetation along runnels and scanty vegetation in small pockets of soil deposition is represented on the vast rocky plateau of Jaisalmer. Salvadora is the dominant species on this rocky areas with negligible soil cover, here euphorbia caducifilia (Thor) form the main associate. Many other lithophytes forming shrubs colonize the tract.

3. Zizyphus Nummularia-Capparis Decidua Type

The exposed and buried pediments and gravelly plains of Jaisalmer with poor sandy loam or loamy and soil cover allow highly adapted species to colonize this habitat. Due to the adaptive capability to the adverse desertic conditions Ziziphus and Capparis are present here. In many regions, largely scattered stunted tree species of Prosopis and Acacia etc are prevalent. The limited associated species of the lithophytic Scrub Desert are also present.

4. Suaeda Fruticosa-Salsola Baryosma Type

This type of vegetation is generally present in low lying saline basin and depressional areas, with one percent slope. In western Rajasthan, these are located at Malhar (Jodhpur), Pachpadra, Uterlai, Thob (Barmer) Lawan, Pokhran etc. mostly Suaeda dominates these tracts.

5. Prosopis-Capparis-Zizyphus Type

The region is dominated by Trees-Prosopis cineraria, Tecomella undulata etc. and Shrubs- Capparis deciduas, Calotropis procera etc. Large part of the flat plains in 25 to 30 cm rainfall zone is dominated by this type which covers the older alluvial plains and buried pediment plains. Due to cultivation the natural vegetation of this zone is highly modified.

6. Prosopis-Tecomella Type

In the sandy undulating plains and ridges of low dunes with very deep soil profile in the district of Barmer, part of Churu and Sikar exhibit the prevalence of this type. As the farmers themselves conserve these plants so species have their predominance in arable lands also. Trees- Prosopis cineraria, Tecomella undulata, Salvadora oleoides, Balanties aegyptiaca. Shrubs- Calotropis procera, Leptadenia pyrotechnica, Lycium barbarum, Acaica jacquemontii, Clerodendrum phlomoides, Ziziphus nummularia, Mimosa Hamata.

7. Prosopis Cineraria-Acacia Nilotica Type

The flat alluvial plains of Shekhawati region having deep to very deep, sandy loam soils support this type present at Nagaur, Sikar, Churu and Jhunjhunu. The low dunes and undulating hummocky plains in the region show the dominance of Prosopis only. The irrigated fields of in flat plains are invariably occupied by both Prosopis Cineraria-Acacia Nilotica species in varying density. Prosopis cineraria show high density. Trees- Prosopis cineraria, Acacia nilotica, Ailanthus excels, Acacia cupressiformis, Ziziphus maruitiana, Tecomella undulata. Shrubs- Zyziphus nummularia, cappairs deciduas, lyceum barbarum, calotropis procera.

8. Salvadora Oleoides-Prosopis Cineraria-Capparis Decidua Type

The flat alluvial plains with heavy soils of sandy clay loam to clay, deep to very deep and moderately saline soils of district Pali, Jalore, southern Barmer, north eastern tract of Churu and Jhunjhunu support this type. Soil column and hardness of the pan is very much reflected by the growth of tree species. Trees- Salvadora sleoides, S. perisca, Tamarix articulate, Acacia nilotica, A. leucphloea, Balanites aegyptiaca. Shrubs- Indigofera oblongifolia, Calotropis proceera, Tamarix ereocoides, Cassia auriculata, Capparis deciduas, Zyziphus nummularia.

9. Acaia Nilotica Type

Deep buried pediment plains, flat alluvial plains with deep sandy clay loam soils but non-saline at the foot hills of Aravalli range in district Pali, Ajmer and part of Udaipur can be witnessed with the predominance of these two species. Trees- Acacia nilotica, Acacia cupressiformis, Azadirachta indica, Salvadora oleoides, S. persica, Tamarix articulate, Ziziphus mauritiana, Moringa oleifera, Albizia lebbeck and Ficus supp. Shrubs- Cassia auriculata, Capparis deciduas, Zyzyphus nummularia, Indigofera oblongifolia.

10. Acacia Senegal-Euphorbia Caducifolia Type

The granite and sand stone hills of Rajasthan are generally well vegetated while limestone and rhyolite hills support very poor vegetation. The protected hills in the rainfall zone of 20 to 35 cm support Acacia Senegal as the dominant forest type. Low hills / rocks, generally unprotected support Euphorbia caducifolia scrub. Trees- Acacia Senegal, Maytenus emarginatus, Anogeissus pendula, Moringa concanesis, Wrightia tinctoria. Shrubs- Euphorbia caducifotia, Capparis deciduas, Ziziphus nummularia, Grewia tenax, Cordia gharaf.

B. VEGETATION OF EASTERN ZONE

The Eastern Zone constitutes the plains, valleys, deeply buried or exposed pediment plains and the Aravalli range. The zone lies in the region which receives rainfall between 50 and 80 cm. The plains lie in the double cropping region and hence the existing vegetation is largely modified due to cultural interferences. Degraded vegetation is present on the rocky wastelands, which is a result of the high biotic impact (Figure 8.1).

1. Acacia Leucophloea-Prosopis Cineraria-Acacia Nilotica Type

Acaica leucophloea dominates the shallow to moderately deep, buried and exposed plains whereas A. nilotica is the main species in the plains with deep soil in western part of Jaipur, Tonk, Alwar and Bhilwara district along with Jaisamand and Ajmer and north western Udaipur. Plains show sparse tree density as the region is extensively cultivated. Prominent Trees are- Acacia leucophlea, A. Senegal, A. nilotica, Aegle moremlos, Azadirachta indica Balanites aegyptiaca, Ficus bengalensis, F. religiosa, Propis cineraria Salvadora perisica, S. oleoides, Phonix sylvestris, Pithecellobium dulco Ziziphus mauritina, Mangifera indica, Madhuca indica. Prominent Shrubs are- Acac jacquemontii, Calotropis procera, Mimosa hamata, Capparis deciduas, zeylanica, C. sepiaria, Dichrostachys cinerea, Cassia auriculata, Lanta camera, Zizyphus nummularia.

2. Acacia Nilotica Type

On the north east, east and south east of the Aravalli range the alluvial plains are largely under double cultivation and dominated by this type of specie. The density of trees is general limited due to cultivation but they exhibit good growth due to good irrigation. Dominate Trees- Acacia nilotica, A. cupressiformis, Azadirachta indica, Mangifra indica, Madhuca indica, Phoneix sylvestris, Ziziphus maauritiana, Ailantl excels, Ficus bengalensis, F. religiosa, Pithecellobium dulce, leucopholea, Cordia dichotoma, Prosopis cineraria. Dominant Shrubs- Zizyp nummularia, Capparis deciduas, C. zeylanica, Ficus palmate, Kirgan reticulate and Lantana camara.

3. Acacia Nilotica-Capparis Decidua Type

This vegetation type is present at dissected ravine terrain along the Chambal Valley covering Dholpur, Sawai Madhopur and Kota districts. This dissected upland has dry situation where as at all lower level, the flat land experiences fast flow of water.

4. Butea Monosperma-Madhuca Indica -Ziziphus Mauritiana Type

Banswara, Dungarpur, South Chittorgarh and Udaipur districts with undulating pediment plains and rocky valleys exhibit this type of vegetation. Here Madhuca indica and Zizphus maruitianna occur in place with deep soil deposition but Butea monosperma (Dhak) dominates the terrain. The dominant Trees here are- Madhuca indica, Azadirachta indica, Zizyphus maruitiana, Mangifera indica, Phoenix sylvestris, Aegle mormelos, Acacia leucphloea, Cassia fuistul and Balanits aegyptica. Dominant Shrubs are- Nyctanthes arbor-tristis, Holerrhena pubescens, Euphorbia neriifalia, E. vinulia, Dendrocalamus, Cassia auriculata, Dichrostachys cinera.

5. Anogeissus Pendula-Boswellia Serrata Type

Alwar, Ajmer, Bhilwara, Jaipur, Tonk, Sawai Madhopur, western Chittorgarh and eastern part of Udaipur are largely dominated with this form of major mixed deciduous vegetation of the Aravalli range. Vegetation at different slope zones is- higher slopes- Anogeissus pendula, Boswellia serrate, Sterculia urens, middle slopes- Acacia Senegal, Bauhinia racemosa, Cassia fistula, Capparis sepiaria, Dichrostachys cinerea, Diospyros melanoxylon, Lannea cormandelia, Wrightia tinctoria. Angle mormelos, Adina cordifolia. Lower slopes and valleys- Butea monosperma, Cassia auriculate, Acacia leucophloea,

Dendroclamus strictus, Euphorbia caducifolia, Holoptelia integrifolia.

6. Tectona Grandis Type

This tract is largely dominated by tectona grandis (Teak) and its main associated species- Bosellia serrata and dalbargia latifoila. It is present in the hilly areas of south-eastern part of Udaipur, southern Chittorgarh and Banswara district. The regions with gentle slope mostly exhibit a good density of trees and shrubs species. Vegetation type changes with change in slope zones. On the Higher slopes- Bauhinia racemosa, Boswellia serrata, Emblica officinalis; Middle slopes- adina cordifolia, aegle mormelos; and Lower slopes - Holerrhena pulsescens, Euphorbia caducifolia etc. are seen.

7. Mangifera Indica-Syzygium Cumini Type

Some important species of this type distributed in this zone are: Butea monosperma, Anogeissus, Moringa concanensis, Anogeissus latifolia, Magnifera indica, Sterculia urens etc. This type of vegetation is present at Mount Abu and the composition of this vegetation is the reflection of the change in them due to the increase in altitude. The region supports dry deciduous, semi-deciduous and evergreen species of plants.

FORESTS

The floral and faunal wealth of forests in the state of Rajasthan is very rich and varied. According to Principal Chief Conservator of Forest, Government of Rajasthan, Jaipur (2016), Rajasthan has 32,828.37 sq. km. of forest area which is about 9.10 per cent of the total geographical area of the state.

The western half is of desert terrain, while most of the area under forests is restricted to eastern and southern parts of the state. The forests are unevenly distributed in vanous districts. Most of the forests are over the hilly area i.e. in Udaipur, Rajasmand, Kota, Baran, Sawai Madhopur, Karaulli, Chittorgarh, Pratapgarh, Sirohi, Bundi, Alwar, Jhalawar, Dungarpur and Banswara districts, which make up for about 80 per cent of the forests of the state.

Dense natural forests are found in protected patches, mostly confined to various national parks and wild-life sanctuaries. Rest of the remaining forests of the state are of varying density. The forests of the state can be divided into three broad forest types; (1) Tropical Thom forests, (2) Tropical Dry Deciduous Forests, and (3) Central Indian Sub-tropical hill forests.

1. Tropical Thorn Forests

Tropical thorn forests are found in arid and semi-arid regions of western Rajasthan. These extend from western Indo Pak border and gradually merge with the dry deciduous mixed forests of the Aravalli hills and the south-eastern plateau. The general composition of vegetation here is the desert and scrub type. These can be further classified as:

(I) Sandy Plains-Under this head following types of forests have been recognized. (i) Desert thorn forest, (ii) Zizyphus scrub forest, (iii) Tropical Euphorbia scrub forest, (iv) Acacia Senegal forest, (v) Rann saline thorn scrub forest, (vi) Salvadora scrub, (vii) Cassia auriculata scrub and (viii) Desert dune scrub.

(II) Saline Flat-Salvadora-Tamarix scrubs Older alluvial plains which have developed alkalinity/salinity and are left uncultivated for a longer period of time are generally dominated by Salvadora oleoides-Tamarix community. Acacia nilotica is also found in the outer periphery of such lands.

Species composition

Tree Canopy- Salvadora oleoides, S.persica, Tamarix articulata, T. aphylla, Prosopis cineraria etc.

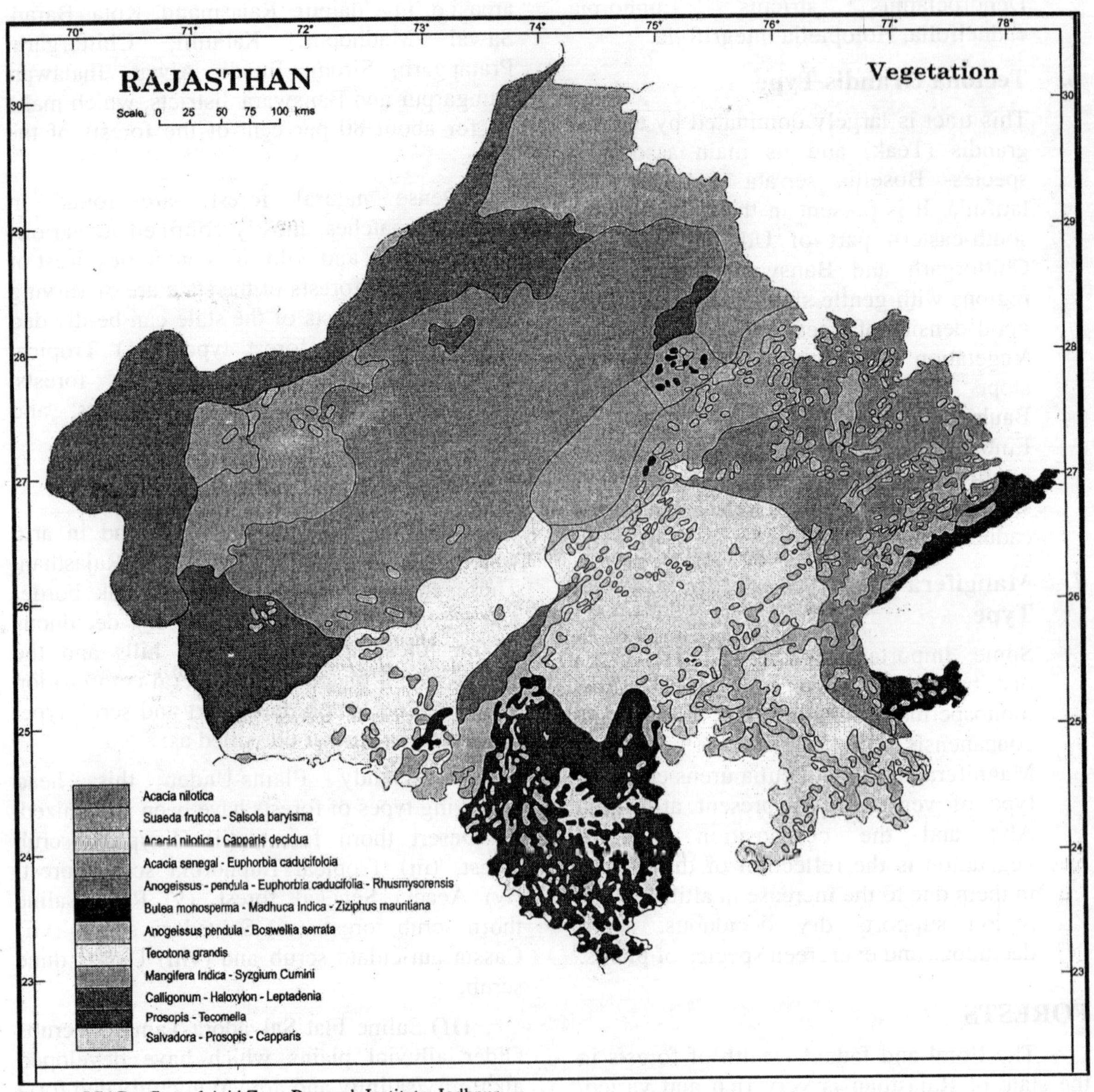

Source: ICAR - Central Arid Zone Research Institute, Jodhpur

Figure 8.1 Vegetation of Rajasthan

Shrub layer Capparis decidua, Tamarix dioca, Zizyphus nummularia, Calotropisprocera and Randia spp etc.

Under Shrubs Suaeda fruticosa, Haloxylon salicornicum, Salsola baryosma, Argemone mexicana etc.

Grasses Sporobolus marginatus, Eleusine compressa, Chloris virgata, Dactyloctenium aegyptium, Tragus biflorus, Aleuropus lagopoides etc.

2. Tropical Dry Deciduous Forests

These forests are mostly found in small patches in few parts of the state. The northern and eastern slopes of Aravalli ranges, mostly in Alwar, Bharatpur and Dhaulpur districts, are covered with this type of forests. Sporadic growth of certain species of dry deciduous forests is found along the dry river beds of Jalor, Nagaur, Ganganagar, Hanumangarh and Bikaner districts. These forests are further divided into following seven sub-groups.

Acacia nilotica sub. Sps Indicaforests

The younger alluvial plains and depressions along the banks ofriver and rivulets are dominated by A. nilotica. The tree assumes gregarious growth under rich soil moisture conditions. Being the most common tree species of the tract, it is mostly lopped for top-feed. The mature trees are harvested or cut for timber and fuel purposes.

Tree canopy Acacia leucophloea, Azadirachta indica, Balanites aegyptiaca etc. Shrub layer Capparis decidua, C. grandis, Zizyphus nummularia etc.

Grasses Heteropogon contortus, Chrysopogonfulvus, Dichanthium annulatum, Bothriochloa pertusa etc.

Butea monosperma forests

Rocky plateau, piedmont plains, rocky out-crops, valley and gravelly plains above 40 cm rainfall zone, specially in the district of Sirohi, Udaipur, part of Pali, Ajmer, Alwar and Chittorgarh, have sqattered, stunted maltreated trees of Butea .

In many regions, it forms small thickets which are almost pure.

1. Shallow, moderately heavy to heavy soils are generally occupied by this species. Species Composition

Tree canopy Butea monosperma, Acacia leucophloea, Balanites aegyptiaca, Zizyphus mauritiana etc.

Shrub layer Zizyphus nummularia, Mimosa rubicaulis, Indigo/era oblongifolia, Dichrostachys cinerea, Maytenus emarginatus, Carissa congesta etc.

Herbaceous species Tephrosia purpurea, Indigo/era cordifolia, Pulicaria wightiana etc.

Grasses Apluda mutica, Heteropogon contortus, Iseilema laxum, Bothriochloa pertusa, Dichanthium annulatum etc.

Anogeissus pendula forests

Anogeissus pendula (commonly known as 'Dhok', 'Dhokra', 'Kalia, Dhav) trees generally form pure stands, often quite well stocked, 6 m high and with a 70 to 80 cm girth. The branching takes place at a height of 1 to 1.5 m height. They provide red hue in winters between November and February but by March, the trees become leafless. Shrubs, undershrubs and ground flora are usually abundant in these forests. The hills of Ajmer, Udaipur, Rajsamand, Chittaurgarh, Bundi, Sawai Madhopur, Karaulli, Kota, Baran, Alwar, Jaipur and parts of Pali districts are dominated by these forests. The hill slopes are generally gradual and well occupied by this species.

Species composition

Tree can opy Anogeissu s pend ula, Aca cia leucophloea, Diospyros cordifolia, Sapindus emarginata, Acacia catechu, Butea monosperma, Azadirachta indica, Ficus glomerata, Lannea coromendelica, Sterculia urens, Boswellia serrata, Tamarindus indica, Albizia odoratissima and Wrightia tinctoria, etc.

Shrub layer Dichrostachys cinerea, Euphorbia caducifolia, Zizyphus mauritiana, Grewia tenax, Rhus mysorensis, Securinega leucopyrus, Mimosa rubicaulis, Pavonia spp. etc.

Undershrub layer Adathoda zeylanica, Capparis sepiaria etc.

Herbaceous species Cassia tora, Indigo/era cordifolia, Borreria hispida, Argemone mexicana, Leucas aspera, Tridaxprocumbens, Sida cordifolia etc.

Grasses Eragrostis spp., Apluda mutica, Chrysopogon fulvus, Cymbopogon martini, Eremopogon foveolatus, Heteropogon contortus, Sehima nervosum, Themeda quadrivalvis.

Mixed Miscellaneous Forests

These forests occupy most of the eastern gravelly/rocky plateau region and Aravalli ranges of Rajasthan. They occur in Alwar, Banswara, Bhilwara, Sirohi, Pali, Bundi, Chittaurgarh, Dungarpur, Kota, Sawai Madhopur, Karaulli and Udaipur districts of the state. A few pockets of these types exists in Udaipur and Dungarpur districts on high elevations in warm moist climate.

Species composition

Open, very dry and degraded teak forests occur on shallow, clayey soils in southern and eastern part of Rajasthan. Where, the annual rainfall varies from 75 to 90,cm. In these areas, teak trees are in various proportions but rarely in pure stands. The under-growth is generally patchy due to heavy grazmg.

The floristic composition of dry teak forest (mixed species) in protected situations show distinct stratification and canopy formation . The top canopy consists of Tectona grandis, Anogeissus p endula and A. latifolia. The forests shed their leaves in the beginning of the winter season and remain leafless till June until the rainy season starts.

Tree canop y-Anogeissus latifolia (Dhavadal Bakali); Lannoa coromendelica (Semut); Holoptelea integrifolia (Kanju/ Pitpapri); Albizia odoratissima (Sirus); Sterculia urens (Kalu gum); Diospyros melanoxylon (Tendu); Mitragyna parvifolia (Haldi); Emblica officinalis (Aonla); Haldina cord ifolia (Haldu); Acacia catechu (Katha/Khair); Peltophorum ptero carpum (Copper pod tree); Sapindus laurifolious (Reetha); Bauhinia racemosa (Asundro); Salmalia malabarica (Semal); Hardwickia binata (Anjan); Tamarindus indica (Imli); Terminalia bellerica (Baheda).

Shrub layer-Lagerstroemia parviflora (Landi); Derris indica (Karanj); Bauhinia purpurea; B. variegata (Kanchan/Kachnar) ; Carrisa congesta; Helicteres isora, Securingea leucopyrus; Rhus mysorensis (Dansri); .May tenus emarginatus (Kankera); Balanites aegyptiaca (Hingota); Wrightia tinctoria (Kerno, Bhakar-Aak) Zizyphus nummularia (Beri/Bordi); Capparis sepia ria; Grewia jlavasc ens; Carissa spinarum; Woodfordia fruticosa; Euphorbia nerifolia (Danda thor); Euphorbia nivulia; Dichrostachys cinerea (Kolai); Lantana camara (Lantana) etc.

Undershrub layer- Solanum incanum; Mimosa hamata;Adhatoda zay lanica etc.

Climbers Cissus repanda, C. quadrangularis, Cissampelospareira, Pueraria tuberosa etc.

Herbaceous layer-Cassia tora, Indigo/ era spp., Atylosia spp., Borreria articularis, Ocimum americanum, Sida Tridaxprocumbens, etc.

Grasses-Themeda quadrivalvis, Apluda mutica, Dichanthium annulatum, Heteropogon contortus, Eragrostis spp., Andropogon pumilus, Aristida spp. etc.

Dry Bamboo Brakes

Bamboo brakes show a marked preponderance in some patches of south-eastern Rajasthan which are at some places biotically disturbed but due to copious rains, ranging from 80 to 90 cm, they grow in dense patches, mixed with grasses and thorny shrubs. Usually, bamboos are found in Banswara, Sirohi, Udaipur, Chittaurgarh and Alwar districts. Here, occurrence of Bamboo (Dendrocalamus strictus) is purely secondary, i.e., it is found mixed with other dominating species. The low hill patches and rocky piedmont plains, merging with undulating terrains, support this vegetation with widely scattered trees and shrub species like Anogeissus pendula, Acacia leucophloea, Tectona grandis, Maytenus emarginatus, Mimosa hamata and Capparisgrandis.

Aegle marmelos Forests

This forest type is usually associated with mixed miscellaneous forests having rich moist conditions and comparatively well-drained soils.

Species composition

Tree canopy-Aegle marmelos, Acacia leucophloea, Wrightia tinctoria etc.

Shrub layer-Nyctanthes arbor-tristis

Herbaceous specie s-Indigo,fera spp., Borreria articularis, Leucas spp., Evolvulus alsinoides, Ipomoea spp., etc.

Grasses-Heteropogon contoruts, Dichanthium annulatum, Apluda mutica etc.

3. Central Indian Sub-tropical Hill Forests

These forests which are most abundant in central India, as in Madhya Pradesh, parts of Gujarat and Maharashtra, are found in Sirohi district of Rajasthan also, mostly on the hills girdling Mt. Abu. These forests have semi-evergreen and some evergreen species of trees. The vegetation of Mt. Abu consists of many plants which are similar to the sub-tropical region of Himalayas. Around Mt. Abu, they are well represented between 700 to 800 m altitudes .

The dominating species are of Bauhinia racemosa (Kachnar), Ma/lotus phillipensis, Mangifera indica, syzigium cuminil, Anogeisus latifolia, Emblica officinalis, Salmalia malabarica, Dalbergia sissoo, Aegel marmelos, Mitragyna parvifolia, Ougeinia oojeinensi and Ixora arborea Ficus glomerata and Dendrocalamus spp. Erythrina suberosa, Anogeissus sericea, etc.

Shrub Layer-Carissa spinarum, Rosa muashcata. Grasses-Cymbopoga martini, Sehima nervosum, Apluda mutica.

Boswellia serrata Forests

On the high hills of Banswara, Sirohi, Pali, Ajmer, Alwar, Udaipur, Chittaurgarh, Boswellia dominates the forest tract. It is associated with stunted trees and shrubs of deciduous nature. At some places, pure stands can be seen showing white shining poles of 12 to 15 m height and 80 to 120 cm gilth. Sterculia urens, though few in number, gain similar height. Both these species become leafless by February. Boswellia is fire resistant and is generally left alone when other trees are removed. It does not coppice well. It occupies higher zones in Anogeissus forests.

Species composition

Tree canopy Boswellia serrata, Lan nea coromendelica, Sterculia urens, Anogeissus latifolia (stunted) and Anogeissus pendula.

Shrub layer Helicteris isora (Maror phali) and Grewia flavescens, etc.

Grasses-Apluda mutica, Heteropogon contortus, Aristida spp., Eleusine spp.

Vegetation and its Adaptations

Inspite of extreme climatic conditions and immense biotic pressure, the vegetation of the state has a lot of biological diversity. Even individual species indicate its variability in different bioclimatic zones of Rajasthan . The climatic conditions and the edaphic features, when combined, play decisive role and hence the vegetation has the existing status. The perennial plants are capable of adapting themselves according to seasonal changes while other try to evade the inhospitable period by shedding their leaves and virtually becoming dormant.

FOREST TYPES

Forest type means a unit of vegetation which possesses broad characteristics in physiognomy and structure sufficiently pronounced to permit of its differentiation from other such units (Figure 8.2 and Table 8.1).

Classification

Mathur (1972), described vegetation type for Rajasthan but the most useful classification for management of the forests is still the Forest Types recognized by Champion and Seth (1968) in their book titled "A Revised Survey of the Forest Types of India". According to this, the forest obtained in Rajasthan fall in :

Major Group II- Dry Tropical Forests

Group 5- Tropical Dry Deciduous Forests

Group 6- Tropical Thorn Forests

Major Group III- Montane Subtrpical Forests

Group 8- Sub-tropical Broadleaved Hill Forests (only in Mt. Abu)

Table 8.1 Forest Type in Rajasthan

Code	Description Type
5A/C1a	Very dry teak forests
5A/C1b	Dry teak forests
5B/C2	Northern dry mixed deciduous forests
5/E1	Anogsissus pendula forests
5/E1/D S1	A.pendula scrub
5/E2	Boswellia forests
5/E5	Butea forests
5/E6	Aegle forests
5/E9	Dry bamboo brakes
5/DS1	Dry deciduous scrub
5/DS2	Dry savannah forests
5/DS3	Euphorbia scrub
6B/DS2	Euphorbia scrub
6/E1	Euphorbia scrub
5/DS4	Dry grassland
5/IS1	Dry tropical riverain forests
5/ID3	Inundation babool
5/2SI	Secondary dry deciduous forests
6B/C1	Desert thorn forests
6B/DS1	Zizyphus scrub
6/E2	Acacia Senegal forests
6/E4	Salvadora scrub
6/E4/DS 1	Cassia auriculata scrub
6/1S1	Desert dune scrub
8A/C3	Central Indian subtropical Hill forests

Source: Champion and Seth - 1968.

Abbreviations:

A - Southern, B - Northern

C - Climatic climax E - Edaphic climax

DS - Degradation stage

Valuable Forests

From the point of view of Forestry Management, following Forest types are important in the context of Rajasthan:

Teak Forests

Teak forests extend over an area of about 5120.0 Sq. Kms., mainly in Banswara and Dungarpur and sporadically in Chittorgarh, Udaipur, Jhalawar, Baran Divisions, representing the northern most limit (24°42' N. Latitude in Udaipur and 25°12' N. Latitude in Baran) of natural teak forests of India. It is found on geological formations consisting of Aravalli system, pre-Aravalli banded gneissic complex and the Deccan trap on elevations varying between 242 to 485 mts., in yearly rainfall zone of 730mm to 1100mm, in localities not affected by severe frost. The teak is often pure (50 to 90% in density) and is commonly associated with Terminalia Crenulata (Sadad), Diospyros Melanoxyplon (Tendu), Anogeissus Latifolia (Dhavra), Lannea Coromandelica (Gurjan of Gol), Boswellia Serrata (Salar); And Occasionally Soymida Febrifuga (Rohan), Miliusa Tomentosa (Umb), Schrebera Swietenioides (Mokha), Zizyphus Xylopara (Ghatbor), Dalbergia Paniculata (Dhoban), Emblica Officinale (Aonla), Sterculia Urens (Kadaya), Dolichandrone Atrovirens (Mendgi), Mitragyna Parvifolia (Kalam), Butea Monosperma (Dhak) and Aegle Marmelos (Bel) are found in tree canopy. The Underwood and undergrowth consists of Dendrocalsmus Strictus (Bans, In Moister Situations), Flacourtia Indica (Kakon), Holarrhena Antidysenterica (Dudhi), Nyctanthes Arbortristis (Tamat) and in Moist Locations Helicteres Isora (Marorphali), Wood-Fordia Fruticosa (Laldhavri) etc. The Ground-flora consists of Cassia Tora (punwar), Ocimum americanum (bepchi), Acanthaceae spp. And grasses like Themeda Spp. (Ratedi, Bhedi), Apluda Varia (Bhangta), Andropogon Pumilus (Masuri), Heteropogon Contortus (Soorwala), Eragrostis (Phuli) Aristida (Lapla) etc. along nalas are found Sorghum Halepense (Barwari), Ischaemum Rugosum (Toil), Echinochloa Colonum (Sawan). Among Climbers Are Acacia Pennata (Chillati), Ampelocissus Spp., Discorea spp. But not profusely. Dendropthoe falcata (benda) parasitises salar trees.

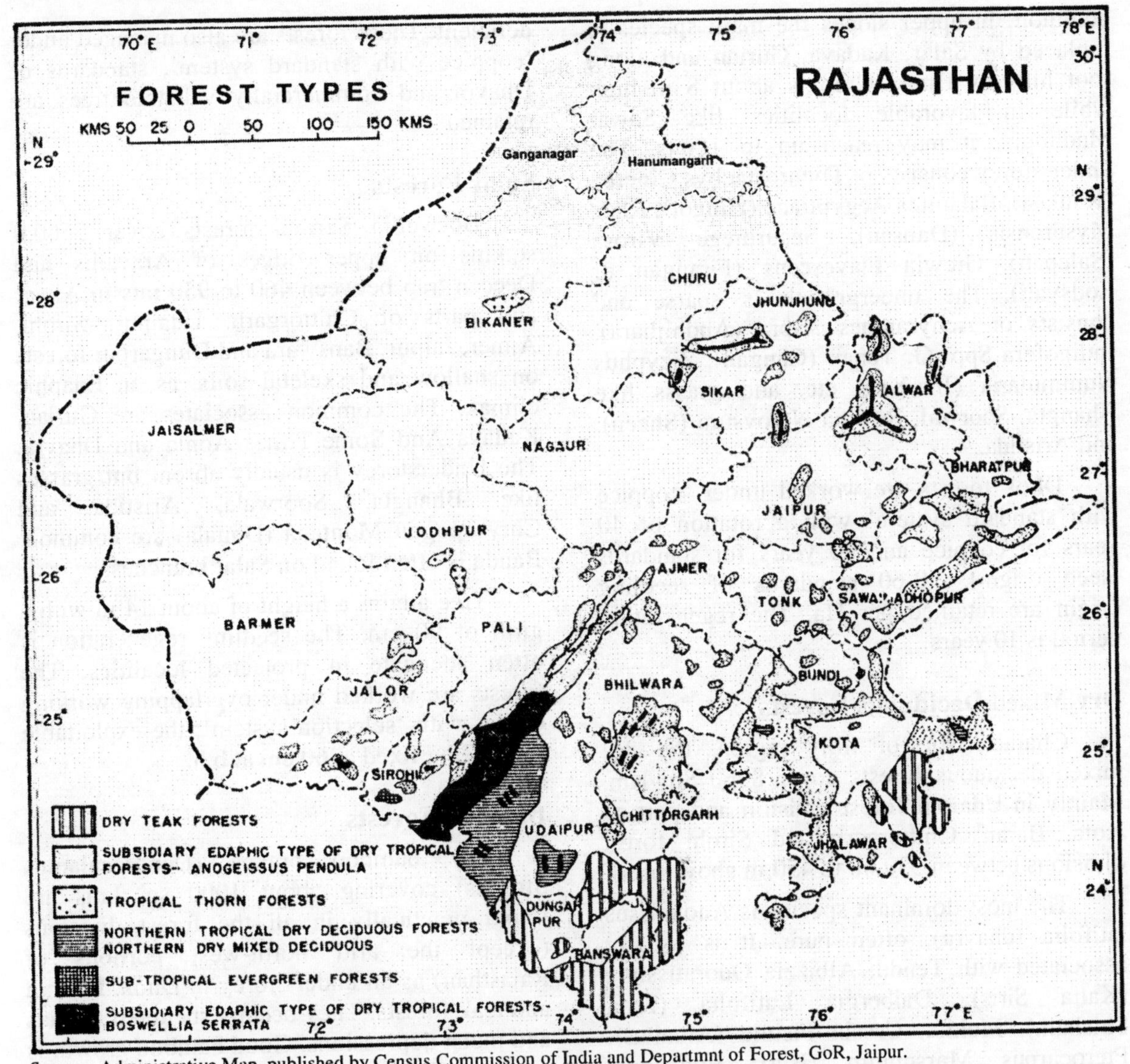

Source : Administrative Map, published by Census Commission of India and Departmnt of Forest, GoR, Jaipur.
Note : The external boundary and the coastlines of India appearing on the map are neither verified nor authenticated by the Survey of India.

Figure 8.2 Forest Types

The teak forests are managed under 'coppice with reserve system' with a rotation of 40 years. Except the reserves of seedling origin, the crop is cut back and tended to yield small timber for agriculture and household purpose.

Dhok (Anogeissus Pendula) Forests

Covering about 21,077 Sq. Km in Kota, Tonk, Bundi, Baran, Jhalawar, Jaipur, Ajmer, Pali, Sirohi, Udaipur, Chittorgarh, Alwar and Bharatpur forest divisions in hilly terrain of Aravalli, Delhi system and Vindhyan system, between elevations of 242 to 760 m, Dhok forms the principal forests.

Dhok generally forms gregarious stands. Its common associates are Acacia Leucophloea (Ronj), A. Catechu (Khair), Bauhinia Rocemosa (Jhinjha), Wrightia Tinctoria (Khirni) and Tendu. In parts of Jaipur, Ajmer, Alwar and Jodhpur; Acacia Senegal (Kumat) is common. In upper slopes the main species is replaced by Salar, Kadaya, Gurjan and along foot hills by Dhak. Crop is about 8 m high while in favorable localities like Sawai Madhopur, it may reach up to 12mts. The under-storey consists of Dichrostachys Cinerea (Kaulasi), Balanites Aegyptiaca (Hingot), Rhus Mysurensis (Dansan), Securinega Virosa (Salepan), Grewia Flavescens (Firangan or Lodsyali). The undergrowth is sparse and consists of Achyranthes Aspra (Andhijhara), Indigofera Spp. G. Tenax (Gangan), Zizyphus Nummularia (Jharberi) etc. and grasss like Bhangta, Soorwal, Sehima Nerovsum (Sheen), and Aristida.

Dhok forests are worked under 'coppice with standard system' with a rotation of 40 years for coppice and 80 years for standards (seed origin). 50-60 standards of seedling origin are retained per Ha. The regeneration period is 10 years.

Dry Mixed Deciduous Forests

Characteristic of hilly region, the dry mixed deciduous forests cover 8960 Sq. Km. Mainly in Udaipur forest division and part of Kota, Baran, Chittorgarh and Sirohi forest divisions between 330 m. to 450 m above m.s.l.

The most dominant species is Anogeissus latifolia (dhavra), often pure. It is usually associated with Tendu, Albizzia Odoratissima (Kalia Siris), Dalbergia Latifolia (Kala Shisham), Mokha, Cassia Fistula (Amaltas), Pterocarpus Marsupium (Bia), Buchannia Latifolia (Chironji), Gurjan, Rohan, Aonla, Bel, Dhak Khair, Cochlospermum Religiosium (Ganiara), Holoptelia Integrifolia, (Churel, Papdi) etc. Along nalla are found Mallotus Philippinensis, (Rohini), Pongamia Pinnata (Karanj), Ficus Glomerata (Gular), Kalam, Terminalia Belerica (Bahera), Madhuca Indica(Mahuwa) and some times Mangifera Indica (Aam), Terminalia Arjuna (Kohda) etc. Under open canopy, grasses are luxuriant consisting of Bhangta, Ratedi, Soorwala, sheen etc. Bamboos are also found in moist situations. The regeneration is generally deficient. These forests are also managed under 'coppice with standard system', standards of Dhavra and commercially valuable trees are retained

Salar Forests

Boswellia Serrata forests cover 2560.0 sq.kms on upper ridges of Aravallis and Deccan trap between 460 to 950 mts in Alwar and parts of Chittorgarh, Udaipur, Sirohi, Ajmer, Jaipur, Banswara and Dungarpur forests on shallow and skeletal soils, as an Edaphic climax. The common associates are Gurjan, Kadaya And Some-Times Aonla and Dhavra. The Understorey is usually absent but grasses like Bhangta, Soorwala, Aristida and Chrysopogon Montana (Dhuala) are common. Banda is often found on Salar branches.

Tree attains a height of about 14 m with a girth of 90 cm. The seedling regeneration is often adequate in protected localities. The forests are worked under overlapping working circles with 'selection system'; the exploitable girth being fixed at 60 cm at b.h.

Bamboo Forests

Dry bamboo brakes (Dendrocalamus Strictus) covering about 1000 sq. kms are found practically in all the forest divisions (except the arid north-west portions of Rajasthan) as an under-storey in Dhok, Dharva and teak forests. They occupy sheltered aspects and moist depressions. Good bamboo growth is found in Chittorgarh, Udaipur, Banswara, Kota and Mt. Abu. The clumps may have 10 to 30 culms in height varying from 5 to 9 m. Their quality is under deterioration year after every year. They need rest.

Khair Forests

Occupying an area of about 1120 sq. kms Acacia Catechu (Khair) forests occur as a facie of pure Khair crop in south eastern region e.g. Baran, Jhalawar, Kota, Chittorgarh, Alwar and Sawai Madhopur and are associated with Ghatbor, Kaulasi, Dhok, Bel, Ronj, Jhinjha etc. It is also found mixed with Dhok and Dhavra forests. The natural regeneration is generally deficient.

Exploited for Katha (Catechin) manufacture by 'Handi' system (country method of boiling Khair heartwood chips in earthen or metallic pots), the Khair trees are worked under 'selection system' with and exploitable diameter of 10 cms at b.h. and felling cycle of 20 years, with reliance on coppice regeneration

Sub-Tropical Evergreen Forests

Restricted over about 55.0 sq. kms in Mt. Abu hills between elevations of 1150.0 to 1830.0 mts, these forests support aam, Jamun (Syzygium Cumini), Kakon, Gadhapals (Erythrina Spp.) Bamboo, Rohini, Karonda (Carissa Congesta), Jasminum Humile, Rosa Moschata, Strobilanthes Callosus, Geradiana Heterophylla, (Bichchhu Buti), Euphorhia Neriifolia (Thor), The Orchids Arides Crispum etc. Anogeisus Sericea (Androk) is also a marked feature at Mt. Abu and Phoenix Sylvetris (Khajur) grows even above 1650 mts.

To preserve aesthetic beauty these forests are not exploited except for some light dry bamboo removal.

Thorn Forests

Spread over 2000 sq.km the thorn forests are very open. They are mainly confined to arid areas in northwest but also on isolated patches of sandy tract and ravenous land in other parts where soil is generally devoid of humus.

The spp. Found are thorny and comprise Prosopis Cineraria (Kherji), Maytenus Emarginatus (Bekal), Caparis Decidua (Kair), Hingot, Acacia Jacquemonti (Baonli), Kumta, Calotropis Procera (Akda), Zizyphus Spp. (Ber, Jharberi), Mimosa Hatata (Jhinjhini) And Euphorbia Nivulia (Thor). In more sandy areas Calligonum Polygonoides (Phog), Crotolaria Burhia (Senia), Leptadenia Pyrotechnica (Khimp) Are Common. The Usual Grasses Are Desmostachya Bipinnata (Dab), Cenchrus Ciliaris(Dhaman, Anjan), C. Setigerus (Kala Dhaman), C. Biflorus (Bhurat), Lasiurus Scindicus (Sewan), Panicum Turgidum (Giramana) and Sporobolus Spp. These forests are managed only as grazing areas.

Flora and Fauna

According to Good (1964), the earth has been divided into six floral Kingdoms, Rajasthan falling under Paletropical or Original floral kingdom. He also recognised 37 Floristic Region or Floristic Provinces, out of which Rajasthan is covered under 9-North African-Indian Desert Region and 17-Indian Region. Phyotogeographically the classification of Chatterjee (1939), is still most widely accepted, who divided India into 8 Botanical Regions; Rajasthan falling in Indus Plains.The characteristic spp. Or floristic elements of this Botanical division are Acacia nilotica, A. Senegal, A. Leucophloea, Anogeissus, Albizzia, Dalbergia, Grewia, Prosopis, Salvadora, Capparis, Euphorbia, and Calotropis.

Flowering Plants

Jacquemont (1832) was the first to botanically explore Rajasthan. He was followed by George King (1868), Blatter and Hallberg (1918-21), Parker (1921) Bhandari (1967) and B. Tigai and S. Sharma (1979). The estimated flowering plants of Rajasthan to belong to 1280 species as follows (Table 8.2):

Table 8.2 : Number of Floral Species

Classification	Dicots	Monocots	Total
Families (After Bentham and Hooker)	92	18	110
Genera	465	128	593
Species	970	210	1280

Source: Jain (1970)

Largest Families

According to number of species the ten largest families in Rajasthan are (Table 8.3)-

Table 8.3 Floral Species

S.No.	Families	Species	Genera
1.	Poaceae	180	74
2.	Leguminosae	177	51
3.	Compositae	86	51
4.	Cyperaceae	63	11
5.	Acanthaceae	46	22
6.	Euphorbiaceae	44	14
7.	Malvaceae	40	11
8.	Labiatae	35	13
9.	Convolvulaceae	35	11
10.	Scrophulariaceae	33	19

Source: BSI and Department of Forest, GoR, Jaipur

Largest Genera

The largest genus is Cyperus with 30 spp. While the next is Indigofera with 20 spp. The Acacia occupies 10th position with 11 spp. in Rajasthan.

Endemic Plants

Rajasthan lies in the zone of poor endemism. Of the 134 genera endemic to India, only six namely; Ouginia, Butea, Caesulia, Glossocardia, Petalidium and Haplanthus are endemic in Rajasthan. Some of the common endemic taxa of Rajasthan are Cleome gynandra var. nana (B1 and Hall) Bhandari, Cleome brachycarpa Vahl ex D.C. var. glausa BI and Hall., convolvulus blatteri Bhandari, Farestia macratha BI. and Hall., Zizyphus truncata BI. and Hall., Tephrosia uniflora sub-spp. Petrosa (Bl and Hall) Gillet and Ali, Lasiurus caudatud, Cenchrus rajasthensis, Tribulus rajasthenisis Bhandari and Sharma Anogeissus sericea brandis var. nana king ex Duthie, Ammannia desertorum BI. and Hall., ex Sant., Euphorbia Jodhpurensis BI. and Hall., Puliciara Rajputance BI. and Hall., Melhania magnifolia BI. and Hall.

Fossiliferous Flora

Fossil evidence is extremely scanty for Rajasthan.

Udaipur and Karauli

Pre-Cambrian plant life of the state is represented by five different types of stromatolites namely Collenia columnaris, C.symnetrica, Baicalica prima and Minjaria calceolate from matoon formation near Udaipur (Banerjee 1971), a filament of Eomycetopsis robusta from Aravalli subground of Udaipur (Banerjee1974) dated more than 2 bilion years and algal forms Gloeocapsomorpha and Palaeoglaucocystis from the vindhyan rocks near Karauli.

Jaisalmer

The lower or Mid Jurassic life recovered from rocks exposed near Jaisalmer (Lathi formation) comprised largely serveral kinds of pteridophytes like Cyathidites australis rimalis Bahme C.minor Couper, Dictyotriletes crateris (Bahme) Srivastva, Cingulatisporites lathiensis Srivastava, Polypodium sporites mathurii Srivastava, and gymnosperms like Ginkgocycadophytes deterius var. majus Der, G.nitidus (Bahme) Srivastava, Monosulcites couperi Der, Cupresacites ramachandrae Srivastava, Araucariacites ghos hii Srivastava, Podocar-pidites spp. Callialasporites dampsieri (Bahme) Der and other spp. of Callialasporites and Classopollis. The plant fossils (upper Jurassic) from Habur village (near Jaisalmer) belong to ferns Gleichenites and Phlebopteris, Bennettitales (Taenopteris, Pterophyllum, Otozamites imbricatus Feistm, Ptilophyllum acutifolium Morris and conifers Elatocladus conferta Oldh and Morris Hall and Pagiophyllum spp. In Mudh and Kotri of Palana (Jaisalmer) Formation Pollen grains of Chelianthes Melia and unidentifiable angiosperms of tertiary period are recorded. Now converted as AKAL wood fossil Park near Jaisalmer, huge gymnospermic fossils of large trees of lower Jurlassic era estimated to be 180 millions years old are lying.

Barmer

Barmer area is also rich in fossil finds. The Cretaceous era plant life recorded is comparable with Polypodium, Lycopodium, Lygodium, Coniopteris, Gnetum, Araucaria, Abies, Pinus, Podocarpus and unidentifiable angiosperms. At kapurdi, Fuller's earth deposits (Barmer) contain the Eocene deposits of evergreen trees like Mesua tertiara, Garcinia borooahii, Calophyllum and fruit of Cocos which are indicative of humid climate with high rainfall

Bikaner

From the Bap-Bikaner area at Mayakor silicified wood and leaf impressions of Ptilophylum acutifolium, Pterophyllum and Equisetites of the same age as that of Lathi formation have been recovered.

Sambhar-Didwana-Ajmer

The mixed xeromorphic Thorn Forests with serveral grasses, sedges, composits and forbes (spp. of Tephrosia, Solanum, Heliotropium, Boerhaavia, Digera and Indigofera) waned about 6000-4000 years ago in the vicinity of Lunkaransar, expanded between 5000-4000 ago in the vicinity of Didwana and appeared at Sambhar between 4500-3000 years ago. Together with this, Psammophytic scrub represented by Calligonum polygon ides and Aerva grew near Lunkaransar. Phog (C.polygonoides) invaded Didwana and Sambhar areas by about 6000 years and 3000 years ago respectively. Ultimately in the northern parts of Rajasthan desert, developed Lithophytic scrub represented by Capparis and Mimosa. Plant life from Pushkar lake near Ajmer is know from about 5000 years. The genus Artemesia which was significantly presented 10,000 years ago is now occasional and even two species A.scoparia and A.nilagirica are now of rare occurrence in the Rajasthan desert.

Threatened Species

Destruction of natural habitats has resulted in decline of many plant species. Although definite information is lacking, the following plants in Rajasthan are either getting rare or threatened with extinction.

FOREST COVER

The forest area of 31,151 sq. km. in the State is only 9.10% of its total area of 342230 sq. km. below the norms of 33% prescribed by National Forest Policy of India 1952. Out of total legal forests (31,151 sq. km.) hardly one third bears good forest cover which can be termed as potentially productive (Figure 8.3 and 8.4). It can be understood by the following statistics:

(a) Potentially productive forests 10,000 sq. km.

(b) Barren hills and grasslands 10,000 sq. km.

(c) Rooted Wastes 11,151 sq. km.

Total: 31,151 sq. km.

LEGAL FOREST AREA

The latest area under legal forests in each of the revenue districts is as follows (Table 8.4)-

Table 8.4 Districtwise Area of Forest According to Legal Status of Rajasthan

(as on 31.3.2016) (In Sq. Km.)

S. No.	Districts	Reserved Area	Protected Area	Un-Classified Forest Area	Total Forest Area
1	Ajmer	194.99	421.69	1.76	618.44
2	Alwar	1010.78	639.73	133.95	1784.46
3	Banswara	0.00	1005.26	1.74	1007.00
4	Baran	0.00	2225.31	14.38	2239.69
5	Barmer	20.30	568.81	36.76	625.87
6	Bharatpur	28.73	393.25	12.97	434.94
7	Bhilwara	438.40	289.62	67.42	795.43
8	Bikaner	0.00	748.49	501.40	1249.88
9	Bundi	853.05	665.03	9.02	1527.09
10	Chittorgarh	1200.15	571.17	0.62	1771.94
11	Churu	7.20	42.39	24.14	73.73
12	Dausa	134.87	149.26	0.36	284.49
13	Dholpur	7.92	597.73	44.03	649.68
14	Dungarpur	257.08	426.43	9.37	692.88
15	Ganganagar	0.00	238.42	395.02	633.44
16	Hanumagarh	0.00	113.25	126.21	239.46
17	Jaipur	678.38	263.17	4.94	946.49
18	Jaisalmer	0.00	238.36	343.23	581.59
19	Jalore	126.13	299.97	83.89	509.99
20	Jhalawar	416.38	924.32	6.21	1346.92
21	Jhunjunu	6.02	399.33	0.00	405.36
22	Jodhpur	4.68	183.78	55.66	244.12
23	Karauli	62.99	1693.13	53.93	1810.05
24	Kota	879.61	430.36	15.59	1325.56
25	Nagaur	0.80	206.28	35.32	242.40
26	Pali	816.56	144.82	2.21	963.58
27	Pratapgarh	907.12	1017.46	0.56	1925.14
28	Rajsamand	277.41	119.20	4.78	401.39
29	Sawai Madhopur	842.87	118.16	22.01	983.04
30	Sikar	9.92	622.40	8.19	640.52
31	Sirohi	614.04	984.72	39.89	1638.65
32	Tonk	101.42	233.17	1.38	335.97
33	Udaipur	2454.98	1434.38	9.80	3899.16
	Total	**12352.78**	**18408.85**	**2066.74**	**32828.37**

Source: Department of Forest, GoR, Jaipur

Forest Cover

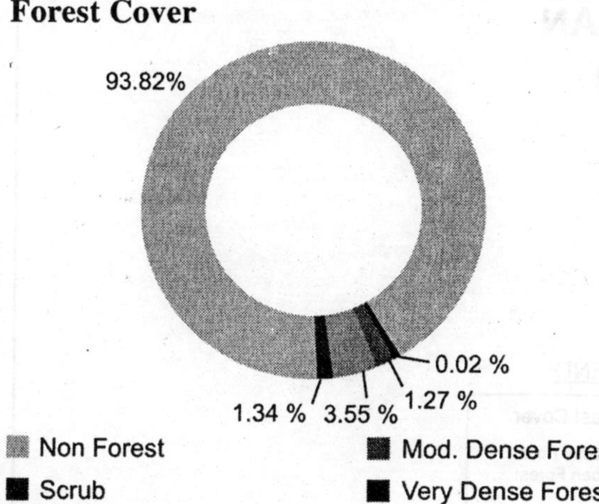

Figure 8.3 Pie Chart Showing Forest Cover of Rajasthan

Based on interpretation of satellite data pertaining to Oct-Dec 2015, the forest cover in the State is 16,572 sq. km. which is 4.84% of the State's geographical area. In terms of forest canopy density classes, the state has 78 sq. km. under very dense forest, 4,340 sq. km. under moderately dense forestand 12,154 sq. km. under open forest.

Forest Cover within and outside Recorded Forest Area

The recorded forest area of the state is 32,737 sq. km. which is 9.57% of its geographical area. The Reserved, Protected and Unclassed Forests are 38.11%, 55.64% and 6.25% respectively of the recorded forest area. However, as the digitized boundary of recorded forest area from the State covers only an area of 23,105 sq. km., the analysis of forest cover within and outside this area is depicted below (Figure 8.5 and Table 8.5).

Table 8.5 Forest Cover within and outside Recorded Forest Area

Forest Cover within Recorded Forest Area (Area in sq km)	
Very Dense Forest (VDF)	14
Moderately Dense Forest (MDF)	1,923
Open Forest (OF)	6,077
Total	**8,014**
Forest Cover outside Recorded Forest Area	
Very Dense Forest	64
Moderately Dense Forest	2,417
Open Forest	6,077
Total	**8,558**
Total Forest Cover	**16,572**
Tree Cover	**8,266**
Total Forest and Tree Cover	**24,838**
Of State's Geographical Area	7.26
Of India's Forest and Tree Cover	3.10 %
Per capita Forest and Tree Cover	0.04 ha

Source: Department of Forest, GoR, Jaipur

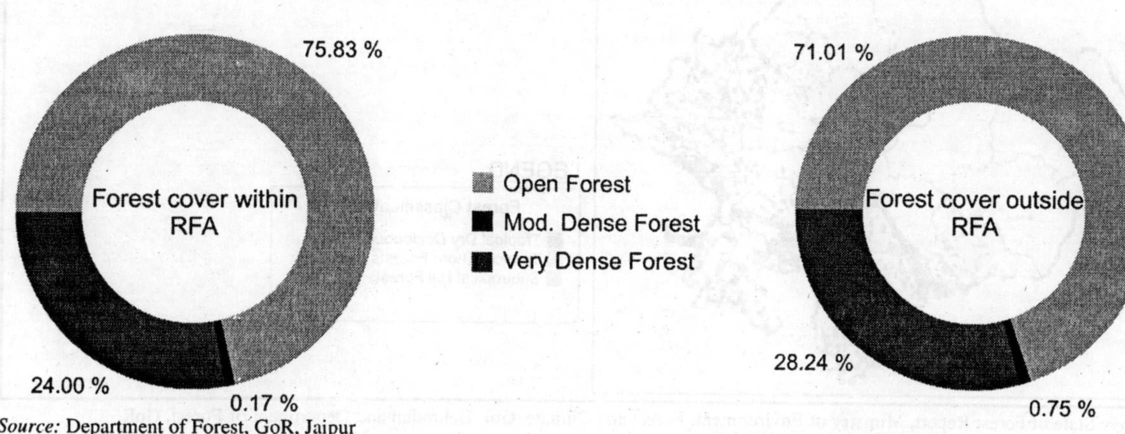

Source: Department of Forest, GoR, Jaipur

Figure 8.4 Pie Chart Showing Forest Cover within and outside RFA

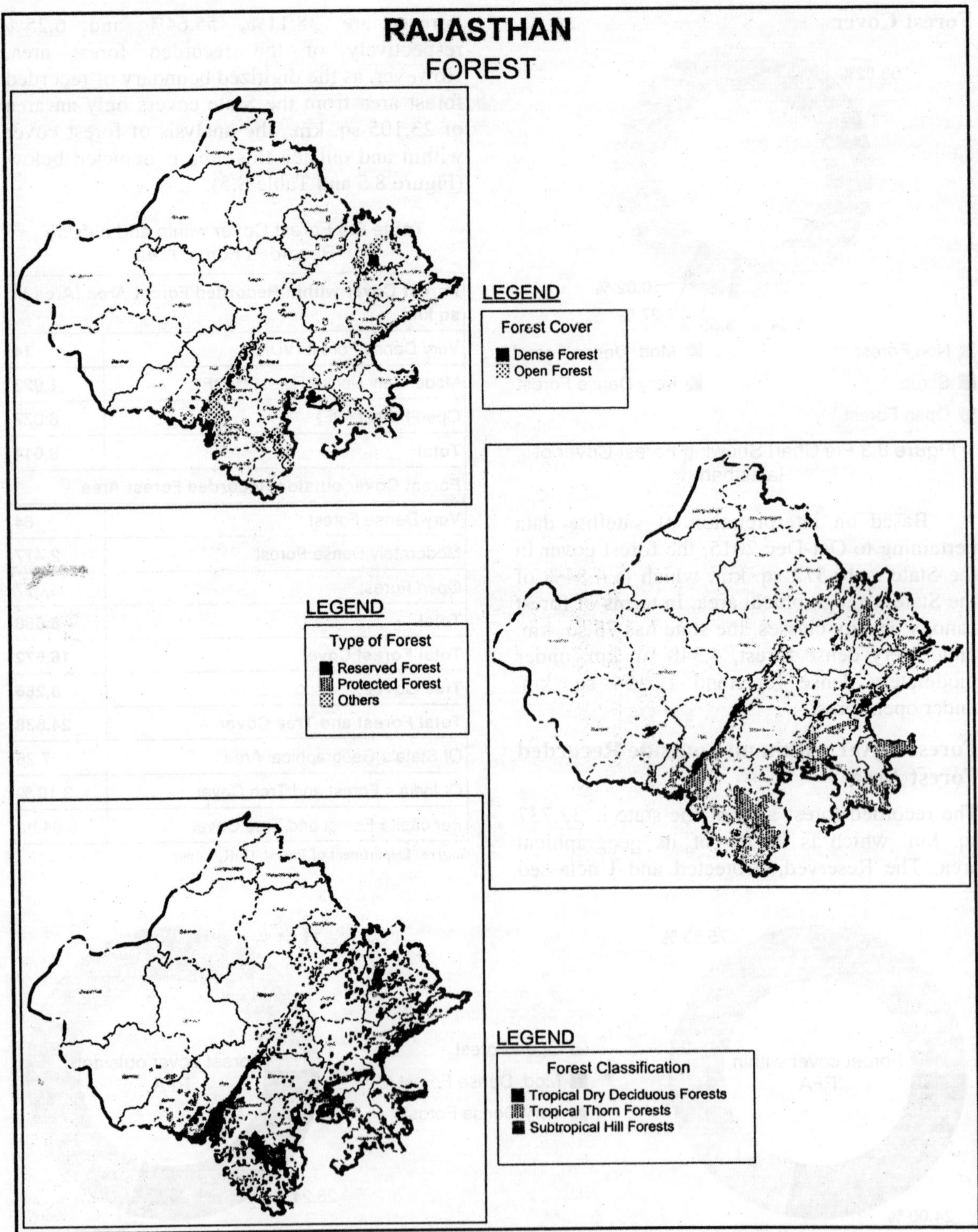

Source: State of Forest Report, Ministry of Environment, Forest and Climate, GoI, Dehradun and Department of Forest, GoR.

Figure 8.5 Rajasthan Forest

Carbon Stock in forest

The total Carbon stock of forests in the State is 89.660 million tonnes (328.753 million tonnes of CO_2 equivalent) which is 1.27% of total forest carbon of the country.

Decadal Change in water bodies within Forest

An increase of 40 sq. km. has been observed in the water body coverage within forest compared to 2005. The graphical and tabular comparison is depicted below (Figure 8.6, Table 8.6 and Table 8.7).

Table 8.6 : Districtwise Forest Cover (area in sq. km.)

District	Geographical Area	2017 Assessment		Open Forest	Total	% of GA	Change*	Scrub
		Very Dense Forest	Mod. Dense Forest					
Ajmer	8,481	0	43	256	299	3.53	13	202
Alwar	8,380	59	335	803	1,197	14.28	-2	246
Banswara[T]	4,522	0	39	222	261	5.77	2	53
Baran	6,992	0	155	858	1013	14.49	-44	103
Barmer	2,8387	0	4	269	273	0.96	86	205
Bharatpur	5,066	0	22	207	229	4.52	13	79
Bhilwara	10,455	0	31	190	221	2.11	-1	158
Bikaner	30,239	1	27	219	247	0.82	28	47
Bundi	5,776	1	138	419	558	9.66	82	151
Chittaurgarh	7,822	0	220	769	989	12.64	-40	86
Churu	13,835	0	3	79	82	0.59	-4	22
Dausa	3,432	0	12	105	117	3.41	2	99
Dhaulpur	3,033	0	80	339	419	13.81	-7	70
Dungarpur[T]	3,770	0	43	248	291	7.72	41	61
Ganganagar	10,978	0	10	103	113	1.03	33	13
Hanumangarh	9,656	1	7	82	90	0.93	-12	1
Jaipur	11,143	12	97	443	552	4.95	49	286
Jaisalmer	38,401	4	53	256	313	0.82	88	227
Jalor	10,640	0	18	257	275	2.58	77	209
Jhalawar	6,219	0	83	356	439	7.06	34	102
Jhunjhunun	5,928	0	21	175	196	3.31	3	188
Jodhpur	22,850	0	4	101	105	0.46	10	167
Karauli	5,524	0	95	775	870	15.75	-7	273
Kota	5,217	0	154	396	550	10.54	-29	135
Nagaur	17,718	0	15	128	143	0.81	22	103
Pali	12,387	0	210	464	674	5.44	-9	308
Pratapgarh[T]	4,449	0	560	484	1,044	23.47	-48	44
Rajsamand	4,655	0	135	376	511	10.98	30	108

District	Geographical Area	2017 Assessment		Open Forest	Total	% of GA	Change*	Scrub
		Very Dense Forest	Mod. Dense Forest					
Sawai Madhopur	4,498	0	154	312	466	10.36	30	114
Sikar	7,732	0	31	161	192	2.48	1	187
Sirohi[T]	5,136	0	302	612	914	17.80	3	206
Tonk	7,194	0	27	138	165	2.29	-4	56
Udaipur[T]	11,724	0	1,212	1,552	2,764	23.58	26	270
Total	**3,42,239**	**78**	**4,340**	**12,154**	**16,572**	**4.84**	**466**	**4,579**

* Change compared to updated 2015 assessment.

Source: State of Forest Report, Ministry of Environment, Forest and Climate, GoI, Dehradun and Department of Forest, GoR.

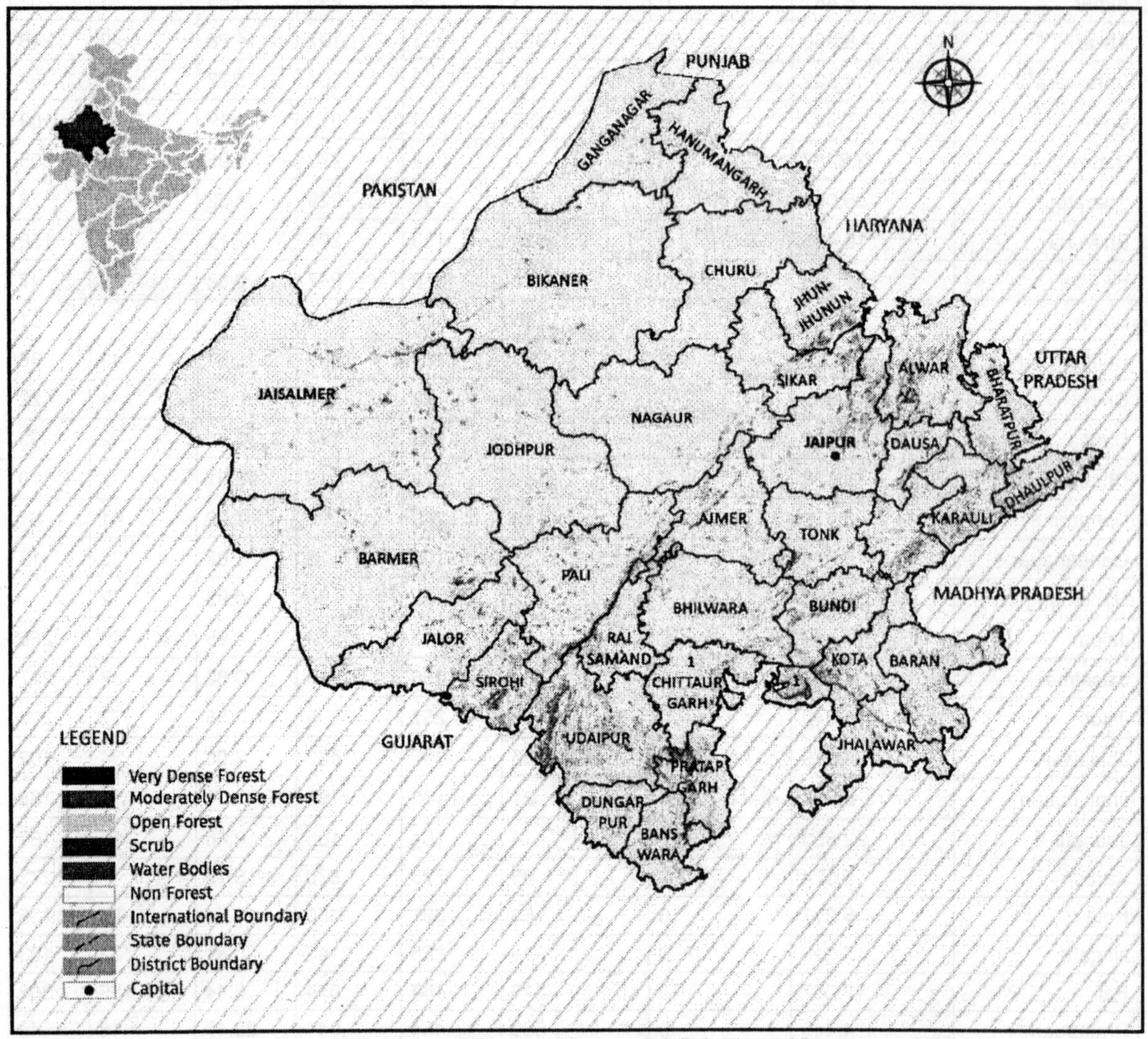

Source: State of Forest Report, Ministry of Environment, Forest and Climate, GoI, Dehradun and Department of Forest, GoR.

Figure 8.6 Forest Cover map of Rajasthan

Table 8.7 Forest Cover Change Matrix (area in sq. km.)

Class	2017 Assessment					Total ISFR 2015 updated
	VDF	MDF	OF	Scrub	NF	
Very Dense Forest	75	1	0	0	1	77
Moderately Dense Forest	3	3,788	500	7	116	4,414
Open Forest	0	474	9,505	217	1419	11,615
Scrub	0	9	432	3,032	807	4,280
Non Forest	0	68	1717	1323	3,18,745	3,21,853
Total ISFR 2017	**78**	**4,340**	**12,154**	**4,579**	**3,21,088**	**3,42,239**
Net Change	1	-74	539	299	-765	

Source: State of Forest Report, Ministry of Environment, Forest and Climate, GoI, Dehradun and Department of Forest, GoR.

Reasons for change detected in 2017 assessment

A net gain of 466 sq km has been observed in forest cover of the State which can be attributed to plantations, regeneration of fast growing species and conservation efforts (Table 8.8).

Table 8.8 : List of Rare and Endangered Plants

S. No.	Latin Name	Local Name	Family	Habit	Tropical locality	Remarks
1.	Ammannia desertorum Blatt. & Hallb.	Moto-Jal-bhangro	Lythraceae	Undershrub	N.W. Rajasthan	Sand stone rocks preferred
2.	Anogessus Sericea Brandis Var. numularia King ex Duthie (=A.rotundifolia Blatt. & Hallb)	Indok	Combretaceae	Tree	Ajmer, Jodhpur, Udaipur	
3.	Anticharis glandulosa Asch. Var. caerulea Blatt. & Hallb. Ex. Sant.		Scrophulariaceae	Herb	Jaisalmer	Found on rocks
4.	Bouchea Marrubifolia Schauer (=Pleurostigma subrotundum Hochst. Nomnud)	Bai	Verbenaceae	Herb	N.W. Rajasthan	
5.	Commiphora wightii (Arnott) Bhandari (=C.mukul Engl. And = Balsamodendron mukul Hook. Ex stock)	Gugul	Burseraceae	Shrub attaining dendroid form.	Aravallis	
6.	Convolvulus scindicus stock	Kaland	Convolvulaceae	Herb	N.W.Rajasthan Barmer	
7.	Crypsis schoenoides Lamk.		Gramineae	Herb (Grass)	Bap, Jaisalmer	
8.	Dignathia hirtella Stapf.		Gramineae	Herb (Grass)	Jalore	For a Short Time After Rains
9.	Dipterygium glaucum Dence		Capparaceae	Undershrub	Jodhpur, N.W. Rajasthan	Sand. Dunes.
10.	Ephedra ciliata Fish & Mey ex C.A. Mey (=E.foliata Boiss)	Suo-Phogaro	Gnetaceae	Climbing Shrub.	N.W. Rajasthan	Seriously Endangered.
11.	Eriolaena hookeriana W.&A.	Bhoti dhaman	Sterculiaceae	Tree	Hill	Very Rare
12.	Euphorbia jodhpurensis Blatt. & Hallb.	Dudheli	Euphorbiaceae.	Herb.	N.W. Rajasthan	
13.	Farsetia Macrantha Blatt. & Hallb.	Motio-hiran-chabbo	Cruciferae	Undershrub	Barmer	Seriously Threatened
14.	Grewia damine Gaertn (=G.salvifolia Weyne ex Roth non Linn.f.)	Phalsa	Tiliaceae	Shrub	Aravalli hills	
15.	Monsonia heliotropoides Boiss		Geraniaceae	Herb	Jaisalmer	Endangered
16.	Melhania denhamii R.Br.		Sterculiaceae	Herb	Barmer	Sand dune

S. No.	Latin Name	Local Name	Family	Habit	Tropical locality	Remarks
17.	M.magnifolia Blatt. & Hallb.		Sterculiaceae	Shrub	N.W. Rajasthan	On rocks
18.	Moringa concanensis Nimmo	Sarguro	Moringaceae	Tree	Aravalli rocks	Getting very rare
19.	Nannea pulla Lamk. Ex. D.C.		Boraginaceae	Herb	N.W. Rajasthan	Wheat fields
20	Pulicaria raputanae Blatt. & Hallb.	Dhola ligru	Compositae	Herb	N.W. Rajasthan	Dry salty bed
21.	Rhynchelytrum villosum (Parl) Chiov		Graminae	Herb(grass)	Sardar Samand (Pali)	Rocky gravel
22.	Rosa lyellii Lindl		Rosaceae	Shrub	Mt. Abu Jaisalmer	
23.	Seetzenia orientalis Decne	Dhakri	Zygophyllaceae	Herb	Jaisalmer	On rocks
24.	Talinum Portulacifolium Aschers ex Schweinf		Portulacaceae	Shrub	Rocks in Jodhpur, Barmer	Very rare
25.	Taverniera cuneifolia Arn.	-	Papilionaceae	Under shrub	Pali	-
26.	Tecomella undulata (Sm) Seem.	Rohida	Bignoniaceae	Tree	Sandy soil	-
27.	Tribulus rajasthanensis Bhandari et Sharma	Gokharu	Zygophyllaceae	Herb	Jodhpur, Jaisalmer & Ajmer	Sand stone rocks preferred.
28.	Tripogon jacquemontii Stapf.	-	Gramineae	Herb (grass)	N.W. Rajasthan (Osian)	Very rare
29.	T. roxburghianus Bhide	-	Gramineae	Herb (grass)	Jodhpur, N.W. Rajasthan	On moist shady rocks
30.	Withania coagulans Dunal	Paneer bandh	Solanaceae	Under shurb	N.W. Rajasthan	Grassy hills tops

Source: Department of Forest, Government of Rajasthn, Jaipur

NATIONAL PARKS AND WILD - LIFE SANCTUARIES

It is the conglomeration of hostile, harsh climatic conditions and the diversity in geological formation imbedded in the Aravalli ranges and lakes that is reflected in the distinctive flora and fauna of Rajasthan. On the western side of the Aravallis, there is the great Thar desert with extreme high temperature during summer with dust storms and erratic rainfall, while on the eastern side, a good forest cover is seen in the villages of Aravalli hill range running parallel of each other from Udaipur to Gujarat state. There are moderate temperatures with rainfall ranging between 25 to 40 cms., providing suitable habitat for wild-life species like the Tiger, Desert cat, Panther, Black buck, Sloth bear, Carcal, Desert fox, Hyaena Wild boar, Percupine and among the birds, Great Indian Bustard, Peacock, Sand grouse and Indian Roller etc. Besides these, snakes, lizards and amphibians are also found in the sanctuaries of Rajasthan (Figure 8.7).

To maintain the ecological balance, two National Parks and twenty five wild-life sanctuaries have been established in suitable forest areas. Wild-life wing of Forest Department maintain these area in a scientific manner, conducive to preservation of wild life in the state. Approximately, 2.7 per cent of the total geographical area of Rajasthan is maintained as National Parks and Sanctuaries in Rajasthan.

The Keoladeo National Park area is partially an artificial lake which gets filled up with water every year during rains and which gets dried up in summer, is a paradise for birds and bird watchers. This is one of the best known bird habitat in the world. Migratory birds like Demoiselle cranes, Pintails, Coots. Painted storks, Pelicans etc. roost and breed in thousands. Siberian crane which is a vanishing species, visits the habitat during winter season although for last few years these birds are not been reported.

Another national park i.e. Ranthambhor, in the dry deciduous forest in Aravalli hills, provides a suitable and healthy habitat to the great Indian Tiger which is at apex of the biotope in most of the Indian Jungles. The National park has been included in Project Tiger Scheme in the year 1973 by Government of India.

Among the wild-life sanctuaries, Sariska is other Tiger reserves in the country. It became a Tiger Reserve under Project Tiger Scheme in 1978. The objective is to maintain a viable population of tiger, which entails improvement of habitat on which the prey species are essentially dependent. In 2004, all Tiger population of the park found missing due to poaching. However, in 2008 one male and one female Tiger have successfully relocated in Sariska by transferring from Ranthombore Tiger Reserve of the State. Besides it, other animals found in the reserve are Panther, Sambher, Blue bull, Wildboar, Wild Cat, Common Langur, Jackal, Hyaena, Porcupine, Mangoose, Palm Civet etc.

Desert National Park (Sanctuary) is a homeland for Great Indian Bustard or locally known as 'Godawan'. It was a common bird a centuary ago. There are now approximately few hundreds birds left in the country which are mostly confined to Rajasthan, parts of Gujarat, Maharashtra and adjoining states. But the last hold and hope for the bird appears to be only Rajasthan, specially in Jaisalmer, Barmer and certain areas of Sonkhlia (Ajmer). The Desert National Park promises preservation and protection of the habitat and the bird to the posterity.

Talchhapar wild-life sanctuary is in a depression land, with abundant salt in its surrounding area. More than 1600 Black Bucks have been counted during last census. A herd of over 100 Black Bucks can be seen at a time in or around the sanctuary area.

Flying Squirrel is found in Sitamata sanctuary, Whereas in the sanctuaries situated at high altitudes area of Mount Abu, Rawati Tatgarh, Kumbhalgarh, Panther, Sambhar, Wild Bear, bear, porcupine, jackal, etc. can be seen, venturing in the protected habitat.

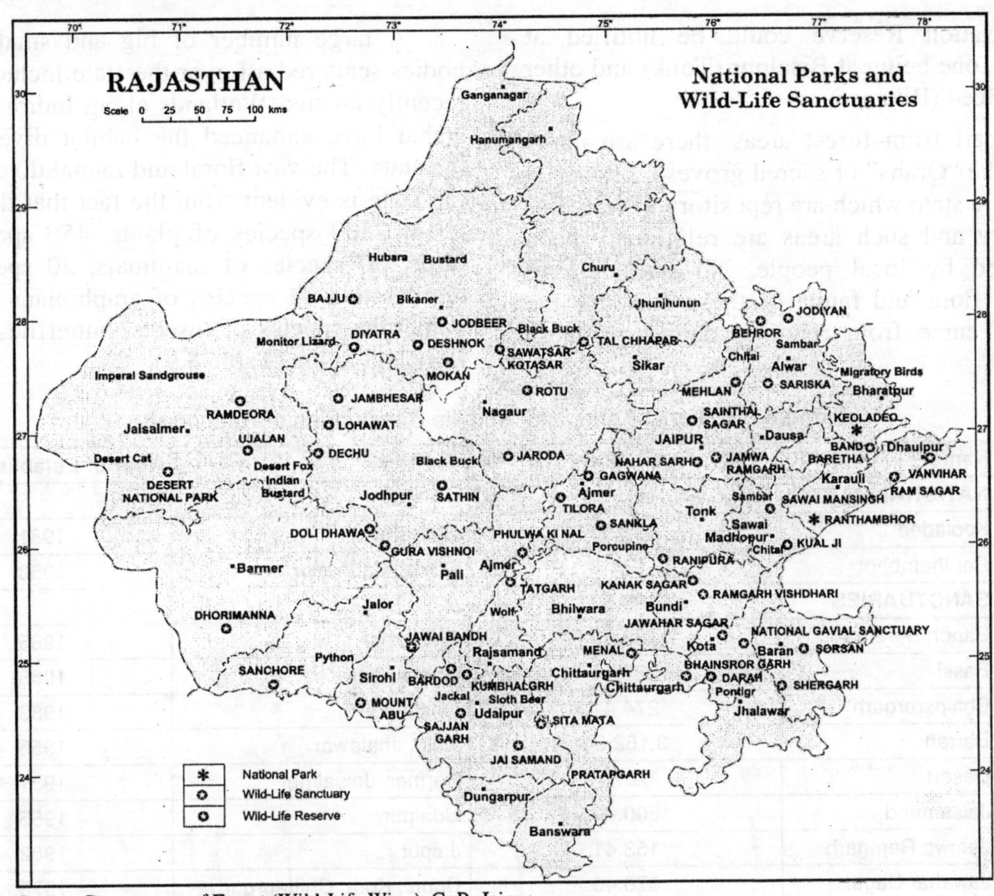

Source: Department of Forest (Wild-Life Wing), GoR, Jaipur

Figure 8.7 National Parks and Wild-Life Sanctuaries

A total of 32 area having fairly a good population of wild-life species but not declared as a national park or a sanctuary, have been declared as a 'CLOSED AREA' under the Wild-Lilfe (Protection) Act, 1972 for a limited time period. Wild-life protection posts have been created in most of these closed areas all over the Rajasthan.

Besides the national parks and sanctuaries in Rajasthan, some traditionally protected areas like Gura Bishnoi (Black buck), Dhava Dolli and Mukam (Chinkara, Blackbuck), Diyatara and Dhorimanna (Chinkara), Sanchor (Jackal, Chinkara), Ramdeora (Spiny Tailed lizard and Chinkara) are of great importance. In the western Rajasthan, Bishnois are the best protector of wild life. There are many instances in which Bishnois sacrificed their lives while protecting the wild-life from hunters. There is awareness amongst the citizens of this state regarding the wild-life and environment protection.

A Bill regarding amendment in wild-life (Portection) Act, 1972, has been passed by the Central Government which has now become Wild-Life (Protection) Amendment Act, 1991. In this Act, provision for heavy penalties are provided which will ensure better protection through law. With deployment of flying squads, number of hunting cases have been reduced considerably in all parts of Rajasthan.

Besides, National Parks and Sanctuaries, the Forest Department is trying to establish Conservation Reserves and Community Reserves for conservation and protection of wildlife as well as entire biodiversity. Two

Conservation Reserve could be notified at present, one being at Bisalpur (Tonk) and other in Jor Beed (Bikaner).

Apart from forest areas, there are many areas like "Orans" or sacred groves scattered in the entire state which are repository of rich bio-diversity and such areas are religiously been protected by local people. No one dare to destroy flora and fauna of these sites for the fear of curse from reigning deities of these sites.

A large number of big and small water bodies scattered all over the state including the recently created Wetlands along Indira Gandhi Nahar have enhanced the habitat diversity in the state. The vast floral and faunal diversity of the state is evident from the fact that there are about 2500 species of plants, 450 species of birds, 50 species of mammals, 20 species of reptiles and 14 species of amphibians besides numerous species of insects, butterflies, micro flora and fauna (Table 8.9).

Table 8.9 National Parks and Wild-life Sanctuaries of Rajasthan

S.No.	Name	Area (sq. Km.)	District	Year of Establishment
	NATIONAL PARKS	28.73		
1.	Keoladeo	392.50	Bharatpur	1981
2.	Ranthambhor		Sawai Madhopur	1980
	SANCTUARIES	199.50		
1.	Bandh Baretha	138.69	Bharatpur	1985
2.	Bassi	229.14	Chittaurgarh	1988
3.	Bhainsrorgarh	274.41	Chittaurgarh	1983
4.	Darrah	3,162.00	Kota, Jhalawar	1955
5.	Desert	52.34	Barmer, Jaisalmer	1981
6.	Jaisamand	300.00	Udaipur	1955
7.	Jamwa Ramgarh	153.41	Jaipur	1982
8.	Jawahar Sagar	676.40	Bundi, Kota, Chittaurgarh	1975
9.	Keladevi	608.57	Karaulli, Sawai Madhopur	1983
10.	Kumbhalgarh	14.76	Udaipur, Rajsamand, Pali	1971
11.	Kesarbagh	112.98	Dhaulpur	1955
12.	Mount Abu	50.00	Sirohi	1960
13.	Nargarh	280.00	Jaipur	1980
14.	National Chambal	492.68	Kota, Sawai Madhopur, Bundi, Dhaulpur, Karaulli,	1979
15.	Phulwari Ki Nal	252.79	Udaipur	1983
16.	Ramgarh Vishdhari	34.40	Bundi	1982
17.	Ramsagar	5.19	Dhaulpur	1955
18.	Sajjangarh	557.50	Udaipur	1987
19.	Sariska	422.94	Alwar	1955
20.	Sitamata	127.76	Chittaurgarh, Udaipur	1979
21.	Sawai Mansingh	98.70	Sawai Madhopur	1984
22.	Shergarh	7.19	Baran	1981
23.	Tal Chappar	463.03	Churu	1971
24.	Todgarh Raoli	25.60	Ajmer, Pali, Rajsamand	1983
	Van vihar		Dhaulpur	1955

Source: Department of Forest (Wild-Life Wing), GoR, Jaipur

WTLDLLIFE PROTECTION

Rajasthan is famous for its splendour and a wide variety of colourful fauna. The jungles, sands, rocks and ravines of Rajasthan provide natural habitat for- various types of wildlife; such as tigers, leopards, chinkaras, sloth bears, wolves, black-hecked storks, sambhars, cheetals, wild neelgais, boars, etc. Herons, geese and storks flutter their wings in the marshes. There are spectacular wildlife sanctuaries, game., reserves, closed areas and national parks which have been developed in the state for the protection and preservation of wildlife. The state has 3 national parks, (Ranthambhoe, Keoladeo and Mukandara Hills), one National Park (Desert Sanctuary), Jaisalmer; 25 wildlife sanctuaries, and 33 closed wildlife areas at present under the Wildlife (protection) Act of 1972, passed by the Government of India. The State Government notified the Wildlife (protection) (Rajasthan) Rules in 1977. These rules are quite comprehensive to ensure the preservation of wildlife in the state. They indicate the procedures for obtaining licenses for trading in wildlife products, permission for shooting and penalties for wildlife offences, etc.

The following spots are important from tire point of view of wildlife preservation and protection in Rajasthan. The three National Parks are as follows:

1. Ranthambhore National Park, Sawai Madhopur

It is spread over 389 sq. kilometres and is regarded as the habitat of tigers. In the recent past there has been some controversy regarding the exact number of tigers in this national park. In the narrow valley of Ranthambhore, tigers, leopards, and bears have their suitable places of hiding. Sambhars, deers, jackals, hyaenas, neelgais, chinkaras, etc., are also found in this park. In the nearby lake there are crocodiles and flocks of birds.

2. Keoladeo (Ghana) National Park, Bharatpur

There are innumerable varieties of geese in this park. Birds make their nests on 'Kikar' trees. Various types of cranes can be seen here. Every year migrant Siberian cranes reach here and take rest for a definite period. Animals like sambhar, wild boar (pig) and neelgai, cheetal, etc., are also found here. UNESCO has given it the status of 'World Natural Heritage'.

3. Mukandara Hills National Park

A National Tiger Sanctuary was sanctioned in March, 2011. Its area lies in Kota and Chittorgarh.

4. National Desert Sanctuary, Jaisalmer

It was set up in 1981. Cheetals, chinkaras, desert cats, fox and rabbits are usually found in this park. Great Indian Bustard, Falcons, Eagles are also seen in it, Great Indian Bustard popularly called Gondawan is fast disappearing from Rajasthan.

There are other places in Rajasthan, where one can find wildlife, particularly near Udaipur. The more important wildlife sanctuaries are:— (i) Kela Devi (Sawai Madhopur); (ii) Kumbhalgarh (Udaipur); (iii) Phulwari Ki Naal (Udaipur); (iv) Todgarh, Rawali (Ajmer); (v) Tiger Project Sariska (Alwar) and (vi) Sitamata (Chittorgarh). Thus, Rajasthan is not devoid of wildlife, even though it is essentially a desert region. Various animals and birds enhance the beauty of nature and help in preserving the environmental and ecological balance in the state. The government is keen to preserve and protect the wildlife as far as possible. But there is need to have strict vigilance to stop illegal hunting or poaching in these areas. Usually we hear about the complaints in this regard. They should be stopped by imparting harsh punishment to the people who indulge in such illegal activities. People's cooperation is necessary for the conservation, preservation and protection of wildlife in Rajasthan. That is essential for promoting eco-tourism in the state.

Chapter 9

DISASTERS MANAGEMENT

Natural hazards have historically displayed an absolute disregard for the national borders established by humankind. Seismic faults, paths of tropical storms and droughts wreak havoc across large areas as determined by the laws of physics rather than those of nations. More often than not, one disaster affects several countries. This is especially true for small island states and other countries of modest territorial holdings.

Natural disasters continue to strike and increase in magnitude, complexity, frequency and economic impact. Whilst, the natural phenomena causing disasters are in most cases beyond human control, vulnerability is generally a result of human activity. Therefore, society must recognize and strengthen traditional methods and explore new ways to live with such risks and take urgent actions to prevent as well as to reduce the effects of such disasters. The capabilities to manage such activities are available and there is need to adopt requisite measures.

The disasters play their role as natural factor, but the least developed countries are the most vulnerable for such events as they are least equipped to mitigate disasters. Developing countries are affected by desertification, drought and other types of natural disasters, are also equally vulnerable and insufficiently equipped to mitigate natural disasters (Figure 9.1).

The poor and disadvantaged groups suffer most from natural disasters and are least equipped to cope with them. In fact, disasters contribute to social, economic, cultural and political disruption in the urban and rural contexts, each of its specific way. Large-scale urban concentrations are parti-cularly fragile because of their complexity and the accumulation of population and infrastructures in limited areas.

Concentrating the issue over the state of Rajasthan, There are various factors responsible for slow pace of development and prevalence of geographic adverse situations, which enhance the severity of disasters in the state. These factors are the hindrances in the acceleration of development process and understanding of these factors is essential to know the reasons striking most.

I. DROUGHT DISASTER

There is no universally accepted definition of droughts. Early geographers defined drought as a prolonged period without rainfall. The Meteorological Office (1936) defined absolute drought as a period as a period of at least 15 consecutive days none of which had rainfall of 0.25 mm or more. According to Ramdas (1960), drought was taken as a situation when the actual seasonal rainfall is deficient by more than twice the mean deviation. In USSR, drought is defined as a period of 10 days with a total rainfall not exceeding 5mm. Gibbs and Mahar (1967) made a study of droughts in Australia by using annual rainfall deciles as drought indicators. Using a network of 100 stations, they draw isolines of annual rainfall deciles and took areas covered by the first docile range as drought areas. Using similar criterion, droughts in Bihar studied by George and Kalyanasundaram (1969). Van Roy (1965) developed a drought analogy index based on rainfall departure and mean of lowest ten values of rainfall in series.

Source: Flood Manual, Rajasthan (Disaster Management and Relief Department.

Figure 9.1 Rajasthan Multi-Hazard Zone

A slightly modified criterion was adopted by Kalyanasundaram and Ramashastry (1969) for study of droughts in Bihar state. According to Bates, drought occurs when the annual precipitation is 75 percent of the normal or when the monthly precipitation is 60 percent of the normal. The Meteorological Department of Government of India, annual rainfall of 75 percent or less of normal is taken as a drought year and 50 percent or less as a severe drought year.

Drought is, however, evaluated in a more meaningful manner balance computa-tion taking in to account the rainfall pattern, evapo-transpiration, moisture characteristics of soil, resting depth and stage of crop growth etc. Van Bavel (1953) found that the amount of moisture deficit is quite sensitive to change in moisture storage capacity of the soil. According to his study, the average number of drought days decreased from 45 to 10 as the moisture storage capacity increased from 25 to 100 mm.

In semi-arid zones like Thar, the specification of drought cannot be divorced from the land utilization. Rainfall too meagre to produce crops may be adequate for pasture. An amount of rain, which would give normal growth of indigenous pasture, may prove inadequate for another area. Thus climatic criteria for drought varies with the type of crop growth, pasture species and the soil type of the locality. The effect of the continued drought in a pasture land would be deterioration of pasture, either by failure of plants to germinate or grow or by reduction of soil moisture to such an extent that even perennial plants cannot maintain themselves. Hence, in the study of the droughts in pasture land, concept of initial effective rainfall is necessary to initiate plant growth following drought and an effective carry over rainfall to maintain growth periods by water balance models during individual years are required. Such analysis would indicate duration of non-growth or drought periods within the growing season (Fitzpatrick *et al.*, 1967, Krishnan, 1969).

Types of Drought

Thernthwaite (1947) has classified droughts in to following four categories:

(i) **Permanent Drought:** which is characteristic of the driest climate, where sparse vegetation growing is adapted to drought and agriculture is possible only by irrigation during the entire crop season.

(ii) **Seasonal Drought:** which is found in climates which have well defined rainy and dry seasons. The natural vegetation is made up of plants that produce seeds during the rainy season and then die and plants that remain alive but become dormant in the dry season. Most of the arid and semi arid zones fall under this category. The planting dates and duration of crop varieties in those regions are so adjusted that their

(iii) growing season falls within rainy season followed by the period having residual moisture storage.

(iv) **Contingent Drought:** which is dependent upon irregularity of rainfall and so does not occur regularity in any definite season.

(v) **Invisible Drought:** Which can occur even when there is frequent rainfall in an area. When rains do not supply enough water to counteract the water loss by evapo-transpiration, the result is a borderline water deficiency in soil resulting in less than optimum crop yields. This occurs usually in humid area.

Barcar (1979) has broadly classified droughts in to following three categories :

(i) **Meteorological Drought:** It arises in a situation, when the actual rainfall is significantly less than the climato-logically expected normal rainfall for an extended period of time over any large area. World Meteorological Organiza-tion (1988) has given following meteorological criterion for identifica-tion of drought situation- "60 per cent less of normal annual precipitation, for over two consecutive years for over 50 percent of the spatial extent of the region."

(ii) **Hydrological Drought:** Such droughts are associated with marked depletion of surface water and consequent drying of ground water resources. These water resources include precipitation, infiltration, lakes, river-systems, reservoirs, and in-flow/ out-flow systems, which may be included in the hydrological water balance equation. A hydrological drought results, if the meteorological drought is sufficiently prolonged. Hydrological drought occurs, when demand by the hydrological system greatly exceeds the water available for the use of the system. Precipitation is the most common natural factor, which usually induces drought. Precipitation has the highest spatial and temporal variability compared with other whether elements.

(iii) **Agricultural Drought:** It is the result of an imbalance between the soil moisture and evapo-transpiration needs of an area over a fairly long period, so as to cause damage to standing crops and a reduction in crop yield. Agricultural drought are defined using the conditions of the crop, livestock, pasture and other agricultural systems.

In Rajasthan, drought is a serious and vulnerable problem and its occurrence is regular in some parts or other. The number of districts and villages affected by drought and human and livestock suffered with the problem (Figure 9.2).

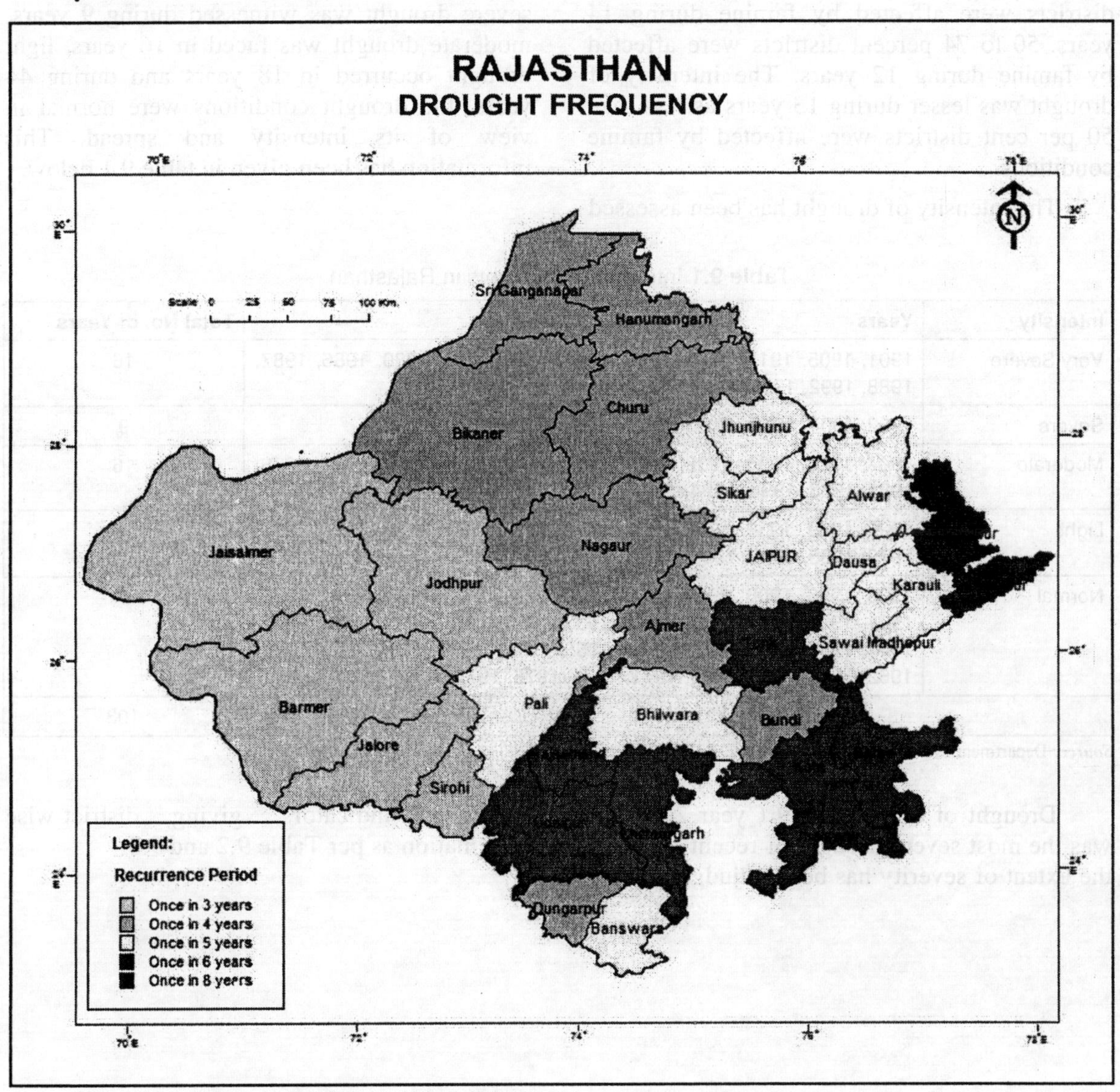

Source: Disaster Management and Relief Department, GoR, Jaipur

Figure 9.2 Drought Frequency in Rajasthan

It is evident from the above information that during the period of last 53 years, there was no famine in any part of the state during six years, i.e., 1959-60, 1973-74, 1975-76, 1976-77, 1990-91 and 1994-95. Similarly during 10 years all the districts were declared famine affected. More than 75 to 99 percent districts were affected by famine during 14 years. 50 to 74 percent districts were affected by famine during 12 years. The intensity of drought was lesser during 13 years as less than 50 per cent districts were affected by famine conditions.

The intensity of drought has been assessed by number of persons affected by famine conditions and for this purpose the drought conditions have been sub-divided in five categories. According to information available with the State Government for past 103 years it is evident that during 16 years the drought conditions were very serious in 16 years, severe drought was witnessed during 9 years, moderate drought was faced in 16 years, light drought occurred in 18 years and during 44 years, the drought conditions were normal in view of its intensity and spread. This information has been given in table 9.1 below.

Table 9.1 Intensity of Droughts in Rajasthan

Intensity	Years	Total No. of Years
Very Severe	1901, 1905, 1911, 1915, 1918, 1951, 1952, 1969, 1980, 1986, 1987, 1988, 1992, 1996, 2001, 2003	16
Severe	1902, 1904, 1913, 1921, 1925, 1939, 1975, 1982, 2000	9
Moderate	1907, 1920, 1936, 1938, 1941, 1953, 1966, 1967, 1970, 1973, 1978, 1981, 1983, 1990, 1994, 1999	16
Light	1922, 1929, 1930, 1932, 1940, 1948, 1949, 1954, 1958, 1964, 1972, 1979, 1985, 1989, 1993, 1997, 1998, 2002	18
Normal	1903, 1906, 1908, 1909, 1910, 1912, 1914, 1916, 1917, 1919, 1923, 1924, 1926, 1927, 1928, 1931, 1933, 1934, 1935, 1937, 1942, 1943, 1944, 1945, 1946, 1947, 1950, 1955, 1956, 1957, 1959, 1960, 1961, 1962, 1963, 1965, 1968, 1971, 1974, 1976, 1977, 1984, 1991, 1995	44
	Total	103

Source: Department of Disaster Management and Relief, Govt. of Rajasthan.

Drought of Samvat 2059 (year 2002-03) was the most severe drought of recent year and the extent of severity has been adjudged on the following indicators, giving district-wise information as per Table 9.2 and 9.3.

Table 9.2 : Extent of Scarcity of Drought

S. No.	District	Total Villages	Affected Villages			Population Affected	Area Sown	Affected Area	Affected Cattle	Value of Damaged Crop	Land Revenue Suspended	Rainfall Deficit %
			50-74 %	75-100 %	Total	(Lacs)	(Lac. Ha.)	(Ha.)	(Lacs)	(Crore Rs.)	(Lac Rs.)	
1	Ajmer	1076	11	1065	1076	11.78	3.06	2.97	7.67	183.62	16.29	65
2	Alwar	1984	123	1857	1980	27.65	2.45	2.16	15.41	129.36	33.46	72
3	Banswara	1524	1510	-	1510	13.49	2.23	1.30	11.82	76.66	8.99	38
4	Baran	1233	688	519	1207	13.76	1.19	0.91	6.73	86.82	16.88	62
5	Barmer	1975	1	1974	1975	14.35	9.77	9.67	42.02	220.09	-	69
6	Bharatpur	1505	-	1504	1504	1.83	1.70	1.70	7.96	84.85	21.97	50
7	Bhilwara	1788	290	1494	1784	14.71	3.20	3.03	19.77	203.46	38.27	56
8	Binaker	889	-	889	889	19.01	1.24	1.01	23.17	10.07	13.10	85
9	Bundi	870	369	501	870	8.86	1.60	1.23	9.62	87.51	10.27	56
10	Chittorgarh	2415	1193	1151	2344	16.34	3.89	2.80	17.16	203.56	21.65	51
11	Churu	907	12	895	907	13.73	4.32	4.24	18.10	121.63	-	64
12	Dausa	1061	117	939	1056	11.93	1.44	1.28	6.56	58.63	8.88	62
13	Dholpur	828	3	825	828	8.87	0.71	0.59	4.53	43.00	5.32	44
14	Dungarpur	872	-	872	872	8.74	1.09	0.93	9.58	9.32	0.87	37
15	Ganganagar	3028	484	1668	2152	7.05	1.53	1.13	8.80	1.77	27.42	76
16	Hanumangarh	1912	314	1525	1839	11.50	3.29	2.79	8.98	-	-	79
17	Jaipur	2340	35	2237	2272	26.16	4.89	4.37	20.36	274.66	51.26	71
18	Jaisalmer	688	36	607	643	4.43	0.24	0.22	12.56	4.34	2.59	83
19	Jalore	728	5	723	728	11.43	4.44	4.30	6.26	260.47	43.88	59
20	Jhalawar	1614	325	1266	1591	10.70	2.68	1.47	8.62	113.91	8.50	46
21	Jhunjhunu	867	-	867	867	16.00	2.32	2.27	12.43	75.85	4.03	75
22	Jodhpur	1082	36	1046	1082	24.06	1.02	-	33.07	279.86	-	76
23	Karauli	829	9	820	829	9.73	1.31	1.23	6.38	81.74	16.62	69
24	Kota	947	521	379	900	9.05	1.28	0.88	3.73	139.52	35.05	53
25	Nagaur	1508	34	1474	1508	19.49	5.87	5.40	19.57	380.65	-	69

S. No.	District	Total Villages	Affected Villages			Population Affected	Area Sown	Affected Area	Affected Cattle	Value of Damaged Crop	Land Revenue Suspended	Rainfall Deficit %
			50-74 %	75-100 %	Total	(Lacs)	(Lac. Ha.)	(Ha.)	(Lacs)	(Crore Rs.)	(Lac Rs.)	
26	Pali	973	8	965	973	14.76	3.23	3.16	28.04	301.56	-	68
27	Rajsamand	1004	10	994	1004	8.03	0.86	0.76	11.93	81.11	10.32	49
28	Sawai Madhopur	800	62	732	800	11.16	1.05	0.93	8.18	50.75	-	71
29	Sikar	1017	6	1.11	1017	18.45	3.80	3.60	16.63	197.99	9.87	67
30	Sirohi	474	-	474	474	6.58	1.16	1.11	9.83	72.10	-	71
31	Tonk	1122	49	1054	1103	10.99	3.02	2.86	9.88	118.75	8.26	70
32	Udaipur	24.06	957	1449	2406	25.77	2.25	1.67	26.28	77.66	15.49	36
	Total	**42266**	**7208**	**33782**	**40990**	**447.80**	**82.26**	**71.97**	**451.63**	**4031.37**	**429.26**	**64**

Source: Department of Disaster Management & Relief, Government of Rajasthan.

Table 9.3 Loss Due To Famine / Scarcity Condition In Rajasthan

Agriculture Year	No. of Districts Affected	No. of Villages Affected	Population Affected (In Lakh)	Land Revenue Suspended ₹ (Lakh)
1	2	3	4	5
1981-82	26	23246	200.12	646.15
1991-92	30	30041	289.00	325.87
2001-02	18	7964	69.70	45.84
2002-03	32	40990	447.80	429.26
2003-04	3	649	5.82	8.80
2004-05	31	19814	227.65	167.77
2005-06	22	15778	198.44	123.21
2006-07	22	10529	136.73	36.49
2007-08	12	4309	56.12	39.86
2008-09	12	7402	100.12	47.69
2009-10	27	33464	429.13	459.04
2010-11	2	1249	13.67	9.53 @
2011-12	11	3739	49.95	30.77 @
2012-13	12	8030	120.90	65.44@
2013-14	17	10225	159.38	101.44
2014-15	13	5841	74.30	15.35
2015-16	19	14487	194.87	171.55@
2016-17	13	5656	90.38	62.00@

* Figures for financial year. @ Likely

Source: Department of Disaster Management and Relief, Govt. of Rajasthan.

Drought Management

The drought is the most disastrous situation and the area affected need immediate relief in terms of wage employment, drinking water support in scarcity affected areas, survival of livestock and other requisite arrangements In such a situation, first and foremost responsibility of the administration is survival of the people with a commitment that not a single person dies from hunger. Similarly, survival of livestock is another area of management being the survival base of the poor and only asset of livelihood. With this motto the administration deputes to manage the situation effectively.

II. FLOOD DISASTER

Flood is a natural climatic phenomenon and occurs in almost all the countries of the World, particularly in tropical areas situated in the zones of receiving heavy rainfall. It is regarded as a most disastrous aspect of human civilization since it destroys every thing whichever comes in its way within moments and there remains no other alternative than to bear its consequences patiently till the cessation of rainfall and lowering down of flood waters in the corresponding areas. The remanent of ancient civilization found out by the Archaeological Survey of India under past investigations in India and other countries reveals that man had started human activities along the river courses particularly in the flood plains with the evolution of life on the earth. It has also proved that water was considered as the main source of life and man's affinity in living along river courses near flood plains even in ancient days has not yet slackened and is continued till today even when floods are experienced every year.

In India, the flood problem arises mainly during the period of south-west monsoon rainfall in the country i.e., 1st June to 30th September since ost of the parts receive about 75 percent of annual normal rainfall or more during this period. Accordingly, almost all rivers receive maximum run-off during these four months whenever heavy to very heavy rainfall occurs in the catchment areas due to which flood of varying intensities are experienced. Flood causes a great loss of human lives, livestock, standing crops in the fields, trees as well as considerable damages to civil properties, roads, railway lines, electricity lines and disruption of all other communication systems all around in the corresponding areas.

Rajasthan does not belong to flood prone states in the country but occurrence of flood is significantly disastrous in view of beyond imagination. There very few strategic areas situated in river catchment areas, but the floods occurrence in such areas are quite rare. Flood situation in Jodhpur and Jaisalmer is beyond imagination and in 1992, when one tourist lady was swept away with the flood waters near Osian tehsil on way from Jodhpur to Phalodi. Similarly, unusual rainfall in July 2000 over Lunkaransar in Bikaner district created severe flood conditions as high sheet of 10-15 feet stood in low lying areas for a month or so (Figure 9.3).

Source: Flood Manual, Rajasthan (Disaster Management and Relief Department)

Figure 9.3 Rajasthan Flood Prone Area

Causes of Floods

Flood is a natural phenomenon and occurs each year in some part or other but with different magnitudes and intensity during the monsoon season. The washing away of Indus Valley Civilization in the pre-Cambrian era is the living examples.

Excessive rainfall on account of unusual storm or cloud burst over an area and consequent discharge or run-off developed from the catchment area create flood conditions when the quantum of run-off so produced does not pass through the available capacity of the corresponding water course, drainage system or river. It is an essential factor; where as other factors are also responsible for developing and intensifying flood conditions.

These important causes are hydro-meteorological factors, climatic factors, physiographical, relief features, geomorphologic conditions, silt load, sand cover, biological vegetation, forest cover, human and live stock population as well as urbanization and other human activities developing in the region are also responsible in intensifying flood conditions of the area.

(i) Hydro-Meteorological Factors

The Hydro-Meteorological Factors are mainly responsible for creating flood conditions in the concerned area. Hydro-meteorological involves hydrological conditions or run-off conditions on one side and meteorological conditions or climatic conditions including rainfall on the other side of the concerning landscape. The quantum of rainfall, its intensity and period whenever exceeds the normal limits over an area, creates abnormal run-off with which flood conditions are developed in those areas, where water courses do not have as much carrying capacity as to pass conveniently without flowing on either side. Under circumstances water passing through as per natural gradients of the landscape may sometimes pass over-flanking the banks of water courses and finding new areas devastating whatever and wherever any property, road, railway line, other communication systems as well as standing crops etc. comes in its way.

The people of Pink city of Jaipur have witnessed floods of July, 1981 in the recent past, which is an example as to how incessant heavy rainfall created flood conditions and intensified further in adjoining areas. Statement showing rainfall data recorded at Jaipur and nearby stations from 18th to 20th July, 1981 is given in Table 9.4.

Table 9.4 Rainfall Recorded in Jaipur and nearby areas (in mms.)

S.No	Stations	18-7-81	19-7-81	20-7-81	Total
1.	Jaipur Airport	162	328	135	625
2.	Jaipur City	236	354	228	818
3.	Kanota	209	589	121	919
4.	Bassi	284	556	15	855
5.	Dausa	52	550	95	697
6.	Chaksu	266	290	155	711

Source: Department of Disaster Management and Relief, Govt. of Rajasthan.

An analysis of the rainfall position reveals that maximum rainfall was recorded in one single day preceded and succeeded caused havoc as the normal rainfall of July in total area was 188 mms. This sort of abnormal rainfall in successive three days did not permit runoff from the regular available systems and the damage was beyond imagination as all the rail, road and air ways were damaged and the city was cut off from every side. The devastating effects and other damages occurred during the period could not be avoided even when the Administration, Military people and all civilians with all sorts of force and equipments were available at the capital city but could no help and remained isolated for three-four days.

In view of above, it is evident that excessive rainfall and other hydrological conditions of particular region are mainly responsible for creating and developing flood conditions in corresponding areas, whenever rainfall crosses normal limits over an area in a short duration.

(ii) Climatological Factors

Climatological factors are mainly responsible for all sorts of catastrophe and other detrimental aspect of human civilization in usual manner. The climate of a region is generally known with the average condition of all weather parameters for which meteorological data are being collected on uniform scientific lines as a regular feature. The Indian Meteorological Department has established a network of meteorological observation stations in whole India for recording all-important meteorological data on uniform scientific lines. It has also adopted similar standards as prescribed by the World Meteorological Organization for maintaining uniform standards the whole world.

The district physiographical conditions with the extension of Aravallis in oblique form from northeast to southwest have played an important role in deciding the climate in Rajasthan. It is a great barrier, which divides the plains of Marwar with the high table land of Mewar. Accordingly, the climate of western Rajasthan is mainly sterile sandy and inhospitable where as the climate of eastern part is somewhat humid and comfortable to a treat extent. Different authors have classified the climate of Rajasthan under different categories.

The India Meteorological Department has divided the whole country in to 35 meteorological divisions for purposes of recording meteorological data and to issue weather forecasts and bulletins. Rajasthan has been classified in to two meteorological divisions i.e., West Rajasthan and East Rajasthan. West Rajasthan comprises of ten districts lying in the west of Aravallis i.e., Barmer, Bikaner, Churu, Ganganagar, Hanumangarh, Jaisalmer, Jalore, Jodhpur, Nagaur and Pali. Other remaining twenty two districts have been placed under second category.

Types of Floods

Cause of the flood is occurrence of rains with high intensity in very short duration, which does not permit regular drainage system to absorb the surface water and overflows in the surrounding areas. In broad term, floods can be categorized as severe and moderate in view of damage in the catchment area.

Analysis of above information reveals that monsoon pattern has been showing live symptoms of variation in the recent past. Twenty one flood years have been identified on the basis of rainfall data collected from all the 32 district headquarters from 1961 to 2000. Only eight flood years – 1961, 1973, 1975, 1976, 1992, 1995 and 1996 had covered more than 50 percent of the geographical area of the state. During 13 flood years, geographical area affected between 20 to 50 percent. Geographical area up to less than 20 percent was affected during 15 years and there was no flood in any part of state during 4 years.

It has therefore, confirmed that the pattern of monsoon is showing live symptoms of a great change particularly excessive rainfall in the western region in the recent past. Lunkaransar flood conditions developed in2000 is the recent example of flood event experienced in the region.

The flood frequency analysis reveals that the frequency is below 20 percent in 27 districts in case severe flood and moderate flood events are evaluated independently. It is however more than 20 percent in 27 districts if total flood events analyzed for all the 32 districts. It has, therefore, been confirmed that flood events occur each year but in different areas with different magnitudes.

It is rather necessary to occur as a regular feature in order to maintain equilibrium in respect of presence of moisture in the environment as well as proportion of water over and below the surface of earth. On the basis of analysis of past 40 years areas have been identified in view if intensity and occurrence of floods:

(a) Highly Sensitive Flood Prone Areas

Areas fall under this category have been identified on the basis of past tract in view of occurrence of rainfall in short spell. This is also one of the geophysical situations and requires proper attention:

(i) Ghaggar flood prone areas in Hanumangarh and Ganganagar districts

(ii) Banganga and Ghambir river systems around Bharatpur city.

(iii) Berach and Gambhiri river systems around Chittorgarh city..

(iv) Chambal river system as a whole including all four reservoirs.

(v) Jakham dam at Anoopura near Dhariawad in Udaipur district.

(vi) Som Kamla Amba dam near Aspur in Dungarpur district.

(vii) Galiakot in Dungarpur district due to Kadana dam.

(viii) Mahi dam near Banswara district.

(ix) Bisalpur dam near Deoli in Tonk district.

(x) Bhim Sagar dam in Jhalawar district.

(xi) Jawai dam in Pali district

(xii) All other Major Irrigation reservoirs.

All above highly sensitive areas require close supervision while passing all floodwater smoothly during the monsoon period. At the time of occurrence of constant rainfall and accrual water, the officers of Irrigation Department put on high alert and to take immediate action in case of any adverse situation.

(b) Sensitive Flood Prone Areas

Areas identified under this category are as under:

(i) Jhalawar city due to Kali Sindh river.

(ii) Pali city due to Bandi river.

(iii) Balotra city due to Luni river.

(iv) Jalore city due to Jawai and Bandi river.

(v) Kot Kasim due to Sahibi river.

(vi) All medium Irrigation reservoirs.

(c) Occasional Flood Prone Areas

Occasional flood conditions have been experienced in densely populated areas particularly because of encroachment along the watercourses for residential purposes in the cities. The administration is helpless in view of humanitarian aspects and resistance of the residents to vacate the low-lying areas. These areas located in cities of Jaipur, Ajmer, Kota, Jaisamler, Jodhpur, Bilara, Samdhari, Lunkaransar, Kolayat, Pilibanga, Jetsar, Anupgarh, Hanumangarh and Suratgarh. Besides these, all minor irrigation reservoirs having capacity of fifty million cubic feet are also covered under this category.

Flood Disaster Management

System for flood disaster management is similar to that of the drought as all these activities are administered by Disaster Management and Relief Department of the State Government. District Collector is the nodal officer at the district level and undertakes all the precautionary and safety measures through the agency of Irrigation Department.

Irrigation Department is responsible to supervise their dams of all the categories and keep constant vigil over the situation. In case of any severe problems envisaged , the District Collector has been authorized to call police and Military to take precautionary measures for safety and security of the human and other causalities.

Preparedness Against Flood

Every year, the preparedness meetings of Government Machinery for proper maintenance of irrigation dams are conducted at all levels. Flood restoration activities are undertaken as part of development activities. Primary responsibility of security of irrigation dams is of Irrigation Department. Regular vigil of rainfall is conducted at field level in view of arrival of rainwater in to the dam and constant monitoring is conducted. Discharge of water from dams is ensured on the basis of water flow level and every precaution is carried out to keep the water level within the safe limits. Boats and helicopters are deployed in the area as per requirement.

Relief Commissioner, Department, District Collectors and Chief Engineers of Irrigation Department conduct regular meetings to assess the situation. All precautionary measures are ensured by respective agencies and regular reporting through written reports, telephonic messages, internet and publicity media are taken to aware the people of respective areas about the possible disasters.

Every effort is made to keep safety and security of people, buildings, infrastructure facilities, roads and railway tracts. All the concerned departments and agencies have delegated the responsibilities on their field officers and all sort of preparedness is ensure before the arrival of monsoon. Emergency measures are taken for the situation caused in view of heavy rainfall and ability of the drainage systems to accommodate the water passes uninterrupted. Serious problems are taken up immediately in view of effective response system.

Response Mechanisms

Monsoon forecasts are being made for each region well in advance for the rainy season as a whole, daily basis on the basis of fluctuations in the sea. Information gathered through satellite is analyzed for the country as a whole and for various regions of each state. Whether forecasts are made for each day on the basis of the findings. In addition, there are 531 rain gauge points, 26 symon rain gauge points and 45 meteorological observatories in the state. Daily rainfall position is recorded and communicated to all concerns for observation of the situation.

During the rainy season, Disaster management and Relief Department set up control rooms at State, District and Block levels and each important event is communicated to all concerns. In addition District Collectors are connected with hot line with State Government dignitaries including Chief Minister, Chief Secretary and other officer and quick communication is made in case of any severe condition. Total administration is on foot to manage the situation.

Response to all the concerns is made immediately through all the communications including telephone for direct information. In addition paper communication is also made simultaneously for official records and immediate remedial measures as per the situation. There are events, when the lives saving efforts have been made even through helicopters, boats, etc.

In case of life casualty, the District Collector is authorized to issue immediate assistance to the family for death rituals as well as one time assistance to the aggrieved family. Immediate assistance is provided to the people marooned in villages due to flood water. Temporary camps are set up for flood-affected people. The total system is part of sanctioned procedure and in case of unwarranted situation immediate assistance measures are taken with the telephonic approval to be confirmed later on papers.

Post Flood Management

Flood restoration activities take long time and cost for bringing the normal situation. This depends on the nature of damage in any area. Post flood management is like emergency plan preparation and implementation. For this purpose, State Government provides all possible assistance and undertakes various activities at its own. Affected people are also associated to manage the situation effectively.

Flood disasters have severe impact on people and area. Irreparable losses are significant and in most of the cases, some of the families are totally ruined. Government assistance is quite meagre in such situations and affected areas and people take long time to mitigate the damage.

III. EARTHQUAKES DISASTER

The sudden slip at the fault causes the earthquake- a violent shaking of the earth when large elastic strain energy released spreads out through seismic waves that travel through the body and along the surface of the earth. After the earth is over, the process of strain build-up at the modified interface between the rocks starts all over again. Earth scientists know this as the Elastic Rebound Theory. The material points at the fault over which slip occurs usually constitute an oblong three- dimensional volume, with its long dimension often running in to tens of kilometers.

Most earthquakes in the world occur along the boundaries of the tectonic plates and are called "Inter-plate Earthquakes" (e.g., 1897 Assam , India earthquake). A number of earthquakes also occur within the plate itself away from the plate boundaries (e.g., Latur, India earthquake); these are called "Inter-plate Earthquakes." In both types of earthquakes, the slip generated at the fault during earthquake is along both vertical and horizontal directions, called "Strike Slip," with one of them dominating sometimes.

Large strain energy released during an earthquake travels as seismic wave in all directions through the Earth's layers, reflecting and refracting at each interface. These waves are of two types- body wave and surface wave; the latter are restricted to near the earth's surface. Body waves consist of Primary Waves (P-waves) and Secondary Waves (S-waves), and surface waves consist of Love Waves and Rayleigh Waves. Under P-waves, material particles undergo extensional and compressional strains along direction of energy transmission, but under S-wave, oscillate at right angles to it. Love waves cause surface motions similar to that of by S-wave, but with no vertical component. Rayleigh wave makes a material particle oscillate in an elliptic path in the vertical plane, with horizontal motion along direction of energy transmission).

P-waves are fastest, followed in sequence by S-, Love and Rayleigh waves. For example, in granites, P- and S-waves have speeds ~4.8 km/sec. and ~3.0 km/sec, respectively, S-wave do not travel through liquids. S-waves in association with effects of Love waves cause maximum damage to structures by their racking motion on the surface in both vertical and horizontal directions. When P- and s-waves reach the earth's surface, most of their energy is reflected back. Some of this energy is returned back to the surface by reflections at different layers of soil and rock. Shaking is more severe- about twice as much, at the earth's surface than at substantial depths. This is often the basis of designing structures buried underground for smaller levels of acceleration than those above the ground.

The instrument that measures earthquake shaking, a Seismograph, has three components- the Sensor, the Recorder and the Timer. The principle on which it works is simple and is explicitly reflected in the early seismograph- a pen attached at the tip of an oscillating simple pendulum – a mas hung by a string from a support, marks on a chart paper that is held on a drum rotating at a constant speed. A magnet around the string provides required damping to control the amplitude of oscillations. The pendulum mass, string, magnet and support together constitute the Sensor; the drum, pen and chart paper constitute the Recorder and the motor that rotates the drum at constant speed forms the timer.

The analog instruments have evolved over time, but today, digital instruments using modern computer technology are more commonly sued. The digital instrument records the ground motion on the memory of the microprocessor that is in-built in the instrument. For knowing the intensity of earthquake, two systems are followed as under:

Richter Scale: Earthquake magnitude is determined by the use of a seismograph, an instrument, which continuously records, ground vibrations. A scale developed by a seismologist named Charles Richter mathematically adjusts the readings for the distance of the instrument from the epicenter, this scale is called Richer Scale and its logarithmic in operation.

MM Scale: MM Scale measures Earthquake intensity. It was developed by an Italian Seismologist, Mercalli in 1902 and modified by Wood and Neuman in 1931. Known as Modified Mercalli Scale (MM Scale). It expresses the intensity of earthquake effects on people, structures and earth's surface in values I to XI.

Causes of Earthquakes

The earth's crust is not one piece but consists of portions called plates, of varying sizes. These plates are not stable but move with three distinctive types of movements along the boundaries of the plates.

Convergent Boundaries: Where plates push each other and one plates slides down the other one.

Divergent Boundaries: Where plates pull away from each other.

Due to continuous movement of the plates, stress accumulation takes place at the boundaries of these plates. When the stress accumulation reaches a point of maximum supportive strain, a rupture takes place generating an earthquake. The earthquake generated this way are most damaging. Approximately, 95 percent of the earthquake activities occur at the plate boundaries. Some earthquakes, however, do occur in the middle of the plates, possibly indicating where earlier plate boundaries might have been. The point of rupture is called the focus.

The intensity of shaking at any place is relating to the amount of energy released, the focal depth, and distance of the place from the focus and the structural properties of the rock or soil on the surface. Some earthquakes are associated with volcanic activities as well.

Characteristics of Earthquake

Ground shaking due to an earthquake is caused by the body and surface waves. The body waves P and S waves, penetrate the body of the earth. P wave travel at a faster rate and provide the initial jolt causing buildings to vibrate in up and down motions. S waves travel at lesser speed than P waves and causes a movement similar to a rope being snapped like a whip. These waves cause a sharper jolt that vibrates the structure side to side and cause more damages.

Surface waves vibrates the ground horizontally and vertically. These are long period waves and cause swaying of tall buildings. Earthquake focus depth is an important factor in shaping the characteristics of the waves and associated damage. The focal depth can be deep (300- 700 kms0 intermediate (60-300 km) or shallow (< 60 km). The shallow focus earthquakes are more damaging due to the close proximity to the ground.

Earthquake Hazards in Rajasthan

Earthquakes continue to be one of nature's most devastating furies known to mankind since time immemorial. India has a very long history of earthquake occurrences. About 56 percent of the total area are vulnerable to seismic damage of buildings in varying degrees. The most vulnerable areas according to the present seismic zone map in India are located in Himalayan and Sub-Himalayan Regions, Kutch and Andaman and Nicobar Islands. Depending on varying degrees of seismicity, the entire country can be divided in to following seismic regions.

1. Kashmir and Western Himalayas- This region covers the states of Jammu and

Kashmir, Himachal Pradesh and sub-mountainous areas of Punjab.

2. Central Himalayas- This region includes the mountain and sub- mountain regions of Uttar Pradesh and the sub-mountainous parts of Punjab.
3. Northern India- This region comprises the whole of Indian Territory to the east of North Bengal.
4. Indo-Gangetic Basin and Rajasthan - This region comprises of Rajasthan, Plain of Punjab, Haryana, Uttar Pradesh and West Bengal (Figure 9.4).
5. Cambay and Rann of Kutch.
6. Peninsular India including the Islands of Lakshwadeep
7. The Andaman and Nicobar Islands

The distribution of magnitudes equal or more than 5.0 in these regions and their average return periods are represented in the Table 9.5.

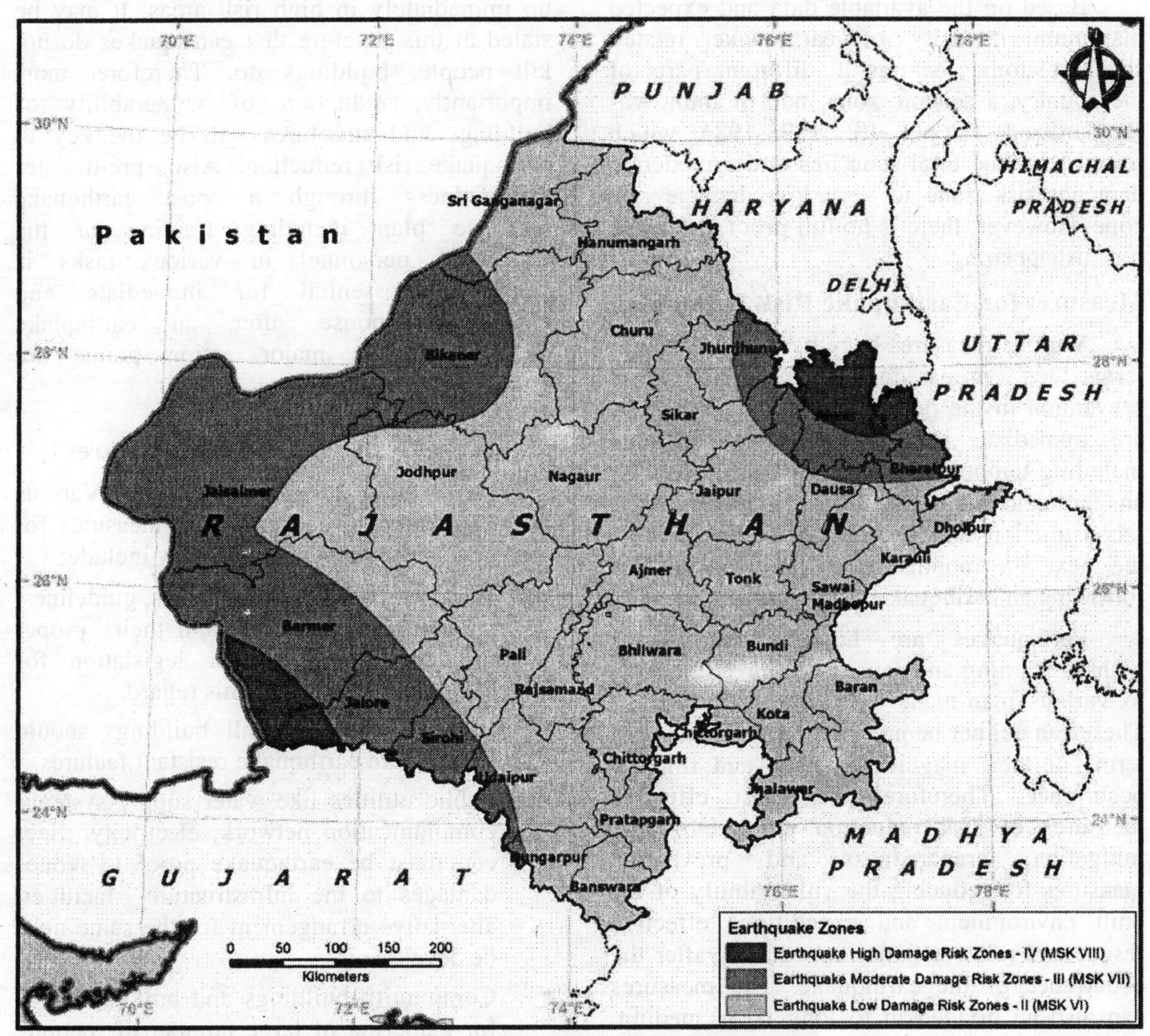

Source: IS 1893: 2001, Flood Manual, Rajasthan (Disaster Management and Relief Department)

Figure 9.4 Rajasthan Earthquake Zones

Table 9.5 Region-Wise Earthquake (M > 5.0) Occurrence in India 1897 to 1993

S. No.	Seismic Region	Number of Earthquakes of Magnitude				Return period
		5.0-5.9	6.0-6.9	7.0-7.9	8.0 +	
1.	Kashmir and Western Himalayas	25	7	2	1	2.5-3 years
2.	Central Himalayas	68	28	4	1	1 year
3.	North East India	200	128	15	4	< 4 months
4.	Indo-Gangetic Basis and Rajasthan	14	6	-	-	5 years
5.	Cambay and Rann of Kutch	4	4	1	1	20 years
6.	Peninsular India	31	10	-	-	2.5-3 years
7.	Andaman and Nicobar	80	68	1	1	< 8 months

Source: Manual on Natural Disaster Management in India.

Based on the available data and expected maximum intensity of earthquake related seismo-tectonic features in different parts of the country, a seismic zone map of India was standardized as per IS: 1893-1984 which reveals that the total state lies under moderate damages risk zone to very low damage risk zone. However, there is no full proof guarantee for mishappening.

Measures for Earthquake Risk Reduction

Various measures have been visualized in terms of preparedness, mitigation and prevention in the pre-earthquake phase. These are immediate rescue and relief measures including temporary sheltering from just after an earthquake event to about 3 months, and reconstruction and rehabilitation measures in the next six months to a three-year period following an earthquake occurrence.

Earthquakes are hazards that strike without warning and cause widespread damage to various man-made structures and systems. These can neither be prevented nor predicted in terms of their magnitude, place and time of occurrence. Therefore, the most effective measures of risk reduction are pre-disaster mitigation, preparedness and preventive measures for reducing the vulnerability of the built environment and expeditious effective rescue and relief actions immediately after the occurrence of the earthquake. The measures can also be divided in to long term, medium term and short term. The long term measures requires five to fifteen years, the medium term one to five years and the short term to be taken up immediately in high risk areas. It may be stated at this juncture that earthquakes do not kill people, buildings do. Therefore, most importantly, reduction of vulnerability of buildings and structures will be the key to earthquake risk reduction. Also, pre-disaster preparedness through a 'post earthquake response plan including training of the concerned personnel in various tasks is considered essential for immediate and effective response after an earthquake occurrence. The major action points are highlighted as follows:

(a) Pre-Disaster Preventive Measures

(i) Long term measures- Various activities taken up as long term measures for earthquake disaster mitigation may include:

- Re-framing the building codes, guidelines, manuals and bye-laws and their proper implementation. Stricter legislation for high seismic areas, in this regard.
- In high-risk areas, all buildings should incorporate earthquake resistant features.
- Public utilities like water supply systems, communication network, electricity lines etc. must be earthquake proof to reduce damages to the infrastructure facilities, alternative arrangement for the same must be developed.
- Community buildings and buildings used for gathering of large number of people, like schools, dharmshalas, hospitals, prayer halls etc. must be made earthquake

resistant in seismic zones of moderate higher intensities.

- Supporting RandD in various aspects of disaster mitigation, preparedness and prevention and post disaster management.
- Improving educational curricula in architecture and engineering institutions and technical training in polytechnics and schools to include disaster related topics.

(ii) Medium Term Measures- The medium term measures for earthquake disaster mitigation may be following:

- Retrofitting of the weak structures in the highly seismic zones.
- Preparation of literature in local language with do's and don'ts for the building construction
- Getting community involved in the process of disaster mitigation by providing proper education and awareness.
- Supporting local technical institutions/ colleges/ schools to conduct research and to organize research and exhibitions for public awareness.
- Networking of local NGOs working in the areas of disaster management.

(b) Post Disaster Preventive Measures

Short term Measures : Urgent measures to be undertaken in the aftermath of a damaging earthquake should include the following:

- Maintenance of law and order, prevention of trespassing, looting etc.
- Evacuation of people
- Recovery of dead bodies and their disposal.
- Medical care for injured
- Supply of food and drinking water.
- Temporary shelters like tents, metal sheds, etc.
- Restoring lines of communication and information
- Restoring transport routes
- Quick assessment of damage and demarcation of damaged areas according to grade of damage.
- Cordoning off severely damaged structures that are liable to collapse during aftershocks.

Coordination between various agencies involved in rescue and relief work is extremely important for success in avoiding gaps and duplication. Pre-disaster preparedness needs to be based on preparing likely damage scenario in probable earthquake occurrences and the estimate of extent of efforts required. Following preparedness measures can be useful:

- Community should be trained to search, rescue and relief at the time of disaster in high risk areas.
- An extensive programme of mass drills can be very useful in high-risk areas for earthquake damage reduction
- Local NGOs should be trained and their capacity and capabilities should be strengthened.
- Introducing earthquake disaster safety do's and don'ts and drills in schools.
- Or organize training of field personnel of the states in the science and art of carrying out post disaster damage survey, for urgent relief purposes and for repair, reconstruction and retrofitting purposes.

Consolidation and Reconstruction

Efforts need to be made in the emergency phase, to involve the effected people to the maximum possible extent so as to create a feeling of self-reliance. They need to be started as quickly as possible so that the period of relief is minimized.

- Detailed survey of building for assessment o damage and decision regarding repair, reconstruction and strengthening or demolition.

- Repair, reconstruction and seismic strengthening or demolition.
- Selection of sites for new settlements, if required.
- Execution of reconstruction programme.
- Review of existing seismic zoning maps and risk maps.
- Review of seismic codes and norms of construction.
- Training of personnel, engineers, architects, builders and masons.
- Statistical and evaluation studies regarding the earthquakes.

IV. DESERTIFICATION AND LAND DEGRADATION HAZARD

Desertification is the diminution or destruction of the biological potential of the land and can lead ultimately to desert-like conditions. It is an aspect of the widespread deterioration of eco-systems and has diminished or destroyed the biological potential, i.e., plant and animal production, for multiple sue purpose at a time when increased productivity is needed to support growing populations in quest of development. Important factors in contem-porary society, struggle for development and the effort to increase food production and to adapt and apply modern technologies, set against a background of population growth and demographic change, interlock in a network of causes and effects. Progress in development, planned population growth and improvements in all types of biological production and relevant technologies must, therefore, be integrated. The deterioration of productive eco-systems is an obvious and serious threat to human progress.

Land degradation is an issue of increasing global concern. It threatens not only the productivity of land but also water quality, human health and the fundamentals of ecosystems on which all life depend. It has also close connection with other major global issues, particularly climate change and biodiversity. It has been estimated that globally around 24 billion tons of fertile soil and 27,000 bio-species are lost each year. While land degradation is acutely felt in the world's arid lands, some 80 per cent is actually occurring outside these areas. For this reason, there is an urgent need to halt and reverse land degradation for ensuring food, water and environment security as well improving living conditions of population residing in such areas.

Desertification, along with climate change and the loss of biodiversity were identified as the greatest challenges to sustainable development during the 1992 Rio Earth Summit which paved the way for the conceptualization and formulation of the United Nations Convention to Combat Desertification (UNCCD). The Convention's 195 parties, including India, work together to improve the living conditions for people in drylands, maintain and restore land and soil productivity and mitigate the effects of drought.

Desertification and land degradation are major threats to agricultural productivity in our country. The process of desertification and land degradation are accelerating due to over-exploitation of natural resources accen-tuated by the impacts of climate change and increasing human and livestock population.

Extent and Magnitude of Desertification

Desertification affects nearly all of the arid regions, to varying degrees, except for the extremely arid climatic deserts. There are very few places, where man has totally destroyed the vegetative cover such as the Sahara, the Alabama and the Tekla Makan, which caused virtually permanent land damages. From slight to high, but not the total, land degradation is very common but equally significant to the vast majority of inhabitants of arid regions.

Extent and magnitude of desertification are qualitatively described as slight, moderate, high and very high. It should be recognized that delineating desertification on the small scale world map could not show those isolated areas of human settlement within the extreme arid regions, oasis dependent on ground water or the floods of exogenous rivers, which faced a distinctive threat of desertification. These areas, small in extent but nevertheless are highly signi-ficant for the economy of the desert regions.

It would be witnessed from Figure 9.5 that slight desertification refers to slight deterioration of plant cover or soil because of man's activities. The moderate desertification means, (a) the plant cover has been degraded to a point, where the range condition would be only fair, or (b) there are indications of accelerated wind and water erosion in the form of hummocks noticeable small dunes and noticeable small gullies, or soil salinity is reducing crop yields moderately. High desertification indicates that (a) forbs and shrubs have replaced desirable grasses or have spread over to such an extent that they dominate the flora, or (b) sheet wind and water erosion have largely denuded the land of vegetation, or (c) Salinity controllable by drainage and leaching has made the land unsuitable for crop production. The very high category of irreversible desertification includes the land covered with large shifting sand dunes, large and numerous gullies or salt crusts on nearly impermeable soils.

The "Very High" category represents the extreme condition that some persons think of it as desertification, where the land is so badly degraded that its utility by man or animal is virtually a zero and economically irreversible, for most purposes. In practice, of course, any land can be made productive again, if enough resources are expended, but it may not be economical to do so (Chouhan, 1994).

Degree of Degradation in Arid and Semi-arid Lands in Rajasthan

On the basis of climate, vulnerability of the land combined with high human-animal pressure, the degree of degradation of arid and semi-arid lands of Rajasthan has been classified in to six categories, e.g., desertized, very high vulnerability, high vulnerability, moderate and slight vulner-ability. Category six indicates the area having negligible vulnerability of deserti-fication. Details about different categories and area covered in given hereunder:

1. Desertized (Already existing desert)

In Rajasthan only a thin strip of western part of Jaisalmer district, which covers 6753 sq. kms area, falls under this category. This constitutes 1.97 percent of the total geographical area. This area is untreatable wasteland not fit for vegetation.

2. Very High Vulnerable Area for Desertification

Nine districts of Rajasthan viz. Bikaner, Churu, Ganganagar, Hanumangarh, Jalore, Jaipur, Jhunjhunu, Nagaur and Sikar fall fully or partly under this category. This area is in the grip of very raid desertification as suitable and concerted efforts can change the situation of total 51952 sq. kms of area i.e., 15.19 percent of the total geographical area.

3. High Vulnerable Area for Desertification

Large area of 88430 sq. kms or 25.84 percent of total geographical area falls under this category. Main districts of this situation are Jaisalmer, Barmer, Jodhpur, Bikaner, Ganganagar and Hanumangarh partly. There are slightly lesser degradation, but serious efforts are required to check desertification.

4. Moderate Vulnerable Area for Desertification

This category consists of parts of Jalore, jodhpur,Nagaur, Barmer, Pali and Sirohi districts and spread over 63731 sq. kms of area i.e., 18.62 percent of the state area.

Though most of the area under this category is desert, but there are comparatively lesser chances of desertification (Figure 9.5).

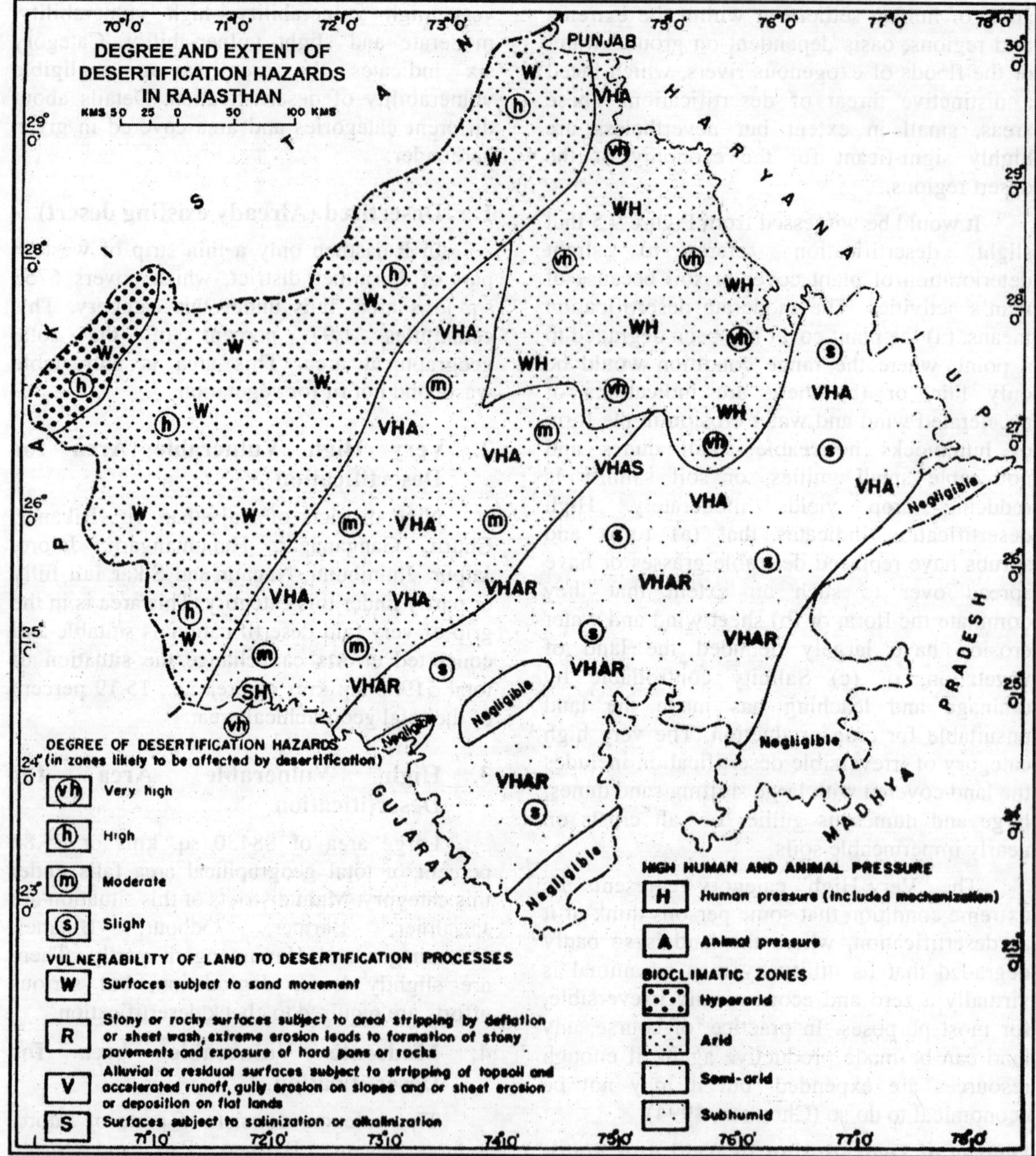

Note: The external boundary of India appearing on the Map are neither verified nor authenticated by the Survey of India. This is for academica purpose only.

Source: Desertification in the World and its Control, Dr. T.S.Chouhan.

Figure 9.5 Degree and extent of Desertification hazards in Rajasthan

5. Slight Vulnerable area for Desertification

Largest area of the state falls under this category. It is spread over in 30.30 percent of the total state geographical area of 103681 sq. km. Main districts fall fully or partly under this category are Sirohi, Udaipur, Dungarpur, Ajmer, Bhilwara, Jaipur, Dausa, Tonk, Bundi, Chittorgarh, Alwar, Bharatpur, Bharatpur, Dholpur, Sawai Madhopur, karauli, Kota, Baran and Banswara. Most of the area is attached to present desert districts and there are various concerns from specialists for spread of desert and desertification, but such fear or concerns have been proved vague.

6. Area of negligible Vulnerability of land to Desertification Hazard

An area of 27692 sq km falls fully or partly consisting of Jhalawar, Banswara, Kota, Baran, Chittorgarh, Sawai Madhopur, Karauli and Dholpur districts. Though the area is only 8.08 percent of the total geographical area of the state but this is a safe zone from the point of view of desertification hazard.

This effort of classification of the total geographical area under different categories from desertification point of view and an indication to State Administration to diversify their efforts to check the problems with suitable mitigation measures.

The total area of the state has been studied on various parameters and revealed that the process of development has not been able to help in the process of desertification but rather the position is deteriorating significantly. In addition negligible vulnerability areas have also been indicated to keep a sequence with the total geographical area of the state. Table 9.6 contains these details.

Table 9.6 : Degree and Extent of Desertification Hazard and Vulnerability of Land to Desertification Processes in Rajasthan

Degree and vulnerability of land to desertification hazard	Area in Sq. kms.	% to total area
1. Desertized (already part of existing desert)	6753	1.97
2. Very High Vulnerability	51952	15.19
3. High vulnerability	88430	25.84
4. Moderate Vulnerability	63731	18.62
5. Slight Vulnerability	103681	30.30
6. Negligible Vulnerability	27692	8.08
Total Area of Rajasthan	**342239**	**100.00**

Source: Desertification in the World and its Control : Dr. T.S.Chouhan, 1994

Processes of Desertification/ Land Degradation Vegetation Degradation (ISRO-SAC Atlas, 2016)

Vegetation degradation is observed mainly as deforestation / forest-blanks / shifting cultivation and degradation in grazing / grassland as well as in scrubland. At places, agriculture is observed within forest lands, this has also been classified under vegetation degradation within forest area. Vegetation is an important factor in the protection of soil and soil fertility. Destruction of vegetation, most often by human activities accelerates soil degradation leading to desertification. When a soil loses vegetation cover, it becomes more susceptible to wind and water erosion. Removal of top soil by water or wind erosion results in loss of organic material leading to decrease in soil aggregation and stability, and hence soil fertility. The water-holding capacity and the nutrient content of the soil are reduced when organic material is lost, which is an additional strain on vegetation survival.

Water erosion

Water Erosion is loss of soil cover mainly due to rainfall and surface runoff water. Water erosion is observed in both hot and cold desert areas, across various land covers and with varying severity levels. The sheet erosion (mostly within agricultural lands) and rills are categorised in slight category, the narrow and shallow gullies are categorized as moderate erosion, while the deep / wide gullies and ravines are classified as severe erosion. Particularly in the context of desertification or land degradation as a whole, water erosion does not refer the river erosion.

Wind erosion

Wind erosion pertains to the Aeolian activities. It denotes the spread of sand by various processes, even up to lofty altitudes of Himalayas. Wind can erode the soil very selectively and intensively in three transporting method, namely Suspension, Saltation and Soil creep. Soil is more vulnerable for wind erosion in conditions such as very sparse or no vegetative cover, increasing wind speed, loose, dry, fine or very light soil, smooth soil surface, large exposed area etc. Wind erosion removes the topsoil, which is rich in all plant nutrients and bacterial activities. Removal of topsoil reduces the capacity of the soil to function and restricts its ability to sustain future uses. Moreover, windblown dust or sand is deposited in cultivated land and buries the lands, thus reducing the production. Various kind of sand cover and their severity are classified based on the depth and spread of sand sheet/dunes and barchans.

Water logging

The undrained land parcels tend to accumulate standing water for longer durations of time on the surface, this condition is called water logging viz ox-bow lakes, low lying areas, and even the shallow water tables. The severity of water logging is determined based on the period of time the water remains stagnant. Several situations can be responsible for the rise in water table like Flood, salt-rich hard pans, excess irrigation, wrong drainage planning etc. Waterlogging may lead to salinization also.

Salinity / Alkalinity

Salinity or Alkalinity is fundamentally the chemical property of the soils. It occurs mostly in cultivated lands, especially in the irrigated areas. At places salinity is clearly observed on satellite images, while the alkalinization is not seen and is mostly inferred based on ground truth and soil sample analysis as well as information/ published maps. Soil salinity refers to the water dissolvable salt present in soil. Salinity can develop naturally, or human-induced. The main causes are excess evapotranspiration, drought, excess irrigation, increase in toxicity, and rise in ground water table. Salinity in dry lands occurs when the water table is between 2 to 3 meters from the soil surface. The salts from the groundwater are raised by capillary action to the surface of the soil. Over time, water evaporates, and the salt remains on the surface. Salinity in irrigated land can occur due to over irrigation and excess use of fertilizers and other chemicals.

Mass Movement

The spontaneous downward movement of soil and rock under the influence of gravity (but without the dynamic action of moving fluids) is included under the general term Mass Movement. The mass movement processes include all forms of down slope movement of soils, overburden, or bedrock under the direct influence of gravity. Mass movement represents the spontaneous yielding of earth materials when gravitational force exceeds the internal strength of the material. It involves sliding, rolling and flowage of masses of soil, overburden and bedrock.

Frost Heaving

Frost heaving is the process of ice lens formation beneath the soil surface during freezing conditions in the atmosphere. The ice grows in the direction of heat loss (vertically toward the surface), starting at the freezing front or boundary in the soil. It requires a water supply to keep feeding the ice crystal growth. The growing ice is restrained by overlying soil, which applies a load that limits its vertical growth and promotes the formation of a lens-shaped area of ice within the soil. The force of one or more growing ice lenses is sufficient to lift a layer of soil, as much as 30 cm or more. Due to coarse spatial resolution this process could not be identified in AWiFS images.

Frost shattering

Frost shattering is the essentially a process of mechanical weathering or break-down of rocks due to regular fluctuation in temperature, around 0°C, in joints or cracks in rocks. At the time of precipitation water enters into the cracks of rock. In winter, it freezes to ice and increases in volume. This creates tremendous pressure on the surrounding rock and generates cracks. This process continues over time, widens the joints/cracks, and causes pieces of rock to shatter from the main rock. The broken pieces of rocks fall down and spread over creating a stony surface called Talus or Scree. This is a regular process in a Periglacial environment. The alternating process of frost shattering slowly widens the joints/cracks, and in time, causes pieces of rock to shatter from the main rock.

Man Made

All those land degradation processes which are induced directly or indirectly by human intervention and are not natural, are categorised as Man Made desertification processes. It includes, Mining/Quarrying, Brick Kiln, Industrial Effluents, City Waste, Urban Agglomeration etc.

Barren / Rocky area

Barren/rocky areas are kind of wastelands which do not have productive capacity. These areas are mostly without or negligible soil cover either due to weathering and erosion of exogenetic processes or due to specific regolith made of hard rock.

Rajasthan is the state with highest area under desertification/ land degradation with respect to country Total Geographical Area (TGA) and second highest area under desertification/ land degradation with respect to state Total Geographical Area. The state is observed with 62.90% of the total geographical area under desertification/ land degradation for the period of 2011-13. The desertification/ land degradation area in Rajasthan has decreased about 0.29% since 2003-05 (Figure 9.6, 9.7 and 9.8).

The most significant process of desertification/ land degradation in the state is Wind Erosion (44.41% in 2011-13 and 44.80% in 2003-05) followed by Vegetation Degradation (7.62% in 2011-13 and 7.59% in 2003-05) and Water Erosion (6.18% in both 2011-13 and 2003-05) (Table 9.7).

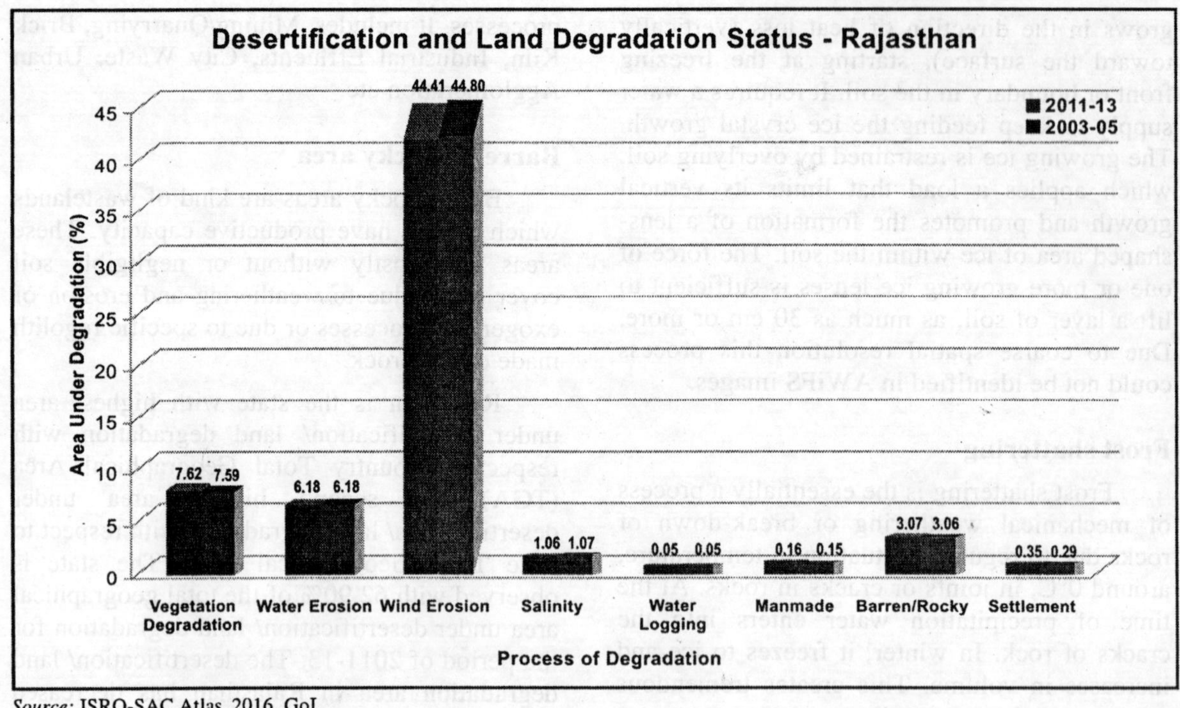

Source: ISRO-SAC Atlas, 2016, GoI.

Figure 9.6 Desertification and Land Degradation Status - Rajasthan

Table 9.7 : Area Under Desertification/ Land Degradation in Rajasthan

S.No.	Desertification / Land degradation Classes		2011-13		2003-05		Change (ha)
	Code	Description (Land Cover, Process, Severity)	Area (ha)	Area (%)	Area (ha)	Area (%)	(2011-13) - (2003-05)
1	Fv1	Forest, vegetation degradation, Low	1479558	4.32	1469518	4.29	10039
2	Fv2	Forest, vegetation degradation, High	252480	0.74	252480	0.74	0
3	Sv1	Land with scrub, vegetation degradation, Low	832578	2.43	832399	2.43	179
4	Sv2	Land with scrub, vegetation degradation, High	41606	0.12	41606	0.12	0
5	Iw1	Agriculture irrigated, water erosion, Low	673640	1.97	673640	1.97	0
6	Iw2	Agriculture irrigated, water erosion, High	4970	0.01	4970	0.01	0
7	Dw1	Agriculture unirrigated, water erosion, Low	334003	0.98	334003	0.98	334003
8	Fw1	Forest, water erosion, Low	83173	0.24	83173	0.24	0
9	Fw2	Forest, water erosion, High	133375	0.39	133375	0.39	0
10	Sw1	Land with scrub, water erosion, Low	658589	1.92	658008	1.92	581
11	Sw2	Land with scrub, water erosion, High	184926	0.54	184926	0.54	0
12	Bw1	Barren, water erosion, Low	15003	0.04	15351	0.04	-349
13	Ew1	Dune / Sandy area, water erosion, Low	28636	0.08	28636	0.08	0
14	Ie1	Agriculture irrigated, wind erosion, Low	2885396	8.43	2908241	8.50	-22845
15	Ie2	Agriculture irrigated, wind erosion, High	89837	0.26	88382	0.26	1455
16	De1	Agriculture unirrigated, wind erosion, Low	5340669	15.61	5426579	15.86	5340669
17	De2	Agriculture unirrigated, wind erosion, High	808076	2.36	822834	2.40	808076
18	Se1	Land with scrub, wind erosion, Low	1136345	3.32	1139509	3.33	-3164
19	Se2	Land with scrub, wind erosion, High	319	0.00	319	0.00	0
20	Be1	Barren, wind erosion, Low	588937	1.72	596828	1.74	-7891
21	Ee1	Dune / Sandy area, wind erosion, Low	2497673	7.30	2510506	7.34	-12833
22	Ee2	Dune / Sandy area, wind erosion, High	1850621	5.41	1838855	5.37	11765
23	Is1	Agriculture irrigated, salinity/alkalinity, Low	31610	0.09	31610	0.09	0
24	Ds1	Agriculture unirrigated, salinity/ alkalinity, Low	154674	0.45	155653	0.45	154674
25	Ds2	Agriculture unirrigated, salinity/alkalinity, High	19058	0.06	19058	0.06	19058

S.No.	Desertification / Land degradation Classes		2011-13		2003-05		Change (ha)
	Code	Description (Land Cover, Process, Severity)	Area (ha)	Area (%)	Area (ha)	Area (%)	(2011-13) - (2003-05)
26	Ss1	Land with scrub, salinity / alkalinity, Low	27716	0.08	27716	0.08	0
27	Ss2	Land with scrub, salinity / alkalinity, High	32360	0.09	32360	0.09	0
28	Bs1	Barren, salinity / alkalinity, Low	5184	0.02	6103	0.02	-919
29	Bs2	Barren, salinity / alkalinity, High	93166	0.27	93166	0.27	0
30	Il1	Agriculture irrigated, water logging, Low	9032	0.03	9032	0.03	0
31	El1	Dune / Sandy area, water logging, Low	8801	0.03	8801	0.03	0
32	El2	Dune / Sandy area, water logging, High	588	0.00	588	0.00	0
33	Fm2	Forest, man made, High	422	0.00	422	0.00	0
34	Tm1	Others, man made, Low	15748	0.05	13417	0.04	2332
35	Tm2	Others, man made, High	36887	0.11	37026	0.11	-139
36	B	Barren	9195	0.03	9195	0.03	0
37	R	Rocky	1043179	3.05	1038623	3.03	4556
38	S	Settlement	118482	0.35	98696	0.29	19786
Total Area Under Desertification/ Land Degradation			**21526512**	**62.90**	**21625604**	**63.19**	**-99092**
39	W	Water body/ Drainage	150463	0.44	150156	0.44	307
40	NAD	No Apparent Degradation	12546925	36.66	12448140	36.37	98785
Total Geographical Area (ha)			**34223900**	**100**	**34223900**	**100**	

Source: ISRO-SAC Atlas 2016, Government of India.

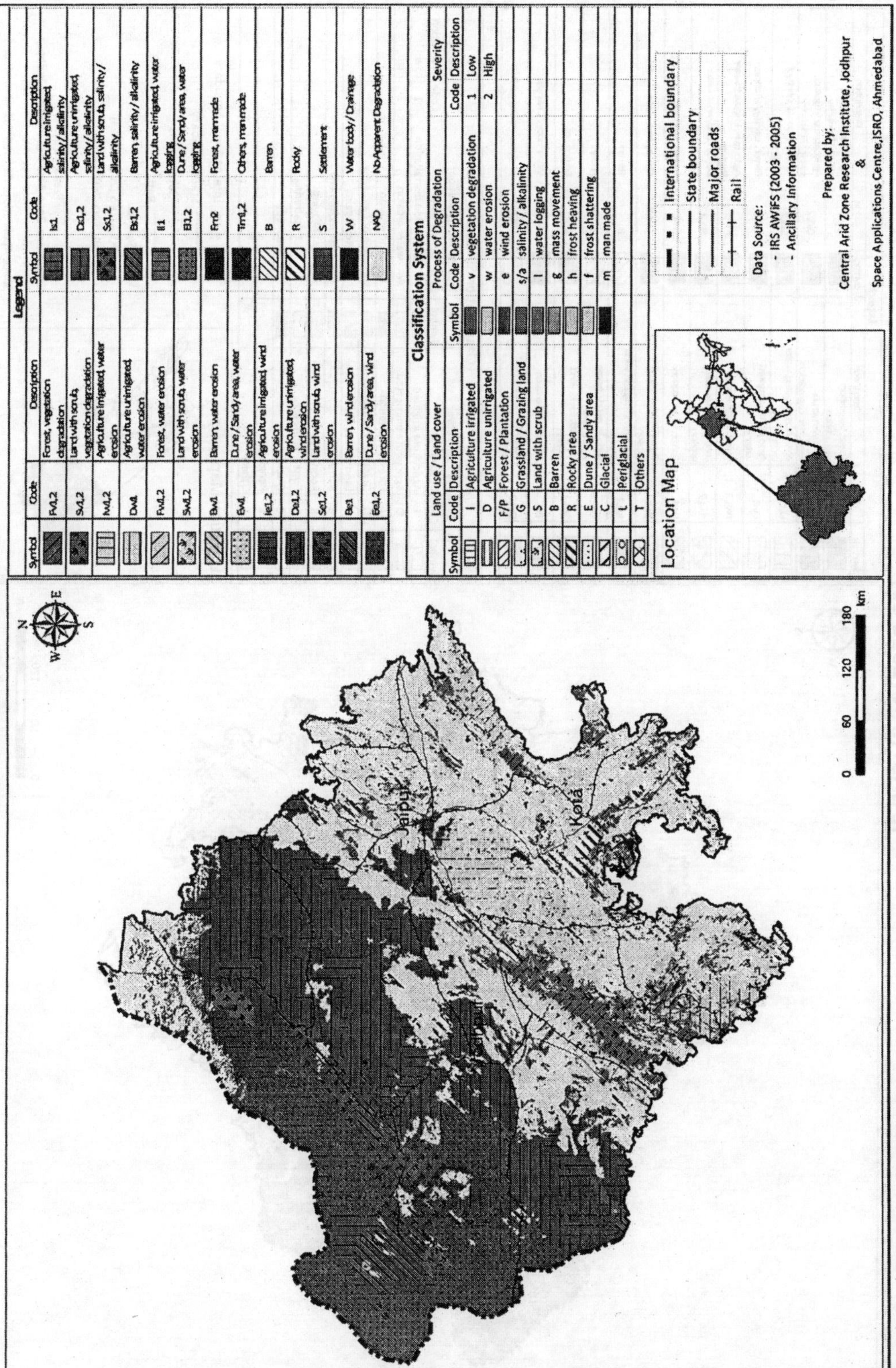

DESERTIFICATION / LAND DEGRADATION STATUS MAP RAJASTHAN (2003-05)

Source: ISRO, CAZRI and SAC 2016, GoI

Figure 9.7 Desertification / Land Degradation Status Map Rajasthan (2003-05)

DESERTIFICATION / LAND DEGRADATION STATUS MAP
RAJASTHAN (2011-13)

Legend

Code	Description	Code	Description
Fv1,2	Forest, vegetation degradation	Is1	Agriculture irrigated, salinity / alkalinity
Sv1,2	Land with scrub, vegetation degradation	Ds1,2	Agriculture unirrigated, salinity / alkalinity
Iw1,2	Agriculture irrigated, water erosion	Ss1,2	Land with scrub, salinity / alkalinity
Dw1	Agriculture unirrigated, water erosion	Bs1,2	Barren, salinity / alkalinity
Fw1,2	Forest, water erosion	Il1	Agriculture irrigated, water logging
Sw1,2	Land with scrub, water erosion	El1,2	Dune / Sandy area, water logging
Bw1	Barren, water erosion	Fm2	Forest, manmade
Ew1	Dune / Sandy area, water erosion	Tm1,2	Others, man made
Ie1,2	Agriculture irrigated, wind erosion	B	Barren
De1,2	Agriculture unirrigated, wind erosion	R	Rocky
Se1,2	Land with scrub, wind erosion	S	Settlement
Be1	Barren, wind erosion	W	Water body / Drainage
Ee1,2	Dune / Sandy area, wind erosion	NAD	No Apparent Degradation

Classification System

Land use / Land cover Code	Description	Process of Degradation Code	Description	Severity Code	Description
I	Agriculture irrigated	v	vegetation degradation	1	Low
D	Agriculture unirrigated	w	water erosion	2	High
F/P	Forest / Plantation	e	wind erosion		
G	Grassland / Grazing land	s/a	salinity / alkalinity		
S	Land with scrub	l	water logging		
B	Barren	g	mass movement		
R	Rocky area	h	frost heaving		
E	Dune / Sandy area	f	frost shattering		
C	Glacial	m	man made		
L	Periglacial				
T	Others				

Location Map

International boundary
State boundary
Major roads
Rail

Data Source:
- IRS AWiFS (2011 - 2013)
- Ancillary Information

Prepared by:
Central Arid Zone Research Institute, Jodhpur
&
Space Applications Centre, ISRO, Ahmedabad

N E S W
Jaipur
Jodhpur
Kota
0 60 120 180 km

Source: ISRO, CAZRI and SAC 2016, GoI

Figure 9.8 Desertification / Land Degradation Status Map Rajasthan (2011-13)

Measures to Combat Desertification

Desertification is like a slow poison, which infested without any warning or visible symptoms. The cultivated or cultivable area is converted in to degraded land of various types such as untreatable wasteland, saline, alkaline etc. The cultivators owning such areas have no voice to spell out their problems and this hazard is completely unnoticed to administrators, planners, geographers and other expert of geo-physical spheres.

The process of desertification is slow and only affected people only visualize this problems being directly affected with the problems. Mitigation measures are very simple but require sincere efforts in well defined implementation and execution systems. Some important issues are detailed below:

(a) Systematic Use of Water Resources: Rajasthan State is significantly facing the water problems and there is huge area, which can yield magnificent production and productivity results. There is necessity to use the available water in the most rationale manner through permanent water harvesting structures, use of water saving devices and adoption of dry farm practices. Such measures are presently being applied in water deficient areas and the results are quite encouraging.

(b) Use of Bio-technology Practices: Substitution of chemical fertilizers and pesticides to bio-technology measures like bio fertilizers and pesticides are able to save the cultivated land from degradation and at least 30 percent saving in the water for irrigation. This ultimately means creation of at least 30 irrigation facilities in other unirrigated areas. The measure is significantly effective to retain the fertility and increase in productivity.

(c) Efforts to Get Flood Water from Rivers: Under the proposed plan of carrying flood river water through deficient areas is a sincere effort, but required to help the state like Rajasthan, where there is large scope of increase of cultivated area, production and productivity. This system needs to be drawn to east of Aravalli to benefit most of the unirrigated area.

(d) Water logging problem is being faced in the command area of Indira Gandhi canal and a significant cultivated land in Ganganagar and Hanumangarh districts has been badly affected with this problem. There is necessity to pump the water through tube wells and pipe line to water deficient areas. This effort is cost effective but could be managed on "Built, Operate and Transfer" (BOT) basis, like in National Highways, where not a single rupee is required to be spent by the State Government.

(e) Tourism is one of the modern measures of social and economic upliftment of the people. Various National Parks, Centuries, Wild Life preservation centers can be developed with the income of tourist inflow. The Triangle Desert Zone had proved a success and need to be redesigned to create interest of the tourists to see glimpses of Thar Desert.

V. DUST STORM DISASTER

Dust Storm is a violent natural phenomenon occurring in the atmospheric blanket enveloping the earth. This problems is faced in Rajasthan during summer months and the dust storm originated in view of severe heat. The hot atmospheric conditions prevail in Thar Desert and create disaster. A vacuum is created with raising high the heated air and to cover the vacuum in the area air moves fast to fill in the vacuum. This process creates havoc as the movement of the air take sand from the surrounding area and sand storm moves to bridge the vacuum area.

The intensity of the sand storm in inversely proportionate to the heating process and creation of vacuum. Sometimes the storm carry large sand dunes of several million tones and moves fast with violent speed. The speed of the sand storm create disaster with the vibrant speed of storm containing huge dust.

The disaster caused through dust storm is loss of human life, livestock and birds, destruction of dwelling units, felling of trees and similar destruction. The duration of the dust storm of quite low say ranging between ten to thirty minutes but the destruction caused in unrepairable.

The speed of dust storm up to speed of 60 kilometers per hour is treated as normal and no disastrous impact in visualized . But the speed exceeding the limit cause devastation significantly. Sometimes the dust storm is followed with the rains, which checks the speed and the atmospheric conditions become pleasant as the impact of dust storm diminishes significantly. In the situation when the dust storm continues, the disaster impact is increased manifold.

Three major devastating effects associated with the dust storm are moderate dust storm , severe storm and very severe storm occur with the atmospheric pressures and their magnitude is classified as :

(a) Moderate dust storm Causes with the speed between 61 to 85 kilometers per hour. The impact of such a disaster is falling of old trees, kutcha dwelling units and causalities due to causatic impact like loss of human life due to displacement of iron roof shed or falling of tree etc.

(b) Severe Dust Storm due to wind velocity between 86-115 kilometers per hour and its significant duration causes significant loss of human life, cattle, birds, katcha buildings etc. The wind velocity and duration are responsible for devastation.

(c) Very Severe Dust Storm: Caused due to storm of more than 116 kilometers per hour wind velocity and longer duration. Once the dust storm is formed, it continues to move and expand, until passes away or the speed is decreased significantly due to atmospheric conditions. In case the speed is accelerated further, the destruction increases manifold.

The areas prone to dust storm is the total state but the highest impact is visualized in significantly extremely hot areas like Dholpur and Churu. Other areas usually face this problem are nearby areas of Thar Desert and Dist Jaipur, Dausa, Sawai Madhopur, Karauli, Bharatpur, Sikar and Jhunjhunu are most vulnerable. Other districts are comparatively face this problem with low severity.

Damages: Dust Storm is the most severe disaster but concentrates in comparatively smaller area specially faced the middle pressure zone. The devastation is caused is related with the wind velocity and intensity impact correlated with the duration before moves forward. The nature and intensity is unpredictable as sometimes the intensity increases while it is situated at particular point. The wind velocity is subject to change in both the directions in terms of speed and change of direction.

The disaster feared with the caution impact of dust storm can be divided in to individual and institution. In case of individual the effects can be loss of life, injuries demanding surgical needs, family disorganization, destruction of houses, personal belongings, livestock, crops and plantation. The institutional damages can be of government properties like partly buildings, vehicles, assets etc. Felling of trees and damages to birds and animals in view of severity are hazardous to environment and the society.

VI. FOREST FIRES DISASTER

Forests face many hazards but the most common hazard is forest fire. Forest trees are as old as the forests themselves. They post threat not only to the forest wealth but also the entire regime of fauna and flora seriously disturbing the bio-diversity and the ecology and environment of a region.

During the summer, when there is no rain for months, the slightest spark litters the forests. The Himalayan forests particularly Garhwal Himalayas have been burning regularly the last few summers, with colossal loss of vegetation cover of that region.

Causes of Forest Fire

Forest fire is treated as the origin of civilized society and man had witnessed the fire and eaten the baked meat, which witnessed the impact of fire to the civic society. But the devastation caused by forest fire is a significant loss and there is necessity to know the causes of severe destruction. The causes can be classified in to two categories as natural causes and man made causes.

(a) Natural Cause

Many forest fires start from natural causes such as lightening, which set trees on fire. Periodic lightening induced fires have been recorded in the history from India, South-Eastern, and Central United States, Australia, Finland and eastern and southern Africa. However, rain extinguishes such fires without causing much damage. High atmospheric temperatures and dryness with low humidity offer fabourable circumstances for a fire to start.

(b) Man Made Causes

Fires are caused primarily due to man made causes, when a source of fire like naked flame, cigarette or bidi, electric spark or any other source of ignition comes in to contact with flammable material. Forest fires also caused by heat generation in the litter and other biomass in summer through carelessness of human neglect. Habitants of the forests are also responsible for forest fire. Carriers of vehicle loaded with highly inflammable material are one of the causes of forests fire which is generally accidental but this is a major source of present time.

Types of Forest Fire

There are two possible types of forest fire-surface fire and crown fire as detailed below:

(a) Surface Fire

A forest fire may burn primarily as a surface fire, spreading along the ground as the surface litter – senescent leaves, twigs and dry grasses etc. on the forest floor are engulfed by the spreading flames.

(b) Crown Fire

The other types of forest fire is a crown fire in which the crown of trees and shrubs burn, often sustained by the surface fire. A crown fire is particularly very dangerous in a coniferous forest because resinous material given off by burning logs burn furiously. On hill slopes, if the fire starts downhill, it spreads up fast as heated air adjacent to a slope tends to flow up the slope spreading flames along with it. If the fire starts uphill, there is less likelihood of it spreading downwards.

Forest Fire in Rajasthan

The total forest area of Rajasthan is 32,828.37 sq. kms (2016), which constitutes 9.10% of the total geographical area of the state. This is quite low in comparison to total forest area of the country constituting of 15.7 percent of the geographical area. By legal status, Reserved forests constitutes 37 percent, Protected forests 53 percent and unclassified forest 10 percent of the total forest area. There are two forest types namely, Tropical Dry Deciduous and Tropical Thorn Forests. Forests are mostly confined in eastern and southern parts of the state. The western part of the state is devoid of forests because of prevailing hot arid conditions. District-wise forest area according to Legal Status of Rajasthan given in Table 9.8.

Analysis of the information given in above table reveals that maximum area under forests of 4,682.20 sq. kms. exists in Udaipur district, which is 37.42 percent of the total geographical area of the district. Churu district has lowest forest area of 80.18 sq. kms which is 0.47 percent of the geographical area of the district.

Table 9.8 : Districtwise Area of Forest According to Legal Status of Rajasthan

(as on 31.3.2016) (In Sq. Km.)

S. No.	Districts	Reserved Area	Protected Area	Un-Classified Forest Area	Total Forest Area
1	Ajmer	194.99	421.69	1.76	618.44
2	Alwar	1010.78	639.73	133.95	1784.46
3	Banswara	0.00	1005.26	1.74	1007.00
4	Baran	0.00	2225.31	14.38	2239.69
5	Barmer	20.30	568.81	36.76	625.87
6	Bharatpur	28.73	393.25	12.97	434.94
7	Bhilwara	438.40	289.62	67.42	795.43
8	Bikaner	0.00	748.49	501.40	1249.88
9	Bundi	853.05	665.03	9.02	1527.09
10	Chittorgarh	1200.15	571.17	0.62	1771.94
11	Churu	7.20	42.39	24.14	73.73
12	Dausa	134.87	149.26	0.36	284.49
13	Dholpur	7.92	597.73	44.03	649.68
14	Dungarpur	257.08	426.43	9.37	692.88
15	Ganganagar	0.00	238.42	395.02	633.44
16	Hanumagarh	0.00	113.25	126.21	239.46
17	Jaipur	678.38	263.17	4.94	946.49
18	Jaisalmer	0.00	238.36	343.23	581.59
19	Jalore	126.13	299.97	83.89	509.99
20	Jhalawar	416.38	924.32	6.21	1346.92
21	Jhunjunu	6.02	399.33	0.00	405.36
22	Jodhpur	4.68	183.78	55.66	244.12
23	Karauli	62.99	1693.13	53.93	1810.05
24	Kota	879.61	430.36	15.59	1325.56
25	Nagaur	0.80	206.28	35.32	242.40
26	Pali	816.56	144.82	2.21	963.58
27	Pratapgarh	907.12	1017.46	0.56	1925.14
28	Rajsamand	277.41	119.20	4.78	401.39
29	Sawai Madhopur	842.87	118.16	22.01	983.04
30	Sikar	9.92	622.40	8.19	640.52
31	Sirohi	614.04	984.72	39.89	1638.65
32	Tonk	101.42	233.17	1.38	335.97
33	Udaipur	2454.98	1434.38	9.80	3899.16
	Total	**12352.78**	**18408.85**	**2066.74**	**32828.37**

Source: Department of Forest, GoR, Jaipur

There are no evidence of major forest fire in the state causing devastation. This is in view of the fact that the area under forest cover is quite low in comparison to all India forest area. Furthermore, with enmass cutting of trees, insignificant protection measures, the dense forests do not exist in the state. However, the smaller events of forest fire were tackled at local level and therefore, no disaster could take place in past fifty years.

Vulnerability

The most vulnerable stretches of the world are the youngest mountains ranges of Himalayas. The forests of Western Himalayas are more frequently vulnerable to forest fires as compared to those of Eastern Himalayas. This is because forests of Eastern Himalayas grow in high rain density. With large scale expansion of Chir (Pine) forests in many areas of the Himalayas, the frequency and intensity of forest fires has increased. The affected area happens to be in the Ganga- Yamuna watershed, the most vital of the country's four watersheds. According to the reports of past events of forest fires, devastation caused by fire in 1999 was very extensive with more than 80,000 hectares of forests turned to ashes. Cumulative figures of Uttar Pradesh are evident that so far 3,75,000 hectares of forest wealth has been destroyed due to forest fires.

Preparedness and Mitigation Measures

Forests are usually seasonal. They usually start in the dry season and can be prevented by adequate precautions. State Governments are aware of the severe damage caused by the fire to forests and ecology areas. Traditional measures of fire control are inadequate and limited in India. The modern methods of fire control are yet to be placed on the ground in the required measures.

During the British period, the fire was prevented in the summer through removal of forest litter all along the forest boundary. This was called "Forest Fire Line." This line used to prevent fire breaking in to the forest from one compartment to another. The collected litter was burnt in isolation. Generally, the fire spreads only if there is continuous supply of fuel (dry vegetation) along the path. The best way to control the forest fire is, therefore, to prevent it from spreading, which can be done by creating fire breaks in the shape of small clearings of ditches in the forest. Use of water is usually last resort, as delivering water on the fire in dense forests on hill slopes, is usually a difficult job. In many developed countries, special aircrafts equipped with water tanks are used to drop tones of water on the burning trees.

Impacts of Fires on Biological Environment

Forest fires also poses serious health hazard by producing smoke and noxious gases, as is evident in Indonesia after forest fires on the islands of Sumatra and Borneo in 1977. The burning of vegetation gives not only carbon dioxide but also a host of other noxious gases such as carbon monoxide, methane, hydrocarbons, nitric oxide and nitrous oxide, that lead to global warming and ozone layer depletion. Consequently, thousands of people suffered from serious respiratory problems due to these toxic gases.

Burning forests and grasslands also add to already serious threat of global warming. Forests play a vital role in keeping the level of carbon dioxide in the atmosphere in check. Forests, grasslands and agricultural lands make up the bulk of the global biomass and biomass burning isa global phenomenon today. Recent measurements suggest that biomass burning may be significant global source of methyl bromide, which is an ozone depleting chemical.

The recent forest fires in the hills of Garhwal may be only a small part of the overall national problem. Already large areas of the Himalayan forests have been cleared indiscriminately for agriculture making them vulnerable to erosion. The only to save the fragile Himalayan ecosystem fro recurring forest fires is to put in place viable disaster management action plan.

VII. LOCUST INVASION

Locust is specific disaster threat to the state, where the locust invasion occurs from the neighboring country Pakistan is a significantly large quantity. This mishappening cause in the monsoon period when the crop in grown in the fields. This attack is sudden without any prior indication and the locust is spread in the cultivated fields in a large area.

The locust lay their eggs and their growth is manifold in short spell. Total crop is damaged by their invasion and the cultivators are helpless to face the situation and have no measures to destroy in view of their large number. This is a natural hazard, which has no cycle of occurrence and there is no system of any prior alarm to the situation.

In the years, when the relations with Pakistan were unfriendly, there was no system of any prior intimation from the neighboring side and therefore, the damage to crop used to be quite significant. The State Government had to approach the Government of India for managing aircraft services to spray pesticides to destroy the insects and used to manage the situation to mitigate crop damage to the possible extent.

Identification of Locust Prone Areas

The area prone to this problem is whole of western Rajasthan but the impact is most in bordering districts of Ganganagar, Bikaner, Barmer and Jaisalmer in the first spell. These locusts used to move very fast and soon spread in the state, in case their growth could not be checked timely and destroying measures are not sufficient. On some occasions the unchecked invasion and growth have witnessed in eastern and southern Rajasthan.

The speed of locust move is so fast that they move with tremendous speed and all the measures to destroy these are proved in vain. In the event of their invasion, the areas of their possible coverage and movement is unable to be judged and the total state and surrounding states are also unsafe with this problem.

Impact of Locust

The invasion of locust is sudden and the quantum of their herd in quite unknown. They move from Pakistan and some of these have their first stoppage in nearby districts, while the movement to other areas is not ruled out. The problem faced by the State Government is movement and direction. The locust moves to ground and attack the standing crops. These lay their eggs in the fields and move forward to other areas for carrying out similar destructive activities. The impact of their arrival and damage is dependent on their number and movement.

The crop damage is quite significant in the affected areas as the mitigation measures could possibly destroy the insects but the eggs laid down in the fields are escaped in view of their being under protected cover and the next generation continue the destruction activities, which is unable to be checked. Sometimes, total crop is damaged in the affected areas.

Revenue officials through enquiry and their own visual estimates calculate the crop damage. The impact is generally limited to some areas where the damage is quite significant.

Control of Locust

With prior warning of the invasion, suitable measures are possible to be taken up. In such cases the spray of pesticides can be quite effective as the insects could be killed before laying the eggs and therefore, the damage could be minimized.

Now the relations with Pakistan are improving and if the process continues to be positive, early warning would be possible to take effective mitigation measures for destruction and moving to other areas.

The crop damage is dealt like similar relief measures applicable in drought and other disasters. The land revenue is suspended till next crop and relief measures are taken up to provide instant support in the form of wage employment measures.

SEVERITY INDEX FOR DISASTERS

Since the dawn of civilization, human society, the natural environment and disasters have been closely interlinked. Natural disasters and increasing environmental degradation play their role in a significant manner and are serious threats to development.

The key elements for development include investment, effective governance and social stability. Unfortunately, disasters lead to exactly the opposite conditions. Loss of manmade and natural capital causes sudden disinvestments, post disaster relief increases both the financial and administrative burden on Government and finally, disasters are socially destabilizing. The social view of sustainable development indicates that vulnerability to natural disasters is definitely a function of human actions and behaviour. Thus, the resilience of Environmental Vulnerability and poverty are mutually reinforcing. Developing countries face the most debilitating consequences in the form of economic and social disruption due to natural disasters. Thus proximity to an extreme natural event combines with low economic or social status to result the deadly consequences.

For many years, the response to disasters and environmental threats was reactive and characterized by increased defensive activities. More recently, our attitude towards such risks has evolved to encompass the move proactive design of projects and policies, the seek of anticipate and limit the damage. In this context, the world is currently exploring the concept of sustainable development- an approach that permits continuing improvements in the present quality of life at a lower intensity of resources use, thereby leaving behind the future generations an undiminished or even enhanced stock of natural resources and other assets.

Disasters hurt people. They injure and kill. They cause emotional stress and trauma. They destroy homes and businesses, cause economic hardships, spell financial ruin to many. But the people hit worst are the poor. A natural disaster can happen anywhere, but for a combination of reasons, political as well as geographic, most large-scale disasters occur in the region, which we call the Third World. For the survivors of a natural disaster, a second disaster may also be looming, for the very aid that is intended to help them recover may be provided in such a manner that it actually impedes recovery, causes further economic hardship and renders the society less able to cope with the next disaster.

The tragic earthquake in Greece and the enormous devastation and suffering wrought by the recent earthquake in Turkey and Taiwan are reminders of one of the most pressing challenges of our time; the extraordinary increase in the number and extent of natural disasters. The facts are startling. The cost of whether related disasters in 1998 exceeded the cost of all such disasters in the decade of the 1980s. Tens of thousands of mostly poor people died during the year, tens of millions have been temporarily or permanently displaced. In the Caribbean, the hurricanes designated George and Mitch killed more than 13,000 people, with Mitch being the deadliest Atlantic storm in 200 years. A much less publicized cyclone in India in June 1998, caused damage comparable to Mitch and an estimated 10,000 deaths. Major floods hit India, Nepal, Bangladesh and much of East Asia with thousands killed. Two-thirds of Bangladesh was inundated for months leaving millions homeless. More than 3,000 people died in China's catastrophic Yangtze River flood, millions were displaced and the financial cost is estimated at an astonishing $ 30 billion. Fires ravaged tens of thousands of square kilometers of forest in Brazil, Indonesia and Siberia, with devastating consequences for human health and local economies. In Afghanistan earthquakes killed more than 9,000 people, while the exact toll of the horrific earthquake in Turkey is still unknown.

The communities will always have to face natural hazards, whether floods, storms or earthquakes. But today's disasters owe as much to human activities as to the forces of nature. Indeed the term "Natural" is increasingly misleading. A wide variation in the number and intensity of natural hazards is normal and to be expected. The occurrences of disasters over the past decades, however, are not nature's variations but a clear upward trend caused by human activities. The upward trend is because of the fact that most of the disasters victims live in developing countries, where poverty and population pressures force growing numbers of people to live in harm's way, on flood plains, in earthquake prone areas and on unstable hillsides. Unsafe building compounds the risks. The vulnerability of those living in risk prone areas is perhaps the single most important cause of disaster causalities and damage. The unsound development and environmental practices exacerbate the problems. Massive logging operations and the destruction of wetlands reduce the soil's ability to absorb heavy rainfall, making erosion and flooding more likely.

It is not just the costs of natural disasters that are exacerbated by human action. Many scientists believe that the recent upsurge of whether related natural disasters is the products of increased global warming much of which is caused by human activity. While the earth has always experienced natural cycles of warming and cooling, the 14 hottest years since measurements first began in the 1860s have occurred in the past two decades. 1998 was the hottest year on record. Given that the pressures of poverty and population growth continues to increase, the disaster trend is likely to worsen if no serious preventive measures adopted in time. Most effective prevention requires better early warning of impending disasters to give vulnerable populations time to move out of harm's way. It also requires more effective disaster response policies. But, above all it requires reducing the vulnerability that cause the damage in the first place.

For years, disasters have been treated as separate events and relief has tended to ignore the implications of disasters for the social and economic development of the society. The most basic issue in disaster are their impact on the poor and the links between poverty and vulnerability to disaster. Nevertheless, the fundamental themes are that the poor, who suffer most in disasters, are more vulnerable in a more complete sense because they are poor. The addressal of their problem in the most crucial factor as their survival and recovery from the problem is only possible with the substantial support.

It is an established fact that damage to and destruction of housing in cataclysmic disasters are the single highest cause of death and injury in the greatest economic loss to the poor in terms of poverty. Housing also illustrates how many of the related developmental issues come in to focus when a disaster strikes. In reality, disaster and development can not be separated, neither can disaster response be divorced from politics. Unfortunately, few government in the Third World are democratic and many regimes perpetuate underdevelopment because it supports the needs of privileged class. Thus many ideas emerge for quantification of impact and vulnerability of natural disasters. Some prefer to obtain foreign assistance in goods keeping in view their vested interest. Other specialists are on the opinion that examination of disasters in the context of development automatically becomes political. Since reducing the vulnerability of the poor is a development issue, these issues need to be tackled politically. But all these aspects do not stop for examining the disasters on the basis of certain well-defined indicators. However, tackling the disaster has certainly becomes political issue and depends on the government to overcome the problem with a political will.

In Indian context, the subject of natural disasters management does not find mention in any of the three lists in the 7th Schedule of the Indian Constitution. The basic responsibility for undertaking rescue, relief and rehabilitation measures in the event of natural disasters, as practiced is that of the concerned State Government. Role of the Central Government is supportive, in terms of supplementation of physical and financial resources and complementary measures in sectors like warning, transport and inter-state movement of food grains etc. On the legal front, neither the Union nor State Government has any enactment to deal with the management of disasters.

Under Centre- State relations, State Governments are not free to take up any issue directly with the external agencies and accordingly require to prepare a Memorandum for submission to the Government of India for providing immediate relief in view of any natural disaster. In this connection the State government is required to pose the gravity of the problem in terms of magnitude and impact and move to the Government of India. Such Memorandums are examined there and Central Study Team is constituted and moved to affected areas to assess the situation and suggest the quantum of relief. In case of most severe problem, assistance from external agencies is envisaged by the Government of India and dealt in response to central-state relations for fund flow mechanisms.

(1) Need For a Severity Plan

Since the State Government is directly responsible for the well being of its people, preparation of the severity plan is the act to be performed by the State Government. First issue for the consideration of the State Government is to prepare of shelf of projects on the basis of past analysis and keep ready the plan for mitigating the natural disaster on the event of its occurrence. Here severity is the most important and crucial factor to be interpreted in a rationale manner. Similarly another issue to draw a plan keeping in view of the severity.

The pre-disaster plan of severity is purely an academic analysis for keeping readiness to manage the problem, but the magnitude and intensity is unforeseen and accordingly the pre-disaster plan need to be drawn in a flexible manner. This can be interpretated as the occurrence of natural disaster in any region or district or any part could not be pre-assessed, what is expected to attempt some concrete proposals where the activities could be initiated in any area in case of any requirement and such plans need to be revised regularly on the basis of anticipated disaster and present status of requirement of an area.

In this connection the Government of Rajasthan has entrusted the task of preparation of shelf of projects to Gram Panchayats and issued instructions to prepare plans for the requirement of activities, which are most essential in the area on the basis of requirement of general people specially the poor. Such activities are of community benefit and prepared and revised in the Gram Sabha meetings with updating and forwarded to District Administration to consider in case of availability of funds for the benefit of the area. These plans are basically the requirement of activities in the area in terms of requirement of the people and generally related to water harvesting measures, drinking water, village road, community buildings, and treatment of wastelands and to some extent activities addressing natural disasters.

Another effort required for preparation of severity plan is directly related to natural disaster and exercise for preparation of plan is carried out for the affected area from the natural disaster. Such plans are prepared on the basis of damage caused by the natural disaster and addressal of related issues is the priority for benefiting affected people.

Other issues are departmental matters, which are addressed in their regular plan preparation measures and respective departments prepare such plans on the basis of requirement of the people and under the agreed pattern of state policies for addressal of developmental issues. In departmental system the policy of the State Government is the most crucial factor. These plans are prepared on the basis of implementation of the policies in a given time frame work and availability of funds.

The issue of severity needs to address first before coming to need for preparation of severity plans. The severity is the situation, caused by any natural disaster and addressal of those issues becomes the overriding priority over the ongoing develo-pmental issues. Severity is assessed in terms of affected area and intensity of the damage. Both the issues are inter-related and in such situation priority is the important factor in terms of providing immediate relief.

Accordingly, the severity of the problem is assessed keeping in view of the magnitude of the problems and in terms of damage caused in view of the intensity of natural disaster. Therefore, the severity is assessed keeping in view the following factors:

(a) Severity is directly related to the disaster caused in the State due to occurrence of an event damaging the people, area and structures. All these aspects are quickly assessed and assessment is taken of the situation at the district level with the help of respective district level authorities.

(b) Severity is also assessed in terms of immediate needs for treating the problem and rescue measures to be taken for mitigating the loss of life and property at the most possible extent.

(c) In case of severity in a larger area, the authorities restrict their activities within each district and assess the problem within the area in view of respective jurisdiction.

(d) Severity is assessed in terms of magnitude of the damage and intensity of the problem. The devastation caused by the disaster is assessed thoroughly. This is assessed by the local authority stationed in nearby area and assigned with the responsibility to handle the situation. The first version of the assessment is generally through telephonic communication and redressal are communication as immediate response to the problem.

(e) In case of the severity restricted to a limited area, the district authorities have authority to take immediate measures and in case of widespread all possible assistance is managed with the help of other authorities who are not directly related but abide with the instructions of the district administration.

(f) The District Collector is the appropriate authority to handle the problems at the district level and the severity of the problem is best adjudged in view of the communication of the local authority and severity is adjudged over verbal communication as the response received from the local people.

(g) Severity of the problem is reciprocated with immediate action required to be taken and similar communication is made to state authorities indicating he nature and magnitude of the problem along with the affected area and population. The counter response received from the state government authorities is abided quickly and local authorities are instructed further with the suitable instructions.

Generally, the quick response for addressal of the situation is taken as immediate action for mitigating the problem to the possible extent and the process is continued till the immediate action required to manage the situation is taken. Therefore, the plan for the severity is prepared for the further actions required to be carried on long term basis as the first task of the addressal of the situation is almost carried out as per necessity and requirement of the situation.

Need for a severity plan is required for managing the situation effectively and in a transparent manner. The need for a severity plan is necessary for following reasons:

(a) Severity plan is needed to assess the situation caused by natural disaster, where quantification of the severity is the most crucial factor to assess. Generally, the Government of India has already framed guidelines for qualifying the severity within the permissible limits and these guidelines form the basis for quantification of the damage.

(b) Severity plan is essential as the primary assessment is based on the visual expressions of the local authority, which need to be quantified in legitimate terms identifying all the related aspects of damage in most authentic terms to the possible extent. The visual expressions are generally based on the crude estimates of the local authority and such assessment is based on the information gathered from the local people.

(c) The severity plan is an estimate of devastated area, damages of different nature and possible further damage in view of the situation aggravated further. An assessment of all sort of damages and possible damages need to be quantified for intensifying the problem and need to be forwarded to state authorities.

(d) Severity plan is required to take appropriate action at the state level about the severity of the problem and taking requisite decisions from every possible manner. This plan covers all the related issues very well known to district authorities in view of past event and is a supplemental to the verbal narration about the event.

(e) Severity plan helps the State Government to know the magnitude of problem and addressal of requisite measures of immediate nature and follow up action to mitigate the problem.

(f) Mitigation measures require financial arrangements, which are possible only on the basis of quantified number of damages and require financial sanction to address all sort of issues. Financial issues require concrete proposal indica-ting the magnitude of problems and the decisions for sanction are taken with the approval of the finance Department.

(g) Natural disasters are eventual occurrences and require additional funds, which have never foreseen in regular development plan and such sanctions can be issued after satisfactory information to the requisite extent about the damages and possible damage further.

(h) Severity plan is a document containing all the aspects of the natural disaster in terms of area and affected people. Such plans can be from more than one districts and uniformity is required in terms of narration of all the aspects to be addressed.

(i) Severity plan is the basis document of quantified damages and forms part of disaster management plan for mitigating the problem. In case the memorandum is required to be prepared for seeking additional assistance from the Government of India, severity plan forms the basis for preparation of such memorandum for qualifying the availability of requisite relief from all the sources including Natural Calamity Contingency Fund and if required, for external assistance by the Government of India, if the problem is of lager magnitude.

(j) This severity plan helps the State Government to submit its claim in logical manner indicating the intensity and magnitude of the problem and also seeking the support of other agencies within the country. Generally communication media and trusts also come forward for assistance in he event of severity of the problem.

(2) Problem of Defining the Severity Index

Every State Government tries to stress its problem of natural calamity for getting maximum possible assistance from the Government of India and try to exaggerate the problem indicating various issues of intensity of the problem. Though the assistance receivable from the Government of India is mix of loan and assistance, but generally State Government try to get maximum possible assistance. But the Government of India has designed a system to constitute Central Team to assess the situation and suggest the magnitude of severity. This team visits the affected areas and later discusses the problems with the officers of State Government and discusses the severity of the problem.

This process is a regular feature between centre and state and the quantum of assistance is recommended by the Central Study Team as the facts narrated in the memorandum are generally treated as exaggerated Therefore, there is necessity of defining the severity index as well as adherence of the State Government to include the logistic information Although the criteria for quantifying the damage from natural disasters has been defined by the Government of India, but it has became a political issue rather than administrative to exaggerate the problem to the maximum possible extent and try to draw maximum possible assistance to witness the success of the State Government in general and the Chief Minister of the State in particular.

Furthermore, this has became the fashion to exaggerate the problem as the State Government feels that the Government of India would cut their proposal significantly and therefore, demanding huge funds in view of convincing the central authorities about their viewpoint. In the specific drought of 2002 the State Government addressed the severity in 136 years and submitted a memorandum for assistance of Rs. 7519.75 crores whereas the amount made available from all sources was Rs. 1332.62 crores. In addition dovetailing of state schemes to supplement requisite contents not admissible under central assistance. This is evident that there is vast variation between the figures quoted by the State Government and made available to State Government to mitigate the disaster.

Therefore, it is necessary to understand the problem in spelling out severity index. For this purpose, there can be some general principles to be sued for all the natural disasters and others for specific nature of problems. The underlying fact lies with the problem that the State Government intends to expose the severity of any natural disaster to the maximum possible extent and intends to include as many issues which could possibly make their representation voluminous and accordingly incorporates the issues under the presumption that the Government of India would never agree their total financial and physical narration. For this purpose, help of media, press and peoples representatives is taken to represent their case boldly at administrative and political levels.

This is true that politics is involved once the natural calamity is evident. The Government of India examines the state proposal on administrative and political basis and ultimately the quantum of assistance is finalized. Even then the State Government tries to stress the gravity of the problem and intends to get as much assistance as possible. Therefore, the administrative issues are surpassed by the systems, which could influence the gravity of the problem at various stages.

It is, therefore, deemed proper to understand the administrative problems being faced in defining the severity index number and later the political and others playing role in manifesting the problems. The administrative problems in defining severity index are as under:

1. First problem being faced by the District authorities in assessment of the area affected by natural calamity. In case of drought, flood, earthquake, desertification hazard, dust storm disaster and locust invasion, area affected from the disaster is difficult to assess. The cultivated area affected with all such problems is directly affecting the cultivators, who are the ultimate looser with the disaster. There are many methods for assessing the severity problems indicating the loss incurred from the disaster. Presently satellite, aerial photography are the latest devices to assess the devastation, but the district administration prefers to get the assessment done with the revenue authorities.
2. Revenue authorities get this assessment done through Patwaries, the filed level

revenue functionaries, who assess the crop damage or area damage with eye view, indicating the loss of crop in terms of up to 25 percent, between 25-50 percent, between 50-75 percent and between 75-100 percent. This eye view is not a perfect method to assess the crop or area damage. Accordingly the physical assessment is to be carried out by the Central Study Team at selected points and incorporated in their recommendations.

3. Loss of human life and livestock is assessed at the time of occurrence of natural disaster and is a crude estimate on the basis of absence of the people. The assessment takes long time to finalize the list of deceased persons, as disappearance for three without recovery of dead body becomes eligibility for claiming the compensation. These lists are prepared earlier on adhoc basis to be revised subsequently.

4. Loss of livestock is not a subject of compensation and individual household has been bearing the loss. Similarly the loss of house is calculated on the basis of full or part damage and the eye view is the only method o calculation of loss. This exercise is carried out by Patwari or Sarpanch who carry out assessment as per his or her own perception. Such severity is to be assessed further by the Central Team on selective basis.

5. Crop damage below 50 percent is not eligible for assistance and damage between 75-100 is serious condition. The marginal cases are in incorporated in the list to attract maximum possible assistance. The revenue authorities usually take liberal view in assessment of crop and area damage. Sometimes, marginal cases are more in number and the Central Team turns down such issues.

6. Regarding wage employment in the affected areas, no material component is admissible for assistance. Without material component, the works of permanent nature could not be taken up and the state Government has to dovetail with the regular departmental activities. Even in case of wage employment the material component is permissible up to 40 percent of the total cost. Under these circumstances manipulations are carried out to cope up the agreed system.

7. The marginal cases of damage of human or cultivated area, affect most as these are the issues could be put in to the damage list to maximize toe losses. The severity index cannot be prepared unless a fool proof system of identification of area and other losses are carried out in a rationale manner.

8. The life casualties of human beings are vague estimates as many people lost their life due to natural death in similar duration are included in the list by the local administration and local residents encourage enlisting those to provide relief to their families.

9. In case of drought, the administration never agrees for any death due to hunger but try to pose it as normal death in view of some disease. Similarly, the temperature figure of 50°C is never witnessed though it occurs in the state at several occasions and locations.

10. The cases of severe hurt or partly hurt are twisted as per requirement of the administration. In case the compensation the State had to pay would try to avoid those cases and when the assistance of Central Government is admissible, those are included in the list.

11. The severity of the problem is adjudged on the political basis to benefit the people of specific area and the issues are interpreted accordingly. Ministers of the State, Members of Parliament and Legislative Assembly try to influence the district administration to show maximum severity in their area to start relief activities.

12. It is funny to note that in the situation of worst famine conditions in the total state the agriculture production is never affected to that extent. This is evident that the figures of severity and damage witnessed by the district administration were not properly assessed.

13. The severity index, which is, formed the basis for witnessing the problem and submitted to Government of India and amount sanctioned to state is significantly varied and the sanctioned amount is used for the areas surpassing the severity.

14. The audit reports prepared by Controller and Auditor General of India include many issues of misappropriation of funds and there are hundreds of audit paras, which witness the severity was manipulated and not real.

(3) Severity Indices for Disasters in Rajasthan

Severity Indices for disasters in Rajasthan need to be prepared on logistic method incorporating the issues of general and specific nature. For this purpose, the Government of India need to formulate the indices taking in to the confidence of the State Government and only agreed issues need to be incorporated and the damage norms must be realistic. There is necessity to treat the issue on humanitarian aspect rather than on political basis.

Natural disasters are serious issues where the devastation is manifold and immediate assistance is required to be provided to the sufferers. In such circumstances, it is very difficult to adhere the financial rules and procedures as procurement of various things become necessary on immediate basis. It is very difficult to adopt the procedure laid down by the Government. It is also necessary the district authorities should not take undue benefit of the situation and try to get the benefit for their personal cause. Therefore, the systems and procedure for immediate assistance need to be attempted afresh from the point of view of the situation.

Other important issue is the spread of area. Sometimes, it is witnessed that the gravity of the problem as assessed in the first instance is aggravated further and the situation continue to worsen. In such a situation, the increasing number of affected persons continues to increase and later during the examination of the accounts, the position seems to be manipulated. The district authorities need to be cautious to keep the record of all such stages and record all the issues for further examination.

Therefore, the severity indices for disasters need to be sub-divided in to three broad categories. In the first category the preliminary assessment must incorporate the area and people affected and the emergency arrangements must be initiated on this basis. Second assessment must be at the time of peak stage of the problem and this assessment must be based on the situation taken in to control and severity assessment need to be attempted on some logistic method incorporating the maximum possible issues. The third and final assessment must be after happening of the disaster and the final assessment must incorporate all the issues of severity of the disaster and carry out relief and rehabilitation measures based on final assessment.

(a) Severity Indices for Drought

Severity indices for drought are primarily related to crop damage. Since most of the cultivated area of Rajasthan is single cropped, crop damage is serious condition to the cultivators specially small and marginal farmers, who suffer most as they do not have substantial irrigation facilities to survive their crops from the damage. In this case the crop loss up to 50 percent is treated as normal situation. This is in view of the fact that productivity up 50 percent and 150 percent is a normal phenomenon in view of unfavourable and most favourable monsoon conditions and this much deviation occurs in the state normally.

Crop damage more than 50 percent and up to 74 percent is treated as major severity as cultivators face many problems like food grain, fodder, drinking water. The crop damage between 75 percent to 100 percent is treated as most hazardous situation and poor cultivators require immediate support of food, fodder, drinking water, medicines and employment for their survival and safety of their livestock, which is a supplemental source of their livelihood.

In case the severity is widespread, the situation aggravated further. In such a situation, the requirement of food and fodder is the immediate need for the larger area. The severity intensifies further when the situation reoccurs during the next year. The severity indices need to be drawn keeping all the factors together and loss of human and cattle population is anticipated most. In such a situation nutritional support and drinking arrangement is the most essential items to be managed for larger area.

The state government is also the most sufferer of the situation as all the developmental activities are seriously affected most and there is most serious financial crisis faced to manage the situation. The state income is affected seriously as the purchasing power of a larger group of population is diminished to the most extent. Business is affected seriously during the severe and widespread disaster.

(b) Severity Indices for Floods

In case of floods, there is some prior indication of heavy rainfall and some predictions of rainfall are visualized earlier. The severity indices are governed by the intensity of rainfall followed by insufficient drainage system to safe clearance of water. The severity occurs in the situation of continuous rainfall of high intensity and widespread. Flood of 1981 in Jaipur city was occurred in view of continuous 3 days downpour followed by insufficient drainage systems allow the huge water to be drained safely.

Loss of human casualties and area affected are the severity indices which include damage of houses, business areas, human and cattle population and submergence of cultivated area under flood. The severity is assessed on the basis of intensity and damage caused. Emergency action required to be taken is safe withdrawal of persons requiring assistance.

(c) Severity Indices of Earthquake

From indices point of view the earthquakes are the most hazardous disasters as their occurrence in unnoticed and unpredicted. The duration of the disaster is very short and nothing is possible even to assess the gravity. The severity is assessed on Richter scale and magnitude less than 4 on Richter scale is very minor disastrous. Magnitude of 3 ~ 3.9 is treated as minor disastrous. Magnitude of 4 ~ 4.9 is light disastrous. Magnitude of 5 ~ 5.9 is moderate, 6 ~ 6.9 is strong, 7 ~ 7.9 is major and 8 and higher is great disastrous. In addition to magnitude, time duration is also important as the severe loss occurs most on the duration factor and magnitude and longer duration make the situation most devastating.

Another aspect is spread of area affected by the earthquake disaster. Casualties is densely populated areas are very high, which aggravated further in case of multiple storied buildings. Dams of major irrigation and multipurpose projects are quite unsafe from earthquake point of view and the large area is submerged in case of any loss to the dam. This type of disaster is multiple as damage through earthquake is followed by floods and severity of the disaster is manifold.

(d) Other Disasters

Locust invasion is one of the serious disasters and indices for assessing the severity are crop damage followed by subsequent problems of the cultivators. The intensity of severity aggravated further in case the situation becomes out of control. A large cultivated area, trees and total vegetative cover are seriously affected due to invasion. Severity indices are similar to drought as the impact if disaster is equally vulnerable in terms of intensity and magnitude.

Desertification is slow processed disaster, where the magnitude of the problem is not assessed fully. Land degradation is slow but permanent and only affected cultivators are aware of the problem. Further there is no hue and cry from any corner as neither politician nor administrators take the problem seriously. The affected cultivators have no voice to express their problem.

(4) Disaster Severity in Different Regions in Rajasthan

Rajasthan is the most disastrous state of the country in view of water deficiency and drought is a recurrent phenomena. Almost total state is water deficient area in view of heavy overdraws of water for cultivation purposes. Water depletion has turned the most parts of the state in to dark zone as recharge is not sufficient to the level of withdrawal.

Most crucial factor is the use of chemical fertilizers and pesticides, which require excess water for irrigation to maintain the productivity level. A large area of the state soil is being degraded in view of this factor and there is close competition between the large and prosperous cultivators to draw maximum possible water for their own cultivation requirements. Drinking water problems are also acute and disaster like situation prevails in the state on regular basis. Drinking water supply through tankers in summer is a regular phenomenon in view of water depletion.

The situation is further aggravated due to failure of monsoon. Cultivation is not a remunerative occupation as the cultivators are unable to get even the farm harvest price from their produce so far as normal monsoon conditions are concerned. The cultivators having single cropped area are the most sufferer of the disaster of any type. The droughts and floods are regular phenomenon as per occurrence of monsoon. Occurrence of monsoon depends on the whether conditions and both the happenings have their own vulnerability and disastrous impact.

So far regional variations are concerned, Aravalli range divides the state in two different geo-physical zones and western zone being Thar Desert is water deficient area. The eastern and southern parts have good moisture and soil conditions but water is the main constraint and in the situation of any natural disaster particularly of drought, the position of any region worsens Disaster severity in different regions has been assessed as under:

(1) Drought Disaster

Natural disaster of drought is the most serious phenomenon and its occurrence is recurrent in one part or other. Total state is vulnerable for drought as the monsoon failure as per whether cycle prevails. Analysis of past fifty years is evident that there were very few years prosperous but the remaining years had witnessed the drought conditions. Region-wise and district-wise analysis of drought severity is given in the Table 9.9, Figure 9.9.

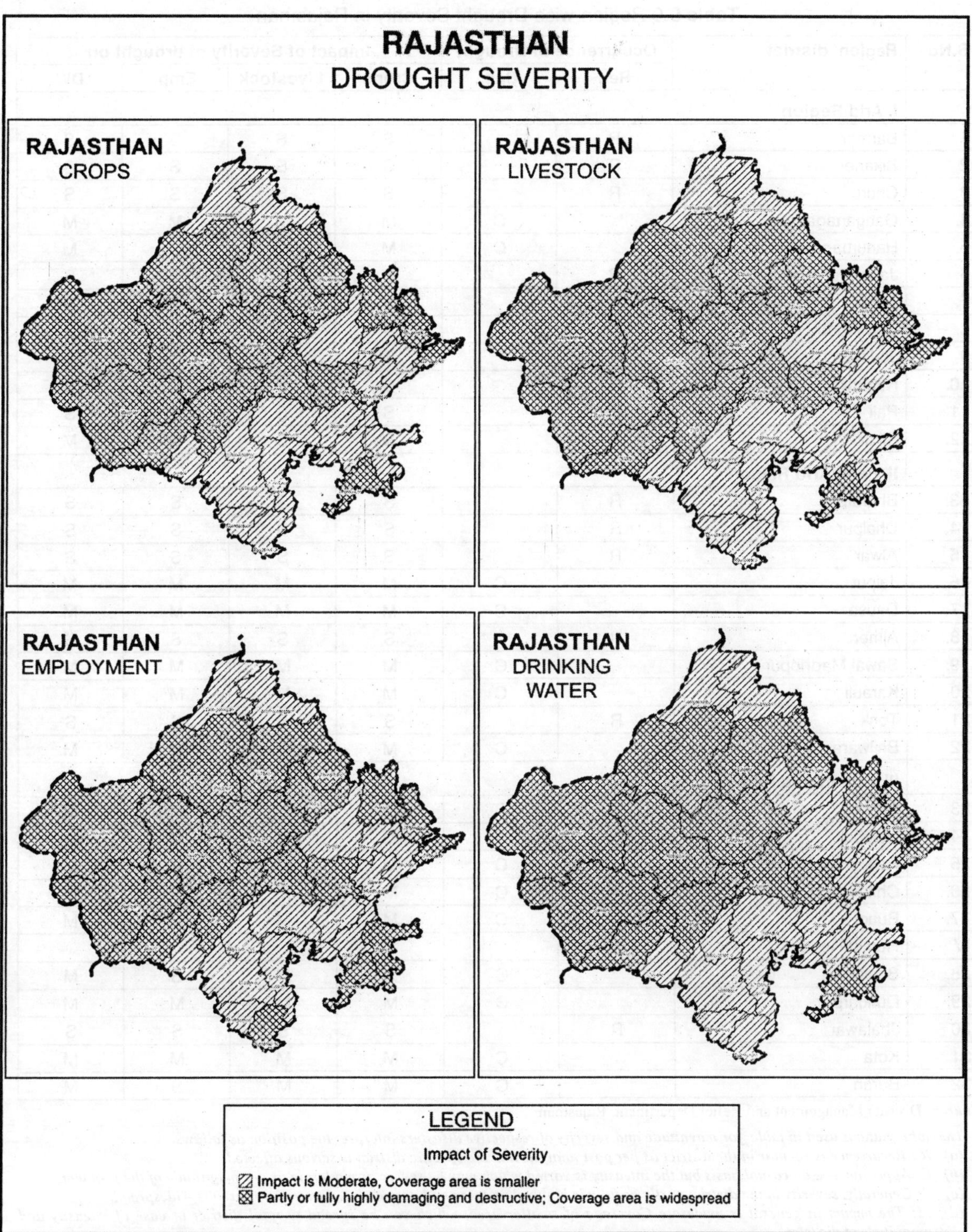

Source: Survey of India (SOI) and Department of Disaster Management and Relief, Government of Rajasthan, Jaipur

Figure 9.9 Drought Severity in Rajasthan

Table 9.9 Region-wise Drought Severity in Rajasthan

S.No.	Region/ district	Occurrence of Drought		Impact of Severity of drought on			
		Reg	Cas.	Crops	Livestock	Emp.	DW
	I. Arid Region						
1.	Barmer	R		S	S	S	S
2	Bikaner	R		S	S	S	S
3.	Churu	R		S	S	S	S
4.	Ganganagar		C	M	M	M	M
5.	Hanumangarh		C	M	M	M	M
6.	Jaisalmer	R		S	S	S	S
7.	Jalore	R		S	S	S	S
8.	Jhunjhunu	R		S	S	S	S
9.	Jodhpur	R		S	S	S	S
10.	Nagaur	R		S	S	S	S
11.	Pali	R		S	S	S	S
12.	Sikar	R		S	S	S	M
	II. Semi Arid Region						
13.	Bharatpur	R		S	S	S	S
14.	Dholpur	R		S	S	S	S
15	Alwar	R		S	S	S	S
16.	Jaipur		C	M	M	M	M
17.	Dausa		C	M	M	M	M
18.	Ajmer	R		S	S	S	S
19.	Sawai Madhopur		C	M	M	M	M
20	Karauli		C	M	M	M	M
21.	Tonk	R		S	S	S	S
22	Bhilwara		C	M	M	M	M
	III. Sub-humid Region						
23	Sirohi		C	M	M	M	M
24.	Udaipur		C	M	M	M	M
25	Rajsamand		C	M	M	M	M
26.	Chittorgarh		C	M	M	M	M
27.	Bundi		C	M	M	M	M
IV	Humid Region						
28.	Banswara		C	M	M	S	M
29.	Dungarpur		C	M	M	M	M
30	Jhalawar	R		S	S	S	S
31	Kota		C	M	M	M	M
32.	Baran		C	M	M	S	M

Source: Disaster Management and Relief Department, Rajasthan

The abbreviations used in table for magnitude and severity of respective disasters interpret the position as follows:

(a) ***R*** *: Recurrence is regular in the district as per past analysis and the specific district is serious affected.*

(b) ***C:*** *Appearance is on casual basis but the intensity is varied which may be severe or moderate as per magnitude of the problem.*

(c) ***S:*** *Generally, severity in the district partly or fully is highly damaging and destructive. Coverage of area is widespread.*

(d) ***M:*** *The impact in general is moderate Coverage of smaller area, which can be spread to total district in case of intensity and magnitude of disaster.*

(e) ***C/S:*** *The occurrence of the disaster is on casual basis, but the disaster in the affected area is severe.*

(f) The analysis is based on the information of past years and severity is set on the basis of disaster over the affected population/ area.

(2) Region -wise Vulnerability of Floods:

Severity of Being water deficient state, chances of flood is very rare and disaster occurs in the event of constant rainfall with high intensity. In the sandy soil, damage to houses, cultivated fields, loss of human life and cattle wealth and damage to community and state assets. The severity intensifies in view of scanty drainage systems for water flow. Region-wise analysis of flood severity and its intensity is given in Table 9.10 and Figure 9.10.

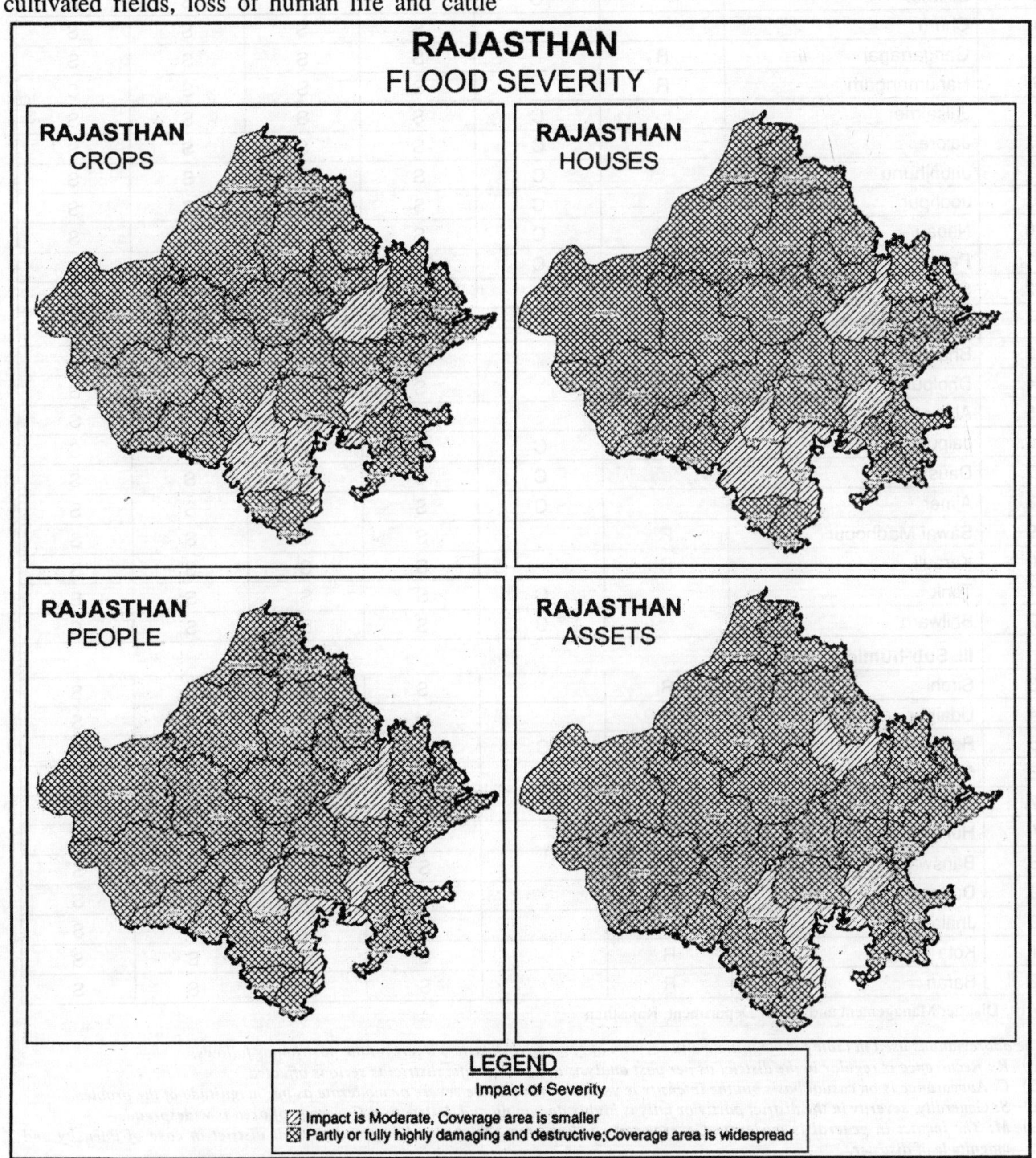

Source: Survey of India (SOI) and Department of Disaster Management and Relief, Government of Rajasthan, Jaipur

Figure 9.10 Flood Severity in Rajasthan

Table 9.10 : Region-wise Flood Severity In Rajasthan

S. No.	Region/ district	Occurrence of Flood		Impact of Severity of Flood			
		Reg	Cas.	Crops	Houses	People	Assets
	I. Arid Region						
1.	Barmer		C	S	S	S	S
2	Bikaner		C	S	S	S	S
3.	Churu		C	S	S	S	S
4.	Ganganagar	R		S	S	S	S
5.	Hanumangarh	R		S	S	S	S
6.	Jaisalmer		C	S	S	S	S
7.	Jalore		C	S	S	S	S
8.	Jhunjhunu		C	S	S	S	S
9.	Jodhpur		C	S	S	S	S
10.	Nagaur		C	S	S	S	S
11.	Pali		C	S	S	S	S
12.	Sikar		C	S	S	S	M
	II. Semi Arid Region						
13.	Bharatpur	R		S	S	S	S
14.	Dholpur	R		S	S	S	S
15	Alwar	R		S	S	S	S
16.	Jaipur		C	M	M	M	S
17.	Dausa		C	S	S	S	S
18.	Ajmer		C	S	S	S	S
19.	Sawai Madhopur	R		S	S	S	S
20	Karauli	R		S	S	S	S
21.	Tonk		C	S	S	S	S
22	Bhilwara		C	S	S	S	S
	III. Sub-humid Region						
23	Sirohi	R		S	S	S	S
24.	Udaipur	R		M	M	S	S
25	Rajsamand		C	M	M	M	M
26.	Chittorgarh		C	M	M	M	M
27.	Bundi		C	M	M	M	M
IV	Humid Region						
28.	Banswara	R		S	S	S	S
29.	Dungarpur	R		S	S	S	S
30	Jhalawar	R		S	S	S	S
31	Kota	R		S	S	S	S
32.	Baran	R		S	S	S	S

Source: Disaster Management and Relief Department, Rajasthan

The abbreviations used in table for magnitude and severity of respective disasters interpret the position as follows:

(a) ***R :*** *Recurrence is regular in the district as per past analysis and the specific district is serious affected.*

(b) ***C:*** *Appearance is on casual basis but the intensity is varied which may be severe or moderate as per magnitude of the problem.*

(c) ***S:*** *Generally, severity in the district partly or fully is highly damaging and destructive. Coverage of area is widespread.*

(d) ***M:*** *The impact in general is moderate Coverage of smaller area, which can be spread to total district in case of intensity and magnitude of disaster.*

(e) ***C/S:*** *The occurrence of the disaster is on casual basis, but the disaster in the affected area is severe.*

(f) The analysis is based on the information of past years and severity is set on the basis of disaster over the affected population/ area.

(3) Region-wise Severity of Earthquakes in Rajasthan

Occurrence of earthquake with high intensity and magnitude is significantly low, but it possibility always exist. Severity of earthquakes feared in high-density areas of the state. The highly vulnerable areas are multi-storied buildings, slate stone roofing houses. Landslides and completely disappearances of cities and villages cannot be ruled out.

Severity of earthquake in whole of the state is quite moderate as evident from past records. The possibility of earthquake always exists as the earthquakes evident during past years had their intensity in Rajasthan, but the magnitude was moderate. However, there are some areas, which may face the problem any time. Severity analysis of earthquake is given in Table 9.11 and Figure 9.11.

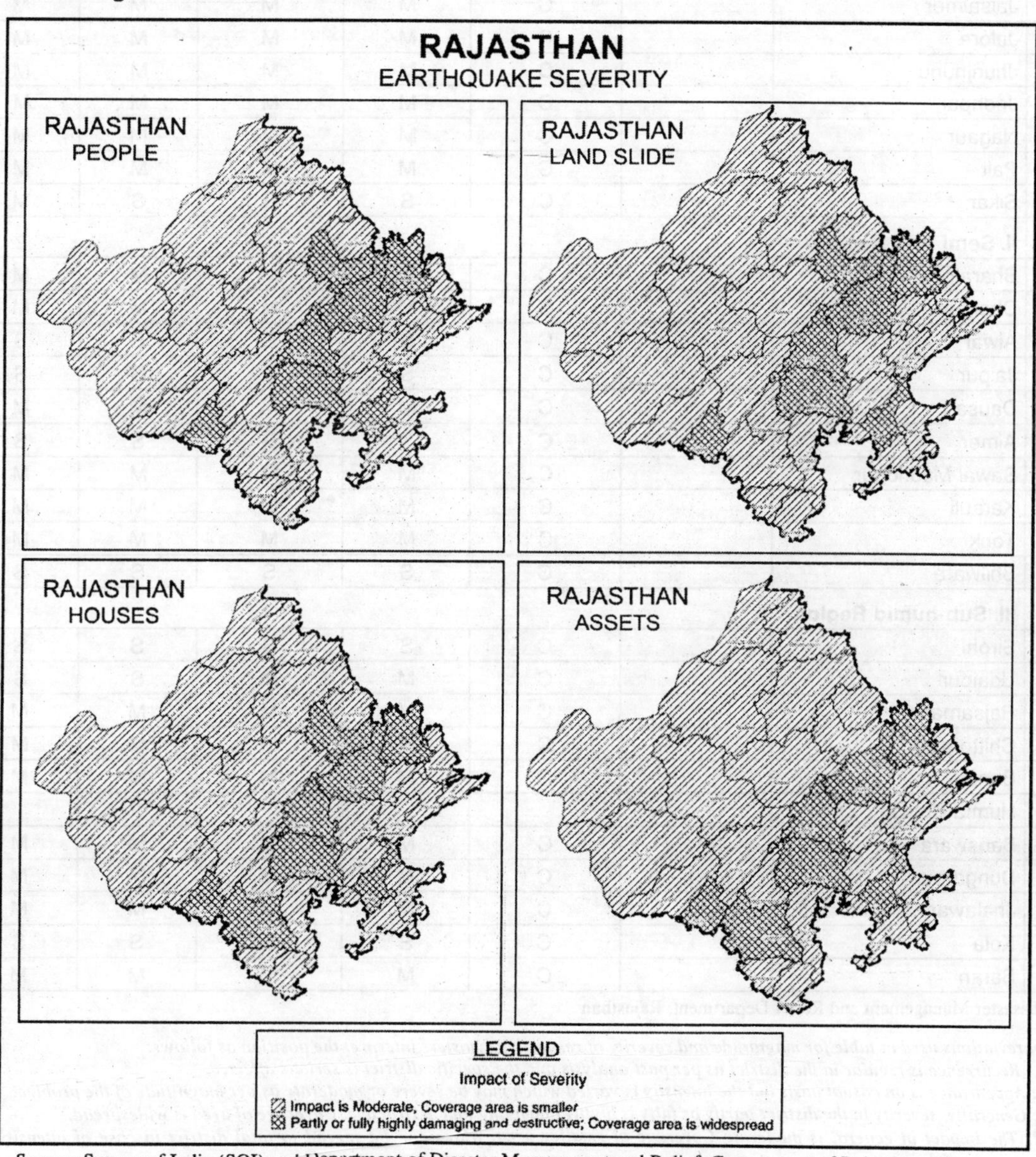

Source: Survey of India (SOI) and Department of Disaster Management and Relief, Government of Rajasthan, Jaipur

Figure 9.11 Earthquake Severity in Rajasthan

Table 9.11 : Region-wise Severity of Earthquake in Rajasthan

S. No.	Region/ district	Occurrence of Earth-quake		Impact of Severity of Earthquake			
		Reg	Cas.	People	Land-slide	Houses	Assets
	I. Arid Region						
1.	Barmer		C	M	M	M	M
2	Bikaner		C	M	M	M	M
3.	Churu		C	M	M	M	M
4.	Ganganagar		C	M	M	M	M
5.	Hanumangarh		C	M	M	M	M
6.	Jaisalmer		C	M	M	M	M
7.	Jalore		C	M	M	M	M
8.	Jhunjhunu		C	M	M	M	M
9.	Jodhpur		C	M	M	M	M
10.	Nagaur		C	M	M	M	M
11.	Pali		C	M	M	M	M
12.	Sikar		C	S	S	S	M
	II. Semi Arid Region						
13.	Bharatpur		C	M	M	M	M
14.	Dholpur		C	M	M	M	M
15	Alwar		C	S	S	S	S
16.	Jaipur		C	S	S	S	S
17.	Dausa		C	M	M	M	M
18.	Ajmer		C	S	S	S	S
19.	Sawai Madhopur		C	M	M	M	M
20	Karauli		C	M	M	M	M
21.	Tonk		C	M	M	M	M
22	Bhilwara		C	S	S	S	S
	III. Sub-humid Region						
23	Sirohi		C	S	S	S	S
24.	Udaipur		C	M	M	S	S
25	Rajsamand		C	M	M	M	M
26.	Chittorgarh		C	M	M	M	M
27.	Bundi		C	M	M	M	M
IV	Humid Region						
28.	Banswara		C	M	M	M	M
29.	Dungarpur		C	M	M	M	M
30	Jhalawar		C	M	M	M	M
31	Kota		C	S	S	S	S
32.	Baran		C	M	M	M	M

Source: Disaster Management and Relief Department, Rajasthan

The abbreviations used in table for magnitude and severity of respective disasters interpret the position as follows:

(a) ***R :*** *Recurrence is regular in the district as per past analysis and the specific district is serious affected.*

(b) C: Appearance is on casual basis but the intensity is varied which may be severe or moderate as per magnitude of the problem.

(c) S: Generally, severity in the district partly or fully is highly damaging and destructive. Coverage of area is widespread.

(d) M: The impact in general is moderate Coverage of smaller area, which can be spread to total district in case of intensity and magnitude of disaster.

(e) C/S: The occurrence of the disaster is on casual basis, but the disaster in the affected area is severe.

(f) The analysis is based on the information of past years and severity is set on the basis of disaster over the affected population area.

(4) Region-wise Severity of Desertification

Desertification has slow but gradual impact. Land degradation is vigilant to the cultivators who realize deterioration of land, soil fertility, salinity and alkalinity in the area. The recent impact is water depletion due to overdraw of excess water than recharged. Eco-system is disturbed constantly. Region-wise severity is given in Table 9.12 and Figure 9.12.

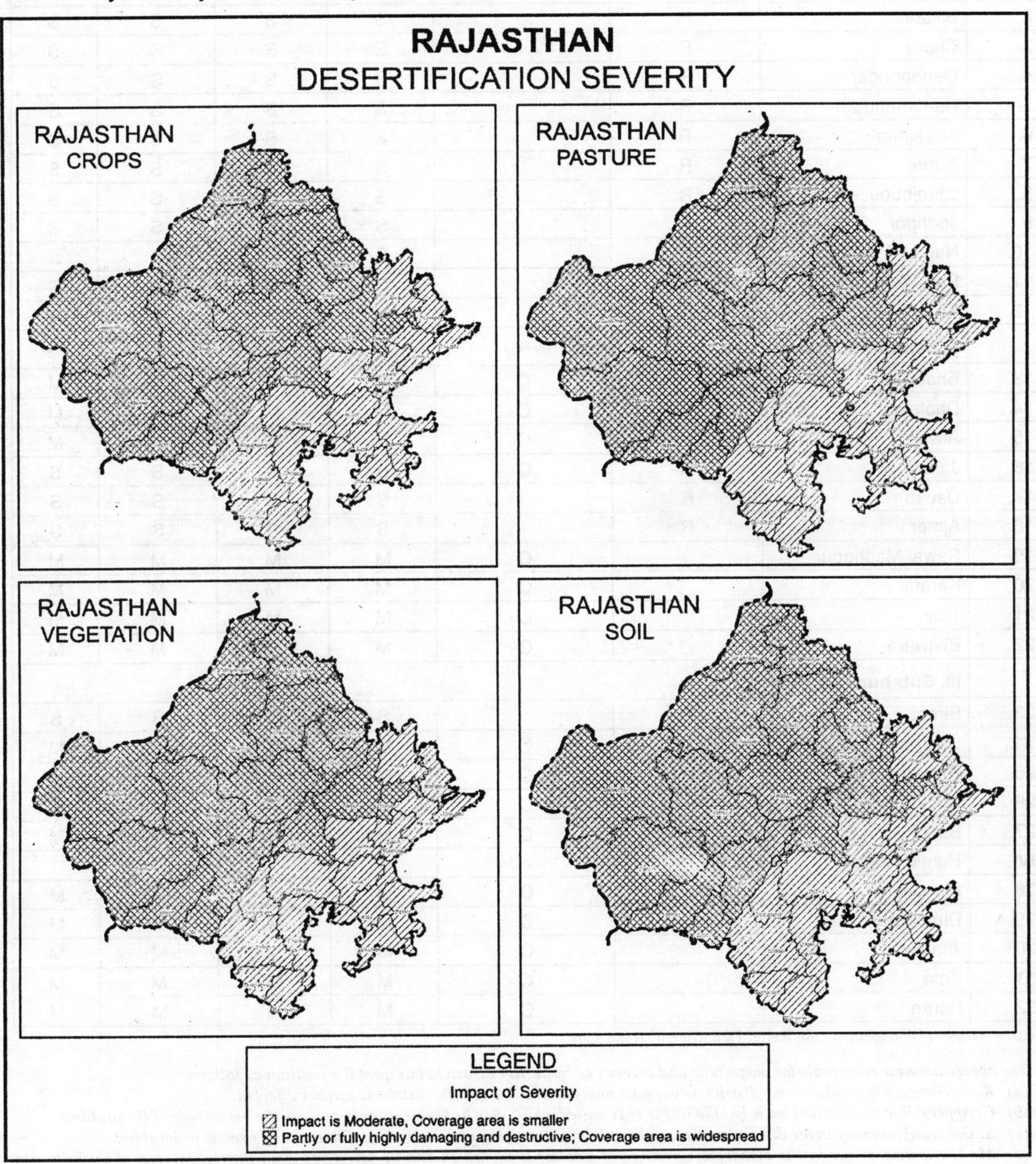

Source: Survey of India (SOI) and Department of Disaster Management and Relief, Government of Rajasthan, Jaipur

Figure 9.12 Desertification Severity in Rajasthan

Table9.12 : Region-wise Severity of Desertification in Rajasthan

S. No.	Region/ District	Occurrence of Deserti fication		Impact of Severity of Deserti- fication			
		Reg	Cas.	Crops	Pasture	Vegetation	Soil
	I. Arid Region						
1.	Barmer	R		S	S	S	S
2	Bikaner	R		S	S	S	S
3.	Churu	R		S	S	S	S
4.	Ganganagar	R		S	S	S	S
5.	Hanumangarh	R		S	S	S	S
6.	Jaisalmer	R		S	S	S	S
7.	Jalore	R		S	S	S	S
8.	Jhunjhunu	R		S	S	S	S
9.	Jodhpur	R		S	S	S	S
10.	Nagaur	R		S	S	S	S
11.	Pali	R		S	S	S	S
12.	Sikar	R		S	S	S	S
	II. Semi Arid Region						
13.	Bharatpur		C	M	M	M	M
14.	Dholpur		C	M	M	M	M
15	Alwar		C	M	M	M	M
16.	Jaipur	R	C	S	S	S	S
17.	Dausa	R		S	S	S	S
18.	Ajmer	R		S	S	S	S
19.	Sawai Madhopur		C	M	M	M	M
20	Karauli		C	M	M	M	M
21.	Tonk		C	M	M	M	M
22	Bhilwara		C	M	M	M	M
	III. Sub-humid Region						
23	Sirohi	R		S	S	S	S
24.	Udaipur		C	M	M	M	M
25	Rajsamand		C	M	M	M	M
26.	Chittorgarh		C	M	M	M	M
27.	Bundi		C	M	M	M	M
IV	Humid Region						
28.	Banswara		C	M	M	M	M
29.	Dungarpur		C	M	M	M	M
30	Jhalawar		C	M	M	M	M
31	Kota		C	M	M	M	M
32.	Baran		C	M	M	M	M

Source: Disaster Management and Relief Department, Rajasthan

The abbreviations used in table for magnitude and severity of respective disasters interpret the position as follows:

(a) ***R :*** *Recurrence is regular in the district as per past analysis and the specific district is serious affected.*

(b) ***C:*** *Appearance is on casual basis but the intensity is varied which may be severe or moderate as per magnitude of the problem.*

(c) ***S:*** *Generally, severity in the district partly or fully is highly damaging and destructive. Coverage of area is widespread.*

(d) ***M:*** *The impact in general is moderate Coverage of smaller area, which can be spread to total district in case of intensity and magnitude of disaster.*

(e) ***C/S:*** *The occurrence of the disaster is on casual basis, but the disaster in the affected area is severe.*

(f) The analysis is based on the information of past years and severity is set on the basis of disaster over the affected population/ area.

(5) Region-wise Severity of Locust Invasion

Locust invasion is sudden and irregular. The insects appear from Pakistan side and the arrival is in very huge quantity, which sometimes becomes unmanageable. First victims of this disaster are the cultivators and landowners of bordering districts. The insect move forward gradually and their unchecked position is significantly destructive. Severity of locust invasion is whole of the state but the districts of arid and semi arid regions are the victims of first stage. Severity position of locust is given in Table 9.13 and Figure 9.13.

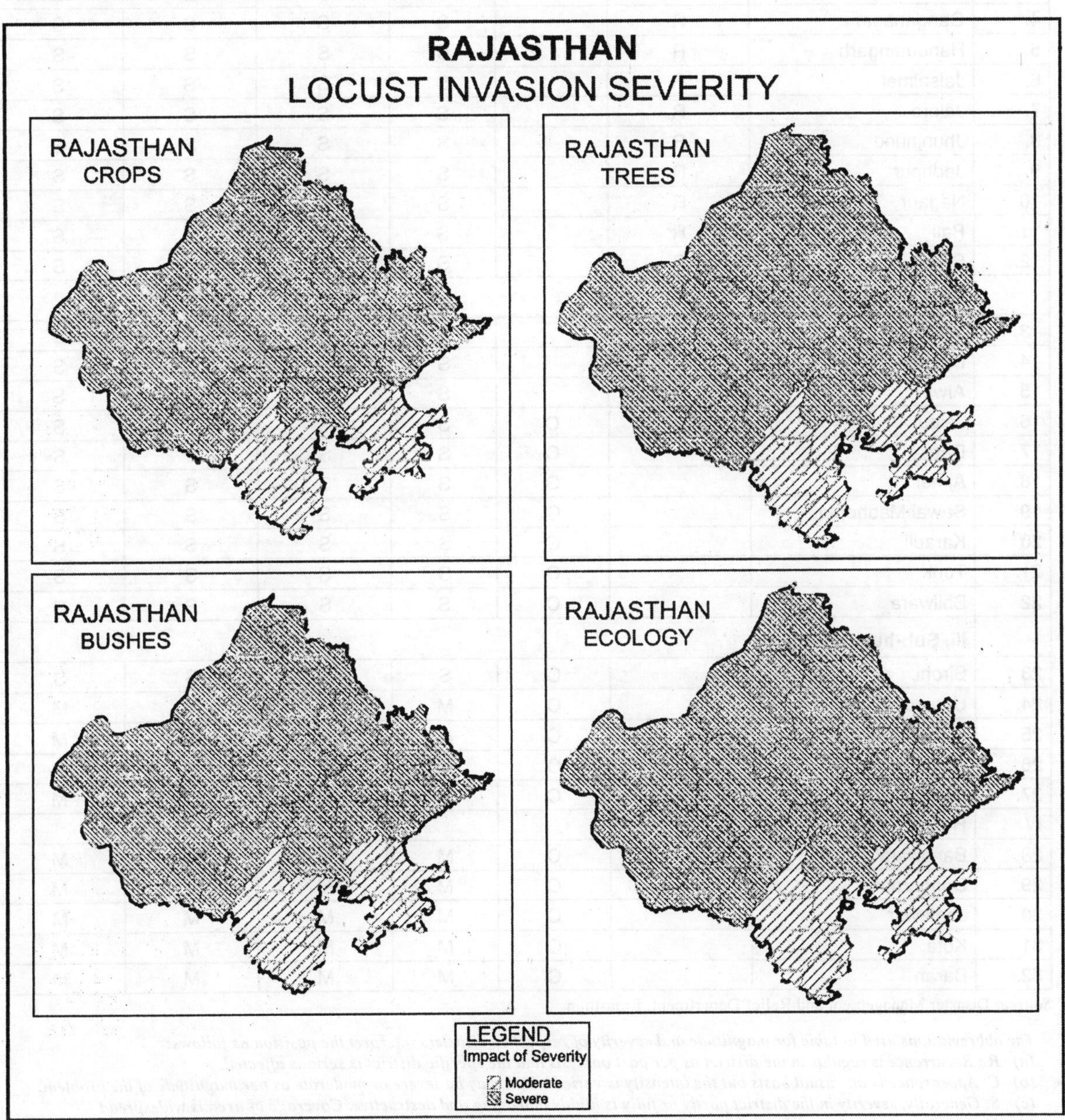

Source: Survey of India (SOI) and Department of Disaster Management and Relief, Government of Rajasthan, Jaipur

Figure 9.13 Locust Severity in Rajasthan

Table 9.13 : Region-wise Severity Of Locust Invasion

S. No.	Region/ district	Occurrence of Locust		Impact of Severity of Locust Invasion			
		Reg	Cas.	Crops	Trees	Bushes	Ecology
	I. Arid Region						
1.	Barmer	R		S	S	S	S
2	Bikaner	R		S	S	S	S
3.	Churu	R		S	S	S	S
4.	Ganganagar	R		S	S	S	S
5.	Hanumangarh	R		S	S	S	S
6.	Jaisalmer	R		S	S	S	S
7.	Jalore	R		S	S	S	S
8.	Jhunjhunu	R		S	S	S	S
9.	Jodhpur	R		S	S	S	S
10.	Nagaur	R		S	S	S	S
11.	Pali	R		S	S	S	S
12.	Sikar	R		S	S	S	S
	II. Semi Arid Region						
13.	Bharatpur	R		S	S	S	S
14.	Dholpur	R		S	S	S	S
15	Alwar	R		S	S	S	S
16.	Jaipur		C	S	S	S	S
17.	Dausa		C	S	S	S	S
18.	Ajmer		C	S	S	S	S
19.	Sawai Madhopur		C	S	S	S	S
20	Karauli		C	S	S	S	S
21.	Tonk		C	S	S	S	S
22	Bhilwara		C	S	S	S	S
	III. Sub-humid Region						
23	Sirohi		C	S	S	S	S
24.	Udaipur		C	M	M	M	M
25	Rajsamand		C	M	M	M	M
26.	Chittorgarh		C	M	M	M	M
27.	Bundi		C	M	M	M	M
IV	Humid Region						
28.	Banswara		C	M	M	M	M
29.	Dungarpur		C	M	M	M	M
30	Jhalawar		C	M	M	M	M
31	Kota		C	M	M	M	M
32.	Baran		C	M	M	M	M

Source: Disaster Management and Relief Department, Rajasthan

The abbreviations used in table for magnitude and severity of respective disasters interpret the position as follows:

(a) ***R :*** *Recurrence is regular in the district as per past analysis and the specific district is serious affected.*

(b) ***C:*** *Appearance is on casual basis but the intensity is varied which may be severe or moderate as per magnitude of the problem.*

(c) ***S:*** *Generally, severity in the district partly or fully is highly damaging and destructive. Coverage of area is widespread.*

(d) ***M:*** *The impact in general is moderate Coverage of smaller area, which can be spread to total district in case of intensity and magnitude of disaster.*

(e) ***C/S:*** *The occurrence of the disaster is on casual basis, but the disaster in the affected area is severe.*

(f) The analysis is based on the information of past years and severity is set on the basis of disaster over the affected population/ area.

Chapter 10

SURFACE WATER RESOURCES

Water is the most critical input for habitation and development of the region. In a development plan, improvement in water resources has an important place in increasing the production of food and fiber, which are basic requirements of human life. As a result of increasing urbanization and other connected developments, there has been shrinkage with the passage of time, in the culturable area available for agricultural development.

Rajasthan being a desert State receives about 1% of the country's water in the form of annual rainfall. A good part of it gets absorbed due to dry soil conditions further restricting surface runoff. The water getting absorbed should normally recharge ground water; however, part of it does not reach the deep ground water table while some part goes to ground water which is already brackish leaving comparatively a small part potable enough to be used as drinking water or fit for agricultural use.

The surface water resources in the State besides being scarce, are mainly confined to South and South-eastern part of the State. There is no perennial river in the State and it has to depend upon its share of water from interstate-river basins and rain water availability in seasonal rivers of State basins during monsoon period.

The arid to semi-arid climatic conditions of Rajasthan are characterized by many rainless months and a high probability of dry years with very little surface water flow. The average annual rainfall in Rajasthan ranges from 200 mm in the North-west to nearly 950 mm in the South-west. More than 90% of rainfall occurs during monsoon, but the distribution is unreliable in terms of space and time. Ground water under these conditions, is a reliable source in many cases also having the advantage of better quality source of water for drinking, industrial as well as for agriculture use.

Out of all the natural resources, water is undoubtedly, most vital for human existence. Adeqate facilities as assured irrigation, imparts confidence to the farmers to adopt improved agricultural practices, helping in achieving much higher level of crop production. Irrigation, no doubt, is a very important input by itself which enhances the effectiveness of all other agricultural inputs like fertilizers, improved seeds, etc.

Rajasthan which is the largest state of the country, in terms of geographical area, is well endowed with land and sunshine but is less fortunate in terms of available water resources. The state has about 256.56 lakhs hectares of culturable area which is nearly 13.89 percent of the country's total culturable area. The surface water potential of the state from internal sources comprising 14 rainfed river basins. The quantity of water flowing in these rivers totally depends on the quantum of precipitation during rainy season.

Although the average precipitation of the state is 530mm, there is marked spatial and temporal variation from one part of the state to another, The bulk of the rainfall generally occurs between four monsoon months of June to September but during this period also, the distribution is uneven. The extreme seasonal and spatial variability of precipitation all over the state is general and in the western half, in particular, is dramatically exemplified by the occurence of drought and floods at about the same time in every monsoon in different parts of the state.

The marked seasonality and variability of rainfall has necessitated construction of water reservoirs like dams, tanks and ponds to store some part of rainfall water to meet the irrigation and drinking water requirements through the rivers, nallahs, etc. and recharge the ground water which is also a major source of water all through the year. Besides, the state gets some water for its use from the inter-state river basins as a part of Inter-state agreements.

To have actual estimates of the available surface water resource in Rajasthan, the entire area in which runoff water flows through rivers and nallahs has been divided into 14 rivers valley catchments. The chambal catchment has got the maximum run off water, folowed by Mahi, Banas, Luni etc. The total surface water potential within the state has been estimated to be 15.80 MAF. To make use of surplus runoff water of the adjoining states of Rajasthan, a number of Inter State Agreements have been made, as a results of which water is already available in Bhakra, Gang, Indira Gandhi Nahar Pariyojana, Chambal Irrigation project and etc.

The state government has accorded highest priority for the development of surface water resources during last serval decades of planned development, as a result of which a large number of irrigation works have been planned and executed in different parts of the state.

At present, there are 23 major irrigation projects in Rajasthan, each with a culturable command area of above 10,000 hectare. These major irrigation projects are depicted in adjoining map. Out of these some projects are multipurpose ones, i.e. They are designed to achieve. The two objectives- power generation and irrigation. Work is in progress on two more major irrigation projects namely. Rajasthan water sector Restructurring project and Narmada.

There are in all, 77 medium irrigation projects which are located almost all over the state, each medium irrigation project has a culturable command area of 2,000 to 10,000 hectare. Surface water and lift schemes having culturable areas less than 2,000 hectare, are categorized as minor irrigation projects and there numbers are 640 spread over different parts of the state.

An irrigation potential of 36.00 lakh hectare has been created from major, medium and minor irrigation projects. Most of the irrigation potential created by major and medium projects has reached the stage of full utilization, actual utilization of the potential of minor irrigation projects have been on the lower side, since a large number of these projects are situated in the desert and drought prone areas where the inflow remains inadequate, due to scanty and erratic rainfall.

To impart trading to in-service personnel of the state department of Agriculture and Irrigation and to the farmers on irrigation management activities and other technological innovations in the field of Agriculture, an institute, the Irrigation Management Training Institute, is functoning at kota since 1984 (Table 10.1).

Table 10.1 India and Rajasthan Comparison

Particular	Unit	India	Rajasthan	%age
Geographical Area	Lakh ha.	3287	342	10.42
Population	Crores	102.86	5.65	5.49
Density of Population	Per Sq.Km.	324	165	-
Culturable Command area	Lakh ha.	1846.67	256.56	13.89
Ultimate Irrigation Potential	Lakh ha.	1156.15 (62% of CCA)	53.50*, (21% of CCA)	-

* Includes the water to Rajasthan from Inter State Agreeme

The total surface water potential for Rajasthan is estimated to be 15.86 million area-feet (MAF). Therefore, the state has to depend on inter-state river basins for water under which 14.50 MAF water has been allotted to the state as follows (Table 10.2) :

Table 10.2 Allotment of Water under River-basins to Rajasthan

	(in MAF) (million acre-feet)
1. Gang Canal	1.11
2. Bhakra Canal	1.41
3. Narmada	0.50
4. Ravi-Beas	8.60
5. Yamuna Water	0.91
6. Mahi Water	0.37
7. Chambal/Kota Barrage	1.60
Total	**14.50**

In inter-state river basins, maximum quantity of water has been allotted from Ravi-Beas, which is 8.60 MAF. Out of this, 7.58 MAF water would be used in Indira Gandhi Nahar Project (IGNP), and the balance 1.02 MAF would be used through Gang, Bhakra, Sidhmukh, Nohar and link Gano Canal-systems (Table 10.3).

Table 10.3 Allocation of water to Rajasthan from Inter-State River Basins (2007)

Particular	Allocation of Water (BCM)*	Developed/ Under Construction
(A) Ravi-BEAS		
(a) Pre partition (Share through Gang canal)	1.37	Developed
(b) Through IGNP Stage-I, II and others	10.60	Developed
2. Bhakra from sutlej	1.74	Developed
3. Yamuna	1.12	Under construction
4. Chambal upto kota Barrage	14.97	Developed
5. Narmada	00.62	Developed
(B) Allocation expected in future-Mahi	1.56	-

* Billion Cubic Meter

Indira Gandhi Nahar Pariyojana

Indira Gandhi Nahar Pariyojana (IGNP) started in 1958 and irrigation started in part area from 1961, was conceived and executed with the basic objective of for converting 19.63 lakh hectare of semi- arid desert wasteland into a Cultural Command Area (CCA). This area spans across 7 districts of Rajasthan, viz. Ganganagar, Bikaner, Hanumangarh, Jaisalmer, Jodhpur,Barmer and Churu. In addition to optimising agriculture production, the project also envisages the regional development and ecological improvement through arresting desertification (Figure 10.1).

IGNP is a grand endeavor to bring the Himalayan 's water to vast stretches of western desertic terrain. The water is diverted from the Hari- Ke- Barrage in Punjab through the 204 km Indira Gandhi Nahar feeder and into 445 km long main canal at Masitawali. The project was divided into stages. Stages-I comprised of irrigating 5.53- lakh hectare white Stage-II covered the remaining 14.10 lakh hectare area. Stage-I area was further sub- divided into two phases for the purpose of execution. Stage-I Phase consisted of all command land whose canal system took off from 0 to 74 km of Main canal white that of phase II comprised of construction from 74 km 189 of main canal. Work has, by and largp, been completed in stage 1 both Phases of and is in progress presently in stage II regions. It is envisaged that the entire construction work would be completed by the year 2012 AD.

Latest estimated cost of IGNP stage-I is Rs. 473 crores and for stage-II, Rs. 3,522 crores. Against this, expenditure incurred upto March 2007is Rs. 434.49 crores and Rs. 2625.73 crores for stage I, and II respectively.

From Indira Gandhi Canal, many urban centers and rural villages of districts of Bikaner, Churu, Ganganagar and Jodhpur, are getting drinking water through lift system on left side of canal.

Formerly it was called Rajasthan Canal Project. On completion of this project it will be regarded as the longest irrigation system in the world. It will convert a large part of the Thar desert into a granary. Churu, Sriganganagar, Bikaner, Jaisalmer, Jodhpur and Barmer districts will be benefited by this project. Its irrigation potential will be 19.63 lakh hectares and culturable command area as 16.17 lakh hectares (stage-I of 5.46 lakh hectares and stage-II of 10.71 lakh hectares) upto March 2017.

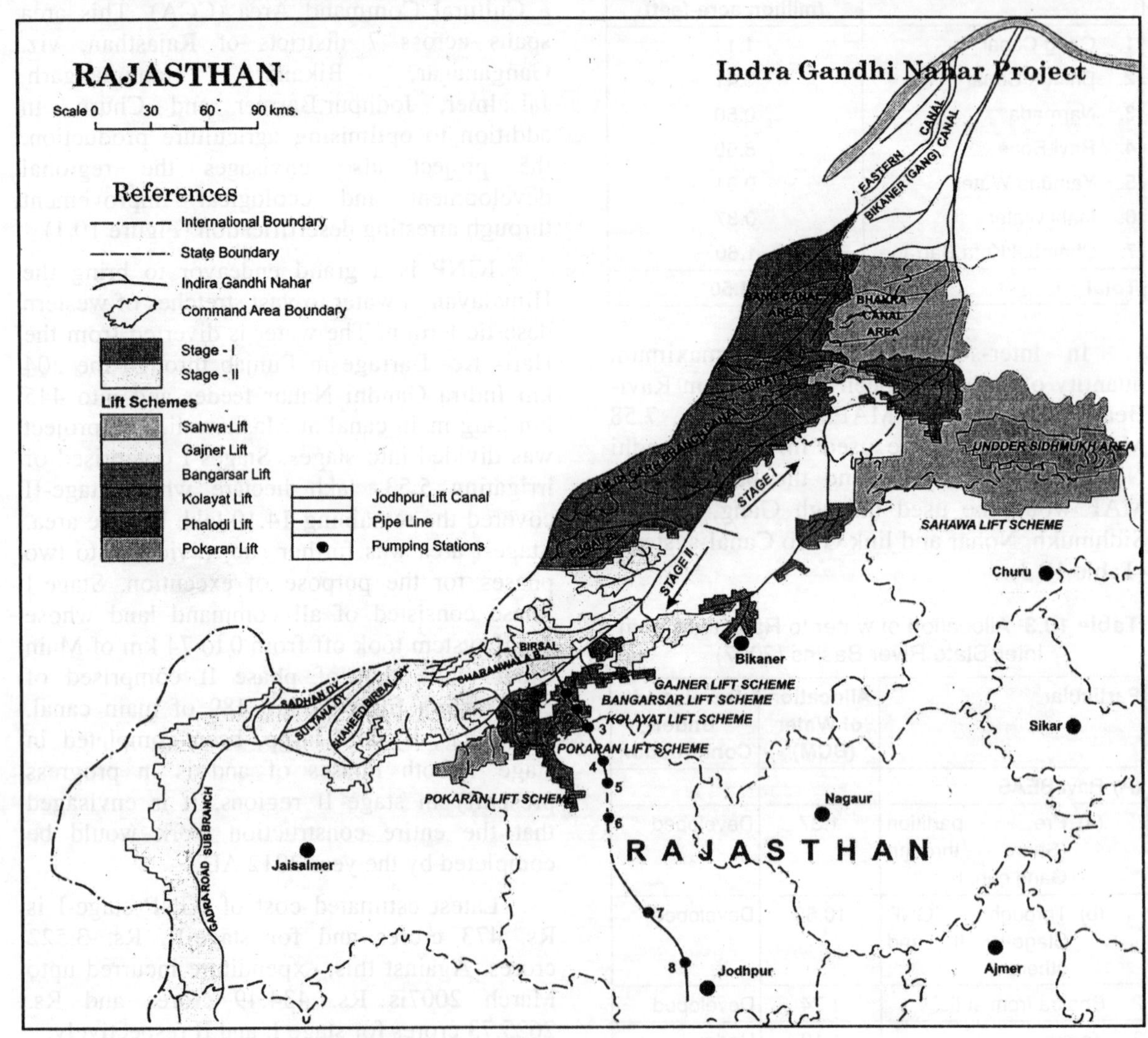

Figure 10.1 Indira Gandhi Nahar Pariyojana

In stage-I, construction works of Rajasthan Feeder of 204 km. (which starts from Harike dam at the confluence of Beas and Sutlej rivers in Punjab and ends at Masitawali village near Hanumangarh), 189 km. long Rajasthan Main Canal and 3454 km. distributaries have been kept, which are nearing completion. In stage-II, construction works of 256 km. long main canal (from a distance of 189 km. to 445 km.), i.e., from Chattargarh to Mohangarh in Jaisalmer district, and 5959 km. distributaries (including culturable command area and areas of seven lift canals) were included. On January 1, 1987, water reached at the last end of the main canal. It was the fulfilment of a happy dream to take the waters of the snow-clad Himalayan mountains to thirsty and hot desert-areas of

Rajasthan. Canal Construction works were completed in 16.17 lakh hectares up to March 2017 outofwhich 16.06 lakh hectares were opened. (5.46 lakh hectares in stage I and 10.60 lakh hectares in stage-II). Agricultural output of several crores is likely to be produced annually with the help of this project.

Lathi Series Area would make important contribution in making Jaisalmer district quite prosperous. Agriculture would develop after the availability of water in this area. 'Sevan' grass grows here under conditions of ordinary rainfall. It is quite nutritious for cattle. Lilwa branch is being developed beyond Mohangarh, the last end of the canal. It would be 90 km. long and irrigate Lathi Series Area. Canal water has reached several kms. far from Mohangarh. The inhabitants, animals and birds of Mohangarh region have got sweet drinking water for the first time, and this dry region has got irrigation facilities too. This project will be extended further up to Gadra Road in Barmer.

Potential Benefits of Indira Gandhi Canal Project

1. The production of wheat, cotton and oilseeds would increase in the state.
2. New industries, new towns and new colonies would be developed as a result of this project. There is a programme of colonisation of lakhs of people in the canal area. A 'master plan' has been prepared for the purpose.
3. For the first time, new lands would be brought under cultivation. The waters of Ravi-Beas would be utilised more intensively.
4. There would be saving in relief-expenditure as a result of the development of irrigation facilities, because the severity of droughts and famines would be reduced. Lilwa branch and Deegha branch would make Jaislmer area economically much better off than at present.
5. Seven lift canals have been drawn from the main canal. Water will have to be lifted for this purpose. At one time, water can be lifted at a height of 60 metres.

These seven lift canals are as follows :

(i) Bikaner-Lunkaransar Lift Canal—Bikaner town and 99 villages will get water from it.

(ii) Gajner Lift Canal (Pannalal Barupal Lift Canal)

(iii) Nohar-Sahwa Lift Canal (Chaudhari Kumbharam Lift Canal)—Apart from several villages, Sardarshahar and Taranagar will also get water from it. 666.50 km. long, this is the biggest Lift Canal.1

(iv) Bangarsar Lift Canal (Vir Tejaji Lift Canal)

(v) Kolayat Lift Canal (Dr. Kami Singh Lift Canal)

(vi) Falodi Lift Canal (Guru Jambheshwar Lift Canal).

Indira Gandhi Nahar Project (CAD Programme), Bikaner

It includes the following six projects :

(i) Indira Gandhi Nohar Project

(ii) Sidhmukh Nohar Irrigation Project

(iii) Aniarsingh Jassana Sub Branch Irrigation Project

(iv) Gang Canal Project

(v) Mandi Development Committee

(vi) Bhakra Nahar Project.

All these projects are controlled by the commissioner, CAD, IGNP, located at Bikaner It includes the following programmes for utilisation of water in arid regions:

(i) Levelling of land,

(ii) Making pucca water-channels,

(iii) Construction of roads and diggies, development of other activities such as education, mandies, rural water-supply, agriculture, cooperation, animal husbandry and fishing, etc.

International Development Association, an affiliate of the World Bank gave assistance for the programmes in the past. Under World Food Programme, free-ration for 24 months and interest-free loan of Rs. 2,000 for each settler were given as assistance to them.)

In 1992-93 afforestation-project was started for greening Indira Gandhi Canal area. It was financed by Overseas Economic Cooperation Fund (OECF) of Japan. Construction of roads, pucca field channels, diggies, etc., were undertaken in Bikaner and Jaisalmer districts. Programmes for the construction of nucca field-channels, roads and new sanitary diggies were undertaken. Tree-plantations were done on the lines of roads and canals. Sand-dune stabilisation work was also undertaken in the canal-area. Efforts have been made to control the problem of sem and salinity in the canal area.

IGNP, Bikaner: As against the expenditure of about Rs. 4.64 crore in 2012-13, a sum of Rs. 4.26 crore has been proposed for 2013-14 for the CAD of IGNP Bikaner.

Farmers are being benefited through agriculture Extension, Adaptive Trials and Soil Survey, Water Logging and Land Reclamation, Abadi Planning, Subsidy on Diggi and Sprinkler, etc.

(vii) Pokharan Lift Canal (Jai Narain Vyas Lift Canal).

Those lift canals will solve the problem of drinking water of several villages and some important towns in Rajasthan.

The location of these seven lift canals can be seen in the previous Map of Indira Gandhi Nahar Project.

Indira Gandhi Nahar Project will help a lot in the development of agriculture and fruit-trees in the large area of the Thar desert.

Thus, Indira Gandhi Nahar Project will increase irrigated area, increase agricultural output, increase the supply of drinking water, stop the onward march of the desert, promote fisheries, develop transportation, help in the construction of grain mandies, encourage animal husbandry and tourism in the state. It will prove to be a boon for the economy of Rajasthan in particular, and India in general.

The first phase of IGNP was completed much earlier, and the second phase by the end of the year 2005. The shortage of cement and coal put obstruction in the completion of the project. The State Government wants that the centre should bear the financial burden of this project like that of the Sutlej-Yamuna Link (SYL) project.

As stated earlier, by the end of March 2016 Irrigation-potential of 16.06 lakh hectares was available under this project. This is being increased every year. As the centre is incapable of financing the IGNP to a greater extent due to paucity of funds, resources will have to be mobilised from external sources to some extent. More benefits should be obtained from the project by developing animal husbandry, fodder development and agriculture, suited to die local conditions.

At several places on Indira Gandhi Canal, a lot of fertile land is going out of cultivation due to increasing salinity in soils, and poisonous grass has erupted on fertile soils due to the problem created by 'sem' water. There is a solid layer of gypsum under the earth, and farmers give excessive water to crops, which has created the problem of 'sem'. If this problem has occurred due to canal, tile-lining over cement-plaster should be done and 'sem' should be stopped. According to late Amrit Nahta, IGN area was more suited for animal husbandry, fruit-trees and horticulture, rather than for normal agricultural crops. Therefore, such activities may be pursued on a priority basis in future in this area. This stoppage of excessive leakage of water on surface soils needs immediate attention, otherwise IGNP will prove to be a bane rather than a boon for the economy of the state.

Thus, the emerging problems of IGNP should be tackled on war footing, otherwise we may not get the benefits from this project to the required extent. Besides irrigation, the project was providing drinking water in the irrigated area. Kanwar Sain Lift Canal is providing drinking water to Bikaner city and 99 villages situated outside the project area. Gandhelisahawa Lift Scheme is providing drinking water to 175 villages of Churu district, while the main canal via Jodhpur Lift Scheme is providing drinking water to Jodhpur city and the enroute towns and villages. In 1999-2000 State budget, Government announced a target of mobilising Rs. 200 crore by selling land in IGN area to speed up the process of colonisation and land allotment in this area. But the actual success achieved was much less than the expectations.

Gang Canal Project

The Gang Canal system is one of the oldest in Rajasthan serving north- western part of Ganganagar. Construction of Gang Canal started in 1920 and canal was opened on 26 October, 1927and the entire work was completed in 1928. Total length of canal system in Rajasthan is 1251 km. The Bikaner Canal feeder to Gang canal provides water to be extent of CCA of 3,07,692 hectares with 65per cent intensity.

Gang Canal-Main canal goes from Ferozpur to Shivpur (Sri Ganganaga). Its length is 137 km. and that of distribtory branches in 1,280 km. It irrigates 1.5 lakh hectares in Ganganagar district. Crop like cotton, wheat, malta etc. are irrigated with its water. This cannal now needs repairs and renovation to a great extent.

In 1984, the work of joining it with 'Gang Canal Link Channel' was started, This link channel can be constructed over a distance of 80 km. Water from Indira Gandhi Nahar would be supplied to it. Link channel originates from Lohgarh in Haryana. It joins Gang Canal at Sadhuwali (Sri Ganganagar). work on modernisation of Gang Canal is under progress with the revise outlay of Rs. 621.42 crore at 2008 price level, out of which an expenditure of Rs. 577.7 crore has been done till March 2014.

Bhakra Project

Pre- independence canals of Sutlej valley project utilised mainly the perennial flow of the Indus river basin and benefitted mostly the area lying in Pakistan white Bhakra Project was under contemplation. The Bikaner State tried to get a share of stored water of Sutlej river, after considerable negotiations, it was decided in 1951, that the Bikaner State and subsequently the state of Rajasthan was permitted (by Government of India as per 1955 Indus treaty agreement) to utilize the CCA of 3,72,469 hectare of land under Bhakra system. Total length of Bhakra Canal System is 1219 km.

This is the largest multi-purpose project of India. Punjab, Haryana and Rajasthan are taking part in it. Rajasthan's share is 15.2 per cent in it. Some land in Ganganagar district has been made cultivable as a result of extension of irrigation from this project Roads, mandies and villages have been developed, apart from concrete lining of the main canal and its branches. Bhakra main canal has irrigation potential of 14.6 lakh hectares, out of which Rajasthan's share is 2.3 lakh hectares, that of Haryana 5.5 lakh hectares and Punjab 6.8 lakh hectares.

Rajasthan has been allotted electricity also at Bikaner and Ratangarh, Churu, Sri Ganganagar, Jhunjhunu and Sikar have benefited from the supply of electricity from Bhakra project (Figure 10.2).

Command Area Development Project

In order to ensure optimum utilization of irrigation water for maximizing agricultral production and productivity through a multi - disciplinary approach, Command Area Development Authorities (CADA) were established for Chambal and Indira Gandhi Nahar Pariyojana (IGNP) in 1974. The basics of IGNP, CADA includes construcction of water courses from the Canal to the fields, roads, sanitation, diggis for drinking water, protective forestry for canal, afforestation, pasture development, fisheries, agriculture research and extension, supply of inputs and

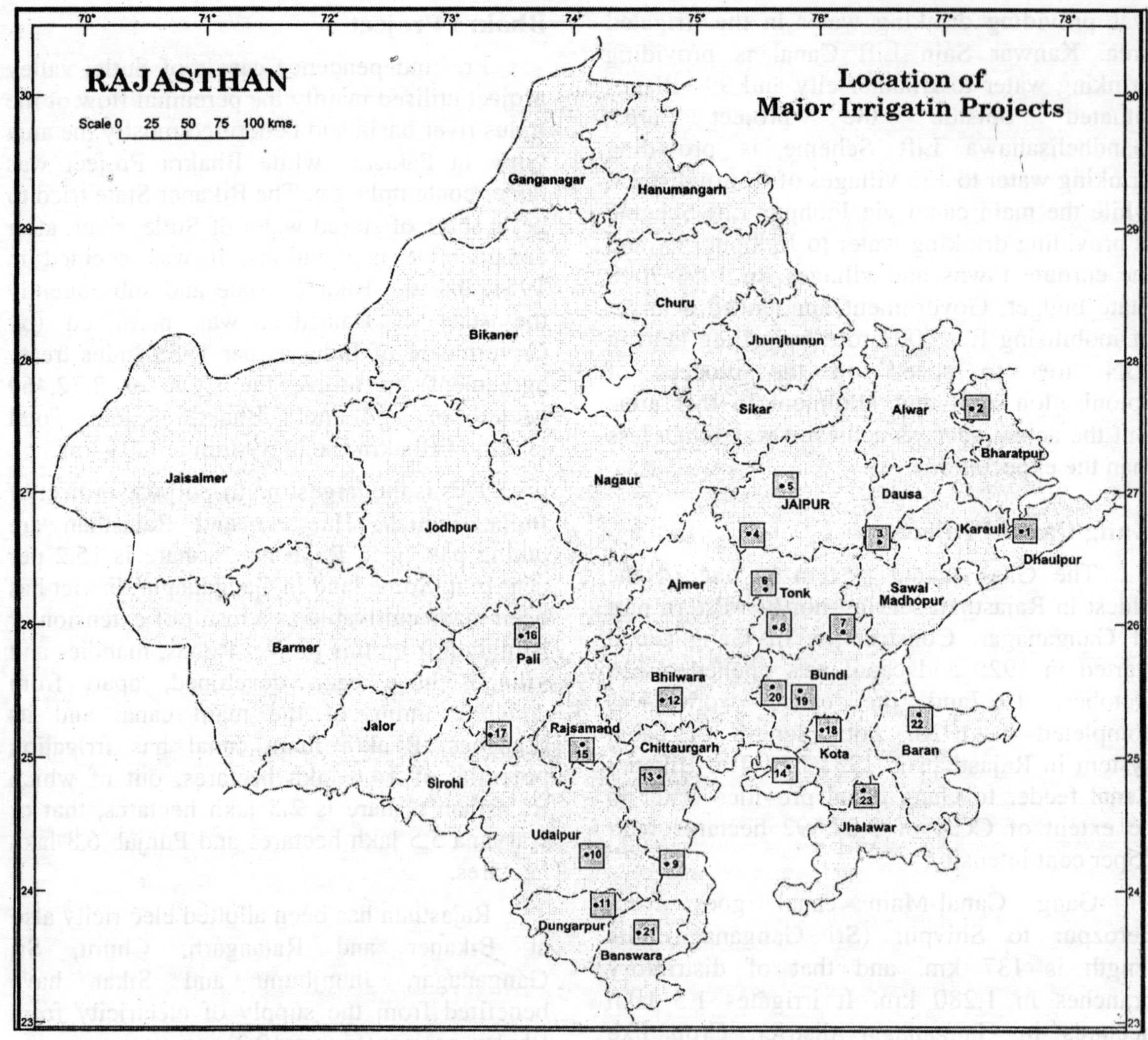

Source: Department of Irrigation, Government of Rajastham, Jaipur

Figure 10.2 Location of Major Irrigatin Projects in Rajasthan

Table 10.4 Name and Number of Major Irrigation Projects with Culturable Command Area in Hectare

SNo.	Name	CCA.	SNo.	Name	CCA.	SNo.	Name	CCA.
1.	Parvati	24,667	9.	Jakham	28,329	17.	Jawai	38,671
2.	Sikrai	10,463	10.	Jaisamand	16,006	18.	Kota Barrage	78,700
3.	Morel	19,393	11.	Som KamlaAmba	17,724	19.	Jawahar Sagar	–
4.	Chhaperwara	11,736	12.	Meja	18,618	20.	Gudha	10,896
5.	Kalak Sagar	10,961	13.	Ghosunda	–	21.	Mahi Bajaj Sagar	80,000
6.	Tordi Sagar	13,960	14.	Rana Pratap Sagar	–	22.	Parwati Pickup Weir	12,550
7.	Galwa Bundh	13,391	15.	Rajsamand	10,441	23.	Harish Chandra Sagar	12,284
8.	Bisalpur Project	81,800	16.	Sardarsamand	10,325			

Source: Department of Irrigation, Government of Rajastham, Jaipur

Table 10.5 Physical status of On-going Irrigation projects and culturable command Area (CCA)

Name of the Scheme	District	Estermated Cost (Rs. in lakhs)	Exp. upto Mar. 2008	CCA (hectare)
(A) Multipurpose Projects - Mahi	Banswara	93000.00	90329.00	80000
(B) Major Projects - Narmada Canal	Jalor	197529.00	122151.00	246000
Rajasthan water sector Restructuring Project	Entire State	73979.00	52403.00	616000
(C) Medium Projects-				
1. Gdrdada	Bundi	14000.00	7502.80	9161
2. Takli	Kota	6969.00	491.53	7386
3. Piplad	Jhalawar	3364.00	632.00	4688
4. Gagrin	Jhalawar	8012.00	1892.40	7943
5. Bandi Sandra		3702.82	2582.01	7135
6. Lahasi		8873.00	2196.70	5755
7. Sukli Selwara		4289.69	3302.04	5000
(D) Modern/Strengthening (Project) Gang Canal	Ganganagar	44579.00	38290.00	96500

Source: Department of Irrigation, Government of Rajastham, Jaipur

services to the farmers, development of abadies and of marketing facilities, drainage system to prevent water - logging and finally to promote settlement in the difficult terrain. CADA work in Chambal included drainage works, On Fram Development (OFD) works, consolidation of land, farm roads, development of nandis, roads in command area, construction of water coureses etc. (Table 10.4).

Mahi Project

Mahi Bajaj Sagar project is a joint multi - purpose venture of the states of Rajasthan and Gujarat for harnessing water of Mahi ri for developmentof irrigation and pow for which an agreement was reach the year 1966 between these sta accordance with this agreement the cost of Mahi Bajaj Sagar Dam being shared between Gujarat and Rajasthan in the ratio of 55:45. As per this agreement, 40 TMC of waterfrom Mahi Bajaj Sagar dam would be released for Gujarat in the interim period till Narmada Canal Project takes shape in the Gujarat and the present Mahi Command in Gujarat is transferred to Narmada water. This agreement further provides that after development of Narmada waters in Gujarat, entire water stored in Mahi Bajaj Sagar Dam and part storage of Kadana dam would be released for use in western arid districts of Rajasthan i.e. Jalor and Barmer.

Accordingly, a dam across river Mahi in Banswara district was constructed and na 72.70 Mahi Bajaj Sagar Dam to impou has been TMCof water. A canal sys of hectares .The constructed to benefit CC Mahi project through power is generated with a installed capacity of two power ho 25 MW at Power House No. land two uni units of 45 MKI at power house II. These th two power house are at a peak load generating stations.

The command area has subsequently been increased from hectares to 1,44,500 hectares of land in Banswara as well as in Dungarpur districts by extending left main canal, named Anandpuri Canal and right main canal, named Mahi -Sagwara Canal.

Irrigation started in 1983 from Mahi - Bajaj Sagar. The increased productivity of the land has brought higher socio- economic status of the population prasticing irrigation. The total latest estimated cost of Mahi Project including Gujarat share is Rs. 898.58 crores and Rs. 863.33 crores expenditure up to Dec 06 has been done . The command area of Mahi Bajaj

Sagar has been increased due to increasing in capacity of reservoir from 72.7 TMC to 77.00 TMC by adding flapongates, subsequently by changing cropping pattern and utilising regeneration of ground water.

It is proposed to provide irrigation for 80000 ha in the second phase of the project as per estimates the cost would be about 898.58 crores out of which expenditure of Rs. 578.32 crores has been done upto March 2006.

It is a joint project of Rajasthan and Gujarat. It will irrigate southern part of Rajasthan and northern part of Gujarat. There was an agreement between the two Govemments in 1966 for the use of water of the river Mahi. As per this agreement, Kadana dam was to be built in Gujarat, whose entire cost was to be borne by that state, and it would take its full benefit. But it was envisaged in the agreement that on the development of dam on Narmada, some amount of water would be given to Rajasthan, which would reimburse appropriate cost of the dam in Gujarat.

The work on Mahi Bajaj Sagar Project has been going on since 1968. Its Unit-I is meant for irrigation in which both Rajasthan and Gujarat have the shares. Main dam is 3,109 metre long. Gujarat w[illegible] share 55 per cent expenditure, and Rajasthan [illegible] per cent. In Unit-II Rajasthan alone has a share [illegible] irrigation and power. In Unit-III also, Rajasthan [illegible] has it own power portion. In Unil-IV, Rajasthan's irrigation, portion is involved. Some expenditure was incurred on Unit-V in the Seventh Plan. This was also meant for irrigation portion of Rajasthan.

Its total power-capacity is 140 MW. First unit of Power House No. 2 was started in February 1986. and the second unit in July 1989. Rajasthan and Gujarat would get irrigation facility for 8.8 lakh hectares. By March 2010-end, 84,707 hectares of land were brought under irrigation, while the preliminary CCA* was estimated at a level of 80,000 hectares. This project will help in the development of the tribal region in several ways.

Mahi Command Area Development Programme

Construction of water-courses, lining in vulnerable reaches, culverts, drop-structures, etc., were to be taken up under the programme. Tribal people who belong to weaker sections of the society will benefit from it. Farmers would benefit due to an improvement in the supply of water. GOI would also provide funds for CAD programme in Mahi region. Earth filling and lining works have been done on this project. Work on Bhikhabhai Sagwada Canal is under progress. It will irrigate 11,818 hectares.

Community Lift Irrigation Programme

Community Lift Irrigation Programme was started in 1980-81 under the Integrated Rural Development Programme (IRDP) for developing irrigation facilities for small and marginal farmers in the Southern and Southeastern parts of the state. A management committee of small and marginal farmers was appointed to monitor the programme. The beneficiary farmers contribute at least 10 per cent cost of the irrigation scheme and the Government gives subsidy for the purpose.

Bisalpur Project

Bisalpur Project is a multipurpose major project being constructed on Banas river. The construction of a dam across river Banas has done in thr year 1999 for drinking water supply to [illegible] town of Ajmer, Beawar.Nasirabad, Jaipur and [illegible]ages in routes etc. and construction of 76.[illegible] canals has been done for providing irrigation faci[illegible] in Tonk and Sawai Madhopur districts in 81,790 [illegible] ha. The total storage capacity of dam is 38.70 TMC,out of which 24.20 TMC is usable, of which 16.20 TMC is for drinking and 8.00 TMC for irrigation. Total Expediture done up to year 2005-06 on the project is Rs. 633.01. crores.

A dam is being constructed near Bisalpur village on river Banas. This village is situated in Tonk district at a distance of 13 km. from Todaraisingh town. Work commenced on it in

1986-87. This project will be completed in two stages.

The main object of the project is to supply drinking water to towns which are Jaipur, Ajmer, Beawar, Kishangarh, Nasirabad and other enroute villages where there is acute shortage of water for domestic as well as industrial uses. The villages of Tonk, Ajmer and Bundi district would get water for irrigation also. Wells and tube- wells would be dug in the catchment area of river Banas at four places which are Sandla, Chhatri, Negadia and Deoli.

The project will irrigate 81,800 hectares of agricultural land of Tonk district.

The total cost of the project was estimated at Rs. 1,100 crore.

Till the end of December 2010, irrigation on 81,500 hectares has been started.

Compensation of the land and properties of villages submerged would be arranged by the government. Rehabilitation and resettlement of the oustees are under progress.

Gurgaon Canal Project

Gurgaon Canal Projact is being constructed of 500 cusecs canal capacity to utilize Yamuna water through Agra Canal during rainy season for irrigating 28,200 hectares of kharif crops in 185 villages of Deeg and Kaman tehsils of Bharatpur District. Present estimated cost of the project is Rs.,3540.00 lakhs. Partial irrigation has been started. Recently an agreement has been reached between the states of Uttar-pradesh, Haryana, Rajasthan and. Delhi regarding sharing of Yamuna water and accordingly Raiasthan share shall be of the order of 0.93 MAF. This all water shall be taken through existing Gurgaon canal system.

It has been drawn from the river Yamuna near Okhla (Delhi). Construction work on it commenced in 1966 and was completed in 1985. The canal enters Jurera village in Kaman tehsil in Bharatpur district. It length is 35 miles in the state. It irrigates 28,200 hectares in Kaman and Deeg tehsils. It is included in the category of major irrigation projects.

Isarda Drinking and Irrigation Project

The project is loacated down stream of Bisalpur project near Isarda Village Hydrology of Project was approved by the Centran Water Commission in July 2000. The estimated cost of the project is 400.12 crore (April 2000). Project report has been sent to Central Water commission New Delhi for approval. On completion of the project irrigation of 22000 ha land in 51 villages will be done. The availability of waterfrom the project will be 10.77 out of which 7.07 TMC will be used for drinking water supply and 2.50 will be used for irrigation purpose after taking account of evaporation and other pounding losses.

Chambal Valley Project

Chambal Valley Project is a multipurpose major project constructed for irrigation as well as Hydel power generation across the river Chambal by the state of Rajasthan and Madhya Pradesh sharing equal costs of construction and storage benefits. The project was completed in 3 stages : Stagel- This first stage comprises of a high masonory gravity dam called Gandhi Sagar Dam, having a gross storage capacity of 6.28 MAF across the Rajasthan- Madhya pradesh boundary. The hydro power station at the dam site consists of four generating unts of 23 MW each and one of 27MW. For the purpose of irrigation, Kota- Barrage was constructed to raise the water level of river flow by about 30m to feed to two canals for commanding areas of both Rajasthan and Madhya Pradesh. The water released after power generation is utilised for irrigation through Kota- Barrage. The first stage construction was started in September 1953 and inaugurated in November 1960. The total anuua! irrigation from Right Main Canal and Left Main Canals is about 11 lakh acres as per original report i.e. of 5.5 lakh acres each in Madhya Pradesh and Rajasthan. The revised cost of Gandhi Sagar and Koat Barrage and

Canals is Rs. 8,035.00 lakhs and Rs. 4,615 lakhs respectively.

Stages II- Under this stage is the second dam in the series of the Chambal Valley Projects, called Rana Pratap Sagar. The dam is 54 m high and 1,143m long at the top. The Power house located on the left side has 4 units of 43 MW each construction was started in September 1960 and all major works completed in 1970. The revised cost of R.P.S. is 3641.00 lakhs.

Stages III- Under this stage is the third dam in the series, called Jawahar Sagar with a height of 44.80 m and lenght at the top 393.20 m long. The power house comprises of 3 units of 33 MW each.The construction was started in September 1962 and most of the works completed in 1973. The revised cost of Jawahar Sagar Dam is Rs. 2,810.82 lakhs.

It is the largest and perennially flowing river of Rajasthan. MP and Rajasthan are jointly working on it. Rajasthan's share is 50 per cent in it. In the first stage, Gandhisagar dam was built. It is the largest barrage. In the second stage Ranapratap Sagar Dam was constructed. In the third stage, there was Jawaharsagar dam or Kota dam, which was a 'pick-up dam' to store water released from the earlier two dams. In the first stage, there is Kota irrigation dam (Kota barrage), which has been built 5 miles north of Kota city.

Chambal project will increase irrigation facilities mainly in Kota and Bundi districts. Problems of saline soils, drainage, etc., have developed in this project, which were tackled to some extent with the assistance from IDA, an affiliate of the World Bank.

In the new programmes related with this project, extension of Bundi branch, raising of the level of Kota barrage and downstream protection works have been included. This project has now been completed. It irrigates 4.58 lakh hectares of land, and generates 386 MW of hydel power. The maximum irrigation capacity under the Chambal Lift Scheme is 47,880 hectares.

Chambal Command Area Development Programme, Kota

This was started in 1974-75. As it was an inhabited area with long period of revenue administration, programmes under CAD here were slightly different from that of IGNP region. Proper drainage system and revival of wild grass have been emphasised for this area. Other programmes include tree-plantations, development of agro-based industries, processing industries, rural godowns and rural housing, etc. The World Bank gave assistance for this project.

In 1991-92, Canadian International Development Agency (CIDA) started Rajasthan Agricultural (Research) Drainage (RAJAD) Project at Kota, which helped in the development of this area. It strengthened the existing irrigation and groundwater development facilities in the region. The canal system was constructed long back and requires maintenance. A project costing ? 1274 crore has been prepared for complete renovation of the irrigation system. It will be financed by NABARD under RIDF.

Narmada Canal Project

As per Narmada Water Disputes Tribunal Award, Rajasthan has been allocated 0.5 MAF of Narmada water. To utilizer this water, Narmada Canal Project, Rajasthan, was formulated at an estimated cost of Rs. 467.53 crores (1989-90), which includes share cost payable to Gujarat for comman works amounting to Rs.280.14 crores and cost of canal works amounting Rs.187.39 crores in Rajasthan territory. According to construction schedule of Sardar Sarovar Project (Gujarat), Narmada water are likely to reach Gujarat - Rajasthan borders by 2007. The Narmada project is likely to be completed by 2009-10.

In the year 2006-07 main canal of 333.25 km length has been constructed and 236.788 km long flow distributaries has been constructed. An expenditure of Rs. 1000.89 crore has been done up to year Dec. 2006.

The Narmada Canal in Gujarat state upto Rajasthan border will have a length of 438.6 km, and in Rajasthan, canal will be 74 km long out of which 65 km in tehsil Sanchore of Jalor district and 9 km in tehsil of Gudamalani of Barmer to cover cutivable area of 1.22 lakh hectares in 76 villages of Jolor and 0.13 lakh hectares in 7 villages of Barmer district, respectively. The rabi irriga-tion is estimated on an area of 74,000 ha.

Narmada Project (Under Accelerated Irrigation Benefit Programme (AIBP).

Sardar Sarovar Narmada Project is a major project of Gujarat. Its revised cost is Rs. 2481.49 crore. An expenditure of Rs. 2120.5 crore has been incurred upto March 2014. A sum of Rs. 150 crore was likely to be incurred during 2012-13 and an outlay of Rs. 180 crore was proposed for 2014-15. Rajasthan will also get irrigation water from it in 76 villages of Jalore district and 7 villagés of Barmer district. This project in both the states would provide water facility to 1107 villages and 2 towns. The water of river Narmada reached Rajasthan in March 2008 and would provide irrigation in 2.46 lakh hectares (for 2.38 lakh hectares created by March 2014) The project is likely to be completed by March 2016.

Sidhmukh Nohar Projects

Sidhmukh Nohar Projects are under execution for providing irrigation facilities to Sidhmukh Nohar areas in nothern part of the state with the assistance of European Economic community to provide irrigation facilities in 1.12 lakh hectares of CCA.The estimated cost of the project is Rs. 197 crores, 0.47. MAF of surplus Ravi-Beas water would be utilized by the implementation of these projects.

This will provide irrigation facilities to Nohar and Bhadra tehsils in Hanumangarh district and Rajgarh (Sadulpur) and Taranagar tehsils in Churu district. With this project, Rajasthan would be able to use surplus water of Ravi-Beas rivers, which was given to it under an agreement of December, 1981, reached between Punjab,' Haryana and Rajasthan. The water which is being made available to Rajasthan will be given from Nangal Headworks. It will flow through Bhakra main canal in Punjab to Fatehabad branch and Kishangarh sub-branch, parallel canal to Haryana.

Nohar tehsil in Ganganagar district will be benefited from it. Both the Sidhmukh and Nohar projects are parts of the same programme. Surplus waters of Rabi-Beas rivers would be used in them. Economic agreement has been reached with European Economic Community (now European Union) to execute the project. Total irrigated area would be 1,14,379 hectares and 99 villages of Hanumangarh district and 14 villages of Churu district would be benefited.

Beas Project

It is a multi-purpose project of Punjab, Haryana and Rajasthan. In this project water from three rivers—Sutlej, Ravi and Beas is being used. It has three units : (i) Beas-Sutlej Link, (ii) Dam on Beas river at Pong and (iii) Beas Transmission System. In the first unit, following works have been included : one dam at Pandoh (in Himachal Pradesh), two tunnels, seven mile-long open hydel channel (from Baggo to Sundar Nagar) and a power plant (at Dehar, with 165 MW capacity).

In the second unit, water was to be stored for Rajasthan at Pong dam (on river Beas). This would irrigate areas in Punjab, Haryana and Rajasthan. A power-plant would also be installed under this unit It should be noted that Rajasthan would not get direct irrigation facility from the Beas project. It would supply water to Indira Gandhi Nahar Project on a permanent basis. It would irrigate 21 lakh hectares in all the three states and Rajasthan would share 150 MW of power (total capacity 240 MW).

There has been Ravi-Beas Water-dispute and Iradi Commission was appointed to resolve it. It submitted its report in May 1987, and raised the share of Punjab from 42.2 lakh acre-feet to 50 lakh acre-feet; Haryana from 35 lakh acre-feet to 38.30 lakh acre-feet, but kept the share of Rajasthan constant at a level of 86 lakh acre-feet. Thus, Rajasthan was discriminated against and its share was not increased, while its need more urgent in this regart. The irrigated areas are already more in the states of Punjab and Haryana. The interests of Rajasthan should not be lost sight of, while settling such river-water disputes, because this state also belongs to the region of Sindh river, like that of the other two states.

Jakham Dam

This has been constructed on river Jakham in Pratapgarh tehsil (district Chittorgarh). Jakham is a tributary of the river Mahi. Construction work on the dam was started in 1962. Irrigation facility would be provided in villages of Dhariawad (Udaipur) and Pratapgarh (Chittorgarh) Panchayat Samiti areas. Tribal people (Bhils) would be greatly benefited by this project.

A pick-up dam has been constructed in Nagaria village, 13 km. below the main dam. This will facilitate irrigation in low-lying productive lands. On right and left banks of the pick-up dam, two canals have been constructed. On main dam, 9 MW power would also be generated. This project wold irrigate 23,505 hectares of land. It pas been constructed under tribal sub-plan of the state.

Bharatpur Canal

This canal was constructed in 1964. It has been drawn from Western Yamuna Canal. Its total length is 28 km. out of which 16 km falls in Uttar Pradesh. It can irrigate eleven thousand hectares of land, out of which 8,500 hectares are irrigated in Bharatpur district alone. It has helped in increasing the production of foodgrains.

Brief Introduction of Some other Dams

Jawai Dam—This has been constructed on river Jawai, which is a tributary of river Luni in western Rajasthan. River Jawai flows on western slope of Aravalli range in Pali district. The dam has been constructed at a distance of 3 km. from Arinpura railway station. The work on it was started in 1946 and was completed in 1951-52. Water from the dam is supplied to Jodhpur, Sumerpur and Pali towns for domestic use. Besides this, 26 thousand hectares in Pali district and 15 thousand hectares in Jalore district get irrigation facility from this project. 176 km. long canal has been constructed from the dam site.

Meja Dam—It has been constructed on river Kothari, at a distance of 8 km. from Mandalgarh town in Bhilwara district. The work commenced in 1957 and was completed in 1972. It irrigates 10 thousand hectares in Bhilwara district. Bhilwara town gets water for domestic use from this dam. Pipeline construction for the same was completed in February, 1985.

Panchana Dam—This earthen dam is being constructed at Gambhiri on the confluence of five rivulets in Sawai Madhopur district near Karauli. The construction work on two canals, etc., has been in full swing. It will irrigate 9,980 hectares in Gangapur, Hindaun, Nadoti and Todabhim tehsils. It comes under medium irrigation projects.

Morel Dam—It is being constructed on river Morel, 16 km. far from Lalsot in Sawai Madhopur district. It irrigates 8.6 thousand hectares. Some of the other main medium irrigation schemes along with their districts are as follows :

Earlier under modernisation category, the irrigation projects were as follows:

Gang Canal (Sri Ganganagar district), Gambhiri (Chittorgarh district), Meja (Bhilwara district), Jaisamand (Udaipur district), etc. It was proposed to start work on medium irrigation projects- Chauli (Jhalawar) and Baithali (Baran) in 1994-95 (Table 10.5, Table 10.6 and 10.7).

Table 10.6 Main Current Medium Irrigation Projects in Rajasthan with the Names of the Projects and their Districts

Name of Medium Projects	Districts
1. Gardada	Bundi
2. Takh	Kota
3. Piplad	Jhalawar
4. Gagrin	Jhalawar
5. Lhasi	Baran
6. Manohar Thana	Jhalawar
7. Hatiyadeh	Baran
8. Rajgarh	Jhalawar
9. Andheri	Baran

Source: Department of Irrigation, GoR, Jaipur

Further details of these projects are given in the following Table 10.7:

Table 10.7 Some New Medium Irrigation Projects

Name	River	Location and District	Estimated Cost	Irrigation Potential
1. Gardada	Mangli Dungari and Ganesh Nallah, Tributary of Chambal	Dam near Village Polaspura, District, Bundi	Rs. 246.16 crore	Target for additional irrigation potential of 9000 hectares in the project.
2. Takli	Takli, Tributary of Arnjar River and Kalisindh River	Dam near Village Dhankya in Kota District	Rs. 131.44 crore (Revised)	will irrigate 7386 hectares and drinking water for 4 villages and 2 towns of Ramganj mandi Tehsil
3. Piplad	Piplad, Tributary of Ahu River.	Dam near Village Sulia Chowki, in Jhalawar District	Rs. 91.20 crore	will irrigate 4688 hectares in 19 villages of Panchpahar Tehsil.
4. Gagrin	Ahu, Tributary of Kalisindh River	Dam site Kalapipal, Jhalawar District	Rs. 185.29 crore,	will irrigate 7943 hectares
5. Lhasi	Earthen Dam on Lhasi River	Khajuria, Baran District	Rs. 106.95 crore (Revised)	will irrigate 4026 hectares
6. Manoharthana Irrigation project	Parwan River (Irrigation Cum Drinking Water	Manoharthana, Jhalawar District	Rs. 93.10 crore	will provide irrigation to 9800 hectares in Jhalawmr district
7. Hatiyadeh	Earthen Dam on Hatiyadeh Nalla	Village Karwari Khurd, Kishanganj Tehsil District, Baran	Rs. 70.66 crore	will irrigate 6885 hectares
8. Rajgarh for Irrigation Cum Drinking Water	Confluence of Ahu and Kanthari Rivers	Rajgarh, Jhalawar District	Rs. 192.13 crore (revised)	An amount of Rs. 20 crore provided for 2015-16
9. Andheri	Earthen Dam on River Andheri	Mimdkiya Village, District Baran	Rs. 87.53 crore	Inter State clearance required, from M.P

1. Draft Annual Plan. GOR. 2016-17. Vol. I. pp. 7.7-7.9

Chapter 11

GROUND WATER RESOURCES

Ground water, plays an important role in the development of water resources in Rajasthan. More area in Rajasthan is irrigated by wells compared to all flow Irrigation Schemes put together. Over exploitation of Ground water is resulting in white, grey zones being converted to dark zones and therefore its impact on ground water quality calls for immediate action to check this problem. Use of ground water and surface water in an integrated manner can help in checking the problem of water logging and salinity in irrigated areas.

Water demand in major part of the state is mainly met from ground water resources because of scarce surface water resources, which endorses ground water to assume the role of supplying water for almost all the development activities. Crystalline, semi consolidated and unconsolidated aquifers are the potential source of water in Rajasthan. In desert region generally the potential aquifer systems often lie adjacent to saline aquifers and receive insignificant replenishment due to water deficit climatic conditions whereas in other parts of the state the rate of withdrawal is so high that the effect of good rainfall is also neglected and the water levels continuously shows a declining trend. Thus in both desertic and non-desertic region trend of water level is towards declining side. Ground water exploitation without adequate recharge has caused depletion in water levels, deterioration in chemical quality of ground water by ingression of saline water and reduction in yield of wells has become common. Water levels vary with rainfall, physiography, drainage and climatic environment. These parameters show wide variation in Rajasthan. Eastern Rajasthan has more rainfall and it decreases westward. From the above analysis it is inferred that in spite of number of wet spells in last 16 years there is a continuous decline in ground water levels in the state which is ultimate result of heavy ground water withdrawals for domestic, irrigation and industrial purposes. If the same trend of exploitation continuous without adequate recharge (natural or artificial) the situation will become so worse that all the productive ground water potential zones will become dry resulting in the permanent damage to the aquifer.

Depth to Water Level

Depth to water table on the eastern side of Aravalli hill range is shallower than that in the western side. The depth to water table in the State varies widely i.e. less than 5m to over 130m. In the areas west of Aravalli hill ranges, deep water table i.e. over 60m is recorded in the zone of inland drainage, in north and southwestern part of Jaisalmer district, southern part of Bikaner district, west central parts of Churu district and north eastern part of Jodhpur district. Locally such areas also exist in the southwestern part of Barmer district, northwestern part of Nagaur district and northwestern part of Jodhpur district. In major portion of western Rajasthan depth to water table is 20 to 60m. Shallow water tables (less than 5m) are noticed locally in the Ghaggar River Basin in Ganganagar district and along Luni River in Jodhpur, Pali and Barmer districts. In the areas on the eastern side of Aravalli hill ranges the depth to water level is less than 5m to over 20m BGL. Shallow water level less than 5m was recorded around Sambhar lake, northern part of Jhalawar district, major part of Kota and Bundi districts, locally in the eastern part of Bhilwara district, in the Mahi-Bajaj canal command area of Banswara district and locally in Dungarpur, Udaipur, Chittaurgarh, Bharatpur and Dhaulpur districts.

Water Level Fluctuation

The fresh water potential aquifers are under tremendous pressure of exploitation. The State is also prone to frequent droughts combating of which promotes construction of more and more ground water abstraction structures. Ever-increasing withdrawal of groundwater for irrigation, industrial and domestic purposes have resulted in declining trends in water levels. The deeper water levels, inferior chemical quality and availability of IGNP water restrict ground water development in westernmost parts of the State and as such the area do not exhibit declining water level trend. The decline in water levels is observed in remaining larger parts (except Mahi and Chambal canal command areas) of the State. Severity of rate of decline in water level also depends on availability of saturated thickness and static ground water resources in the area. In hard rock areas saturated thickness is generally restricted to weathered and fractured zones and hence decline in water level is rapid and impact of which is of serious nature. In alluvial areas having limited saturated thickness (say 10 to 50m) decline in water level of few centimeters to 1m per year may be taken as serious. However, decline in water level of few centimeters or even over 1m may not be considered significant for aquifers having saturated thickness of more than 100 to 500m as found in Lathi basin area of western Rajasthan (Table 11.1).

Table 11.1 Change in Water Level

S. No.	District	Average Rise/ Drop (m)	Max. Drop (m) (Spot Value)
1	Ajmer	-5.59	-19.03
2	Alwar	-5.12	-19.45
3	Banswara	0.00	-5.80
4	Baran	-2.01	-17.90
5	Barmer	-1.33	-19.38
6	Bharatpur	-1.11	-7.45
7	Bhilwara	-4.71	-12.60
8	Bikaner	2.47	-12.85
9	Bundi	-3.30	-18.75
10	Chittorgarh	-4.09	-19.70
11	Churu	0.04	-13.56
12	Dausa	-2.18	-16.50
13	Dholpur	-1.35	-7.75
14	Dungarpur	-2.02	-8.20
15	Ganganagar	5.62	-1.25
16	Hanumangarh	4.42	-3.67
17	Jaipur	-5.74	-21.25
18	Jaisalmer	0.23	-7.98
19	Jalore	-8.00	-21.45
20	Jhalawar	-1.22	-6.60
21	Jhunjhunu	-6.22	-14.81
22	Jodhpur	-6.60	-54.26
23	Karauli	-3.43	-29.75
24	Kota	-1.51	-13.17
25	Nagaur	-6.50	27.50
26	Pali	-6.69	-28.86
27	Rajsamand	-4.69	-16.80
28	Sawai Madhopur	-1.74	-9.00
29	Sikar	-4.77	-16.82
30	Sirohi	-6.49	-24.99
31	Tonk	-2.37	-11.60
32	Udaipur	-2.67	-13.33

Source: GWD, Government of Rajasthan, Jodhpur

Long Term Water Level Fluctuation

Decline in water level is a countrywide phenomenon. Rajasthan witness that decline in water levels by more than 8m in parts of Jodhpur, Nagaur, Jalore, Jhunjhunu, Jaipur and Dausa districts. Decline in water levels between 4m and 8m has been found in parts of Sikar, Jhunjhunu, Jalore, Nagaur, Ajmer, Jodhpur, Jaisalmer, Barmer, Alwar, Jaipur, Dausa, Rajsamand, Karauli, Jhalawar and Bhilwara districts. In major parts of the Bikaner, Churu, Barmer, Baran, Kota, Sawai Madhopur, Dhaulpur, Bhilwara, Udaipur, Bundi and Tonk districts decline in water level is mostly less than 4m.

Annual Water Level Fluctuation

The overall rainfall in the State was more by 22.9% in 01-02 (June – May) as compared. Thirteen districts received normal rainfall (+20% to –20%). Seventeen districts received above normal rainfall (>20%) Out of these, fifteen districts received rainfall in the range 20% to 60% above normal rainfall and remaining two districts rainfall more than 60% above normal rainfall. This has resulted in rise of water levels. The deficit rainfall (-20% to -60%) was observed in two districts.

Decline in Water Level

Decline in water level less than 2m was observed in 33% of total NHS monitored in the State mainly falling in the districts of Hanumangarh, Dausa, Ganganagar, Alwar, Bharatpur, Karauli, Jaipur, Dholpur, Churu, Bundi, Sikar, Pali, Sirohi and Dungarpur. In other districts decline of this range was observed at isolated stations only. In 4% of NHS, decline in water level in the range of 2m to 4m was observed at isolated locations in Alwar, Karauli, Pali, Nagaur, Bharatpur, Sirohi, Kota, Jodhpur, Sawai Madhopur, Dhaulpur, Dausa and Jaipur districts. Fall in water level of more than 4m was observed in 5% of wells monitored and was noticed at isolated patches in Karauli, S. Madhopur, Bhilwara, Barmer, Sirohi, Alwar, Udaipur, Ganganagar, Hanumangarh, Dhaulpur, Jalore, Dausa, Bikaner and Jaipur districts.

Rise in Water Level

Rise in water levels, less than 2m, was observed in 40% of wells monitored in the State falling in all the districts. Rise in water levels between 2m to 4m was recorded in 13% of NHS falling mainly in Banswara, Rajsamand, Baran, Udaipur, Chittorgarh and Dungarpur districts. Isolated locations in Ajmer, Jodhpur, Barmer, Nagaur, Bhilwara, Jhalawar and Jalore districts exhibited situation. In 5% stations, rise in water level more than 4m was observed in isolated stations in the districts of Rajsamand, Ajmer, Bhilwara, Kota, Jodhpur, Udaipur, Jalore and Bundi.

GROUND WATER UTILISATION

The water balance computations use the computed figures of natural ground water replenishment as an input to the ground water storage and additional inputs come as return flows, thus increasing the supply capacity of aquifers. This increase is growing with time, as more return flows come from the growing domestic, industrial and agricultural water utilization. Drafts from the ground water storage are restrained to the limits of aquifer storage and monthly withdrawal capacity.

Thus the water balance computations invariably results in a balanced exploitation of aquifers on average, with over-utilization and under utilization in the high and low dependability climatic states respectively, when higher or lower than average ground water withdrawals were necessary to meet demands, because surface water availability was reduced or increased in these climatic states. This reflects the concept of conjunctive use of surface and ground water which may be encouraged through the development of artificial recharge works in the identified areas.

The assessed mean natural replenishment of all the aquifers in Rajasthan is about 13.14 BCM (10.64 MAF) per year. Acute over exploitation of aquifers is evidenced by constantly falling water levels and increase in ground water salinity in the northern basins (Shekhawati, Ruparail, Sabi), in parts of Luni and Banas basins and in part of the outside basin. Exclusively high ground water levels, water logging of soils and salinity are a growing concern in the parts of IGNP and Chambal Projects, as a result of high percolation from leaking canals and in places the presence of a shallow hard pan beneath the surface.

GROUND WATER DEVELOPMENT

Block wise analysis of development status of ground water when categorized into safe to over exploited stage indicates that 55% of all the blocks are currently being over exploited whereas only 15% of the blocks fall under safe

Redcliff Line

Godawan

Rohira

Chinkara

Khejari

Keoladeo National Park

Machia Park

Ranthambhore-Bangal Tiger

Mukundra hills (darrah) National Park

Sarishka

Mung Crops

Bajra

Soybean

Mango Tree

Mustard

Rose Plantation

Wheat Field

Soil Erosion

Pond

Canal

Tube Well

Jawahar Sagar Dam

Mahi Bajaj Sagar

Rana Pratap Sagar Dam

Bisalpur Dam

Gang Canal

Jakham Dam

Indira Gandhi Canal

Chambal River Rajasthan

Mahi River Rajasthan

Pichola Lake

Sambhar Lake

Jaswantpur Jalore

Mountain Rain

Sandstorm

Mount Forest

Raghunathgarh

Chambal Revins

Gurushikhar

Rajasthan Dry Forest

Jalore Granite Mines

Makrana

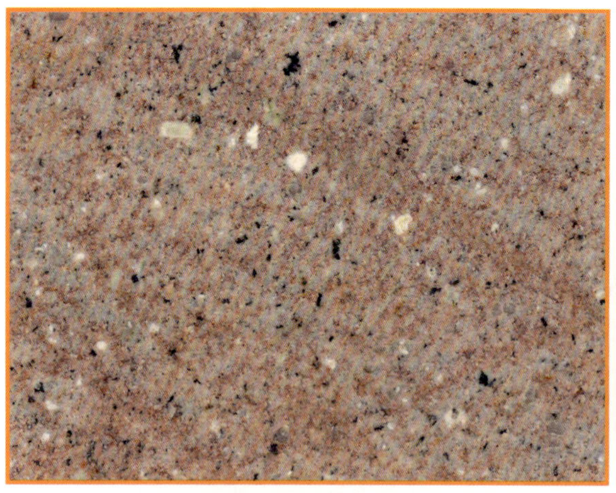

Jalore Granite

State Secretriate

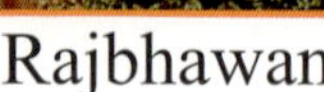

Rajbhawan

Asembly

Fairy Queen-World`s
Oldest Steam Locomotive

Railbus Service in Rajasthan
Merta City-Merta Road

Desert Festival, Jaisalmer

Tribal Women

Sephard

Rajasthan Solar

Jodhpur Sanstone Mines

Watershed

Eco-Friendly Friends

Live Stock Care

Irrigation

Live Stock

Mangala Processing Terminal (Barmer)

Ceramic Glass Factory

Lignite Mines Bikaner

Jodhpur Sandstone house

Sonar Fort (Jaisalmer)

Lodurva Jain Temple (Jaisalmer)

Pushkar Lake

Goram Ghat

Wind Mills (Jaisalmer)

Suratgarh Super Therminal Power Station (Sri Ganganagar)

Cement Industry, Nimbahera(Chittorgarh)

category. About 13% of the blocks have been notified where any further development of ground water is not permitted (Figure 11.1, 11.2 and 11.3).

Categorization on the basis of development of GW	% age of State	Block Name
Safe	13.33%	Anandpuri, Anupgarh, Aspur, Bagidora, Banswara, Bap, Barmer, Bhadra, ChhotiSarwan, Ganganagar, Garhi, Gharsana, Ghatol, Hanumangarh, Karanpur, Lunkaransar, Nohar, Padampur, Peepalkhoont, Pilibanga, Raisinghnagar, Rawatsar, Sadulshahar, Sajjangarh, Sangaria, Sardar Shahar, Shahbad, Suratgarh, Tibi
Semi-Critical	9.05%	Bichhiwara, Churu, Dungarpur, Fatehpur, Itawa, Keshorai Patan, Khajuwal, Kishanganj, Kolayat, Kotra, Kushalgarh, Ladpura, Luni, Pali, Phagi, Ratangarh, Sagwara, Simalwara, Sultanpur, Todaraisingh, Tonk
Critical	14.46%	Sindhari, Chohtan, Kaman, Deeg, Bayana, Nagar, Bari, Sam, Khanpur, Nadoti, Karauli, Nagaur, Ladnu, Rohat, Sumerpur, Dhariawad, Pindwara, Abu Road, Deoli, Kherwara, Rishabhdeo, Jhadol, Sarada, Lasadiya, Salumbar.
Over Exploited	50.40%	Ahore, Alsisar, Amet, Antah, Arain, Arnod, Asind, Atru, Bakani, Balesar, Bali, Balotra, Bamanwas, Bandikui, Banera, Bansur, Baran, Bargaon, Bari Sadri, Baseri, Begun, Bhadesar, Bhainsrorgarh, Bhim, Bhinay, Bhinder, Bhopalsagar, Bikaner, Bonli, Bundi, Chaksu, Chhabra, Chhipabarod, Chhoti Sadri, Chitalwana, Dag, Danta Ramgarh, Dausa, Degana, Deogarh, Desuri, Dhaulpur, Dhorimanna, Didwana, Dudu, Dungargarh, Dungla, Gangapur, Gangrar, Girwa, Gogunda, Hindaun, Hindoli, Hurda, Jahazpur, Jaisalmer, Jaitaran, Jamwa Ramgarh, Jaswantpura, Jawaja, Jayal, Jhalrapatan, Kapasan, Kathumar, Kekri, Khairabad, Khamnor, Khandar, Khandela, Khetri, Kishangarh, Kishangarh Bas, Kotkasim, Kotputli, Kotri, Kumbhalgarh, Kumher, Lachhmangarh, Lalsot, Mahwa, Makrana, Malpura, Mandal, Mandalgarh, Mandawar, Manohar Thana, Marwar Junction, Masuda, Mavli, Nadbai, Nainwa, Neem Ka Thana, Neemrana, Niwai, Nokha, Parbatsar, Phalodi, Piprali, Pirawa, Pratapgarh, Railmagra, Raipur, Rajakhera, Rajgarh, Rajsamand, Ramgarh, Rani, Rashmi, Reni, Reodar, Riyan, Rupbas, Sahara, Sangod, Sankra, Sapotra, Sawai Madhopur, Sewar, Shahpura, Sheo, Sheoganj, Shergarh, Sikrai, Sirohi, Siwana, Sojat, Srinagar, Sujangarh, Suwana, Talera, Taranagar, Thanagazi, Tijara, Umren, Uniara, Viratnagar, Weir
Over Exploited (Notified)	11.66%	Amber, Bassi, Bawari, Baytoo, Behror, Bhinmal, Bhopalgarh, Bilara, Buhana, Chirawa, Chittaurgarh, Dhond, Govindgarh, Jalor, Jhotwara, Jhunjhunun, KuchamanCity, Mandor, Merta, Mundwa, Nawalgarh, Nimbahera, Osian, Peesangan, Rajgarh, Raniwara, Sambhar, Sanchore, Sanganer, Sayla, Shahpura, Srimadhopur, Surajgarh, Todabhim, Udaipurwati
Saline	1.10%	Khajuwal, Taranagar

Source: Ground Water Department, GoR, Jodhpur

Basis for categorization:

Ground water development	Categorization
<= 70%	Safe
70% - 90%	Semi critical
90% - 100%	Critical
> 100% Over exploited	> 100% Over exploited

In **Notified Blocks**, development of ground water is not permitted any more.

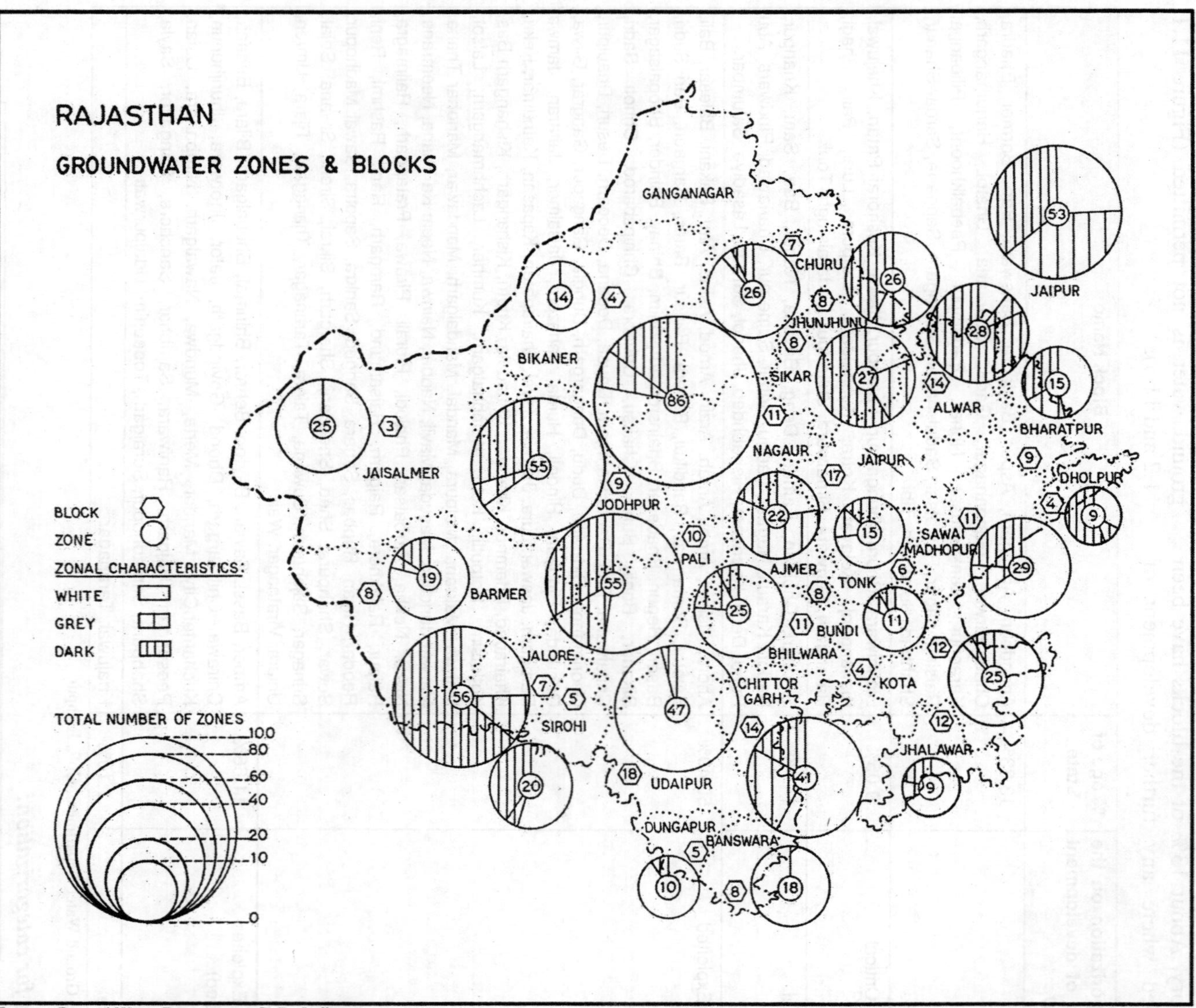

Source: Ground Water Department, GoR, Jodhpur

Figure 11.1 Rajasthan Ground Water Zones and Block

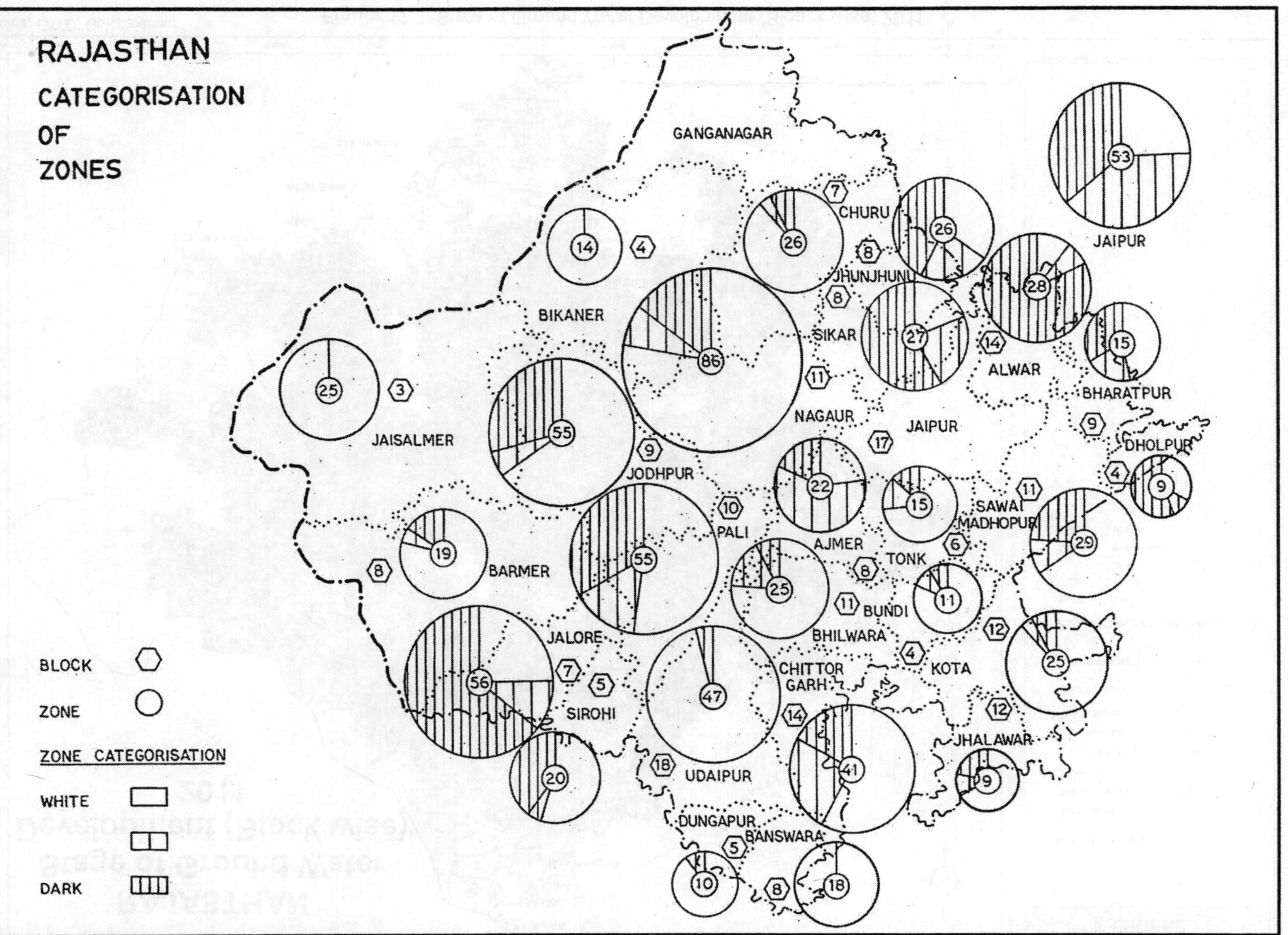

Source: Ground Water Department, GoR, Jodhpur

Figure 11.2 Rajasthan Categorisation of Zones

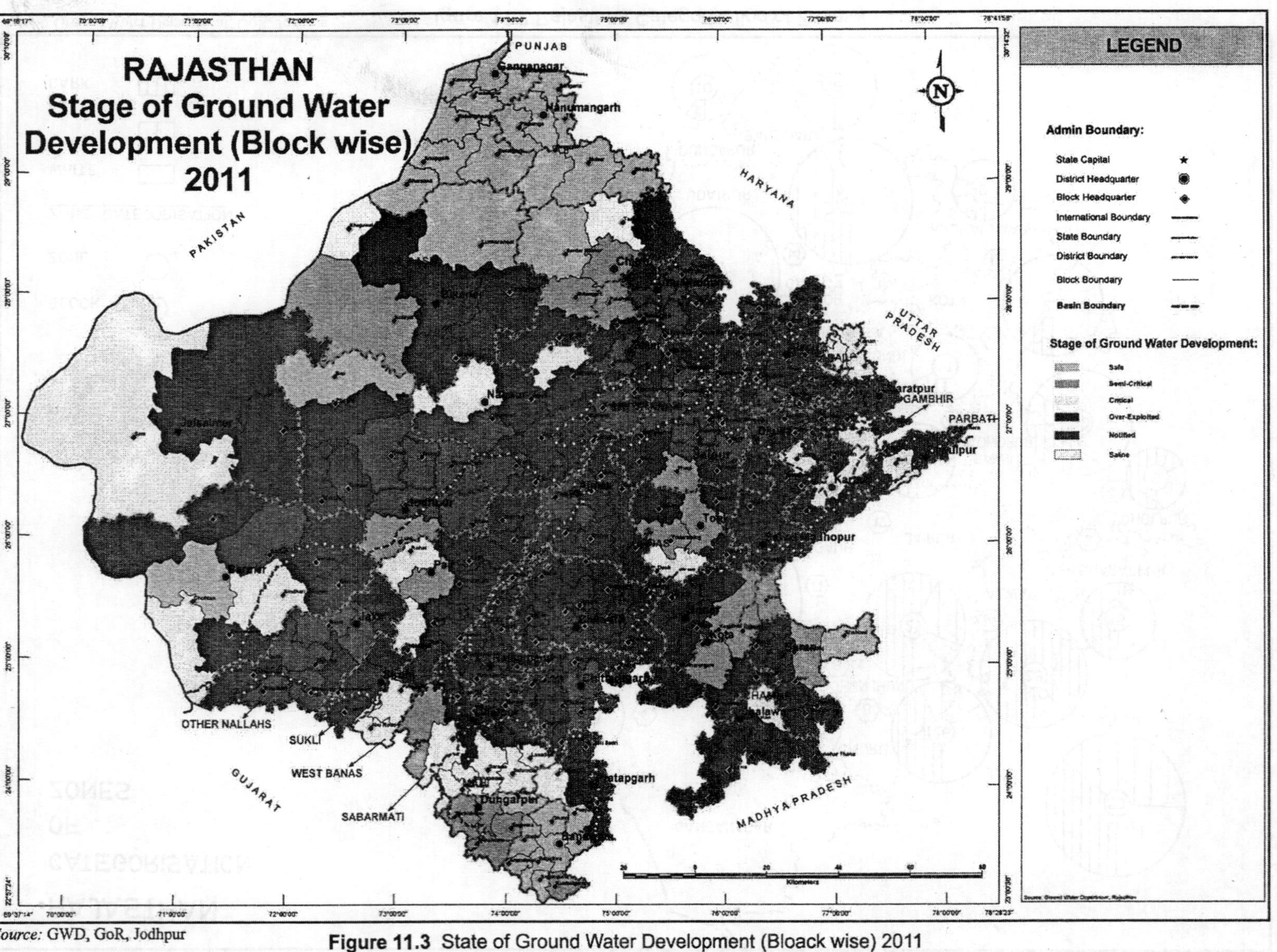

Source: GWD, GoR, Jodhpur

Figure 11.3 State of Ground Water Development (Bloack wise) 2011

DEPTH TO WATER LEVEL (PRE MONSOON – 2010)

Depth to water level in the state varies from less than 10m below ground level to more than 130m. A quick comparison of this Figure 11.4 with rainfall distribution reveals a somewhat direct relationship, i.e., the areas that receive better rainfall have groundwater occurring at shallower depths whereas in rain deficient areas water occurs at far greater depths. In addition, around major cities the water has generally gone down as seen around most of the district head-quarters. East of Aravalli, the general depth to water level is less than 30m, except around major cities; and in the western part most of the area shows deeper water table i.e., more than 50m and reaching upto 130m of depth. The shallower water levels in the west is largely restricted to the vicinity of Indira Gandhi Canal and some isolated pockets around Barmer, east of Jaisalmer and around Pali.

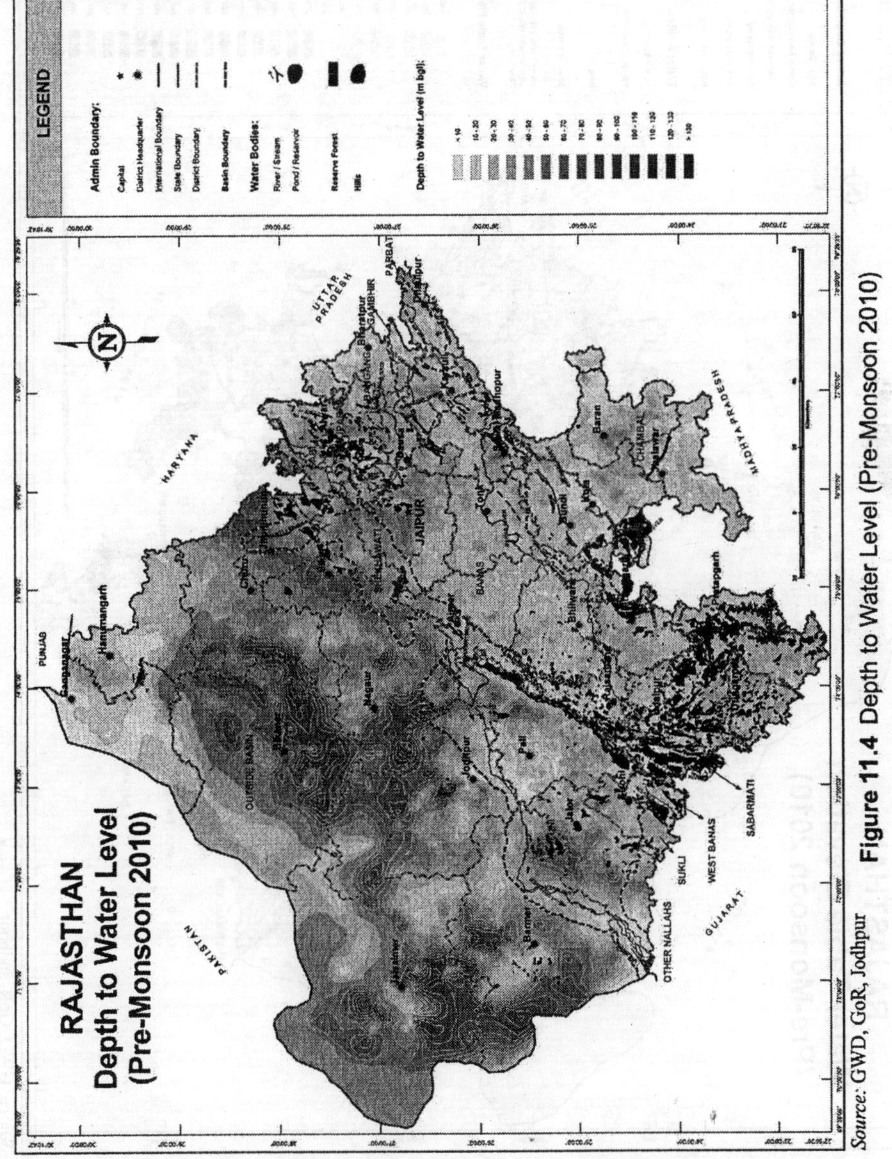

Figure 11.4 Depth to Water Level (Pre-Monsoon 2010)

Source: GWD, GoR, Jodhpur

WATER TABLE ELEVATION (PRE MONSOON –2010)

Water table elevation Figure 11.5 provide a different perspective and groundwater flow directions can be visualized from this map. Aravallis act as regional flow barriers and ground water flows towards east to the east of this range and to the west in the western part. In Mt. Abu area, maximum water table elevation of about 920m amsl was observed and minimum elevation of less than 40 m amsl is found in the western and southwestern part of the state bordering with Pakistan and Gujarat.

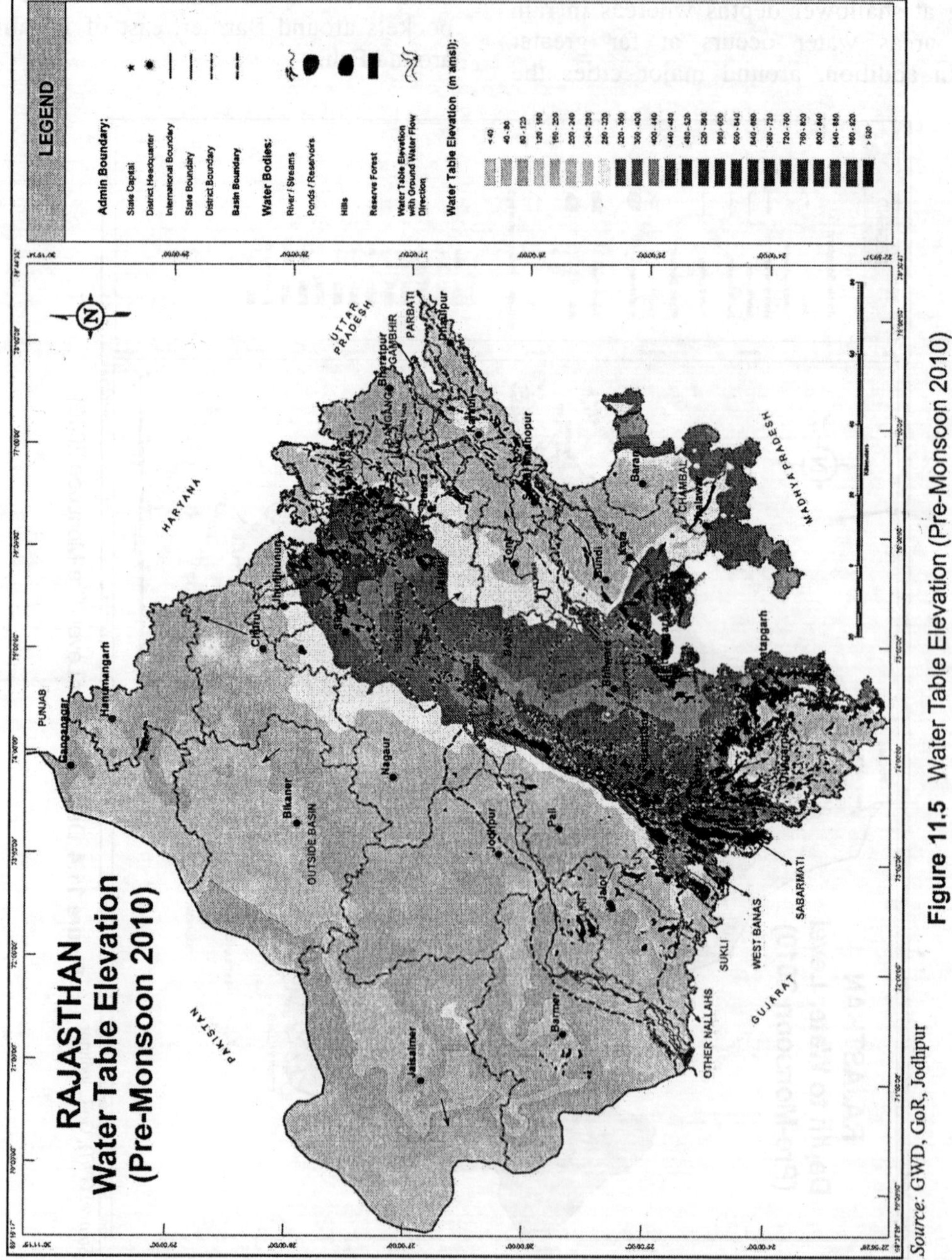

Figure 11.5 Water Table Elevation (Pre-Monsoon 2010)

Source: GWD, GoR, Jodhpur

WATER LEVEL FLUCTUATION (PRE TO POST MONSOON 2010)

Difference in water level between pre-monsoon and post monsoon measurements bring forth areas that had shown (Figure 11.6) increase or decrease in ground water storage, by the end of monsoon period. The –ve fluctuation areas indicate that water level has gone down in post monsoon season i.e., ground water in those areas are being withdrawn from static ground water resources and the situation is alarming. Such situation persists in major part of western Rajasthan and in pockets north of Jaipur and around Alwar, Kota, Baran etc. Significant decline of water table by 6-8m and more is indicated near Bikaner, Barmer, Hanumangarh, Karauli etc. However, a general rise of water level by 2m and upto 6m is observed over the larger part of the areas in the state, and some localized pockets have shown significant rise of more than 10 m and upto 18m west of Bhilwara, Bundi, Tonk and adjacent to the Aravalli range, largely in hardrock aquifers.

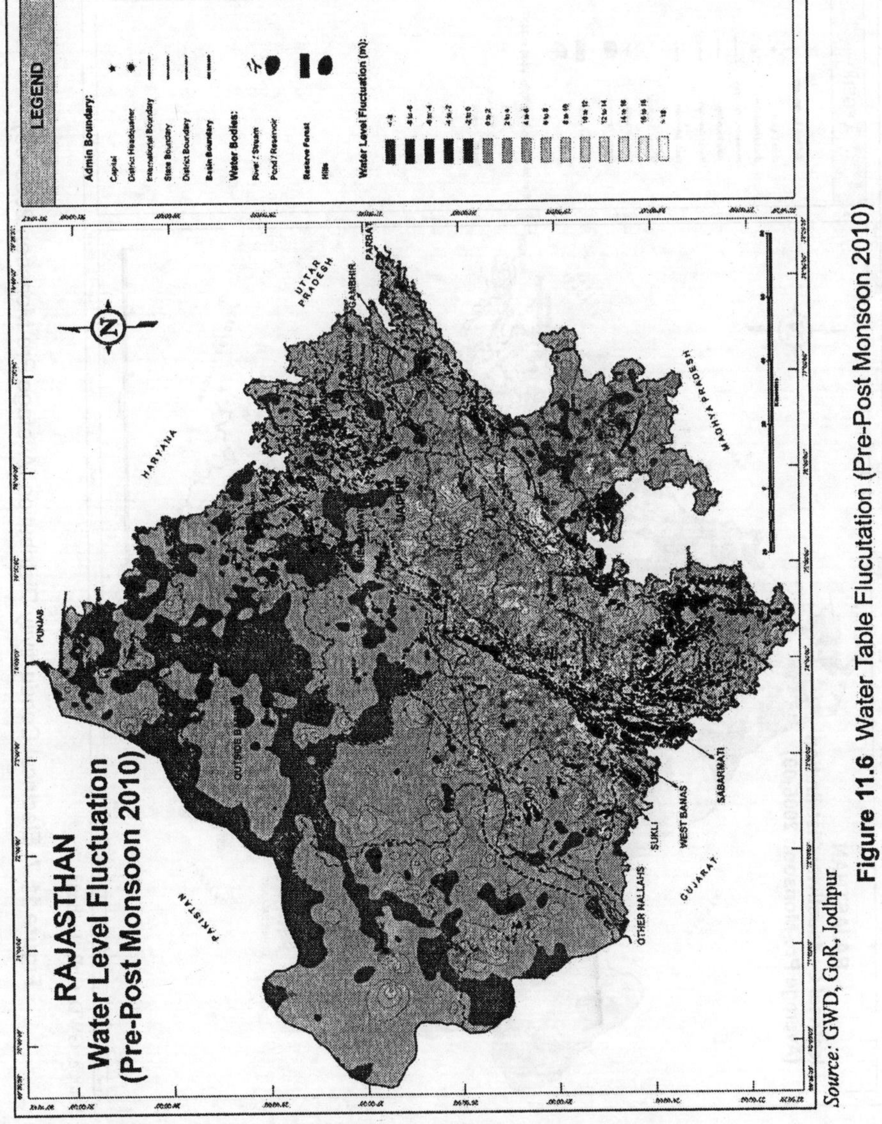

Source: GWD, GoR, Jodhpur

Figure 11.6 Water Table Flucutation (Pre-Post Monsoon 2010)

QUALITY OF GROUND WATER

Ground Water Electrical Conductivity Distribution

High Electrical Conductivity values are seen largely in western Rajasthan especially around Pali, Barmer, Jodhpur, west of Bikaner, Jaisalmer and Chru with some moderately high salinity areas in between. Except for some very high salinity areas around Bharatpur and in the region between Jaipur, Ajmer, Tonk and Bhilwara; areas in the eastern Rajasthan (around Bundi, Kota, Baran and Jhalawar) and southeastern parts around Banswara and Dungarpur indicate the presence of fresh potable groundwater.

Average values of EC (mS/cm at 25°C) from measurements between 2005 and 2009 have been plotted and EC ranges selected for depiction are in line with suitability for human drinking purpose (Figure 11.7).

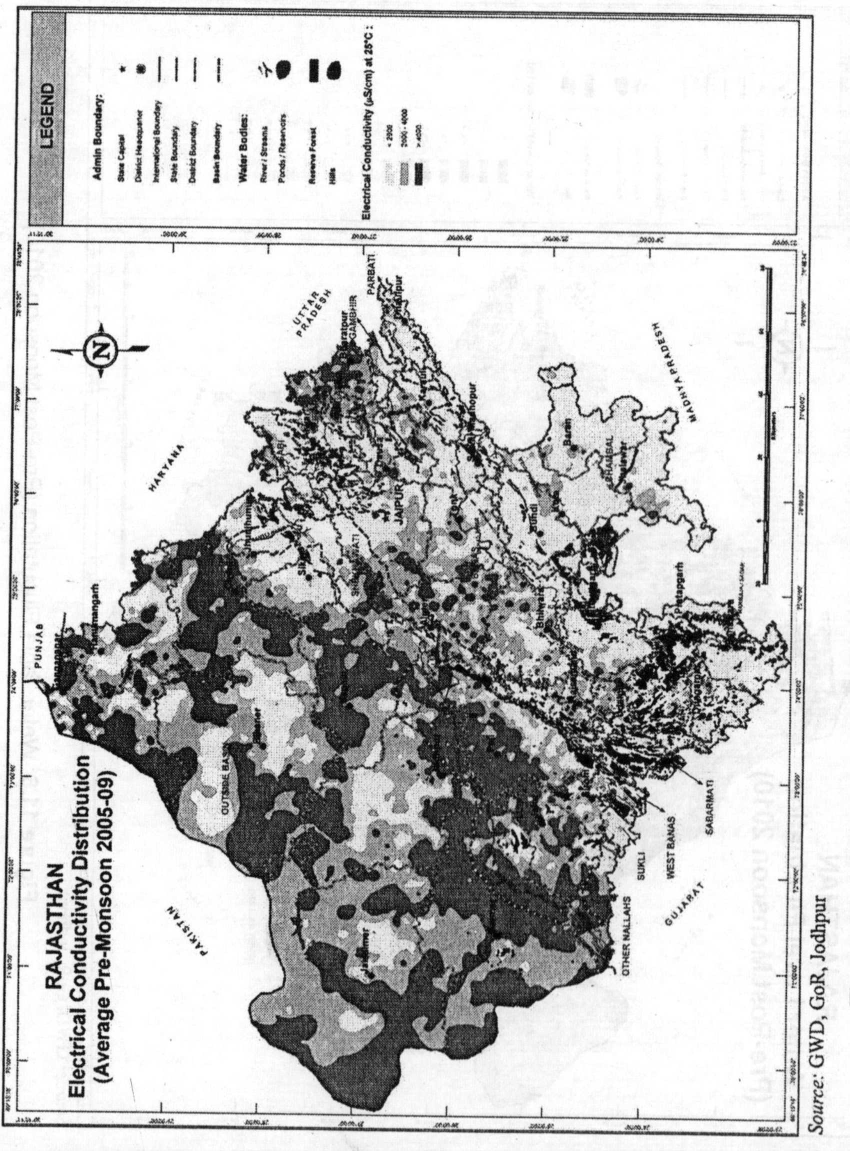

Source: GWD, GoR, Jodhpur

Figure 11.7 Electrical Conductivity Distribution (Avefage Pre-Monsoon 2005-09)

Ground Water Chloride Distribution

The chloride distribution also follows the similar distribution pattern as that of Electrical conductivity in the State. However, ranges of values selected here for presentation are in line with suitability for drinking purpose by humans based on chloride concentration. Banswara, Bundi, Karauli, Kota, Pratapgarh and to a good extent Sikar, Udaipur, Jhalawar and Jhunjhunun have ground water chloride concentration largely within potability limits. Barmer, Bharatpur, Jaisalmer, Churu etc. have very high chloride concentration in ground water (Figure 11.8).

Here also, an average chloride concentration of 2005 – 2009 is used for study.

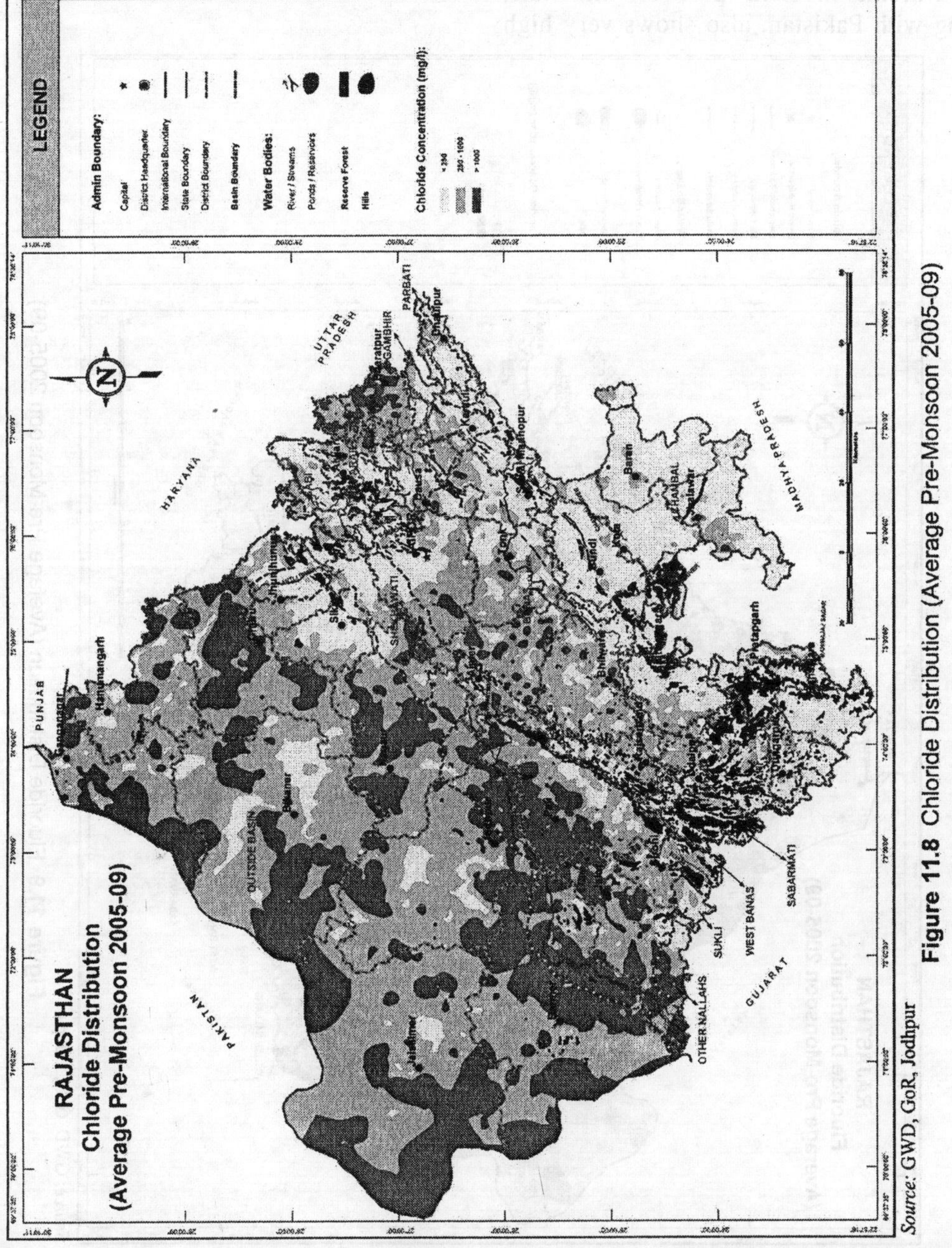

Figure 11.8 Chloride Distribution (Average Pre-Monsoon 2005-09)

Source: GWD, GoR, Jodhpur

Ground Water Fluoride Distribution

Fluoride distribution (Figure 11.9) depicts fluoride concentration as observed in ground water samples of wells in the state. High fluoride concentration can be seen on both sides of Aravalli range in Tonk-Alwar-Bhilwara region in the east, whereas, Nagor-Jodhpur-Pali-Jalor belt in the west of the range. In the extreme western part of the state, bordering with Pakistan, also shows very high concentration of fluoride that renders water unusable for human drinking. Apart from specific localized pockets, in general, the areas around Jhalwar-Kota-Baran-Bundi-Chittaurgarh region, Dhaulpur-Karauli-Dausa-Kirauli belt and that around Churu, Bikaner, Barmer and Jaisalmer have by and large have shown very low concentration of fluoride in ground water.

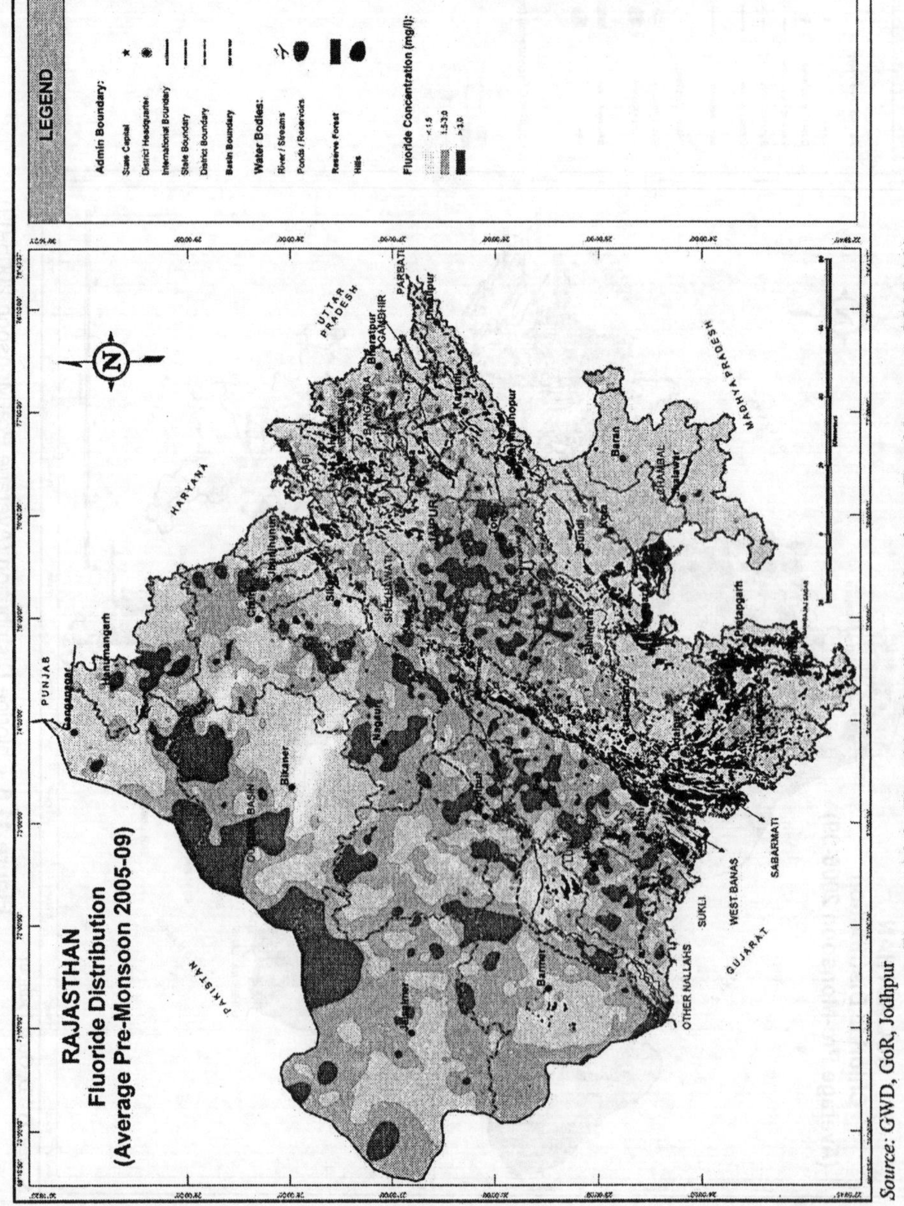

Source: GWD, GoR, Jodhpur

Figure 11.9 Fluoride Distribution (Avefage Pre-Monsoon 2005-09)

Ground Water Nitrate Distribution

The state shows, in general, high concentration of Nitrate in groundwater. The red colour regions represent >100 mg/l of Nitrate in groundwater which is considered unsuitable for agriculture. Major part of western Rajasthan falls under high Nirtrate category and good part of Bhilwara-Rajsamand-Udaipur-Dungarpur belt has indicated presence of high Nitrate concentration. In other parts also, pockets of high Nitrate concentration are found widely distributed all over. The areas around Jhalawar-Bundi-Baran belt, around Bikaner, south of Jaisalmer and scattered areas in rest of eastern Rajasthan, low to moderate concentrations of Nitrate are found (Figure 11.10).

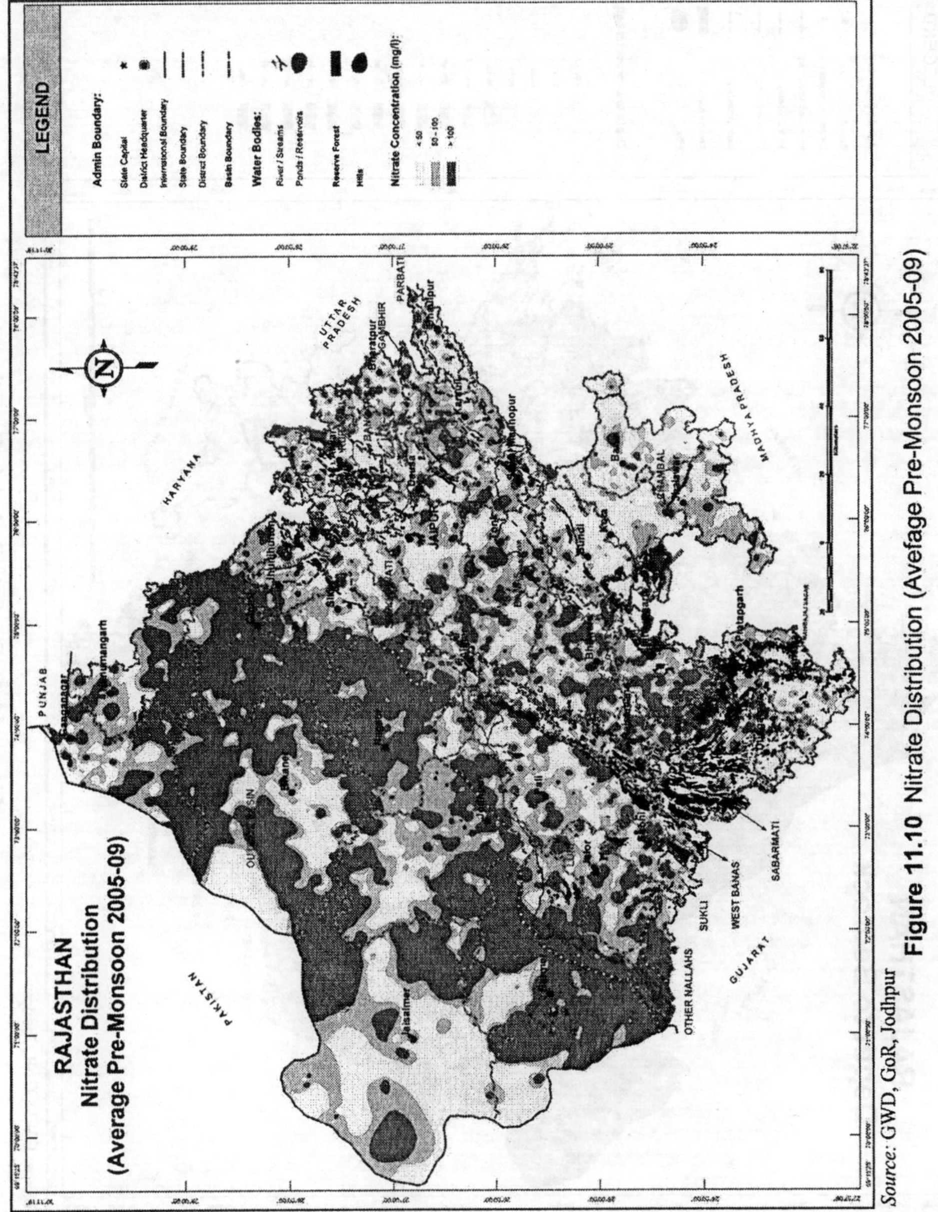

Figure 11.10 Nitrate Distribution (Avefage Pre-Monsoon 2005-09)

Source: GWD, GoR, Jodhpur

DEPTH TO BEDROCK

From hydrogeological perspective, the beginning of massive bedrock has been considered for defining top of bedrock surface. These rocks are overlain by alluvial deposits of sand, clay, silt and admixture of these in different proportions and thicknesses along with the weathered and fractured zones of the underneath massive bedrock. A perusal of map and data reveals that eastern Rajasthan has shallower bedrock depth whereas in the western part the bedrock occurs at greater depths reaching upto 300m bgl (Figure 11.11).

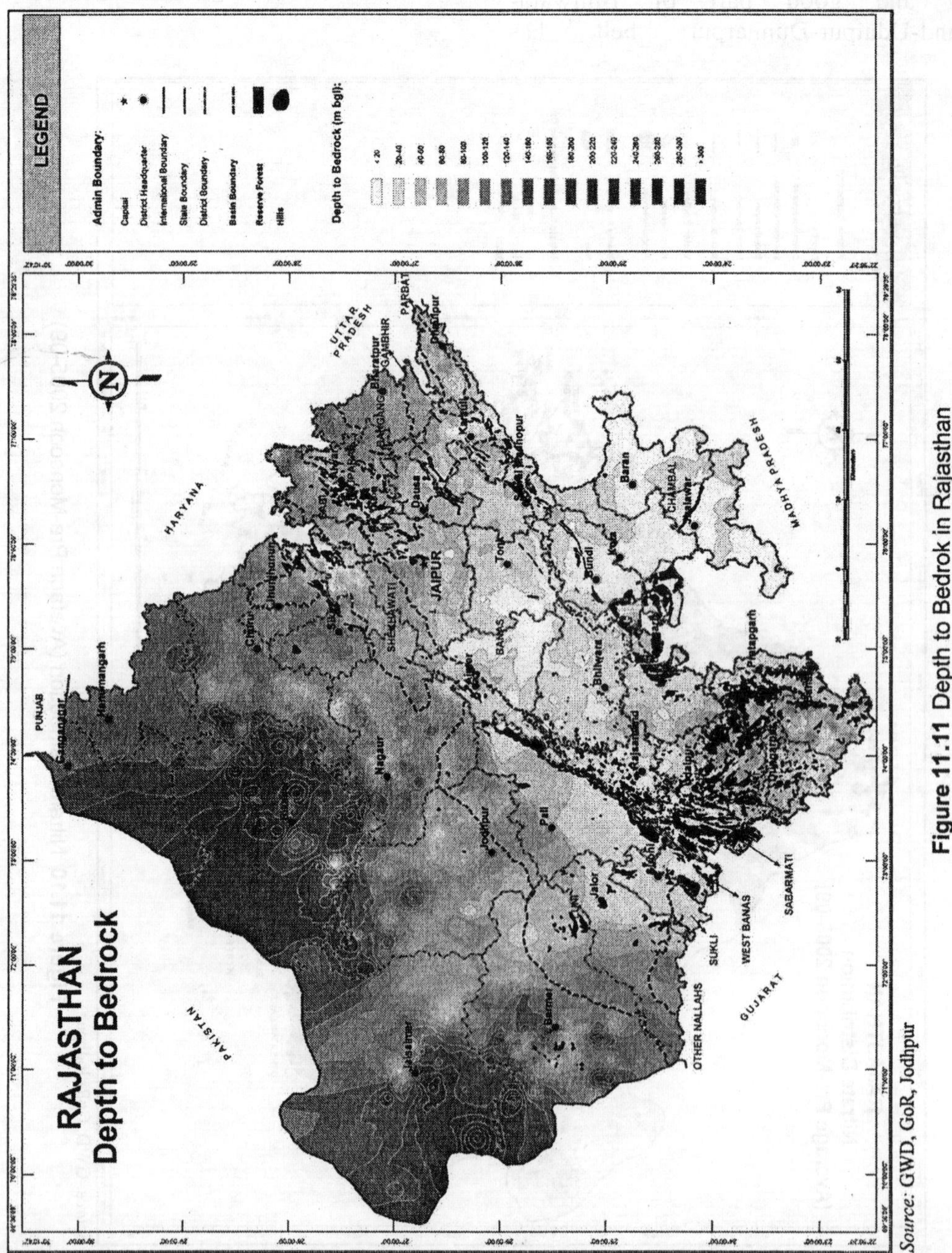

Figure 11.11 Depth to Bedrok in Rajasthan

Source: GWD, GoR, Jodhpur

UNCONFINED AQUIFER

Unconfined aquifer in alluvial areas

Alluvial aquifers occur mostly in north, northeast and southwestern parts of the state. These attain significant thickness in Hanumangarh, Jalor, Bikaner and Barmer districts whereas spatially they are most prominent aquifers in Bikaner, Barmer, Jaisalmer, Ganganagar, Hanumangarh, Jalor and Churu where their total area is around 9,000 sq kms or more reaching upto 15,400 sq kms (in Bikaner) (Figure 11.12).

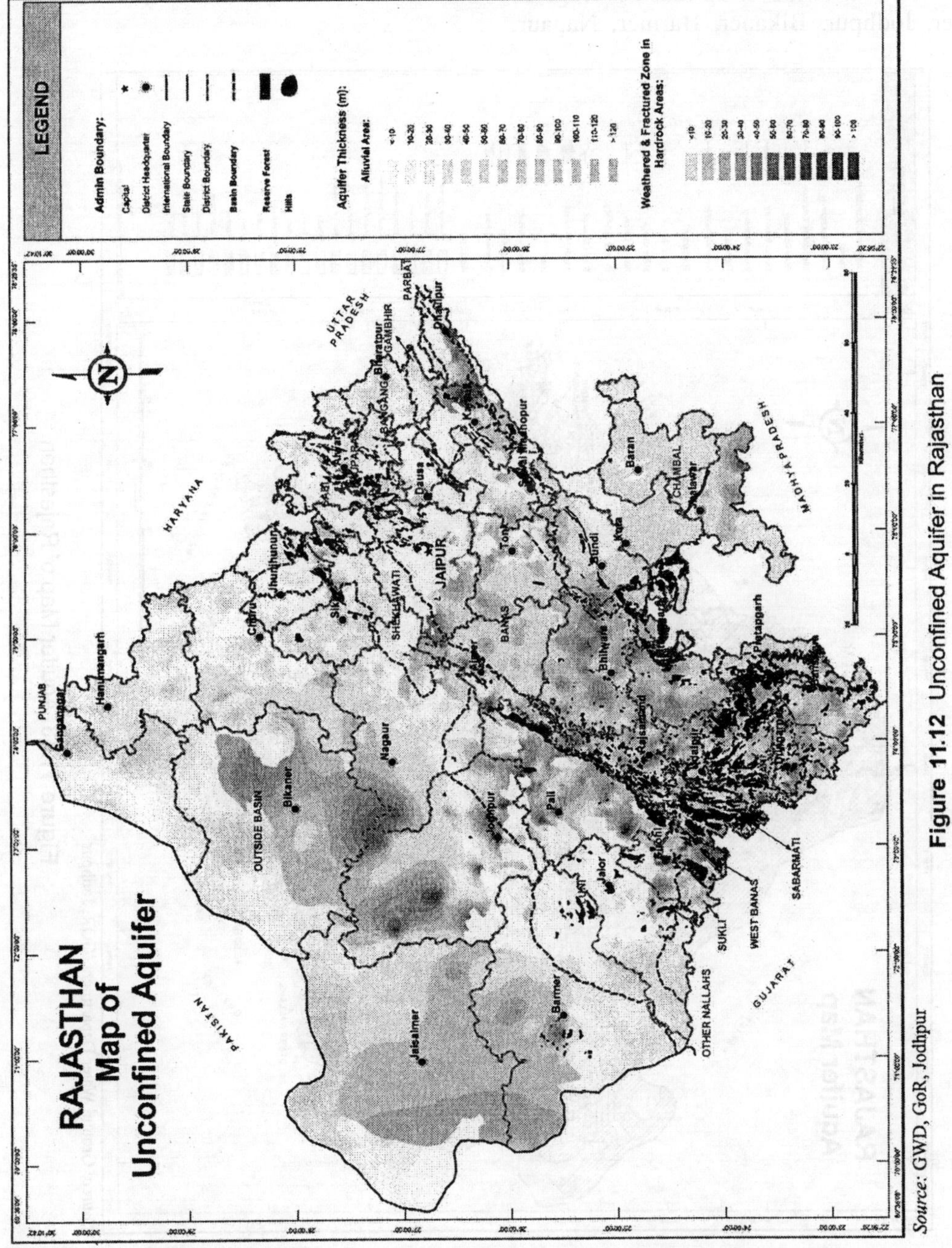

Figure 11.12 Unconfined Aquifer in Rajasthan

Source: GWD, GoR, Jodhpur

Unconfined Aquifer Hardrock Areas

The weathered and fractured hardrocks (both consolidated and semi-consolidated) also constitute good aquifers in Central, southeast and western parts of the state. While Pali (90-100m thickness), Dhaulpur, Rajsamand seem to have significant thickness of such zones, they are significant in terms of spatial coverage in Jaisalmer, Jodhpur, Bikaner, Barmer, Nagaur, Bhilwara etc. having more than 9,000 sq km area and reaching upto 24,300 sq km (in Jaisalmer). It is also important to note that Bhilwara, Udaipur, Chittaurgarh, Jhalwar, Banswara, Rajsamand, Dungarpur etc. do not have any significant alluvial aquifers and ground water is mostly drawn from hardrock aquifers only (Figure 11.13).

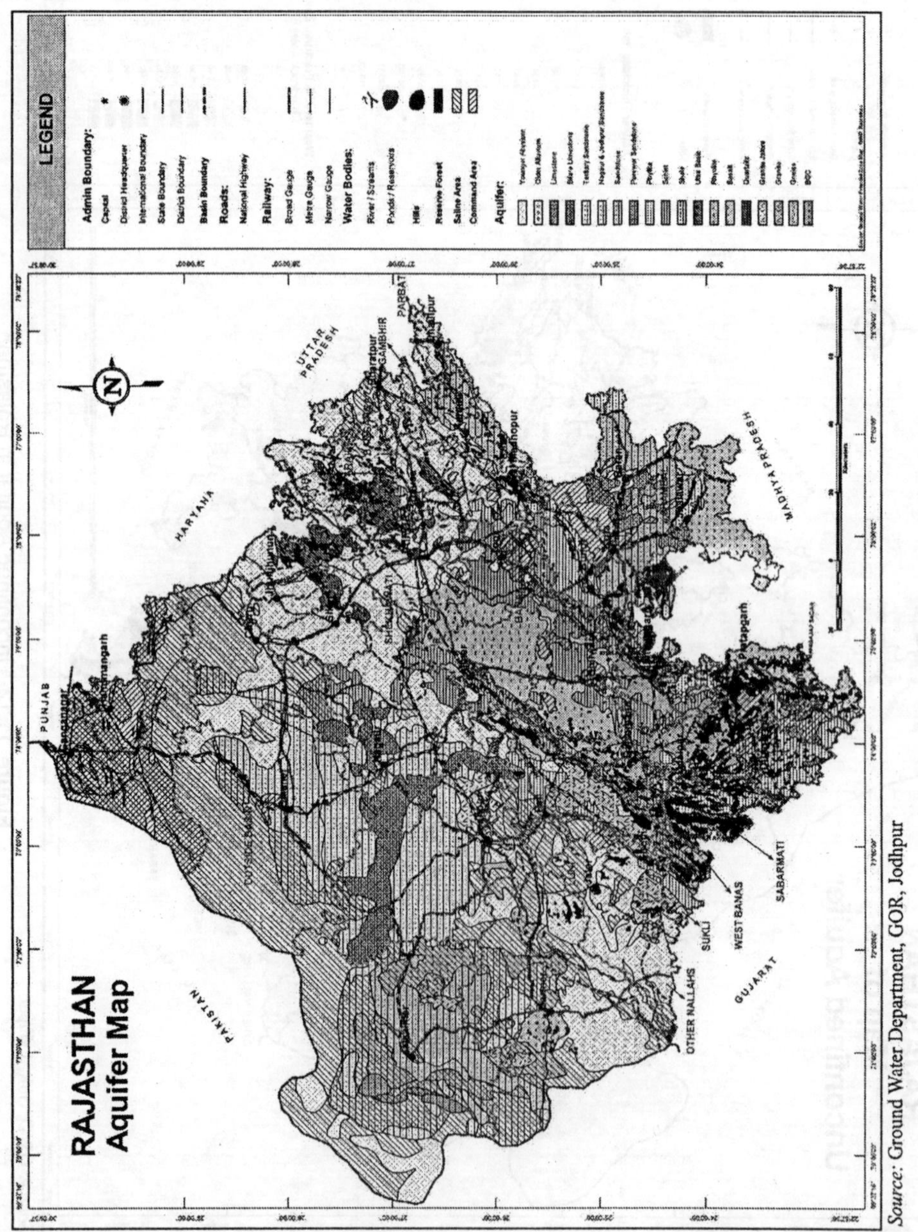

Source: Ground Water Department, GOR, Jodhpur

Figure 11.13 Aquifer Map of Rajasthan

Present Ground Water Scenario

Systematic and regular monitoring of ground water levels brings out the changes taking place in the groundwater regime. The maps so generated are of immense help for regional groundwater flow modelling which serves as a groundwater management tool to provide the necessary advance information to the user agencies to prepare contingency plans in case of unfavourable groundwater recharge situation. The data also has immense utility in deciding the legal issues arising out of conflicting interests of groundwater users.

Water level data of the collected during the year 2016-2017 has been utilized to prepare various maps showing depth to water level and fluctuation of water level. Depth to water level maps is useful in dealing with problems of water logging and artificial recharge, where the relative position of water level with reference to the ground surface is of critical importance. Water level fluctuation maps (rise or fall) are indispensable for estimation of change in storage in the aquifer.

The data is analysed for each set of measurement, and report prepared which include following maps to understand the groundwater regime in the State.

- Depth to water level maps
- Seasonal fluctuation maps- water level fluctuation in comparison to pre-monsoon.
- Annul fluctuation maps - water level fluctuation in comparison to same month in the previous year.
- Decadal fluctuation maps - water level fluctuation in the month of measurement with reference to the decadal average for the same month.

Depth to Water Level (Unconfined Aquifer)

An analysis was done to understand the water level behaviours of the NHS stations in the different categories of the water levels during every monitoring period and the same is depicted in Figure 11.14. Shallow water level 0 to 2 m bgl observed in 3% to 15% of majority of stations whereas deeper water level recorded in 18% to 19% of the stations in the year 2016-17.

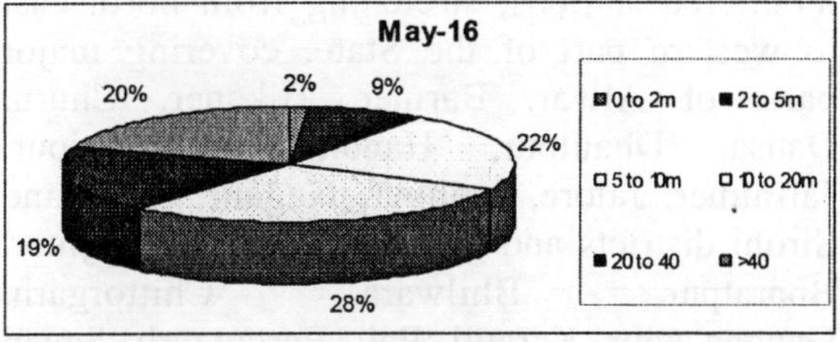

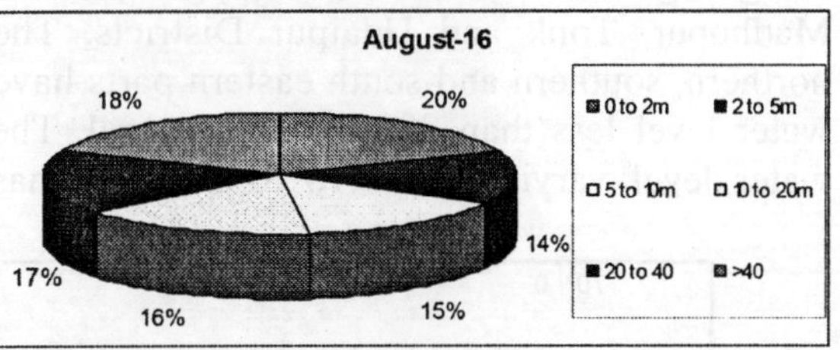

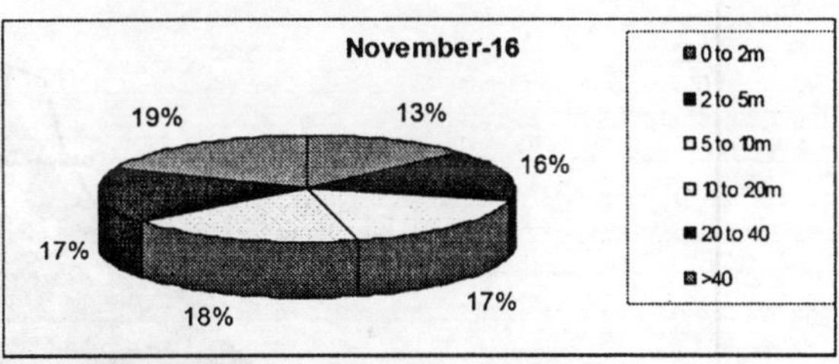

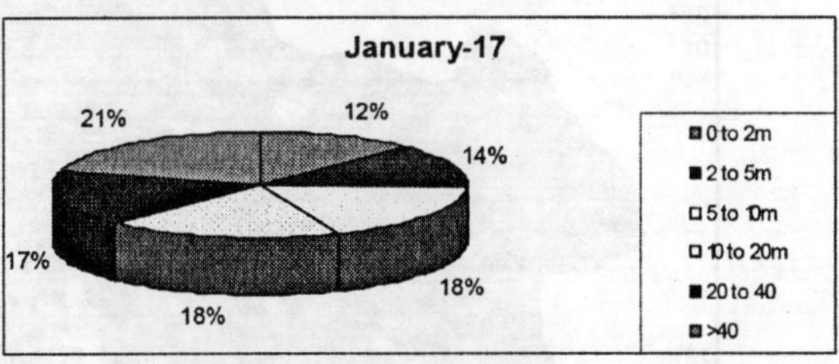

Source: CGWD, Jaipur

Figure 11.14 Distribution of Wells In Different Categories of Water Levels (2016-17)

Pre-Monsoon Period

Total number of wells analysed are 917. (Figure 11.15) displays a wide variation of ground water level during the monitored period of May, 2016 in Rajasthan State. It varies from 0.35 mbgl (at Rawatbhata in Chittorgarh district) to 120.45mbgl (at Kathora in Jaisalmer district). Water levels of more than 40 m bgl in 19.63% of stations fall mostly in the districts of

Barmer, Bikaner, Churu, Jaipur, Jaisalmer, Jalore, Jhunjhunu, Sikar and Jaipur representing north-central and western part of the state. Depth to water level between 20 to 40 mbgl has been recorded in 18.97% of monitored stations, stretching from north east to western part of the State, covering major parts of Alwar, Barmer, Bikaner, Churu, Dausa, Dhaulpur, Hanumangarh, Jaipur, Jaisalmer, Jalore, Jodhpur, Nagaur, Sikar and Sirohi districts and in small patches in Ajmer, Bharatpur, Bhilwara, Chittorgarh, Ganganagaur, Karauli, Pali, Partapgarh, Sawai Madhopur, Tonk and Udaipur Districts. The northern, southern and south eastern parts have water level less than 20 mbgl in general. The water level varying from 10 to 20 mbgl has been recorded at 29.01% of monitored stations falling in all the districts. The depth to water level ranging from 5 to 10 mbgl has been recorded at 22.36% of monitored stations in patches in Ajmer, Banswara, Baran, Bharatpur, Bhilwara, Bundi, Chittorgarh, Dhaulpur, Dungarpur, Ganganagar, Jhalawar, Kota, Pali, Rajsamand, Tonk, Pratapgarh and Udaipur. The water level ranging between 2 to 5 mbgl has been observed at 8.51% of monitored stations covering south and south-east and in isolated patches in northern part of State. Shallower water level of less than 2 mbgl have been seen at 1.53% of stations occurring as isolated patches only in Ajmer, Banswara, Ganganagar, Hanumangarh, Pali, Tonk and Udaipur districts.

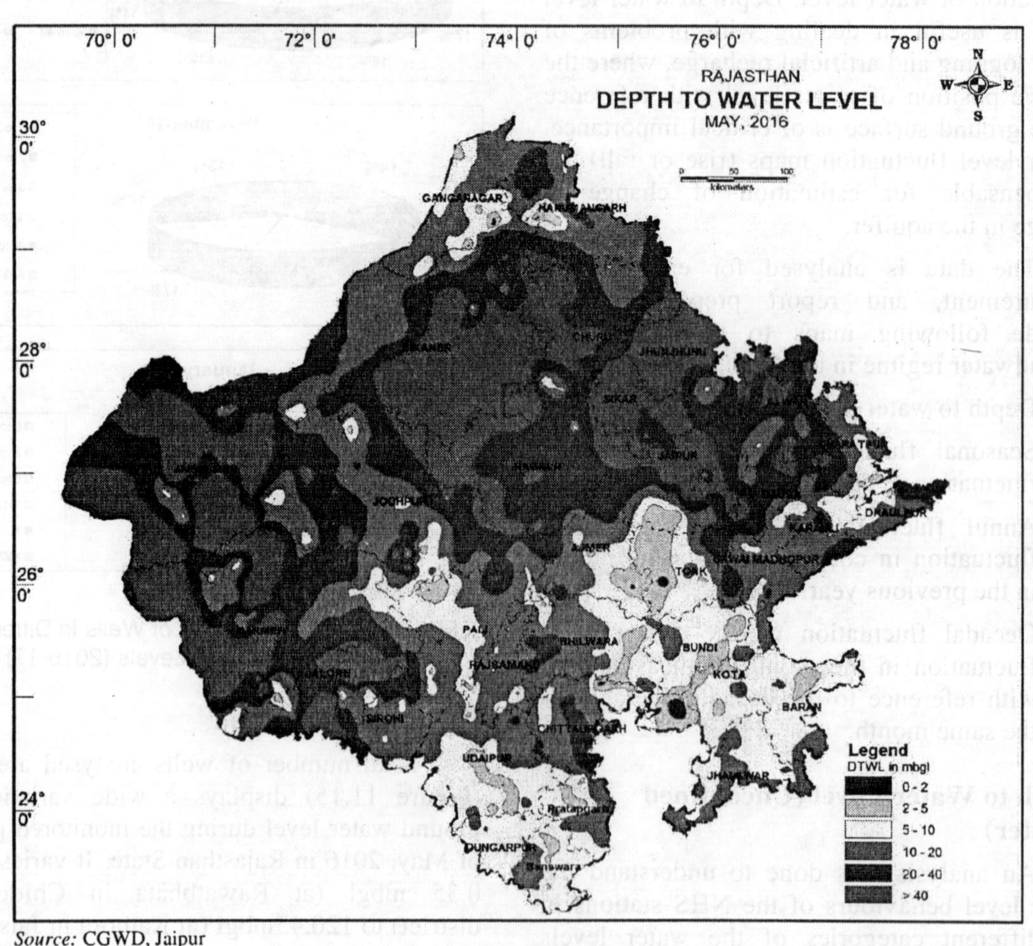

Source: CGWD, Jaipur

Figure 11.15 Depth to Water Level (May 2016)

Monsoon Period

Total number of well analyzed are 1018. A perusal of map in Figure 11.16 reveals a wide variation of ground water level during the monitored period of August, 2016 in Rajasthan State. It varies from 0.03 mbgl (Chandujikaguda in Banswara district) to 120.46 mbgl (at Kathora in Jaisalmer district). The water levels of more than 40 mbgl have been recorded at 18.37% of monitored stations in large patches extended from north east to western and upper central part, covering major parts of Jhunjhunu, Sikar, Bikaner, Jalore, Jaipur, Churu, Jaisalmer, Barmer, Alwar and Jodhpur Districts. Depth to water level between 20 to 40 mbgl has been recorded in 17.49% of monitored stations, stretching from north east to western part of the State, covering major parts of Nagaur, Alwar, Barmer, Dausa, Churu, Jaisalmer, Bikaner, Barmer, Jalore, , Bharatpur and Ganganagar Districts. The water level vary from 10 to 20 mbgl in 16.11% of monitored stations spread in all the districts except Banswara, Dungarpur, Udaipur, Bundi, Bikaner, Kota, Jhalawar and Pratapgarh district. The depth to water level ranging from 5 to 10 mbgl has been recorded at 14.83% of monitored stations falling in all the districts of southern and south-eastern part i.e east of Aravllis. The water level ranging between 2 to 5 mbgl has been observed at 14.34% of monitored stations covering south, south east and isolated patches in northern part of State. Shallower water level of less than 2 mbgl have been seen at 18.86% of stations occurring in Banswara, Dungarpur, Pratapgarh, Udaipur, Rajsamand, Pali, Baran, Kota and Tonk districts.

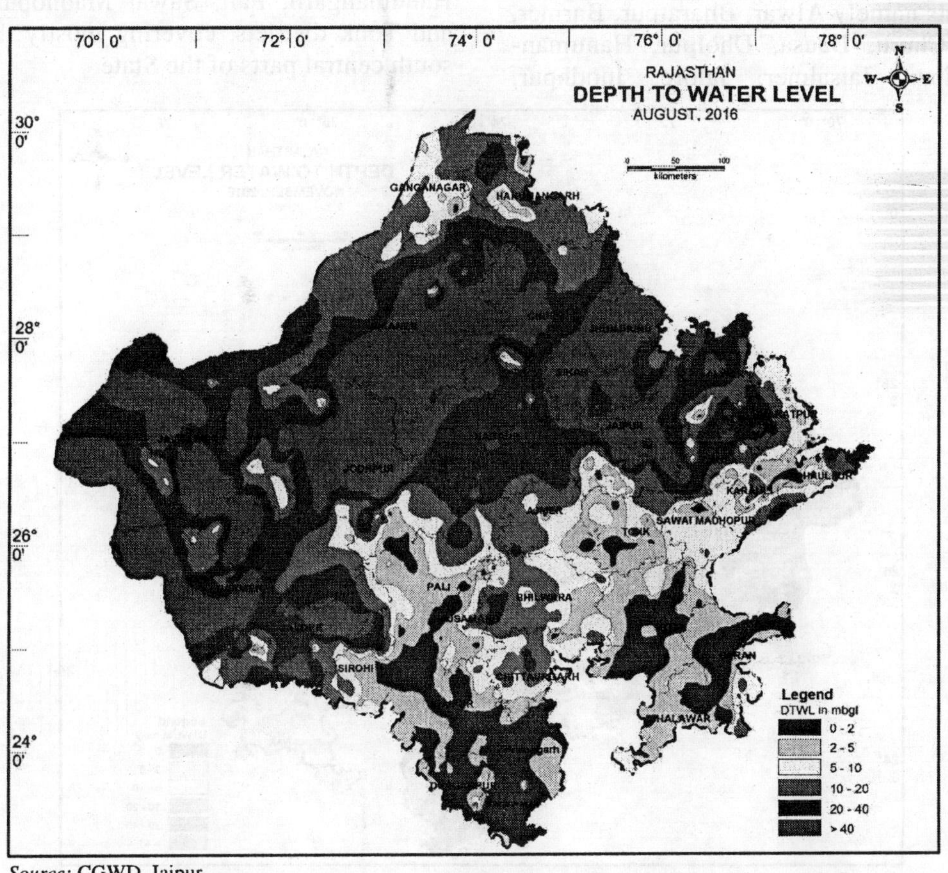

Source: CGWD, Jaipur

Figure 11.16 Depth to Water Level (August 2016)

Post Monsoon Period

Total number of wells analysed are 998. Perusal of the map (Figure 11.17) shows wide variation of ground water level during the monitored period in the State of Rajasthan, which varied from 0.12 m at Chirwa in Udaipur district to 122m at Sadhsar in Bikaner district. Large patches of water levels of more than 40 mbgl have been recorded in 18.54% stations representing north east-central and western part of the state, extending from north east to western part covering major parts of the districts of Barmer, Bikaner, Churu, Jaipur, Jaisalmer, Jalore, Jhunjhunu, Nagaur, Sikar and in small patches in Alwar, Bharatpur and Jodhpur districts. Depth to water level between 20 and 40 mbgl. have been recorded in 16.73% stations in upper central half and north east to western part of the state in stretches in most of the districts namely Alwar, Bharatpur, Barmer, Bikaner, Churu, Dausa, Dholpur, Hanumangarh, Jaipur, Jaisalmer, Jalore, Jhodhpur, Nagaur, Sikar and in small scattered patches in Ajmer, Bhilwara, Chittorgarh, Ganganagar and Karauli districts. Depth to water level between 10 and 20 mbgl. have been recorded in 18.74% stations falling mainly in all the districts except Banswara, Dungarpur, Udaipur, Jhunjhunu, Kota, Baran, Jhalawar and Pratapgarh districts. Depth to water level between 5 to 10 m bgl have been recorded in 17.33% stations in patches in all the districts except Jhunjhunu, Nagaur and Sikar district. Depth to water level between 2 and 5 mbgl have been recorded in 15.53% stations in most of the districts in small isolated pockets except Bikaner, Churu, Dausa, Jhunjhunu, Nagaur and Sikar districts. Water level less than 2 m bgl have been recorded in 13.13% stations in isolated pockets in Ajmer, Banswara, Dungarpur, Udaipur, Baran, Chittorgarh, Dhaulpur, Ganganagar, Hanumangarh, Pali, Sawai Madhopur, Sirohi and Tonk districts, covering mostly the lower south central parts of the State.

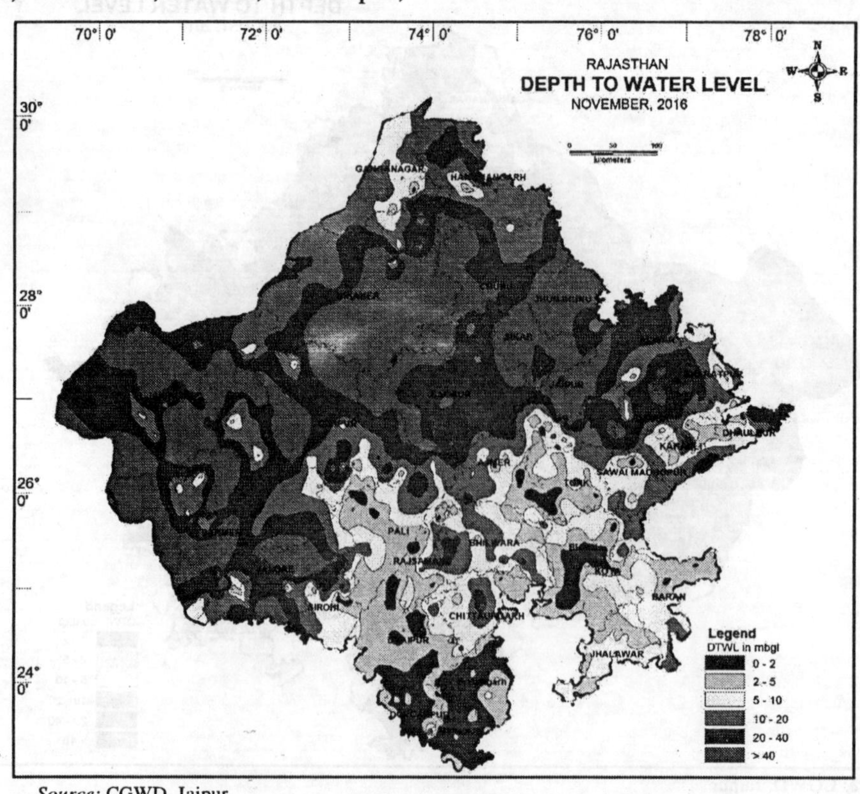

Source: CGWD, Jaipur

Figure 11.17 Depth to Water Level (November 2016)

Total number of wells analysed are 1066. The Figure 10.18 show a variation of ground water level during the monitored period of January, 2017 in Rajasthan State. It varies from 0.06mbgl at Bagidora in Banswara district to 122.1mbgl at Sadhsar in Bikaner district. The water levels of more than 40 mbgl have been recorded in 20.64% of monitored stations in large patches extending from north east to west direction, covering major parts of Jhunjhunu, Sikar, Nagaur, Churu, Bikaner, Jaipur, Jaisalmer, Barmer and Jodhpur districts. The depth to water level between 20 and 40 mbgl has been recorded in 17.26% of monitored stations, stretching from north east to western part of the State, covering major parts of Alwar, Nagaur, Churu, Jaisalmer, Barmer, Hanumangarh, Dholpur, Dausa, districts and small patches in Ajmer, Bharatpur, Chittorgarh, Ganganagar, Karauli,Sikar and Sirohi districts. The water level vary from 10 to 20 mbgl has been recorded at 18.20% of monitored stations falling in all the districts except Jhunjhunu and Banswara. Depth to water level ranging from 5 to 10 mbgl has been recorded in 18.20% of monitored stations falling in southern Rajasthan and in isolated patches in Ganganagar and Hanumangarh districts. The water level ranging between 2 to 5 mbgl has been observed in 13.60% of monitored stations covering south, south east and isolated patches in northern part of State. Shallower water levels of less than 2 mbgl has been observed in 12.10% of monitored stations falling in Banswara, Pratapgarh, Dungarpur, Udaipur, Kota and Bundi districts and in isolated patches in Rajsamand and Pali districts.

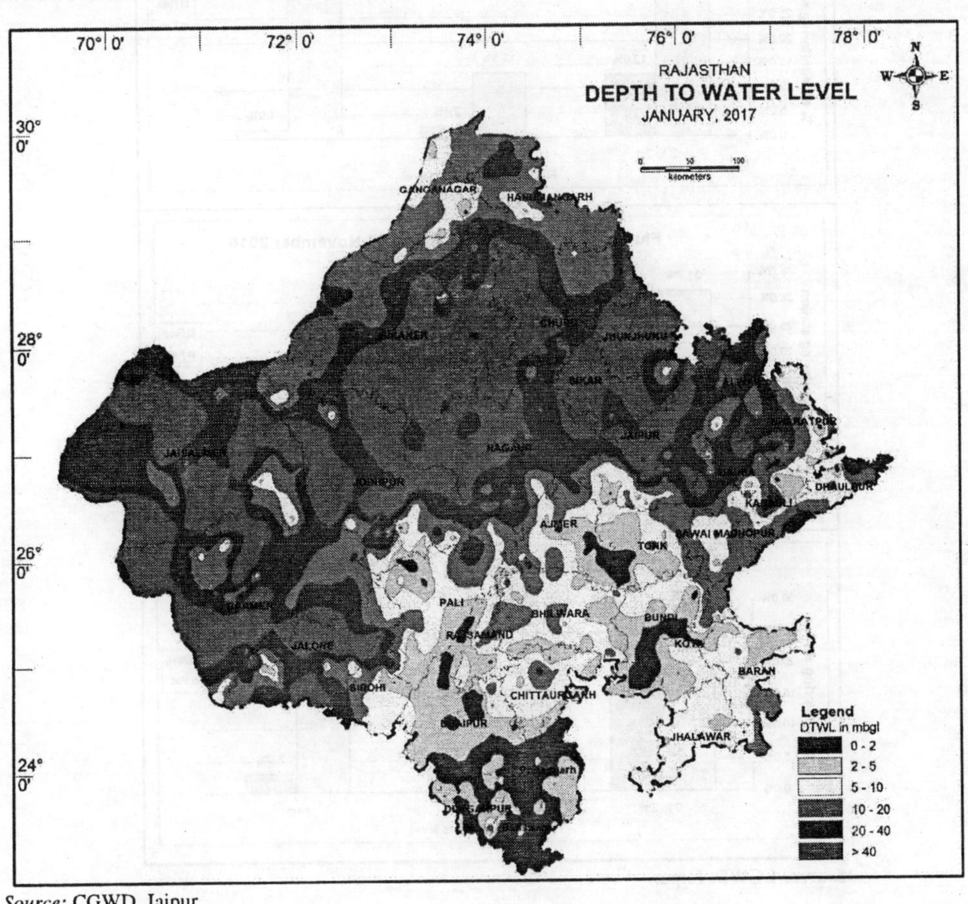

Source: CGWD, Jaipur

Figure 11.18 Depth to Water Level (January 2017)

Seasonal Water Level Fluctuation

To study effect of monsoon on the groundwater regime and subsequent utilisation of groundwater for various needs like agriculture, irrigation, Domestic etc., changes in depth to water levels with respect to May data are studied. The change in groundwater in the region over different periods is presented graphically in Figure 11.19.

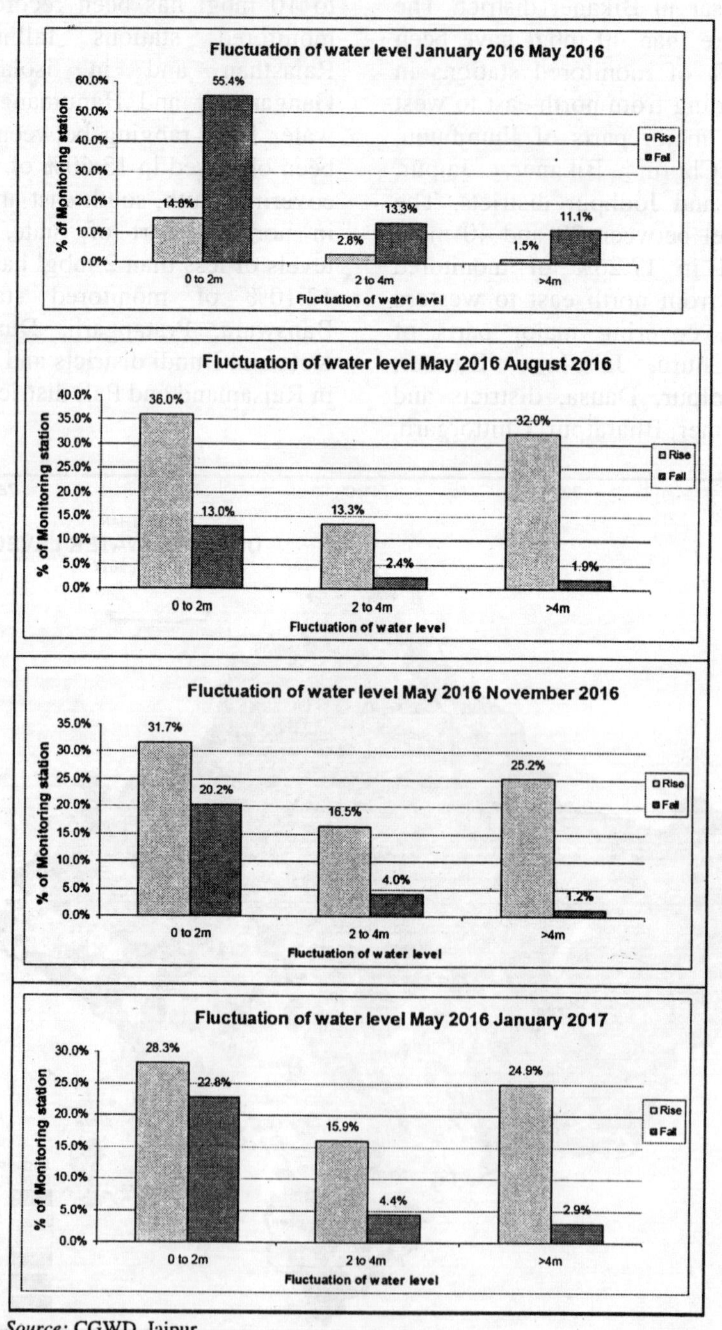

Source: CGWD, Jaipur

Figure 11.19 Seasonal Fluctuation of Water Level (2016-2017)

Pre Monsoon to Monsoon Period

Total number of wells analysed are 848. A perusal of (Figure 11.20) of seasonal fluctuation of water level from May, 2016 to August, 2016 shows that about 81.30% of the stations scattered in most of the districts in small patches have shown rise in water level during this period. The water level rise ranging between 0 to 2m has been observed at 36% of the stations, occurring in small to large patches in all the districts except Bikaner district. The water level rise of 2 to 4m has been observed at 13.3% of the stations scattered in districts falling in northern, western and south western parts in isolated patches in all the districts, except Ajmer, Bhilwara, Sikar and Nagaur districts. The rise of water level more than 4m has been observed at 32% falling in all the districts except Ajmer, Ganganagar, Jaipur and Nagaur districts. The water level fall of 0 to 2 m has been observed at 13% except Banswara, Baran, Bundi, Dhaulpur, Dungarpur, Jhalawar, Karauli, Kota, Pali, Pratapgarh, Rajsamand and Sirohi districts. The water level fall of 2 to 4 m has been observed at 2.4% of stations falling in only Alwar, Barmer, Bhilwara, Bikaner, Chittaurgarh, Churu and Jhunjhunu districts. The water level fall of more than 4 m has been recorded in small isolated patches at 1.9% of stations falling in Barmer, Bikaner, Chittaurgarh Churu Jhunjhunu,Jodhpur ,Sikar and Tonk District. In this period, a maximum rise of 23.10m has been recorded at Choti Sadri in Pratapgarh district, whereas minimum rise of 0.01 m at Ramthala in Tonk, Malawala in Jaipur and Bamboo in Churu district. A maximum fall of 11.86m has been recorded at Kapasan1 in Chittorgarh district and minimum fall of 0.01 m at Gomanwali in Gaganagar; Tigaria in Jaipur; Kathora in Jaisalmer district. Thirteen stations show no change.

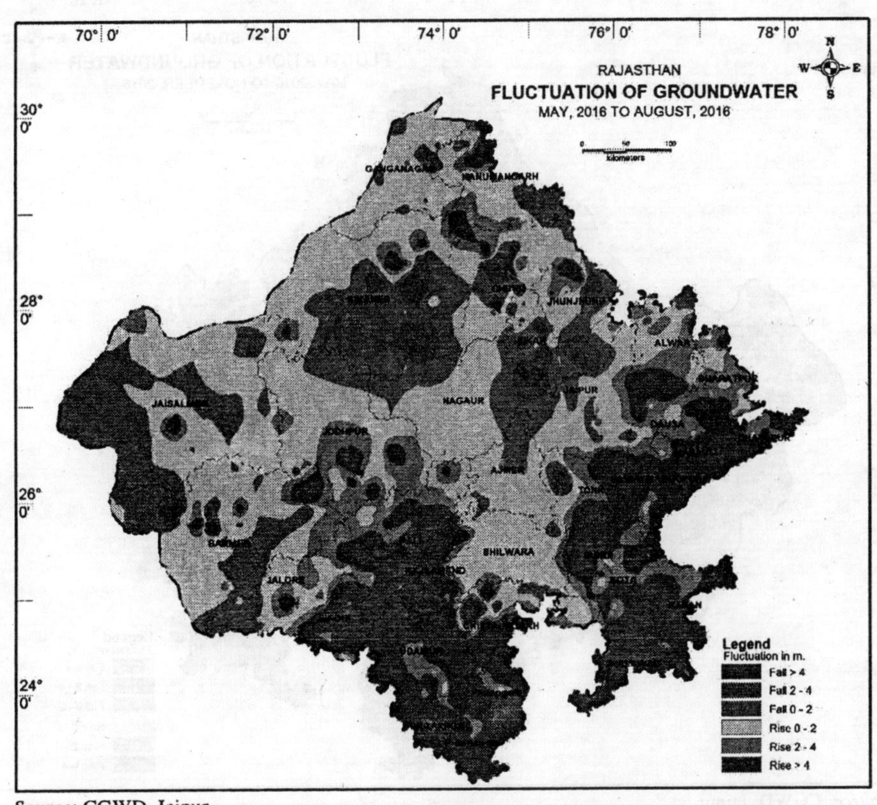

Source: CGWD, Jaipur

Figure 11.20 Water Level Fluctuation – May 2016 to August 2016

Pre to Post Monsoon

Total number of wells analysed are 826. Figure 11.21 shows about 73.4% of the stations observed rise of water level between the period of May, 2016 and November, 2016. South to south-eastern and northwest central parts of the state shows rise in water level. Rise in the fluctuation of water level between 0 to 2m are observed in 31.7% of the stations scattered in all the districts. Rise of 2 to 4 m are observed in 16.5% of the stations scattered in isolated patches in all the districts in the state except Hanumangarh and Nagaur districts. Rise of more than 4m are recorded in 25.2% of the stations falling in isolated patches in most of the districts in the State except Ajmer, Hanumangarh, Nagaur and Sikar districts. East-west and central parts in the state mostly experienced the Fall in Water Level. About 20.2% of the stations shows fall of 0 to 2m scattered in most of the districts except Banswara, Baran, Bundi, Dungarpur, Jhalawar, Pratapgarh and Sirohi districts. Fall of 2 to 4 m are observed in 4% of stations in small patches in Barmer, Bharatpur, Bikaner, Dausa, Ganganagar, Hanumangarh, Jaisalmer, Jhunjhunu,Jhodhpur,Nagaur and Sikar districts. Fall of more than 4 m occurs in the isolated patches in 1.2% of stations and scattered only in Alwar, Bharatpur, Bikaner, Churu, Hanumangarh, Jodhpur, Sikar and Tonk districts. A maximum rise of 27.32m is recorded at Posaliya in Sirohi distirct, whereas minimum rise of 0.01 m is recorded at Chhipa Barodi in Baran, Bikaner, Churu and Sikar district. A maximum fall of 11.06m is recorded at Sabalpura in Sikar district and minimum fall of 0.01m is recorded at Baishakhi in Jaisalmer district. Ten stations show no change.

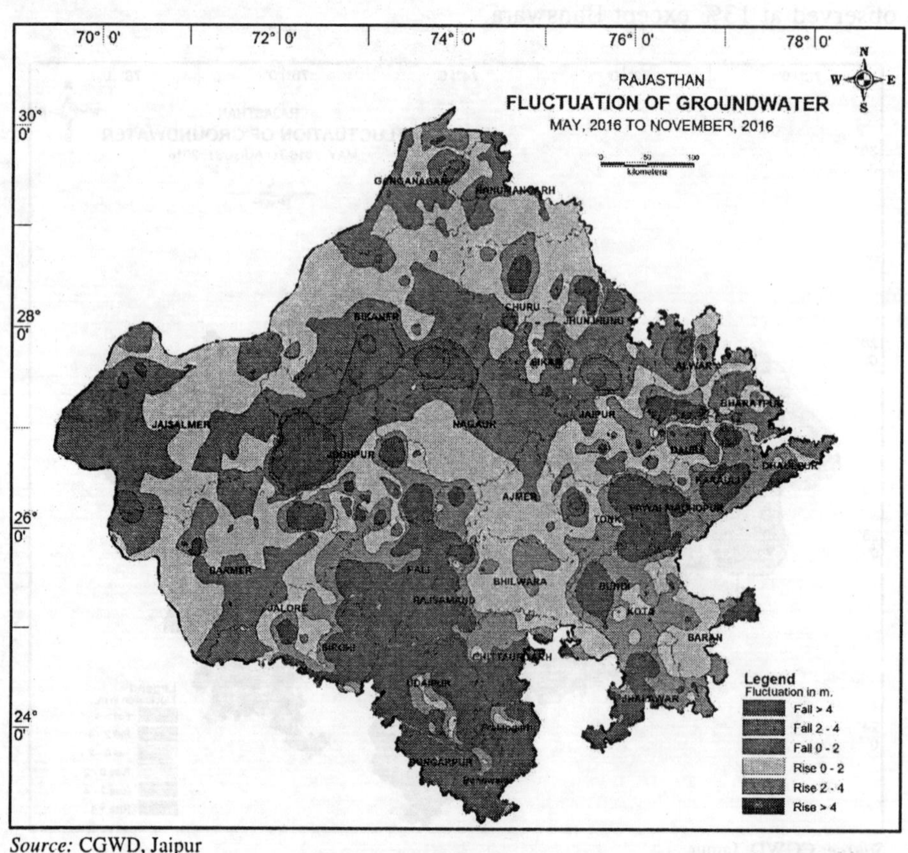

Source: CGWD, Jaipur

Figure 11.21 Water Level Fluctuation – May 2016 to November 2016

Total number of wells analysed are 859. A perusal of (Figure 11.22) of seasonal fluctuation of water level from May, 2016 to January, 2017 depicts that about 69.10% of the stations scattered in most of the districts have shown rise in water level during this period. The water level rise ranging between 0 to 2m has been observed at 28.3% of the stations, occurring in large patches in Bundi, Hanumangarh, Churu, Kota, Pali, Ganganagar, Tonk, Sikar, Baran, Bikaner and Bharatpur districts. The water level rise of 2 to 4m has been observed at 15.9% of the stations scattered in all the districts except Sikar districts. About 24.9% of stations scattered in all the districts except Dhaulpur, Ganganagar and Nagaur districts, rise of water level more than 4m has been seen during this period. The water level fall ranging between 0 to 2m has been observed at 22.8% of the stations scattered in all the districts except Ajmer, Banswara, Dungarpur, Pratapgarh and Sirohi district. The water level fall of 2 to 4 m has been observed at 4.4% of stations scattered in patches in all districts except Banswara, Baran, Bundi, Chittaurgarh, Dhaulpur, Dungarpur, Jalore, Jhalawar, Karauli, Kota, Pali, Pratapgarh, Rajasamand, Sawai Madhopur and Udaipur districts. The water level fall of more than 4 m has been recorded in small isolated patches at 2.9% of stations, majorly in Bikaner, Barmer, Churu, Dausa Ganganagar Hanumangarh Jaipur, Jhunjhunu, Jodhpur, Sikar and Sirohi districts. In this period, a maximum rise of 22.64m has been recorded at Choti Sadri in Pratapgarh district, whereas minimum rise of 0.02 m at Bai in Sikar district. A maximum fall of 20.95 m has been recorded at Bau in Sikar district and minimum fall of 0.01 m at Sewada in Jalore district during this period. Six stations show no change.

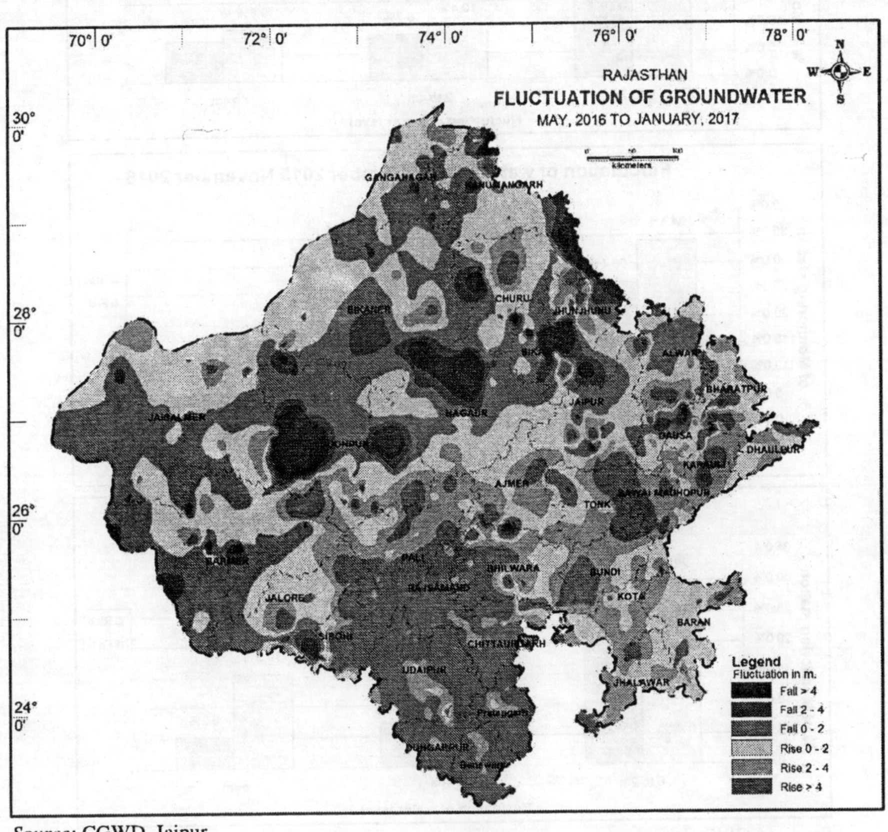

Source: CGWD, Jaipur

Figure 11.22 Water Level Fluctuation – May 2016 to January 2017

Annual Water Level Fluctuation

Annual Fluctuation in the water levels of the different monitoring periods were analysed graphically and depicted in Figure mostly fall is dominant over the rise in all annual periods (Figure 11.23).

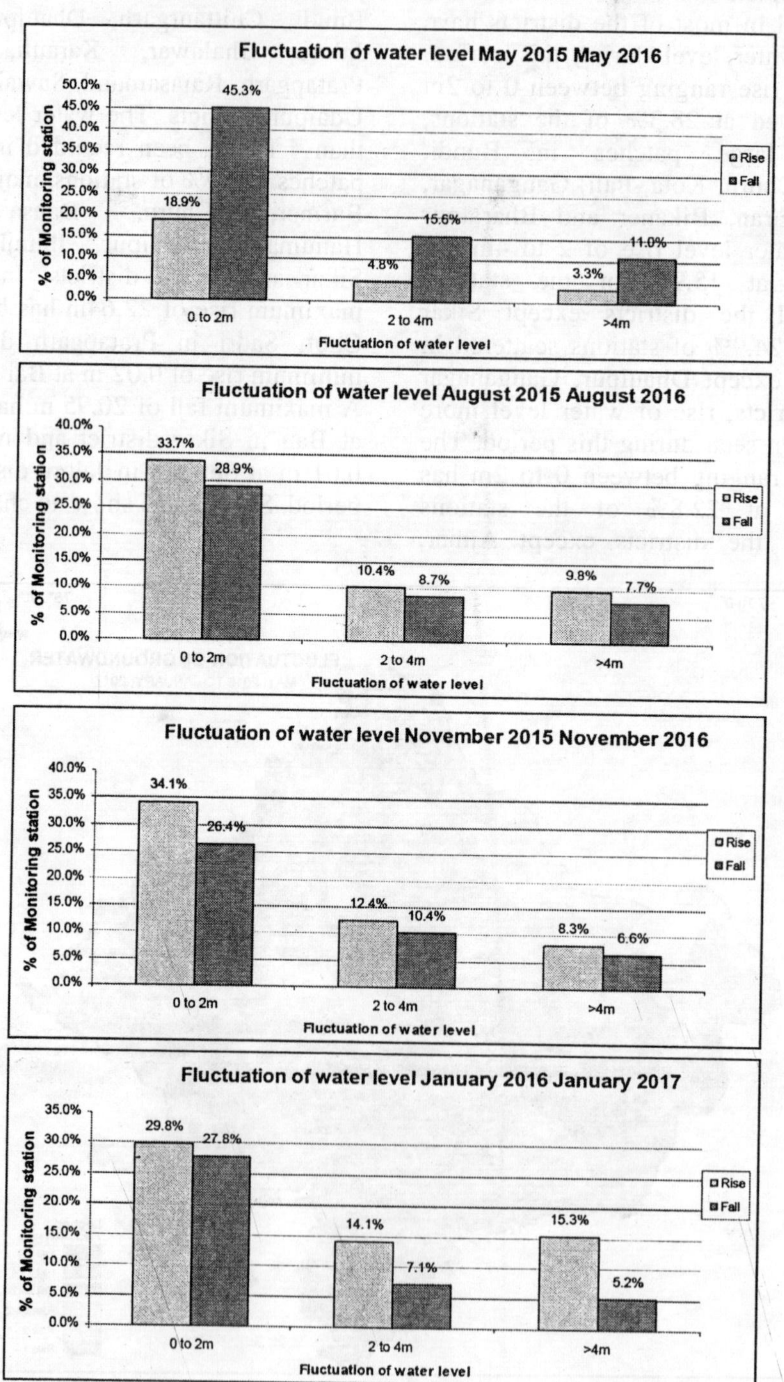

Source: CGWD, Jaipur

Figure 11.23 Annual Fluctuation in the water levels - different monitoring periods

Pre Monsoon Period

Total number of wells analysed are 748. A perusal of (Figure 11.24) for annual fluctuation of water level from May, 2015 to May, 2016 reveals that about 27% of the stations scattered mostly in south to south-western, north western, western and west central parts of the State have shown rise in water level during this period. The water level rise <2 m has been observed at 18.9% of the stations scattered in most of the districts, majorly in Baran, Barmer, Bhilwara, Bikaner, Chittorgarh, Churu, Dhaulpur, Ganganagar, Hanumangarh, Jaisalmer, Jalore, Jhalawar, Jodhpur, Nagaur, Pali, Pratapgarh, Rajsamand, Sikar, Sirohi and Tonk districts. The rise in water level between 2 and 4m in 4.8% of the stations scattered in isolated patches only in Ajmer, Banswara, Barmer, Bhilwara, Bikaner, Ganganagar, Hanumangarh, Jaipur, Jaisalmer, Jalore, Jhalawar, Jhunjhunu, Jodhpur, Karauli, Pali, Rajsamand, Sikar and Udaipur districts. The rise of water level more than 4m has been recorded at 3.3% of the stations falling mainly in Barmer, Bhilwara, Bikaner, Chittorgarh, Churu, Dholpur, Dungarpur, Ganganagar, Jalore, Nagaur and Sirohi districts. The maximum rise of 16.01 m has been recorded at Dhanwara in Jalore district and minimum rise of 0.01 m at Kasba-thana in Baran district. About 45.3% of the stations scattered in all the districts, majorly in eastern, southern, south eastern and central parts, have shown fall in water level ranging from 0 to 2 m during this period. The water level fall of 2 to 4 m has been recorded at 15.6% of stations occurring in patches in almost all the districts. The water level fall of more than 4 m has been shown by 11% of stations occurring in all the districts in isolated patches except Ganganagar, Hanumangarh, Jaisalmer, Jalore and Kota districts. The maximum fall of 22.42 m is recorded at Neema in Churu district and minimum fall 0.01m at Dungargarh in Bikaner and Gaindoli in Bundi district. Eight station show no change.

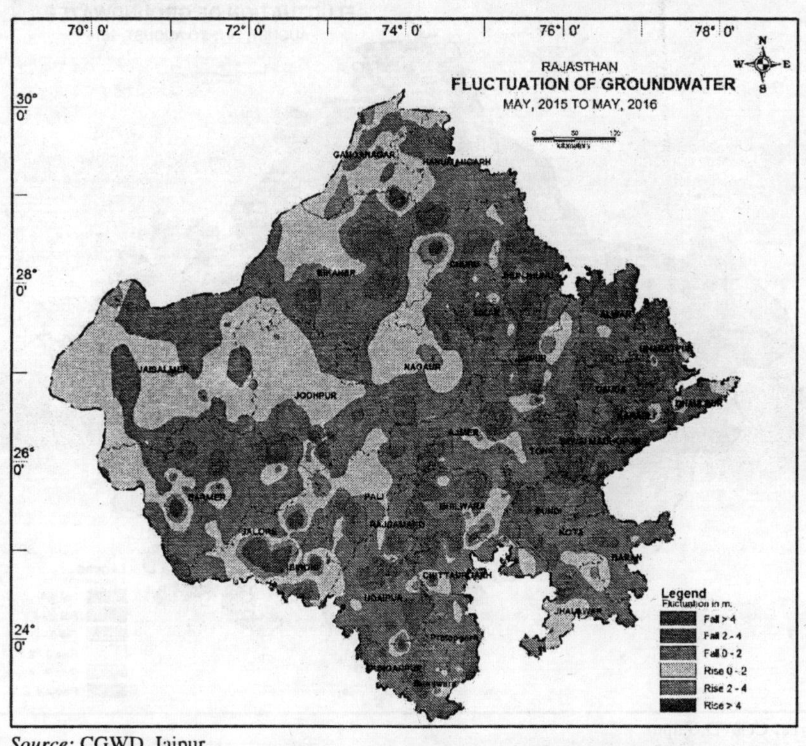

Source: CGWD, Jaipur

Figure 11.24 Annual Water Level Fluctuation May 2015 to May 2016

Monsoon Period

Total number of wells analysed are 789. A perusal of (Figure 11.25) for annual fluctuation of water level from August, 2015 to August, 2016 reveals that about 53.9% of the stations scattered mostly in south to south-western, north western, western and west central parts of the State have shown rise in water level during this period. The water level rise of 0 to 2 m has been observed at 33.7% of the stations scattered in most of the districts mostly in Pratapgarh, Bundi, Pali, Kota, Rajsamand, Bharatpur, Jhalawar, Banswara, Tonk, Udaipur, Churu and Ganganagar districts. The rise in water level ranges from 2 to 4 m in 10.4% of the stations falling in all the districts except Ajmer and Sirohi districts. The rise of water level more than 4m has been recorded at 9.8% of the stations falling in all the districts except Baran, Bhilwara, Jhunjhunu and Sirohi districts. The maximum rise of 20.9 m has been recorded at Dhanwara in Jalore district and minimum rise of 0.02 m at Rookansar and Sekhiwar in Sikar and Rajgarh in Jailsalmer district. During the period, about 28.9% of the stations, scattered all over State mainly in eastern, southern, south eastern and central parts, shows fall in water level ranging from 0 to 2 m. The water level fall of 2 to 4 m has been recorded at 8.7% of stations occurring in patches in all the districts except Banswara, Bundi, Dhaulpur, Pali, Sawai Madhopur, Tonk and Udaipur. The water level fall of more than 4 m has been shown by 7.7% of stations covering major parts of Bhilwara, Ajmer, Chittaurgarh, Jhunjhunu, Sikar and Tonk districts. The maximum fall of 19.70 m is recorded at Pipli in Jhunjhunu district and minimum fall 0.01m at Surwania in Banswara, Odan in Rajsamand and Suranwali in Ganganagar district. Six stations show no change.

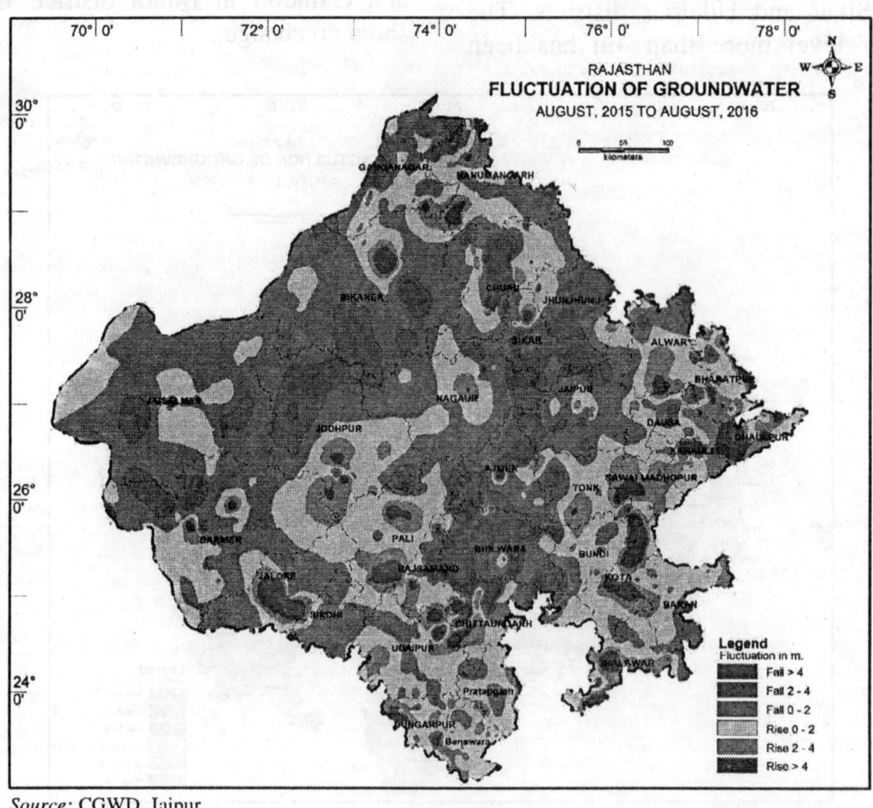

Source: CGWD, Jaipur

Figure 11.25 Annual Water Level Fluctuation August 2015 to August 2016

Post Monsoon Period

Total number of wells analysed are 800. A perusal of (Figure 11.26) exhibits that about 54.8 % of the stations shows rise in water level in patches scattered in almost all the districts. Rise in the water level between 0 to 2m are observed in 34.1% of the stations scattered in isolated patches in all the districts. Rise of 2 to 4m are observed in 12.4% of the stations scattered in most of the districts except Ajmer, Dausa, Nagaur, Sikar, Barmer, Jodhpur and Jalore districts. Rise of more than 4m are observed in 8.3% of the stations scattered in patches in Banswara, Barmer, Bharatpur, Bhilwara, Bikaner, Bundi, Chittorgarh, Churu, Dausa, Dhaulpur, Dungarpur, Ganganagar, Hanumangarh, Jaipur, Jaisalmer, Jalore, Jhalawar, Jodhpur, Karauli, Kota, Pali, Pratapgarh, Rajsamand, Sawai Madhopur, Sirohi, Tonk and Udaipur districts. The maximum Rise in water level of 15.85m is recorded at Purabsur in Hanumangarh district and minimum Rise of 0.01 m at Bhopalpura in Ganganagar district. Fall in water level are mainly in the range of 0 to 2 m and observed in 26.4% of the stations scattered in all the districts in the State. Fall of 2 to 4m are observed in 10.4% stations shown in isolated patches scattered mostly in all the districts except in southern and south-eastern part of the State. Fall of more than 4m are recorded in small isolated patches in 6.6% stations falling in Ajmer, Nagaur, Dausa, Sikar, Jhunjhunu, Bikaner and Churu districts. The maximum decline in water level of 18.28m is recorded at Pipli in Jhunjhunu district and minimum decline of 0.03m is recorded at Dholipal in Jaisalmer district. Fifteen stations show no change.

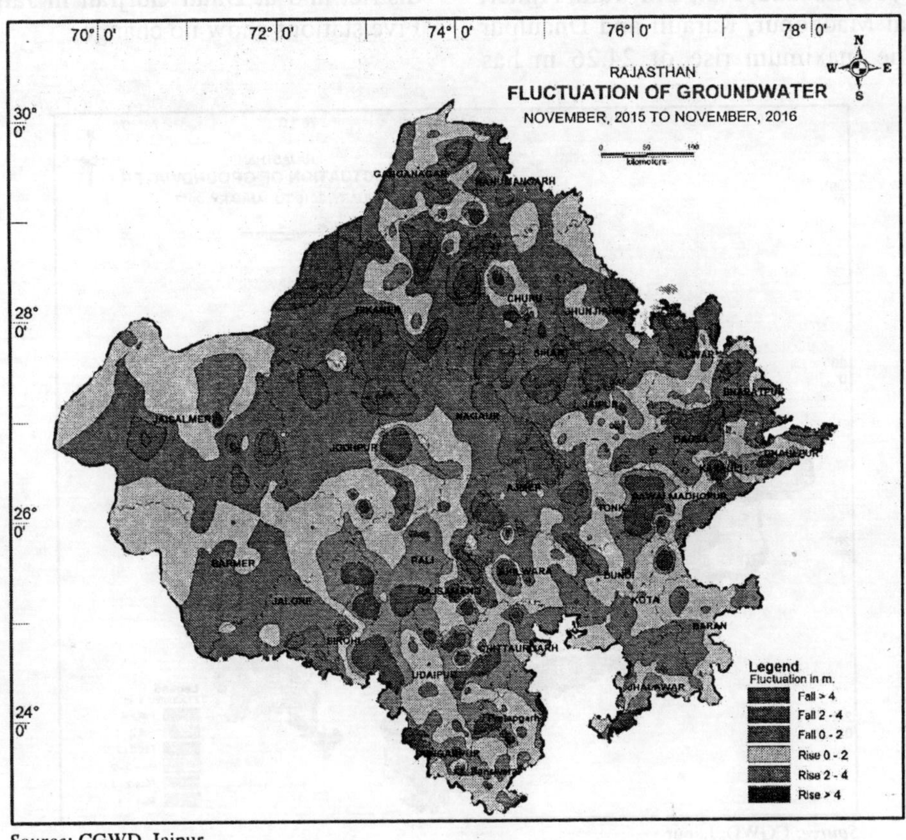

Source: CGWD, Jaipur

Figure 11.26 Annual Water Level Fluctuation November 2015 to November 2016

Total number of wells analysed are 828. A perusal of (Figure 11.27) representing annual fluctuation of water level from January, 2016 to January, 2017 reveals that about 59.20% of the stations scattered mostly in south to south-western, north western, western and west central parts of the State have shown rise in water level during this period. The water level rise of 0 to 2 m has been observed in 29.8% of the stations falling in east of Aravallis in southern and south-eastern half. The water level rise ranging from 2 to 4 m has been shown by 14.1% stations scattered all over the districts, mostly in Banswara, Dungarpur, Pratapgarh, Udaipur, Chittorgarh, Rajsamand, Pali, Bhilwara, Ajmer, Tonk, Sawai Madhopur, Karauli and Dhaulpur districts. The rise of water level more than 4m has been recorded at 15.3% of the stations falling in all the Banswara, Dungarpur, Pratapgarh, Udaipur, Chittorgarh, Rajsamand, Pali, Bhilwara, Ajmer, Tonk, Sawai Madhopur, Karauli and Dhaulpur districts. The maximum rise of 24.26 m has been recorded at Purabsar in Hanumangarh district and minimum rise of 0.01 m at Rewali in Barmer district and 6PB in Bikaner district. About 27.8% of the stations scattered in all the districts, mostly in eastern and central parts, have shown fall in water level ranging from 0 to 2 m during this period. The water level fall of 2 to 4 m has been recorded at 7.1% of stations occurring in patches in all the districts except Banswara, Bundi, Chittorgarh, Dungarpur, Jalore, Kota, Pali, Pratapgarh, Rajsamand, Sawai Madhopur, Sirohi and Udaipur districts. The water level fall of more than 4 m has been shown by 5.2% of stations occurring in patches in Sirohi, Barmer, Jaisalmer, Jodhpur, Nagaur, Bikaner, Sikar, Jhunjhun, Alwar and Dausa districts. The maximum fall of 28.25 m is recorded in Alwar district and minimum fall 0.02m at Dhirera_Pz in Bikaner, at Bolanwali in Hanumangarh district and at Datal Gurjran in Jaipur district. Five stations show no change.

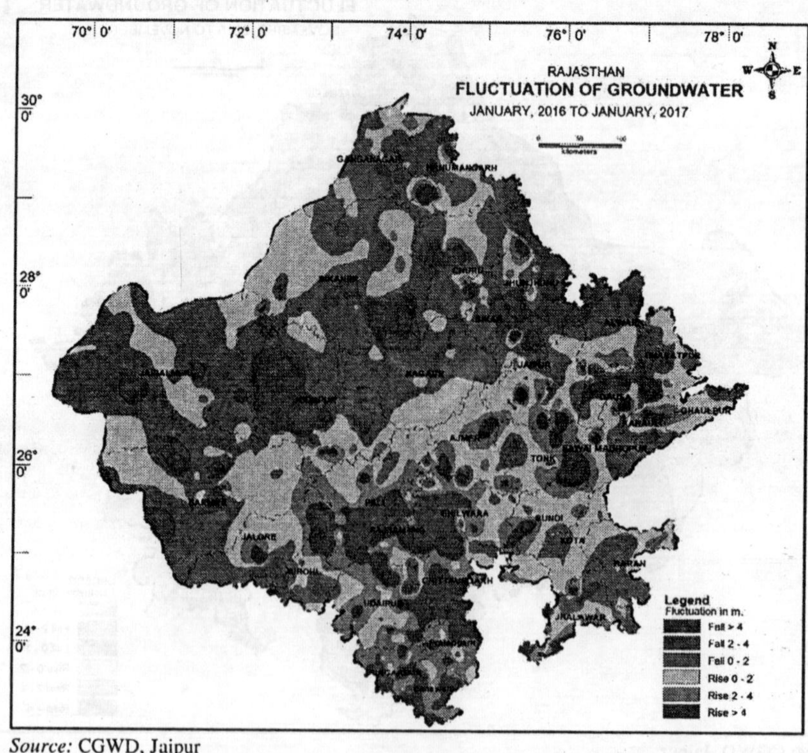

Source: CGWD, Jaipur

Figure 11.27 Annual Water Level Fluctuation January 2016 to January 2017

Decadal Variations

Decadal Fluctuation in the water levels of the different monitoring periods were analysed graphically and depicted in Figure 11.28. This figure shows a comparison of the changes of the water levels during different seasons with their respective decadal averages. It is noticed that the rise mostly dominates over the fall.

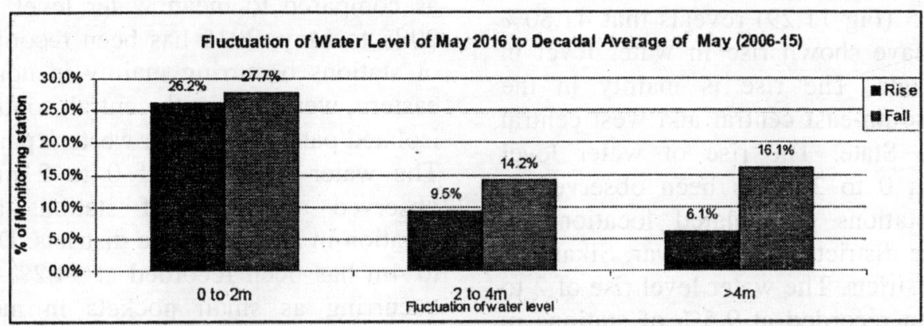

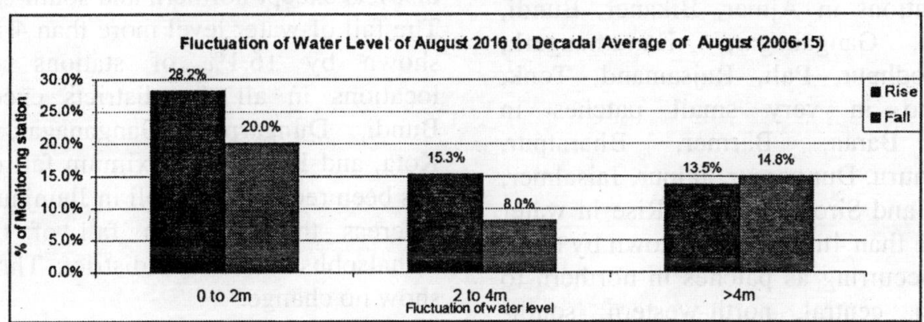

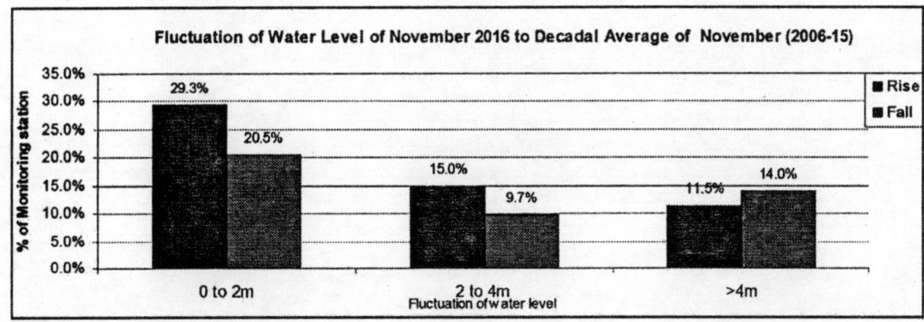

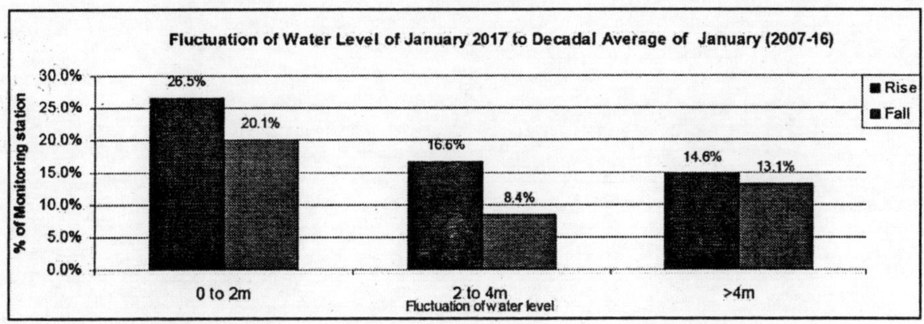

Source: CGWD, Jaipur

Figure 11.28 Decadal Water Level Fluctuation

Decadal average of May (2006 to May 2015) to May 2016

Total number of wells analysed are 846. A comparison of the water level of May, 2016 with the mean of water levels from May, 2006 to May, 2015 (Fig 11.29) reveals that 41.80% of stations have shown rise in water level in Rajasthan State. The rise is mainly in the northern to south-east central and west central parts of the State. The rise of water level ranging from 0 to 2m has been observed at 26.2% of stations in isolated locations in almost all the districts except Alwar, Sikar and Jhunjhunu districts. The water level rise of 2 to 4 m has been recorded at 9.5% of stations in isolated locations in Ajmer, Bikaner, Bundi, Chittorgarh, Ganganagar, Hanumangarh, Jhalawar, Jodhpur, Pali, Rajsamand, Tonk, Udaipur and in very small patches in Banswara, Baran, Barmer, Bharatpur, Bhilwara, Churu, Dungarpur, Jaipur, Jaisalmer, Jalore, Kota and Sirohi districts. Rise in water level of more than 4m has been shown by 6.1% of stations occurring as patches in northern to south-eastern, central, north-western, south-western and south eastern parts. The maximum rise of 32.13 m has been recorded at Lalera in Bikaner district and minimum rise of 0.02 m at Urpuria in Baran and Kuchera in Nagaur districts. The fall in water level in May, 2016 as compared to mean water level from May, 2006 to May, 2015 has been recorded at 58% of stations occurring mainly in north eastern, eastern, west and south central, southern and in isolated patches in south western parts of State. The water level fall of 0 to 2 m has been observed at 27.7% of stations at isolated location in almost all the districts. Decline of 2 to 4m has been recorded at 14.2% of stations occurring as small pockets in most of the districts except northern and south-eastern part. The fall of water level more than 4 m has been shown by 16.1% of stations at isolated locations in all the districts except Baran, Bundi, Dungarpur, Ganganagar, Jhalawar, Kota, and Pali. The maximum fall of 28.35 m has been recorded at Pipli in Jhunjhunu district, whereas the minimum fall of 0.02 m at Rohalsobhsar in Sikar district. Three stations show no change.

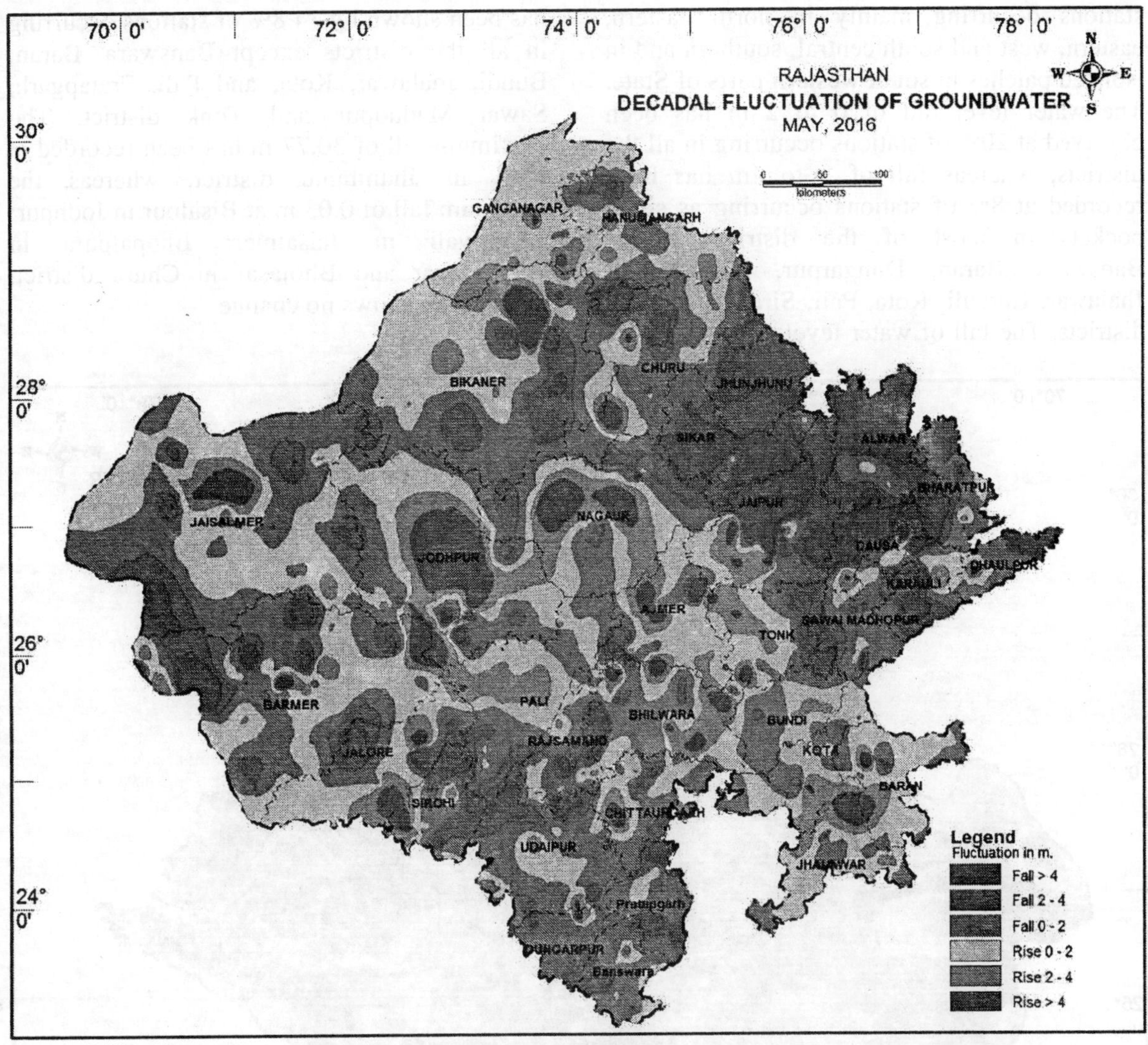

Source: CGWD, Jaipur

Figure 11.29 Decadal Water Level Fluctuation –Average May to May 2016

Decadal average of August (2006 to 2015) to August 2016

Total number of wells analysed are 871. A comparison of the water level of August, 2016 with the mean of water levels from August, 2006 to August, 2015 (Figure 11.30) reveals that 57% of stations have shown rise in water level in Rajasthan State. The rise is mainly in the northern, southern, south-eastern, and west central parts of the State. The rise of water level ranging from 0 to 2m has been observed at 28.2% of stations at isolated locations in all the districts except in central part of the state. The water level rise of 2 to 4 m has been recorded at 15.3% of stations occurring in all the districts except Bhilwara, Jalore, Jhunjhunu, Jodhpur and Sikar districts whereas water level rise of more than 4m has been shown by 13.5% of stations at isolated patches in all the districts. The maximum rise of 26.15 m has been recorded at Lalera in Bikaner district and minimum rise of 0.01 m at Dara in Kota and Baroda in Dungarpur district. The fall in water level in August, 2016 as compared to decadal mean has been recorded at 42.8% of stations occurring mainly in north eastern, eastern, west and south central, southern and in isolated patches in south western parts of State. The water level fall of 0 to 2 m has been observed at 20% of stations occurring in all the districts, whereas fall of 2 to 4m has been recorded at 8% of stations occurring as small pockets in most of the districts except Banswara, Baran, Dungarpur, Ganganagar, Jhalawar, Karauli, Kota, Pali, Sirohi and Tonk districts. The fall of water level more than 4 m has been shown by 14.8% of stations occurring in all the districts except Banswara, Baran, Bundi, Jhalawar, Kota, and Pali, Pratapgarh, Sawai Madhopur and Tonk district. The maximum fall of 30.77 m has been recorded at Pipli in Jhunjhunu district, whereas the minimum fall of 0.03 m at Bisalpur in Jodhpur; Ghantiyali in Jaisalmer; Bhopalpura in Ganganagar and Bhojasar in Churu district. One station shows no change.

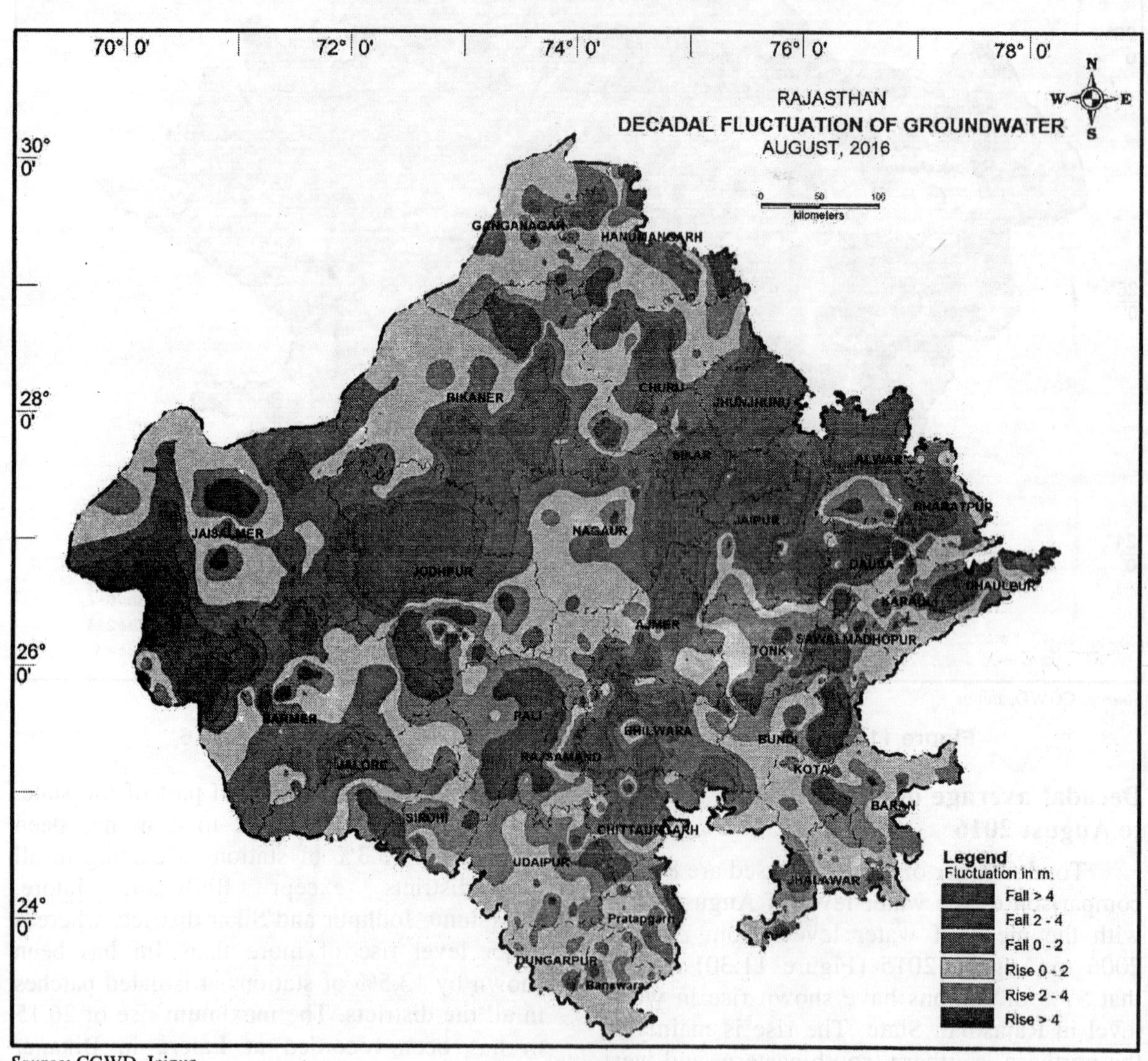

Source: CGWD, Jaipur

Figure 11.30 Decadal Water Level Fluctuation –Average August to August 2016

Decadal average of November (2006 - 2015) to November 16

Total number of wells analysed are 873. A comparison of the water level of the November, 2016 with the average water level of the November for last decade (Figure 11.31) reveals that rise in water level is experienced in 55.8% of stations analysed in the State. Rise is mainly in the northern to south-east central and west central parts of the state. Rise is mostly in the range of 0-2m (29.3%) 2 to 4m (15%) and more than 4m (11.5%). Rise of 0 to 2m are observed in all the districts except Jhunjhunu, Sikar disricts. Rise of 2 to 4m are observed in all the districts in patches except Bharatpur, Jhunjhunu and Sikar distritct. Whereas rise of more than 4m are observed in patches in most of the districts, except Baran, Churu, Jhunjhunu and Sikar district. The maximum rise of 32.54m is recorded at Lalera in Bikaner district, whereas minimum rise of 0.04m is recorded at Sawaipur in Bhilwara district. Fall in water level are mainly recorded in the north-east to lower and upper west central parts in the State. Fall is experienced in 44.1% of stations analysed in the State. Fall is mostly in the range of 0-2m (20.5%) 2 to 4m (9.7%) and more than 4m (14.0%). Fall of 0 to 2m is observed in all the districts except Dungarpur and Jhunjhunu districts. Whereas fall of 2 to 4m is observed in small pockets in most of the districts except Banswara, Bundi, Dhaulpur, Dungarpur, Jalore, Kota, Rajsamand and Udaipur districts. Fall of more than 4 m are observed in patches in all the districts in the State except Banswara, Baran, Bundi, Ganganagar,Jhalawar, Pali, Pratapgarh, Sawai Madhopur and Sirohi districts. The maximum fall of 31.08m is recorded at Pipli in Jhunjhunu districts whereas the minimum decline of 0.01 m is observed at Mandapia RS in Bhilwara and Mayathiwala in Jaisalmer districts.

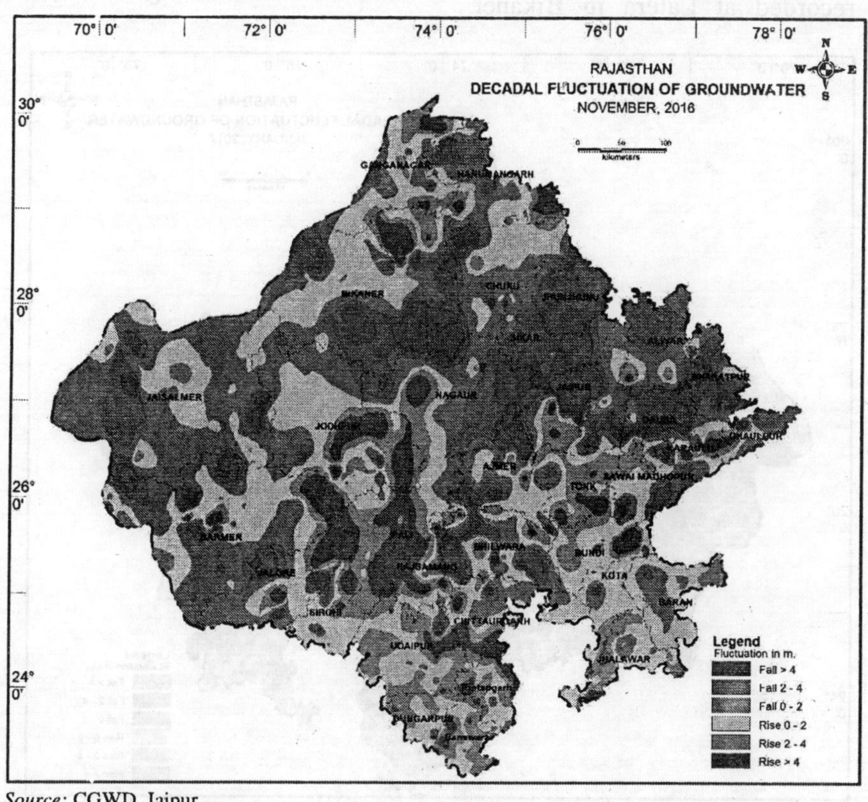

Source: CGWD, Jaipur

Figure 11.31 Decadal Water Level Fluctuation –Average November to November 2016

Decadal average of January, 2007 to January, 2017

Total number of wells analysed are 937. A comparison of the water level of January, 2017 with the mean of water levels from January, 2007 to January, 2016 (Figure 11.32) reveals that 57.7% of stations have shown rise in water level in Rajasthan State. The rise of water level ranging from 0 to 2m has been observed at 26.5% of stations occurring in all the districts, mostly in Ganganagar, Ajmer, Kota, Baran, Bikaner, Jodhpur and Sawai Madhopur districts. The water level rise of 2 to 4 m has been recorded at 16.6% of stations at isolated locations in all the districts, mostly in Dungarpur, Banswara, Ajmer, Pali, Tonk, Udaipur and Jhalawar, whereas water level rise of more than 4m has been shown by 14.6% of stations occurring in patches in all the districts except Alwar, Baran, Dausa, Jhunjhunu and Sawai Madhopur. The maximum rise of 25.38 m has been recorded at Lalera in Bikaner district and minimum rise of 0.01 m at Ramgarh2 in Ajmer district and Rewali in Barmer district. The fall in water level in January, 2017 as compared to mean water level from January, 2007 to January, 2016 has been recorded at 41.6% of stations. The water level fall of 0 to 2 m has been observed at 20.1% of stations occurring in all the districts except in Dungarpur district, whereas fall of 2 to 4m has been recorded at 8.4% of stations occurring as small pockets in most of the districts except Banswara, Baran, Bundi, Churu, Dungarpur, Kota, Pali and Udaipur. The fall of water level more than 4 m has been shown by 13.1% of stations occurring in all the districts except Banswara, Baran, Bundi, Chittorgarh, Dungapur, Jhalawar, Kota, Nagaur, Pali, Pratapgarh and Rajsamand. The maximum fall of 37.09 m has been recorded at Neemrana in Alwar district, whereas the minimum fall of 0.03 m at Girdhapura in Kota district. Six stations show no change.

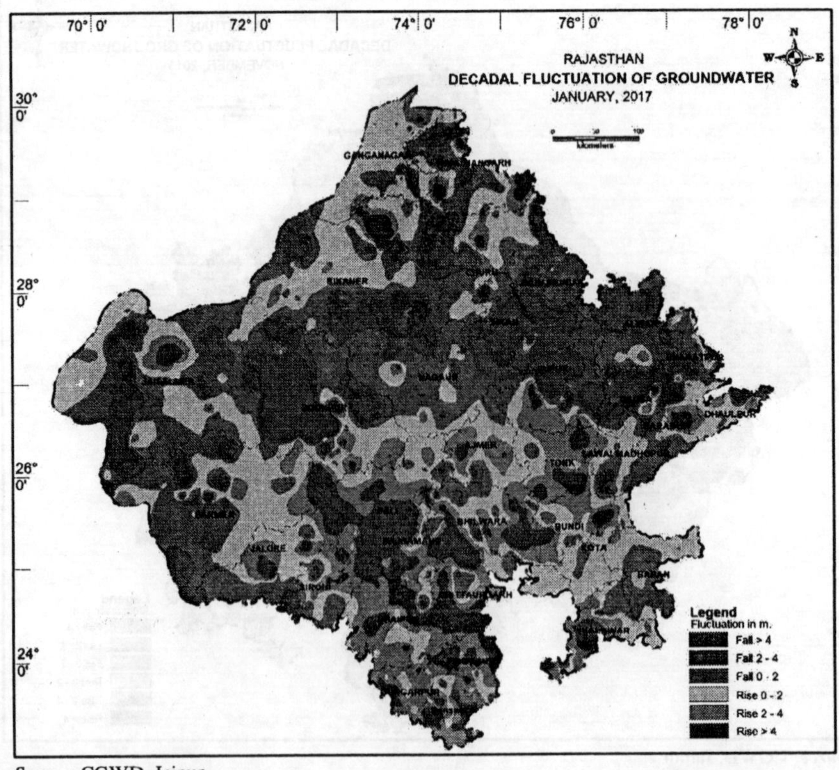

Source: CGWD, Jaipur

Figure 11.32 Decadal Water Level Fluctuation –Average January to January 2017

Hydrochemistry

For the evaluation of Hydro-Chemical status and distribution of various chemical constituents in Ground water of Rajasthan state, 661 water samples were collected from NHS during pre-monsoon, 2016, out of these 646 samples were chemically analyzed in chemical laboratory. Distribution of major constituents in percentage in the State where the principal chemical constituents are (1) within accep-table limit (2) permissible limit (3) beyond permissible limit. District wise percentages of samples where the major chemical constituents are beyond permissible limit for drinking water. District wise distribution of major constituents in (1) within Acceptable limit (2) permissible limit (3) beyond permissible limit has been shown in Table 11.2.

Standards for drinking use

The presence of various chemical constituents in excess quantity in water affects the human health adversely. In our country Bureau of Indian Standard has prescribed standard limits for drinking purpose (IS-10500: 2012). Acceptable and permissible limits for major constituents and their probable effects on human body are as follows:

Table 11.2 Water Quality Standards for drinking use

S. No.	Constituents	Acceptable limit (ppm)	Permissible limit (ppm)	Probable effects
1	TDS	500	2000	Beyond limit water bitter in taste and can cause stomach disorder.
2	Chloride	250	1000	Indigestion, bitter taste
3	Sulphate	200	400 (if Mg does not exceeds 30ppm)	Causes stomach disorder.
4	Nitrate	45	-	Methemoglobinemia in bottle fed infants and Gastro-intestinal problems.
5	Fluoride	1	1.5	Above permissible limit causes skeletal and dental fluorosis and non skeletal fluorosis
6	Total Hardness	200	600	Calcification of arteries, urinary concretions, diseases of kidney or bladder, stomach disorder.
7	Calcium	75	200	Insufficiency causes rickets, excess causes stones in kidney or bladder, essential for human health.
8	Magnesium	30	100	Its salts are cathartic and diuretic, excess is laxative.
9	Iron	0.3	No relaxation	Bitter sweet taste, staining of laundry, trace is essential for nutrition.

Source: CGWD, Jaipur

Quality of Ground Water

For the beneficial use of water its purity is essential otherwise it may affect human health adversely. The quality of water depends on its physical and chemical properties. Physical properties include colour, odour and turbidity which can be determined by our senses. The chemical properties depend on the nature and quantity of various chemical constituents individually or jointly. The possible sources, effect on human health and distribution of some major Chemical constituents are described in following chapter.

Electrical Conductance (EC)

Electrical Conductance is the ability of a substance to conduct an electric current. Chemically pure water in liquid form has a very low conductance. The presence of dissociated ions in solution renders the solution conductive (Figure 11.33).

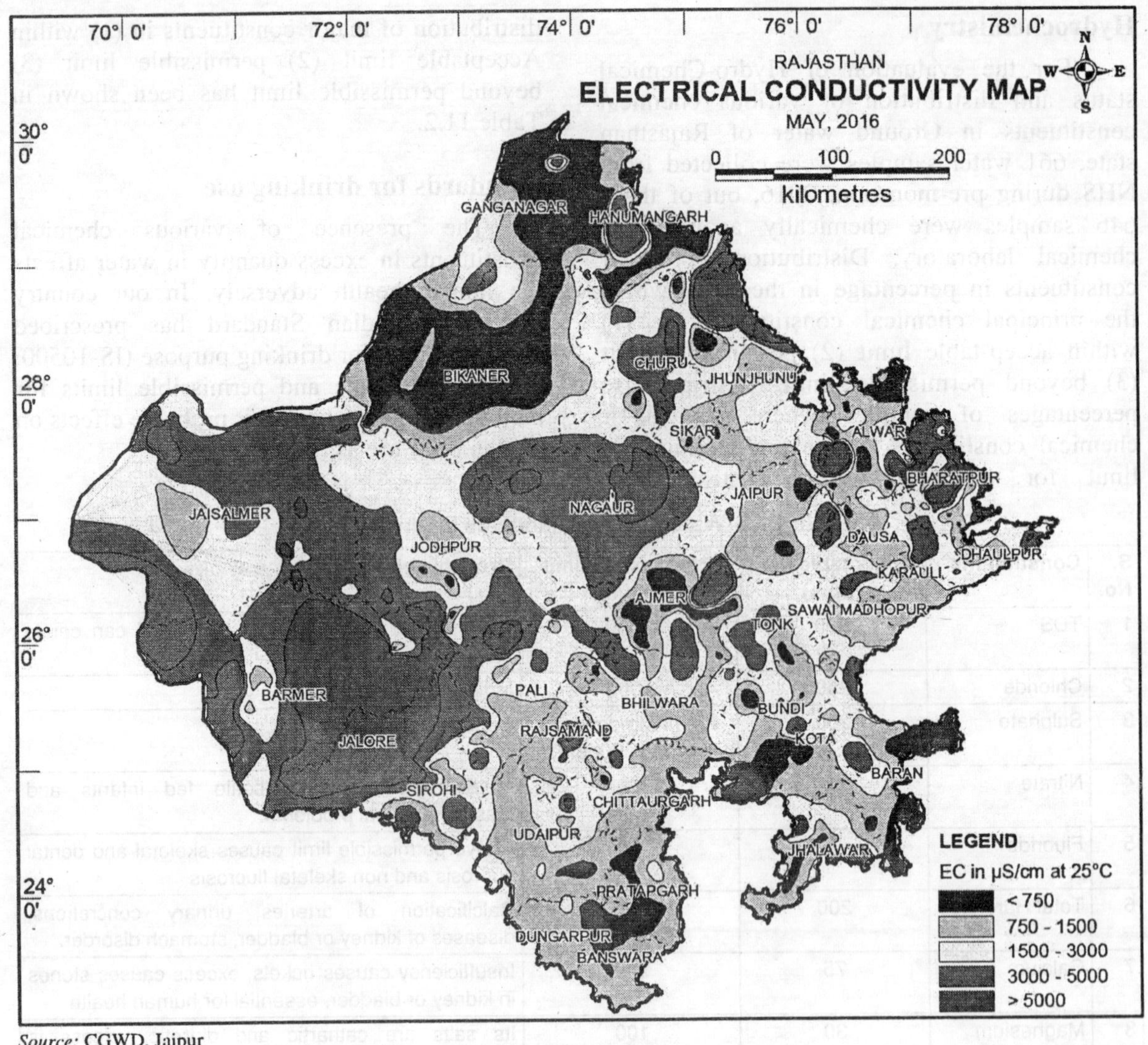

Source: CGWD, Jaipur

Figure 11.33 Distribution of Electrical Conductance –May 2016

Therefore, EC of a solution gives an idea about the quantity of ions or dissolved solids present in it. In western, central and some eastern parts of the state higher EC values of water (>3000 μS/cm) have been observed thus making the ground water saline and non-potable. In southern and some eastern part of the state water is fresh as the EC values are within 1500 μS/cm.

Out of 646 water samples analysed, 19.05% samples have EC values less than 750 μS/cm, 31.55 % in the range of 750-1500 μS/cm, 26% in the range of 1500-3000 μS/cm, 23.39% have more than 3000 μS/cm. Distribution of Electrical Conductivity of Ground Water is shown in Figure 11.33. A perusal of map and data, shows that in the districts of Jalore 100%, Nagaur 81.25%, Barmer 53.84%, Churu 47.82%, Jaisalmer 44.74% Pali 38.89%, Jodhpur 37.93%, Ajmer 30.77 and Dausa 30.77% samples have EC values in the range of 3000-26900 μS/cm i.e. Beyond Permissible Limit. Minimum value of EC has been found at Satur (Bundi) as 190 μS/cm and maximum value as 26900 μS/cm Dasuk in Ajmer district.

Chloride (Cl)

It is one of the most common constituent present in natural water and remains soluble in water unaffected by biological processes therefore reducible by dilution. Natural mineral origin can also be a cause of high chloride content. Industrial effluents (galvanizing plants, water softening plants, oil wells, refineries and paper works) may also leach into ground water.

Sewage effluents contain a larger concentration of Chlorides. Chloride ions have some functions in the body. The tolerance limits of chloride vary with climate and excretion. It is the cation associated with chloride that has usually harmful effects on human body. Individual affected by heart and kidney disease should restrict water consumption with a high chloride concentration.

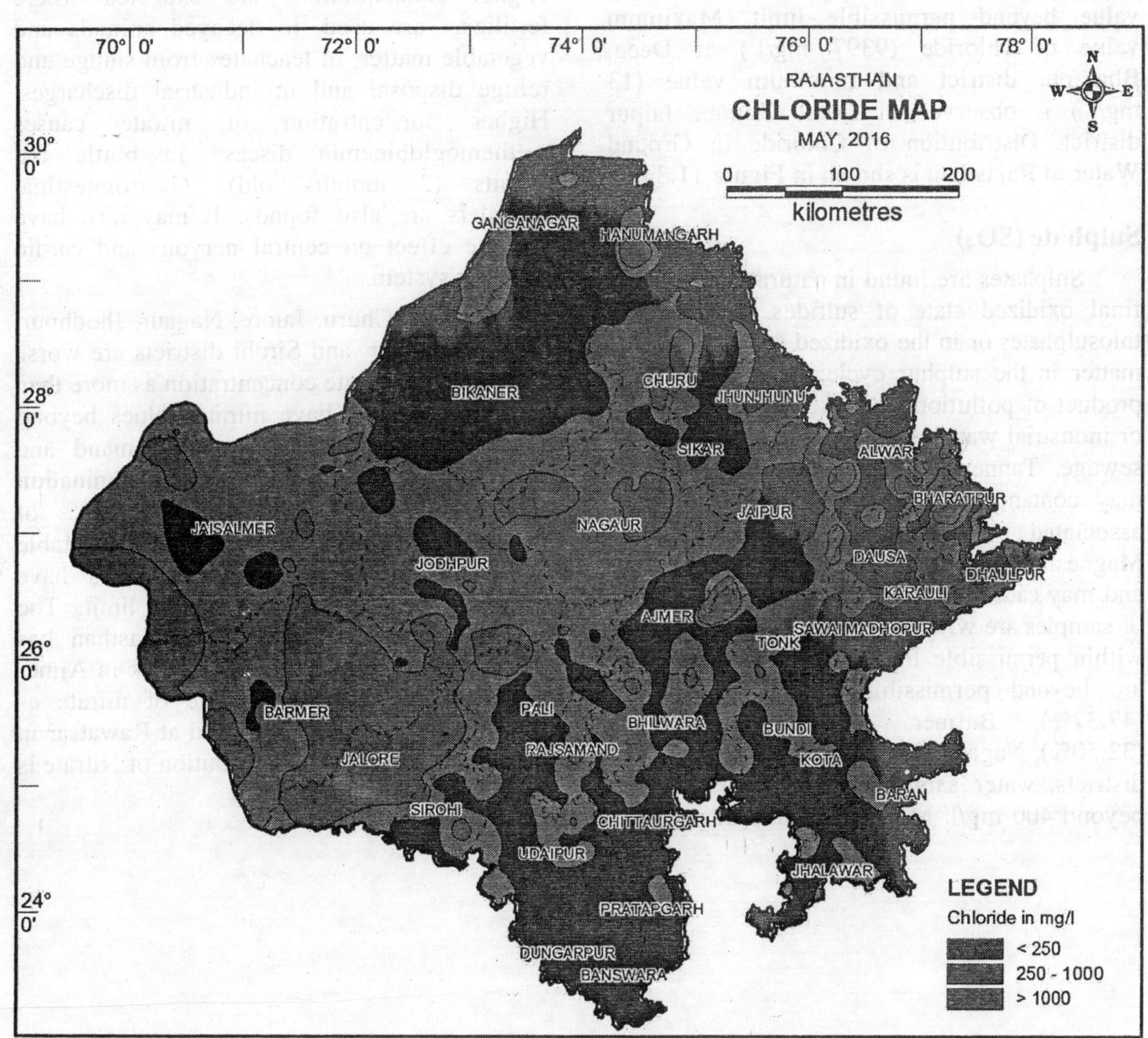

Source: CGWD, Jaipur

Figure 11.34 Distribution of Chloride May 2016

Out of 646 sample analysed, 11.22% have chloride value beyond permissible limit of 1000 mg/L, 28.54% within permissible limit and 60.24% samples are within Acceptable limit. In Jalore (100%) Nagaur (43.75 %) Barmer (38.46%) and Bharatpur (37.50) districts, chloride value is beyond permissible limit. In Chittorgarh, Jhunjhunu Banswara, Baran, Bundi, Dungarpur, Ganganagar, Kota, Pratapgarh, Bikaner, Dholpur and Sikar districts, none of the sample shows chloride value beyond permissible limit. Maximum value of chloride (9397 mg/L) at Deeg, Bharapur district and minimum value (13 mg/L) is observed at Datal Gujran, Jaipur district. Distribution of Chloride in Ground Water of Rajasthan is shown in Figure 11.34.

Sulphate (SO_4)

Sulphates are found in natural water in the final oxidized state of sulfides, Sulfites and thiosulphates or in the oxidized stage of organic matter in the sulphur cycle; in all cases as a product of pollution sources related to mining or industrial waste. Detergents add Sulphate to sewage. Tanneries, steel mills, textile plants may contaminate water. Sulphate ions when associated with high concentration of Magnesium and sodium ions, acts as laxative and may cause gastric disorders. In all, 71.55% of samples are within Acceptable limit, 16.30% within permissible limit and 12.15% samples are beyond permissible limit. In Jaisalmer (47.37%), Barmer (38.46%), Bharatpur (32.50%), Nagaur (31.25%) and Pali (22.22%) districts, water samples have Sulphate value beyond 400 mg/l. Minimum value of sulphate (3.0 mg/L) at Gajwara in Jhalawar district and maximum (4380 mg/L) has been observed at Majri khurd in Alwar district.

Nitrate (NO_3)

Sources of Nitrate are mineral deposits (sodium and potassium nitrates), soils, sea water and atmosphere. Nitrate is used as a fertilizer, as a food preservative and as an oxidizing agent in the chemical industries. Higher concentrations are expected where fertilizers are used, in decayed animals and vegetable matter, in leachates from sludge and refuge disposal and in industrial discharges. Higher concentration of nitrate causes methemoglobinemia disease in bottle fed infants (3 months old). Gastrointestinal disorders are also founds. It may also have adverse effect on central nervous and cardio vascular system.

Barmer, Churu, Jalore, Nagaur, Jhodhpur, Sikar, Jaisalmer, and Sirohi districts are worse affected with nitrate concentration as more than 50 % of samples have nitrate values beyond permissible limit. Bhilwara, Rajsamand and Karauli Districts have nitrate contamination between 40 and 50 %. Around 65.53% of samples have nitrate values within Acceptable limit and remaining 34.47% samples have nitrate value beyond permissible limit. The minimum value of nitrate in Rajasthan has been observed as 0.0 mg/L in Dasuk in Ajmer districts. The maximum value of nitrate as 3640 (mg/L) has been observed at Rawatsar in Hanumangarh district. Distribution of Nitrate is shown in Figure 11.35.

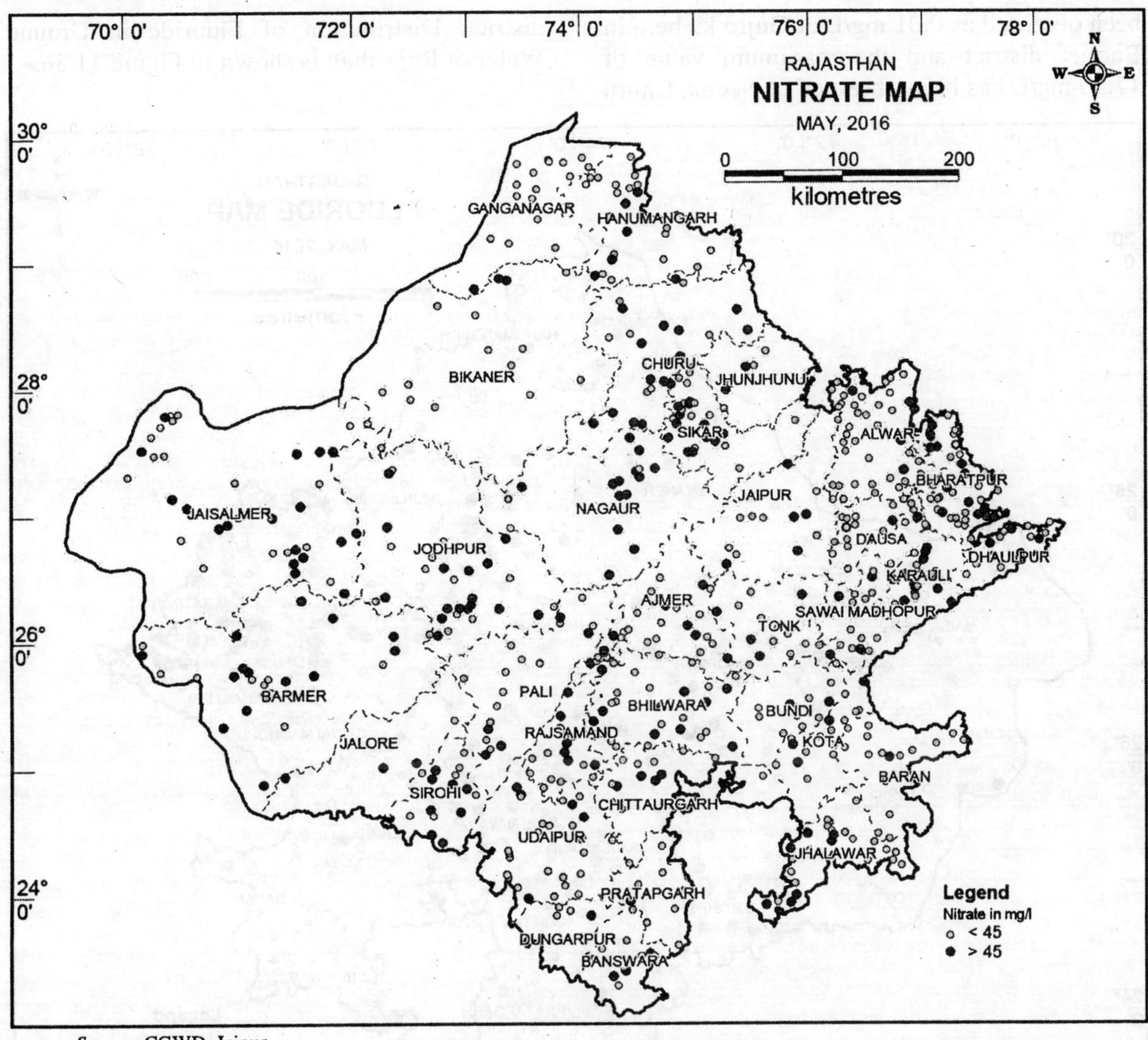

Source: CGWD, Jaipur

Figure 11.35 Distribution of Nitrate May 2016

Fluoride (F)

Fluoride is as inherent component of igneous rocks. The main sources of fluoride in natural water are fluorite (CaF_2), Cryolite (Na_2AlF_2), Fluorapatite. In minerals like mica, amphilaiboles and topaz etc, the fluoride ions are bound on the mineral surfaces. Food in the diet is the major source of fluoride. Tea contains high fluoride concentration. Fluoride reduces dental carries, very high concentration may cause crippling skeletal fluorosis in human body. Less than 1.0 mg/L is essential.

Occurrence of high fluoride in the ground water of Rajasthan is a great concern as 24.64% of 650 ground water samples contain fluoride value beyond Permissible limit i.e. 1.5 mg/L whereas 58.49% and 16.47% of samples are within Acceptable and permissible limit respectively. Ajmer, Sirohi, Sikar and Churu districts are worst affected where more than 50% samples shows fluoride value more than 1.5 mg/L and Bhilwara, Bikaner, Dausa, Bharatpur and Jaipur districts are moderately affected where samples shows fluoride between 40% - 50%. Minimum value of fluoride has been observed as 0.01 mg/L at Gujro ka bera in Barmer district and the maximum value of 17.50 mg/L has been observed at Neema Churu district. Distribution of Fluoride in Ground Water of Rajasthan is shown in Figure 11.36.

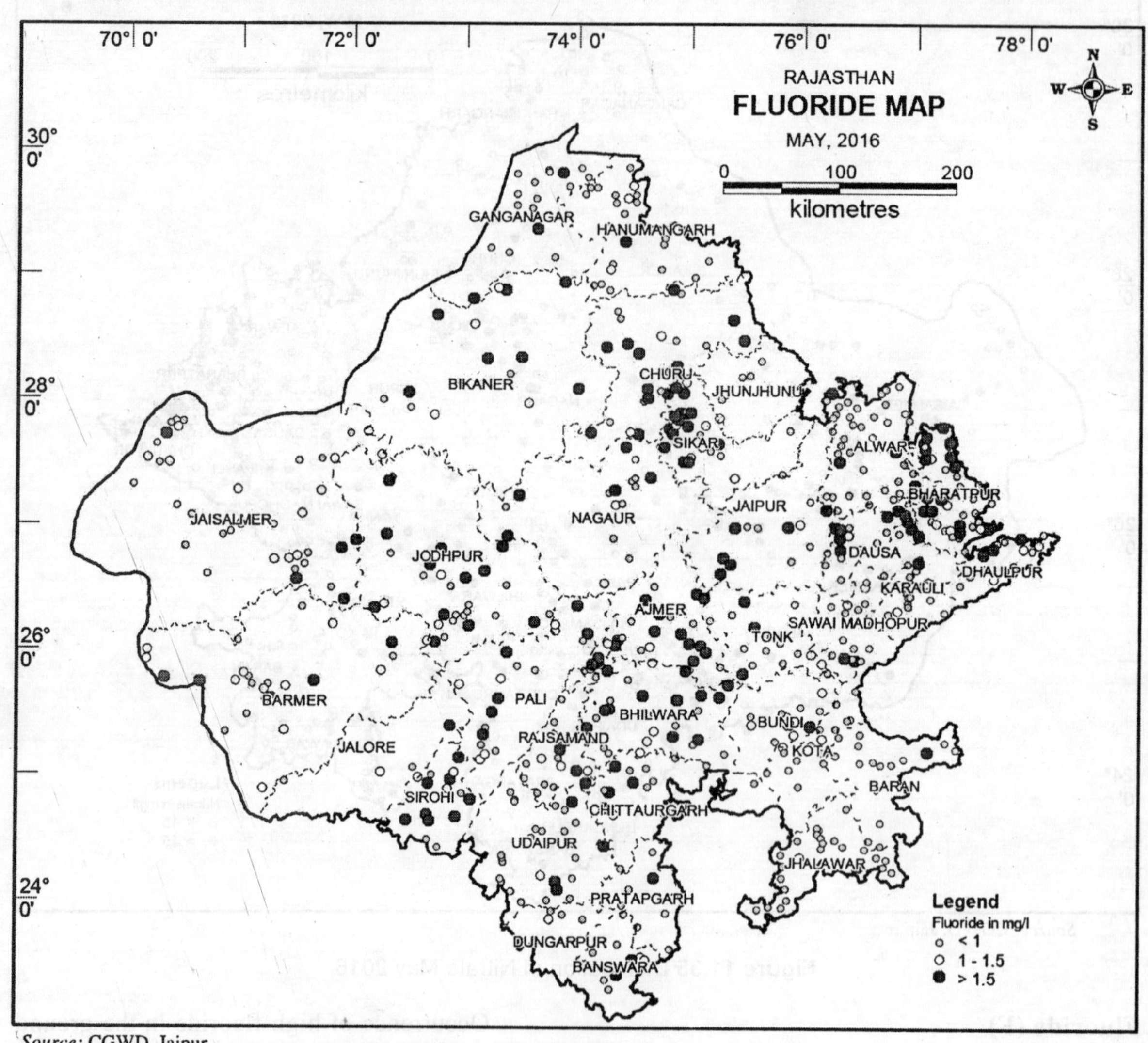

Source: CGWD, Jaipur

Figure 11.36 Distribution of Fluoride May 2016

Total Hardness (TH)

It is primarily determined by sum of calcium and magnesium ions expressed as calcium carbonate. Other substances such as iron, manganese, aluminum, strontium, zinc may also contribute to a very small extent due to low solubility. An inverse correlation between hardness of water and cardiovascular diseases (Heart, hypertension and stroke) has been shown. High values may cause calcification of arteries, urinary concretions and stomach disorder.

Out of 646 samples analysed, 23.20% samples are within Acceptable limit of 200 mg/L. Only 17.41% of samples have value beyond permissible limit of 600 mg/L whereas 59.38% samples are within permissible limit. Jalore is the worst affected district where more than

80.0% samples shows Total Hardness value beyond permissible limit. In Pratapgarh, Jhunjhunu, Sikar, Kota and Chittorgarh districts none of samples have Total Hardness value beyond permissible limit. The minimum value of hardness as 20 mg/L has been found at Minksar of Bikaner district. The maximum value has been observed as 7500 mg/l at Dasuk of Ajmer district (Figure 11.38 and 11.39).

Calcium (Ca)

It is always found in combination in limestone, marble and chalk. Its most common compounds are limestone, gypsum, fluorite; also calcium carbide, chloride, hypochlorite. Calcium is essential for human body. Its low content in soft water has been linked with rickets and defective teeth. Its excess may cause stones in kidney or bladder. Gout, Rheumatism etc. are also linked with its high concentration.

There is no cause of concern about the calcium hazard as only 4.82% samples are beyond the permissible limit of 200 mg/L. Banswara, Baran, Jalore, Chittorgarh, Jaipur, Sawai Madhopur, Jhalawar, Jhunjhunu, Kota, Sikar, Bikaner, Jodhpur, Bhilwara, Pratapgarh, and Sirohi districts have Calcium within permissible limit (30.26%) and 64.92% samples are within acceptable limit. The minimum value of calcium has been observed as 2.0 mg/L at Minksar of Bikaner district. The maximum value as 2800 mg/L has been found at Dasuk of Ajmer district.

Magnesium (Mg)

It is never found as a free element. It constitutes a large deposit as magnesite and common rock forming dolomite.The presence of magnesium is beneficial for heart and nervous system. However higher concentrations have laxative and diuretic effect. Only 13.13% of samples have magnesium value beyond permissible limit of 100 mg/L and 35.54% samples are within Acceptable limit and 51.33% samples within permissible limits. No sample in Banswara, Chittorgarh, Sikar, Bikaner, Pratapgarh, Sirohi and Jhunjhunu districts have magnesium value beyond permissible. In Bharatpur (47.50%) Samples, Churu (30.43%) and Barmer (31.03%) districts have Mg value beyond permissible limit. The minimum value of Mg as 2.0 mg/L has been found at Athmeel of Dholpur district and maximum value 1209mg/L at Deeg of Bharatpur district.

Iron (Fe)

Common ores of iron are Hamatite, Magnetite, Limonite, and Pyrite. Leaching of iron salts (acid mine drainage) and iron products industrial waste may be a pollution source. Iron is an essential element in human nutrition. About 35.12% of samples have iron value beyond the permissible limit of 1.0 mg/L and 64.88% samples are within Acceptable limit of 0.3 mg/L. Dungarpur and Rajsamand are worst affected districts where 75.00% and 65.00% samples have Iron value beyond permissible limit respectively. Minimum value of iron as 0.0 mg/L has been observed at in Hasanpur in Alwar district and maximum value of 13.08 mg/l at Deeg in Bharatpur district (Figure 11.37, 11.38 and 11.39).

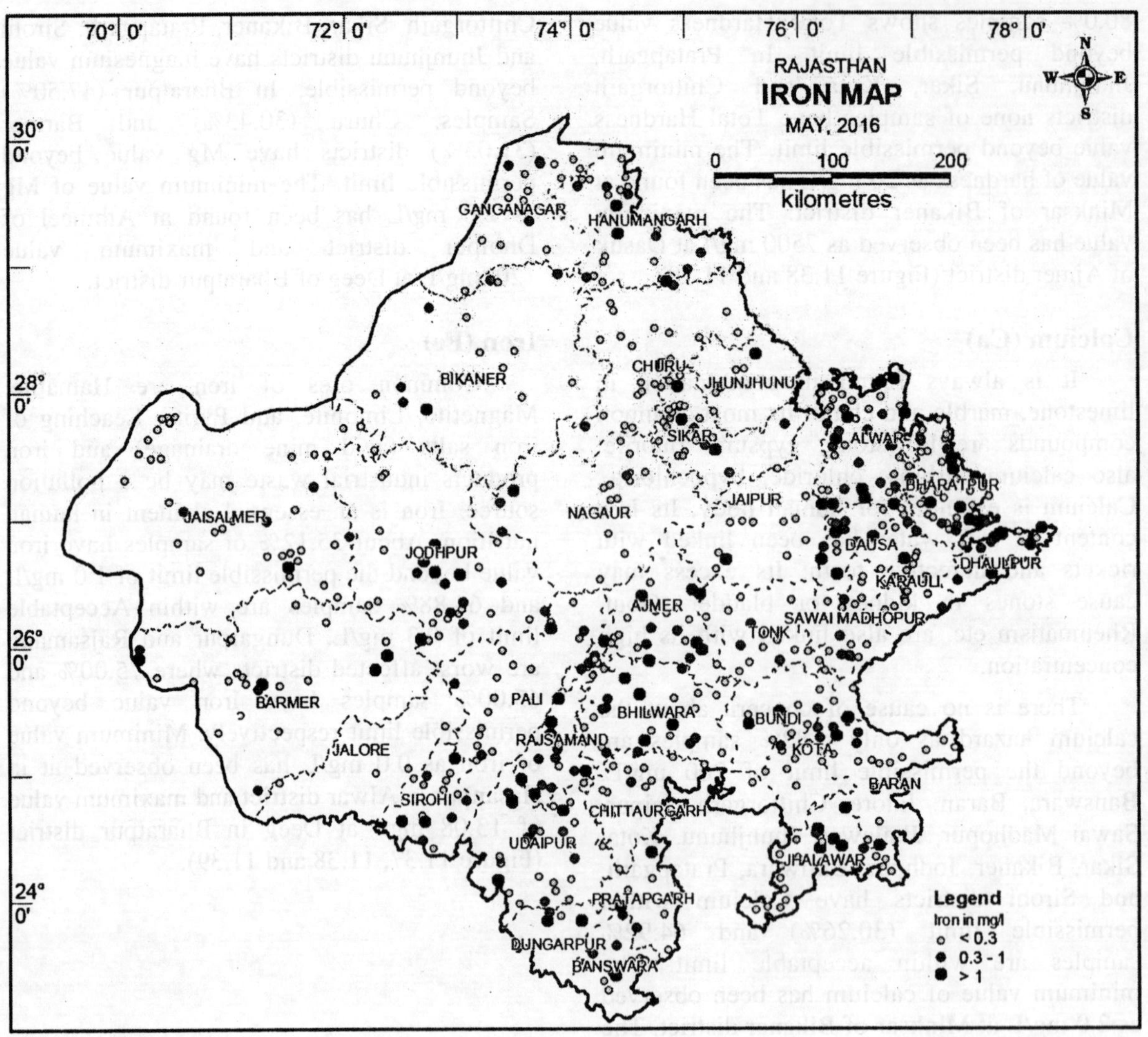

Source: CGWD, Jaipur

Figure 11.37 Distribution of Iron May 2016

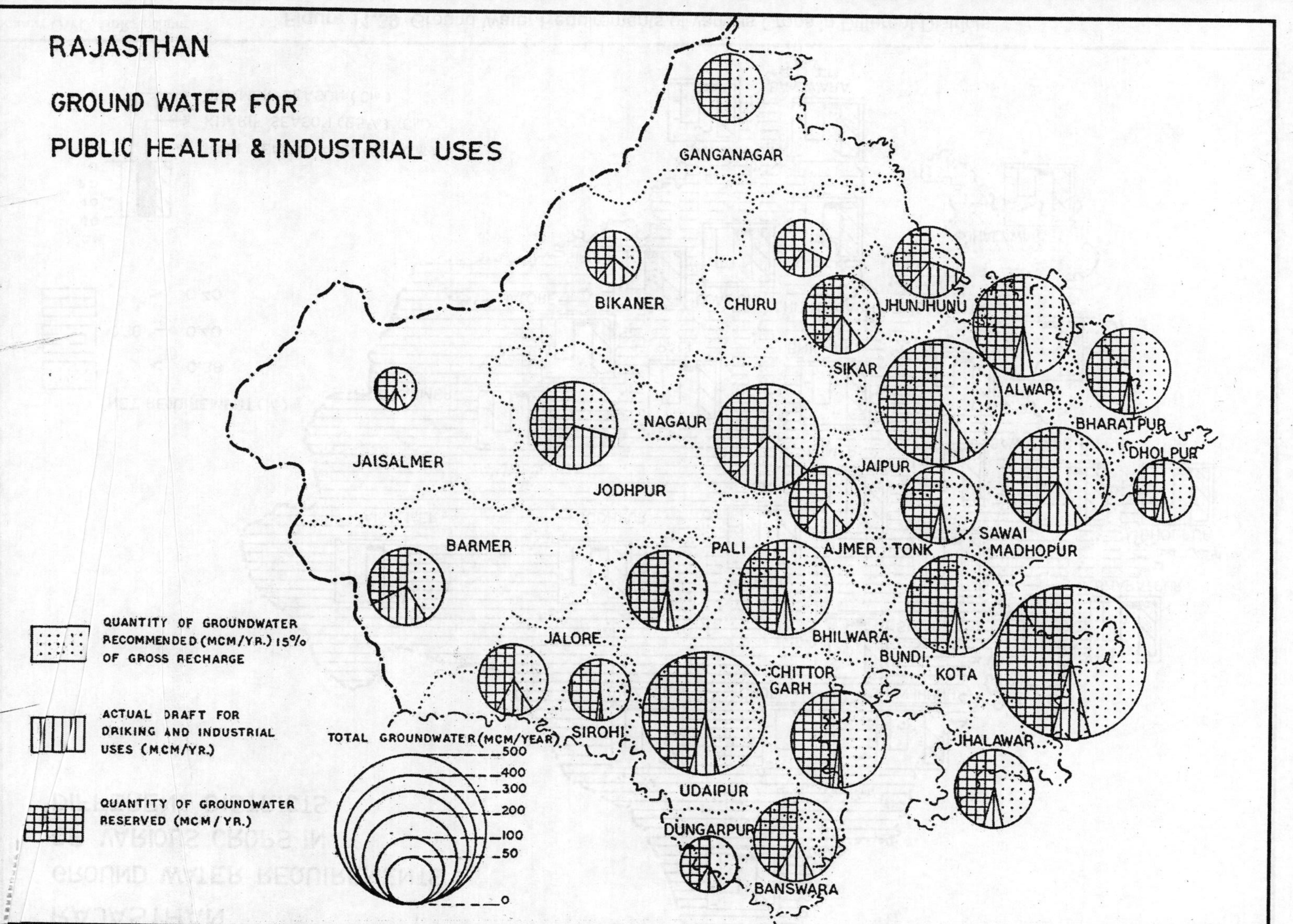

Sorce: GWD, GoR, Jodhpur

Figure 11.38 Ground Water for Public Health and Industrial Uses

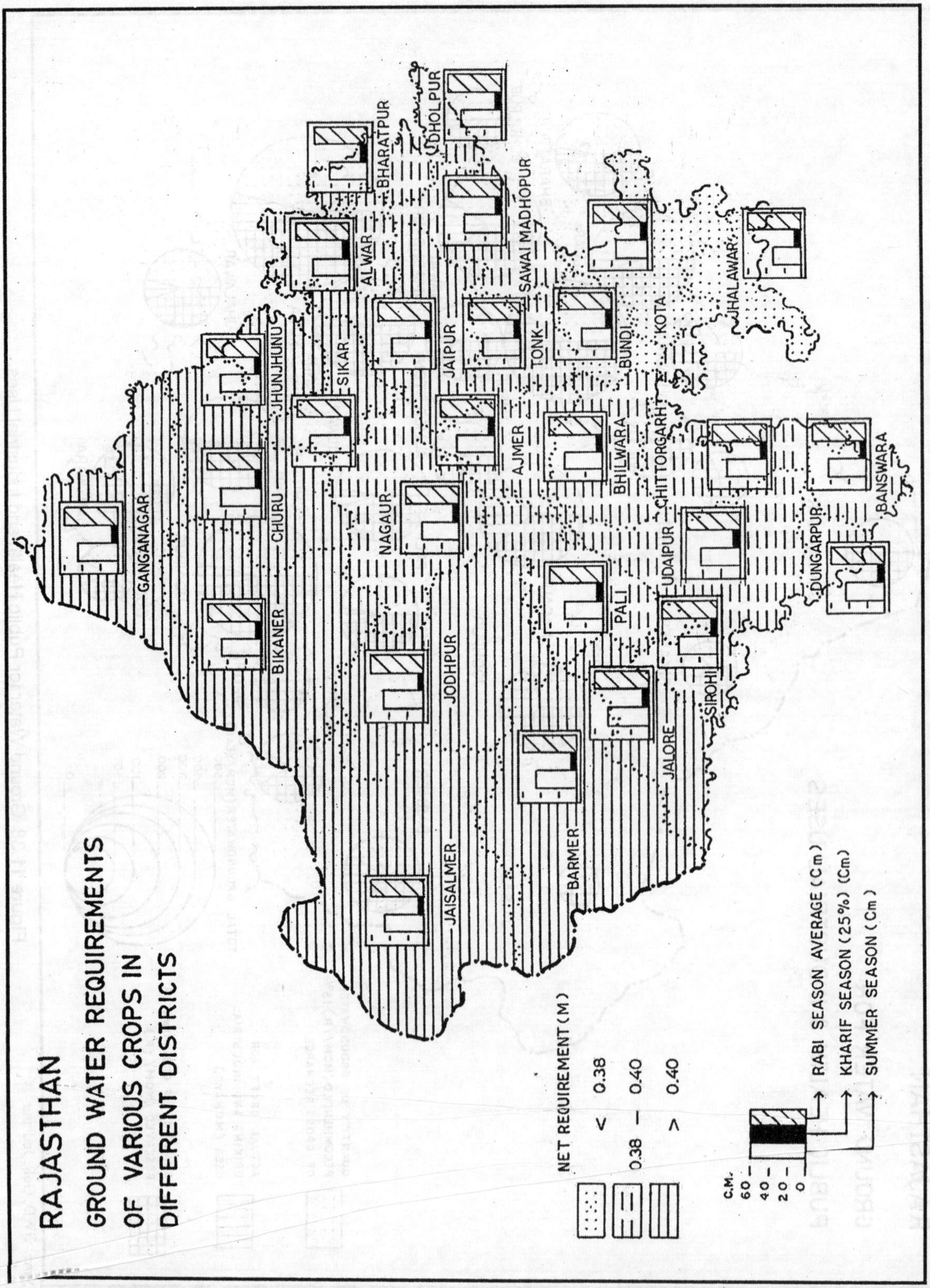

Figure 11.39 Ground Water Requirements of various Crops in Different Districts

Source: GWD, GoR, Jodhpur

Chapter 12

LAND USE AND AGRICULTURE DEVELOPMENT

The various parts of he State present a variety of land use patterns which exhibits, to a large extent, the availability of soil and water resources in the area and human endeavours to harness them. In the areas having greater concentration of population, the significance for agricultural cropping of the land increases and therefore net sown area in these districts is relatively high. In 1952 in the state net area sown is 93.13 lakh ha. which increases to 180.2 lakhs ha. In the year 2015-16. Whereas on the other hand, in the districts where the soil is less fertile, extensive farming is needed to feed even a smaller population, so that the net area sown again increases.

The common nine-fold classification of land use has been generalized here into following broad groups. Forest, Not available for cultivation (Area under non-agriculture uses and Barren and unculturable land), other uncultivated land excluding fallow lands (Permanent pasture/ other grazing land, land under miscellaneous trees, crops and groves and culturable wasteland) Fallow land (Fallow lands other than current fallow and current fallows) and net area sown.

Total culturable area in the state is 180.24 lakh ha. (2015-16). In the western districts of Jaisalmer, Bikaner and Barmer, the percentage of available culturable wasteland and fallow land is the highest an account of the stretch of sandy soils in these areas. The net area sown in these districts is much less than the other types ofland uses.

In the semi-arid zone and sub humid parts of the state, the net area sown is higher on account of extensive agriculture and growing demands of increasing population. However, in the mid western districts of arid and semi arid zone, mainly in Churu, Nagaur, Jodhpur, Pali and Jalor districts the. percentage of fallow land is quite high for the reason that only rainfed crops are grown in this area.

The percentage of cultivable waste land the land put to non agricultural use is highest in parts of Bhilwara, Chittaurgarh, Udaipur, Kota and Baran districts, owing to their hilly terrain. The percentage of forest areas are noticeable high in Udaipur, Dungarpur, Banswara, Sirohi, Kota, Jhalawar, Baran, Bundi, Sawai Madhopur and Karauli districts. Forest areas in general, are conspicuously absent in the entire western Rajasthan. Higher percentage of pasture land, which are often seasonal, exist Barmer, Jodhpur, Jalore and Nagaur districts.

The land use pattern of a region determines the ecological balance in the region and helps to understand the environmental status as well. It includes the management and modification of natural environment in to build environment such as settlements, cultivated land etc. Approximately half of the area of the Rajasthan state is under cultivation with an average Cropping Intensity of 125%. 12.6% is put to Non agricultural Uses i.e. not available for cultivation. The other Land Use classes include 13.27% of land As Culturable Waste, 7.76% under Forest, 10.75% Fallow Land, 5 % under Pastures and Grazing Land. The percentagewise classification is presented in Figure 12.2 and 12.3.

During 2015-2016, the maximum area under forest was in Ajmer, Alwar and Udaipur districts, whereas the lowest three districts were Nagaur, Jodhpur and Churu. Udaipur, Jaisalmer and Bikaner districts lead in area that is not available for cultivation and Sikar, Dausa and Jhunjhunu are the bottom three districts.

Under the use of Permanent pasture and other grazing lands Bharatpur, Hanumangarh and Ganganagar have the least area whereas Barmer, Jodhpur and Bhilwara have the largest area under this category. Churu and Rajsamand districts have no land under miscellaneous trees, crops and groves. Districts Nagaur, Jalore and Hanumangarh cover the least whereas Dungarpur, Ganganagar and Jhalawar cover the maximum area under the same category.

Culturable Wasteland is highest in Ajmer, Alwar and Jaisalmer and minimum at Hanumangarh, Jhunjhunu and Bharatpur. Fallow lands other than current fallow cover the highest area in Alwar, Ajmer and Jodhpur whereas they are lowest at Karauli, Dholpur and Bharatpur. The current fallow land covers the highest area in Ajmer, Alwar and Barmer whereas the lowest are at Karauli, Banswara and Rajsamand.

Under the category of Net Area Sown and Net Area Sown more than once Ajmer and Alwar stand on the top two positions with Barmer and Hanumangarh at the third positions respectively. Dhoplur, Dungarpur and Rajsamand are the lowest three in the category of Net Area Sown and Dungarpur, Sirohi and Rajsamand are the last three of Net Area Sown more than once.

It can be seen that the western districts of Jaisalmer, Bikaner and Barmer have large stretches of sandy soils therefore a very high percentage of cultivable waste and fallow land is present in the region. The net sown area is also less than the other types of land uses here. Higher percentage of seasonal pasture lands exist in Barmer, Jodhpur, Jalore and Nagaur districts. Generally, Forest areas are evidently absent in the entire western Rajasthan.

Churu, Nagaur, Jodhpur, Pali and Jalore districts lying in the mid western districts of arid and the semi arid zone support only rain-fed crops in this area and because of low rainfall the percentage of fallow land is quite high. The net sown area is the highest in the semi-arid zone and the sub-humid parts of eastern Rajasthan which also accounts for extensive agriculture. Due to the presence of hilly terrain in parts of Bhilwara, Chittorgarh, Udaipur, Kota and Baran districts the percentage of cultivable waste and the land put to non agricultural use is highest. The percentage of forest areas in Kherwara (Udaipur), Dungarpur, Banswara, Sirohi, Kota, Jhalawar, Baran, and Sawai Madhopur districts are noticeably high.

LAND-USE CLASSIFICATION

The common nine-fold classification of land use has been generalized here into six broad groups:

- Forest,
- Land put to non-agricultural uses,
- Permanent pastures,
- Net sown area,
- Fallow land(including current and other fallow lands) and
- Cultivable wasteland (including barren and uncultivable land, and land under miscellaneous tree crops).

The percentage of cultivable waste and the land put to non-agricultural use is highest in parts of Bhilwara, Chittaurgarh, Udaipur, Kota and Baran districts, owing to their hilly terrain. The percentage of forest areas is noticeably high in Kherwara (Udaipur), Dungarpur, Banswara, Sirohi, Kota, Jhala-war, Baran and Sawai Madhopur districts. Forest areas, in general, are conspicuously absent in the entire western Rajasthan. Higher percentages of pasture land, which are often seasonal, exist in Barmer, Jodhpur, Jalore and Nagaur districts of state (Figure 12.1).

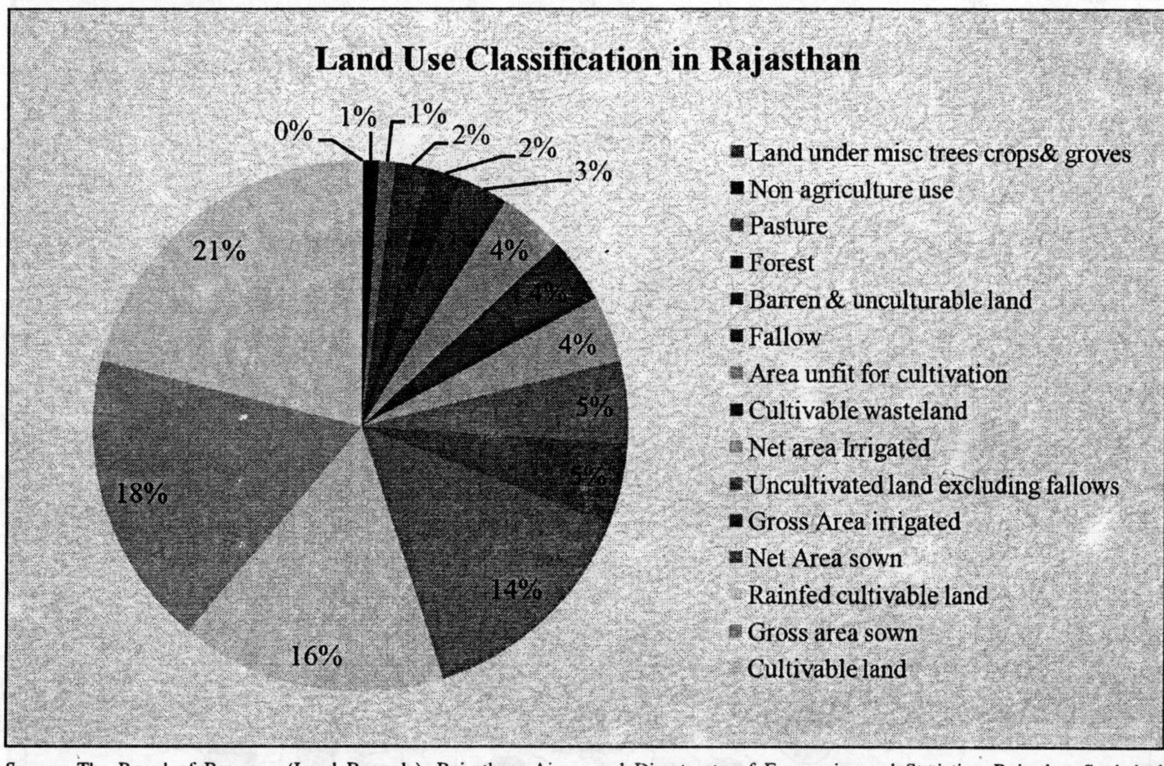

Source: The Board of Revenue (Land Records), Rajasthan, Ajmer and Directorate of Economics and Statistics, Rajasthan Statistical Abstract, Rajasthan, Jaipur

Figure 12.1 Land Use Classification in Rajasthan

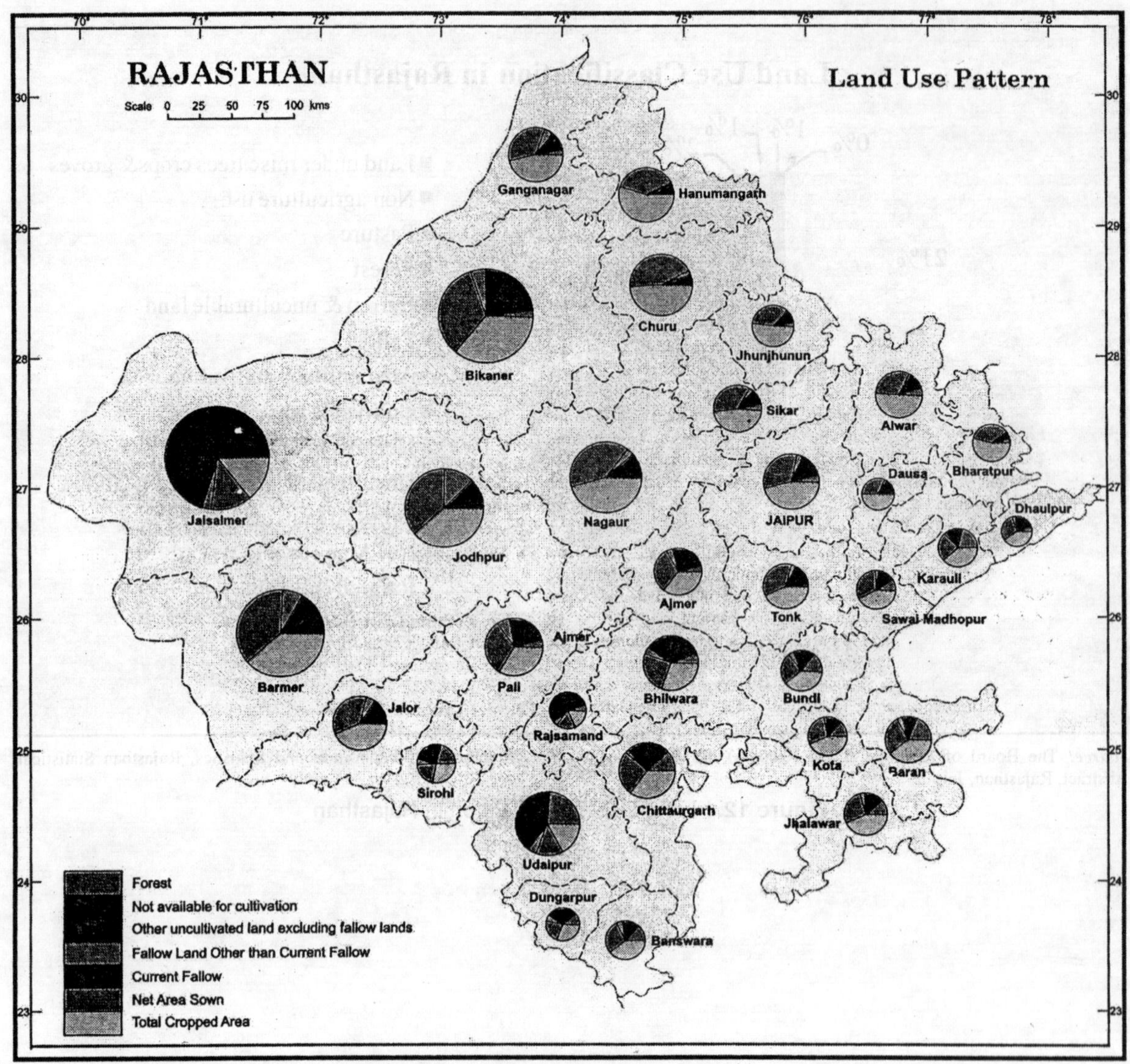

Source: Rajasthan Agricultural Statistics at a Glance Year 2016-17, Commissionerate of Agriculture, Rajasthan, Jaipur

Figure 12.2 Land Use Pattern in Rajasthan

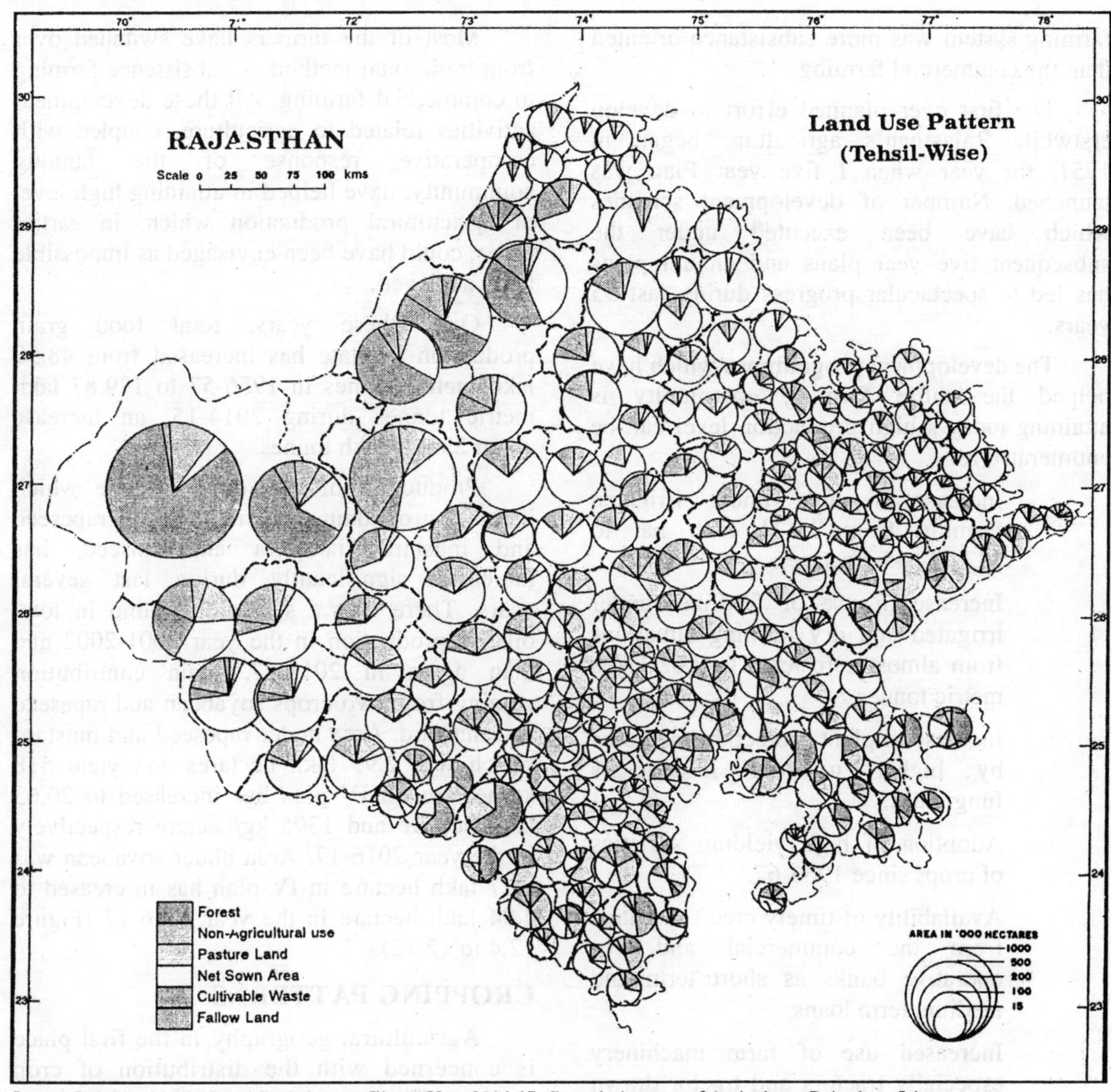

Source: Rajasthan Agricultural Statistics at a Glance Year 2016-17, Commissionerate of Agriculture, Rajasthan, Jaipur

Figure 12.3 Land Use Pattern (Tehsilwise) in Rajasthan

AGRICULTURAL DEVELOPMENT

Rajasthan is basically an agrarian economy. Most of its population lives in small villages and dhanies. It has a wide range of agro climatic regions from very low rainfall in western part to high rainfall in south and south-eastern part of the state. Major part of the state is, however, covered by arid and semi arid climatic conditions which has a characteristic low, erratic and uneven distribution of rainfall associated with lack of other water resources for irrigated farming. Growing only one rainfed crop in kharif season, that too associated with high risk, has led to dependence of village community on livestock. More harsher the climatic conditions, higher was the shift towards animal component. Due to low capital investement capacity, lesser availability of agriculture credit and lack of adequate infrastructural facilities like road, power etc., farming system was more subsistance oriented than the commercial farming.

The first ever planned effort to develop erstwhile Rajasthan's agriculture began in 1951, the year when I five year Plan was launched. Number of development schemes which have been executed under the subsequent five year plans and annual plans has led to spectacular progress during last 55 years.

The development programmes which have helped the entire farming community is attaining today's high production level can be enumerated as:

- Increase in area under irrigated farming from 11.71 lakh ha. to 132.04 lakh hectares.
- Increase in use of fertilizer under irrigated and dry farming situations from almost zero level to 7.62 lakhs metric tonnes.
- Increase in plant protection measures by higher use of insecticides fungicides.
- Adoption of high yielding varieties of crops since 1966-67.
- Availability of timely credit facilities from the commercial and co-operative banks as short term and medium term loans.
- Increased use of farm machinery especially tractors and tractor drawn implements.
- Rural electrification.
- Extensive network of rural roads.
- A good network of Krishi Upaj Mandis for marketing of agricultural produce.
- Warehousing facilities to store farm produce.
- TV programme of Agriculture Extension and Research for adoption of improved agricultural practices.

Most of the farmers have switched over from traditional method of subsistence farming to commercial farming. All these development activities related to agriculture, coupled with co-operative response of the fanning community, have helped in attaining high level of agricultural production which, in earlier times, could have been envisaged as impossible to be achieved.

Over these years, total food grain production in state has increased from 48.50 lakh metric tonnes in 1956-57 to 179.87 lakh metric tonnes during 2014-15, an increase about 231.04 lakh tonnes.

Production of oil seeds in state which includes groundnut, sesamum, castor, rapeseed and mustard, taramira and linseed, has increased significantly during last several years. There was a substantial jump in total oilseed production in the year 2001-2002 and than again in 2016-17, main contribution coming from two crops soyabean and rapeseed and mustard. Area under rapeseed and mustard which was 2.99 lakh hectares and yield 518 kg/hectare, in IV plan has increased to 20.63 lakh hectare and 1306 kg/hectare respectively in the year 2016-17. Area under soyabean was 0.17 lakh hectare in IV plan has in creased to 4.94 lakh hectare in the year 2016-17 (Figure 12.4 to 12.12).

CROPPING PATTERNS

Agricultural geography in the first place is concerned with the distribution of crop systems. The crop system of Rajasthan is so varied and complex that the distribution of particular types can only be shown on very large scale maps of land use. Such maps would be very detailed and the pattern would be complicated. Various maps, which are prepared after incorporating all the conceivable facts about different crops, do not build up into simple, large- scale patterns. When dealing with such a large area as Rajasthan, it would be appropriate to formulate and select as criteria, some main characteristics common to farming of large areas. The variations may be indicated within the major regions.

In the following pages the study of various agricultural crops has been attempted on a statistical basis. The crops have been grouped under the food crops like cereals and pulses, and commercial crops like oilseeds, sugarcane and cotton. The various crops have been described on the basis of acreage devoted to them in each district and then each crop area as the percentage of the total cropped area in the district. Both these criteria clearly indicate the heavy concentration of different crops in different parts of the State.

Agriculture in Rajasthan is marked by certain characteristics which are lacking in industrialised regions. Its main features are as follows :

1. Aridity and topographic features have played important roles in the agriculture of this State. 25 cm rainfall line divides the State into two major divisions, (*i*) the western arid and semi arid sandy plains, and (*ii*) the eastern Aravalli hilly region, the Banas Basin and Plateau area where the rainfall is up to 200 centimetres.
2. Nearly two-fifths of the total area of the State is devoted to farming.
3. A large part of the area devoted to farming, *i.e.* about 90 per cent is under food crops. Of this nearly 65 per cent is devoted to cereals and 25 per cent to pulses. Bajra alone occupies about 30.8 per cent of the area under crops.
4. About 45 per cent area of the State is covered by the districts of Bikaner, Jaisalmer, Barmer, Jodhpur, Nagaur and Churu, all lying in the western half of the State. The total cropped area in all these districts is about 26 per cent, while almost an equal area is classified as culturable waste. About seven-tenths of the total cropped area in these districts is under bajra.
5. Bikaner and Jaisalmer districts lie in the driest part and cover about one-fifth of the total area of the State. About 8.8 per cent of this area is under crops of which three-fifths is under bajra cultivation. About 54.2% of the area of these two districts is classified as culturable waste.
6. No separate specific crop is grown as fodder in this State.
7. The yield per acre is low and in most of the crops, the average yield is much lower than that of India.
8. Over large parts of India, the fields produce two, or more crops in a year. But here, due to aridity, high summer temperatures and large sandy stretches in the west, the double cropped areas are only about 6.8 per cent of the net sown area.
9. In farming, there is little use of fertilizers and manures. Animal refuse which could be used as good manure is mostly burnt.
10. Very little cropped area is devoted to commercial crops; cotton and sugarcane occupy about 1.7 per cent and 0.6 per cent of the total cropped area respectively.
11. The district of Ganganagar, located in the north of the State, mainly because of the facilities for irrigation, grows nearly 32 per cent of the gram acreage, 11 per cent of wheat and barley, 16.6 per cent of sugarcane and about 30 per cent of the cotton acreage of the State.

Cotton is another important cash crop and its production has increased from 1.93 lakh bales to 10.98 lakh bales from I Five Year Plan period to 12 plan period.

Guar seed is also an important crop, specially of desert districts. The state ranks first in guar seed production in the country, contributing about 60 per cent of the central pool.

Rajasthan constitutes about 11 per cent of the geographical area with about 5.49 per cent of population, but has only about one per cent of the irrigation resources of the country. Today, it is contributing 8.48 per cent of food grain production, 15.89 per cent of oil seeds, 5.14 per cent of cotton and 14.91 per cent of total pulses of the country's total production of these crops. Agriculture in the state, being the primary sector and mainstay of over 80 per cent of the state's population, is contributing about 50 per cent of the state's total income. Besides, it provides raw material to a large number of agriculture based industries which is providing employment to a large number rural as well as urban population.

Based on national norms of 160 kg. of requirement of its food grain per adult per annum, Rajasthan inspite of its inhospitable climate and lack of resources, especially irrigation, is one of the few elite states of the country which is surplus in food grains. Present per capita food grain production per annum at the state level is 205. However, inspite of this perceptible development in the field of agriculture, it has to take long strides to fulfil its commitments for still better growth on a sustainable basis in future.

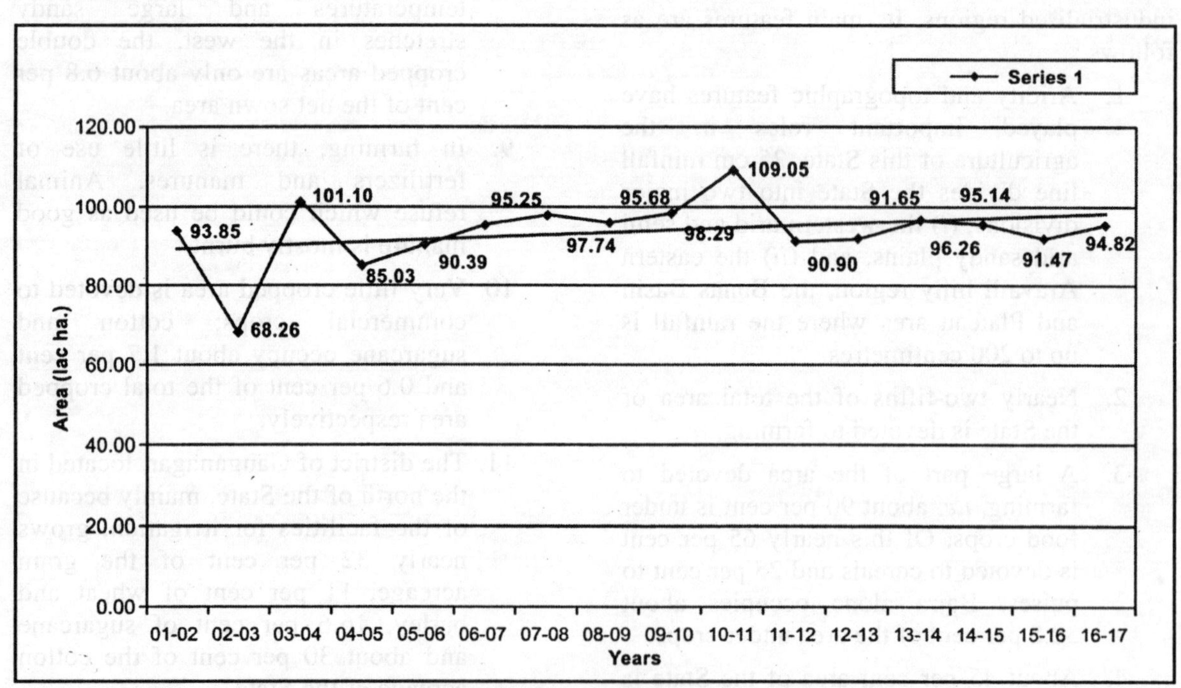

Source: Rajasthan Agricultural Statistics at a Glance Year 2016-17, Commissionerate of Agriculture, Rajasthan, Jaipur

Figure 12.4 Area Under Total Cereals

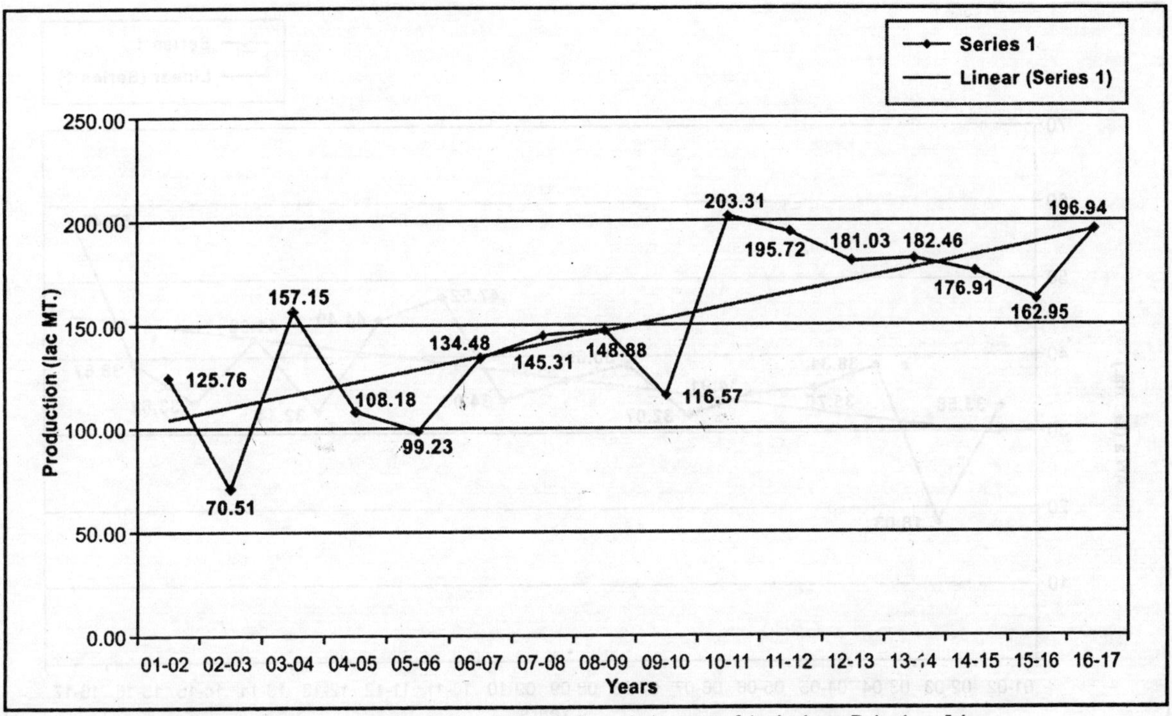

Source: Rajasthan Agricultural Statistics at a Glance Year 2016-17, Commissionerate of Agriculture, Rajasthan, Jaipur

Figure 12.5 Production Under Total Cereals

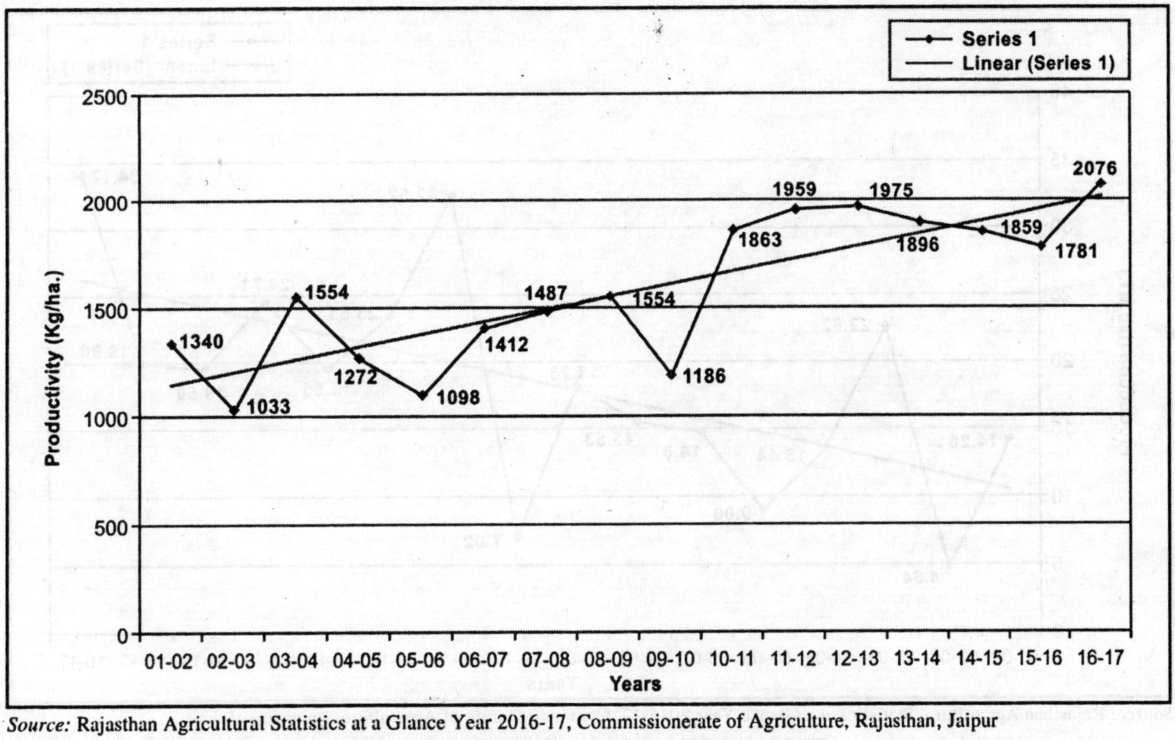

Source: Rajasthan Agricultural Statistics at a Glance Year 2016-17, Commissionerate of Agriculture, Rajasthan, Jaipur

Figure 12.6 Productivity of Total Cereals

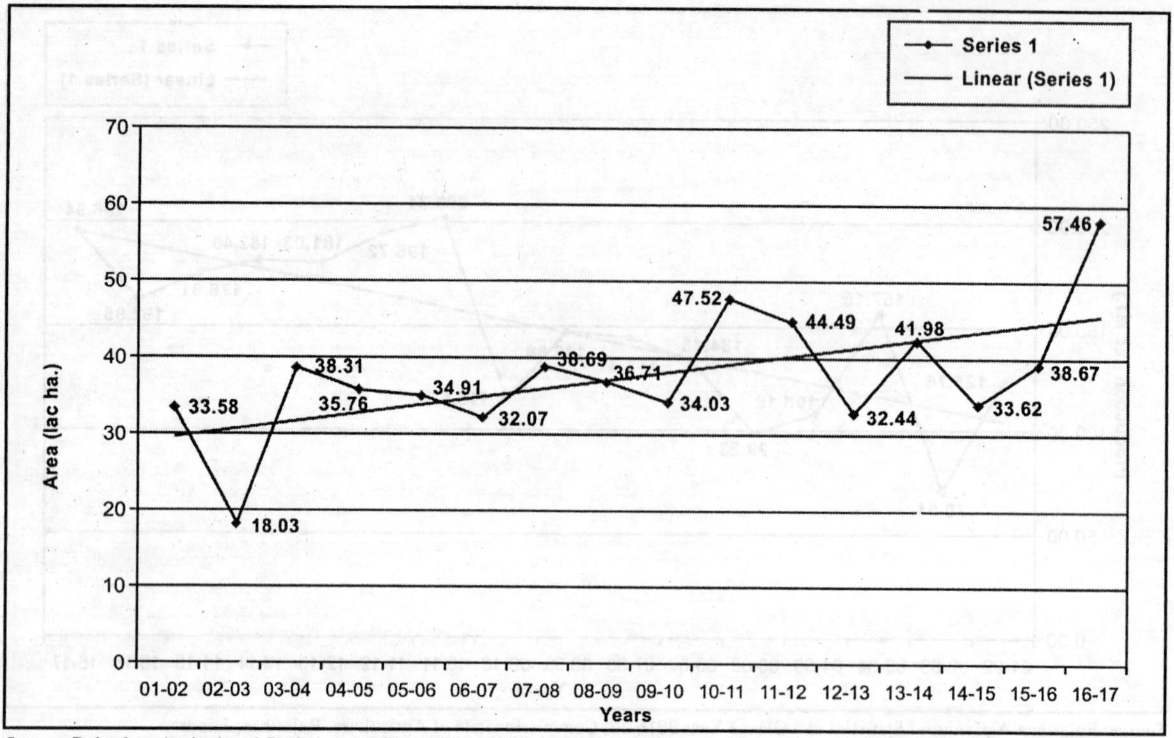

Source: Rajasthan Agricultural Statistics at a Glance Year 2016-17, Commissionerate of Agriculture, Rajasthan, Jaipur

Figure 12.7 Area Under Total Pulses

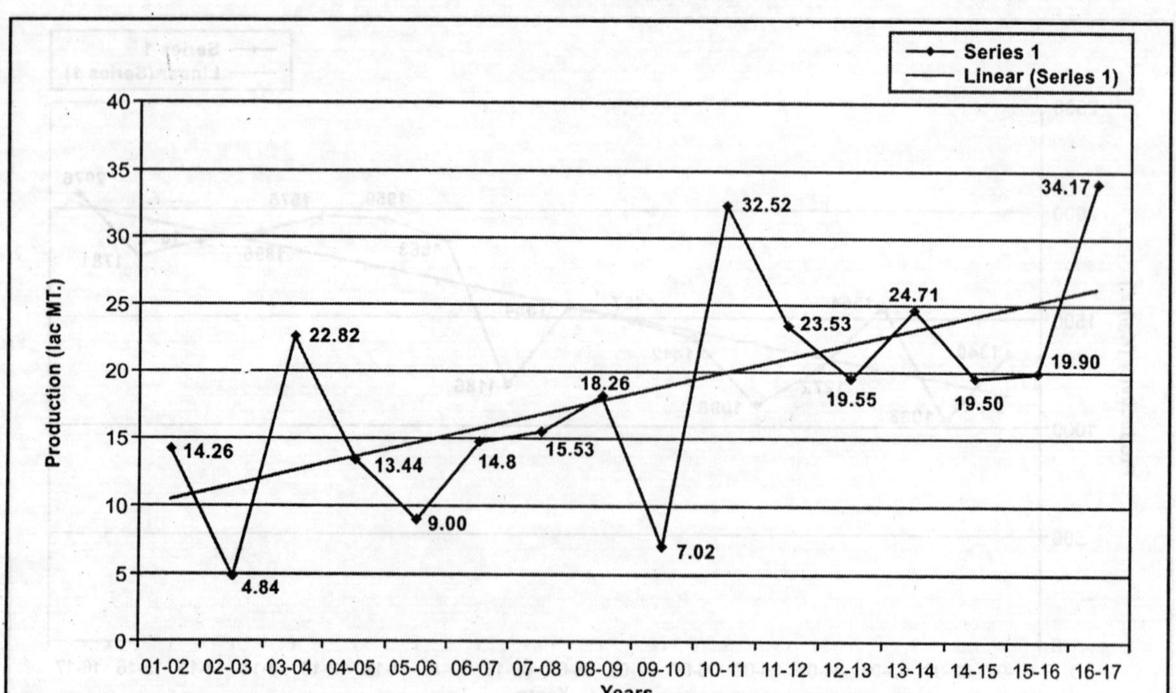

Source: Rajasthan Agricultural Statistics at a Glance Year 2016-17, Commissionerate of Agriculture, Rajasthan, Jaipur

Figure 12.8 Production Under Total Pulses

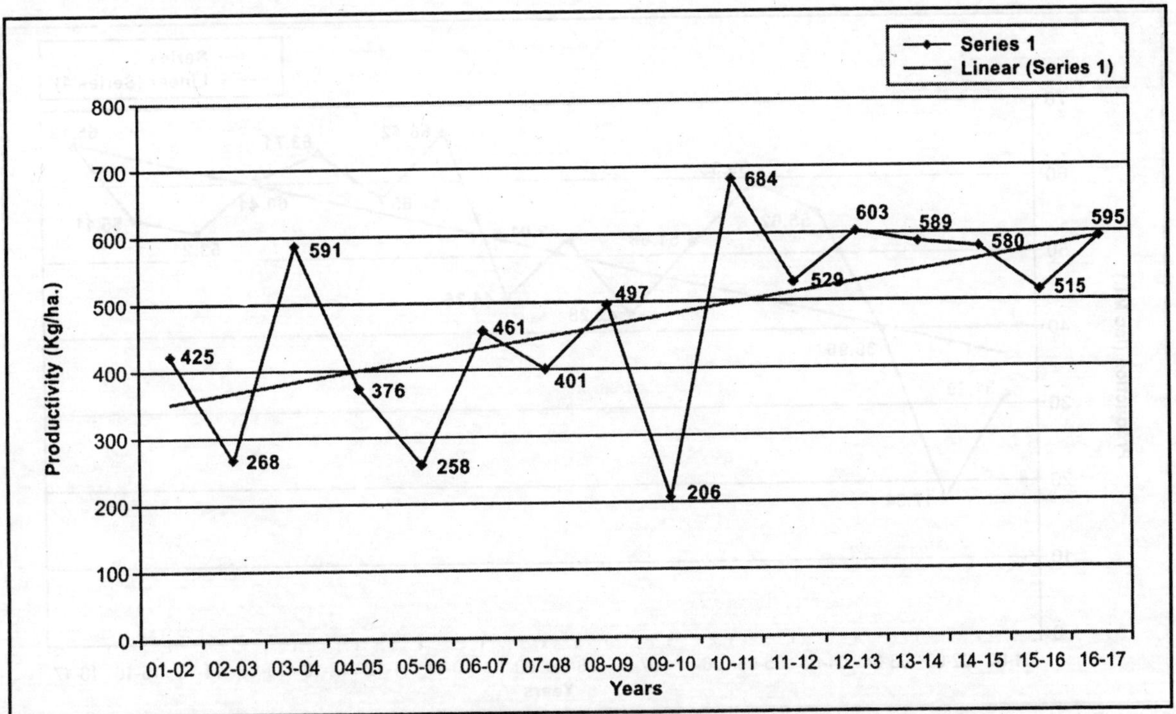

Series 1
Linear (Series 1)
Productivity (Kg/ha.)
800
700
600
500
400
300
200
100
0
425
268
591
376
258
461
401
497
206
684
529
603
589
580
515
595
01-02 02-03 03-04 04-05 05-06 06-07 07-08 08-09 09-10 10-11 11-12 12-13 13-14 14-15 15-16 16-17
Years

Source: Rajasthan Agricultural Statistics at a Glance Year 2016-17, Commissionerate of Agriculture, Rajasthan, Jaipur

Figure 12.9 Productivity of Total Pulses

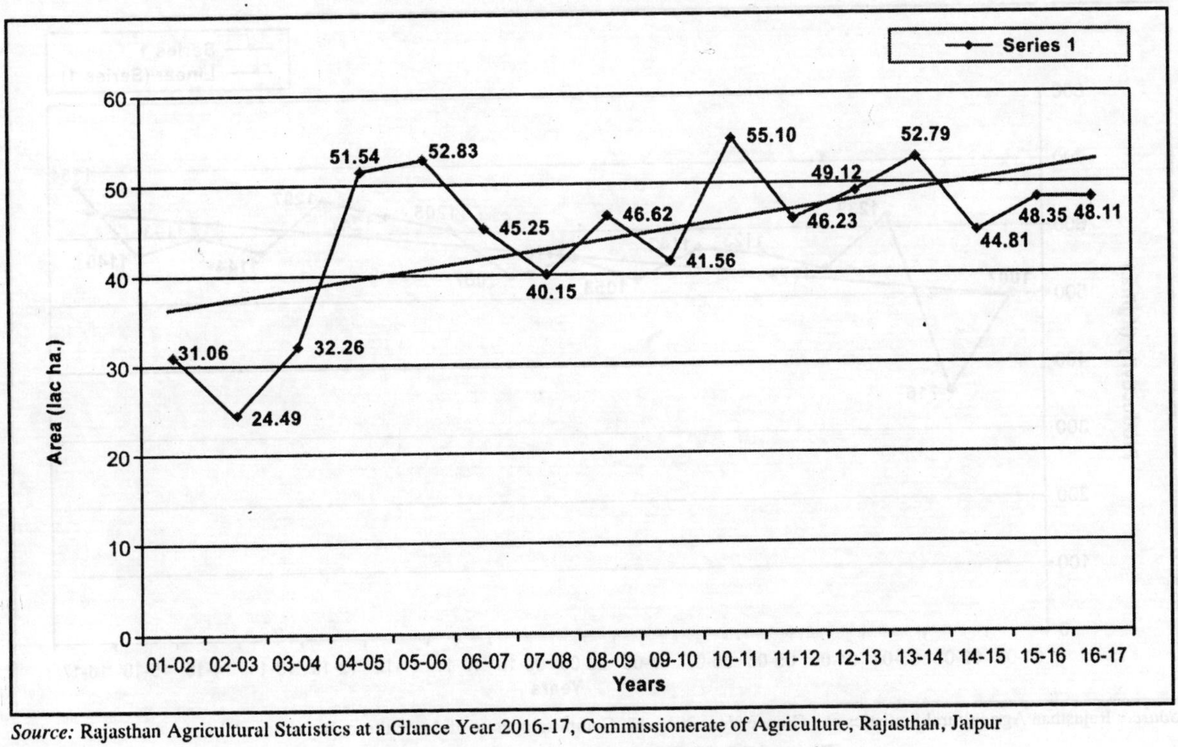

Series 1
Area (lac ha.)
60
50
40
30
20
10
0
31.06
24.49
32.26
51.54
52.83
45.25
40.15
46.62
41.56
55.10
46.23
49.12
52.79
44.81
48.35
48.11
01-02 02-03 03-04 04-05 05-06 06-07 07-08 08-09 09-10 10-11 11-12 12-13 13-14 14-15 15-16 16-17
Years

Source: Rajasthan Agricultural Statistics at a Glance Year 2016-17, Commissionerate of Agriculture, Rajasthan, Jaipur

Figure 12.10 Area Under Total Oil-seed

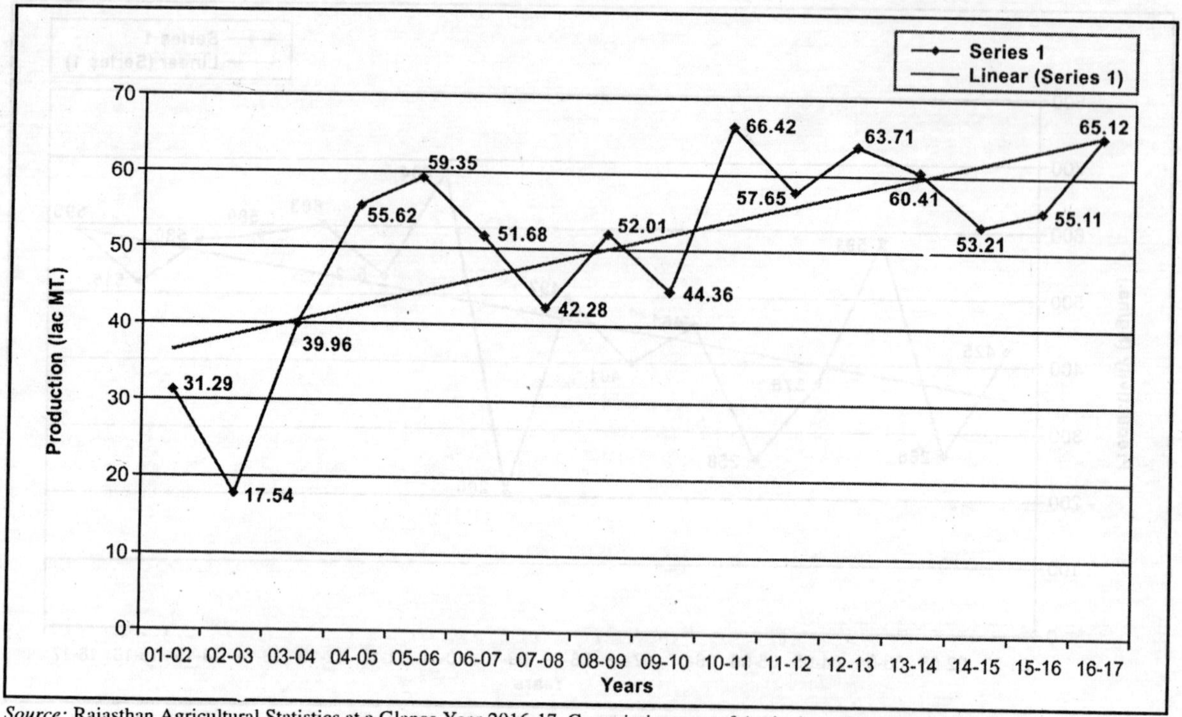

Series 1
Linear (Series 1)
Production (lac MT.)
Years

Years	01-02	02-03	03-04	04-05	05-06	06-07	07-08	08-09	09-10	10-11	11-12	12-13	13-14	14-15	15-16	16-17
Production (lac MT.)	31.29	17.54	39.96	55.62	59.35	51.68	42.28	52.01	44.36	66.42	57.65	63.71	60.41	53.21	55.11	65.12

Source: Rajasthan Agricultural Statistics at a Glance Year 2016-17, Commissionerate of Agriculture, Rajasthan, Jaipur

Figure 12.11 Production Under Total Oil-seed

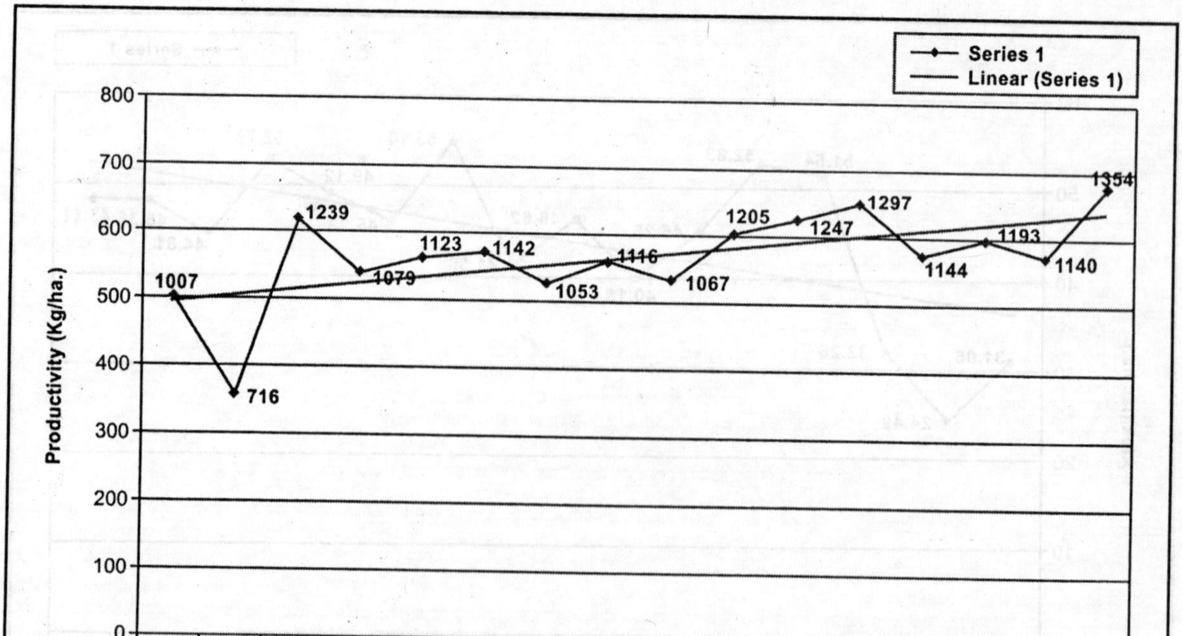

Series 1
Linear (Series 1)
Productivity (Kg/ha.)
Years

Years	01-02	02-03	03-04	04-05	05-06	06-07	07-08	08-09	09-10	10-11	11-12	12-13	13-14	14-15	15-16	16-17
Productivity (Kg/ha.)	1007	716	1239	1079	1123	1142	1053	1116	1067	1205	1247	1297	1144	1193	1140	1354

Source: Rajasthan Agricultural Statistics at a Glance Year 2016-17, Commissionerate of Agriculture, Rajasthan, Jaipur

Figure 12.12 Productivity of Total Oil-seed

BAJRA

Bajra is the major food crop of Rajasthan. This millet can be grown in sandy soils under rainfed conditions and hence assumes importance in arid region of Rajasthan. This food grain crop is also grown abundantly in the arid tract of Gujarat, Uttar Pradesh, Karnataka, Maharashtra and Andhra Pradesh. It has multiple uses besides being a staple food, its fodder is an important feed for milch animals.

Bajra grows mostly in those areas, which have high diurnal temperatures. The temperatures during growing season remains between 25°C to 35°C. The optimum rainfall requirements range between 35 to 50 cm. Bajra can be grown in areas which receive less than 35 cm of annual rainfall but prolonged spells of warm, rainless weather may be detrimental and may led to crop failures, At harvest time, dry warm weather is most suitable.

Although several strains of new hybrid of Bajra have been developed, there has not been a substantial increase in its productivity, chiefly for the reason that this crop is still a rainfed crop. Failure of monsoon or its uneven distribution leads to crop failures. The productivity of Bajra in past decades has thus varied from 89 kg/ha (1979-80) 790 kg/ha (2016-2017) in Rajasthan. Because of its tolerance of high temperatures and drought, amongst competing other kharif crops, its cultivation is predominantly high in low and erratic rainfalls areas.

Rajasthan accounts for 55.29 percent of total area of 10.58 millions hectare in India under bajra, other predominant bajra growing states are Maharashtra, Gujarat, Uttar Pradesh, Haryana and Karanataka. Looking to the high production efficiency of Bajra crop, supplemental irrigation has also been provided to this cop, particularly to the high yielding strains in Gujarat and Haryana state. In Rajasthan, the major districts which are providing irrigation to bajra crops are Jaipur, Sikar, Ganganagar, Jodhpur, Nagaur and Jhunjhunun.

Bajra is grown in over 58.58 lakh hectares of land in Rajasthan. The major areas lies in western Rajasthan, the highest (above 10.20 lakh ha) is in Barmer district, alone. In Jodhpur, Jalore and Nagaur districts crop grown over 7.09 lakh ha, 3.92 lakh ha and 5.61 lakh ha respectively. Other districts where this crop is significant are Churn, Jhunjhunun, Sikar, Bikaner, Jaipur and Dausa. In Ganganagar, as also in the entire southern and south-eastern Rajasthan, bajra cultivation is less. The area under this croop, especially in western districts fluctuates from year to year, as this crop is totally dependent on the rains.

The production of the crop is highest in Jodhpur district 8.27 lakh metric tones. Total production in the state is 66.57 lakh metric tones and contributes 27.00 per cent of the total production of food grains crops in Rajasthan (2016-17). In predominantly bajra growing districts productions ranges from 1.00 lakh metric tones to 3.00 lakh tones.

Productivity is markedly high in eastern as compared to western Rajasthan. The state productivity is 1136 kg/ha, while in Karauli 2,264 kg/ha, Sawai Madhopur 1,780 kg/ha, Dhaulpur 1,778 kg/ha, Alwar 1,728 kg/ha, Bharatpur 1,673 kg/ha, Dausa 1,856 kg/ha and Jaipur 1,678 kg/ha in other district like Barmer, Bikaner and Jaisalmer the yield is lower ranging from 5,00 to 1,000 kg/ha. Yearly variations in productivity of this crop especially in western parts are very high. While in good rainfall years it is fairly high, but their drastic reduction during drought years.

Among the newly developed hybrids, short duration hybirds will be more beneficial for cultivation in western arid tracts while medium maturing types are suitable for rest of the areas. Besides, the hybrids, cultivation of composite varieties has an added advantage (Figure 12.13 and 12.14).

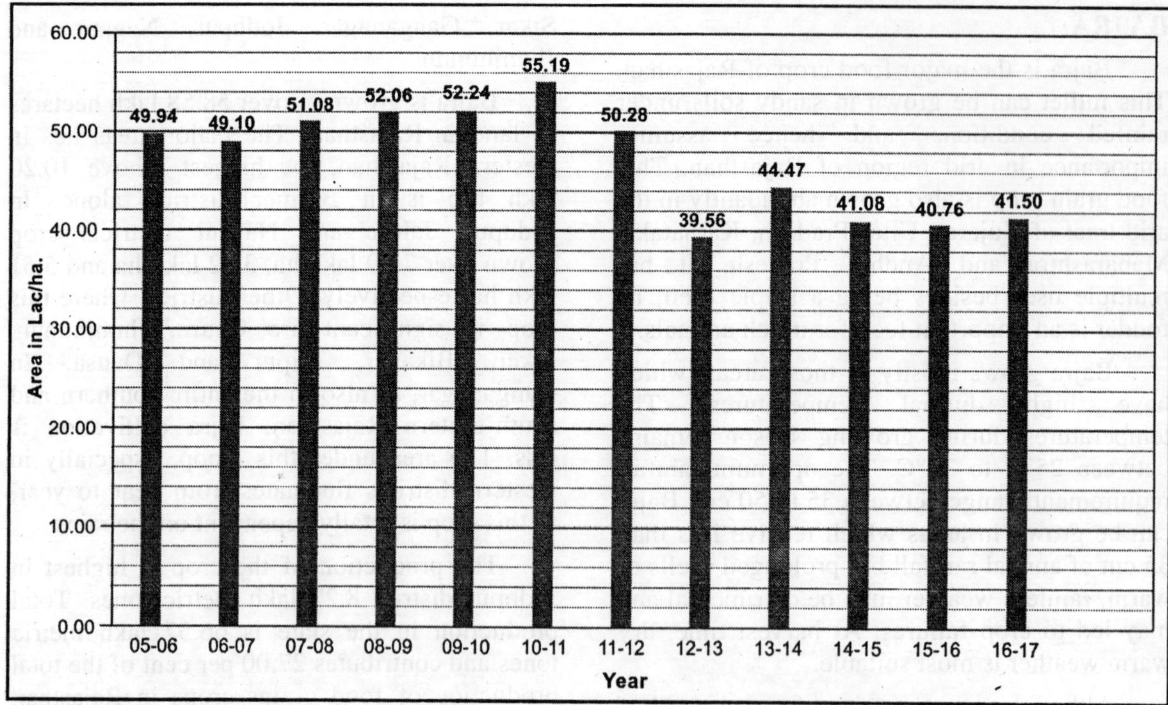

Source: Rajasthan Agricultural Statistics at a Glance Year 2016-17, Commissionerate of Agriculture, Rajasthan, Jaipur

Figure 12.13 Yearwise Area under Bajra Crop in Rajasthan (2005-06 to 2016-17)

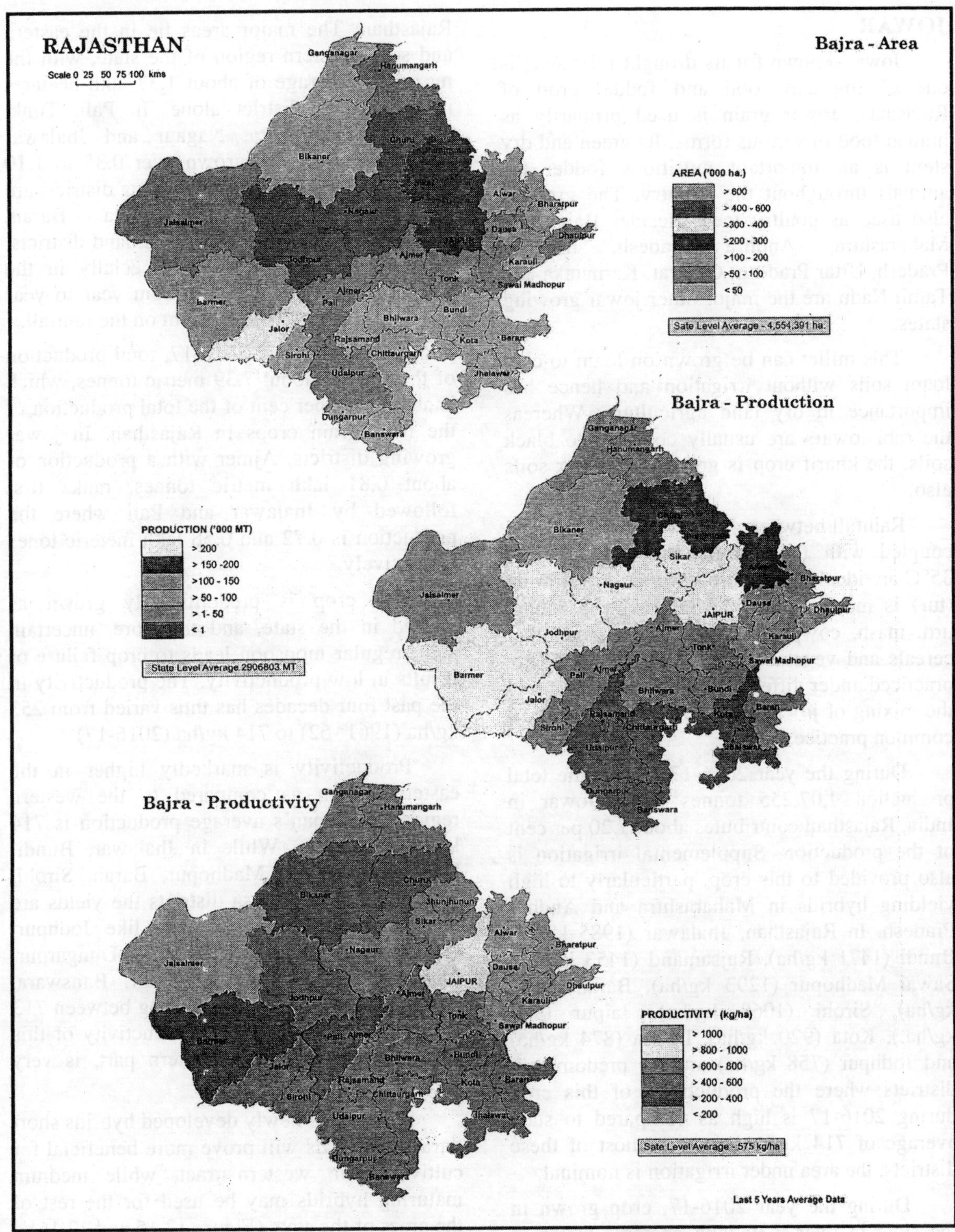

Source: Rajasthan Agricultural Statistics at a Glance Year 2016-17, Commissionerate of Agriculture, Rajasthan, Jaipur

Figure 12.14 Crop - Bajra Area in Rajasthan

JOWAR

Jowar, known for its drought tolerance, is one of imprtant food and fodder crop of Rajasthan. Jowar grain is used primarily as human food in various forms. Its green and dry stem is an important nutritious fodder for animals throughout the country. The grain is also used as poultry feed. Besides Rajasthan, Maharashtra, Andhra Pradesh, Madhya Pradesh, Uttar Pradesh, Gujarat, Karnataka and Tamil Nadu are the major other jowar growing states.

This millet can be grown on loam to clay loam soils without irrigation and hence has importance in dry land agriculture. Whereas the rabi jowars are usually confined to black soils, the kharif crop is grown on lighter soils also.

Rainfall between 40 to 100 cm per annum, coupled with temperature between 25°C to 35°C are ideal for this crop. Mixed jowar-arhar (tur) is most common. Mixtures with mung, urd, mash, cowpea and even bajra and other cereals and vegetables, etc. during kharif, are practiced under different situations. During rabi the mixing of jowar with safflower is the most common practice.

During the year 2016-17, out of the total production 4,07,255 tonnes under Jowar in India, Rajasthan contributes about 7.20 per cent of the production. Supplemental irrigation is also provided to this crop, particularly to high yielding hybrids in Maharashtra and Andhra Pradesh. In Rajasthan, Jhalawar (1985 kg/ha) Bundi (1471 kg/ha), Rajsamand (1453 kg/ha), Sawai Madhopur (1293 kg/ha), Baran (1115 kg/ha), Sirohi (1068 kg/ha), Jaipur (947 kg/ha.), Kota (920 kg/ha), Dausa (874 kg/ha) and Jodhpur (758 kg/ha) are the predominent districts where the productivity of this crop during 2016-17 is high as compared to state average of 714 kg/ha but even most of these districts, the area under irrigation is nominal.

During the year 2016-17, crop grown in the state in over 7.39 lakh hectares which is 5.59 per cent of the gross cropped area of Rajasthan. The major areas lie in the eastern and south eastern region of the state, with the maximum average of about 1.37 lakh hectares in the Ajmer district alone. In Pali, Tonk, Bharatpur, Bhilwara, Nagaur and Jhalawar districts, the crop is grown over 0.35 to 1.10 lakh hectares. Other jowar growing districts are Jaipur, Jodhpur, Alwar, Kota, Baran, Chittaurgarh, Udaipur and Rajsamand districts. The area under this crop, especially in the western districts, fluctuates from year to year as the crop is totally dependent on the rainfall.

During the year 2016-17, total production of the state is about 7.39 metric tonnes, which is about 4.11 per cent of the total production of the food grain crops in Rajasthan. In jowar growing districts, Ajmer with a production of about 0.81 lakh metric tonnes, ranks first followed by Jhalawar and Pali where the production is 0.72 and 0.58 lakh meteric tones respectively.

The crop is predominently grown as rainfed in the state, and therefore, uncertain and irregular monsoon leads to crop failure or results in low productivity. The productivity in the past four decades has thus varied from 252 kg/ha (1961 -62) to 714 kg/ha (2016-17).

Productivity is markedly higher in the eastern region as compared to the western region. Rajasthan's average production is 714 kg/ha (2016-17). While in Jhalawar, Bundi, Rajsamand, Sawai Madhopur, Baran, Sirohi, Jaipur, Kota and Dausa districts the yields are more than 874 kg/ha, district like Jodhpur, Ganganagar, Banner, Bikaner, Dungarpur, Jaisalmer, Jalore, Karauli, Sikar, Banswara, and Dhaulpur have yields ranging between 713 to 758 kg/ha. Variation in productivity of this crop, especially in the western part, is very high.

Among the newly developed hybrids short duration hybrids will prove more beneficial for cultivation in western tract, while medium maturing hybrids may be used for the rest of the areas of the state (Figure 12.15 and 12.16).

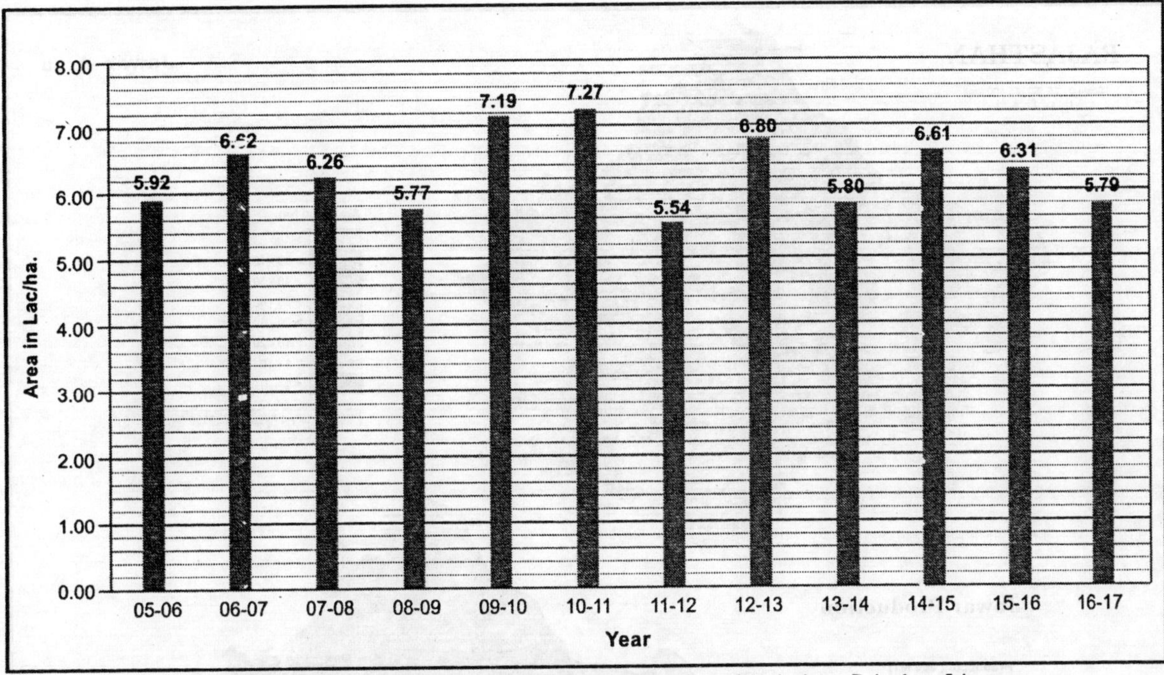

Source: Rajasthan Agricultural Statistics at a Glance Year 2016-17, Commissionerate of Agriculture, Rajasthan, Jaipur

Figure 12.15 : Yearwise Area under Crop Jowar in Rajasthan (2005-06 to 2016-17)

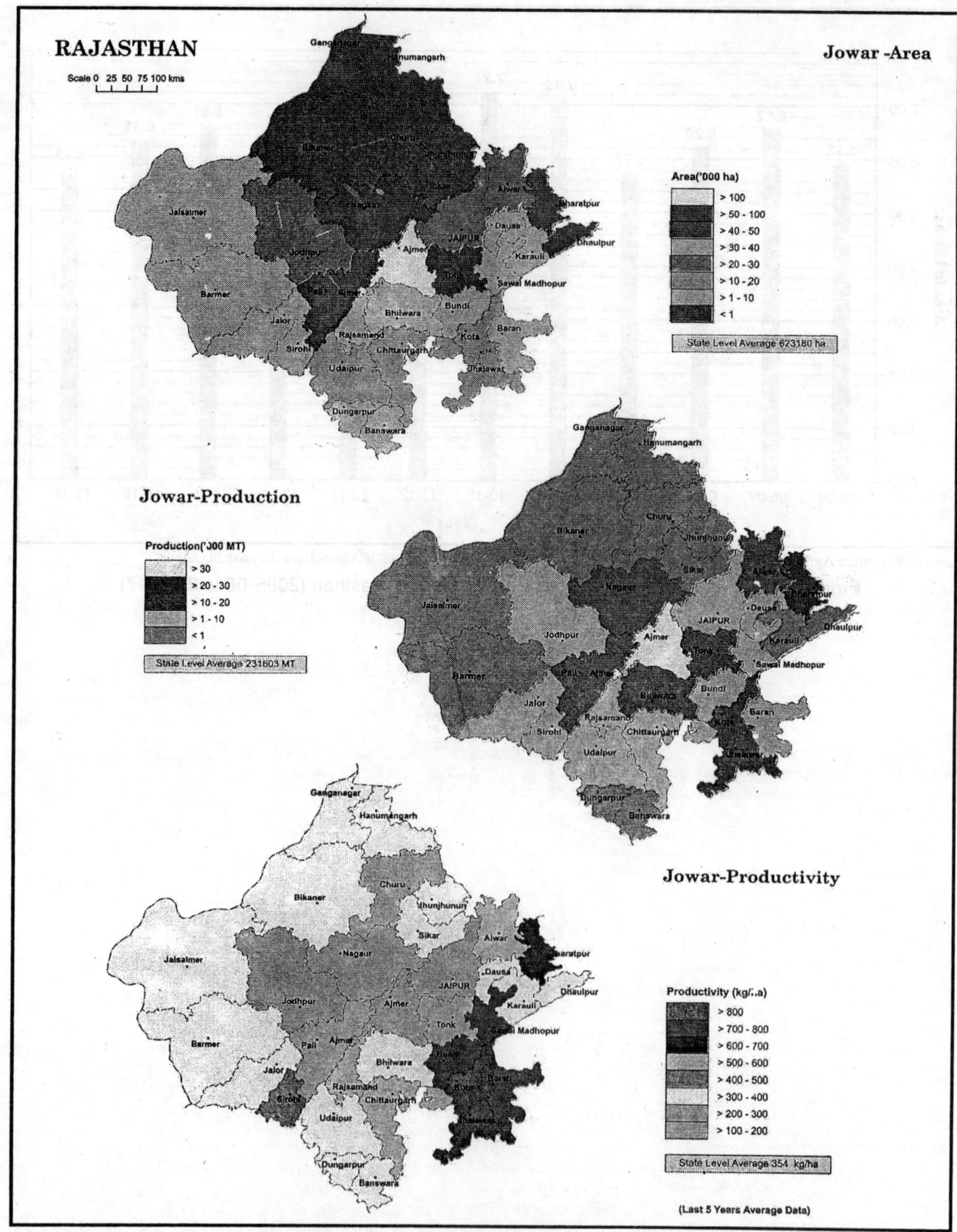

Source: Rajasthan Agricultural Statistics at a Glance Year 2016-17, Commissionerate of Agriculture, Rajasthan, Jaipur

Figure 12.16 Crop - Jowar Area in Rajasthan

MAIZE

Maize, also called com, is predominantly a kharif season crop. It is the staple food grain of southern districts of Rajasthan, namely; Udaipur, Rajsamand, Banswara, Dungarpur, Chittaurgarh, Bhilwara and Jhalawar. These districts have good well distributed rainfall over the crop season. They also have sandy loam to loamy soils with good drainage and therefore are most suited to this crop. However, its area under cultivation in rabi/summer season is also increasing in the southern parts of the state.

It is grown in areas where the average maximum and minimum temperatures remain about 35°C and 12°C respectively. It is mostly a rainfed crop and is grown in areas receiving 50 to 100 cms of average rainfall. Mostly sown in July with the commencement of the first monsoon showers, the maize crop requires 3 to 4 months to ripen. Harvesting is done in Oct.-November. The crop yields higher when grown in fertile loamy soils. Stagnation of water in the field causes decay of plants and so the crops are mostly grown in well drained soils.

In India, this crop is mostly grown in parts of the states of Uttar Pradesh, Uttaranchal, Rajasthan, Madhya Pradesh and Chattishgarh. About 15-20 per cent of the country's total area under this crop is covered in Rajasthan. Although it is mainly a rainfed crop, supplemental irrigation is also given to some areas. Maize is also commonly grown in Andhra Pradesh, Bhiar, Himachal Pradesh, Gujarat, Karnataka, Punjab and Jammu and Kashmir.

The area sown in Rajasthan state under this crop is about 7.52 per cent of the area covered by kharif crop. The highest average, of more than 1.81 lakh hectares is in Bhilwara district followed by Udaipur 1.80 lakh hectares. In Chittaurgarh, Banswara, Dungarpur, Jhalawar, Rajsamand districts, maize in grown in areas ranging from 0.60 to 1.75 lakh hectares. The other districts which grow maize are Bundi, Baran, Ajmer, Sirohi, Pali, Tonk, Kota, Alwar, Jaipur and Dausa in 2016-17.

While the production of this crop in the state is about 20.69 meteric tonnes, its share in the total food grain production of the state is about 11.50 per cent. Major producers of this crop are Chittaurgarh, Udaipur, Bhilwara, Banswara, Jhalawar and Rajsamand districts, each producing more 1.41 lakh meteric tonnes. Production is also high in Dungarpur, Bundi, Ajmer, Sirohi, Baran and Pali districts ranging from 0.45 to 0.89 lakh meteric tonnes. Tonk, Kota, Alwar and Jaipur districts have production ranging from 0.11 to 0.23 lakh meteric tonnes in 2016-17. During low rainfall years, production is adversely affected.

Productivity of the crop is variable, depending largely upon the quantum of annual rainfall and its distribution at the time of sowing and growth of the plants. In Rajasthan, the productivity during the last 40 years ranged from 377 kg/ha as in (1987-88) to 1863 kg/ha in (2016-17), but the average productivity has usually remained around 800-900 kg/hect over the last four decades. Per hectare yield of this crop in 2016-17 varies in different districts from 1179 to 2285 kg. The yields are more than 1900 kg/ha in Jhalawar, Rajsamand, Chittaurgarh, Banswara, Churu, Udaipur and Dhaulpur districts. Sikar, Ganganagar, Barmer, Nagaur, Karauli, Dausa, Sawai Madhoupr, Jalore, Hanumangarh, Bharatpur, Jodhpur, Sirohi, Bundi and Pali districts have a range from 1800 to 1882 kg/ha. In The rest of the districts, except Bikaner, Jaisalmer and Jhunjhunun the yields are either around or lower than state average.

New strains in this crop are being developed, particularly with a view to stabilize production. Short duration hybrids and composites whill go a long way in giving a fillip to production in the state (Figure 12.17 and 12.18).

Normally about 1/8th output of maize in the country comes from Rajasthan. It is produced in Udaipur, Chittorgarh, Bhilwara districts to a greather extent. Its production also fluctuates widely from year to year 2014-15 had reached a level of 15.6 lakh tonnes.

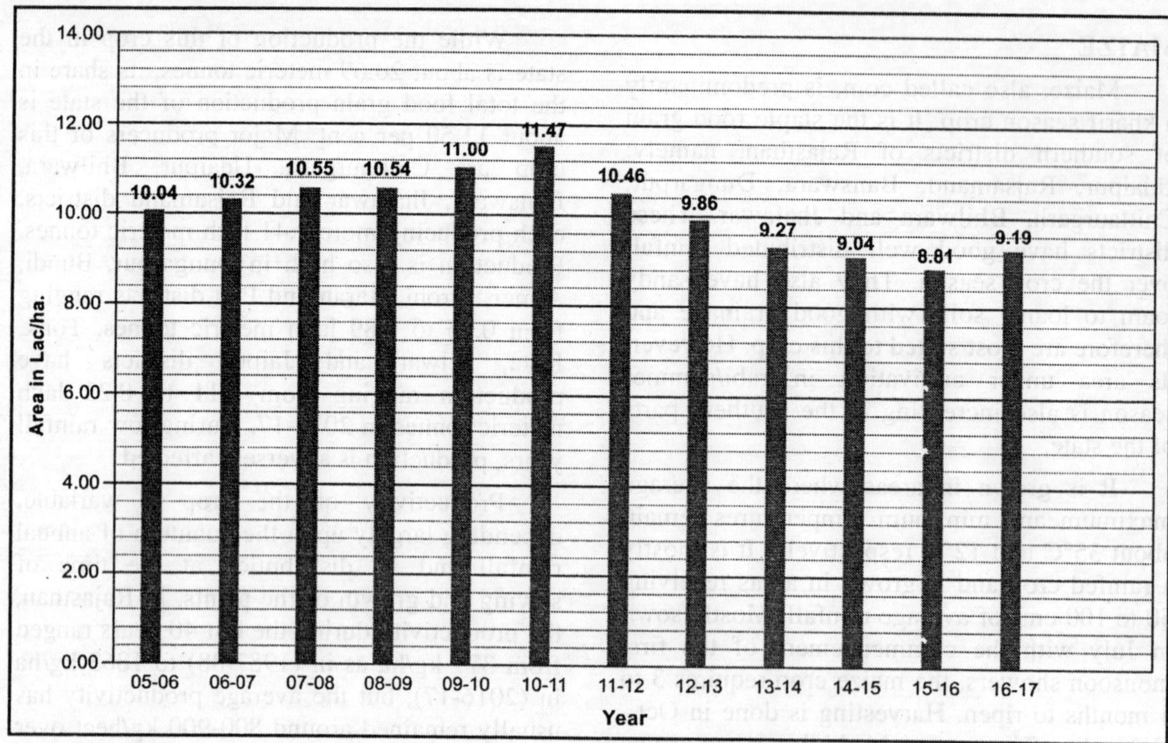

Source: Rajasthan Agricultural Statistics at a Glance Year 2016-17, Commissionerate of Agriculture, Rajasthan, Jaipur

Figure 12.17 Yearwise Area under Crop Maize in Rajasthan (2005-06 to 2016-17)

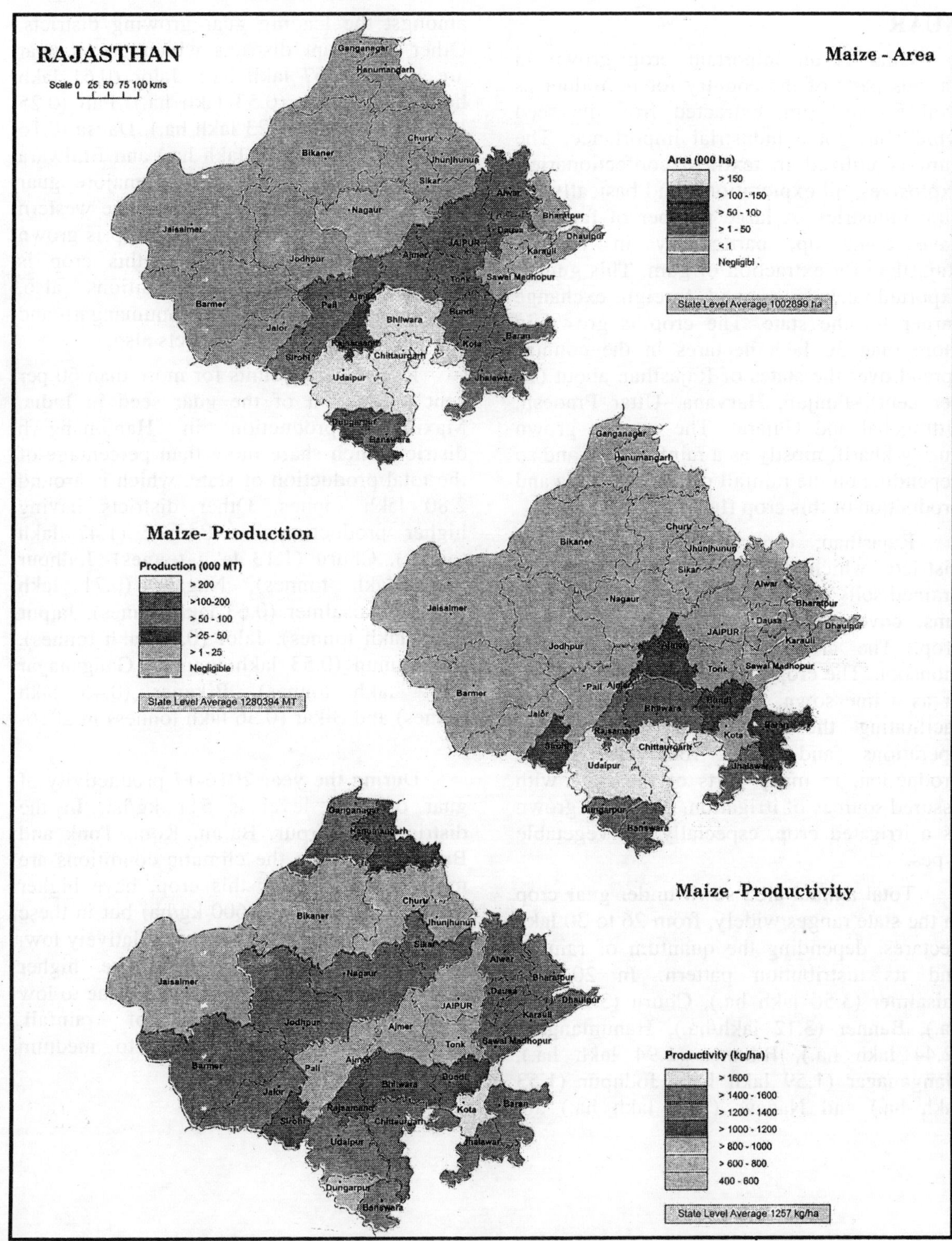

Source: Rajasthan Agricultural Statistics at a Glance Year 2016-17, Commissionerate of Agriculture, Rajasthan, Jaipur

Figure 12.18 Crop - Maize Area in Rajasthan

GUAR

Guar is an important crop grown in various parts of the country for its fodder as well for its gum, extracted from its seed which has got a industrial importance. The gum is utilized in textiles, confectionaries, explosives, oil explorations, and basically the gum industries. A large number of factories have come up, particularly in western Rajasthan for extraction of gum. This gum is exported and is a good foreign exchange earner for the state. The crop is grown in more than 30 lakh hectares in the country spread over the states of Rajasthan about (76 per cent), Punjab, Haryana, Uttar Pradesh, Uttranchal and Gujarat. The crop is grown during kharif, mostly as a rainfed crop and so depending on the rainfall pattern, the area and production of this crop fluctuates annually.

Rajasthan, especially in the western districts which have light textured, well drained soils and rainfall in range of 10 to 30 cms. cover the maximum area under this crop. The crop is sown with onset of monsoon. The crop is sown both as broadcast or as a line sown, latter practice helping in facilitating the process of inter-cultural operations and give relatively higher production. In many parts of the state with assured sources of irrigation, its is also grown as a irrigated crop, especially the vegetable types.

Total annual area sown under guar crop in the state ranges widely, from 26 to 30 lakh hectares, depending the quantum of rainfall and its distribution pattern. In 2016-17 Jaisalmer (3.56 lakh ha.), Churu (3.27 lakh ha.), Banner (3.12 lakh/ha.), Hanumangarh (2.44 lakh ha.), Bikaner (1.94 lakh ha.), Ganganagar (1.59 lakh ha.), Jodhpur (1.53 lakh ha.) and Nagaur (1.41 lakh ha.) are amongst the leading guar growing districts. Other important districts which grows guar are Jaipur (0.67 lakh ha.), Jalor (0.61 lakh ha.), Jhunjhunun (0.53 lakh ha.), Pali (0.25 lakh ha.), Alwar (0.23 lakh ha.), Dausa (0.16 lakh ha.), Tonk (0.12 lakh ha.) and Bhilwara (0.10 lakh ha.). Most of the majore guar growing districts are situated on the western arid part of the state since this crop is grown mostly as rainfed. However, this crop is grown under irrigated conditions also, especially, in Ganganagar, Hanumangarh and to some extent in other distircts also.

Rajasthan accounts for more than 60 per cent production of the guar seed in India. Maximum production in Hanumangarh district which share more than percentage of the total production of state, which is around 2.80 lakh tonnes. Other districts having higher production are Banner (1.47 lakh tonnes), Churu (1.13 lakh tonnes), Jodhpur (0.83 lakh tonnes), Nagaur (0.71 lakh tonnes), Jaisalmer (0.63 lakh tonnes), Jaipur (0.56 lakh tonnes), Jalor (0.54 lakh tonnes), Jhunjhunun (0.53 lakh tonnes), Ganganagar (0.52 lakh tonnes), Bikaner (0.46 lakh tonnes) and Sikar (0.36 lakh tonnes) in 2016-17.

During the year 2016-17 productivity of guar of state level is 511 kg/ha. In the districts-Dungarpur, Baran, Kota, Tonk and Bundi etc. where the climatic conditions are more favourable for this crop, have higher productivity (999 to 1600 kg/ha) but in these districts area under this crop is relatively low. In western districts which have higher proportion of area under this crop, due to low and uneven distribution of rainfall, productivity of guar is low to medium (Figure 12.19 and 12.20).

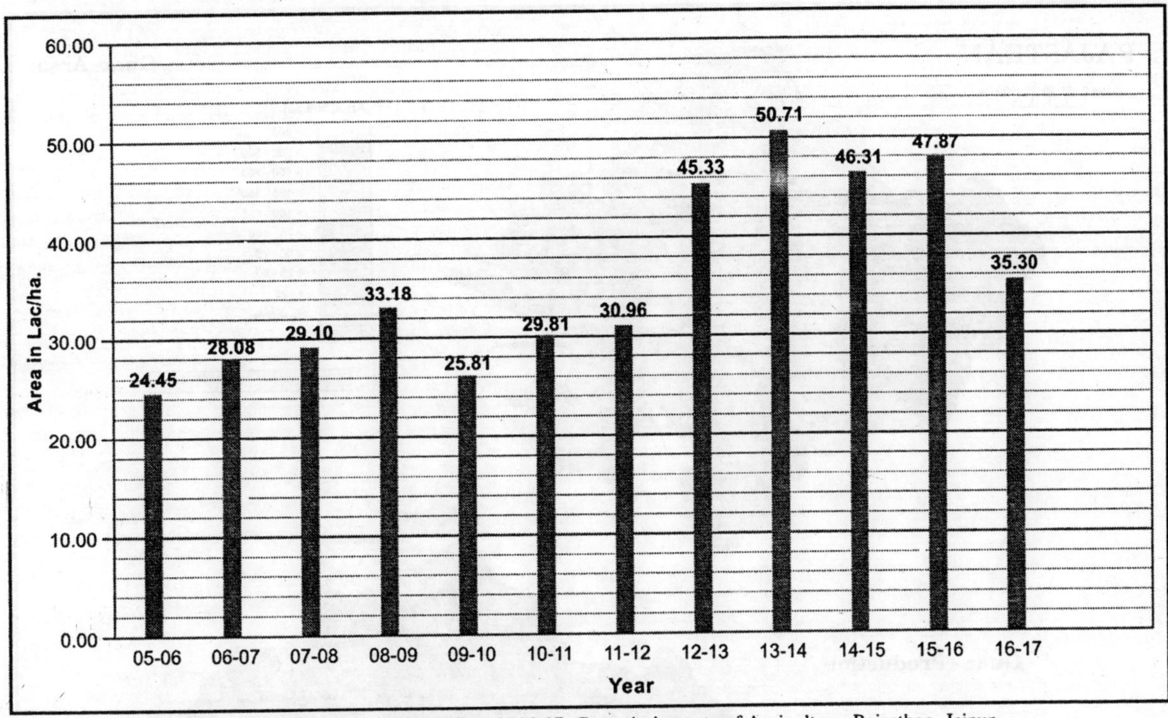

Source: Rajasthan Agricultural Statistics at a Glance Year 2016-17, Commissionerate of Agriculture, Rajasthan, Jaipur

Figure 12.19 Yearwise Area under Crop Guar in Rajasthan (2005-06 to 2016-17)

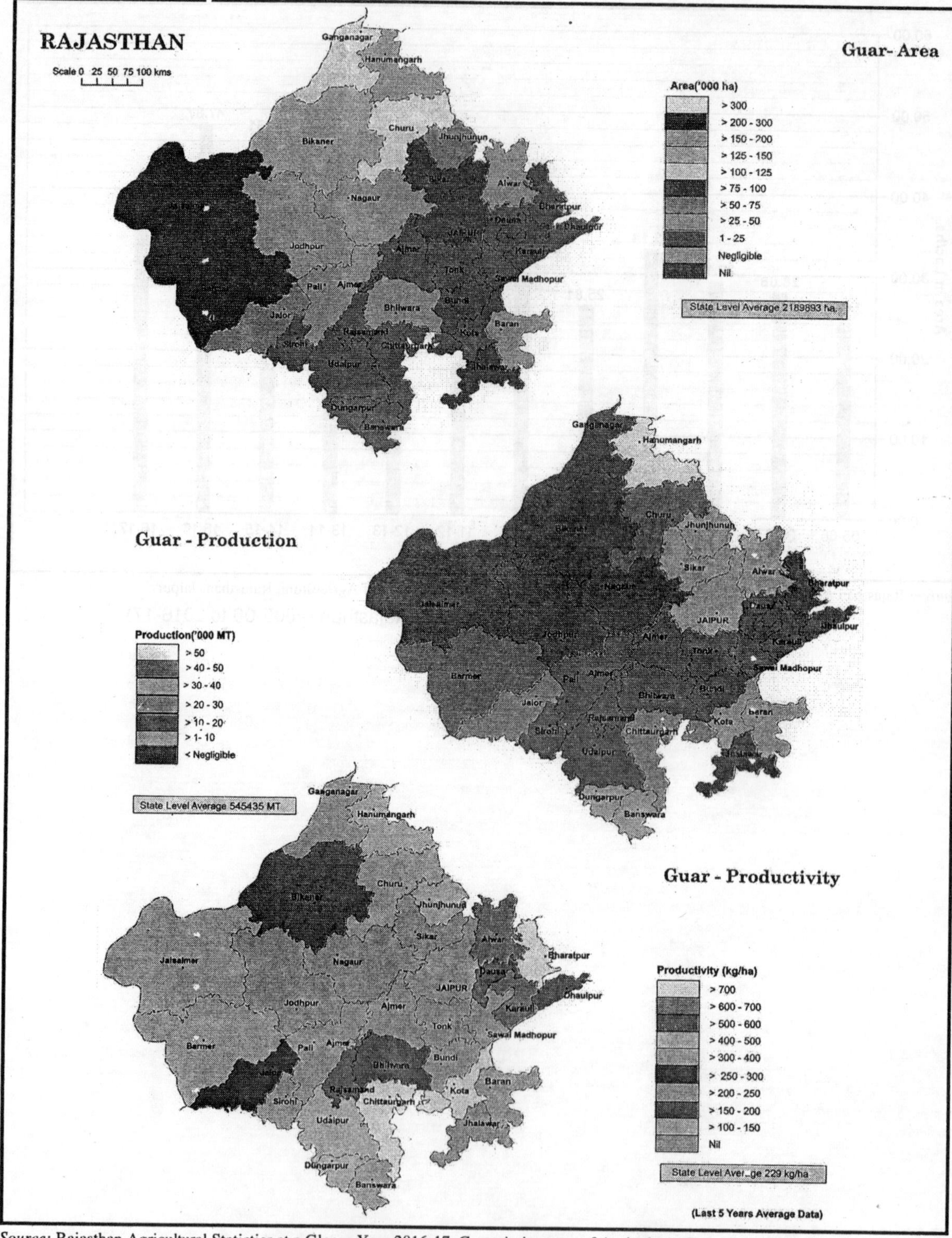

Source: Rajasthan Agricultural Statistics at a Glance Year 2016-17, Commissionerate of Agriculture, Rajasthan, Jaipur

Figure 12.20 Crop - Guar Area in Rajasthan

COTTON

Cotton is an indigenous plant of India and its cultivation has been going on since ancient times. India is the one of the main cotton producing country in the world today.

Cotton is a tropical plant. It is grown in the kharif season (June-October). For healthy growth of plants the optimum temperatures are between 20°C to 30°C. The rainfed crops need an average precipitation of 80 cm to have optimum yields. At the time of flowering and fruit formation, the plants need dry weather with high temperatures. During flowering, excessive rainfall with humid weather and overcast skies usually affects ball formation and its opening, resulting in poor yields.

There has been an appreciable advance in the development of high yield varieties of cotton. More than 14 hybrids which are resistant to pests and weather hazards, are available for cultivation. Cotton fibres of hybrids are long-stapled, having good strength and better shine. High yielding varieties of cotton are grown on black soils of the Deccan trap which are spreads over extensive areas in Maharashtra, Gujarat, Karnataka, Andhra pradesh and southern parts of Rajasthan.

Cotton is a cash crop and is grown both under irrigated and rainfed conditions in the state. The latter category of cotton is sown only in those areas which have balck cotton soils and copious rainfall, which is generally, well distributed over the crop season. While the irrigated crop is taken by advanced sowings a few weeks earlier than the monsoon rains, rainfed crop is sown only with the onset of monsoon. Cotton crop is slightly salt tolerant and hence can be grown in saline lands of irrigated areas also.

During last 50 years of planned economic development in Rajasthan, there has been significant increase both in the area and production of cotton. In the 1950-51, at the begining of the I five year plan area under cotton was 1.93 lakh hectares which by the end of the IX plan roses to 6.23 lakh hectares. The production in the corresponding period also increases from 1.31 lakh bales to 10.80 lakh bales (one bale = 170 kg).

Currently (2003-04) cotton is grown in about 4.84 lakh hectares, which is about 2.60 per cent of the gross cropped area of the state. It is the main cash crop in Ganaganagar, Hanumangarh, districts where its area exceeds 2.59 lakh hectares. It is also an important cash crop in Alwar, Banswara, Bikaner, Jodhpur, Nagaur, Bhilwara and Ajmer districts each of them having an acreage of 0.06 to 0.19 lakh hectares. The crop is also grown in Chittaurgarh and Sirohi districts.

The annual production of the state in 2016-17 is 14.57 lakh bales of cotton, the highest over 2.66 lakh bales, is from Hanumangarh district alone followed by Ganganagar district 2.63 lakh bales. In Alwar, Banswara, Nagaur, Jodhpur, Bikaner and Pali districts annual production ranging from 0.10 lakh bales to 0.40 lakh bales.

Cottoon yields are responsive to the use of fertilizers and pesticides. During the year 2016-17 Nagaur districts has the highest yields of 527 kg/ha followed by Sirohi and Banswara 520 and 507 kg/ha respectively.

New high yielding strains have been developed recently which are of short duration with high quality lint. These are suited well in double cropping systems, thus helping in increasing the gross productivity (Figure 12.21 and 12.22).

Its output was 15.27 lakh bales in 2015, which was higher than 2013-14. Sri Ganganagar district is the leading district for the production of cotton in Rajasthan due to better irrigation facilities. There are three varieties of cotton that are grown in the state: (i) Deshi Cotton is mostly grown in Udaipur, Chittourgarh and Banswara districts; (ii) American cotton is mostly grown in Sri Ganganagar district; (iii) Malwi cotton is grown in Kota, Jhalawar and Tonk district.

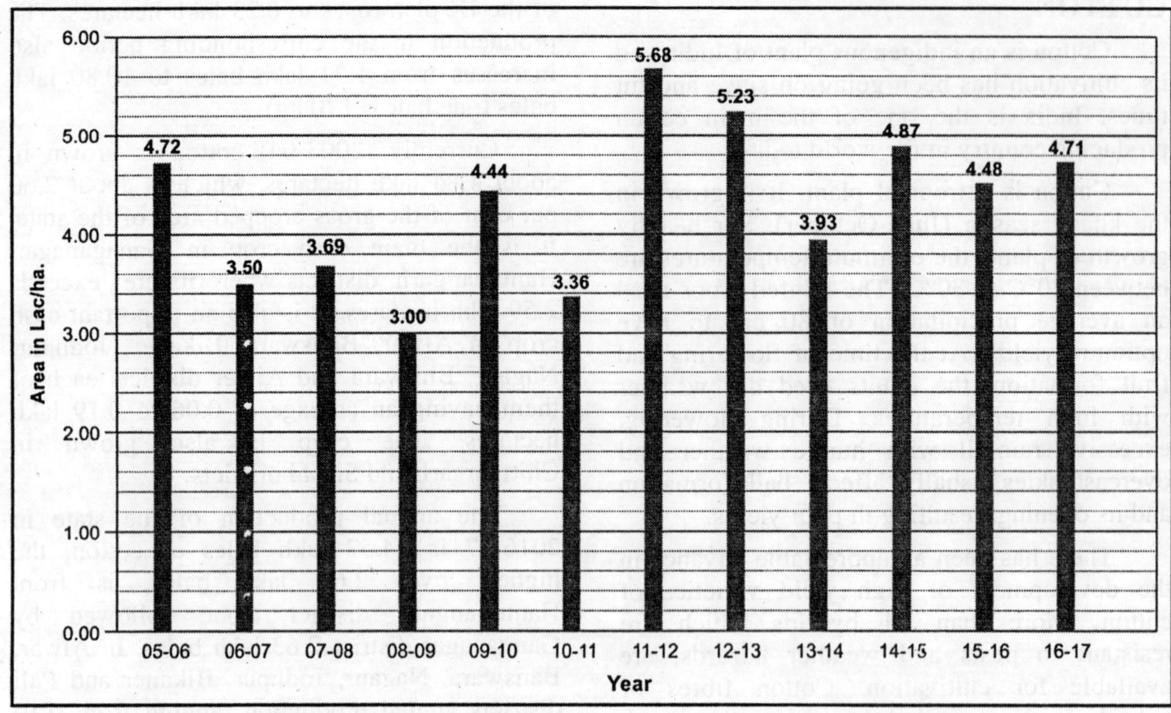

Source: Rajasthan Agricultural Statistics at a Glance Year 2016-17, Commissionerate of Agriculture, Rajasthan, Jaipur

Figure 12.21 Yearwise Area under Crop Cotton in Rajasthan (2005-06 to 2016-17)

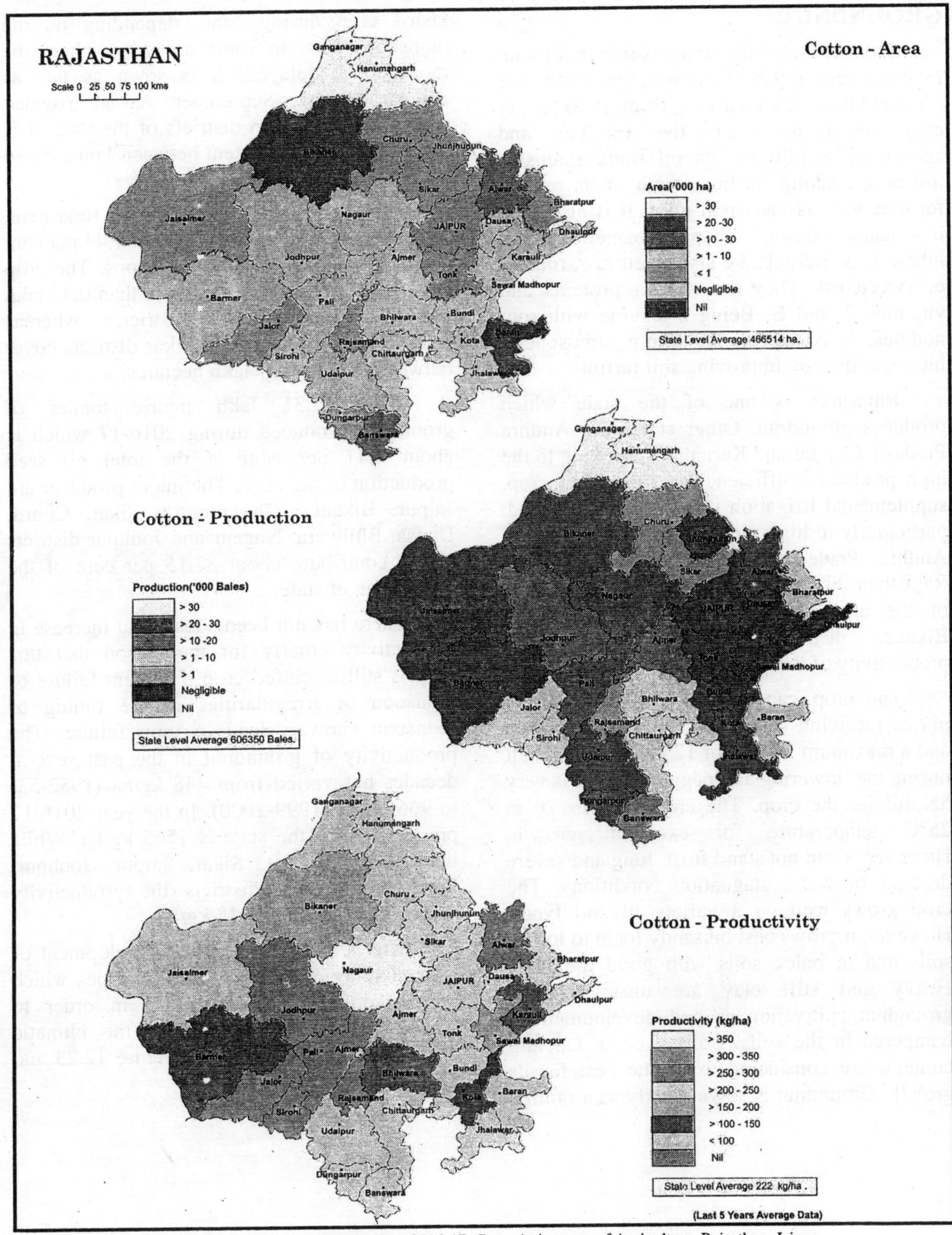

Source: Rajasthan Agricultural Statistics at a Glance Year 2016-17, Commissionerate of Agriculture, Rajasthan, Jaipur

Figure 12.22 Crop - Cotton Area in Rajasthan

GROUNDNUT

India occupies the first position, both with regard to area and production in the world. The oil contents of the seed vary from 44 to 55 per cent, depending upon the varieties and agronomic conditions. Its oil finds extensive use as a cooking medium, both in its refined form as well as vanaspati ghee. It is also used in manufacturing soap, cosmetics and lubericants. Kernels are also eaten raw, roasted or sweetened. They are rich in proteins and vitamins A and B. Being a legume with root nodules, it is capable of fixing atmospheric nitrogen, thereby improving soil fertility.

Rajasthan is one of the state which produces groundnut. Other states are Andhra Pradesh, Gujarat and Karnataka, Looking to the high production efficiency of groundnut crop, supplemental irrigation is also being provided, particularly to high yielding strains in Gujarat, Andhra Pradesh and Karnataka states. In Rajasthan this crop is grown over 61 per cent of the area under irrigation in Jaipur and Bikaner districts (2016-17) where the productivity of this crop is high.

The crop can be grown sucessfully in places receiving a minimum rainfall of 50 cm and a maximum rainfall of 125 cm. The rainfall during the flowering and pegging stages is very helpful for the crop. The crop requires 20 to 25°C temperature for excellent growth. However, it can not stand frost, long and severe drought or water stagnation conditions. The crop grows well on a variety of soil types. However, it grows best on sandy loam to loamy soils and in balck soils with good drainage. Heavy and stiff clay are unsuitable for groundnut cultivation as pod development is hampered in the soils. Soils, rich in calcium contents are considered to be the best for its growth. Groundnut is sown mostly as a rainfed kharif crop during June, depending on the monsoon rains. In some areas, or where the monsoon is delayed, it is sown as late as August or early September. As an irrigated crop, in the southern districts of the,state, it is grown, to a limited extent between January and March as a summer crop.

During 2016-17 area sown in Rajasthan, 2.12 lakh hectares, which is about 1.44 per cent of the area covered by kharif crops. The crop covers the highest area i.e. more than 0.44 lakh hectares in Jaipur district, whereas Chittaurgarh, Bikaner and Sikar districts cover between 0.16 to 0.37 lakh hectares.

About 3.31 lakh metric tonnes of groundnut produced during 2016-17 which is about 8.41 per cent of the total oil seed production of the state. The major producer are Jaipur, Bikaner, Chittaurgarh, Sikar, Churu, Dausa, Bhilwara, Nagaur and Jodhpur distircts which contribute about 87.15 per cent of the production of state.

There has not been substantial increase in productivity chiefly for the reason that this crop is still a rainfed crop. Frequent failure of monsoon or irregularities in the timing of monsoon showers lead to crop failure. The productivity of groundnut in the past several decades has varied from 448 kg/ha (1952-53) to 966 kg/ha (1999-2000). In the year 2016-17 productivity of the state is 1565 kg/ha. While in Churu, Bikaner, Sikar, Jaipur, Jodhpur, Barmer and Jalor districts the productivity ranges from 1570 to 2413 kg/ha.

Efforts are underway for development of high yielding, short duration genotypes which possess drought tolerance also in order to reduce the adverse effects of the climatic variations particularly rains (Figure 12.23 and 12.24).

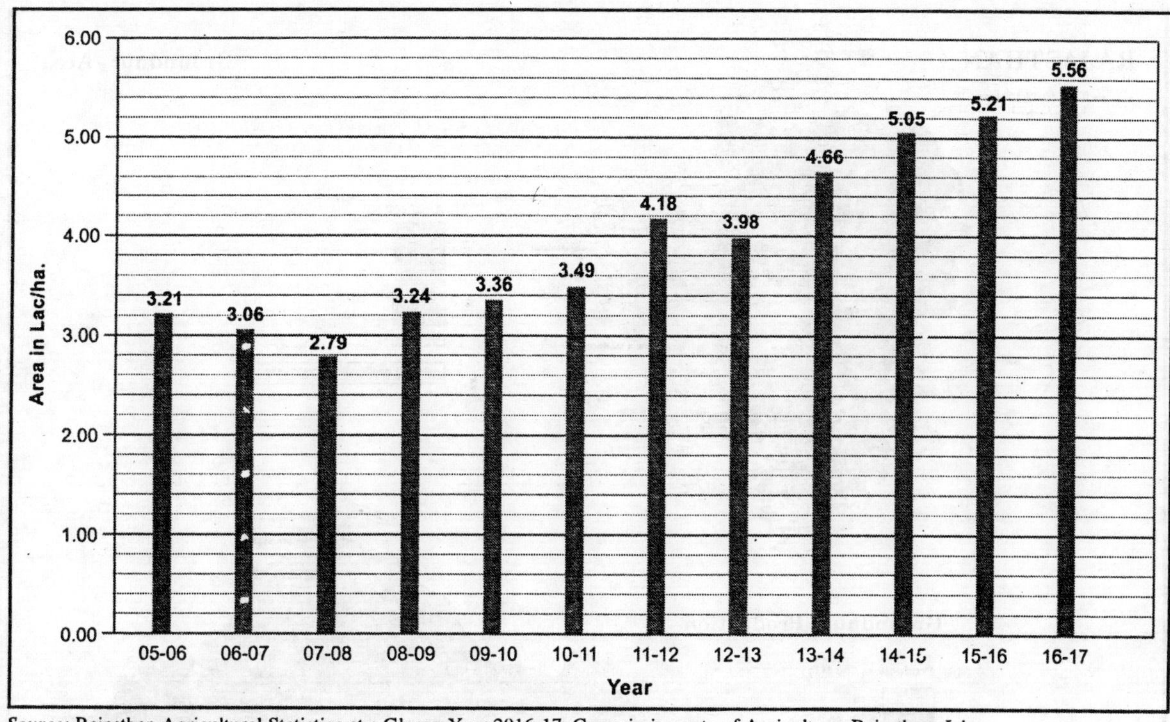

Source: Rajasthan Agricultural Statistics at a Glance Year 2016-17, Commissionerate of Agriculture, Rajasthan, Jaipur

Figure 12.23 Yearwise Area under Crop - Groundnut in Rajasthan (2005-06 to 2016-17)

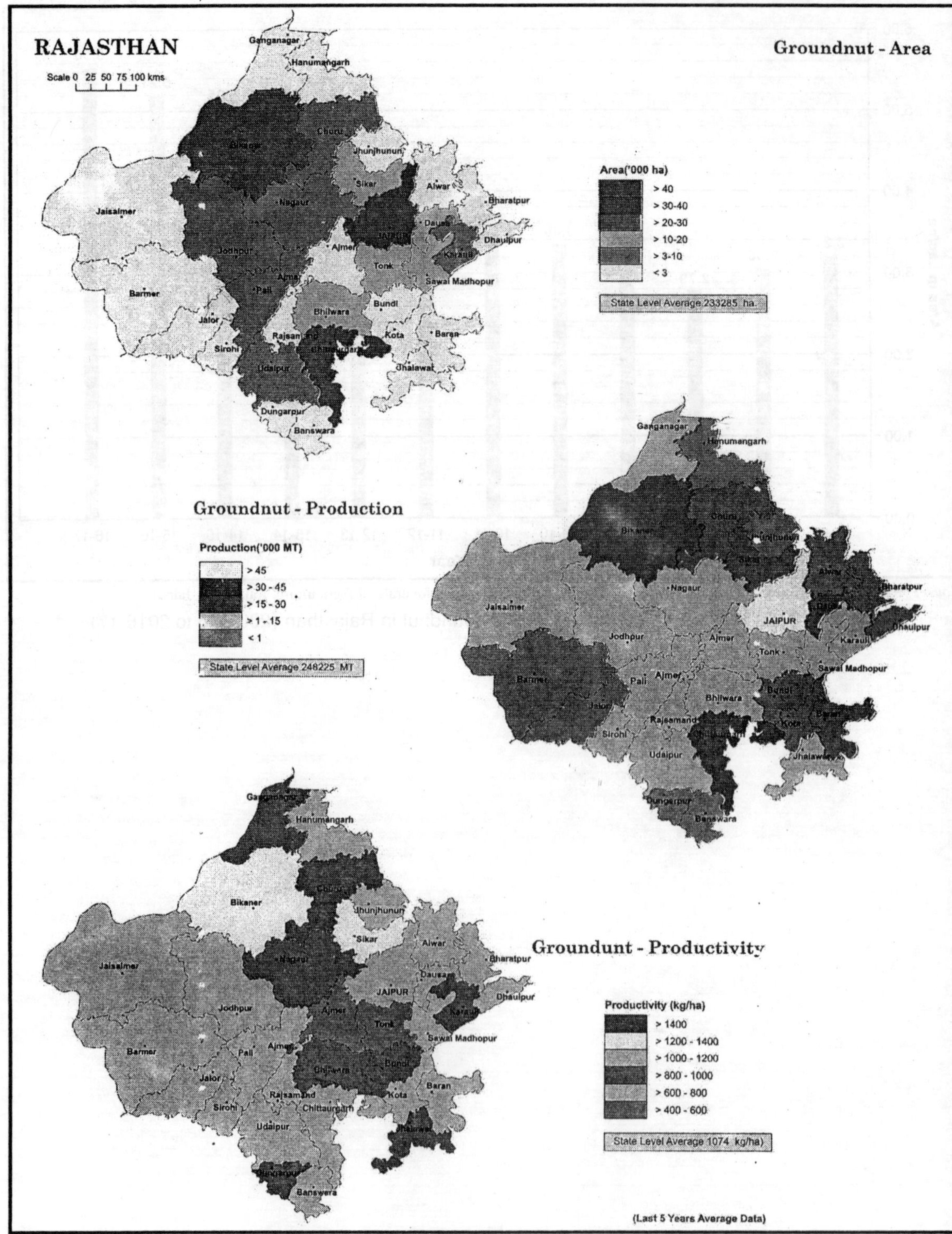

Source: Rajasthan Agricultural Statistics at a Glance Year 2016-17, Commissionerate of Agriculture, Rajasthan, Jaipur

Figure 12.24 Crop - Groundnut Area in Rajasthan

PADDY

Paddy is the staple food crop of humid areas. It prefers low-lying and water logged areas where none of the other cereals can be grown. However, evolution of new plant has made it possible to grow paddy in areas which have relatively low rainfall and heavier soil types in parts of Punjab, Haryana, Western Uttar Pradesh and Rajasthan etc. It has been observed that paddy yield in newly acquired areas are much higher than the traditional growing areas. Higher yield of paddy are attributed to evolution of high yielding dwarf plant types, better soil and water management practices, alongwith an efficient nutrient management schedule etc. In Rajasthan, Paddy is more grown in irrigated areas which have soil that can retain water for longer duration as it has the highest requirements of water compared to the other cereals of the same season and duration. However, rainfed or upland paddy is also taken in southern Rajasthan.

Marked progress has been achieved in increasing rice production chiefly on account of the development of a large number of high yielding varieties, which are resistant to drought, lodging, diseases and pests, and their responsivencess to high fertilizer levels.

Rajasthan, being an arid state and having meagre means of canal irrigation, is not a paddy producing area in the country. It is only in parts of Ganganagar, Hanumangarh, Bundi, Kota, Baran, Banswara and Dungarpur districts where paddy cultivation is being practised. Whereever canals have not yet appeared, upland paddy continues to be grown which is coarse and of inferior quality, and is used locally by inhabitants in Kota, Baran, Banswara and Dungarpur districts.

Paddy in grown in about 1.00 lakh hectares (2016-17) in Rajasthan. The major areas lie in the south-eastern part of Rajasthan with the highest of above 0.28 lakh hectares being in Banswara district, followed by Hamumangarh (0.21 lakh hectares) and Dungarpur (0.19 lakh hectare). Other districts which has an acreage of about 0.02 to 0.07 lakh hectares each are Bundi, Kota, Udaipur, Ganganagar, Karauli and Bharatpur.

In 2016-17 total production in state is 4.57 lakh tonnes. This is 2.11 per cent of the total production of cereals in Rajasthan. Hamumangarh district has the highest production 0.67 lakh metric tonnes annually. Followed Banswara (0.24 lakh metric tonnes) Dungarpur (0.20 lakh metric tonnes), Bundi (0.15 lakh metric tonnes) and Ganganagar (0.10 lakh metric tonnes).

Productivity is markedly higher in district of Hanumangarh and Ganganagar 3213 and 2583 kg/ha., which is higher than the state average 1653 kg/ha. during 2003-04, due to availability of canal irrigation and adoption of improved technology of crop production. In Baran, Bundi, Kota, Sirohi, Ajmer, Alwar, Rajsamand, Bhilwara, Chittaurgarh, Dhaulpur, Jhalawar, Sawai Madhopur, Bikaner and Tonk, the average yield ranges from 1600-1944 kg/ha. In Banswara, Udaipur and Dungarpur districts, the yields are low, being a rainfed crop (Figure 12.25 and 12.26).

Many dwarf, high fertilizer-responsive genotypes have been developed, raising the production in this crop considerably, but compared to other paddy growing states, it is still much lower. Development of short duration high yielding varieties is the demand of the day so that they fit in the sequential cropping system, thereby increasing gross production per unit area.

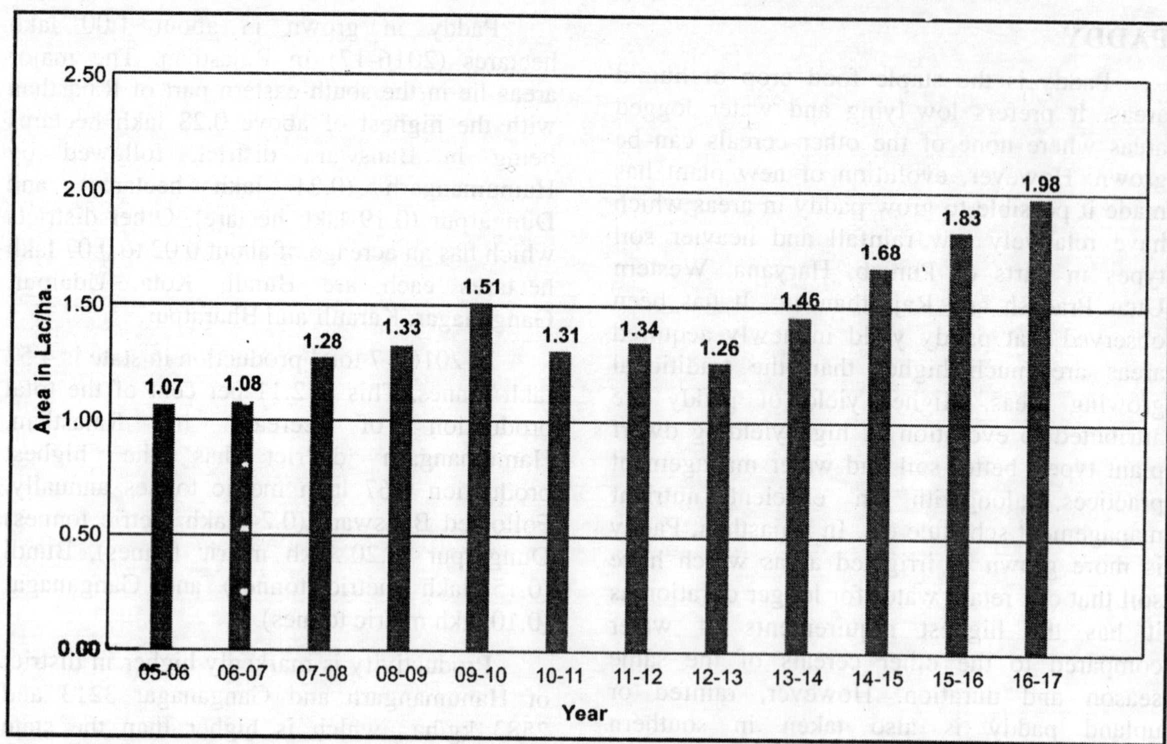

Source: Rajasthan Agricultural Statistics at a Glance Year 2016-17, Commissionerate of Agriculture, Rajasthan, Jaipur

Figure 12.25 Yearwise Area under Crop Paddy in Rajasthan (2005-06 to 2016-17)

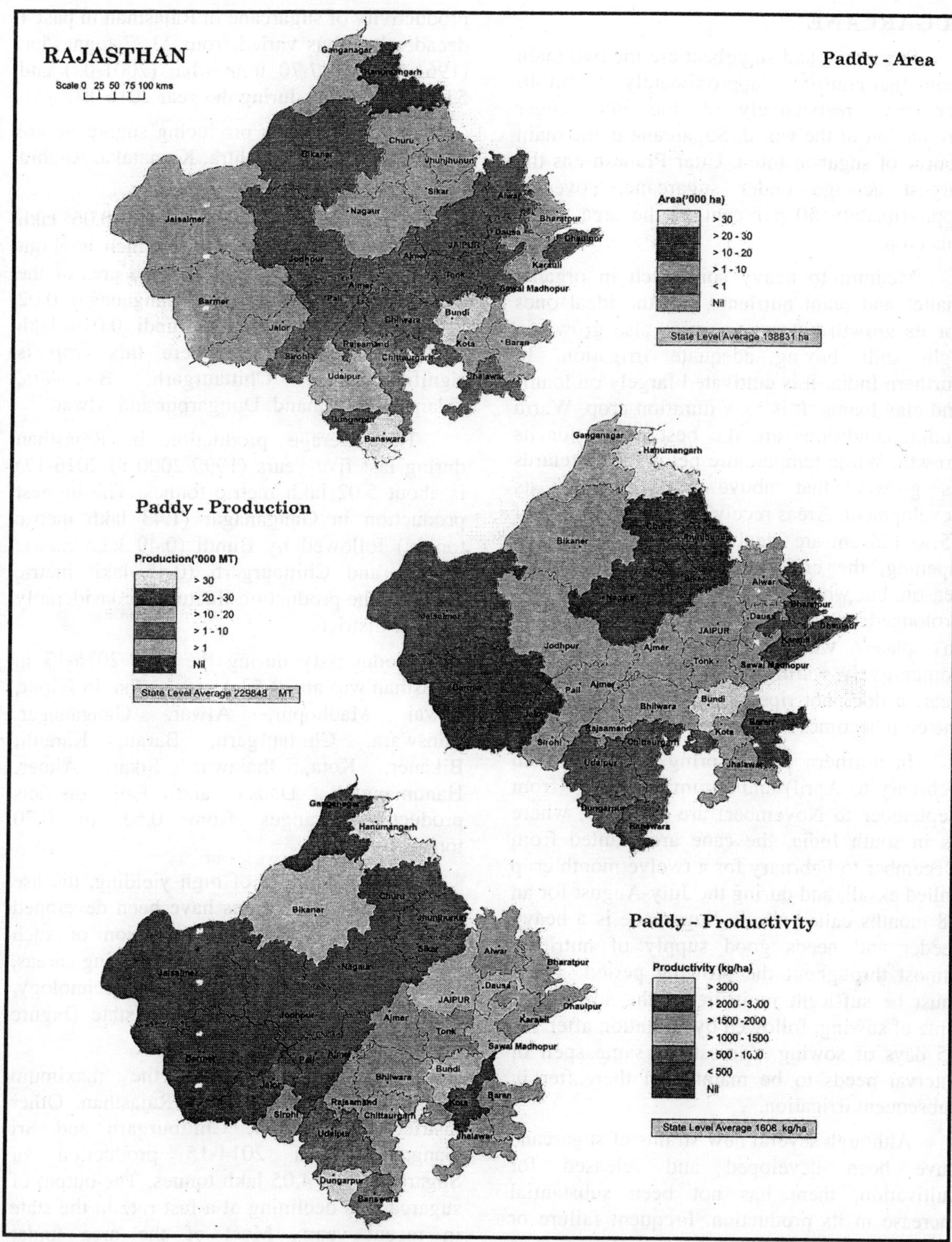

Source: Rajasthan Agricultural Statistics at a Glance Year 2016-17, Commissionerate of Agriculture, Rajasthan, Jaipur

Figure 12.26 Crop- Paddy Area in Rajasthan

SUGARCANE

Sugarcane and sugarbeet are the two main crops that contribute approximately 56 and 46 per cent, respectively of the total sugar production in the world. Sugarcane is the main source of sugar in India. Uttar Pradesh has the largest acreage under sugarcane, covering approximately 50 per cent of the area under this crop.

Medium to heavy soils, rich in organic matter and plant nutrients, are the ideal ones for its growth but sugarcane is also grown in light soils having adequate irrigation. In northern India, it is cultivated largely on loams and clay loams. It is long duration crop. Warm humid conditions are the best suited for its growth. While temperature below 20°C retards its growth, that above 50°C inhibits its development. Areas receiving annual rainfall of 75 to 125 cm are ideal for its growth. For its ripening, the crop requires a cool and dry season, but where the rainfall is too heavy and prolonged, the quality of juice tends to be low. In place where the weather remains compartively warm and moist throughout the year, it does not ripen well. In drought years, the crop becomes a little fiberous.

In northern India, spring planting (from February to April) and autumn planting (from September to November) are common, where as in south India, the cane are planted from December to February for a twelve month crop called eksali, and during the July-August for an 18 months called adsali. Sugarcane is a heavy feeder and needs good supply of nutrients almost throughout the growing period. There must be sufficent moisture in the soil at the time of sowing, followed by irrigation after 12-15 days of sowing. Almost the same spell of interval needs to be maintained thereafter in subsequent irrigation.

Although several new strains of sugarcane have been developed and released for cultivation, there has not been substantial increase in its production. Frequent failure or irregualr monsoons lead to crop failure and consequently productivity declines. The productivity of sugarcane in Rajasthan in past 4 decades has thus varied from 23.27 tonnes/ha. (1961-62) to 47.70 tonnes/ha. (2001-02) and 53.41 tonnes /ha. during the year 2016-17.

Prominent states producing sugarcane are Uttar Pradesh, Maharashtra, Karnataka, Andhra Pradesh and Rajasthan.

Sugarcane is grown in over 0.06 lakh hectares of land in Rajasthan, which is about 0.04 per cent of the gross cropped area of the state. The major areas are Ganganagar 0.02 lakh hectare followed by Bundi 0.014 lakh hectare other districts where this crop is significant are Chittaurgarh, Banswara, Udaipur, Rajsamand, Dungarpur and Alwar.

The average production in Rajasthan during last five years (1999-2000 to 2016-17) is about 5.02 lakh metric tonnes. The highest production in Ganganagar (1.43 lakh metric tonnes) followed by Bundi (0.49 lakh metric tonnes) and Chittaurgarh (0.45 lakh metric tonnes). The production fluctuates considerably in other districts.

Productivity during the year 2016-17 in Rajasthan was about 53.41 tonnes/ha. In Jaipur, Sawai Madhopur, Alwar, Ganganagar, Banswara, Chittaurgarh, Baran, Karauli, Bikaner, Kota, Jhalawar, Sikar, Ajmer, Hanumangarh, Dausa and Pali districts productivity ranges from 0.53 to 0.70 tonnes/ha.

A large number of high yielding, disease and pest resistant strains have been developed in the country. With the adoption of such strains in the conventional growing areas, coupled with high production technology, production can be increased in state (Figure 12.27).

Bundi district gives the maximum production of Sugarcane in Rajasthan. Other district are Udaipur, Chittourgarh and Sri Ganganagar. In 2014-15 production of Sugarcane was 4.05 lakh tonnes. The output of sugarcane is declining at a fast rate in the state in recent years. Most of the area under sugarcane is irrigated.

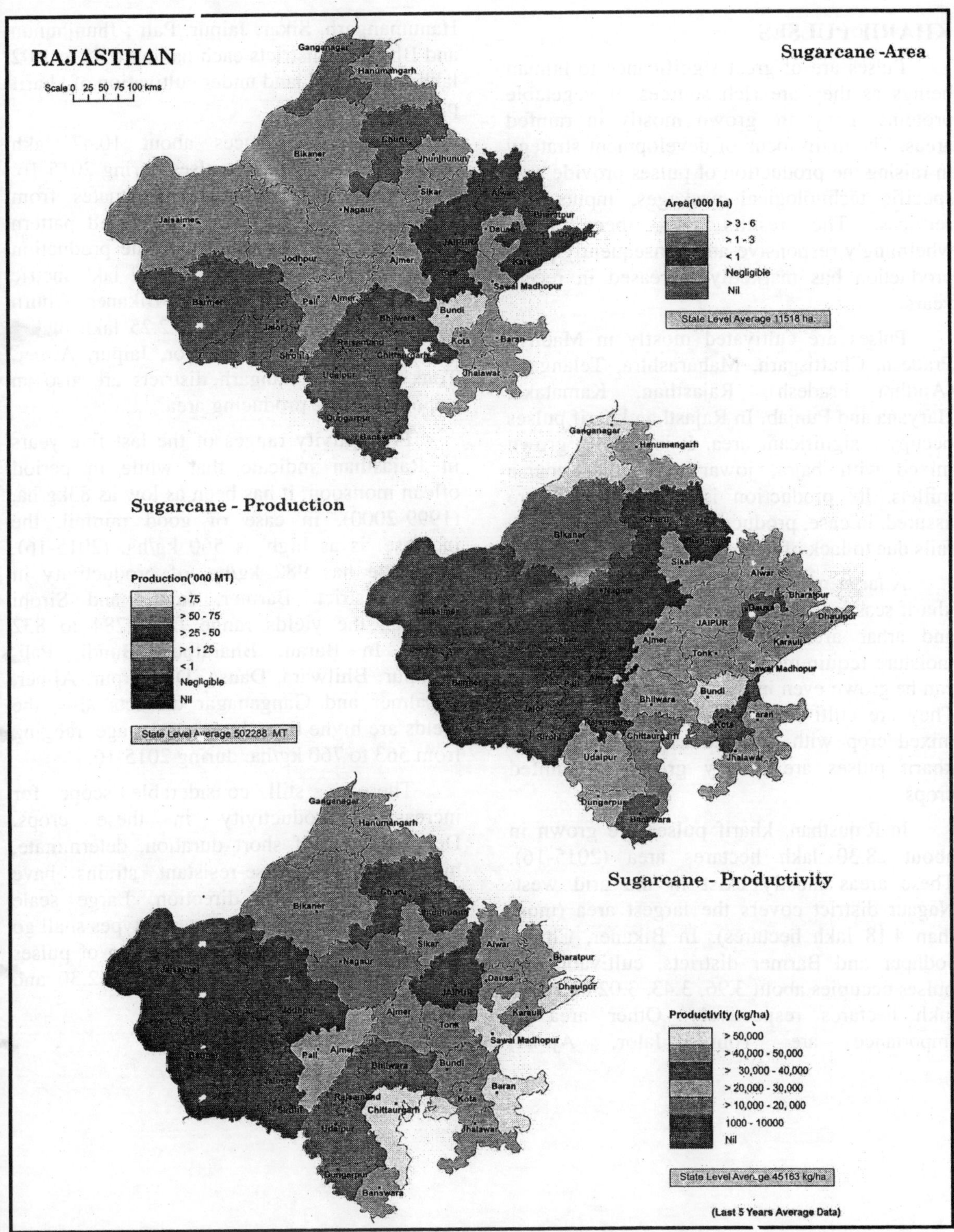

Source: Rajasthan Agricultural Statistics at a Glance Year 2016-17, Commissionerate of Agriculture, Rajasthan, Jaipur

Figure 12.27 Crop - Sugarcane Area in Rajasthan

KHARIF PULSES

Pulses are of great significance to human beings as they are rich sources of vegetable proteins. They are grown mostly in rainfed areas. The main focus of development strategy in raising the production of pulses provide area specific technological packages, inputs and services. The results have been overwhelmingly responsive and consequently pulse production has markedly increased in recent years.

Pulses are cultivated mostly in Madhya Pradesh, Chattisgarh, Maharashtra, Telangana (Andhra Pradesh), Rajasthan, Kamataka, Haryana and Punjab. In Rajasthan kharif pulses occupy a significant area. It is usually grown mixed with bajra, jowar and other coarse millets. Its production is regarded as more assured, in case, production of main food crops fails due to lack of adequate rams.

A large number of pulses is grown in the kharif season. Among them moong, moth, urad and arhar are important. Pulses haye less moisture requirements for their growth and so can be grown even in areas having low rainfall. They are cultivated as individual or also as mixed crop with other kharif crops. All the kharif pulses are mostly grown as rainfed crops.

In Rajasthan, kharif pulses are grown in about 28.30 lakh hectares area (2015-16). These areas mostly exist in the arid west. Nagaur district covers the largest area (more than 4.18 lakh hectares). In Bikaner, Churn, Jodhpur and Barmer districts, cultivation of pulses occupies about 3.96, 3.43, 3.02 and 3.01 lakh hectares respectively. Other area of importance are Tonk, Jalor, Ajmer, Hanumangarh, Sikar, Jaipur, Pali; Jhunjhunun and Bhilwara districts each having 0.55 to 1.02 lakh hectares of land under cultivation of kharif pulses.

Rajasthan produces about 10.47 lakh metric tonnes of kharif pulses during 2015-16. Total production, however, fluctuates from year to year depending upon rainfall pattern during monsoon. In Rajasthan, the production is highest in Barmer district 2.52 lakh metric tonnes followed by Nagaur, Bikaner, Churn and Jodhur districts (0.96 to 2.25 lakh metric tonnes during 2015-16). Jalor, Jaipur, Ajmer, Tonk and Hanumangarh districts are also an important pulse producing area.

Productivity ranges of the last five years in Rajasthan indicate that while in period oflean monsoon, it has been as low as 83kg/ha. (1999-2000), in case of good rainfall, the increase. is as high as 560 kg/ha. (2015-16). The state has 982 kg/ha. of productivity in Alwar district. Barmer, Jaipur and Sirohi districts, the yields range from 784 to 832 kg/ha. In Baran, Bharatpur, Bundi, Pali Udaipur, Bhilwara, Dausa, Dungarpur, Ajmer, Jaisalmer and Ganganagar districts also, the yields are highe than the state average, ranging from 563 to 760 kg/ha. during 2015-16.

There is still considerable scope for increasing productivity in these crops. Developments of short duration, determinate, high-yielding disease-resistant strains have paved way in this direction. Large scale avialability of seeds of these genotypes shall go a long way in increasing production of pulses in the state (Figure 12.28, 12.29, 12.30 and 12.31).

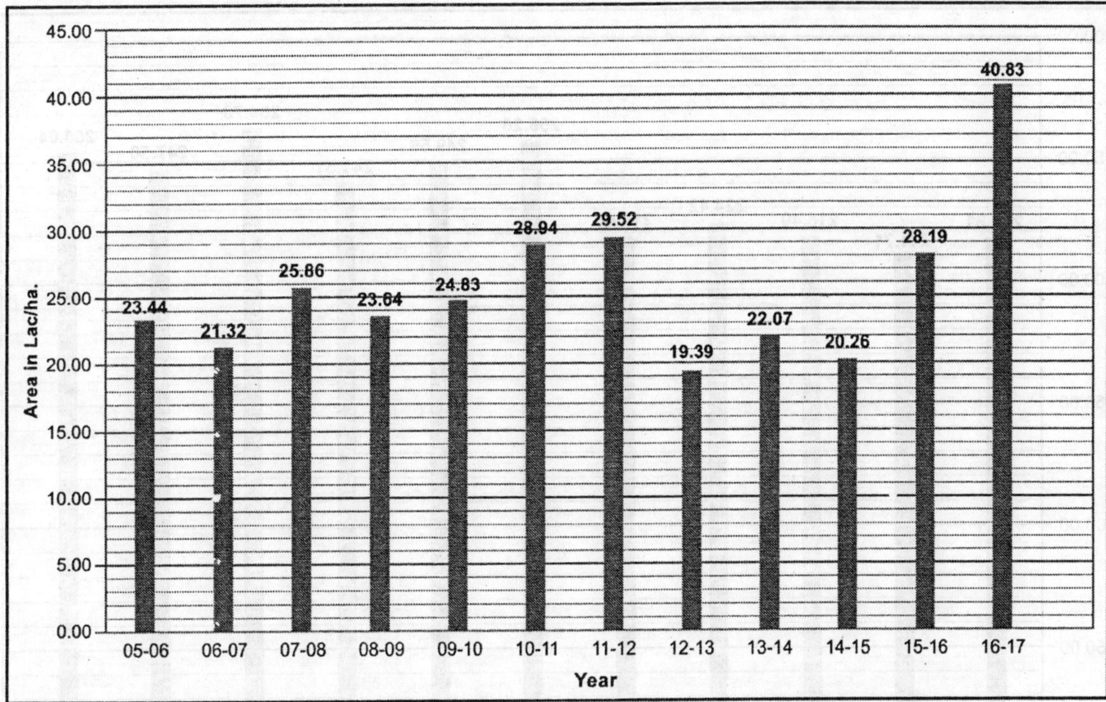

Source: Rajasthan Agricultural Statistics at a Glance Year 2016-17, Commissionerate of Agriculture, Rajasthan, Jaipur

Figure 12.28 Yearwise Area under Crop- Kharif Pulses (2005-06 to 2016-17)

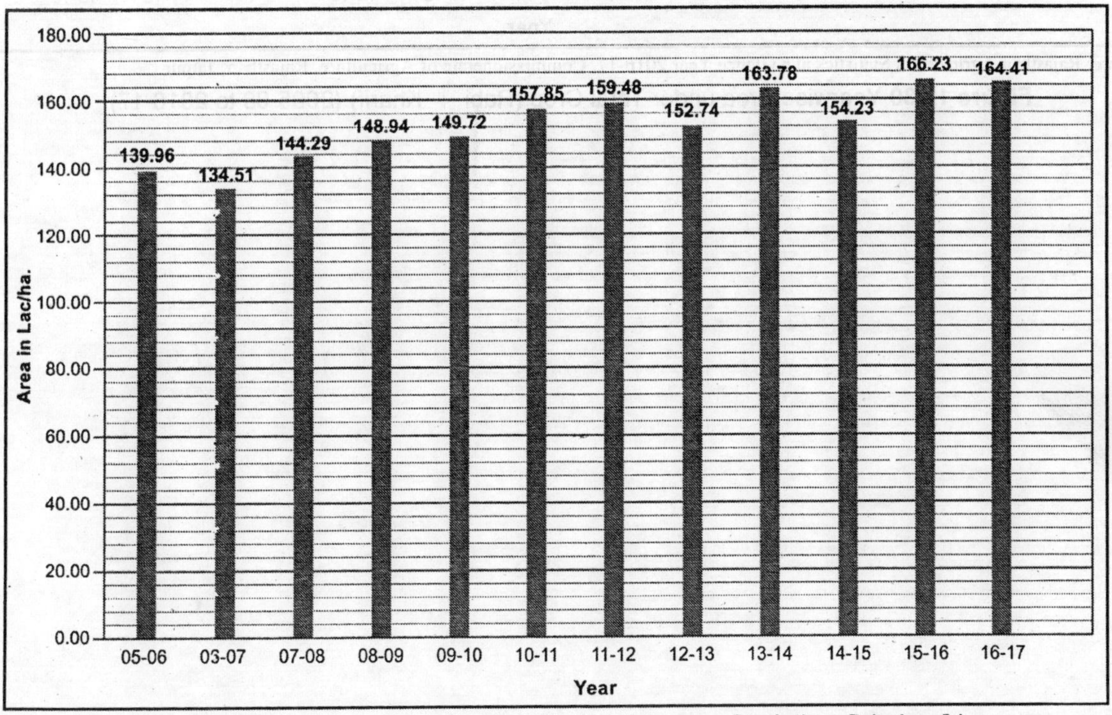

Source: Rajasthan Agricultural Statistics at a Glance Year 2016-17, Commissionerate of Agriculture, Rajasthan, Jaipur

Figure 12.29 Yearwise Area under Crop- Total Kharif (2005-06 to 2016-17)

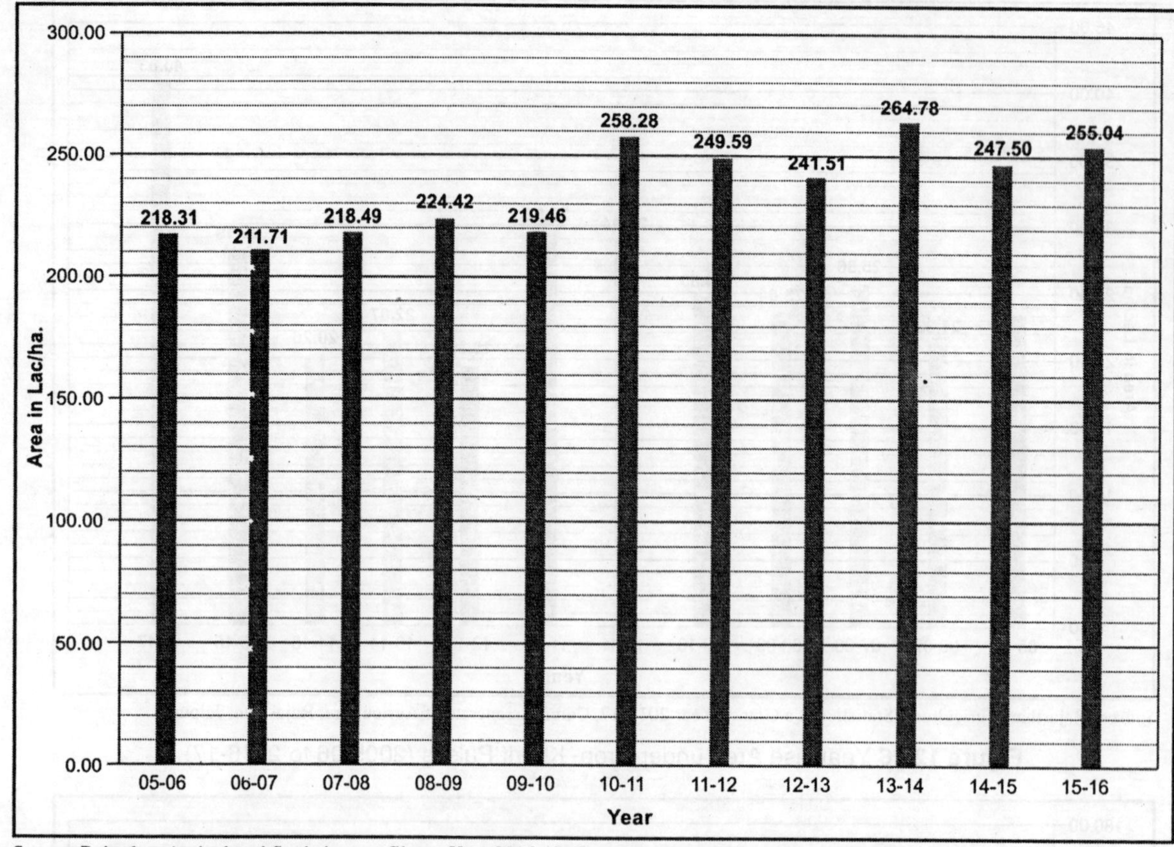

Source: Rajasthan Agricultural Statistics at a Glance Year 2016-17, Commissionerate of Agriculture, Rajasthan, Jaipur

Figure 12.30 Yearwise Area under Total Crop (Rabi + Kharif) (2005-06 to 2016-17)

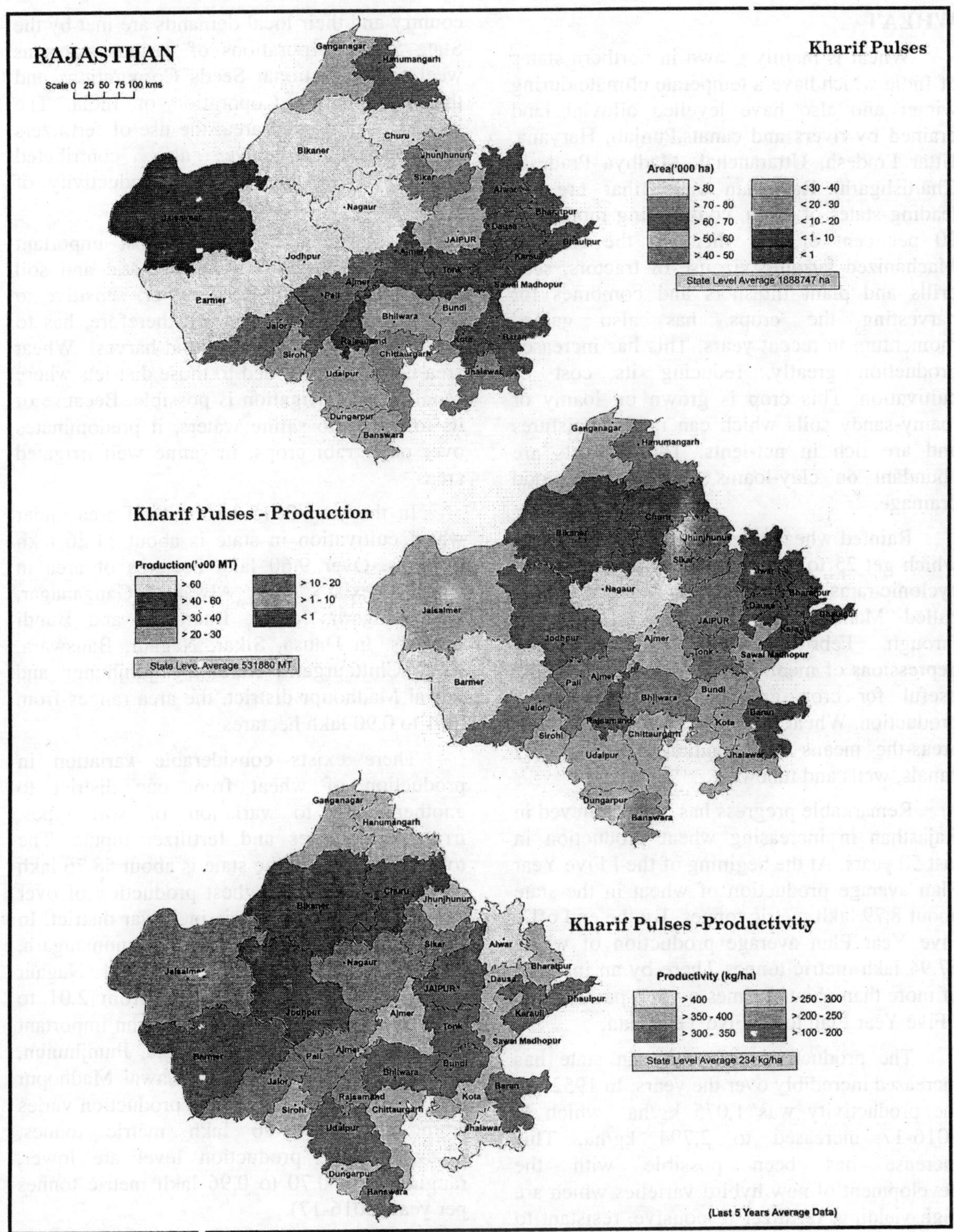

Source: Rajasthan Agricultural Statistics at a Glance Year 2016-17, Commissionerate of Agriculture, Rajasthan, Jaipur

Figure 12.31 Kharif Pulses Area in Rajasthan

WHEAT

Wheat is mainly grown in northern states of India which have a temperate climate during winter and also have levelled alluvial land drained by rivers and canals.Punjab, Haryana, Uttar Pradesh, Uttaranchal, Madhya Pradesh, Chattishgarh, Rajasthan and Bihar are the leading states of India, contributing more than 90 per cent of the wheat of the country. Machanized farming viz.,use of tractors, seed drills and plant thrashers and combines for harvesting the crops, has also gained momentum in recent years. This has increased production greatly, reducing its cost of cultivation. This crop is grown on loamy or loamy-sandy soils which can retain moistures and are rich in nutrients. The harvests are abundant on clay-loams soils having good drainage.

Rainfed wheat crop can be grown in areas which get 25 to 75 cm of rainfall. The winter cyclonic rains in northern parts of India, locally called "Mahawat", which occure in December through February due to secondary depressions of mediteranean region, are highly useful for crop growth, resulting in high production. Wheat is usually grown in irrigated areas-the means of irrigation being usually canals, wells and tubewells.

Remarkable progress has been achieved in Rajasthan in increasing wheat production in last 50 years. At the begining of the I Five Year Plan average production of wheat in the state about 8.79 lakh metric tonnes. By the end ofIX Five Year Plan average production of wheat 67.94 lakh metric tonnes There by an increase of more than about 8 times o-ver a period from I Five Year Plan to IX Five Year Plan.

The productivity of wheat in state has increased incredibly over the years. In 1952-53, the productivity was 1,075 kg/ha., which in 2016-17 increased to 2,794 kg/ha. This increase has been possible with the development of new hybird varieties which are high-yielding, fertilizer-responsive, resistant to pest-diseases and dry conditins. Today, not less than 400 varieties of wheat are avilable in the country and their local demands are met by the State Seed Corporations of various states as weel as the National Seeds Corporations and the State Farms Coporation of India. The increase in irrigated area, the use of fertilizers and pesticides have also contributed significantly to increase the productivity of wheat throughout the country.

In Rajasthan, wheat is most important cereal crop. It has a wide climatic and soil adaptation range. It is, however, sensitive to soil moisture deficiency and therefore, has to be irrigated for obtaining good harvest. Wheat area is mostly confined to those districts where canal or well irrigation is possible. Because of its tolerance to saline waters, it predominates over other rabi crops, in saline well irrigated areas.

In the year 2015-16 the total area under wheat cultivation in state is about 31.26 lakh hectares. Over 9.60 lakh hectares of area in wheat exist in Alwar, Ganganagar, Hanumangarh, Jaipur, Bharatpur and Bundi districts. In Dausa, Sikar, Nagaur, Banswara, Kota, Chittaurgarh, Karauli, Jhjunjhunun and Sawai Madhoupr district, the area ranges from 0.61 to 0.90 lakh hectares.

There exists considerable variation in production of wheat from one district to another, due to variation of soil types, irrigation facilites and fertilizer inputs. The total production in the state is about 58.76 lakh metric tonnes. The highest production of over 6.49 lakh metric tonnes is in Alwar district. In Bharatpur, Ganganagar, Jaipur, Hanumangarh, Dausa, Bundi, Karauli, Sikar and Nagaur districts, the production varies from 2.01 to 5.35 lakh metric tonnes. It is also an important rabi crop in · Chittaurgarh, Kota, Jhunjhunun, Dhaulpur, Banswara, Baran, Sawai Madhopur and Tonk districts, where the production varies from 1.24 to 1.96 lakh metric tonnes. Elsewhere, the production level are lower, ranging form 0.70 to 0.96 lakh metric tonnes per year (2016-17).

The average yield of wheat per hectare for the state is 2,794 kg/ha (2016-17). Higher levels of yield are observed in Bharatpur (3,764), Karauli (3,639), Dhaulpur (3,475), Alwar (3,449), Dausa (3,163), Chittaurgarh (3,025), Jaipur (2,981), Kota (2,866), Churn, Jaisalmer and Rajsamand (2,794) kg/ha. In the semi-desert districts like Jhunjhunun and Sikar, the yields also higher 2,763 and 2,640 kg/ha. respectively. Inthe western Rajasthan, the yields are lower because of lack of irrigation and also due to sandy texture of soils, coupled with high temperatues prevailing at ripening stage.

Wheat is one of those crops which has shown spectacular increase its area. production and productivity especially after the year 1966-67 when its high yielding varieties were introduced (Figure 12.32 and 12.33).

Wheat is a rabi crop. Rajasthan is the fifth largest producer of wheat in India. Sri Ganganagar, Jaipur, Alwar, Kota and Sawai Madhopur are the major wheat producing districts in the state. In 2013-14 the production of wheat was 110.2 lakh tonnes, which was 10 percent of all India a wheat output in 2014-15, Rajasthan obtained the highest place in the production of wheat in India and was given a special award by Prime Minister Narendra Modi.

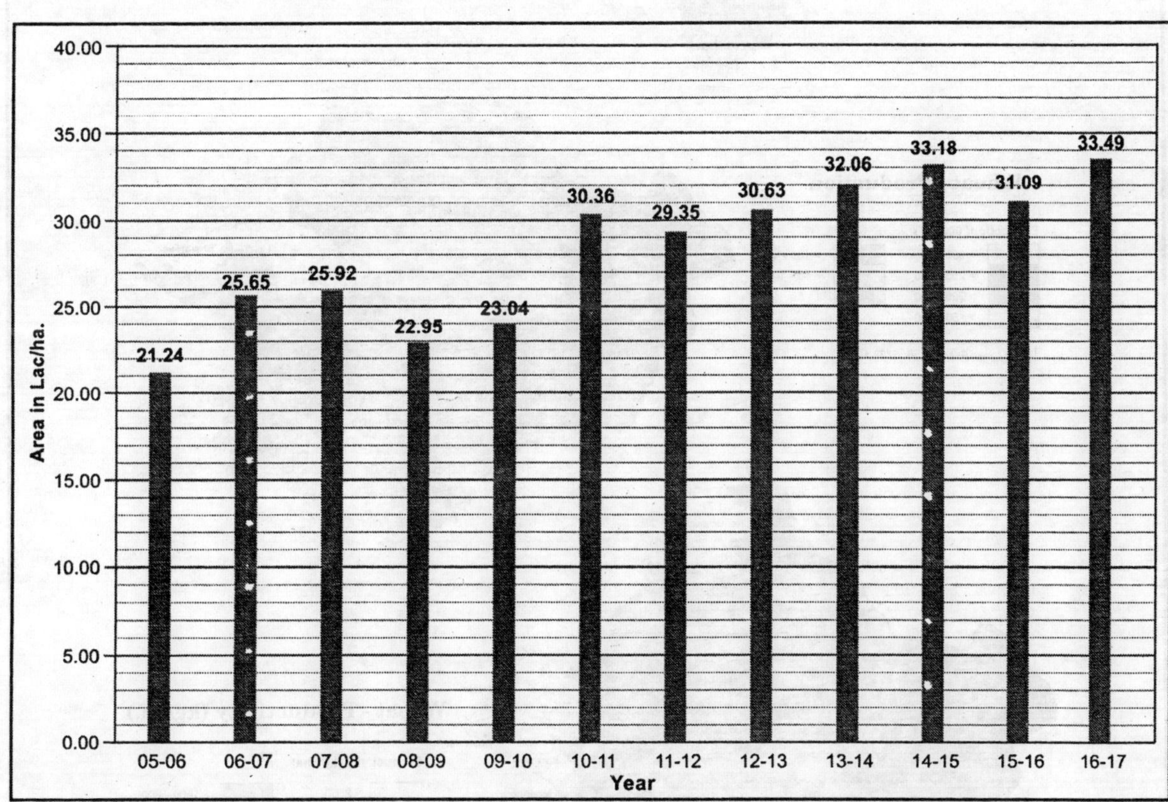

Source: Rajasthan Agricultural Statistics at a Glance Year 2016-17, Commissionerate of Agriculture, Rajasthan, Jaipur

Figure 12.32 Yearwise Area under Total Crop Wheat (2005-06 to 2016-17)

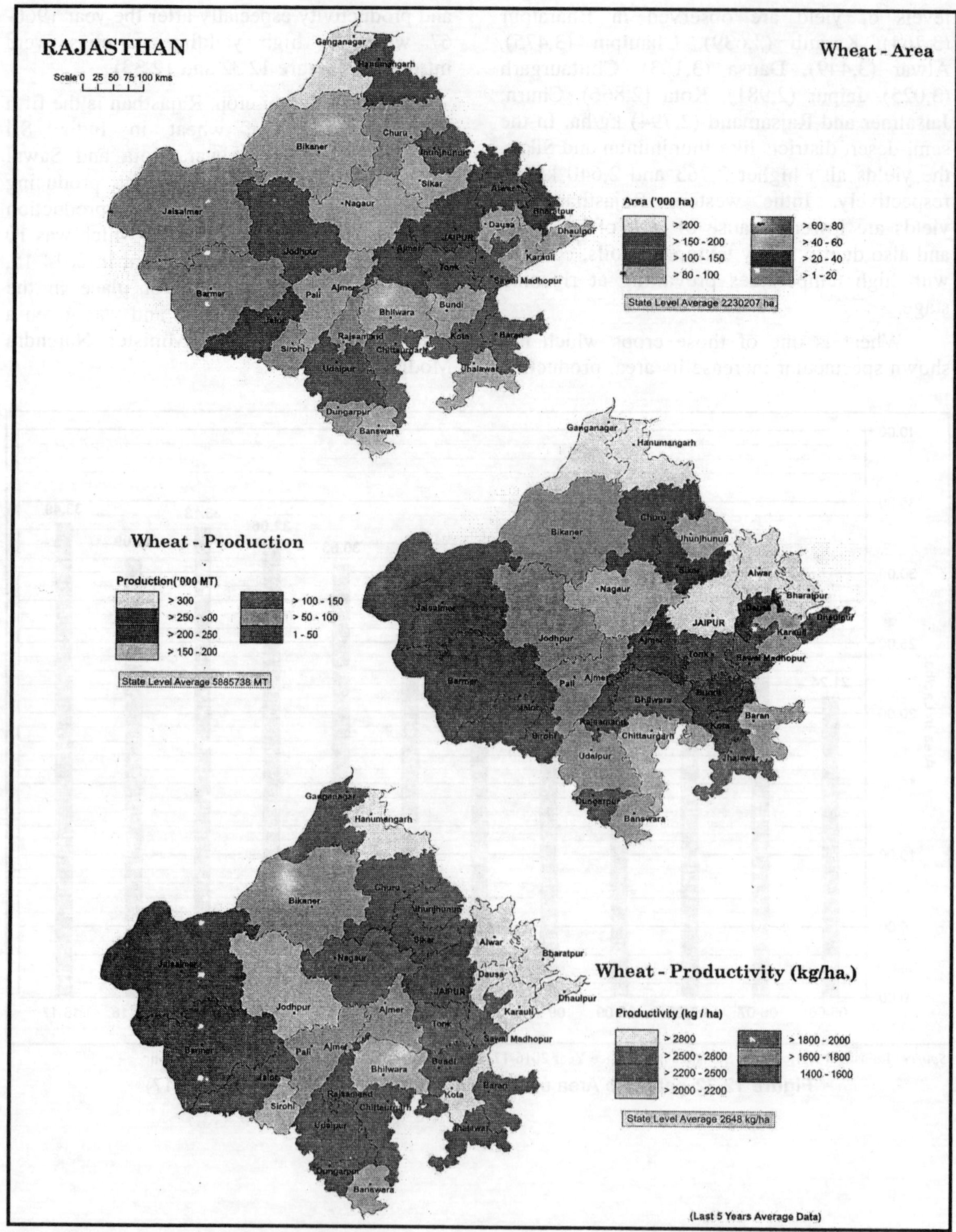

Source: Rajasthan Agricultural Statistics at a Glance Year 2016-17, Commissionerate of Agriculture, Rajasthan, Jaipur

Figure 12.33 Crop Wheat Area, Production and Productivity in Rajasthan

BARLEY

Barley is an important cereal of the state. The major portion of this crop is consumed by people, either as flour (Pure or mixed with that of wheat or gram) for chapati making or as parched grain to make ·sattu. Itis also use to prepare malt for manufacturing beer, whisky and other industrial products such as alcohol and vinegar. Malt syrup is also used in making candies. While surplus grain provides feed for cattle and horeses, the straw too being fed to cattle.

The crop is usually grown on loamy or loamy sandy soils. The unirrigated barely crop can be grown in areas which receive 25 to 75 cm of rainfall on conserved soil moisture. Because of its low water requiremetns, the crop is well suited for drought prone areas. The crop prefers bright cold winter days and nights, free from frost during flowering. The crop can be grown sucessfully on saline lands or on marginally fertile lands with inadequate water supply where wheat can not give economic results. In any given region cultivated barley matures 2 or 3 weeks earlier than wheat and thus needs comparatively less irrigation. Barley is grown pure or mixed with other rabi crops, such as wheat, gram, peas and lentil. Sometimes rapeseed and mustard, taramira and linseed are also grown mixed with barley or as an inter crop.

The prductivity of barley in the state has increased remarkably over the years. In 1952-53 the productivity was 1,335 kg/ha., which is 2016-17, increased to 2,956 kg/ha. This increase has been possible with development of new hybrid varieties which are fertilizer responsive, resistant to diseases and pests and are of short duration. The increase in irrigated area and use of fertilizers have also contributed significantly to the increased productivity of barley throughout the country.

Besides Rajasthan, the other barley producing states are Uttar Pradesh, Haryana, Bihar and Madhya Pradesh. States like Punjab, West Bengal, Himachal Pradesh and Jammu and Kashmir have limited area under this crop.

In Rajasthan, the crop occupies maximum acreage in Jaipur (0.51 lakh ha.) followed by Skiar (0.22 lakh ha.), Alwar (0.16 lakh ha.), Nagaur (0.11 lakh ha.) and Ganganagar (0.10 lakh ha.) while areas with lesser coverage are in the districts of Hanumangarh (0.09 lakh ha.), Dausa (0.075 lakh ha.), Ajmer (0.073 lakh ha.), Jhunjhunun (0.062 lakh ha.), Tonk (0.061 lakh ha.), Bhilwara (0.049 lakh ha.) and Udaipur (0.045 lakh ha.) in 2016-17. As irrigation is predominantly through wells and their recharging depends upon the rainfall during the monsoon season, its area in the western districts of the state fluctuates considerably.

Barley contributes about 2.27 per cent of the total production of food grain crop in Rajasthan. Jaipur district stands first in production (11.56 lakh metric tonnes). Next in importance are the districts of Sikar and Alwar which produces 0.56 and 0.32 lakh metric tonnes, respectively. In other districts which produce barley, the production ranges from a minimum 36 metric tonnes (Jaisalmer) to a maximum 0.25 lakh metric tonnes (Nagaur).

Newly developed genotypes which are not only high fertilizer-responsive, dwarf varieties, but also are disease resistant, which are available and are required to be propagated more in the farming community to attain high levels of barley production (Figure 12.34 and 12.35).

Rajasthan comes after Uttar Pradesh in the production of barley. About 1/4th output of barley in the country comes from Rajasthan. It is mostly produces in Jaipur, Udaipur Alwar, Tonk and Bhilwara districts. In recent years new varieties like Jyoti RS-6 etc. have come into vogue. In 2014-15 its production was 9.6 lakh tonnes.

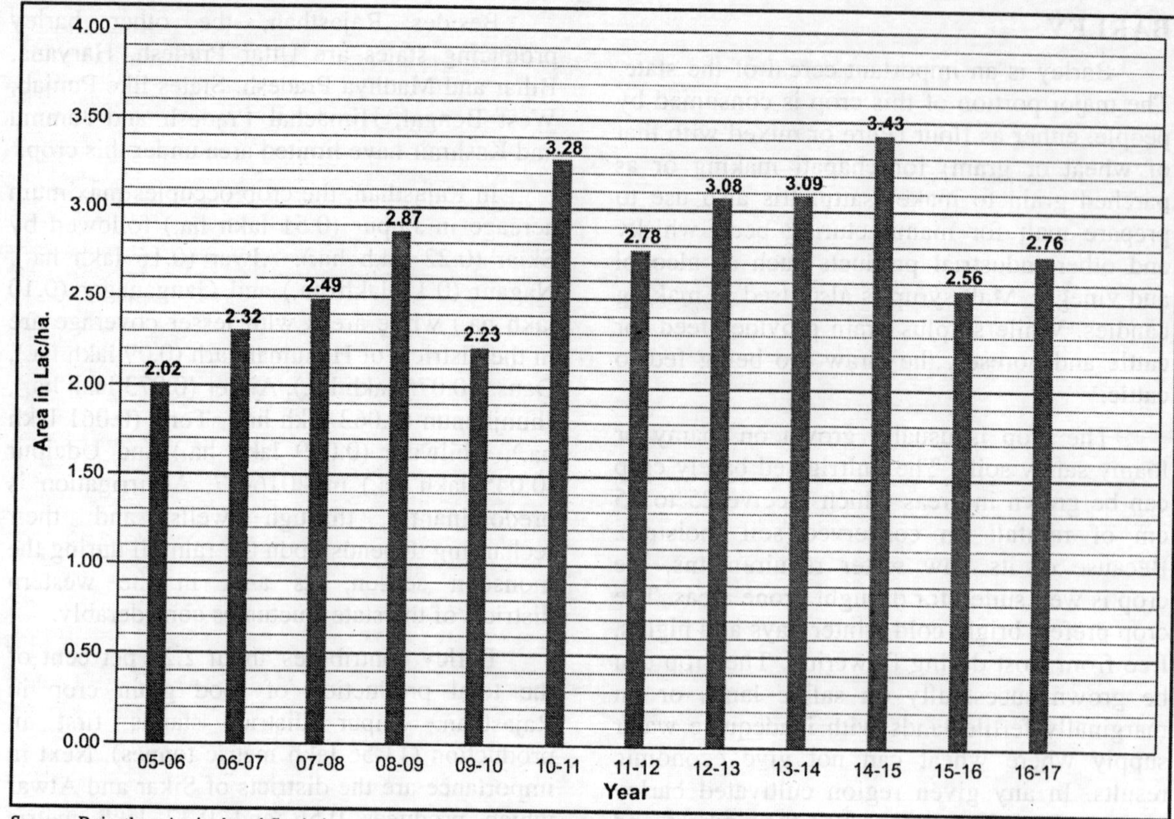

Source: Rajasthan Agricultural Statistics at a Glance Year 2016-17, Commissionerate of Agriculture, Rajasthan, Jaipur

Figure 12.34 Yearwise Area under Crop Barley in Rajasthan (2005-06 to 2016-17)

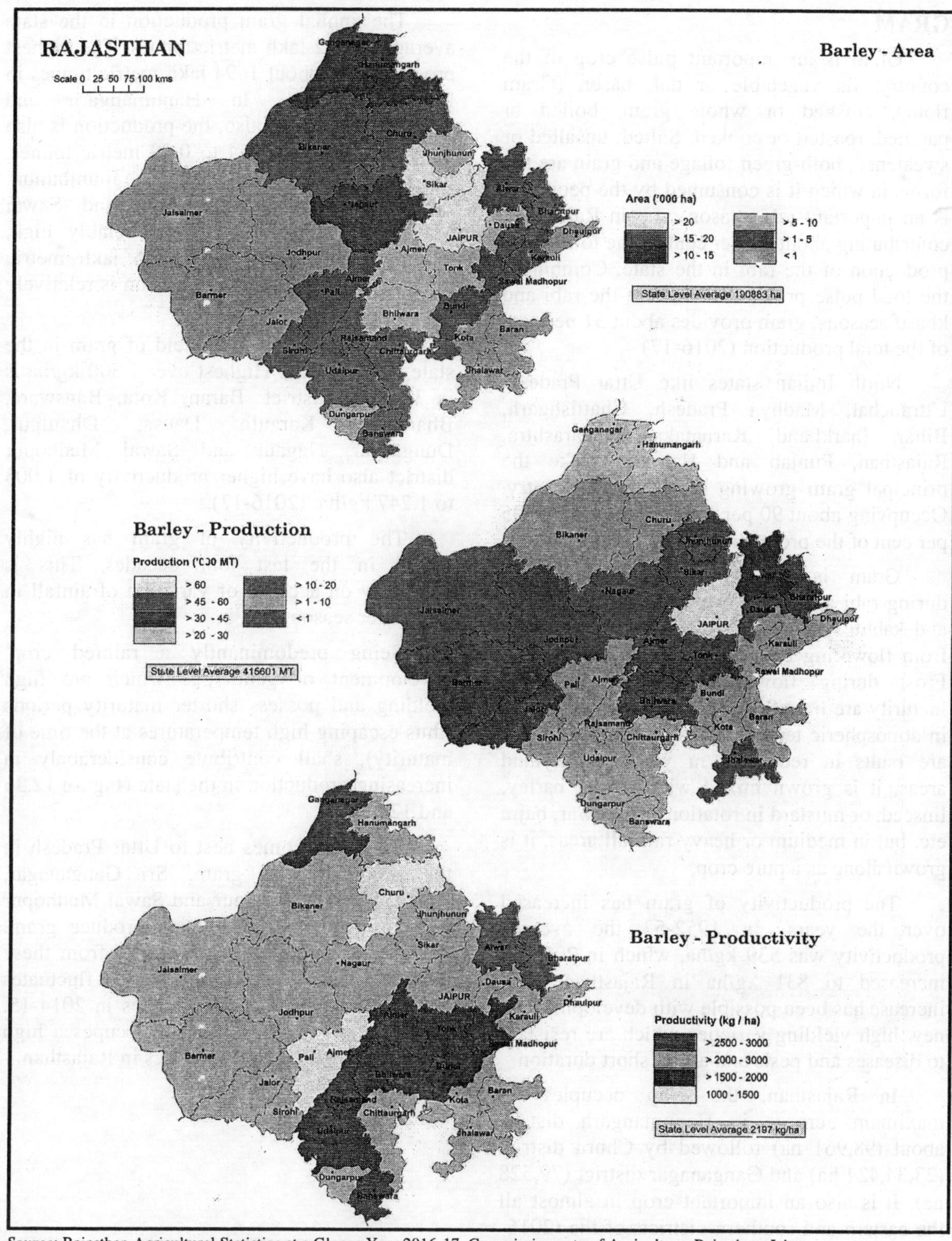

Source: Rajasthan Agricultural Statistics at a Glance Year 2016-17, Commissionerate of Agriculture, Rajasthan, Jaipur

Figure 12.35 Crop Barley Area, Production and Productivity in Rajasthan

GRAM

Gram is an important pulse crop of the country. Its vegetable or dal, basen (Gram flour), crushed or whole gram, boiled or parched, roasted or cooked, Salted, unsalted or sweatened, both-green foliage and grain are the forms in which it is consumed by the people..It is an important rabi season crop in Rajasthan, contributing about 93 per cent of the total pulse production of the rabi in the state. Combining the total pulse proudction of both-the rabi and kharif seasons, gram provides about 31 per cent of the total production (2016-17).

North Indian states like Uttar Pradesh, Uttranchal, Madhya Pradesh, Chattishgarh, Bihar, Jharkhand, Karnataka, Maharashtra, Rajasthan, Punjab and Haryana make the principal gram growing areas in the country. Occupying about 90 per cent of the area and 95 per cent of the production of the nation.

Gram is generally grown unirrigated during rabi season. However, present varieties and kabuli types requires irrigation, especially from flowering to the grain formation period. Frost during flowering and rains during maturity are harmful to the crop. An early rise in atmospheric temperature during grain filling are rsults in reduction in yield. In dryland areas, it is grown mixed with wheat, barley, linseed, or mustard in rotation with jowar, bajra etc. but in medium or heavy rainfall areas, it is grown alone as a pure crop.

The productivity of gram has increased over the years. In 1952-53 the average productivity was 539 kg/ha, which in 2016-17 increased to 831 kg/ha in Rajasthan. This increase has been possible with development of new high yielding varieties which are resistant to diseases and pests and are of short duration.

In Rajasthan, the crop occupies its maximum acreage in Hanumangarh district about (98,961 ha) followed by Churn district (23,34,424 ha) and Ganganagar district (79,528 ha). It is also an important crop in almost all the eastern and southern districts of the (2016-17).

The annual gram production in the state averages 11.32 lakh metric tonnes. The highest production of about 1694 lakh metric tonnes in Bikaner district. In Hanumangarh and Ganganagar district also, the production is also high ranging from 0.54 to 0.59 metric tonnes. In the districts Sikar, Karauli, Jhunjhunun, Alwar, Nagaur, Jaipur, Churn and Sawai Madhopur, production is appereciably high, varying from 82,017 to 60,226 lakh metric tonnes. In other district production is relatively low (2016-17).

Average per hectare yield of gram in the state is 633 kg/ha. Highest over 1,300 kg/ha is in Udaipur district. Baran, Kota, Banswara, Bharatpur, Karauli, Dausa, Dhaulpur, Dungarpur, Nagaur and Sawai Madhopur district also have higher productivily of 1,003 to 1,247 kg/ha. (2016-17).

The productivity of gram has highly varied in the last four decades. This is obviously on account of variation ofrainfall in the winter season.

Being predominantly a rainfed crop, development of genotypes which are high yielding and possess shorter maturity periods (thus escaping high temperatures at the time of maturity), shall contribute considerabaly in increasing production in the state (Figure 12.36 and 12.37).

Rajasthan comes nest to Uttar Pradesh in the production of gram. Sri Ganganagar, Alwar, Bharatpur, Jaipur and Sawai Madhopur are the main districts, which produce gram. About $3/4^{th}$ output of gram comes from these districts. The output of gram fluctuates drastically. It was 9.1 lakh tonnes in 2014-15. It comes under rabi pulses. It occupes a high place in the production of pulses in Rajasthan.

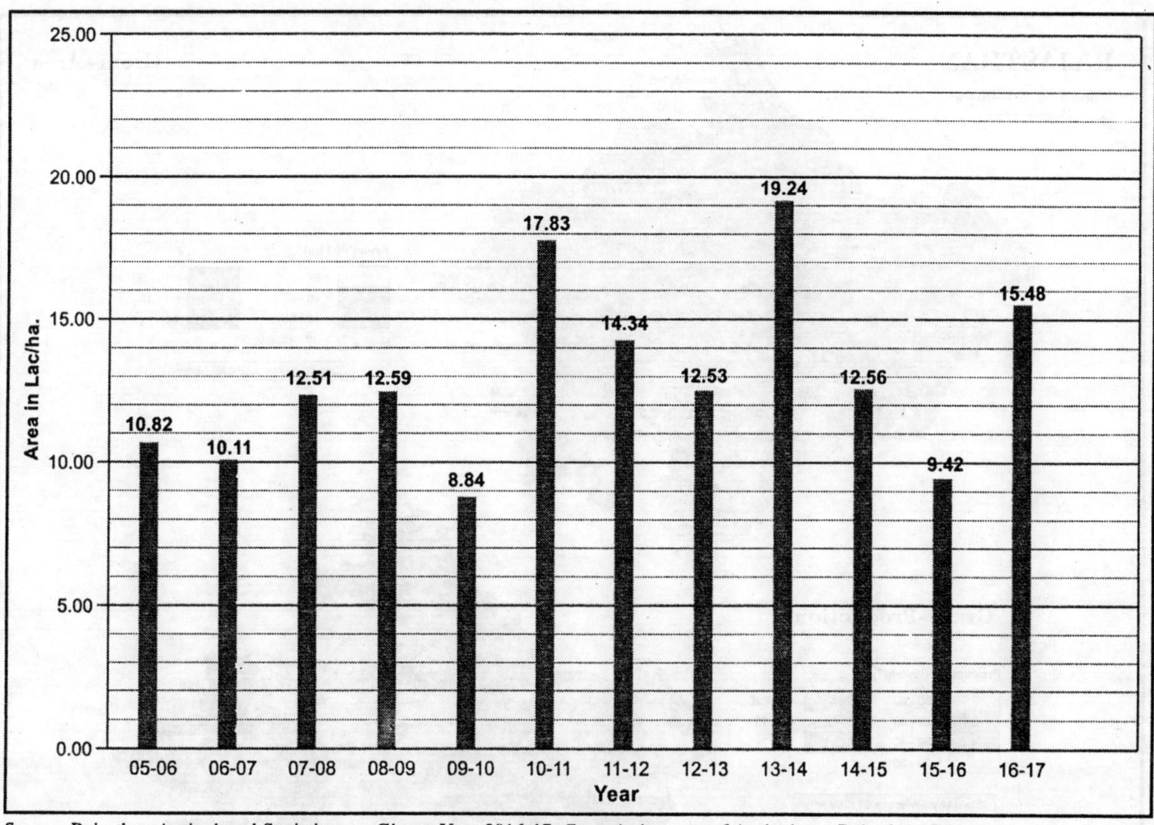

Source: Rajasthan Agricultural Statistics at a Glance Year 2016-17, Commissionerate of Agriculture, Rajasthan, Jaipur

Figure 12.36 Yearwise Area under Crop Gram in Rajasthan (2005-06 to 2016-17)

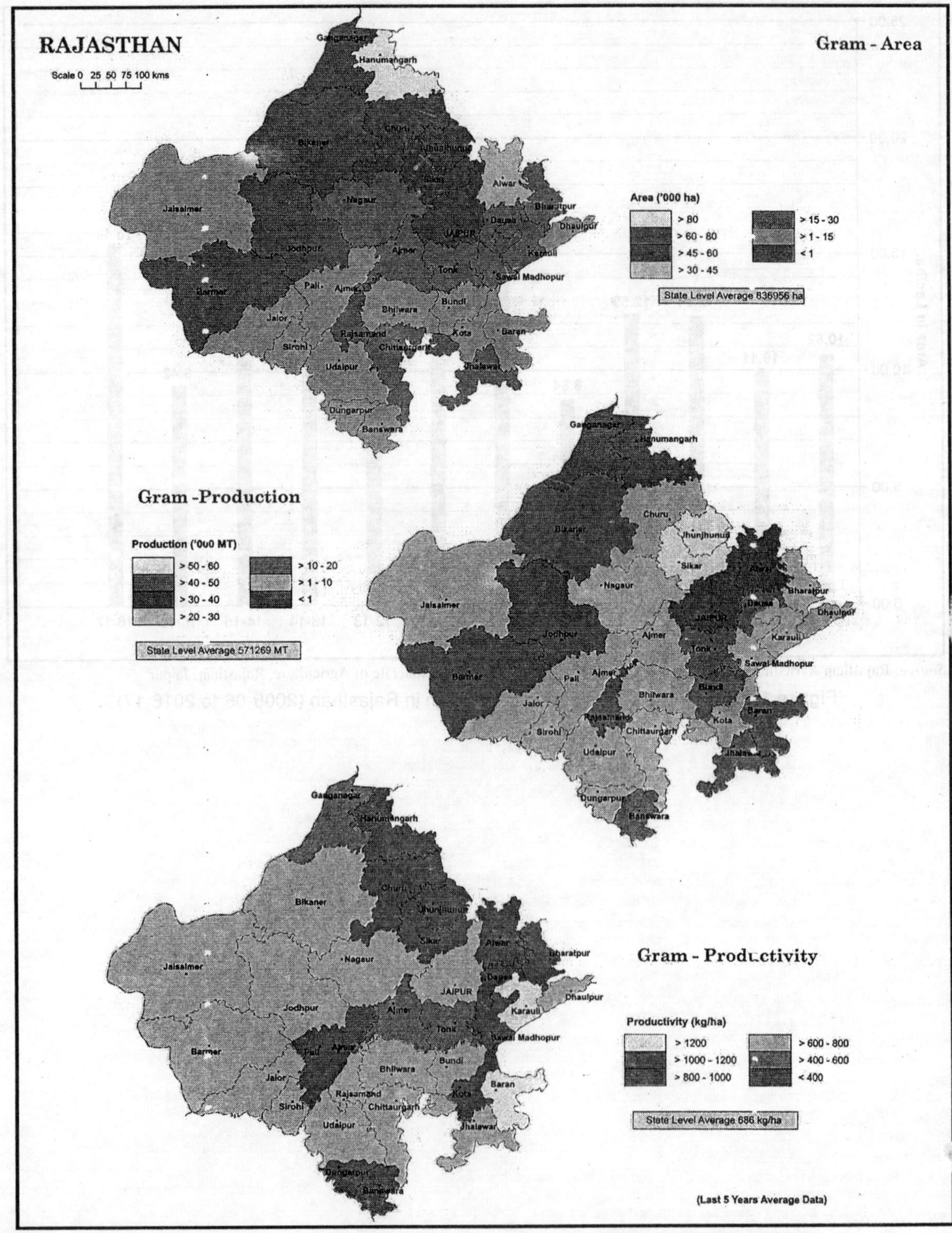

Source: Rajasthan Agricultural Statistics at a Glance Year 2016-17, Commissionerate of Agriculture, Rajasthan, Jaipur

Figure 12.37 Crop Gram Area, Production and Productivity in Rajasthan

RAPESEED AND MUSTARD

On the oil seed map of the world, India occupies a prominent position, both with regard to acreage and production. Rapeseed and mustard are the most important source of edible oil. The oil content of the seeds of different form ranges from 30 to 50 per cent. The oil is mainly used as a cooking medium in northern India and can not be replaced by other edible oils. The seeds and oil are used as a condiment in preparation of pickles and for flavouring curries and vegetables. The oil cake of sarson is mostly used as cattle feed.

It is a rabi season crop which requires cool, dry, weather and bright sunlight. The crop may grown in rainfed conditions but higher yields are obtained under irrigated areas. The crop grows well in areas where the annual rainfall ranges between 75 to 100 cm.

Allthough several strains of new high yielding varieties of rapeseed and mustard have been introduced in the state, there has not been a substantial increase in its productivity in the rainfed areas. Failure of monsoon or uneven distribution in the rainfall leads to lower yields and consequently productivity declines. Thus the productivity of rapeseed and mustard in the past five decades has thus, varied from 333 kg/ha. (1952-53) to 1,306 kg/ha. (2016-17) in Rajasthan.

Owing to its commercial importance, this crop has replaced other rabi crops significantly in major parts of the state. Each year the area under this crop is increasing considerably. Nearly 20.62 lakh hectares are being covered by this crop which is about 32.60 per cent of the area covered by rabi crops (2016-17). This crop contributes about 98 per cent of the oil seed production (rabi) and 68 per cent of the total oil seed production (Kharif + Rabi) in the state (2016-17).

Besides Rajasthan, other predominant states producing rapeseed and mustrad are Uttar Pradesh, Madhya Pradesh and Gujarat.

Ganganagar district covers the highest area under this crop (2.60 lakh hectares). Whereas Alwar, Bharatpur, Baran and Sawai Madhopur districts cover more than 1.23 lakh hectares. Hanumangarh, Nagaur, Jhunjhunun, Tonk, Kota, Dhaulpur, Jalor, Jaipur, Dausa, Bundi, Karauli and Jodhpur districts, have acreage ranging from 0.50 to 1.09 lakh hectares (2016-17). Increasing coverage in eastern Raj asthan has open a new era for oil industry in the state.

Total production in the state is nearly 32.93 lakh metric tonnes. Production of this crop is highest in Ganganagar district about 3.29 lakh metric tonnes followed by Alwar (3.46 lakh metric tonnes) and Bharatpur (3.22 lakh metric tonnes) (2016-17).

Productivity is markedly higher in northern and eastern Rajasthan as compared to the western part of the state. The state average yield is nearly 1,275 kg/ha. While in Gangangar, Kota and Dhaulpur districts yield is about 1344, 1452, 1326 kg/ha. respectively, in other districts like Bharatpur, Karauli, Jhunjhunun, Banswara and Banner yield ranges from 1274 to 890 kg/ha. (2016-17). Production and Productivity of this crop, especially in western Rajasthan are high in good rainfall years but is adversly affected during drought conditions.

In order to increase its production with high disease tolerance and to lower the effect of hannful ingredients in its oil, intensive efforts are underway in the direction of exploitation of hybird vigour. Thus, high yielding hybird of this crop will boost production in the state (Figure 12.38 and 12.39).

The state occupies first place in the production of Rapeseed and mustural in the country. Earliear, mustrad was mainly produce in Alwar, Bharatpur, Jaipur and Sri Ganganagar districts. But after the implementtation of agricultureal extension programmes, its production has storted in Jalore, Sirohi, Udaipur, Chittourgarh, Kota and Bundi districts also. The production of all the nine oilseeds was 55.1 lakh tonnes in 2015-16. Now Rajasthan is a leading state in the production of mustard and rape-seed in India. Its output was 28.8 lakh tonnes in 2014-15.

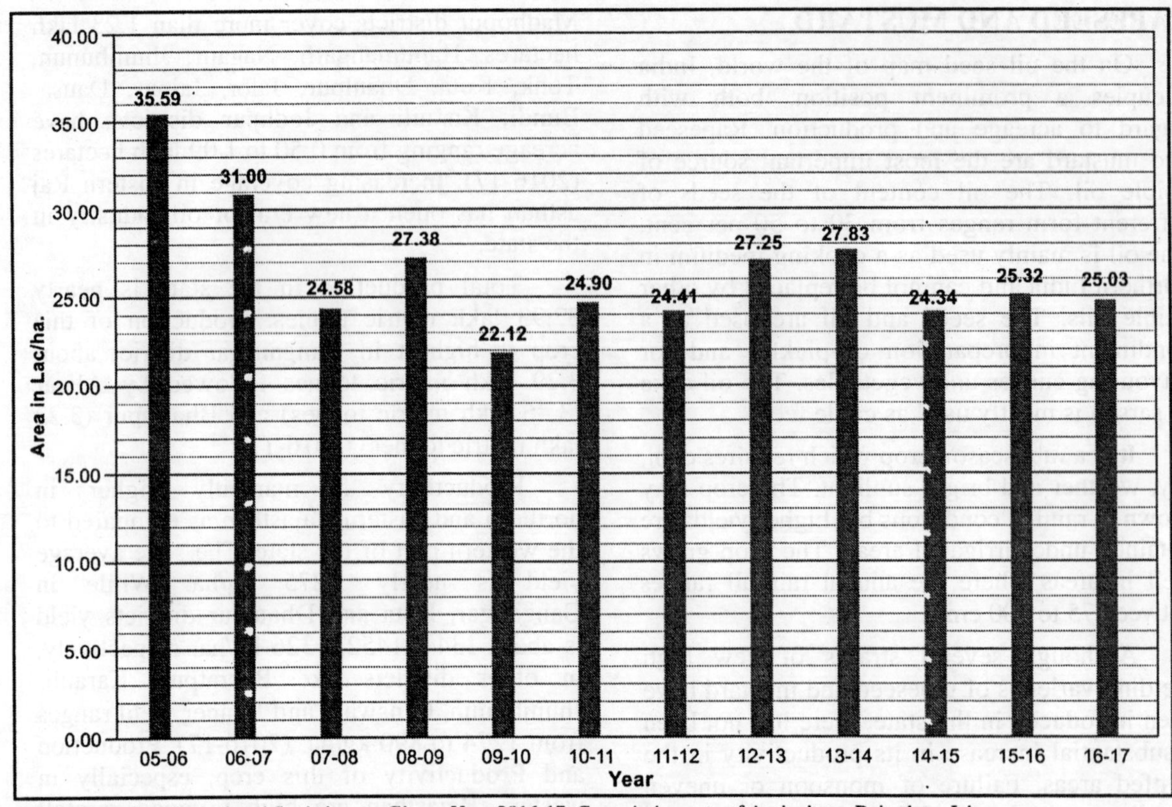

Source: Rajasthan Agricultural Statistics at a Glance Year 2016-17, Commissionerate of Agriculture, Rajasthan, Jaipur

Figure 12.38 Yearwise Area under Crop Rapeseed and Mustrad in Rajasthan (2005-06 to 2016-17)

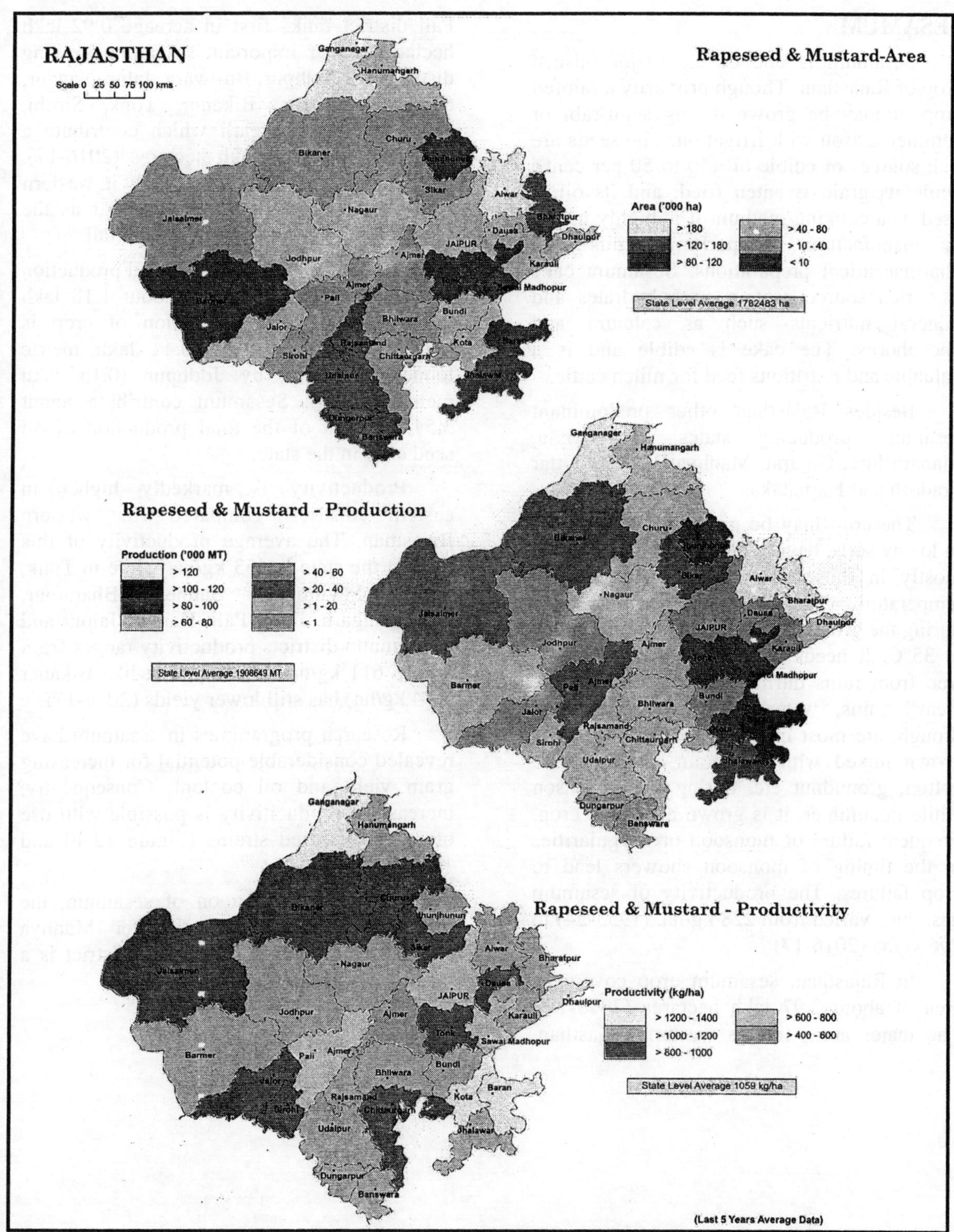

Source: Rajasthan Agricultural Statistics at a Glance Year 2016-17, Commissionerate of Agriculture, Rajasthan, Jaipur

Figure 12.39 Crop Rapeseed and Mustard Area, Production and Productivity in Rajasthan

SESAMUM

Sesamum is one of the major oilseed crop of Rajasthan. Though primarily a rainfed crop, it may be grown during semi-rabi or summer season with irrigation. The seeds are rich sources of edible oil (40 to 50 per cent), while its grain is eaten fried, and its oil is used as a cooking medium it is highly useful in manufacture of perfumed oils and pharmaceutical preparations. Sesamum cake is a rich source protein, carbohydrates and mineral nutrients such as calcium and phosphorus. The cake is edible and is a valuable and nutritious feed for milch cattle.

Besides Rajasthan other predominant sesamum producing states are Orissa, Maharashtra, Gujarat, Madhya Pradesh, Uttar Pradesh and Karnataka.

The crop may be grown on light sandy to loamy soils, having PH between 5.5 to 8.2, mostly in those areas which have diurnal temperature. The favourable temperatures during the growing season range between 25° to 35°C. It needs warm and humid climate, free from rains during its flowering period. Heavy rains, waterlogging and prolonged drought are most injurious for the crop. It is grown mixed with red gram, bajra, jowar, cotton, groundnut etc. during kharif season while in summer, it is grown as a pure crop. Frequent failure of monsoon or irregularities in the timing of monsoon showers lead to crop failures. The productivity of sesamum has, thus, varied from 228 kg/ha. (1953-54) to 296 kg/ha (2016-17).

In Rajasthan, sesamum crop covers an area of about 3.97 lakh hectares (2016-17). The major areas lies in western Rajasthan. Pali district ranks first in acreage 0.92 lakh hectares. Other important sesamum growing districts are Jodhpur, Bhilwara, Jalor, Nagaur, Sawai Madhopur, Bikaner, Tonk, Sirohi, Bundi and Chittaurgarh which contribute a total of about 1.69 lakh hectares (2016-17). The area under this crop especially in western district, fluctuates from year to year as the crop is totally dependent on the rainfall.

During the year 2016-17 total production of seasmum in the state is about 1.18 lakh metric tonnes. The production of crop is highest in Pali district (0.44 lakh metric tonnes) followed by Jodhpur (0.15 lakh metric tonnes). Sesamum contribute about 3.54 per cent of the total production of oil seed crop in the state.

Productivity is markedly higher in eastern district compared to western Rajasthan. The average productivity of this crop in the state is 453 kg/ha. while in Tonk, Sawai Madhopur, Banner, Bharatpur, Hanumangarh, Jalor, Pali, Karauli Jaipur and Jhunjhunun districts productivity ranges from 455 to 611 kg/ha., other district like Bikaner (174 kg/ha) has still lower yields (2016-17).

Research programmes in sesamum have revealed considerable potential for increasing grain yield and oil content. Consequently, increase in producitivity is possible with use of newly released strains (Figure 12.40 and 12.41).

In the production on of sesamum, the place of Rajasthan comer after Madhya Pradesh and Uttar Pradesh. Pali district is a leading district for this crop.

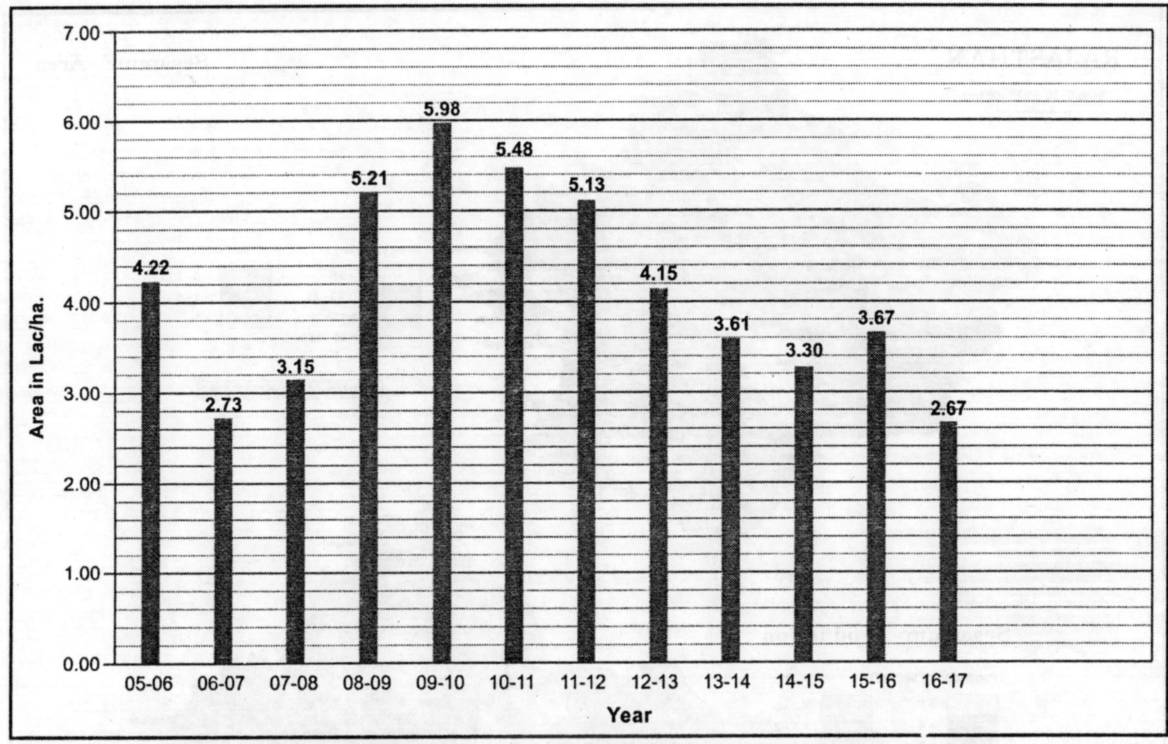

Source: Rajasthan Agricultural Statistics at a Glance Year 2016-17, Commissionerate of Agriculture, Rajasthan, Jaipur

Figure 12.40 Yearwise Area under Crop Sesamum in Rajasthan (2005-06 to 2016-17)

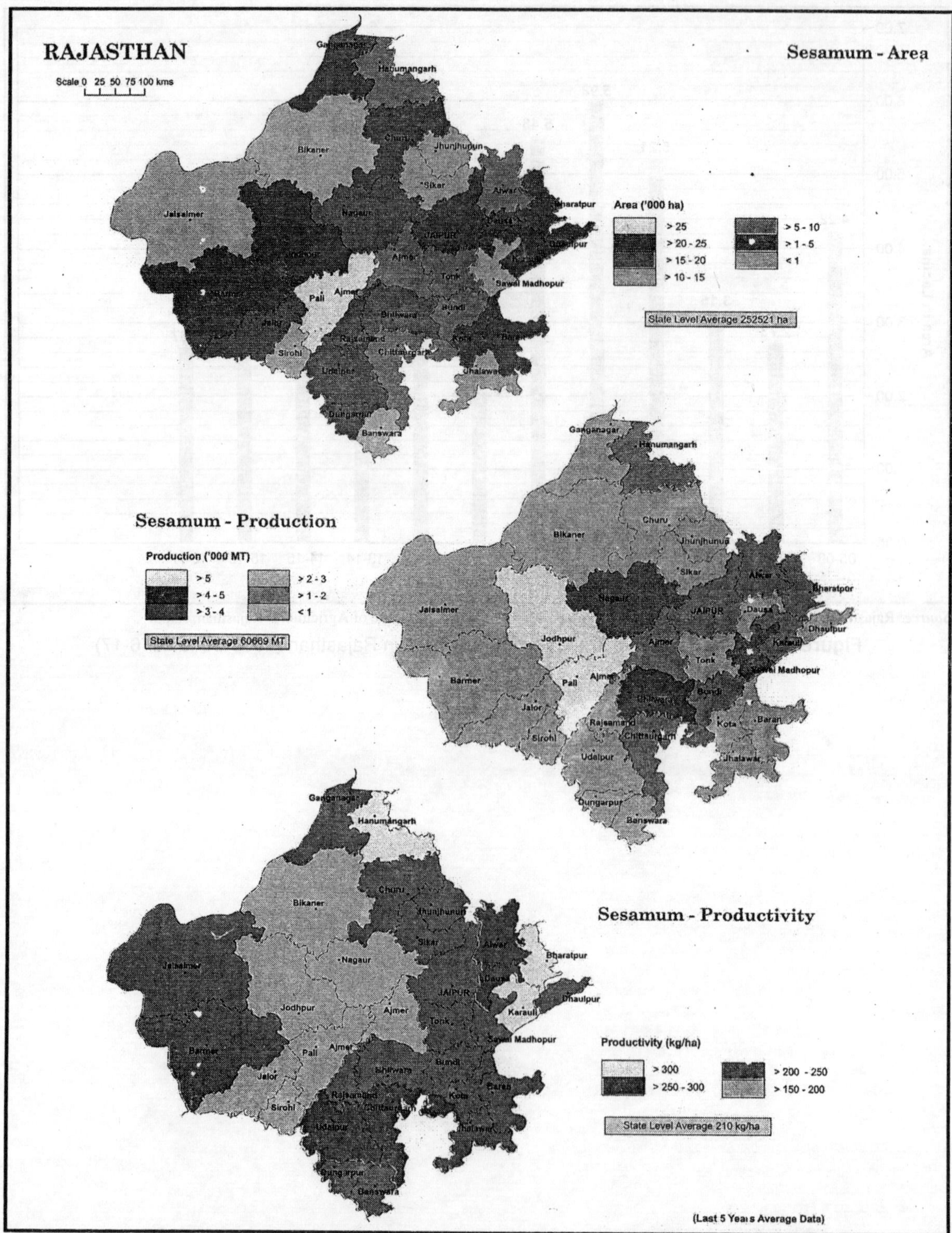

Source: Rajasthan Agricultural Statistics at a Glance Year 2016-17, Commissionerate of Agriculture, Rajasthan, Jaipur

Figure 12.41 Crop Sesamum Area, Production and Productivity in Rajasthan

LINSEED

This crop is under cultivation from ancient times for fibre (flax) and for its seeds which are rich in oil. Linseed has greater importance among other oil seeds owing to its various uses and special qualities. The crop is mainly grwon in the state for its seeds, which is used for extracting oil. The oil contents of the seed varies from 33 to 47 per cent. The oil is used in manufacturing paints and varnishes, oil-cloth, waterproof fabrics and linoleum. It is also used as an edible oil in some area. Linseed produces fibre of good quality and has importance in paper and plastic industries. It is an important rabi crop in Rajasthan, Uttar Pradesh, Madhya Pradesh, Maharashtra, Bihar, Karnataka and West Bengal.

Linseed grows mostly in those areas where the annual rainfall is 70 to 100 cm. The crop for seed purposes does well under moderate cold, but the fibre crops grows best in cool moist conditions. The optimum temperatue requirements range between 20-30°C. The crop can be grown on different kind of soils, except sandy, and poorly drained heavy clays, but it does best on clay loams. September and October period are best for sowing this crop, which may somtimes extend upto November.

The crop is mainly grown as unirrigated. Its productivity is directly related to intensity and timing of the monsoon precipitation during the rainy season. Frequent failure of monsoon and irregularties in the timing of monsoon results in decrease in productivity. The productivity of linseed in the past three decades has thus varied from 310 kg/ha (1973-74) to 805 kg/ha. (2016-17).

Other predominating states producing linseed are Madhya Pradesh, Uttar Pradesh, Maharashtra and Bihar, Looking to the high production effeciency of linseed crop in recent past, farmers have realised importance of providing supplemental irrigation to this crop, particularly to the high yielding strains.

Rajasthan covers an area of 0.02 lakh hectares under this crop. The major linseed producing districts are : Kota (414 hectares) Nagaur (284 hectares), Baran (275 hectares), Bhilwara (259 hectares), Jhalawar (240 hetares), Chittaurgarh (230 hectares), Bundi (228 hectares) and Tonk (156 hectares). These districts cover more than 90 per cent area of the total linseed crop (2016-17).

The total production in the state is nearly 1853 metric tonnes. It contribute about 0.05 per cent of the total oilseed production of the country. Production of this crop is highest in Baran district (311 metric tonnes), followed by Kota (307 metric tonnes) and Nagaur (229 metric tonnes).

State average yield is 805 kg/ha. In Baran, Dhaulpur, Bundi, Bikaner, Banner, Churu and Bhilwara yield is more than state average ranging from 807 to 1131 kg/ha. In other districts Ajmer, Nagaur, Sirohi and Udaipur the yield ranges from 800 to 806 kg/ha. (2016-17).

Use of available high yielding, short duration strains with disease-resistance, will boost the average productivity in the state. For specific areas, varieties suited for unirrigated conditions have already been released and may be popularised (Figure 12.42).

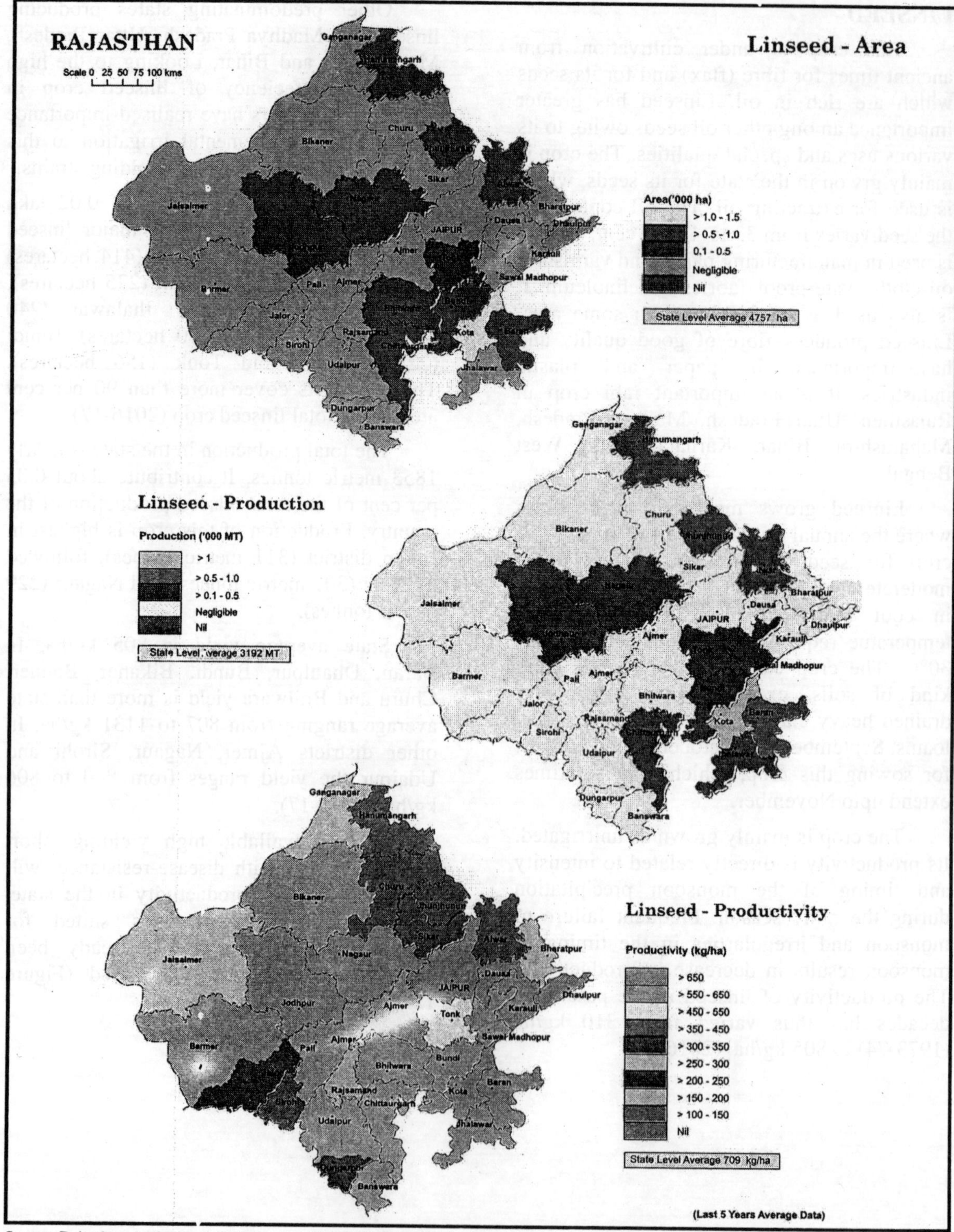

Source: Rajasthan Agricultural Statistics at a Glance Year 2016-17, Commissionerate of Agriculture, Rajasthan, Jaipur

Figure 12.42 Crop Linseed Area, Production and Productivity in Rajasthan

GROSS CROPPED AREA

The gross cropped (cultivated) area varies from one region to another due to variation in land capability and irrigation facilities. There has been 10.78 per cent increase in gross cropped area in the state within last two decades. It was 47.83 per cent in 1979-80 which in 2016-17 increased to 73.00 per cent of the total geographical area of the state. This may be attributed to the increase in the irrigation facilities.

The gross cropped area out of the total area of the district varies from district to district. In Alwar, Bharatpur, Dausa, Hanumangarh, Jaipur, Jhunjhunun and Sikar districts it ranges from more than 70-98 per cent of the total geographical area of the district. In Banswara, Dhaulpur, Ganganagar, Jalor, Jhalawar and Tonk districts it ranges from more than 50 to 70 per cent. In other part of the state it is less than 50 per cent. In Jaisalmer district it is only 2.37 per cent due to sandy soils, poor precipitation (2016-17) (Figure 12.43).

NET CROPPED AREA

The net cropped area (cultivated) also varies from one region to another region due to variation in soil fertility and availability of irrigation facilities. In Rajasthan, it is higher in the northern and north eastern parts where canal water and well water irrigation facilities is available. In Alwar, Bharatpur, Dausa, Ganganagar, Hanumangarh, Jaipur, Jalor, Jhunjhunun and Sikar districts it ranges from more than 50 to 76 per cent of the total geographical area of the district. In Ajmer, Banswara, Barmer, Bhilwara, Bundi, Chittaurgarh, Churu, Dhaulpur, Jhalawar, Kota, Nagaur, Sawai Madhopur and Tonk districts it ranges more than 30 to 50 per cent. In other parts of the state it is less than 30 per cent. In Jaisalmer district it is only 20.90 per cent (2016-17) (Figure 12.43).

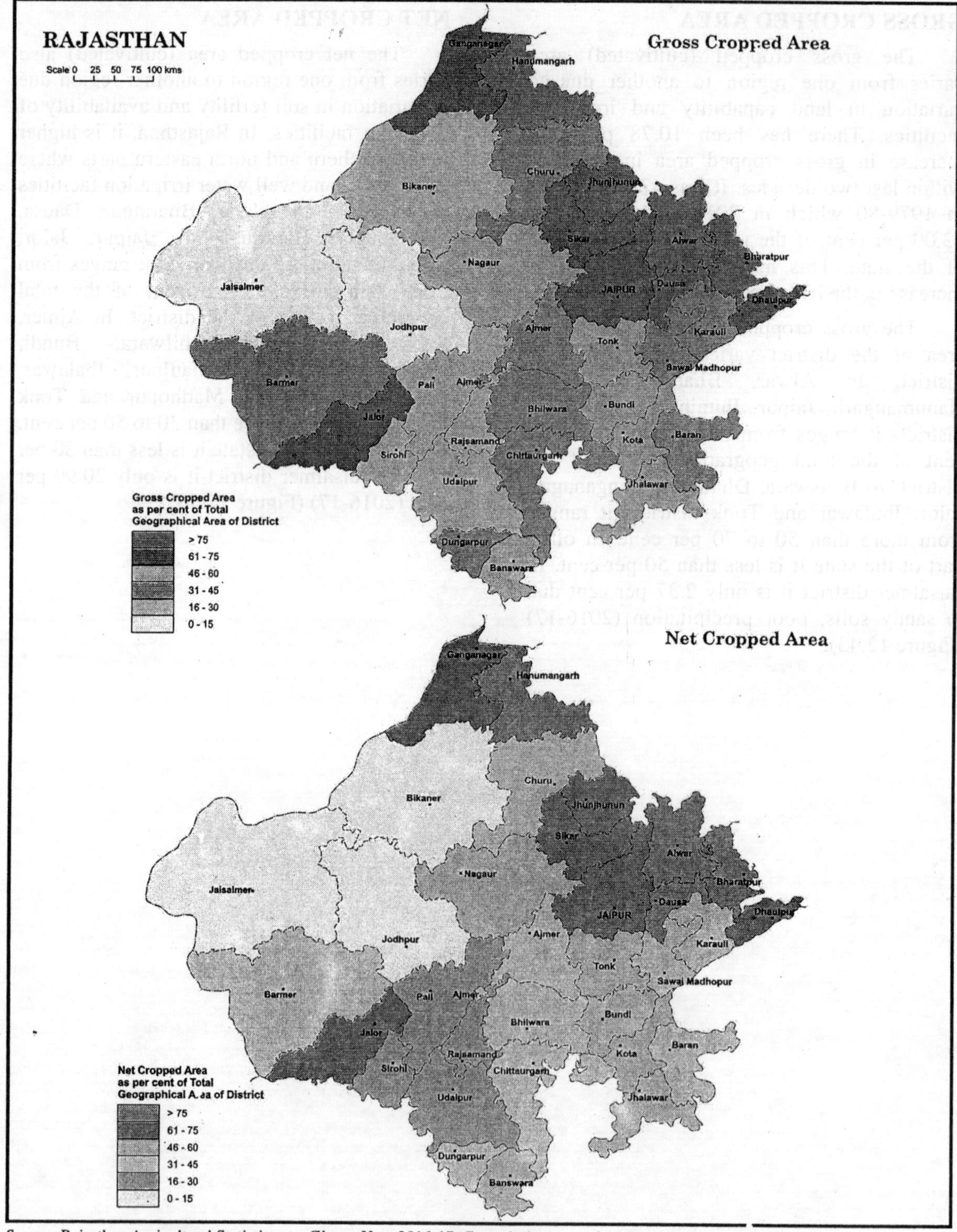

Source: Rajasthan Agricultural Statistics at a Glance Year 2016-17, Commissionerate of Agriculture, Rajasthan, Jaipur

Figure 12.43 Gross Cropped and Net Cropped Area in Rajasthan

LAND HOLDINGS

Agricultural productivity, to a large extent, is dependent upon certain socio-economic factors such as the size of land-holdings and land-tenure systems. Prior to the independence of the country, large sized land-holdings existed in Rajasthan due to 'zamindari' system of land-tenancy. After the abolition of zamindari system the land property rights were conferred upon persons who were either real cultivators of the land or were tenant cultivators. Thus, large-sized land holdings got fragmented in a number of smaller units.

These land-holdings of the farmer, over the ages, have today become divided and sub-divided by division of property among heirs and distant relatives, by expansion of cultivation irregularly over wastelands, by purchase and sale of small plots of land, or due to the new immigrants in villages. The law of succession in India results in the sub-division and fragmentation of holdings.

The disadvantages of fragmented and scattered holdings are many. It puts a large proportion of land outside the possibility of effective cultivation and economic development. The small plots are difficult to work with modem machinery like tractors and threshers. There is, thus, a considerable waste of labour in ploughing, weeding and harvesting of the fields. The input costs of human labour, fertilizer, irrigation and transport of farm produce increase many times in small land-holdings. Road linkages from smaller plots of land to grain markets are seldom available and so animal-driven carts are used. Smaller land-holdings are essentially, therefore, a type of subsistence farming.

The size of land holdings according to their area, for efficient and economically operational cultivation, is only one criterion to improve agricultural production. Rather more significant is the land efficiency to increase farm produce. Thus, sterile barren lands, even of larger hectarage, may produce less and become economically inefficient in utilization of modem farm technology, while on the other had, smaller holdings of fertile, loamy soil with adequate means of irrigation may be economically viable for use of agricultural innovations. However, they are inefficient units, even to sustain a family and the farm livestock, and therefore, the need is to consolidate smaller fragmented land units to make them economically viable.

Out of the total land of the state (34.2 million hectares), about 19.9 million hectares (58 per cent) are operated as individual, joint or institutional holdings which number more than 4.48 million. Though the per capita average size of land holdings is 4.44 hectares, the majority (29 per cent) are very small, of less than 1 hectare each. These small holdings account for only 3 per cent of the total area of the holdings. On the other hand, there are some large holdings of 10 hectares or more each, which account for 11 per cent of the total number and 50 per cent of the total area (Table 12.1).

In fertile areas like Alwar, Bharatpur, Dhaulpur, Sawai Madhopur, Karauli, Ajmer and Tonk districts, large number of land holdings are less than 1 hectare, showing a greater intensity of agricultural cropping and demographic pressure on the land. Similarly, in the southern tribal areas of Udaipur, Sirohi, Dungarpur, Banswara and Chittaurgarh, the size of the land holdings of less than 1 hectare is larger. On the other hand, Jaisalmer, Barmer, Bikaner and Jodhpur districts have larger land holdings, exceeding 10 hectares. In all these western districts, the land, however, is mostly devoid of irrigation. In Churu and Nagaur districts also, the number of larger land holdings is high (more than 10 hectares) because of the sandy soil and unirrigated lands (Figure 12.44).

Table 12.1 DISTRICTWISE AND CATEGORY-WISE NUMBER AND AREA OF OPERATIONAL HOLDINGS FOR ALL SOCIAL GROUPS

(Holding in Number)
(Area in Hectare)

S. No.	District	Marginal		Small		Semi Medium		Medium		Large		Total	
		Number	Area	Number	Area	Number	Area	Number	Area	Number	Area	Number	Area
1	Ajmer	95,407	40,090	44,941	65,362	39,160	109,404	28,163	167,427	7,740	120,195	215,411	502,478
2	Alwar	136,239	66,640	73,093	102,522	52,545	147,224	28,355	163,190	3,164	47,701	293,396	527,277
3	Banswara	78,457	36,705	35,804	51,375	27,089	75,213	12,242	69,906	1,089	18,311	154,681	251,510
4	Baran	35,387	18,194	28,050	41,042	29,038	82,426	21,728	130,499	4,162	60,801	118,365	332,962
5	Barmer	7,126	3,421	9,949	15,268	24,894	75,152	62,597	424,199	78,233	1,755,706	182,799	2,273,746
6	Bharatpur	115,585	53,796	49,568	71,041	39,967	112,915	23,938	138,911	2,196	30,216	231,254	406,879
7	Bhilwara	113,400	57,945	77,379	109,801	61,254	170,989	31,843	185,812	4,428	66,739	288,304	591,286
8	Bikaner	1,903	1,148	4,623	6,921	18,126	55,701	67,398	434,588	53,848	1,082,383	145,898	1,580,741
9	Bundi	37,273	19,611	28,388	41,405	26,114	73,065	16,451	97,440	2,450	36,350	110,676	267,871
10	Chittaurgarh	80,079	42,382	63,982	91,252	53,269	149,302	32,707	192,737	4,555	65,974	234,592	541,647
11	Churu	2,469	1,511	7,841	12,183	29,054	87,875	68,064	452,515	54,501	994,124	161,929	1,548,208
12	Dausa	34,269	17,166	25,353	36,093	22,885	65,052	14,913	88,835	2,674	40,149	100,094	247,295
13	Dhaulpur	55,359	26,479	23,826	33,993	16,989	47,444	7,916	44,916	808	11,809	104,898	164,641
14	Dungarpur	78,771	31,317	29,604	42,597	22,453	61,648	7,907	43,420	507	11,842	139,242	190,824
15	Ganganagar	2,599	1,554	8,466	12,597	23,258	70,243	61,717	393,117	24,069	402,231	120,109	879,742
16	Hanumangarh	9,990	5,887	18,917	27,777	37,070	110,474	56,391	358,999	23,566	375,256	145,934	878,393
17	Jaipur	74,400	37,375	59,370	85,939	64,038	184,477	46,596	281,864	12,563	205,150	256,967	794,805
18	Jaisalmer	2,938	2,114	3,617	5,371	4,897	14,586	17,202	115,482	22,614	533,868	51,268	671,421
19	Jalor	11,270	6,332	20,294	30,806	36,085	103,187	42,756	273,538	23,002	391,063	133,407	804,926
20	Jhalawar	48,093	23,990	34,108	49,287	32,407	91,740	24,229	145,588	4,417	62,874	143,254	373,479
21	Jhunjhunun	32,746	19,758	47,283	68,852	51,218	144,391	30,763	179,703	3,693	50,859	165,703	463,563
22	Jodhpur	15,209	8,458	22,343	33,815	43,142	125,324	68,716	449,088	57,867	1,191,924	207,277	1,808,609
23	Karauli	----- Included in Sawai Madhopur District -----											

S. No.	District	(Holding in Number) (Area in Hectare)											
		Marginal		Small		Semi Medium		Medium		Large		Total	
		Number	Area	Number	Area	Number	Area	Number	Area	Number	Area	Number	Area
24	Kota	25,237	12,996	21,421	31,488	22,757	64,950	19026	115,084	3,779	56,284	92,220	280,802
25	Nagaur	25,845	12,184	34,906	52,667	62,780	183,360	84,455	539,136	43,389	711,433	251,375	1,498,780
26	Pali	56,475	30,374	47,136	68,767	45,726	126,185	36,913	229,217	18,791	351,076	205,041	805,619
27	Pratapgarh	----- Included in Chittaurgarh District -----											
28	Rajasmand	69,668	29,022	29,493	41,697	21,209	59,169	11,386	66,602	1,781	28,485	133,537	224,975
29	Sawai Madhopur	111,174	50,499	61,714	87,045	48,899	139,064	30,554	180,569	4,862	72,008	257,203	529,185
30	Sikar	40,687	23,283	50,037	72,402	58,276	165,398	42,275	251,952	7,016	94,244	198,291	607,279
31	Sirohi	27,094	14,543	19,371	28,079	15,803	44,480	11,031	67,751	3,134	51,132	76,433	205,985
32	Tonk	40,352	20,222	32,434	46,810	37,466	108,141	34,118	210,938	8,986	134,318	153,356	520,429
33	Udaipur	145,168	65,110	71,702	101,297	49,210	136,087	21,739	124,508	2,797	43,390	290,616	470,392

Marginal < 1.00 ha., small 1.00-2.00 ha., semi medium 2.00-4.00 ha., Medium 4.00-10.00 ha., Large >10.00 ha.

Source: Agriculture Census Rajasthan At a glance 2010-11, Revenue (Agriculture Census) Department, Rajasthan

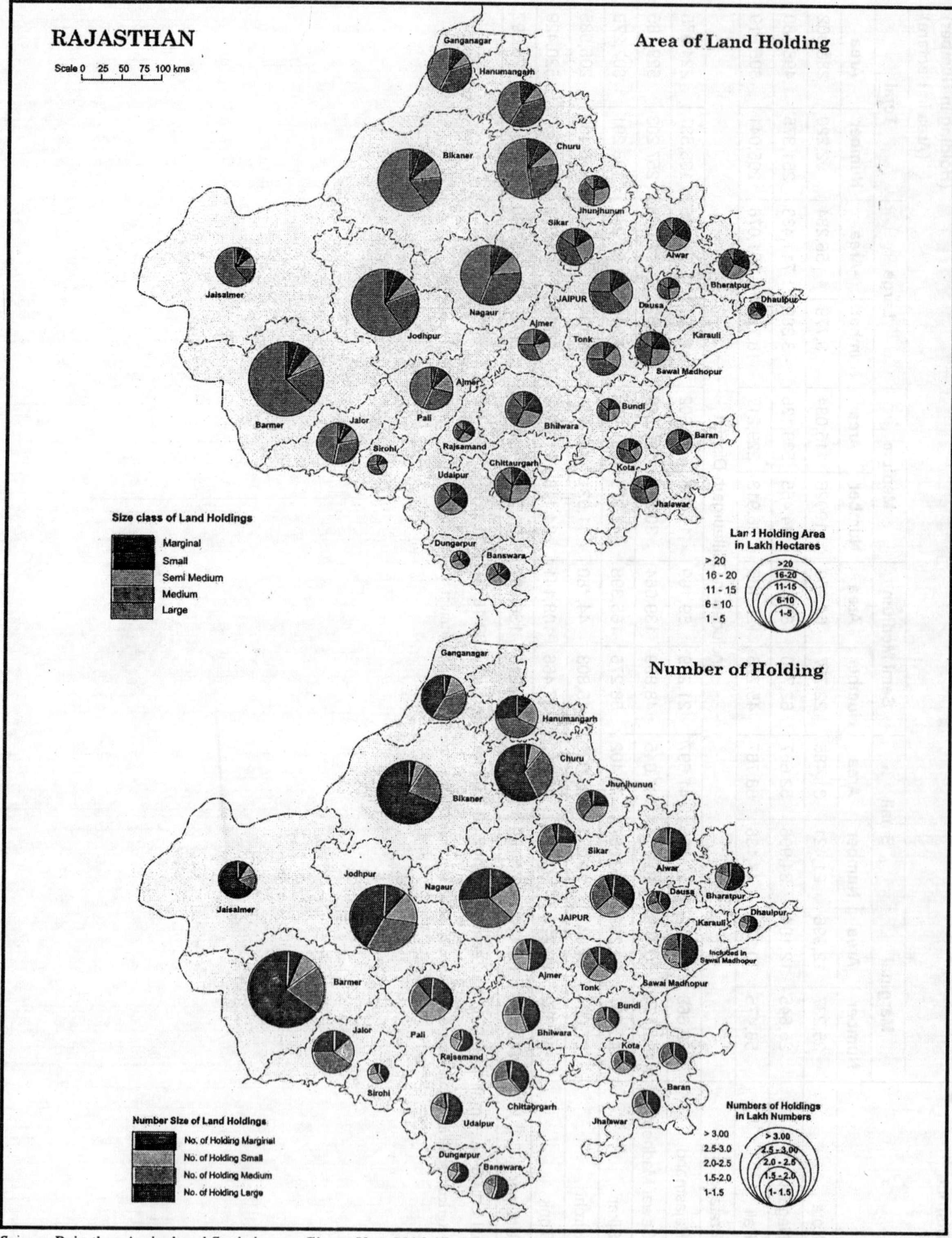

Source: Rajasthan Agricultural Statistics at a Glance Year 2016-17, Commissionerate of Agriculture, Rajasthan, Jaipur

Figure 12.44 Area of Land Holding and Number of Holding in Rajasthan

FERTILIZER CONSUMPTION

Plants require sixteen elements for their natural growth. Out of the these, carbon and oxygen are absorbed by plants directly from air while hydrogen is absorbed from water. All remaining elements namely-nitrogen, phosphorous, potash, calcium, magnesium, sulphur, iron, manganese, zinc, boron, molybodenium and chloride are absorbed from the soil and so their availability in the soil is essential for better crop production. Amongst these first six elements are absorbed in large quantities and so are usually referred as macro-nutrients while rest are also essential but are required in minute quantities and so are commonly referred to as micro-nutrients.

Extent of crop production from any field depends on the fertility of the soil. Virgin-soils are usually rich in all requisite quantities of plant nutrients. Continuous cultivation without any replenishment of the nutrients absorbed by the plants or those lost by other means like leaching etc. leads to reduction in the fertility of soil. Use of organic manures-Farm Yard Manure or Green Manures is the practice followed from times immemorial for maintaining the fertility levels of the soil. Non-availability of requisite quantities of organic manures results in low level of crop production. Application of fertilizers helps to make up the deficient nutrient levels in the soil and thus augment the soil fertility.

Fertilizer is the key input for obtaining higher crop production and thus occupies a prestigious place in the agricultural development stragegy of any region. They are used alone or in addition to organic manures for promoting intensive cultivation of food and cash crops, Fertilizer industries, which are established in the countiy nearly four decades ago, have come up in a planned way to put India in the forefront among the fertilizer producing countries.

Amongst the macro elements, soils in Rajasthan, in general, are rich in calcium and magnesium carbonates and so plants do not feel their deficiency but available contents of other macro elemetns namely nitrogen, phosphorous, potash and sulphur vary widely from area to area and at times even in field to field. Some crops on particular soil types also exhibit deficiency symptoms of micro nutrients which are required proportionately in higher quantities and their availability greatly governs the overall production of crop.

Traditionally, Rajasthan's farmers have been using organic manures to increase soil fertility. Initially, they were sceptical about the use of fertilizers to replenish soil nutrients. Concerted efforts of the government through demonstrations, from early fifties onwords, exemplified the benefits of the use of fertilizers. Fertilizers are now finding an increasing use for production of all major crops and especially cash crops, particularly in those areas where means of irrigation are adequate.

Soils of Rajasthan, in general, are low to medium in availability of nitrogen. Amongst all the nutrient elements maximum research work has been done on nitrogen and continues to receive special attention today. The reason for this that plant require it in much more quantities than it is usually available in the soils. Sandy to sandy loam soils of western part usually have a lower level of available nitrogen than loam to clay loam soils of eastern, south eastern and southern parts of the state. Soils which are impoverished due to extensive cropping and also areas having assured sources of irrigation require more nitrogenous fertilizers for better crop growth. A wide variety of nitrogenous fertilizers are available in various forms amongst which urea, calcium ammonium nitrate and diammonic phosphat are most common. Per hectare use of nirogenous fertilizers vary widely from district to district and from area to area within district also. Total consumption of nitrogenous fertilizer in the state is 9,98,276 m.t. with the highest per hectare consumption of 285.60 kg in Karoli district followed by 8.04 kg in Barmer district (2016-17).

Rajasthan soil also varies widely in available contents of Phospohorous. Usually all these soils are medium to high in total phsophate content but due to its property to remain in various fixed froms in the soil, it is not readily availlable to plant roots. Plants, therefore show its deficiency symptoms or at times even without attaining this stage, adversely affects the crop growth and its production. Under such circumstances, based on soil tests, phosphatic fertilizers are required to be applied, especially in good rainfall areas or in fields with assured irrigation. All phosphatic fertilizers are used as a basal dose and applied at some depth, depending on the crop grown. Phosphatic fetilizers are available in a wide variety of form : Super phosphate, diammonium phosphate and ammonium sulphate phosphate. Total consumption of phosphatic fertilizers is about 3,40,708 metric tonnes in Rajasthan with highest per hectare conumption of 48.60 kg in Kota followed by 34.33 kg in Baran district (2016-17).

Soils of Rajasthan are medium to high in available potassium contents. As compared to nirogenous and phosphatic fertilizers, proportionately veiy low quantities of potasic fertilizers are made use of in the state. Common form of fertilizers used are muriate of potash and sulphate of potash. Total consumption of potasic fertilizers in the state is about 15,685 metric tonnes. Use of potash per hectare is less than 1 kg in most of the distirct. It is the highest in 4.55 kg in Udaipur district followed by 2.83 kg in Sirohi distirct (2016-17).

Use of fertilizers in Rajasthan has remarkably increased in last 65 years. The total use of fertilizers (NPK) which was 0.01 lakh tonnes 1960-61 has increeased 13.45 lakh tonnes in 2016-17 at the state level. There is a great disparity in the use of fertilizers in different part of the state. The maximum quantity of fertilizer used in Ganganagar district about 8.44 lakh metric tonnes which has an assured source of irrigation by canals. It is followed by Kota 6.85 lakh metric tonnes and Flanumangarh 0.51 lakhs (2016-17) where there is also a well developed system of canal irrigation. In desert districts, consumption of fertilizers is still relatively low compared to eastern district as there is low rainfall and lack of other resources for irrigation (Figure 12.45).

In 2010-11, the average size of operational holdings was 3.07 hectare in Rajasthan, while it was only 1.16 hectares for the country as a whole. Thus, the average size of operational holding is quite high in Rajasthan. It had second place amongst 17 states of India.

In 2016-17, out of total of 1,35,460 tonnes of fertilizers, 5,91,177 tonnes was used in kharif season and the rest in rabi season.

Fertilizers in Rajasthan are available both as single fertilizers and as mixed in different proportions of the major nutrients. The use of sulphur is mostly in the oilseed crops.

Sulphur is also required to be applied especially in those soils on which crops show it deficiency or its application gives higher crop production. There are some crops like groundnut, rapeseed and mustard etc., which respond well to its application. Gypsum is the common form in which it usually applied as abasel does.

Efficient use of fertilizers can be made by finding out the deficiency of soils nutrients through soil testing. The state has 30 soil testing laboratories, out of which 20 are stationary and 20 are mobile (2016-17). They operate to advise farmers about the application of does of fertilizers in their fields. They are all government-run institutions. In addition, there are laboratories for testing of fertilizer samples for their quality. Fertilizer samples collected by the Agriculture Department officials under fertilizer Act are sent to these laboratories for quality control.

Farmers of the State generally are using more of nitrogenous fertilizer as compared to phosphatic and potasic fertilizers. The use of micro-nutrients is almost negligible. Thus the practice of use of balanced fertilizers is not usually followed. Whatever is the total quantity of fertilizers used, the major share is under irrigated forming, while is dry land farming areas, its use is still proportionately very small.

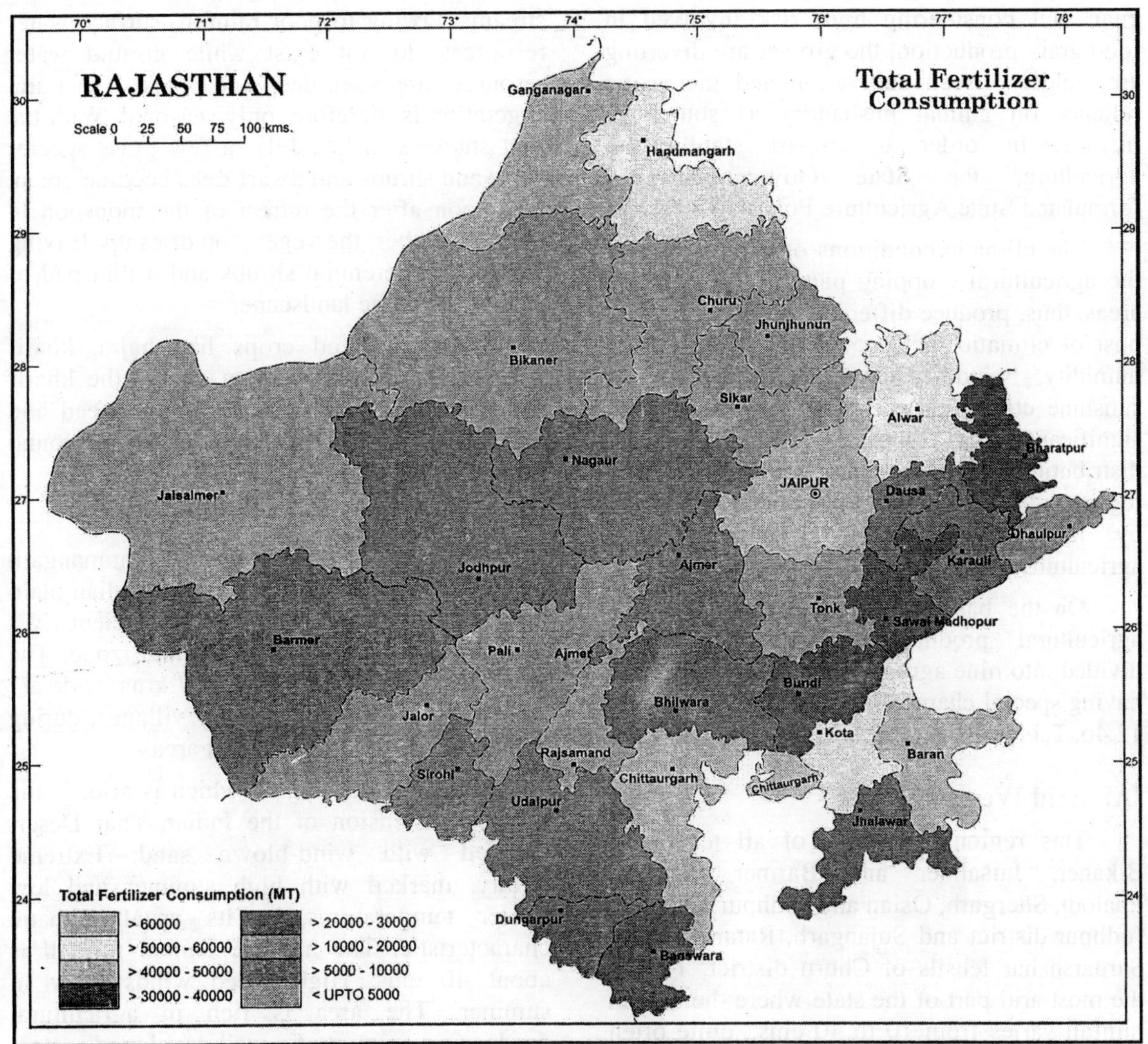

Source: Rajasthan Agricultural Statistics at a Glance Year 2016-17, Commissionerate of Agriculture, Rajasthan, Jaipur

Figure 12.45 Total Fertilizer Consumption in Rajasthan

AGRO-CLIMATIC ZONES

The state of Rajasthan has extreme geo-climatic conditions. It had 48 drought years in 100 years for the century ending 2002. It has about 11 per cent of the country's land mass, but only 1 per cent of total surface water of the country is available in the state. Of the 14 river basins, Chambal river is perennial. Even rainfall is erratic and scanty as average rainfall is 57 cm. Out of the total 237 blocks in the state, only 30 blocks are in safe category and balance are either "over exploited" or "critical". Apart from water resource limitations, the weather is severe as the temperature ranges from as low as 3 °C to as high as 49° C. Although it has 10 agro-climatic zones, about two-third is arid or semi-arid area with tropical desert climate. Weather uncertainties and unfavourable climatic conditions lead to high level of fluctuation in crop productivity as well as agricultural production which results into wide fluctuation in Gross State Domestic Product (GSDP). It is the largest producer of mustard, bajra, coriander, cumin, cluster beans, isabgul and guar. But considering high risk involved in food grain production, the grower are diverting area under oilseed cultivation and increasing reliance on animal husbandry as source of income. In order to impart stability to agriculture, the State Government has formulated State Agriculture Policy in 2012.

The climatic conditions of a region affect the agricultural cropping pattern and different areas, thus, produce different crops. Amongst a host of climatic factors, rainfall, temperature, humidity, wind velocity and duration of sunshine etc. affect the cropping pattern in a significant way. Annual rainfall and its distribution over the entire year, and the regimes of diurnal and annual temperatures are, by far, the prominent factors affecting agriculture and the life style of the people.

On the basis of climatic conditions and agricultural produce, Rajasthan has been divided into nine agro-climatic zones, each one having special characteristics ofits own (Figure 12.46, Table 12.2 and 12.3).

IA. Arid Western Plain

This region comprises of all tehsils of Bikaner, Jaisalmer and Barmer districts, Phalodi, Shergarh, Osian and Jodhpur tehsils of Jodhpur district and Sujangarh, Ratangarh and Sardarshahar tehsils of Churu district. This is the most arid part of the state where the annual rainfall varies from 10 to 40 ems., quite often erratic, so much so, that the entire rainfall of the year may fall on a single day and the rest of the year may be dry.

Summer temperatures are always high and the diurnal range exceeds even 20° C. During the day, the summer temperatures may as high as 49° C but in the night, the temperatures may fall, to less than 20° C. In winters, the day temperatures are higher but the night temperatures may be near freezing point. Winters are of short duration, not exceeding two months-December and January.

This is an area of not-so-developed drainage system where there are no flowing streams. Owing to poor rainfall, surface water resources do not exist while ground water resources are often deep and brackish. Natural vegetation is therefore, only seasonal. With the first showers in late July, a few grass species grow and shrubs and dwarf trees become green. But, soon after the retreat of the monsoon in mid September, the vegetation dries up, leaving only a few perennial shrubs and a thin pad of pale grass on the landscape.

Mostly rainfed crops like bajra, kharif pulses, guar etc. are grown during the kharif season. Rabi crops like wheat, rape-seed and mustard are grown only in areas where ground water is available for irrigation.

I B. Irrigated North-Western Plains

The entire Ganganagar and Hanumangarh districts which are an alluvial and aeolian plain formed by the river Ghaggar (the ancient river Saraswati) forms this agro-climatic zone. The Ghaggar channel, even though 7 krns. wide at_ some places, inundates nearby villages, during heavy rainfall in its catchment areas.

A part of this region which is arid, is the northern extension of the Indian Thar Desert covered with wind-blown sand. Extreme aridity, marked with high summer and low winter temperatures, is its usual climatic characteristic. The average annual rainfall is about 40 ems. High speed winds blow in summer. The area is rich in agricultural produce on account of a well developed system of canal irrigation. Today, a large network of Gang Canal and Bhakhra Canal which along with the Indira Gandhi Canal, has made the entire area green and productive.

Due to abundance of canal water irrigation, this region has today become the greenary of Rajasthan. Amongst the kharif crops cotton, Sugarcane and pulses are of importance. In the rabi season, wheat, mustard, gram, vegetables and fruits are produced. The total production as well as productivity levels of all crops are relatively much higher in this zone as compared to other parts of the state.

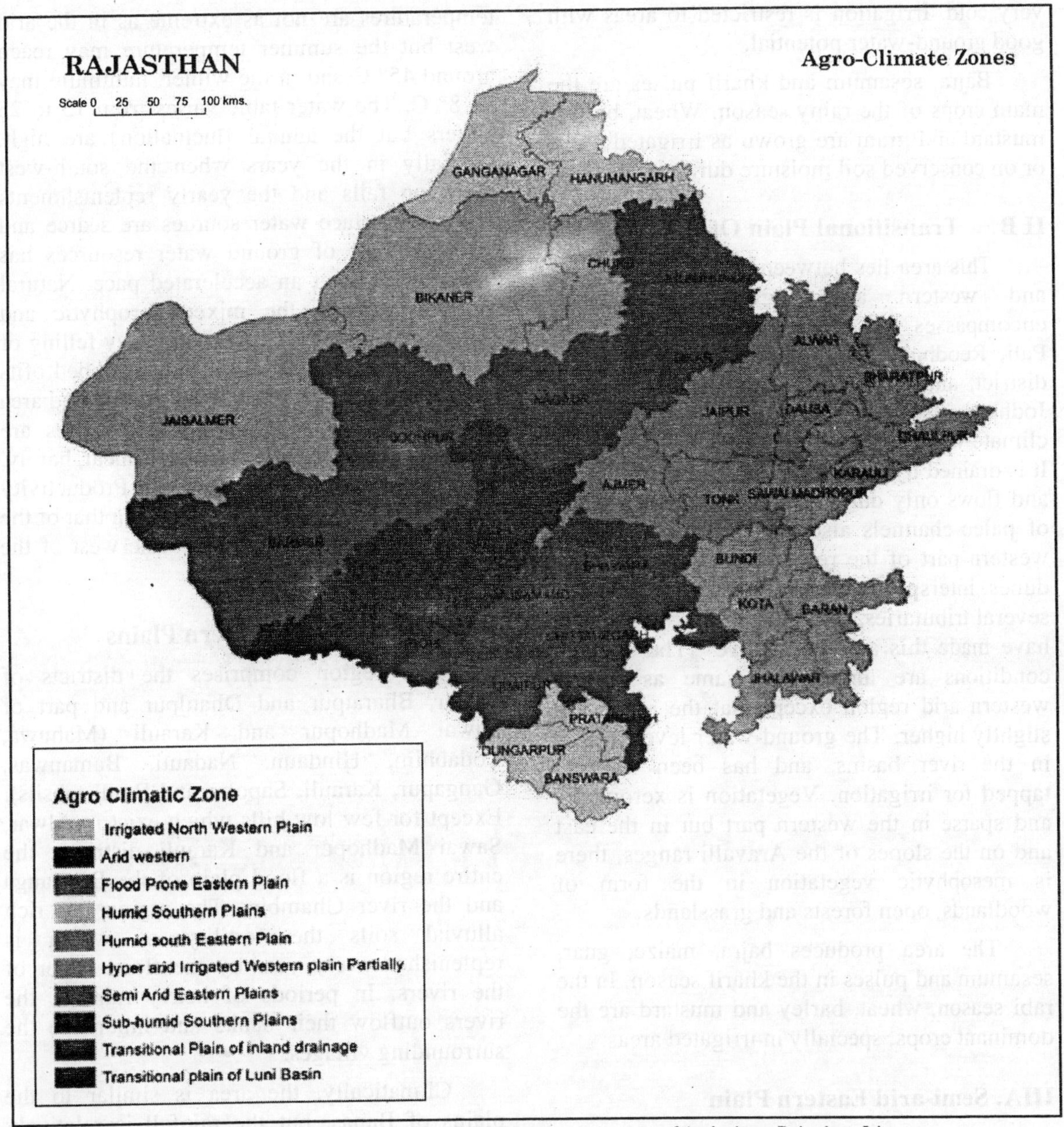

Source: Rajasthan Agricultural Statistics at a Glance Year 2016-17, Commissionerate of Agriculture, Rajasthan, Jaipur

Figure 12.46 Agro-Climatic Zones of Rajasthan

IIA. Transitional Plain Oflnland Drainage

This zone comprises all tehsils of Nagaur, Sikar and Jhunjhunun districts and Taranagar, Churu and Rajgarh tehsils of Churu district. The area is covered with sand dunes and inter-dunal sandy plains. Drainage is not well developed and streams which flow in the rainy season, disappear in sandy fields after covering some distance. Climatically, this zone is slightly better as compared to the adoining zone of the Arid Western Plain. Rainfall is slightly higher, temperatures in summer months do go very high but the winters are very cold. Irrigation is restricted to areas with good ground-water potential.

Bajra, sesamum and kharif pulses are the main crops of the rainy season. Wheat, barley, mustard and gram are grown as irrigated crops or on conserved soil moisture during rabi.

II B. Transitional Plain Of Luni Basin

This area lies between the Aravalli ranges and western arid region. The region encompasses the entire districts of Jalor and Pali, Reodhar and Sheoganj tehsils of Sirohi district, and Bilara and Bhopalgarh tehsils of Jodhpur district. The region has semi-arid climate with an annual rainfall of 30 to 50 ems. It is drained by the river Luni which is seasonal and flows only during rainy season. A number of paleo-channels also exist in this area. The western part of the region is dotted with sand dunes, interspersed in alluvial soil. Luni and its several tributaries, like Sukri, Mithri and Jawai have made this area productive : The climatic conditions are almost the same as in the western arid region except that the rainfall is slightly higher. The ground-water level is high in the river basins, and has been usefully tapped for irrigation. Vegetation is xerophytic and sparse in the western part but in the east and on the slopes of the Aravalli ranges, there is mesophytic vegetation in the form of woodlands, open forests and grasslands.

The area produces bajra, maize, guar, sesamum and pulses in the kharif season. In the rabi season, wheat, barley and mustard are the dominant crops, specially in irrigated areas.

IIIA. Semi-arid Eastern Plain

This region comprises four districts namely, Jaipur, Dausa, Tonk and Ajmer. Bans, with its several tributaries, forms a rich fertile plain. On the western side, the region is flanked by the low Aravalli hills which extend from the south-west to the the north-east.

The animal rainfall of the region varies from 50 to 60 ems. with an increasing trend towards the east. Summer and winter temperatures are not as extreme as in the arid west but the summer temperature may reach around 45° C and in the winter, minimum may be 8° C. The water table varies from 15 to 25 meters but the annual fluctuations are high, specially in the years when the south-west monsoon fails and the yearly replenishments are low. Surface water sources are scarce and so harnessing of ground water resources has been going on at an accelerated pace. Natural vegetation is of the mixed xerophytic and mesophytic type, but owing to heavy felling of trees, the surface mantle has been robbed ofits natural wealth. In the total gross cultivated area of this zone, bajra, sorghum and pulses are grown in the kharif season, and · wheat, barely, gram mustard in the rabi season. Productivity of all crops in this zonę is better than that of the agro-climatic zones that are to the west of the Aravalli range.

III B. Flood Prone Eastern Plains

This region comprises the districts of Alwar, Bharatpur and Dhaulpur and part of Sawai Madhopur and Karauli (Mahuwa, Todabhim, Hindaun, Nadauti, Bamanwas, Gangapur, Karauli, Sapotra and Bonli tehsils). Except for few low hills which exist in Alwar, Sawai Madhopur and Karauli districts, the entire region is a flood plain of the Banganga and the river Ghambiri. The region has rich alluvial soils the fertility of which is replenished every year by the flood water of the rivers. In periods of heavy rainfall, the rivers outflow their banks and innundate the surrounding villages.

Climatically, the area is similar to the plains of Banas, but the rainfall is relatively higher in the east, the annual average being about 75 ems. Natural vegetation exists on mountain slopes, wetland areas, and protected zones but the excessive plundering of forest wealth has dwindled the natural cover. The region produces a variety of crops because both surface water and ground water irrigation sources exist. A network of canals drawn from the upper Yamuna Canal and the Panchana

Dam irrigate this area. The ground water aquifers vary from 5 to 15 meters and therefore, well water irrigation is also practiced. The region produces bajra, sorghum, maize, sugarcane, sesamum and a variety of pulses in the kharif season. Wheat, barely, gram and mustard are the dominant crops during rabi season.

IVA. Sub-humid Southern Plains and TheAravalli Hills

Bhilwara district, all tehsils of Udaipur and Rajsamand districts, except Dharyiawad, Salumber and Sarada, all tehsils of Chittaurgarh and Pratapgarh distircts, except Chotti Sadri, Pratapgarh, Arnod and Bari Sadri and Abu Road and Pindwara tehsils of Sirohi district form this agro-climatic zone. For the most part, this is a region of low Aravalli hills with the inter-mountain plateau, deeply dissected by streams and rivers. The region has a moderately warm climate in summers and with mild winters. The annual rainfall varies from 50 to 95 ems. Th highest precipitation in the state is recorded in Abu hills (Sirohi district). There are number of surface water streams like Ghambiri, Sabarmati, Banas and its tributaries but they are all ephemeral. The surface rocks are granitic and highly metamarphosed and so the depressions on the surface are filled with rain water in the form of natural tanks.

The area is rich in natural vegetation which grows on the slopes of the Aravallis and in the wetland areas but exeessive felling of trees has degraded these open forests. Tank water irrigation is most common. The area produces maize as the chief food crop of the kharif season but in irrigated areas, paddy is also grown. In the rabi season, wheat, gram and oil seeds are the maiu crops. In area of black soil, cotton and opium are also cultivated.

IVB. Humid Southern Plains

The districts of Dungarpur and Banswara, Paiis of Udaipur (Dhariyawad, Salumber, and Sarada tehsils), Pratapgarh and Chittaurgarh (Chotti Sadari, Bari Sadari, Pratapgarh and Arnod tehsils) are included in this region. The area comprises oflow Aravalli hills with intrusions of black lava rocks. This is mostly a tribal area where Bhils, Girasiyas and Damors live amidst forests and hills. The area has a humid climate with an average rainfall of more than 70 ems per year. The temperature regimes do not fluctuate much in summer and winter so the area has mild winters and mild summers. The humidity is always high and all these factors, combined together, promote a profuse growth of natural vegetation. The mountain slopes have an abundance of natural forests.

There are a number of surface water streams. Mahi and its tributaries likeAnas,Arau and Jhakham have made this area very fertile. The commissioning of Mahi Bajaj Sagar multipurpose project has provided this area with canal water irrigation and electirc resources. Cotton and sugarcane.are the chief cash crops grown in the black soil region. Maize, sorghum and paddy are the Qhief crops of the kharif season. Groundnut, mustard, sesamum and rapeseed also grown.

V. Humid South-Eastern Plains

Popularly known as the Hadauti plateau, this region includes the districts of Kata, Baran and Jhalawar and two tehsils of Sawai Madhopur namely Khandar and Sawai Madhopur. The black soil region of this plateau is fe1iile and is used for cultivation of sugarcane, cotton and opium. The region has low hills of the Gwalior series, interspersed with broad plateaus of Vindhyan rocks. The Chambal is the main river along with its tributaries like Parvati, Kali Sindh, Parwan and Banas. The development of canal irrigation system with a series of dams and barrages on the chambal, has made this area rich in agriculture production. Gandhi Sagar, Rana Pratap Sagar and Jawahar Sagar dams togehter with Kata Barrage have generated enough resources of electircity and canal water for irrigation.

The region has warm summers but mild winers. Summers temperatures sometimes touches 45°C. The relative humidity is generally high in this zone. The annual rainfalls varies from 60 to 85 ems.

Natural vegetation exists in the form of woodlands, Parklands and open forests which have, however now degraded. Paddy and sorghum are the chief food crops grown in kharif season. This area is suitable for soyabeen crop also, Wheat, barley, gram and mustard are grown in winter.

Table 12.2 Agro-Climatic Zones of Rajasthan

Zone No.	Zone Name	District Covered	Major Crops		Soils
			Kharif	Rabi	
I-A	Arid Western plain	Barmer and Jodhpur	Pearlmillet, Mothbean, Sesame	Wheat, Mustard, Cumin	Desert soils and sand dunes aeolian soil, coarse sand in texture some places calcareous
I-B	Irrigated North Western plain	Sriganganagar, Hanumangarh	Cotton, Clusterbean	Wheat, Mustard, Gram	Alluvial deposits calcareous, high soluble salts and exchangeable sodium
I-C	Hyper arid partial irrigated Western plain	Bikaner, Jaisalmer, Churu	Pearlmillet, Mothbean, Clusterbean	Wheat, Mustard, Gram	Desert soils and sand dunes aeolian soil, loamy coarse in texture and calcareous
II-A	Transitional plain of Inland drainage	Nagaur, Sikar, Jhunjhunu, Part of Churu	Pearlmillet, Clusterbean, Pulses	Mustard, Gram	Sandy loam, shallow depth red soils in depressions
II-B	Transitional plain of Luni basin	Jalore, Pali, Sirohi, Part of Jodhpur	Pearlmillet, Clusterbean, sesame	Wheat, Mustard	Red desert soils in Jodhpur, Jalore and Pali, Sierozems in Pali and Sirohi
III-A	Semi arid Eastern plains	Jaipur, Ajmer, Dausa, Tonk	Pearlmillet, Clusterbean, Sorghum	Wheat, Mustard, Gram	Sierozems, eastern part alluvial, west north west lithosols, foot hills, brown soils
III-B	Flood prone Eastern plain	Alwar, Dholpur, Bharatpur, Karoli, S.Madhopur	Pearlmillet, Clusterbean, Groundnut	Wheat, Barley, Mustard, Gram	Alluvial prone to water logging, soils of recently alluvial calcareous nature
IV-A	Sub-humid Southern plains	Bhilwara, Rajasamand, Udaipur, Chittorgarh, Parts of Sirohi	Maize, Pulses, Sorghum	Wheat, Gram	Soil are lithosols at foot hills and alluvial in plains
IV-B	Humid Southern plains	Dungarpur, Banswara, Pratapgarh, Parts of Chittorgarh and Udaipur	Maize, Paddy Sorghum Blackgram	Wheat, Gram	Predominantly reddish medium texture, well drained calcareous, shallow on hills, deep soils in valleys
V	Humid south Eastern plain	Kota, Jhalawar, Bundi, Baran, Parts of Sawai Madhopur	Sorghum Soyabean	Wheat, Mustard	Black Soils of alluvial origin, clay loam, groundwater salinity

Source: Water Resources Department, Govt. of Rajasthan.

Table 12.3 Details of Agro-Climatic Zones

Name of Zone	Arid western	Irrigated North Western Plain	Hyper arid Irrigated Western plain Partially	Transitional Plain of inland drainage	Transitional plain of Luni Basin
Districts Covered (Blocks)	Part of Barmer district (Barmer, Balotra, Baytod, Chohtan, Dhorimanna, Sindhari, Sheo) and Jodhpur district (Balesar, Bap, Part of Bhopalgarh, Luni, Mandore, Osian, Phalodi, Shergarh)	Ganganagar, Hanumangarh	Bikaner, Jaisalmer and Part of Churu (Ratangarh, Sardarshahar, Sujangarh)	Nagaur, Sikar, Jhunjhunun and Part of Churu (Churu, Rajgarh, Taranagar)	Jalor, Pali, Part of Sirohi (Reodhar, Sheoganj, Sirohi), Barmer (Siwana) and Jodhpur (Bilara, Part of Bhopalgarh)
Total area (Million Ha)	4.56	2.06	7.70	3.71	3.17
Net Sown Area (Million Ha)	2.34	1.60	2.44	2.68	1.93
Rainfall (mm)	216-496	183-344	192-420	266-627	139-1,273
Soil Type	Desert soils and sand dunes aeolian soil, coarse sand in Texture and some places Calcareous	Alluvial deposits calcareous, high soluble salts and exchangeable sodium	Desert soils and sand dunes aeolian soil, Loamy coarse in texture and calcareous	Sandy loam, shallow depth red soils in depressions	Red desert soils in Jodhpur, Jalor and Pali sierozems in Pali and Sirohi
Major Crops	Bajra, Moth bean, sesame, Wheat, Mustard, Cumin	Cotton, Guar, Wheat, Mustard, Gram	Bajra, Moth bean, Guar, Wheat, Mustard, Gram	Bajra, Guar, Pulses, Wheat, Mustard, Gram	Bajra, Guar, Sesame, Wheat, Mustard

Source: Water Resources Department, Govt. of Rajasthan.

Name of Zone	Semi Arid Eastern Plains	Flood Prone Eastern Plain.	Sub-humid Southern Plains	Humid Southern Plains	Humid south Eastern Plain
District Covered (Blocks)	Ajmer, Jaipur, Tonk, Part of Dausa (Bandikui, Dausa, Lalsot, Sikrai)	Alwar, Bharatpur, Dhaulpur, Karauli, Part of Dausa (Mahuwa) and Sawai Madhopur (Bamanwas, Bonli, Gangapur)	Bhilwara, Chittaurgarh, Rajsamand, Part of Sirohi (Abu Road, Pindwara), Udaipur (Bargain, Bhinder, Girwa, Gogunda, Kherwara, Kotra, Jhadol, Mavli, Sarada) and Pratapgarh (Choti Sadri, Pratapgarh)	Dungarpur, Banswara, Part of Udaipur (Salumber) and Pratapgarh (Dhariyawad, Peepalkhunt, Arnod	Baran, Bundi, Jhalawar, Kota, Part of Sawai Madhopur (Khandar, Sawai Madhopur)
Total Area (Million Ha)	2.95	2.44	3.76	1.20	2.69

Name of Zone	Semi Arid Eastern Plains	Flood Prone Eastern Plain.	Sub-humid Southern Plains	Humid Southern Plains	Humid south Eastern Plain
Net Sown Area (Million Ha)	1.77	1.41	0.92	0.57	1.27
Rainfall (mm)	313-694	140-736	311-1,470	511-1,043	452-985
Soil	Sierozems, eastern part alluvial, west north west lithosols, foot hills, brown soils	Alluvial prone to water logging, nature of recently alluvial calcareous has been observed	Soil are lithosol at foot hills and alluvial in plains	Predominantly reddish medium texture, well drained calcareous, shallow on hills, deep soil in valleys	black, of alluvial origin, clay loam
Major Crops	Bajra, Guar, Jowar, Wheat, Mustard, Gram	Bajra, Guar, Groundnut Wheat, Barley, Mustard, Gram	Maize, Pulses, Jowar, Wheat, Mustard Gram	Maize, Paddy, Jowar, Black gram, Wheat, Mustard, Gram	Jowar, Soybean, Wheat, Mustard,

Source: Water resources Department, Govt. of Rajasthan.

KHADIN AGRICULTURE

About 8% area in the Western Rajasthan is rocky/gravely, generating run-off in resopnse to intense rainfall. This *in-situ* rainwater harvesting practice is known as khadin cultivation in local parlance. It is somewhat similar to 'Haveli' cultivation in the Bihar state. This practice dates back to several centuries and is being followed in 100-200 mm rainfall zone in Wesetern Rajasthan. Entire run-off and sediment is conserved *in-situ*. The run-off is harvested in contignous farmlands by constructing earthen, bunds across the gentle slop of the land in the valley bottom to facilitate spreading and standing of harvested water on the farm soil. Water is thus made to accumulate to a depth of 60 to 90 cm. This water seeps into the ground and recharges the adjoining wells as well as ground water. Besides, on conserved moisture cropping is done- during the Kharif/Rabi seasons, depending upon the depth/amount of accumulated run-off. Cropping is started from upper reaches to lower reaches on receding situation. The Khadins can be replicated on a vast scale at appropriate places perferably in heavier soil area-sandy loam onwards. With this technique the present intensity/single Kharif crop cand be doubled in many situation. In most of the Khadins local varieties of crop is grown and the fertilizers used is negligible.

There is, thus, considerable scope to improve upon the system to increase production by agro-horticulture and agro-forestry system. This technique requires a cropland in proximity to rocky catchments and is therefore site specific. The catchments to command area ratio varies from 11:27, 8:22 to 2:22 in 217 mm, 240 mm and 310 mm rainfall zones, respectively (Figure 12.47).

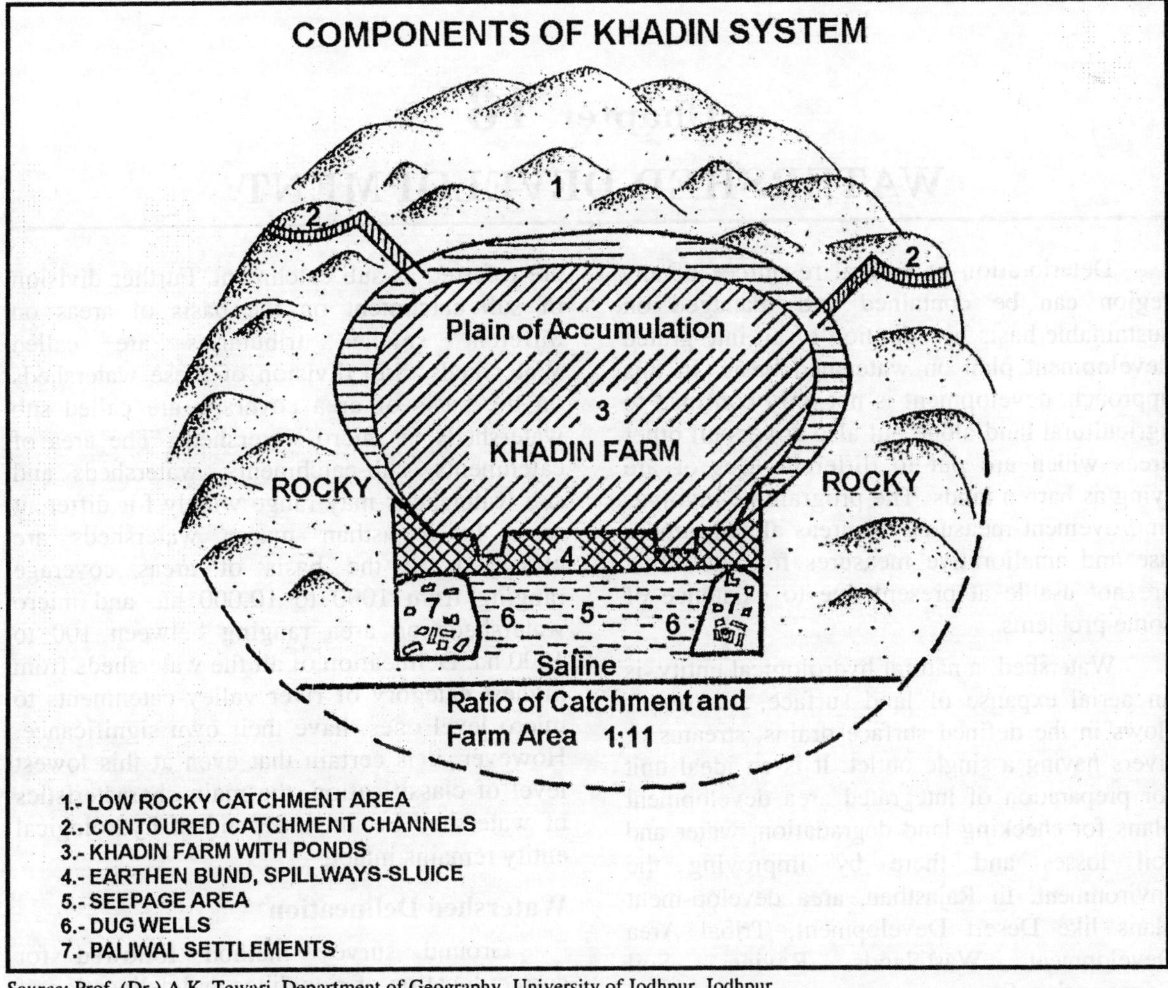

Source: Prof. (Dr.) A.K. Tewari, Department of Geography, University of Jodhpur, Jodhpur

Figure 12.47 Components of Khadin System in Rajasthan

Chapter 13

WATERSHED DEVELOPMENT

Deterioration of natural resources of any region can be contained and managed on sustainable basis by adoption of an inte-grated development plan on watershed basis. In this approach, development is not only confined to agricultural land alone but also covers all other areas which are put to different uses or are lying as barren lands. The programme involves improvement measures for areas already under use and ameliorative measures for lands that are not usable at present due to existence of some problems.

Watershed, a natural hydrological entity, is an aerial expanse of land surface, its run-off flows in the defined surface drains, streams or rivers having a single outlet. It is an ideal unit for preparation of integrated area development plans for checking land degradation, water and soil losses and there by improving the environment. In Rajasthan, area develop-ment plans like Desert Development, Tribal Area Development, Wastelands, Ravines, Soil Conservation Programmes, etc. are planned on watershed approach, so it is essential to have a "frame work of watersheds" in the form of suitable maps and the supporting statements. The "Watershed Atlas of Rajasthan" is an attempt to provide a uniform delineation, codification, priotisation of the watersheds that could be followed by all concerned with watershed approach on a common basis.

The term watershed, is synonymous to catchments, sub-catchments, watersheds, sub-watersheds, macro watersheds and micro watersheds depending upon the size of the drainage area. Catchment is referred to the entire drainage area of a river, through which its runoff is drained. Division of the catchments on the basis of different tributaries of the river, are referred as sub catchment. Further division of sub catchment on the basis of areas on different smaller tributaries are called Watersheds. Sub division of these watersheds on the basis of area coverage are called sub watersheds or micro watersheds. The area of catchments, sub-catchments, watersheds and sub watersheds may range widely for different area. In Rajasthan macro watersheds are classified on the basis of areas coverage ranging from 1000 to 10,000 ha. and micro watersheds on area ranging between 100 to 1000 ha. Delineation of all the watersheds from highest category of river valley catchments to micro level ones, have their own significance. However, it is certain that even at this lowest level of classification, the main characteristics of watershed i.e. an independent hydrological entity remains intact.

Watershed Delineation

Ground survey method followed for correct location of ridgelines and delineation of the watersheds is time consuming and very difficult especially in non accessible terrains. Remote sensing technique has proved to be a boon for quick and precise delineation of watersheds.

Conventional Method

In this method, a field surveyor studies 1:50,000 scale Survey of India toposheets of the project area and based on the terrain features, attempts, the possibility of delineation of watersheds. Then, he reconnoitrers the area to demarcate ridgelines or water dividing lines depending on the drainage pattern of each watershed. For this purpose he has to traverse on foot, spending lot of time, and energy. After compilation of this information, in the form of maps, detailed soil and topographical surveys are conducted, which form the basic ingredients for preparation of development plans for the watershed.

Remote Sensing Based Approach

Satellite data because of its synoptic view, is of greater assistance in delineation of the ridgelines and study of drainage, land use pattern, etc. on catchments, sub catchments and watersheds. Study of aerial photographs, because of the stereoscopic vision, is helpful in mapping of land use pattern, drainage characteristics and precise location of ridge lines of the macro and micro watersheds. Thus, by adoption of standard methods of visual interpretation satellite, false colour composite imagery of 1:50,000 scale and aerial photographs of 1:25,000 scale, delineation of watersheds over a wide range from catchments to micro level is possible with more accuracy, and in less time.

Codification

Uniform codification system of the watershed helps to understand the entire watershed network over large area and avoid usage of the same symbol for different watersheds causing ambiguity system of codification of watersheds for any region may be exclusively based on digits or use of alphabets and digits. The all India Soil and Land Use Survey Organisation (AISLUS). Ministry of Agriculture, Government of India, New Delhi a pioneer organisation in Countrywide delineation of watersheds have made use of alphabets and digits in watershed codification and is outlined in brief as follows:

1. Region- The Country has been divided info 6 regions, disignated as 1 to 6. Rajasthan is covered under region 1 and 6.
2. Basins- The regions have been further divided into 35 basins under the six regions and denoted with Capital alphabet A to Z.
3. Catchments- These are further bifurcations of basins and denoted by digit 1, 2, 3 ....
4. Sub catchments- Sub catchments of rivers are denoted by small alphabets a to z.
5. Watersheds- Sub division of sub catchment is made into watersheds denotes by digits 1, 2, 3 ....
6. Sub watersheds- are denoted by small alphabets a to z.

Thus a sub watershed in Luni basin of Pali district of Rajasthan as per AISLUS, New Delhi may be represented as "6A2d1a".

State Remote Sensing Application Centre (SRSAC) Proposed Addition

The minimum unit of sub watershed is of the order of more than 15,000 ha., where as in Rajasthan execution of watersheds are planned in watersheds ranging from hundreds to thousands of hectare and termed as micro (less than 1000 ha) and macro (1000 to 10,000 ha.). The macro watersheds can be represented by digit 1 to 99 where as micro watershed by small alphabets a to z.

The system based purely on digits, may offer possibilities of computerisation, while the other system which makes use of both alphabets and digits is more understandable in its application in programmes of local and regional level interest.

Taking into considerations the possibilities of making its use in Watershed Management Information System (WMIS), digital system is proposed for codification of watersheds in Watershed Atlas. The delineation of macro and micro watershed has been covered pahchayat samiti wise so that modified digit based codification will accordingly hold good for panchayat samitiwise. Though the hierarchy wise the system will be extension of AISLUS system. Looking to limitations of digital codifications, order of streams in rivers flowing in Rajasthan (in most cases not more then four), preparation of panchayat samiti based maps and micro level execution of watershed requirement, the attributes have been limited to 4 variables. Accordingly in

"Watershed Atlas of Rajasthan" the codification system is adopted as follows

1. Catchment

Catchment area is denoted by the first two digits. All the river valley catchments have been codified onthis basis and the code numbers given to each of them are given Table 13.1.

Table 13.1 River Catchment in Rajasthan

S.No.	River Catchment	Code No.
1.	Kantli	01
2.	Dohan	02
3.	Sahibi and Sota	03
4. .	Bar	04
5.	Mendha Riveer (Sambhar Lake)	05
6.	Luni	06
7.	Banas	07
8.	Banganga	08
9.	Chambal	09
10.	Sukal	10
11.	Banas West	11
12.	Sabarmati (Wakal)	12
13.	Mahi	13

Source: Watershed Atlas of Rajasthan

2. Watershed

All the watersheds within each of the above river valley catchment have been given next two digits. Thus first watershed of Luni catchment will be 0601 and second one will be denoted as 0602 and so on.

3. Macro Watershed

All the macro watersheds within each of the watershed are given next two digits. The code number for first macro watershed in first watershed of Luni catchment will thus be 060101 and second one will be 060102 and so on.

4. Micro Watershed

All the micro watershed within each of the macro watershed are given still further next two digits after the macro watershed code number. Thus first micro watershed in first macro watershed, of the first watershed of Luni catchment will be 06010101 and second one as 06010102 and so on. This codification stands for Panchayat Samitiwise watersheds (Figure 13.1).

The above codification system can be summarised as following (Table 13.2):

Table 13.2 Watershed Codification

S. No.	Category	Codification	Remarks
1.	Catchment	Arabic numericals from 1 to 13	For 13 River Valley Catchment of Rajasthan
2.	Watershed	Arabic numericals from 1 to 99	For each watershed with in each catchment
3.	Macro watershed	Arabic numericals from 1 to 99	For each macro watershed with in each watershed
4.	Micro watershed	Arabic numericals from 1 to 99	For each micro watershed within each macro watershed.

Source: Watershed Atlas of Rajasthan

The above codification has been done keeping in view the panchayat samitiwise delineation of watersheds. Drainage order in the State also limits to eight digit codification. Therefore catchment, watershed, macro and micro watersheds have been included in this Codification System. At number of times sub-catchment and watershed will have some extent. However in few cases of two adjoining panchayat samiti having same catchment and watershed will have same code number of different macro and micro watersheds.

Due to deciphering nature of drainage, catchments and watershed code are not followed as per River Valley Catchment. In such cases code number for catchment denoted by "00" and for watershed by "0000". This holds good for desertic terrain where scattered drainage exist.

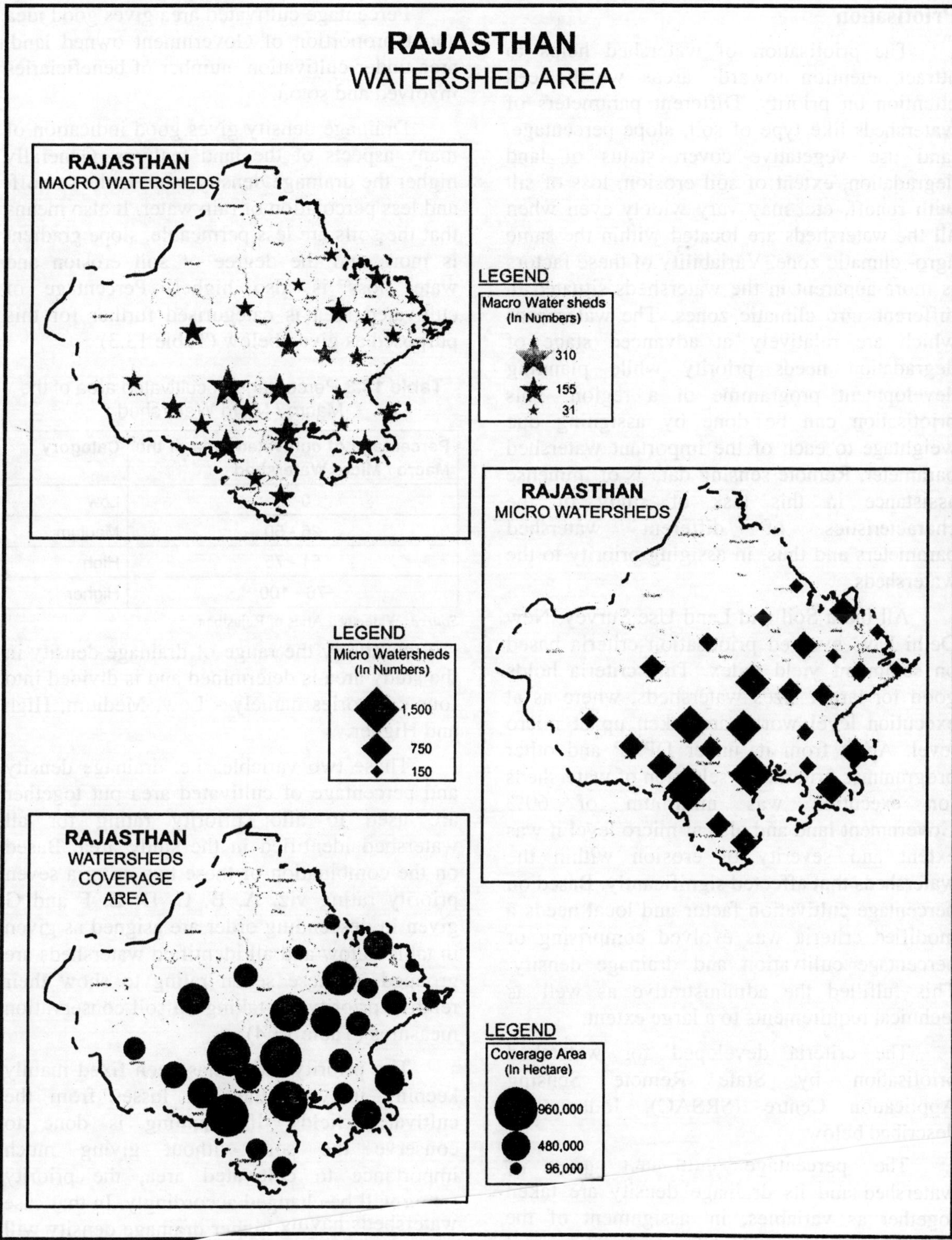

Source: Rajasthan Watershed Atlas.

Figure 13.1 Watershed Area in Rajasthan

Priotisation

The priotisation of watershed helps to attract attention towards areas which need attention on priority. Different parameters of watersheds like type of soil, slope percentage, land use, vegetative cover, status of land degradation, extent of soil erosion, loss of silt with runoff, etc. may vary widely even when all the watersheds are located within the same agro- climatic zone. Variability of these factors is more apparent in the watersheds situated in different agro climatic zones. The watersheds which are relatively at advanced stage of degradation needs priority while planning development programme of a region. This priotisation can be done by assigning due weightage to each of the important watershed parameter. Remote sensing data is of immense assistance in this task of study of the characteristics of different watershed parameters and thus, in assiging priority to the watersheds.

All India Soil and Land Use Survey, New Delhi have evolved priotisation criteria based on sedimemt yield index. This criteria holds good for larger sizes watersheds, where as at execution level works are taken up at micro level. Apart from it, under DPAP and other programmes criteria for selection of watersheds for execution was minimum of 60% Government land and also at micro level it was extent and severity of erosion within the watersheds that affected significantly. Based on percentage cultivation factor and local needs a modified criteria was evolved comprising of percentage culitvation and drainage density. This fulfilled the administrative as well as technical requirements to a large extent.

The criteria developed for watershed priotisation by State Remote Sensing Application Centre (SRSAC), Jodhpur is described below

The percentage cultivated area of watershed and its drainage density are taken together as variables, in' assignment of the priority to the watershed for undertaking soil conservation work.

Percentage cultivated area gives good idea about proportion of Government owned land, area under cultivation, number of beneficiaries involved and so on.

Drainage density gives good indication of many aspects of the land features. Generally higher the drainage density more is the run off, and less percolation of rain water. It also means that the soils are less permeable, slope gradient is more and the degree of soil erosion and water loss is also higher. Percentage of cultivated area is categorised further for this purposed as given below (Table 13.3)::

Table 13.3 Percentage of cultivated area of the Macro / Micro Watershed

Percentage of cultivated area of the Macro / Micro Watershed	Category
0-25	Low
26 - 50	Medium
51 - 75	High
76 - 100	Higher

Source: Watershed Atlas of Rajasthan

Similarly the range of drainage density in the study area is determined and is divided into four categories namely - Low, Medium, High and Higher.

These two variables i.e. drainage density and percentage of cultivated area put together are used to allot priority rating for all watershed identified in the study area. Based on the combination of these two criteria seven priority rating viz. A, B, C, D, E, F and G given in descending order are asigned as given in table below and all identified watersheds are grouped in these seven rating to show their relative priority for taking up soil conservation measures (Table 13.4).

The priority rating has been fixed mainly keeping in view the soil losses from the cultivated fields. If planning is done to conserve the soil without giving much importance to cultivated area, the priority rating will be changed accordingly. In that case watersheds having higher drainage density will have to be given top priority.

Table 13.4 Soil Conservation Measures

Cultivated area group	Drainage density group	Priority rating
Higher	Higher	A
Higher	Higher	
Higher	High	B
High	High	
Higher	Medium	C
High	Medium	
Medium	Higher	D
Medium	High	
Higher	Low	
Medium	Medium	E
Medium	Low	
Low	Higher	F
Low	High	
Low	Medium	G
Low	Low	

Source: Watershed Atlas of Rajasthan

Watershed Coverage in Rajasthan

Rajasthan occupies an area of about 3.42 lakh Sq. Km. and is covered by 237 Panchayat Samities. Well developed drainage exists in eastern half and part of the western half of the state covered by Luni basin area. Rest of the western Rajasthan has a inland drainage system (Table 13.5).

Based on the drainage characteristics as studied by interpretation of 1:250,000 scale IRS LISS-I FCC data superimposed on the 1:1 M Survey of India toposheet, the entire well drained area of the state in the first instance was grouped into the thirteen River Valley Catchments. Based on the drainage area of tributaries of the rivers the catchment areas were further sub divided into 103 watersheds. Further sub-division of these watersheds has lead to grouping of the drainage areas at the macro watershed and furhter to the micro watershed using 1:50,000 Satellite imagery and black and white aerial photographs. Details of the identified catchments and watersheds are given below (Figure 13.2).

Table 13.5 Number of identified catchments and watershed in Rajasthan

S. No.	Name of River	River Code	Number Catchment	Watershed
1.	Kantli	01		03
2.	Dohan	02	1	02
3.	Sahibi and Sota	03	1	04
4.	Bar	04	1	04
5.	Mendha River (Sambhar Lake)	05	1	05
6.	Luni	06	1	14
7.	Banas	07	1	12
8.	Banganga	08	1	10
9.	Chambal	09	1	19
10.	Sukal	10	1	03
11.	Banas West	11	1	05
12.	Sabarmati (Wakal)	12	1	08
13.	Mahi	13	1	14
	Total	**-**	**13**	**103**

Source: Watershed Atlas of Rajasthan

River Valley Catchments

The rivers of Rajasthan, except for the Chambal, are ephemeral and flow only during the rainy season. The river catchment areas, which are largely influenced by local topography and amount of precipitation, vary largely showing an indelible impact upon the drainage system. In the absence of any formidable mountain barrier in the state and the low quantum of annual rainfall, the river catchement areas are mostly small. The flat, rugged and undulating rolling lands of the low hills and plateaus, which are the origins of the rivers and streams, are highly dissected and devoid of deep channels to add stream density and so the catchments are of low .stream density.

The Aravalli ranges which run across the state from the south-west to the north-east direction largely influence the drainage system of the state. While a major part of western Rajasthan has an inland drainage system, the southern, the southeastern and the eastern parts have a well developed drainage system. The river system of the state can be conveniently put into tow major groups —

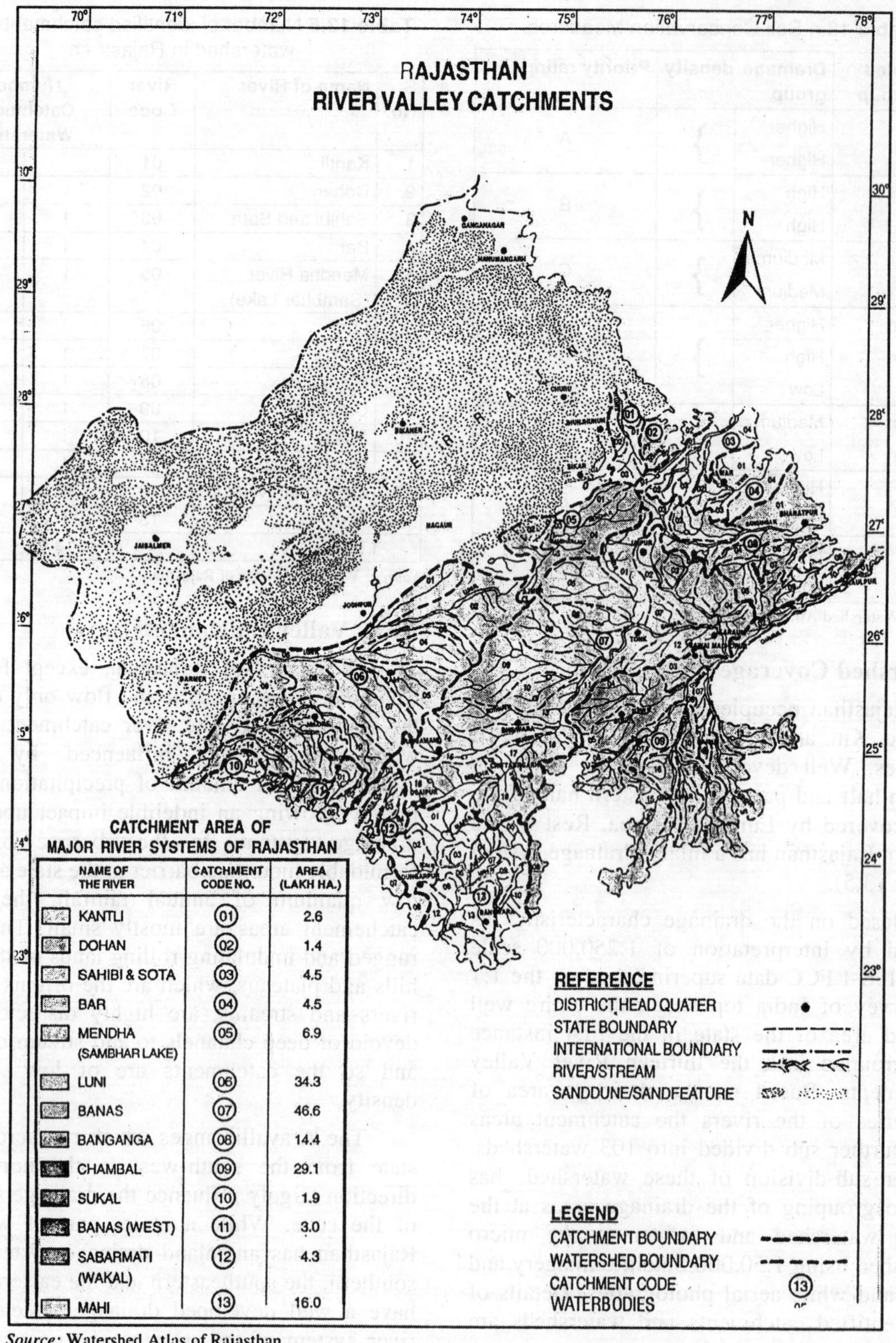

Source: Watershed Atlas of Rajasthan

Figure 13.2 River Valley Catchments in Rajasthan

(i) draining its run off ultimately to the Bay of Bengal, and (ii) flowing to reach the Arabian Sea. While all the rivers flowing on the eastern side of the Aravalli ranges fall in the first group, rivers flowing in the southern and the western part of the state belong to the second group.

The river Chambal, the only perennial river of the state, which forms the boundary with Madhya Pradesh, has its origin south of Mhow, near Manpur from the northern parts of the Vindhyans. It enters Rajasthan near Chaurasigarh Fort and flows though Kota District. Later, it flows to form the district boundary between Kota and Bundi and also between Kota and Sawai Madhopur. This river covers a distance of about 135 kms in Rajasthan out of its total length of about 965 kms. and has therefore, a smaller share of drainage system in the state. It has a catchment of 75,680 sq. km. (including Banas river) in Rajasthan. The river collects the major part of drainage discharge from Jaipur, Dausa, Sawai Madhopur, Karauli, Ajmer, Tonk, Bhilwara, Bundi, Kota, Baran, Jhalawarr, Chittaurgarh, Udaipur anti Rajsamand districts. Its main tributaries are Banas, Kali Sindh. Parvati etc. The river Chambal joins the river Jamuna near Etawah in Uttar Pradesh.

The river Banas originating from the Khamnor hills of the Aravalli ranges (about 5 km from Kumbhalgarh) is one of the major rivers of the state which, in its entire course, flows through Rajasthan. It flows from Kumbhalgarh towards the south upto the Gogunda plateau and after cutting the Aravalli ranges at right angles, it flows through Nathdwara, Rajsamand and Railmngra. The river collects the major run-off of Udaipur, Chittaurgarh and Bhilwara districts by the tributary, Berach. The Run meets the river Chambal near Khandar. This river usually dries up during the summer season. The total length of this river is about 480 kms and its main tributaries are Berach, Kothari, Khari, Dai, Mashi, Dundh and Morel.

The river Banganga which originates from the low hills of Bairath (Jaipur district), flows towards the east, entering Sawai Madhopur district and finally in Bharatpur district where its water spreads over a large area. The total length of this river is about 380 kms. The river joins the river Jamuna near Fatehabad of Agra district.

The Gambhiri which originates from Karauli hills in Karauli/Sawai Madhopur districts, flows through these districts to a greater length and innundates several areas of Bharatpur district before joining the river Jamuna. Its water has now been bounded at Panchna and Sikri Dams, and the problem of floods has been reduced to a great deal.

Luni is the only major stream on the west of the Aravalli ranges Which meets the sea in the Rann of Kutch. It originates in Nag Pahar in Ajmer and flows south-west through Jodhpur, Barmer and Jalor over a distance of about 320 kms, covering a total catchment area of 34,250 sq. km. Its catchment area has spread over some parts of Nagaur district also. It flows only during the rainy season and even at this time, the river is choked with advancing sands at many places. Even in the monsoon months when the water flow rate is high and at its peak, the flowing water is not able to cut these sand deposits on the river bed, as a result of which, the run off spreads to the adjoining fields on both the banks of the river braiding it. This results in sand deposition on arable and non-arable lands, thus, adversely affecting their productivity. A unique characteristic of this river is that its water is good upto Balotra, but it becomes brackish thereafter till it drains in the Rann of Kutch. A number of streams and tributaries like Bandi, Sukri, Mithri, Jawai, Sagi, Jojri, Lilri, Guhiya, etc., flows on the western side of the Aravalli range and drain the ran off of some parts of Pali, Jalor, Sirohi and Barmer district areas.

The river Mahi, which orginates from the Mhow ranges of Madhya Pradesh, and after flowing for some distance towards north, enters Rajasthan state near Khandu village of

Banswara district. Its catchment area in Rajasthan largely spreads over the districts of Udaipur, Banswara, Chittaurgarh and Dungarpur, covering a total geographical area of about 16,030 sq. kms. Important tributaries of this river are Som, Jakham, Anas, Chap and Moren.

The West Banas originates from the western slope of the Aravalli ranges and drains some parts of Sirohi district, finally draining its ran off in the Rann of Kutch.

The Sabarmati, originating from the low hills of Gogunda (Udaipur district) and called Wakal in its initial course, covers a catchement area of 4,300 sq. km. in the state and drains part of its water from Udaipur and Sirohi districts before finally traversing across Gujarat to meet the Gulf of Cambay.

The major part of western Rajasthan has sandy terrain with relatively low rainfall. There are number of samll rivers/ nallahs and seasonal streams which flow in this area for some distance, and then disappear in the vast expanse of sandy terrain.

The river Kantli originates from the Khandela hills of Sikar district. It flows during the rainy season only. After flowing over a distance of 100 Kms. towards the north, through Sikar and Jhunjhunun districts, it enters Churu district where its water is lost in the sandy terrain. The river may flow with gushing waters during heavy rainfall, causing lateral cutting of the banks and a braided channel in some parts during its course.

Sahibi river has its origin from the Sewar hills (Jaipur district). It flows through Bansur, Bahror, Mandawar and Tijara tehsils to enter Gurgaon district of Haryana state where, after flowing for some distance, its water is either lost in the adjacent areas or spread over vast stretches of land to cause flood havoc.

Chapter 14
IRRIGATION

Rajasthan has a coverage of 53.50 lakh hectare irrigated area which is propor-tionately much less compared to areas of the adjoining States like Uttar Pradesh, Punjab and Haryana. Out of the total culturable area of the 273.24 lakh hectares; the area under net irrigated agriculture is only 43.72 lakh hectares. This irrigated area is 16.00% of the total cultivable area of the State. Surface water irrigation has made good progress all these years. The total irrigated area has increased from 12 lakh hectares in 1951-52 to 105.62 lakh hectares in 2015-16. At present it covers about 22% of the total irrigated area of the State. Since 1951, the area under irrigated by wells and tube wells have remarkably been increasing, and at present 77.24% of the total area exist under irrigated farming under this system in the State.

GROSS IRRIGATED AREA

Gross irrigated area includes area irrigated more than once. This icreased from abobe 12 lakh hectares in 1950-51 to 101.7 lakh hectates in 2014-15. The gross irriaged area as a ratio of gross cropped area increased from 12 percent in 1950-51 to 42 percent in 2014-15. This means that even today about 60 percent (about 3/5) cropped area depends on rainfall. Dry farming area essential for thidexeloprnent of agriculture in Rajasthan. Whete sprinkle and drip-irrigation system become absolutely essential for increasing crop output (Figure 14.1).

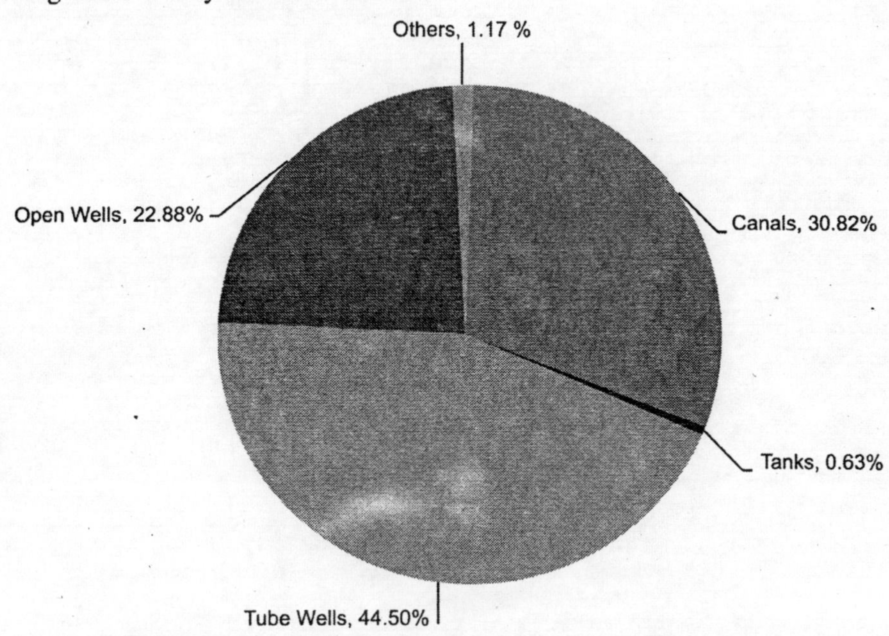

Source: Department of Water Resources, GoR, Jaipur

Figure 14.1 Source Wise Gross Irrigated Area in Percentage, 2015-16

Gross Irrigated area includes the double – cropped or triple -- cropped area, which is irrigated in Kharif, Rabi and summer-seasons in the same field. Such areas are found more in canal irrigated and well irrigated regions. Irrigation in the Kharif season is usually required for cotton, sugarcane, paddy and oil seeds. In Rabi-season, irrigation is mostly used for wheat, mustard, barley, gram, vegetables and fruit plants. The gross irrigated area in the State varies from year to year, depending upon the quantum and pattern of precipitation.

The gross irrigated area varies from district to district. It is highest in Ganganagar district. In Kota, Bundi, Jaipur, Dausa and Alwar also, the percentage of irrigated area is observed more. In Jaisalmer and Churu districts irrigation by wells and canals is on a much lower scale. In Dungarpur district also, it is on the lower side, owing to poor aquifers in the hard metamorphic rocks but the situation is much better in Banswara district, where the Mahl-Bajaj canal has now started covering more area under irrigated agriculture.

Gross irrigated area includes area irrigated more than once. This increased from about 12 lakh hectares in 1950-51 to 101-7 lakh hectated in 2014-15 and 105.62 lakh in 2015-16 (Figure 14.2 and 14.3) The table shows that the gross irrigated area as a ration of gross cropped area increased from 12% in 1950-51 to 42% in 2014-15 and 68.55@ in 2015-16 (Figure 14.4).

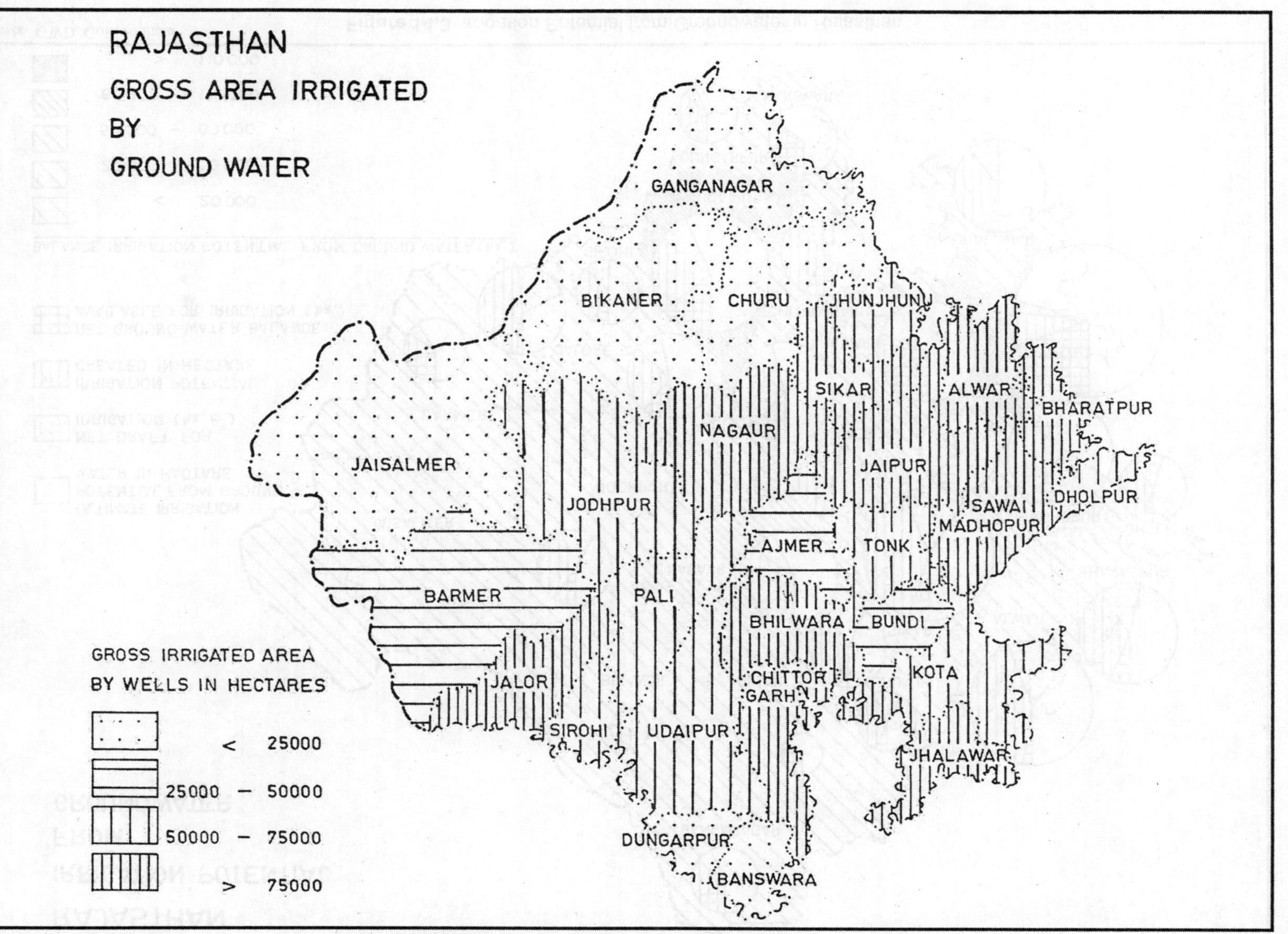

Source: GWD, GoR, Jodhpur

Figure 14.2 Gross Area Irrigated by Ground Water in Rajasthan

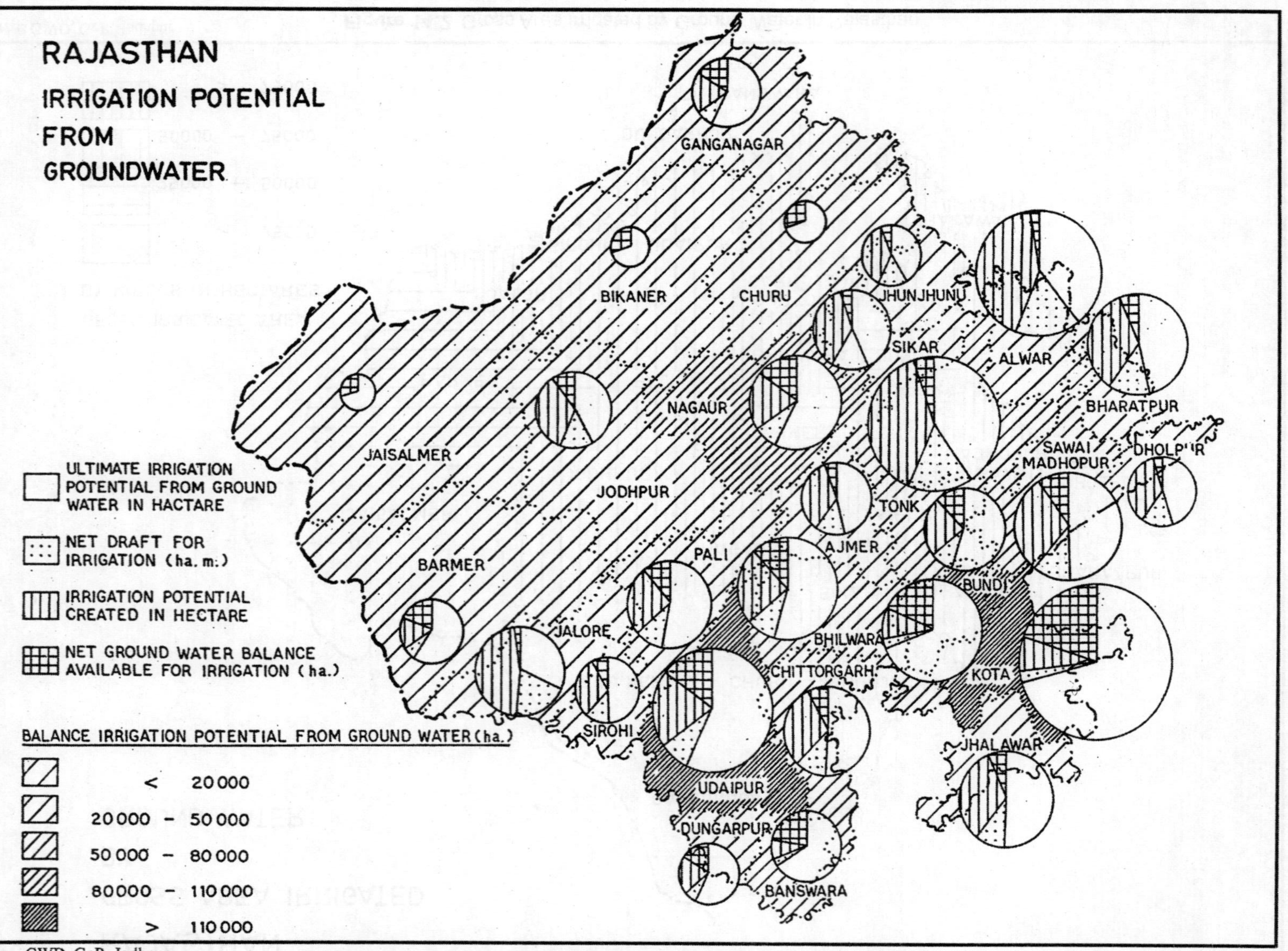

Source: GWD, GoR, Jodhpur

Figure 14.3 Irrigation Potential from Groundwater in Rajasthan

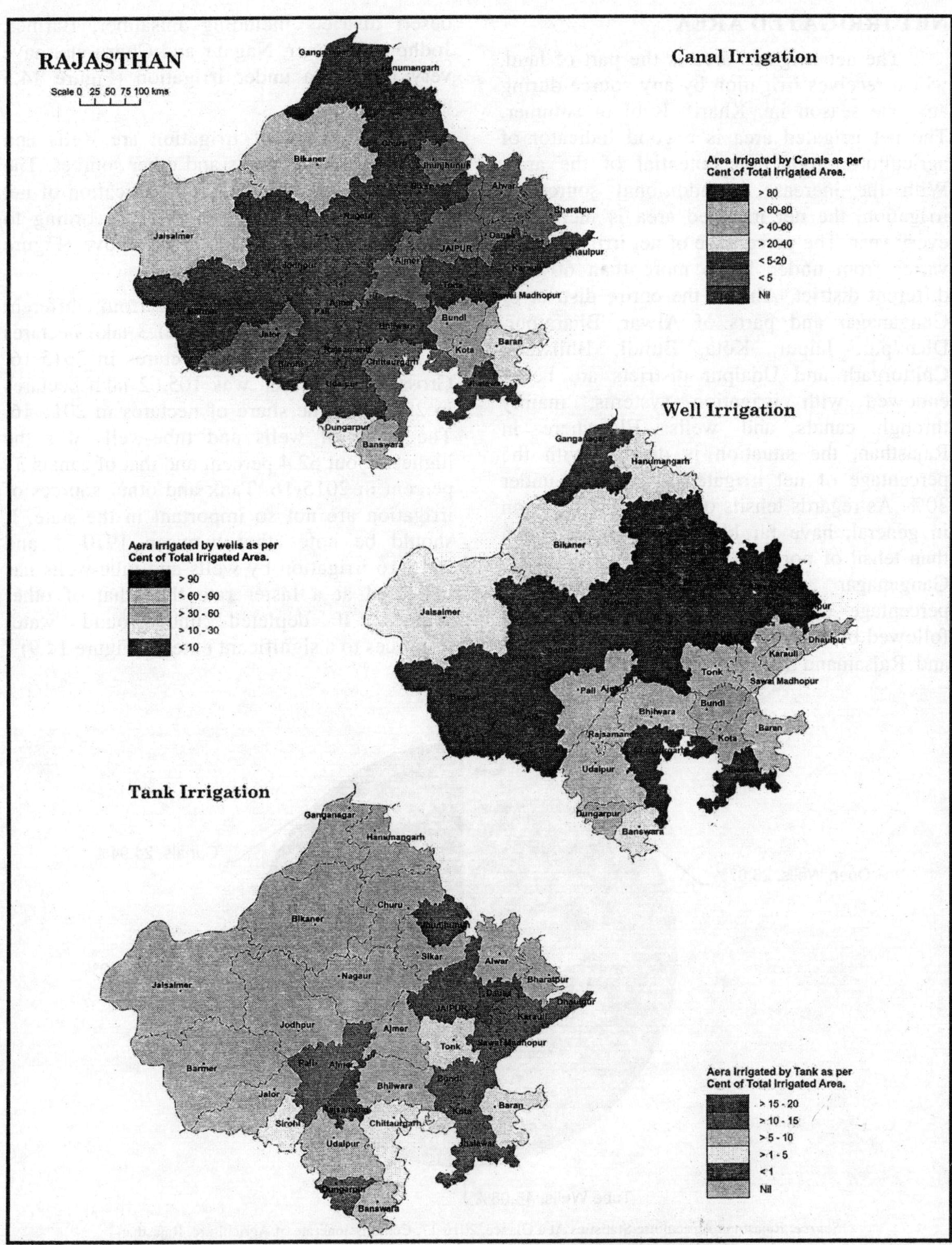

Source: Rajasthan Agriculture Statistics At a Glance, 2016-17, Commissionerate of Agriculture, Rajasthan

Figure 14.4 Canal, Well and Tank Irrigation in Rajasthan

NET IRRIGATED AREA

The net irrigated area is the part of land, which receives irrigation by any source during any one season i.e. Kharif, Rabi or summer. The net irrigated area is a good indicator of agriculture production potential of the area. With the increase in additional source of irrigation, the net irrigated area is increasing every year. The percentage of net irrigated area varies from under 15 to more than 60% in different district. Almost the entire district of Ganganagar and parts of Alwar, Bharatpur, Dhaulpur, Jaipur, Kota, Bundi, Bhilwara, Chittorgarh and Udaipur districts are better endowed with irrigation systems, mainly through canals and wells. Elsewhere in Rajasthan, the situation is dismal, with the percentage of net irrigated area being under 30%. As regards tehsils of the southern region in general, have far better net irrigated area than tehsil of northern and western Rajasthan. Ganganagar tehsil has the maximum percentage of net cultivated area (83.5%) followed by Anoopgarh (75.3%), Jaipur (75.2) and Rajsamand (71.4%). Many tehsils of the desert districts including Jaisalmer, Barmer, Jodhpur, Bikaner, Nagaur and Churu etc. have very little area under irrigation (Figure 14.5 and 14.6).

The sources of irrigation are wells and tube-well, canals, tanks and other sources. The following table shows the classification of net irritaged ared for year 2015-16 according to different sources as mentioned above (Figure 14.7 and 14.8).

The net area irrigated from different sources has increased from 21.3 lakh hectares in 1970-71 to 79.4 lakh hectares in 2015-16. Gross irrigated area was 105.62 lakh hectares in 2015-16. The share of hectares in 2015-16. The share of wells and tube-wells was the highest about 62.4 percent and that of canals 31 percent in 2015-16. Tank and other sources of irrigation are not so important in the state. It should be notes that between 1970-71 and 2015-16 irrigation by wells and tube-wells has increased at a faster rate than that of other sources. It depleted underground water resources to a significant extent. (Figure 14.9)

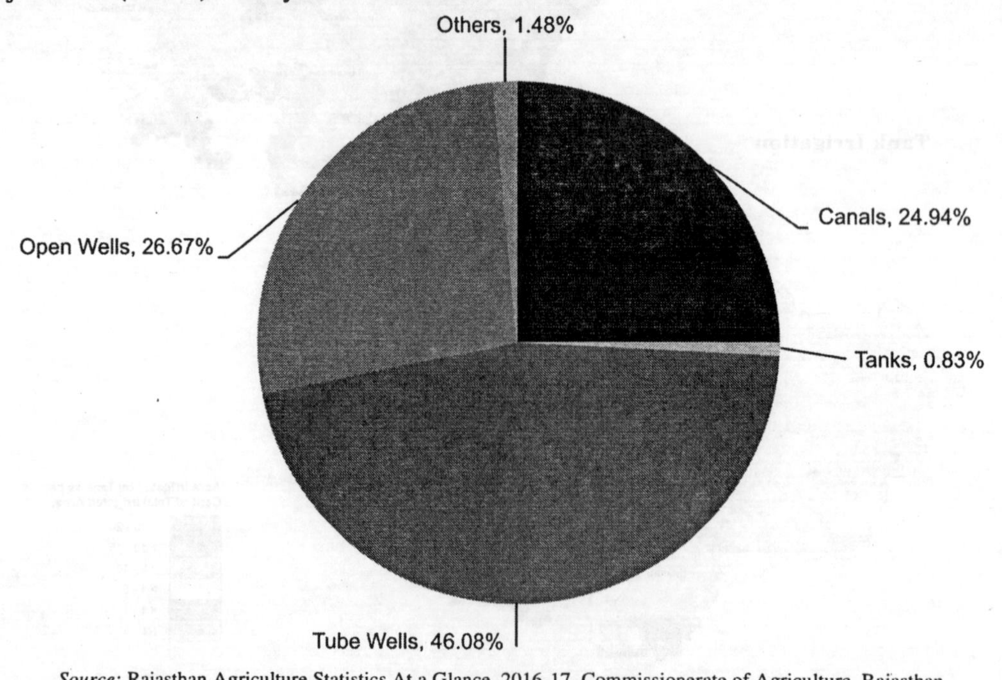

Source: Rajasthan Agriculture Statistics At a Glance, 2016-17, Commissionerate of Agriculture, Rajasthan

Figure 14.5 Source Wise Net Irrigated Area in Percentage, 2015-16

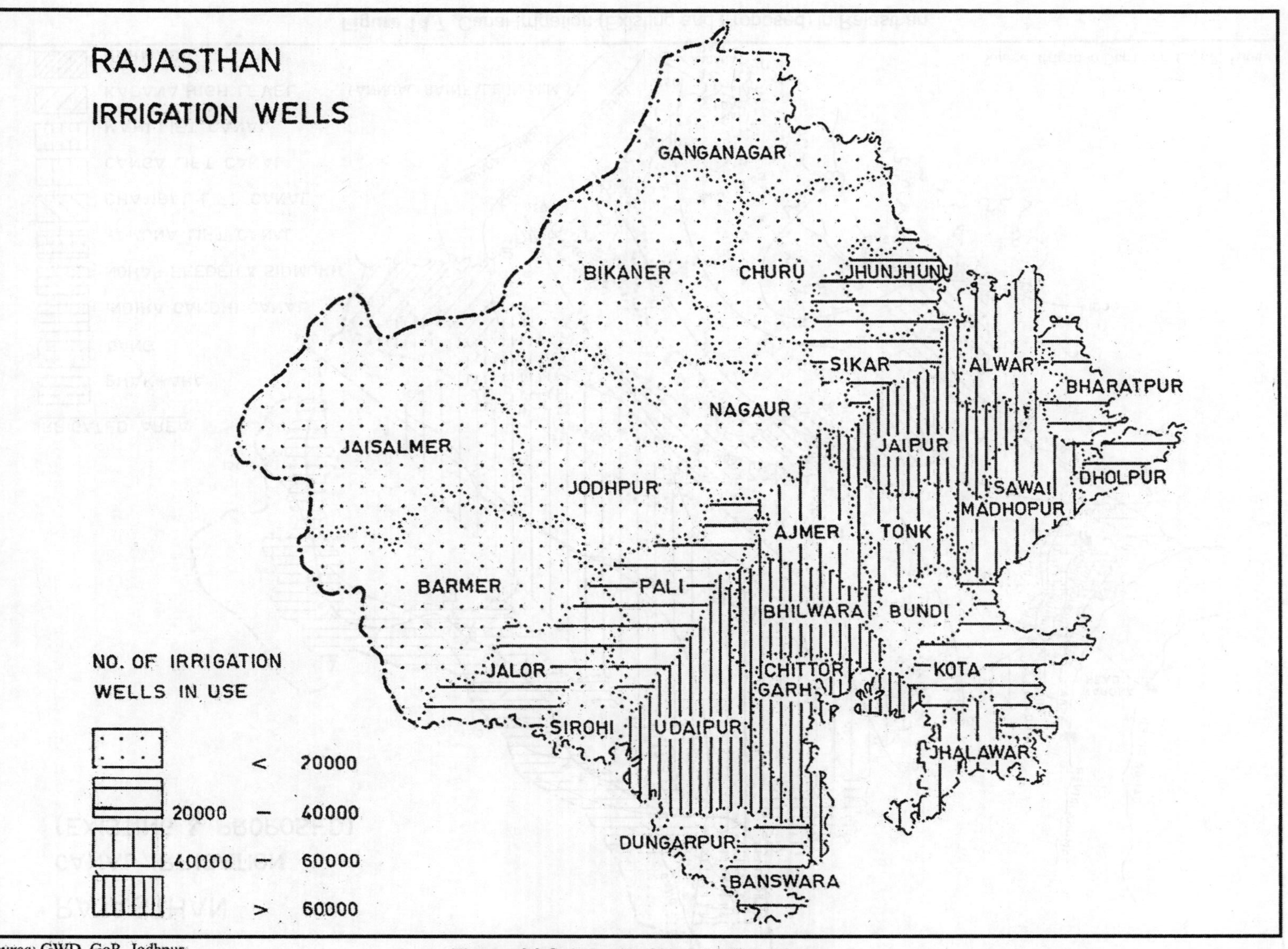

Source: GWD, GoR, Jodhpur

Figure 14.6 Irrigation Wells in Rajasthan

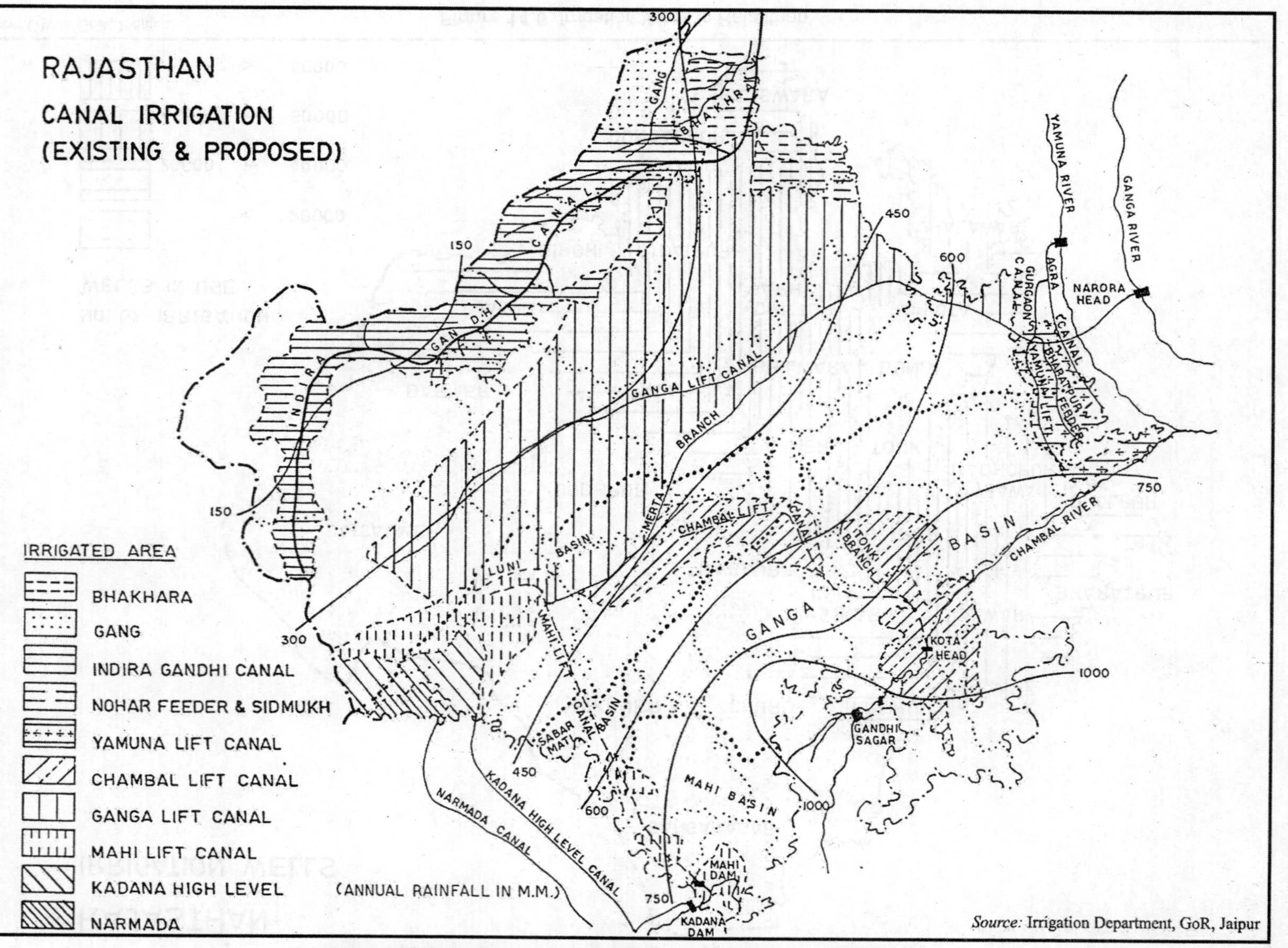

Source: Irrigation Department, GoR, Jaipur

Figure 14.7 Canal Irrigation (Existing and Proposed) in Rajasthan

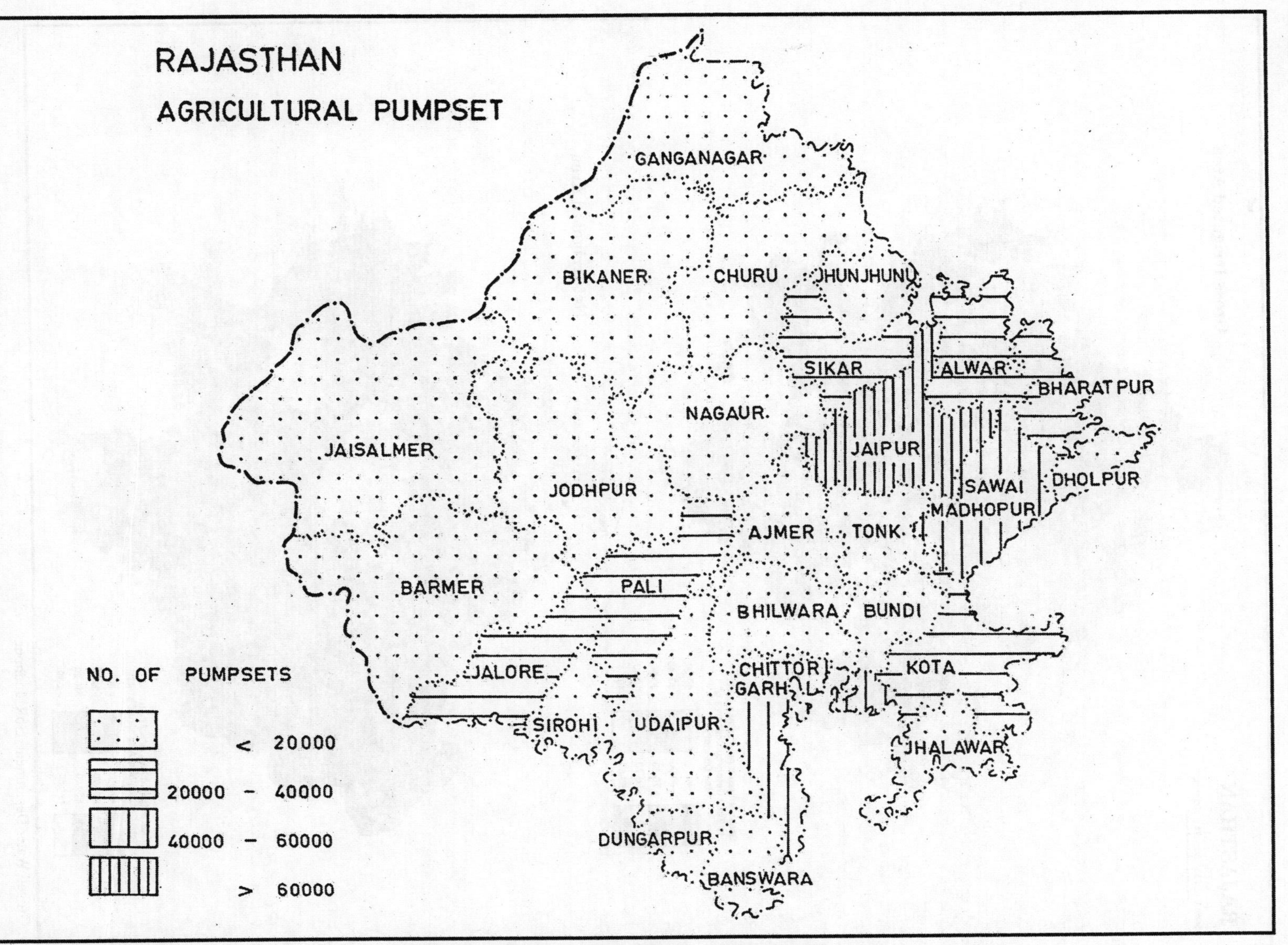

Source: GWD, GoR, Jodhpur

Figure 14.8 Agricultural Pumpset in Rajasthan

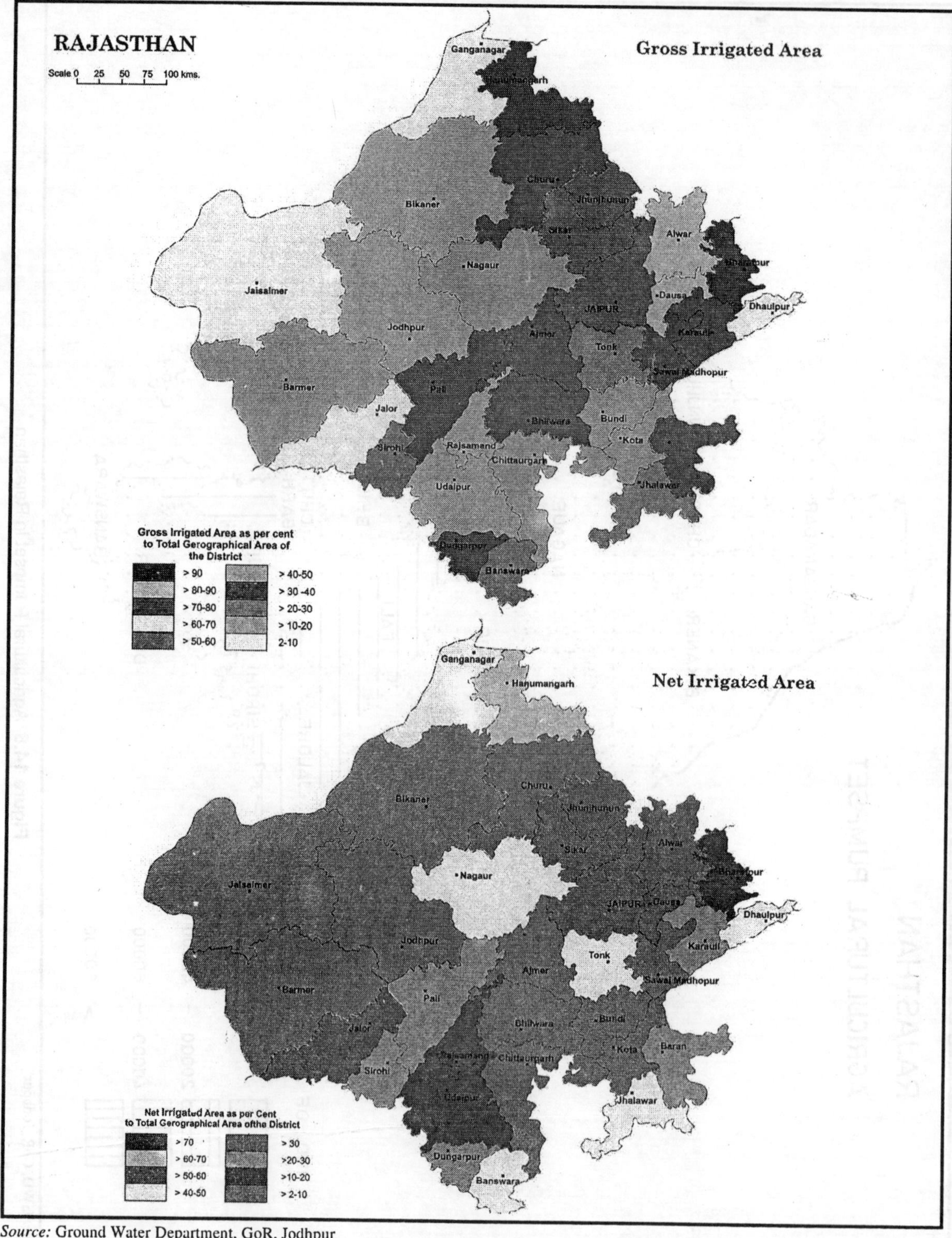

Source: Ground Water Department, GoR, Jodhpur

Figure 14.9 Gross and Net Irrigated Area in Rajasthan

IRRIGATION RESOURCES

Irrigation Commission (1972) had assessed the ultimate Irrigation potential of the country as 113.5 million ha comprising 58.5 million ha by Major and Medium Irrigation schemes, about 15 million ha by Minor surface water schemes and about 40 million ha by ground water. Indications are that minor surface irrigation potential could be about 17 million ha and ground water potential could be around 64.5 million ha taking the ultimate Irrigation Potential to about 140 million ha. This potential does not include additions due to inter basin transfers.

The mean annual surface and ground water resources of the State are as under:

	Mcum	MAF
(a) States internal resources:-	19580	15.86
I. Already harnessed	8210	6.65
II. To be harnessed by on going schemes	2100	1.70
Sub Total (I +II)	**10310**	**8.35**
III. Balance to be harnessed	9270	7.51
(b) External surface resources:-		
i. Surface water allocated under various inter states agreements	17901	14.50
ii. Contemplated allocations from other inter state sources viz Ganga water 45800 Mcum (37.10 MAF), Mahi (ex-kadana) water 1926 Mcum (1.56MAF)	47726	38.66
Sub Total (I+II)	**65627**	**53.16**
Total long term available surface water resources from internal and external sources	85207	69.02
Ground Water Resources	13142	10.64
Grand Total	**98349**	**79.66**

The total culturable area in the State is about 157 lac ha (out of total geographical area of 342 lac ha) which can be used for agriculture production if water required for irrigation is available. The State of Rajasthan is a water deficit State and its present total surface (both external as 14.5 MAF and internal as 15.86 MAF) water resources is estimated as 37481 Mcum (30.36 MAF).

An integrated and comprehensive water resources planning study for the State for a 50 year plan horizon i.e. upto 2045 has been under taken through consultants. As per the preliminary report of the study, the likely utilizable resources will be only 16432 Mcum (13.31 MAF) out of the total internal surface resources of 19580 Mcum (15.86 MAF) of the State.

Thus, the total utilizable water including all losses in the State with external allocated surface water of 17901 Mcum(14.5 MAF) shall be 34333 Mcum (27.81 MAF). The demand for various sectors has also been tentatively estimated by the consultants, though they are to be further refined during the course of the study. According to the present study the demand of all other sectors combined, except the Agriculture sector, for the year 2045 works out to 8864 Mcum (7.18 MAF) out of which about 2628 Mcum i.e.2.13 MAF (20% of 13142 Mcum) is assumed to be from ground water sources, while the balance 6236 Mcum (5.05 MAF) shall be from surface water resources. This means that the water available for agricultural shall be only 28097 Mcum (22.76 MAF) as per details given below.

	Mcum	MAF
(a) Surface Water		
Internal sources (Utilizable as per Consultants)	16432	13.31
Inter State Agreement	17901	14.50
Sub Total	**34333**	**27.81**
(b) Ground Water	**13141**	**10.64**
(c) Demand for other user Departments except Agriculture	8864	7.18

	Mcum	MAF
(d) Demand met from ground water (20% of b)	2628	2.13
(e) Balance to be met from surface water (c-d)	6236	5.05
(f) Net surface water available for Agriculture (a-c)	28097	22.67

The ultimate irrigation potential with the utilizable surface water (28097 Mcum) for the State with the the constraint of water availability is estimated as 47 lac ha (considering required water depth as 0.60 m including storage and all other losses).

The ultimate irrigation potential through contemplated allocations from other inter basin sources viz. Ganga water (45800 Mcum) and Mahi water at Kadana (1926 Mcum) is estimated to be 41.5 lac ha and 2.20 lac ha respectively, which add up to 43.70 lac ha. The ultimate irrigation potential from ground water resources has been estimated as 29.13 lac ha (Table 14.1).

The combined ultimate irrigation potential through surface water and ground water is likely to be as under:-

i. With existing internal and external resources 76.13 lac ha

ii. With existing internal, external and future external resources 119.83 lac ha

Table 14.1 : Major Irrigation Projects of the State are as follows

Name	Districts	Maximum Irrigation-potential on completion of other details (in hectares)
1. Jakham	Udaipur	23505
2. Yamuna Water Project (2 big projects)	For Bharatpur For Jhunjhunu & Churu districts	Yamuna Water of 3195 cusecs to be used
3. Parwan Irrigation & Drinking Water Project	To be constructed in Jhalawar district and to irrigate Kota, Baran and Jhalawar districts	131400
4. Kali Sindh	Jhalawar	Cost : Rs. 416.76 crore
5. Dholpur Lift	Dholpur	34665
6. Narmada (under AIBP)	Jalore	2.46 lakh
7. Sidhmukh-Nohar	Hanumangarh & Churu	33620
8. Bisalpur	Tonk	81500
9. Rajasthan Water Sector Restructuring Project (RWSRP)	Declared by World Bank to reform irrigation system (surface) in different districts	Started in 2002-03

Source: Irrigation Department, GoR, Jaipur

BASIN-WISE WATER RESOURCES AND IRRIGATICN DEVELOPMENT

The State has fourteen major river basins which are further sub-divided in 59 sub-basins. Irrigation Department, Rajasthan. According to Master Plan the State has been divided into 14 River Basins viz Chambal, Banas, Luni, Mahi Shekhawati, Sabarmati, Parwati, Sabi, Banganga, Ruparail. Gambhiri, West Banas, Sukli and other nallahs of Jalore Districts.

Surface Water- According to the total Surface Water Resources of Rajasthan. The total surface water petential for Rajasthan is estimated to be 15.86 million area-fee (MAF). Therefore, the state has to depend on inter state river basins for water under which 14.50 MAF water has been allotted to the state.

Ground Water- Based on hydro-geological investigations, the ground water recharge i.e. replenishable ground water resources in all river basins is estimated as 13142 Mcum/year (10.64 MAF). The average ground water resources for irrigation is 10957.68 Mcum/year (8.875MAF), and the net irrigation draft is 6336.38 Mcum/year (5.375 MAF).

The ultimate ground water irrigation potential is estimated as 2.913 m.ha. and the irrigation potential created up to December, is 1.807 m.ha. Leaving a balance of 1.106 m.ha.

A brief summary of the State of surface and ground water resources, irrigation and culturable land use from existing, ongoing and proposed surface water schemes, basin wise so far identified, is described as under-

Shekhawati Basin

The mean annual surface water potential has been estimated as 183.49 Mcum (0.148MAF), and mean gross recharge from ground water is 275.90 Mcum (0.224 MAF). While the net irrigation draft from ground water is about 385.33 Mcum (0.313 MAF). The total area of basin is 1.269 mha and culturable area of 0.356 mha.

Ruparail Basin

The mean annual surface water potential has been estimated as 183.49 Mcum (0.148 MAF), and mean gross recharge from ground water is 270.05 Mcum (0.219 MAF). While the net irrigation draft from ground water is about 285.88 Mcum (0.232 MAF). The total area of basin is 0.302 mha and culturable area of 0.212 mha of the basin.

Banganga Basin

The mean annual surface water potential has been estimated as 448.25 Mcum (0.148 MAF), and mean gross recharge from ground water is 843.95 Mcum (0.219 MAF). While the net irrigation draft from ground water is about 1034 Mcum (0.840 MAF). The total area of basin is 0.972 mha and culturable area of 0.652 mha of the basin.

Gambhir Basin

The mean annual surface water potential has been estimated as 531.36 Mcum (0.430 MAF), and mean gross recharge from ground water is 468.20 Mcum (0.379 MAF). While the net irrigation draft from ground water is about 394 Mcum (0.320 MAF). The total area of basin is 0.439 mha and culturable area of 0.263 mha of the basin.

Parbati Basin

The mean annual surface water potential has been estimated as 248.53 Mcum (0.201 MAF), and mean gross recharge from ground water is 223 Mcum (0.181 MAF). While the net irrigation draft from ground water is about 131 Mcum (0.106 MAF). The total area of basin is 0.248 mha and culturable area of 0.132 mha of the basin.

Sabi Basin

The mean annual surface water potential has been estimated as 213.85 Mcum (0.173 MAF), and mean gross recharge from ground water is 313 Mcum (0.253 MAF). While the net irrigation draft from ground water is about 421 Mcum (0.342 MAF). The total area of basin is 0.505 mha and culturable area of 0.243 mha of the basin.

Banas Basin

The mean annual surface water potential has been estimated as 3451.50 Mcum (2.795 MAF), and mean gross recharge from ground water is 2792 Mcum (2.261 MAF). While the net irrigation draft from ground water is about 2003 Mcum (1.626 MAF). The total area of basin is 4.705 mha and culturable area of 2.183 mha of the basin.

Chambal Basin

The mean annual surface water potential has been estimated as 8059 Mcum (6.528 MAF), and mean gross recharge from ground water is 3381 Mcum (2.739 MAF). While the net irrigation draft from ground water is about

1141 Mcum (0.926 MAF). The total area of basin is 3.136 mha and culturable area of 1.478 mha of the basin.

Mahi Basin

The mean annual surface water potential has been estimated as 4529.52 Mcum (3.669 MAF), and mean gross recharge from ground water is 988 Mcum (0.800 MAF). While the net irrigation draft from ground water is about 326 Mcum (0.265 MAF). The total area of basin is 1.661 mha and culturable area of 0.588 mha of the basin.

Sabarmati Basin

The mean annual surface water potential has been estimated as 513 Mcum (0.415 MAF), and mean gross recharge from ground water is 145 Mcum (0.117 MAF). While the net irrigation draft from ground water is about 47 Mcum (0.038 MAF). The total area of basin is 0.419 mha and culturable area of 0.093 mha of the basin.

Luni Basin

The mean annual surface water potential has been estimated as 790 Mcum (0.640 MAF), and mean gross recharge from ground water is 1601 Mcum (1.297 MAF). While the net irrigation draft from ground water is about 1727Mcum (1.402 MAF). The total area of basin is 3.827 mha and culturable area of 1.419 mha of the basin.

West Banas Basin

The mean annual surface water potential has been estimated as 280 Mcum (0.227 MAF), and mean gross recharge from ground water is 91 Mcum (0.073 MAF). While the net irrigation draft from ground water is about 59 Mcum (0.048 MAF). The total area of basin is 0.187 mha and culturable area of 0.069 mha of the basin.

Sukli Basin

The mean annual surface water potential has been estimated as 87 Mcum (0.070 MAF), and mean gross recharge from ground water is 53 Mcum (0.043 MAF). While the net irrigation draft from ground water is about 42 Mcum (0.034 MAF). The total area of basin is 0.096 mha and culturable area of 0.025 mha of the basin.

Other Nallah of Jalore District-Basin

The mean annual surface water potential has been estimated as 55 Mcum (0.045 MAF), and mean gross recharge from ground water is 57 Mcum (0.046 MAF). While the net irrigation draft from ground water is about 57 Mcum (0.047 MAF). The total area of basin is 0.186 mha and culturable area of 0.022 mha of the basin.

Area falling out-side Basin

This area is situated west of Aravali and occupies about half the area of the State. There are no streams and rivers in this area, and except for external sources of surface water of 13728 Mcum (11.22MAF) being proposed to be utilized through Gang Canal System, Bhakara canal System and Indira Gandhi Nahar Project, and meager ground water sources estimated as 2300 Mcum (1.865 MAF), it is wholly dependent on rainfall. Out of the total area of 16.273 mha of this region, the total culturable area is 8.009 mha. The prestigious Indira Gandhi Nahar Project has been taken up for utilization of the allocated share of the Inter – State Ravi Beas river to Rajasthan which shall be creating to a command of 1.269 mha (Figure 14.10).

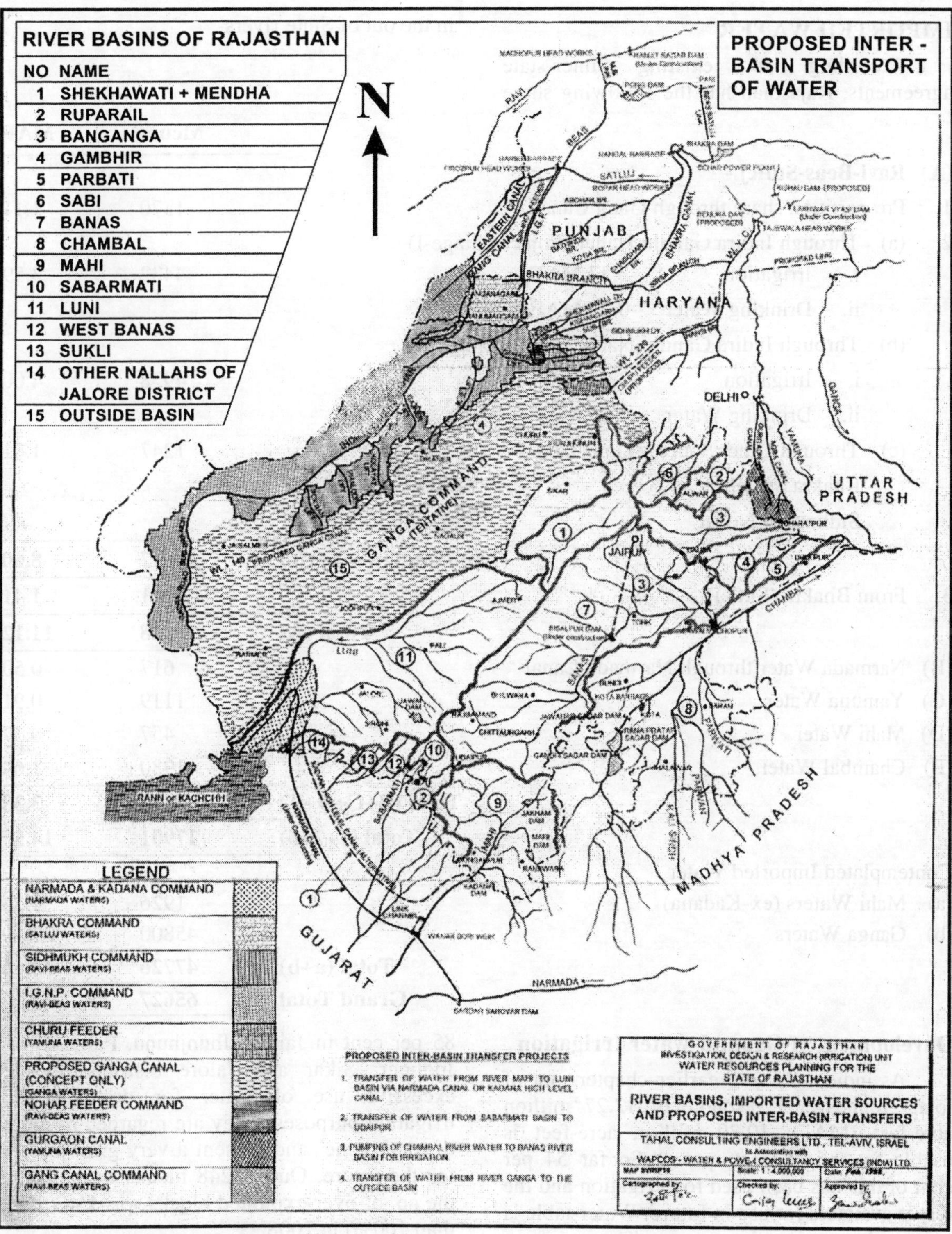

Source: Irrigation Department, Govenment of Rajasthan, Jaipur

Figure 14.10 Proposed Inter-Basnin Transport of Water

IMPORTED WATER

According to existing Inter-state agreements, Rajasthan has the following share in the out of State rivers-

		Mcum	MAF
(A) Ravi-Beas-Sutlej -			
1. Pre-partition share through Gang Canal		1370	1.11
2. (a) Through Indira Gandhi Nahar Project (Stage-I)			
i. Irrigation -3.37 MAF		4432	3.59
ii. Drinking Water -0.22 MAF			
(b) Through Indira Gandhi Nahar Project (Stage-II)			
i. Irrigation -3.35 MAF		4938	4.00
ii. Drinking Water -0.65 MAF			
(c) Through Other Sources Gang Canal	-0.30 MAF	1247	1.01
Bhakra System	-0.24 MAF		
Sidhmukh-Nohar	-0.47 MAF		
	Sub total (a)+(b)+(c)	**10617**	**8.60**
3. From Bhakra (Sutlej)		1741	1.41
	Total (A)	**13728**	**11.12**
(B) Narmada Water through Narmada Canal		617	0.50
(C) Yamuna Water		1119	0.91
(D) Mahi Water		457	0.37
(E) Chambal Water		1980	1.60
	Total (B+C+D+E)	**4173**	**3.38**
	Total (1+2+3)	**17901**	**14.50**
Contemplated Imported Water			
(a) Mahi Waters (ex-Kadana)		1926	1.56
(b) Ganga Waters		45800	37.10
	Total (a+b)	**47726**	**38.66**
	Grand Total	**65627**	**53.16**

Development of Ground-water Irrigation

As indicated in an earlier chapter, out of total ground-water resource of 12.27 million acre-feet (MAF), 10.80 million acre-feet is usable for irrigation purposes. So far 54 per cent of this has been used for irrigation and the rest is available for use in future. Water-table is going down in the state. The use of groundwater has reached a level of more than 85 per cent in Jaipur, Jhunjhunu, Pali, Alwar, Jodhpur, Sikar and Jalore districts due to excessive use of water for drinking and irrigation purposes. They are regarded as areas of 'dark zone' and present a very grim picture for the future. Out of 248 blocks in the state, the no. of over-exploited blocks was 168 (more than 100%) in 2008.

Chapter 15

LIVESTOCK AND DAIRY DEVELOPMENT

The importance of livestock in Rajasthan agriculture is of great significance. The entire field operation from ploughing to the harvesting of crops is carried out by the draught animals mainly bullock and camel. They are the keystone in farming. In addition to crops, draught and milch stock are raised and maintained by the individual farmer. They are his constant companions in the field by day, and live beside his house or even under his roof. The maintenance of livestock and their proper care and treatment closely affect the prosperity of the Rajasthan cultivator.

India supports the largest cattle population, which including buffaloes, is 20% of the world's estimated 1115 million bovine populations. In spite of this large number of cattle, the average per capita daily consumption of milk in India is only 130 gram as against 741 grams in Switzerland etc. It is considerably small on account of the very low milk yield capacity of the Rajasthan cattle. The Rajasthan agriculture differs fundamentally from that in the western countries in that it is firmly dependent on the use of draught force. Paradoxically enough, no branch of Rajasthan agriculture is so much neglected as animal husbandry.

In Rajasthan animal husbandry occupies an important role in the state economy. Since more than half of the area is arid or semi-arid and is not suitable for traditional crop farming. Livestock raising has superiority over crop cultivation with respect to growth, stability and resource conservation. The natural forces like soil climate-compeis have less impact on livestock farming. Animal husbandry is not a mere adjunct to crop agriculture, but a major economic activity contributing over 12 percent of the total net cosmetic income. The state produced 2.8 million tones of milk and over 10 thousand tones of wool annually which account for over 10 percent and 40 percent of the total milk and wool produced respectively in the country. The State has only 7 per cent and 21 per cent respectively of the total cattle and sheep population of India. There are significant regional differences with respect to animal husbandry practices and the role of livestock farming in the State economy. The livestock in the State are of better quality in comparison to those found in other parts of the country.

There are seven well defined breeds of cattle namely Haryana, Gir, Nagauri, Kankrej, Tharparker, Rathi and Malvi in Rajasthan. Out of these Tharparker and Rathi are generally classified as dairy breeds, Narauri and Malvi for producing draught animals and Haryana, Gir and Kankrej for producing bullock and milch animals. The important breeds of sheep in the State are Chokle. Magra, Marwari, Nali, Pugal, Jaisalmeri, Malpura and Sonadi. Chokla sheep produce medium apparel and excellent carpet quality wool while Magra, Marwari, Nali Pugal and Jaisalmeri produce medium to good carpet quality wool. The Sonadi and Malpura produce poor quality wool but are suitable for mutton production. The importand breeds of goats are Marwari, Thakrana, Sirohi, Jamnapari, and Barbari. The Barbari is highly prolific and produce twins and triplets. Thakrana is reputed for high mild production. Jamnapari is a hardy and heavy animal and can adapt to almost all types of climate and food.

LIVESTOCK AND CATTLE REGIONS

There are significant regional differences with respect to animal husbandry practices and the role of livestock farming in the economy of the state.

Arid Region

The arid and semi-arid areas covering over half the State are agro-climatically not suitable for crop farming. Here animal husbandry is evidently the only source of regional economy which not only support the natives of the area but also constitute an important item of sale to other regions. The natural forces which reduce the productivity and lead to instability in agriculture have much less impact on livestock farming. This is primarily due to the ability of the desert animals to withstand the extreme conditions and the trans-migratory system of husbandry followed which imparts greater adjustability. Further, the region is endowed with some of the best breeds of cattle, sheep and draught animals. Hardy perennial grasses like Sewan (Lasiurus sindicus), Ajjan (Cenchrus setigerus), Dhaman (Cenchrus ciliaris) and also abundantly available. A comparatively low investment and marginal improvement in management and conservation of fodoer and water resources can bring remarkable change in returns from animal husbandry whereas in returns from animal husbandry whereas in agriculture even higher input may not have desired impact. Stability in the case of crop farming would need adequate irrigation facilities, soil conservation and other capital investment methods which are precarious in arid areas.

Arid regions are characterized by a stagnant economy and absence of industrial infrastructure. The existing cropping pattern in these areas indicates that it is mainly designed to serve the subsistence needs of the human and cattle population. Low yield coupled with recurrent crop failures necessitates hoarding of the production achieved during a favorable year. This reduces the chances of marketable surplus from crop products. On the other hand, in the case of livestock farming the situation is different. Because local needs of the livestock products are limited, the major share of the produce is surplus. The yields are relatively profitable and are not unstable. About half the milk production and almost the entire quantity of wool and mutton produce in the area are surplus to local requirement. Livestock farming, therefore, in the region by its very nature offers scope for commercialization. This also provides scope for livestock based industries. Unfortunately, the importance of livestock farming vis-a-vis crop farming has not been fully realized and attempts were made to bring increasing areas under crop farming. Since the area was unsuitable for such activity, stress on crop farming brought about faster deterioration of soil as vegetation cover and wasteful use of scarce water resources. It is felt that the resource depletion trend can be checked by reverting all the sub-marginal land to natural plant cover of grasses and shrubs, which will support livestock and cattle development.

Eastern and Southern Regions

In the eastern districts of the State small and marginal farmers predominate. Most of them derive part of their income from products of animal origin. But in the absence of any organized market for fluid mild, they do not have any incentive for increasing its production. Here the number of animals is quite high as compared to the western Rajasthan but the production of milk, wool and mutton is frustratingly low. No surplus has ever been achieved due to high local consumption.

The livestock in the southern districts of the State are of poor quality. The fodder resources though available in plenty are nutritionally poor. However, the number of stock is so large that small improvement in their production is likely to raise the overall production of the State considerably. The areas coming under the command of the Chambal and Mahi projects offer great potential for livestock farming along with cultivation.

Growth of Livestock and Cattle Population

In Rajasthan the livestock and cattle population is already high and has shown a further increase during the present century. At present there are more than 57.7 million and 13.3 million of total livestock and cattle population in the State respectively. The growth rate of different categories of domestic animals is not uniform. The differential growth rate of domestic animals is and indication of changing emphasis of cattle and livestock in the rural economy of Rajasthan.

Animal Husbandry is a major economic activity of the rural peoples, specially in the arid and semi-arid regions of the Rajasthan. Development of livestock sector has a significant beneficial impact in generating employment and reducing poverty in rural areas.

As per the 19th Livestock census 2012, there are 577.32lacs Livestock (which include Cattle, buffalo, Sheep, Goat, Horse and Ponies, Mules, Donkeys, Camel, Pig) and Poultry 80.24 lacs.

The livestock population fluctuates bared on drought condition the state. There is drastic/dicline in livestock population in years of actue scarcity of rainfall (perticularly of sheep).

There was a decline of about 19% in the population of sheep in 2012, while the population of cattle and buffaloes increased by about 10% and 17% respectively during 2007-2012.

As pointed out earlier, 25 percent of the total sheep population of India is fond in Rajasthan; More than 2 lakh families are engaged in sheep husbandry and also a signifcant number is involved in wool-processing activities. Annual production of wool exceeds 170 lakh ka and several lakh sheep and goat-heads are used for meat every year.

LIVESTOCK DISTRIBUTION PATTERN

The Livestock population details are given in Table. Districtwise population distribution is represented in Figure 15.1. It is clear from the figure that Barmer, Jodhpur and Nagaur districts leads in the total live stock population while Sirohi and entire eastern Rajasthan is having least population of the livestock. Brief Descrip-tion of the individual major component of Livestock population is given in Table 15.1 to 15.5.

CATTLE

The percent share of cattle to total Livestock population has decreased from 35.02% in 1966 to 23.08% % in 2012 (Figure 15.4 and 15.5). The very low and low cattle distributions were observed in Sawai Madhopur, Dausa Karauli, and Jhunjhunu and Bharatpur districts. Medium ranges were registered in Kota, Jalore, Sirohi, Alwar and Sikar districts. High ranges were found in Ajmer, Nagaur, Bikaner, Barmer and Banswara district, while very high ranges were located in Udaipur, Jodhpur, Bhilwara and Chittaurgarh district of Rajasthan. However, during 1972, ten districts namely Barmer, Bikaner, Churu, Jaisalmer, Jalore, Jhunjhunu, Jodhpur, Nagaur, Sikar and Sirohi indicated change in lower categories. This might be due to severe drought during that interval of livestock census. None of the districts indicated change in higher level of category.

BUFFALOES

The total number of buffaloes was arranged in five ranges namely, very low, low, medium, high and very high. Very low categories was found Jaisalmer, Barmer and Bundi district. Low ranges were identified in Baran, Kota, Dungarpur, Bikaner districts while Medium ranges were traced in Udaipur, Jalore, Pali, Bhilwara, Chittaurgarh, Bhilwara, Ajmer, Nagaur, Sikar, Dausa, Karauli, Jhunjhunu, Hanuman-garh and Ganganagar district of Rajasthan. Very high ranges were found in Jaipur, Alwar and Bharatpur district of Rajasthan.

Rajasthan is rich in animal wealth, it has the distinction of having the largest population of camels, sheep and goat of different breeds. According Animal Husbandry Department, Govt. of Rajasthan.

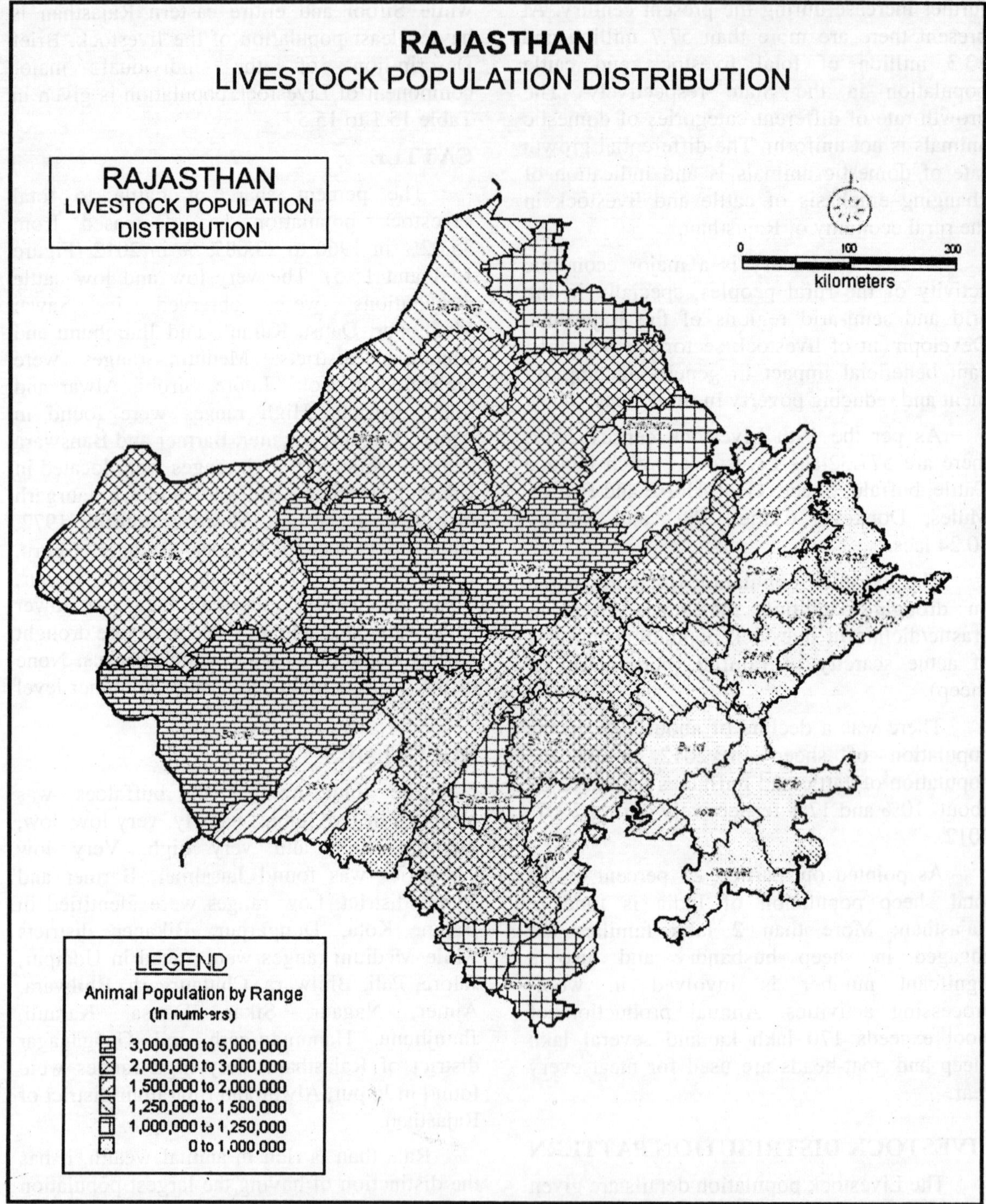

Source: Survey of India (SOI), Board of Revenue, Ajmer and Department of Animal Husbandry, Government of Rajasthan, Jaipur

Figure 15.1 Livestock Population Distribution in Rajasthan

Important breeds of Camel, Sheep and Goat in Rajasthan :

Camel	Sheep	Goat
Bikaneri	Chokla	Marwari
Jaisalmeri	Magra	Sirohi
Mewari	Marwari	Jhakrana
Mewat	Sonadi	Parbatsar
-	Malpura	-
-	Nali	-
-	Jaisalmeri	-
-	Pugal	-

SHEEP

The sheep population was arranged and mapped in group of five different categories namely very low, low, medium, high and very high. Very low category was found in Easter region namely Dausa, Karauli, Dhaulpur, Baran, Kota, Jhalawar and Banswara. Medium category is found in Udaipur, Rajsamand, Tonk, Jaipur, Jhunjhunu, Sikar, Hanumangarh and Ganganagar districts. Highest concentration of population were found in Desertic regions of Rajasthan like, Jaisalmer, Bikaner, Barmer, Jodhpur, Nagaur, Pali district.

These animals are reared mostly for wool and meat, and can subsist on grasses. The sheep population is higher in almost all the western districts particularly in Nagaur, Bikaner, Jaisalmer, Pali, Jodhpur and Barmer all of which have more than ten lakh population of sheep each. A slightly lower population, ranging from 3 to 8 lakhs is found in Ajmer, Bhilwara, Churu, Ganganagar, Hanumangarh, Jaipur, Jalor, Jhunjhunun, Rajsamand, Sikar, Sirohi, Tonk and Udaipur districts. The other areas of concentration of sheep population is Bharatpur, Bundi, Dausa and Sawai Madhopur districts. In southern and south eastern Rajasthan like Banswara, Baran, Dhaulpur, Jhalawar and Kota districts their concentration is low (Fig. 15.2A and 15.2B).

GOAT

The percent share of goat in was worked out and grouped into five different categories. Very low category was observed in Eastern districts like Bharatpur, Karauli, Dausa, Dhaulpur and Hanumangarh districts.

Highest Goat concentration was found in Barmer district of Rajasthan followed by Jodhpur and Nagaur districts.

Goats are reared both for meat, fur and milk. Some of the goat breeds may yield upto 2 kg of milk per day. These animals can survive on meagre quantities of fodder and can, therefore, be reared in every household. They are found all over Rajasthan, but their highest population exceeding eighteen lakhs, is concentrated in Barmer district followed by above twelve lakhs in Jodhpur district and above ten lakhs in Nagaur district.

Figure 15.2A Sheep and Goat Breads in Rajasthan

Nali
Marwari
Jhakrana
Barbari
Pugal
Jaisalmeri

Figure 15.2B Sheep and Goat Breads in Rajasthan

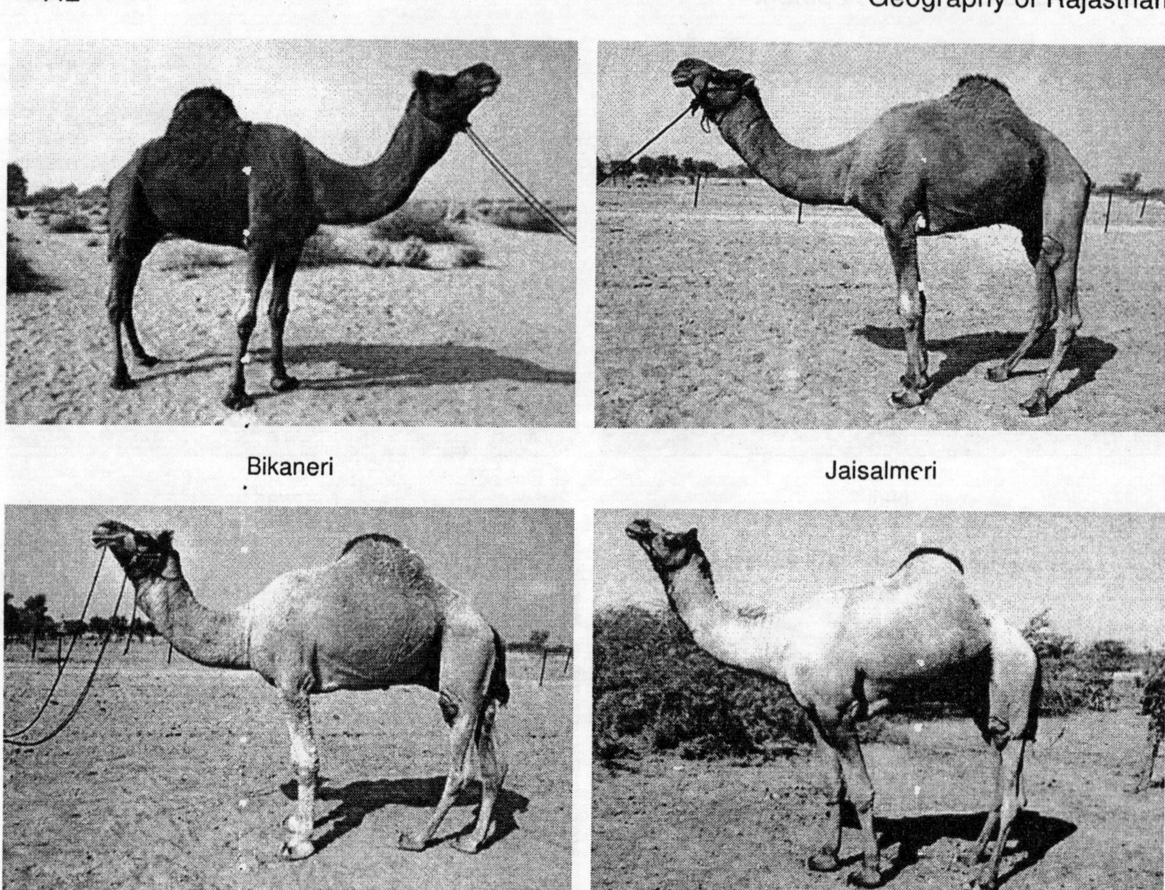

Bikaneri Jaisalmeri

Mewari Jalori

Figure 15.3 Camel Breeds in Rajasthan

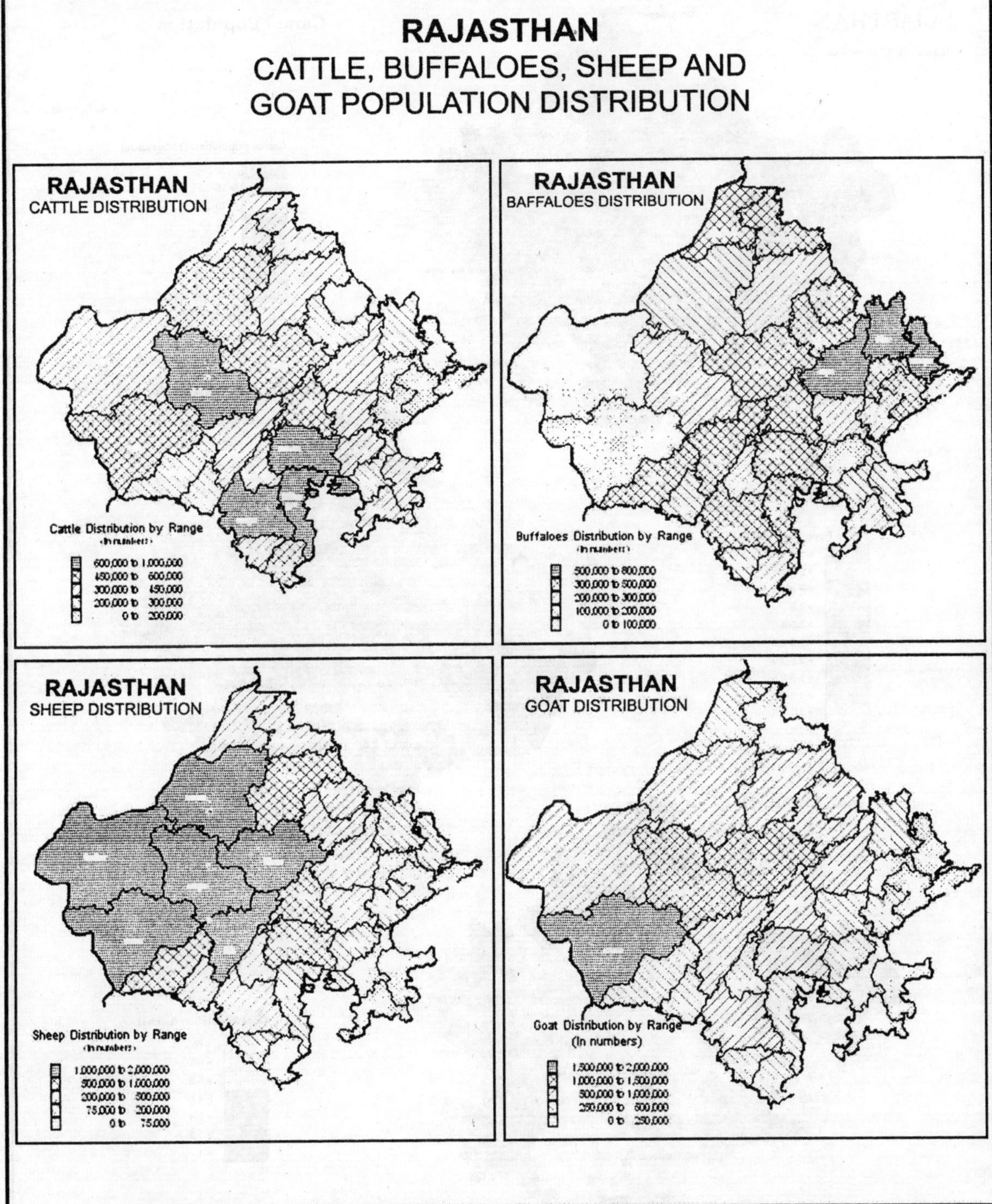

Source: Survey of India (SOI), Board of Revenue, Ajmer and Department of Animal Husbandry, Government of Rajasthan, Jaipur

Figure 15.4 Cattle, Buffaloes, Sheep and Goat Population Distribution in Rajasthan

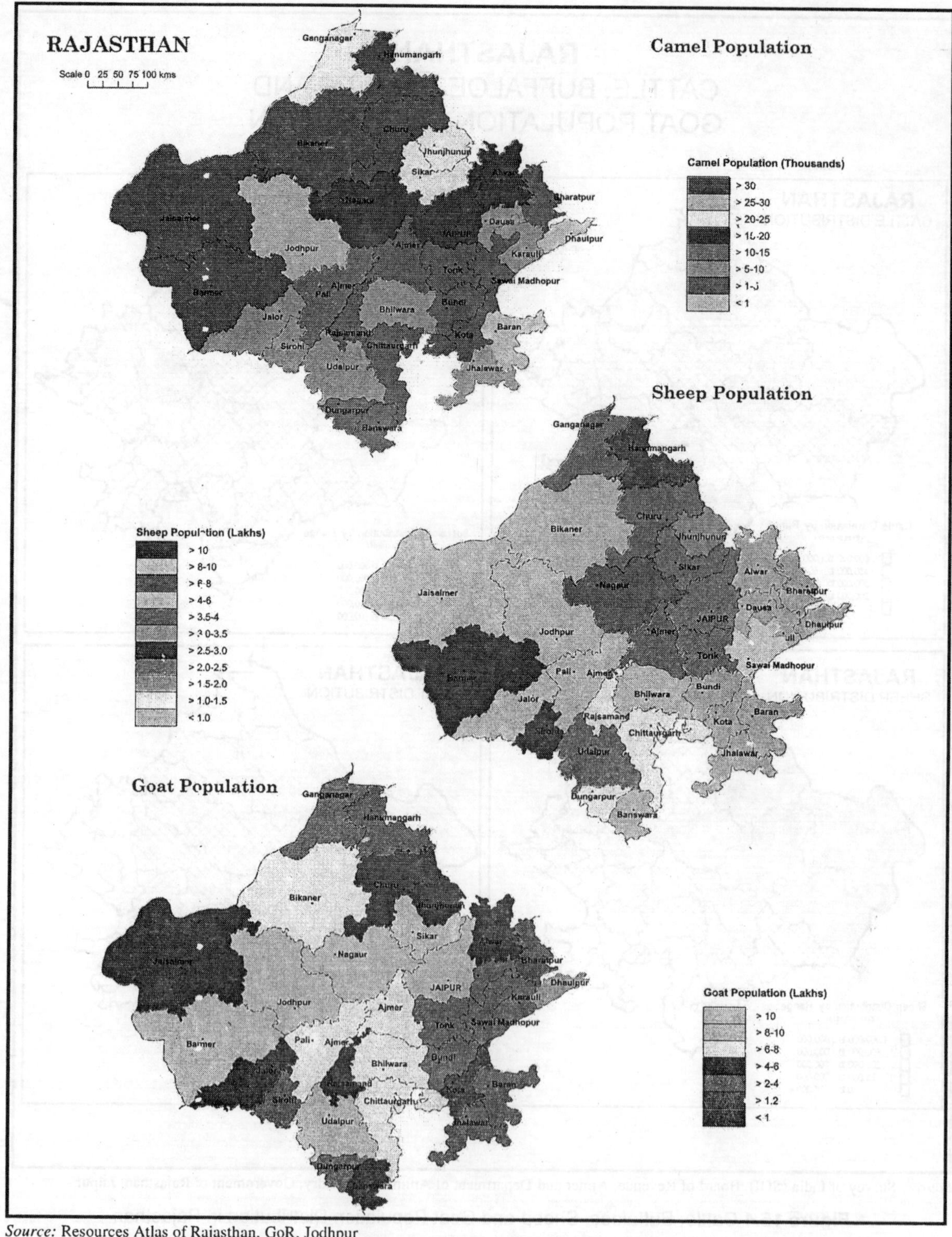

Source: Resources Atlas of Rajasthan, GoR, Jodhpur

Figure 15.5 Cattle, Buffaloes, Sheep and Goat Population Distribution in Rajasthan

Other districts with higher goat population, ranging from 4 to 10 lakhs are Churu, Sikar, Bhilwara, Bikaner, Udaipur, Pali, Ajmer, Alwar, Chittaurgarh and Rajsamand districts.

Animals such as goats and sheep are used for milk, meat and wool. Some other animals like horses and camels are used for drought purposes. The concentration of these animals in eastern and south eastern parts of the state is greater as compared to the western half. Due to availability of natural pasture lands for grazing these animals and enough left-overs of the farm produce are available for feeding them.

CAMEL

The contribution of camel to Rajasthan's livestock population was worked out and arranged into five different categories namely very low, low, medium, high and very high. It indicated that camel population is maximum in desert region of Rajasthan. It is lowest in eastern region of the state. Nagaur, Pali, Jaipur and Alwar districts along with Sikar and Ganganagar is having second highest population of camel.

The camel is draught animal used for transport of goods and human beings. It can walk with ease on sandy soils and can remain without water for days together. Out of total camel population 77.81 per cent found in the state and remaining 22.19 per cent in rest of India. Although camels are reared all over Rajasthan, their concentration is greater in the western arid region The districts have camel population of above 25,000 are Banner (43,172), Churu (31,226), Bikaner (46,209), Hanumangarh (33,959), Jodhpur (16,749), Jaisalmer (49,917), Jhunjhunun (12,813), Nagaur (7,085), Ganganagar (12,310) and Sikar (27,552). Other areas where the camel population is higher are Alwar, Bhilwara, Dausa, Jalor, Bharatpur, Bundi, Chittaurgarh, Jaipur, Karauli and Sawai Madhopur districts. They are fewer in number in Dhaulpur, Jhalawar, Kota, Banswara, Baran, Dungarpur and Tonk districts.

Camel the "Ship of the Desert" due to its ability to sustain water scarcity and excellent adoptive mechanism under harsh conditions will continue its existence as a single or few animals for agriculture and transport for rural population. Camel carts are being used for transportation of various goods like grains, fuel, wood, fodder, construction material, water and LPG cylinder etc. Bikaneri and Jaisalmeri breed of camel are used in "camel .safari". The region is prone to frequent drought and camel is the only source which provides the source of livelihood in rural as well as urban areas (Figure 15.3).

Rajasthan, the species wise population (2012) is as follows Camel 325,713, Sheep 90,79,702 and Goat 21,665,939 (Figure 15.4 and 15.5).

Camel hair is being used under village cottage industries in making Shawls, blankets, Ropes, Carpets and other items of daily use. Camel bones and leather are used for preparation of ornaments and Handicraft items. Camels are intensively used in desert part of border for military/border surveillance purposes.

PACK ANIMAL

The pack animal consists of horses, ponies, mules and donkeys. The population contribution of pack animals was worked out and grouped into five categories namely very low, low, medium, high and very high. It is observed that population of pack animals was very low in Dausa, Dhaulpur, Karauli and Kota districts. The highest concentrations were observed in Barmer and Jaisalmer districts followed by Jalore, Jodhpur, Bikaner and Churu districts. The entire Aravalli belt was found to have medium to low concentration of Pack Animals in the state of Rajasthan (Figure 15.6).

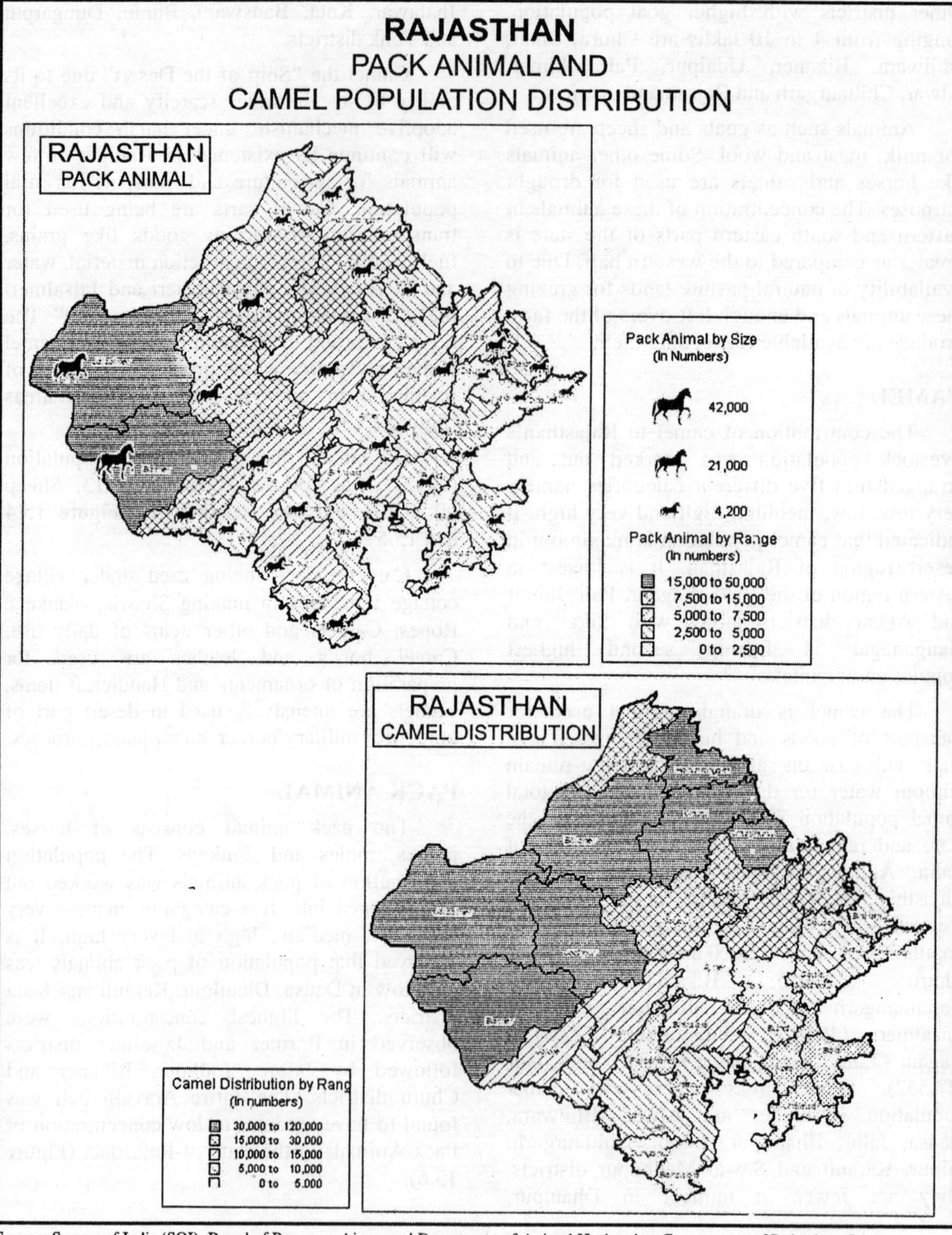

Source: Survey of India (SOI), Board of Revenue, Ajmer and Department of Animal Husbandry, Government of Rajasthan, Jaipur

Figure 15.6 Pack Animal and Camel Population Distribution in Rajasthan

Table 15.1 Livestock Population of Rajasthan - 2012

S. No.	District	Cattle			Buffalo	Sheep	Goat	Horses and Ponies
		Exotic/ C.B. Cattle	Indigenous	Total Cattle				
1	2	3	4	5	6	7	8	9
1	Ajmer	42798	361928	404726	438804	365108	730758	1849
2	Alwar	63326	143539	206865	1060734	52166	379776	952
3	Banswara	9906	588547	598453	282438	7207	504758	165
4	Baran	5263	337748	343011	250281	9631	183499	493
5	Barmer	2301	786065	788366	214168	1404031	2896620	2591
6	Bharatpur	37420	129474	166894	836220	65049	172391	754
7	Bhilwara	121893	596001	717894	452320	405505	846696	2256
8	Bikaner	57421	848654	906075	193433	653028	961907	3047
9	Bundi	14010	228947	242957	316926	54027	329366	1390
10	Chittorgarh	33066	423805	456871	393268	43941	474799	1717
11	Churu	32018	315452	347470	292571	348522	820043	824
12	Dausa	33934	104087	138021	488365	55908	306342	719
13	Dholpur	6349	53337	59686	359012	11807	89652	622
14	Dungarpur	2736	372287	375023	232133	62652	416729	138
15	Ganganagar	172768	463652	636420	270992	273476	384446	1048
16	Hanumangarh	97918	404153	502071	389303	189210	212993	1223
17	Jaipur	344908	290033	634941	1073386	229948	837094	1175
18	Jaisalmer	1637	432986	434623	4057	1185150	1513386	1089
19	Jalore	2248	293591	295839	469162	385416	469640	1920
20	Jhalawar	3126	376949	380075	310513	11329	313475	1327
21	Jhunjhunu	183475	49215	232690	384902	126047	521668	1241
22	Jodhpur	56479	791864	848343	305238	731229	1681913	1616
23	Karouli	9029	92032	101061	476583	56677	282980	582
24	Kota	9083	213524	222607	236499	18302	149610	334
25	Nagour	72611	428979	501590	556598	584841	1485051	2086
26	Pali	8988	346549	355537	311691	850596	767468	1193
27	Pratapgarh	15712	314338	330050	150493	21917	259899	292
28	Rajsamand	26105	234730	260835	222293	100488	536901	1002
29	Sawai Madhopur	1859	108144	110003	318184	83155	272341	1000
30	Sikar	200989	130045	331034	554076	146283	1073389	824
31	Sirohi	3089	191486	194575	186218	205736	307708	616
32	Tonk	9494	218180	227674	391599	200694	375827	953
33	Udaipur	53113	919069	972182	553635	140626	1106814	738
	Total	**1735072**	**11589390**	**13324462**	**12976095**	**9079702**	**21665939**	**37776**

Table 15.1 *Continous.....*

District	Mules	Donkeys	Camel	Pig	Livestock	Dogs	Rabbits	Poultry At Backyard	Poultry At Farm	Total Poultry
1	10	11	12	13	14	15	16	17	18	19
Ajmer	4	2132	1516	20552	1965449	19206	381	98324	2639518	2737842
Alwar	592	1284	5928	15142	1723439	16713	639	94192	74521	168713
Banswara	1	1713	558	125	1395418	16831	124	267610	1097	268707
Baran	197	906	775	12013	800806	6818	176	50785	623	51408
Barmer	28	17495	43172	261	5366732	32208	193	17348	10	17358
Bharatpur	81	1442	2287	24297	1269415	26972	666	147542	25806	173348
Bhilwara	264	1160	4655	14542	2445292	7408	303	65861	51144	117005
Bikaner	45	8712	46209	859	2773315	29520	288	22762	1000	23762
Bundi	8	734	2737	13595	961740	11914	107	22341	10594	32935
Chittorgarh	15	440	2166	4052	1377269	3906	133	66373	45	66418
Churu	179	5063	33959	1202	1849833	11341	178	65213	484394	549607
Dausa	21	431	2269	10816	1002892	6564	309	29650	11543	41193
Dholpur	155	822	304	7141	529201	6115	20	9233	8900	18133
Dungarpur	1	1214	1672	38	1089600	13625	86	177647	160	177807
Ganganagar	162	4609	12310	1781	1585244	86322	2080	44152	13550	57702
Hanumangarh	407	3370	31226	1462	1331265	49647	973	59223	77204	136427
Jaipur	54	1300	4896	21203	2803997	24666	975	185503	152202	337705
Jaisalmer	16	5846	49917	1129	3195213	2871	22	20170	11	20181
Jalore	7	3334	5114	743	1631175	14585	242	17038	4	17042
Jhalawar	28	827	259	7096	1024929	7294	181	55932	28250	84182
Jhunjhunu	257	1601	12813	3438	1284657	15676	938	463929	1114605	1578534
Jodhpur	178	4176	16749	822	3590264	25765	567	30083	35241	65324
Karouli	113	849	3394	11484	933723	4255	216	10707	99	10806
Kota	61	454	1839	14760	644466	9791	440	19720	1578	21298
Nagour	40	1771	10578	7456	3150011	15155	613	74245	548	74793
Pali	10	2066	8353	5290	2302204	11366	144	49352	22092	71444
Pratapgarh	67	320	109	2410	765557	7462	314	138149	0	138149
Rajsamand	3	938	1572	3137	1127169	2669	53	30043	6127	36170
S.Madhopur	25	1240	3687	13867	803502	6975	236	22331	707	23038
Sikar	65	1164	7085	4126	2118046	7620	440	85241	216441	301682
Sirohi	285	1402	4121	530	901191	9548	737	52209	27	52236
Tonk	0	268	789	10820	1208624	13533	393	35951	13171	49122
Udaipur	6	2385	2695	1485	2780566	45234	375	504353	0	504353
Total	**3375**	**81468**	**325713**	**237674**	**57732204**	**569575**	**13542**	**3033212**	**4991212**	**8024424**

Source: Department of Animal Husbandry, Government of Rajasthan

Table 15.2 Demographic Characteristic of Districtwise Livestock Population and Density in Rajasthan 2012

S.No.	DISTRICT	AREA (SQ. KMS.)	LIVESTOCK CENSUS 2012	LIVESTOCK DENSITY (PER SQ. KM.)
1	Ajmer	8481	1965449	232
2	Alwar	8380	1723439	206
3	Banswara	5037	1395418	277
4	Baran	6992	800806	115
5	Barmer	28387	5366732	189
6	Bharatpur	5066	1269415	251
7	Bhilwara	10455	2445292	234
8	Bikaner	27244	2773315	102
9	Bundi	5776	961740	167
10	Chittorgarh	10856	1377269	127
11	Churu	16830	1849833	110
12	Dausa	3432	1002892	292
13	Dholpur	3033	529201	174
14	Dungarpur	3770	1089600	289
15	Ganganagar	10978	1585244	144
16	Hanumangarh	9656	1331265	138
17	Jaipur	11143	2803997	252
18	Jaisalmer	38401	3195213	83
19	Jalore	10640	1631175	153
20	Jhalawar	6219	1024929	165
21	Jhunjhunu	5928	1284657	217
22	Jodhpur	22850	3590264	157
23	Karouli	5524	933723	169
24	Kota	5217	644466	124
25	Nagour	17718	3150011	178
26	Pali	12387	2302204	186
27	Pratapgarh	-	765557	-
28	Rajsamand	3860	1127169	292
29	S.Madhopur	4498	803502	179
30	Sikar	7732	2118046	274
31	Sirohi	5136	901191	175
32	Tonk	7194	1208624	168
33	Udaipur	13419	2780566	207
	Total	**342239**	**57732204**	**169**

Source: Department of Animal Husbandry, Government of Rajasthan

Table 15.3 Percentage Change in Livestock Population in Rajasthan 2012

S.No.	Species	2007	2012	Change	% of Change
1	Cattle	12119512	13324462	1204950	9.94
2	Buffalo	11091974	12976095	1884121	16.99
3	Sheep	11189855	9079702	-2110153	-18.86
4	Goat	21502996	21665939	162943	0.76
5	Horse and Ponies	25438	37776	12338	48.5
6	Mules	886	3375	2489	280.93
7	Donkeys	102130	81468	-20662	-20.23
8	Camel	421836	325713	-96123	-22.79
9	Pigs	208556	237674	29118	13.96
	Total	**56663183**	**57732204**	**1069021**	**1.89**
1	Rabbits	9301	13542	4241	45.6
2	Dogs	1246036	569575	-676461	-54.29
3	Poultry	4993620	8024424	3030804	60.69

Source: Department of Animal Husbandry, Government of Rajasthan

Table 15.4 Growth of Livestock Population in Rajasthna (1951-2012)
(Percentage of Total Livestock Population of the year.)

SPECIES	1951	1956	1961	1966	1972	1977	1983	1988	1992	1997	2003	2007	2012*
Cattle	107.82	120.73	131.36	131.23	124.7	128.96	135.04	109.21	116.66	121.41	108.54	121.2	133.24
% of total livestock pop.	42.26%	37.73%	39.20%	35.01%	32.07%	31.18%	27.20%	26.69%	24.08%	22.37%	22.08%	21.38%	23.08%
Buffalo	30.45	34.3	40.19	42.23	45.92	50.72	60.43	63.44	77.75	97.7	104.14	110.92	129.76
% of total livestock pop.	11.93%	10.57%	11.99%	11.27%	11.81%	12.26%	12.17%	15.50%	15.98%	17.95%	21.28%	19.58%	22.48%
Sheep	53.87	73.73	73.6	88.06	85.56	99.38	134.31	99.32	124.91	145.85	100.54	111.9	90.80
% of total livestock pop.	21.11%	22.73%	21.97%	23.49%	22.01%	24.03%	27.05%	24.24%	25.78%	26.33%	20.41%	19.75%	15.73%
Goats	55.62	87.3	80.52	103.23	121.62	123.07	154.8	125.78	152.85	169.71	168.09	215.03	216.66
% of total livestock pop.	21.80%	26.92%	24.03%	27.54%	31.28%	29.76%	31.18%	30.79%	31.55%	31.16%	34.17%	37.95%	37.53%
Camel	3.41	4.36	5.7	6.54	7.45	7.52	7.56	7.19	7.46	6.69	4.98	4.22	3.26
% of total livestock pop.	1.34%	1.34%	1.70%	1.74%	1.92%	1.82%	1.52%	1.76%	1.53%	1.22%	1.01%	0.75%	0.56%
Pigs	0.55	0.73	0.71	0.83	1.17	1.3	1.79	2.07	2.53	3.03	3.37	2.09	2.38
% of total livestock pop.		0.22%		0.22%					0.52%	0.55%	0.68%	0.37%	0.41%
Others	3.44	3.13	3.01	2.64	2.36	2.64	2.57	2.16	2.29	2.16	1.7	1.27	1.23
% of total livestock pop.	1.56%	0.94%	1.11%	0.70%	0.90%	0.95%	0.88%	1.02%	0.47%	0.39%	0.39%	0.22%	0.21%
Total	255.16	324.28	335.09	374.76	388.78	413.59	496.5	409.17	484.45	546.55	491.36	566.63	577.32
Dogs	0	0	0	0	0	0	14.33	12.54	15.02	16.74	23.71	12.46	5.70
Poultry	2.54	4.57	7.25	8.65	12.35	15.9	22.13	25.85	30.13	44.06	61.92	49.94	80.24
Rabbits	0	0	0	0	0	0	0	0	0	0.19	0.23	0.09	0.14
* Provisional													
Percentage of total livestock population of the year.													

Source: Department of Animal Husbandry, Government of Rajasthan

Table 15.5 Districtwise Density of Total Livestock and Bovine Population in Rajasthan (1999-2012)

Districts	**Livestock** (No. per sq km)				**Bovine** (No. per sq km)			
	1997	**2003**	**2007**	**2012**	**1997**	**2003**	**2007**	**2012**
Rajasthan	**159**	**144**	**166**	**169**	**64**	**62**	**68**	**77**
Ajmer	248	190	239	232	94	69	86	100
Alwar	194	199	240	206	115	131	140	151
Banswara	259	310	309	309	178	203	204	195
Baran	117	112	124	115	85	74	83	85
Barmer	145	116	158	189	22	24	28	35
Bharatpur	182	207	183	251	127	159	119	198
Bhilwara	259	203	194	234	108	88	85	112
Bikaner	84	81	85	92	22	25	27	36
Bundi	169	155	172	167	95	88	96	97
Chittaurgarh	221	238	256	176	139	141	158	109
Churu	154	104	137	134	39	30	34	46
Dausa	233	243	283	292	143	147	160	182
Dhaulpur	149	158	174	174	110	120	123	138
Dungarpur	273	303	309	289	153	168	175	161
Ganganagar	130	120	151	144	67	61	79	83
Hanumangarh	129	120	140	138	66	67	76	92
Jaipur	208	221	255	252	108	117	131	153
Jaisalmer	64	46	74	83	8	6	9	11
Jalor	170	154	179	153	57	57	66	72
Jhalawar	149	167	182	165	106	111	117	111
Jhunjhunun	210	200	237	217	80	85	92	104
Jodhpur	172	116	146	157	38	31	40	50
Karauli	144	140	177	169	89	84	94	105
Kota	126	126	139	124	86	81	91	88
Nagaur	183	149	176	178	52	44	50	60
Pali	223	172	187	186	60	46	54	54
Pratapgarh				172				108
Rajsamand	265	233	249	242	114	97	110	104
Sawai Madhopur	176	166	196	179	93	86	82	95
Sikar	232	241	293	274	87	91	100	114
Sirohi	188	189	188	175	67	65	72	74
Tonk	175	141	169	168	81	62	77	86
Udaipur	216	253	233	237	122	134	125	130

Source: 16th, 17th, 18th and 19th Livestock Census, Department of Animal Husbandry, Dairying and Fisheries, Ministry of Agriculture and Farmers Welfare, Govt. of India.

MILCH ANIMALS

Cows and buffaloes are the common milch animals reared in the state. Livestock raising is an important occupation of fanners to augment farm income. The state has more than 600 to 1594 milch animals per 1,000 hectares in the north and north-eastern tracts covering Ganganagar (639.29 per 1,000 ha.), Hanumangarh (673.40 per 1,000ha), Alwar (1311.95 per 1,000 ha), Bharatpur (1594.54 per 1,000 ha.), Jaipur (1168.17 per 1,000), Dausa (1466.53 per 1,000 ha.) and Dhaulpur (1200.17 per 1,000 ha.) districts. The southern districts of Udaipur (1168.99 per 1,000 ha.), Rajsamand (1171.11 per 1,000 ha.), Dungarpur (1680.02 per 1,000 ha.), Banswara (1823.73 per 1,000 ha.) and Jhalawar (1109.84 per 1,000 ha.). All these districts are hilly and have comparatively thick forests. The western part of the state, comprising of Jodhpur division Jodhpur (306.37 per 1,000 ha.), Sirohi (647.08 per ha.), Jalor (567.14 per 1,000 ha.), Pali (463.53 per 1,000 ha.), Barmer (235.36 per 1,000 ha.) and Jaisalmer district (63.92 per 1,000 ha.) has a comparatively lower population of milch animals. The area is a frequently drought prone and has less fodder to feed the milch animals. However, The Rathi cows of Bikaner, Tharparkar cows of Jodhpur and Kankrej (Sanchori) cows of Barmer provides a good breed of milch animals.

Buffaloes are reared mostly in eastern Rajasthn, specially in Bharatpur, Alwar, Jhunjhunun, Jaipur and Sikar districts. Buffaloes require more feed per day and the yield of milk is also higher than that of cows. Availability of green fodder and forage is higher in the eastern parts of Rajasthan and so is the concentration of these animals (Figure 15.7).

The population of cattle and Buffalo species in the state is about 108.53 lakhs of cattle and 104.14 lakhs buffaloes respectively.

The state has the following important breeds of cattle:

Rathi	Tharparkar
Kankrej (Sanchori)	
Malvi	Nagauri
Haryana	Gir

Following breeds of buffaloes are found in the state:

Murrah	Mehsana
Suriti	

NON-MILCH ANIMALS

Rajasthan is rich in animal wealth. According to Livestock Census 2003, the specieswise Non-Milch Animals population excluding Camel, Sheep and Goat is as follow:

Horse and Ponnies	> 0.25 Lakhs
Pigs	> 3.37 Lakhs
Donkeys	> 1.42 Lakhs
Dogs	>23.71 Lakhs

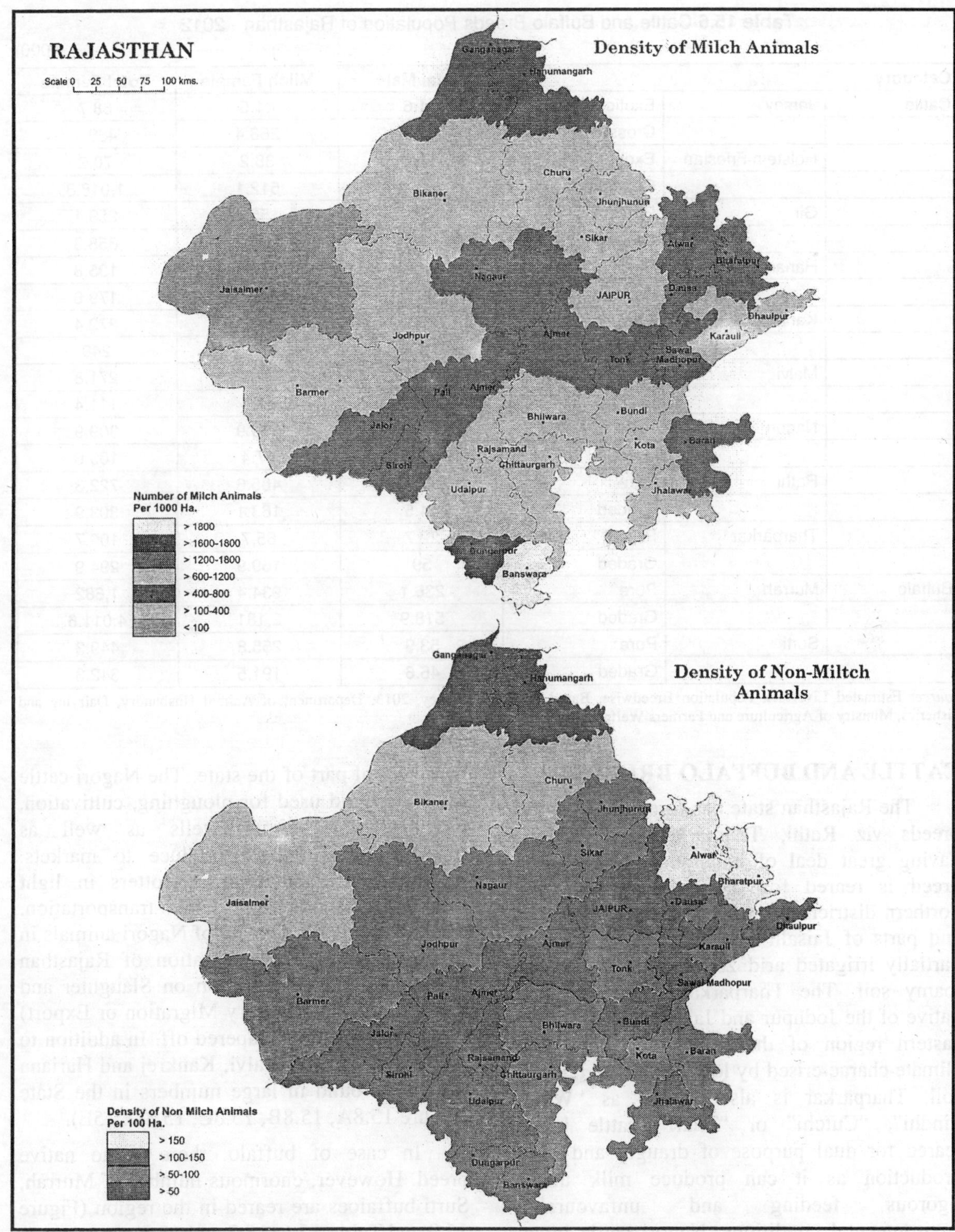

Source: Department of Animal Husbandry, GoR, Jaipur

Figure 15.7 Density of Milch and Non-Milch Animals in Rajasthan

Table 15.6 Cattle and Buffalo Breeds Population of Rajasthan - 2012

('000)

Category			Total Male	Milch Female	Total Female
Cattle	Jersey	Exotic	11.6	41.5	68.7
		Crossbred	85.4	268.4	469.1
	Holstein Friesian	Exotic	11.2	38.2	70.2
		Crossbred	130.3	512.1	1,018.8
	Gir	Pure	70	155.4	259.1
		Graded	85.9	205.2	358.3
	Hariana	Pure	34.9	117.5	195.8
		Graded	33.1	102.3	179.6
	Kankrej	Pure	79.3	218.1	372.4
		Graded	75.9	135.6	240
	Malvi	Pure	203.8	164.2	271.8
		Graded	83.2	67.4	111.4
	Nagori	Pure	59.7	190.9	309.9
		Graded	23.9	67.4	109.6
	Rathi	Pure	137.5	405.6	722.3
		Graded	54.5	163.1	303.9
	Tharparkar	Pure	22.7	65.7	109.7
		Graded	59	169.9	294.9
Buffalo	Murrah	Pure	236.1	934.4	1,682
		Graded	518.9	2,181	4,011.6
	Surti	Pure	53.9	255.8	449.3
		Graded	46.6	191.5	342.3

Source: Estimated Livestock Population Breedwise, Based on Breed Survey, 2013, Department of Animal Husbandry, Dairying and Fisheries, Ministry of Agriculture and Farmers Welfare, Govt. of India.

CATTLE AND BUFFALO BREEDS

The Rajasthan state has three native cattle breeds viz Rathi, Tharparker and Nagori, having great deal of endurance. Rathi cattle breed is reared for dairy purposes in the northern districts of Shri Ganganagar, Bikaner and parts of Jaisalmer which are irrigated or partially irrigated arid zones with alluvial or loamy soil. The Tharparkar cattle breed is native of the Jodhpur and Jaisalmer districts in eastern region of the state which has arid climate characterised by low rainfall and desert soil. Tharparkar is also known as "White Sindhi", "Cutchi" or "Thari" cattle breed reared for dual purpose of draught and milk production as it can produce milk under rigorous feeding and unfavourable environmental conditions. Nagori cattle breed has been named after the Nagaur district which is in central part of the state. The Nagori cattle are sturdy and used for ploughing, cultivation, drawing water from wells as well as transportation of field produce to markets. Earlier they were used as trotters in light ironwheeled carts for quick transportation. There was a good demand of Nagori animals in Bihar but after implementation of Rajasthan Bovine Animal (Prohibition on Slaughter and Regulation of Temporary Migration or Export) Act, the demand has tapered off. In addition to native breeds, Gir, Malvi, Kankrej and Hariana cattle are found in large numbers in the State (Figure 15.8A, 15.8B, 15.8C, 15.8D, 15E).

In case of buffalo, there is no native breed. However, enormous number of Murrah, Surti buffaloes are reared in the region (Figure 15.9 and Table 15.6, 15.7 and 15.8).

Table 15.7 Distribution of Cattle Breeds Population in Rajasthan

Breed	Breeding Tract	Utility	Distribution
Rathi	Bikaner, Ganganagar and Jaisalmer districts of Rajasthan	Milch	Mainly distributed in Bikaner, Ganganagar and Hanumangarh districts
Tharparkar	Jodhpur, Barmer, Jaisalmer districts of Rajasthan and Kutch district of Gujarat	Milk and Draught	Distributed in Jaisalmer, Jodhpur and Barmer districts
Nagori	Nagaur, Bikaner and Jodhpur districts of Rajasthan	Draught	Mainly distributed across Nagaur, Jodhpur, Bikaner districts

Source: Animal Genetic Resources of India (AGRI-IS), National Bureau of Animal Genetic Resources, ICAR and 18th Livestock Census, Department of Animal Husbandry, Dairying and Fisheries, Ministry of Agriculture and Farmers Welfare, Govt. of India

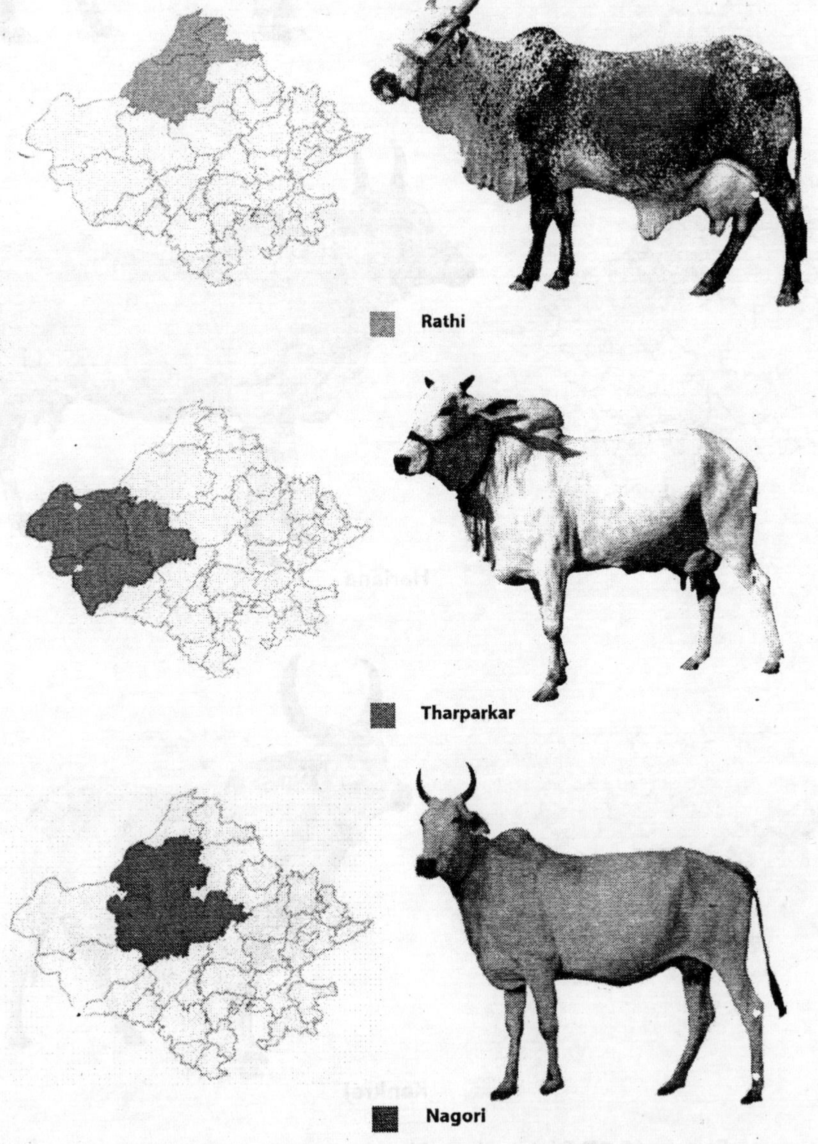

Figure 15.8A Distribution of Indigenous Cattle Breeds

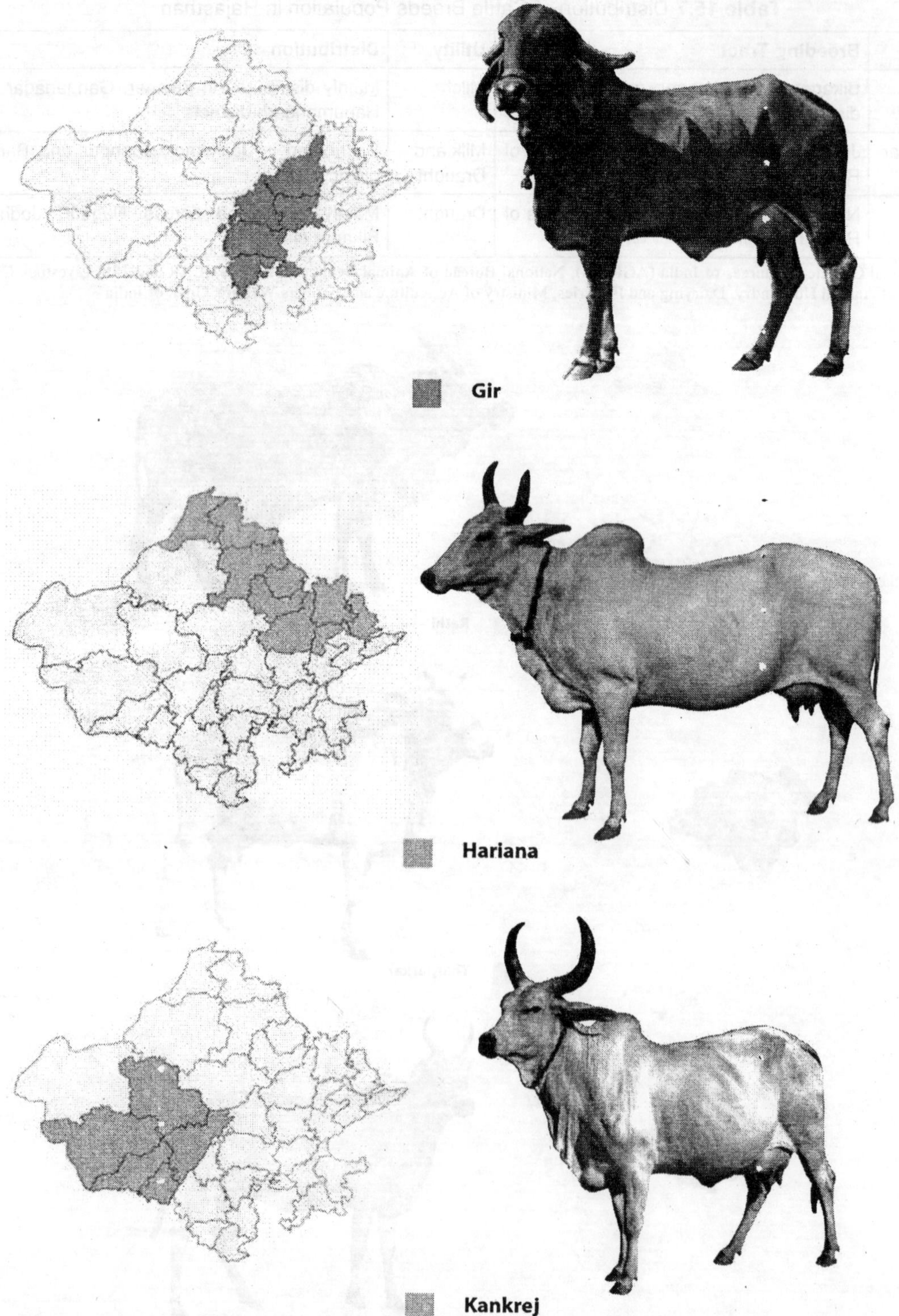

Figure 15.8B Distribution of Indigenous Cattle Breeds

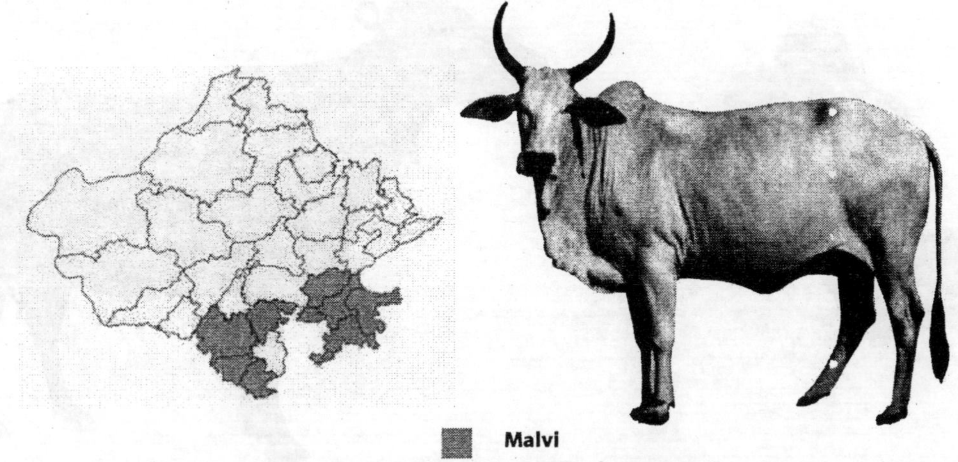

Figure 15.8C Distribution of Indigenous Cattle Breeds

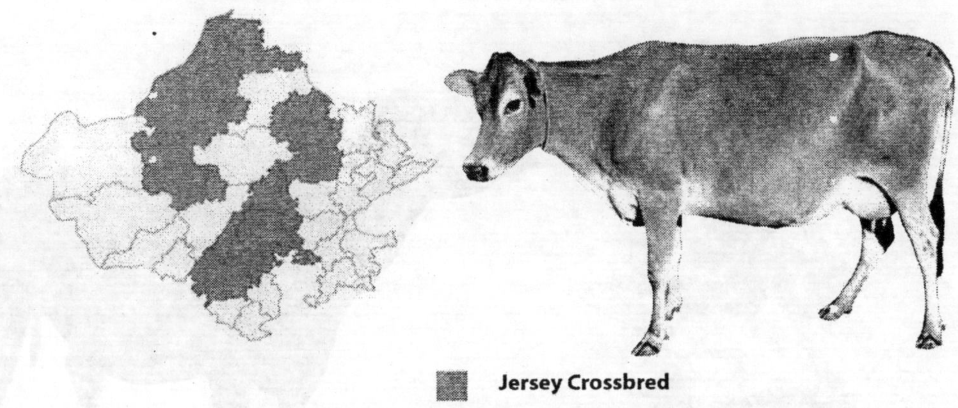

Figure 15.8D Distribution of Jersey Crossbred

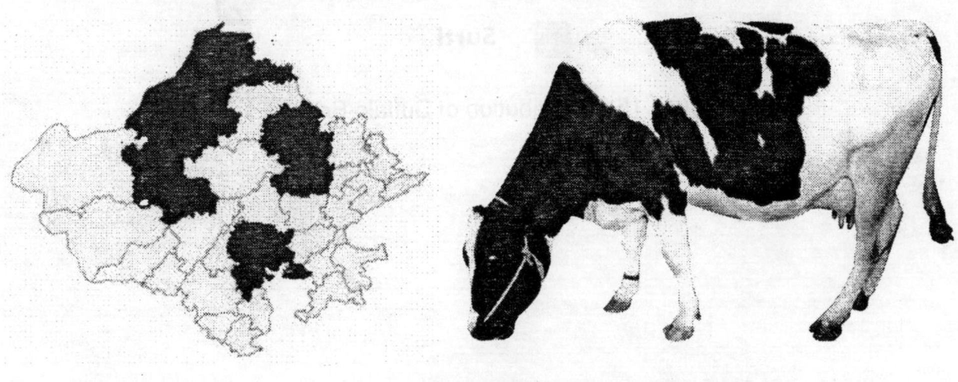

Figure 15.8E Distribution of Holstein Friesian Crossbred

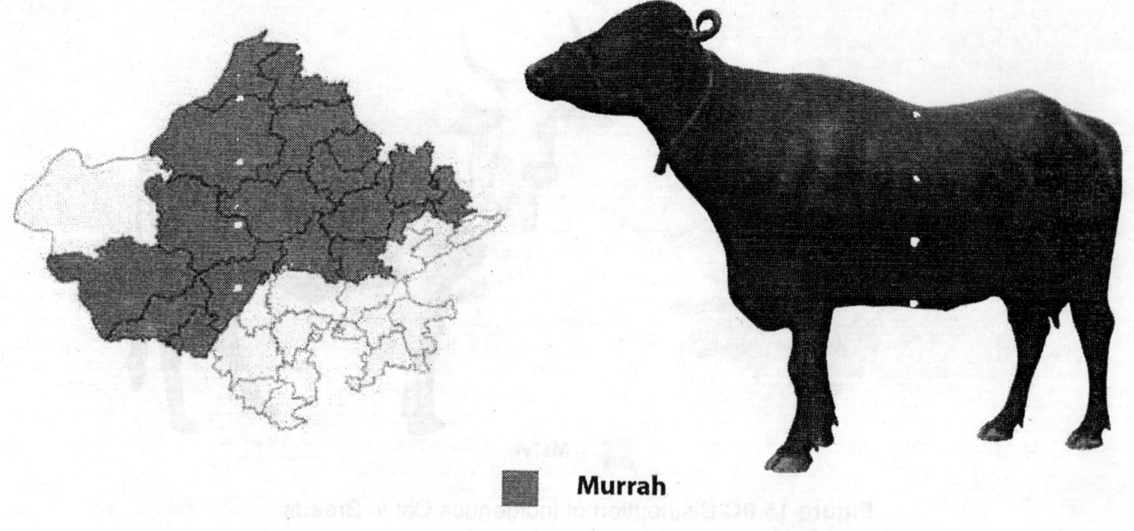

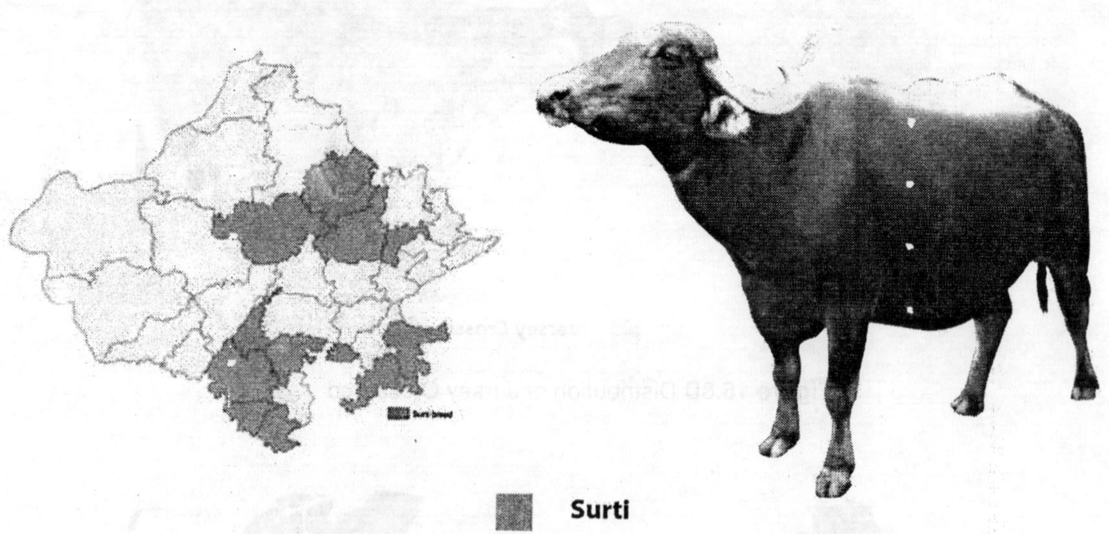

Figure 15.9 Distribution of Buffalo Breeds

PERFORMANCE OF MAJOR BREEDS

Table 15.8 Performance of Native Cattle Breeds

Parameters	Cattle		
	Rathi	Tharparkar	Nagori
Breed Population as on 2012	1,218,294	486,339	503,193
Colour	The animals are usually brown with white patches all over the body, but animals having completely brown, or black coat with white patches are often encountered.	Animals are white or light grey. Face and extremities are of a darker shade than the body.	Generally white or light grey. In some cases head, face and shoulder are slightly greyish.
Average Adult Body Weight (Kg)	Female : 295	Male : 475 Female : 295	Male : 363 Female : 318
Lactation Yield (Kg)	1,560 (1,062 - 2,810)	1,749 (913 - 2,147)	603 (479 - 905)
Lactation Length (days)	336	285	267 (237 - 300)
Calving Interval (days)	512 (420 - 600)	425 (403-565)	455 (420 - 540)
Age at First Calving (days)	1,392 (1,080 - 1,560)	1,231 (1,101 - 1,575)	1,421 (1,260 - 1,500)
Average Milk Fat (%)	3.7 - 4.0	4.9 (4.7 - 4.9)	5.8

Source: Animal Genetic Resources of India (AGRI-IS), National Bureau of Animal Genetic Resources, ICAR and estimated Livestock Population Breed Wise, Based on Breed Survey 2013, Department of Animal Husbandry, Dairying and Fisheries, Ministry of Agriculture and Farmers Welfare, Govt. of India. AE Nivsarkar *et al*, (2000), Animal Genetic Resources of India, Cattle and Buffalo, ICAR publication.

MILK PRODUCTION

In Rajasthan, livestock sector play major role in improving socio-economic status and fulfilling nutritional needs of rural masses. The state is second highest in milk production in the country. Of the total milk produced, 53 per cent is bufalo milk, 36 per cent is cattle milk and 11 per cent is goat milk. Per capita availability of milk is highest in Jaisalmer district having 1,085 grams of milk per day per person. Approximately, 50 per cent of the milk produced is sold in the market and only 25 per cent is consumed at home as liquid milk. Remaining 25 per cent of the milk produced is converted into milk products. Highest livestock density is in Rajsamand and Dausa district having 292 livestock per square kilometer.

Animal Husbandry is a major economic activity of the rural people of Rajasthan, especially in the arid and semi-arid regions. The State faces frequent droughts and famine which causes frequent crop failure as most of the agriculture is rain fed. In this climatic scenario, dairy sector provides sustainable year-round income to large number of farmers. Over a decade, there has been an incremental increase of 3 per cent in number of in-milk bovine with maximum increase in Jaisalmer and Pali districts (Figure 15.10 to 15.25).

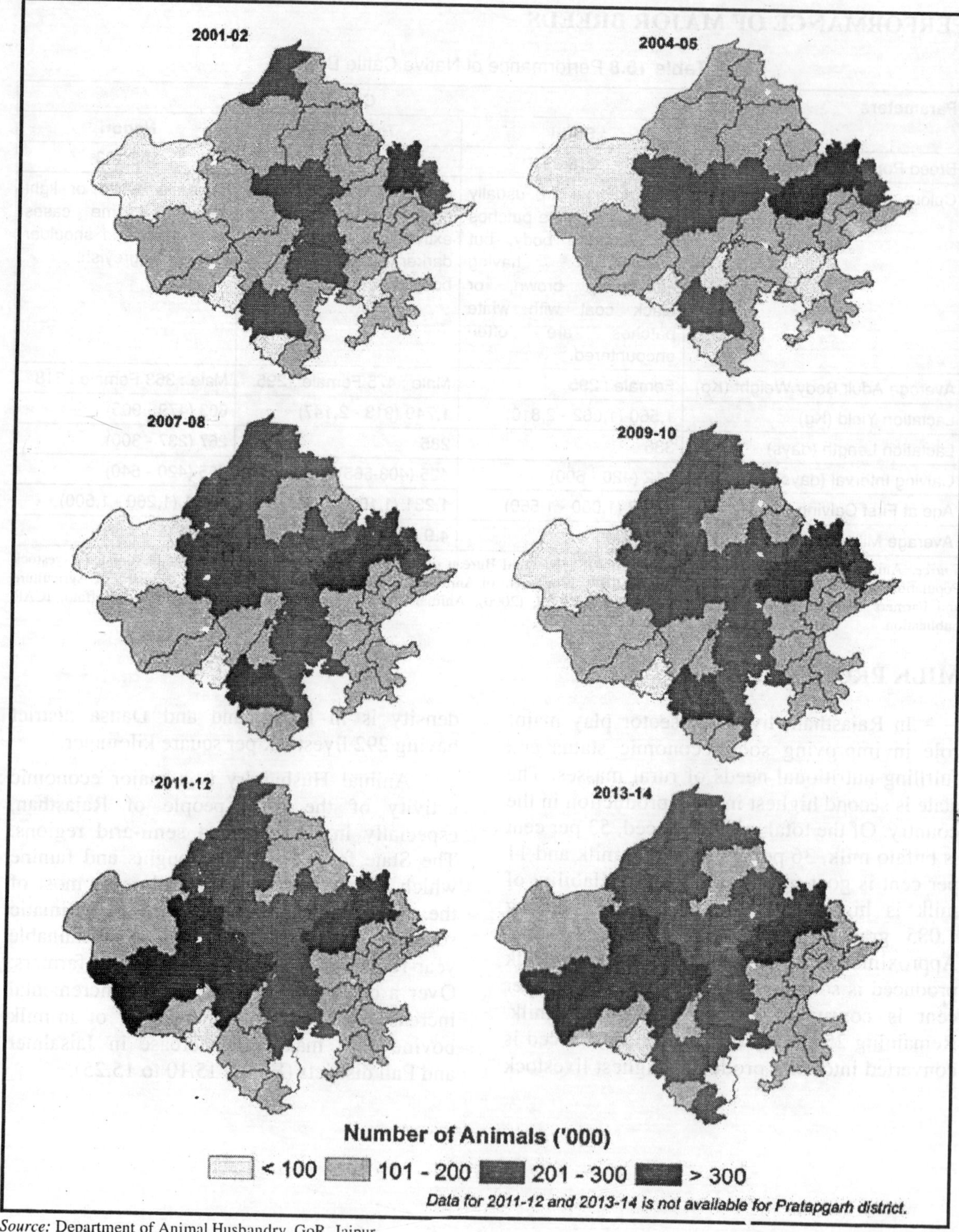

Source: Department of Animal Husbandry, GoR, Jaipur

Figure 15.10 Districtwise changing scinerio in-milk bovine Population in Rajasthan (2001-02 to 2013-14)

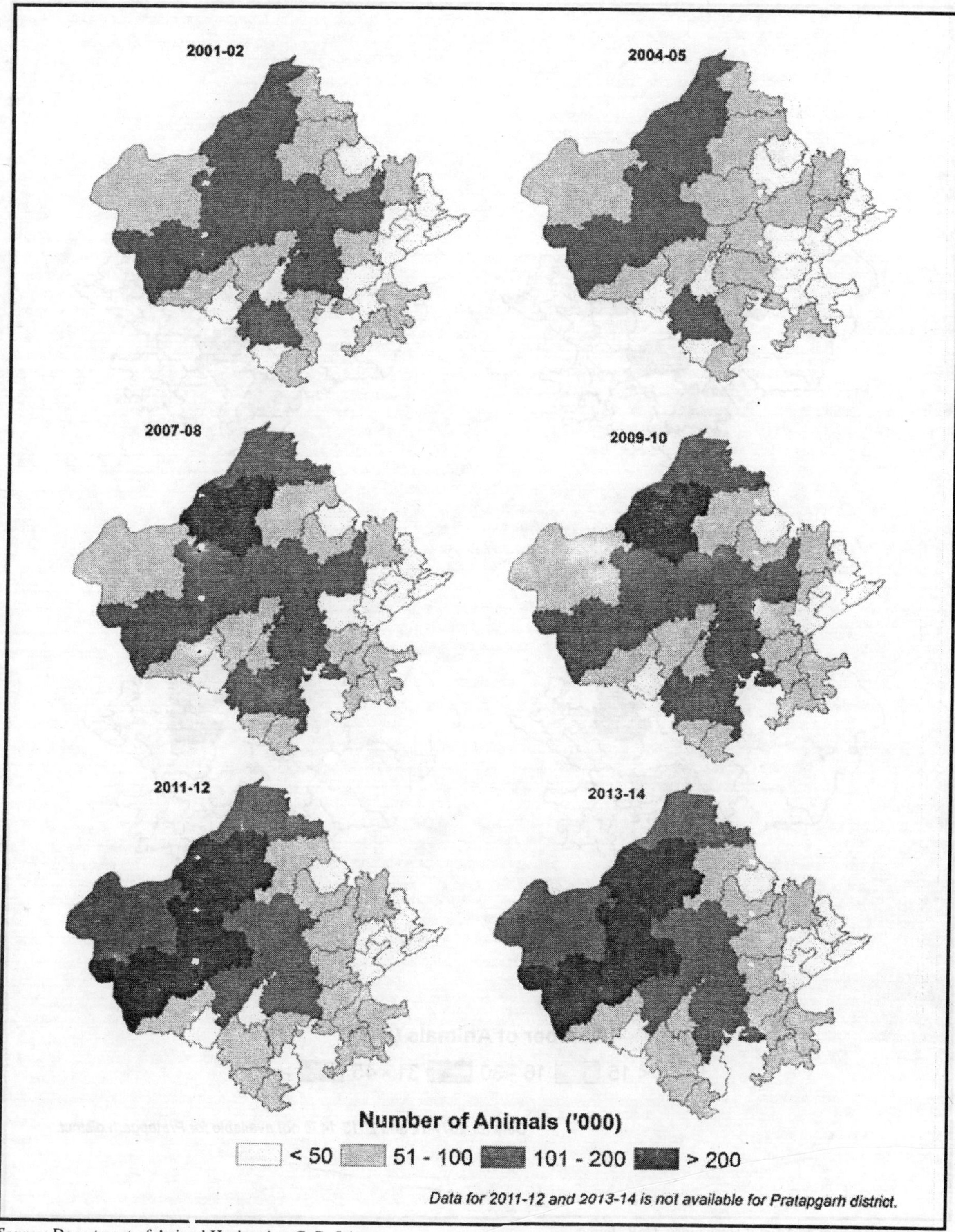

Source: Department of Animal Husbandry, GoR, Jaipur

Figure 15.11 Districtwise In-milk Indigenous Cattle Population in Rajasthan (2001-02 to 2013-14)

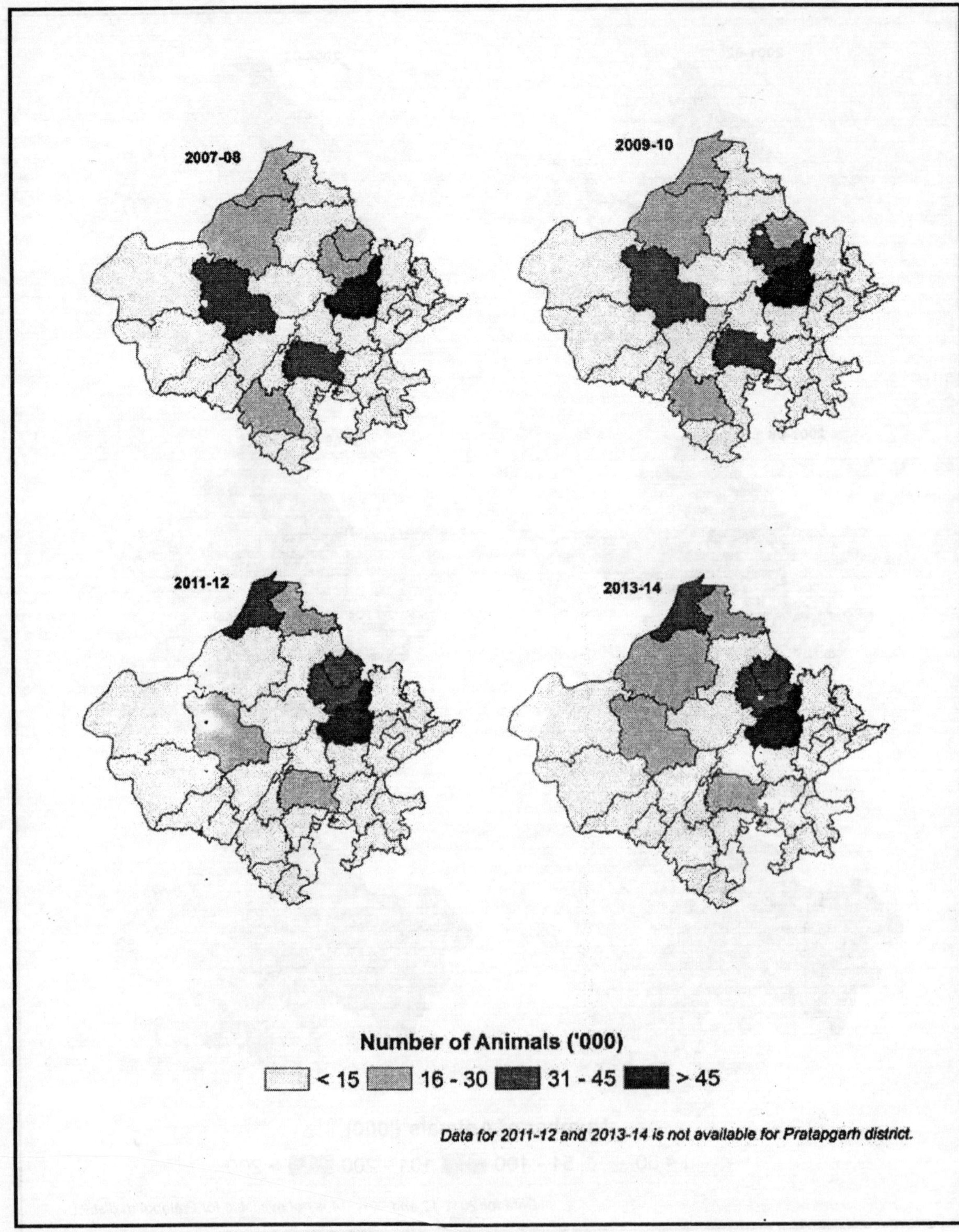

Source: Department of Animal Husbandry, GoR, Jaipur

Figure 15.12 Districtwise In-milk Crossbred Cattle Population (2007-08 to 2013-14)

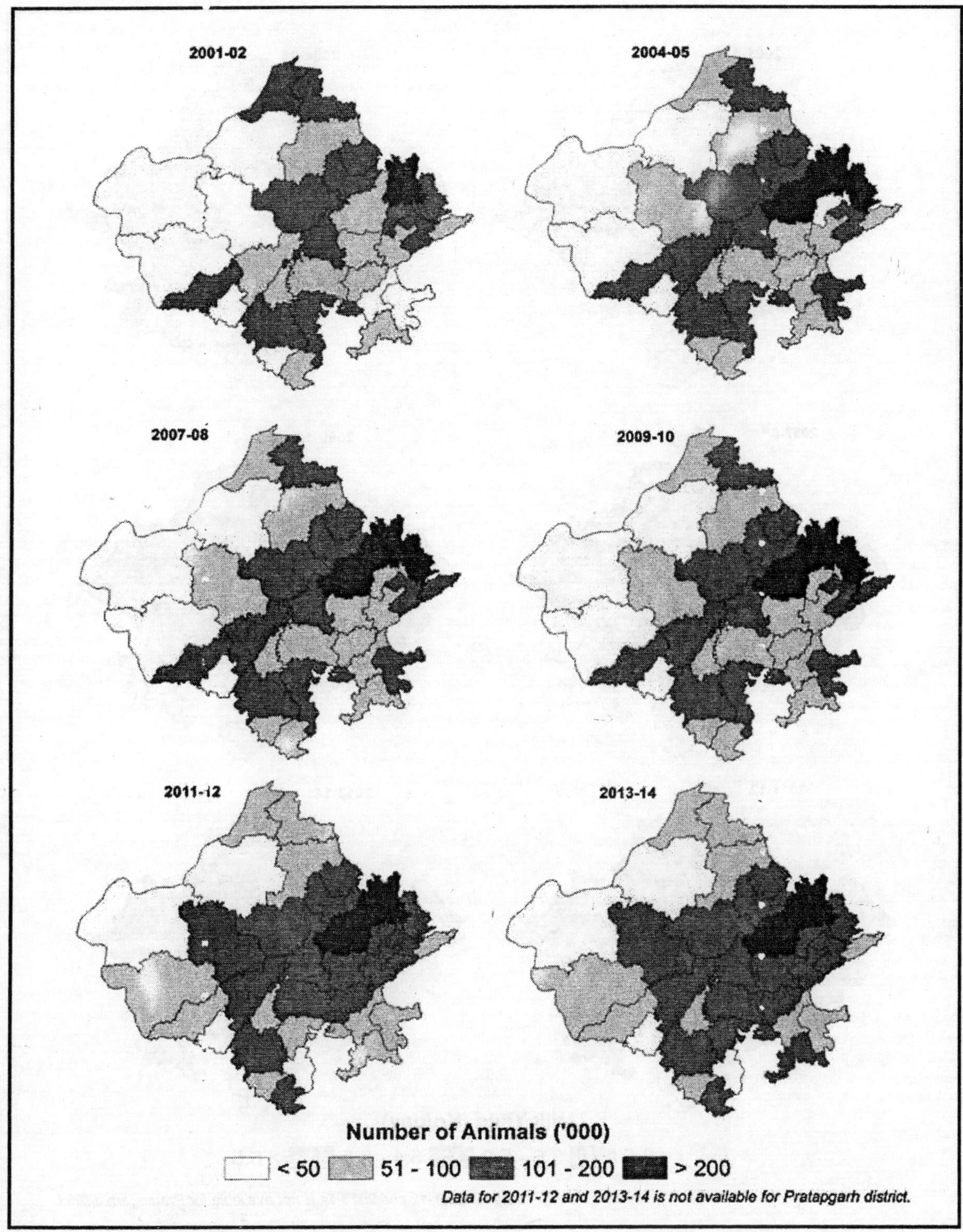

Source: Department of Animal Husbandry, GoR, Jaipur

Figure 15.13 Districtwise In-milk Buffalo Population in Rajasthan (2001-02 to 2013-14)

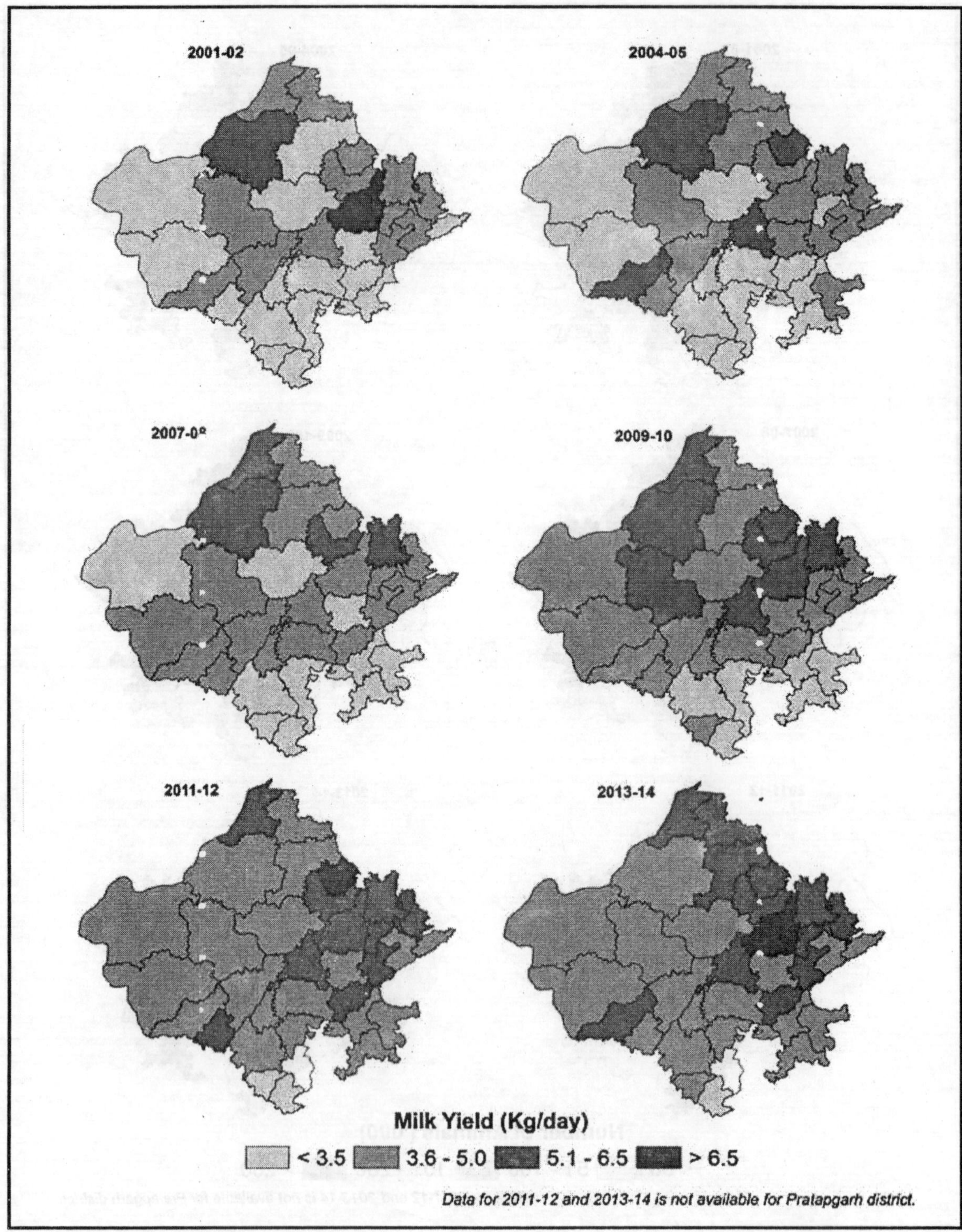

Source: Department of Animal Husbandry, GoR, Jaipur

Figure 15.14 Districtwise In-milk Bovine Milk Yield in Rajasthan (2001-02 to 2013-14)

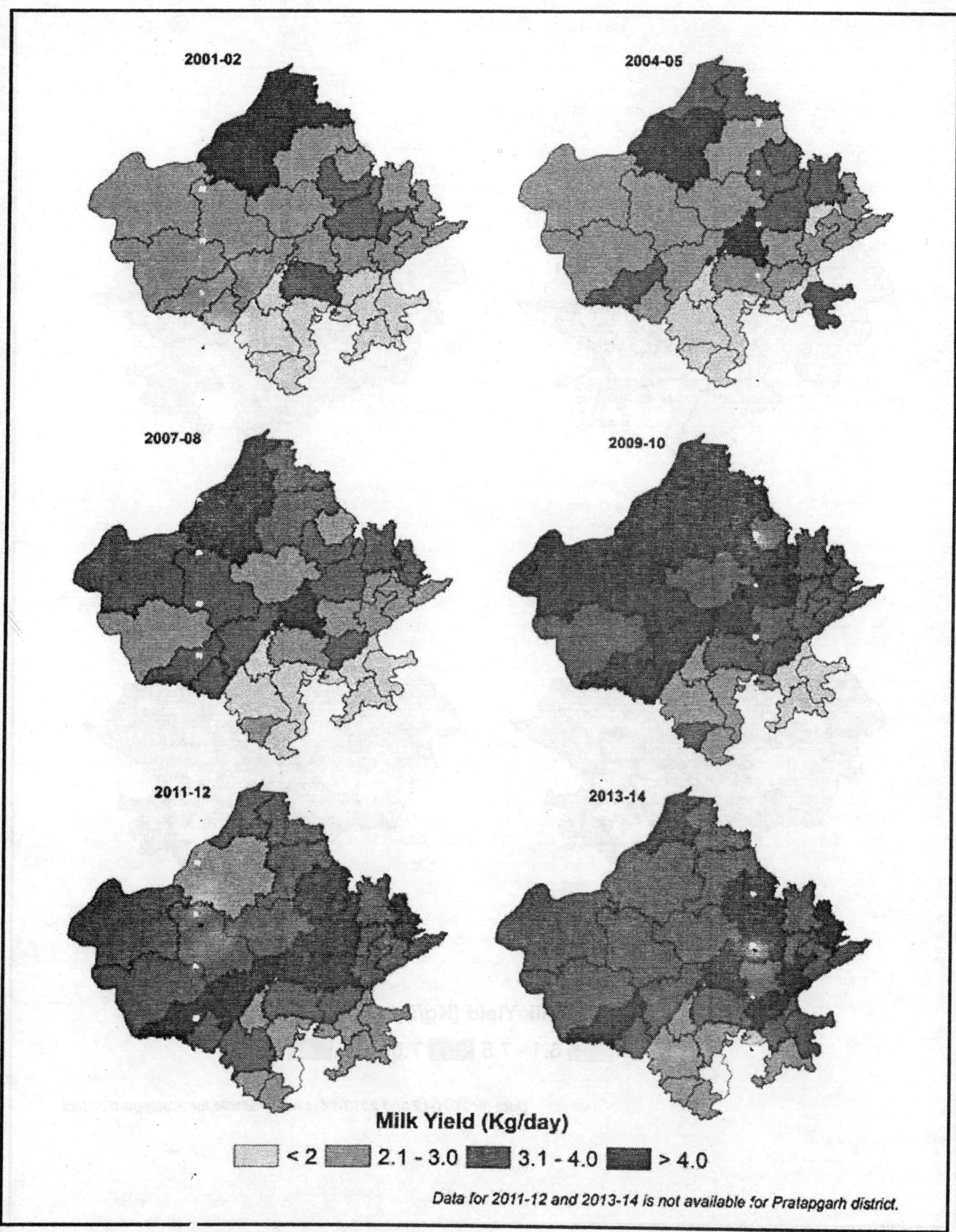

Source: Department of Animal Husbandry, GoR, Jaipur

Figure 15.15 Districtwise Indigenous Cattle Milk Yield in Rajasthan (2001-02 to 2013-14)

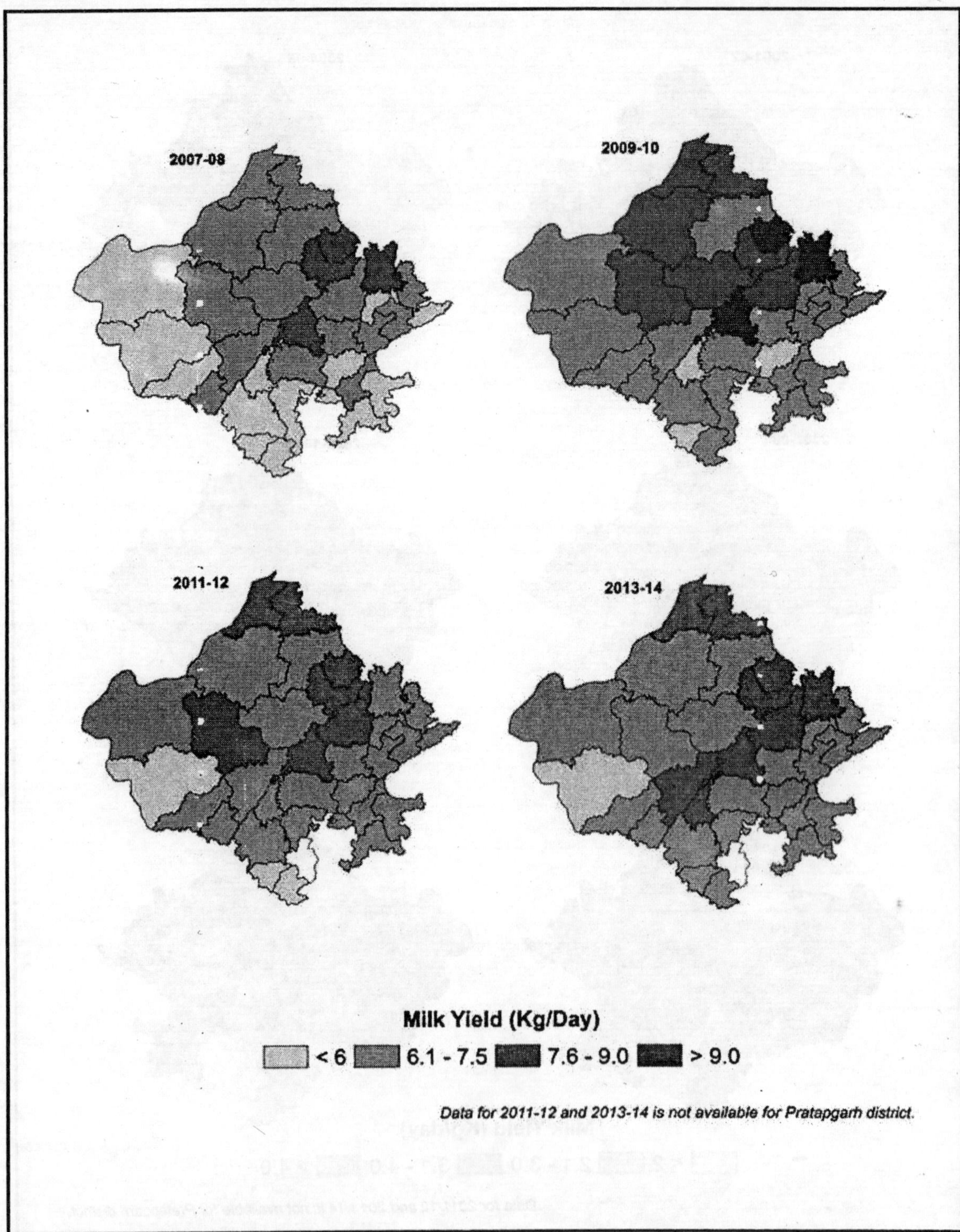

Source: Department of Animal Husbandry, GoR, Jaipur

Figure 15.16 Districtwise Crossbred Cattle Milk Yield in Rajasthan (2007-08 to 2013-14)

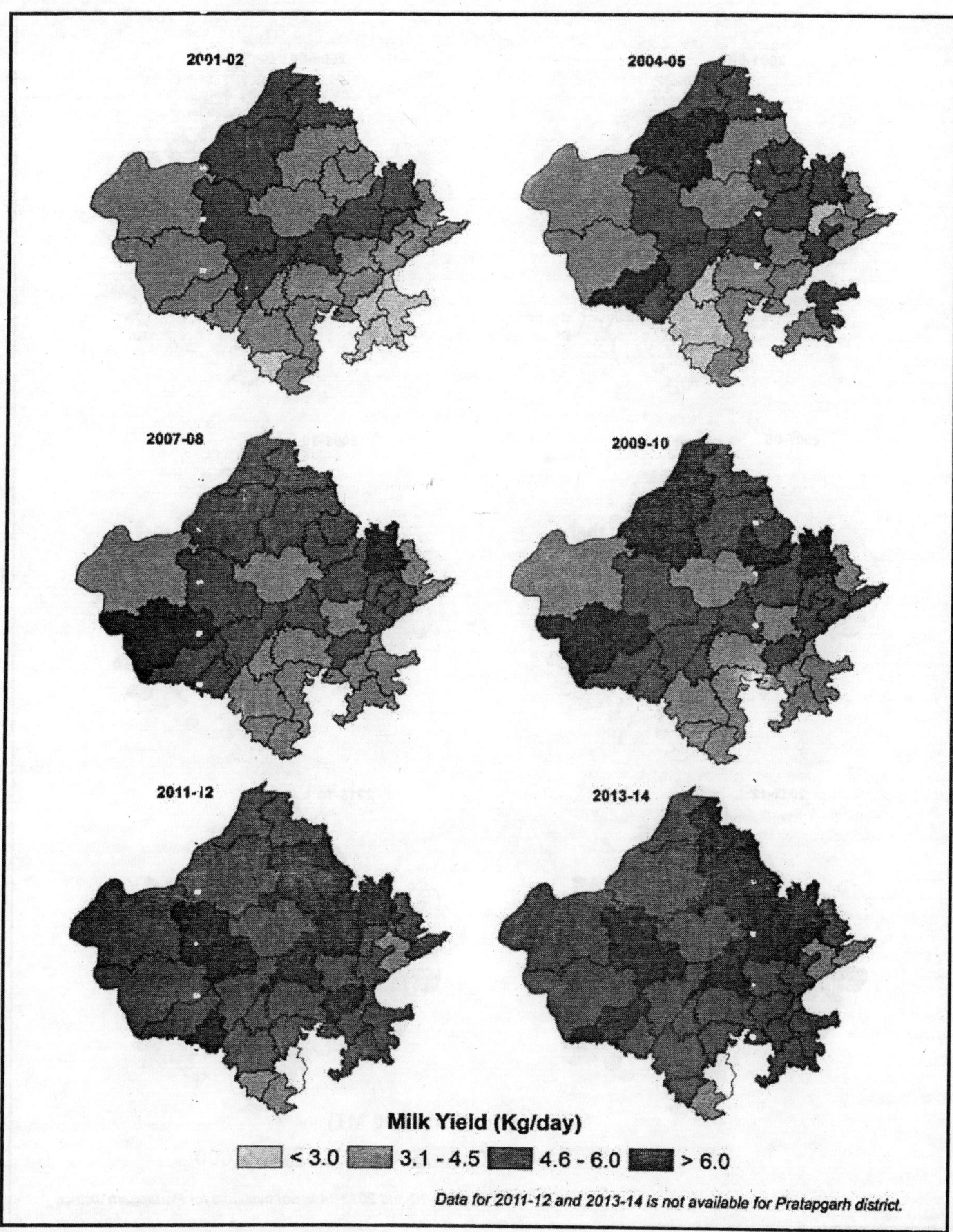

Source: Department of Animal Husbandry, GoR, Jaipur

Figure 15.17 Districtwise Buffalo Milk Yield in Rajasthan (2001-02 to 2013-14)

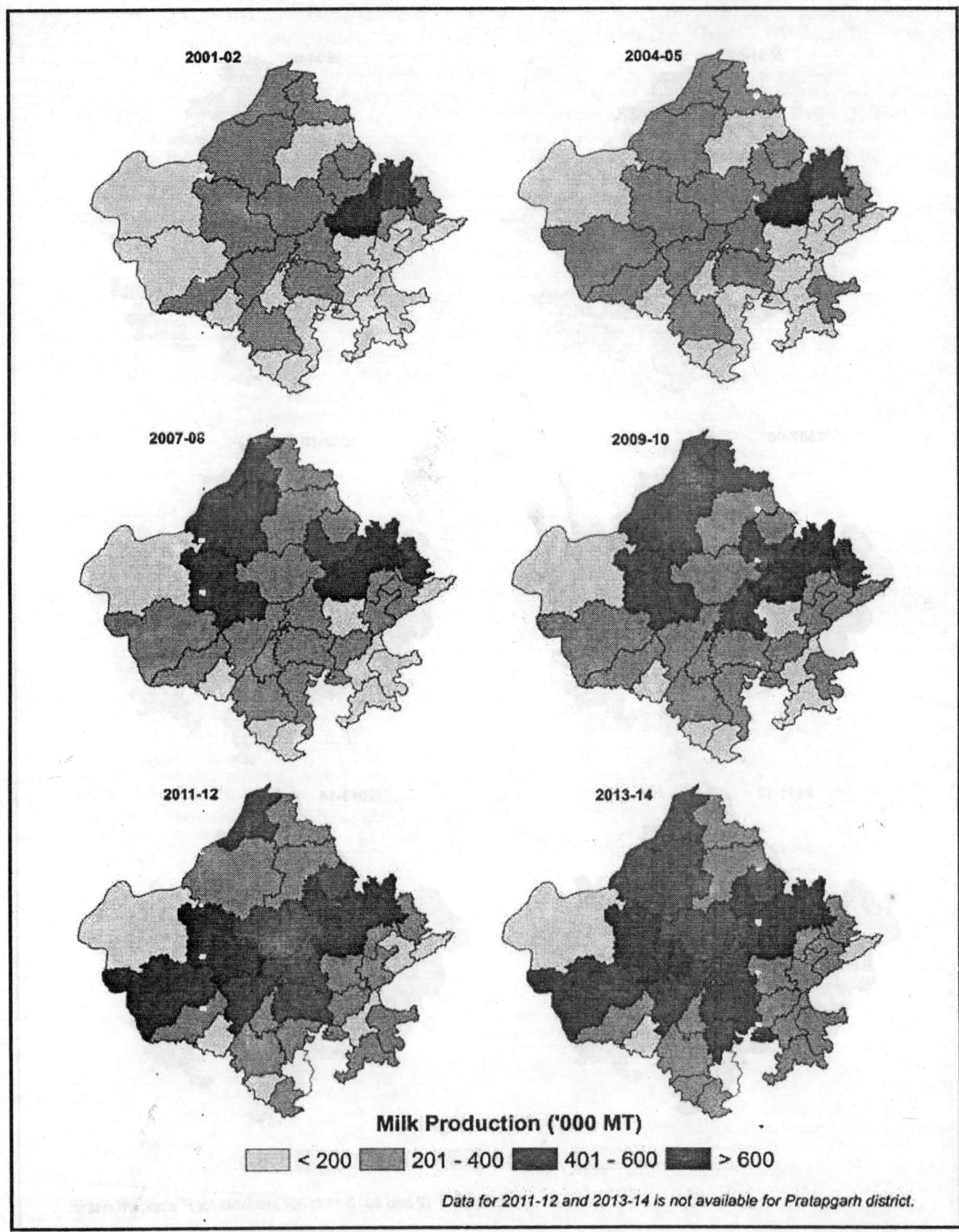

Source: Department of Animal Husbandry, GoR, Jaipur

Figure 15.18 Districtwise Bovine Milk Production in Rajasthan (2001-02 to 2013-14)

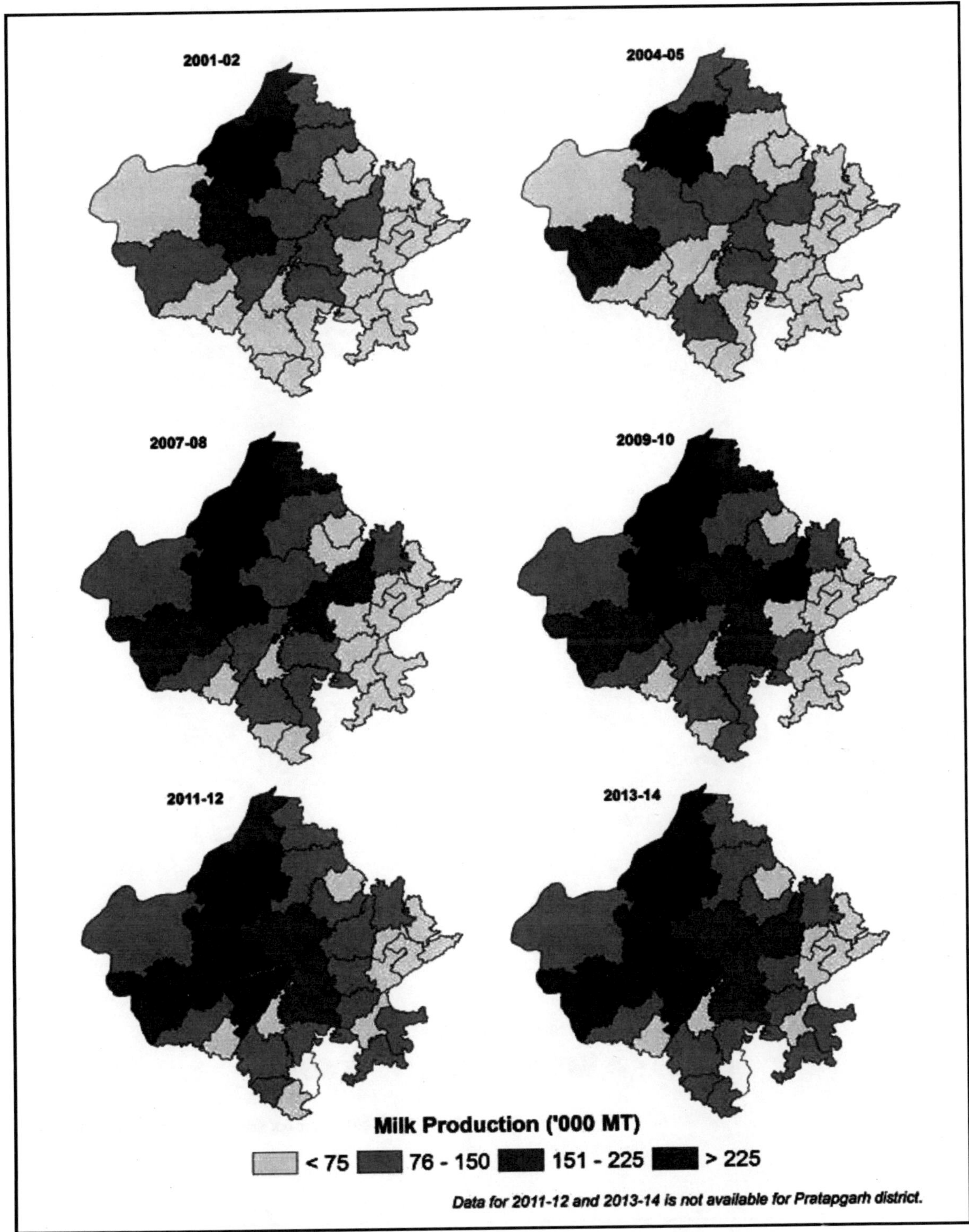

Source: Department of Animal Husbandry, GoR, Jaipur

Figure 15.19 Districtwise Indigenous Cattle Milk Production in Rajasthan (2001-02 to 2013-14)

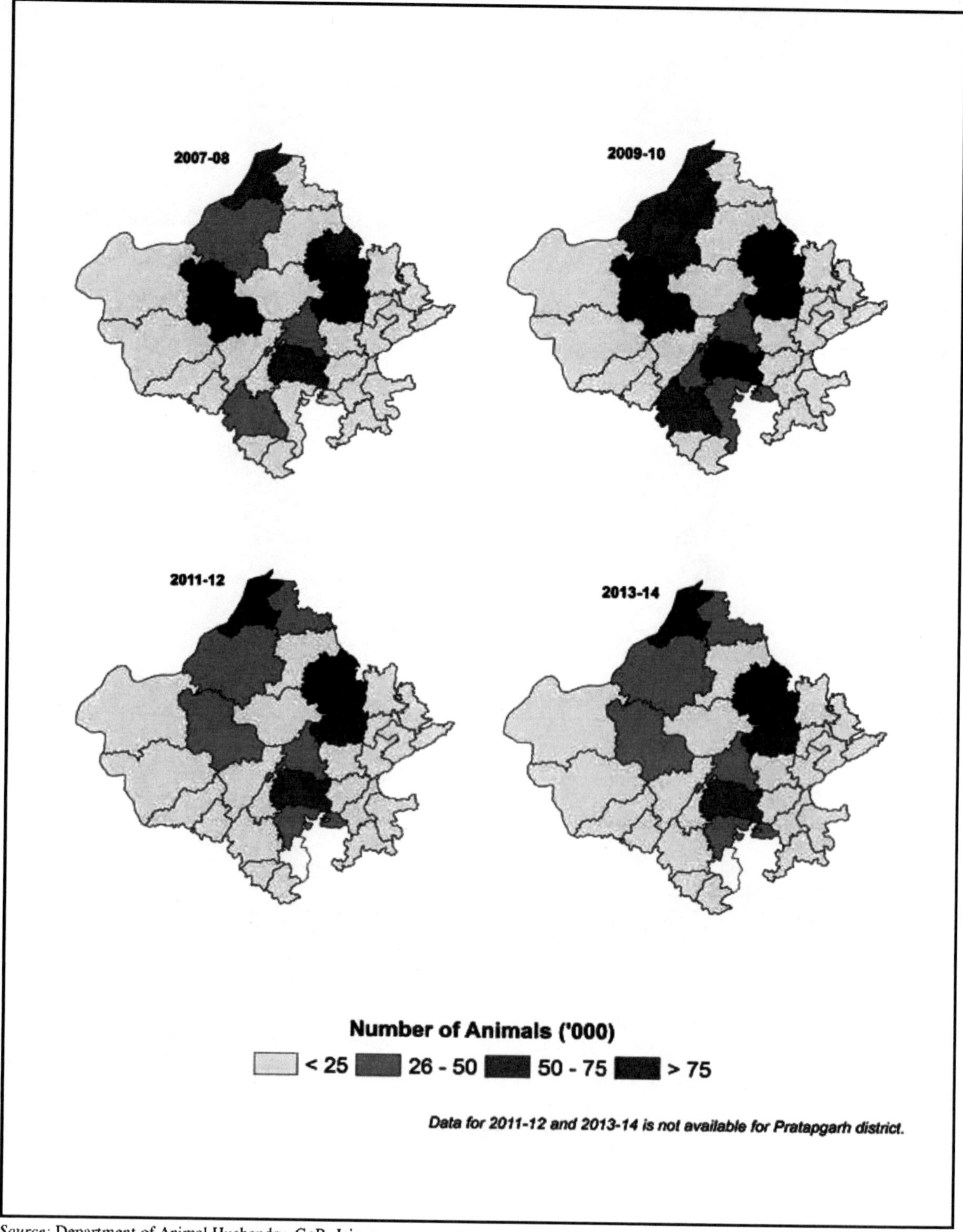

Source: Department of Animal Husbandry, GoR, Jaipur

Figure 15.20 Districtwise Crossbred Cattle Milk Production in Rajasthan (2001-02 to 2013-14)

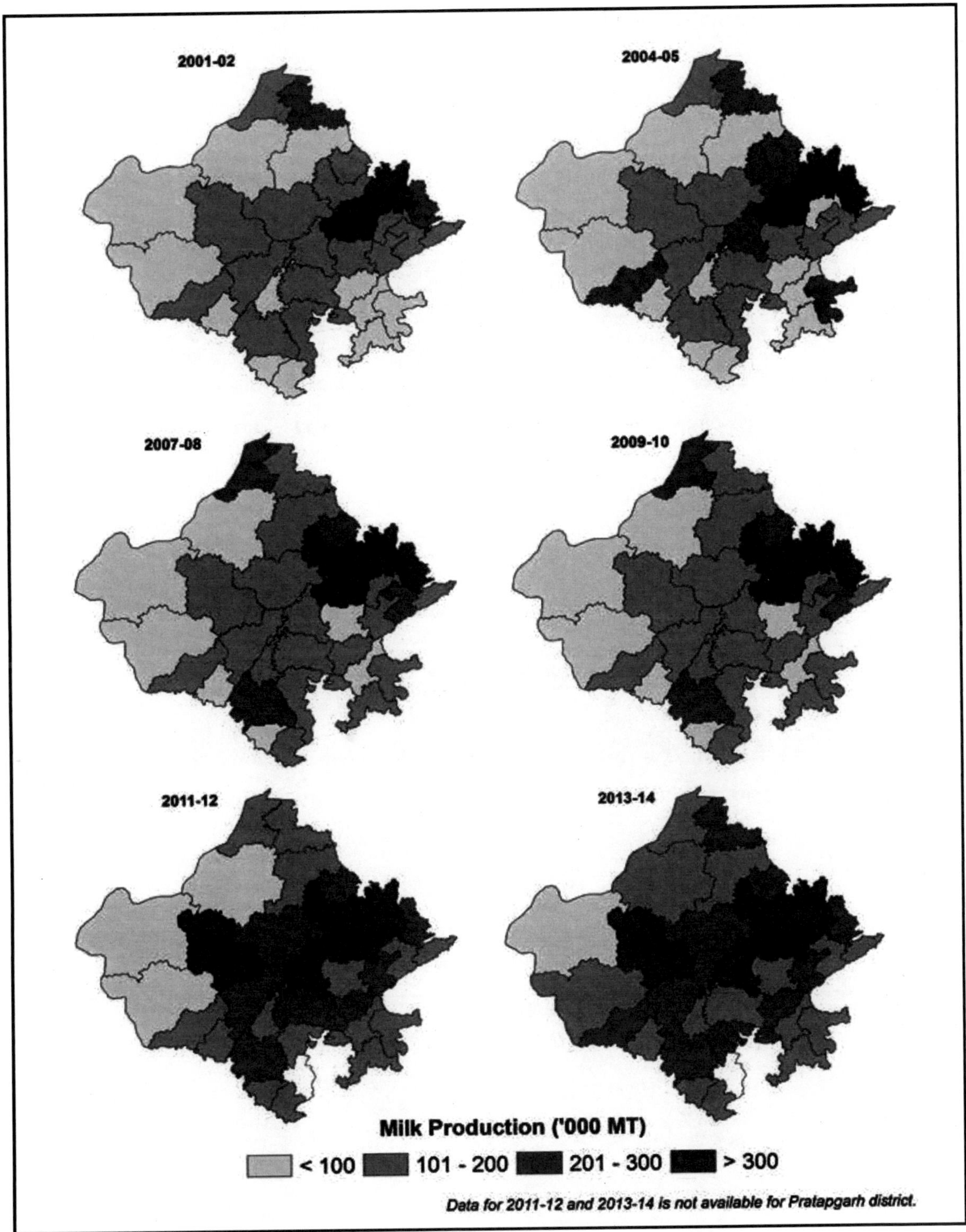

Source: Department of Animal Husbandry, GoR, Jaipur

Figure 15.21 Districtwise Buffalo Milk Production in Rajasthan (2001-02 to 2013-14)

Rajasthan's Milk Production (MMT) (2007-08 to 2013-14)

(Change due to No. of In-milk Animals and Average Yield)

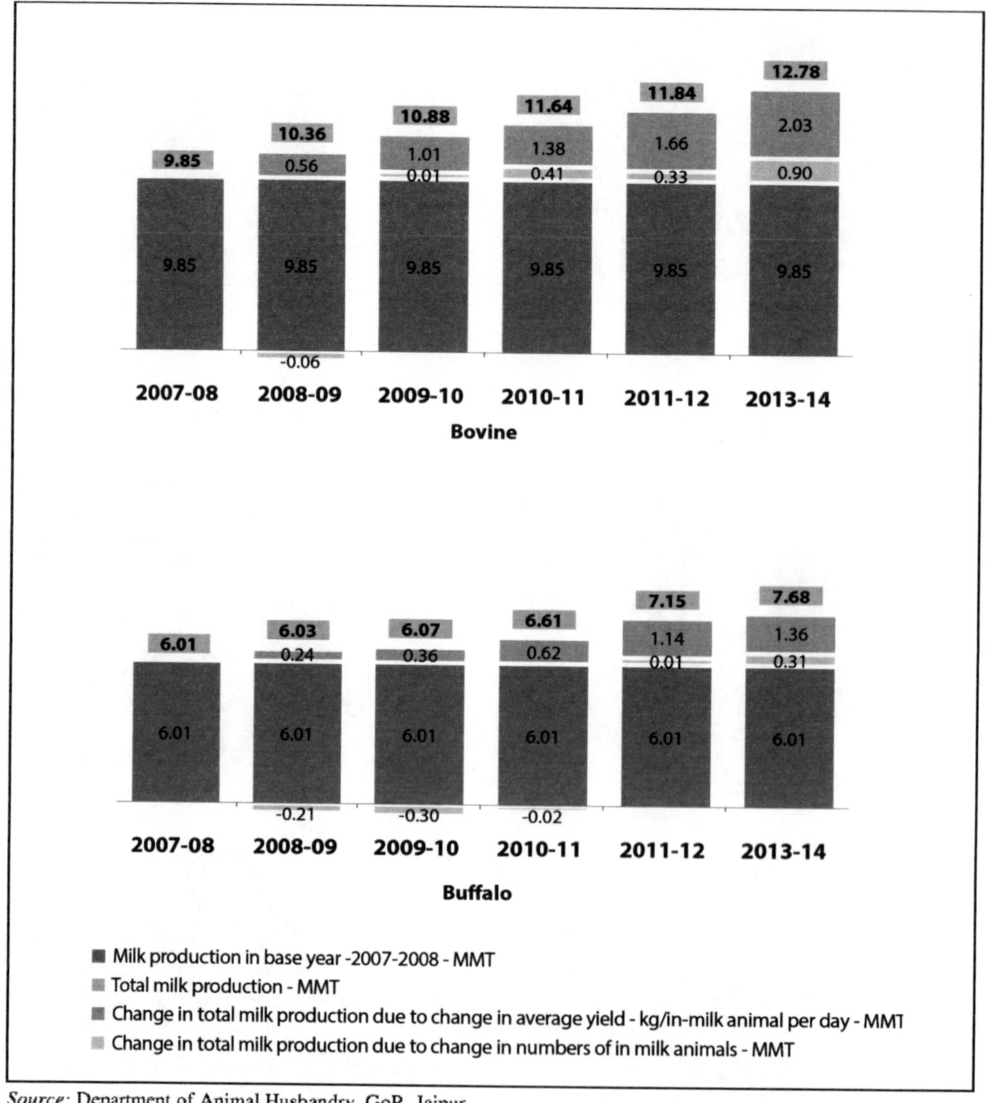

Source: Department of Animal Husbandry, GoR, Jaipur

Figure 15.22 Rajasthan Milk Production (MMT) (2007-08 to 2013-14)

Rajasthan's Milk Production (MMT) (2007-08 to 2013-14)

(Change due to No. of In-milk Animals and Average Yield)

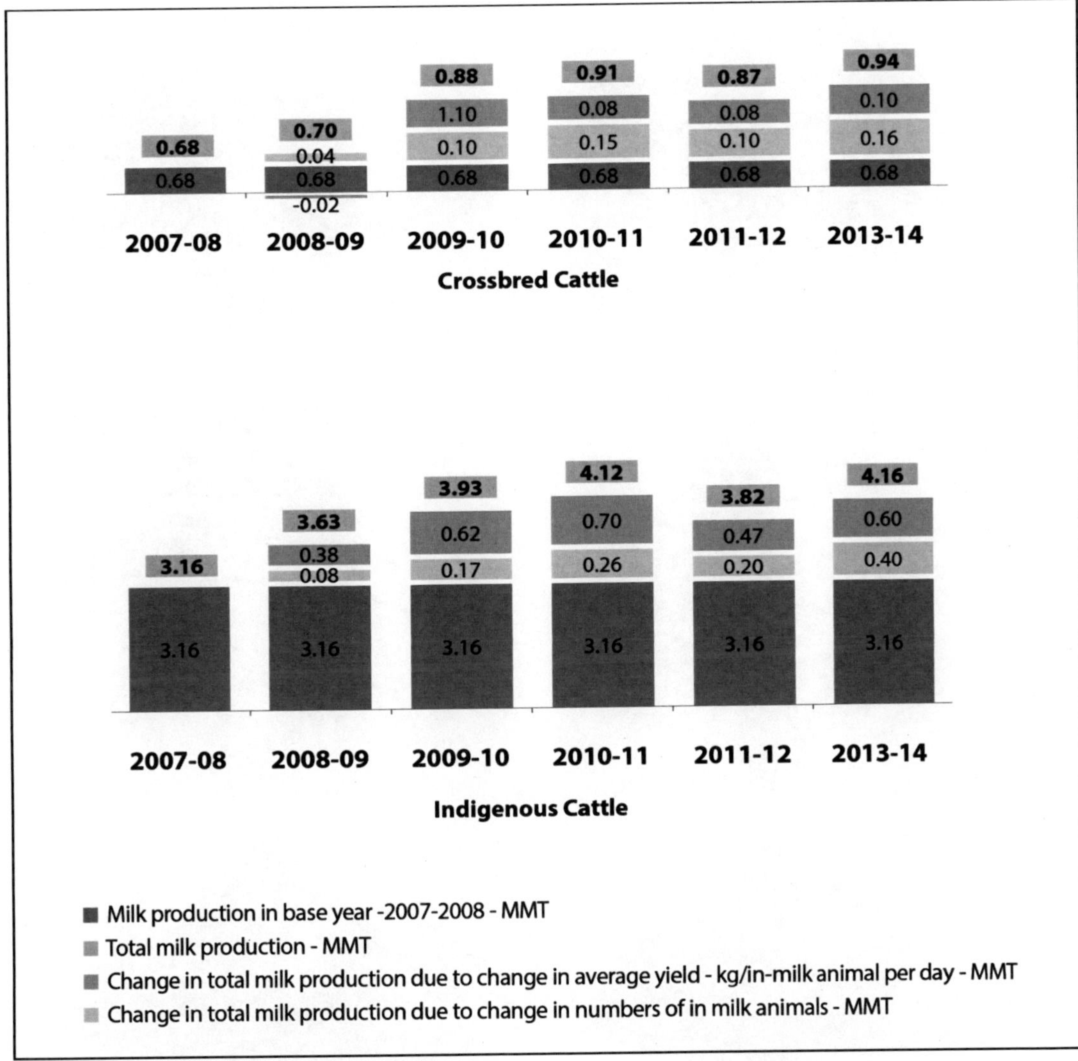

Source: Department of Animal Husbandry, GoR, Jaipur

Figure 15.23 Rajasthan's Milk Production (MMT) (2007-08 to 2013-14)

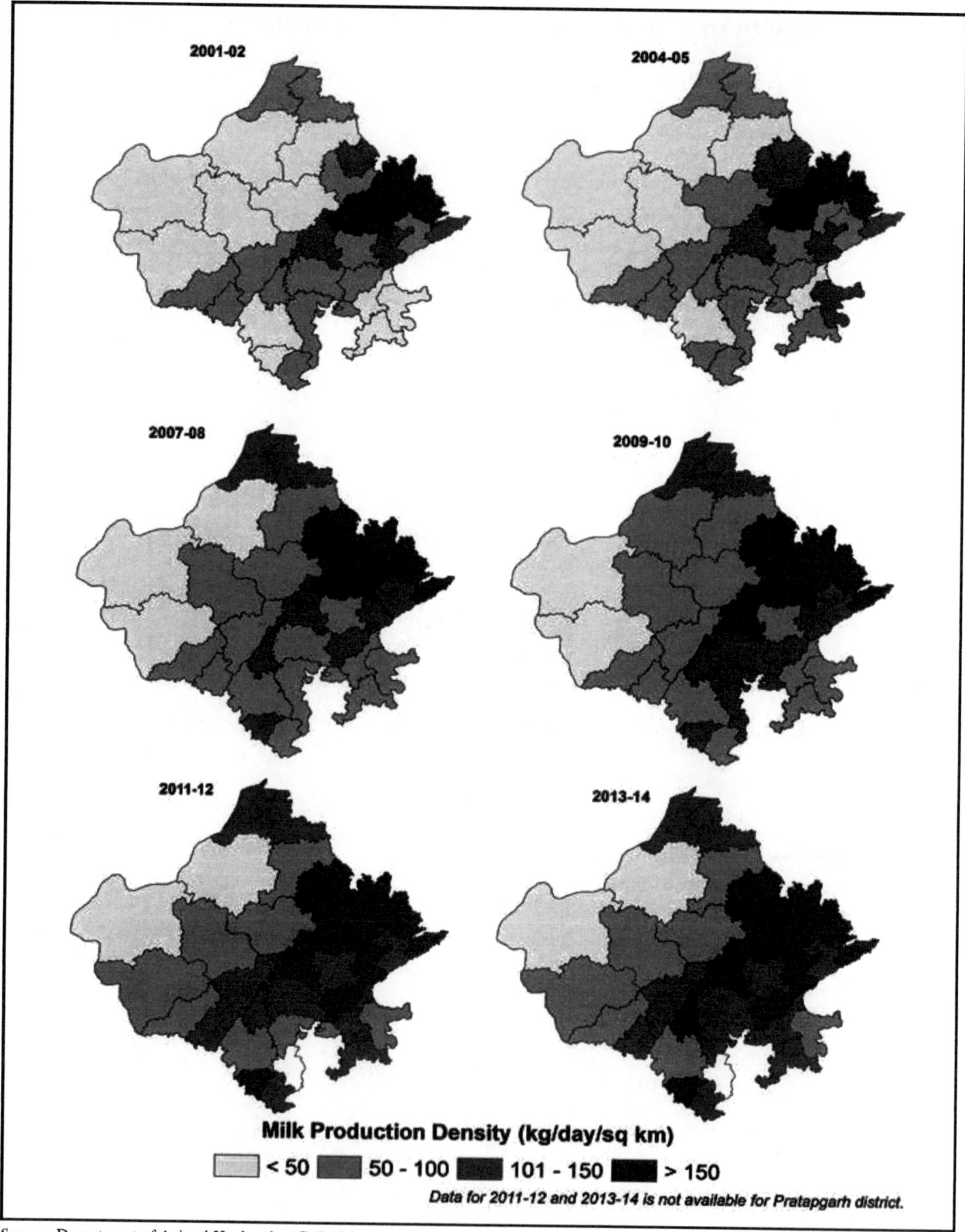

Source: Department of Animal Husbandry, GoR, Jaipur

Figure 15.24 Districtwise Milk Production Density in Rajasthan (2001-02 to 2013-14)

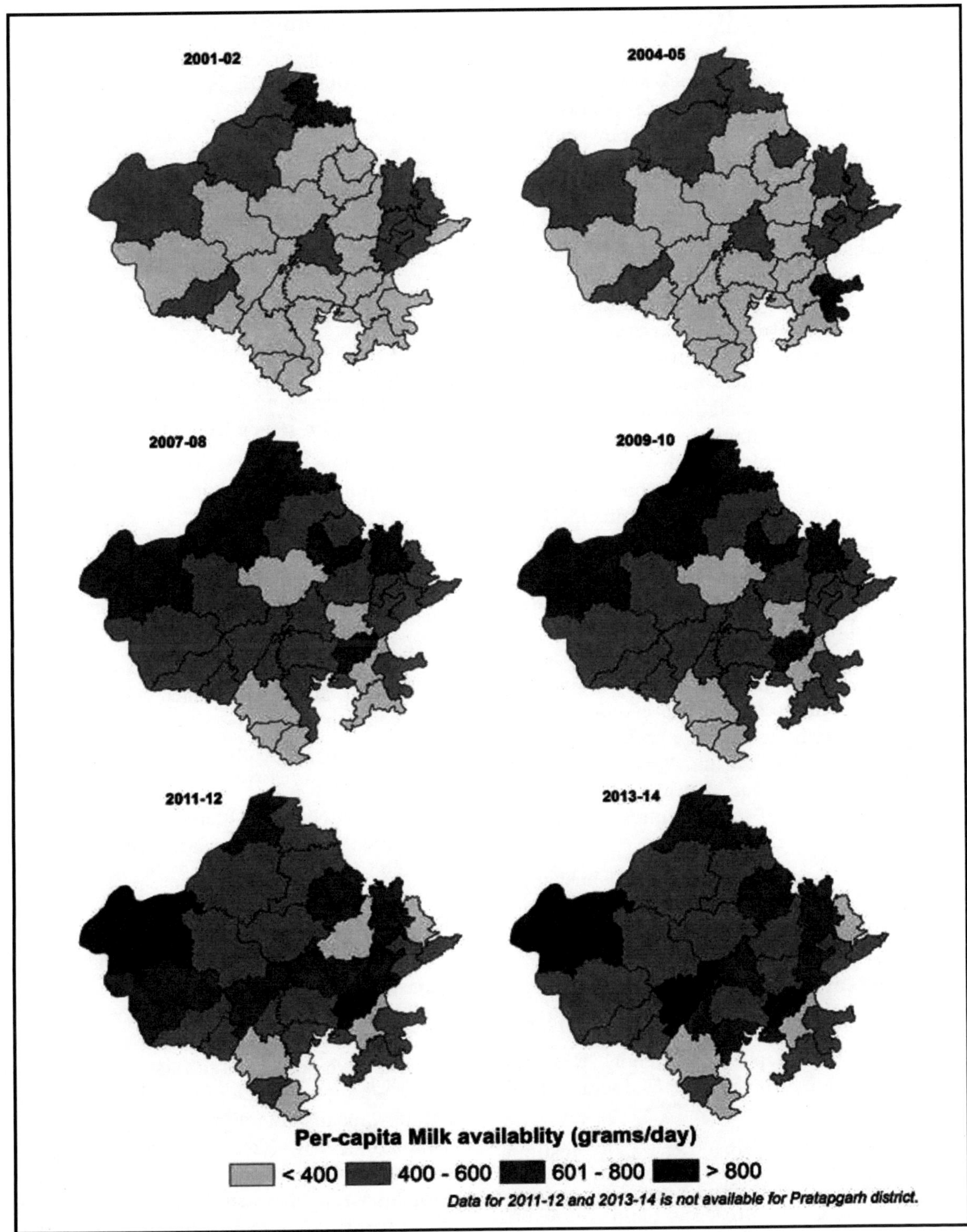

Source: Department of Animal Husbandry, GoR, Jaipur

Figure 15.25 Districtwise Per Capita Milk availability in Rajasthan (2001-02 to 2013-14)

ANIMAL HUSBANDRY FACILITIES

In order to accelerate the pace of the "White Revolution", livestock breeds are to be improved and the existing cattle wealth has to be cared for. This can be achieved by providing good animal husbandry facilities.

In 2012 state had a total 1,425 veterinary hospitals, 285 animal husbandry dispensaries, 1,720 veterinary sub centres, 4 cattle breeding, 10 Horse breeding and 1 Sheep breeding farms and 2,311 artificial insemination units available to provide adequate health coverage to the existing livestock. The main function of these institution is to provide treatment for the ailing animals, to vaccinate against prevalent infections and contagious diseases and to castrate the non-desi bull. Besides these available artificial insementation centres are used to improve cattle breeds (Figure 15.26).

Rajasthan, has abundant wealth of sheep, and therefore, the sheep and wool department has opened number of sheep extension centres distributed throughout the state. The number of such centres is greater in western Rajasthan where sheep are major species of livestock rared by people to sustain the economy of the area. The state has four cattle breeding and bull rearing farms located in Ajmer, Bharatpur, Jhalawar and Nagaur districts. In addition, there are number of sheep centres available in the state.

Some of the important animal husbandry research, breeding and training institutes which functioning in Rajasthan are:

Central Government

- Central Camel Research Institute, Bikaner
- Central Sheep and Wool Research Institute, Avikanagar, Malpura (Tonk district)
- Central Arid zone Research Institute (CAZRI), Jodhpur
- Central Cattle Breeding Farm, Suratgarh (Ganganagar)

State Government

The Cattle breeding and bull rearing farms for different breeds of animals are located at

- Kumher (Bharatpur) Bull breeding farm
- Nagaur Nagauri breed of cattle
- Ramsar (Ajmer) Sirohi and Parbatsar breed of goats
- Dug (Jhalawar) Malvi breed of cattle

▪ Facilities for training livestock assistants exist at Jaipur, Jodhpur, Kota and Udaipur. Besides this, one in-service training centre at Jaipur provides training to officers and staff of the department in the frozen semen technoloay.

▪ The Sheep and Wool Department has a training school at Jaipur.

▪ The Dairy Federation has training facilities at the major dairy plants where training is provided to the field staff and members of the primary Dairy cooperative societies.

▪ Veterinary ploycIinics are functioning at Jaipur, Jodhpur, Bikaner, Ajmer Kota and Udaipur.

▪ A Forzen Semen Laboratory is functioning at Bassi (Jaipur).

▪ Liquid Nitrogen Plants are functioning at Bassi (Jaipur), Nagaur, Banswara, Bundi and Jhalawar.

DAIRY DEVELOPMENT

Dairy Development programme in Rajasthan was initiated during the year 1971 for marketing surplus milk available with the rural milk producer, so that they get the remunerative prices and to raise their economic status. The state is rich in good breeds of milch animals, localised in different region like Rathi, Tharparkar, Gir, Kankrej, Sanchori, Haryana and Malvi. The Dairy. Development Programme in the state started, initially, as a department of State Government and later on, as a corporation named, Rajasthan State Dairy Development Programme and finally as a federation and renamed Rajasthan Cooperative Dairy Federation (RCDF). The dairy development programme and its activities in the state are now governed by this federation on "Amul" pattern, having three-tier cooperative system. This envisages milk producers cooperative societies at the village level federated into distirct milk cooperative unions and the federation of all the unions in the state into an apex body, the Rajasthan Cooperative Dairy Federation (RCDF).

Source: Department of Animal Husbandry, GoR, Jaipur

Figure 15.26 Animal Husbandary Facilities in Rajasthan

All the major activities of the federation and union are aimed at implementation of the "Operation Flood" programme in the state which includes (Figure 15.27):

- Organisation of a dairy cooperative supplemented with all the required related italicised programme activities.
- Setting up of infrastructure for milk processing, manufacturing of cattle feed, raising of exotic Jersey Bulls for breeding, freezing of semen of pedigreed exotic and indigenous bulls and buffalo bulls far artificial insemination, production of improved variety of fodder seeds and its distribution, implementation of programmes and activities.
- Marketing of milk and milk products within and outside the state to fulfill the requirements of consumers as well as to pay back remunerative milk price to milk producers.
- Presently, 18 milk unions in the state have got infrastructure of 6,308 registered dairy (milk) cooperative societies in the villages with more than 4.58 lakhs as its registered members. With these milk unions, there are number of dairy plants distributed in the state having capacity of handling more than 10.00 lakhs liters milk per day, along with milk powder plants of 10.00 M.T. capacity per day. Plants at Raniwara, Ajmer, Alwar, Jaipur, Hanumangarh and Bikaner has 5 tonnes capacity milk powder plant. Alongwith these plants, there are 24 chilling centres in the milksheds to process and chill the milk received from the remote villages, with the capacity of handling 4.0 lakh litres of milk per day. Alongwith the milk, the milk products like Ghee, Skimmed milk powder, Whole milk powder, Butter, Paneer, Mawa, Chach, Lassi, Srikhand etc. are also manufactured and marketed by the dairy plants.
- Four cattle feed plants, of 100 metric tonnes capacity each located at' Ajmer, Jodhpur, Bikaner and Nadbai (Bharatpur district) to manufacture the balanced cattle feed to be made available to the farmers in the villages through the milk cooperative societies of respective unions.
- To raise man power to run the dairy cooperative societies for all the activities and services to be rendered. Fodder development is also an activity to be taken up by the farmers in the villages which is very important, not only for raising milk production but also from animal health point of view and therefore, RCDF established a seed mullification farm at Rojari, in Bikaner district. The quality fodder seed is being produced for distribution to the milk producers of the dairy cooperative societies.
- To raise the Jersey breeding bulls for intensive cross breeding programme in the remote and inaccessible villages where artificial insemination programme has not been taken up. Likewise, there is a frozen semen bank with the RCDF located at Bassi (Jaipur district) which collects and processes the semen of valuable pedigreed, exotic and indigenous bulls and buffalo bulls, for its supply alongwith the latest technology to the Dairy cooperative societies in the milksheds for artificial insemination, which is done by the trained personnel in the villages. It also supplies semen to the Department of Animal Husbandry.
- For improving packing system, RCDF has also asceptic milk packaging plant in the state. The processed milk is packed in the such card sheets of tetra packs so as to sustain its quality and life upto 90 days without refrigeration.
- To have the quality control there is one, central quality control laboratory with RCDF so as to regular monitor and check the quality of milk and milk products as per I.S.I. specifications and provision of P.F.A. It also regularly checks the packaging material and cattle feed samples.

Due to lack of requisite irrigation facilities and lower precipitation over a mjor part of the state, rural economy is more dependent on rearing of cattle and animals. Dairy Development Programme, thus has an important role to play in rural areas in raising over all socio-economic status of the farmers.

Presently, the milk cooperative societies with their federal body in the milkshed i.e. milk unions, are being managed by the elected boards, representing the milk producers. Likewise, the board of federation is nominated by the government comprising mostly Chairmen of the milk unions. To strengthen the cooperative network at the grass root level i.e. Village Dairy Cooperative Societies, cooperative development programme has also been taken up in all the milk unions, with assistance from National Dairy Development Board. The basic purpose is to develop the consciousness of among the milk producers to manage and ran their milk cooperative societies in the villages with elected management committees.

Women Dairy Development Programme (STEP) has been sponsored by the Ministry of Women and Child Welfare, Government of India in number of districts of the state. It is a well known fact that all the activities related to management of dairy animals are mostly being managed and looked after by rural women. Sofar, around 495 women Dairy Cooperative Societies have been organised with around 18,000 women members in the state. They are being managed by elected women managing committee.

For valuable stock of cross breed livestocks available with the small and marginal farmers and landless labourers, cattle feed at the subsidised rates, is being provided to raise them to become high yielding milch animals. This activity is being taken up under the special livestock breeding programme. This scheme is being covered through the milk unions in different districts of the state.

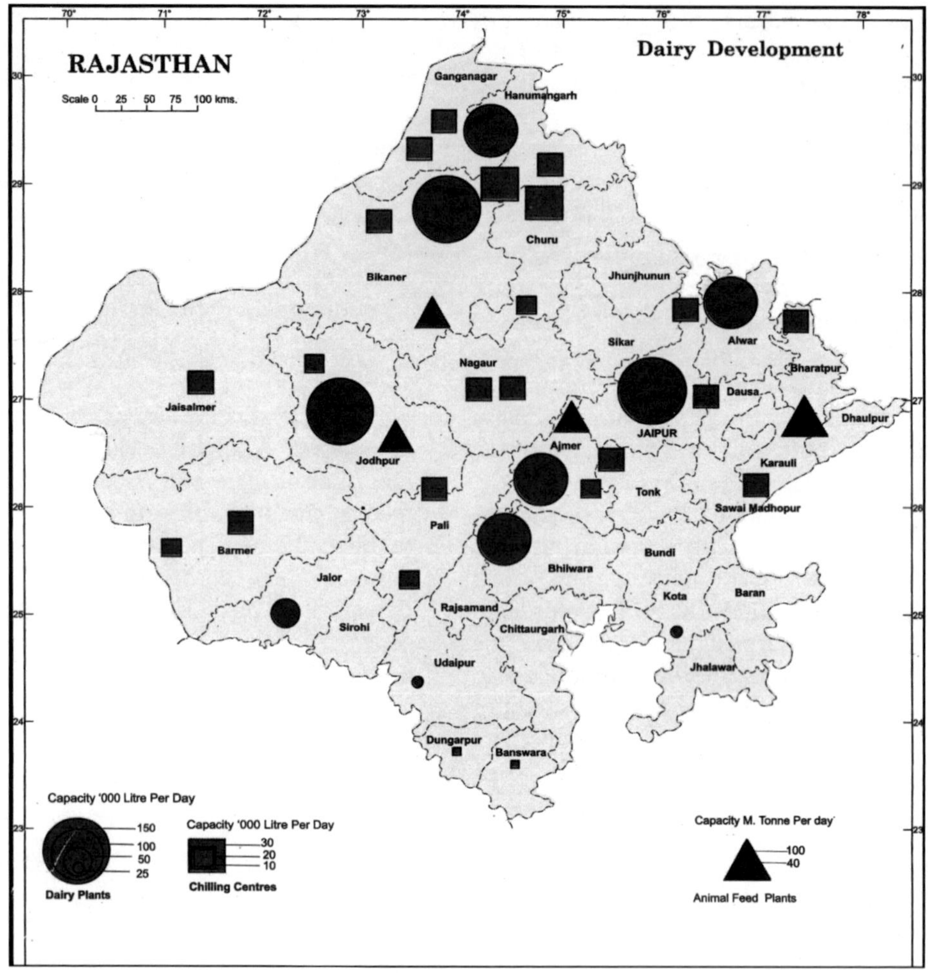

Source: Department of Animal Husbandry, GoR, Jaipur

Figure 15.27 Dairy Development in Rajasthan

CATTLE FAIRS

A unique of its kind in the country, cattle fairs of Rajasthan have a colourful tradition of over six hundred years. They are organised periodically in different parts of the state with the aim of bartering and selling cattle, horses, buffaloes, bullocks, ponies, camels, goats, sheep and donkeys. The special feature is that no cow is sold in these annual fairs. The bullocks, horses, and camels for sale are brought by traders from the adjoining states like Punjab, Haryana, Uttar Pradesh, Madhya Pradesh, Gujarat and Maharashtra. Ten state level cattle fairs are organised by the State Animal Husbandry Department under a special Act. A number of other cattle fairs are also organised by Panchayat Samitis, Gram Panchayats and Municipalities. The total annual turn over of these fairs exceeds crores of rupees and forms the major source of income to the animal breeders of the state.

These fairs also provide opportunities to the traders to sell all types of household items and agricultural implements. Most of these fairs are organised in the memory of famous warriors, lokdevtas, saints, social leaders, Gods and Goddesses.

The State Animal Husbandry Department organises the following ten cattle fairs (Table 15.9, Figure 15.28) namely:

Table 15.9 Cattle Fairs in Rajasthan

S.No.	Cattle Fairs	Place	Period
1	Shri Mallinath Cattle Fair	Tilwara (Barmer district)	Chaitra Krishana 11 to Chaitra Sukhla 11.
2	Baldev Cattle Fair	Merta City (Nagaur district)	Chaitra Sukhla 15 to Baisakh Krishana 6.
3	Gomti Sagar Cattle Fair	Jhalrapatan (Jhalawar district)	Baisakha Sukhla 13 to Jaisth Krishana 5.
4	Gogameri Cattle Fair	Gogameri (Ganganagar district)	Shrawan Sukhla 15 to Bhadrapad Sukhla 15.
5	Shri Veer Tejaji Cattle Fair	Parbatsar (Nagaur district)	Shrawan Sukhla 15 to Bhadrapad Krishana 15.
6	Jaswant Cattle Fair	Bharatpur (Bharatpur district)	Ashwini Sukhla 5 to Ashwini Sukhla 15,
7	Pushkar Cattle Fair	Pushkhar (Ajmer district)	Kartik Sukhla 8 to Marg Shish Krishana 2.
8	Chandra Bhaga Cattle Fair	Jhalrapatan (Jhalawar district)	Kartik Sukhla 11 to Marg Shish Krishana 5.
9	Shri Ramdeo Cattle Fair	Nagaur (Nagaur district)	Magh Sukhla 10 to Phalgun Krishana 15.
10	Shivaratri Cattle Fair	Karauli (Karuli district)	Phalgun Krishana 6 to Phalgun Krishana 14.

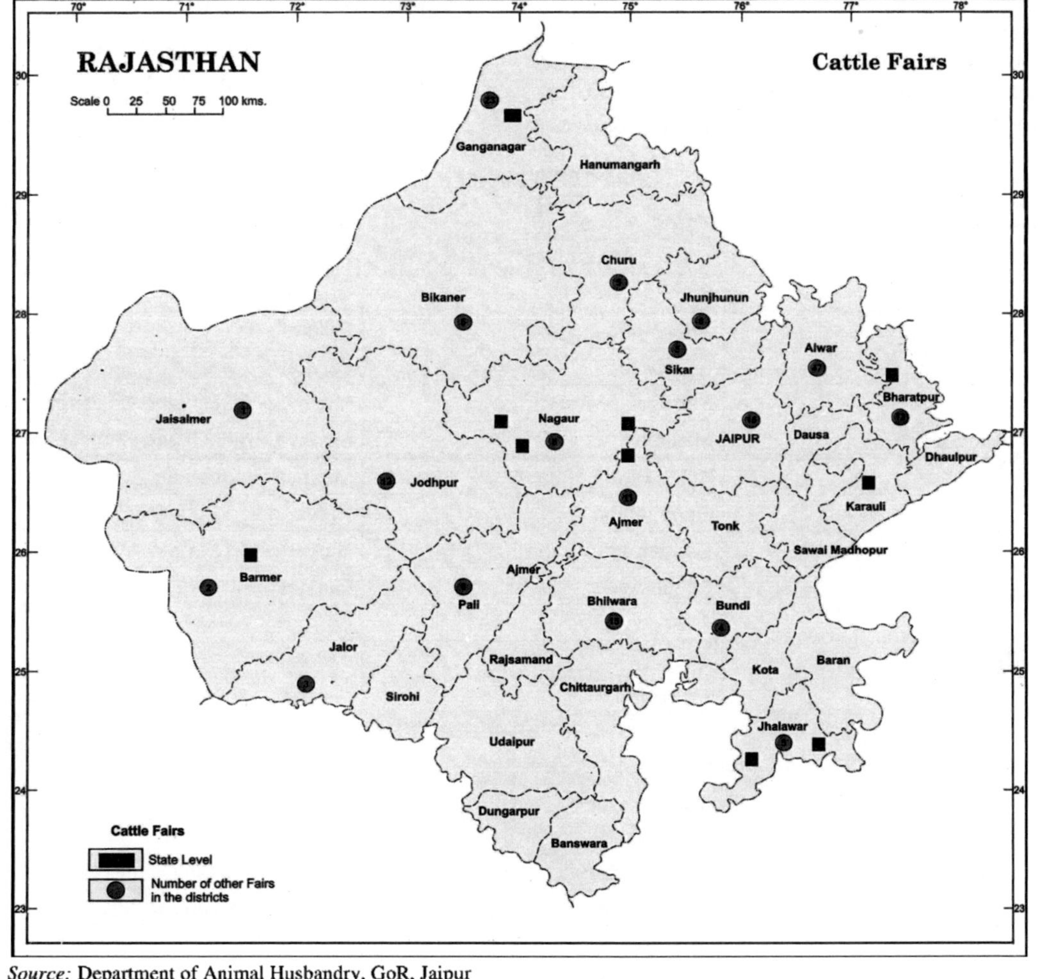

Source: Department of Animal Husbandry, GoR, Jaipur

Figure 15.28 Cattle Fairs in Rajasthan

LIVE STOCK MIGRATION ROUTES

Rajasthan State a typical example of climatic variation, arid in west sub humid and humid in the east with annual rainfall varying from 100 mm in Jaisalmer to 750 mm in Banswara district. Due to low and erratic rainfall, high temperature, deep ground water, lack of vegetation and salinity the entire western part has been classified as arid region, part of great "Thar Desert". More than 60 per cent of the entire geographical area of state faces frequent droughts and only few districts of south eastern part of state classified as sub humid and humid climates. As a consequence of this situation, the state faces frequent droughts, leading to cattle migration.

Western part of Rajasthan, being a rain deficit area, can offer fodder for livestock only during monsoon season. In the event of failure of monsoon, the situation worsens, and the cattle breeders do not find enough pasture lands to graze their livestock. Under these adverse circumstances, the cattle and sheep breeders have developed a definite pattern of migration with their livestock to areas where sufficient grazing, ground is available. These migratory routes are followed year after year by the livestock raisers. They take away their domestic animals from their native places to such distant areas as Madhay Pradesh, Uttar Pradesh, Punjab, Haryana and Gujarat where enough pasture is available. Sometimes, the cattle and sheep breeders remain away from their native places for several years and return to the home land only when favourable conditions prevail for grazing of their animals (Figure 15.29 and 15.30).

The traditional migration routes for sheep and cattle are given in Table 15.10.

Table 15.10 Migration Routes

S.No.	Breeders From Districts	Route	Destination
		(A) CATTLE MIGRATION	
1	Jaisalmer	Barmer	Gujarat
	Barmer	Dhorimana	Gujarat
	Jodhpur	Sanchore	Gujarat
	Nagaur	Beawar	Madhya Pradesh and Chattisgarh
	Jodhpur	Masuda	Madhya Pradesh and Chattisgarh
2	Barmer	Kekri, Bundi, Kota	Madhya Pradesh and Chattisgarh
3	Churu	Ganganagar	Haryana and Punjab
	Bikaner	Ganganagar	Haryana and Punjab
	Jaisalmer	Sanchore	Gujarat
4	Pali	Jalor	Gujarat
	Jalor	Sirohi	Gujarat
	Sirohi	Abu Road	Gujarat
		(B) SHEEP MIGRATION	
1	Jaisalmer	Bikaner	Punjab
	Bikaner	Ganganagar	Punjab
	Churu	Rajaldesar	Haryana
	Jaisalmer	Jodhpur, Ajmer, Jaipur	Madhya Pradesh Chattisgarh and Haryana
	Jodhpur	Ajmer, Tonk, Bharatpur	Uttar Pradesh and Utranchal
2	Barmer	Sanchore	Gujarat
	Pali	Bhilwara, Bundi, Kota	Madhya Pradesh and Chattisgarh
	Jaisalmer	Nagaur, Sikar, Alwar	Haryana

S.No.	Breeders From Districts	Route	Destination
3	Jaisalmer	Barmer, Sanchore	Gujarat
	Jodhpur	Pali, Sirohi, Abu, Rajsamand, Dungarpur, Banswara	Gujarat
	Nagaur	Bhilwara, Chittaurgarh, Banswara	Madhya Pradesh and Chattisgarh
	Jalor	Sanchore, Pali, Rajsamand	Gujarat
	Pali	Bhilwara	Madhya pradesh and Chattisgarh
	Sirohi	Abu	Gujarat
	Pali	Ajmer, Jaipur, Dausa, Bharatpur	Uttar Pradesh and Uttranchal
	Pali	Ajmer, jaipur, Tonk	Uttar Pradesh and Uttranchal
	Jodhpur	Beawar, Bhilwara, Kota	Madhya pradesh
(C) CAMEL MIGRATION			
1	Jaisalmer	Ganganagar	Punjab and Haryana
		Barmer	Jalor
		Jodhpur, Nagaur, Sikar	Alwar
2	Bikaner	Nagaur	Ajmer
		Sikar	Alwar
		Churu, Jhunjhunun	Alwar
3	Barmer	Jodhpur	Ajmer
		Ajmer	Jalor

Source: Department of Animal Husbandry, GoR, Jaipur

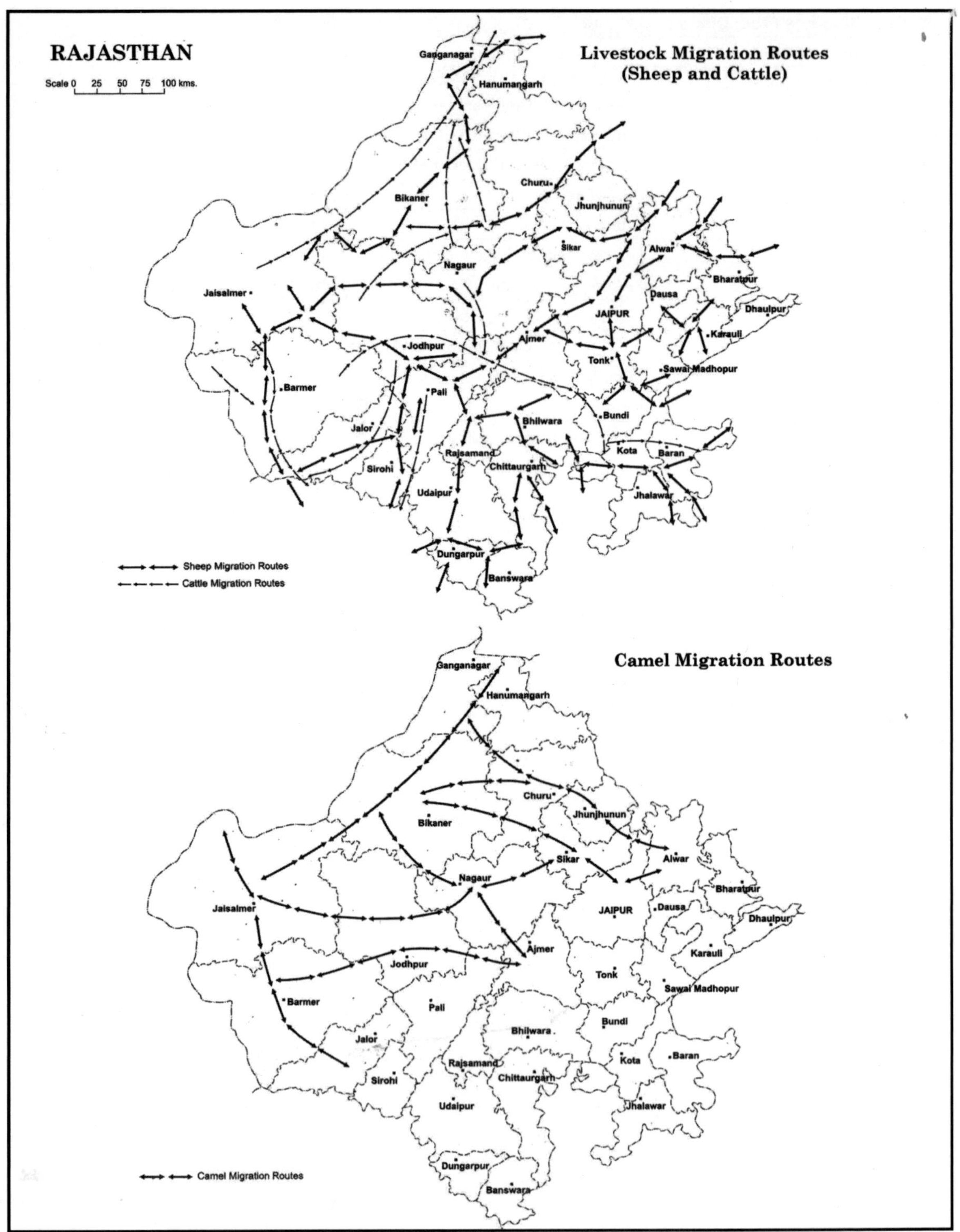

Source: Department of Animal Husbandry, GoR, Jaipur

Figure 15.29 Livestock Migration Routes of Sheep, Cattle and Camel

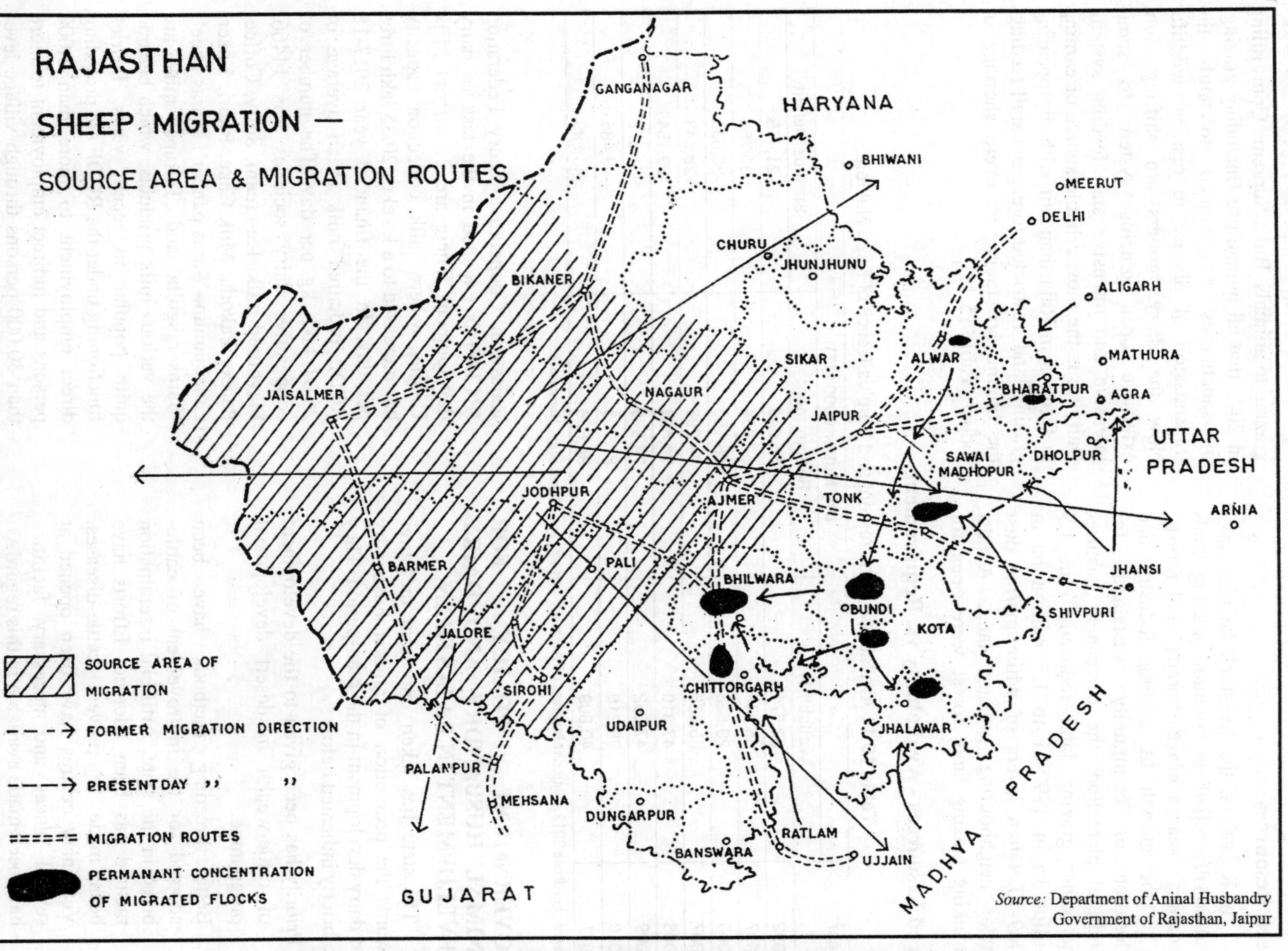

Figure 15.30 Sheep Migration (Source Area and Migration Routed) in Rajasthan

Feed Resources

In Rajasthan, the livestock keepers have traditionally relied on common grazing lands "gochars", scared groves "orans" and forests. With the growth of mining industry and allocation of community wastelands for biodiesel plantation, the permanent pastures and other grazing land reduced from 1.9 million ha in 1990-91 to 1.7 million ha in 2009-10. Often layers of white marble dust choke neighbouring grazing land. Amidst shrinking grazing land, forest were open for grazing to animals. But a Supreme Court ruling in 2002 that all human use (including grazing) in sanctuaries be banned to curb the transmission of disease to resident wildlife. Now the dairy farmers are shifting from extensive open grazing system to semi-intensive and intensive stall feeding system. Dairying is the most reliable source of earning to farmers in Rajasthan but with disappearing grazing land, restricted forest and stall feeding, the bovine are facing a severe shortage of fodder (Table 15.11).

Feed Resources - Availability vs. Requirement

Table 15.11 Availability, Requirement and Surplus/Deficit of Feed Nutrient

Year	Dry Matter ('000 MT)		
	Availability	Requirement	Surplus/Deficit
1992	33,571	55,046	-21,475
1997	35,848	66,634	-30,786
2003	29,523	66,153	-36,630
2007	45,655	74,298	-28,643
2008	47,310	76,464	-29,154
2009	47,052	78,929	-31,877
2010	38,218	81,703	-43,485
2011	40,809	84,808	-43,999

Source: Feedbase 2012, National Institute of Animal Nutrition and Physiology, Bangalore.

GOVERNMENT POLICY FOR ANIMAL HUSBANDRY AND DAIRY DEVELOPMENT IN RAJASTHAN

The state has taken important measures towards the promotion of animal husbandry and dairy development in the plan-period. They are briefly indicated below :

1. Priority has been given to the development of cattle-wealth in desert development programme.
2. Better breeding methods have been adopted for the improvement of cattle-breed in the state. Artificial insemination method has been adopted. Efforts have been made to prevent cattle-diseases. Veterinary centres have been opened at several places and necessary facilities have been made available in this regard.
3. Rajasthan Cooperative Dairy Federation (RCDF) is engaged in the task of dairy development. There are 21 district Milk Unions, where milk collection capacity has increased to a level of 20.35 lakh litres per day. In the financial year 2017-18 RCDF procured milk at an average of 26.07 lakh kg per day. The number of cooperative dairy societies was 14466. Flavoured milk, Ice-cream, Softy, Coffee, Paneer Pakoda, Milk Cake, Rabri, Kheer, Gulabjamun, 'Saras paneer', 'Saras ghee', 'Saras rasgulla' and 'Tetrapak milk' are the various milk products, which became quite popular in recent years. Dairies functioning under the RCDF are providing direct employment to more than 3,000 persons and indirect employment to more than 30,000 persons through village level cooperative societies.

4. Various cross-breeding farms/centres have been started (target of 50 centres) with the help of Bharat Agro Industries Foundation (BAIF). They are being developed at Bhilwara, Kota, Bundi, Udaipur, Chittorgarh, Dungarpur and Banswara, Bassi (Jaipur) Cattle-breeding farm (Exotic Nucleus Farm) is already known for raising exotic Jersy Bulls.

5. Cattle-fairs are organised at several places in Rajasthan. More famous are the cattle-fairs of Parbatsar and Piplu.

Thus, efforts are going on in the state towards the improvement of cattle- breeding, reducing and controlling cattle-diseases, increasing productivity of cattle and improving the living conditions of the cattle-breeders by these measures. Planning for cattle development would go a long way in improving employment opportunities in the state. Therefore, more attention should be given to this sector than what has been attempted so far.

Chapter 16

MINERAL RESOURCES

Mineral resources are the backbone for the industrial and socio-economic development of a country. Rajasthan ranks seconds only to Jharkhand in mineral deposits. Some minerals produced like Wollastonite Jaspar are found only in Rajasthan. Important minerals produced in the State are basemetals viz. copper, lead and zinc; and other industrial minerals like soapstone, calcite, baryte, wolastonite, clays, silica sand, fluorite. cement grade limestone, garnet, gypsum, rock phosphate etc. The State also produces a variety of dimensional and decorative stones viz. granite, marble, slate sandstone, quartzite etc. In fuel minerals, the State has vast resources of lignite suitable for power generation. Oil and Natural Gas has also been found at some places in Western Rajasthan.

The State has virtual monopoly in country in the production of wolastonite. Jasper and zinc while it is a producer of gypsum, asbestos, soapstone, rock phosphate, ball clay, calcite, feldspar, lead etc. accounting for more than 70 per cent of country's production of these minerals. The state occupies an important place in the field of dimensional and decorative stones. 95 per cent of country's production of marble is from the State of Rajasthan. There has been a steady increase in production of sandstone, slate and granite in recent years.

Rajasthan, endowed with a rich mineral wealth, occupies a significant position on the mineral map of the country. In the area of metallic and non metallic minerals, Rajasthan, by possessing 19% of working mines in the country, leads the country. It possesses a wide range of mineral resources and offers considerable potential for the future. The important metallic minerals with which the name of the State is intimately associated are: lead, zinc, copper and tungsten. It is also richly endowed with a variety of non metallic minerals. Rajasthan has also enormous deposits of building stones like marble, granite, sandstone, limestone, slate etc. It is a leading producer of asbestos, soapstone/steatite, gypsum, rock phosphate, calcite, feldspar, clay etc. Presently, the State enjoys a monopoly in the production of wollastonite, emerald, jasper and semi precious garnet. In addition, the State has good deposits of lignite in Bikaner, Nagaur and Barmer districts.

TYPE OF MINERALS

Occurrences of minerals in Rajasthan is shown in Figure 16.1. Brief descriptions of minerals with their occurrence areas in Rajasthan are described below:

Apatite

Apatite occurs as a primary constituent of igneous rocks, but only as accessory mineral. It also occurs as a vein mineral. It is a natural phosphate bearing mineral having a definite chemical composition. Apatite is mainly of two types. One is chlorapatite $3Ca_3(PO_4)_2CaCl_2$ and the other is fluorapatite $3Ca_3(PO_4)_2$ CaF_2. Its main use is as fertilizer. In Rajasthan, occurrences of Apatite have been reported from Sikar and Udaipur districts.

Asbestos

Rajasthan accounts for about 96% of the country's total production of asbestos, whereas Andhra Pradesh and Karnataka, the other producing States contribute the remaining 4%.

The asbestos deposits are mainly located in the southern parts of the State, confined to the altered ultramafic intrusives within the rocks of the Bhilwara, Aravalli and Delhi Supergroups. Rajasthan possesses crysotile and amphibole varieties of asbestos.

Barytes

Barytes, Barite or the 'Heavy Spar' are different names of the same mineral having the composition $BaSO_4$. The mineral is very important and vital to the petroleum industry which consumes more than 80° of the world's Barytes production. Barytes finds it principal utility in drilling mud for exploration of because of its high specific gravity, inertness to acids insolubility in water and above all its lower cost compared to many other available heavy materials.

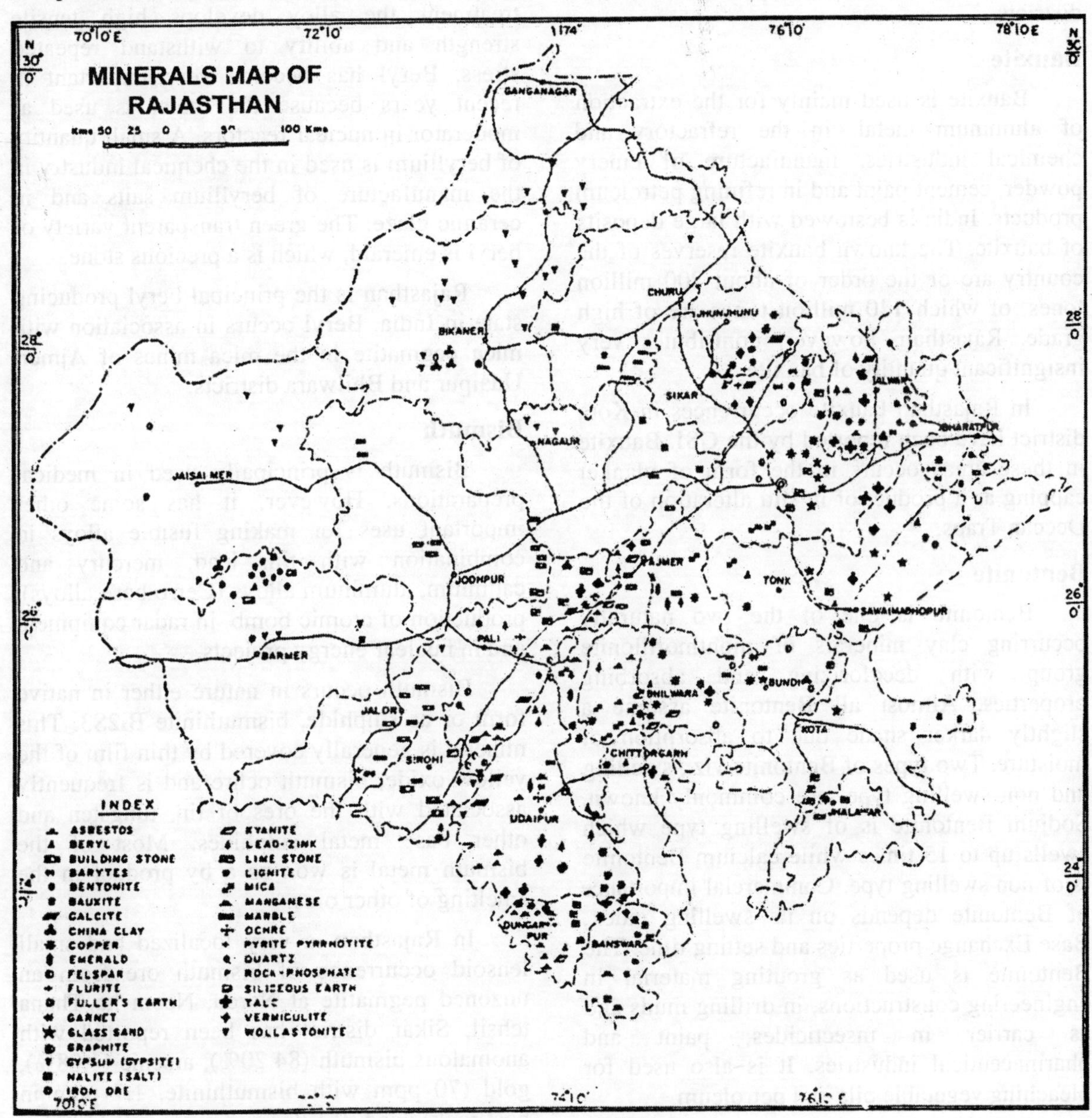

Source: Geological Survey of India, Department of Geology and Mines, GoR, Udaipur

Figure 16.1 Minerals Map of Rajathan

In Rajasthan, Barytes mineralization is associated with igneous, metamorphic and sedimentary rocks. It occurs as veins, cavity fillings, bedded and residual deposits. Occurrences of Barytes are known in the districts of Alwar, Bharatpur, Bhilwara, Bundi, Chittorgarh, Jalore, Pali, Sikar, Sirohi and Udaipur, whereas important deposits are limited to Alwar, Bharatpur, Pali and Udaipur districts.

Bauxite

Bauxite is used mainly for the extraction of aluminum metal, in the refractory and chemical industries, manufacture of emery powder, cement paint and in refining petroleum products. India is bestowed with large deposits of bauxite. The known bauxite reserves of the country are of the order of about 200 million tones, of which 140 million tones are of high grade. Rajasthan, however, contributes very insignificant quantity of bauxite.

In Rajasthan bauxite occurrences in Kota district have been reported by the GSI. Bauxite in these areas occurs in the form of blanket capping as a product of in situ alteration of the Deccan Traps.

Bentonite

Bentonite is one of the two naturally occurring clay minerals of montmorillonite group with decclorizing and absorbing properties. Almost all Bentonite assume a slightly darker shade due to absorption of moisture. Two types of Bentonite viz., swelling and non swelling type, are commonly known. Sodium Bentonite is of swelling type which swells up to 15 times while calcium Bentonite is of non swelling type. Commercial importance of Bentonite depends on its swelling index, Base Exchange properties and setting time. The Bentonite is used as grouting material in engineering constructions, in drilling muds and as carrier in insecticides, paint and pharmaceutical industries. It is also used for bleaching vegetable oils and petroleum.

Rajasthan possesses inexhaustible reserves of both swelling and non swelling types of Bentonite. Deposits of Bentonite are mostly located in Barmer and Jaisalmer districts.

Beryl

Beryl, a silicate of beryllium and aluminum, is the principal source of beryllium metal which is lighter than aluminum. When alloyed with copper and after suitable heat treatment the alloy develops high tensile strength and ability to withstand repeated stress. Beryl has become very important in recent years because the metal is used as moderator in nuclear reactors. A small quantity of beryllium is used in the chemical industry in the manufacture of beryllium salts and in ceramic glaze. The green transparent variety of beryl is emerald, which is a precious stone.

Rajasthan is the principal beryl producing state in India. Beryl occurs in association with mica pegmatite in the mica mines of Ajmer, Udaipur and Bhilwara districts.

Bismuth

Bismuth is principally used in medical preparations. However, it has some other important uses for making fusible alloys in combination with tin, lead, mercury and cadmium, aluminum alloys (Cerro base alloys), production of atomic bomb, in radar equipment and in nuclear energy projects.

Bismuth occurs in nature either in native form or as sulphide, bismuthinite Bi2S3. This mineral is generally covered by thin film of the yellow oxide, bismuth ochre and is frequently associated with the ores of tin, tungsten and other base metal sulphides. Most of the bismuth metal is won as a by product in the smelting of other ores.

In Rajasthan, a very localized and small lensoid occurrence of bismuth ore from an unzoned pegmatite at Narda, Neem ka Thana tehsil, Sikar district has been reported with anomalous bismuth (84.20%), arsenic (1.08%), gold (70 ppm with bismuthinite, 1.7 ppm in host rock), lead (0.17%) and copper (0.11°%).

Basalt / Rhyolite

South eastern parts of Rajasthan are partly occupied by the Deccan Traps, which constitute the main building stone in this area. The rhyolite is mostly used for railway ballast and other construction purposes in western Rajasthan.

Flaggy Stone (Kotà Stone)

Upper Vindhyan limestone, quarried in Kota and Chittorgarh districts, occurs in shades of yellow, green and blue. The quarries are located at Ramganj Mandi, Chechat, Suket, Manpura, Morak and Mur-ka Kheri in Kota district. Flaggy limestone also occurs near Aroutiya, Paroliya and Kishanpura in the Jhalawar district.

Granite

Granite and associated rocks form excellent dimensional and decorative stones. Durability, hardness, capability of taking mirror polish and fascinating colours makes the granite most distinctive stone among marble, limestone and sandstone. It is widely used as polished tiles, slabs and monumental stone all over the world.

The granite occurrences in Rajasthan are distributed over wide areas in Ajmer, Alwar, Banswara, Barmer, Bhilwara, Chittorgarh, Dungarpur, Jaipur, Jaisalmer, Jalor, Jhunjhunun, Jodhpur, Nagaur, Pali, Rajsamand, Sawai Madhopur, Sikar, Sirohi, Tonk and Udaipur districts. A district wise brief account of important localities along with nature, colour and commercial names of the granite is given below

Marble

In Rajasthan, marble and marble based industries are very well developed. The marble and marble products are used in house construction, statues, marble chips, marble/dolomite powder and plaster making; as tyre detergent; and in gur, sugar, and paper industries.

Marble is mainly mined from Makrana, Raj Nagar, Rikhabdev, Ajmer, Andhi and Bhaislana areas. Abur limestone and Jaisalmer limestone take good polish and are used as decorative stones. These are marked as 'Supari stone' and 'Yellow marble' respectively.

Morwar in Rajsamand district has a single mine (out of numerous mines) which produces 300,000 metric tones of blocks slabs of white marble per year creating a world record which find a place in the Guinness Book of World Records.

Rajasthan possesses the largest reserves exceeding 1,100 million tones accounting for about 95% of the total production of marble in India.

Sandstone

Sandstone is mainly found in the Vindhyan and the Marwar Supergroup sequences of Bharatpur, Tonk, Sawai Madhopur, Bundi, Chittorgarh districts and Jodhpur, Bikaner, Pali, Jaisalmer districts respectively. Rajasthan has been the main center of export of sandstone to other parts of the country. The important occurrences of sandstone are found in Bharatpur, Bikaner, Bundi, Jaipur, Jodhpur, Kota, Tonk, Sawai Madhopur districts.

Slate / Schist

Slate finds its use as material for roofing, paving and shelves. Slate deposits are found mainly in the Delhi and Aravalli rocks in parts of Ajmer, Alwar, Bharatpur, Bundi, Pali, Sawai Madhopur, Tonk, Udaipur and Chittorgarh districts. Quarries in Jaipur, Udaipur, Alwar, Dungarpur and Banswara are providing schist and phyllities as roofing material.

Alwar slate has acquired a reputation in the world market for its quality and has left China and Spain far behind in the international market. Alwar district has enormous reserves of slates and has monopolized the Indian export. Alwar slates are used in the preparation of school slates, as roof tiles, and tabletop of the Billiard table.

The slates are available around Bahrod, Maton and Neemrana in huge quantity. Mines are also situated around Jakharana, Bassai, Kalor, Adind, Bhaiyasar, Mahatavas and Khundrot villages. RSMDC is engaged in slate mining at Maton.

Calcite

Calcium carbonate (CaC03) is an important constituent of sedimentary rocks, next to quartz and clay minerals. In the form of limestone it is extensively used in the manufacture of lime. cement, ceramic products, calcium carbide and bleaching powder. In the pulverized form, it is used in paint, rubber, textile, paper, leather and sugar industries. Limestone of high purity is used for imparting glaze in pottery and as a flux in metallurgical operations. Pure, crystallized, transparent variety of calcium carbonate, known as "Iceland Spar" is used in optical instruments.

The total reserves of calcite in India are estimated at 10,573 thousand tones. Rajasthan is the leading producer of calcite in the country. In Rajasthan, occurrences of calcite have been recorded from Sikar, Pali, Udaipur, Sirohi and Jaipur districts.

China Clay

Extensive deposits of china clay are found in India. With the increasing demand from industries like ceramic, refractory, textile, paper, rubber and pesticides, the production of the mineral has increased to a very large extent. Almost the entire production of china clay is consumed indigenously. A small quantity of superfine variety is also imported for specialized uses.

Rajasthan contributes about 50% production of china clay, 40% of ball clay and 10% of fire clay in the country. China clay is found to occur in Bhilwara, Bundi, Chittorgarh, Jaipur, Jhalawar, Sawai Madhopur, Sikar, Pali districts etc. Ball clay occurs in Alwar, Bikaner and Bundi districts.

Copper

Copper is the most essential metal used by the man since the 'Bronze age'. In spite of its part substitution by aluminium, the demand of copper is ever increasing. At present the production of copper metal in the country is only 15,000 tones as against the annual demand of about 0.5 million tones. A major part of the demand for this metal is, therefore, met by imports causing heavy drain on our foreign exchange.

Substantial reserves of copper ore have been established in Rajasthan, which now hold a significant place in the country in this important base metal resource. The main occurrences of copper ore are located along the prominent mineralized belts viz, Khetri Copper Belt and Alwar Copper Belt in the northeast part of the State. Besides, a number of small copper deposits associated with other base metal occurrences, which are under various stages of exploration, have been brought to light in other parts of the State.

Corundum

Corundum, due to its hardness, is used mainly in manufacturing abrasive material. In sintered form, it is used for special refractory crucibles. Synthetic corundum is in great demand as the production of natural corundum is insufficient to meet the demand.

Ruby and sapphire are the transparent gem verities of corundum. Ruby is rose red, carmine or purple. True sapphire is blue. Other colours in sapphire include white, yellow, green or pink.

Dolomite

Dolomite occurs in massive or bedded form. Its main industrial application is as a flux material in the iron and steel industry, which accounts for 91% of the total consumption in the country. Besides, it is used in glass, ferroalloys, alloy steel, fertilizer, ceramics, chemicals, electrode, rubber, building construction etc.

Though there are many occurrences of dolomite in the various geological formations of Rajasthan, production of dolomite is very small as there are no metallurgical and other industries in the State.

The dolomite deposits occur in Ajmer, Alwar, Bhilwara, Jaipur, Jaisalmer, Jhunjhunun, Jodhpur, Nagaur, Pali, Sawai Madhopur and Udaipur districts. The total reserves in the State are placed at 1,62,644 thousand tones.

Feldspar

Feldspar is an alumino silicate mineral of potassium, sodium and/or calcium. Though feldspar is among the most widely distributed rock forming minerals and occurs as a constituent of most of the rocks, its commercial deposits are mainly confined to pegmatite. Feldspar is either quarried from the pegmatite veins or is recovered as by product to the mining of mica and beryl from the pegmatite. Feldspar is chiefly used in the ceramic and glass industries and in insulator making.

Rajasthan is the principal feldspar producing State in the country and its reserves are about 10.13 million tones which constitute 72% of the total Indian reserves.

In Rajasthan, feldspar occurrences have been reported from Ajmer, Alwar, Jaipur, Pali and Sikar districts (Figure 16.2).

Fluorite

Fluorite (CaF2) is generally colourless to white and purple in colour. The finest grade of fluorite is used for enameling iron, manufacture of opaque and opalescent glasses and for the production of hydrofluoric acid. The inferior grade fluorite is used as a flux in foundry work for the manufacture of steel. Transparent fluorite is used in the manufacture of lenses. Use of fluorite in the steel metallurgy and in the manufacture of artificial chryolite makes it an important mineral.

Rajasthan has a large number of occurrences and small deposits of fluorite and is only next to Gujarat. Occurrences of fluorite are reported from Ajmer, Alwar, Bhilwara, Dungarpur, Jaipur, Jalor, Jhunjhunun, Nagaur, Sikar and Sirohi districts. Workable deposits occur in Dungarpur, Jalor, Jhunjhunun, Sikar and Sirohi districts. The total reserves of fluorite from.

Fuller's Earth

Fuller's earth, also known as Multani Mitti is a clay of sedimentary origin composed essentially of attapulgite, montmorillonite and kaolinite. It is used as cosmetic and for cleaning woolen clothes. It has absorbent properties which enable it to remove greasy and oily matter. It is also used in refining oils and animal fats.

Major production of fuller's earth in India comes from Rajasthan. Annual production is about 5,0006,000 tones. Barmer is the main producing district contributing about 77% of total production of the State. The rest of the production comes from Bikaner (18%) and Jaisalmer (5%) districts. With the increase in the output of mineral oil and vegetable oil products in India, a progressive increase in demand for fuller's earth is apprehended.

Aquamarine

Aquamarine, a beryllium aluminium silicate, is a transparent variety of beryl. Aquamarine is mainly recorded from Toda Raising and Hanotia in the Tonk district. It occurs in association with mica and pegmatite. Aquamarine is usually cut in pendant form.

Amethyst

Amethyst, a purple to violet coloured transparent variety of quartz, occurs in well crystallized form. When heated strongly it looses its colour. Under moderate heating it turns yellow and is sold as citrine. Though faceted, the steep brilliant cuts are common. It is also carved into attractive forms and shapes. Amethyst is reported to occur near the Kishangarh railway station in Ajmer distrct and from Bidera, Girota, Didwana and Choup in Jaipur district.

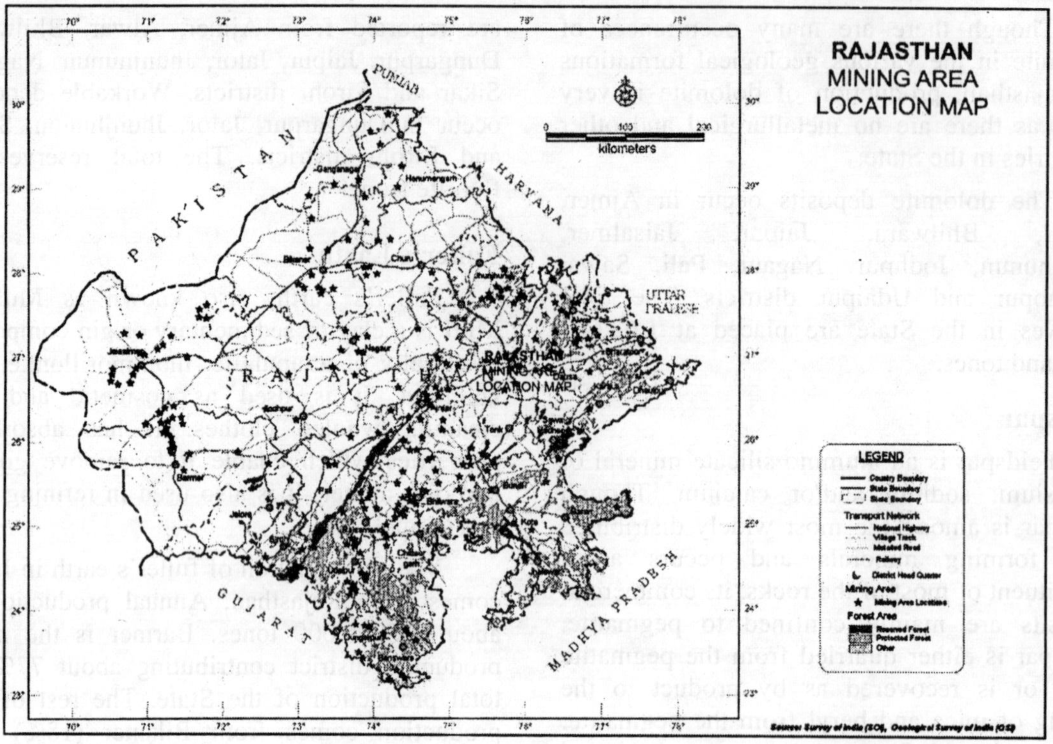

RAJASTHAN
MINING AREA
LOCATION MAP

kilometers

LEGEND
Country Boundary
State Boundary
Transport Network
National Highway
Village Track
Metalled Road
Railway
District Head Quater
Mining Area
Mining Area Locations
Forest Area
Reserved Forest
Protected Forest
Other

PAKISTAN
PUNJAB
HARYANA
UTTAR PRADESH
MADHYA PRADESH
GUJARAT
RAJASTHAN

Source: Survey of India (SOI), Geological Survey of India (GSI)

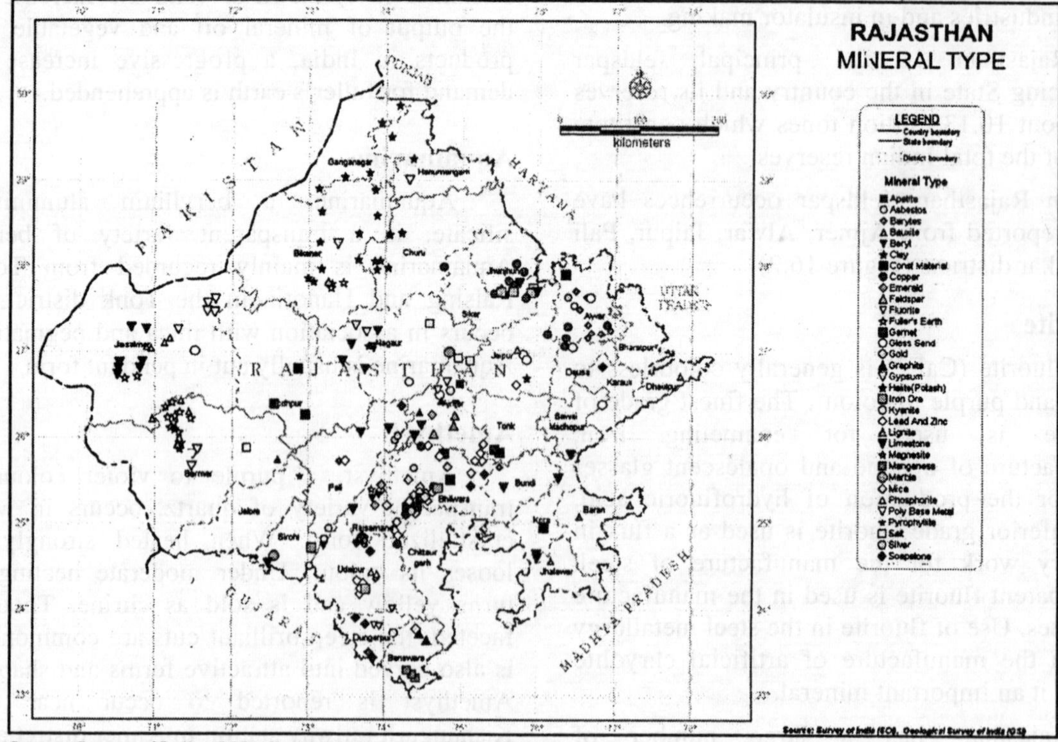

Source: Survey of India (SOI), Geological Survey of India (GSI)

Figure 16.2 Rajasthan Mining Area Location and Mineral Type Map

Chrysoberyl

Chrysoberyl (oxide of beryllium and aluminium) is green in colour and differs from beryl in its composition. Chrysoberyl occurs in association with mica and aquamarine bearing pegmatites of Ajmer and Bhilwara districts.

Dumortierite quartz (blue aventurine)

A massive opaque blue variety of crystalline quartz of semi precious quality is reported to occur at Dewal near Shahpura, Jaipur distrct. The intergrown crystals of dumortierite impart blue colour to quartz.

Emerald

Transparent variety of beryl possesses attractive colours and lustre. The colour varies from colourless to green, blue and yellow. Greeen coloured stones are known as emerald (Panna), while aquamarine has yellow colour. Emerald is a very highlypriced precious stone. In India emeralds are available mostly in Ajmer and Udaipur districts of Rajasthan.

Fuchsite quartzite (green aventurine)

Fuchsite quartzite is reported from Bhinder area Bhilwara district. It occurs in high grade psammites of the Pur Bunera Group. It is used in statue building.

Garnet

Garnet is used in the manufacture of coated abrasive like paper, cloth and discs for grinding and finishing glass, wood, hard rubber etc. In powder form, it is used for making lapping and buffing compounds. Garnet is generally obtained from small pits and shallow underground mines. It is also recovered as co product in processing of beach sands. Rajasthan is the only State producing gem variety. The garnet occurrences in Rajasthan are recorded from Ajmer, Bhilwara and Tonk districts.

Rock Crystal

Transparent crystal of quartz is used for necklaces and other ornaments. Associated with pegmatite it is available in Ajmer and Tonk districts.

Ruby / Corundum

Pegmatites intruding ultrabasic rocks give rise to development of corundum by desilication of feldspars. A small deposit of corundum/ruby at Jual in Tonk district and an occurrence associated with muscovite at Galwa in Bhilwara district are reported. The corundum is mainly used for manufacture of emery cloth.

GOLD

Gold, the 'king of metals', has always attracted mankind for its untarnishable yellow glitter for making the brilliant ornament par excellence and its relative scarcity and physical properties render it to this day the acknowledged international measure of material values. The usages of gold are as the standard for monetary system in the form of coinage or as bullion, for ornamentation especially in jewellery, in making gold leaf, gold lace and various fabrics, gold plating and in some chemical and photographic preparations. It also has some medicinal values.

Incidences of gold in multiple environments in several part of Rajasthan are well known. The occurrences are found in the Bhilwara, Aravalli and Delhi Supergroups of rocks of Archaean to Proterozoic age. Gold mineralization is hosted by a variety of rocks types. It is invariably associated with massive to semimassive, disseminated sulphides comprising pyrrhotite, pyrite, chalcopyrite, arsenopyrite, sphalerite, bismuthinite etc.

GRAPHITE

Graphite, also known as 'black lead', is a soft crystalline form of naturally occurring carbon. Graphite crucible industry and foundry units are the major consumers of natural graphite. Refractory industries consumes graphite for manufacturing magnesium carbon bricks for use in steel making in basic oxygen and electric arc furnaces.

There are numerous reported occurrences of graphite deposits in the country. The graphite mined is not of high grade. About 50% of our requirements are met by import. Domestic graphite is mostly used in the metallurgical industry for steel foundry and graphite crucibles. The mineral occurs in lump, flaky or powdery forms which together with the total fixed carbon (F.C.) content, determines the uses.

In Rajasthan, graphite occurrences have been located in Ajmer, Alwar, Banswara and Jodhpur districts.

Gypsum

Gypsum, a hydrous calcium sulphate containing 79% CaS04 and 21% water, is a widely distributed mineral and is generally transparent or white in colour. It occurs both as massive type or as fibrous lumps and transparent plates. The varieties of gypsum are selenite (finely crystalline), stainspar (fibrous with silky lustre), alabastar (fine grained, massive and compact) and gypsite porous gypsum of inferior quality mined with sand and grit. Gypsum is mainly used in Portland cement to retard and control the time of setting and is also used in the manufacture of ammonium sulphate for artificial fertilizer, plaster of Paris, distemper, 'burtonizing' water for beer making, potteries and as a flux in smelting nickel ores. Its uses in paper industry and as filler in insecticides are also important. Selenite is used for making gypsum plates for petrological microscopes. Alabaster is used for ornamental purpose.

Rajasthan accounts for about 90% of total production of gypsum in the country having a total reserves of 1,013.70 million tonnes as estimated by GSI. This constitutes a lion's share of the total proven reserves of 1,150 million tonnes of gypsum in the country. Production of gypsum in Reserves of gypsum have been located mainly in Bikaner, Nagaur, Barmer, Jaisalmer, Pali and Ganganagar districts (Table 29). The most important deposits occur in Nagaur district.

Ilmenite

Ilmenite is an ore from which titanium metal is derived. It is used in the manufacture of pigments. India possesses rich deposits of beach sands of ilmenite mainly along the coast of Kerala and at a few places along the east coast of India.

In Rajasthan, ilmenite is found to occur as scattered disseminations in the country rocks in various parts of Ajmer, Nagaur, Pali, Jodhpur and Sikar districts but these are of only academic importance.

Iron Ore

Iron is extracted from its oxide ores such as hematite and magnetite which when smelted with coke and limestone yield metallic iron. Iron ores are needed to meet the demands of modern iron and steel industry. Iron deposits of Rajasthan are of minor importance to meet the demands of any large scale iron and steel industry within the State. The iron ore deposits are located mainly at Morija, Rampura, Neemla Raialo, Kotputli and Dabla in Jaipur district; at Tonda, Sior, Zamalpur and Kali Pahari in Jhunjhunun district and at Narda, Manowas, Bagoli and Sarai in Sikar district. These occurrences are found associated with the Delhi Supergroup of rocks.

Kyanite

Kyanite is an important high alumina refractory mineral. Kyanite, sillimanite and andalusite are a closely related trio of aluminium silicate (A1203. Si02) minerals having the same chemical composition but differing slightly in physical properties.

In Rajasthan, occurrence of coarse bladed crystals of kyanite and sillimanite in small veins and lenses are reported from Ajmer, Bhilwara, Dungarpur, Jaipur, Pali and Udaipur districts.

Lead and Zinc

In India, Rajasthan and Gujarat have been centers for Pb and Zn mining since centuries

Zawar, Udaipur many mines bear ancient world Zawar is the place recognised as first ancient smelting site in the world.

In south eastern Rajasthan Pb Zn deposits o within the Bhilwara Supergroup of Archaean to Lo Proterozoic age (e.g. Agucha, Dariba Rajpura, Banera), Lower Proterozoic Aravalli Supergroup (zawar) and Lower to Upper Proterozoic Delhi S pergroup (e.g. Kayar Ghugra). In terms of tonnage a d grade the deposits occurring within the Archaean b sement rocks far exceed those occurring in the main ravalli and Delhi basins.

Lignite

Lignite, a low grade fossil fuel, represents an intermediate stage in the alteration of wood into coal. It is dark brown to blackish brown in colour carries a good amount of moisture and often crumbles into powder on drying. In Rajasthan, large deposits of lignite occur in Tertiary formations of Middle Eocene age in the Barmer, Jaisalmer and Nagaur basins falling in Barmer, Jaisalmer and Bikaner Nagaur districts, respectively.

Limestone

Limestone is one of the most important industrial mineral required in the manufacture of lime, cement, chemicals (soda ash, caustic soda, bleaching powder, calcium carbide etc.), fertilizer (ammonium nitrate etc.) and as a flux in iron and steel, ferro alloy and other metallurgical industries.

Limestone occurs in almost all the districts of the State but important deposits are located in Ajmer, Bundi, Chittorgarh, Jodhpur, Nagaur, Pali, Sirohi and Jaisalmer districts.

Megnesite

Megnesite, a hard white mineral, is the chief source of magnesium. Its importance lies in its property to withstand high temperature. It is extensively used in the manufacture of refractory bricks used in steel making furnaces, in the manufacture of some special types of cement (sorel cement) and in glass and rayon industry.

Megnesite deposit near Sarupa Chhaja in Ajmer district is known to have reserves of about 1.5 million tones of magnesite with an average grade of 50% MgCO3 (24% MgO). Occurrences have also been reported from Bhimana in Pali district and Lewaka Gurha area in Udaipur district.

Manganese

Manganese finds its main use in the manufacture of pig iron and steel, Ferro manganese (a chief ingredient for production of manganese steel), chemicals and in dry battery cells. India possesses vast resources of manganese ores and is ranked third in the world in the production of manganese. It is second to iron ore in terms of export. In Rajasthan manganese ore deposits of significance occur in Banswara district.

Mica

Mica is one of the important and indispensable minerals used in electrical and electronic industry due to its excellent dielectric strength and insulating properties. India, once enjoyed the status of being a leading mica exporter but with mica substitution coming in a big way, the export oriented mica mining industry is facing a great slump forcing many mica mines to close down.

In Rajasthan, five distinct mica pegmatite belts have been delineated : Beawer Ajmer belt, Bhilwara belt, Dungarpur Banswara belt, Tonk belt and Kaunthal belt. Mica is generally of ruby variety but mostly stained and spotted due to mineral inclusions.

Molybdenum

Molybdenum is mainly used as an alloying constituent in steel industry for imparting strength, elasticity and resistance to shock and fatigue. Metal molybdenum finds application in electrical industry where it is used for contact points, thermocouples and electrodes. Molybdenum tungsten alloy is used as filament support, in incandescent electric lamps, wireless receivers and transmitter valves. In India the entire requirement of this metal is met by imports.

Molybdenum does not occur in nature in its free or native state. In India molybdenum is generally found associated with Pb, Zn and Cu ores and is recovered as by product concentrates. All India recoverable reserves are placed at 8.4 million tonnes of ore.

In Rajasthan molybdenite (the sulphide of molybdenum) is found in the sodalite cancrinite pegmatite traversing the nepheline syenites at Manderia near Kishangarh in Ajmer district. Minor occurrences of academic interests are also reported from Khandela area of Sikar district. Molybdenum associated with the copper sulphide ores is also reported from the Khetri Copper Belt of Jhunjhunun district.

Nickel

Nickel is an important ingredient in the production of stainless steel. It is also used as a catalyst in making alloys and coins and in the storage of batteries.

Nickel compounds are widely used in electroplating, chemical and ceramic industry. The requirement of the country in respect of nickel is met solely by imports.

Ochre

Ochre are mineral pigments with iron as the colouring material either in the form of hematite or limonite or both. These earth pigments are insoluble and are used directly as paint material or to give colour, body and opacity to paint, cement, linoleum, rubber, glasses, enamels, plastics etc.

The maximum production is reported from Chittaurgarh district (69.5%) followed by Bikaner district (17.9%).

Oil and Natural Gas

The Eocambrian Hanseran Evaporite Group (HE of the Marwar Supergroup contains foetid dolomite a dolomitic limestone at four levels below halite beds N 2,3,4 and 5. It is gray to dark and black carbonaceous fine grained, massive to vuggy dolomite. It stains pap and smells strongly of petroleum.

Foetid dolomite from Lakhasar contains 30.40 m kg of ethane and hydrocarbon, and is considered t source rock for the petroleum. The hydrocarbon migrated and got trapped in structural traps around Devikot Nachana subsurface high separating at Jaisalmer basin from the Nagaur Bikaner basin.

Potash

Potash is one of the key elements used in fertilizers. India produces a meager quantity of potash in Gujrat as by product in manufacture of salt from sea water.

In Rajasthan, GSI has established by detailed geophysical investigation aided by extensive drilling, occurrences of subsurface halite bearing evaporates over an area of 30,000 km z covering parts of Ganganagar, Hanumangarh, Bikaner, Churu and Nagaur districts.

Pyrite – Pyrrhotite (Sulphur)

India does not possess any significant deposit of native sulphur. The entire sulphur requirement of the country is met from imports. The uncertainty of obtaining native sulphur in the international market and its exceedingly high prices has necessitated the search for alternate indigenous sources such as pyrite (iron sulphide). The economic value of pyritepyrrhotite lies in its being a source of sulphur. The sulphur as well as the pyrite itself constitutes the principal raw materials for the manufacture of sulphuric acid, which is use mainly in the manufacture of superphosphate fertilizer and also in several other chemical industries. Pyrite pyrrhotite deposit near Saladipura in Sikar district is of paramount importance.

Rock Phosphate

Rock phosphate, popularly known as Phosphorite, is an essential ingredient in the manufacture of super phosphate, a fertilizer, whose paramount importance in boosting the green revolution in our country needs no emphasis. Rajasthan contributes a major share in the production of rock phosphate in India.

In Rajasthan, phosphorites occur in diverse lithological formations ranging in age from Lower Proterozoic to Tertiary. Of these, the discovery of the Precambrian phsophorites in the Aravalli rocks of Udaipur district, in particular, has opened up new vistas in the sphere of exploration of rock phosphate in similar metasediments in India.

The phosphorite deposits in Udaipur and Banswara districts in Rajasthan are located within the rocks of the Aravalli Supergroup of Lower Proterozoic age while those of Alwar and Jaipur districts occur in the rocks of Delhi Supergroup of Lower to Middle Proterozoic age and that of the Chittorgarh district occur in the rocks of the Lower Vindhyan of Middle Proterozoic age. Phosphorite occurrences are also reported from the Mesozoic Formations of Jaisalmer.

Salt

Salt is one of the most common substances used in our daily life and is an essential constituent of human food. Salt is mainly prepared by evaporating seawater. Salt also occurs in nature as beds of rock salt, which are in fact, evaporates derived from ancient basins of highly saline seas. In dry regions like Rajasthan, salt is precipitated from saline ground water at the ground water surface in the form as efflorescence. Extraction of salt from efflorescence forms local cottage industries. Considerable quantities of common salt are prepared from the brines of saline lakes in Rajasthan. Salt lakes of Rajasthan, vary in characters are important sources of salt production in the country. The chief salt producing areas are in Nagaur, Barmer and Bikaner districts.

Silica Sand (Quartz and Glass Sand)

Quartz is an important constituent of almost all rock forming minerals. Free silica occurs in forms such as pure quartz (crystalline or massive), opal, agate and chalcedony. The main source for silica sand and quartz are sandstones, quartzite, pegmatite, quartz veins etc.

In Rajasthan, glass sand occurrences suitable for utilisation in glass industries are known from Jaipur, Bundi, Dausa, Sawai Madhopur, Tonk, Jaisalmer, Bikaner and Barmer districts. The other use of silica sand is in acidic Remi mass in foundry units.

Silver

Silver is valued as a medium of exchange and also for its use in jewellery. In recent years, its application in industry has increased rapidly

In Rajasthan silver is associated with the Pb Zn ores of Zawar, Rajpura Dariba and Bharak areas of Udaipur, Rajsamand and Bhilwara districts respectively and is extracted during the smelting of Pb from the Pb concentrates.

Siliceous Earth

Siliceous earth composed mainly of colloidal silica and other siliceous matters contains chiefly 69 81% Si02 and 4 12% Al203 with very little CaO, MgO and iron oxides. It is homogeneous, porous, easily breakable and light in weight. It has high absorption power. Its main use is as filters and refractory silica bricks. Siliceous earth occurrences are known from Barmer and Jaisalmer districts.

Steatite

Steatite, talc and soapstone are the three terms used in trade for a hydrous silicate of magnesium. A purer variety of compact and massive talc is steatite.

Rajasthan is the prime producer of soapstone in India. Rajasthan with 6.3 million tones of reserves accounts for about 41% of the all India total reserves of all categories. It produces about 86% of the national output. Occurrences of soapstone have been reported from Jaipur, Bhilwara, Dausa, Sawai Madhopur, Udaipur, Dungarpur and Banswara districts.

Tin

Tin is recovered from its oxide cassiterite. Metal tin is widely used in alloys and in the chemical industry. Tin oxide is used as polishing material fcr marble and granite. India does not possess workable deposits of tin and the entire requirement is met by imports. In Rajasthan, Bhilwara district holds some promise of holding a deposit.

Tungsten

Tungsten is a metal of strategic importance. It is used in the manufacture of high speed alloy and special steels, and is essential for machine tools, Armour plates and other military equipment. It is also used in various electrical appliances. Tungsten carbide, much harder than the hardest steel, is used in making drill bits, cutting tools etc. The best known deposit of the country is that of Degana in Rajasthan.

Uranium

Uranium is a valuable metal for atomic energy and holds important place both in war and peace.

Occurrences of uranium ores are known to occur at several places in Sikar, Bhilwara and Udaipur districts.

Vermiculite

Vermiculite, yellow and brown in colour, is an alternation product of mica and retains the micaceous cleavage. When vermiculite is heated, its volume increases considerably (12 times or more). The mineral is used where some extreme lightness and low heat conductivity are required.

Deposits of vermiculite have been recorded near Gudas in Ajmer district'

Wollastonite

Wollastonite, a calcium metasilicate (CaSi03) is one of the latest to enter in the field of industrial minerals, due to its fast growing industrial applications. It is used in the ceramic, enamel, glass, matches, paints, paper, plastics and plywood industries; in the preparation of art ware, ceiling tiles, floor tiles, insulators, as an extender of short fibred asbestos or as a replacement for non fibrous materials. Rajasthan is the only wollastonite producing State in India. Large deposits of wollastonite have been located at Pali, Sirohi, Udaipur and Ajmer districts in Rajasthan.

Status of Rajasthan in Occurrence of Mineral and its Production

The status of Rajasthan with regard to category wise recoverable reserves of minerals and production of minerals is shown in following Table 16.1 and Table 16.2.

Table 16.1 Major Mineral Statistics in Rajasthan (2016-17)

S. No.	Mineral	Leases (No.)	Area (in Hec.)	Production (Lac' Tons)	Sale Value (Crore' Rs.)	Revenue (Lac' Rs.)	Employment (Nos.)
A. Metallic Minerals							
1	Cadmium	0	0.00	0.001	1.28	19.12	0
2	Copper Ore	3	706.75	10.55	211.06	1582.93	1890
3	Iron ore	18	2240.10	35.63	743.82	3225.28	899
4	Lead Zinc	8	6964.97	61.36	1456,315	158476.09	7093
5	Silver	0	0.00	0.0046	1766.88	12229.46	0
6	Manganese	1	18.90	0.03	0.41	6 89	70
B. Other Minerals							
7	Asbestos	0	0	0	0	0	0
8	Beryl	3	14.3346	0	0	0.76194	0
9	Emerald Crude	1	46.32	0	0	5.59773	8
10	Epidote	1	5	0	0	0	0
11	Bauxite	1	123.5	0.0000	0.0000	1.232	1
12	Fluorite	7	992.95	0.00	0.00	0.12	2
13	Garnet (Abr.and Crude)	17	79.32	0.02	0.18	2.57	29
14	Kyanite	5	163.91	0.00	0.00	0.20	0
15	Lignite	7	14634.23	85.18	1137.28	7049.57	330
16	Limestone	41	22219.70	663.06	1374.74	56984.98	5143
17	Magnesite	4	19.75	0.00	0.00	0.00	0
18	Rock-Phosphate	3	1673.38	7.88	157.69	3783.82	621
19	Selenite	4	1016.30	0.13	1.29	31.68	28
20	Siliceous Earth	44	313.95	0.30	1.98	43.08	142
21	Vermiculite	3	34.63	0.00	0.00	1.61	0
22	Wollastonite	16	302.41	1.66	15.28	217.87	177
	Miss Income	0	0	0	0	0.3675	0
	Total	**187**	**51570.40**	**865.802**	**6868.19**	**243662.87**	**16433**

Source: Department of Mines and Geology, GoR, Udaipur

Table 16.2 Minor Mineral Statistics in Rajasthan (2016-17)

S. No.	Mineral	Leases (No.)	Area (in Hectors)	Production (Lac' Tons)	Sale Value (Crore' Rs.)	Revenue (Lac' Rs.)	Employment (Nos.)
1	Ball Clay	116	5362.75	34.37	240.61	2223.33	700
2	Barytes	2	69.85	0.03	0.32	1.63	15
3	Bentonite	28	53.31	1.41	6.49	179.80	162
4	Brick Earth	10	936.05	155.87	1069.78	4194.78	11530
5	Calcite	39	774.43	0.93	5.86	111.08	234
6	China Clay	265	4060.10	28.55	110.68	1350.16	2932
7	Chert	2	1.71	0.0049	0.0173	0.4900	0
8	Chips Powder/Limestone	33	33.55	0.86	2.14	57.45	41
9	Dolomite	15	1163.07	4.76	19.52	387.51	1469
10	Felspar	1332	6791.27	56.11	181.01	2153.24	5730
11	Fire Clay	3	59.99	0.00	0.00	0.00	0
12	Fuller's Earth/Kharia Mitti	20	206.21	0.19	0.82	23.22	63
13	Jasper	1	25	0.004	0.004	0.245	5
14	Granite	962	2327.03	38.43	1061.79	8149.60	7447
15	Gypsum	63	12782.44	18.17	97.93	3126.62	2485
16	Kaolin	2	8.00	0.01	0.04	0.70	30
17	Kankar-Bajri	108	98074.48	568.41	974.78	23594.22	21600
18	Limestone (Burning)	438	8368.99	85.85	170.66	11046.05	7434
19	Limestone (Dimnl.)	466	2056.50	39.26	458.24	6544.09	5311
20	Marble	1772	3015.49	135.24	1776.25	28054.71	17469
21	Masonary Stone	5576	6181.05	929.84	1501.50	28533.16	52308
22	Mica	15	208.11	0.03	0.42	1.50	22
23	Mill Stone	6	989.50	0.01	0.03	0.92	27
24	Mitti	7	7.00	66.97	36.06	825.57	1255
25	Murram/Gravel/Gitti	0	0.00	8.41	11.53	341.12	1925
26	Ochres	96	1257.65	17.06	37.55	827.38	569
27	Phylite-shist/Patti Katla	55	123.59	57.62	26.87	1020.62	3831
28	Phyrophilite	8	58.66	0.04	0.16	3.21	18
29	Quartzite	45	56.22	0.14	0.34	13.88	20
30	Quartz	1234	6969.68	15.00	46.38	1096.28	6782
31	Rhyolite	217	830.41	14.99	12.00	778.76	837
32	Salt Petre	0	0.00	0.00	0.00	0.00	0
33	Sandstone	1176	6354.11	120.11	625.95	17148.20	48397
34	Serpentine	274	337.58	9.66	144.92	2319.41	1665
35	Silica Sand	156	1950.89	8.56	38.79	697.23	909
36	Slate Stone	15	33.74	0.00	0.00	8.79	3
37	Soapstone	188	8400.96	12.01	99.89	1973.30	2754
	Inc. from Govt. Deptt.	0	0.00	0.00	0.00	9271.18	0
	Misc. Income	0	0.00	0.00	0.00	12431.68	0
	Total	**14745**	**179928.85**	**2428.90**	**8759.337**	**168491.10**	**205979**

Source: Department of Mines and Geology, GoR, Udaipur

MINERAL DEVELOPMENT AND MANAGEMNT

The National Mineral Policy announced in 1993 has opened the doors for both domestic and foreign direct investment in the Indian Mineral Industry. This is expected to give a fillip to the domestic mineral industry in furthering its growth and making the entire gamut of minerals and mineral based products domestically and internationally competitive. Sweeping changes in the national Mineral Policy has been recently made by Government of India making it adequately liberalised investor-friendly through amendments to MMRD act (now MMDR act). The economic liberalisation, which has been followed by promulgation of the new economic policy and subsequent General Agreement on Tariffs and trade (GATT) pose a new challenge before the nation's mining sector.

The mining operations in the State are carried out either by opencast method or buy underground method. The underground mining in the State is mostly in the organised sector where public sector organisations like M/s Hindustan Zinc Limited. M/s Hindustan Copper Limited, M/s RSMM Limited and M/s RSMDC are operating where due consideration is given to the protection of environment. However in the open-cast mining that too, mostly in stone mining, by and large are in unorganised sector, are discriminately dumping the waste and are not maintaining the proper environmental problem of stone industry in the state is waste generation (both during quarrying and processing).

Fuel Minerals

The State Directorate of petroleum has identified large of lignite at depths varying from 200-600 m. in Barmer-Sanchore basin. Recently a team from Australia namely M/s Line Energy has shown interest to develop underground Coal Gassification (USG) project in koslu area of Barmer district.

The State Government has also undertaken RandD project for Coal bed Methane (CBD) in Barmer and Bikaner districts. New Delhi has sanctioned Rs.4027 Crores for drilling. The exploration works has been entrusted to M/s Mineral exploration Corporation Ltd. (MECL).

Recently, Government of India has released 7 blocks of Coal Bed Methane for competitive bidding in which one blocks of Rajasthan (north of Barmer-Sanchore basin) has been included.

Although lignite deposits have been proved at 33 locations, favourable geological environment spread over a large area in parts of Bikaner, Nagaur and Barmer districts provides further scope for lignite exploration both shallow as well as deeper levels.

Gem Minerals

A long belt has been identified for exploration of emerald in parts of Udaipur, Rajsamand and Ajmer districts. Emeralds of Rajasthan has a good market value. Besides, State also has potential of semi-precious stones like Garnet, Aquamarine and precious stone like Diamond.

Lime Stone

This mineral has wide ranging applications and it has been located over 26 districts of the State. About 9000 MT reserves have been established. At present 15 major and 4 medium cement plants with an installed capacity of 16 MT of cement per annum are already in operation and about 10 cement plants each of over 1 MT capacity are likely to come up in near future on the areas already explored by the department in various districts. Apart from cement industry, the State is also producing large quantity of SMS grade limestone in Jaisalmer districts which otherwise would have been imported.

Arial Prospecting for Basemetals and Noble Metals

With the liberalisation of economy, the department look lead in inviting foreign collaboration in the mineral sector. Rajasthan is the first State in the country to grant reconnaissance permits over large areas for aerial survey for basemetals. In all 54 permits have been sanctioned to multinational companies covering almost 74000 sq. km area. The data generated by these surveys will be very useful for planning detailed ground prospecting for noble metals and basemetals.

New Prospecting Policy

For promoting scientific mining, bringing transparency, the State Government has farmed separate policies for granite, marble and gypsum.

Mineral's Waste

The dimensional stone industry has high percentage of waste generation. Solid stone waste, powder and slurry are the major source of environmental degradation in the mining belts of marble, sandstone, granite, flaggy limestone, state stone etc. The marble waste generation varies widely from 30 per cent by weight in mechanised quarrying to 65 per cent by weight in manual mining. In granite mines, waste generation is even on higher side.

Reclamation of Mined Out Land

Reclamation of the broken land concurrent with the excavation is an integral part of a properly planned and implemented mining strategy. The Indian Bureau of mines has laid down regulations in this regard and every lessee has to submit short term and long term plans for the land reclamation before commencing the mining activities. The IBM monitors the implementation of the proposed plan periodically.

The large scale mines owners and the government undertaking like M/sHZL, HCL, RSMML, RSMDC, Wolkem (India) Ltd., ASD, ASI, ACC and the major cement looks plants have a functional environmental management department. Which looks after land reclamation in their respective mining leases.

The mines and geology department in collaboration with the District Collector and District Industries Department has initiated the work of reclamation. The department has carried out a study to suggest measures for land reclamation in the Kotastone mining belts of the Kota and Jhalawar district using remote sensing data.

Dumping of Marble Slurry

A large quantity of marble slurry is being generated by the thousands of the marble cutting and polishing units installed in various parts of the State. Unplanned disposal of this slurry results in chocking of the natural drainage, corruption of the fertile top soil and silting in the water bodies, besides several other deleterious effects to the natural environment like air pollution.

In districts like Udaipur and Rajsamand, the District Collectors have identified lands(s) for dumping marble slurry in the industrial areas. The mineral based industry owners also have a growing awareness to pour this fluid waste to dry up in their factory premises and the resulting solid waste is being dumped by them in the nearby low lying areas. There is a need to identify such lands in all the industrial areas where stone cutting and polishing units are under operation.

Structural use of Marble Powder Slurry: At present, 7-8 hundred marble cutting industries in Rajasthan are producing 2000 MT of marble powder slurry per day as waste product. Set up of more such industries is increasing at a fast rate.

The slurry is being dumped in open and big heaps and dumps lying adjacent to the road side are spread in several Km. Disposal of marble powder slurry is a big problem, causing serious environmental hazard and pollution problem.

Malaviya National Institute of Technology (MNIT), Jaipur has studied the physical and chemical properties of marble powder slurry and concluded that marble powder slurry can be used for preparing compressed bricks, tiles and decorative tiles. To disseminate the outcome of the study, a workshop was organised in August, 1998 which was attended by the number of entrepreneurs, scientists and concerned officers of various departments.

Studies for the Solid Waste Management

The biggest challenge in the environment management of the mining belts in Rajasthan involves waste rock disposal. An unprecedented growth in the mining activities in last two decades has resulted in generation of huge quantities of waste rock in the mining belts, most of which is being dumped at the quarry-head or in the close proximity, without any scientific considerations. This cam be potentially hazardous from both environmental and mineral conservation points if view. The unplanned and unscientific dumping of the waste rock results in disturbing the regional environment and natural water regimes. An additional problem is posed by the land degradation by the mine waste, which at places, is dumped directly over the cultivable lands. Moreover, in absence of adequate dumb-stabilisation measures, the waste rock spills over the cultivable land and the slurry generated by the rainwater deposit fine-grained mud over the soil in the vicinity of the waste dumps. This results in disturbing the chemical and physical balance of the natural soil. Expedient dumping of the waste rock over the quarry head leads to repeated handling of the waste rock, and occasionally the process becomes too expensive to mine out the mineral economically, leading to the wastage of economically viable minerals.

The problems are especially acute in the dimensional and decorative stones mining belts, where the waste rock problem is exaggerated by the small size of the leasehold areas, which gives the mines owners very little space for systematic dumping of the mining waste.

In view of the above stated problems, the department carried out several studies using remote sensing data in ecologically the worst affected mining belts, viz. the marble mining belts of the Rajsamand district, limestone (Kota stone) mining belts of the Kota and Jhalawar districts and the sandstone mining belt of the Jhalawar districts. In these belts, the remote sensing data were interpreted to prepare land-use maps, which were subsequently used for identifying non-mineralised fallow land use in vicinity of the mining areas, which could be used for dumping the mining waste. Several such areas have been tentatively identified in the mining belts of the Rajsamand district, Work is now continuing using GIS techniques for delineating such areas.

Similar studies have been carried out in the Bijoliya sandstone mining belts also. However, in these parts dumping in the mined out parts of the quarries is not possible because of almost horizontal dips of the sandstone horizon. In this belt, the non-mineralised wastelands need to be identified in the vicinity. However, the waste rock in this belt usually comprises fractured sandstone, which can be used as building stone and road material.

Afforestation in Mining Clusters

The department has established an environment and development cell in the Directorate, Udaipur in August, 1995. Since then, all the mining officers in different districts of the state have been instructed to encourage and monitor afforestation and plantation activities in all the mining leases falling their jurisdiction. This has yielded encouraging results and the leases have started plantation and afforestation activities in their leases.

THE NEW MINERAL POLICY 2015

Rajasthan is the largest State in India. It boasts of three natural heritage facets: the Aravalli hill range, the Thar Desert and the Minerals. The State is a virtual treasure trove of minerals of indescribable qualities and potential – 79 varieties of minerals are established, of which 57 are being produced. The State enjoys 9% of the country's mineral production. So far, 34,991 Leases have been allotted, of these 3,403 are of Major Minerals and 11,861 of Minor Minerals, including 39 PL and 1 RP. The revenue from minerals during 2013 – 14 was Rs 3,088.66 crores. The Crude Oil alone contributed Rs 5,953 crores during 2013 - 14. Nearly four lakh direct and about 20 lakh indirect jobs are provided.

The State now recognizes minerals as the economic-power-house. The New Mineral Policy 2015 views that the mineral resources are finite. Therefore, the emphasis should now be on introducing cutting-edge technology in exploration, exploitation, minimizing wastage; the waste be re-cycled, widespread job-creation, especially to workers drawn from scheduled castes and backward communities, and greater revenue to the Government. The private sector is the fulcrum of this sector. Therefore, the Policy aims at:

(i) the development of economically sound and stable mining, minerals, metal and mineral reclamation industries,

(ii) the orderly and economic development of mineral resources, reserves, and reclamation of metals and minerals to help assure satisfaction of industrial, security and environmental needs,

(iii) the study and development of methods for the disposal, control, and reclamation of mineral waste products, and the reclamation of mined land, so as to lessen any adverse impact of mineral extraction and processing upon the physical environment that may result from mining or mineral activities.

MINERAL EXPLORATION

Mineral Sector

- Mineral exploration to embrace modern technologies, invite and facilitate the national and international organizations as well as entrepreneurs.
- Induction of latest technology by procuring modern equipments like Portable XRF, DGPS, UV Lamp, Core splitter, Bore hole camera, Bore hole Scanner, Analyser and Logger, Satellite Imageries, Camp Equipments, etc. to expedite the exploration and reducing the cost.
- Renovation of the chemical laboratory and equipped, obtaining sophisticated and advanced analysing instruments such as XRF (4 KW), Bomb Calori Meter, Digital Reflectance Meter, etc.
- For deep seated mineral deposits like base metal, noble metal, lignite and Potash, appropriate technology to be adopted for underground mining protocols.

Hydrocarbon Sector

- The Petroleum Exploration License (PEL) was granted by the State Government to Shell India and ONGC in 1995. Shell India carried out exploration but in 2001 Cairn became its shareholder in place of Shell India, of this Block. Cairn India carried out exhaustive exploration and discovered oil.
- In march, 2010, the resource potential from 3.7 - 4.1 billion barrels (i.e. 480-500 million tonnes) were revised to 5 billion barrels (i.e. 690 million tonnes) of oil equivalent Oil in-place have been proved alone in Barmer-Sanchore Basin.
- Subsequently, in March, 2012, these reserves were again revised to 6.5-7.2 billion barrels. (i.e. 900-1000 million tonnes) of oil equivalent in-place.
- About 30 billion cubic meter of rich and lean gas reserves have been discovered in Jaisalmer Basin in the gas fields namely Manhera tibba (1994), Tanot, Dandewala area (1996) and SGL Field (Shahgarh) (2009-10). Natural gas is being produced to the tune of 9.5 cubic meter per day from Jaisalmer basin. It is being supplied presently to Ramgarh Power Plant (110MW+160 MW).
- Oil India Limited has undertaken trial production (2009) of heavy oil in

Baghewala, Jaisalmer area, for the exploitation of proved in-place heavy oil reserves of 25.00 million tonnes and bitumen reserves of 53.00 million tonnes and has produced about 64 metric tonne of heavy oil.

- The State will regularly monitor exploration and development activities in these blocks as envisaged under Minimum Work Programme (MWP) of Production Sharing Contract (PSC) signed with GOI.
- Exploration is underway in four petroliferous basins like Jaisalmer Basin, Barmer-Sanchore Basin, Bikaner-Nagaur Basin and Vindhyan Basin covering 15 districts such as Barmer, Jaisalmer, Bikaner, Sriganga-nagar, Hanumangarh, Jalore, Kota, Jodhpur, Jhalawar, Baran, Bundi, parts of Chittorgarh and Sawai Madhopur.

Socio-economic development of the state and nation is directly related to availability of indigenous minerals, their proper exploitation and utilization. Fortunately Rajasthan is renowned for its mineral wealth. It occupies a significant position on the mineral map of the country.

In the area of "Major Minerals" Rajasthan is possessing 1,361 working mines and stands in front line of the countries major mineral production. The value of mineral production in Rajasthan in the year 2004-05 at Rs. 2,910 crores increased by 2 per cent from that in the previous year. Its Share in total value of mineral production of the country in the year 2004-05 was about 4 per cent. The value of "Minor Mineral" produced during the year 2004-05 was worth of Rs. 16,762,624,000 and Rajasthan ranked among leading state of the country in production of minor minerals with a total of 8,243 mining lease and 14,396 quarry licences for minor mineral excavations.

State possesses a wide range of mineral resources and offers considerable potential for future. The important metallic minerals with which name of the state is intimately associated are lead, zinc, copper, silver, cadmium and iron ore. Tungeston exploitation is presently not in operation and deposit is under reassessment for future mining. Gold occurences in association with copper have also got success in located in state. In recent past oil exploration agencies also got success in locating natural gas and petroleum deposits in the state and now the state has emerged as a repository of wealth of crude- oil, natural gas and lignite as well. The extensive occurrences of petroliferous basins in western Rajasthan have made it a large potential region for hydro- carbons. As a result of exploration activities of various agencies in Rajasthan, lignite deposits/ occurrences have been proved at 36 locations in Bikaner, Nagaur, Barmer and Jaisalmer districts. It is also richly endowed with the variety of non-metallic minerals distributed throughout its territories and significantly contributing to the industrial growth of the country. Rajasthan has also enormous deposits of building stones like marble, granite sandstone, limestone, slate, shale, phyllite, schist etc.

The state produces 50 varieties of minerals with the occurrence of 79 kind of minerals. Rajasthan was the sole producer of garnet (Gem), jasper, selenite and wollastonite and presently state also enjoys a monopoly in production of these minerals. Almost the entire production of calcite gypsum, lead concentrate, zinc concentrate and amphibole variety of asbestos in the country was reported from state and it is leading producer of these minerals in the country. Besides, Rajasthan was the leading producer of ball clay accounting 93 per cent, silver 96 per cent, and stealite 77 per cent production in the country. It was also the second largest producer of fluoride (graded) 23 per cent, felspar 20 per cent, copper concentrate 44 per cent and silica sand 18 per cent in the country. Further increase in production was reported is respect of natural gas (utilized) by 27 per cent, lead concentrate 12 per cent, zinc concentrate 13 per cent, felspar 10 per cent, gypsum 28 per cent, dolomite 70 per cent, from that of previous year.

Metallic Minerals

The basic metal mineralistion in Rajasthan is, perhaps, the most extensive compared to other states as indicated by the abundance of ancient working representing past mining activities. The copper, lead and zinc deposits in the state are confined to a stretch from north east i.e. Alwar district to south west up to Banswara, Pali, Rajsamand, Udaipur, Sirohi, and Dungarpur districts for a strike length of 260 kms and a width of 60 to 90 kms. These deposits are having associations of gold, silver and cadmium too. Besides these metals manganese deposits of Banswara and iron ore deposits of Jaipur, Dausa, Sikar, Jhunjhunun, Udaipur, Bhilwara, districts are also attractions of the entrepreneurs.

A total of 10 large area reconnatory permits have been granted over an area of 11,190.47 sq. kms in the state of Rajasthan to private and multinational companies. Further 46 application for the grant of large area permits for the exploration of the base metal deposits in Rajasthan are under processing. Which it self shows the importance of the metallic mineral deposits zone in the state (Figure 16.3).

Rich deposits of zinc and lead are extensively mined from Rampura- Agucha (Bhilwara), Zawar (Udaipur), Rajpura and Dariba (Rajsamand). New lead and zinc deposits have been proved near Ghughra in Ajmer district and mining lease have been granted to Hindusthan Zinc Ltd. for exploitation. Exploration for copper and multi metals was carried out by MEC around Bhagal copper deposit Chittaurgarh district and Devtalai multi metal deposit in Bhilwara district. Investigation for base metals are either completed or under progress at Sindeswar Khurd and Kalian, Dariba Bethumbi belt Rajsamand district, Sawar and Srinagar Ajmer district, Dhani- Based in Dausa district etc. Copper is mainly mined in the Khetri- Sighana belt in Jhunjhunun district. Smaller deposits of copper are also found in Kho Dariba, the Bhagoni area in Alwar district, Deri-Basantgarh, Azari, Pipela, Goliya, Sampganj in Sirohi district, Anjani in Udaipur district, Padar ki Pal in Udaipur district and Pur- Dariba Banera in Bhilwara district. Silver and Cadmium ore recovered as by product form the lead zinc ores of Zawar, Rajpura Dariba and Rampura Agucha deposits. Gold associations are proved with Khetri Singhana copper deposits, Dhani Based copper deposit, Srinagar copper prospect and a lot more. Prospecting have also been carried out for gold near village Bhukia, district Banswara, proved a deposit with workable parameters under present economic senerio.

Wolframite and scheelite are ore of tungsten which is a mineral of strategic importance. Wolframite is found near Degana in Rawat hill in Nagaur district and near Baida in Sirohi district. Scheelite occurance have been locaked near Nana and. Bar in Pali district while Bekla and Khila in Sirohi district. Low grade manganese is found is Banswara and Udaipur districts. Deposits of low grade iron are also occurs ip parts of Jaipur, Dausa, Udaipur, Jhunjhunun, Sikar, Ajmer and Nagaur district. Base metal deposits along with nobel metal deposits of state are being explored throughout in state.

Mineral Fuels

Energy is universally recognised as one of the most significant inputs for economic growth and human development. Oil, natural gas and coal are the three primary commercial sources of energy. The share of commercial sources of energy is almost 70 per cent in the country. Rajasthan has vast tract of rock formations spread over to 1,50,000 sq. km in Barmer - Sanchor basin, Jaisalmer basin and Bikaner Nagaur basin which has the potential of hydro carbon and lignite depsoits.

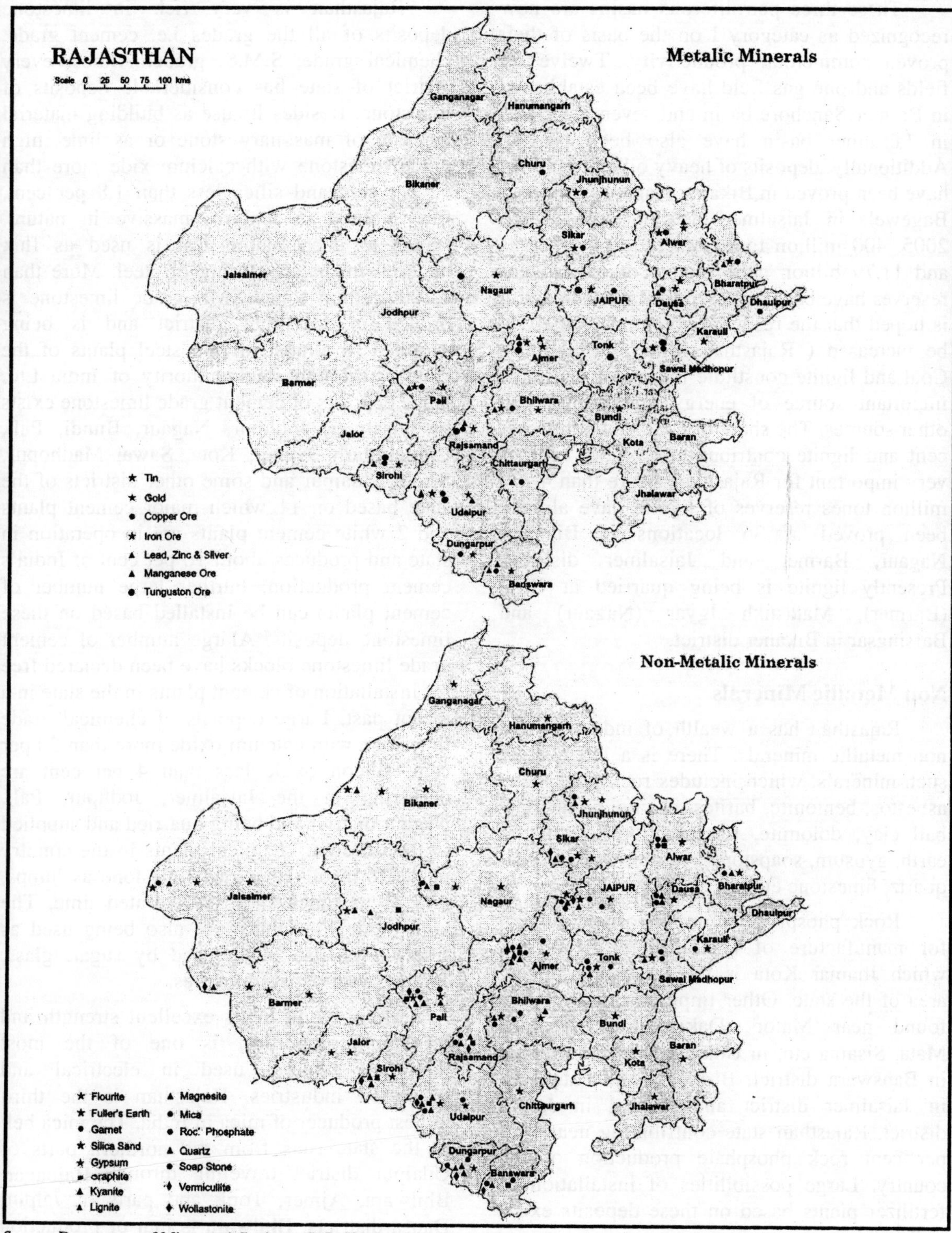

Source: Department of Mines and Geology, GoR, Udaipur

Figure 16.3 Metalic Minerals in Rajasthan

These three petroliferous basins are now recognized as category I on the basis of their proven commercial productivity. Twelve oil fields and one gas field have been established in Barmer Sanchore basin and seven gas fields in Jaisalmer basin have also been proved. Additionally, deposits of heavy oil and bitumen have been proved in Bikaner Nagaur basin near Bagewela in Jaisalmer district. Upto March 2005, 400 million tones of crude oil reserves and 11.79 billion cube meters of natural gas reserves have been assessed in Rajasthan and it is hoped that the resource potential is likely to be increased (Rajasthan Mineral Bulletin). Coal and lignite constitute the major and most important source of energy and outrank all other sources. The share of coal is about 96 per cent and lignite contribution is 4 per cent, is very important for Rajasthan. More than 4,270 million tones reserves of lignite have already been proved at 36 locations in Bikaner, Nagaur, Barmer and Jaisalmer districts. Presently lignite is being quarried at Giral (Barmer), Matasukh Igyar (Nagaur) and Barsingsar in Bikaner district.

Non Metallic Minerals

Rajasthan has a wealth of industrial and non metallic minerals. There is a long list of such minerals, which includes rock phosphate, asbestos, bentonite, barites, calcite, china clay, ball clay, dolomite, felspar, fluorite, fuller's earth, gypsum, soapstone kyanite, wollastonite, quartz, limestone etc.

Rock phosphate is an important mineral for manufacture of chemical fertilizers, for which Jhamar Kota is the largest producing area of the state. Other important deposits are found near Mator, Dakankotra, Neemach Mata, Sisama etc, in Udaipur district. Sallopat in Banswara district, Birmania and Fatehgarh in Jaisalmer district and Achrol in Jaipur district. Rajasthan state contributing nearly 75 per cent rock phosphate production of the country. Large possibilities of installation of fertilizer plants based on these deposits exists in the state.

Rajasthan is very rich in limestone deposits of all the grades i.e. cement grade, chemical grade, S.M.S. grade. Almost every district of state has considerable deposits of limestone. Besides its use as building material in from of masonary stone or as lime, high grade limestone with calcium xide more than 52 per cent and silica less than 1.8 per cent, fine grained in texture, massive in nature, stands to decripitation test is used as flux material in manufacturing of steel. More than 1,100 million tones SMS grade limestone is proved in Jaisalmer district and is being supplied to almost all the steel plants of the country through steel authority of India Ltd. Huge deposits of cement grade limestone exists in Jaisalmer, Jodhpur, Nagaur, Bundi, Pali, Chittaurgarh, Ajmer, Kota, Sawai Madhopur, Sirohi, Udaipur and some other districts of the state based on 11 which major cement plants and 2 white cement plants are in operation in state and produces about 16 per cent of India's cement production. Further large number of cement plants can be installed based on these limestone deposits. Alarge number of cement grade limestone blocks have been declared free for installation of cement plants in the state in a recent past. Large deposits of chemical grade limestone with calcium oxide more than 50 per cent, silicon oxide less than 4 per cent are occurring in the Jaisalmer, Jodhpur, Pali, Nagaur district and being quarried and supplied to the different chemical plants in the country in the form of either raw limestone as lumps, gitti or as quick lime or hydrated lime. The high grade limestones are also being used as filler. It is also being used by sugar, glass, ceramic, and paper industries.

Mica owing to its excellent strength and insulating properties, is one of the most important mineral used in electrical and insulation industries. Rajasthan is the third largest producer of mica in India. The mica belt in the state runs from the northern parts of Udaipur district traveling through Rajnagar, Bhilwara, Ajmer, Tonk and parts of Jaipur, Dausa districts. Bhilwara is maj or production centre of mica.

The state rich in asbestos deposits, is found in Jharol and Kherwara of Udaipur district and Deogarh belts in Rajsamand district, Arjunpura, Mangliwas, Hatundi and Rajgarh area in Ajmer District, Dewal, Khemaru in Dungarpur district and Asan, Sendra, Manpura in Pali district contributing about 89 per cent of country's production.

Silica sand is the main mineral for production of glass and the state of Rajasthan is the second leading producer of silica sand in the country. This is mainly found in Bundi, Sawai Madhopur, Bharatpur, Jaipur, Dausa, Sikar, Bikaner and Barmer districts (Figure 16.4).

Wollastanite is one of the latest minerals added to the list of industrial minerals of Rajasthan. The state is the sole producer of this mineral in the country. Significant deposits of wollastonite are fond in Belka Pahar near Khilla of Sirohi district and Khera Uperla in parts of Pali district and Sirohi districts. Recently large deposits of wollastonite has been located in Gola Alipura, Pratappura Kranti area of Ajmer district and Anna, Asan area of Pali district.

Rajasthan contributes 95 per cent of the country's gypsum production. Gypsum is widely used as a fertilizer and for making plaster of Paris. Gypsum is widely distributed through out of the north western part of the state, particularly in the districts of Hanumangarh, Bikaner, Ganganagaer, Nagaur, Churu, Banner, Jaisalmer, Pali and Jalor.

Bentonite which possesses inherent swelling properties is an important clay of great commercial importance being used in oil well drillings, cleaning and moulding etc. Rajasthan is the second largest producer of this mineral. Rich deposits of bentonite are found in Barmer and Jhalawar districts besides smaller deposits in Bikaner, Jaisalmer and Nagaur districts.

Fluorite also known as fluorspar is a strategic mineral of great importance. It is used in manufacture of hydrofluoric acid, artificial cryolite ceramics, glass, enamels and welding rods etc. It is also used as flux in metallurgy in iron and aluminum. It is found in district of Jalor, Sirohi, Bhilwara and Ajmer.

Rajasthan is also second largest producer of barites in the country, it is mainly found in Alwar and Udaipur district. It is used in preparing heavy muds for oil well drilling, paints, lithophone industries and source material for barium chemicals. It is also used as filler in paper, rubber, leather and textile industries. Barites is also located as strtiform deposit in association with magnetic deposits of Rajsam and and Bhilwara district and intersected in drilling.

Soapstone is very important raw material in various industries like cosmetics, paper, textile, pesticides, insecticides, paints and rubber depending on the grade. Rajasthan is the main producer of various grades of soapstone and contributes about 87 per cent of country's production. Important production centres are the districts of Ajmer, Alwar, Banswara, Bharatpur, Bhilwara, Dungarpur, Jaipur, Dausa, Jhunjhunun, Sawai Madhopur, Tonk, Pali, Rajsamand and Udaipur.

The china clay and ball clay are mainly used for ceramic, paper, textile, and rubber omdistries, where as fire clay is mainly used for manufacture of firebricks, blocks, retorts and crucibles etc. Rajasthan contributes about 50 per cent production of china clay, 71 per cent of ball clay and 10 per cent of fire clay in the country. China clay is found to occur in Bhilwara, Jaipur, Chittaurgarh, Sawai Madhopur, Sikar, Pali, Jhalawar, Bundi etc, ball clay occurs in Bikaner, Alwar, Bundi and Nagaur districts.

Rajasthan enjoys monopoly of gem variety of garnet in different colors red, deep red, lilacs, pink. Tonk, Ajmer and Bhilwara are the main producing districts of gem varieties of garnet.

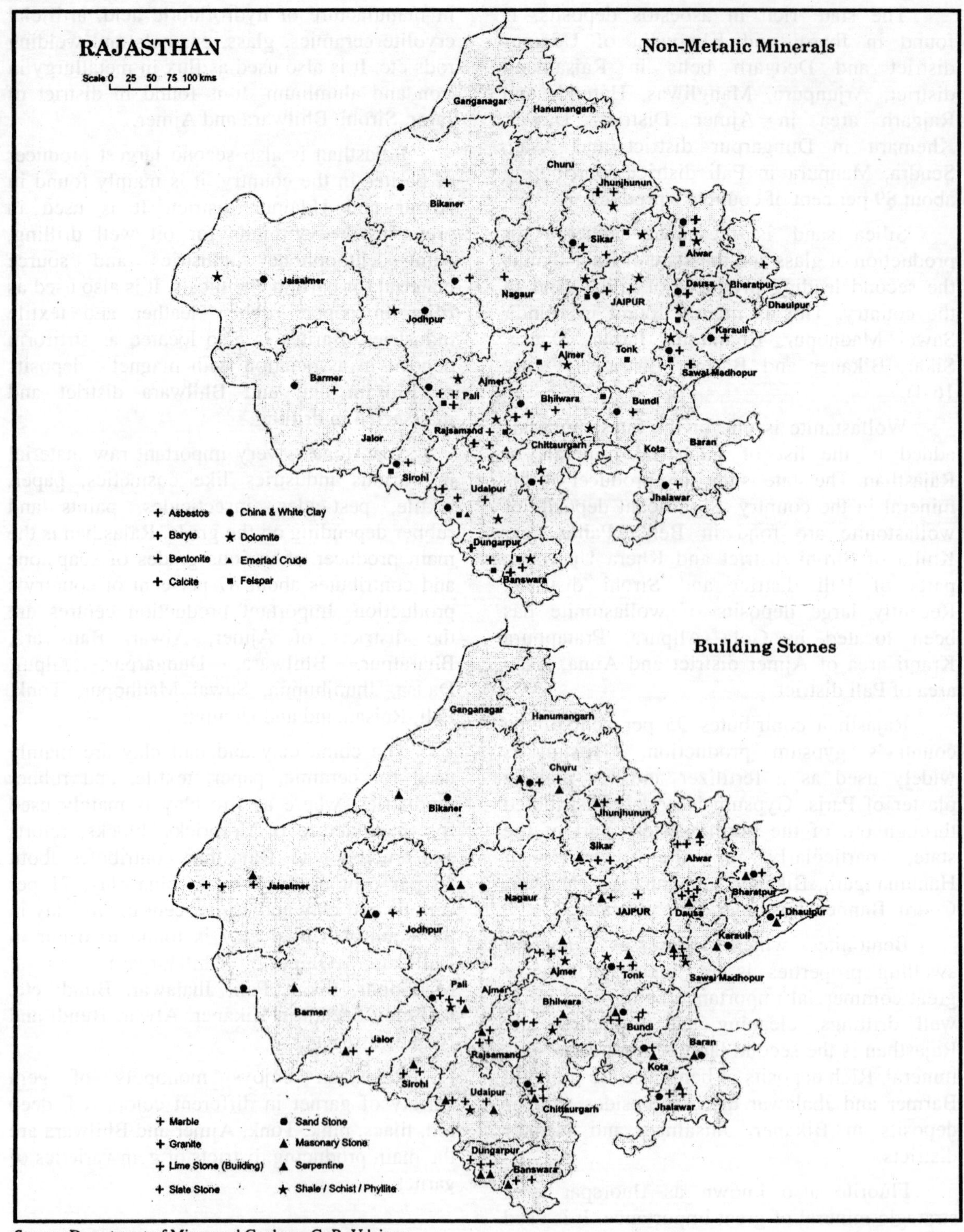

Source: Department of Mines and Geology, GoR, Udaipur

Figure : 16.4 Non-Metalic Minerals in Rajasthan

Felspar is mainly used in glass making and in ceramics. Rajasthan is contributing about 70 per cent of country's production of felspar. Felsper occurrences are mainly in Bhilsara, Ajmer, Pali, Tonk, Rajsamand, Alwar and Jaipur districts.

Quartz is produced in Bhilwara, Ajmer, Sikar, Sirohi, Pali, Jhunjhunu, Bundi, Alwar, Rajsamand, Udaipur and Tonk districts. It is an essential raw material of ceramic industry. High grade quartz is also important raw material of optical industries and electronic components.

Building Stones

The state is famous for its varieties of ornamental, building and decorative stones such as marble, granite, sandstone, limestone, etc., white marble of Makrana, red and buff colored sand stone of Jodhpur and Dhaulpur, yellow limestone of Jaisalmer, flaggy limestone of Kota, Nimbhahera and Chittaurgarh and pink granites of Jalor, red granites of Lakha, Jaisalmer have won legendary fame. Besides Makrana (Nagaur district) sizeable resources of different kinds of marble are found in Udaipur, Ajmer, Sirohi, Banswara, Chittaurgarh, Dungarpur, Alwar and Rajsamand districts. Sandstone is extensively quarried in Jodhpur, Bundi, Dhaulpur, Sawai Madhopur, Nagaur and Kota districts. Kota blue stone is found in parts of kota districts near Ramganj Mandi, and Chechat areas. In Rajasthan large tract of granites are found in Jalor, Barmer, Sirohi, Pali, Tonk, Ajmer, Alwar, Bhilwara, Jaipur, Sikar and Rajsamand districts. The schist, phyllite and slate stone of Chittaurgarh, Bhilwara, Tonk, Ajmer, Alwar are also having their place in world market.

Chapter 17

WASTELANDS

Wastelands, in general, are those lands which are degraded and at present lying unutilised or which are not being used to its optimum potential due to different constraints. The word "Wasteland" is a very broad term and is being defined in a number of ways by different authors. The description and classification of Wastelands are suggested by the National Wastelands Development Board (NWDB), New Delhi (1986) is now usually followed by all concerned. The Board has described Wastelands as "Degraded land which can be brought under vegetative cover, with reasonable efforts and which is currently under utilised and land which is deteriorating for lack of appropriate water and soil management or on account of natural causes. Wastelands can result from inherent/imposed disabilities such as by location, environment, chemical and physical properties of the soil or financial or management constraints".

These lands are of two broad classes, the culturable and the non-culturable Wastelands. Culturable Wastelands are capable of or have the potential for development for agricultural or pasture purposes or even can be afforested. They are not being used, at present, due to such constraints as lack of water for irrigation, salinity or alkalinity in the soil, soil erosion, water-logging, and unfavourable physiographic position or human neglect. All these lands can be put to some productive use after proper treatment and reclamation measures.

Non-culturable Wastelands, on the other hand, are barren lands that can not be put to any porductive use, either for agriculture or to develop forest cover. Barren rocky outcrops are one fo the example of such lands.

Availability of a reliable database on the wastelands of the country is pre-requisite before embarking upon planning strategies for their development. Information on the extent, nature and location wastelands helps in designing interventions targeted at the resource-poor sections of the rural population. It is towards this end that DoLR initiated an exercise for carrying out the mapping of wastelands in entire country in phases, using remote sensing technology.

Management of natural resources, especially land resources, is the key to attain food, water and environmental security. This has special relevance since per capita availability of agricultural land in India is decreasing rapidly due to population growth, industrialization and urban expansion. Amongst the various options available for improving land productivity, development of wastelands is one of the most viable options. This calls for preparing comprehensive spatial database on wastelands, their types and spatial distribution so that such lands can be put under appropriate use like greening, reclamation so that more land is available for productive use.

In India, this widening gap had led to over-utilization of land resources regardless of their potential and limitations, resulting in the creation of vast stretches of wastelands affected by soil salinity, water logging, desertification, soil erosion etc., and thus progressive decrease in per capita cultivable land. Ecological imbalance is also another manifestation of excessive pressure on land resources. To arrest this negative trend and to bring the wastelands back to their productive capabilities, for sustaining the current production levels, adoption of appropriate and innovative technologies for wasteland amelioration has to be planned on a scientific basis, which hinges on the information base to be generated on wastelands, i.e., their spatial distribution, severity of the problem, current utilization status etc.

Remote Sensing data has an important role to play in wastelands inventory, planning and monitoring. The experience of utilization of remote sensing data for the above stated purposes has been highly rewarding in the fields of agriculture, water resources, ground water targeting, watershed development etc. The information obtained from satellite platform is timely, reliable, accurate and cost-effective. Also, due to its availability on a repetitive mode, it is useful for monitoring the status periodically.

TYPES OF WASTELAND

Gullied / Ravine Land

Gullies are formed as a result of localized surface run-off affecting the unconsolidated material resulting in the formation of perceptible channels causing undulating terrain. Gullies develop from rills which are tiny water channels with a few centimeters deep, formed as a resultant impact of heavy rainfall and wearing action of run-off generated there from. These are the first stage of excessive land dissection followed by their networking. Ravines are basically a extensive systems of gullies developed along river courses.

They are commonly found in sloping lands, developed as a result of concentrated run-off. They appear in light yellow to bluish green in color depending on the surface moisture and depth of erosion. They vary in size, shape with irregular broken network pattern. Further classification of this category is possible based on the depth, width, bed slope, frequency and morphology of bed material of the ravines. Two categories are recognized in ravine lands.

1. Gullied / Ravine Land - Medium

Medium ravines that have a depth ranging from 2.5 to 5.0 metres

2. Gullied / Ravine Land - Deep

Deep ravines that have a depth of more than 5 metres,

Land with scrub

These areas possess shallow and skeletal soils, at times chemically degraded, extremes of slopes, severely eroded and lands subjected to excessive aridity with scrubs dominating the landscape. They have a tendency for intermixing with cropped areas. They appear in light yellow to brown to greenish blue depending on the surface moisture cover and vary in size from small to large having either contiguous or dispersed pattern. Two sub-classes are possible to be delineated based on the presence of vegetation cover on such lands.

3. Land with scrub - Dense

The vegetal cover is often more than 15 percent. Scrublands are associated with moderate slopes in plains and foot-hills and are generally surrounded by agricultural lands.

4. Land with scrub - Open

This category has a similar description as mentioned in the earlier class excepting that they possess sparse vegetation, generally less than 15 percent or devoid of scrub. This land is generally prone to deterioration due to erosion and has a thin soil cover.

Waterlogged / Marshy Land

Waterlogged land is that land where the water is at/or near the surface and water stands for most of the year. Marsh is a land, which is permanently or periodically inundated by water and is characterized by vegetation, which includes water and reeds.

These lands are distinctly seen in light to dark blue tone of varying size and shapes on satellite image. By virtue of their location in flood plains, coastal tidal flats, lagoons and river plains, these can be delineated with good accuracy. Subtle variations in tones depending on the presence of aquatic vegetation, irregularity in shape, textural behavior of the vegetation - smooth or mottled depending on the duration of waterlogging may sometimes share common spectral properties with that of crops. Data from other sources including ground data has to be used for its proper delineation.

Depending on duration of water logging period/seasonality, two subclasses viz., seasonal waterlogged and permanent water-logged areas are possible to be delineated.

5. **Waterlogged/Marshy Land -Permanent**

The duration of water logging is for at least 6 months or more in a year.

6. **Waterlogged / Marshy Land -Seasonal**

The duration of water logging is for less than 6 months in a year.

Land affected by salinity/alkalinity

Lands that are affected by salinity/alkalinity portray the qualities that have adverse effect on the growth of most plants due to action or presence of excess soluble salts (saline) or high exchangeable sodium.

Salt affected lands appear in different tones of dull white to white on the satellite data in different geographical conditions. In the irrigated alluvial plains of Punjab, Haryana, Uttar Pradesh and northeastern Rajasthan, they appear in bright white to white tone, while in inland coastal areas. like Gujarat, they appear in light blue tone. The size and shape of these land units vary from small to big, contiguous to non-contiguous. In the coastal belt like parts of Surat, Eastern part of Tamilnadu, they are seen in small patches midst irrigated crop areas. Coastal saline soils may be with or without ingress or inundation by sea water.

Considering the degree of salinity, two sub-classes are possible to be delineated, which are given below.

7. **Land affected by Salinity/Alkalinity - Strong**

These patches appear bright white in colour. These lands do not support any kind of vegetation during any season. Chemically, their Electrical Conducti-vity (EC) levels (dS/m) are more than 30, pH more than 9.8 and Exchangeable Sodium Percentage (ESP) is more than 40.

8. **Land affected by salinity/alkalinity - Medium**

They appear dull white to light red in tone and are often associated with some scrub vegetation or salt-resistant crop during kharif season. Chemically, their Electrical Conductivity (EC) levels (dS/m) are between 8 and 30, pH 9.0 and 9.8 and Exchangeable Sodium Percentage (ESP) is between 15 and 40.

Shifting Cultivation

It is the traditional practice of growing crops on forested / vegetated hill-slopes by 'slash and burn' method. Known by different names in different regions like bush-fallow jhuming, podu cultivation etc, this is a commonly observed practice in the North-Eastern states, parts of Orissa, Andhra Pradesh and the tribal areas of Chattisgarh. They appear in light yellow or greenish tone in color in small, scattered patches, irregular in shape, non-contiguous and dispersed and located on hill slopes. These are associated with mountainous / hilly areas midst forest cover and forest- cleared areas. Two sub-categories viz., current Jhum and abandoned Jhum are possible to be delineated under this class.

9. **Current Jhum**

These are the areas that are used for cultivation by the process mentioned above which are clearly perceptible on the current season satellite image that are in pre-burnt /post-burnt condition.

10. **Abandoned Jhum**

These are the areas that were under shifting cultivation, left idle for more than one year but less than 5 years thereby giving a scope for the regeneration of secondary vegetation, especially bamboo or grasses. This category has a tendency to get mixed with forested areas.

11. Under-utilized/degraded notified forest land

Lands within Notified Forest boundaries, with various types of forest cover, in which vegetative cover is less than 20% are classified as degraded/ underutilised. These lands are generally confined to the fringe areas. Such lands appear in dark gray to light red tone during the maximum green period. The tonal variations are subject to change with the foliage cover and the season of data acquisition.

12. Agricultural land inside notified forest land

These are the areas within Notified Forest boundaries, where regular agricultural activities are practiced with Kharif, Rabi or Zaid seasonal crops.

13. Degraded pastures/grazing land

These are the areas in non-forest areas, whether or not they are permanent pastures or meadows, which have become degraded due to lack of proper soil conservation and drainage measures. They appear in light brown to light red in tone, which may vary, depending on the amount of grass cover associated with the feature. At times they exhibit dull red tone and never reach the bright tone during any season. They are small in size, irregular in shape, scattered and associated with village fringes and flood plains.

14. Degraded land under plantation crop

These are the degraded areas that sustain the growth of plantation crops located outside the notified forest areas. This category can be better delineated using multi-season satellite data in conjunction with ground data. They exhibit dark brown to light red tone, especially during peak growing season; however, the tonal variation is controlled by the foliage cover percentage vis-a-vis the season of satellite data acquisition. These are small in shape exhibiting a regular pattern and scattered.

Sand (Desert / Coastal / Riverine)

These are the areas that have stabilized accumulation of sand in coastal or inland areas that can be either desertic or riverine. They appear as white to light yellow/bluish depending on moisture content and at times light red when vegetation is associated with the class, vary in size, with regular to irregular shape with contiguous to linear pattern Mostly they are located in deserts, riverbeds and along the shores.

15. Coastal sand

Coastal sands are the sands that are accumulated as a strip along the seacoast. Very high reflectance exhibited by this category especially in the NIR region of the spectrum enable their separation with the salt affected land.

16. Riverine sand

Riverine sands are those that are seen as accumulations outside the main course of the river(outside the river banks), usually in the active flood plain as sheets, which are the resultant phenomena of river flooding.These are generally covered by either scrub or crops.

17. Desertic sand (dunes less than 15 m high and interdunal areas)

Desertic sands are those confined to arid environment where the rainfall is scanty. These lands are characterized by accumulation of sand developed in situ or transported by Aeolian processes.

18. Desertic sand - moderately high dunes

These are essentially sandy areas with stabilized to semi-stabilised dunes which are 15 - 40m high. Hardy and sparse vegetation is present in these areas.

19. Desertic sand - high dunes

These are active dunal areas with semi-stabilised to shifting sand dunes which are more than 40 m high. Their shapes and sizes vary, depending upon the prevailing wind conditions. These areas rarely support any vegetation.

Mining/Industrial wasteland

These are the lands where large-scale mining operations bring about the degradation of land and resultant mine dumps, industrial wastelands, which are subject to degradation caused by large scale industrial effluent discharges. The features exhibit dark gray (coal mining areas) to light bluish to back (iron ore waste) tone subject to the color of the mine dump, small to medium in size, depending on the extent of mining area, irregular in shape with mottled texture, located at or near active mining areas and industrial complexes. These areas are generally associated with lines of transportation especially railway and road. The features of this category need detailed ground truth for better classification. These areas are subjected to removal of different earth material (both surfacial and sub- surfacial) by manual and mechanized operations. Large scale quarrying and mechanization results in mining and mine dumps. It includes surface rocks and stone quarries, sand and gravel pits, brick kilns, etc.

20. Industrial wasteland

These are areas of stockpile of storage dump of industrial raw material or slag/effluents or waste material or quarried/mixed debris from earth's surface. These are the lands which have been deteriorated due to large scale industrial effluent discharge. These areas are seen conspicuously around urban areas and other areas where industrial activity is prominent.

21. Mining wasteland

Mining wastelands are those areas where waste debris is accumulated after extraction of required minerals. Generally these lands are confined to the surroundings of the mining area.

22. Barren/Rocky/Stony Waste

These are rock exposures of varying lithology often barren and devoid of soil and vegetation cover. They occur amidst hill-forests as openings or as isolated exposures on plateau and plains. Such lands can be easily discriminated from other categories of wastelands because of their characteristic spectral response. They appear in greenish blue to yellow to brownish in color depending on the rock type. They vary in size with irregular to discontinuous shape with a linear to contiguous or dispersed pattern. They are located in steep isolated hillocks/hill slopes, crests, plateau and eroded plains associated with barren and exposed rocky/ stony wastes, lateritic out-crops, mining and quarrying sites These areas appear in light gray to black tone due to hill shadow on one side and light red on the other side due to vegetation, the tonal variation is subject to degree of soil erosion.

23. Snow Covered/Glacial Area

These are the areas under snow cover confined to the Himalayan region. They appear in bright white to white in color depending on the moisture and thick-ness of the snow spread in large areas. They possess irregular shape with a contiguous pattern, located in mountain peaks and slopes and high relief areas.

Out of these, first nine categories are treated as culturable and remaining four categories as non culturable wastelands.

A significant reduction in total wasteland area (10264 Sq. Km.) was noticed in Rajasthan state during 2005-06 and 2008-09 (Table 17.1 and 17.2). Most of the change was noticed in the three classes 'desertic sands' (4,500 Sq. Km.), which have been converted into 'cropland'. Marked decrease was also recorded in the spatial extents of 'dense scrubland', 'gullied land', 'degraded forest', 'degraded pastures' and 'barren rocky' classes. Correspondingly, increase in non-wasteland classes of 'plantations' and 'forest plantations' were recorded, indicating the trend of change. A few of these changes are shown in Figures 17.1 and 17.2.

From the results of the study, the following conclusions were drawn regarding Remote Sensing and GIS Applications in Wastelands Change Analysis.

Use of 3 – season Resourcesat-1 LISS-III data along with limited field checks has facilitated better discrimination of various categories of wastelands.

Geographical Information System (GIS) enabled Updation of previous mapping cycle wasteland vector layer with current season satellite data and identification of change if any, in their status and spatial extent.

The spatial analysis of wasteland change in different states of the country shows that about 32,000 Sq. Km. of wastelands got changed into non-wasteland class, while 27,000 Sq. Km. of non-wasteland was converted into wasteland category during the period of study. The temporal analysis of the change in the total area of wastelands between 2008-09 and 2005-06 showed an aggregate reduction of 5,240 Sq. Km. (0.52 million ha), which is equivalent to decrease of 0.17 % of the wastelands in 2005-06.

Most of the area that showed change from wastelands to non-wastelands has been converted into cropland (including fallow land) during 2008-09. Considerable extent of wasteland was recorded converting into forest and forest plantations and other plantations classes.The database will serve as a good reference on wastelands for studying their temporal behaviour and for evaluating the effectiveness of various wasteland development programmes, being funded by the Department of Land Resources (DoLR) under the Ministry of Rural Development (MRD), Government of India.

Based on the study of data obtained by NRSC, Hyderabad, under National Project of Wasteland mapping on 1:50,000 scale using satellite data, Rajasthan state has about 1.01 lakhs sq.km. of wastelands of differnet categories. Land with or without scrub cover the maximum with 36.32 per cent of total wastelands in the state followed by desertrc; sands (29.10 per cnet) and u nder utilised/degraded notified forest land (9.48 per cent). Other waste.lands categoreis identified in the state are degraded pastures/grazing lands (8.64 per cent), gullied and or/ravinous land (6.5 per cent), barren rocky/stony waste/sheet rock area (4.97 per cent), land affected by salinity/alkalinity (3.71 per cent), waterlogged and marshy land (0.26 per cent), mining wastelands (0.23 per cent), steep slopping land (0.21 per cent) and degraded land under plantation crop (0.18 per cent).

Different categories are unevenly distributed in 33 districts of the state (Figure 17.3). While all the western desertic districts have desertic sands as dominating category of wastelands, districts falling along Aravalli hill ranges and Vindhyans have degraded forest or scrub lands and barren rocky areas. Ravines are restricted to both bank of different rivers, flowing in the districts of Kota, Bundi, Baran, Dhaulpur, Jaipur, Tonk, Dausa and Alwar, Salt affected (land affected by salinity and or alkalinity) occur in area near by to all the salt pans situated in different parts of western Rajasthan as well as agricultural fields which have inherent problem of salinity or are irrigate9 with saline waters.

To meet the increased demand of fuel and fibre of the ever increasing human population, it is essential that more land is brought under plough and productivity of the existing arable land is maintained on a optimum level. Culturable wastelands cover a large area in the state. Due to inherent limitations it is not possible to bring all the area under arable use but a large chunk have the potential for improvement. All attemmpts may be made with the help of wasteland maps of each district to sort out such lands and explore the possiblity of their improvement and management.

District wise and Category wise Wasteland available in State (Table 17.1 and 17.2)

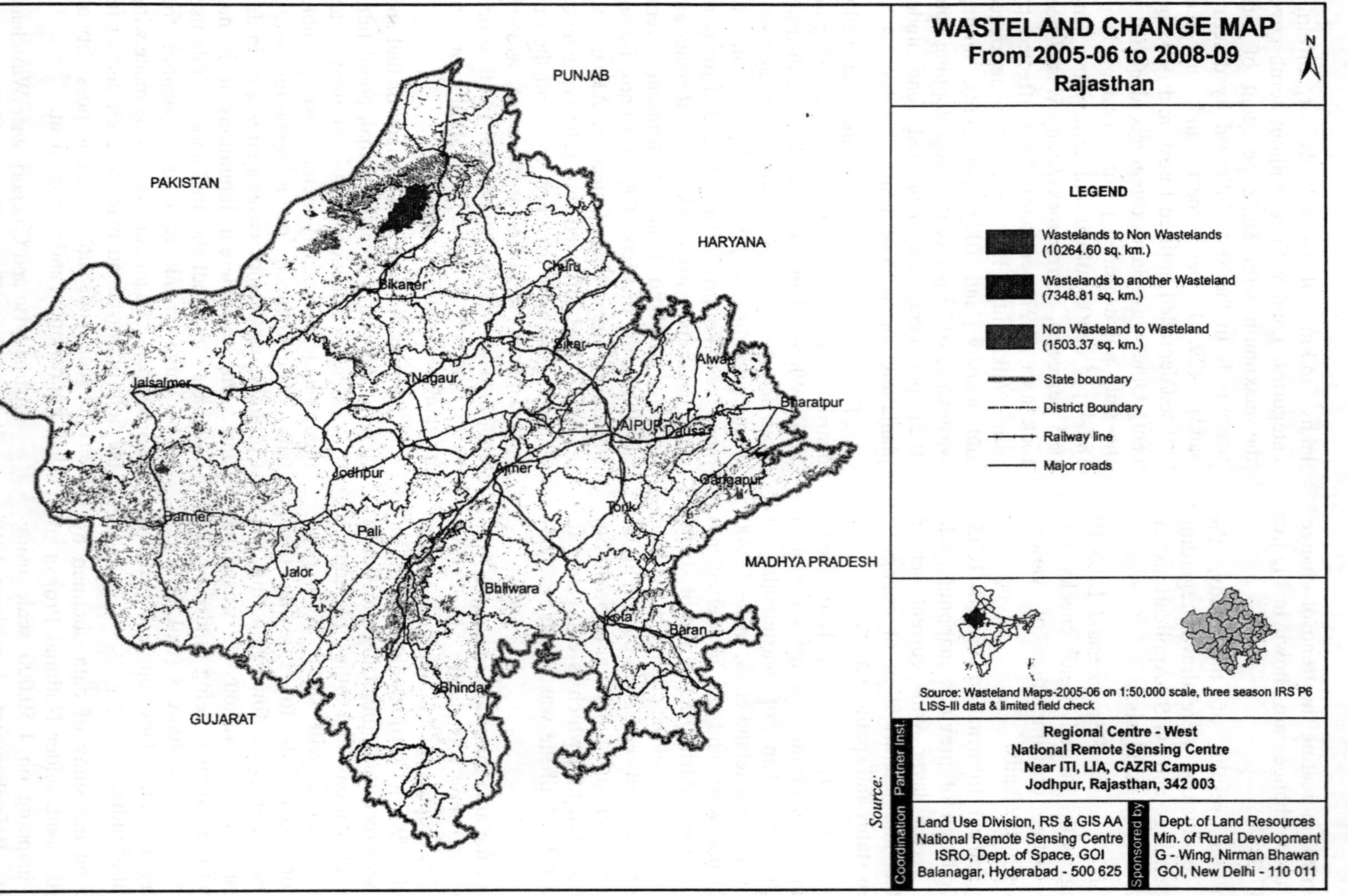

Figure 17.1 Rajasthan Wasteland Change Map from 2005-06 to 2008-09

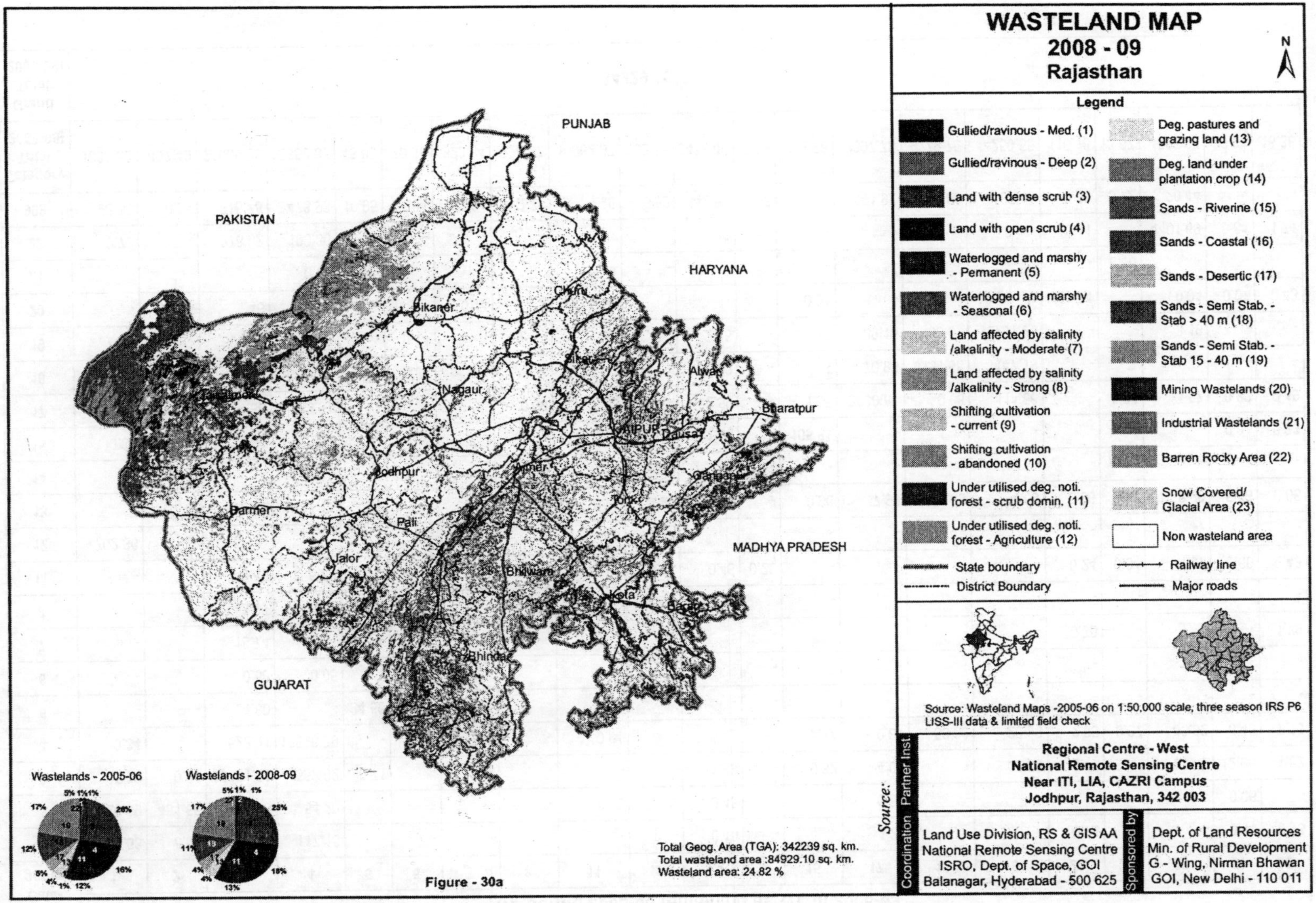

Figure 17.2 Rajasthan Wasteland Map (2008-09)

Table 17.1 Change matrix showing inter-class area changes in different wasteland categories in Rajasthan during 2005-06 and 2008-09

WL Cate-gories	WL during 2005-06 remaining as WL in 2008-09																				
	1	2	3	4	5	6	7	8	11	12	13	14	15	17	18	19	20	21	22	24	25
1	253.53	5.07	551.56	117.16	-	-	-	-	-	-	0.10	-	-	1.25		-	0.23	-	-	0.41	0.16
2	26.89	431.66	315.33	54.20	-	-	-	-	0.03	-	0.45	-	-	0.81	-	-		-		0.06	-
3	5.06	0.04	18331.94	1552.92	0.71	-	1.07	-	-	-	11.99	-	0.32	4.08	-	39.35	6.17	-	85.46	12.06	1.82
4	0.34	-	522.21	12596.79	-	-	0.42	-	0.94	-	0.49	-	2.27	0.81	58.30	28.93	2.05	0.87	43.62	6.49	7.15
5	-	-	1.80	-	34.56	-	-	-	-	-	-	-	-	-	-	-	-	-	-	-	-
6	-	-	0.58	0.05	-	43.84	-	-	-	-	-	-	-	-	-	-	-	-	-	-	-
7	-	-	18.74	14.81	-	-	284.10	-	-	-	-	-	-	-	-	-	0.70	-	-	-	1.79
8	-	-	3.07	0.61	1.34	-	-	224.00	-	-	-	-	-	-	-	-	-	-	-	-	-
11	-	-	2.89	6.89	-	-	-	-	10934.90	38.44	0.46	0.22	-	-	-	-	0.24	0.04	1.07	0.60	2.43
12	702.30	-	-	-	-	-	-	-	15.10	114.52	-	-	-	-	-	-	-	-	-	-	-
13	-	0.48	112.32	117.11	-	-	0.47	0.30	8.96	0.90	3274.54	-	0.96	7.99	0.83	4.05	0.96	0.13	0.20	5.16	0.06
14	-	-	-	-	-	-	-	-	-	-	-	-	-	-	-	-	-	-	-	-	-
15	0.35	-	35.06	28.02	-	-	-	-	-	-	1.68	-	105.81	-	-	-	-	-	-	0.04	0.33
17	-	-	111.58	83.57	-	-	-	0.06	0.03	-	2.59	-	1.05	3030.81	0.08	111.66	-	0.05	1.57	0.52	0.49
18	-	-	179.15	59.48	0.43	5.34	-	-	-	-	0.90	-	-	70.93	9151.68	1147.19	-	-	-	-	-
19	-	-	303.56	61.76	-	0.14	-	-	-	-	2.45	-	0.18	89.16	66.99	12751.96	-	-	3.19	-	-
20	-	-	1.46	1.71	-	-	0.14	-	0.91	-	-	-	0.04	-	-	-	101.74	-	0.04	0.07	0.43
21	-	-	-	0.07	-	-	-	-	0.14	-	-	-	-	-	-	-	-	8.86	-	-	-
22	0.72	-	278.02	190.30	-	-	-	-	-	-	0.13	-	-	-	-	-	0.41	-	4401.69	0.81	1.71
999	98.33	0.24	316.78	473.93	10.95	21.04	37.64	2.21	1.96	5.82	142.62	0.67	2.15	191.92	1.88	167.41	3.68	3.72	20.44	-	-
Category Total (2008-09)	1087.52	437.49	21086.05	15359.37	48.00	70.36	323.84	226.57	10962.97	159.68	3438.40	0.89	112.78	3397.75	9279.75	14250.55	116.18	13.67	4557.27	26.24	16.38
Grand Total (2008-09)	84929.10																				

Table 17.1 Cond...

WL Cate-gories	WL 2005-06 becoming non-WL in 2008-09							(2005-06)	
	26	27	28	29	30	31	32	Category Total	Grand Total
1	46.61	42.42	-	0.53	0.24	-	0.90	**1020.17**	
2	18.87	13.76	-	0.57	1.78	-	0.64	**865.03**	
3	983.86	2376.10	23.16	1.90	13.92	0.06	3.67	**23661.70**	
4	562.51	758.94	13.97	0.22	1.21	-	10.88	**14619.38**	
5	0.76	2.48	-	-	-	-	25.28	**64.88**	
6	1.73	8.06	-	-	-	-	0.67	**54.94**	
7	17.54	9.09	-	-	-	-	0.34	**347.12**	
8	5.84	12.90	-	-	-	-	21.62	**269.39**	
11	1.17	2.20	1.73	239.66	126.10	-	7.04	**11366.10**	
12	0.03	0.61	-	21.58	-	-	0.20	**854.34**	**93690.33**
13	181.77	191.04	4.09	-	0.44	-	5.70	**3918.42**	
14	-	-	-	-	-	-	-	-	
15	8.80	13.19	0.10	-	-	-	3.29	**196.69**	
17	805.52	503.30	-	0.13	2.78	-	0.09	**4655.88**	
18	357.73	211.88	2.27	-	0.29	0.94	-	**11188.21**	
19	1663.52	609.07	6.63	-	27.83	-	-	**15586.44**	
20	0.11	0.20	-	-	-	-	-	**106.86**	
21	-	-	-	-	-	-	-	**9.06**	
22	10.18	19.49	1.19	-	0.96	-	0.10	**4905.72**	

WL Category	Reduction from 0506 to 0809		Increase from 0506 to 0809	
	Area	No of polygons	Area	No of polygons
1	91.27	672	98.33	46
2	35.67	149	0.24	3
3	3622.59	8655	316.78	749
4	1361.36	6731	473.93	768
5	28.52	23	10.95	7
6	10.46	24	21.04	8
7	28.76	69	37.64	12
8	40.36	49	2.21	12
11	380.93	1998	1.96	23
12	22.42	433	5.82	17
13	388.25	1390	142.62	185
14	-	-	0.67	1
15	25.76	129	2.15	3
17	1312.84	3140	191.92	186
18	573.11	679	1.88	10
19	2307.05	3192	167.41	250
20	0.81	13	3.68	28
21	-	-	3.72	7
22	34.45	233	20.44	135
Total	**10264.60**	**27579**	**1503.37**	**2450**

Source: Land Use Division, RS & GIS AA National Remote Sensing Centre, ISRO, Dept. of Space, GOI, Balanagar, Hyderabad; Dept. of Land Resouces, Min. of Rural Development, GOI, New Delhi

Table 17.2 District-wise and Category-wise Distribution of Wastelands (sq.km.) During 2008-09 vis-a-vis 2005-06

WL class	Ajmer			Alwar			Barmer			Banswara			Baran			Bharatpur		
	2005-06	2008-09	Change	2005-06	2008-09	Change	2005-06	2008-09	Change	2005-06	2008-09	Change	2005-06	2008-09	Change	2005-06	2008-09	Change
1	0.09	33.97	33.88	56.69	44.98	-11.72		0.53	0.53				78.30	73.22	-5.07	59.41	2.69	-56.72
2	-	-	-	67.94	39.75	-28.19	20.22	19.47	-0.75	-	-	-	100.47	39.22	-61.25	-	0.16	0.16
3	654.08	558.86	-95.22	449.93	496.08	46.15	1032.84	619.49	-413.34	169.50	171.89	2.40	191.44	290.52	99.08	193.16	277.02	83.85
4	910.43	1076.74	166.31	83.24	51.27	-31.96	384.68	719.52	334.84	164.26	176.66	12.40	131.87	165.89	34.01	58.70	134.84	76.14
5	8.31	8.63	0.31	-	-	-	-	0.17	0.17	-	-	-	-	-	-	-	-	-
6	-	-	-	-	-	-	-	21.31	21.31	-	-	-	-	-	-	-	7.61	7.61
7	11.98	10.82	-1.15	0.20	0.30	0.10	123.12	105.29	-17.83	-	-	-	-	-	-	0.25	0.26	0.01
8	19.90	17.63	-2.27	-	-	-	9.93	3.49	-6.44	-	-	-	-	-	-	-	-	-
9	-	-	-	-	-	-	-	-	-	-	-	-	-	-	-	-	-	-
10	-	-	-	-	-	-	-	-	-	-	-	-	-	-	-	-	-	-
11	210.79	200.39	-10.40	111.10	97.69	-13.41	124.69	113.15	-11.54	629.12	628.11	-1.01	1197.68	1160.31	-37.38	57.07	39.92	-17.15
12	-	-	-	40.59	10.33	-30.25	0.75	-	-0.75	25.97	15.03	-10.94	130.61	7.34	-123.27	0.19	0.21	0.02
13	128.01	87.74	-40.27	-	0.05	0.05	279.09	156.45	-122.64	23.57	19.39	-4.18	106.82	95.55	-11.27	-	-	-
14	-	-	-	-	-	-	-	-	-	-	-	-	-	-	-	-	-	-
15	0.08	-	-0.08	0.35	0.05	-0.29	-	2.30	2.30	-	-	-	46.52	39.71	-6.81	6.94	-	-6.94
16	-	-	-	-	-	-	-	-	-	-	-	-	-	-	-	-	-	-
17	19.14	13.19	-5.95	0.08	-	-0.08	402.27	160.04	-242.23	-	-	-	-	-	-	6.04	0.77	-5.27
18	-	-	-	-	-	-	1728.70	1313.17	-415.53	-	-	-	-	-	-	-	-	-
19	-	-	-	-	-	-	2598.27	1387.10	-1211.17	-	-	-	-	-	-	-	-	-
20	3.37	5.27	1.91	0.75	0.64	-0.11	1.95	3.09	1.14	3.71	2.80	-0.91	-	-	-	1.41	0.99	-0.42
21	-	1.05	1.05	-	-	-	-	0.15	0.15	0.34	0.19	-0.15	-	-	-	-	-	-
22	182.53	143.93	-38.59	273.70	228.05	-45.65	299.06	394.79	95.73	102.90	91.01	-11.89	35.27	27.51	-7.75	196.08	91.06	-105.02
23	-	-	-	-	-	-	-	-	-	-	-	-	-	-	-	-	-	-
Total	2148.69	2158.22	9.53	1084.56	969.20	-115.37	7005.56	5019.52	-1986.04	1119.37	1105.09	-14.28	2018.97	1899.26	-119.71	579.25	555.51	-23.74
TGA	8481.00			8380.00			28387.00			5037.00			6955.00			5092.00		

1. Gullied and/ or ravinous land (Medium)
2. Gullied and/ or ravinous land (Deep)
3. Land with Dense Scrub
4. Land with Open Scrub
5. Waterlogged and Marshy land (Permanent)
6. Waterlogged and Marshy land (Seasonal)
7. Land affected by salinity/alkalinity (Medium)
8. Land affected by salinity/alkalinity (Strong)
9. Shifting Cultivation - Current Jhum
10. Shifting Cultivation -Abandoned Jhum
11. Under-utilised/degraded forest (Scrub domin)
12. Under-utilised/degraded forest (Agriculture)
13. Degraded pastures/grazing land
14. Degraded land under plantation crop
15. Sands-Riverine
16. Sands-Coastal
17. Sands-Desertic
18. Sands-Semi Stab.-Stab>40m
19. Sands-Semi Stab.-Stab 15-40m
20. Mining Wastelands
21. Industrial wastelands
22. Barren Rocky/Stony waste
23. Snow covered /Glacial area

Table 17.2 Cont. ...

Area in sq.km.

WL class	Bhilwara			Bikaner			Bundi			Chittaurgarh			Churu			Dausa		
	2005-06	2008-09	Change	2005-06	2008-09	Change	2005-06	2008-09	Change	2005-06	2008-09	Change	2005-06	2008-09	Change	2005-06	2008-09	Change
1	12.32	11.61	-0.71	59.16	57.71	-1.45	99.29	80.99	-18.29	-	0.09	0.09	-	4.13	4.13	73.49	9.47	-64.01
2	-	-	-	37.25	35.47	-1.78	210.29	191.08	-19.21	-	-	-	-	-	-	33.23	2.80	-30.43
3	1618.22	1406.31	-211.91	1430.39	132.49	-1297.90	247.19	255.08	7.89	866.63	771.24	-95.39	52.82	150.47	97.65	129.78	191.60	61.81
4	768.28	814.28	46.00	265.73	398.29	132.56	37.20	82.93	45.73	442.76	388.50	-54.26	343.38	240.95	-102.43	93.33	63.95	-29.37
5	-	-	-	0.37	0.37	-	-	-	-	-	-	-	1.32	1.32	-	-	-	-
6	-	-	-	-	-	-	-	-	-	-	-	-	-	-	-	-	-	-
7	-	-	-	-	-	-	-	-	-	2.54	2.54	-	14.71	14.53	-0.18	0.17	-	-0.17
8	-	-	-	-	-	-	-	-	-	-	-	-	0.39	0.01	-0.38	0.15	0.06	-0.09
9	-	-	-	-	-	-	-	-	-	-	-	-	-	-	-	-	-	-
10	-	-	-	-	-	-	-	-	-	-	-	-	-	-	-	-	-	-
11	363.48	351.70	-11.77	70.87	70.06	-0.81	770.48	717.47	-53.00	892.10	901.72	9.62	-	-	-	75.97	63.51	-12.46
12	9.11	3.24	-5.87	-	0.81	0.81	-	1.05	1.05	183.16	5.05	-178.11	-	-	-	1.24	0.75	-0.48
13	324.26	313.52	-10.74	485.71	466.05	-19.66	27.20	26.45	-0.74	50.57	49.28	-1.29	128.12	130.00	1.89	15.10	11.04	-4.06
14	-	-	-	-	-	-	-	-	-	-	-	-	-	-	-	-	-	-
15	-	-	-	-	-	-	-	-	-	-	-	-	-	-	-	0.90	0.48	-0.43
16	-	-	-	-	-	-	-	-	-	-	-	-	-	-	-	-	-	-
17	0.08	-	-0.08	1111.56	951.66	-159.90	0.35	0.35	-	-	-	-	58.42	14.36	-44.05	0.48	0.25	-0.23
18	-	-	-	1278.94	1013.70	-265.24	-	-	-	-	-	-	-	-	-	-	-	-
19	-	-	-	5907.50	5301.61	-605.89	-	-	-	-	-	-	109.99	30.79	-79.20	-	-	-
20	2.65	2.58	-0.07	-	-	-	0.41	0.53	0.12	4.99	5.80	0.81	0.04	0.04	-	0.72	0.25	-0.47
21	-	-	-	-	2.62	2.62	-	-	-	-	-	-	-	-	-	-	0.05	0.05
22	206.16	205.17	-0.98	19.63	.20.06	0.43	88.81	56.88	-31.94	98.87	88.62	-10.25	4.94	2.64	-2.30	26.70	7.70	-19.01
23	-	-	-	-	-	-	-	-	-	-	-	-	-	-	-	-	-	-
Total	**3304.55**	**3108.42**	**-196.13**	**10667.12**	**8450.91**	**-2216.21**	**1481.21**	**1412.80**	**-68.40**	**2541.62**	**2212.84**	**-328.78**	**714.14**	**589.25**	**-124.88**	**451.25**	**351.93**	**-99.33**
TGA	**10455.0**			**27244.00**			**5550.0**			**10856.00**			**16830.00**			**3432.00**		

1. Gullied and/ or ravinous land (Medium)
2. Gullied and/ or ravinous land (Deep)
3. Land with Dense Scrub
4. Land with Open Scrub
5. Waterlogged and Marshy land (Permanent)
6. Waterlogged and Marshy land (Seasonal)
7. Land affected by salinity/alkalinity (Medium)
8. Land affected by salinity/alkalinity (Strong)
9. Shifting Cultivation - Current Jhum
10. Shifting Cultivation -Abandoned Jhum
11. Under-utilised/degraded forest (Scrub domin)
12. Under-utilised/degraded forest (Agriculture)
13. Degraded pastures/grazing land
14. Degraded land under plantation crop
15. Sands-Riverine
16. Sands-Coastal
17. Sands-Desertic
18. Sands-Semi Stab.-Stab>40m
19. Sands-Semi Stab.-Stab 15-40m
20. Mining Wastelands
21. Industrial wastelands
22. Barren Rocky/Stony waste
23. Snow covered /Glacial area

Table 16.2 Cont. ...

Area in sq.km.

WL class	Dhaulpur			Dungarpur			Ganganagar			Hanumangarh			Jaipur			Jaisalmer		
	2005-06	2008-09	Change	2005-06	2008-09	Change	2005-06	2008-09	Change	2005-06	2008-09	Change	2005-06	2008-09	Change	2005-06	2008-09	Change
1	68.06	74.67	6.61	-	-	-	-	-	-	-	-	-	3.31	199.23	195.92	-	-	-
2	120.08	30.94	-89.13	-	-	-	-	-	-	-	-	-	17.89	17.86	-0.04	-	-	-
3	315.96	343.94	27.98	371.09	399.55	28.47	21.95	76.29	54.35	13.66	24.01	10.35	181.24	215.81	34.56	8352.40	7748.97	-603.42
4	83.08	95.94	12.85	552.60	551.04	-1.56	89.39	81.19	-8.20	33.61	41.32	7.71	874.92	816.74	-58.18	3870.31	3565.04	-305.27
5	-	-	-	-	-	-	-	7.49	7.49	5.00	7.18	2.18	10.10	-	-10.10	-	-	-
6	-	-	-	-	-	-	2.97	7.41	4.44	1.40	1.04	-0.36	-	-	-	15.33	4.76	-10.57
7	-	-	-	-	-	-	-	-	-	-	-	-	19.12	15.40	-3.72	11.36	11.76	0.40
8	-	-	-	-	-	-	-	-	-	-	-	-	39.87	32.33	-7.54	40.67	38.77	-1.89
9	—	-	-	-	-	-	—	-	-	—	-	-	-	-	-	—	-	-
10	-	-	-	-	-	-	-	-	-	-	-	-	-	-	-	-	-	-
11	297.98	294.25	-3.73	324.88	311.04	-13.84	-	-	-	23.12	23.12	-	385.51	380.31	-5.20	145.87	110.99	-34.87
12	56.96	1.44	-55.52	18.22	18.03	-0.20	-	-	-	-	-	-	-	-	-	0.04	0.04	-
13	-	-	-	27.97	26.01	-1.96	-	-	-	21.46	10.95	-10.52	55.98	42.09	-13.89	85.98	86.74	0.76
14	-	-	-	-	-	-	-	-	-	-	-	-	-	-	-	-	-	-
15	0.15	-	-0.15	-	-	-	-	-	-	-	-	-	23.23	12.86	-10.36	-	-	-
16	-	-	-	-	-	-	-	-	-	-	-	-	-	-	-	-	-	-
17	-	-	-	-	-	-	154.59	169.79	15.20	330.54	102.84	-227.70	97.69	7.63	-90.06	528.99	546.69	17.70
18	-	-	-	-	-	-	1088.21	-	-1088.21	20.44	-	-20.44	-	-	-	6905.69	6815.99	-89.70
19	-	-	-	-	-	-	171.83	926.54	754.72	-	118.69	118.69	-	-	-	5058.99	4902.85	-156.14
20	0.22	0.15	-0.07	2.23	2.40	0.17	-	-	-	-	0.16	0.16	3.20	2.60	-0.60	0.84	0.65	-0.19
21	-	-	-	0.14	0.14	-	2.57	2.57	-	-	-	-	-	-	-	-	-	-
22	20.50	20.31	-0.20	52.80	43.33	-9.47	-	-	-	-	-	-	80.79	70.80	-9.98	1228.13	1226.28	-1.85
23	-	-	-	-	-	-	-	-	-	-	-	-	-	-	-	-	-	-
Total	**963.00**	**861.64**	**-101.36**	**1349.93**	**1351.54**	**1.61**	**1531.49**	**1271.27**	**-260.22**	**449.23**	**329.30**	**-119.93**	**1792.85**	**1813.66**	**20.82**	**26244.58**	**25059.53**	**-1185.05**
TGA	**3008.00**			**3770.00**			**10978.00**			**9656.00**			**10636.00**			**38401.00**		

1. Gullied and/ or ravinous land (Medium)
2. Gullied and/ or ravinous land (Deep)
3. Land with Dense Scrub
4. Land with Open Scrub
5. Waterlogged and Marshy land (Permanent)
6. Waterlogged and Marshy land (Seasonal)
7. Land affected by salinity/alkalinity (Medium)
8. Land affected by salinity/alkalinity (Strong)
9. Shifting Cultivation - Current Jhum
10. Shifting Cultivation -Abandoned Jhum
11. Under-utilised/degraded forest (Scrub domin)
12. Under-utilised/degraded forest (Agriculture)
13. Degraded pastures/grazing land
14. Degraded land under plantation crop
15. Sands-Riverine
16. Sands-Coastal
17. Sands-Desertic
18. Sands-Semi Stab.-Stab>40m
19. Sands-Semi Stab.-Stab 15-40m
20. Mining Wastelands
21. Industrial wastelands
22. Barren Rocky/Stony waste
23. Snow covered /Glacial area

Table 17.2 Cont. ...

Area in sq.km.

WL class	Jalor			Jhalawar			Jhunjhunun			Jodhpr			Karauli			Kota		
	2005-06	2008-09	Change	2005-06	2008-09	Change	2005-06	2008-09	Change	2005-06	2008-09	Change	2005-06	2008-09	Change	2005-06	2008-09	Change
1	51.89	47.83	-4.07	19.71	4.98	-14.72	1.42	18.75	17.33	11.72	10.02	-1.70	134.53	33.28	-101.25	96.61	13.38	-83.23
2	20.42	7.14	-13.28	0.26	-	-0.26	27.46	22.12	-5.34	-	-	-	26.95	1.92	-25.03	156.24	17.61	-138.63
3	233.51	279.12	45.61	731.81	782.93	51.12	30.23	70.42	40.19	439.60	364.84	-74.76	310.24	316.28	6.04	238.47	401.63	163.16
4	80.84	91.67	10.83	374.63	297.80	-76.83	324.37	291.17	-33.20	945.42	910.64	-34.78	60.69	87.94	27.25	19.84	96.08	76.24
5	-	-	-		-	-	-	-	-	-	-	-	-	-	-	-	-	-
6	-	0.67	0.67	-	-	-	-	-	-	27.57	27.57	-	-	-	-	-	-	-
7	11.91	11.60	-0.31		-	-	-	-	-	48.97	49.77	0.80	-	0.06	0.06	-		-
8	36.66	34.26	-2.39		-	-	-	-	-	6.18	5.98	-0.20	-	-	-	-	-	-
9	-	-	-	-	-	-	-	-	-	-	-	-	-	-	-	-	-	-
10	-	-	-	-	-	-	-	-	-	-	-	-	-	-	-	-	-	-
11	161.06	152.02	-9.04	604.36	532.07	-72.30	169.99	168.40	-1.59	57.53	59.90	2.37	1121.07	1096.01	-25.06	521.43	533.84	12.41
12	-	0.13	0.13	12.83	0.73	-12.10	-	0.22	0.22	-	0.49	0.49	243.21	7.09	-236.12	-	-	-
13	286.88	270.22	-16.67	16.15	15.74	-0.41	56.71	50.68	-6.03	661.95	643.43	-18.51	1.38	2.09	0.70	6.08	5.15	-0.93
14	-	-	-	-	-	-	-	0.89	0.89	-	-	-	-	-	-	-	-	-
15	0.03	0.13	0.10	-	-	-	15.48	11.38	-4.10	4.10	0.24	-3.86	-	-	-	0.34	0.36	0.02
16	-	-	-	-	-	-	-	-	-	-	-	-	-	-	-	-	-	-
17	285.73	239.58	-46.15	-	-	-	19.21	14.18	-5.03	1084.12	1002.53	-81.58	0.05	0.10	0.05	-	-	-
18	20.57	16.30	-4.27	-	-	-	-	-	-	145.66	120.59	-25.07	-	-	-	-	-	-
19	132.30	112.70	-19.60	-	-	-	-	-	-	1349.97	1375.47	25.50	-	-	-	-	-	-
20	0.91	0.91	-	3.30	3.57	0.27	-	0.31	0.31	20.59	20.19	-0.41	0.56	0.51	-0.06	0.53	0.36	-0.17
21	0.61	0.65	0.04	-	-	-	-	-	-	0.99	1.12	0.13	-	-	-	-	-	-
22	173.30	173.70	0.41	13.90	11.99	-1.92	32.69	27.03	-5.67	500.21	502.12	1.91	55.44	34.61	-20.82	11.32	5.98	-5.34
23	-	-	-	-	-	-	-	-	-	-	-	-	-	-	-	-	-	-
Total	**1496.61**	**1438.64**	**-57.97**	**1776.94**	**1649.81**	**-127.14**	**677.56**	**675.54**	**-2.02**	**5304.56**	**5094.87**	**-209.69**	**1954.13**	**1579.89**	**-374.24**	**1050.85**	**1074.38**	**23.53**
TGA	**10640.00**			**6219.00**			**5928.00**			**22850.00**			**5524.00**			**5481.00**		

1. Gullied and/ or ravinous land (Medium)
2. Gullied and/ or ravinous land (Deep)
3. Land with Dense Scrub
4. Land with Open Scrub
5. Waterlogged and Marshy land (Permanent)
6. Waterlogged and Marshy land (Seasonal)
7. Land affected by salinity/alkalinity (Medium)
8. Land affected by salinity/alkalinity (Strong)
9. Shifting Cultivation - Current Jhum
10. Shifting Cultivation -Abandoned Jhum
11. Under-utilised/degraded forest (Scrub domin)
12. Under-utilised/degraded forest (Agriculture)
13. Degraded pastures/grazing land
14. Degraded land under plantation crop
15. Sands-Riverine
16. Sands-Coastal
17. Sands-Desertic
18. Sands-Semi Stab.-Stab>40m
19. Sands-Semi Stab.-Stab 15-40m
20. Mining Wastelands
21. Industrial wastelands
22. Barren Rocky/Stony waste
23. Snow covered /Glacial area

Table 16.2 Cont. ...

Area in sq.km.

WL class	Nagaur			Pali			Rajasamand			Sikar			Sirohi			Sawai Madhopur		
	2005-06	2008-09	Change	2005-06	2008-09	Change	2005-06	2008-09	Change	2005-06	2008-09	Change	2005-06	2008-09	Change	2005-06	2008-09	Change
1	-	33.70	33.70	14.35	9.17	-5.18	-	-	-	21.43	87.06	65.64	42.62	42.08	-0.54	114.95	128.83	13.89
2	-	-	-	-	-	-	-	-	-	2.64	1.87	-0.77	10.74	7.86	-2.88	9.36	-	-9.36
3	291.82	288.26	-3.56	1188.03	1203.95	15.92	1478.24	779.53	-698.70	50.38	112.71	62.33	526.85	523.03	-3.83	163.49	165.68	2.19
4	289.53	477.90	188.36	361.50	412.79	51.29	328.08	669.65	341.56	392.10	312.84	-79.26	400.07	408.39	8.32	156.77	126.13	-30.64
5	38.05	22.74	-15.31	-	-	-	-	-	-	0.09	0.09	-	1.62	-	-1.62	-	-	-
6	-	-	-	7.01	-	-7.01	-	-	-	0.67	-	-0.67	-	-	-	-	-	-
7	18.09	9.69	-8.40	77.78	73.39	-4.39	-	-	-	6.93	9.31	2.38	-	9.11	9.11	-	-	-
8	102.56	85.76	-16.80	7.72	6.59	-1.13	-	-	-	4.48	1.24	-3.24	-	-	-	0.62	0.45	-0.17
9	-	-	-	-	-	-	-	-	-	-	-	-	-	-	-	-	-	-
10	-	-	-	-	-	-	-	-	-	-	-	-	-	-	-	-	-	-
11	127.64	126.85	-0.79	197.16	188.91	-8.24	41.21	26.60	-14.61	309.74	287.94	-21.80	504.43	504.73	0.31	388.69	341.39	-47.29
12	-	-	-	4.16	2.87	-1.29	10.66	-	-10.66	-	0.10	0.10	14.84	9.35	-5.48	15.14	3.93	-11.21
13	528.26	419.44	-108.82	395.64	338.81	-56.83	30.73	26.47	-4.26	68.08	46.20	-21.88	50.28	44.78	-5.51	11.68	10.84	-0.84
14	-	-	-	-	-	-	-	-	-	-	-	-	-	-	-	-	-	-
15	-	-	-	69.59	36.53	-33.06	0.09	0.04	-0.05	19.85	1.56	-18.29	4.51	3.51	-1.00	2.59	1.94	-0.65
16	-	-	-	-	-	-	-	-	-	-	-	-	-	-	-	-	-	-
17	327.84	90.32	-237.52	8.87	3.03	-5.84	-	-	-	178.00	41.65	-136.35	33.16	30.36	-2.80	8.54	8.08	-0.45
18	-	-	-	-	-	-	-	-	-	-	-	-	-	-	-	-	-	-
19	248.36	85.74	-162.62	-	-	-	-	-	-	-	-	-	9.24	9.07	-0.17	-	-	-
20	4.27	6.50	2.24	5.85	6.25	0.40	12.52	18.17	5.66	0.12	0.35	0.23	4.97	4.81	-0.16	0.36	0.37	0.01
21	-	-	-	1.37	1.37	-	-	0.72	0.72	-	-	-	2.08	2.08	-	-	-	-
22	92.67	63.70	-28.97	289.24	265.71	-23.53	392.05	337.81	-54.24	28.33	26.38	-1.94	117.27	102.02	-15.25	12.33	21.45	9.12
23	-	-	-	-	-	-	-	-	-	-	-	-	-	-	-	-	-	-
Total	**2069.09**	**1710.60**	**-358.50**	**2628.27**	**2549.39**	**-78.88**	**2293.56**	**1858.98**	**-434.58**	**1082.84**	**929.31**	**-153.53**	**1722.69**	**1701.17**	**-21.52**	**884.51**	**809.10**	**-75.41**
TGA	**17718.00**			**12387.00**			**4689.00**			**7732.00**			**5136.00**			**5003.00**		

1. Gullied and/ or ravinous land (Medium)
2. Gullied and/ or ravinous land (Deep)
3. Land with Dense Scrub
4. Land with Open Scrub
5. Waterlogged and Marshy land (Permanent)
6. Waterlogged and Marshy land (Seasonal)
7. Land affected by salinity/alkalinity (Medium)
8. Land affected by salinity/alkalinity (Strong)
9. Shifting Cultivation - Current Jhum
10. Shifting Cultivation -Abandoned Jhum
11. Under-utilised/degraded forest (Scrub domin)
12. Under-utilised/degraded forest (Agriculture)
13. Degraded pastures/grazing land
14. Degraded land under plantation crop
15. Sands-Riverine
16. Sands-Coastal
17. Sands-Desertic
18. Sands-Semi Stab.-Stab>40m
19. Sands-Semi Stab.-Stab 15-40m
20. Mining Wastelands
21. Industrial wastelands
22. Barren Rocky/Stony waste
23. Snow covered /Glacial area

Table 17.2 Cont. ...

Area in sq.km.

WL class	Tonk			Udaipur			Rajasamand		
	2005-06	2008-09	Change	2005-06	2008-09	Change	2005-06	2008-09	Change
1	0.83	65.12	64.29	-	-	-	1020.17	1087.52	67.35
2	3.32	2.22	-1.10	-	-	-	864.75	437.49	-427.27
3	324.15	324.88	0.73	1362.60	1343.16	-19.44	23661.70	21086.05	-2575.65
4	269.22	269.72	0.50	1428.55	1441.58	13.02	14619.38	15359.37	739.99
5	-	-	-	-	-	-	64.88	48.00	-16.88
6	-	-	-	-	-	-	54.94	70.36	15.42
7	-	-	-	-	-	-	347.12	323.84	-23.28
8	-	-	-	-	-	-	269.12	226.57	-42.55
9	-	-	-	-	-	-	-	-	-
10	-	-	-	-	-	-	-	-	-
11	102.92	104.56	1.64	1377.87	1376.00	-1.87	11365.78	10962.97	-402.81
12	-	-	-	86.67	71.46	-15.21	854.34	159.68	-694.66
13	4.18	2.44	-1.75	40.60	40.81	0.21	3918.42	3438.40	-480.03
14	-	-	-	-	-	-	-	0.89	0.89
15	1.94	1.69	-0.25	-	-	-	196.69	112.78	-83.90
16	-	-	-	-	-	-	-	-	-
17	0.16	0.36	0.21	-	-	-	4655.88	3397.75	-1258.13
18	-	-	-	-	-	-	11188.21	9279.75	-1908.46
19	-	-	-	-	-	-	15586.44	14250.55	-1335.89
20	0.17	0.17	-	26.21	25.75	-0.46	106.86	116.18	9.32
21	-	-	-	0.97	0.97	-	9.06	13.67	4.61
22	22.49	15.47	-7.02	247.63	251.18	3.55	4905.72	4557.27	-348.45
23	-	-	-	-	-	-	-	-	-
Total	**729.38**	**786.63**	**57.25**	**4571.11**	**4550.91**	**-20.20**	**93689.47**	**84929.10**	**-8760.37**
TGA	**7194.00**			**12590.00**			**342239.00**		

1. Gullied and/ or ravinous land (Medium)
2. Gullied and/ or ravinous land (Deep)
3. Land with Dense Scrub
4. Land with Open Scrub
5. Waterlogged and Marshy land (Permanent)
6. Waterlogged and Marshy land (Seasonal)
7. Land affected by salinity/alkalinity (Medium)
8. Land affected by salinity/alkalinity (Strong)
9. Shifting Cultivation - Current Jhum
10. Shifting Cultivation -Abandoned Jhum
11. Under-utilised/degraded forest (Scrub domin)
12. Under-utilised/degraded forest (Agriculture)
13. Degraded pastures/grazing land
14. Degraded land under plantation crop
15. Sands-Riverine
16. Sands-Coastal
17. Sands-Desertic
18. Sands-Semi Stab.-Stab>40m
19. Sands-Semi Stab.-Stab 15-40m
20. Mining Wastelands
21. Industrial wastelands
22. Barren Rocky/Stony waste
23. Snow covered /Glacial area

Source: Land Use Division, RS & GIS AA National Remote Sensing Centre, ISRO, Dept. of Space, GOI. Balanagar, Hyderabad; Dept. of Land Resouces, Min. of Rural Development, GOI, New Delhi

RAJASTHAN
CATEGORY-WISE WASTELANDS

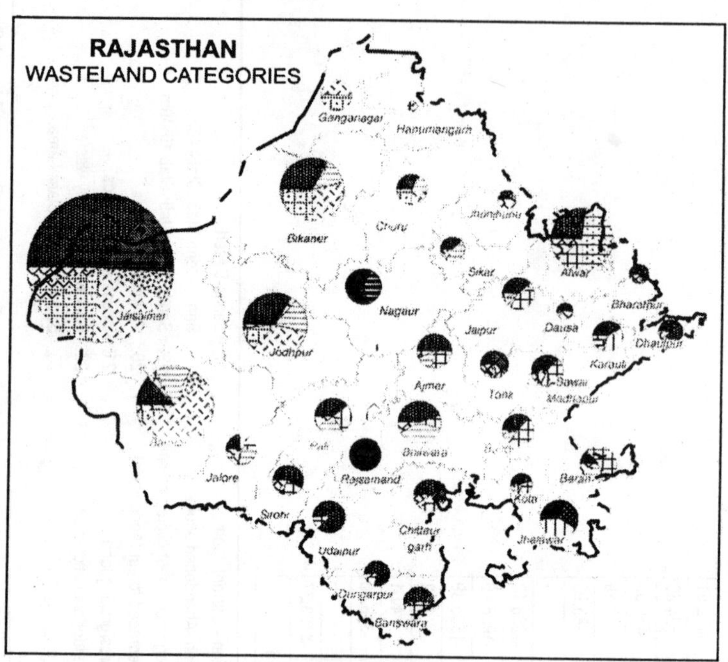

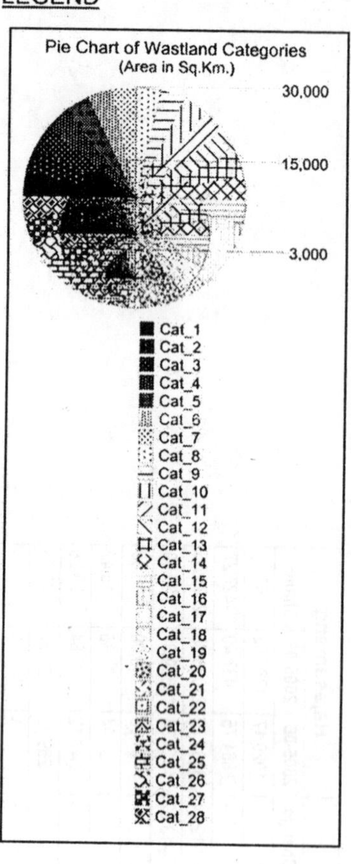

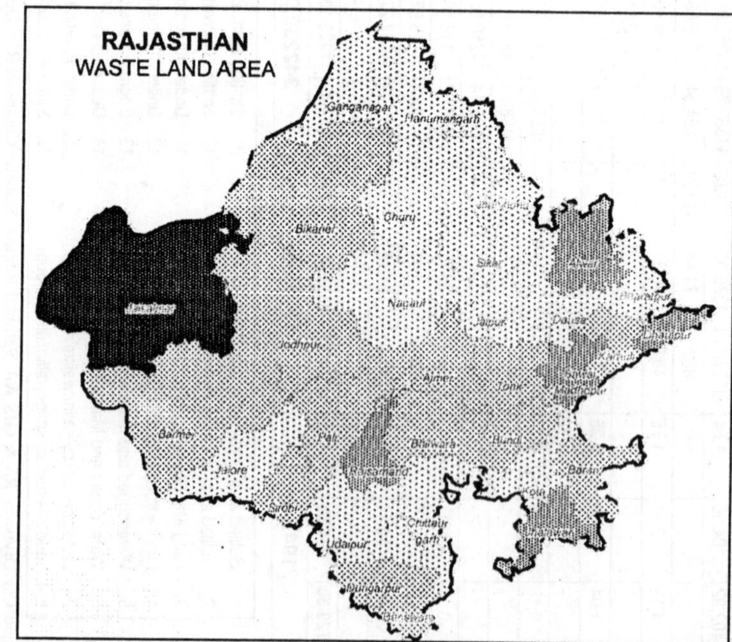

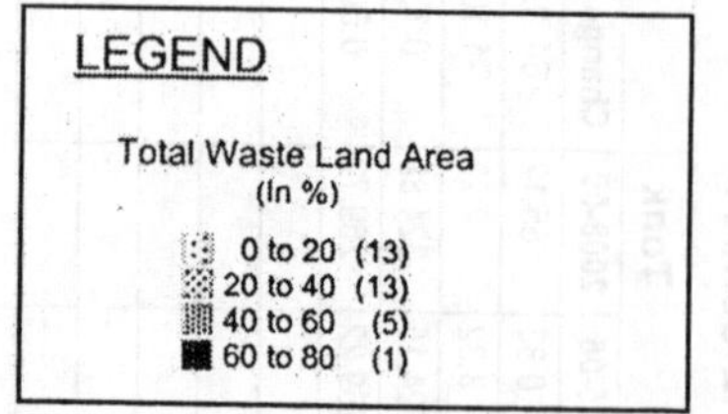

Source: Wasteland Atlas of India, National Remote Sensing Agency

Figure 17.3 Category-wise Wastelands of Rajasthan

Chapter 18
ENERGY DEVELOPMENT

Sufficent power availability plays an important role in industrial, agriculture development and other sectors of the country and state. About 75 per cent of the energy, sold by the state, is consumed by industrial and agriculture sectors alone while 15 to 16 per cent is consumed by domestic and commercial sectors. The balance electricity meets the requirement of drinking water supply, public lighting system and other miscellaneous sectors.

Electricity generation in India commenced more than 110 years back, with the first commercial generating unit being commissioned in 1895 by the Darjeeling Municipal Council. Electricity was recognised by the Govermment of India as a potential industry in 1910 by enacting the Indian Electricity Act. The Electricity (Supply) Act 1948 provided for constitution of Central Electricity Authority and State Electricity Boards and their responsibility to promote generation, transmission and distribution of electricity. The Rajasthan State Electricity Board (RSEB) came into existence on July 1, 1957. Later on for better and reform programmes in power sector. RSEB has been unbundled into five.

The development of power occupies a central place in the process of economic development. The regular availability of electricity in sufficient quantity and at reasonable rates affects the growth of agriculture and industry. Electricity enjoys the highest place in infrastructure development. In the post-economic liberalisation period from July 1991 onwards, there has been a lot of emphasis on the growth of installed capacity of power in different states, and for this purpose, private investment is being promoted from domestic as well as foreign sources. The demand for power was increasing faster than its supply in different states, therefore, it was absolutely essential to increase the supply of power without undue delay, so that the development process was not held up for a longer period. There was, of course, need for full transparency in power agreements and the national interest should never be compromised in any way. The development of power is a pre-condition for the speedy and sustainable industrial development of a region.

There are two types of sources of power:

1. **Conventional Sources of Energy–** Under it we include hydel power, thermal power (generated from coal, gas and oil) and electricity generated from nuclear power.

2. **Non-conventional Sources of Energy–** This includes power generated from wood and bio-gas, solar energy, smokeless chulha, wind power, etc. These are also called renewable sources of energy, because it is possible to renew and regenerate them, unlike conventional sources, which once exhausted, can't be replaced unless new sources are discovered.

In 2014-15 per capita consumption of electricity in Rajasthan was 1122.82 kwh, while it was 1009.61 kwh for the country as a whole. The state had 11th rank among 17 states of India in per capita consumption of electricity. Gujarat with 2104.69 kwh, had the highest place in per capita consumption of electricity', while Punjab had second place with per capita consumption of 1857.88 kwh in the same year.

At the end of March 2017, the total installed capacity of power in Rajasthan was

18677.18 MW. In 1951-52 it was 13 MW only. Thus, during the plan-period, the installed capacity has increased substantially. But the gap between demand and supply is increasing continuously, and this should be bridged up by increasing the installed capacity in future.

At the end of March, 2017, the total installed capacity of electricity was 18677.18 MW, out of which 6810.79 MW (36.5%) was under state owned and shared projects. It is absolutely essential to develop power resources in future to have a better balance between the demand and supply of power in the state.

We indicate below the various power-projects that supply power to Rajasthan at present.

1. Projects in which Rajasthan has been given its shared-capacity. In these projects, Rajasthan has been given its share in power; they are:

(i) Bhakra-Nangal Project.

(ii) Beas Unit (Dehar) and Unit II (Pong).

(iii) Chambal Project.

All the three projects as stated above, are hydro-electric projects.

(iv) Satpura Thermal Power Project (thermal power house) (Madhya Pradesh).

2. The state has got allotted-capacity in the following power projects :

(i) **Singrauli Super-thermal Power Project (Uttar Pradesh)**—Its total capacity was 2,050 MW and Rajasthan was allotted a share of 15 per cent in it. This is being operated as a central project by National Thermal Power Corporation (NTPC).

(ii) **Rihand Super-thermal Power Project (UP)**—It is also being managed by NTPC. It installed power capacity is 1,000 MW, and Rajasthan has been allotted a share of 9.5 per cent in it.

(iii) **Anta Gas Power Station**—This is also being managed by NTPC. Its total capacity is 413 MW and Rajasthan has been allotted a share of 19.8 per cent in it.

(iv) **Auraiya Gas Project (UP)**—Its total capacity was 652 MW and Rajasthan was allotted a share of 9.2 per cent in it. This project is also under NTPC.

(v) **Narora Nuclear Power Projects (UP)**—Its total capacity was 470 MW and Rajasthan's allotted share is 9.6 per cent.

(vi) **Rajasthan Atomic Power Project (RAPP)**—It was set up at Rawatbhata. near Ranapratap Sagar Powerhouse in 1973 with the cooperation of Canada in the central sector by the Nuclear Power Corporation (NPC). With two units of 235 MW each, its capacity initially was 470 MW. It was hundred per cent meant and designed for Rajasthan. The third and fourth units (total capacity 440 MW) were to be commenced from July and December 1999 respectively. Four units (each of 500 MW) were to be installed later sometime in 2000 AD. In the past the closed units were restarted by the efforts of Indian engineers and experts in atomic power. As this could be done without mechanical assistance from Canada is a notable achievement of the indigenous technique. There was an agreement with Nuclear Power Corporation (NPC) at a later date under which RAPP would get full electricity (440 MW) from the third and the fourth units.

3. The projects which would be under state's own capacity are as follows :

(i) Kota Thermal Power Station (KTPS)

Stage I (2 × 110) = 220 MW (Commenced in 1983)

Stage II (first unit) = 210 MW (on September 25, 1988)

Stage II (second unit) = 210 MW (on May 1, 1989)

Stage III (first unit) = 210 MW (on April 11,1994). Its cost has been estimated to be Rs. 480 crore.

Sixth unit of 195 MW capacity was to be started in August 2003. Seventh unit of

195 MW capacity was to commence production sometime in 2008. By the end of November 2010, total capacity of KTPS had reached a level of 1,240 MW and 6 units.

(ii) Mahi Hydel Project.

(iii) Mini-hydel Schemes of Rajasthan— Under Indira Gandhi Nahar Project—Anoopgarh branch, Suratgarh branch, Mangol, Charanwala and Pugal branches.

Others—Right main Canal Mahi I & II, Itwa Bisalpur and Jakliam projects (total 10 mini schemes).

GENERATION

Devoid of natural resources essential for power generation like river water and fossil fuels, Rajasthan had to depend on partnership with adjoining states to share their potential namely, Bhakra and Beas Hydel projects with Punjab and Himachal Pradesh, Chambal Hydro and Satpura Thermal projects with Madhya Pradesh. Subsequently, the Government of India set up a nuclear based thermal power plant at Rawat Bhata(Kota). The first unit of220 MW capacity was commissioned in 1972-73, the second in 1980-81, third and fourth units were commissioned in 1999-2000 and 2000-2001 respectively. The first coal-based power plant in the state sector came up in 1983 at Kota with a capacity of 2×110 MW. Later on, an additional capacity of 210 MW each was installed during 1988, 1989 and 1994.The Sixth unit of KTPS with a capacity of 195 MW was commissioned on August 2003. Now existing installed capacity of KTPS is 1045 MW. The second coal based power plant came up in 1998 at Suratgarh with a capacity of 1 x 250 mw. After that another 4 units of 250 MW was commissioned from the year 2000-03. In addition to that two units of 37.5 MW each of Ramgarh gas based thermal project was commissioned in August 2002 and June 2003.

State has an attractive policy for setting up power projects based on non-conventional energy source (wind and bio mass based projects). 367 MW capacity of power projects have already been commissioned upto March 2006 under this catagory.

Over the years power was also allocated from CPSU's coal based power station at Singarauli, Rihand and Unchahar and gas based power stations at Anta (Kota) and Auriya (UP) and Dadri.

Generation Mix

With the gradual increase in generation capcity there has been a distortion in hydro-thermal mix. In 1980, the hydro-thermal mix which was 65:35 has now changed to 25:75. The advantages in having higher hydro-thermal mix is that the generation can be regulated with much more flexibility to meet the demand. Thermal/Nuclear power stations generally operates as base load station. The load varies from time to time. The peak load normally occurs either in early hours of the morning or in the evening when there is a general lighting load. Higher hydro-thermal mix helps in meeting these peak demands. With the distortion in hydro-thermal mix after 1980, the state has no choice except to restrict the supply of power to agricultural consumers in early morning and evening peak hours and to industrial consumers during evening lighting load hours.

The demand of energy has been increasing day by day but the increase in generation capacity could not match this increasing pattern of load. The state had, therefore, to take some concrete steps within its means to meet the demand. These were:

(a) Increasing plant load factor of Kota Thermal Power Station (KTPS) and STPS

(b) Reducing transmission and distribution losses.

(c) Private Sector participation in power sector.

In 1980-81, the transmission losses in the state were 26.10 per cent but these have gradually reduced to 22.80 per cent but slowly increases to 28.75 per cent in the year 2000-01 and now in the year 2005-06 these are 40.94 per cent. The sudden increase in loss is due to change in calculation pattern of energy accounted for flat rate consumers.

KTPS has earned a productivity incentive award from the Government of India five times for having achieved a plant load factor of 57.32 per cent in 1984, 71.46 per cent in 1987, 51.76 per cent in 1989 and 66.35 per cent in 1992. It would be worthwhile to mention that Gold Shield Award has been given to KTPS and STPS for best performance in the year 2001 to 2004. Also 90.60 and 90.98 per cent load factor has been achieved by 2001 to 2004. Also 90.60 and 90.98 per cent load factor has been achieved by KTPS and STPS respectively during 2005-06.

Government owned companies incoiporated under companies Act 1956. These companies came into operation on 19-7-2000 as per Notification issued by government of Raj asthan and these companies are:

- Rajasthan Rajya Vidyut Utpadan Nigam Ltd., (RVUN)
- Rajasthan Rajya Vidyut Prasaran Nigam Ltd., (RVPN)
- Jaipur Vidyut Vitran Nigam Ltd., (JVVNL)
- Ajmer Vidyut Vitran Nigam Ltd., (AVVNL)
- Jodhpur Vidut Vitran Nigam Ltd., (JdVVNL)

The history of power development in Rajasthan dates back to the formation of Rajasthan state in 1949. Electricity power, till then, was confined to royal palaces and a few towns. The totat number of towns and villages electrified at the time were only forty-two. The installed generating capacity was only 13.27 MW.

The allocation of state plan budget to the power sector has gradually increased. In the year 2000-01 it was 24.23 per cent where as in the year 2003-04 it was 38 per cent and now in the year 2005-06 it was per cent.

Additional Generation Capacity

The criteria of allocation of power from central generating power stations, is based on the weightaed average of central plan assistance received and energy consumed in past years. Presently it has been decided to treat the present "formula" as "guidelines" for new central sector power stations for allocation of power from central sector generating stations to the states/union territories. The implication of this change are enumerated below:

- It will not distrub the allocation already made under the "formula". There will be no change in the contents of fonnula.
- Power from new central sector power stations will be made in accordance with power purchase agreements (PPAS) to be signed between the CPSU and the states/union territories or any of their authorised agency/board.
- First offer for purchase of power shall be made by the CPSU to each constituent (state/union territories or their authorised agency) of the region as per their entitlement.
- In case any constituent of the region does not buy its share or part thereof, the CPSU shall have the right to sell that power to any other state/union territories in accordance with the PPA to be entered into by them. However, such offer by the CPSU shall first be made to the state ($)/UT (S) within the region (where power station is located) before diverting the power to state ($)/UT ($) outside the region.
- Where there are more than one claimants to the surplus power, so offered weightage in allocation shall be given to the power sector reforming state ($)/UT(s).
- This does not affect allocation of 10% of the power to the state where the central thermal power plant is located and the 12% free power from Central hydel power

stations to the state(s)/UT(s) of the regions (including the state where the hydel project is located).

The peak demand in Rajasthan is expected to go up from the present level of about 5,000 MW to 7,500 MW by the end of Five Year Plan. In order to meet the increased power demand and energy requirement, the state government has drawn up comprehensive programme for development of additional capacity of about 4,500 MW. Out of this, 1,500 MW capacity will be added through State Sector Projects. The work on State Sector Projects viz Giral Lignite TPS (250 MW), Dhaulpur Gas TPS (330 MW) and Chhabra Coal based TPS (500 MW) has already been taken up. In addition, the capacities of existing thermal power plant are also proposed to be enhanced by installing additional units viz 195 MW at Kota Super Thermal Power Station and 250 MW at Suratgarh Super Thermal Power Station. About 1,500 MW capcity is proposed to be added through allocation from the Central Sector Projects which are under various stages of implementation. Balance 1,500 MW capacity is proposed to be developed through Private Sector Projects.

Lignite is the only fuel source available in Rajasthan. We are encouraging exploitation of lignite for power generation by State, Central and Private Sector companies. Rajasthan Vidyut Utpadan Nigam (RVUN) is setting up a 250 MW TPS based on lignite reserves of Giral in Barmer district. M/S Neyveli Lignite Corporation (NLC) is developing 250 MW project based on lignite reserves in Bikaner district. Another 1,000 MW project based on lignite reserves at Kapurdi and Jalipa in Barmer district is being developed in private sector. A captive power plant of 70 to 120 MW capacity is also coming up based on lignite reserves of Gudha east in Bikaner district. Other lignite reserves in the state are also proposed to be exploited for power generation (Figure 18.1).

The state has also announced a policy for grant of attractive concessions to the private entrepreneurs desirous for setting up of power plant in the state subject to selling of 50% of power within the state. Many entrepreneurs have expressed interest in setting up power projects.

ENERGY CONSUMPTION

In 2016-17, approximately 2/3rd of the energy was consumed by the industrial sector, while the proportion of agriculture was only 2.79 per cent. Energy consumption in agriculture development received an impetus from 2016-17 onwards (Table 18.1, 18.2 and 18.3 and Figure 18.2).

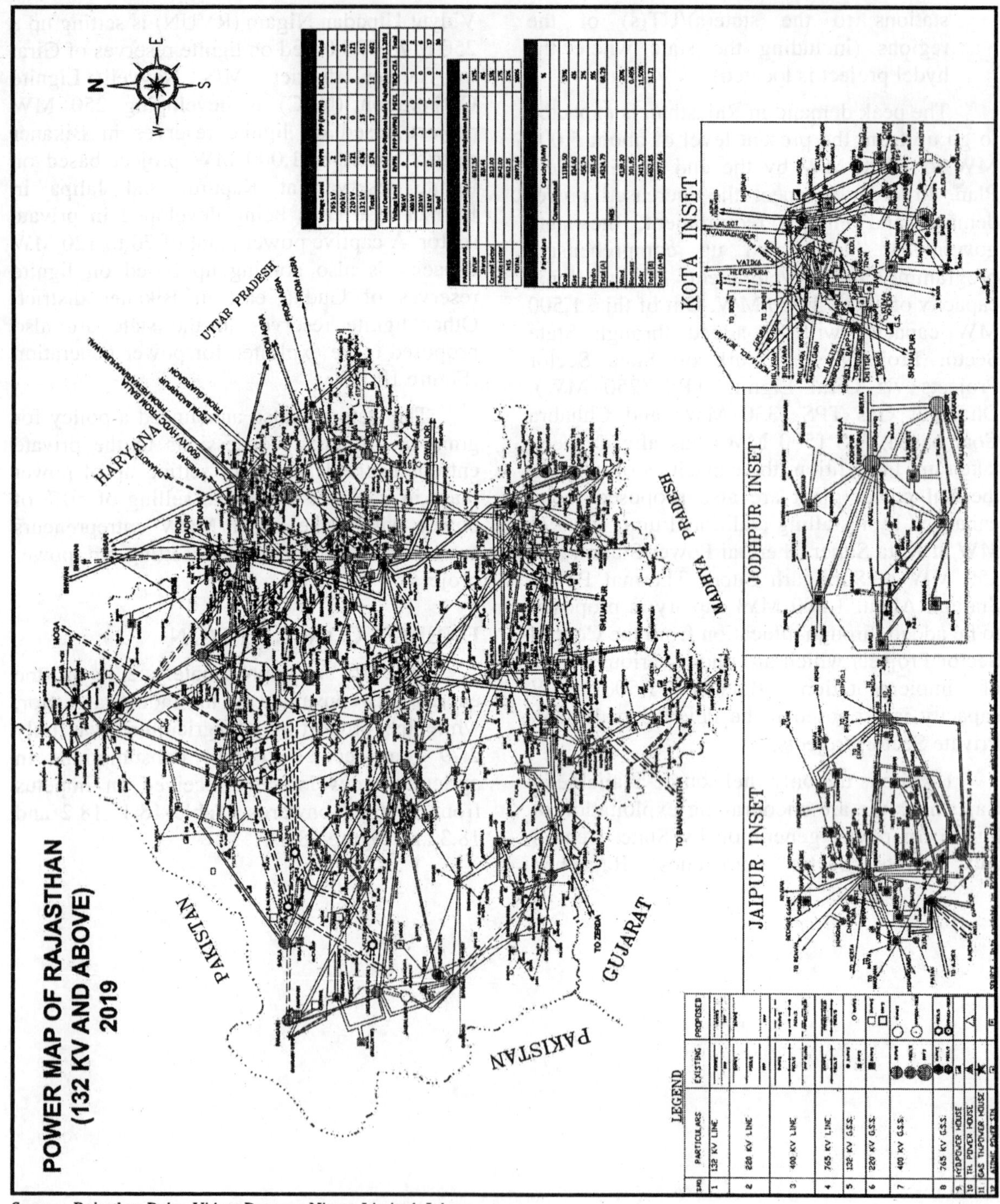

Source: Rajasthan Rajya Vidyut Prasaran Nigam Limited, Jaipur

Figure 18.1 Power Map of Rajasthan

Table 18.1 Electricty 2016-17 and 2017-18

Item	Unit	2016-17(P)	2017-18(P)
A. CONSUMPTION OF ELECTRICITY	**Lakh Unit**	**507623**	**557807**
1. Domestic	,,	114618	119983
2. Commercial(Non-Domestic)	,,	42855	45766
3. Agriculture	,,	213919	228334
4. Industrial (Small +Medium + Large)	,,	103882	131676
5. Public street light	,,	3638	3477
6. Public water works (Small +Medium + Large)	,,	19642	
7. Others	,,	9069	7271
B. TOTAL ELECTRICITY AVAILABLE	**Unit**		
1. Total Electricity Available for distribution (sale)	M. U.	71153.052	74197.172
2. Sale to / drawl by common pool consumers and other States	M. U.	-147.708	147.161
3. Total Electricity Available for supply within the State	M. U.	73923.886	76823.356
C. CONSUMERS SERVED	**Unit**	**2016-17**	**2017-18**
1. Domestic	Nos.	9228166	11375212
2. Commercial(Non-Domestic)	,,	952048	1375081
3. Agriculture	,,	1211546	1394271
4. Industrial (Small +Medium + Large)	,,	165088	271168
5. Public street light	,,	13574	20471
6. Public water work (Small +Medium + Large)	,,	49439	76763
7. Others	,,	37034	48359
TOTAL	,,	**11656895**	**14561325**

ELECTRICITY (TRANSMISSION LINES)

D. TRANSMISSION LINES			
1. 765 KV (D.C.)	CKt/ Kms.	425.5	425.5
2. 400KV (D.C)	CKt/ Kms.	2223.74	2223.74
3. 220 KV (D.C.)	CKt/ Kms.	7793.66	7908.39
4. 132 KV (D.C.)	CKt/ Kms.	13145.76	13456.34
5. 765 KV (S.C.)	CKt/ Kms.	-	-
6. 400 KV (S.C.)	CKt/ Kms.	1713.59	2717.99
7. 220 KV (S.C.)	CKt/ Kms.	6627.26	6996.9
8. 132 KV (S.C.)	CKt/ Kms.	3625.29	3732.96
TOTAL		**35554.80**	**37461.82**

S.C.-Single Circuit, D.C. Double Circuit, F- Final, P- Provisional

Source: Statistical Year Book of Rajastham , 2018

Table 18.2 Consumption Pattern

S. No.	Name of District	Electricity Consumers Served (Numbers) 31.03.2018							
		Domestic	Commercial (Non-Domestic)	Public Street Light	Agriculture	Industrial (Small +Medium-I Large)	Public water works (Small + Medium + Large)	Others	Total
1	2	3	4	5	6	7	8	9	10
1	Ajmer	517995	81994	1176	31217	18605	1278	2864	655129
2	Alwar	562289	61481	778	115380	18475	4901	1692	764996
3	Banswara	233493	16274	231	15864	3036	408	1151	270457
4	Baran	181725	14540	125	36101	2567	386	829	236273
5	Banner	310981	32431	267	45254	4344	2884	2303	398464
6	Bharatpur	329918	30601	366	30002	6903	2250	1033	401073
7	Bhilwara	435199	47255	1190	55182	12040	1840	2024	554730
8	Bikaner	352311	48083	1286	41656	9057	2155	1359	455907
9	Bundi	180279	16235	238	29173	2958	688	949	230520
10	Chittaurgarh	262427	25534	681	71279	8208	1145	1555	370829
11	Churu	347841	33277	562	22194	5997	3259	1994	415124
12	Dausa	208306	21332	272	44191	4419	1200	331	280051
13	Dhaulpur	116298	9141	59	7879	4402	234	323	138336
14	Dungarpur	225120	14934	313	26370	2599	425	1285	271046
15	Garrgarragar	368380	46607	554	15781	7163	770	2191	441446
16	Hanumangarh	330952	30140	853	27971	4737	564	1349	396566
17	Jaipur	1342263	279239	2810	129847	38544	14779	3091	1810573
18	Jaisalmer	95249	10478	66	9866	1073	663	728	118123
19	Jalor	254455	23552	363	62974	5321	1949	827	349441
20	Jhalawar	219342	14888	204	39622	4355	316	1112	279839
21	Jhunjhunu	397150	44864	438	58226	6781	5783	2076	515318
22	Jodhpur	621818	94186	1941	72047	19729	5092	1771	816584
23	Karauli	168044	13635	114	30086	3355	2124	704	218062
24	Kota	365680	54987	820	30087	7834	778	1212	461398
25	Nagaur	514474	47952	686	57115	13635	7384	2513	643759
26	Pah	471090	51100	924	38391	10510	2028	1887	575930
27	Pratapgarh	128707	7920	179	36675	1567	314	723	176085
28	Rajsamand	237353	20864	548	20020	9835	752	1511	290883
29	Sawai Madhopur	191104	16585	211	27593	3303	1121	901	240818
30	Sikar	481018	61268	550	73608	10288	5570	2068	634370
31	Sirohi	182046	23750	468	26836	5011	1021	708	239840
32	Torrk	230064	19483	255	17809	3418	1329	1160	273518
33	Udaipur	511841	60471	943	47975	11099	1373	2135	635837
	Rajasthan	**11375212**	**1375081**	**20471**	**1394271**	**271168**	**76763**	**48359**	**14561325**

Source: Statistical Year Book of Rajastham , 2018

Table 18.3 Consumption of Electricity

S. No.	Name of District	Consumption of Electricity (In Lacs Unit)31.03.2018							
		Domestic	Commercial (Non-Domestic)	Public Street Light	Agriculture	Industrial (Small + Medium+ Large)	Public water works (Small + Medium + Large)	Others	Total
1	2	11	12	13	14	15	16	17	18
1	Ajmer	5948.49	2245.96	130.59	1652.73	9637.51	1346.43	342.42	21304.13
2	Alwar	5998.95	1987.08	190.93	16686.05	31409.06	617.48	235.51	57125.06
3	Banswara	1785.28	332.98	23.07	786.92	726.10	90.66	37.59	3782.60
4	Baran	1304.07	324.08	40.27	6149.13	265.13	125.90	66.78	8275.36
5	Banner	2164.75	929.95	40.29	9029.81	1993.63	493.77	279.74	14931.93
6	Bharatpur	2718.78	710.68	145.19	3979.25	1447.68	263.18	210.51	9475.27
7	Bhilwara	3722.32	1191.96	108.63	3264.57	12323.38	300.12	48.69	20959.67
8	Bikaner	3951.83	1443.55	180.77	25447.35	2606.22	1214.63	733.17	35577.50
9	Bundi	1575.32	343.33	84.81	3745.77	826.03	189.61	147.27	6912.14
10	Chittaurgarh	2098.53	617.98	51.46	6865.12	3768.50	505.43	94.63	14001.66
11	Churu	3106.90	778.44	74.85	9810.47	730.62	1156.44	61.25	15718.98
12	Dausa	1616.51	463.80	49.69	4927.69	506.02	183.55	28.97	7776.23
13	Dhaulpur	996.63	234.38	17.98	1208.17	888.74	204.79	19.97	3570.66
14	Dungarpur	1586.67	323.53	19.95	1250.77	535.46	76.54	22.21	3815.13
15	Ganganagar	5137.58	1252.66	61.35	1667.45	1301.43	551.89	783.07	10755.44
16	Hanumangarh	3996.73	773.68	39.15	4741.52	812.63	554.34	49.42	10967.46
17	Jaipur	24809.36	14581.80	760.08	16475.33	21009.06	2276.67	1027.50	80939.81
18	Jaisalmer	724.99	506.50	21.33	5386.67	654.70	547.87	501.17	8343.23
19	Jalor	1494.09	557.15	25.75	12740.08	2285.28	522.07	25.38	17649.80
20	Jhalawar	1640.92	383.00	45.06	4444.83	1886.16	165.06	56.64	8621.67
21	Jhunjhunu	3639.09	1160.85	67.36	9585.74	1157.03	992.89	82.43	16685.39
22	Jodhpur	8077.47	3525.03	421.33	29865.48	6559.91	2834.14	1227.15	52510.50
23	Karauli	1475.81	356.23	99.65	3771.93	234.08	380.49	58.38	6376.57
24	Kota	5723.79	2282.10	217.26	4152.10	4633.71	535.57	496.35	18040.88
25	Nagaur	3998.95	1085.52	70.65	13069.97	2730.61	1215.01	67.21	22237.92
26	Pali	3401.00	1031.73	80.54	2954.53	3087.71	405.23	75.13	11035.88
27	Pratapgarh	775.37	148.39	10.14	2457.06	64.85	43.89	8.25	3507.95
28	Rajsamand	1642.26	541.45	58.35	877.72	5455.02	231.15	29.60	8835.56
29	Sawai Madhopur	1887.88	469.36	36.93	3428.68	243.15	172.55	47.49	6286.04
30	Sikar	4542.52	1473.59	72.80	11790.00	2449.90	862.75	46.73	21238.30
31	Sirohi	1382.29	624.23	44.85	2630.66	2635.84	180.63	56.73	7555.23
32	Tonk	1972.60	483.83	57.47	1346.82	891.06	1590.66	38.54	6380.98
33	Udaipur	5085.57	2600.96	127.98	2143.70	5920.19	468.69	264.91	16612.00
	Rajasthan	**119983.30**	**45765.77**	**3476.49**	**228334.07**	**131676.41**	**21300.09**	**7270.77**	**557806.89**

Source: Statistical Year Book of Rajastham , 2018

The agriculture load has picked up at a faster rate than the other sectors specially industrial sector. Agriculture load demands extensive transmission and distribution system which is also one of the maj or factor higher transmission losses. The agriculture load is generally at its peak during winter months with the onset of Rabi season. From the year 1980 to meet the agriculture load in cropping season, power cuts on industrial has been introduced.

RURAL ELECTRIFICATION

In 1951, there were only 42 electrified localities (26 towns and 16 villages). Power supply was then confined to only domestic, commercial and public lighting services. The Pace of rural electrification Picked up in third decade.

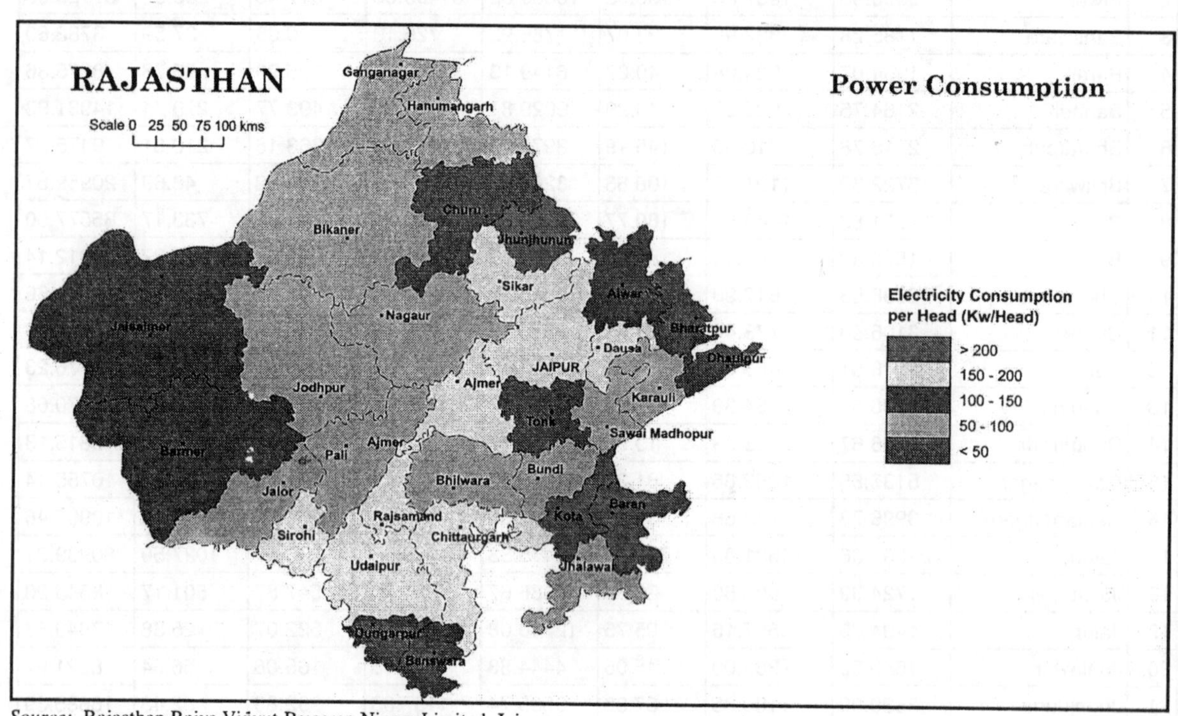

Source: Rajasthan Rajya Vidyut Prasaran Nigam Limited, Jaipur

Figure 18.2 Power Consumption in Rajasthan

NON-CONVENTIONAL SOURCES OF ENERGY

State Government created a autonomous agency in 1985 with the name of Renewable Energy Development Authority (REDA), since then most of the REDA programmes were only subsidy programmes though REDA installed variety of systems but their long term impact was not seen. From 1999 onwards agency modified its approach and an effort was made to off-load the subsidy and make these programmes people's programme.

The State of Rajasthan has a strong resource of Non-conventional Energy comparing the lesser availability of electricity, coal, petroleum products and other energy sources.

State has adopted a policy with a view to promote generation of grid-grade power through Non-conventional energy sources in the context of non polluting renewable and decentralised availability of inputs like solar, wind, mini-small hydel, biomass and other wastes. However, in view of immense potential of power generation through exploitation of wind resources in Rajasthan and special incentives required because of special nature of wind as resource, it has been felt necessary to adopt a scientific promotion policy for generation of power by use of wind resources. The objectives of this policy is to support a programme of creating 100 MW power generation capacity based on exploitation of wind resource in the state, largely in private sector.

Rajasthan is endowed with plenty of sunshine, wind and biomass, the basic ingredients of non-conventional energy sources. The town Jodhpur is renowed as "Sun City" since it gets maximum solar radiation almost throughout the year. Besides, western Rajasthan particularly Jaisalmer is one of the biggest tourist attraction, not only on account of "Desert Havelis" but for the availability of enough sunlight suitable for "Sun Bath" for those foreign tourists, coming from temperate regions. High wind speed, almost throughout the year, is a great natural gift to the state. The state, particularly western Rajasthan has rich biomass.

Non-conventional energy is obtained from the continuous or repetitive currents of energy occurring in the natural environment. It is renewable but low in intensity, approximately 300 Wm^{-2} or less. The dispersed nature of non-conventional energy sources makes it most appropriate to serve as decentralized energy system. However, its fluctuating nature requires special concentration, storage and control consideration. There is vast variety of non-conventional energy sources which can be broadly grouped under three categories i.e. (a) The sun, (b) The Tidal energy i.e. the motion and gravitational potential of the sun, moon and earth and (c) chemical sources of energy such as hydrogen energy (fuel cell). So far, technically feasible and financially viable technology have been developed for harnessing only solar and wind energy while the chemical sources are at Research and Development level, At present, non- conventional energy is contributing hardly one per cent of the total demand but has the potential to meet upto 25 per cent demand in near Future. A Summary of non- conventional energy sources vis-a-vis end use demand is as Table 18.4.

Solar Radiation

The state has plenty of solar radiation and minimum of cloudy days. The total amount of radiation falling on a unit horizontal surface is called Global Solar Radiation which is measured by an instrument, called Pyranometer. The portion of radiation from the sky and clouds is called the diffused solar radiation. The difference between the two is called direct solar radiation which is measured by an instrument called Pyrheliometer. The Indian Meteorological Department maintains a network of radiation stations which measure various parametes, like quantity of solar radiation, daily duration of sunshine etc. Most parts of Rajasthan receive solar radiation of 5.8 to 6.4 $KWhm^{-2}$ with 300 clear sunny days in a year.

Table 18.4 Non-conventional Energy Sources vis-a-vis end Uses

Sector	End Use	Sources	Devices
Domestic Sector	I. Cooking (Meal)	I. Solar II. Biogas III. Energy Plantation IV. Briquettes	Cooker Gas burner Improved chulhas
	II. Lighting	i. Solar ii. Biogas iii. Producer gas iv. Wind	Photovoltaic modules and fluorescent lamps Incandescent lamps/electrical generator Electrical generator Wind electricity generator
	III. Water (general Purpose)	i. Solar ii. Wind iii. Biogas iv. Producer gas	Photovoltaic pumps Wind convertor Biogas engine Producer gas engine

Sector	End Use	Sources	Devices
	IV. Drinking water	i. Solar ii. Biogas	Solar stills Heat operated
Agriculture Sector	I. Irrigation	i. Solar ii. Wind iii. Biogas iv. Producer gas	Photovoltaic pumps Wind turbine Biogas engine Producer gas engine
	II. Mechanical work (with modern farms)	i. Methanol ii. Ethanol	Internal Combeviton engine Internal Combeviton engine
	III. Drying	i. Solar	Dryer
	IV. Refrigerators	i. Solar ii. Wood iii. Biogas iv. Producer gas	Cooling devices Cold storage Cold storage
Industries	I. Small scale (milling, handloom, etc.)	i. Solar ii. Wind iii. Biogas iv. Producer gas	Photovoltaic Wind generator Biogas engine Producer gas engine
	II. Cootage (pottery, bricks, leather etc.)	i. Biogas ii. Biomass	Biogas furnace Wood or biomass briquettes and furnace
	III. Street lightings	i. Solar ii. Biogas iii. Producer gas	Photovoltaic Biogas generator Producer gas generator
	IV. Community Centre	i. Solar ii. Biogas iii. Producer gas	Photovoltaic Biogas generator Producer gas generator
Transport	I. Transport (mobile mechanical work)	i. Methnnol ii. Ethanol	Internal Combeviton Engine Internal Combeviton Engine

Source: REDA

Solar Thermal

The main grades of heat which can be provided by solar energy and the possible application are as follows (Table 18.5).

Table 18.5 Solar Thermal Energy

Type of Thermal Energy	Application
Low grade (below 100 °C)	Water heating, air heating, crop drying, refrigeration, space heating, water desalination, etc.
Medium grade (100-300 °C)	Cooking, steam generation for industrial application, drying, refrigeration, power generation, water desalination, water pumping, etc.
High grade (above 300°)	Power generation

Source: REDA

The device used for collecting solar radiation and transferring it to a fluid, passing in contact with it is called 'solar collector'. They are of two types namely (a) flat-plate solar collector and (b) focussing type collectors. Flat-plate collectors are used where temperatures below 90 °C are required and concentrating collectors are suitable for higher temperatures like power generation.

The most popular solar cooker is box-type collector. For distillation process, glass roof-type solar stills have been used in many places in Rajasthan state for producing distilled water. The state also has the old tradition of constructing buildings based on solarpassive design approach.

SOLAR PHOTOVOLTAIC POWER PLANTS

Solar Photovoltaic technology is being utilised in the State for providing stand alone domestic and street lighting systems in the rural areas and have proved to be very successful as this has obviated drawings long transmission/distribution lines which would have called for massive investment. MNES provides 50 per cent of financial assistance for the projects. The main advantage of grid interactive power plants being their maintenance free operating capacity. Three grid interactive power plants of 55 Kwp capacity have been installed so far.

Direct conversion of solar energy into electricity is achieved through Photovoltaic technology. Silicon water photo electric cells, popularly called as "Solar cells" are usd. They have theoretical efficiency of about 25 per cent but a practical value is around 15 per cent. Rajasthan Electronices and Instruments Ltd., Jaipur, is one of the five companies in the country which is manufacturing and assembling photovoltaic modules, Stand-alone modules are available for lighting, pumping, communication, railway signals, etc. Recently, a programme of solar lantern has been promoted, particularly, for border areas. At prsent, chief problem is the high cost. With the expected breakthrough in the large scale production of solar cells, made of amorphous silicon, this technology will have wider applications. Solar powered wind pumps upto 1300 watts capacity have been developed in the country and are being installed in the state.

SOLAR POWER PROJECT

140 MW Integrated Solar Combined Cycle Power at Mathania is in the advanced stage of implementation with an investment of Rs.871.74 Crores. $ 45 million GEF grant from World Bank will be available as 100 per cent grant for the project. It is expected that Ministry of non-conventional Energy Sources. Government of India and Government of Rajasthan will provide Rs.50 Crores each as equity. M/s KFW, Germany have completed the appraisal of the project. Loan agreement in respect of KFW composite loan of dm 250 Million has been signed on 29.10.2001 between Government of India and KFW Germany, besides this, Global Environment Fund (GEF) is also providing $7.50 lac for conducting various studies and technical guidance. It is expected to finalise EPC cum O and M contract upto August/September, 2002. The ground work at site may start in the mouth of October, 2002.

Rajasthan in blessed with maximum solar radiation intensity of about 6-7 kwh/m2 day and maximum sunny days (more than 325 days) in a year, capable of making the state leader in Solar Power Generation. It can become the biggest hub of solar power due to the gift of nature, and investment friendly and effective gift of nature.

Firm Capacity of Solar Power at the end of December 2017 was 1656.70 MW in Rajasthan. The development of solar power in the state would save a lot of gas and fuel. The first solar energy fridge was set up at Balesar higher primary health centre in Jodhpur district. A panel of glass plates was made on the roof. The solar energy was stored in the fridge-battery which operates the fridge.

Earlier it was proposal to start an Integrated Solar Combined Cycle (ISCC) plant of the capacity of 140 MW at Mathania in Jodhpur. Its revised cost was Rs. 700 crore which was to be shared between the World Bank, Germany, Raj. Government and the Centre. Solar energy would be used for street tube-lights, solar-cookers, water-heaters, and solar-pumps. These pumps would be used for drawing water at Banner, Nagaur and Churu. Colour TV sets would also be installed in border areas.

There is a big programme to develop solar energy in Rajasthan, for which various centres have been selected from districts, suitable for the purpose. State Government announced Solar Energy Policy in 2011 and also in 2014. Solar plants will be set up in the state under

National Solar Mission and under the policy of 2014. Government would instal 25,000 MW additional solar power capacity with private investment (detailed discussion has been given later under Solar Energy Policy 2014).

WIND POWER GENERATION

Rajasthan Energy Development Agency had carried out monitoring of wind in past 13 years under a MNES sanctioned project in association with Indian Institute of Tropical Methodology, Bangalore. It has been assessed that State has a gross potential of 1200 MW power generation from wind with the available technology, out of which 400 MW can be harnessed immediately. Agency has already installed 2 MW wind power demonstration farm at Jaisalmer, 2.25 MW at Devgarh and 2.10 MW Phalodi with the help of MNES.

In private sector four wind farms has been set up with a total capacity of 9.76 MW. It is expected that by 31.3.2004, State will be able to achieve its target of 100 MW power from wind with capital investment from private sector. Agency has received proposals for projects for a combined of 237 MW from available industry in the wind energy sector. Investments from profit making companies avails 100 per cent depreciation from its profits IT Act.

There are number of Wind Monitoring and Wind Mapping Stations are available in the state. The data from the wind survey are utilized to identify windy areas having mean annual wind speed greater than 18 km/hr, which is suitable for electricity generation.

The velocity of wind in Rajasthan is in the range of 20 to 40 km. per hour. There is a programme to develop production of fodder and grazing grounds in Indira Gandhi Nahar Project Area by installing 10 wind-mills with simple and low cost equipment. Wind air generators would be procured for the development of desert in Rajasthan.

A wind-based project of 2 MW was started on 14th August, 1999 in Jaisalmer and a second wind-power-demonstration-project was commissioned on 6th March, 2001 at Deogarh in Chittorgarh district. Till December 2017, 4121.20 MW Capacity Wind Power Plants were established against technical potential of 5400 MW. Rajasthan Renewable Energy Corporation Ltd. has established three wind power plants of total capacity of 6.35 MW at Jaisalmer, Phalodi and Devgarh. In addition to this, Wind Power Plant of 25 MW, 10.2 MW and 10.2 MW have been set up in 2004, 2006 and 2010 respectively.

BIOMASS POWER PROJECTS

There are now technologies available where surplus biomass such as mustard stalk, rice husk, wheat stalk and other agro wastes can be run power plants. REDA has got studies done for such availability of biomass in ten blocks in the State. It is estimated that power plants based on biomass totalling to a capacity of about 300MW can materialise. REDA have already received applications from private parties proposing power plants based on biomass totalling to a capacity 30 MW. Ten more blocks have been taken up for assessment of surplus biomass.

Biomass constitutes nearly 65 per cent of the total energy consumption in the state. Firewood, agro-residues and cattle-and camel-dung are essentially used for cooking in a very efficient manner. Biogas technology and Improved Chulhas have become very popular to recycle cattle-dung and conserve firewood, respectively. Gasification of wood and agro-residues for mechanical and electrical applications upto 100 KW has started becoming economically viable. The state has over 6 million cultivable wasteland, potentially suitable for energy plantations. Renewable Energy Centre at College of Technology and Agricultural Engineering, Udaipur, has identified trees for energy plantation in arid areas. These are: Prosopis juliflora, Acacia auriculiformis, Eucalyptus, Leucenia, etc.

Gobar-gas plants are expanded in the villages of Rajasthan. There is a lot of saving of kerosene oil and fuelwood by extending the use of bio-gas plants. There is an emphasis on the maintenance of the existing plants rather than starting new ones. Surplus biomass such as mustard stalk rice husk & other agro wastes as well as Juliflora can be utilised to run power plant. Biomass Power Generation Plants of 101.95 MW capacity have been set up by March 2017 in the state.

INTRODUCING POWER PROJECTS OF RAJASTHAN

Rajasthan Rajya Vidyut Utpadan Nigam Ltd. owns and operates the following thermal, gas and hydel power stations in the state sector.

(A) Installed Power Capacity in the State—As on 31 March 2016 was as follows (Table 18.6):

(A) Thermal Power Projects	
1. Suratgarh Thermal Power Station	1500
2. Kota Thermal Power Station	1240
3. Chabbra Thermal Power Plant Units 1 to 4	1000
4. Kalisindh Thermal Power Plant	1200
5. Giral Lignite Thermal Power Units 1 to 2	250
(B) Gas Based Power Projects	
1. Dholpur Combined Cycle Gas Project	330
2. Ramgarh Gas Thermal Plant	273.50
(C) Hydro Projects	
1. Mahi Hydel Power Station	140
2. Mini Micro Hydel Schemes	23.85
Total	**5957.35**

Source: REDA

(B) Nine New Power Projects—Following projects were to be commissioned in the 12th Five Year Plan (2012-17):

		Capacity (MW)
1.	Chhabra Super Critical Thermal Power Project Stage II, Unit 5 & 6 (2 × 660 MW)	1320
2.	Suratgarh Super Critical Thermal Power Project Stage VI Unit 7 & 8 (2 × 660 MW)	1320
3.	Dholpur Gas Thermal Power Project Stage II (3 × 110 MW)	330
4.	Ramgarh Gas Thermal Power Project Stage-IV	160
5.	Suratgarh Super Critical TPS Project Stage VI (Unit 9 & 10) (Sri Ganganagar) (2 x 660) MW	1320
6.	Kalisindh Super Critical TPS Stage II (Unit 3 & 4) (Jhalawar) (2 × 660 MW)	1320
7.	Banswara Super Critical TPP Unit 1 & 2 (2 × 660 MW)	1320
8.	Kota Gas Thermal Project (3 × 110)	330
9.	Chhabra Gas Thermal Project (3 × 110)	330
	Total	**7750**

Source: REDA

Projects to be developed by private parties were as follows :

1. Jalipa Kapurdi Lignite Mining cum Power Generation Project, issued letter of intent to M/S West Power to set up 1,000 MW lignite mining cum power generation project.
2. 1320 MW Banswara Thermal Power Project (2 × 660 MW), approved on January 13, 2009.
3. 70 MW Gurha Thermal Power Project in South West of Bikaner, lignite to be supplied by Raj. State Mines & Minerals Ltd. and water from Indira Gandhi Canal. Company obtained commencement of business certificate on 13 November, 2009.

Rajiv Gandhi Gramin Vidyutikaran Yojana (RGGVY)—The Govt. of India launched this scheme in April 2005. Under this scheme all the villages were to be electrified during the next 5 years to provide electricity to BPL families. Connections to BPL families were to be given free of cost and centre was to give 90 per cent of the cost as grant and 10 per cent as loan. Rural Electrification Corporation (REC) was the nodal agency for implementation of this scheme. Around 11.77 lakh BPL households have been electrified under this scheme. Under BJP rule, this scheme has now been subsumed in the Deen Dayal Upadhyaya Gram Jyoti Yojana (DDUGJY) to which 6 remaining schemes of districts of RGGVY have been handed over.

Twelfth Five Year Plan Outlays on Power Development A sum of Rs. 47,246 crore was allotted in the Twelfth Five Year Plan (2012-17) for different projects, which is shown in the following table 18.6 :

	Project	Outlay in XII Plan (Rs. Crore)
1.	Chhabra Thermal Power Project (Unit 1 & 2)	300
2.	Chhabra TPP, Unit 3 & 4	568
3.	Chhabra Super - Critical, Unit 5 & 6	7605
4.	Suratgarh Super - Critical TPP, Unit 7 & 8	7625
5.	Suratgarh Super - Critical TPP, Unit 9 & 10	7920
6.	Ramgarh Gas Thermal Power Project Stage III	169
7.	Ramgarh Gas Thermal Power Project Stage IV	640
8.	Kalisindh TPP, Unit I & II	2719
9.	Kalisindh Super - Critical TPP, Unit 3 & 4	7920
10.	Banswara Super - Critical TPP, Unit I & II	7880
11.	Survey Investigation Schemes & Carried over Liabilities	50
12.	Dholpur Gas CCPP Stage II	1210
13.	Kota Gas Project	1320
14.	Chhabra Gas Project	1320
	Total	47246.00

Source: REDA

Thus, vast funds were to be spent on power development in the 12th Five Year Plan in Rajasthan.

Rajasthan Energy Development Agency and Rajasthan State Power Corporation Ltd. were merged in 2002 to form a new corporation, namely, Rajasthan Renewable Energy Corporation (RREC) for the development of non-conventional energy sources in the state.

The aforesaid discussion indicates that the power scenario is changing quite fast in Rajasthan. It has the potential for generating additional power, based on conventional sources like lignite, gas, diesel and water, and also on non-conventional sources like solar-energy, bio-gas and wind-power, etc. With proper planning and policy measures in the power sector, the state can definitely solve the problem of power-shortage in future and can hopefully become a power surplus state in the country.

NEW POLICY DEVELOPMENT IN THE POWER SECTOR

1. Deen Dayal Upadhyaya Gramin Jyoti Yojana (DDUGJY): New BJP Government has launched this yojana and the earlier RGGVY has been subsumed under it. Remaining 6 schemes of RGGVY, relating to the districts namely Jaipur, Banswara, Hanumangarh, Udaipur, Dholpur and Banner have been transferred to it for implementation; The funding pattern for DDUGJY would be 60% grant by the centre and the balance 40% by the State Govt./ implementing agencies.

Rajasthan Solar Energy Policy, 2014

Earlier Rajasthan Solar Energy Policy was released in 2011, it has been decided to review it in 2014.

New policy would reduce the dependence on conventional sources of energy and obtain self sufficiency in its energy needs. The aim is to install 25,000 MW of solar power through state or private enterprises or through Public-Private-Partnership or through individual efforts.

Rajasthan Renewable Energy Corporation (RREC) will act as Nodal Agency for Clearance of projects. It will facilitate projects of capacity above 20 MW. It will set up Solar Power Plants in Rajasthan for sale to DISCOMS.

State will promote development of Solar Parks by Private Sector Developers. Solar Power Producer shall contribute towards

Rajasthan Renewable Energy Development Fund of Rs. 1 lakh/MW every year for the entire life of the project.

It was hoped that Solar Energy Policy of 2014 and Wind Energy Policy 2012 would promote renewable energy sources in Rajasthan. State was on the threshold of rapid power development in the near future.

Power Development Programme in the Budget for 2018-19

Rajasthan State Power Transmission Corporation was rewarded for the best performance during the last 3 years. Power generating capacity was added by about 6700 MW during the last 4 years, and 3161 MW under Renewable Energy sources. Power Generation Corporation has entered profit zone by December 2017 to the extent of Rs. 400 crore. Organisational structure of power distribution companies has been reformed. Farmers would be provided with 2 lakh agriculture-connections during 2018-19. New sub-stations would be set up (400 KV-one, 132 KV-15 & 33 KV-200). 7 lakh new domestic connections would be allotted. State would promote energy development in 2018-19.

MICRO HYDEL PROJECTS

MNES has now designed small hydel projects also as non-conventional energy sources based power projects. State Government has authorised RSPCL for executing projects based on non-conventional Energy Sources, to facilitate setting up of following small hydel projects:-

- Jakham Dam (Chittorgarh)
- Alina RMC at head (Kota)
- Bisalpur dam RMC (Tonk)
- Som Kamala Amba, RMC head (Dungarpur)

There are other places also on the IGNP where suitable water falls are available for putting up small hydro power stations. It is estimated that if all such falls on the canals in the State are utilised a further 10MW electricity generation capacity can be added and local area can be served. At such falls small capacity plants can be taken up with short gestation period.

Small Hydel Power

Small hydel power systems can provide energy for rural and remote villags in a decentralized and cost effective manner. Projects of capacities less than one MW, are called Micro-hydel, 1-5 MW as Mini-hydel and 5-15 MW as small hydel systems. The state has the potential in this field, including water falls in the main and minor canals.

PROMOTIONAL PROGRAMME FOR CONSUMER PRODUCTS

With the modified strategy of creating mass awareness about promotion of non-subsidised non-conventional energy consumer devices. REDA has found a huge market for promotion of varied products some can be listed as Solar Photovoltaic Home Lighting System, Solar Water Heating System, Solar Cookers, Solar Lantern, SPV Pumps, Compact Fluorescent Lamp, Pressure Cookers, Nutan Stoves etc. Agency has authorised about 180 Aditya Solar Shops in private sector who are doing necessary supporting work for demonstration off the shelf availability, sale, repair and providing guarantees for customer satisfaction without any subsidy.

MINERAL OIL EXPLORATION

Mineral fuels which include crude petroleum and natural gas are the key to the pace of development of any country. Attempts to explore hydrocarbons in Rajasthan dates back to 1956. During 1962 and 1967, Oil and Natural gas Commission (ONGC) in a joint venture with Campagnie Generale De Geophysique (CGG) and International Forasol Petroleum (IFP) of France, carried out seismic data acquisition, followed by exploratory drilling. During 1981, a major policy decision assigned a high priority to Rajasthan which imparted requisite thrust in exploration activities. In 1982, ONGC was allotted 14,000 sq kms area for exploration, while 28,600 sq kms were alloted to Oil India Limited (OIL), a sister organization. Currently, both ONGC and OIL are operating in Rajasthan (Figure 18.3).

Sedimentary tract to the west and north-west of Aravalli up to International border, known as the Great Rajasthan Desert, covers an area of about 1,20,000 sq kms The area has been divided into four geolgic basins, namely, Jaisalmer Basin, Bikaner-Nagaur Basin, Barmer Basin and Sanchor Basin. Jaisalmer Basin has a thick sedimentary section with positive evidence of hydrocarbon accumulations. Bikaner-Nagaur Basin is also rated as prospective and shows positive evidence of oil accumulations. The Barmer and Sanchor Basins have commercial accumulations of oil. Jaisalmer and Bikaner-Nagaur Basins come under category-II basins, where oil and gas occurrences are known. Commercial production of gas by OIL and ONGC has commenced since 1996. The main aim is to upgrade these basins to the category-I i.e. a basin having commercial production of hydrocarbon.

Geo-scientific efforts carried out during last 30 years have enabled delineation of a large number of prospective areas within the Jaisalmer Basin and Bikaner-Nagaur Basin. Nine gas fields, viz. Manhera Tibba, Ghotaru, Bakhri Tibba, Khartar, Bankia, Tanot, Dandewala and Bagıtibba have been proved as a result of exploratory efforts made by ONGC and OIL so far. A major find of heavy crude oil has been made at Baghewala in Bikaner-Nagaur Basin. In addition to these gas fields, good hydrocarbon indications have been observed in other structure as Jalalwala, Bhuana, Vikharan Nai, Shahgarh etc. Some 25 more prospective sites have been identified which are under process of further detailing by seismic survey.

Gas has been discovered in Jasalmer Basin in multiple reservoirs, particularly in Ghotaru sandstones of Cretaceous age, and Khuiala limestone and sandstone of Eocene age. Gas along with minor oil exists in Pariwar sandstone of Cretaceous age. Heavy, balck oil has been found in two reservoirs within Bilara and Jodhpur formations in Bikaner-Nagaur Basin. The prognosticated resources in Jaisalmer area is 180 MMT (oil + oil equivalent gas), while the discovered reserves in this Basin are of the order of 8.6 billion m^3 of gas.

Plans are being made to commercially exploit the established resources of heavy oil and bitumen from Baghewala field in Bikaner-Nagaur Basin and crude oil from Barmer-Sanchor basin by Caim-ONGC.

PETROLIUM AND NATURAL GAS

As indicated eariliear, reserves of natural gas and crude oil have been indicated mostly in the Jaisalmer and Barmer districts. The State Government should try to increases the production of natural gas and oil so that the related development may follow in due course. Private investment domestic and foregin will have to be tapped to develop the mineral and power resources of the state. The industrial growth in the mining as well as power sectors. These should be a simultaneous development in these three sectors alongwith the infrastructureal development so that Rajasthan may be in a position to tackel problems of poverty and unemployment in future. The Report f the S.C. Tripathi committeee says "Discovery of hydro-carbons in Rajasthan is very important event.

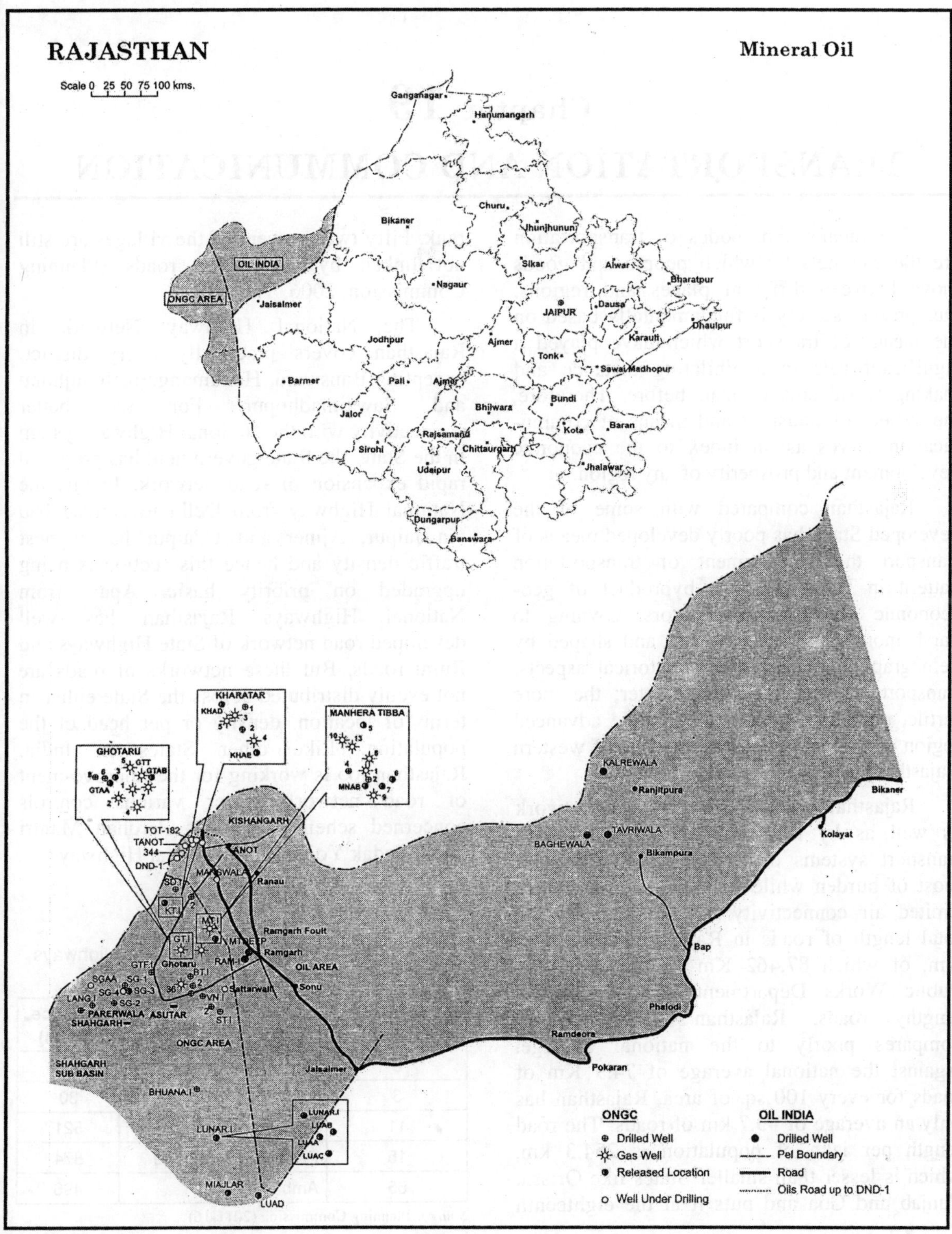

Source: ONGC, GoI

Figure 18.3 Mineral Oil in Rajasthan

Chapter 19

TRANSPORTATION AND COMMUNICATION

The means and modes of transportation are the channels by which people and goods move between different places and regions. Our present society is fundamentally based on the means of transport which have played a significant role in annihilating distance and making world smatter man before. Therefore, the system of transport and means of communication serves as an index to the economic development and prosperity of any region.

Rajasthan compared with some of the developed State, has poorly developed means of transport the development of transportation pattern in Rajasthan is a byproduct of geo-economic and historical factors. Owning to harsh morphology and climate and shaped by demographic, economic and historical aspects, transport network is serving better; the more fertile, populous, economically more advanced region of eastern Rajasthan than that of western Rajasthan.

Rajasthan is well served by road network as well as by railways. The rail and road transport systems of Rajasthan have to bear most of burden while only a few towns have limited air connectivity. As of 2000-01, the total length of roads in Rajasthan is 151,865 Km, of which 87,462 Km is maintained by Public Works Department. Despite of the lengthy roads, Rajasthan's road density compares poorly to the national average. Against the national average of 74.9 Km of roads for every 100 sq. of area, Rajasthan has only an average of 43.7 km of roads. The road length per lakh of population is 254.3 km, which is lesser than smaller States like Orissa, Punjab and Goa and puts it at the eighteenth rank. Fifty two per cent of the villages are still not linked by all-weather roads (Planning Commission, 2006, Table 19.1).

The National Highway Network in Rajasthan covers practically every district, excepting Banswara, Hanumangarh, Jhunjhunu and Sawaimadhopur. For the better connectivity with the National Highway system in the State, the State government has proposed rapid expansion of road network. In this the National Highway from Delhi to Ahmadabad via Jaipur, Ajmer and Udaipur has highest traffic density and hence this section is being upgraded on priority basic. Apart from National Highways Rajasthan has well developed road network of State Highways and Rural roads. But these networks of roads are not evenly distributed across the State either in terms of location, density or per head of the population. Like other States in India, Rajasthan too is working for the improve-ment of road network under various controls concerned schemes such as Pardhan Mantri Gram Sadak Yojana and National Highways.

DEVELOPMENT PROJECT

Table 19.1 Rajasthan: Major National Highways, 2011

National Highway No.	Between Cities	Distance (in Km)
8	Delhi-Ahmadabad	678
3	Agra-Mumbai	30
11	Agra-Bikaner	521
15	Pathankot-Sanchore	874
65	Ambala-Pali	496

Source: Planning Commission (2011-16)

Source: RSRTC, Jaipur

Figure 19.1 Rajasthan Transport Network

A State's sinews are roads, railways and communications. The effective transport and communication systems accelerate the growth of the economy by integrating markets, reducing the cost of production, enhancing the overall productivity and also helping in attracting large scale industrial investment.

ROADS

The State has a net work of 1.51 lakh km (151914 km) of roads consisting 2964 km as national highway, 9966 km as state highway, 5947 km as district main roads, 66395 km as other district and village roads and 2239 km of border roads constructed by PHED and 64403 km by other departments in the state respectively. The road density as measured by length of roads per 100 sq. km of geographical area of the State was 44.42 km (Figure 19.1).

The State had accounted for 6.12% of road net work at the all India level of 24.66 lakh km. Total network of roads is encompasses 11 national highways and 519 state highways covering the entire state. Rajasthan State Road Transport Corporation is the main state government department for providing transportation facilities with about 4258 vehicles operating through 47 depots on 2624 routes (including 519 state highways) in the state of Rajasthan.

From the Table and Figure it is clear that Jaipur has the highest National Highway Road length followed by Bikaner and Ajmer district. Sawai Madhopur, Jhunjhunu, Banswara and Hanumangarh are the districts having negligible National highway length in their districts. Jodhpur has the highest all type road length while Dhaulpur district hast the lowest share in the Rajasthan state.

Ganganagar district has the highest road connectivity to its villages followed by Jaipur and Alwar districts. Lowest village connected to roads were found in Jaisalmer and Dhaulpur district of Rajasthan (Table 19.2 to 19.7).

Table 19.2 Roads (Total Road Length)

Classification	Unit	2015-16	2016-17
(A) PWD Roads			
(a) National Highways	Km.	3466.54	3354.09
(b) State Highways	,,	15188.36	15019.16
(c) Major District Roads	,,	7237.57	8053.46
(d) Other District Roads	,,	15332.84	15391.52
(e) Villages Roads	,,	107854.33	112470.29
Total (A)		**149079.64**	**154288.52**
(B) Other Departmental Roads			
(a) National Highways	Km.	4701.66	4848.11
(b) State Highways	,,	418.69	418.69
(c) Major District Roads	,,	408.64	408.64
(d) Other District Roads	,,	14980.39	16039.65
(e) Villages Roads	,,	48118.23	50850.25
Total (B)		**68627.61**	**72565.34**
(C) Total Roads			
(a) National Highways	Km.	8168.2	8202.2
(b) State Highways	,,	15607.05	15437.85
(c) Major District Roads	,,	7646.21	8462.1
(d) Other District Roads	,,	30313.23	31431.17
(e) Villages Roads	,,	155972.56	163320.54
Total (A) + (B)		**217707.25**	**226853.86**

Source: Statistical Year Book of Rajasthan, 2018

Table 19.3 Road Length (PWD) - 31.03.2017

S. No.	Name of District	Total Road Length (PWD) (in Km.)	Area in Sq. Kms.	Projected Population ('000) (As on 1st Oct,2016)	Road Length per 100 Sq. Km. of area	Road Length per Lakh of Population
1	2	3	4	5	6	7
1	Ajmer	4737.99	8481	2855	55.87	165.95
2	Alwar	5975.00	8380	4064	71.30	147.02
3	Banswara	4135.75	4522	1994	91.46	207.41
4	Baran	2880.51	6992	1353	41.20	212.90
5	Banner	10968.82	28387	2895	38.64	378.89
6	Bharatpur	3672.69	5066	2819	72.50	130.28
7	Bhilwara	5796.04	10455	2663	55.44	217.65
8	Bikaner	7049.47	30239	2623	23.31	268.76
9	Bundi	2457.47	5776	1228	42.55	200.12
10	Chittaurgarh	3750.74	7822	1704	47.95	220.11
11	Churu	4552.24	13835	2257	32.90	201.69
12	Dausa	3395.51	3432	1813	98.94	187.29
13	Dhaulpur	2086.10	3033	1336	68.78	156.15
14	Dungarpur	3916.08	3770	1539	103.87	254.46
15	Ganganagar	4707.09	10978	2165	42.88	217.42
16	Hanumangarh	3488.14	9656	1964	36.12	177.60
17	Jaipur	6758.81	11143	7389	60.66	91.47
18	Jaisalmer	5323.24	38401	747	13.86	712.62
19	Jalor	4874.77	10640	2029	45.82	240.25
20	Jhalawar	3727.97	6219	1560	59.94	238.97
21	Jhunjhunu	4636.87	5928	2355	78.22	196.89
22	Jodhpur	9632.05	22850	4088	42.15	235.62
23	Karauli	2587.85	5524	1613	46.85	160.44
24	Kota	2572.58	5217	2160	49.31	119.10
25	Nagaur	9010.04	17718	3656	50.85	246.45
26	Pali	5347.65	12387	2244	43.17	238.31
27	Pratapgarh	2478.12	4449	961	55.70	257.87
28	Raj Samand	3083.72	4655	1279	66.25	241.10
29	Sawai Madhopur	2478.78	4498	1479	55.11	167.60
30	Sikar	4862.48	7732	2955	62.89	164.55
31	Sirohi	2064.54	5136	1147	40.20	179.99
32	Tonk	3240.83	7194	1569	45.05	206.55
33	Udaipur	8038.58	11724	3396	68.57	236.71
	Rajasthan	**154288.52**	**342239**	**75899**	**45.08**	**203.28**

Source: Statistical Year Book of Rajasthan, 2018

Table 19.4 Road Length in Rajasthan

S. No.	Districts	Type of Roads					Total
		National Highway (N.H.)	Painted (B. T.)	Metaled (W.B.M.)	Graveled (G.R.)	Fair weather & dressed up tracks (F.R.)	
1	Ajmer	242	2490	81	31		2844
2	Alwar	35	3344	105	156		3640
3	Banswara		1476	110	121		1707
4	Baran	232	886	52	315	11	1496
5	Barmer	189	3998	238	793		5218
6	Bharatpur	71	1869	53	6		1999
7	Bhilwara	129	2551	357	838	13	3888
8	Bikaner	325	2908	87	316	10	3646
9	Bundi	107	1121	95	187		1510
10	Chittorgarh	202	2299	178	162		2841
11	Churu	239	2513	203	145		3100
12	Dausa	88	1148	83	273		1592
13	Dhaulpur	28	978	83	49		1138
14	Dungarpur	34	1265	156	63		1518
15	Ganganagar	124	3101	11	92		3328
16	Hanumangarh		2460	23	44		2527
17	Jaipur	401	3455	164	44	42	4106
18	Jaisalmer	196	2887	42	731		3856
19	Jalore	41	2153	180	268	18	2660
20	Jhalawar	126	1002	116	144		1388
21	Jhunjhunu		2069	267	140		2476
22	Jodhpur	218	4490	226	1004		5938
23	Karauli	4	1227	164	105	2	1502
24	Kota	120	1299	78	139	2	1638
25	Nagaur	330	3444	533	731		5038
26	Pali	218	2959	440	1028		4645
27	Rajsamand	155	1283	214	231		1883
28	S. Madhopur		976	86	149		1211
29	Sikar	191	1842	168	308		2509
30	Sirohi	143	1131	78	97		1449
31	Tonk	109	1343	37	185	8	1682
32	Udaipur	250	2929	197	106	7	3489

Source: RSRTC, Jaipur

Table 19.5 Road Transport Route Nationalized

S. No.		Depot (Districts)	No. of Routes	Route Distance (Km.)	Average Route Distance (Km.)
1		Ajmer	67	8520	127
	A	Ajaymeru	71	8896	125
	B	Beawar	78	11645	149
2		Alwar	78	8985	115
	A	Matsya Nagar	53	5675	107
	B	Tizara	33	4630	140
3		Banswara	50	10874	217

S. No.		Depot (Districts)	No. of Routes	Route Distance (Km.)	Average Route Distance (Km.)
4		Barmer	62	12844	207
5		Bharatpur	51	9656	189
	A	Lohagarh	61	8796	144
6		Bhilwara	51	8452	166
7		Bikaner	42	12285	292
8		Bundi	62	11765	190
9		Baran	29	5160	178
10		Chittorgarh	93	15722	148
11		Churu	12	2411	201
	A	Sardarsahar	44	8980	204
12		Dausa	77	7953	103
13		Dholpur	39	7132	183
14		Dungarpur	70	12322	176
15		Ganganagar	79	15826	200
16		Hanumangarh	66	13408	203
	A	Anoopgarh	56	13001	321
17		Jaipur	67	19775	295
	A	Delux	34	13534	398
	B	Jhalana Dunagri	74	3810	51
	C	Sanganer	70	2429	35
	D	VaishaliNagar	71	7697	108
	E	Vidhyadhar Nagar	90	5038	56
	F	Kotputli	38	5094	134
18		Karauli			
	A	Hindaun	45	7054	157
19		Jaisalmer			
20		Jalor	41	10634	259
21		Jhalawar	60	9872	165
22		Jhunjunu	39	6724	172
	A	Khetri	45	7835	174
23		Jodhpur	64	15808	247
	A	Phalodi	28	7193	257
24		Kota	65	13751	212
25		Nagaur	41	9475	231
26		Pali	50	6685	134
	A	Falana	41	7494	183
27		Rajsamand			
28		Sawai Madhopur	33	5283	160
29		Sikar	103	16251	158
30		Sirohi	40	7307	183
	A	Aburoad	43	7982	186
	B	Rewdar	43	9360	218
31		Tonk	37	9530	258
32		Udaipur	87	17927	206

Source: RSRTC, Jaipur

Table 19.6 Villages and Panchayat H.Q. connected by Roads - 31.03.2017

S. No.	Name of District	Total Villages (As per Census-2011)	Villages connected by Roads	Villages not connected by Roads	Total Panchayat H.Q.	Panchayat H.Q. connected by Roads
1	2	3	4	5	6	7
1	Ajmer	1099	1081	18	282	282
2	Alwar	2021	1912	109	512	512
3	Banswara	1485	1332	153	346	344
4	Baran	1114	925	189	221	221
5	Barmer	2452	1989	463	489	486
6	Bharatpur	1432	1354	78	374	374
7	Bhilwara	1790	1662	128	384	384
8	Bikaner	857	780	77	290	290
9	Bundi	867	787	80	183	183
10	Chittaurgarh	1581	1306	275	290	290
11	Churu	873	854	19	254	254
12	Dausa	1079	975	104	231	231
13	Dhaulpur	801	781	20	171	171
14	Dungarpur	972	902	70	291	289
15	Ganganagar	2855	2198	657	336	336
16	Hanumangarh	1831	1212	619	251	251
17	Jaipur	2126	2038	88	532	532
18	Jaisalmer	756	557	199	140	137
19	Jalor	791	770	21	274	274
20	Jhalawar	1488	1321	167	252	252
21	Jhunjhunu	926	912	14	301	301
22	Jodhpur	1836	1480	356	466	464
23	Karauli	851	797	54	227	226
24	Kota	805	759	46	155	155
25	Nagaur	1575	1544	31	467	467
26	Pali	1017	981	36	321	321
27	Pratapgarh	970	742	228	165	163
28	Rajsamand	1037	906	131	207	207
29	Sawai Madhopur	758	673	85	200	200
30	Sikar	1162	1119	43	343	342
31	Sirohi	470	451	19	162	162
32	Tonk	1116	1027	89	230	230
33	Udaipur	2471	2041	430	544	541
	Rajasthan	**43264**	**38168**	**5096**	**9891**	**9872**

Source: Statistical Year Book of Rajasthan, 2018

Table 19.7 Villages Connected with Roads

S. No	District	Villages connected with Road				Villages not connected with Painted Road
		Painted	Metalled	Gravelled	Total	
1	Ajmer	568	104	154	826	159
2	Alwar	1020	43	155	1218	728
3	Banswara	596	52	146	794	637
4	Baran	329	23	97	449	621
5	Barmer	570	29	385	984	641
6	Bharatpur	640	27	75	742	603
7	Bhilwara	527	66	325	918	647
8	Bikaner	364	7	54	425	155
9	Bundi	299	19	55	373	453
10	Chittorgarh	669	83	234	986	1186
11	Churu	518	33	19	570	356
12	Dausa	348	20	72	440	569
13	Dholpur	254	20	51	325	226
14	Dungarpur	419	63	69	551	295
15	Ganganagar	1364	4	17	1385	1523
16	Hanumangarh	785	4	1	790	740
17	Jaipur	1105	64	117	1286	845
18	Jaisalmer	208	3	80	291	227
19	Jalore	405	50	141	596	69
20	Jhalawar	298	32	70	400	1048
21	Jhunjunu	479	37	91	607	217
22	Jodhpur	648	29	129	806	54
23	Karauli	335	56	19	410	331
24	Kota	331	22	40	393	418
25	Nagaur	730	92	186	1008	366
26	Pali	571	60	155	786	118
27	Rajsamand	383	55	129	567	306
28	Sawai Madhopur	245	22	32	299	424
29	Sikar	467	100	90	657	274
30	Sirohi	237	19	71	327	119
31	Tonk	304	21	66	391	628
32	Udaipur	750	87	256	1093	1213

Source: RSRTC, Jaipur

ROAD NETWORK

Since man first traced a path across a landscape, his routes of travels have played a vital rote in political and cultural exchange, commerce and military conquests. Among the different transportation links such as roadways, railways, waterways and airways, roods were the earliest channels known to man. For about 3,000 years, his mode of transportation was by foot or on the back of animals and as such there were hardly any problems felt by him.

Increase in population, together with industrial revolution and technological advancement has changed the scenario over a period of time. Increase in population has resulted in large scale movement of people from one region to another for earning their livelihood. The quick transportation of agricultural products, vegetables, finished goods, fuel, commericial product etc. are key factor for economic growth of the country. The challanges of drought, floods, earth-quakes and other natural calamities, are effectively met, when a proper network of roads is available for quick and better transport to provide relief to the vicitims. Roads play and imprtant role in wars also to meet the challanges of the enemy by quick transport of manpower, equipments, arms and amunition and other materials etc., the sucess of war being largely dependent on constant flow along its supply lines to the front.

The improved speed and increased carrying-capacity of vehicles, have added new dimensions to road, which are now not simply meant to provide a link between various regions but along with quicker movement of passengers and freight, are intended to provide a mode that is efficient, safe without any obstruction and danger to vehicle as well as to other users including pedestrians and animal traffic. Addition of new roads, widening of the roods to accomodate more traffic, improvement of the road suface to make it suitable for higher speeds and tonnage, are the needs of the time of for any developing country.

Rajasthan state does not have a good network of roads due to its large desert area. The district wise raod is determined in terms of length of roads and their type of construction, and also density of roads in the state i.e. length of roads per 100 sqkm of area. The total length of road in the state is 87,462 km including all types. The length of painted roads (Bitumen and without Bitumen) is however, the largest (73,443 km) followed by gravelled roads (9,043 km.) and fair weather road travels have played a vital rote in political and cultural exchange, commerce and military conquests. Among the different transportation links such as roadways, railways, waterways and airways, roods were the earliest channels known to man. For about 3,000 years, his mode of transportation was by foot or on the back of animals and as such there were hardly any problems felt by him.

Increase in population, together with industrial revolution and technological advancement has changed the scenario over a period of time. Increase in population has resulted in large scale movement of people from one region to another for earning their livelihood. The quick transportation of agricultural products, vegetables, finished goods, fuel, commericial product etc. are key factor for economic growth of the country. The challanges of drought, floods, earthquakes and other natural calamities, are effectively met, when a proper network of roads is available for quick and better transport to provide relief to the vicitims. Roads play and imprtant role in wars also to meet the challanges of the enemy by quick transport of manpower, equipments, arms and amunition and other materials etc., the sucess of war being largely dependent on constant flow along its supply lines to the front.

The improved speed and increased carrying-capacity of vehicles, have added new dimensions to road, which are now not simply meant to provide a link between various regions but along with quicker movement of passengers and freight, are intended to provide a mode that is efficient, safe without any obstruction and danger to vehicle as well as to other users including pedestrians and animal traffic. Addition of new roads, widening of the roods to accomodate more traffic, improvement of the road suface to make it suitable for higher speeds and tonnage, are the needs of the time of for any developing country.

Rajasthan state'does not have a good network of roads due to its large desert area. The district wise raod is determined in terms of length of roads and their type of construction, and also density of roads in the state i.e. length of roads per 100 sqkm of area. The total length of road in the state is 87,462 km including all types. The length of painted roads (Bitumen and without Bitumen) is however, the largest

(73,443 km) followed by gravelled roads (9,043 km.) and fair weather road (113km.) Thee are twelve National highways which passes through Rajasthn state. The total length of National highways in the state is 4,547 km. National highways pass through most of the districts except Banswara, Haumangarh, Jhunjhunun and Sawai Madhopur. There are state highways with a total run of 8,714 km (Figure 19.2).

Various districts of Rajasthan have a raod density varying form 10.04 km/100 sq. km. (Jaisalmer) to more than 47 km/100 sq.km. Rajsamand (40.15 km 100 sq. km.), Dungarpur (40.26 km/100 sq.km.), Jhunjhunun (41.77 km/100 sq. km.), Alwar 43.44 km/100 sq. km.) and Dausa (47.02 km/100 sq. km.) districts have the higher road density, while the lowest road density is in Jaisalmer distrit (10.04 km/100 sq.km.). In the most of the districts of the state road density varies from (15 to 30 sq. km.)

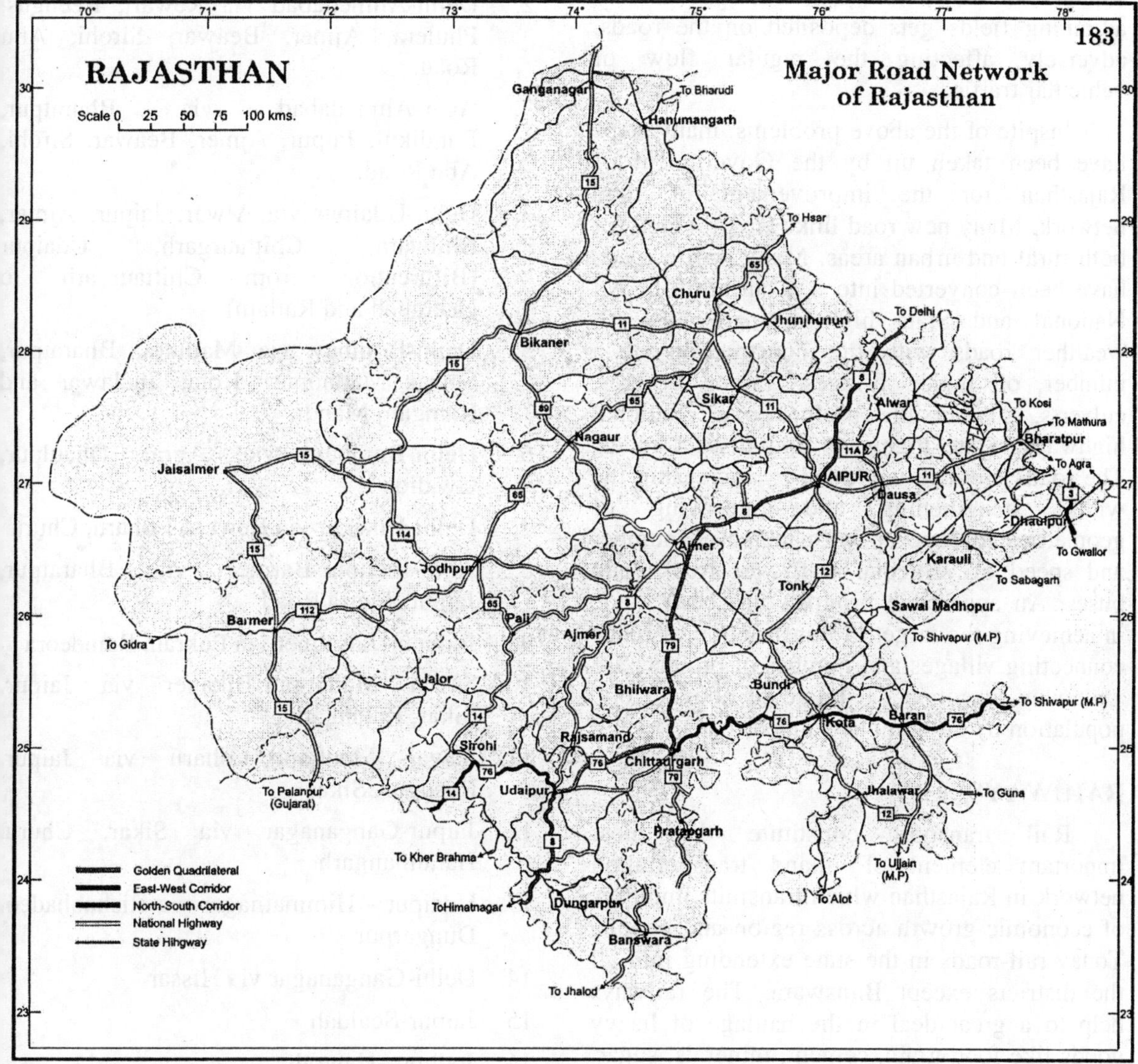

Source: RSRTC, Jaipur

Figure 19.2 Major Road Network of Rajasthan

There are some typical problems in construction and maintenance of roads in some parts of the state. As in other parts of the country, roads in the state especially in the eastern region are subject to road cuts, due to excessive rains and floods. A major part of western Rajasthan has typical problem of its own owing to its sandy terrain. Due to noncohesive nature of soil or absence of stabilised strata, difficulties are faced in construction of roads to tackle traffic of high speed and heavy vehicles. Further, during summer months, moving sand from the adjoining fields gets deposited on the roads, adversely affecting the regular flow of vehicular traffic.

Inspite of the above problems, many steps have been taken up by the Government of Rajasthan for the improvement of road network. Many new road links are provided for both rural and urban areas. Many traffic roads have been converted into a 4-lane traffic. All National and State highways are now all wealther roads with the construction of a number of major bridges, minor bridges, culverts, submersible bridges over National highways, State highways and district roads. The roads are also upgraded by increasing the width, stengthening and improving of geometries to cope up with increased number and speed of vehicular traffic of trucks and buses. An appercible progress has been made in achieving the target of construction of roads connecting villages and number of projects are underway to connect habitate of above 1,000 population by bituminous painted roads.

RAILWAYS

Rail transport constitute the most important element of inland transportation network in Rajasthan which transmits impulses of economic growth across region and people. Today rail-roads in the state extending into all the districts except Banswara. The railways help to a great deal in the haulage of heavy goods like food grain, cement, minerals, sugar, salt, coal, cotton and textiles. Besides thousands of passengers also travel daily both for short and long distance (Figure 19.3).

The railways of the state are managed by number of Railway zones of the Country. A rail Bus has been also introduced in the state to carry passengers from merta city to merta road and back.

Some of the important rail-road of Rajasthan are:

1. Delhi-Ahmedabad via Alwar, Jaipur, Ajmer, Beawar, Sirohi, Abu-road
2. Delhi-Ahmedabad via Rewari, Reengus, Phulera, Ajmer, Beawar, Sirohi, Abu Road.
3. Agar-Ahmedabad via Bharatpur, Bandikui, Jaipur, Ajmer, Beawar, Sirohi, Abu Road.
4. Delhi-Udaipur via Alwar, Jaipur, Ajmer, Bhilwara, Chittaurgarh, Udaipur (Bifurcation from Chittaurgarh to Neemuch and Ratlam)
5. Delhi-Bombay via Mathura, Bharatpur, Sawai Madhopur, Kota, Jhalawar and Ramganj Mandi.
6. Delhi-Bombay via Agra, Dhaulpur, Gwalior
7. Delhi-Bikaner via Rewari, Loharu, Churu
8. Agra-Jodhpur-Barmer via Bharatpur, Jaipur, Merta Road
9. Jodhpur-Jaisalmer via Pokran, Ramdeora
10. Sawai Madhopur-Bikaner via Jaipur, Sikar, Ratangarh
11. Sawai Madhopur-Loharu via Jaipur, Reengus, Sikar
12. Jaipur-Ganganagar via Sikar, Churu, Hanumangarh
13. Udaipur - Himmatnagar via Rishabhadeo, Dungarpur
14. Delhi-Ganganagar via Hissar
15. Jaipur-Sealdah
16. Jodhpur-Hawrah
17. Jodhpur-Bandra (via Ahmedabad)

18. Jodhpur-Chennai
19. Jodhpur-Secunarabad and many more.

Most of the meter gauge in the state has already been converted into Broad gauge in the state for remaining conversion is in progress.

The main nodal centres of rail traffic in the state are Jaipur, Kota, Jodhpur, Sawai Madhopur, Ajmer, Udaipur, Bharatpur. Shortage of railway wagons for goods transport and the passenger coaches for passengers are the main constraints for efficient services of railways. Morever, the competition between rail and road traffic has also increased in recent years. Thus there is a great need to extend the railway services to those areas which are uncovered so far and to reduce the freight traffics to compete with the road transport system (Table 19.8).

Table 19.8 Railway Rout and Zone in Rajasthan

S.No.	Particulars-Gauge/Route Kms.	Railway Zone falling in Rajasthan	Unit	2015-16	2016-17
1	Broad Guage (B.G.)	Northern	Km.	31.17	31.17
		Western	,,	52.55	52.55
		West Central	,,	758.22	758.22
		North Central	,,	266.49	266.49
		North Western	,,	4069.48	4164.91
		Total	,,	**5177.91**	**5273.34**
2	Metre Guage (M.G.)	Western	,,	0.00	0.00
		North Western		628.43	533.83
		Total	,,	**628.43**	**533.83**
3	Nerrow Guage (N.G.)	North Central	,,	86.76	86.76
		Total		**86.76**	**86.76**
	Total Route Kms. Of All Zone Railway	**Grand Total (a+b+c)**	,,	**5893.1**	**5893.93**

Source: Statiscal Year Book of Rajasthan, 2018

On the front to rail networks, Rajasthan has the largest rail network after Uttar Pradesh, but the majority of the length of this railways system belongs of rules of the former States of Jaipur, Jodhpur, Bikaner and Mewar. The railways of the state are managed by three railway zones of the country viz., the Northern, Western and the Central railways. The total length of rail network in state is over 6227 km, extending in all the districts except few districts.

Most of areas of present day Rajasthan remain without the facility of railways. The district headquarters of Jhalawar, Banswara, Karauli, Sirohi, Rajsamand and Tonk remained unconnected with any railways line even today. The eastern Rajasthan has well developed rail network as this region happens to be on the main lines of Delhi to Mumbai (Planning Commission, 2006). Main rail routes in Rajasthan are as follows:

(i) Delhi – Ahmadabad
(ii) Jodhpur – Howrah rail route
(iii) Jodhpur – Delhi
(iv) Delhi – Kota – Mumbai
(v) Ajmer – Sialdah
(vi) Jaipur – Jammu Tavi route

Railways, the artery of the economic system, have a supra-regional feature of integrating various States in India. Development of railway system is shaped by interplay of factors such as laying of new lines, conversion of meter gauge into broad gauge, renewal and replacement of tracks and electrification of lines. The railways are functioning as a commercial undertaking as well as public utility service. Though the National Transport Policy Committee suggested a model split in which the share of long distance freight for railways should be 72 per cent and the rest for road transport system. Total rail track length is of 5917 km which is about 9.42 % of the all India rail track length. 17.29 According to the statistics Railway in Rajasthan has a density of 17.29 km per 1000 sq. km. of the geographical area of the state.

AIRWAYS

Air transport is the fastest means of transport 'today for passengers, cargo and mail. The Indian Airlines and Vayudect services which form a network of internal air transport in India, have important destina-tions at a few cities in Rajasthan as well an account of the fast growing passenger traffic.

The tourist centres of the state like Jaipur, Jodhpur and Udaipur have put a heavy demand on increasing domestic air services in Rajasthan. The Indian Airlines, setup under the Air Corporation Act 1953, operate its regular flights for the three tourist centres of international importance (Jaipur, Jodhpur, Udaipur) connecting them with international airport of Bombay and Delhi, Air transport in Rajasthan, as in other parts of India, is not a substitute to other modes of transport rather it is a means of utmost importance to the people and for transhipment of cargo where speed and time are concerned.

Airways have provided decisive boost to exports and imports. The main exports of Rajasthan, which are precious stones, diamonds, cotton textiles, indigenous wollen babrics, handicrafts of blacquer - work, brasswares and marbles, are today sent by air to various parts of the world. Therefore beside the tourist industry, air cargo also plays a significant role in promoting the state revenue.

Despite the spectacular progress shown by the development of airway, the state still lacks in this mode of transport. Certain important tourist centres like Ajmer, Chittaurgarh, Mound Abu, Sawai Madhopur and Bharatpur, which are favourite tourist spots of religious importance and scenic beauty are still unconnected by air services and foreign tourist find difficult to cover long distances to reach them by surface transport means from the nearest airports. There is also great need to increase the passenger seating capacity in the existing air services in view of the mounting pressure of air- traffic at Jaipur, Jodhpur and Udaipur.

On account of tourist importance of Jodhpur, Jaisalmer, Jaipur and Udaipur, there is large traffic of foreigners and Indian travellers by airlines. To cope up with this demand private air taxi services came forward and are already in operation in between Bombay Jodhpur-Jaipur-Bombay etc.

In terms of Air transport, Rajasthan has four airports in Jaipur, Jodhpur, Kota and Udaipur connected with regular air service from Delhi and Mumbai. The prime cause of air transport development is Rajasthan is that State has many world famous tourist destination and Air service act as connecting service for tourist both from abroad and home. Presently there are five airports (Figure 19.3) working in Rajasthan and these are as follows:

Table 19.9 Rajasthan: Major Airports, 2011

Name of Airport	Place	Nature(National/ International)
Sanganer Airport	Jaipur	International
Dabok Airport	Udaipur	Domestic
Ratjanada Airport	Jodhpur	Domestic
Kota Airport	Kota	Domestic
Jaisalmer Airport	Jaisalmer	Domestic

Source: Planning Commission (2011; 63)

At present 5 airports at district head quarters of Jaipur, Jodhpur, Jaisalmer Kota and Udaipur districts are in operation and are linked to almost all the metro cities of India. The Sanganer Airport in Jaipur has got recognition as International Airport and is going to play a main role in the state's economy in coming years. Twelve Airstrips and 128 Helipads are already available in

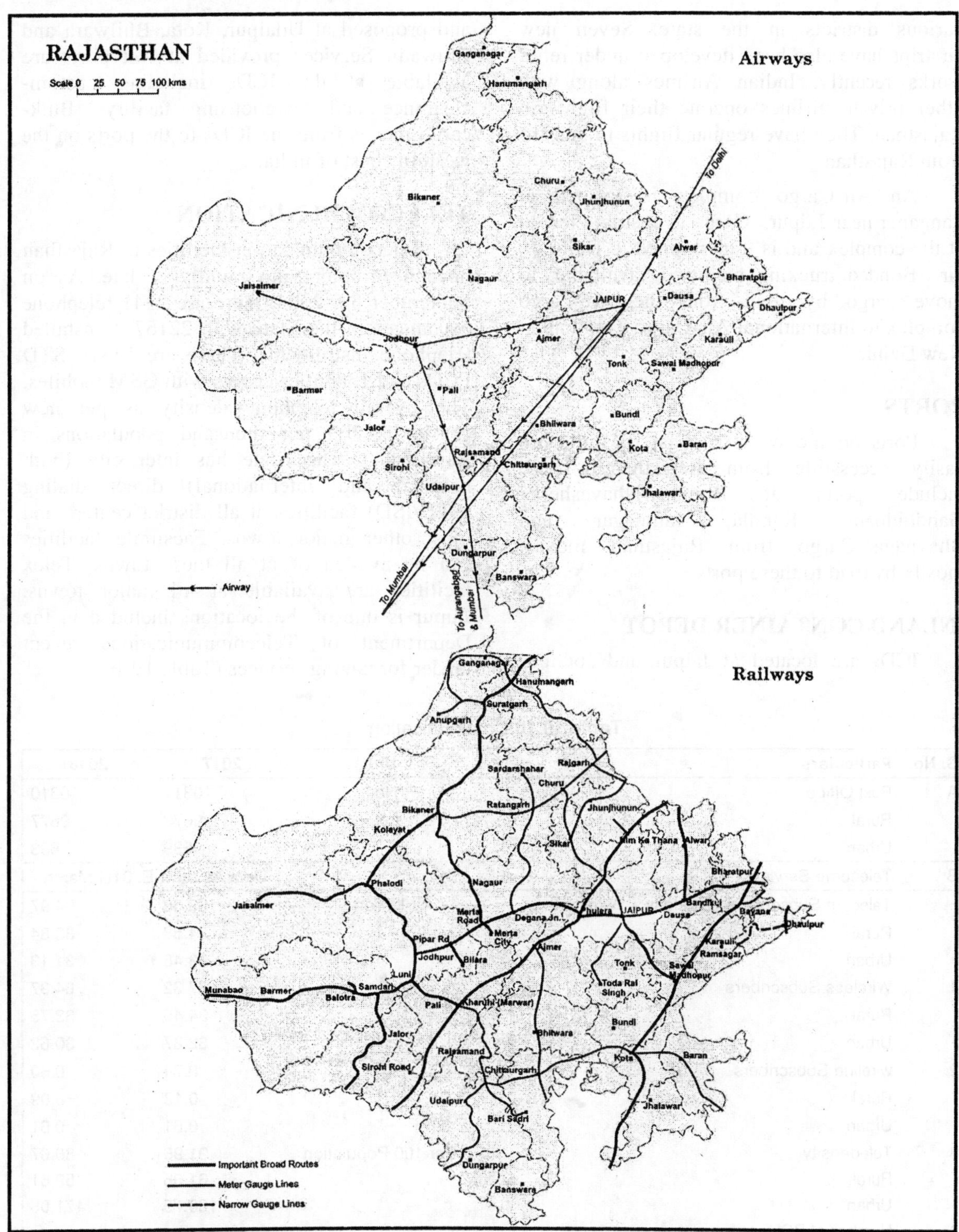

Source: Statistical Year Book of Rajasthan, 2018

Figure 19.3 Airways and Railways Network in Rajasthan

various districts in the state. Seven new airstrips have also been developed under relief works recently. Indian Airlines along with other private airlines operate their flights in Rajasthan. They have regular flights to and fro from Rajasthan.

An Air-Cargo complex is located at Sanganer near Jaipur. Cargo is custom-cleared at the complex and is transshipped to Delhi by air. Bonded trucking facility is available, to move cargo by road from the Air-Cargo complex to International Air-Cargo Terminal at New Delhi.

PORTS

Ports on the western coast of India are easily accessible from Rajasthan. These include ports at Bombay/Nhavashera. Gandhidham/ Kandla, Jamnagar and Bhavnagar. Cargo from Rajasthan moves, mostly by road to these ports.

INLAND CONTAINER DEPOT

ICDs are located at Jaipur and Jodhpur and proposed at Udaipur, Kota, Bhilwara and Bhiwadi. Services provided at wet ports are available at the ICDs including custom-clearance and warehousing facility. Bulk-cargo moves from the ICDs to the ports on the western coast of India.

TELECOMMUNICATION

Telecommunication facilities in Rajasthan have been improving rapidly of late. As on September 30, 2004, there are 2341 telephone exchanges in the State with 2216751 installed telephone instruments. There are 53879 STD PCOs and 136 cities covers with GSM mobiles. The overall telephone density as per new census is 269 per thousand populations in Rajasthan. The State has inter city (both national and international) direct dialing (STD/ISD) facilities at all district centers and most other major towns. Facsimile facilities can be availed of at all these towns. Telex facilities are available at all major towns. Jaipur is one of the locations included in the Department of Telecommunications recent tender for paging services (Table 19.10).

Table 19.10 Communication

S. No.	Particulars	Unit	2017	2018
A	Post Office	Nos.	10311	10310
	Rural	"	9672	9677
	Urban	"	639	633
B	Telecome Services		Data As on Q.E. 31st March	
	Telecom Subscribers (Wireless + Wireline)	Million	68.06	64.97
	Rural	"	34.58	33.84
	Urban	"	33.48	31.13
1	Wireless Subscribers	"	67.32	64.37
	Rural	"	34.45	33.75
	Urban	"	32.87	30.62
2	wireline Subscribers	"	0.74	0.60
	Rural	"	0.13	0.09
	Urban	"	0.61	0.51
3	Teledensity	Per 100 Population	91.96	86.67
	Rural	"	61.66	59.61
	Urban	"	186.75	171.08
4	Number of PCOs (BSNL+Bharti Group+Sistema Shyam+Tata Group)	Nos.	17799	6204*

5	Number of VPTs (Village Public Telephones) - BSNL	Nos.	16879	7857
6	**GSM Service:-**			
	% of Pre-paid in total Subscribers	%	97.15	97.27
7	**CDMA-** Full Mobility Service:-			
	% of Pre-paid in Total Subscribers	%	96.10	97.56
	* -Sistema Shyam is excluding			

Tele-density figures are derived from the subscriber data provided by the operators and the population projections, the Office of the Registrar General & Census Commissioner, India.

Source: Statistical Year Book of Rajasthan, 2018

POST AND TELEGRAPHS

Total numbers of post office in urban areas are 929 while in rural the number is 9359. They are catering about 21.70 lakh persons of the state with 0.30 numbers per 10 sq. km. density in the state of Rajasthan.

From the Table 19.10 and Figure 19.4 it is clear that Jaipur being the capital of the state has the highest number of Post Offices, Telegraph offices, Telephone exchange but in PCO category Bhilwara leads the tally (Table 19.11).

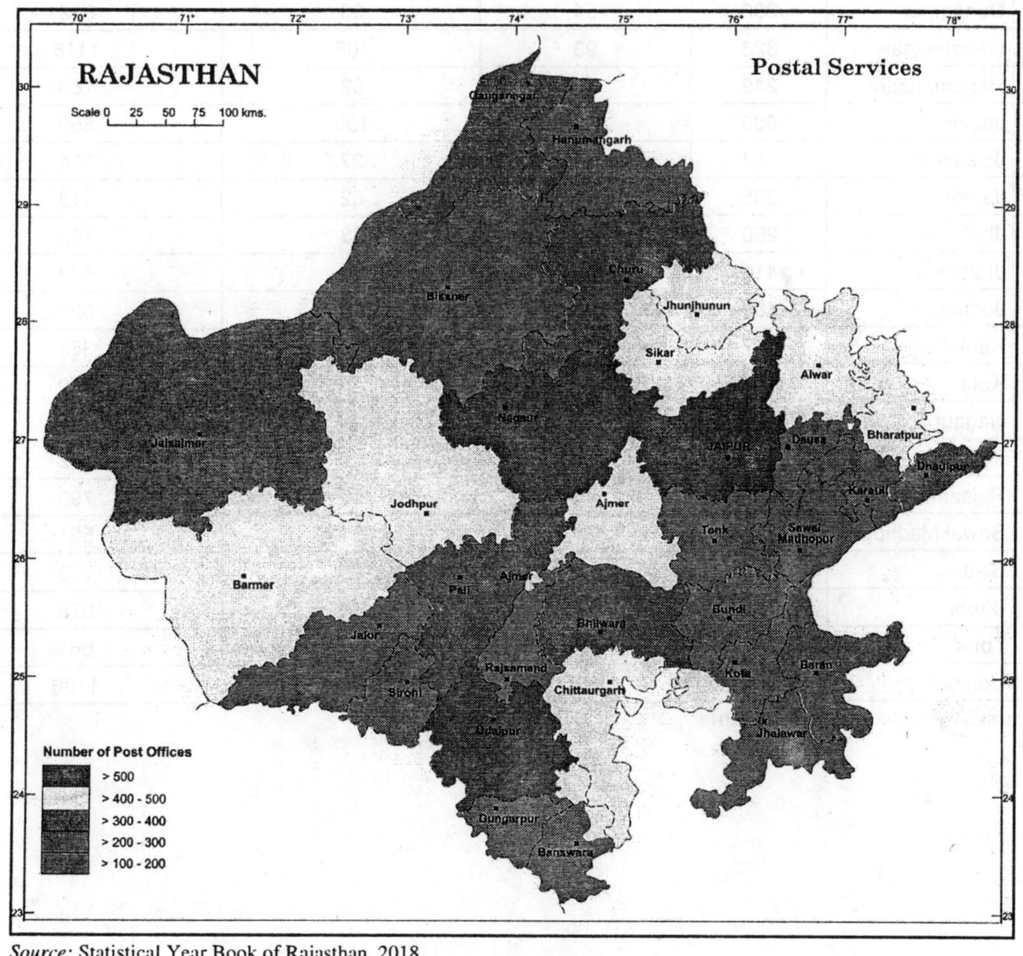

Source: Statistical Year Book of Rajasthan, 2018

Figure 19.4 Postal Services in Rajasthan

Table 19.11 Post, Telegraph and Telephone Offices

S. No	District	Post office	Telegraph Offices	Telephone Exchanges	Public call Offices (Rural)
1	Ajmer	430	61	84	989
2	Alwar	485	82	102	1073
3	Banswara	277	58	34	815
28	Baran	197	53	33	724
4	Barmer	479	104	68	896
5	Bharatpur	415	79	57	880
6	Bhilwara	393	95	75	1180
7	Bikaner	222	97	69	537
8	Bundi	176	31	38	512
9	Chittorgarh	403	61	61	780
10	Churu	397	89	85	682
29	Dausa	244	19	41	638
11	Dholpur	188	31	14	340
12	Dungarpur	298	54	36	532
13	Ganganagar	328	93	105	1118
31	Hanumangarh	249	22	62	760
14	Jaipur	600	143	150	859
15	Jaisalmer	154	27	27	374
16	Jalore	265	60	62	513
17	Jhalawar	250	69	33	702
18	Jhunjunu	414	106	70	839
19	Jodhpur	400	86	93	822
32	Karauli	247	27	26	373
20	Kota	196	59	45	699
21	Nagaur	542	165	101	1026
22	Pali	378	95	133	772
30	Rajsamand	210	33	50	790
23	Sawai Madhopur	214	71	34	551
24	Sikar	470	153	93	773
25	Sirohi	180	40	49	378
26	Tonk	221	62	47	664
27	Udaipur	494	85	82	1136

Source: Statiscal Year Book of Rajasthan, 2018

Chapter 20

TOURISM AND CULTURAL HERITAGE

The rich cultural and historical heritage of Rajasthan attracts people from all over the world. The richness and diversity of this culture creates natural infrastructure for tourism promotion and development. Rajasthan is fortunate to have rich treasure of natural beauty, flora and fauna, accompanied by places of religious, historical and archaeological importance. In a nutshell, except sea shores and snow peaks, the state is endowed with everything which attracts a tourist. The Department of Tourism, Art and Culture (Government of Rajasthan) is working for promotion, development and regulation of tourism throughout the state. For the same aim, Rajasthan Tourism Development Corpora-tion has developed a newtwork of good hotels and midways through out the state for domestic and foreign travellers (Figure 20.1).

The Tourist centres of the state can be broadly categorised into six types:

Hill Resort

Mount Abu is the only Hill resort in the state located under shadow of highest mountain peak of Aravalli range called "Guru Shikar" (1,772 m above MSL). It is an exhilarating dreamland with fresh environ-ment, charming and beautiful scenery, salubri-ous atmosphere and the treasure of art and nature. The famous Jain temples of Delwara and Achalgarh are an added attractions.

Centres of Wild-Life and Natural Spelendour

The lush green jungles, the vast streches of golden sand, the sprawling rocks and ravines in the state has created some of the most spectacular wild-life sanctuaries of the nation, besides world heritage sight Keoladeo National Park (Bharatpur) and Ranthambore National Parkk of "Project Tiger" fame (Sawai Madhopur). Other Places of interest are Sariska (Alwar), Jaisamand (Udaipur), Darrah (Kota), Dhawa (Jodhpur), Van Vihar and Ramsagar (Dhaulpur), Tal Chappar (Churu), Kumbhalgarh (Rajsamand), etc. The Desert National Park (Jaisalmer) is famous for fossils and the protection of "Godawan" the Great Indian Bustard.

Historical Centres

The famous historical sites have been the seats of ancient and medieval culture which offers golden opportunity to get acquainted with opulent past of Rajasthan. The magnificent forts of Chittaurgarh, Kumbhalgarh, Jodhpur, Nagaur, Ranthambore, Jaisalmer, Bikaner, Bundi, Amber and many more, are indeed the embodiments of architecultural excellence and rightly be called "treasure houses" of rich cultural and historical heritage. Most of the fort house popular deities which make them regligious centres too, like those at Sawai Madhopular and Amber.

The state has a rich cardie of historically important temples and monuments which present best examples of art and architecture of bygone era. The temple of Osian (Jodhpur), Arthuna (Banswara), Jhalrapatan (Jhalawar), Bijolia (Bhilwara), Deshnok (Bikaner), Jagat Shiromani (Jaipur), Menal (Bundi), Jagat temple (Udaipur) etc. are renowned for their architecure, and serenity.

Archeological Sites

Rajasthan is famous for its glorious past whose ancestores displayed their art in not only

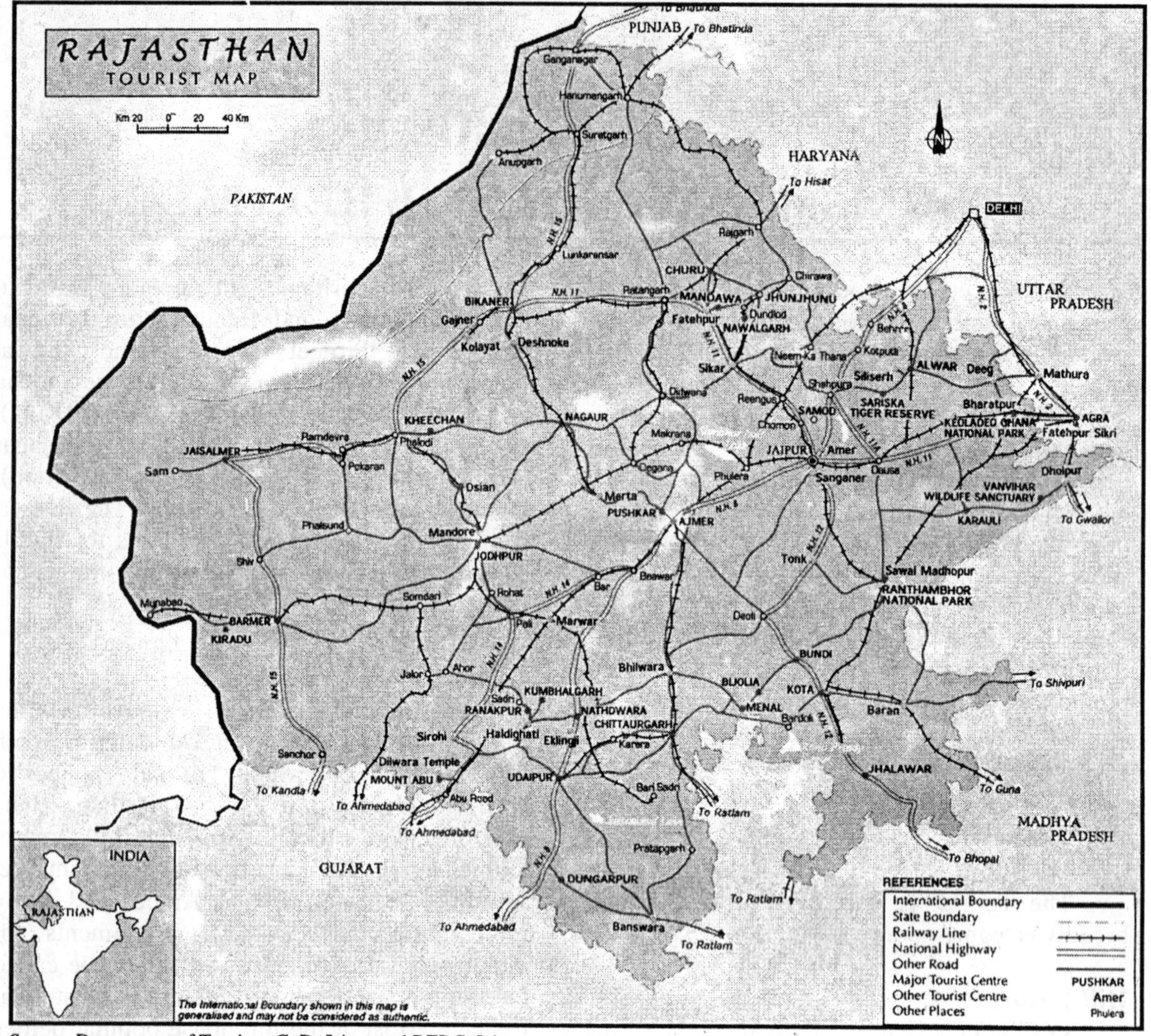

Source: Department of Tourism, GoR, Jaipur and RTDC, Jaipur

Figure 20.1 Tourist Map of Rajasthan

in battle fields, but also in building magnificent forts, palaces and other settlements. The following places have been excavated by archaeologist in the state which throw light on the ancient art, architeculture and socio cultural life of the people : Bagor, Aharr, and Gilund (Rajsamand), Tilwara (Barmer), Kalibangan, and Rangmahal (Ganganagar), Noh (Bharatpur), Nagaland Rairh (Tonk), Jodhpura, Bairath or virat and Sambhar (Jaipur), Sunari (Jhunjhunun), Ganeshwar (Sikar), Chittaurgarh and Nagari (Chittaurgarh), Mandore (Jodhpur), Bhinmal (Jalor) andNadol (Pali).

Religious Centres

Rajasthan is a unique state where different faiths unite. Places of worship enjoy reputation on national level and thousands of pilgrims and visitors across the borders of the state regularly visit them. The temples of Pushkar (Ajmer), Ranthamabhore and Kela Devi (Sawai Madhopur), Ramdeora (Near Pokaran, Jaisalmer), Deshnok (Bikaner), Nathdwara (Rajsamand), Eklingi (Udaipur), etc are famous among hindus. The Dargah of Ajmer, the holy shrines of Bohras at Galiyakot (Dungarpur), the Jain Shrines at Mahaveerji (S. Madhopur), Lodherva (Jaisalmer), Ranakpur (Pali), Keshriya ji (Udaipur), Nakoda ji (Barmer) and Delwara (Sirohi) are famous places where pilgrims and tourists pay their homage.

Recreational Centres

Occurance of natural recreational centre is a God-gift to the state, like wild-life sanctuaries, national parks, natural lakes on one hand, and vast streches of golden glittering sand on the other hand. It is land of festivals and fairs, each marking historical, religious or seasonal event. Traditional festivals like Gangaur and Teej, Tribal fair like Baneshwar (Dungarpur), cattle fairs like Nagaur and Tilwara (Barmer), are very popular. Department of Tourisn has gone a long way to design a number of festivals on different occasions. Thousands of domestic and foreign tourist flock each year to attend these festivals (Figure 20.2).

TOURIST CIRCUIT

(1) Desert Circuit: Jaisalmer-Barmer-Jodhpur-Bikaner Churu - Ganganagar - Hanumangarh

(2) Mewar Circuit: Udaipur-Rajsamand-Chittaurgarh Bhilwara

(3) Vagad Circuit: Dungarpur-Banswara

(4) Dhundhat Circuit Jaipur-Dausa-Tonk

(5) Godwar Circuit: Pali-Jalor-Sirohi

(6) Merwara Mewar Circuit: Ajmer-Nagaur

(7) Brij - Mewat Circuit Alwar Bharatpur Karauli -Sawai Madhopur Dhaulpur

(8) Shekhawat Circuit: Sikar-Jhunjhunun

(9) Hadoti Circuit: Kota - Bundi - Baran - Jhalawar.

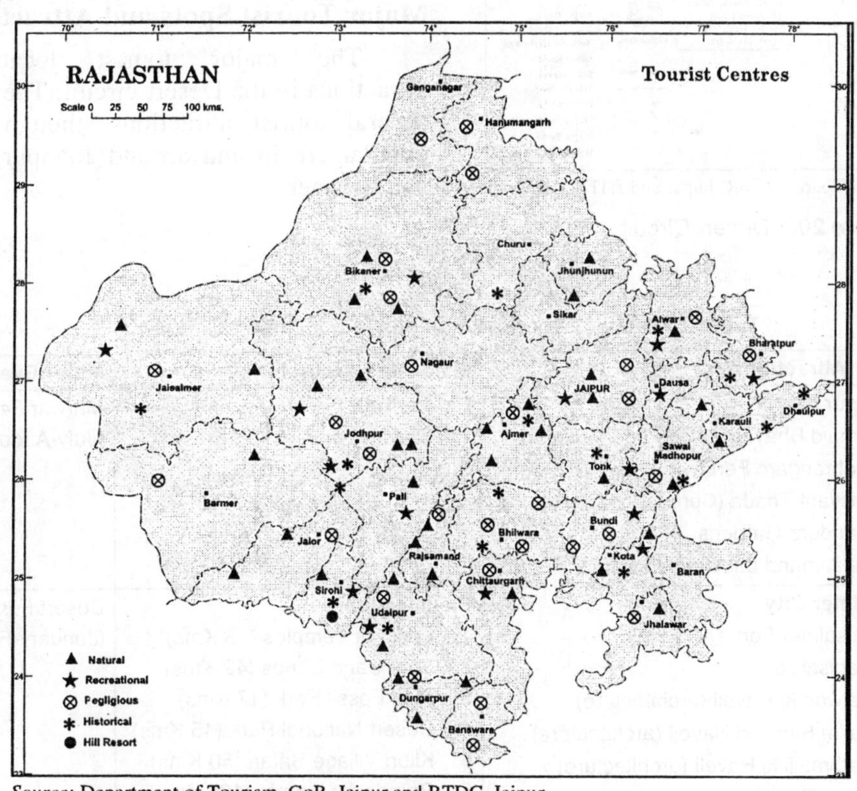

Source: Department of Tourism, GoR, Jaipur and RTDC, Jaipur

Figure 20.2 Tourist Centres in Rajasthan

(1) DESERT CIRCUIT

Overview of the Circuit Background. The desert circuit includes three districts in western Rajasthan rising out of the golden sands of the Thar Desert. The circuit is woven around the principal tourist locations in these districts, namely, Jodhpur, Jaisalmer and Bikaner (commonly referred to as the Desert Triangle) and now increasingly the desert town of Barmer. A map of the desert region showing the main tourist locations is provided in Figure 20.3.

Source: Department of Tourism, GoR, Jaipur and RTDC, Jaipur

Figure 20.3 Desert Circuit

Geography

The arid and forbidding Thar Desert lies in north-west Rajasthan and is characterised by shifting sand dunes and high summer temperatures of upto 45°C. Jodhpur lies on the edge of this arid tract and is the link between the desert on the west and the semi-arid but cultivable regions to the East. The region receives very little rainfall- 90% of it in the monsoon (July-September) 6.2.4 Jodhpur with a population approx. 8.5 Lakhs is the largest city in the region. It is also the second largest city in the state. Other important cities in the region include Bikaner (population 5.3 Lakhs), Barmer (84,000) and Jaisalmer (48,000). The Desert circuit is the largest though the most sparsely populated region in the state. 6.2.5 Given the climatic profile of the Desert circuit the tourist season is limited to the winter months and almost the entire tourist traffic comes between October to March (Table 20.1).

Major Tourist Spots and Attractions

The major tourist locations and attractions in the Desert circuit. The region has several tourist attractions, though the most popular are in and around Jodhpur, Jaisalmer and Bikaner.

Table 20.1 Major Tourist Locations in the Desert Circuit

District	Major Attractions	Major Day-excursions	Major Events
Jodhpur	**Jodhpur City** 1. Umaid Bhavan Palace and Museum 2. Mehrangarh Fort 3. Jasvant Thada (Cenotaphs) 4. Mandore Gardens 5. Balsamand Lake and Gardens	**Ex-Jodhpur** 1. Osiyan Temples (58 Kms)	Marwar Festival (July-August)
Jaisalmer	**Jaisalmer City** 1. Jaisalmer Fort 2. GadisarLake 3. Patwon ki Haveli (architecture) 4. Salim Singh ki Haveli (architecture) 5. Nathmalji ki Haveli (architecture) 6. Tazia Tower	**Ex-Jaisalmer** 1. Lodurva Temples (16 Kms) 2. Sam Sand Dimes (42 Kms) 3. Akal Fossil Park (17 Kms) 4. Desert National Park (45 Kms) 5. Kliuri Village Safari (50 Kms) 6. Ramdeora temple (100 Kms)	Desert Festival (January-February)

District	Major Attractions	Major Day-excursions	Major Events
Bikaner	**Bikaner City** 1. Junagarh Fort and Museums 2. Lalgarh Palace 3. Camel Research Farm 4. Devi Kund	**Ex-Bikaner** 1. Gainer Palace and Wildlife Sanctuary (32 Kms) 2. Shri Kolayat Shrine (50 Kms) 3. Deshnok Kami Mata temples (30 Kms) 4. Katriyasar Desert Village (55 Kms)	Camel Festival (January) Kolayat Fair (November) Gangaur Festival (April)
Barmer	**Barmer Town** 1. Ancient fort and temple ruins 2. Sun Temple	**Ex-Barmer** 1. Kiradu Temples (30 Kms) 2. Khed and Jasol Temples (70 Kms)	Tilwara Cattle Fair (March-April) Khed Fair (August-September)

Source: Department of Tourism, GoR, Jaipur and RTDC, Jaipur

Main Tourist Locations and Attractions

Unlike most other circuits in the state, the Desert circuit does not have one particular 'hub' per se though Jodhpur is the most popular entry point. Given the large geographical area covered, the major cities in the region are hubs for the tourist undertaking local excursions. Some salient observations with respect to the main tourist locations and the type of tourists they attract are as follows:

Jodhpur: Jodhpur, one of the principal cities in the circuit, has the massive 15th century Mehrangarh fort. The city also houses the grand Umaid Bhavan palace (and hotel), the only 20th century palace. Other attractions include the beautiful white marble royal cenotaphs of Jasvant Thada and the ancient capital of Marwar, Mandore with its cenotaphs and gardens. Another emerging destination, around an hour's drive from the city is the Osiyan village, which houses 15 beautifully sculpted Jain and Brahmanical temples belonging to five different centuries.

Jaisalmer: Jaisalmer, the golden city of Rajasthan, is one of the favourite destinations for both domestic and foreign tourists. The main attraction is the Jaisalmer fort, which houses an entire living area within its ramparts and is known for its beautifully carved havelis and its ancient Jain temples. Attractions outside the fort, are the intricately carved havelis such as the Patwon ki Haveli and the Salim Singh ki Haveli and the Gadisar Lake. Other famous attractions ex-Jaisalmer include the Lodurva Jain temples, Sam sand dunes, the Desert National Park and the Khuri village safari. The popular tourism 'products' of the region include the desert village camps, desert safaris and the desert festival.

Bikaner: Bikaner lies at the northern most point of the desert circuit, and is famous for its forts, palaces and havelis. Places of tourist interest include the Junagarh fort and museum and the Lallgarh palace. Other attractions are the Camel Research Farm and the Bhandeshwar and Sandeshwar temples (visited by the domestic tourists). The popular excursions from this city include the famous Karni Mata temple (at Deshnok), the Gajner Palace and wildlife sanctuary and Shri Kolayat temple.

Barmer: A small desert town, Barmer is famous for its carved wooden furniture and hand block printing. It is best visited during the annual fairs held nearby. The places of interest in this region include the ruins of Juna Barmer and the Kiradu temples. Other places in the district like Khed and Jasol are also known for their temple architecture.

(2) MEWAR CIRCUIT

Mewar region encompasses beautiful hills, lakes and deep valleys. The circuit offers a combination of religion and history and is particularly popular amongst the domestic tourist. This circuit primarily includes locations like Udaipur, Chittaurgarh and Nathdwara. The Mewar circuit showing the main tourist locations in the circuit (Figure 20.4).

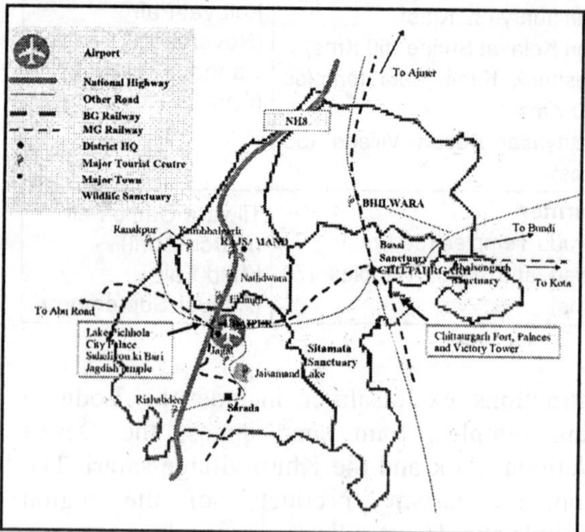

Source: Department of Tourism, GoR, Jaipur and RTDC, Jaipur

Figure 20.4 Mewar Circuit

Geography

Mewar area lies to the north of the Tropic of Cancer. The climate in this area is hot during the summer months of April to June. The winters are mild and pleasant and hence the peak tourist season is from September to March. The area is a mixture of plains and hills and has large tracks of land covered by forests. With a large number of sanctuaries and lakes in the area. In general, the Mewar region in a hilly forested and lake filled region in sharp contrast to the harsh desert on its west. Udaipur with a population of 3.9 Lakhs is the largest city in the circuit.

Major Tourist Spots and Attractions

The major tourist locations in the Mewar circuit. There are a large number of tourist spots in the Mewar area, though the most popular are the ones in and around Udaipur which is the tourism hub for the region (Table 20.2).

Table 20.2 Major Tourist Locations in Mewar Region

District	Major Attractions	Major Excursions	Major Events
Udaipur	**Udaipur City**	**Ex-Udaipur**	Mewar Festival (March-April)
	1. City Palace	1. Chittaurgarh Fort (120 Kms)	
	2. Jagdish Temple	2. Eklingji Temples (22 Kms)	Gangaur Festival (March-April)
	3. Sahelion ki Bari (Gardens)	3. Nagda (temple ruins) (23 Kms)	
	4. Pratap Memorial	4. Haldighati (battle site) (40 Kms)	Teej (July-August)
	5. Bhartiya Lok Kala Museum	5. Nathdwara Temple 48 Kms)	
	6. Fateh Sagar Lake	6. Kankroli Temple (65 Kms)	
	7. Pichhola Lake	7. Rajsamand Lake (66 Kms)	
	8. SajjangarhFort	8. Jagat (Temple) (58 Kms)	
	9. Ahar Cenotaphs	9. Jaisamand Lake (48 Kms)	
	10. Gulab Bagh (Rose Garden)	10. Gogunda (Historic site) (26 Kms)	
		11. Chavand (Historic site) (83 Kms)	
		12. Rishabdeo (Jain Temple) (58 Kms)	
		13. Kumbhalgarh Fort (84 Kms)	

Source: Department of Tourism, GoR, Jaipur and RTDC, Jaipur

Main Tourist Locations and Attractions

Some salient observations with respect to the main tourist and the type of tourists they direct are as follows:

Udaipur: Udaipur is the main tourist centre in Mewar. The city is full of lakes and palaces. The main attractions are the City Palace, Lake Pichhola, Fateh Sager Lake, Jag Niwas Palace, Jagdish Temple, Sahelion Ki Bari, Ahar, etc. are all located within the city. Sajjangarh fort and the adjoining small wild life sanctuary are another attraction located around 8 Kms from the city. On the north of Udaipur lies Nathdwara (famous Krishna temple) Nagda (temple ruins), Eklingji (temple complex), Haldighati (battleground) visited mainly by the domestic tourists for both religious and historical purposes. Further north lies Rajsamand Lake and Kankroli. Both these tourist spots are less popular and visited only by a few domestic tourists.

Chittaurgarh: Chittaurgarh, around 120 Kms east of Udaipur is very popular as a place of historical interests to domestic tourists particularly with the Bengali tourist. The fort is also gaining in popularity with foreign tourists.

Kumbhalgarh: Kumbhalgarh, around 84 Kms west of Udaipur is an important location for tourists enroute Ranakpur. The fort is particularly popular with domestic tourists and known for its camping sites and horse safaris.

(3) VAGAD CIRCUIT

Vagad region has a varied topography that encompasses wild and rugged terrain in the northeast to fertile plains of alluvial soil in the southwest. The circuit offers a combination of tribal culture, religion and history and is particularly popular amongst the domestic tourist. This circuit primarily includes Dungarpur and Banswara. A map of the Mewar circuit showing the main tourist locations in the circuit (Figure 20.5).

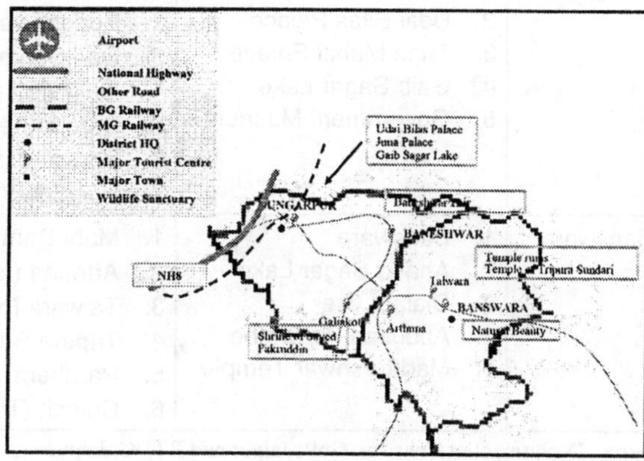

Source: Department of Tourism, GoR, Jaipur and RTDC, Jaipur

Figure 20.5 Vagad Circuit

Geography

Vagad area lies to the south of the Mewar Circuit bordering Gujarat. The climate in this area is hot during the summer months of April to June. The winters are mild and pleasant and hence the peak tourist season is from September to March. The area is a mixture of plains and hills and has large tracks of land covered by forests. With a large number of sanctuaries and lakes in the area. The region consists of the two districts of Dungarpur and Banswara separated by the Mahi river. In general, the Vagad region is the tribal belt of Rajasthan. Banswara with a population of 86,000 is the largest city in the circuit.

Major Tourist Spots and Attractions

The major tourist locations in the Vagad circuit. Although, there are a large number of tourist spots in the Vagad area, the most popular are the Dungarpur and Banswara (Table 20.3).

Table 20.3 Major Tourist Locations in Vagad Region

District	Major Attractions	Major Excursions	Major Events
Dungarpur	1. Dunaarour 2. Udai Bilas Palace 3. Tuna Mahal Palace 4. Gaib Sagar Lake 5. Government Museum	1. Ex-Dunaarrrar 2. Beneshwar Temple (60 Kms) 3. Deo Sonmath Temple (24 Kms) 4. Galiyakot (Sufi shrine) (58 Kms) 5. Baroda (temples) (41 Kms) 6. Bhuvaneshwar (temple) (9 Kms) 7. Poonjpur (temple) (37 Kms)	Beneshwar Fair (Febraary-March) BarBij Fair (October- November)
Banswara	1. Banswara 2. Anand Sagar Lake 3. DialabLake 4. Abdullah Pir Shrine 5. Madareshwar Temple	1. Mahi Dam (18 Kms) 2. Arthuna (temple ruins) (55 Kms) 3. Talwara (ruins) (15 Kms) 4. Tripura Sundari (temples) (19 Kms) 5. Parahera (Temple) (22 Kms) 6. Chinch (Temple) (18 Kms)	

Source: Department of Tourism, GoR, Jaipur and RTDC, Jaipur

Main Tourist Locations and Attractions

Some salient observations with respect to the main tourist and the type of tourists are as follows:

Dungarpur: Dungarpur is famous for its unique architectural style, which is reflected in the Udai Bilas Palace and Juna Palace. Another interesting place to visit is the famous shrine dedicated to Shrinathji on the banks of the Gaib Sagar Lake. Tourists also visit Dungarpur to see the large variety of migratory birds that frequent the water bodies at Dungarpur during winters.

Banswara: Banswara is known for its scenic beauty, especially the Mahi Bajaj Sagar Dam. Mahi river has number of islands inside and thus Banswara is also known as 'City of Hundred Islands". Others places worth visit in the town are Anand Sagar Lake, Dialab Lake, Madareshwar and the shrine of Abdulla Pir.

Baneshwar: Baneshwar, which lies 60 Kms east of Dungarpur at the confluence of the Som and Mahi rivers, is famous for its Shiva Temple. This temple is much revered by the Bhils (tribals) and the annual fair (held in February-March) attracts hordes of tribals from the Rajasthan and the adjoining states of Gujarat and Madhya Pradesh.

Arthuna: Arthuna lies at a distance of 55 Kms from Banswara. Arthuna is famous for its ruins of the Hindu and Jain temples dating back to the 11th, 12th and 15th centuries.

(4) DHUNDHAR CIRCUIT

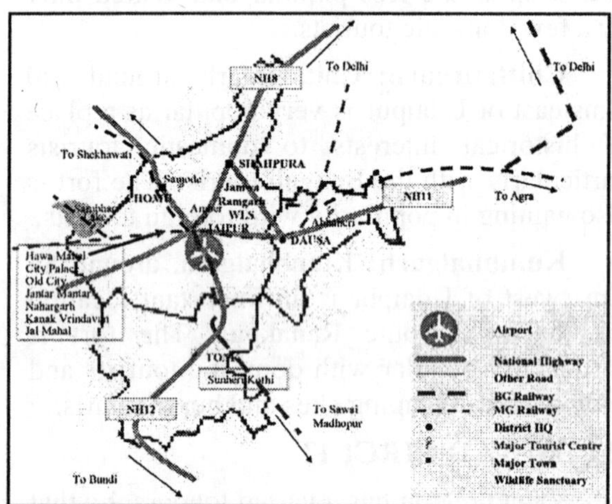

Source: Department of Tourism, GoR, Jaipur and RTDC, Jaipur

Figure 20.6 Dhundhar Circuit

The Dhundhar circuit constitutes of tourist locations around Jaipur and is known for its blend of culture, history, architecture and religion. It includes tourist locations in Jaipur, Dausa and Tonk districts, mainly the Jaipur City and others like Samode, Abhaneri and Ramgarh. Jaipur the capital city of Rajasthan, is the hub for this circuit. Other locations in the circuit are usually covered by taking daily excursions from Jaipur. A map of the

Dhundhar tourism circuit is provided in Figure 20.6.

Geography

The Dhundhar circuit lies in the central part of the state between the northern Aravalli range and close to the Sambhar Basin. The climate in the region is extremely hot during the peak summer months of April–June. As a result, the tourist season extends from September to March. The region gets an annual rainfall of around 640 mm, most of it in the monsoon months of June to September.

Major Tourist Spots and Attractions

Dhundhar region has a large inventory of attractions, which are diverse as well as unique. Hence the region is popular amongst domestic as well as foreign tourists. The main attractions in the region are given in Table 20.4.

Table 20.4 Major Tourist Locations in Dhundhar Circuit

District	Major Attractions	Major Day-excursions	Major Events
Jaipur	**Jaipur** 1. The City Palace 2. Jantar Mantar Observatory 3. Hawa Mahal Palace 4. Govind Devji Temple 5. Galtaj i Pilgrim C entre 6. Moti Doongri, Lakshmi Narayan Temple 7. Statue Circle 8. Ram Niwas Gardens with Zoo 9. BM Birla Planetarium 10. Old City **Others** 1. Maharani ki Chhatri Cenotaphs 2. Jal Mahal Palace 3. Kanak Vrindavan Temple Complex 4. Gaitore Cenotaphs 5. Ghat ki Gimi gardens (Sisodia Rani gardens) 6. Amer fort (Sheesh Mahal and other palaces) 7. Jaigarh fort 8. Nahargarh fort 9. Sanganer palaces and temples	1. Bagru Fort (35 Kms)- on Ajmer Road 2. Ramgarh Lake, temples and fort ruins (32 Kms NE) 3. Samode Elaveli and Bagh (40 Kms NW) 4. Bairath ruins (86 Kms on Shalipura-Alwar road) 5. Abhaneri well and temple (128 Kms NE off Agra road) 6. Sambhar Lake - (94 Kms W) 7. Madhogarh Tunga (40 Kms on Bassi-Lalsot Agra road)	Gangaur Festival (March-April) Elephant Festival (March) Teej (July-August)
Dausa	Abhaneri		
Tonk	**Tonk** Sunheri Kothi		

Source: Department of Tourism, GoR, Jaipur and RTDC, Jaipur

Main Tourist Locations and Attractions

Although termed as the Dhundhar circuit, the circuit is mainly concentrated at Jaipur. As seen in the Exhibit, the Dhundhar circuit has a large inventory of attractions covering history, religion, culture and leisure. Jaipur also offers other products such as convention facilities, health centres, etc. Jaipur has some unique attractions such as the Hawa Mahal, the Jantar Mantar Observatory, etc.

Other places in the circuit are within a radius of 80 Kms from Jaipur. Most tourists stay in Jaipur and visit other places in the circuit taking day excursions. Thus, the Dhundhar circuit is usually completed within 2 days from Jaipur.

(5) GODWAR CIRCUIT

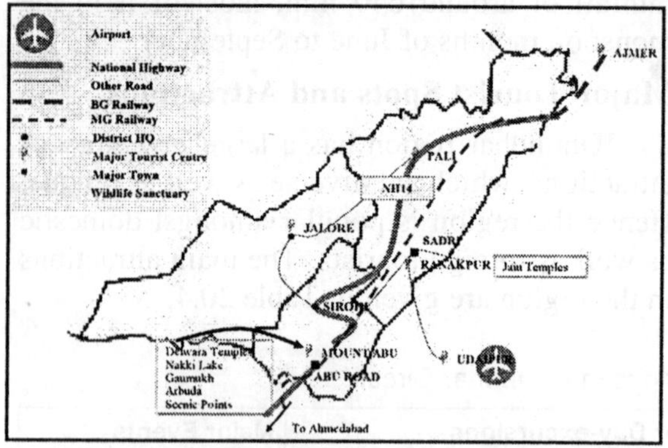

Source: Department of Tourism, GoR, Jaipur and RTDC, Jaipur

Figure 20.7 Godwar Circuit

The Godwar circuit consist of the only hill station in Rajasthan- Mount Abu. The beautiful temple town Ranakpur is the other important attraction in the circuit. The former is very popular leisure destination with the tourist coming from Gujarat, whereas the latter is a pilgrim centre, popular with both domestic as well as foreign tourist. Other location falling under the circuit, which is relatively unexplored by the tourist, is the simple Rajasthani town of Jalore. Thus, the Godwar circuit is essentially a two location circuit, of which Mount Abu is often covered as a standalone circuit. A map of the Godwar tourism circuits provided in Figure 20.7.

Geography

Mount Abu lies 1720 m above the sea level, and forms the south-western tip of the Aravalli range. The hill station stretches along a 20 Kms long plateau and is approached from Abu Road in the south east. The area is well wooded with flowering trees, and a good variety of birdlife. Mount Abu benefits from lower temperatures and cooling breezes. The summer temperatures in Mount Abu are 27-28°C when those in the plains are 32-34°C. Mount Abu lies in the Sirohi district of Rajasthan.

Ranakpur lies in the Pali district of Rajasthan close to the Pali and Rajsamand district border. Surrounded by hills on all sides the Ranakpur temples lie in the protec-ted area of Kumbhalgarh wildlife sanctuary.

Major Tourist Spots and Attractions

Godwar region has relatively lesser number of tourist attractions. However, these attractions are very unique, and hence the region attracts many domestic as well as foreign tourists. The main attractions in the region are given in Table 20.5.

Table 20.5 Major Tourist Locations in Godwar Circuit

District	Major Attractions	Major Day-excursions	Major Events
Sirohi	1. Mount Abu 2. Dilwara Jain Temples 3. NakkiLake 4. Gaumukh Temple 5. Arbuda 6. Adhar Devi Temple 7. Sunset Point 8. Honeymoon Point 9. Guru Shikhar (Highest peak) 10. Achalgarh Fort	1. Ranakpur (181 Kms)	Summer Festival (June) Winter Festival (December)
Pali	**Ranakpur** 1. Jain Temple Complex	Kumbhalgarh (48 Kms) Mount Abu	
Jalore	**Jalore** 1. Jalore Fort		

Source: Department of Tourism, GoR, Jaipur and RTDC, Jaipur

Main Tourist Locations and Attractions

Tourist activity is mainly concentrated around two major locations- Mount

Abu and Ranakpur: Very few tourists go to Jalore and other places in the region.

Mount Abu: Mount Abu is the popular hill station of Rajasthan. It offers two main tourism products- leisure and religion. Many domestic tourist come to Mount Abu for its pleasant climate and scenic beauty. Mount Abu has the beautiful Nakki Lake, around which most of the tourist activity is centered. Other interesting sites in the place such as the Sunset point and the Honeymoon point also attract crowds of domestic tourists. Mount Abu also receives many foreign tourists, who come to see the famous Dilwara temples, famous for their intricate architecture. Various other Jain and Hindu temples around the place, notable amongst them being the Gaumukh, Arbuda temple and Adhar Mata temple, are also visited by the tourists arriving in Mount Abu.

Mount Abu is usually covered as a standalone destination. Many tourists also cover Mount Abu on short overnight excursion from Udaipur coming via Kumbhalgarh and Ranakpur.

Ranakpur: The other important tourist location in the region Ranakpur is a small temple town known for its Jain temple complex. This place is popular among the domestic as well as the foreign tourists. It is one of the five holy places for the Jains. The beautiful temple architecture and carving attract many foreign travellers who take a short excursion ex-Udaipur.

(6) MERWARA-MARWAR CIRCUIT

The desert circuit primarily includes pilgrim towns of Ajmer and Pushkar, and relatively unexplored places like Merta and Nagaur. The circuit covers the two central districts of Rajasthan, namely, Ajmer and Nagaur. A map of the Merwara-Marwar circuit is provided in Figure 20.8.

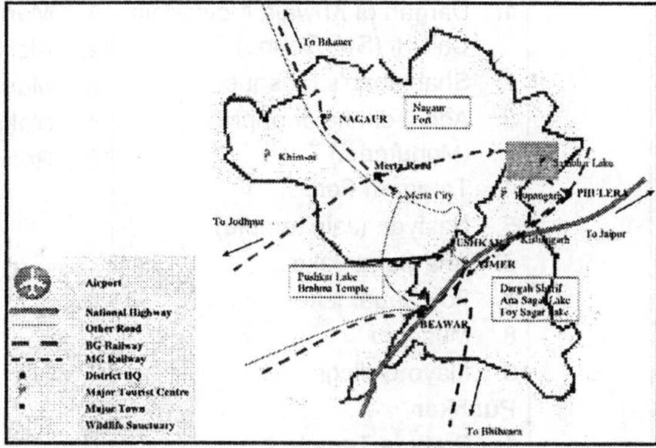

Source: Department of Tourism, GoR, Jaipur and RTDC, Jaipur

Figure 20.8 Merwara-Marwar Circuit

Geography

Ajmer and Pushkar are located 132 Kms west of Jaipur, in a scenic valley encircled by the Aravalli hills. The valley is interspersed with several lakes forming important attractions for the two holy towns. Surrounded by hills on three sides, Pushkar is 11 Kms from Ajmer and situated on the banks of the Pushkar lake.

Ajmer with a population approx. 4.9 Lakhs is the fifth largest city in the state. Other important cities in the region include Kishangarh and Nagaur. The Merwara–Marwar circuit falls in one of the most densely populated region in the state (Table 20.6).

Major Tourist Spots and Attractions

The major tourist locations and attractions in the Merwara–Marwar circuit. The region has relatively less number of tourist attractions, though some of the most visited ones, as compared to other circuits in the state (Table 20.7).

Table 20.7 Major Tourist Locations in Merwara-Marwar Circuit

District	Major Attractions	Major Day-excursions	Major Events
Ajmer	**Aimer City** 1. Dargah of Khwaja Moinuddin Chishti (Sufi Shrine) 2. Shahjahan's Mosque 3. Adhai-din-ka-Jhonpra (Monument) 4. Taragarh Fort 5. Nasiyan (Jain temple) 6. Ana Sagar lake 7. Foy Sagar lake 8. Museum 9. Mayo College **Pushkar** 1. Brahma Temple 2. Savitri Temple 3. Pushkar lake and ghats	**Ex-Aimer / Pushkar** 1. Merta (Temples and Ruins) (75 Kms) 2. Kishangarh (Paintings) (27 Kms) 3. Mangliyawas (Kalpa Vriksha) (26 Kms) 4. Makrana (Marble Quarries) (108 Kms.) 5. Roopangarh (Fort) (25 Kms)	· Pushkar Fair (October-November) · Annual Urs at Ajmer (according to lunar calendar)
Nagaur	**Nagaur City** 1. Nagaur Fort 2. Moti Mahal (Palace) 3. Badal Mahal (Palace) 4. Hawa Mahal (Palace) 5. Baradari (Courtyard) 6. Temples 7. Tarkinka Dargah (Mosque)	1. Merta (Temples and Ruins) (75 Kms)	· Nagaur Cattle Fair (January-F ebraury)

Source: Department of Tourism, GoR, Jaipur and RTDC, Jaipur

Main Tourist Locations and Attractions

Although termed as a circuit, the Merwara-Marwar circuit is mainly concentrated around the Ajmer-Pushkar hub. Other locations in the circuit are of very little significance and hence less visited by both, the domestic and the foreign tourists. The theme of the circuit is really 'pilgrim tourism'. Some salient observations with respect to the main tourist locations and the type of tourists they attract are as follows:

Ajmer: Ajmer, one of the main cities in the circuit, has the famous shrine or dargah of the sufi saint Khwaja Moinuddin Chishti. The annual Urs at the dargah attracts pilgrims of all faiths form around the country and also from abroad. Other attractions in Ajmer include the marvelous Shahjahan's mosque inside the dargah, Adhai-din-ka-Jhonpra and the lakes.

Pushkar: Pushkar houses the only temple in India dedicated to Lord Brahma and is an important pilgrimage for the Hindus. Located on the banks of the Pushkar Lake, the temple town is famous for the colourful Pushkar fair held every year during October-November. It attracts large number of domestic as well as foreign tourist during the 5-day fair.

(7) BRIJ MEWAT CIRCUIT

The Brij Mewat circuit covers locations bordering Delhi, Uttar Pradesh, Madhya Pradesh, Sariska, and Bharatpur. Other places in the circuit include mainly Alwar, Deeg, Karauli and Sawai Madhopur. The circuit is most famous for its Keoladeo National Park, Sariska National Park and Ranthambhor National Park and is very popular with wildlife enthusiasts. The region also has a fair share of places of historic and archaeological importance in the areas. Although termed as a circuit these tourist locations are covered as daily excursions from Jaipur or enroute Jaipur from either Delhi or Agra. The Brij Mewat region is constituted of three separate sub-circuits, namely, Alwar-Sariska, Bharatpur-Deeg and Sawai Madhopur.

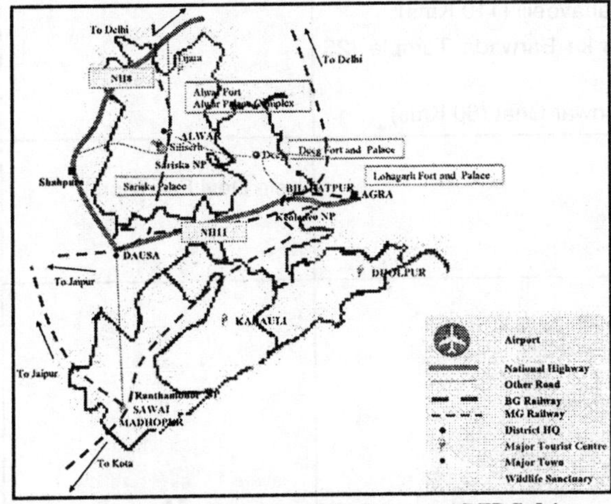

Source: Department of Tourism, GoR, Jaipur and RTDC, Jaipur

Figure 20.9 Brij Mewat Circuit

Dholpur and Karauli in the Brij Mewat circuit are particularly known for their fairs and attract a large number of domestic pilgrimage tourists. A map of the Brij Mewat tourism circuit is provided in Figure 20.9.

Geography

The circuit is nestled between cluster of small hills of the Aravali range. The southern and eastern part of the circuit, bordering Uttar Pradesh has large areas under wetlands, locally called as Orans. The tourist arrivals at the locations in the circuit are largely during the winter months from October to February.

The Alwar-Sariska sub-circuit is towards the North bordering Delhi. Bharatpur-Deeg is towards the East (54 Kms from Agra), whereas Sawai-Madhopur-Ranthambhor lies towards the South. Karauli and Dholpur lie between the Bharatpur and Sawai Madhopur sub-circuits.

Major Tourist Spots and Attractions

The main attractions in the Brij Mewat Circuit are the National Parks at Sariska, Bharatpur and Ranthambhor. The Brij Mewat Circuit (especially Sariska and Bharatpur) is also popular as a weekend tourist destination for Delhi and the National Capital Region (NCR).

The main attractions in the region are given in Table 20.8.

Table 20.8 Major Tourist Locations in Brij Mewat Circuit

District	Major Attractions	Major Day-excursions	Major Events
Alwar	**Alwar** 1. Bala Qila- The Alwar fort 2. Alwar Palace Complex-Palace, Centotaphs, Tanks 3. Vijay Mandir Palace 4. Purjan Viliar Gardens 5. Hope Circus-architectural monument **Sariska** 1. Sariska Palace 2. Tal Vrakhsha Temples **Siliserh** 1. Siliserh Lake Palace	1. Jaisamand Lake 2. Sariska wildlife sanctuary and Tiger reserve 3. Kankwari, Tehla and Kushalgarh forts in the Sariska WLS	Holi, Gangaur and others Alwar Utsav (February)
Bharatpur	**Bharatpur** 1. Lohagarh fort- Palaces, Museum, Towers	1. Keoladeo Ghana National Park 2, Deeg Palace (35 Kms)	

District	Major Attractions	Major Day-excursions	Major Events
Sawai Madhopur	**Sawai Madhopur** 1. Ranthambhor Fort- Palaces, Cenotaphs 2. Ranthambhor National Park 3. Amareshwar Mahadev Temple 4. Bhairon Temple 5. Chamatkar Jain Temple	**Ex-Sawai-Madhopur** 1. Tonk (90 Kms) 2. Kakod fort (22 Kms) 3. Hati Bhata (32 Kms) 4. Shri Mahaveeiji (110 Kms) 5. Chauth ka Barwada Temple (25 Kms) 6. Rameshwar Ghat (60 Kms)	Mahaveerji Fair (March-April)
Karauli	**Karauli** 1. Palace, Lake, Temples (Madan Mohanji and Kalyanji) 2. Kela Devi Temple- Karauli		Kela Devi Fair (April)
Dholpur	**Dholpur** 1. Shergarh Fort 2. Ghanta Ghar 3. Van Viliar- Ramsagar wildlife sanctuary 4. Talab Shahi Lake 5. Machkund (pilgrim place)		

Source: Department of Tourism, GoR, Jaipur and RTDC, Jaipur

Main Tourist Locations and Attractions

Brij Mewat circuit is most popular for its wildlife experience. The national parks at Sariska, Bharatpur and Ranthambhor figure on the tour itinerary of most foreign tourists visiting Agra/ Jaipur/ Golden Triangle. Besides the three national parks, the circuit also has other attractions, most notable amongst them being the palaces at Sariska and Deeg.

Alwar-Sariska sub-circuit also has some beautiful palaces and lakes. Siliserh lake palace is a weekend getaway for tourist from Delhi. Similarly, Lake Jaisamand near Alwar is also a popular picnic destination during the rains.

In Bharatpur, the Lohagarh fort, which lies in the town, is usually visited alongwith Keoladeo National Park. It houses some beautiful palaces and the Iron Tower (Lohastambh). Similarly, Sawai Madhopur has some interesting tourist attractions such as the Kakod fort and the Hati Bhata (huge elephant carved out of one stone).

(8) SHEKHAWATI CIRCUIT

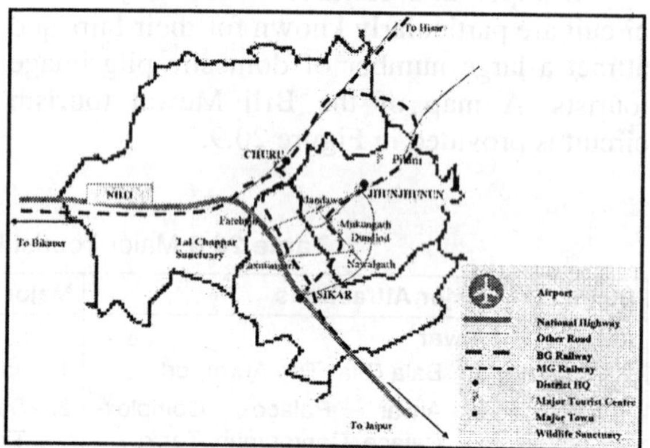

Source: Department of Tourism, GoR, Jaipur and RTDC, Jaipur

Figure 20.10 Shekhawati Circuit

The Shekhawati region includes the three districts of Churu, Sikar and Jhunjhunun in the north-eastern part of Rajasthan bordering Haryana. The region, known as the 'open art gallery of Rajasthan' is famous for its painted havelis. A map of the Shekhawati region is provided in Figure 20.10.

Geography

Shekhawati is a semi-desert area bounded on two sides by the Delhi-Jaipur and Jaipur-Bikaner highways. Sikar lies north of Jaipur, while Jhunjunun in the north touches Churu district and the state of Haryana. Most of the painted towns and villages are concentrated in the west and the north of Shekhawati. Sikar with a population approx. 1.8 Lakhs is the largest town in the region. Other important cities in the region include the district headquarters of Jhunjhunun and Churu. The Shekhawati region is one of the most thickly populated regions in the state.

Major Tourist Spots and Attractions

Table shows the major tourist locations and attractions in the Desert circuit. The region has several tourist small towns famous for their painted havelis, though most of them are within lose distance of each other (Table 20.9).

Table 20.9 Major Tourist Locations in Shekhawati Circuit

District	Major Attractions	Major Day-excursions	Major Events
Sikar	**Sikar City** 1. Biyani Haveli 2. Murarka Haveli 3. Somani Haveli 4. Sagarmal Sodhani Haveli 5. Madhav Niwas Kothi 6. Jubilee Hall 7. Fort 8. Jain temples	**Lachhmangarh** 1. Sawant Ram Chokhani Haveli 2. Mirijamal Kyala Haveli 3. Bansidhar Rathi Haveli 4. Kedia Haveli 5. Sanganeria Haveli **Fatehpur** 1. Ramgopal Mahavirprasad Haveli 2. Goenka Haveli 3. Hukumsichand Choudhri Haveli 4. Jalan Haveli 5. Bhartiya Haveli	Teej (July / August) Gangaur Festival (March/April)
Jhunjunun	**Jhuniunun Town** 1. Nar Singh Das Tibrewal Haveli 2. Khaitan Haveli 3. Ishwar Das Modi Haveli 4. Khatri Mahal (Wind Palace) 5. Bihariji Temple	**Nawalaarh** 1. Aatli (eight) haveli complex 2. Jodhraj Patodia Haveli 3. Bansidhar Bhagat Haveli 4. Chokhani Haveli 5. Hotel Roop Niwas Palace **Churi Ajitgarh** 1. Shiv Narain Nemani Haveli 2. Kothi Shiv Datt Haveli 3. Rai Jaganlal Tibrewal Haveli 4. Ram Rampratap Nemani Haveli **Chirawa** 1. Rai Bahadur Tulsian Haveli 2. Nemani Haveli 2. Mangalchand Dahnia Haveli 3. Rarachand Keshardev Dahnia Haveli 4. Dulichand Kakrania Haveli 5. Bhola Ram Kakrania Haveli 6. Manoharlal Vaidh Haveli 7. Kakrania Well 8. Sehksaria Well	Teej (July / August) Gangaur Festival (March/April)

District	Major Attractions	Major Day-excursions	Major Events
		Dundlod 1. Dundlod Fort 2. Goenka Haveli **Mukundgarh** 1. Mukimdgarh Fort 2. Kanoria Haveli 3. Ganeriwala Haveli 4. Textiles and brasswork **Khetri** 1. Raghunath Temple 2. Bhopalgarh Fort 3. Pamia lal Shall ka Talab (water tank) 4. Sukli Mahal (Palace) 5. Hari Singh Temple 6. Baghore Fort **Mandawa** 7. Mandawa Castle 1. Chokhani Haveli 2. Saraf Haveli 3. Goenka Haveli 4. Ladia Haveli **Parasrampura** - Shardul Singh's Chhatri (Cenotaph) **Baggar** - Reservior **Ramgarh** - Havelis and temple **Surajgarh** - Fort **Alsisar and Malsisar** - Havelis, Temples, Fort **Bissau** - Havelis **Mehansar** - Fort and temples	
Churu	**Churu** 1. Churu Fort 2. Kanhaiyalal Bagla Haveli 3. Kothari Haveli 4. Surana Haveli 5. Poddar Haveli 6. Nagarshree Chandmal Bhantia Haveli 7. Sagarmal Vaidya Haveli 8. Bajrangmal Mantri Haveli 9. Jai Dayal Goenka Haveli 10. Parakh Haveli 11. Jain Temple 12. Balaji Temple		Teej (July / August) Gangaur Festival (March/April)

Source; Department of Tourism, GoR, Jaipur and RTDC, Jaipur

Main Tourist Locations and Attractions

Unlike most other circuits in the state, where there are a few unique attractions, Shekhawati circuit itself is unique from the remaining circuits. The main attraction in the circuit are the painted havelis, which are more or less identical, are spread out over the entire region and as such the uniqueness of the region lies in its homogeneity. As a result the tourist usually does not visit more than 2 - 3 places in the circuit. Amongst the more popular ones are Mandawa, Nawalgarh, Mukundgarh, Sikar, Jhunjhunun and Dundlod.

Mandawa: Mandawa is an important stop over for groups travelling from Delhi enroute Bikaner. The place is famous for its fort, the Mandawa Castle, now converted into a heritage hotel. Some havelis worth visiting in this small town are Chokhani, Saraf, Goenka and Ladia

Nawalgarh: Nawalgarh has one of the finest havelis in the region. Some of the havelis are maintained by the trusts formed by respective owners and are converted into museums. Nawalgarh also has a fort, which now lies in a dilapidated condition.

Mukundgarh: Mukundgarh town built around a temple square is known for its fort, which is now a heritage hotel. Like other towns in the region, Mukundgarh also has painted havelis. It has good handicrafts market, selling textiles, brassware and iron scissors.

Sikar: Sikar, the largest town in the region, is known for its havelis and temples. Tourists also take small half-day excursions to Harshnath temples and Jeen Mata temples located close to Sikar

Dundlod: Dundlod is known for its fort and the horse safaris. It also has the famous Goenka Haveli

Jhunjhunun: Jhunjhunun, the capital of Shekhawati also has splendidly painted havelis. Other interesting attractions include Khatri Mahal, Badalgarh, Jorawargarh, Mertani Baori, Ajit Sagar and Jain temple.

(9) HADOTI CIRCUIT

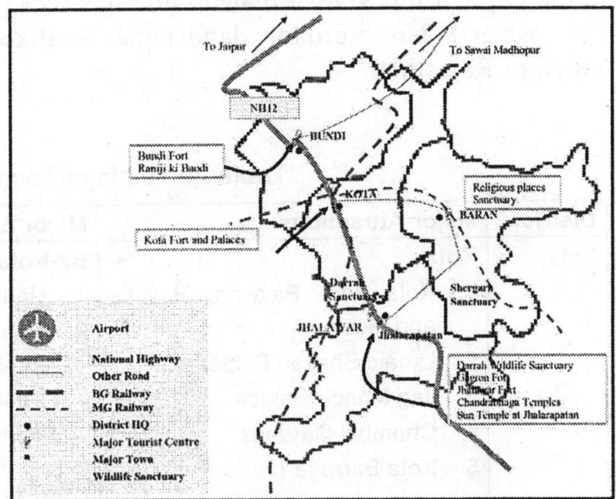

Source: Department of Tourism, GoR, Jaipur and RTDC, Jaipur

Figure 20.11 Hadoti Circuit

Hadoti circuit is the least explored regions of Rajasthan. The region is known for its beautifully sculpted temples and forts and palaces. This circuit includes locations like Bundi, Kota and Jhalawar. Places of tourist interest in Baran district are also covered as a part of this circuit.

Kota, Bundi, Jhalawar and Baran compose the backbone of the Hadoti region, which has a history that goes back several centuries. Surprisingly unexplored, the Hadoti region of Rajasthan has some splendid treasures for the tourist. Its impregnable fortresses, sprawling palaces, exquisitely wrought temples and lovely waterways act as a magnificent foil to exotic wildlife and delicate fresco paintings. A map of the Hadoti tourism circuit is provided in Figure 20.11.

Geography

The Hadoti region falls in the Harawati plains, which is interspersed with the Bundi hills and Kota plateau. The region has beautiful hills, valleys and lakes, which form a picturesque countryside.

Kota is a city strung out along the eastern bank of River Chambal, just below a rocky gorge. Bundi is surrounded by the Aravalli hills on three sides and is circumscribed by a massive wall with four gateways. Jhalawar lies at the edge of the Malwa plateau and has rocky but water-laden verdant landscape, unlike much of Rajasthan.

Major Tourist Spots and Attractions

Hadoti has a large number of attractions. The main attractions in the region are given in Table 20.10.

Table 20.10 Major Tourist Locations in Hadoti Circuit

District	Major Attractions	Major Excursions	Major Events
Kota	**Kota** 1. Kota Fort- Palaces, Havelis and Museum 2. Umaid Bhavan Palace 3. Jag Mandir Palace 4. Chambal Gardens 5. Kota Barrage	**Ex-Kota** 1. Badoli Temple Complex (48 Kms)- Baran 2. Rana Pratap Sagar Dam and Bhainsrogarh (50-55 Kms) 3. Bhanddeora Temple (110 Kms)- Baran district 4. Garhgachh Temple ruins (110 Kms)- Baran district 5. Sitabari (120 Kms)- Baran district 6. Shergarh Fort (125 Kms)- Baran district 7. Darrah Wildlife Sanctuary (50 Kms)	Kota Dussehra (October)
Bundi	**Bundi** 1. Taragarh Fort 2. The Palace 3. Chhattar Mahal or Palace of Towers 4. Ratan Daulat Monument 5. Chitrashala: Gallery of Miniatures and Murals 6. Nawal Sagar Lake 7. Raniji ki Baori Wells 8. Sukh Mahal Palace 9. Phool Sagar Palace 10. Shikar Burj 11. Keshar Bagh 12. 84 Pillared Cenotaph	**Ex-Bundi** 1. Jait Sagar Lake (3 Kms) 2. Rameshwaram Cave Temple (20 Kms) 3. Keshavraipatan Temples (45 Kms) 4. Ramgarh Sanctuary (45 Kms) 5. Bijolia Fort (50 Kms) 6. Menal Temples (70 Kms) 7. Talwas Fort and Temples (53 Kms) 8. Dugari Fort and Paintings (65 Kms) 9. Indragarh Fort (77 Kms)	Kajli Teej (July-August)
Jhalawar	**Jhalawar** 1. Jhalawar Fort 2. Government Museum 3. Bhavani Natya Shala Theatre	**Ex-Jhalawar** 1. Rain Basera Cottage (6 Kms) 2. Jhalarapatan Temples (6 Kms) 3. Chandrabhaga Temples (7 Kms) 4. Gagron Fort and Buddhist Caves and Stupas 5. Dag Temples (100 Kms) 6. Kakuni Temples and Fort (65 Kms) – Baran district 7. Bhimsagar Dam (24 Kms) 8. Atishey Jain Temple (35 Kms) 9. Dalhanpur Temples (54 Kms) 10. Fort of Gangadhar (120 Kms)	Jhalawar Fair (October-November)

Source: Department of Tourism, GoR, Jaipur and RTDC, Jaipur

Main Tourist Locations and Attractions

As seen in Table, the Hadoti region has a vast inventory of attractions. The region is truly, the best kept secrets of Rajasthan, and has many beautiful and unexplored forts, palaces and old temples and temple ruins.

Kota: Kota is a major industrial town of the state. Yet, memories of its ancient links linger strongly in form of various structures, ruins, paintings and murals. The most notable among these are the Kota fort and the various palaces and havelis within the fort, Jag Mandir and the Chambal Gardens. There are many small yet interesting attractions scattered in the city. A day's excursion from Kota is Baran district, which has many interesting attractions such as forts and temples.

Bundi: 36 Kms from Kota is a tiny picturesque town of Bundi. Interesting monuments including impressive medieval forts, palaces, havelis, temples with beautiful stone idols and chhattris with carved pillars, along with a picturesque lake in the heart of the town, add to its charm. Bundi is very famous for its intricate carvings and murals.

Jhalawar: Jhalawar boasts of rich historic as well as natural wealth with some exquisite pre-historic cave paintings, massive forts and thickly wooded forests and exotic wildlife variety. One can spot countless species of birds as one drives past the lush countryside. The countryside is all the more fascinating and colourful during winters.

RAJASTHAN TOURISM GROWTH

Tourism is ragarded as an important industry in the entire world. India has also earns a lot of foreign exchange every year from tourism, Rajasthan contributes a lot in this respect. On the basis of the available information it has been estimated that out of three foreign tourists one definitely visits Rajasthan. Thus, the state helps in increasing foreign exchange earning of the country, currently several lakh tourists visit Rajasthan every year, out of whom a large number is from diferent states, While the rest are foregn tourists, who constitute about 1/3 of the tourist who come from abroad every year. It is that in 2017 about 16,09,963 lakh tourists visited Rajasthan out of whom 16,09,963 lakh were foreign tourists and 4,59,16,573 lakh were Indian tourists. Tourists continue to come in all seasons these days (Table 20.11 to 20.15).

Tourists who visit Rajasthan spend a lot of money year in the state. There is some amount of contribution of tourism in the economich development (Table 20.16 to 20.18).

The foreign tourist triffic in Rajasthan is a part of the larger triffic in India. It was only after 1949 that much attention has been given to tourism in India and thus recorded a continuos growth in tourist traffic. In 1971 only 8.80 lac home domestic tourist and 42,500 foreign tourists visited Rajasthna while this number increased in 2017 upto 45916573 lac domestic tourists and 1609963 lac foreign tourists respectively (Figure 20.12)

Rajasthan is a leading tourism State in the country. Its glorious heritage, colorful living traditions and vibrant culture are special attractions for both, domestic and foreign tourists. The tangible and intangible tourism products of the State offer immense potential for growth of the tourism industry. Development of tourism resources and increasing both, domestic and foreign tourist arrivals is a high priority area for the Union and State Governments. Therefore, there is a need for expanding tourist centric infrastructural facilities to keep pace with the present trends and for potential tourism growth in the State. Department of Tourism had announced a Rajasthan Tourism Unit Policy in 2007. This Policy will now be replaced by Rajasthan Tourism Unit Policy, 2015 in order to extend more support and incentives for establishment of Tourism Units in the State.The new Policy has been framed keeping in view the guidelines under the 'Suraj Sankalp' Policy document, new emerging trends in the tourism sector, representations given by various stakeholders including the tourism and trade organisations and also the suggestions received from other departments.

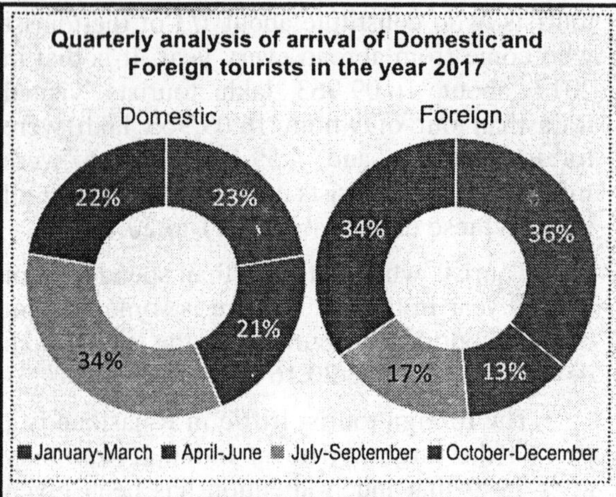

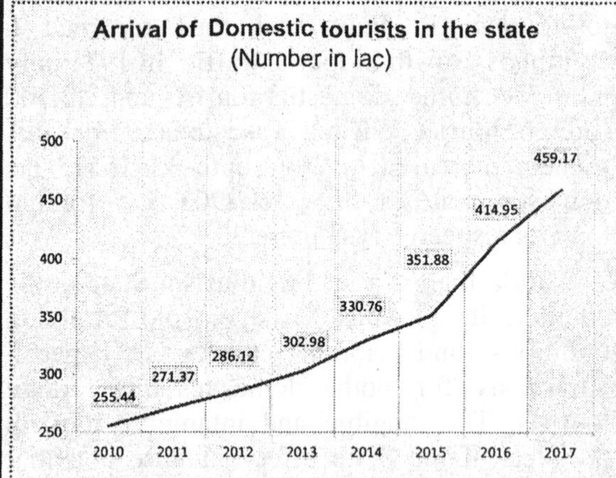

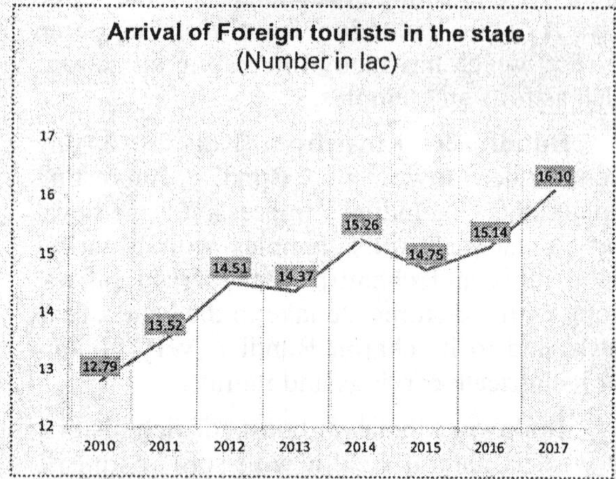

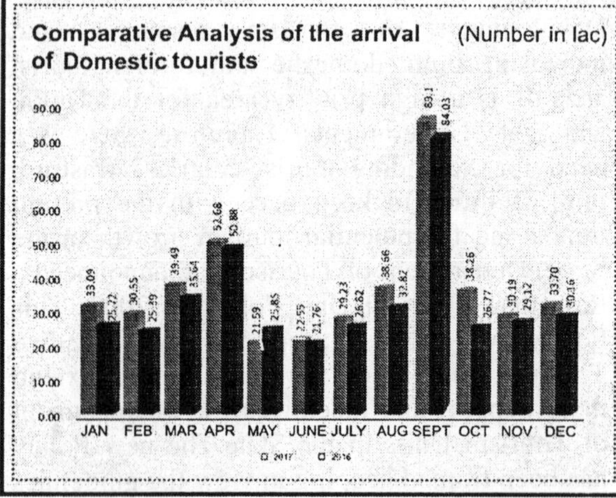

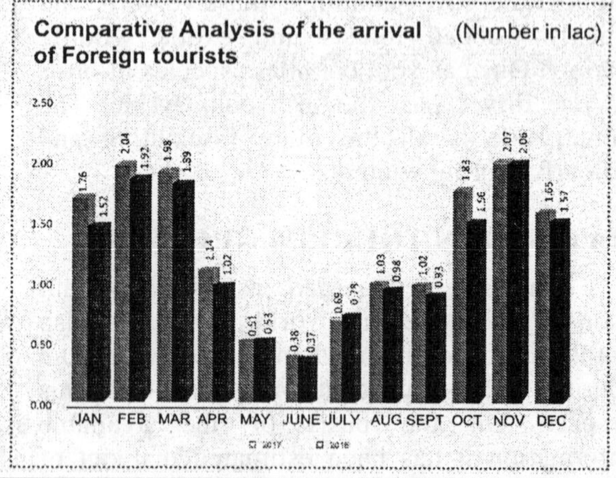

Source: Department of Tourism, GoR, Jaipur and RTDC, Jaipur

Figure 20.12

The Rajasthan Tourism Unit Policy, 2015 primarily addresses issues relating to time bound conversion of land for tourism units including new hotels and heritage hotels, time bound approval of building plans, grant of Patta to heritage hotels, allotment of land for tourism units on DLC (District Level Committee) rates, applicability of Rajasthan Investment Promotion Scheme, 2014 (RIPS-2014) for tourism units and smooth and speedy implementation of the provisions of related departments like Revenue, Urban Development and Housing (UDH) and Local Self Government (LSG), Panchayati Raj, etc. It is expected that this Policy will strengthen the existing infrastructure, will foster infrastructure development, income and employment generation and increase the much needed availability of hotel rooms for the tourists. By allowing Heritage Hotels in Rural Abadi/ Panchayat areas Rural Tourism is likely to increase manifold.

The definition of Tourism Unit has been expanded to cover various types of tourism units and activities including budget hotels, heritage hotels, resorts, golf courses, adventure sports, etc. It is expected that these steps will bring speedy investment in the State.

Tourism Sector

Enterprises of the tourism sector covered under sub-clause (a) and (b) of clause 2 (xxxviii) of the Scheme shall be granted the following benefits:

(a) Investment Subsidy of 50% of VAT and CST which have become due and have been deposited by the enterprise for seven years;

(b) Employment Generation Subsidy up to 10% of VAT and CST which have become due and have been deposited by the enterprise, for seven years;

(c) Reimbursement of 25% of amount of VAT paid on purchase of plant and machinery or equipment for a period up to seven years from the date of issuance of the entitlement certificate;

(d) Exemption from payment of 50% of Entertainment Tax for seven years;

(e) Exemption from payment of 100% of Luxury Tax for seven years;

(f) Land allotment in urban and rural areas at DLC rates;

(g) 25% additional exemption from payment of stamp duty chargeable on the instrument of purchase or lease of more than 100 years old heritage property in the State, for the purpose of hotel development under the Scheme declared by the Tourism Department as provided in Notification no. F.12 (20) FD /Tax / 2005-219 of 24.03.2005;

(h) 50 percent additional exemption from payment of conversion charges; and

(i) 100 percent exemption from payment of development charges.

Notwithstanding anything contained in the Scheme, an enterprise making investment in the tourism sub-sector defined under sub-clause (c) of clause 2 (xxxviii) of the Scheme shall be granted the benefits, as may be provided to it by the State Empowered Committee.

Kota stone, Marble and Granite Sector: Enterprises making a minimum investment of twenty five lakh rupees in the sector shall be granted the following benefits:

(i) Investment Subsidy of 55% of VAT and CST which have become due and have been deposited by the enterprise for seven years; and

(ii) Employment Generation Subsidy up to 10% of VAT and CST which have become due and have been deposited by the enterprise, for seven years.

Table 20.11 Centrewise details of Tourists visiting Major Tourist Destinations of The State (2014-17)

S. No.	Tourist Place	2014		2015		2016		2017	
		Domestic	Foreign	Domestic	Foreign	Domestic	Foreign	Domestic	Foreign
1.	Mount Abu	2376831	4376	2017636	1598	1983435	1672	2682871	2579
2.	Udaipur	720120	166936	727266	165525	756440	183964	830784	190521
3.	Jaipur	1170152	568234	1201152	596756	1544730	565978	1702665	633990
4.	Pushkar	3234750	70603	3786360	69494	3961130	97651	4636005	101673
5.	Jodhpur	520198	139640	598967	126772	972337	138558	894085	144363
6.	Ajmer	4245710	33069	4546300	36423	4896070	41112	4651920	59405
7.	Jaisalmer	250716	91759	266175	84533	359497	90937	493755	122851
8.	Nathdwara	702801	21	637722	10	678006	0	713431	0
9.	Chittourgarh	612587	48374	640688	38879	791840	52340	449091	3417
10.	Bharatpur	69225	40386	66322	39608	72701	26368	111796	18398
11.	Bikaner	347294	67098	348772	60767	325244	60300	356094	61930
12.	Ranakpur	555951	122403	532039	102994	525455	114806	494174	105272
13.	Kota	51467	3516	90598	2574	89546	1778	202298	1860
14.	S. Madhopur	77800	61495	85200	67935	106000	51265	139428	55190
15.	Jhunjhunu	126184	46828	86555	37420	127687	24477	174883	35999
16.	Banswara	114040	141	113410	82	121487	139	129925	117
17.	Alwar	104418	18650	95787	10634	119815	8524	147653	12296
18.	Sariska	13087	237	14487	150	10932	198	10069	639
19.	Rishabdeo	12489	0	25800	0	26022	0	30419	0
20.	Bundi	49925	15063	54574	15290	59864	15420	65021	16442
21.	Sikar	116905	0	48305	16	102282	826	152416	1941
22.	Siliserh	1910	0	2071	0	1752	0	1862	2
23.	Bahroad	4402	2	3650	31	2533	8	1802	0
24.	Jhalawar	93344	108	92426	114	88440	130	103102	166
25.	Others	17504185	26635	19105311	17706	23771870	37278	26741024	40912
	Total	**33076491**	**1525574**	**35187573**	**1475311**	**41495115**	**1513729**	**45916573**	**1609963**

Source: Department of Tourism, GoR, Jaipur and RTDC, Jaipur

Table 20.12 Countrywise details of Foreign Tourists Visiting the State in the Year 2017

S. No.	Tourist Place	U.K. 1		France 2		Italy 3		Canada 4		U.S.A. 5		Germany 6		Australia 7		Switzerland 8		Japan 9		Sri Lanka 10	
		Tourist	Night halt	Tourist	Night halt	Tourist	Night halt	Tourist	Night halt	Tourist	Night halt	Tourist	Night halt	Tourist	Night halt	Tourist	Night halt	Tourist	Night halt	Tourist	Night halt
1	Mount Abu	321	346	293	344	337	338	150	150	221	308	220	222	285	292	162	211	69	69	0	0
2	Udaipur	22343	38344	27829	48521	9308	16805	8261	15650	22423	41154	16692	30179	12777	24377	2679	5003	1272	2259	239	469
3	Jaipur	62935	94408	45101	67657	21666	32505	24136	36212	76439	114663	35620	53436	19844	29773	5014	7527	23935	35908	1841	2766
4	Pushkar	10481	10924	15036	16601	5504	5789	3540	3636	5154	5641	8618	9010	5292	5723	1533	1697	668	883	33	38
5	Jodhpur	13442	15895	25681	30768	8906	10532	5526	6291	12018	14115	12063	13815	8950	10333	1680	1913	2036	2258	117	131
6	Ajmer	2102	2302	1945	2176	338	388	737	771	1504	1544	1977	2149	999	1078	194	209	118	136	191	215
7	Jaisalmer	3888	8026	5194	10400	739	1494	963	1918	4505	9166	2546	5158	1819	3656	757	1519	797	1708	15	30
8	Nathdwara	0	0	0	0	0	0	0	0	0	0	0	0	0	0	0	0	0	0	0	0
9	Chittourgarh	491	491	705	705	79	79	154	154	205	205	164	164	406	406	80	80	49	49	0	0
10	Bharatpur	1611	1703	1483	1599	479	513	609	638	738	785	4010	4257	617	627	217	225	375	390	1	1
11	Bikaner	1795	1834	21878	22555	5348	5460	1873	1924	1118	1134	11870	12146	2605	2637	1090	1096	85	85	8	8
12	Ranakpur	9497	9497	25879	25879	7591	7591	3110	3110	8251	8251	14488	14488	6219	6219	1230	1230	370	370	40	40
13	Kota	122	122	129	129	3	3	44	44	187	187	328	328	33	33	68	68	91	91	6	6
14	S. Madhopur	15558	15558	1987	1987	1143	1143	2937	2937	10022	10022	5274	5274	3586	3586	650	650	192	192	84	84
15	Jhunjhunu	1738	1785	13699	13866	2429	2454	837	859	500	502	6328	6358	1137	1190	485	491	44	67	12	12
16	Banswara	8	8	2	8	4	9	3	3	23	33	2	2	19	24	2	2	3	9	0	0
17	Alwar	1207	1207	877	877	429	429	596	596	480	480	738	738	389	389	322	322	1913	1913	6	6
18	Sariska	11	11	29	29	6	6	22	22	3	3	125	125	8	8	0	0	0	0	0	0
19	Rishabdeo	0	0	0	0	0	0	0	0	0	0	0	0	0	0	0	0	0	0	0	0
20	Bundi	1850	2594	5078	7205	909	1413	721	1214	680	1136	1325	2126	1326	2148	408	728	151	286	10	20
21	Sikar	390	390	1064	1064	43	43	49	49	24	24	120	120	36	36	14	14	2	2	0	0
22	Siliserh	0	0	0	0	0	0	0	0	0	0	0	0	0	0	0	0	0	0	0	0
23	Bahroad	0	0	0	0	0	0	0	0	0	0	0	0	0	0	0	0	0	0	0	0
24	Jhalawar	41	81	8	24	6	30	0	0	22	92	2	3	11	22	0	0	3	27	0	0
25	Others	6852	6897	8713	8738	1251	1256	1917	1917	3066	3077	4351	4357	2462	2462	837	837	112	112	22	22
	Total	**156683**	**212423**	**202610**	**261132**	**66518**	**88280**	**56185**	**78095**	**147583**	**212522**	**126861**	**164455**	**68820**	**95019**	**17422**	**23822**	**32285**	**46814**	**2625**	**3848**

Source: Department of Tourism, GoR, Jaipur and RTDC, Jaipur

Table 20.13 Countrywise details of Foreign Tourists Visiting the State in the Year 2017

S. No.	Tourist Place	Pakistan 11		Bangladesh 12		Singapore 13		Iran 14		U.A.E. 15		Saudi Arab 16		Malasia 17		Others 18		Total 1 to 17		Total 1 to 18	
		Tourist	Night halt	Tourist	Night halt	Tourist	Night halt	Tourist	Night halt	Tourist	Night halt	Tourist	Night halt	Tourist	Night halt	Tourist	Night halt	Tourist	Night halt	Tourist	Night halt
1	Mount Abu	0	0	0	0	0	0	0	0	0	0	0	0	0	0	521	542	2058	2280	2579	2822
2.	Udaipur	19	43	651	1148	804	1632	218	446	197	393	127	256	939	1844	63743	119019	126778	228523	190521	347542
3	Jaipur	138	209	16827	25241	2792	4191	6848	10278	461	694	843	1267	6929	10399	282621	423857	351369	527134	633990	950991
4	Pushkar	2	2	272	305	166	189	59	74	9	11	21	26	254	404	45031	55849	56642	60953	101673	116802
5	Jodhpur	226	226	410	445	529	572	132	157	41	52	50	57	594	653	51962	61051	92401	108213	144363	169264
6	Ajmer	557	3345	40146	50193	73	93	65	85	60	87	14	21	105	146	8280	9748	51125	64938	59405	74686
7	Jaisalmer	7	14	24	48	75	150	0	0	4	8	2	4	33	66	101483	202337	21368	43365	122851	245702
8	Nathdwara	0	0	0	0	0	0	0	0	0	0	0	0	0	0	0	0	0	0	0	0
9	Chittourgarh	0	0	0	0	8	8	0	0	0	0	0	0	1	1	1075	1075	2342	2342	3417	3417
10	Bharatpur	0	0	2	3	2	2	1	1	0	0	5	5	4	4	8244	8619	10154	10753	18398	19372
11	Bikaner	2	2	48	48	83	86	10	10	0	0	3	3	124	124	13990	14644	47940	49152	61930	63796
12	Ranakpur	2	2	48	48	288	288	26	26	78	78	30	30	309	309	27816	27816	77456	77456	105272	105272
13	Kota	0	0	10	10	31	31	27	27	2	2	7	7	11	11	761	768	1099	1099	1860	1867
14	S. Madhopur	0	0	4	4	178	178	112	112	40	40	1	1	9	9	13413	13413	41777	41777	55190	55190
15	Jhunjhunu	0	0	4	4	24	24	8	8	2	2	2	6	28	28	8722	10016	27277	27656	35999	37672
16	Banswara	0	0	0	0	0	0	0	0	0	0	0	0	0	0	51	1555	66	98	117	1653
17	Alwar	0	0	3	3	7	7	0	0	0	0	0	0	29	29	5300	5300	6996	6996	12296	12296
18	Sariska	0	0	0	0	0	0	0	0	0	0	0	0	4	4	431	431	208	208	639	639
19	Rishabdeo	0	0	0	0	0	0	0	0	0	0	0	0	0	0	0	0	0	0	0	0
20	Bundi	0	0	26	52	26	50	0	0	4	10	0	0	17	32	3911	5774	12531	19014	16442	24788
21	Sikar	0	0	25	25	0	0	0	0	4	4	1	1	1	1	168	168	1773	1773	1941	1941
22	Siliserh	0	0	0	0	0	0	0	0	0	0	0	0	0	0	2	2	0	0	2	2
23	Bahroad	0	0	0	0	0	0	0	0	0	0	0	0	0	0	0	0	0	0	0	0
24	Jhalawar	0	0	10	92	1	8	0	0	4	56	0	0	0	0	58	473	108	435	166	908
25	Others	0	0	13	13	93	93	13	13	16	16	4	4	43	43	11147	11161	29765	29857	40912	41018
	Total	**953**	**3843**	**58523**	**77682**	**5180**	**7602**	**7519**	**11237**	**922**	**1453**	**1110**	**1688**	**9434**	**14107**	**648730**	**973618**	**961233**	**1304022**	**1609963**	**2277640**

Source: Department of Tourism, GoR, Jaipur and RTDC, Jaipur

Table 20.14 Yearwise Comparative details of Foreign Tourist arrivals in India

Year	Foreign Tourist Visited in India	Change of Percentage	Foreign Tourist Visited in Rajasthan	Change of Percentage
1990	1707158		417641	
1991	1677508	-1.74	494150	18.32
1992	1867651	11.33	547802	10.86
1993	1764830	-5.51	540738	-1.29
1994	1886433	6.89	436801	-19.22
1995	2123683	12.58	534749	22.42
1996	2287860	7.73	560946	4.90
1997	2374094	3.77	605060	7.86
1998	2358629	-0.65	591369	-2.26
1999	2481928	5.23	562685	-4.85
2000	2649378	6.75	623100	10.74
2001	2537282	-4.23	608283	-2.38
2002	2384364	-6.03	428437	-29.57
2003	2726214	14.34	628560	46.71
2004	3457477	26.82	971772	54.60
2005	3918610	13.34	1131164	16.40
2006	4447167	13.49	1220164	7.87
2007	5081504	14.26	1401042	14.82
2008	5282603	3.96	1477646	5.47
2009	5167699	-2.18	1073414	-27.36
2010	5775692	11.77	1278523	19.11
2011	6309222	9.24	1351974	5.74
2012	6577745	4.26	1451370	7.35
2013	6967601	5.93	1437162	-0.98
2014	7679099	10.21	1525574	6.15
2015	8027133	4.53	1475311	-3.29
2016	8804411	9.68	1513729	2.60
2017	10177248	15.59	1609963	6.36

Source: Department of Tourism, GoR, Jaipur and RTDC, Jaipur

Table 20.15 Month-wise Comparative details of Foreign Tourist visited in Rajasthan

Month	Foreign Tourist Visited in India		Change in Percentage	Foreign Tourist Visited in Rajasthan		Change in Percentage
	2016	2017		2016	2017	
	844533	983413	16.44	151543	176102	16.21
	848782	956337	12.67	191777	203584	6.16
	809107	904888	11.84	189309	198441	4.82
	592004	740275	25.05	102473	114417	11.66
	527466	630438	19.52	53046	51276	-3.34
	546972	669989	22.49	37489	38050	1.50
	733834	788377	7.43	78280	68696	-12.24
	652111	724067	11.03	98150	102563	4.50
	608177	722567	18.81	92788	102186	10.13
	741770	875838	18.07	156417	182691	16.80
	878280	1004826	14.41	205534	207452	0.93
	1021375	1176233	15.16	156923	164505	4.83
Total	**8804411**	**10177248**	**15.59**	**1513729**	**1609963**	**6.36**

Source: Department of Tourism, GoR, Jaipur and RTDC, Jaipur

Table 20.16 Yearwise Comparative Statement of Indian and Foreign Tourist Visited in Rajasthan (1971-2017)

S. No.	Year	Tourist arrivals			Changes % Compared to last year		
		Domestic	Foreigner	Total	Domestic	Foreigner	Total
1.	1971	880694	42500	923194			
2.	1972	902769	48350	951119	2.51	13.76	3.02
3.	1973	1157959	54611	1212570	28.27	12.95	27.49
4.	1974	998227	55781	1054008	-13.79	2.14	-13.08
5.	1975	1117663	66207	1183870	11.96	18.69	12.32
6.	1976	1303633	92272	1395905	16.64	39.37	17.91
7.	1977	1618822	125112	1743934	24.18	35.59	24.93
8.	1978	2042586	160134	2202720	26.18	27.99	26.31
9.	1979	2306550	195837	2502387	12.92	22.30	13.60
10.	1980	2450282	208216	2658498	6.23	6.32	6.24
11.	1981	2600407	220440	2820847	6.13	5.87	6.11
12.	1982	2780109	237444	3017553	6.91	7.71	6.97
13.	1983	2932622	266221	3198843	5.49	12.12	6.01
14.	1984	3040197	259637	3299834	3.67	-2.47	3.16
15.	1985	3120944	268774	3389718	2.66	3.52	2.72
16.	1986	3214113	291763	3505876	2.99	8.55	3.43
17.	1987	3424324	348260	3772584	6.54	19.36	7.61
18.	1988	3495158	366435	3861593	2.07	5.22	2.36
19.	1989	3833008	419651	4252659	9.67	14.52	10.13
20.	1990	3735174	417641	4152815	-2.55	-0.48	-2.35
21.	1991	4300857	494150	4795007	15.14	18.32	15.46
22.	1992	5263121	547802	5810923	22.37	10.86	21.19
23.	1993	5454321	540738	5995059	3.63	-1.29	3.17
24.	1994	4699886	436801	5136687	-13.83	-19.22	-14.32
25.	1995	5248862	534749	5783611	11.68	22.42	12.59
26.	1996	5726441	560946	6287387	9.10	4.90	8.71
27.	1997	6290115	605060	6895175	9.84	7.86	9.67
28.	1998	6403310	591369	6994679	1.80	-2.26	1.44
29.	1999	6675528	562685	7238213	4.25	-4.85	3.48
30.	2000	7374391	623100	7997491	10.47	10.74	10.48
31.	2001	7757217	608283	8365500	5.19	-2.38	4.60
32.	2002	8300190	428437	8728627	6.99	-29.57	4.34
33.	2003	12545135	628560	13173695	51.14	46.71	50.92
34.	2004	16033896	971772	17005668	27.81	54.60	29.09
35.	2005	18787298	1131164	19918462	17.17	16.40	17.13
36.	2006	23483287	1220164	24703451	25.00	7.87	24.02
37.	2007	25920529	1401042	27321571	10.38	14.82	10.60
38.	2008	28358918	1477646	29836564	9.41	5.47	9.21
39.	2009	25558691	1073414	26632105	-9.87	-27.36	-10.74
40.	2010	25543877	1278523	26822400	-0.06	19.11	0.71
41.	2011	27137323	1351974	28489297	6.24	5.74	6.21
42.	2012	28611831	1451370	30063201	5.43	7.35	5.52
43.	2013	30298150	1437162	31735312	5.89	-0.98	5.56
44.	2014	33076491	1525574	34602065	9.17	6.15	9.03
45.	2015	35187573	1475311	36662884	6.38	-3.29	5.96
46.	2016	41495115	1513729	43008844	17.93	2.60	17.31
47.	2017	45916573	1609963	47526536	10.66	6.36	10.50

Source: Department of Tourism, GoR, Jaipur and RTDC, Jaipur

Table 20.17 Month wise Comparative Statement of Indian and Foreign Tourist Visited in Rajasthan

Month	Year 2017			Year 2016			Percent change compared to past period		
	Domestic	Foreign	Total	Domestic	Foreign	Total	Domestic	Foreign	Total
1	2	3	4	5	6	7	8	9	10
January	3308560	176102	3484662	2572165	151543	2723708	28.63	16.21	27.94
Feburary	3055214	203584	3258798	2538731	191777	2730508	20.34	6.16	19.35
March	3949266	198441	4147707	3564473	189309	3753782	10.80	4.82	10.49
April	5268458	114417	5382875	5087658	102473	5190131	3.55	11.66	3.71
May	2159261	51276	2210537	2584850	53046	2637896	-16.46	-3.34	-16.20
June	2254567	38050	2292617	2175542	37489	2213031	3.63	1.50	3.60
July	2922797	68696	2991493	2681950	78280	2760230	8.98	-12.24	8.38
August	3866195	102563	3968758	3281802	98150	3379952	17.81	4.50	17.42
September	8917186	102186	9019372	8402851	92788	8495639	6.12	10.13	6.16
October	3825926	182691	4008617	2676946	156417	2833363	42.92	16.80	41.48
November	3019317	207359	3226676	2911900	205534	3117434	3.69	0.89	3.50
December	3369826	164505	3534331	3016247	156923	3173170	11.72	4.83	11.38
Total	**45916573**	**1609963**	**47526536**	**41495115**	**1513729**	**43008844**	**10.66**	**6.36**	**10.50**

Source: Department of Tourism, GoR, Jaipur and RTDC, Jaipur

Table 20.18 List of Fair-Festivals and Cultural programs organized by the Department (2017)

S.No.	Day, Fair and Festival	Place	Date and Year
1	Camel Festival	Bikaner	13-14 January, 2017
2	Kite Festival	Jaipur	14 January, 2017
3	Shekhawati Handicraft Fair	Jhunjhunu	18-26 January, 2017
4	Sawai Madhopur Foundation Day	Sawai Madhopur	19 January, 2017
5	Vintage Car Rally	Jaipur	25-26 January, 2017
6	Nagaur Fair	Nagaur	1-4 Feburary, 2017
7	Baneshwar Fair	Dungarpur	7-10 Feburary, 2017
8	Maru Festival	Jaisalmer	8-10 Feburary, 2017
9	World Music Festival	Udaipur	10-12 Feburary, 2017
10	Jalor Festival	Jalor	15-17 Feburary, 2017
11	Bharatpur Foundation Day	Bharatpur	19 Feburary, 2017
12	Photo Festival	Jaipur	22-25 Feburary, 2017
13	Brij Holi Festival and Bird Fair	Bharatpur	7-8 March, 2017
14	Dhulandi Festival	Jaipur	13 March, 2017
15	Johar Fair	Chittaurgarh	22-24 March, 2017
16	Rajasthan Day	Jaipur	27-30 March, 2017
17	Gangaur Fair	Jaipur	30-31 March, 2017
18	Mewar Festival	Udaipur	30-31 March and 1 April, 2017
19	Mahavirji Fair	Karauli	9 April, 2017

S.No.	Day, Fair and Festival	Place	Date and Year
20	Nadbai (Bharatpur) Cultural Programme	Bharatpur	19 April, 2017
21	Summer Festival	Mount Abu	9-10 May, 2017
22	Jodhpur Foundation Day	Jodhur	12 May, 2017
23	Prathaviraj Jayanti	Ajmer	23 May, 2017
24	Pratap Fair	Rajsamand	28-30 May, 2017
25	Jagannath Fair	Alwar	3-5 July, 2017
26	Teej Festival	New Delhi	22-26 July, 2017
27	Teej Festival	Jaipur	26-27 July, 2017
28	Meera Mahotsav	Chittourgarh	29 July to 5 August, 2017
29	Kajjali Teej	Bundi	9-10 August, 2017
30	Kaman Festival	Bharatpur	15-16 August, 2017
31	Maharaja Daharsen Jayanti	Ajmer	24 August, 2017
32	Gogamedi Fair and Cultural Programme	Hanumangarh	30 August, 2017
33	Diggi Fair	Tonk	31 August to 2 September, 2017
34	Abhaneri Festival	Dausa	21-23 September, 2017
35	Dol Fair and Cultural Programme	Baran	27 September, 2017
36	Kota Dussehra Fair	Kota	30 September to 17 October, 2017
37	Marwar Festival	Jodhpur	4-5 October, 2017
38	Tourism Festival	Rajasthan	5-25 October, 2017
39	Saarc Sufi Festival	Rajasthan	6-9 October, 2017
40	Khalkhani Mata Fair	Jaipur	7-10 October, 2017
41	Bharat Festival	Delhi	23-25 October, 2017
42	Pushkar Fair	Ajmer	28 October to 4 November, 2017
43	Dungarpur/Vagar Foundain Day	Dungarpur	29-31 October, 2017
44	Kolayat Fair	Bikaner	2-4 November, 2017
45	Chandrabhaga Fair	Jhalawar	3-5 November, 2017
46	Bundi Festival	Bundi	6-8 November, 2017
47	Lake Festival	Udaipur	18-20 November, 2017
48	IITF Rajasthan Day	New Delhi	20 November, 2017
49	Matasya Festival	Alwar	25-26 November, 2017
50	Kumbhalgarh Festival	Rajsamand	1-3 December, 2017
51	Mangarh Dham Fair	Banswara	2-3 December, 2017
52	Music in the Park	Jaipur	Jan. to May, 17 and Oct. to December
53	Bhakti Festival	Pushkar	16-17 December, 2017
54	Ranakpur Festival	Pali	21-22 December, 2017
55	Winter Festival	Mount Abu	29-30 December, 2017
56	Bhilwara Festival	Bhilwara	24-26 December, 2017

Source: Department of Tourism, GoR, Jaipur and RTDC, Jaipur

CULTURAL HERITAGE

World is gifted with different civilisations, each one of them having different identity of its culture, which depicts the values, faiths, beliefs of people, their way of life, social behaviour, creativity, literature, arts, sculputre, architecture, music, religions, Gods and Goddesses. India is proud of its rich culture which has persistently retained its identity, despite various aggressions and being ruled by other people. Ours, being a vast country, has a lot of diversity inspite of common cultural links and values. There is unity in diversity. Different states in the country have distinct characteristics, Rajasthan is also in no way, exception to it.

Culture of Rajasthan can be traced back to the pre-historic age. In this region Rigveda was composed in the valley where Sarasvati-Drishadvati once flowed. Reference to Tirtharaj Pushkar in the Epic reveals the development of a creative cultural environment that existed here. Mythological allusions thread an uninterrupted cultural heritage that is discernible even today in the rituals associated with life-cycle of the rural and the tribal folk. This age is supposed to have initiated polyandry, elaborate ritualistic worship of Gods and Goddesses, belief in rebirth and vindication of the significance of penance in religious practice. Besides these influences, the dominance of the priesthood in the social fabric can be traced even in the Reformation Movement in Rajasthan (Figure 20.13).

ANCIENT CULTURAL LINKS

Archaeological Excavations and Explorations

Recent archeaological excavation at different sites in Rajasthan further strengthen the claims of the state's richess in ancient cultural heritage. 'Hand axes' found at Jaipur and Indergarh (preserved in Calcutta Museum) relate to the stone age and further evidences found in Jhalawar consolidate this claim. The second link is provided by equipments like 'scraper' and 'point' found in the basin of river Berach in Chittaurgarh. Later, stone age equipments found at Bagore (Udaipur) and Tilwara (Marwar) provide links with the Later Stone Age.

Excavations at Ganeshwar (Sikar), Gilund (Rajsamand), Ahar (Udaipur), and Kalibangan (Ganganagar) provide ample evidences of Rajasthan being an important cultural centre around 2,800 B.C.

Excavation at Kalibangan reveal a Pre-Harappan civilization, rich in agriculture with 'grid-pattem' ploughing which yielded two crops. Remains of town planning and architecture, with square, well founded houses, bricks and earthen pots, whitewash and the patterns on the pottery, help to piece together the picture of a developed civilization and culture. Several centres of culture have also been found along the basins of the Bearch and Banas rivers.

Explorations at Kurada (Nagaur) and Mallah (Bharatpur) establish cultural links with the West Asia. A meter and a half deep deposit of copper Age pottery was found at Jodhpura (Jaipur district). Besides, roofs covered with tiles, floors and brick walls have been discovered. A dish-on-stand has been unearthed from the same site. Noh (Bharatpur) revealed evidences of the Iron Age. Other sites offering similar evidences are Viratnagar (Jaipur) and Sunari (Jhunjhunun). Excavations have been carried out at Rairh (Tonk), Sambhar (Jaipur), Nagar (Tonk), Naenwa-Nagari (Chittaurgarh) and Bhinmal (Jalor), Nadol (Pali), Balathal (Udaipur), Rangmahal (Ganganagar), Jayal (Nagaur) reveal the cultural links of Rajsathan with the rest of India.

Sculptures

Proto-historic clay and metal sculptures, found in Kalibangan excavations, are the most ancient evidences of the art in Rajasthan. Icons of the deities of different sects of Shung Period (184-72 B.C.), Shak-kushan Period (1st century B.C. 3rd A.D.) and then Gupta period (3rd A.D. 600 A.D.), mark a linear development in

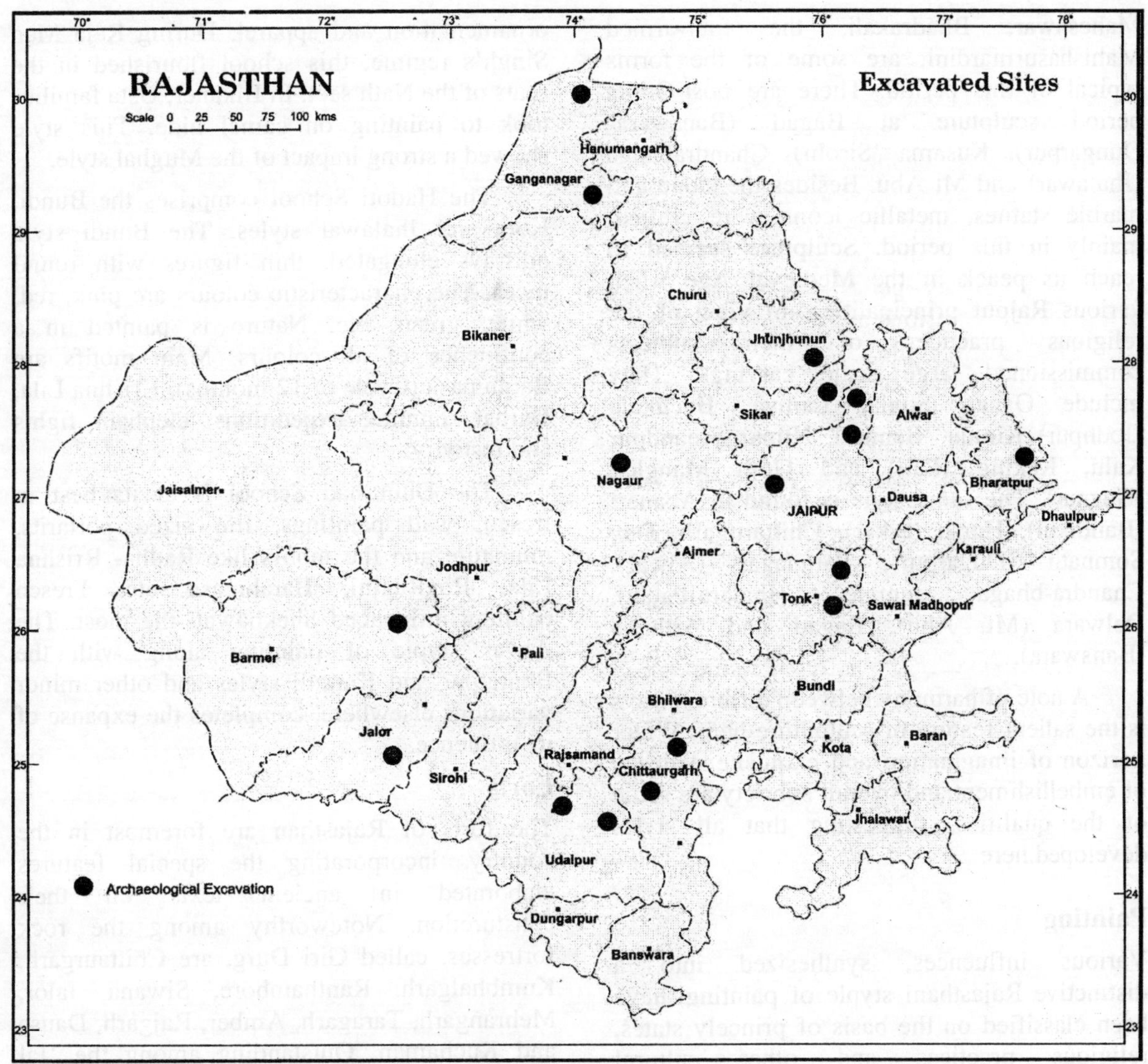

Source: Department of Archological and Museums, GoR, Jaipur

Figure 20.13 Excaveted Sites in Rajasthan

this art. The 12,h century Samaddhiswar temple and Kumbh Meru or the victory tower built during the reign of Maharana Kumbha (1490-1525) at Chittaur have the peculiar sculpture of the Mandans. The Osian temples of8th, 10"'andthe 12th centuries reveal the peculiarities of Mahamaru style.

Suculptures at Gamadi (Bharatpur), Nand (Ajmer), Malavnagar (Tonk) and Hawa Mahal (Jaipur) reveal different forms of Lord Shiva in different museums ir. Rajasthan.

Vaishnava sculpture at Nagar (Malavnagar, Tonk) and Sambhar (Jaipur) is made of Keolin. The icons depict Vishnu, Akanesha, Vasudev and Balram.

Buddhist sculpture are found at Bairath (Jaipur), Lalsot 'Dausa) and Noh (Bharatpur). The cultural grandeur of the Gupta leriod has been expressed at Mukandara and Charchoma in Hadoti, Nagari (Chittaur), Gangdhar (Jhalawar), Mandore 'Jodhpur) and the Saraswati basin (Ganganagar). Uma-

Maheshwar, Bhadrakali, the four-armed Mahishasurmardini, are some of the forms typical of this period. There are post-Gupta period sculpture at Bagad (Banswara-Dungarpur), Kusama 'Sirohi), Chandrabhaga (Jhalawar) and Mt Abu. Besides the stone and marble statues, metallic icons were crafted, mainly in this period. Sculpture seemed to reach its peack in the Medieval Age when various Rajput principalities, in view of the religious practices of their subjects commissioned large scale carvings. They include Osian, Soyla, Lamya, Buchkala (Jodhpur) Kiradu 'Banner), Nimaj, Anandpur Kalu, Keking (Pali) and Goth Manglod (Nagaur). The same style is found in Abaneri (Bandikui), Harsh (Sikar), Chittarugarh, Dev Somnath (Dugarpur), Paranagar (Alwar), Chandra-bhaga temple (Jhalara-Patan), Delwara (Mt. Abu), Bagad and Arthuna (Banswara).

A note of harmony between different sects is the salient feature of sculputure here. Wider horizon of imagination and exquisite delicacy of embellishment and overall sobriety are some of the qualities, suggesting that all styles developed here.

Painting

Various influences, synthesized into a distinctive Rajasthani styple of painting, have been classified on the basis of princely states, religious practices and other cultural organisations.

The Mewar School of painting, an amalgamation of the Jain, Apbhransha and Malava styles, had got the name Chavanad from the capital of Mewar during Maharana Pratap's time and thus came to be classified as the Udaipur school. The most important form of this school is the Nathadwara style.

The Marwar School of painting includes Jodhpur; Bikaner and Kishangarh styles with minor variations in Jaisalmer, Nagaur, Ghanerao and Ajmer. Paintings based on the hymns of Soorsagar and Rangpriya reveal impressive colour schemes, depiction of posh ornamentation and apparel. During Raja Man Singh's regime, this school flourished in the seats of the Nath sect. In Bikaner, Usta families took to painting on camel hide. This style showed a strong impact of the Mughal style.

The Hadoti School comprises the Bundi, Kota and Jhalawar styles. The Bundi style portrays elongated, thin figures with round faces. The characteristic colours are pink, red, white, green, etc. Nature is painted in a confluence of all colours. Main motifs are Barahmasa (Cycle of 12 months), Krishna Lila, Darbar, hunting expeditions, elephant fights and festivity.

The Dhunrhad School is at its best in fresco, wall paintings, life sized portarits, miniature and the motifs like Radha- Rrishna Lila, RagRagini, Barahmasa etc. Fresco painting influenced Shekhawati the most. The Alwar styple of painting along with the Bharatpur and Karauli styles and other minor variations elsewhere, completes the expanse of its influence.

Forts

The forts of Rajasthan are foremost in the country, incorporating the special features elaborated in ancient texts on their consturction. Noteworthy among the rock fortresses, called Giri Durg, are Chittaurgarh, Kumbhalgarh, Ranthambore, Siwana, Jalor, Mehrangarh, Taragarh, Amber, Rajgarh, Dausa and Kuchaman. Outstanding among the 'Jal Durgs' is Gagron fort near Jhalawar, built on the confluence of the rivers Ahu and Kali Sindh. Besides, there are some well-known forts constructed in plains the forts of Jaisalmer,'Junagarh (Bikaner), Lohagarh (Bharatpur). It is not only the architectural beauty, the imposing structure and reputed invicibility of the forts, but also the legends of chivalry associated with them that attract the tourists (Figure 20.14).

Temples

Alongwith Hinduism, Jainism has also remained the backbone of the culture of Rajasthan. Jainism flourished here towards the

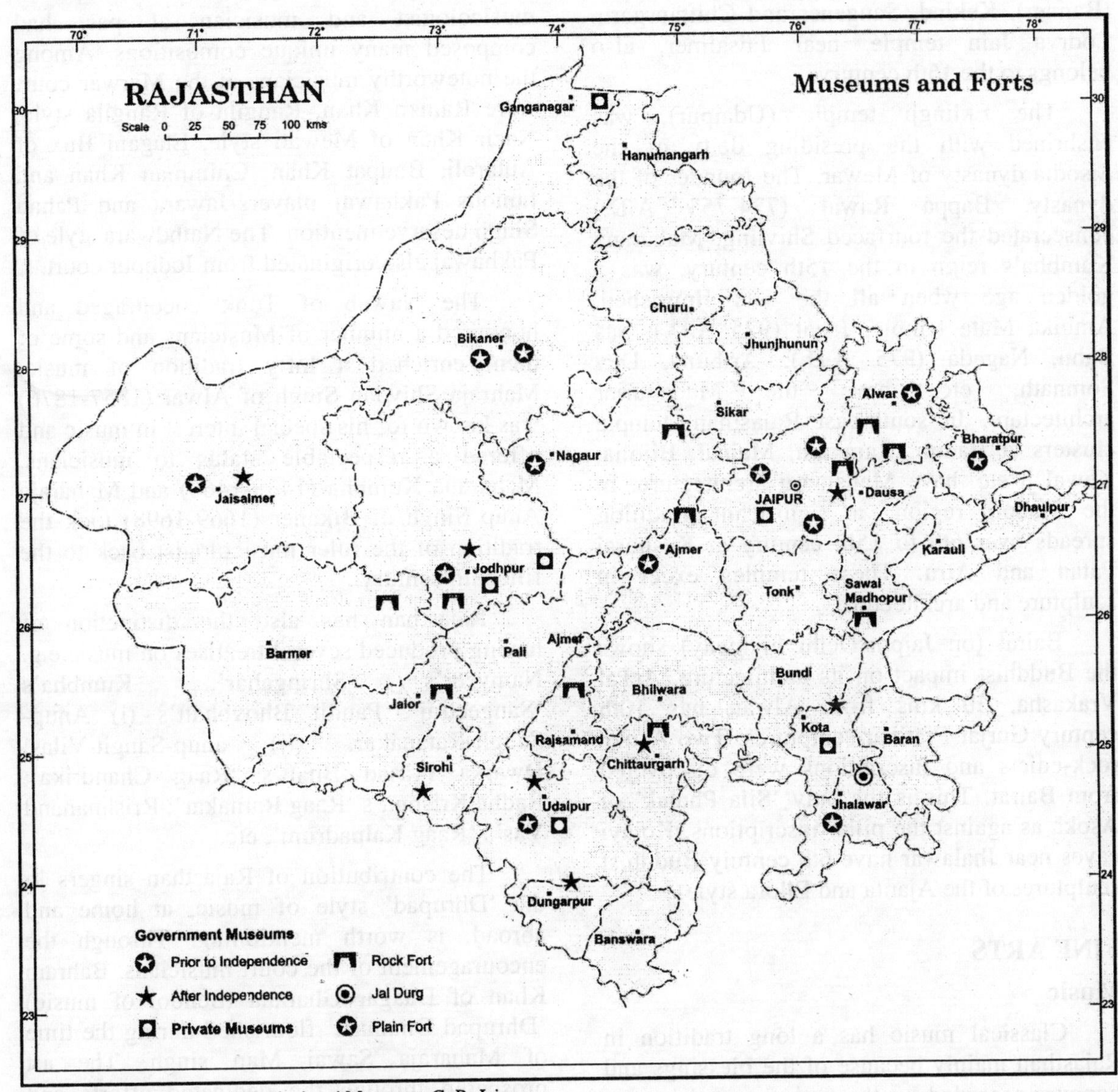

Source: Department of Archological and Museums, GoR, Jaipur

Figure 20.14 Museums and Forts in Rjasthan

close of the 8th century and the oldest Jain temple of Mahaveerji, in Osian (Jodhpur District), was built during the reign of the Pratihar King Vatsraj (783-785 A.D.). Other Jain temples in western Raj asthan, built between 81'1 and 12th Century are Mahaveer ji temple, Ghanerao; Parasvanath temple, Sadri; Navlakha Parasvanath temple, Pali, Adinath temple, Narlai and Nadol (Pali District). Besides, there are important Jain temples in Bhinmal, Jalor, Sirohi and Mt. Abu.

Delwara Jain temples of Mt. Abu remain the most exquisite examples of stone (marble) carving. Adinath temple (1031 A.D.) built by Vimalshah and the Neminath temple (1131 A.D.) built by Vastupal-Tejpal, have the rarest and the finest marble carvings, revealing an astounding aesthetic sense and breathtaking beauty. More Jain temples in Aravalli range are found at Ranakpur, Barkana (Pali), Nakoda

(Banner), Kekind, Sanganer and Chittaurgarh. Lodrva Jain temple, near Jaisalmer, also belongs to the 15th century.

The Eklingji temple (Udaipur), was enshrined with the presiding deity of the Sisodia dynasty of Mewar. The founder of the dynasty, Bappa Rawal (734-753 A.D.) consecrated the fourfaced Shivling. Maharana Kumbha's reign in the 15th century, was a golden age when all the arts flourished. Ambika Mata temple, Jagat (925 A.D.); Sas Bahu, Nagada (975 A.D.); Arthuna, Dev Somnath, etc. have the Mahaguriar architecture. In south-east Rajasthan, temple clusters at Badoli, Ramgarh, Mainal, Bijolia, Kawal ji etc. have Mahamaru architecture. In the Hadoti region, a significant tradition spreads over 6th to 15lh century at Keshorai Patan and Atru. These temples excel in sculpture and architecture.

Bairat (on Jaipur-Delhi highway) shows the Buddhist impact on its architecture. AtTal Vrakasha, 30 kms from Alwar, has 10th century Gurjar-Pratihar sculptures. Two Asoka rock-edicts and inscriptions were discovered from Bairat. This is the only 'Sila Phalak' of Asoka as against the pillar inscriptions. Kolavi caves near Jhalawar have 6th centuiy Buddhist sculptures of the Ajanta and Ellora style.

FINE ARTS

Music

Classical music has a long tradition in Rajasthan mainly because of the blessings and honours accorded by the various royal houses of Jaipur, Jodhpur, Tonk and Bikaner from time to time. Maharaja Pratap Singh of Jaipur (1779-1803) was himself a great musicologist. Ustad Chand Khan, his court musician, wrote 'Swar Saagar'. Later musicians in Gunijan Khana were Amrit Sen (Sitar), Mubark Ali, Behram Khan, Munammad Ali Khan, Amir Baksh, Munna Khan and Rajjab Khan. Several musicians from Uttar Pradesh also sought patronage in Jaipur. The rulers of Jodhpur patronised musicians for centuries together. Man Singh (1803-43) Who himself was a musicologist and musician at par, had composed many unique comositions. Among the noteworthy musicians at the Marwar court were Ramzn Khan, Rangila of Rangila style, Nazir Khan of Mewati style, Elagani Bux of Autaroli, Bhupat Khan, Chimman Khan and famous Pakhawaj players-Jawara and Pahad Singh deserve mention. The Nathdwara style of Pakhawaj also originated from Jodhpur court.

The Nawab of Tonk encouraged and honoured a number of Musicians and some of them enriched a lofty tradition of music. Mahraja Shivdan Singh of Alwar (1857-1876) was known for his special interest in music and offered a respectable status to musicians. Maharana Kumbha (1433-1468) and Maharaja Anup Singh of Bikaner (1669-1698) took the tradition of the ruler musicologist back to the fifteenth century.

Rajasthan has also the distinction of having produced several treatises on music e.g. Namveer's Shringahar', Kumbha's 'Sangeetraj', Pandit Bhovbhatt's (i) Anup-Sangit-Ratanakar, (ii) anup-Sangit-Vilas; Dwarka Prasad Bhatt's 'Raag Chandrika'; Radha Krishna's 'Raag Ratnakar'; Rrishnanand Vas's 'Raag Kalpadrum', etc.

The contribution of Rajasthan singers in the 'Dhrupad' style of music, at home and abroad, is worth mentioning. Through the encouragement of the court musicians, Bahram Khan of Daagar Gharana (School of music) 'Dhrupad Dhamars' flourished during the time of Maharaja Sawai Man singh. The art prospered through the dedicated efforts and innovative variations of several Gharanas of Music like the Allahdiyan Khan Gharana, Sainiya Gharana, Agra Gharana, Rangila Gharana, Qawaal Bachhon Gharana and others.

The melodious Raga 'Maand', originally a folk tune of Jodhpur, of infinite sonorous significance, is today sung and played by all music masters of classical compositon. Allahillai Bai of Bikaner and Govri Devi of Jodhpur each, have added charm of a freshness and individuality in their exposition of Maand.

In Rajasthan, Dholi, Damami, Nagarchi, Langa, Manganiar, Dhadi, Kalawant, Bhaat, Jaga, Rao, Jogi, Kamad, Vairagi, Gandharva, Bhopa, Bhavai, Rana, Doom, etc. are castes of professional 'Musicians and singers'.

In the temples of Nathdwara, Kankroli, Kishangarh, Jaipur, Jodhpur, Kota etc. 'Haveli Sangeet', with all its traditional distinctive featurs, is also still alive.The couplets of devotees have been set to diferent Rags according to time theory of music.

PERFORMING ARTS

Drama

The development of modem drama in Rajasthan runs parallel to the trends at the national and international levels. Nineteenth century marks a beginning in all these cases. Professional and progressive theatres flourished between 1879 and 1895. During the reign of Sawai Ram singh II (1800-1835), Ram Prakash Theatre was built in Jaipur. Persian theatre style was in vogue and the playwrights like Husseini Mian, 'Zarif Munshi, Nanak Prasad Luckhnavi, Hafiz Mohammed Abdula Fatehpuri, Mirza Nazir Beg Akbarabadi and 'Raunak Banarasi' are well known for popularising theatre. Rana Nihal Singh Lokendra of Dhaulpur patronished Indian Imperial Theatrical Company during his rule (1873-1901). Maharana Sir Bhawani Singh of Jhalawar (1886-1929) was a great theatre enthusiast. He established the Shakespear's Dramatic Society, translated five of Shakepearean plays and patronised regular performances. His successor built 'Bhawani Natyashala (a theatre in memory of his father). Maharaja Kishan Singh of Bharatpur commissioned a theatrical company, 'The Bharatpur Dramatic Society' in the last decade of the nineteenth century. In Tonk, Nawab Ibrahaimm Ali Khan and then Nawab Saadat Ali Khan (1930-1947), patronised dramatic activity. In Bundi and Kota states, too, drama flourished. During the reign of Mahraja Jaswant Singh (1873-95), the Rajputana Malwa Theatrical company initiated dramatic activities which gained momentum later.

Dance

Rajasthan has a long histoiy and tradition in the field of dance. Kathak, the most popular dance form, even today, can be traced to have its origin in Rajasthan. Nayaks in the Vaishnav and the Bhojak in the Jain temples practised this art. The Mughal emperors in Delhi and other places, had a number of artists in their courts during 16lh to the 18"' centuries. Jaipur state alone patronised a large number of dancers. Jaipur Gharana, particularly that of Sanwal Das, made a significant contribution in develop-ment of this dance form. It represents the pure Indian tradition while the experts and scholars believe that the other Gharanas have imbibed Mughal and other influences.

The Ghoomar is popular all over Rajasthan. There are other regional folk dances too : the Gindar and Chang nritya of Shekhawati, the Gair and Loor of Marwar-Mewar, the 'Dandia' of Marwar, the Dhol nritya of Jalor, the Agni nritya of Jasnathis of Bikaner and the Bum nritya of Alwar-Bharatpur. The Chari dance, Bhawai, Teratali and Kachchi ghori are the enchanting dances of some of the social groups.

FOLK ARTS

FolkTheatre

Various folk theatre styles with their regional peculiarities form a vital cultural force that pulsates with a live tradition, entertaining masses, even in the midst of urbanising trends.Turra-Kalagi of Chittaur, Rammats of Bikaner and Jaisalmer, Rasdhari of Mewar and Marwar, Tamasha of Jaipur Bhatt Brahmins, Nautanki of Bharatpur-Kama, Gavari of the Bheel tribe in the Aravalis, Khyal styles of Kishangarhi, Chidavi, Shekhawati, Hathrasi, Nagauri, Mewari and Kuchamamani, Allabakshi Khyal of Alwar and Ramlila of Kota and Jhalawar have all been developed in theatrical styles, with an uninterrupted tradition of about three centuries.

Gopiji Bhatt is veteran player of Tamasha style. In the folk theatre forms, male actors play female roles, improvise contemporary comments and create biting satires in these forms which are very popular with the masses.

Puppetry

It is believed by great connoisseurs of folk art that the nats acrobat nomads took up the art of puppetry about four or five centuries back from the Mughal Puppeteers. They are classified as nomads, camping in different towns of northern India and travelling as far as Burma, In Rajasthan, some families have settled in Nagaur and Chum districts. About 5,000 puppeteers play the well-knit legendary tale of Amar Singh Rathore all over the country. Some puppeteers have developed new programmes to meet the requirements of national policies and programmes under the auspices of the Union Ministry of Information and Broadcasting, Envrionment, Family Welfare etc. Some selected groups have also given performances abroad. Innovative experiments through workshops are on the cards.

Pad

Singing devotional and legendary themes with simultaneous showing of and explaining of a scroll by Bhopa and Bhopi is a great popular folk tradition. Legends of the Bagdawat and Pabuji are sung, following the depiction of the tale on the printed cloth spread at the spot. These scrolls have invoked an interest in the inland and foreign tourists alike, to make them a part of interior decoration. The tradition is still patronised by the rural folk.

Folk Music and Instruments

Folk music of Rajasthan tmely represents its culture. Folk songs are commonly related to rituals, social ceremonies, festivals, seasons, Gods and Goddesses, love legends and other occasions, sung to melodious traditional tunes. The large tradition of flok songs is available with the women folk of Rajasthan. Every caste has its own lore of songs. These songs are sung during life-passage ceremonies, festivals and seasons. The men folk of eastern Rajasthan have contributed to folk songs. Devotional folk songs are sung in every village. Langa, Monganiyar, Dholi, Mirasi Damami, Jogi, Kalawant, Bhaat, Rao, Kanad Vairagi, Dhadhi etc. are castes of performing artistes who have developed special styles over centuries with their own musical instruments.

Folk instruments devised on simple acoustic principles of sound, reveal the ingenuity of the creators. Wide variety of the instruments have been adopted by different professional caste groups. These can conveniently be grouped as under:

1. Membranophonic Instrument - Mridang, Dholak, Nagara, Dhol or Drum, Nagara, Nobat, Madal, Chang, Khanjari, Duff, Bum, Damama, Dhoonsa, Bhen, Kamat, Deru, Dhak.
2. Idiophonic Insrtruments - Manjira, Tali, Janjh, Thali, Khartal Ghante, Chimta, Jhaler, Ghunghru, Morchary, Lsindhi, Gujrala, Jogia.
3. Chorodophonic Instruments - Sarangi, Jantar, Rvavan Hattha, Ravaz, Tandara, Iktara, Chikara, Kendra, Kamaicha, Surinda, Surindi.
4. Aerophonic Instruments - Bansuri, Algoja, Peti, Pungi, Murali, Shahnai, Nafiri, Tots, Satara, Narh, Pavari, Tarpo, Mashak.

Membranophonic instruments include special folk Mridang played by the Ravalia artistes. Dholk is the most popular accompaniment to devotional and stage music. The Dhol or the drum has variations of shape and size and the style of playing according to the ethnic groups which traditionally play them. The Nagara is an accompaniment for dance and theatre. The Nobat is a festive rhythm instrument. The male and female drums are played with sticks of sheesam or babul. The Madal is a flok instrument like Mridang, which is made of clay instead of wood. The Chang is a poplar frame drocm instrument associated with the festival of colours of Holi, groups of revellers sing in chours to its rhythm. The Khanjari, made of mango wood, mounted with leather, is played by nomadic Nath tribes.

Idiophonic instruments like Cymbals (Manjira), Janjh, Thali and Khartal are played for devotional music sung by the pople.

The string (Chordophonic) instruments are very refined and sophisticated Kamaicha, Sindhi and Gujaratan sarangis are two varieties in use. Janta is a primitive of the Veena. Ravan Hattha is the most popular string instrument played by flok and tribal musicians, Ravaz is played with the help of gajand strokes of nails Tandvra, Chautara and Nishan are the different names of the same Tanpura-a-type instrument. Iktara is made out of pumpkin or gourd and Sadhus play it with devotiona songs. Bhapang is made out of bottle-gourd with string tied to a key.

Source: Department of Archological and Museums, GoR, Jaipur

Figure 20.15 Akademis and Cultural Organisation in Rajasthan

With the adjustment of the key, different nots are played. Kamayicha is like Sarangi but with a large round blly, mounted with leather to achieve resonance. It is played with bow.

Aerophonic instruments cover-Bansuri (flute), Algoja, Pungi, Shanai, Satara etc. Morchang (the Jew's harp) is an iron tong presged between the teeth and manoeuverad by striking a wire. Akademis and Cultural Organisations (Figure 20.15).

Various Akademis in Rajasthan are established with the main objectives of preservation, promotion and propogation of literature/art. These Akademis are registered autonomous bodies. Most of the language Akademis are under the Ministry of Education while Rajasthan Sangeet Natak Akademi is supervised and aided by the Minsitry of Culture.

1. Rajasthan Sahitya Akademi, with its headquarters at Udaipur, publishes creative and critical literature, recognises literary institutions and publication for support, presents awards to outstanding creative writers every year and publishes monographs of renowned writers.
2. Rajasthan Sangeet Natak Akademi came into existence for enrichment of music, dance and drama.
3. Lalit Kala Akademi, Jaipur, promotes all kinds of painting experiments by organising camps, workshops, exhibitions and by providing grants to organisations and individual painters.
4. Rajasthani Bhasha Sahitya Evam Sanskrit Akademi, Bikaner was established for upliftment of Rajasthani literature. Sanskrit Akademi was established to preserve and promote Sanskrit literature.
5. Arabic and Persian Research Centre, Tonk was established to preserve valuable manuscripts and to conduct research in Arabic and Persian literature.
6. Urdu Akademi was established to promote creative writing in Urdu.
7. Brij Bhasha Akademi came up to publish, promote and evaluate literature wirtten in Brij Bhasha.
8. Sindhi Akademi, Jaipur came into being to publish and promote literature and performing arts in Sindhi.
9. Kathak Kendra, Jaipur promotes kathak dance of Jaipur style.
10. Rajasthan Sangeet Sansthan, Jaipur, aims at imparting training in classical music and dance.
11. Rajasthan School of Arts, Jaipur gives training to upcoming artists in the field of painting etc.
12. Rajasthan Oriental Research Institute, Jodhpur came into existence for preservation and publication of old manuscripts.
13. West Zone Cultural Centres, Udaipur is one of the seven cultural centres which were established in different zones of the countiy, to present to the masses, an integrated image of Indian culture, was established in 1986. It comprises Rajasthan, Gujarat, Maharashtra, Goa, Daman and Diu as member states. Shilpa Gram has been developed near Udaipur as a cultural complex with representative settings of the member states.
14. Jawahar Kala Kendra, Jaipur has a beautiful sprawling cultural campus, designed by Charles Correa, is situated at Jaipur and has an auditorium, exhibition gallery, performing spaces, library, Shilp Gram, studios and the infrastructure required to document, preserve and promote visual and performing arts.

Chapter 21

INDURISTAL DEVELOPMENT

Prior to independence, the country was divided into a number of feudal units. During the princely administration, the programme of industrial development was on low priority. Consequently, after independence and with the integration of the princely states, Rajasthan had to start with almost a negligible industrial base and infrastructure.

The state is gifted with many natural resources. It has a profusely rich mineral base, large livestock population, agriculture production and forest wealth, along with people having entrepreneurial capability and business acumen. The entrepreneurial capabilities of the state have been accepted throughout the country. There were only 13 large scale units and 207 registered factories in 1949. Subsequently, industrial develop-ment in all sectors seemed to have gained momentum. Many fields of activity recorded a significant growth, e.g. mineral processing, textiles, engineering, plastics, chemicals and electronics. Various supportive measures, extended by the government, in the form of proper development of infrastructure, fiscal incentives, marketing support and the inherent availability of natural wealth in the state set the movement towards industrialization.

Large and Medium Scale Industries

There are 657 medium and large scale industries in the state out of which 389 units are presently operative. The Rajasthan Industrial Development and Investment Corporation Limited (RIICO) has financed 913 units, with an assistance of Rs. 1820.75 crores.

Today, there are more than 500 large and medium scale industrial units in the states. During 2016-17 registered Micro, Small and Medium industries units were 5.53 lakh, with investment level of about Rs. 44546.2 crore in them (Figure 21.1).

Predominant sectors among the medium and large scale industries are Agro, textiles, mineral processing, engineering, chemical and electronics.

Textile Industry

Rajasthan has a dynamic textile sector with the most modem technology in spinning, weaving and processing. The state has 590 working textile mills, which produce cotton, synthetic and woolen yam and textile. There are 16 process houses, processing mostly fabrics made out of man made fibre.

Cotton textile industry occupies an important place in large scale industries in Rajasthan At the time of the formation of Greater Rajasthan in 1949, there were 7 cotton textile mills in the state. In 2001 their number was 23, out of which 17 were in private sector, 3 in public sector (two at Beawar and one at Vijainagar) and 3 in cooperative sector (at Gulabpura, Gangapur and Hanumangarh). Cotton textile mills are located at Beawar (3), Bhilwara (3), Jaipur (2), Kishangarh (2), Udaipur, Pali, Gangapur (Bhilwara), Hanumangarh, Kota, Bhawani-mandi, Vijainagar, Ganganagar, Gulabpura (Bhilwara), and other centres.

The first cotton textile mill—'The Krishna Mills Ltd.' was set up at Beawar in 1889 in the private sector. The second mill—'Edward Mills Ltd.' was also set up in 1906 at Beawar. Third mill—'Maha Laxmi Mills Ltd.' was set up in 1925 in the state. Thereafter in 1938, Mewar Textile Mills was established at

Bhilwara, and in 1972 Maharaja Umaid Mills was started at Pali. In 1946 'Sardul Textile Mills' was started at Ganganagar. Later on, due to sickness Krishna Mills and Edward Mills were taken over by the National Textile Corporation (NTC), and thus they were transferred from the private sector to the public sector.

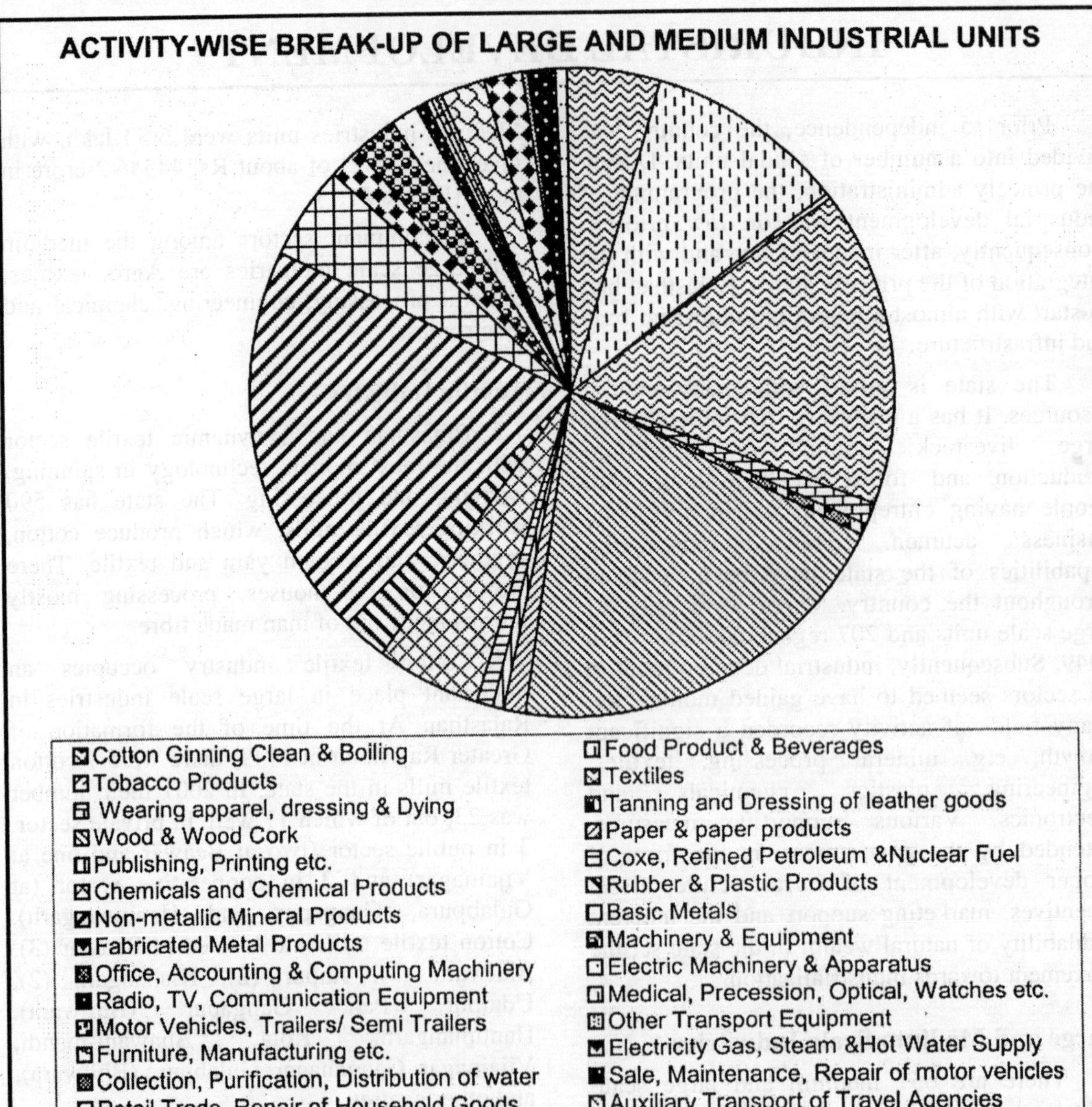

Source: Department of Industris and Commissionerate of Industries, GoR, Jaipur

Figure 21.1 Activitywise Breakup of Large and Medium Industrial Units

Factors affecting the Localisation of Cotton Textile Industry

Localisation of this industry is influenced more by nearness to market for finished product than by nearness to raw material, i.e., cotton; because cotton can be sent to far-off places, provided other factors are favourable for setting up cotton textile mills. Following factors have influenced the localisation of this industry in Rajasthan:

1. Availability of Raw Material—This has helped in setting up cotton textile mills at Ganganagar, because cotton is available there. Cotton is grown in Ajmer, Bhilwara, Jhalawar, Chittorgarh and Jaipur district as well. In Beawar also, cotton has been made available from inside as well as outside the state.

2. Nearness of the Market and Availability of Labour—Labour is available in the surrounding areas. Market for finished product, i.e., textile cloth, is near the production-centres.

3. Climate, water-supply, availability of land, etc., have also influenced the locali-sation of cotton textile industry in Rajasthan.

4. Coal is imported from outside the state. Electricity is available in production- centres and permission has been given for the setting up of diesel-generating sets also.

The annual production of cotton fluctuates from year to year. In 2011-12 it was 17.3 lakh bales, while it was 8,6 lakh bales in 2010-11. (each bale of 170 kg.)

The production of all kind cotton cloth was 4.28 crore metre in 2015. While the production of cotton yam was 13.26 crore kg. in 2015.

1. Rajasthan Cooperative Spinning Mills Ltd., Gulabpura (Bhilwara)—This was set up in 1965. It purchases cotton from the member-famiers. It participates in activities like ginning, spuming, weaving, dyeing, etc. Its main objective is to arrange remunerative price from cotton growers by selling cotton yam. By amalgamating the three State Cooperative Spinning Mills Ltd. at Gulabpura, Gangapur and Hanumangarh and ginning mills at Gulabpura into Rajasthan Rajya Sahkari and Ginning Mills Federation Ltd., (SPINFED) has come into existence with effect from 1.4.1993. Due to old machinery its total loss had reached a level of Rs. 117.9 crore till December, 2014.

2. Gangapur Cooperative Spinning Mills Ltd.—It was set up in 1981. It is also engaged in ancillary industries for the benefit of members. It was amalga-mated with SPINFED from April 1, 1993.

3. Sriganganagar Cooperative Spin-ning Mills Ltd.—It was set up in 1978. Its office is situated at Hanumangarh junction. It uses the cotton grown in the district and supplies raw material to powerlooms and handlooms. It was running in loss during the recent past. It has also been amalgamated with SPINFED as indicated above from April 1, 1993.

Sugar Industry

For the past several years three sugar mills have been operating in Rajasthan. They are—(1) The Mewar Sugar Mills, Bhopal Sagar (District Chittorgarh), which was set up in 1932, (2) The Ganganagar Sugar Mills Ltd. which was under the possession of Bikaner Industrial Corporation Ltd. in 1945, and later transferred to Sriganganagar Sugar Mills Ltd. as state-enterprise, (3) Shri Kesoraipatan Cooperative Sugar Mills Ltd. was set up in 1965 in Bundi district. Thus, three sugar mills represent three different organisations, viz., private, public and cooperative in this sector. Sugar mills are usually set up near sugarcane producing areas so that the inconvenience of transporting it to long distances may be reduced. The location of sugar mills in Rajasthan has been influenced by the availability of sugarcane in the aforesaid three districts.

1. Sugarcane Production—It fluctu-ates from year to year to a significant extent. In 1977-78 it was 28.3 lakh tonnes, which declined to a level of 5.31 lakh tonnes in 2015-16 (final) and likely to the only 3.65 lakh tonnes in 2016-17 (Advance).

2. Trend in the production of Sugar—In 1978, production of sugar was 41 thousand tonnes, which came down to 9 thousand tonnes in 1988. It increased to 26.3 thousand tonnes in 1993 and declined to 12 thousand tonnes in 1994. In 1999 it was 23.4 thousand tonnes and declined sharply to a low level of 6,448 tonnes in 2003. Thus the production of sugar is highly uncertain in Rajasthan.

The Ganganagar Sugar Mills Ltd.— It was organised as a public enterprise in July 1956. It is operating the following units: Sugar factory, Sriganganagar; Distillery units at Sriganganagar and Atru and some wine-centres at other places for the production of rectified spirit from molasses, licensed-shopkeepers to sell indigenous wine (in tribal areas of Kota and Udaipur) and Hi-tech Glass Factory at Dholpur to manufacture bottles, glass-products, etc. (now a sick unit). Ganganagar Sugar Mills Ltd. has incurred financial losses in the past.

Shri Kesoraipatan Cooperative Sugar Mills Ltd. (Bundi District)— It was set up in 1965 to help the sugarcane farmers of the surrounding areas. Due to continuous losses, crushing of sugar-cane was stopped in 2000-01. It is closed at present. Sugar industry in Rajasthan needs to be reorganised so as to make it more efficient and competi-tive in future. Problems related with the availability of sugarcane, machinery, finance, management, etc., need to be tackled properly in years to come.

Cement Industry

Rajasthan is regarded as a loading state in cement industry. Ample quantity of cement-grade limestone is found here. Gypsum is also available in the state. Coal is imported from outside the state. In 1988 there were 9 large scale cement factories in Rajasthan, viz., ACC Ltd., Lakheri Jaipur Udyog, Sawai Madhopur; Birla Jute, Chittorgarh; Hindustan Sugar, Udaipur, J.K. Cement, Nimbaheda; Manglam Cement, Modak; Straw Products, Sirohi district, Sri Cement, Beawar and Sriram Cement, Sri Ramnagar, Kota. There are several mini-ccment plants at present in Rajasthan.

Production of cement (all types) has increased phenomenally in Rajasthan in recent years. During the year 2014, the level of production was 264.4 lakh tonnes and is likely to be about 277.1 lakh tonnes in 2015. With large reserves of cement-grade limestone available in Rajasthan, the state can increase the production of cement in future to a significant extent.

Rajasthan is emerging on the map of India as a leading cement producer. There are mini-cement plants at Abu Raod, Neem-ka-Thana, Banswara, Hindaun City and Kotpuli. They provide more employment to the people. The location of cement industry has been influenced by the availability of raw material and market-access.

The capacity of cement production has increased in the state in recent years. Cement manufacturers got the benefit of rising cement prices during the recent past.

Salt Industry

Rajasthan possesses natural conditions favourable for the production of salt due to the existence of salt water lakes in the state. The biggest salt water lake is the Sambhar Lake, besides that of Deedwana and Pachpadra lakes. The salt deposits of Sambhar Lake are inexhaustible with large annual production and salt of this lake is also of good quality. Production of salt on small scale is one in private sector in salt units at Falodi, Kuchaman City, Pokaran and Jabdinagar (Navan tehsil in Nagaur district). We shall describe salt-sources first and then the factories based on them.

State Salts, Deedwana— This source is spread over an area of 1910 acres. At present the traditional indigenous people and the department of the state are operating is this area. The rain water is stored in bunds on both sides of his source. After seepage, water goes to salt-production-area. This water is called 'brine'. As sodium sulphate quantity is higher in brine as compared to salt, it is unfit for consumption. Therefore, this source produces

80-85 per cent quantity of non-edible salt, and there-are difficulties in selling it in the market.

State Salts, Pachpadra— It is spread over an area of 32 sq. miles. Here the annual production-capacity is 6 lakh quintals. Pachpadra is situated at a distance of 128 km in south-west of Jodhpur. This area has been under operation since 1964.

Both these sources of salt are looked after by the State Government, while the Government of India manages the production of salt at Sambhar. Sambhar Salts Ltd., a subsidiary company of Hindustan Salts Ltd. is managing it. Sambhar salt is famous for its quality.

The department is developing a new salt source at Jabdinagar near Sambhar. State Enterprises Based on Salt-

Rajasthan State Chemical Works, Deedwana (Sodium Sulphide Factory)— It was set up in 1966. It produces sodium sulphide which is used in leather and tanning industry. Due to some difficulties its production was stopped in September 1988. It could be started again after making the necessary arrangements, in the joint sector. At present this unit is lying closed.

Rajasthan State Chemical Works, Deedwana (Sodium Sulphate Works)—It was set up in 1964 It produces sodium sulphate. It has suffered financial losses in the past.

Rajasthan Government Salt Works, Deedwana—It was set up as a departmental enterprise in 1960. It produces edible, non-edible, industrial and iodised salt. Pure salt is manufactured from sodium sulphate, which is extracted from brine, In September 1981, it was given on lease to M/s. Deedwana Chemical Private Ltd., but there were some disputes and court-proceedings were resorted to in that regard.

Rajasthan Government Salt Works, Pachpadra—This was set up in 1960 Here also edible, non-edible, industrial and iodised salts are manufactured At Pachpadra and Deedwana plants have been set up to manufacture iodised salt Due to lack of iodine, the goiter disease is very common in hilly areas. Iodine-salt is very helpful in making up this deficiency.

The production of salt was 18 lakh tonnes in 2001. The production of Sodium Chloride (salt) was 5.6 lakh kg. in 2005.

The position of salt works in Rajasthan is quite uncertain and management reforms in state enterprises are the need of the hour.

Glass Industry

Apart from sand, several chemicals and coal arc used in the glass industry. Sand stones, silica sand, sodium sulphate, etc., arc available in sufficient quantity for the development of this industry. Limestone is also available in abundance. There are two glass factories at Dholpur. One is Dholpur Glass Works in private sector, whose annual production is 1,000 tonnes, and the second is Hitech Glass Factory under Ganganagar Sugar Mills Ltd. As mentioned earlier, this unit became sick due to lack of availability of good quality coal and its old furnace. Glass industry may be developed at Jaipur, Sawai Madhopur, Bikaner, Bundi and Udaipur in future.

The aforesaid discussion shows that there has been some development of cotton textile, sugar, cement, salt and glass industries in Rajasthan in the past. There are bright prospects of increasing cement production in the state. Other industries need innovative measures to modernize them so as to make them viable and competitive. Under new industrial policy as adopted by the State Government, industrial diversification is taking place in the state. More units are coming up in mineral- based industries, environment-friendly electronics industry, chemical-based industries and agro-based industries. Private investment proposals from home and abroad have been made in recent years. Rajasthan faces a bright industrial scenario in the next decade, provided infrastructural facilities in terms of physical facilities like power, telecommuni-cations, transport, etc., are enhanced and social infrastructure in terms of education and health facilities is made available in new industrial centres.

Mineral based Industries

Rajasthan has large deposits of important minerals and stands next only to Bihar in the country. Among the non metallic minerals predominant in Rajasthan are asbestos, emerald, felspar, garnet, gypsum, limestone, mica, quartz, silica sand, clays and soapstone.

Limestone reserves of the state are estimated to be about million tones. There are 19 large and medium scale portaland cement plants, (3 large scale white cement plants) and 115 mini cement plants in the states. The reserves have sufficient raw materials for establishing more cement plants to meet the country's growing needs. Rajasthan is the leading producer of good quality felspar accounting for 70 per cent of the country's production. It is the second largest producer of silica sand and clay. Despite being the ideal home for glass and ceramic raw materials, there are only 8 medium scale industries based on the these raw materials in the state besides a few SSI units. There is only one large scale unit based on asbestos in the state. This unit manufactures asbestos pressure pipes and is located at Bhilwara, consuming mostly imported asbestos.

The principal metallic minerals are copper, zinc wolframite (tungsten ore) and iron. There are two zinc smelter plants, located at Udaipur and Chanderia (Chittaurgarh) and one copper smelter plant at Khetri (Jhunjhunun).

The state has also inherited large deposits of various building, decorative and ornamental stones. These include marble, dolomite, sandstone, Kota stone, Jodhpur stone, Dhaulpur stone, Karauli stone, Jaismalmer stone, Slate stone and granites. Only a small portion of these deposits have been exploited sofar for industrial use.

The marble reserves of the state are estimated to be over million tonnes and a good number of medium scale industries, producing marble tiles, slabs and flooring tiles are running. Similar is the case with granite based units. Units based on marble and granite in the small scale sector are over in number. Units in medium and SSI sector based on marble, dolomite and granite have further scope for expansion.

The state has almost a monopoly in the country with respect to minerals like lead zinc ore, asbestos, copper, calcite, felsper, clays, gypsum and soapstone. Under the Constitution of India, mineral rights and administration of mining laws are vested with the State Government.

Setting up of Mineral Based Industries

Looking to the availability of Limestone Asbestos, Clay, Quartz, Feldspar, and other economic minerals a large number of lime, asbestos based industries are operating in the State. Pulverizing units to grind quartz, feldspar, soapstone etc., are also operating in the state. The status of the state with regard to existing mineral-based industries is given in Table 21.1.

Table 21.1 Existing Mineral-based Industries

S.No.	Industry	Number
1	Cement Plants(Major)	21
2	Cement Plants(White)	2
3	Copper Smelter	1
4	Fertilizer (Single super phosphate/urea)	8
5	Glazed Tiles and Ceramic	4
6	Granite Cutting and Polishing units	About 800 (Including 8 Gang Saw Plants)
7	Insulators (L.T. and H.T.)	1
8	Lead Zinc Beneficiation Plants	3
9	Marble Cutting and Polishing units	1500 Gang Saw 2000 Tiling Plants
10	Sanitary Ware	1
11	Sulphuric Acid	3
12	Super Zinc Smelter	1
13	Zinc Smelter	1
14	Quartz and Feldspar Grinding and Pulvarizing units	1000
15	Plster of paris	500
16	Gitti Crusher	3000
17	Lignite based Power plants	13
18	Power plant coal based	6
19	Iron Puetisation Plants	1
20	Gas based Power thermal plants	3

Source: Department of Mines and Geology ,Udaipur

A vast mineral resource provides innumerable opportunity for taking up of a variety of mineral-based industries in the state. There are 8869 mineral-based industries located in the state out of which maximum 819 units are situated in Ajmer district followed by Nagaur. 4300 units of stone granite cutting and polishing constitute the largest portion of the mineral-based units. More than 70% of the mineral based units are based on stones and related industries alone. Following table shows the figures of various minerals production, area, sale value and total revenue earned and no. of employment generated in the sector (Table 21.2).

Table 21.2 Various Minerals, Production, Area, Sale Value, Total Revenue and Employment

S. No.	Mineral	Leases (No.)	Area (in Hec.)	Production (Lac'Tons)	Sale Value (Crore'Rs.)	Revenue (Lac'Rs.)	Employment (Nos.)
Metallic Minerals							
1	Cadmium	0	0	0.002	3.989	421.320	0
2	Copper Ore	3	706.75	10.030	200.602	1365.050	1890
3	Iron ore	16	2214.73	29.832	275.830	2202.321	992
4	Lead Zinc (ROM)	8	6581.63	72.825	1724.232	95432.394	4434
5	Lead	0	0.00	1.514	0.000	1837.920	0
6	Zinc	0	0.00	13.988	0.000	0.000	0
7	Silver	0	0.00	0.003	1281.579	9365.309	0
8	Manganese	1	18.90	0.039	0.966	6.330	70
Other Minerals							
9	Asbestos	4	228.80	0	0	0	0

S. No.	Mineral	Leases (No.)	Area (in Hec.)	Production (Lac'Tons)	Sale Value (Crore'Rs.)	Revenue (Lac'Rs.)	Employment (Nos.)
10	Ball Clay	105	5353.75	31.113	149.343	1245.402	700
11	Barytes	2	34.59	0.026	0.155	2.210	25
12	Calcite	43	938.05	0.935	3.393	64.453	640
13	China Clay	204	3840.43	18.812	110.050	657.191	1591
14	Dolomite	18	1707.58	3.678	17.262	231.889	404
15	Emrald	1	46.32	0.000	0.000	0s	0
16	Fire Clay	4	59.99	0.000	0.000	0.340	0
17	Fluorite	6	1084.95	0.000	0.000	0.911	15
18	Garnet (Abr. and Crude)	13	59.53	0.008	0.075	1.960	10
19	Gypsum	59	13645.20	30.878	143.926	3273.838	871
20	Jasper	7	232.88	0.015	0.162	0.840	15
21	Kyanite	3	154.00	0.000	0.000	0.408	5
22	Lignite	7	11470.00	76.857	709.842	4807.633	700
23	Limestone	37	17223.25	566.723	1219.832	36510.977	2961
24	Magnesite	3	14.40	0.000	0.000	0.430	0
25	Mica	14	116.41	0.073	0.918	4.122	106
26	Ochres	80	1553.05	27.460	47.945	514.568	607
27	Phyrophilite	6	44.71	0.083	0.207	2.488	14
28	Quartz	1304	10164.55	17.891	45.450	959.554	6497
29	Felspar	1134	6954.49	16.685	46.385	947.605	6706
30	Rock-Phosphate	5	1998.87	11.176	223.515	4160.256	1155
31	Selenite	3	920.85	0.000	0.000	8.104	1
32	Silica Sand	75	1472.29	9.965	45.239	331.123	1164
33	Siliceous Earth	19	199.71	0.272	1.856	22.469	75
34	Soapstone	206	9353.14	8.096	70.769	1405.057	3087
35	Vermiculite	2	9.33	0.000	0.000	0.100	0
36	Wollastonite	11	126.88	1.906	15.633	209.770	258
	Misc. Income	'0	'0.00	'0.000	'0.000	0.412	'0
	Total	**3403**	**98529.99**	**950.883**	**6339.1558**		**34993**

Source: Department of Mines and Geology, Udaipur

Scope of Setting up of Mineral Based Industries

Looking at the availability of Minerals in the State, there is tremendous scope for setting up of new minerals based industries. Various minerals mined in Rajasthan are being sent to other States in raw form. The State is therefore, deprived of the value addition in minerals as well as employment avenues to rural and backward classes. The endeavor therefore would be to encourage the establishment of mineral based industries in the State. While contributing 4.4% to the State's GDP, the mining sector has affected every segment of life, like improvement in infrastructure, health and medical, education, skill development, providing sustainable livelihood to a large number of people belonging to weaker sections of the society. The mining activity has benefitted in the local area development which is visible in the form of social infrastructure like school buildings, health centers, children playgrounds, availability of drinking water etc. Greenery is an added benefit to society through mining sector. Most of the mine operators plant trees in and around their sites and also at the over-burden areas. The State is proud to possess huge reserves of Lignite, Crude Oil and high quality Gas. It is also renowned for its deposits of Marble, Sand Stone and some unique decorative stones. Mining is not only a major source of employment in the rural and tribal areas of the State, but also a major source of revenue to the Government, playing an important role in the development of the State.

Prospecting and Exploration

The Department of Mines and Geology initiated systematic Mineral Survey and Prospecting activities in the year 1967. They are being continued and cover about 57% (2,00,000 sq.km.) area of Rajasthan under Regional Mineral Survey (RMS); 22,000 sq.km. under Regional Geological Mapping (RGM); 4,500 sq.km. under Detailed Geological Mapping (DGM) and about 5,30,000 meterage of drilling. The planned efforts by the Department resulted in discovering world class mineral deposits like Jhamar Kotra Rock Phosphate, Rampura-Agucha multi metal deposits, Deri-Basantgarh base-metal, etc. Widespread deposits of SMS and Cement grade Limestone, Lignite, Gypsum, etc. have been located by the Department. A number of mineral deposits like Fluorite, Barytes, Magnesite, Wollastonite, Calcite, Granite, Marble, Slate, etc. have also been discovered by the Department. As a result, 3 Smelters, 23 Cement plants, a number of thermal power plants, a float glass manufacturing unit and thousands of mineral processing units and other mineral based industries have been set up in the State. Some notable global companies have also commenced their mining and industrial operations in the State. There is a huge potential for setting up new Cement plants, Lignite based thermal power units, glass-ceramic industries, and fertilizer industries in the State.

Mineral Concession, Production and Revenue

The number of mineral concessions in the State in the year 1960-61 was 5,713 (500 major mineral leases, 806 minor mineral leases and 4,407 quarry licences),which has increased to 33,375 (138 major mineral leases, 15,136 minor mineral leases and 18,249 quarry licences) in 2014-15. There are 74 Prospecting Licences (PL) in operation currently. Mineral production in 1960-61 was 76.12 lakh tonnes (27.29 lakh tonnes from major minerals and 48.83 lakh tonnes from minor minerals), which increased to 5,694 lakh tonnes (958 lakh tonnes from major minerals and 4,736 lakh tones from minor minerals) in 2014-15. The State's revenue from minerals was Rs. 92.4 lakh in 1960-61, which increased to Rs. 3,635 crore in 2014-15. Crude Oil has emerged as an outstanding resource in the State in recent years. It contributed revenue of Rs. 20,843 crore to the State till 2014-15. Petroleum drilling operations are going on in all 4 petroliferous basins i.e. Barmer-Sanchor basin, Jaisalmer basin, Bikaner- Nagaur and Vindhyan Basin. The mining sector provides direct employment to about 7-8 lakh persons and indirect employment to about 22-25 lakh persons in the State. Some of the industries which may come up in the State are given Table 21.3.

Table 21.3 Some of the Industries which may come up in the State

S.No.	Mineral/Metal	Industries
1.	Copper	Wire drawing and alloys, defence, die casting , railway, chemicals.
2.	Lead	White lead and Chrome lead, Storage batteries, pipe industries.
3.	Zinc	Zinc chemicals,dry batteries. Storage batteries, gulvanizing Industries.
4.	Cement grade limestone	Cement
5.	Chemical grade limestone	Calcium ammonium nitrate, Calcium carbide, Precipitated calcium carbonate, Hydrated lime
6.	Rock Phosphate	Single super phosphate, Triple super phosphate, Diammonium phosphate and Phosphoric acid, elemental phosphorous.
7.	China Clay	Ceramics,cosmatics,paint, paper,cement,pottery.
8.	Ball Clay	Ceram ics,pottery,pesticides,insulators sheets, detergent,medicines.
9.	Fire clay	Refractories, sanitary ware,mortas,foundary, crucibles, insecticides.
10.	Calcite	Glazed Tiles.glass and ceramics, textiles,filter paint and mortar.
11.	Quartz and Silica	Boilers and Glass lumps and Fluorescent tubes, TV Glasses, Maruti and tata car glasses, fire glasses
12.	Bentonite and Fullers earth	Pulverizing units, Earth activation units,edible oil industries.
13.	Soapstone	Insecticides, Cosmetics, Micronizing units,kitchen sinks, electronics and insulation.
14.	Gypsum	Plaster of Paris, Gypsum Board,cement industries,fertilizers, conditioners.
15.	Fluorspar	Cryolitic, Hydro fluoric acid,Ceramic,welding electroder,glass industries, artificial teeth
16.	Garnet	Abrasive, Cutting and Polishing,gem industries,finishing glass.
17.	Lignite	Briquetting , Liquid lignite, Power plants,in all other industries as fuel.
18.	Potash	Murates of Potash, fertilizers, explosives, match box.
19.	Granite	Decorative and dimensional stone industries.
20.	Marble	Decorative industries.

Source: Department of Mines and Geology, Udaipur

Prospects of Agriculture based industries

A vast agriculture produce is the encouraging factor for taking up some agro-based industries in the State. State is dominating in production of many crops like bajra, cumin, rape and mustered guar and various spices. This sector also throw plenty of opportunity for export of slew of products which includes spice, Dry roses and rose products, Medicinal plant produce etc. industries that could be set-up in the State and export potential of agriculture sector is discussed in ensuing paragraphs.

Details of various minerals and metals available in the different region as per area production, sale value, revenue generated and total employment is given in the below Table 21.4.

Table 21.4 Prospects of Agricultur based Industries

Crop	Area	Possible Industries
Maize	Udaipur, Dungarpur, Banswara, Chittorgarh, Ajmer, Jhalawar and Bhilwara District	Corn Flake, Starch
Wheat	Ganganagar, Kota, Bundi, Chittor, Bhilwara, Jaipur, Tonk, Alwar, Bharatpur, Sawai Madhopur, Pali and Hanumangarh District	Flour, Noodles, Sapghetti Sooji, Maida, Bread and Biscuits.
Barley	Jaipur, Udaipur, Alwar, Tonk, Ajmer, Bhilwara, Sikar and Nagaur	Alcohol and Processed Food
Soya-been	Kota, Bundi, Jhalawar, Chittorgarh and Baran.	Soya Meal
Mustard	Almost in all districts of the State	Oil and Solvent Extraction.
Cotton	Ganga Nagar, Bhilwara, Ajmer, Nagaur, Pali and Hanumangarh	Cotton seed oil
Guar	Hanumangarh, Churu, Nagaur, Jalore, Sikar, Jhunjhunu, Bikaner, Jaisalmer and Barmer.	Guar Gum and animal food
Moth	Churu, Nagaur, Bikaner, Jodhpur and Barmer	Papad, Bhujila
Gram	Jaipur, Ganga Nagar, Churu, Alwr and Hanumangarh	Pluse, Flour
Spices	Kota, Jaipur, Swaimadhopur, Chittorgarh, Jalore and Barmer	Spices, Spices Olerosin

Source: Department of Industris and Commissionerate of Industries, GoR, Jaipur

Engineering Industries

The state has a wide base of engineering industries. There are about 60 medium and large scale industries in engineering products, which include ball bearings, roller bearings, precision machines, machines tools, precision instruments, railway wagons, truck chasis, tractors, earth moving machinery, scooters, mopeds, mini steel plants, automobile ancillaries, transmission towers, electric meters, LPG gas cylinders, railway coaches etc.

Chemical Industries

Important chemical based product in the state are caustic soda, glyoxal, sulphuric acid, hydrochloric acid, fertilizers, pesticides and nylon.

Rajasthan is an important inland source of common salt. The brine of Sambhar Lake predominantly contains sodium chloride. Edible salt is manufactured by Sambhar Salts Ltd. Besides Sambhar, salt is also manufactured at Pachpadra, Phalodi, Kuchaman, Sujangarh, etc. from the saltish water available there.

Electronic Industries

Rajasthan is a late entrant in electronic industry but is keen to develop it rapidly. The industry is being offered certain incentives over and above those available to other new industries. As per estimates prepared by the Rajasthan State Industrial Development and Investment Corporation Limited, the existing investment in this industry is Rs. 150 crores. More than 170 electronic units have been established. There is a growing inclination of entrepreneurs taking up varied electronic projects.

Small Scale Industries

Small Scale Industries (SSI) has emerged as the most dynamic sector of the country's industrial economy. With the growing spirit of entrepreneurship, financial and technical help given by the government, including various concessions and fiscal incentives, the SSI sector has exhibited an impressive growth.

In the present definition of small scale industry, the investment limit in plant and machinery is Rs. 1 crore. Earlier it was Rs. 60 lakhs, raised to Rs. 3 crore; and again reduced to Rs. 1 crore in December 1999. At the end of 2015-16 the total number of small scale units registered with the State Government was 4.54 lakh, with investment of Rs. 29495.3 crore and employment to 21.03 lakh persons. The data for factory sector units are collected regularly, but the information about non-factory sector units is very scanty in nature. Small scale units are found in different types of industries. They are described below :

Small Scale Industries Based on Agro-products

Under this head are included vegetable oil and vegetable ghee units, gur and khandsari units, cotton ginning and pressing factories, dari and niwar units, etc. Vegetable oil units are found in Jaipur, Bharatpur, Sawai Madhopur, Sri Ganganagar, Kota, Bundi, Ajmer and Pali districts, where oilseeds are produced in sufficient quantity. Some vegetable oil factories are as follows : 'Vir Balak' in Vishwakanna (Jaipur), at Khairthal (Alwar), at Dausa, Niwai, Bharatpur (Engine-brand mustard oil), Gangapur City, Sawai Madhopur, Jalore and other places. Vegetable ghee factories are found in several places, such as in Vishwakarma (Jaipur—'Maharaja Vanaspati'—Premier Vegetable Products:

Ajmer Vanaspati P.V.P. Ltd. in Jhotwara Industrial Area, "Kesri Vanaspati" (At Niwai), Rohitash in Durgapura and some units at Chittorgarh and Bhilwara). There are "daf units related with arhar, moong, urad and moth, etc., and printing of cloth are done at several places. Sugarcane is used for the production of gur and khandsari.

Livestock-based Small Scale Units

Woollen cloth, hides and skins, bones, milk and milk-products come under it. Woollen mills were started at Bikaner, Churu and Ladnu in the small scale sector earlier, but their financial performance has not been satisfactory. Leather factory at Tonk has also been closed down. But units related with milk-products can be more viable in the state.

Mineral-based Small Scale Units

Marble and granite stones are found in various parts of Rajasthan. Processing units have been started to use these resources. Brass and other types of utensils are prepared at Jaipur, Pali, Jodhpur, Bharatpur, Kishangarh, etc. Products of various types based on gold and silver are manufactured at Jaipur. Agricultural implements are made at several places in Rajasthan. An implement called 'cultivator' is manufactured, which is a component for tractor. Agricultural implement-producing units have been set up at Gajsinghpur (Sri Ganganagar) and Jhotwara (Jaipur).

Forest-based Small Scale Units

There are wooden toy-units at Udaipur, Sawai Madhopur and Jodhpur. Bamboo products are also made in these places. There is a straw' board factory at Kota. Bidi-manufacturing is done with the help of tendu leaves. Gum, Kattha, lac and other minor forest products are also important. Furniture units are also found at several locations. Match-box units exist at Ajmer and Alwar.

Thus, several types of resource-based small scale units are found in different parts of Rajasthan.

Starting from a small number of SSI units in 1949, the number of registered small scale industries as on 31 st March, 2007 reached 29,1,305 with an investment of Rs. 5,91,862.58 lakhs and providing employment to 11.51 lakh persons.

The number of powerlooms, are 21,206 (31st March, 2007). There are 15,807 SSI units doing dyeing and printing work while 11,800 units are engaged in ginning and carding, and there are 22 units for sizing cotton yam.

It is worth mentioning that fabric weaving out of man made fibre in the small scale has started very recently and it is the achievement of the last 8 to 10 years only. The power loom sector has shown remarkable progress in Bhilwara, Ajmer, Jaipur, Pali, Sirohi, Udaipur, Alwar and Jodhpur districts. Twenty five years ago, only four units consisting of thirty two power looms were working in Bhilwara for weaving synthetic fabric, though there were a few units of cotton power looms too. Today, Bhilwara has 13,336 registered power looms of which about 13,000 powerlooms work on synthetic yam. Bhilwara has emerged as the number one textile producing centre in the state. In Ajmer district, only cotton power looms, have developed where Kishangarh is already a good cotton power loom centre. Other places, where synthetic fabric is produced is Alwar, Banswara and Udaipur. There are over 20,000 SSI units based on minerals of different varieties. They are marble cutting and polishing units, marble and granite tiles manufacturing, mineral grinding, clay washing, asbestos cement pipes and fittings making, cement products of various types, ornamental stone, carving, plaster of paris, lime manufacturing, hydrated lime, lime products, distemper, building bricks, cement hollow blocks, emery stone wheels, soil pipes etc. manufacturing units.

The development of SSI units in the engineering sector is equally substantial. Foundries with engineering workshops, medium level engineering workshops, welded items and steel furniture units, small workshops, repair shops and fabricated item manufacturing units have been established. There are other units manufacturing ACSR conductors, utensils, flour mills, wire nails, screws, hinges, building fittings, nuts and bolts, construction equipments, masonary tools and equipments etc. which are prominent in this sector.

Cottage, household and village industries, normally, employ family members as workers and skilled workers with a low level of capital investment. Cottage/ household industries are mainly confined to traditional arts and crafts and are not markedly affected by modernization. On the contrary, SSI units are more adaptive, generally upgrading and reflecting modem technological advancement. These, in a sense, have to compete, to some extent, with the more sophisticated organised sector (MSI and LSI). However, some of the SSI units are at a level with those in the organised sector. The cottage and household industries sector is linked with the SSI units. This sector is part and parcel of the SSI sector or as per the new definition, is known as the tiny sector. However, the cottage/ village and household industries may be classified into (i) Handlooms, (ii) Khadi and Village Industries and (iii) Handicraft groups.

(i) Handlooms

According to a survey conducted by the Government of Rajasthan, the scenario of the handloom sector emerged as under:

Handlooms are operating in a threefold framework. Firstly, those enrolled as members of co operative society, their interest and working are looked after by the Apex Body of the co operative societies viz. the Rajasthan Rajya Bunkar Sahkari Sangh Ltd. The interest of individual weavers are looked after by the Rajasthan State Handloom Development Corporation, a state government undertaking. The handlooms, working as SSI units are being taken care of the entrepreneurs themselves.

The Rajasthan State Handloom Development Corporation imparts training to traditional weavers in the modern technique of weaving, modernization of handlooms, apprising the weavers of the fashion conscious consumer market demand, providing proper raw material and marketing assistance, supplying latest designs in weaving, etc. The corporation has several outlets within and outside the state for marketing. The needs of weavers under the cooperative fold are catered by the Apex Society. The entrepreneur has himself to be alive to the weavers need and the market demand.

The popular handloom products of Rajasthan are Kota doria sarees and the mulmul of Mathania. Designs in tie and dye from Jodhpur and Sikar are rare specimen of creativity Alongwith this ıs the style of dyeing and designing fabrics in lahariya and mothada. Rajasthan chunnis and safas of dyed mulmul are popularly used to enhance the ethnic look. Bed sheets, bed covers, towels, turkish towels, bedspreads, woolen blankets, khes, janta cloth, cotton quilts, tapestry, printed cotton cloth (especially that of Bagru, Sanganer and Banner) are the famous products of Rajasthan. In recent years, cotton and polyester blended fabrics, viz.

shirtings, kurta, pyjama and trouser pieces are also being produced on handlooms.

(ii) Khadi and Village Industries

The activities of Khadi and Village Industries (KVI) are monitored by the Rajasthan Khadi and Village Industries Board incorporated in the year 1955. All India Khadi and Village Industries Commission is the highest body in the country, which spells out he various programmes to be adopted for the upliftment of Khadi and Village Industries in the State. The State Board bears the responsibility of planning, organizing and implementing the programmes of the commission. The objective is to enhance avenues of employment in rural areas for the benefit of un employed and under employed persons to augment their income and, thus to give a boost to the village economy. KVI programmes are implemented through registered institutions, co-operative societies and individual artisans and trained entrepreneurs. During the year 2006-07, village industries produced goods worth Rs. 924.25 crores and employment to 1,34,201 persons. One hundred thirteen types of industries are covered under the village industries programme, which include mineral, forest, agriculture, chemicals, engineering, textile and service based industries. Under Khadi, four types of Khadi viz. cotton, woolen, silk and polyester khadi are covered.

The Khadi and Village Industries Commission (KVIC) is a statutory body established by an Act of Parliament (No. 61 of 1956, as amended by act no. 12 of 1987 and Act No. 10 of 2006. In April 1957, it took over the work of former All India Khadi and Village Industries Board. The broad objectives that the Khadi and Village Industries Commission has set before it, are...

- The social objective of providing employment.
- The economic objective of producing saleable articles.
- The wider objective of creating self-reliance amongst the poor and building up of a strong rural community spirit.

Some of the major functions of Khadi and Village Industries Commission are,

- The Khadi and Village Industries Commission is charged with the planning, promotion, organisation and implementation of programs for the development of Khadi and other village industries in the rural areas in coordination with other agencies engaged in rural development wherever necessary.

Its functions also comprise building up of a reserve of raw materials and implements for supply to producers, creation of common service facilities for processing of raw materials as semi-finished goods and provisions of facilities for marketing of Khadi and Village Industries products apart from organization of training of artisans engaged in these industries and encourage-ment of co-operative efforts amongst them. To promote the sale and marketing of khadi and /or products of village industries or handicrafts, the Khadi and Village Industries Commission may forge linkages with established marketing agencies wherever feasible and necessary.

The Khadi and Village Industries Commission is also charged with the responsibility of encouraging and promoting research in the production techniques and equipment employed in the Khadi and Village Industries sector and providing facilities for the study of the problems relating to it, including the use of non-conventional energy and electric power with a view to increasing productivity, eliminating drudgery and otherwise enhancing their competitive capacity and arranging for dissemination of salient results obtained from such research.

- Further, the Khadi and Village Industries Commission is entrusted with the task of providing financial assistance to institutions and individuals for develop-ment and operation of Khadi and village industries and guiding them through supply of designs, prototypes and other technical information.

- In implementing Khadi and Village Industries activities, the Khadi and Village Industries Commission may take such steps as to ensure genuineness of the products and to set standards of quality and ensure that the products of Khadi and village industries do conform to the standards.

The Khadi and Village Industries Commission may also undertake directly or through other agencies studies concerning the problems of Khadi and/or village industries besides research or establishing pilot projects for the development of Khadi and village industries.

Rajasthan has been famous for its cottage and rural industries and handicrafts and they-occupy an important place in the economy of the state. Usually members of a household work in them, but sometimes work may be arranged through some hired labour also. Gems and jewellery work, dyeing and printing of textiles, carpet-making, etc., are quite important from the point of view of employment, income and export- earning s.

They may give whole time employment or part time employment. They may be conducted in rural areas or in urban areas. Electricity may or may not be used in them. But most of the work is of manual nature.

This is a traditional household industry of Rajasthan. It provides whole time as well as part time employment to the people. Women can also find employment in them to some extent. Cotton and woollen khadi both come under it. At present more than 1.5 lakh persons have got partial or full-time employment in them. Under woollen khadi, Barrfi of Jaisalmer, woollen blankets and some other products of Bikaner, 'khes' of Chomu and 'reji' of some other places are very famous. Profit-range is higher in woollen khadi as compared to cotton khadi.

During 2011-12, the value of Khadi output was about 40.66 crore, out of which cotton khadi was valued at Rs. 19.13 crore, woollen khadi at 13.6 crore, silkan khadi 0.04 crore and poly khadi at Rs. 8.13 crore.

Government subsidises the sale of woollen, cotton and silk khadi every year to promote their sales.

Those who have studied the working of khadi industry quite intensively, have pointed out various problems facing this industry; such as low returns to produccrs of wool, spinners and weavers, who do a lot of hard work; non-payment of even minimum wages to khadi employees, and irregularities in the purchase of dyes, etc. Thus, khadi workers are victims of exploitation, while Khadi Institutes are earning commercial profits to some extent. The management of these institutes should be improved and the interests of workers in the production of khadi should be given greater attention in future.

Rajasthan Khadi and Village Industries board manages the following rural industries also, apart from the sector; ghani oil, gur, kliandsari, hand-made paper, soap from non-edible oils, leather, pottery, bee-keeping (honey) hand-pounding of rice. Leather and ghani oil occupy' higher place in these eight industries from the point of view of value of production and sales. There has been an improvement in village industries in terms of output in the past. They give employment to rural people in off-season as well. These industries have to face marketing problem for their products. The rural artisans face difficulties in getting raw materials at reasonable prices. There are other issues also related with production-technology, training of workers, subsidy, working conditions, finance, etc., which need to be resolved in future.

Khadi and village industries taken together provide jobs to a large number of people, which is more than double the employment in the factory sector The productivity of this sector needs to be increased so as to improve the economic conditions of the people engaged in them. The value of output in khadi and village industries should also be enhanced beyond the present level, for which technology-upgradation and quality-improvement in products would be quite essential (Figure 21.2).

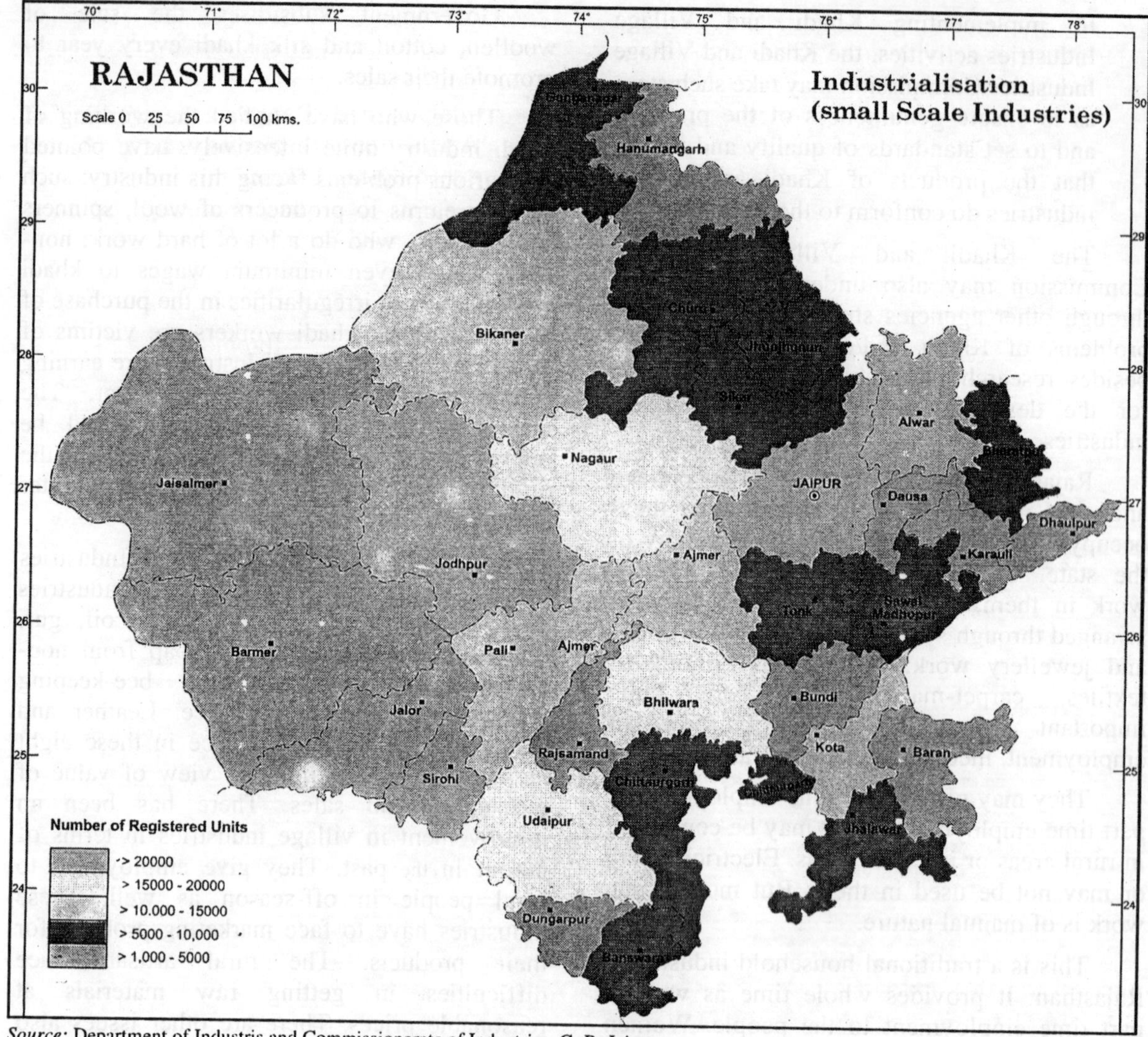

Source: Department of Industris and Commissionerate of Industries, GoR, Jaipur

Figure 21.2 Industrialisation (Small Scale Industries)

(iii) Handicrafts

Rajasthan is famous for its handicrafts in and outside the country. The patronage to artisans and craftsmen provided by the princely states and later continued by the central and state governments after independence, has kept the arts and crafts of the state alive. The workmanship has excelled in the country and its artisans/ craftsmen have won many national awards.

The Rajasthan Small Scale Industries Coiporation looks after the inerest of craftsmen engaged in the manufacture of handicrafts. Their products are marketed through a chain of well established handicrafts emporia in almost important cities of the country and places of tourist interest.

The handicrafts of Rajasthan represent vivacious variety in art and culture of the state. The artisans show their craftsmanship and high class talent by working on brass, stones, clay, leather, textiles, wood, etc. R. Natarajan has given a beautiful description of the lively nature of Rajasthani handicrafts in the following words, - "There is a profusion of exuberant colours and fascinating forms in the handicrafts of Rajasthan. They are, in fact, a tribute to human refinement and sensibility and also a challenge to the stark desert environment. Each period of history has displayed its own steak of ingenuity, contributing to the aesthetic heritage of this land. The erstwhile kings and nobles were lovers and patrons of arts and crafts. This has led to the tendency of the Rajasthan people obsessively decorating almost everything, be it a camel or a kitchen wall, their clothes, to themselves even. No wonder each Rajasthani handicraft piece never ceases to amaze."

The hand-printing and dyeing of textiles at Sanganer, Pali, Bagra, etc., have been very famous. 'Ajrak print' of Banner, 'Pichwaiyan' (in the back side of sculpture) of Nathdwara near Udaipur, in which firstly the cloth is dyed with black colour and then the activities of Lord Krishna pertaining to his childhood, etc., are depicted on it, and the life-sketch of some great man is represented on a typical piece of cloth. Various types of beautiful clothings for ladies such as chunaria , laharia and odhania, etc., are famous which are made available mostly at places like Jodhpur and Jaipur.

250 gram 'rajai' of Jaipur is very famous and the same is purchased by tourists with great interest for the winter season. The precious and semi-precious gems and artistic jewellery of gold and silver, brass artware and utensils with designs, lac bangles, marble-statues, light salma-sitara embroi-dery work on clothing and shoes, blue pottery products (with Persian origin and made of ground quartz stone flourished in Jaipur under Maharaja Ram Singh's patronage), clay and wooden toys, sandal wood products woollen carpets of Jaipur and Bikaner, products made from camel-skin, etc., are some of the examples of handicrafts of Rajasthan. Carpets, gems and jewellery' are important foreign exchange earners, and need to be developed systematically in future for export promotion purposes. Silk industry can be developed from mulberry-plantation (for silk-worm rearing) in Kota, Udaipur, Bharatpur, Bundi and Chittorgarh districts; Tassar (artificial silk) can also be developed in Kota. Udaipur and Banswara districts, for which 'Arjun' trees are planted which help in restoring ecological balance to some extent. Artificial silk can be produced through chemical process. Jaipur has been : famous for its engraved brassware which is usually enamelled.

Government is promoting various handicrafts in the state by starting craft design development centres and encouraging exports by providing freight-subsidy on exports though Inland Container Depots (ICDs). Thus, the state possesses a rich cultural heritage of the past, and handicrafts reflect the greatness of Rajasthan in several ways. There is fairly range of handicrafts produced in the state. Each area is identified for its specific craft, for example, Jaipur is known for its exquisite gems, Jewellery, woollen carpets, designs, portraying of myths, legends and romances, rangoli patterns, folk paintings and with exquisite colour combinations.

Amongst fabric, falling under the category of handicrafts, block printing, indigo algerian printing, resist printing, and calico sand block printing are done at Jaipur, Bagru, Sanganer, Banner, Jaisalmer, Jodhpur, Udaipur, Tonk and Balotra. Embroidery and zari work on cloth and on leather, studded with glass, druggets and namda, camel hide work, silver ornaments with enamelling, metalwares, brass articles with artistic engraving and enamelling, bronze items, blue pottery, filigree cutting on silver, gold ornaments with enamelling, applique work, quilts, wooden lattice work, miniature carving on wood, cutting of precious stones, gauntlet work, embroidered leather shoes, tie and dye work, different types of paintings, miniature painting on marble, paper, silk and ivory, etc. are some of the items of export. The list is very large as Rajasthan is the home for handicrafts.

Notwithstanding the above progress, the state lags behind many other states in the sphere of industrialization. The contribution of the industrial sector is barely 8 per cent of the net domestic product of the state. Obviously, the industrial scenario needs strengthening. Keeping this objective in view, the state government announced Industrial Policy in the year 1990 which seeks to encourage investment in the industrial sector and in tourism which has also been declared as an industry. The industries which are labour oriented and have a limited requirement of water and power, are to be given top priority, so much so that special incentives are being offered to such undertakings. Fiscal incentives and other facilities announced in the industrial policy are encouraging and conducive to industrial growth. This policy changed the industrial scenario of the state, by creating an industrial climate. In order to give further impetus to the industrial development.

The State Government announced RIPS 2003 in the year 2003 to encourage investment in the state. The State Government announced Policy Package to Micro, Small and Medium Enterprises 2008 on 15.02.2008 to help micro, small and medium enterprises achieve globally competitive level (Table 21.5).

Table 21.5 Investment and Employment in Small Scale Industries

S.No.	Name of the District	Registered Units (Nos.)	Employment (Nos.)	Investment (Rs. in lakhs)
Ajmer Division				
1	Ajmer	8	2325	5737.17
2	Tonk	5	1089	286.90
3	Bhilwara	59	32413	3510.28
4	Nagaur	5	732	466.48
Bharatpur Division				
1	Bharatpur	1	245	172.22
2	Dhoplur	3	877	65.58
3	Alwar	91	32791	20228.27
4	Karoli	0	0	0
Bikaner Division				
1	Bikaner	1	750	94.00
2	Churu	1	70	25.25
3	Hanumangarh	2	1486	82.90
4	Sriganganagar	7	4090	7448.13
Jaipur division				
1	Jaipur	28	14178	3197.09
2	Dausa	0	0	0
3	Jhunjhunu	0	0	0
4	Sikar	6	1903	425.52
Udaipur Division				
1	Udaipur	21	7549	742.91
2	Chittorgarh	11	6490	9491.24
3	Raj samand	1	1904	248.18
4	Dungarpur	3	2237	241.14
5	Banswara	7	16000	2617.40
6	Pratapgarh	0	0	0

S.No.	Name of the District	Registered Units (Nos.)	Employment (Nos.)	Investment (Rs. in lakhs)
Jodhpur Division				
1	Jodhpur	18	3233	1007.00
2	Pali	6	5342	7114.41
3	Sirohi	7	4306	4794.28
4	Jalore	0	0	0
5	Barmer	3	2807	28490.67
6	Jaisalmer	0	0	0
Kota Division				
1	Kota	12	7345	4914.37
2	Bundi	5	1352	810.31
3	Jhalawar	1	3594	353.61
4	Baran	4	4449	7755.12
5	Sawai Madhopur	0	0	0
	Grand Total	**316**	**162237**	**110320.40**

Source: Commissionerate of Industries. GOR, Jaipur

Large Scale Industries.

As on 31.3.2014, Rajasthan had 316 Large scale units, in which Rs. 110320.40 Crore were invested and 1.62 Lakhs of employment was generated from these industries. The information about number of units, employment and investment as on 31st March 2014 region wise is given in the Table 21.6.

There are 221 Medium scale industries are registered from 2007-08 to 2014-15 Alwar and Bhiwadi with 91 units' accounts for 29% of large scale industries followed by Bhilwara and Jaipur district 59 and 28 respectively. 316 large units having investment of Rs. 110320.40 Lakhs, reported as on 31.3.2014.

Micro, Small and Medium Industries

Table 21.6 Year-wise inforamtion of Micro, Small and Medium and Medium Industries.

Year	Registered Micro, Small and Medium and Medium Industries	Investment (Rs. in Crores)	Employment (persons in No.)
2012-13	15363	2788.06	94562
2013-14	17601	2778.89	98791
2014-15	18655	2513.64	91831

Source: Department of Industris and Commissionerate of Industries, GoR, Jaipur

It is evident from the numbers inscribed into the table above that number of registered units; employment and investment are growing over the year. Total 18655 no of registered Micro Small and Medium Industries during the 2014-15 with total 91831 no of employment and Rs. 2513.64 crore of total investment in these enterprises. Jaipur district with 27069 units has maximum number of Micro Small and Medium Industries in the state. District-wise status of Micro Small and Medium Industries is given in the Table 21.7.

Table 21.7 District-wise Status of Micro Small and Medium Industries

S.No.	Name of DIC	Micro, Small and Medium and Medium InJustries (Numbers)
1.	Ajmer	5341
2.	Alwar	4035
3.	Banswara	2040
4.	Baran	990
5.	Barmer	1430
6.	Bharatpur	2653
7.	Bhilwara	5655
8	Bhiwadi	5007
9	Bikaner	5280
10	Bundi	1681
11	Chittorgarh	2513
12	Churu	1719

S.No.	Name of DIC	Micro, Small and Medium and Medium Industries (Numbers)
13	Dausa	2148
14	Dholpur	1252
15	Dungarpur	1275
16	Hanumangarh	1359
17	Jaipur (Urban)	21066
18	Jaipur (Rural)	6003
19	Jaisalmer	1851
20	Jalore	1520
21	Jhalawar	1719
22	Jhunjhunu	2379
23	Jodhour	7710
24	Karoli	2044
25	Kota	5427
26	Nagaur	4061
27	Pali	4530
28	Pratapgarh	1074
29	Rajsamand	3295
30	Sawai Madhopur	1878
31	Sikar	2416
32	Sirohi	1975
33	Sri Ganganagar	2586
34	Tonk	2170
35	Udaipur	5965
36	Phalodi	233
	Total	**124280**

Source: Department of Industris and Commissionerate of Industries, GoR, Jaipur

Mineral Based Industries

Mineral Potentiality: The State is geologically so endowed that it becomes a veritable repository of minerals. Rajasthan is fortunate enough to have a wide spectrum of mineral deposits. There are about 64 different major and minor minerals produced in the State, contributing annual revenue of more than 600 crores.

Rajasthan is the sole producer of garnet (gem variety), jasper and wollastonite. Almost entire production of zinc (concentrate), calcite, asbestos and gypsum in the country was reported from Rajasthan. Besides, Rajasthan is the leading producer of ball clay (40%), feldspar (70%), fluorite (graded) (59%), Kaolin (44%), lead concentrate (80%), ochre (72%), phosphorite (79%), silver (54%), steatite (85%), barytes (53%), copper (34%), quartzite (33%) and silica sand (21%).

The State Government welcomes new investments in the mineral sector and is pledged to provide all necessary facilities to potential investors. Given the availability of a large variety of metallic and non-metallic industrial minerals, there is a huge potential of mineral-based industries in Rajasthan. In addition, there is well-developed transport, communication and power infrastructure in the State. The mineral policy of the State has been formulated to facilitate investment in mineral-based industrial sector. The State offers an investor-friendly environment and welcomes Indian and multinational investment in mineral-based industries.

Exports

Exports hold the key not only for a healthy balance of payments positions and macro economic, but also for a planned and sustained economic development. In the absence of better performance on the economic front, the balance of payments position of the country comes under strain and retards economic progress. Rajasthan has traditionally been an exporter of primary goods. The significance of exports from the state lies not only in earning foreign exchange for the country's exchequer but also in indirect benefits to the state such as:

1. Expansion of market opportunities for its produce;
2. Improvement in product quality and subsequent handling techniques;
3. Technological up gradation in terms of plant, machinery and manufacturing process;
4. Greater employment opportunities
5. Exposure to the international marketing and technological environment etc. The state government has identified exports as

one of the thrust areas. Export promotion cell has been set up in the state commissionerate of Industries. The RSIC ltd. has been recognized as an export house. Promotional efforts for export growth are also being made by the Bureau of Industrial Promotion and the RIICO ltd. the manufacturer and exporter have given an excellent account of themselves in export efforts, which is evident from the sharp increase in the volume of exports in the recent years.

Government of Rajasthan has provided several incentives like over-riding priority in release of power connections to units set up in EPIP / Export processing zone, exemption from power cuts, purchase tax exemption depending on the level of investment and entry tax exemption to help promote export-oriented units. The policy of encouraging exports has already started paying rich dividend and in turn exports have led to the overall industrial development of the state. State has also declared its SEZ policy and act lot of concessions and facilities has been incorporated.

The entrepreneurs of Rajasthan are exporting a wide variety of products, which includes Handicrafts, carpets, readymade garments, printed fabrics, makeup's, leather goods, chemicals, minerals, stones viz. marble, granite, slate, sand stone, engineering goods, precious and semi precious stones and jewellery etc. the details of product wise and year wise is as below (Table 21.8).

Table 21.8 Export Product Year-wise

(Rs. in Crore)

S.No.	Product	2010-11	2011-12	2012-13	2013-14
1	Textile	2500.13	3402.74	3800.08	4139.65
2	Agro and Food Products	2869.72	1078.38	14517.17	9080.75
3	Gem and Jewellery (including diamond)	2686.24	2970.03	3023.36	4564.15
4	Engineering	1920.17	2698.21	2914.81	3517.26
5	Metal : 1.Ferrous	483.13	441.80	683.21	759.40
	2. Non- Ferrous	4498.52	2779.21	1124.94	923.79
6	Dimensional Stones, Marbles, Granites and articles of stone mica etc.	933.25	1812.18	2028.89	2539.42
7	Mineral fuels, Mineral oil and Products etc.	2915.90	162.29	1052.91	171.65
8	Electrical, Electronics, and computer Software	120.38	421.00	1171.82	1500.00
9	Wool and Woollen	16.92	34.26	35.25	39.75
10	Chemical and Allied	1347.43	2743.14	3779.69	4123.64
11	Drugs and Pharmaceuticals	186.52	281.22	293.76	359.90
12	Plastics and Linoleums	136.39	252.74	517.22	407.71
13	Handicrafts	1875.95	2150.65	2363.83	3029.18
14	Leather and Leather Products	81.45	83.47	98.26	137.16
15	Readymade Garments	933.25	1281.45	1288.20	1543.17
16	Carpets (Durries)	237.60	370.62	394.36	442.61
17	Handlooms	1.20	2.10	NA	NA
18	Others	9.05	77.11	18.94	7.52
	Total	**23753.20**	**32749.60**	**39106.70**	**37286.71**

Source: Department of Industris and Commissionerate of Industries, GoR, Jaipur

Export Infrastructure

RIICO plays a catalytic role in industrial development of Rajasthan. Keeping in view the specific requirements of export-oriented units, RIICO has given emphasis for development of special parks for setting up of such units. RIIICO on behalf of Govt. of Rajasthan is providing and developing these basic infrastructure facilities for the exporting units.

1. Export Promotional Industrial Parks
2. Special Economic Zones
3. ASIDE Scheme (Assistance to state for developing export Infrastructure and allied activities)
4. AIR Cargo Complex (ACC) and inland container Depot (ICD)

Export Promotional Industrial Parks

The EPIPs have been developed / are being developed under the scheme of department of Commerce, Govt. of India. In EPIPs only those units that undertake to export a minimum of 33% of their production, are allotted land. Brief details of these EPIPs are as under:

(a) EPIP at Sitapura, Jaipur- The first EPIP of the state was developed by RIICO at sitapura, Jaipur over an area of 361 acres (409 plots). The park is functional since march-1997 with 208 units (230plots) under production and 62 plots (28 units) are under construction.

(b) EPIP at Boranada, Jodhpur- The second EPIP of the state has been developed by RIICO at Boranada, Jodhpur over an area of 207 acre on which an expenditure of approximately Rs. 18 crore has already been incurred. Plots are allotted mainly to handicrafts units. All basic infrastructure facilities are available since July 2003.

(c) EPIP at Neemrana, Alwar- The third EPIP of the state is being developed at Neemrana located on NH-8 (Delhi-Jaipur Road). The park companies of 218 saleable industrial plots of which 210 plots have been allotted. This EPIP is being developed with an estimated project cost of about Rs. 35 Crore and expenditure of Rs. 24 Crore has already been incurred.

Special Economic Zone (SEZs) : The Wings of Change

SEZs are specially delineated duty free enclaves treated as a foreign territory for the purpose of industrial, service and trade operations, with exemption from custom duties and a more liberal regime in respect of other levies, foreign investment and other transactions. Domestic regulations, restrictions and infrastructure inadequacies are sought to be eliminated in the SEZ for creating a hassle free environment. The SEZ scheme seeks to create a transparent system by introducing simplified procedures for enhancing productivity and making it easier to do business. SEZs envisages key role for the state government in export promotion and creation of infrastructure, investment and employability. Benefits derive from investment and additional economic activity in the SEZ and the employment generated has a multiplier effect on the economy. SEZs have been driving industrial growth in India. This is an established fact today that for a business to be globally competitive and eco system has to evolve where all the key elements like infrastructure, resources, logistics, hassle free environment etc. need to be provided. SEZs act as a catalyst in providing these benefits to business. Rajasthan government has been very proactive for the growth of industries in the state. It has network of support organization like Rajasthan state industrial development and investment corporation (RIICO), Bureau of Investment Promotion (BIP), and the Rajasthan Finance Corporation (RFC) amongst other. The state has adopted the Public private partnership model very successfully for the implementation of the SEZ projects in the state. Presently there are four operational SEZs in Rajasthan, while three of these are state owned by Rajasthan state industrial development and investment corporation ltd. (RIICO). The other one is joint venture between the RIICO and Mahindra and

Mahindra. Among those in the pipeline are six SEZs at advanced stages of formal approval from the Govt of India and nine that have in principal approval. The details of the these four SEZs are as below, of these three SEZs have been developed / are being developed byz RIICO,The brief details of these SEZs are as under:

(a) SEZ for Gems and Jewellery units at Sitapura, Jaipur (Phase-I)- The SEZ for Gems and jewellery units (phase I), has been set up at sitapura, Jaipur. It is planned over 21.50 acres of land. All of the 51 Plots planned have already been allotted. Roads, drainage, boundry wall work and power supply/ street light work and other basic infrastructure facilities have been developed.

(b) SEZ for Handicrafts units at Boranada, Jodhpur- The SEZ for handicraft units at boranada, Jodhpur has been planned over 180.94 acres of land (291) plots. Basic infrastructure facilities have been provided in this SEZ.

(c) SEZ for GEMS and jewellery units at Sitapura, Jaipur (Phase-II)- This SEZ for GEMS and jewellery units at sitapura (phase-II), Jaipur has been planned over 89.39 acres of land (190 plots). Basic infrastructure facilities have been provided in this SEZ.

(d) Public-Private Partnership for SEZ-RIICO on behalf of the state government, is also promoting its first multi product SEZ in PPP mode on Jaipur-Ajmer National Highway. The SEZ shall be developed on 2500 acres of land and the domestic tariff area will be developed on 550 acres of land. In the phase–I government of India has approved IT/ITes SEZ, Handicrafts SEZ and light engineering SEZ and also accorded in principal approval for multi product SEZ. RIICO has also developed theme based parks such as special economic zones at Jaipur, Jodhpur and Neemrana (being developed) , IT parks at Jaipur, Jodhpur, Kota and Udaipur; EPIP at Jaipur and Jodhpur: Agro food parks at Kota, Jodhpur and Sri Ganganagar; Apparel Park at Jaipur; Bio-Tech Parks at Jaipur , Jodhpur and Bhiwadi besides growth enters and other special parks.

CLUSTER INDUSTRIES

Rajasthan is famous for its rich heritage - its history, the Maharajas and their palaces. But the state is more than just a tourist destination, state is now host to a number of small and medium enterprises alongside some major industrial units that have done well in the desert state are textiles, agro and food products, gems and jewellery, marble, handicrafts, readymade garments, carpet and durries and gem stones.

Most of these sectors are export oriented and have shown high levels of growth. SME across the various sectors have registered a good growth in exports during 2014-15. A number of enterprises promoted by pioneers have over the years become major industry clusters and the unprecedented growth of scattered industries has now created a need for ensuring organized development. State has specialty of having several pockets ranging from very sophisticated items to traditional products throughout its territory. These concentrations are either due to skill/ craftsmanship find in that particular area or availability of particular raw material or the skills are traditionally cherished and developed there. Cluster Industries have been identified for development in the State. 44 handicraft cluster Industries have been identified for intensive credit planning and focused development by the MSMEs up to the 2013-14, out of these activities in 32 clusters are completed. For Sectoral development of handicraft industries in the State, following cluster industries have been identified by the state government and developed by various agencies, details of some of these are as given in next Table.

A brief detail about some of these clusters is as follows:

Gems and Jewellery- With a history of dating back more than two and half centuries, the internationally recognized pink city of Jaipur is still a hot spot in the gems and Jewellery sector. From manual processing to value added branded products, the industry has grown, keeping pace with the changing trends. With over 2,00,000 peoples engaged in the industry, Rajasthan accounts for Rs. 1650 Crore of the total $17 billion worth of Indian export of gems and Jewellery. With more than 1300 exporters in the SME segment alone the sector has been witnessing a growth rate of 15-20% (Table 21.9).

Table 21.9 Cluster Industries in Rajasthan

S.No.	Name of the Cluster	Location of the cluster
1	Dying and Printing Cluster	Akola- Chittorgarh
2	Handloom Cluster	Darib- Churu
3	Murtikala Cluster	Talwara-Banswara
4	Charm Juti Cluster	Bhinmaal- Jalore
5	Gota Loom Cluster	Ajmer
6	Aari-Tari Cluster	Nayla-Jaipur
7	Murtikala Cluster	Gola Ka bas-Alwar
8	Honey Cluster	Bharatpur
9	Pottery and Terakota Cluster	Ramgarh- Alwar
10	Abharayan Utpad(Forest Product) Cluster	Baran
11	Stone Art ware cluster	Dungarpur
12	Leather Product Cluster	Jodhpur
13	Kashth kala Cluster	Bassi-Chittorgarh
14	Kashida kari Cluster	Pugal- Bikaner
15	Stone cluster	Pindwara-Sirohi
16	Terra-Cotta Cluster	Siyava-Sirohi
17	Auto-Component Cluster	Alwar
18	Handloom Cluster	Rajpura Patalwas -Jaipur
19	Marble Artizan Cluster	Chhitoli-Jaipur
20	Sand Stone Carving Cluster	Pichupada-Dausa
21	Brass ware Carving Cluster	Balahedi-Dausa
22	Embroidery and Handicraft Cluster	Shiv ,Badmer
23	Tye and dye Cluster	Jodhpur
24	Carpet Cluster	Thanagazi, Alwar
25	Glass Embroidery Cluster	Dhanau ,Badmer
26	Kota Doria Cluster	Kaithoon, Kota
27	Leather product and Leather Tannig Cluster	Bansoor, Alwar
28	Bamboo Cluster	Malikheda , Banswara
29	Leather Product Cluster	Sawarda, Dudu ,-Jaipur
30	Abharayan Utpad(Forest Product) Cluster	Talchapar (Churu)
31	Lac Cluster	Jaipur
32	Terra-Cotta Cluster	Sagwara. Dungarpur
33	Terra-Cotta Cluster	Pokran , Jaisalmer
34	Terra-Cotta Cluster	Molela ,Rajsamand
35	Hand Block Cluster	Bagru -Jaipur
36	Jaisalmer Stones Cluster	Jaisalmer
37	Leather Cluster	Dhiroda; Alwar

Source: O/o Commissioner of Ind., GOR, Jaipur

Handicrafts- Another traditional industry that Rajasthan is well known for is the export of handicrafts. The industry that started about 25-30 years ago today boasts of an annual turnover of approximately 1400-1500 crore rupees. There are over 500 handicraft manufacturing and exporting units providing employment to about two lakhs people in Jaipur and jodhpur, the two main handicraft centers in the state. In order to promote the industry an exclusive handicraft center is being set up in the Boranada indusrtrial area, jodhpur, besides setting up industrial parks at Basni and Boranada.

Garments- Garment manufacturing and exports is yet another important industrial cluster in the state. After the end of the quota system, the garment industry witnessed a sudden spurt lof growth of 25-30%. Huge export orders kept all the exporters busy even in the off season for which tailors and fashion professionals were hired from neighboring states. Over 2502 garment manufacturing and exporters today account for annual exports worth Rs. 850 Crore, of which Rs. 500 Crore is from direct exports. The industry provides employment to over two lakhs people. The industry is now eyeing the growing domestic market too. The industry needs organized textile parks and apparel centers which have sound printing and dyeing facilities of international standards.

Printing and Dyeing- Rajasthan has long been known for its printed and dyed textiles. But lately the hand block printing and dyeing units that employing over two Lakhs people, have been in thick of controversy over creating an ecological imbalance due to excessive use of dyes. With a total turnover of Rs. 500 Crore, the industry is getting orders from exporters across the country and especially from Delhi, Mumbai and Kolkata. Over a century old the cluster is asking for industry status as it has invested about Rs. 600 Crore in machinery.

Carpets- The state is also famous for heritage carpet weaving. With over 170 manufactu-ring and exporting units, Jaipur is one of the largest processing centers of handmade woolen carpets after Bhadoi in UP. Providing livelihood to over 2.5 lakhs people, the industry accounts for annual exports worth Rs. 450 Crore. A hand made carpet with fine quality work generally costs between $6 and $30 per sq. ft. in the international market. The labour intensive natures of the industry and frequent droughts have helped the industry flourish in the state. Carpet manufacturing needs no power or water, which have always been scarce in the state. The industry emerged as regular source of income for rural artisans and masses during drought in the state.

Agro Industry- With about 20,000 units in the state, the agro industry has an export market of over Rs. 500 Crore. The size of the domestic market could be several times that of exports, but details are not available. There 2164 agro units in nagore,1496 in Sawai Madhopur, 1391 in Jaipur and 1192 in jodhpur. The industry has helped bring prosperity to the villages and increase production and income as well as absorb surplus labour in the area. In a way it helped solve the problem of unemployment. The major agro products from the state include oil seeds (Ground nut , Caster seeds), horticulture (fruits and Vegetables), tobacco, spices, cotton, diary and guargum. In broad sense, this category may include food-products, milk-products and meat-produce by But in a narrow sense, this includes industries based on agricultural raw materials only such as cotton ginning and pressing factories, cotton textiles (khadi, handloom, powerloom and mill-loom), silk industry, vanaspati ghee and edible oils, soap, gur, khandsari, sugar, bakery-products, confectionery, 'methi\ mehandi, papad- bhujiya, citrus fruits, vegetables and flowers, etc. In the recent past there were 23 cotton textile mills in Rajasthan—17 in the private sector, 3 in the public sector and 3 in the cooperative sector. There are three sugar mills and some vegetable oil factories in the state.

Stones- Rajasthan is home to exquisite natural stones like granite, marble stone, slate and sand stone and it has a long tradition of stone usage. Of the total exports from India, rajasthan accounts for about Rs. 800 Crore of stones, of which sandstone and marble are the main items.The kishangarh cluster about 100 KM. from Jaipur, is a major market for stones quarried in Makarana and Rajsamand. Huge marble deposits, a well developed quarrying base, availability of cheap labour and indigenously developed machinery, a large domestic and export market and a well established distributaion network are the main reasons for the business doing well in the state.

Auto Cluster- Auto cluster at Bhiwadi is yet another important industrial desti-nation in the state. The presence of Honda Siel cars, at India's second manufacturing plant at the tapukara industrial area in a Bhiwadi indicates that Rajasthan could be the next big automotive destination in north India. Large numbers of auto component units have also started their operations in Bhiwadi, Chopanki and Alwar industrial areas. Having prepared a blueprint to attract investment of Rs. 5000 Crore, the RIICO has also signed MOU with the Japanese External Trade Organization (JETRO) for Japanese investment in Neemrana. Ashoka Leyland (truck chasis assembly plant), Eicher, caparo fasteners, climate system, federal Mogul, Hi-tech gears, Amtek India and Balkrishana Tyres are some of the other players operating from bhiwadi region.

Plastic Hub- Jaipur is the hub of the plastic industries in the state, with over 800 manufacturers. In all, there are about 1300 plastic manufacturing units in the state, providing direct employment to about 2.5-3 lakh people and indirect employment to about 4 lakh. With an annual production of 1.5 lakh tonnes, the industry has an annual market of Rs.900 crore alongside exports of Rs. 50-60 crore. Rajasthan is the seventh largest producer of plastic in the country. The entire industry runs around a profit margin of 10-12 %. Some of the clusters is earlier selected by the state government was given Table 21.10.

Table 21.10 Cluster Industries

S.No.	Item	Location of Cluster
1	Hand made paper	Sanganer (Jaipur)
2	Marble	Makrana, Udaipur, Rajasamand and and Kishangarh
3	Oil Mills	Alwar, Bharatpur and Sawai Madhopur.
4	Food Processing (Rasgulla, Bhujia and Papad)	Bikaner
5	Mineral Grinding	Beawar
6	Gem and Jewellery	Jaipur
7	Statue Making (Murtikala)	Jaipur
8	Welding Electrodes	Kota
9	Hand Tools	Nagaur
10	Wollen Carpets	Jaipur
11	Screen Printing	Barmer and Pali
12	Textile Hand printing	Jaipur
13	Slate Industry	Hindaun (Karauli)
14	Umbrella Assembly	Falna (Pali District)
15	Wooden Toys	Udaipur
16	Stainless Steel Re-rolling	Jodhpur
17	PVC Wires and Cables	Jaipur
18	Cement based	Beawar, Sirohi and Behror

Source: Department of Industris and Commissionerate of Industries, GoR, Jaipur

Number of units, investment and employment in some the cluster industries taken for the developmental activities are given as follows:

Handmade Paper- Handmade paper industries are located in Sanganer near Jaipur. There are around 13 units manufacturing handmade paper with an investment ranging from Rs.10 lacs to Rs.50 lacs per unit. Few units are 100% export oriented units and their export is to the tune of Rs.1 crore per year. These units provide employment to more than 700 persons. In Rajathan, the literacy is around 31.03% with more and more emphasis given for increasing the literacy in the State, there exists good scope for the development of

Handmade paper industry in the State. The demand of handmade paper is increasing day by day in domestic as wellas in intnl. market. Kumarappa National Handmade Paper Institute, Sikarpura Road, Sanganer, Jaipur is functioning in the State to cater to the needs of Handmade Paper Industry.

Mineral Grinding- A cluster of mineral grinding is located in Beawar. There are 150 units of mineral grinding, which include 120 of ball mills and 30 hammer mills of quartz and feldspar in the cluster and 50 new units are in pipe line. 85% of the mineral grinded goes outside the State for industrial use in glass and ceramic, rubber, paints industry. Good demand of the grinded mineral beckoning to bright future of this cluster.

Marble Cutting- Rajasthan is richest with regard to marble deposits. It has the reserve of 1000 million tonnes of good quality marble. 17 districts in the State have so far shown presence of marble. Some important places are Makrana, Udaipur, Rajsamand, Banswara and Ajmer. Kishangarh is the main mandi for marble cutting. As per the statistics data available with department of Mines and Geology, Government of Rajasthan, the State produced 16,41,000 M.T. Marble having a sale value of Rs. 202 crores. Nearly 1 lac persons are employed directly or indirectly in this industry. Annual export is estimated to be worth Rs. 1075.31 crores.

Oil Mills- Oil mill units are mainly concentrated in Sawai Madhopur, Alwar and Bharatpur. An estimated 700 number of units are manufacturing oil with an investment of Rs. 1400 lacs providing employment to 2-3 persons per unit.

Food products- Food products like manufacture of Papad, Mangori, Namkin and Rasgulla are localised in Bikaner district. There are 842 registered SSI units in the district employing 2,053 persons with a total investment of Rs. 1,034 lacs. The annual export of Agro and Food product is estimated to Rs. 1320.07 crore.

Cement- Cement manufacturing units are found in Sirohi district and Behror in Alwar district. There are about 25 units in Behror and 5 units in Sirohi district. Their investment is around Rs.1 crore per unit with a capacity of 50 TDP and the production is estimated to be 40 TDP. This industry provides employment to 60-65 persons per unit. Mineral Grinding Industry in Beawar is a good cluster having about 100 units manufacturing mineral powder.

Statue Making- Statue making is also located in and around Jaipur. There are about 1000 units manufacturing Statue/Murti using traditional skill. Investment is Rs. 50,000/- to Rs. 1 lac per unit and employment 3-4 persons per unit and the production is around Rs.6 lacs per year.

Welding Electrodes- There are about 15 units manufacturing this item in Kota district. The total production capacity of this item in the State is about 500 M.T. employ-ing about 200 persons. The units are utilising about 60% of the production capacity.

Hand Tools- Plier, Spanner, Hammer and Goldsmith Tools are mainly manufactured in Nagaur District. There are more than 1270 units including artisan manufacturing these items with traditional skill.

Woollen Carpets- Manufacturing of Woollen Carpets is localized in Jaipur. Most of the units are exporting their products. The annual export of woolen carpets was Rs. 273.16 crores .

Screen printing- Screen printing industrial units in the form of clustered are in existence at Barmer and Pali of the State.

Textile hand printing- The cluster of this industry is located at Sanganer and Bagru towns of Jaipur district. The annual export of these items was Rs. 2023.16 crores.

Slate Industry- The Slate industry as a cluster is functioning at Hindaun, District Karauli in Rajasthan State.

Umbrella Assembly- Falna of Pali district has a good concentration of Umbrella manufacturing units.

Wooden Toys- Udaipur has a good concentration of wooden toys units which functions as a household industry.

Stainless Steel Re-rolling- Jodhpur has a concentration of more than 100 units engaged in the manufacture of stainless Re-rolling.

PVC Wires and Cables- A good number of units manufacturing PVC Wires and Cables are located at Jaipur.

Other Cluster Industries- Other cluster industries identified are lacquer craft in Jaipur, Screen Printing in Barmer and Pali, Textile Hand Printing in Jaipur, Spinning and Processing in Bhilwara and Textile Processing and White Metal Industry in Jodhpur, Plaster of Paris in Bikaner, Slate Industry in Hindaun, district Karauli, Umbrella assembly in Falna (Pali district) Electrical Transformers and ACSR conductors in Jaipur and HDPE Pipes and Sprinklers in Jaipur having 70 units in operation.

AGRO FOOD PARKS

Rajasthan is one of the most fascinating States in India and perhaps the richest in agriculture heritage. It is a common belief that it is a land of desert, few people know and even less is aware that Rajasthan has vast tracts of land that are irrigable. It is predominantly agrarian State where agriculture contributes 25-40% of the State's Domestic Product and 70% of its produce is engaged in agriculture and allied activities. Thar desert that runs, predominantly in the western part of the State and covers 250.000 Sq. Kms.has for centuries been a non irrigable tract of land. The district of Barmer,Bikaner, Jaisalmer and Jodhpur provide the co-ordinates of the desert to a common travels. Indira Gandhi Canal has resulted a significant increase in the State.This can covers the seven district of Rajasthan namely Barmer, Bikaner, Churu, Hanumangarh, Jaisalmer Jodhpur and Sriganganagar. Crops of mustard, cotton and wheat now flourish in the semi-arid western reason of the State where nothing but sand ruled the roost for several years. The canal is slowly but surely changing the face of Rajasthan. Government of Rajasthan is actively engaged in promoting various new techniques of modern day farming and bringing the technology to the farmer by organizing several fares and on site demonstrations. The governments massive be evinced by the number of subsidies and financial support that it is extending to the customer.

Government of India is providing grant-in-aid for development of Agro Food Park, so that all requisite information like Common Effluent Treatment Plant, Warehousing, Testing Laboratory etc. are available at place and Agro units located in these parks can derive advantage of large scale. Congenial at place and Agro units located in these parks can derive advantage of large scale. Congenial environment, hassle free labour law and best-suited infrastructure are key-elements, on which these parks have been designed. Rajasthan being primarily an agrarian economy and leader in producing Bajra, Barely, Pulses, Coriander and host of industrial ventures of this sector, Looking to the current need of the hour, developing of Agro Food Park at Ranpur, Kota and Boranada, Jodhpur has been undertaken in the State. The details of the Parks are as under (Table 21.11).

Table 21.11 Agro Food Parks in Rajasthan

Description	Agro Food Park, Ranpur, Kota	Agro Food Park, Boranada, Jodhpur	Agro Food Park, Sriganganagar
Area Acres	136.74	192.47	81.14
No. of Plots	100	183'	77
No. of Plots for Dry Process	73	146'	-
No. of Plots Wet Process	18	28	
No. of Plots for Green House	9	9	-

Source: Department of Industris and Commissionerate of Industries, GoR, Jaipur

These parks have been approved by the Ministry of Food Processing Industries. These Parks will have requisite infrastruc-ture including Common Effluent Treatment Plant and warehousing. The Agro Food park a Sri Ganganagar is in planning state.

POTENTIAL INDUSTRIES

The strength and weakness of Potentiality and hindrance of in industrali-zation in the State. Varieties of products are manufactured and exported from the State. The State Capital is connected with Broad gauge Railway line with the rest of the country. It has developed good infrastructure facilities. The State is providing about 35% of the total wool production in the country. The population of the State is growing at a faster rate as compared to the average growth rate of other States. Of the total population 44% constitutes as main workers as per 2011 Census.

There is abundant scope for setting up of MSMEs in the State. Based on available resources in different districts in the State, a good number of Potential Industries have been visualized for development as Table 21.12.

Table 21.12 Potenital Industries in Rajasthan

(Rs.in lacs)

S. No.	Name of the item	Investment in P and M
1	Refined Edible Oil (50 MT Per day)	100.00
2	Rice and Dall Mill	15.00
3	Bread and Biscuits	15.00
4	Confectionery (Milk & Sugar based)	10.00
5	Cattle Feed	15.00
6	Roller Flour Mill (40 MT per day)	60.00
7	Milk Chilling Plant (600 MT)	20.00
8	Egg Powder	15.00
9	Mineral Water	15.00
10	Chips and Wafers	02.00
11	Garlic Powder	10.00
12	Isabgul processing	15.00
13	Hosiery (Knitted)	5.00
14	Sewing thread Reel	04.00
15	Crape Bandage	05.00
16	Zip Fasteners (Nylon)	70.00
17	Readymade Garments	15.00
18	Woolen Jersy (for defence)	08.00
19	Surgical Cotton	15.00
20	Nylon Socks	05.00
21	Elastic Tape	15.00
22	Under Garments (Hosierv)	15.00
23	Wooden Boards	15.00
24	Wooden Panel-Doors and Windows	12.00
25	Wooden Ply	25.00
26	Cycle Chain	15.00
27	Wire Mesh and Netting	20.00
28	Agricultural Implements	10.00
29	Aluminum Furniture	05.00
30	Sanitary Fittings (metallic)	10.00
31	Building Hardware	15.00
32	Paper Pins and Gem clips	10.00
33	Ball Bearing (small)	75.00
34	Staple pins	10.00
35	Cycle Free Wheel	35.00
36	Nuts and Bolts	35.00
37	Offset Printing	20.00
38	Wooden Furniture	25.00
39	Steel furniture	05.00
40	Tin Containers	10.00
41	G.I. Wire Drawing	20.00
42	Copper Wire Drawing	15.00
43	S.S.Tubes and Pipes	100.00
44	Stainless Steel Wire	50.00
45	Aluminum Die-casting	15.00
46	Auto Leaf Spring	35.00
47	Stainless Steel Wire	15.00
48	Welding Electrodes	35.00
49	Copper Powder	20.00
50	Aluminum Powder	20.00
51	Electric Motor (1 to 10HP)	15.00
52	Ceiling Fan	10.00
53	Domestic Electric Appliances	05.00
54	PVC Wires and Cables	15.00

S. No.	Name of the item	Investment in P and M
55	G.L.S. Bulbs	10.00
56	Compact Florescent Tubes	60.00
57	Syringe Disposable	50.00
58	Distribution Transformers	25.00
59	Synthetic Detergent Powder	03.50
60	Disposable Cups and Plates	06.00
61	Corrugated Paper Cotton	10.00
62	Zinc Sulphate	15.00
63	P.P. Films	10.00
64	Computer Stationery	50.00
65	Plastic Folders Laminated	10.00
66	Handmade paper	50.00
67	Fibre Glass products	10.00
68	Plastic Moulded items	10.00
69	Paper Envelopes	5.00
70	Phenyl	15.00
71	Plastic Containers	15.00
72	Guar Gum Powder	100.00
73	Rigid PVC Pipes	40.00
74	Waxed Paper	15.00
75	Cycle Tubes and Tyres	30.00
76	Plastic Water Storage Tank	25.00
77	PVC Doors and Windows	60.00
78	Canvass Rubber Foam	60.00
79	HDPE Woven Sacks	60.00
80	Ceramic Glazed Flooring Tiles	45.00
81	Marble Cutting and Polishing	45.00
82	S.W.G. Pipes	60.00
83	R.C.C. Pipes	40.00
84	Granite Tiles and Slabs	40.00
85	Sanitary Ware	60.00
86	Refractory Bricks	50.00
87	Bone china Crockery	30.00
88	Surgical Grade Plaster of Paris	50.00
89	Fly Ash Bricks	40.00
90	Marble Slurry Bricks	15.00
91	Marble Slabs/cutting	40.00
92	Diamond Cutting and Polishing	05.00
93	Gem Cutting and Polishing	05.00
94	Toughened Glass	20.00
95	Sodium Silicate	20.00
96	Sanitary Wares	45.00
97	Mineral Grinding	15.00
98	Plaster of Paris	15.00
99	A.C.Pipes and Fittings	10.00
100	PVC Footwear	15.00
101	Leather Shoe Upper	10.00
102	Chrome Tanned Hides and Skins	25.00
103	Vegetable Tanned Hides and Skins	25.00
104	Fancy Leather items	02.00
105	Leather Garments	05.00
106	Hand Gloves (Industrial use)	05.00
107	Safety Shoes (Industrial use)	05.00
108	Mojari Shoes	05.00
109	Leather Belt	02.00
110	Bone Crushing/Mill	10.00

Source: O/o Commissioner of Industries, GoR, Jaipur

Chapter 22
POPULATION

The distribution of population in various regions is closely associtated to a number of factors pertaining to climatic conditions, fertility of soils, availability and development of means of transport and communication, growth of trade, infrastructural facilities and other secondary and tertiary activities. In Rajasthan, a close relationship between the size of population and area of various districts is clearly discernible. The most populous districts like Bharatpur, Jaipur, Alwar, Dhaulpur and Kota lie on the eastern fringe of the state, while least populous districts like Jaisalmer, Barmer and Bikaner lie in the west, as compared to total geographical area of the district. Thus Jaipur district having only 3.11 per cent of the total area of the state, is inhabited by 9.29 per cent of the population while Jaisalmer district, occupying 11.22 per cent of the area, has only 0.90 per cent of the population.

These gross disparities in distribution of population are undoubtedly related to geographical factors which work in situ. The availability of means of sustenance, regardless of the areal extent of the district, could account for the inequitable distribution of population in various districts. The dispersal of population closely follows the pattern of annual rainfall regime which exhibits a consistent decreasing trend from east, south-east to west and north-

The fertile plains of the east, drained by number of ephemeral rivers and streams which have deposited fertile alluvial soil over the years, provide rich arble soils to sustain the people. Coupled with this, are the climatic factors-the moderate climate in the east, providing a comfortable zone of temperature, humidity and precipitation, and accentuating better living conditions. The levelled plains of the east have also promoted development of surface transport systems in the form of roads and rail-road, facilitating the growth of industries, trades and business centres and hence in all the eastern districts, the size of population is invariably higher. On the other hand, the inhospitable climatic conditions of western Rajasthan along with limited resources of surface and ground water, lack of fertile soils and the inadequacy of means of transport have restricted population distribution in this part of the land. Hence, the population of the districts of western half of the state in far smaller in related to their area. It only in the central part of Rajasthan, as the the districts of Chum, Nagaur, Pali, Jodhpur and Jalor, that the area and population are proportionate.

The population of the State according to 2011 Census is 6.85 crores. Males constitute 3.55 crores (52%) and the remaining 3.3 crores (48%) are females. 75% of the total population in the State lives in rural areas and 25% live in urban areas. The sex ratio of population was 928 females per 1000 males. The population decadal (2001-11) growth rate of the state is higher than as compared the country's decadal growth rate, whereas it shows decrease as compared to1991-2001.The following Table (Table 22.1) depicts the population of Rajasthan since 1951 as compared to the average in India.

Table 22.1 Population trend in Rajasthan

S.No	Year	Total Population (in crores)		Population decadal Growth Rate (in%)	
		Rajasthan	India	Rajasthan	India
1	1951	1.60	36.01	15.2	13.3
2	1961	2.02	43.91	26.2	19.3
3	1971	2.57	54.70	27.63	24.80
4	1981	3.41	68.38	32.36	24.74
5	1991	4.40	84.40	28.44	23.50
6	2001	5.65	102.86	28.41	21.52
7	2011	6.85	121.05	21.3	17.7

Source: Census of India, 2011

Factors Affecting Density of Population

Topography: The relief features play an important part in determining the density and distribution of population in this State. The larger aggregations of population are found in the level tracts with better water supply either by rain or by irrigation. In such areas along with adequate water supply, fertility of the soil and food production are the main factors. There is a grater concentration of population in the river valleys in the eastern plain area and to some extent in the Luni basin and the sub-montane zone in the west. The high density of population in these plains in general and Trans-Yamuna in particular indicates that this area is a better food producing region. The soils are renewed by the detrital matter brought down by the rivers from the hills and this action enriches the low lying tracts.

In the southeastern part of the State, there are a few areas of dense population. Such areas are well watered, low lying tracts where rainfall is heavier and more regular. The soil in this part is more fertile. There are extensive areas of black soil which produce good crops of wheat and barley.

West of the Aravallis, except the districts of Sirohi, Pali, Sikar and Jhunjhunu, the whole of this area lies in the Western Sandy Plains which includes the arid and semi-arid areas. In a large part of this area, extreme arid conditions prevail, temperature is very high during the greater part of the year, irrigation facilities are almost absent (except the area served by Indira Gandhi Canal) and the land is covered by vast stretches of sands which has resulted in a very low density of population. The river Luni is the only stream in the west and even that is non-peremial. The Luni basin drained by this river and some smaller tributaries occasionally floods this basin. This results in the deposition of alluvial material which produces some wheat. This area supports a larger population than that lying further west.

Rainfall: There is a relationship between the density of population and the rainfall. It is generally stated that rainfall along with the fertility of the soil and the surface terrain combine to control the density of population. This is however, a wide generalisation which has numerous exceptions and is not always applicable. In Rajasthan, the density increases from west to east and southwest to northeast. Jaisalmer, the driest part of the State with less than 100 mm of annual rainfall, has the lowest density while Alwar and Bharatpur in the northeast have a density of about 438 and 503 persons per sq kilometre, the highest in the State. But some of the districts in the southeast like Jhalawar with greater rainfall than the northeast region, have a lower density of 227 persons per sq kilometre. Dungarpur located on the eastern edge of the Aravallis in the south receives about 900 mm of rainfall. Although this,area is more hilly and rugged in the west than Udaipur district, it has a density of 368 persons per sq kilometre, much higher than in Udaipur where the density is only 262 persons per sq kilometre. This is due to the smaller area of Dungarpur and better and fertile land available for cultivation. The population here consists of tribals like '*bhils*' and '*menas*' who have a lower standard of living.

Settlements: Western Rajasthan includes the whole of the arid plain of Marwar and part of Jaisalmer, Barmer and Bikaner. The lower density of less than 25 persons per sq kilometre is in the districts of Bikaner, Barmer and Jaisalmer, the lowest being in Jaisalmer where it is only 7 persons per sq kilometre. The whole of this region is deficient in rainfall, climatic conditions are inhospitable, the surface topography is marked by wind blown sand ridges and the drainage is characteristically internal. In spite of these, man has succeeded in settling down even in these remoter parts of the State. Many settlements like Nokh, Bap, Bikampur, Girasar, Sihar and many others have grown up in this region. Water supply is made available from wells which are sometimes about 50 metres or even more deep in the northwest of Phalodi where the annual rainfall is less than 250 mm. In this part, chains of crescentric sand dunes cover up the ground with an average elevation of over 30 metres. Further northwest in Ganganagar, part of the land has been reclaimed in areas where irrigation is available. The extreme western parts of Jaisalmer, Bikaner and northwest Barmer receive rainfall ranging between 100 mm to 250 mm which is completely inadequate for any agricultural operations. The road from Gadra Road to Phalsund is the dividing line between the northern Jaisalmer desert and the southern Barmer desert. The northern desert has sief or longitudinal type of dunes and the southern part has sand dunes of the barchan type or the transverse dunes. In the south the dunes are much higher averaging between 500 to 1000 metres above the ground level. They are more stable as they cannot be shifted by winds. These fixed dunes have resulted in a higher density of population, *i.e.* an average of 17 persons per sq kilometre as compared to 17 persons per sq kilometre in Jaisalmer area. The railway line that runs from Barmer to Bhimarlai traverses through the country interspersed with shallow basins (blowouts) which are occupied by rural settlements. Barmer with a population of 26,03,751 (2011) is the largest commercial town in this area. It is located on the eastern edge of a degradational piedmont surface carved out of Jurrasic sandstone hills, and dotted with residual hills. The formation of these steep sided, isolated and residual hills is characteristic of hot dry lands in the western part of the State. The nature of settlements are determined to a great extent by the size and the location of sand dunes which rise to a height of about 100 metres. East of the river Luni the sand dunes are of fixed type. The density of population is 92 persons per sq kilometre, which is the highest in this region.

The Jaisalmer desert is more desolate and covered mostly by the shifting sand locally known as '*dhrians*'. 'West of Shahgarh the *dhrians*' run in chains and are of very complex type. It is, however, only in the blowouts of this desert that man could get some foothold. The number of such settlements of precarious existence does not exceed 100. The largest number is to be found in the Jaisalmer pediplain, which is also carved out of Jurassic sandstones. A number of short, intermittent and discontinued streams traverse the Jaisalmer pediplain, the dry beds and banks of which could more easily be tapped for groundwater; and wherever the water supply is more or less assured settlements could spring up. To the north of Jaisalmer a number of playa lakes, designated as ranns, occur in basins more or less rimmed by low scarps. These lakes, though fed by centripetal drainage, remain dry for the greater part of the year, but their importance is mainly due to the fact that they yield salts. A number of settlements, including Sri Mohangarh, the largest village with a population of about 9,928 (2011) owe their existence to the salt deposit in these ranns.' Near the Aravallis, the rainfall increases to 500 mm in the north and to 650 mm in the south and there are a large number of streams coming down the hills. This has resulted in the concentration of population in the foothill zone especially in the triangular area bounded in the north by the river Luni and in the southeast by the Jawai-Sukri rivers respectively. There are several towns in this fertile tract including Pali town. Settlements have also sprung up around salt lakes and marble quarries. Didwana in the northeast and Pachpadra in the southwest are the typical salt towns. Makrana is the well known centre of marlble quarries. Jodhpur, the largest town in this region, is located on a rocky surface around a large ancient fort. South of Luni is characterised by isolated rocky hills, bare and unpopulated whereas at the base of the Aravallis are a large number of villages which receive their water supply from hill streams during the monsoon season.

The Aravalli range, stretching for a distance of about 550 kilometres from Khetri to Khed Brahma from northeast to southwest across the state provides another inhospitable environment for human settlements. The hills consisting of hard quartzite rocks rise abruptly from the adjoining plains. 'In the Aravalli range the villages are few and far between and in many cases settlements consist of a few homesteads, erected either in forest clearings or in depressions between rugged quartzite ridges. Some concentration of population is, however, found where an extensive flat surface intervenes between rugged ridges, as in Bhorat Plateau between Kumbhalgarh and Gogunda.' There are many gap towns north of Beaver, and the largest town being Ajmer. The Sambhar lake occupies one such gap in the north, around which a number of settlements have grown up. In the south is located the artificial lake Dhebar (Jai Samand) and the lake city of Udaipur. The northern part of the Aravallis is much more dissected and contains few settlements though Alwar is located on the eastern edge of the hill.

The eastern Rajasthan though slightly bigger in area than the western part contains more than four times the population of the western part. The Mewar Plain covers the eastern and southern part of eastern Rajasthan. This area is drained by the river Banas and its two tributaries Kothari and Khari. Here the rainfall is higher, the soils are fertile and better irrigation facilities from wells and tanks are possible. Hence such areas have larger dispersal of population. The rural population is concentrated along the river courses and around irrigation tanks. The famous Nathdwara temple town is located in this area. The Jaipur Plain occupies the northern part of eastern Rajasthan and the soil consists of alluvium and sand drift. Jaipur the capital of this State is located on an outlier of the Aravallis at the northern edge of the alluvial plain. In eastern Rajasthan, the plateau and wooded ridges are sparsely populated. These plateau and ridges are formed of sandstone rocks of the Vindhyan age. Bundi town is located on a sandstone ridge and Kota town at the edge of the flat land drained by the river Chambal, Kali Sindh and Parbati.

POPULATION DENSITY

Rajasthan occupies nearly 10.42 per cent of the country's total area. Yet its population is only abut 5.49 per cent of the total population of the country. In the present era/century, the state's population density has consistenly increased from 30 to 165 persons/sq.km. However, at present, it is much lower than to the countiy's average of 324 persons/sq.km. (Table 22.2)

Table 22.2 Density of Popuiation (Persons/sq.km.)

Year	India	Rajasthan
1901	77	30
1911	82	32
1921	81	30
1931	90	34
1941	103	41
1951	117	47
1961	142	59
1971	177	75
1981	221	100
1991	267	129
2001	324	165
2011	382	200

Source: Census of India, 2011

Regional disparities are markedly discernible amongst various districts of Rajasthan. Jaipur district is most thickly populated district with a population 471 persons/sq. km. followed by Bharatpur (414 persons/sq.km.), Dausa (384 persons/sq.km.) and Alwar (357 perosns/sq.km.). The lowest density of 13 persons/sq.km. is recorded for Jaisalmer district. Inspite of this record of the lowest density, Jaisalmer has registered the highest growth rate of population amongst all the districts of the state, an increase which in absolute numbers is altogether unexpected. Interestingly, the situation with regard to population density has remained unchanged in the last decade except for Sirohi which in 1991, recorded a density lower than the state's average, while in 2001, it is just one point above the state average.

Despite the relatively high growth rate of population, the density of population in Rajasthan, as per the 2011 census is 200 persons per sq. km. The Table below reflects the trends of density of population of state as compared to the India. Population density in the state is approximately half of the national density which is 382 in 2011.

The density of population in Rajasthan, i.e. the number of people per sq. km. was 200 in 2011. It was 165 in 2001. India the density of population was 382 in 2011 and 325 in 2001. The highest density was found in the state of Bihar, which was 1106; and lowest was in Arunachal Pradesh, which was 17 only. There are inter-density defferences in density levels of 33 districts of Rajasthan. Highest density was 595 in the Jaipur district, while the lowest was 17 in the Jaisalmer district. In 19 districts, the density was more than the state-average; while it was less inrest of the 14 districts.

DECADAL GROWTH OF POPULATION

The population of Rajasthan has increased from 10,294,090 in 1901 to 68,54,8437 in 20011, registering a population growth of 448.60 percent in 100 years. While analysing the growth of popuiation in last 100 years in rural and urban segments of the state, it is found that the growth of population in urban areas is much higher than that of rural areas which stood at 751.60 percent and 394.86 percent respectively.

The growth of population in rural areas of the state since 1901 has shown a fluctuating trend. However, it has registered a higher growth rate of 27.49 percent during the decade 1991-2001 as against 25.46 percent registered during 1981-91. Similarly it has shown a fluctuating trend in urban areas also since 1901. It registered the highest growth rate of 58.69 percent during the decade 1971-81. Contrary to the growth rates in rural areas it has shown a declining trend from 39.62 percent registered in 1981-91 to 31.17 percent during 1991-2001. Decline in the decadal growth rate of urban population is mainly attributed to declassification of 7 Munici-palities into Gram Panchayats (Table 22.3).

Table 22.3 Percentage Decadal Variation of Population from 1901 to 2011 by Rural-Urban Composition in Rajasthan

Year	Total	Rural	Urban
1901	-	-	-
1911	+6.70	+8.74	-4.83
1921	-6.29	-7.26	-0.03
1931	+14.14	+13.63	+17.21
1941	+18.01	+17.25	+22.43
1951	+15.20	+10.80	+39.59
1961	+26.20	+29.65	+11.04
1971	+27.83	+25.77	+38.47
1981	+32.97	+27.47	+58.69
1991	+28.44	+25.46	+39.62
2001	+28.33	+27.49	+31.17
2011	21.30	19.0	29.0

Source: Census of India, 2011

Percentage Decadal Growth at District level

Jaisalmer, Bikaner, Barmer, Jaipur and Jodhpur are the districts which have attained the highest percentage growth rates of 47.45, 38.18, 36.83, 35.10 and 33.77 respectively during the decade 1991-2001. Whereas, contrary to it, Rajsamand, Jhunjhunun, Chittaurgarh, Pali and Jhalawar are the districts which have the lowest percentage growth rates of 19.88, 20.90, 21.46, 22.39 and 23.34 respectively during the same period.

Percentage Decadal Growth in Rural areas

The state has registered percentage decadal growth rate of 27.49 during the decade 1991-2001. There are 11 such districts which have attained higher growth rates and the remaining 21 have attained lower than that of state's growth rate. The following table shows the districts which have registered higher/lower growth rates than that of 27.49 i.e. average of the state.

Highest growth rate in rural areas have been attained by the districts adjoining to international border i.e. Jaisalmer, Bikaner and Barmer. Apart from these, Jodhpur and Sirohi have also attained higher growth rates. On the other hand, the five districts which have recorded the lowest percentage growth rates in rural areas are Rajsamand, Jhunjhunun, Chittaurgarh, Kota and Hanumangarh (Table 22.4).

Table 22.4 First Five and last Five districts of Rajasthan as per their rural decadal growth rate.

First Five Districts		Last Five Districts	
Jaisalmer	(47.99%)	Rajsamand	(18.38%)
Bikaner	(47.82%)	Jhunjhunun	(20.74%)
Barmer	(40.85%)	Chittaurgarh	(20.84%)
Jodhpur	(37.41%)	Kota	(21.02%)
Sirohi	(32.97%)	Hanumangarh	(21.08%)

Source: Census of India, 2011

Percentage Decadal Growth in Urban areas

As compared to the decadal growth rate of 27.49 in rural areas of the state, a higher growth rate of 31.17 in urban areas has been recorded during 1991-2001. At the district level, 14 have recorded higher decadal growth rate than that of state average whereas 18 have recorded lower rate. The following are the five districts each having highest and lowest decadal growth rates in urban areas (Table 22.5).

Table 22.5 First five and last five districts of Rajasthan as per their Urban decadal growth rate.

Highest			Lowest		
1.	Jaipur	(46.17)	1.	Barmer	(0.75)
2.	Jaisalmer	(44.54)	2.	Jhalawar	(11.43)
3.	Baran	(40.73)	3.	Sirohi	(18.15)
4.	Sawai Madhopur	(40.17)	4.	Churu	(20.10)
5.	Hanumangarh	(39.33)	5.	Banswara	(20.32)

Source: Census of India, 2011

Jaipur district has the highest decadal growth rate of 46.17 in urban areas of the state whereas Barmer district has recorded the lowest growth rate of 0.75. Jaisalmer is the district which has higher growth rates in both rural and urban areas of the state as compared to other districts.

SEX RATIO

Sex ratio, the number of females per thousand males in the population, is an important demographic indicator which helps in protraying in the birth rate, fertility rate, death rate, and availability of man-power and also points to the future growth of population. While a higher rate of population growth rate is considered to be detrimental to the economy, The rise of population with an imbalanced sex-ratio is also equally detrimental and any serious imbalance between the two sexes may create several social problems. Ideally, the proportion of men and women should be balanced but infact, it is not usually so, as male and female births can never be equalized.

In the state, there has always been a preponderance of males and females, i.e. the sex ratio has always been adverse to women. During the last hundred years (i.e. 1901 to 2001), the sex ratio has never remained uniform in the state. In 1901 and 2001 sex ratio of 905 and 922 respectively, declined to 896 in 1921. Then in 1931, the ratio again rose to 907. However, it declined marginally and stood 906 in 1941 but again rose to 921 in 1951. It steeply declined to the level of 908 in 1961 and then, moved upto 911 and 910 in 1971 and 1991, respectively. Finally, in the 2001 Census, the ratio has again increased to 922, recording a increase 12 females for every 1,000 males. The sex ratio of the state is lower compared to the national figure of 933 (Census 2001).

The sex ratio denotes the number of females per thousand males. A high sex ratio would mean excess of females on males and a low sex ratio would mean excess of males over females.

Table 22.6 Sex ratio in India and Rajasthan

Year	1951	1961	1971	1981	1991	2001	2011
Rajasthan	924	908	911	919	910	921	928
India	946	941	930	934	927	933	943

Source: Census of India, 2011

The sex ratio is defined as the number of females per thousand males. In Rajasthan, as in other parts of the country the sex ratio has been adverse to females i.e. the number of females per thousand males is significantly low. As per the provisional results of Census 2001, 922 females have been recorded per thousand males in Rajasthan (Table 22.6). A look at the sex ratio recorded for India and Rajasthan over a period of last 100 years shows that it has always been on a lower side in Rajasthan as compared to that of India.

Rajasthan has registered a some increase in the sex ratio, which is 921 in 2001 increased to 928 in 2011. Table showing the population distribution, percentage of growth rate, sex ratio and population density (district wise) is given in Table 22.7

Table 22.7 Sex ratio for Rural-Urban areas at state level for 1901-2011 (Rajasthan)

Census year	Total	Rural	Urban
1901	905	898	947
1911	908	904	936
1921	896	896	897
1931	907	908	903
1941	906	907	897
1951	921	919	958
1961	908	913	882
1971	911	919	875
1981	919	930	877
1991	910	919	879
2001	922	932	890
2011	928	933	914

Source: Census of India, 2011

The above table shows that the sex ratio in urban areas was higher than that of rural areas upto 1921. After that, the sex ratio in rural areas was higher than that of urban areas till date except in 1951.

Sex Ratio in Rural and Urban areas at District level

The sex ratio for rural and urban areas of Rajasthan has been recorded as 932 and 890 respectively in 2001. The sex ratio at district level in Census 2001 is universally higher in rural areas of all the districts than in urban areas except in Bharatpur, Dhaulpur, Karauli, Tonk and Baran districts. The sex ratio in the rural areas of the districts varies from 821 in Dhaulpur district to 1,035 in Dungarpur district. Rajsamand (1,013) and Pali (1,003) are also the districts besides Dungarpur of which the sex ratio favours females. In 17 districts sex ratios in rural areas are lower than that of state's average of 932.

None of the districts in the state is such where sex ratio favours females in the urban areas. The sex-ratio in urban areas of the districts of Rajasthan varies from the lowest of 764 in Jaisalmer to the highest of 945 in Tonk. In urban areas of 15 districts of the state, sex ratio is lower than the state average of 890.

The five districts each for highest and lowest sex ratio in rural and urban areas are shown to the following Inset Table 22.8.

Table 22.8 Highest and Lowest Sex Ratios in rural and urban areas of Rajasthan.

Rural				Urban			
Highest		**Lowest**		**Highest**		**Lowest**	
1.Dungarpur	(1035)	Dhaulpur	(821)	1. Tonk	(945)	Jaisalmer	(764)
2.Rajsamand	(1013)	Jaisalmer	(832)	2. Churu	(936)	Alwar	(838)
3.Pali	(1003)	Bharatpur	854)	3. Dungarpur	(935)	Ganganagar	(848)
4. Udaipur	(988)	Karauli	(855)	4. Banswara	(930)	Dhaulpur	(857)
5..Banswara	(982)	Ganganagar	(881)	5. Rajsamand	(927)	Barmer	(858)

Source: Census of India, 2011

Sex Ratio at Tehsil level

All the 241 tehsils of the state have been classified into 5 ranges of sex ratio. These ranges are (i) Below 850 (ii) 850 to 899 (iii) 900 to 949 (iv) 950 to 999 and (v) 1,000 and above. While classifying these tehsils into above five ranges of sex ratio we find the maximum number of tehsils i.e. 94 are those which have sex ratio in between 900 - 949.1. It is followed by 63 and 39 in the ranges 950-999 and 850-899 respectively. It is a matter of great satisfaction that 29 tehsils in the state are such of which sex ratio favours females i.e. above 1000. Sex ratio varies from 788 in Jaisalmer tehsil of Jaisalmer district to 1,077 in Aspur tehsil of Dungarpur district.

LITERACY

In Rajasthan, the literacy rate has risen remarkably in last forty years. From a percentage of 18.12 in 1961 the percentage of literates has, now, risen to 60.40 per cent. Looking into the progress of literacy in the last decade (1991-2001), the growth in percentage of literate population at 21.85 per cent, though laudable, is yet below the national literacy percentage of 65.38.

In the 2001 Census, all children below 6 years of age, have been treated as illiterate. The Census Department, therefore, estimates the literacy percentage of the entire state to be 60.40 in 2001. In Rajasthan, out of the total population of 56,507,188, literate persons are 27,702,010, of which 18,047,157 are male and 9,654,853 are female. This means that we have 28,805,178 illiterate persons, if absolute numbers are taken into consideration.

The districtwise percentage of the literate in the total population shows marked variation, the lowest being in Banswara (44.60 per cent) and the highest in Kota (73.50 per cent). The Census of 2001 shows that thirteen districts of the state, namely Kota, Sikar, Jhunjhunun, Jaipur, Churu, Ganganagar, Ajmer, Bharatpur, Karauli, Hanumangarh, Dausa, Alwar and Dhaulpur have literacy rate in between above 60 to 70 per cent, sixteen districts namely Baran, Barmer, Udaipur, Jhalawar, Nagaur, Bikaner, Jodhpur, Sawai Madhopur, Rajsamand, Bhilwara, Bundi, Pali, Chittarugarh, Sirohi, Tonk and Jaisalmer have literacy rate in between above 50 to 60 per cent and lowest rate ranging from 40 to 50 per cent is observable in Jalor, Dungarpur and Banswara districts of the state.

The decadal growth of literacy in various districts shows that it has been remarkably high in Barmer (36.02 per cent but very low in Ajmer (12.26 per cent). Other districts where literacy percentage above state level are Baran, Bundi, Churu, Dausa, Dhaulpur, Hanumangarh, Jalor, Jhalawar, Jhunjhunun, Karauli, Nagaur, Rajsamand, Sikar, Sirohi and Udaipur.

There is a high positive relationship between illiteracy and poverty, hence no programme of poverty eradiction can be successful without a simultaneous programme of eradication of illiteracy. It gives rise to a 'culture of poverty'. The document 'Challanges of Education' published by ministry of Education (1985) states, 'appropriatelly fashioned Adult Education programmes could serve as effective instruments for converting the country's population from being a drag on development into an engine of development'.

Literacy, progress and development go had in hand, hence a literacy drive must be launched on war footing.

For eradication of illiteracy governmental efforts alone would not be sufficient. Voluntary agencies, employers in organised sector and the student community should be invited or urged to share their responsibility. Then only this colossal problem can be tackled.

As per the Census 2011 state recorded literacy rate of 66.1 which is quite significant as it was only 60.4 in 2001. Rajasthan is still an educationally backward State. The situation of female literacy is particularly alarming as 52.10% of the female population is literate as compared to 79.20 % of male population of the state. A wide gap exists between the rate of literacy in urban and rural areas and among male and female (Figure 22.1).

Rajasthan has marched ahead in literacy by achieving literacy rate of 61.03 in 2001 as against 38.55 in 1991. Rajasthan has got a distinction in achieving the highest decadal difference of literacy rate of 22.48 among the other states of India. Chhatisgarh is near to it with a difference of 22.27.

If we consider the literacy rates by rural-urban composition of the state, we find that literacy rate for urban areas is much higher i.e. 76.89 percent as against 55.92 percent for rural areas of the state.

Literacy at District level (Rural-Urban)

At the district level for rural areas, following are five districts, each the best and worst performers in literacy (Table 22.9).

Table 22.9 Literacy at District Level (Rural and Urban)

Best performers (Rural)		Worst performers (Rural)	
Name of District	**Literacy Rate**	**Name of District**	**Literacy Rate**
1. Jhunjhunun	73.24	1. Banswara	40.78
2. Sikar	70.39	2. Bhilwara	44.59
3. Kota	67.34	3. Jalor	44.81
4. Churu	65.29	4. Dungarpur	45.69
5. Hanumangarh	63.65	5. Bikaner	46.33

Source: Census of India, 2011

Likewise following are the five districts, each for the best and worst performers on literacy in urban areas (Table 22.10).

Table 22.10 Best and Worst Performers on Literacy (Urban and Rural)

Best performers (Urban)		Worst performers (Rural)	
Name of district	**Literacy rate**	**Name of district**	**Literacy rate**
1. Udaipur	86.19	1. Jalor	66.33
2. Banswara	84.80	2. Dhaulpur	67.48
3. Alwar	82.27	3. Nagaur	69.37
4. Ajmer	81.69	4. Tonk	69.57
5. Chittaurgarh	81.01	5. Karauli	70.22

Source: Census of India, 2011

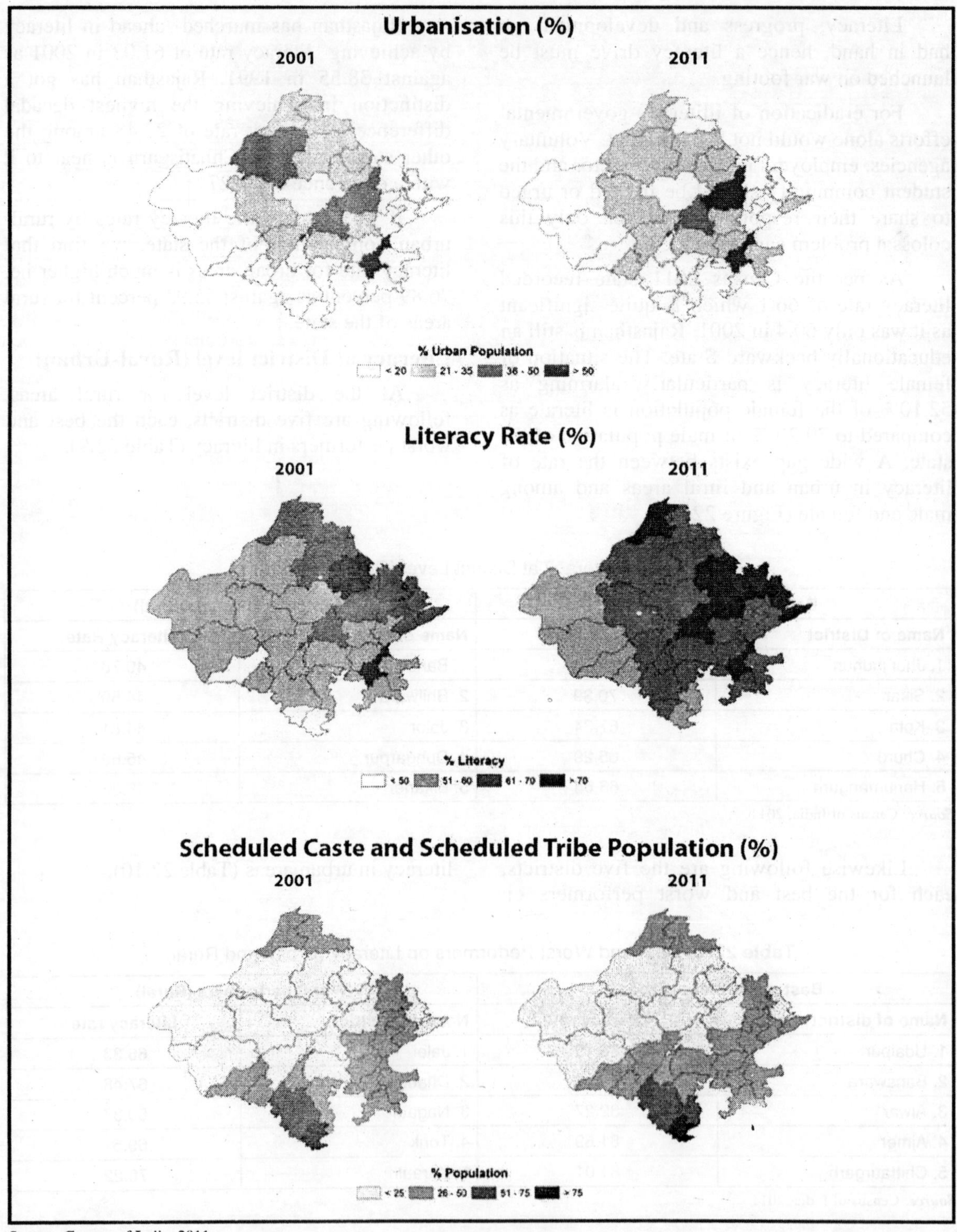

Source: Census of India, 2011

Figure 22.1 Scheduled Caste and Scheduled Tribe Urbanisation, Litracy Rate and Population

Literacy by sex

The male literacy rates have always been higher than the female literacy rates at all the levels i.e. district/tehsil for rural and urban areas. as can be seen from Table 22.11 given at the end of this volume.

Table 22.11 Literacy Rates by Sex at State Level for Rural and Urban Areas In 1991 And 2001

Year	Total			Rural			Urban		
	P	M	F	P	M	F	P	M	F
1991	38.55	54.99	20.44	30.37	47.64	11.59	65.33	78.50	50.24
2001	61.03	76.46	44.34	55.92	72.96	37.74	76.89	87.10	65.42

Source: Census of India, 2011

Among males, the literacy rate has increased from 54.99 percent in 1991 to 76.46 percent in 2001, which is higher than that of all India average i.e. 75.85. In case of females, it has increased more than double from 20.44 to 44.34 percent from 1991 to 2001 respectively, which is still lower than all India average of 54.16 .

It is a matter of great significance for Rajasthan that the female literacy rate of 11.59 percent in rural areas which was the lowest at the national level in 1991, has increased more than threefold i.e. 37.74 percent in 2001. The male literacy rate has also progressed well, rising from 47.64 percent to 72.96 percent during last decade.

POPULATION DISTRIBUTION

According to 2011 Census the total population of Rajasthan is 68548437. Out of Total Population 51500352 persons were living in rural areas and the rest in urban areas of the state. The proportion of the population living in rural and urban areas comes to 75.1 per cent and 24.9 per cent respectively.

Rural-Urban Composition since 1901

The urban component of population of Rajasthan i.e. the proportion of urban population to its population had been higher than that of India from 1901 to 1951. After that the proportion of urban population of Rajasthan had been lower than that of India. As regards to the trend, the proportion of urban population after declining from 15.06 percent in 1901 to 13.44 percent in 1911, indicated an increasing trend upto 1951 reaching at 18.50 percent. Again, after declining to 16.28 percent in 1961 it registered a continuous increase. As regards to India, after declining from 10.84 percent in 1901 to 10.29 percent in 2011 it has shown a continuous increasing trend which can be seen from the Table given 20.12.

Table 22.12 Rural-Urban Composition of Population in India and Rajasthan (1901-2001)

Year	Proportion of Population (%)			
	India		Rajasthan	
	Rural	Urban	Rural	Urban
1901	89.16	10.84	84.94	15.06
1911	89.71	10.29	86.56	13.44
1921	88.82	11.18	85.67	14.33
1931	88.01	11.99	85.28	14.72
1941	86.14	13.86	84.73	15.27
1951	82.71	17.29	81.50	18.50
1961	82.03	17.97	83.72	16.28
1971	80.09	19.91	82.37	17.63
1981	76.69	23.31	78.95	21.05

1991	74.27	25.73	77.12	22.88
2001	72.22	27.78	76.62	23.38
2011	68.86	31.14	75.13	24.87

Source: Census of India, 2011

It is correct to visualise in the context of Rajasthan that 'India lives in villages'. Out of 32 districts, the percentage of rural population to total population in 25 districts is higher than that of the state average of 76.62. Kota, Jaipur, Ajmer, Bikaner, Jodhpur, Churu and Ganganagar are the districts where the percentage of rural population to total population is lower than that of state average. Kota is the only district in Rajasthan which can claim to be an urban as only 46.58 percent of population lives in rural areas of the district.

Jalor, Barmer, Dungarpur and Banswara are the districts where more than 90 percent of the population lives in villages. On the other hand, Kota, Jaipur and Ajmer are the districts where the proportion of the rural population to total population is less than 60 percent.

The proportion of urban population to total population has decreased in 11 districts viz Ajmer, Bikaner, Jodhpur, Churu, Pali, Udaipur, Sirohi, Jaisalmer, Jhalawar, Barmer and Banswara. In other words, the proportion of rural population to total population has increased in these districts. It is very interesting to note that inspite of being Divisional Headquarters and the top cities as their constituents, the proportion of rural population has increased in Ajmer, Jodhpur, Bikaner and Udaipur districts during the decade 2001-2011.

Conversely at the urban front, Jaipur tehsil has the highest proportion of urban population of 96.31 percent to its total population in the state. It is followed by Ladpura tehsil (83.49 percent), Jodhpur tehsil (80.40 percent), Bikaner tehsil (75.07 percent) and Ajmer tehsil (73.22 percent). All these tehsils have major cities as their urban constituents.

WORKMAN PARTICIPATION

The assessment of human resources is largely related to the proportion of its population actually engaged in gainful economic activities. These activities can be classified into different occupational groups. Each occupational group has its own vital role to play in the regional planning and development.

The percentage of workers in the total population is a significant parameter of the dependency of the people. Usually, people of the age group 18-55 years form the working group of the population. Young Childern and the old are dependent upon them. Therefore, the larger percentage of working population in a district, lesser is the dependency ratio.

In Rajasthan, the percentage of working population varies from 34.51 (Kota district) to (51.58 Chittaurgarh district). Work participation is higher than state average in Alwar, Banswara, Baran, Barmer, Bhilwara, Bundi, Chittaurgarh, Dhaulpur, Dungarpur, Jalor, Jhalawar and Tonk districts. Remaining districts have less workman participation than the state average.

On the basis of work participation, human resources can be grouped into three categories, viz. main workers, marginal workers and non-workers. The main workers include cultivators, agricultural labourers, household industrial workers and other workers like engaged in government jobs, commerce, factories, transport, social work, artists, teachers, etc. The marginal workers are those who work during some part of the year. The non-workers are those who have not worked at any time at all in the year e.g., persons performing houshold activities, students, infants, dependent children, beggars, convicts, unempolyed youths, etc.

The proportion of main workers varies from 51.58 per cent in Chittaurgarh district to 34.51 per cent in Kota district. In general, in Alwar, Banswara, Baran, Barmer, Bhilwara, Bundi, Chittaurgarh, Churu, Dhaulpur, Dungarpur, Jalor, Jhalawar, Sawai Madhopur and Tonk districts shows higher level of work participation (Main + Marginal Workers). The porportion of marginal workers is relatively higher in the districts of Alwar, Baran, Bharatpur, Bundi, Dhaulpur, Dungarpur, Jalor, Jhunjhunun, Rajsamand and Udaipur, the highest being for Rajsamand (55.44 per cent) followed by Dungarpur (48.76 percent).

OCCUPATION AGRICULTURAL WORKERS

Agricultural occupation forms the mainstay of employment in state because a large proportion of the population lives in rural areas. The predominace of agriculture in some areas shows that urabanization and industrialization have moved at a slower pace in those areas. Agricultural workers may, therefore, be found in a large percentage in those areas where land is fertile, means of irrigation are available and industrial activities have not yet developed to the extent of absorbing growing population in secondary and tertiary activities.

Agricultural workers have their main concentration in rural areas. This group, though constituting a major proportion of the main workers in the state, generally falls into the unorganised sector. The group of agricultural workers includes both the cultivators and the agricultural labourers. Cultivators cultivate the land owned by themselves or the land leased or mortaged to them by the owners, while the agricultural labourers work on wages for land owners and cultivators.

On the national level, cultivators constitute 31.65 per cent of the total workers. In Rajasthan state, they constitute 55.28 per cent of the total workers. Rajasthan is not counted among the agriculturally developed states, for the reason that agricultural workers are more and the farm produce is less.

Rajasthan has the highest ratio of cultivators to agricultural workers on a national level which indicates that the proportion of landless labour is lowest in the state. Agricultural workers constitute only 10.62 per cent of the total workers.

The Percnetage of agricultural workers as compared to main workers, varies from less than 40 per cent to more than 80 per cent of total workers in various tehsils. The porportion of agricultural workers is the least (less than 40 per cent) in the tehsils having big urban areas like. Ajmer, Beawar, Nasirabad (Ajmer district), Baran (Baran district), Bharatpur (Bharatpur district), Bhilwara (Bhilwara district), Bikaner (Bikaner district), Karanpur, Ganganagar, Sadulshahar, Padampur (Ganganagar district), Hanumangarh (Hanumangarh district), Sanganer, Jaipur (Jaipur district), Jaisalmer (Jaisalmer district), Jalor (Jalor district), Jodhpur (Jodhpur district), Ladpura, Ramganj (Kota district), Pali, Desuri, Sumerpur, Bali (Pali district), Amet, Rajsamand, Nathdwara, (Rajsamand district), Abu Road (Sirohi district), Tonk (Tonk district), Mavli, Girwa (Udaipur district). The minimum percentage is in Jaipur tehsil (4.20 per cent) followed by Ladpura (Kota) (8.80 per cent), Ajmer (11.90 per cent) and Jodhpur (13.00 per cent)respectively, showing a greater impact of urbanization or urban functions. The tehsils which are mainly rural, have maximum proportion of Agricultural workers.

Industrial Workers

Industrialization has been very slow in the state and therefore, the proportion of industrial workers is also low. It ranges from 1 to 9 per cent in most of the tehsils. Sporadic industrial growth is noticeable in:

1. Some tehsils of Jaipur district and one tehsil of Tonk district.
2. In Some tehsils of Pali district and one tehsil of Rajsamand district.

Most of the industrial workers are engaged in cottage industries or low or medium scale industries. Large scale industries are mostly concentrated in Ajmer, Khetri, Zawar, Bhilwara, Bikaner, Kota and Jaipur and hence these tehsils exhibits a better proportion of industrial workers in the state. The proportion of industrial workers engaged in house hold industry, manufacturing, processing, servicing and repairs, etc., varies from 0.80 per cent in Chattargarh Tehsils of Bikaner district to 7.80 per cent in Tonk tehsil of Tonk district.

Other Workers

This class of workers consits of persons who are engaged in secondary and tertiary activities like government jobs, commerce, trade, transport, social work, art, teaching etc. Such workers are mainly concentrated in urban areas, industrial towns and commercial centres. These workers, though in a Small minority in the frame work of total workers, are largely a part of the organised sector.

The percentage of such workers varies from less than 10 per cent to 70 per cent in various tehsils of the state. In addition to tehsils having large urban centres like Jaipur, Jodhpur, Kota, Ajmer, Bikaner and Udaipur the some tehsils of Alwar and Bhilwara, districts, too, have a large percentage of such workers. The maximum percentage is found in Jaipur (89.60 per cent) followed by Ladpura (Kota) (80.30 per cent), Jodhpur (79.00 per cent), Ajmer (77.60 per cent), Jaisalmer (71.40 per cent), Bikaner (69.80 per cent), Girwa (Udaipur) (69.30 per cent) and Bhilwara (63.00 per cent). The tehsils of lowest proportion of such workers are those scattered in various parts of the state, the minimum percentage being for Amaud of Chittaurgarh district (5.10 per cent) followed by Ghatol (5.90 per cent), Bagidora (6.30 per cent), Kushalgarh (6.50 per cent) of Banswara district, Chattargarh of Bikaner district (7.90 per cent), Chohtan of Banner district (8.50 per cent) and Sanchore of Jalor district (9.70 per cent).

URBAN CENTRES

The urban centres are those areas which lie in Municipal limits, the Cantonment boards, and notified town areas. In addition to this, areas which are designated as towns fulfil the criteria of (a) a minimum population of 5,000, (b) having aleast 75 per cent of male working population engaged in non-agricultural and allied activities, and (c) a density of population that is atleast 400 persons per square kilometer.

The state had 216 urban centres in 2001. The percentage of urban population to the total population was (23.38 per cent) which was less than the national average of urban population (27.82 per cent), the number of towns in the state has increased from 135 (1901) to 222 including census towns (2001). In the Census year 1951, the number of towns was enumerated to be 227 but after independence, due to the change in the definition of towns, certain urban centres were disqualified and the number of towns declined steeply to 145 in 1961. Thereafter, while adhering to the revised definition of urban centres, the number has now gone to 222.

During the past 100 years, the urban population has increased from 1.6 million (1901) to more than 13.21 million (2001). In the first fifty years (from 1901 to 1951), the net increase in the urban population over all the towns in state was only 1.4 million but during the last 50 years, the addition has been extent of 9.93 million which is nearly 10 times the number of persons added to the urban population. In past ten years alone, about 31.74 lakhs persons were added to the urban centres, showing an increasing trend of urbanization. Thus, the rural urban demographic balance is under going a process of significant change.

The state has six types of towns and cities based upon the size of their population. In 2001, the number of class I towns was 20. These had a population of more than one lakh. There were 26 class II towns (population 50,000-99,999); 90 class III towns (population 20,GOO- 49,999); 59 class IVtowns (population 10,000-19,999); and 17classV towns (population 5,000-9,999) and 4 class VI Towns (population less than 5,000). It is interesting to find that in big cities of the state viz. Ajmer, Jodhpur, Bikaner and Udaipur divisional Head Quarters ratio of rural population to total population is more (Table 22.13).

Table 22.13 Decadal Urban Population

Year	Urban Population	Decadal Variation	Decadal per cent variation	Percentage Urban Population of Total Population
1901	1,550,656	-	-	15.06
1911	1,475,829	-74,827	-4.83	13.44
1921	1,475,335	-494	-0.03	14.33
1931	1,729,205	+253,870	+17.21	14.72
1941	2,117,101	+387,896	+22.43	15.27
1951	2,955,275	+838,174	+39.59	18.50
1961	3,281,478	+326,203	+11.04	16.28
1971	4,543,761	+1,262,283	+38.47	17.63
1981	7,210,508	+2,666,747	+58.69	20.93
1991	1,004,0118	+2,829,610	+39.24	22.88
2001	13,214,375	+3,174,257	+31.62	23.38
2011	17,048,085			24.9

Source: Census of India, 2011

In the decade, 1991-2001, Jaipur (59.37 per cent), Bhilwara (52.30 per cent), Ganganagar (37.99 per cent), (Pali 37.07 per cent), Tonk (35.34 per cent), Kota (31.14 per cent) have recorded high growth rates among the class I towns of the state. Low growth rates are observed in Beawar (17.99 per cent) and Ajmer (21.71 per cent) (Table 22.14).

Table 22.14 Population, Growth Rate and Sex Ratio of some important Urban Agglomeration/ Cities with a Population of 1,00,000 and above

S.No.	UA*/Cities	Population (2001	Growth Rate 1981-91	1991-2001	Sex Ratio 2001
1	Jaipur	2,324,319	+49.18	+59.37	875
2	Jodhpur	856,034	+28.10	+28.48	878
3	Kota	704,731	+49.74	+31.14	881
4	Bikaner	529,007	+44.36	+27.08	873
5	Ajmer	490,138	+7.01	+21.71	912
6	Udaipur	389,317	+32.29	+26.17	896
7	Alwar UA	265,850	+44.83	+26.51	856
8	Bhilwara	280,185	+49.88	+52.30	885
9	Ganganagar	222,833	+30.47	+37.99	828
10	Bharatpur	205,104	+48.99	+30.74	862
11	Sikar	185,506	+43.96	+25.11	919
12	Pali	187,571	+49.39	+37.07	890
13	Beawar UA	125,923	+18.57	+17.99	920
14	Tonk UA	135,663	+29.00	+35.34	934

*UA = Urban Agglomeration

Source: Census of India, 2011

TRIBLS

The population structure and demographic processes vis-à-vis hysical environmental, economic, bio-social and cultural context of major Scheduled Tribes of demographically backward state of Rajasthan, namely, Sahariya, Mina, Bhil, Kathodi, Damor and Garasia belonging to the districts of Baran, Sawai Madhopur, Udaipur, Dungarpur and Sirohi have been attempted.

POPULATION : SCHEDULED CASTES AND SCHEDULED TRIBES

Scheduled Castes

Raj asthan had 17.16 per cent of Scheduled Castes population in 2001. The distribution of this population is uneven in various tehsils and districts. The highest percentage of Scheduled Caste population is found in Karanpur tehsil of Ganganagar district (43.24 per cent) while the lowest percentage is in Kotra tehsil of Udaipur district (1.02 per cent).

The distribution pattern of Scheduled Caste population shows that the entire Ganganagar districts has highest (33.72 per cent) followed by Hanumangarh district (26.13 per cent) of this tye of population. Higher percentages are also observed in Bharatpur, Churu, Dausa, Dhaulpur and Karauli districts (20-25 per cent). In the southern districts specially in Banswara and Dungarpur districts, the percentage of this type of population is very low (less than 5 per cent). In other districts of Rajasthan, the percentage vary from 6 to 24 per cent.

Scheduled Tribes

The tribal population of Rajasthan is mostly concentrated in southern, south-eastern and eastern part of the state. Rajasthan has 12.56 per cent of Scheduled Tribes population scattered in the entire state. The percentage of male Scheduled Tribes population is (12.40 per cent) less than the female population (12.72 per cent) in the state.

The distribution pattern of this population shows in Ghatol (78.64 per cent), Bagidora (80.36 per cent), Kushalgarh (89.24 per cent) tehsils of Banswara district, Dungarpur teshil (71.55 per cent) of Dungarpur district and Kotra (89.32 per cent), Dhariawad (78.76 per cent), Sarada (75.30 per cent) tehsils of Udaipur districts have the highest population.

In the state composition of Rural-Urban Scheduled Tribes population is 15.52 and 2.88 per cent respectively. The distribution pattern of this type population shows that Banswara district is having highest ST population (72.27 per cent) of the total population of the district. Dungarpur and Udaipur districts have also higher percentage of this population 65.14 and 47.86 per cent of the total population of the area.

In Baran, Bundi, Chittaurgarh, Dausa, Karauli, Rajsamand, Sawai Madhopur and Sirohi districts concentration of this population ranges from 13.09 to 26.82 per cent. In Ajmer, Bharatpur, Bikaner, Churu, Dhaulpur, Ganganagar, Hanumangarh, Jhunjhunun, Jodhpur, Nagaur, Sikar districts, the Scheduled Tribes population is less than 5%.

As per 2011 Census 1,22,21,593 persons i.e. 17.8 % of the total population of the State belong to SC and 92,38,534 persons i.e. 13.5 % of the total population is belong to ST community. Following table shown the comparative details of the SC/ ST population in 2001 and 2011 (Table 22.15 and 22.16).

Table 22.15 Scheduled Caste/Scheduled Tribe Population 2001 and 2011

Category	Year 2001		Year 2011	
	SC	ST	SC	ST
Rajasthan	17.2	12.6	17.8	13.5

Source: Census of India, 2011

Table 22.16 Districtwise SC and ST Population - 2001 and 2011

Districts	**SC Population ('000)**		**ST Population ('000)**	
	2001	**2011**	**2001**	**2011**
RAJASTHAN	9,694.5	12,221.6	7,097.7	9,238.5
Ajmer	386.3	478.0	52.6	63.5
Alwar	539.0	653.0	239.9	289.2
Banswara	64.3	80.1	1,085.3	1,373.0
Baran	181.1	221.2	216.9	276.9
Barmer	309.0	436.4	118.7	176.3
Bharatpur	455.9	557.3	47.1	54.1
Bhilwara	316.5	407.9	180.6	229.3
Bikaner	334.2	493.6	5.9	7.8
Bundi	174.3	210.8	194.9	228.5
Chittaurgarh	250.8	250.2	388.3	201.5
Churu	407.2	451.7	10.1	11.2
Dausa	279.4	354.3	353.2	433.3
Dhaulpur	197.9	245.7	47.6	58.6
Dungarpur	46.0	52.3	7,21.5	983.4
Ganganagar	603.4	720.4	14.7	13.5
Hanumangarh	396.6	494.2	10.0	14.3
Jaipur	777.6	1,003.3	412.9	528.0
Jaisalmer	74.1	99.1	27.8	42.4
Jalor	261.3	357.2	126.8	178.7
Jhalawar	184.6	243.6	141.9	182.2
Jhunjhunun	309.2	360.7	36.8	41.6
Jodhpur	456.4	608.0	79.5	118.9
Karauli	280.1	354.5	270.6	325.0
Kota	300.6	405.4	152.0	183.8
Nagaur	545.2	699.9	6.5	10.4
Pali	323.5	398.1	105.8	144.6
Pratapgarh		60.4		550.4
Rajsamand	122.5	148.2	129.2	160.8
Sawai Madhopur	223.2	278.8	241.1	285.8
Sikar	339.8	418.8	62.5	75.3
Sirohi	163.0	201.9	210.8	292.5
Tonk	233.1	287.9	145.9	178.2
Udaipur	158.3	188.5	1,260.4	1,525.3

SC: Scheduled Caste, **ST:** Scheduled Tribe

Source: Census 2001 and 2011, Office of the Registrar General and Census Commissioner, Govt, of India. New Delhi.

As per 2011 Census 1,22,21,593 persons i.e. 17.8 % of the total population of the State belong to Scheduled Caste and 92,38,534 persons i.e. 13.5 % of the total population is belong to Scheduled Tribe community. Following table shown the comparative details of the SC/ ST population in 2001 and 2011 (Table 22.17).

Table 22.17 Comparative SC/ ST population in 2001 and 2011.

Category	Year 2001		Year 2011	
	SC	ST	SC	ST
Rajasthan	17.2	12.6	17.8	13.5

Source: Census of India, 2011

Census of India 2001 As per the 2001 census, the Scheduled Tribe (ST) population of Rajasthan State is 7,097,706 constituting 8.4 percent of the total ST population of India. The Scheduled Tribes of the State constitute 12.6 percent of the total population (56,507,188) of the State. It holds 12th position among all States and UTs in respect of the percentage share of ST population to total population. The ST population has registered a growth rate of 29.6 per cent during 1991-2001 which is 1.2 per cent higher than the growth of the total population.

The Scheduled Tribes are overwhelmingly rural as 94.6 per cent of them inhabit in rural areas whereas 76.6 per cent of the total population of the state live in villages.

District wise distribution of tribal population shows that they have their highest concentration in Banswara district (72.3 per cent), followed by 65.1 percent and 47.9 per cent in Dungarpur and Udaipur districts respectively. Nagaur (0.2 per cent) preceded by Bikaner (0.4 per cent) has the lowest share of tribal population in the total population. Population-Size and Distribution

Out of twelve (12) tribes scheduled for the State, Mina is the most populous tribe, having a population of 3,799,971 constituting 53.5 per cent of the total ST population followed by Bhil (2,805,948). Mina and Bhil together constitute 93 per cent whereas Garasia, Damor, Dhanka and Saharia combine to form 6.6 per cent of the total ST population. Six tribes, Bhil Mina, Naikda, Kathodi, Patelia, Kokna and Koli Dhor along with the generic tribes constitute the residual 0.3 per cent of the total tribal population. Among the twelve STs, Koli Dhor is the smallest tribe with a population below 100 preceded by Kokna (405), Patelia(1,045), Kathodi (2,922) etc.

Mina account for 99 - 99.8 per cent of all ST population in the five districts namely Dhaulpur, Bharatpur, Sawai Madhopur, Karauli, and Dausa. The tribe share more than 90 per cent of the total tribal population in Alwar, Jhunjhunun, Sikar, Tonk and Churu districts. The second major tribe, Bhil have a share of 99 per cent of all ST population in Barmer district followed by Jaisalmer, Jodhpur, Rajsamand (95 per cent) districts.

Sex Ratio

The Scheduled Tribes of the State have registered considerably lower sex ratio both in total as well as in the 0-6 yrs. population in comparison to those recorded by all STs at the national level. The over all sex ratio of the ST population is 944 females per 1000 males which is lower than the national average of 978 in respect of all ST population.

Among the major tribes, Mina and Dhanka have overall sex ratio lower than the national as well as state averages whereas Bhil, Garasia, Seharia tribes have returned overall sex ratio higher than that of the State average. Damor have registered the overall sex ratio significantly higher than the national average.

The sex ratio among ST children in the age group 0-6 yrs. is 950. Among the individual tribes, Damor have female children outnumber the male children whereas among Garasia both male and female children are in equal number in the corresponding age group. The child sex ratio among Bhil is higher than the state as well as national level. The lowest child sex ratio has been reported by Mina (Table 22.18).

Table 22.18 The Sex Ratio

Age Group	All STs (India)	All STs (State)	Damor	Garasia (excluding Rajput Garasia)	Bhil	Seharia	Mina	Dhanka
All ages	978	944	997	971	970	961	924	923
0-6 yrs.	973	950	1011	1000	975	960	925	907

Source: Census of India, 2011

Literacy and Educational Level

The Scheduled Tribes of the state have a literacy rate of 44.7 per cent which is lower than the national average (47.1 per cent) in respect of all STs. There has been a significant improvement in the rate of literacy during the decade 1991-2001. The over all literacy rates which was merely 19.4 per cent at 1991census has increased to 44.7 per cent at 2001, higher by 25.3 percentage points. Male literacy has gone up from 33.3 per cent in 1991 to 62.1 per cent which is higher than that of national average of 59.2 per cent. On the other hand, though the female literacy has increased six times from meagre 4.4 per cent at 1991 to 26.2 per cent at 2001, it is still lower than the national average of 34.8 per cent recorded by all tribal females (Table 22.19).

Among the larger groups, Dhanka and Mina have registered literacy rate higher than that recorded by all STs at the national level whereas Damor, Bhil, Seharia and Garasia have lower literacy rate than those of national as well as State averages. The given below brings out variations in total and female literacy rates among the major tribes.

Table 22.19 Literacy Rate

Literacy Rate	All Scheduled Tribes (India)	All Scheduled Tribes (State)	Dhanka	Mina	Damor	Bhil	Seharia	Garasia (excluding Rajput Garasia)
Persons	47.1	44.7	61.8	52.2	41.2	35.2	34.2	30.7
Females	34.8	26.2	45.9	31.8	23.2	19.1	18.7	15.4

Source: Census of India, 2011

So far as the levels of education are concerned, slightly more than half (50.8 per cent) of the literate persons are either without any educational level or have attained below primary level of education. The primary level literates constitute 24.2 per cent followed by literates upto middle level (13.2 per cent). The persons educated up to matric/ secondary/ higher secondary constitute 9.1 per cent whereas 2.5 per cent only are graduates and above. Non-technical and technical diploma holders form negligible percentage (0.1).

There has been increase in the drop out rate after the primary level of education. The percentage of matriculate persons is almost one fourth of the persons educated up to primary level.

Graduate persons constitute approximately one third of the matriculates. The major tribes have shown the similar trend in respect of the level of education (Table 22.20).

Table 22.20 Educational level among the major Scheduled Tribes

Names of STs	Literate without educational level	Below primary	Educational levels attained				
			Primary	Middle	Matric/Secondary Higher Secondary/ Intermediate etc.	Non-technical and Technical diploma or certificate not equal to degree	Graduate and above
All STs	9.5	41.3	24.2	13.2	9.1	0.1	2.5
Mina	7.9	37.7	24.6	15	11.2	0.1	3.5
Bhil	12.6	48	23.4	9.8	5.3	0.1	0.9
Garasia (excluding Rajput Garasia)	16.2	52.8	19.4	7.3	3.6	0.1	0.6
Damor	10.1	40.8	25.5	13.6	8.1	0.2	1.7
Dhanka	4.5	33.6	29.7	18.8	10.9	0.1	2.4
Seharia	7.9	59	26.2	4.8	2	0.1	0.1

Source: Census of India, 2011

Little more than half of the total 20 lakh tribal children in the age group 5 -14 go to school. It is noted that around 9.2 lakh (46 per cent) children of this age group are not attending school. Mina, the numerically largest tribe have 62 per cent children in the corresponding age group go to school. Dhanka have the highest percentage of school going children in the same age group. Bhil, Seharia, Garasia lag behind as they have less than half of the total children in the corresponding age-group attend school. Work Participation Rate (WPR)

The work participation rate of the ST population in Rajasthan is 47.6 per cent which is lower than that of the national average (49.1 per cent). The male work participation rate has declinedslightly from 51.8 per cent to 50.1 per cent whereas female WPR has increased from 40.6 per cent to 44.9 per cent during 1991-2001. Among the total workers, 66.8 per cent are main workers, which is nearly equal to that recorded for all STs at the national level (68.9 per cent).

Individually, Bhil (48.4 per cent) and Mina (47 per cent) have WPR lower than the national average whereas Damor (55.2 per cent) and Garasia (49.4 per cent) have WPR higher than the national average. Damor tribe also have the highest female WPR (54.5 per cent) followed by Garasia (47.1 per cent) and other tribes. Category of the Workers

Agriculture is the main economic activity of the tribes of Rajasthan. While 69 per cent of total workers are 'Cultivators' which is significantly higher than the national average of 44.7 per cent, 'Agricultural Labourers' constitute only 14 per cent which is less than half of that recorded by of total STs at the national level (36.9 per cent). 'Other Workers' constitute 16.3 per cent and workers in 'Household Industry' account for only 0.7 per cent.

Expectedly, majority of the workers are 'Cultivators' among Mina, Damor, Bhil and Garasia tribes whereas Seharia have maximum proportion of 'Agricultural Labourers' followed by 'Cultivators'. Dhanka have the highest proportion of 'Other Workers' (Table 22.21).

Table 22.21 Percentage Distribution of Workers in four Economic Categories

Economic Category	All Scheduled Tribes	Mina	Damor	Bhil	Garasia (excluding Rajput Garasia)	Seharia	Dhanka
Cultivators	69.0	74.8	73.1	64.3	60.3	24	15.1
Agricultural Labourers	14.1	9.8	14.8	17.7	18.4	66.4	15.1
HHI Workers	0.7	0.7	0.9	0.7	0.5	0.5	2.3
Other Workers	16.3	14.7	11.2	17.2	20.8	9.1	67.5

Source: Census of India, 2011

Marital Status

The data on marital status show that the 'never married' persons (51.7 per cent) constitute more than half of the total population. 'Married' persons account for 44.5 per cent while 3.5 per cent are 'widowed'. A negligible proportion (0.2 per cent) of STs are 'divorced and separated'.

Like other communities of the State, practice of the marriage of girls and boys below the legal age is prevalent among the tribes also. It may be discerned from the proportions of married tribal girls below 18 years (4.3 per cent) and tribal boys below 21 years (5.7 per cent) are considerably higher than those at national level (2.1 per cent and 2.8 per cent respectively). The STs namely, Mina, Seharia have registered the higher percentage of married boys and girls below the legal age in comparison to those of the state average (Table 22.22).

Table 22.22 Percentage of married girls and married boys below the stipulated

Age group	All Scheduled Tribes (State)	Seharia	Mina	Bhil	Dhanka	Damor	Garasia (excluding Rajput Garasia)
Married girls (less than 18 yrs.)	4.3	5.9	5.0	3.6	3.5	3.0	1.2
Married boys (less than 21yrs.)	5.7	8.6	6.0	5.5	4.1	4.5	3.2

Source: Census of India, 2011

The mean number of children ever born per ever married tribal woman (age-group 45-49 yrs.) is 5 which is higher than that of all STs at the national level (4).

Religion

The Scheduled Tribes professing Hinduism form 99.5 per cent, whereas those professing Islam and Christianity constitute a meagre 0.1% and 0.3% respectively. Tribal following Sikhism, Budhism, Jainism, Other religion and persuations together constitute a very small (7,174) population.

Composition of Population

Eighty-nine percent of the state's population comprises Hindus. Scheduled Castes and Scheduled Tribes form 17.29 percent and 12.44 per cent, respectively, of the total population.

About 8 per cent of population is Muslim, most of whom are Sunnis. There is a small affluent community of Shia Muslims in southeastern Rajasthan, known as Bohras.

Caste system integral to Hinduism, dominates Rajasthan, as it does in other parts of India. However, there are some local variations of the caste system. Instead of the Brahmins, the Rajputs are at the top of the social hierarchy. The Rajputs comprise various clans (or Khamps) according to their dynastic families. Due to the integration of the princely kingdoms into the state of Rajasthan and the abolition of the system of *Jagirdari*, the Rajputs were nudged from their positions of power by the Brahmins and other castes. In Rajasthan, the Brahmins are subdivided into two groups – Channatiyas and non-Channatiyas – and intermarriage between them is traditionally forbidden.

Below the Rajputs are the Vaisyas, who are divided into two groups: those who profess Jainism, and those professing Vaishnavism.

Below the Vaishyas are Scheduled Castes who are scattered throughout the state. Apart from them, the communities of Jats and Gujjars are traditionally engaged in farming and animal husbandry.

Scheduled Castes

In 1991, Scheduled Castes comprised 17.3 percent of the population of Rajasthan compared to 16.5 percent in the whole country. The highest percentage of Scheduled Castes to total population in 1991 has been noticed in the district of Ganganagar (29.0 percent), whereas the lowest percentage has been recorded by Dungarpur and Banswara (about 5 percent by each).

The Scheduled Castes in the state are: Chamar (leather workers), Meghwal, Balai, Khatik, Bola, Bairwa, Dḥed, Jingar, Dabgar and Pasi. In addition, there are castes traditionally associated with scavenging: Bhangi, Mehtar, Valmiki and Chura. In addition, the artisan castes, like Koli (Kori or Koria), Sargara, Mahar or Mohar, Gavaria, Salvi, Gancha (Garancha Mehtars), Tirgar (Kamangers) and Koochband (Kuchband) inhabit different parts of the State. The castes associated with music and entertainment are, Dom (Dome), Kalbelia (Sapera), Nat (Nut), Dholi, Bhand, Kamad (Kamadia), Badi, Bazigar (Madari), and Bansphor. Moreover, there are castes traditionally associated with criminal activities, such as, Bawaria or Baori or Moghias, Bagri, Sansi (including Kapadia Sansi), Kanjar (Kunjar) and Bedia (Beria). Other miscellaneous castes inhabiting the state of Rajasthan are : Dhanak (Dhankia), Garuda/Garada, Dhobi (washerman), Aheri (hunter), Santia, Chandal, Singiwala, Sarbhangi (Aghoris), Thori, Majhabi, etc. Apart from these, there are those who are not associated with traditional occupations but have separate names. These are Birgis/Vargi /Bargi, Bajigar, Adi Dhasni, Bidakias, Godhi, Khangar and Marg-Gradi.

Scheduled Tribes

According to 1991 Census (Rajathan), the highest concentration of Scheduled tribes is noted in Banswara and Dungarpur districts, where their proportion to the total population of the district is 72.63 percent and 64.44 percent, respectively. Other districts with higher proportions are Udaipur (36.8 percent), Sirohi (23.4 percent), Sawai Madhopur (22.6 percent), Bundi (20.8 percent), Chittorgarh (20.3 percent), and Kota (14.2 percent). On the other hand, the districts – Ganganagar, Bikaner, Churu and Nagaur have less than 1 percent of Scheduled Tribes. According to the Scheduled Castes and Scheduled Tribes Orders (Amendment) Act, 1976, the Scheduled Tribes of the State are given in Table 22.24. Of the twelve Scheduled Tribes, the Minas and the Bhils, along with their sub-groups, Damor, Garasia, and Saharia, constitute the major tribal groups of the state. These population groups have been considered for studying in detail in the present paper along with Kathodis. Based on the recorded history as well as the current socio-economic status, the Minas, who constitute a sizeable population amongst the tribes, do not qualify to be 'scheduled' as a community for compensatory discrimination. The Minas are the descendants of tribes, which ruled over most of the later princely state of Jaipur before the Kachachwahas established their rule nearly a thousand years ago. A Mina chief was always associated with the coronation ceremonies of Kachachwaha rulers. A significant percentage of Minas were employed as watch and ward staff by the Jaipur durbar and his Jagirdars. Another group of Minas became skilled agriculturists, cultivating substantial landholdings, thus making the entire Mina community well off. Today, they are availing of the special facilities available to

Scheduled Tribes to improve their condition still further.

By contrast to the Minas are the Bhils, the largest Scheduled Tribe, who live in the hilly forests of southwestern Rajasthan. Despite their high socio-political status on account of the military support they gave to the Sisodia rulers of Mewar in general and Maharana Pratap in particular, they lead a miserable existence, mostly devoid of the amenities of education, communication, transport and other benefits of modernization (Table 22.23).

Table 22.23 Tribal groups of Rajasthan

S. No.	Name of the Tribal/ Sub-tribal group	Population	Percentage to total population
1.	Bhil, Bhil Garasia, Dholi Bhil, Dungri Bhil, Dungri Garasia, Manasi Bhil, Rawal Bhil, Tadvi Bhil, Bhagalia, Bhilala, Pavira, Vasava, Vassava	18,61,502	44.500
2.	Bhil Mina	18,687	0.446
3.	Damor, Damari	30,603	0.732
4.	Dhanka, Tadvi, Teleria, Valvi	16,238	0.388
5.	Garasia (exclude Rajput Garasia)	1,21,939	2.915
6.	Kathodi, Kattcari, Dhar Kattkari, Dhar Kattari, Son Kathadi, Son Kattkari	2,541	0.061
7.	Konka, Kokai, Kukna	165	0.004
8.	Koli, Dhan, Tokre Koli, Kolcha, Kolgha	1,881	0.045
9.	Mina	20.69,456	49.471
10.	Naikda, Nayaka, Choli vale Kapadia, Nayaka, Mota Nayaka, Nana Nayaka	10,287	0.246
11.	Patelia	1,724	0.041
12.	Saharia, Seharia, Sahariya	41,487	0.990
13.	Others (unspecified)	1,674	0.161
	Total	**41,83,124**	**100.00**

Source: Census of India, 1991, Rajasthan

The Bhils and Minas account for roughly 50 and 44 percent of the tribal population of Rajasthan. Of the rest, Garasias and Sahariyas constitute roughly three and one percent, respectively. The southern districts of Rajasthan - Banswara, Dungarpur, Udaipur, and Chittorgarh have concentration of tribal groups of Bhils and Damors, while Garasias are mainly concentrated in Sirohi and Pali. Minas inhabit Jaipur, Sikar and Alwar, whereas Tonk, Bundi, Kota and Jhalawar have both Bhils and Minas. Sahariyas inhabit the tehsils of Kishanganj and Shahbad of Baran (erstwhile Kota) district.

Apart from the Scheduled Tribes, there are some other tribal communities inhabiting different parts of the state. They are broadly classified as Denotified Tribes and include Baori, Kanjar, Sansi, Bogri, Nat, Bhat, etc.; and Nomadic Tribes, which include Banjara, Gadulia Lohar, Kalbelia, Shikkaligar and few other groups; and Seminomadic tribes comprising Rabari, Jogi, Masani and others.

The Gujjars, who profess Hinduism, dwell in eastern Rajasthan, including Jaipur, Udaipur, Alwar, Kota and Bharatpur. They are divided nomadic Rabari or Raika are also Hindu. They are divided into two groups, the Marus, who breed camels, and the Chalkias, who breed sheep and goat (Table 22.24).

Table 22.24 Tribal population of Rajasthan according to Census, 1981

Tribal groups	Population	Percentage to total Tribal population
Mina	2,969,456	49.47
Bhil	1,861,502	44.50
Garasia	121,939	2.91
Saharia	41,427	0.91
Damor	30,603	0.73
Others (Unspecified)	58,197	1.40

Source: Census of India, 2011

The affluent Oswals hail from Osiyan near Jodhpur, and are successful in trade and commerce. They are predominantly Jains, although a few profess Vaishnavism. Oswal women observe strict purdah or seclusion. Another community of Rajasthan, which deserves a special mention because of its currently relevant tradition, is Bishnois. They are famous because of their conservationist religious beliefs that restrict taking of animal and vegetable life. The environment protectionist beliefs of Bishnois are rooted in the religious teaching of their spiritual mentor, Jambhoji. Today, the Bishnois are a wealthy farming orthodox about observing the twenty-nine (bishnao in Hindi) principles laid down by their guru Jambho-ji Maharaj (Figure 22.2).

Sahariyas: They are the most backward of all the Scheduled Tribes of Rajasthan. They are also considered as a primitive Tribe. Their customs and manner bear great resemblance with caste Hindus with whom they live in their present habitat (mostly known as 'sehranas' at the periphery of the main villages), but who consider them untouchables. There are 41,427 Sahariyas residing in Rajasthan forming 0.91 percent of the total Scheduled Tribe population. They cultivate land and work as agricultural/casual labourers and raise livestock. They reside in district of Baran (erstwhile Kota District), Jhalawar, Udaipur, Dungarpur, Jaipur, Sawai Madhopur and Churu with their main concentration (99.2 percent) in Baran. They do not have their own language; around 61 percent of them speak *Khariboli,* 23 percent *Brijbhasha* and 15 percent speak *Hadoti*, while others have taken to local dialects.

Minas: They held power over a large part of Rajasthan before the advent of the Rajputs. Those loyal to the new rulers were granted land and other sources of production. Those who could not reconcile to their defeat formed organised bands and adopted the path of resistance, crime and violence. Attempts were made to rehabilitate them by offering the job of village watchman. That is why there are two divisions –the Zamindars and the Chowkidars, – which continue till date. There are around 2,969,456 (Census of India, 1981) Minas in Rajasthan who constitute the largest part (49.47 percent) of the total Scheduled Tribe population of the State.

They are spread over the whole state, but are mainly concentrated in Jaipur, Sawai Madhopur and Udaipur where over 51 percent of their population reside. They cultivate land, and both men and women participate in agricultural activities. They are Hindus but have faith in certain animistic traits as well. The Minas are not a Scheduled Tribe in Ajmer.

Bhils: They derive their name from the Dravidian word, Billu, meaning a bow. They live in scattered hutments separated widely, or parched on hilltops, depending upon the topography.

There are 1,861,502 Bhils (Census of India, 1981) in Rajasthan; they constitute 44.50 percent of the total population of Scheduled Tribes. They inhabit the villages of the state where they practise agriculture, wherein both men and women participate. They are spread all over the state but are mainly concentrated in the districts of Udaipur, Banswara and Dungarpur. About 40 percent of them speak *Bhilli* and its allied dialect called *Wagdi*, which is spoken in Banswara and Dungarpur. Bhils living in other parts have adopted local dialects.

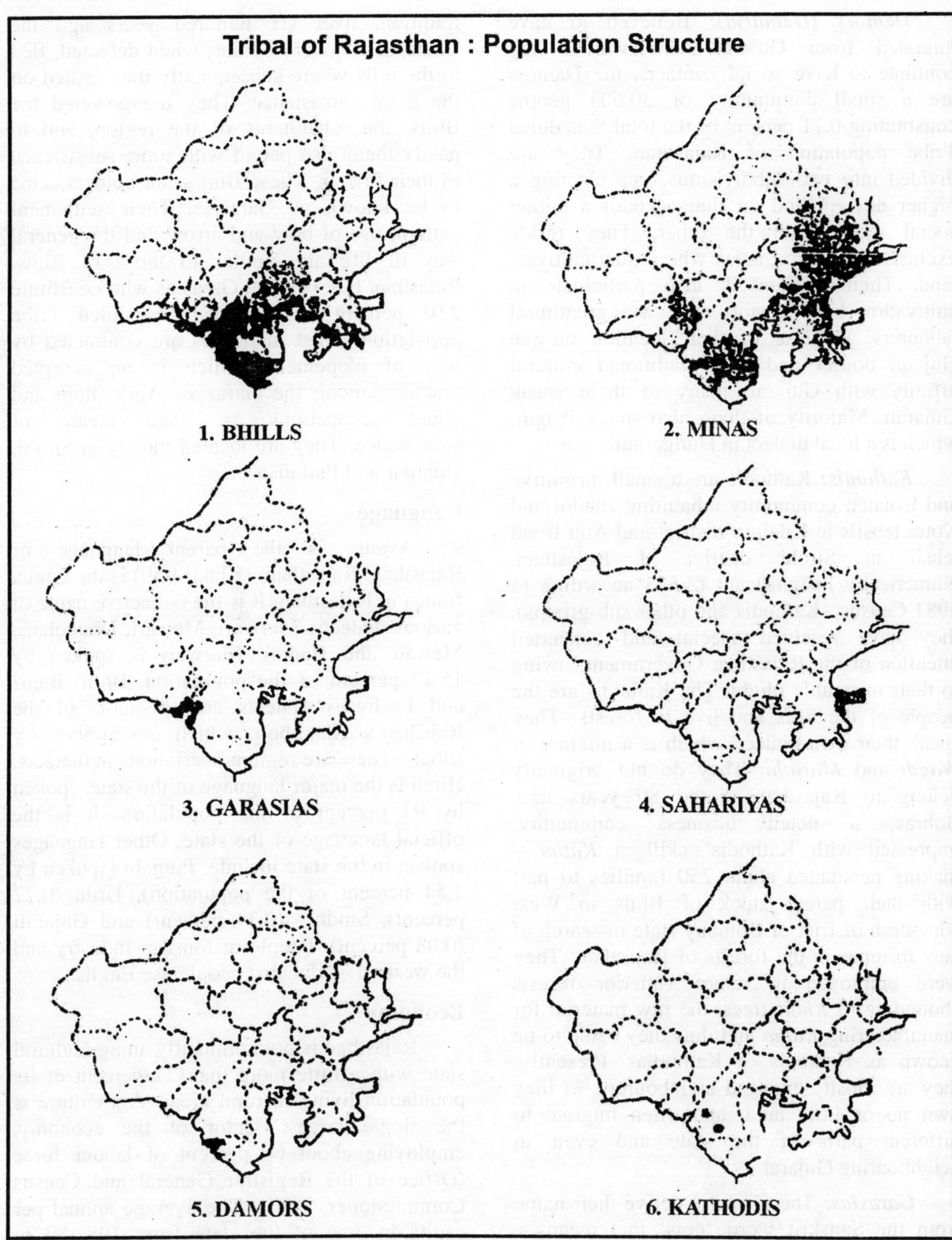

Source: Census of India, 2011

Figure 22.2 Dispersal of Scheduled Tribes in Rural Areas of Rajasthan State

Damors (Damaria): Believed to have migrated from Gujarat with which they continue to have social contacts, the Damors are a small community of 30,603 people constituting 0.71 percent of the total Scheduled Tribe population of Rajasthan. They are divided into two subdivisions, one claming a higher descent and on that account a higher social status than the other. They reside exclusively in rural areas where they cultivate land. Their womenfolk also participate in cultivation but do not work as agricultural labourers. Because of their location on the Gujarat border and their traditional cultural affinity with Gujarat, many of them speak Gujarati. Majority of them also speak *Wagdi*, which is a local dialect in Dungarpur.

Kathodis: Kathodis are a small, primitive and isolated community inhabiting Jhadol and Kotra tehsils in Udaipur district and Abu Road tehsil in Sirohi district of Rajasthan. Numerically insignificant (2,533 according to 1981 Census –Kathodis and other sub-groups), they have received special and concerted attention of the Rajasthan Government, owing to their miserable plight. The Kathodis are the people of the hills and live in forests. They speak their own dialect, which is a mixture of *Wagdi* and *Marathi*. They do not originally belong to Rajasthan. Some 80 years ago, Bohras, a noted business community, impressed with Kathodis' skill at *Katha* – making persuaded about 250 families to part with their parent stock of Bhils in West Khandesh district of Bombay state in search of new fortunes in the forests of Rajasthan. They were employed in remote interior forests abounding in *Khai*r trees, the raw material for manufacturing *Katha* and thus they came to be known as Kathodis or Kathodias. Presently, they are mostly engaged as labourers, as they own no or little land, and often migrate to different parts of the state and even in neighbouring Gujarat.

Garasias: The Garasias derive their name from the Sanskrit word 'gras' that means a morsel or subsistence. According to their tradition, over six hundred years ago the Chauhan Rajputs of Jalor, when defeated, fled to the hills where subsequently they settled on the grant subsistence. They overpowered the Bhils, the inhabitants of the region, and to pacify them also parted with some subsistence in their favour. These Bhil grant-holders came to be known as Garasias. Their settlement pattern, use of bow and arrow and the general way of life are similar to those of Bhils. Rajasthan has 121,939 Garasias who constitute 2.91 percent of the total Scheduled Tribe population. Most marriages are contracted by way of elopement, which is an accepted practise among the Garasias. Agriculture and allied occupations are their means of subsistence. They are located mostly in Sirohi, Udaipur and Pali districts.

Language

Avanti is the parent language of Rajasthan. Rajasthani (Khadi-boli) is the lingua franca in Rajasthan. It is the collective name of various dialects: Marwari, Mewari, Dhundhari, Mewati and Hadoti. Marwari is spoken by 15.27 percent of the population. Bhili, Bagri and Pachwari dialects are offshoots of the Rajsthan's Khadiboli, which are spoken by tribals. There are regional variations in dialects. Hindi is the major language of the state, spoken by 93 percent of the population. It is the official language of the state. Other languages spoken in the state include, Punjabi (spoken by 1.54 percent of the population), Urdu (0.77 percent), Sindhi (0.34 percent) and Gujarati (0.08 percent). People of tourism industry and the western – educated people use English.

Economy

Rajasthan is predominantly an agricultural state with a little more than 77 percent of its population living in rural areas. Agriculture is the single largest sector of the economy, employing about 60 percent of labour force (Office of the Registrar General and Census Commissioner, 1991). The average annual per capita income of the state from 1982-83 to 1984-85 was Rs. 1,818 (Centre for Monitoring

Indian Economy, 1991). During 1987-88, 26 percent of the rural population and 19 percent of the urban population were estimated to be below the poverty line.

Agriculture

Agriculture and animal husbandry form the mainstay of the state's economy, representing 44.79 percent of the state's revenue in 1994-95, as against only 31.74 percent of the country as a whole. There are two agricultural seasons –Kharif and Rabi, of which Kharif season particularly depend on the quantum of rain as well as its proper distribution over a reasonable time span and its intensity. Similarly, Rabi crop is adversely affected if the skies are overcast continuously in the winter season. Cereals, pulses, food grains, oil seeds, chillies, tobacco, sugarcane and cotton, are grown during the two agricultural seasons. The state ranks sixth in India in terms of the per capita production of food grains and tenth in the growth rate of the production of food grains. The average annual per capita food grain production of the state from 1987-90 was 188 kilograms. Irrigation is an essential input for agricultural production in the state. Surface water resources are scarce as there are no perennial rivers traversing the southeastern region of the state. Therefore, to a great extent agriculture is dependent upon the vagaries of monsoon. There are four major sources of irrigation, viz., canals, tanks, wells and tube-wells. Out of the total area cultivated in the State, during 1995-96, only 32.33 percent area (on an average) was under irrigation. Out of the total irrigated area, 62.76 percent was covered by wells and tube wells, 33.24 percent by canals and 4.00 percent by other sources.

The recharge and sustainability of wells and tube wells depend primarily on rainfall, so the whole area barring the 33 percent that is canal irrigated, depends on rainfall. Irrigation facilities are available for an area of over 9.88 million acres (4 million hect.). Agricultural production is inadequate to meet local needs, in three out of five years due to insufficient or untimely rainfall. Consequently, there is a severe shortage of food, fodder, fuel and drinking water. Frequent droughts lead to temporary out migration of human and cattle population.

Animal Husbandry

Despite ecological limitations, agriculture and animal husbandry constitute the primary occupation in Rajasthan, with livestock management having been refined into a productive art to suit the arid and semiarid lands. The minimal vegetation found in the region has been skilfully exploited to sustain large herds of sheep, goats and camels. In conditions of low rainfall and high levels of evaporation and transpiration of moisture, the people of Rajasthan have evolved varieties of cattle, which are amongst the best in the country. Today, welladapted packages of agriculture and animal husbandry are being threatened by new policies of agricultural development, which have little relevance to the ecological specialities of Rajasthan. The traditional ecology-specific life style of people is being modified without taking into account the possible long-term impact of current changes. These eco-developmental considerations are crucial to the understanding of the transhumant way of life of pastoral people of Rajasthan who lead a nomadic or semi-nomadic existence. The settled livestock farmers produce a great quantity of milk, which is collected by public sector dairies for pasteurisation and consumption.

A large number of small and marginal farmers, agricultural labourers and other local poor depend upon livestock for gainful employment. There were 484.46 lakh livestock and 30.13 lakh poultry in the State (1992 Livestock Census). Under the Integrated Dairy Development Programme, 16 milk producers Co-operative unions are operational in all the districts of the State. Sheep husbandry is popular and it provides employment opportunities to weaker sections. According to Livestock Census of 1992, there were 124.91 lakh sheep in the State, which is 25 percent of the country's sheep population. 170-lakh kg wool is produced in the state, which is nearly 40 percent of the country's wool production.

Industry

During the last 35 years, Rajasthan has become a major producer of synthetic yarns, cement, zinc, copper, trucks, tractors, scooters, tyres, cords and cables, railway wagons, ballbearings, water and power metres, automobiles parts, instrumentations, electrical equipments and electronic goods like, copper foils, copper clad laminator, television sets, picture tubes, milk testers, wireless equipments, sugar, marble, sandstone and a number of handicrafts. In 1992-93, there were 30 textile mills in the state, employing around 50,000 people. Almost half of the national output of polyester viscose yarn is produced here, contributing around Rs. 400 million to the national economy. The other big industry of Rajasthan is the cement industry, employing about 15,000 people. The productions of sugar, as well as mining of marble and sandstone, also contribute to the state's economy.

Handicrafts and Small Scale Industries

Handicrafts of Rajasthan are very popular and contribute to its economy. It is estimated that more than 600,000 people are employed in this sector. In 1994, there were 164,000 small scale industries, including cotton and wool spinning, carpet making, block printing, gem cutting and polishing, ivory carving, pottery and brassware production, leather goods production, marble carving and enamelling, among others. Rajasthan is also famous for its vibrantly coloured ethnic clothes with tying-dyeing, block printing and numerous forms of embroidery and appliqué. In the village of Kaithone, in Kota district, the most highly prized *Masuria* (woven from both cotton and silk) is made into beautiful sarees.

Chapter 23

GEOGRAPHICAL REGION

Rajasthan is largest state of country spread over on 10.41 percent of area, while population of state as per 2011 census is 6,86,21,012 persons, which is 5.67 percent of country. Sex ratio of state is 926 females per thousand males, while density of population is 201 persons per sq. kms. Regional variation of state can be accessed from the fact that normal rainfall of state is 57.51 centimeters, varying from 18.55 centimeters in Jaisalmer to 95.03 centimeters in Banswara. Rainfall is erratic creating disturbed conditions of famine and floods in state. Geographical formations cover one percent high altitude, 51 per cent plain area, 31 per cent uplands and 11 per cent mountains and hills. In administrative system, state is divided into thirty three districts.

Criteria for delimitation of regional boundaries are determined by purpose and objective for which the regional formation is necessitated. According to Prof. Herbertson, geographical region is land or territory, where factors influencing common human life. It focuses on specific area of earth surface and studies through gamut of geographical factors, physical, biotic, human, working in unison and intersecting among themselves and with other areas for a long period of time and provide relative homogeneity or distinct character. Hartshorne regarded hierarchical method as basis for demarcating various regions, which is helpful in demarcation of various regions. It is helpful in demarcating physiographic, climate and vegetation region.

Methodology for determination of geographical regions has undergone various conceptual changes to associate cultural peculiarities while studying any specific area from regional point of view There are several objectives for demarcation of geographical regions comprising of climate, river valley, topography, soil and vegetations. Cultural cohesion comprise of common language, history, religion, ideology or political limitations. Distinctive economic factors include industrial development, nature of agriculture and source of minerals. Technical and welfare considerations include transport, incidence of diseases, pest control or illiteracy. There are many criteria included in demarcations of geographical regions, though many have no relevance from such aspects.

Criteria for delimitation of regional boundaries are determined by purpose and objective of regional systems. Determination of regions of state can be done on geographic factors comprising of climate, river valleys, topography, soils and vegetations. Cultural factors of division can be common language, history, religion or ideology and similarity of political institutions. Distinctive economic factors for regional demarcation would cover industrial development, nature of agriculture and sources of minerals. Various technical and welfare considerations comprising of air transport, incidence of diseases, pest control and illiteracy. Keeping various factors in view, regional classification of state is determined on geographical concepts which have been covered and analyzed hereunder.

Usually physiography of various regions, climatic conditions, vegetation, types of soil, agricultural practices, availability of minerals, industrial growth and demographic factors have been taken as basis of geographical regions, where sex ratio, expectations of life is among few criteria for such determination. Quality of man-land relations has been established in region, nature of development can be specific feature transforming any geographical region into live region, where prosperity of people is visible from geographical factors. Hall had pointed out the limit of region, as not critical factor but qualitative aspects have significance with visible variations.

District is treated as revenue unit of any state, but such unit has significance of regional system as homogeneity features, but such situation is also not significant for Rajasthan, where Jaisalmer has geographical area of 38401 sq kms spread over 11.22 per cent of state area, while Dholpur district with 3033 sq kms of area is 0.89 per cent of state area. These factors have another relevance, where livelihood measures are significant factors, where population density of Jaisalmer is 17 persons per sq kms, while Dholpur has 398 persons per sq km and Jaipur with 598 sq kms. People of state had preferred to reside in area where survival of people is possible and even in high density locations, people prefer to reside instead of shifting to low density areas.

V.C. Mishra (1966 and 1968) has given scheme of seven geographical regions and described them briefly in general. The second atempt was made by A.K. Sen (1968), who has given scheme of four geographical regions based on climatic conditions and natural regions. Third attempt was made by R.L. Singh (1971) while dividing India into regions, He suggested the division of Rajasthan into two middle order regions. Then he further divided these two regions into four first order and thirteen second order regions. In this scheme due attention was paid to local differentioations in addition to natural and cultural similarities, fourh attempt was made by B.K. Roy (1981) of Census of India based on phyio-geographical regions including physiography, geology, vegitation, climate, soil and population. While dividing India into several regions, he suggested two divisions for Rajasthan based on phisiographical regions- (1) The Great Plains and (2) The Decan Plateau. These geographical regions has distinct characteristics in form of landforms, relief, population and settlement, these regions were further divided into meso, micro and sub-micro regions.

Fifth attempt was made by T.S. Chouhan (1987) in Agricultural Geography: A Study of Rajasthan State on the basis of preceding investigation, Four well idefined Geo-agriculture regions were identified in Rajasthan. The division is based on the natural, cultural and local homogeneity factors of the region. They may be considered as an important planning frame for development in Rajasthan—

(1) The Western Geo-Agricultural Planning Region.

(2) The Northern Geo-Agricultural Planning Region.

(3) The Central and Geo-Eastern Agricultural Planning Region.

(4) The South - Geo-Eastern Agricultural Planning Region.

Various features of area and population are determined on factors have relevance, which have been categorized as manmade factors comprising of irrigation, industrial development, means of communication, while disparity in geo-graphical factors and man made factors have relevance from different points of views. Keeping these aspects in view, Rajasthan has been divided into seven distinct geographical regions given in Table 23.1.

Each geographical region has distinct characteristics in terms of physiographic features, climatic conditions, type of soils, vegetation growth, agricultural practices, minerals deposits, population concentration and means of communications. In addition to natural features, people manage various facilities for their survival base through availability of resources. Western arid region comprises of Thar Desert, while presence of canal has changed geographical features of area to vital extent. Similar situation is of plateau region of state, where inter-state river water has changed physical features to significant extent. Accordingly geographical features have changed conditions for developmental impacts (Figure 23.1).

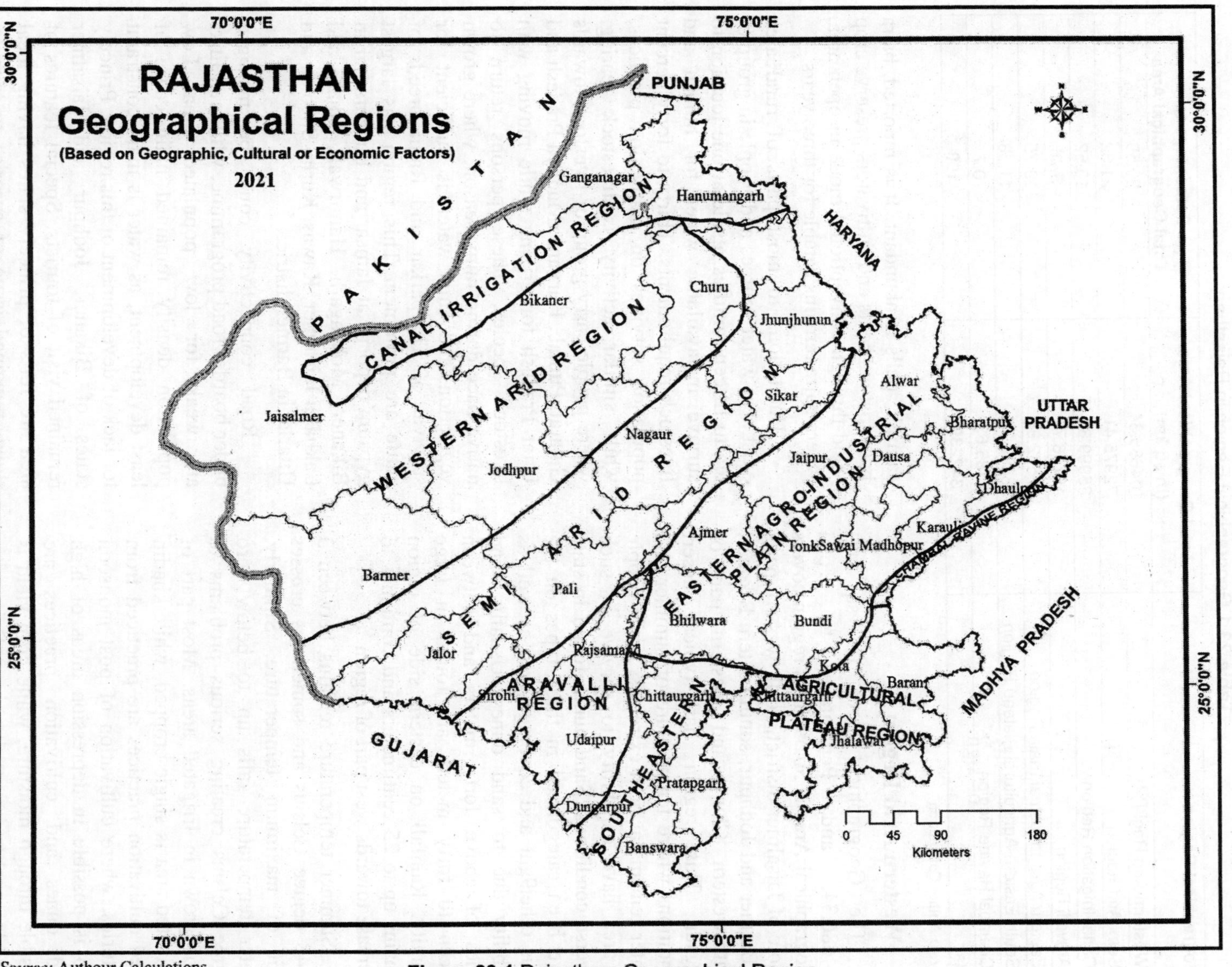

Source: Authour Calculations

Figure 23.1 Rajasthan: Geographical Region

Table 23.1 Geographical Region in Rajasthan

Geographical Region	Total Geographical Area (in sq km)	Percentage to Total Geographical Area
1. Westerm Arid Region	104686.34	30.59
2. Semi Arid Region	72975.41	21.32
3. Canal Irrigation Region	35999.86	10.52
4. Aravalli Region	19716.72	5.76
5. Eastern Agro-Industrial Plain Region	67582.66	19.75
6. South Eastern Agricultural Plateau Region	38892.38	11.36
7. Chambal Ravine Region	2385.63	0.7
Total	**342239**	**100**

Source: Author Calculation

1. Western Arid Region

The Geographical Area (in sq km) 104686.34, and Percentage to Total Geographical Area is 30.59. This region covers whole of Jaisalmer district, north-western parts of Barmer and Jodhpur, south-eastern Bikaner, south-western Churu and western part of Nagaur. This region has typical desert conditions where temperature of summer and winter remains at both extremes, while people of area have capability to survive in such adverse conditions. Sand dunes found in state have been categorized in three types on the basis ofhei9ght and coverage area as well as mobility due to sand dunes. Usually, sand dunes of above forty meters and between fifteen to forty meters are found in large quantity. Rainfall on eastern side of region remains up to 25 centimeters and continues to decrease towards west part of region.

Summer temperature remains between 3 to 48 degree Celsius and sometimes crosses limit of maximum temperature. Similarly, winter temperature falls up to below zero degree Celsius creating serious problems to crops grown in irrigated areas. Most part of cultivated area is single cropped, while khadin based cultivation practices are practiced from centuries, where cultivation of post monsoon crop is possible in depression areas of high sand dunes and cultivation practices are possible through moisture, while crop yield is possible equal to irrigated regions. Bajra crop is possible in large area, where crop damage is recurrent feature, while crop yield in regular rainfall spells is abundant. It is reported, bajra has most nutrients and cultivators preserve crop yield through conventional mode in open area, where bajra remains eatable for three years.

Habitations in most parts of rural area exist near water sources and people in group of ten to fifteen families reside at one location to survive in available water for human and livestock requirements. Thirty to forty percent population survives on cultivation practices, while similar activity is livestock rearing, where people carry group of livestock towards Gujarat, Punjab, Haryana, Madhya Pradesh and Uttar Pradesh for grizing while moving with livestock species. These persons return to native pace during rainy season, where grown vegetation and rain water is sufficient for survival of Tharparkar and Rathi breeds of cattle are prominent This region has largest gypsum deposits at Jamsar and Lunkaransar in Bikaner, Mohangarh, Hamirwali, Dhani and Lakha in Jaisalmer, Kawas, Kurla, Sheokar and Uttarlai in Barmer district.

Road connectivity could be managed under boarder road programme, while rail links are weak for low populated area. Low population density remained main cause of least development, as water is main constraint for lack of development of this region. Princely states of Bikaner, Jodhpur and Jaisalmer remained visit of tourists. Special features of area are forts of princely states having local stone, remained issue of surprise to tourists. This region reported to be prosperious area during existence of ancient Saraswati river, which was disappeared before millions of years creating adverse conditions for survival of people. It is also fact that desert region remained under sea before million years and region recovered back creating salty lakes and arid conditions.

2. Semi Arid Region

The Geographical Area (in sq km) 72975.41 and Percentage to Total Geographical Area is 21.32. This region falls west of Aravallis and reported to be controlling spread of desert area towards east. This area has normal rainfall above 25 cms and less than 45 cms comprising of Jalore, Pali, south-eastern Jodhpur and Nagaur, Sikar, Jhunjhunu and north-eastern part of Churu district. Southern part of region is drained by Luni river and its small tributaries, which remain functional during occurrence in catchment area. Flow of water has helped in recharge of ground water, while people of area have dug large number of tube wells in Kantli seasonal river originates from hills of Sikar districts and spreads towards Jhunjhunu and disappearing while touching boundary of Churu district. Southern part of region is drained by Luni river, which becomes saline after crossing Barmer.

Northern part of region possesses interior drainage due to presence of interior drainage due to presence of scattered hills of low heights in comparison to Aravalli region. Temperatures of low and high are of similar extremes of arid region. Fatehpur of Churu district and Pilani of Jhunjhunu district have low and high temperatures more than arid region. This region has up to sixty percent cultivated area, where monsoon crops are grown with help or rainy water and post monsoon crops are possible in drainage area of hills, where ground water recharge is possible, but cultivators have installed tube wells to draw more than two hundred percent ground water than annual recharge created serious problems to those water sources. People try to grow all possible crops in their cultivated fields to every possible extent.

Region is rich in livestock wealth comprising of Harayana bred in north, Nagauri breed in middle and Sanchore breed in south. Livestock breed is dependable source of livelihood in case of crop damage due to failure of rainfall. Dairy Federations are collecting milk from village and provide money of milk and nutritious fodder for high milk yield. Presence of medium and low sand dunes are helpful for moisture resistance for crop growth through moisture cultivation. Region is rich in non-metallic minerals and high quality gypsum in Jodhpur, Pali, Barmer and Nagaur. Rail-road connectivity is better for passenger and goods transport, but watr is main constraint in development of region.

3. Canal Irrigation Region

The Geographical Area (in sq km) 35999.86, Percentage to Total Geographical Area is 10.52. Canal irrigated region is converted from arid and semi arid region of state and area covered under this category covers whole Ganganagar and Hanuman districts, north-eastern part of Bikaner and part of Jaisalmer district. Process of canal irrigation remained slow for complete cemented canal and drainage system of feasible area. Presence of sand dunes on solid structures created problem in canal construction in systematic mode. In beginning, availability of canal water in Ganganagar district helped in high production and state remained in surplus producing area in country. Now total canal region of arid and semi-arid region has been gripped in serious land degradation problems comprising of water logging, uncontrolled growth of weed and high chemical water flown by Punjab degraded total cultivated area of these districts (Figure 23.2).

Rajasthan state has been benefited under Indus Water Treaty, where state was allotted 8.6 million acre feet water for irrigation and other requirements of drinking water besides irrigation in villages of Ganganagar, Hanumangarh, Bikaner and Jaisalmer districts. Drinking water support to hard hit villages of Jodhpur and Churu was also covered through canal system. Problem in irrigation canal

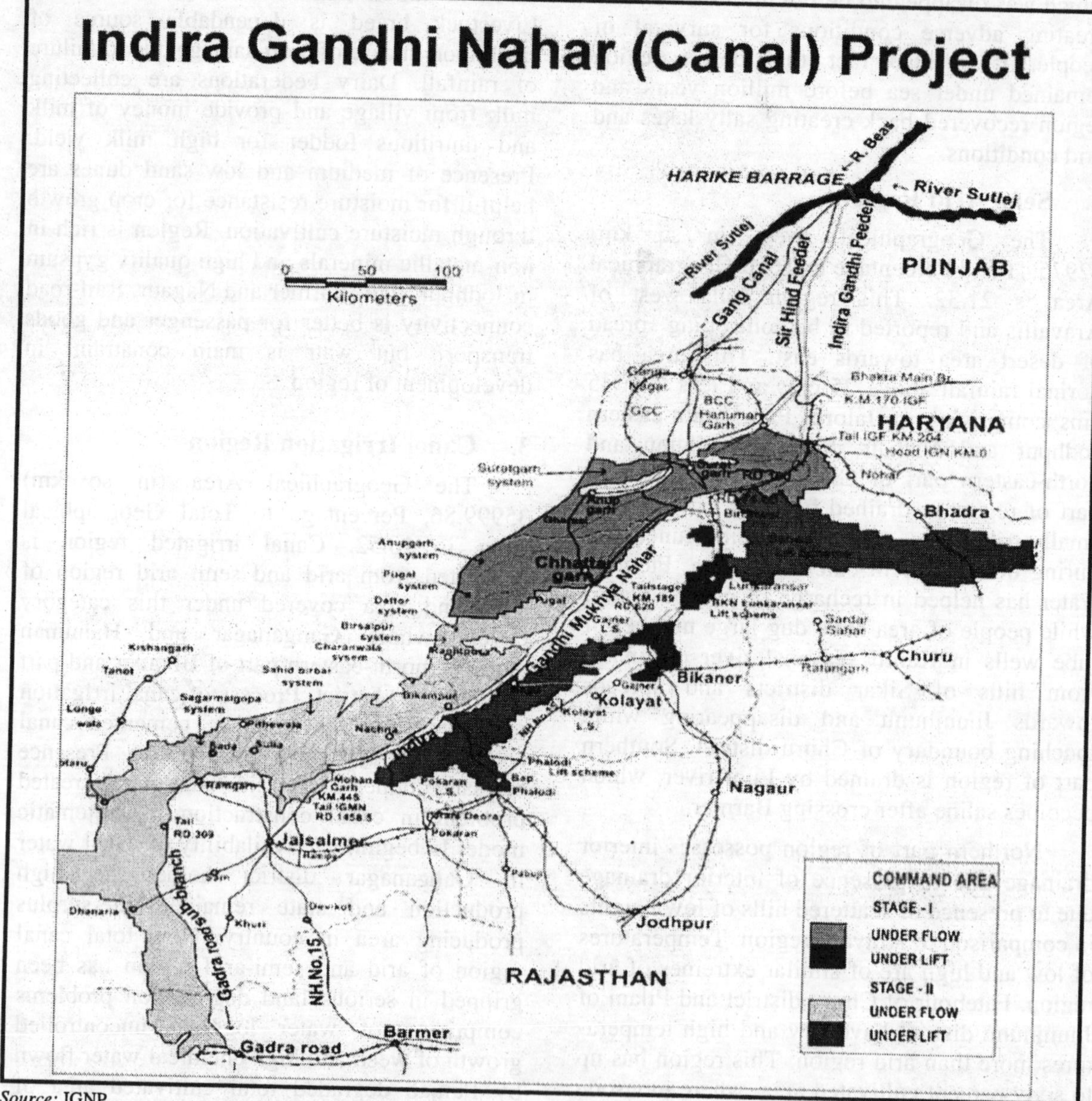

Source: IGNP

Figure 23.2 Index Map of Indira Gandhi Canal Project

region started with excessive use of water without understanding consequences to create water logging problem In addition, cultivators started providing excessive chemical fertilizers to maintain crop level. Both situations created serious land degradation problem due to water logging and poisonous impact in crop yield due to excessive use of chemical fertilizers.

Another problem created by Punjab farmers to flow chemical infested water into canal water, which disturbed total cultivation system, as water flown in canals contain highest pollution of chemicals and most of land of cultivators has become waste for no cultivation and high poisonous contents of chemicals. Cultivators of Punjab flow infested water as there is no mode to flow such contaminated water in any other source than canals carrying water for Rajasthan. Even after serious impact on land and crops, state government of Rajasthan continued to protest for acts of Punjab instead of evolving measures to purify water for drinking by human being and livestock as well as irrigation as source of livelihood. Now cultivators have no alternate to continue cultivation in their fields to eliminate cultivation.

Total area irrigated through canal irrigation in Ganganagar, Hanumangarh, Bikaner and Jaisalmer is more than 11.54 lakh hectare besides other problems being faced by people. In addition, no efforts were initiated to start bio-fertilizers to use in cultivation for safety of cultivated land from soil degradation and contamination impact of chemicals. This problem was faced by cultivators from more than two decades, but no efforts were made to eradicate problem by government by deploying expert to eliminate all problems created by cultivators of region and by cultivators of Punjab. Government of Rajasthan could not handle problem being faced by cultivators.

4. Aravalli Region

The Geographical Area (in sq km) 19716.72, Percentage to Total Geographical Area is 5.76. Aravalli mountain range is one of the oldest mountain system dividing state into two distinct regions, where western part of Aravalli hills is desert and semi-desert region, while eastern part comprises of plain and plateau areas of state having unique features. This region covers almost whole of Udaipur, south-western part of Pali and Sirohi and western parts of Udaipur district. This is most distinctive region of state, predominantly largely by Aravalli ranges and hilly region of south-west. Western boundary of region is marked by edge of hills and covered by Banas and Chhappan plains on eastern side. Southern part of region is covered by about fifty km wide hills and ranges branch towards south and south-west part of state. South-eastern part of Sirohi is occupied by principal Aravalli ranges.

Guru Shikhar with 1727 meter height is the highest peak of Aravalli ranges in Rajasthan. Other highest peaks of this region lie around Gogunda and Kumbhalgarh, north-west of Udaipur, This part of region is known as Bhorat Plateau with average elevation of 1225 meters. Rainfall normally occurs in region ranges from fifty to one hundred centimeters mostly during monsoon season. Mountains and hills of this region have well set drainage systems during rainy season but no systematic approach for rain water harvesting and use of water could be managed by state. Government of India floated watershed scheme for rainwater harvesting, but investment of more than four thousand crores could not yield success in any part of this region.

Under watershed approach, state government was required to add money for rain water harvesting structures to flow water to lower side and yield irrigation facilities for larger area of state, but agriculture experts preferred to flow water from upper reaches to stop soil erosion in hilly and mountain regions, but they failed to understand reality that this region had no soil cover for constant flow of water from centuries. Some water drains to form well set drainage systems, but no efforts could be made to use water in judicious mode to manage irrigation facilities on assured basis. Irrigation is possible from natural process during rainy season. In Udaipur, about seventy percent land is barren, where cultivation is not possible, In addition, no efforts could be made by government to convert barren land into fertile land through expert devices.

No efforts could be made to treat waste, barren and fallow lands, which can be treated with treatment of soil in systematic manner. Similarly, use of rain water for judicious use in crop cultivation remained failure for no interest of government machinery. Evan fallow lands and decreased productivity lands have never been treated by experts, though agriculture department claims its expertise to treat such lands in systematic mode. Similarly, waste land, barren land and degraded pasture lands have never been used for raising plants of various species to decrease impact of pollution of various categories. Such efforts could help in assured employment through horticulture practices in this region, where resources are available but no efforts could be made to manage systematically.

This region is rich in mineral deposits requiring judicious use through expertise support for getting maximum part of deposits. Aravalli region has highest deposits of lead and zinc in Jawar mines Beryllium mines are situated at Bari Sikarbari, Selaka Gudha and Champa Gudha in north and south of Amet, mica in northern Udaipur, emerald between Deogarh in the north to Kankroli. Similarly, soap stone talk and steatite in Udaipur district. Feldspar and asbestos are also found in this region. Main problem of systematic excavation remained problem as many minerals of high cost and deposits were given to private sector for political benefit and state could not get requisite money from mismatch mode of beneficiation.

5. Eastern Agro-Industrial Plain Region

The Geographical Area (in sq km) 19716.72, Percentage to Total Geographical Area is 5.76. Eastern agro-industrial region is comparatively plain region, where cultivation practices are possible in maximum part of coverage area. This region is spread over in districts of Jaipur, Dausa, Ajmer, Sawai Madhopur, Bhilwara, Bundi, Alwar, Bharatpur, Dholpur and north-western part of Kota. Western part of this region is demarcated by 50cm isohyt, northern by Uttar Pradesh and Haryana, eastern and southern part by ravines of Chambal and south-eastern part by agricultural region. River Banas and its tributaries flow through this region, where Bisalpur dam is catering drinking water requirements of cities of Jaipur, Ajmer, Nasirabad and adjoining areas, though dam was constructed for meeting irrigation requirements of farmers in catchment area.

Soil strata of this region is widely varied from alluvium, red and yellow in larger part of Ajmer, while eastern Bhilwara, Bundi and south-western soils are mixed red and black soil. Northern part of region is largely covered with alluvium soils except in southern pat of Alwar and in part of Bharatpur, where detached Aravalli hills are more prominent. Rainfall of this region ranges between fifty to ninety centimeters and fertile soils helps in growth of all crops subject to availability of water. Main economic activity of region is cultivation and industrial activities are associated with processing and manufacturing activities. Cultivated area is proportionately decreases from north to south and west mostly for land degradation and inability of cultivators to revive fallow lands. Cultivation practices in Alwar and Bharatpur are carried out in seventy to seventy five percent of total reported area. Crops grown in this region are bajra, jowar, maize, pulses, wheat, gram, rape and mustard.

Crop cultivation is also substantial activity in Jaipur, Sawai Madhopur, Dausa and Tonk between sixty to seventy percent of reported area, as cultivators try to use every cultivable land for livelihood and productive use of land. Cultivable land is thirty to forty percent of reported area in Ajmer, Bundi and Bhilwara. Presence of hilly tracts in all these district are responsible for meager cultivable land. In cultivated lands, where irrigation facilities are available from any source, cultivators prefer to grow high cost crops instead of Bajra. In kharif crops season, crop yield depends on regular spells of monsoon, while in rabi crop season, cultivators remain fully aware of water availability from known sources. Out of more than seventy percent rural population, about forty to fifty percent persons of rural population remain associated with land based activities as cultivators and twenty percent usually function as agriculture labour.

In view of serious water problem for cultivation practices, larger area of state is single cropped area, where cultivation practices are based in occurrence of rainfall in regular time bound spells, while crop failure is regular feature in one or two years in rainfall cycle of five years. Pressure of human population is high in this region in view of high density of population comprising of 598 persons per s km in Jaipur, 503 in Bharatpur, 476 in Dausa, 438 in Alwar, 399 in Bhilwara, 398 in Dholpur against state average of 201 persons per sq km. Rail and road connectivity has been increased in various parts of state, while many district headquarters are still lacking rail connectivity for better transport facilities.

6. South Eastern Agricultural Plateau Region

The Geographical Area (in sq km) 38892.38, Percentage to Total Geographical Area is 11.36. Southern agriculture region lies east of Aravalli region and south of agro-industrial region of state. Districts are covered under this region are Banswara, Chittorgarh, Pratapgarh, Baran, Jhalawar and Kota. The Vindhyan Plateau has an elevation 350 to 580 meters and generally formed if sand stones and shales. Part of this region in Kota, Baran and Jhalawar consists of stony uplands and partly includes Bundi district. This area has dark deep black soil having high fertility. High elevations and hard structure beneath small soil cover of les than one feet, cultivators remain effortful to preserve soil cover for crop growth. Even after high rainfall above state average, hard structure of land does not allow water to be charged underground.

Chambal river and its tributaries have formed alluvial basin in Kota, which helped in construction of dams for irrigation facilities in Kota and surrounding area. Soil capability is known from growth of Basmati rice exported to developed nations for high cost to cultivators. Jhalawar has become orange cultivation area and fruit variety is better than Nagpur, which helped in purchase of fruits from fields of farmers. Agriculture is main activity of region, where fifty to sixty percent of reported area in districts of Kota, Baran and Jhalawar is used for cultivation with more than average yield. Cropped area is comparatively low in Baran, Jhalawar, Banswara, Dungarpur, Chittorgarh nd Pratapgarh districts. Chittorgarh and Pratapgarh districts are famous for opium cultivation, where smuggling of product remained recurrent feature of region.

Banswara and Dungarpur districts are declared as tribal districts for more than fifty percent tribal population and funds for development of tribal population are received from Central Government, but most of the money received for socio-economic development of tribal people is drained for lack of awareness among tribal people for their own rights and benefits being provided to them. Banswara has highest rainfall of 93 centimeters, but total water is drained and cultivated fields are situated on highlands where irrigation facilities are meager. In addition, tribal families have not been provided livelihood sources after denudation of forests and they are surviving in most pity situation, though funds for their social and economic development are provided by central and state governments.

7. Chambal Ravine Region

The Geographical Area (in sq. km.) 2385.63, Percentage to Total Geographical Area is 0.70. Ravinous land is network of gullies formed generally in deep alluvium and ente4ring nearby river flowing much lower than surrounding table lands. The ravines are extensive system of gullies developed along river courses. Three categories of ravinous lands have been made comprising of deep ravines of more than five meter deep, medium ravines of depth between 2.5 to 5.0 meters depth and shallow ravines blow 2.5 meter depth.

In reference to Chambal ravinous region, area under Bharatpur, Dholpur and Sawai Madhopur districts is 2385.63 sq kms, which is severely problematic due to shelter-home of dacoits and anti-social elements hide them to escape from police administration. In addition, main geographical problem is further deepening of ravinous area due to cut by rainfall flowing water in ravinous areas in catchment of Chambal river. Topography of Chambal valley consists of hills and plateau and whole of the valley consists of new alluvial deposits. General slope of valley is towards north-east.

REFERENCE

Administrative Atlas-Rajasthan (2011), Census of India 2011, Directorate of Census Operations, Rajasthan, Jaipur.

Air Ministry (1936), British Rainfall 1935, Meteorological Office, London.

Annual Reports of RIICO and RFC for 2016-17.

Assam Earthquake (1897), The Great Assam Earthquake of 1897, Assa Portal

Barcar (1979), India Drought is Worst in Decades., The New York Times and T.S. Chouhna in Space Technology and GIS Disaster (2018), Scientific Publisher, Jodhpur

Bhasin, M.K. and Shampa Nag (2007), Demography of the Tribal Groups of Rajasthan : Population Structure, Anthropologist, 9(1) : 1-3 (2007)

Bhushan, S.K., (2001). Geology and gecochemistry of the magmatic rocks of the Malani Igneous Suite and Tertiary alkaline province of western Rajasthan. *Mem. Geol. Surv. Ind.* 126.

Budget Speech of CM, Vasundhara Raje. 12 February 2018 for Budget 2018-19.

Budget Study and Budget at a Glance,bothfor 2018-19, DES, Jaipur.

Chamipion, Harry, G. and Seth, S.K. (1968), A Revised Survery of Forest Types of India, Manager of Publication, New Delhi.

Chouhan T.S. (1987) Agricultural Geography: A Study of Rajasthan State, Academic Publishers, Jaipur, India.

Chouhan T.S. (1988), Integrated Area Development of Indian Desert, Geo-Environ Academia and Divyajyoti Prakashan, Jodhpur, India

Chouhan T.S. (1991). Environmental Monitoring and Pollution Measures, Academic Publishers, Jaipur, India

Chouhan T.S. (1992), Desertification in the World and its Control, Scientific Publishers, Jodhpur, India

Chouhan T.S. (1992), Readings in Remote Sensing Applications (ed.), Scientific Publishers, Jodhpur, India

Chouhan T.S. (1993), Natural and Human Resources of Rajasthan (ed.), Scientific Publishers, Jodhpur, India

Chouhan T.S. (1993), Rajasthan Atlas (Hindi), Vigyan Prakashan, Jodhpur.

Chouhan T.S. (1993), Studies in Arid Land Management (ed.), Scientific Publishers, Jodhpur, India

Chouhan T.S. (1995), Indian Desert: Resources and Perspective of Development (ed.), Printwell Publishers, Jaipur, India

Chouhan T.S. (1995), Irrigation in Arid Lands (ed.), Printwell Publishers, Jaipur, India

Chouhan T.S. (1996), Applied Remote Sensing and Photo-Interpretation, Vigyan Prakashan, Jodhpur

Chouhan T.S. (1996), Encylopaedia of Rajasthan (ed.) Vo. I-2, Vigyan Prakashan, Jodhpur.

Chouhan T.S. (1998), BIO STATISTICS: Concepts and Applications, Universal Scientific Publisher, Jaipur.

Chouhan T.S. (1998), Remote Sensing and Photogrammetry (Principles and Applications) Vol. I and II, Vigyan Prakashan, Jodhpur

Chouhan T.S. (2003), Geographical Information System and Remote Sensing for Sustainable Development, Vol. I and II, Universal Scientific Publishers, Jaipur.

Chouhan T.S. (2003), Remote Sensing for Natural Resources Management, Universal Scientific Publishers, Jaipur.

Chouhan T.S. (2013), Biogeography: Theoritical and Applied (Vol. 1 and 2 Set), Shruti Publications, Jaipur.

Chouhan T.S. (2018), Combating Desertification, Land Degradation and Climate Change:

Management of Drylands, Sceintific Publishers (India), Jodhpur.

Chouhan T.S. (2018), Geo-Informatics for Combating Land Degradation and Desertification, Sceintific Publishers (India), Jodhpur.

Chouhan T.S. (2018), Remote Sensing and GIS GPS Based Resources Management, Sceintific Publishers (India), Jodhpur.

Chouhan T.S. (2018), Space Technology and GIS for Disaster Monitoring and Mitigation, Sceintific Publishers (India), Jodhpur.

Chouhan T.S. (2019) Disaster Perception And Management, Scientific Publishers, Jodhpur, India

Chouhan T.S. (2019), Drought Proofing and Water Resource Management, Sceintific Publishers (India), Jodhpur.

Chouhan Tej Singh (2018), Geography of Rajasthan (Hindi, IInd Edition), Published by: Scientific Publishers (India), Jodhpur.

Climate of Rajasthan State (1988), India Meterologica Department, Government of India, Pune.

Coulson, Al., (1928), The geology of Bundi State, Rajputana. *Rec. Geol. Surv. Ind.* 44(1)

Dairying in Rajasthan: A Statistical Profile (2016), National Dairy Development Board, Anand-388 001 Gujarat.

DeMartonne, E. (1928), Areismeet indice and aridite C.R. Acad Sci., 128

Desertification and Land Degradation Atlas of India (Based on IRS AWiFS data of 2011-13 and 2003-05), (2016), Published by Space Application Centre, ISRO, Ahmedabad.

Draft Annual Plan (2016-17), Volume I; Planning Department, Government of Rajasthan.

Dutta, B.K. and Joshi D.C. (1989), Forms and availabilit of Potassium in the medium textured aluvial soils of arid Rajasthan. Ann. Arid Zone, 28, 101-111.

Dutta, B.K. and Joshi, D.C. (1989), Status and availability of potassium in sand soils of arid Rajasthan, Internal Symposium on Managing Sandy Soils. Abstract, Vol. 264-269.

Economic Review (2017-18), English, Directorate of Econimics and Statistics, Department of Planning, Rajasthan, Jaipur.

Economic Survey)2017-18), GOI (Vol. I and II, January 2018, printed by OUP, and released by end March 2018) (composite volume, this time) and Vol. II in August 2017.

Emberger, L. (1955), Afrique dunold ouest. In: Plant ecology reviews fo reseach, UNESCO, Paris, Arid Zone Research, 60

Geology and Mineral Resources of Rajasthan (2011), Geological Survey of India, Jaipur.

George and Kalyanasundaram (1969), Drought in Bihar, Wikipedia.

Ghose, B., Singh, S and Kar A. (1977), Geomorphology of the Rajasthan Desert in Desertification and its Control, ICAR, New Delhi.

Gibbs, W.J. and J.V. Manthur (1967), Rainfall deciles as drought indicators by Government Publication.

Greygory, S. (1969), Rainfall reliability. *In* Environment and Landure in Africa. Editors, Thomas, M.F.G.W. Methuen, London and Washington.

Ground Water Atlas (1999), Department of Ground Water, Government of Rajasthan, Jaipur

Ground Water Year Book (2016-17), Rajasthan State, Central Ground Water Board, Western Region, Jaipur

Gupta, B.C. (1934). The Geology of Central Mewar. *Mem. Geol. Surv. Ind.* 65(2) 107-169

Hacket C.A., (1881), On the geology of the Aravalli region, central and eastern. *Rec. Geol. Surv. Ind.* 14(4).

Hacket, C.A., (1877), Note on Aravalli series in north-eastern Rajputana. *Rec. Geol. Surv. Ind.* 10(2).

Herbertson A. J. (1899). An Introduction to Human Geography, A. and C. Black, London

Herbertson A. J. (1905). The Major Natural Regions: An Essay in Systematic Geography. Published by The Royal Geographical Society, London.

Heron A.M., (1917), The geology of north-eastern Rajputana and adjacent district. *Mem. Geol. Surv. Ind.* 45(1) : 1-128.

Heron A.M., (1936), The geology of south-eastern Mewar, Rajputana. *Mem. Geol. Surv. Ind.* 68 (1) : 1-120.

Heron A.M., (1953), The geology of central Rajputana. *Mem. Geol. Surv. Ind.* 79 : 1-389.

Hydrogeological Atlas of Rajasthan (2013), Ground Water Department, Government of Rajasthan, Jaipur. (European Union State Partnership Programme), Jaipur.

India: Economy and State Forest Report (2017), Forest Survey of India (Ministry of Environment Forests and Climate Change), Kaulagarh Road, P.O-IPE, Dehradun -248 195.

Indian Economy, Twelfth Five Year Plan, Vol. I, II and III, PC, GOI and Daily Financial Newspapers like Business Standard, the Financial Express. Indian Express and EPW (leading articles by experts)

Joshi D.C. (1993), Soil Resources of Rajasthan in Natural and Human Resources of Rajasthan (Ed.), T.S. Chouhan, Published by Scientific Publishers (India), Jodhpur.

Joshi, D.C. and Dhir R.P. (1983), Distribution of Micronutrient forms along dune land scape. Ann. Arid Zone 22, 135-141.

Joshi, D.C. and Dhir, R.P. (1983), Available forms of manganese and iron in some arid soils and their ralation with soil properties. Ann. Arid Zone 22, 7-14.

Joshi, D.C., Choudhari, J.S. and Jain, S.V. (1973), Distribution of Sulphur fractions in relation to forms of phosphorus in soils of Rajasthan. Journal of the *Indian Society of Soil Science*, 289-294.

Joshi, D.C., Dhir, R.P. and Gupta, B.S. (1981), Forms of Iron and Manganese in Some Soils of Arid Regions of Rajasthan. Journal of the *Indian Society of Soil Science*, 462-468.

Joshi, D.C., Singh, N. and Kolarkar, A.S. (1988), Forms of iron and manganese and their availability in the salt affected soil of Guhiya catchment. Tropical Agriculture 64, 41-48.

Kalyanasundaram and Ramasharer (1969), Drought in Bihar State. (2018) T.S. Chouhan, Space Technology and GIS Disaster Monitoring and Mitigations, Scientific Publisher , Jodhpur.

Karunakaran, C., (1950), Report on the geological mapping in parts of T.S. 45 J/16, 45 K/9 and 13, in Ajmer-Merwara (F.S. 1948-49). 13281-82

Krishan, A. and Thanvi (1977), "Drought in Rajasthan", Indian Journal of Geography, 11-12, No. 1.

Livestock Census Rajasthan (2012), Department of Animal Husbandry, Government of Rajasthan, Jaipur.

Lodha P.S. and Baser, B.L. (1971), Available micronutrients in sandy soils of Rajasthan. Ann. Arid Zone 10. 99-104.

Mathur, C.M. (1970), Forest Types of Rajasthan, Unpublished Ph.D. thesis, University of Rajasthan, Jaipur.

Mathur, C.M. (1972), Selection and Management of Tree Spp. for Fuel, Fodder and Small timber in desert area with reference to Rajasthan. Semi. on Management of Forests and Pastures, I.G.F.R.I, Jhansi.

Mehta, K.M., Shan Karriaryan, H.S. and Jaisinghani, C.J. (1962), Study of Pedo Genesis of soils of Bundi district (Rajasthan). *Soil Sci and Plant Nutr.* 8, 32-38.

Miller, A.A. (1961), Climatology, Methuen and Co. Ltd., London.

Ministry of Agriculture (1944), Governemnt of India, Delhi.

Mishra V.C. (1966), "Geographical Regions of Rajasthan", The Indian Journal of Geography, Vol. No. 1, Jodhpur, January, 1966.

Mishra V.C. (1967), Geography of Rajasthan, National Book Trust, New Delhi.

Population Census 2011, Government of India and Government of Rajasthan, Jaipur.

Pramanik S.K., etc. (1952), Indian Journal Meteor and Geophys 3(3).

Rajasthan Agricultural Statistics at a Glance (2016-17), Commissionerate of Agriculture, Rajasthan, Jaipur

Rajasthan Development Report (2006), Planning Commission, Government of India, New Delhi, Published by Academic Foundation, New Delhi.

Rajasthan Mineral Policy (2015), Department of Mines and Gelogy, Government of Rajasthan, Udaipur-313 001

Rajasthan State Action Plan on Climate Change, Government of Rajasthan.

Rajasthan, Twelfth Five Year Plan (2012-17) Volume I, Planning Department Government of Rajasthan, Jaipur.

Rajasthan: Statistical Abstract of Rajasthan (2015), Directorate of Econimics and Statistics, Department of Planning, Government of Rajasthan, Jaipur.

Ramdas, L.A. (1960), Crops and Weather in India, ICAR, New Delhi.

Ramdas, L.A. and A.K. Malik (1948), Agricultural Situation in India, Technical Bulletin, ICAR, New Delhi, Ministry of Food and Agricultural, Directorate of Economics and Statistics and Co-operatioon, Delhi.

Rao Raja, C.S. and Gupta, B.D., (1971), Geology around Jaipur, Rajasthan, Rec. Geol. Surv. Ind. 98(2) : 122-131.

Report on Annual Survey of Industries, Rajasthan, (2013-14), March 2017 (DES).

Resource Atlas of Rajasthan (1994) and up-dated Editon, 2009), State Remote Sensint Application Centre, Deparment of Science and Technology, Government of Rajasthan, Jaipur.

Richard Hartshone (1959). Perspective on the Nature of Geography, Published for the Assoication of American Geographers, Publisher: Rand McNally and Company, Chicago. John Murray.

Sen, A.K. (1968), Agricultural Atlas of Rajasthan, CAZRI (ICAR), Jodhpur.

Singh R.L. (1971), India-A Regional Geography, Published by: N.G.S.I., Varanasi, U.P.

Soil Resource Atlas of Rajasthan (2010), State Remote Sensint Application Centre, Deparment of Science and Technology, Government of Rajasthan, Jaipur.

State Disaster Management Plan (SDMP) (2014), Disaster Management and Relief Department, Government of Rajasthan, Jaipur.

State Finances: A Study of Budgets (2017-18), (RBI) (Latest).

State Industrial Profile (2014-15), Ministry of Micro, Small and Medium Enterprises, Government of India, MSME-Development Institute, Jaipur.

Statistical Year Book of Rajasthan (2017 and 2018), Director of Economics and Statistics, Department of Planning, Rajasthan, Jaipur.

Thornthwaite, C.W. (1948), An approach towards a rational classification of climate. Geogr. Rev. 38.

Thornthwaite, C.W. and Mather, J.R. (1955) The Water Balance. Drexel Institute of Technology—Laboratory of Climatology, Publications in Climatology, Centerton, 104 p.

Trewartha, G.T. (1947), An Introduction to Weather and Climate. Amazon.com

Van Bavel C. H. M. (1954), "Water deficits and irrigation requirements in the southern United States", Journal of Geophysics Research, 64.

Wastelands Atlas of India (Change Analysis Based on Temporal Satellite Data of 2005-06 and 2008-09), Department of Land Resources, Ministry of Rural Development, Government of India, New Delhi and National Remote Sensing Centre, Indian Space Research Organisation, Department of Space, Government of India, Hyderabad-500 625.

Watershed Atlas (1997), State Remote Sensing Application Centre, Department of Science and Technology, Government of Rajasthan, Jaipur

World Bank. Changing Wealth of Nations, 2018 (Trends in Wealth by Country Income Group).

World Development Indicators, 2017. World Bank Group.

World Meterological Organization (1988), Geneva, Switzerland.

■

Randhawa, M.S. (1960), Crops and Weather in India, ICAR, New Delhi.